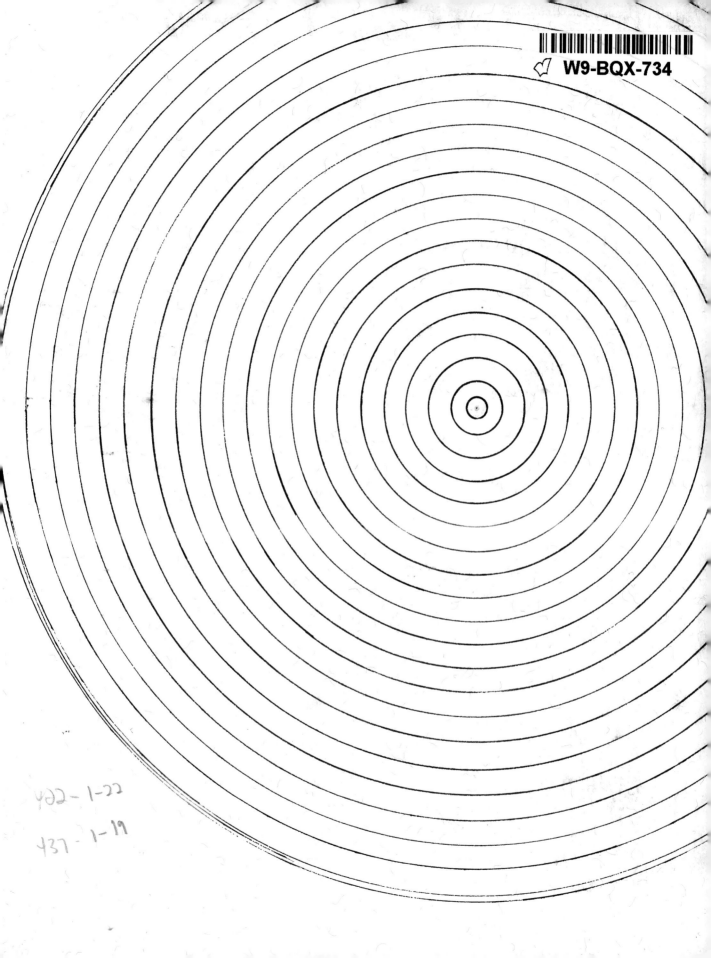

Second Edition

# TECHNICAL MATHEMATICS WITH CALCULUS

Paul Calter

Professor of Mathematics
Vermont Technical College

PRENTICE HALL, Englewood Cliffs, NJ 07632

**Library of Congress Cataloging-in-Publication Data**

Calter, Paul.
Technical mathematics with calculus / Paul Calter . — 2nd ed.
p. cm.
ISBN 0-13-902834-X
1. Engineering mathematics. 2. Calculus. I. Title.
TA330.C344 1990b
510—dc20 89-22979
CIP

Editorial/production supervision: Ed Jones and Mary Carnis
Interior and cover design: Meryl Poweski
Development editor: Roberta Lewis
Manufacturing buyer: Dave Dickey
Cover photo: Michael Haynes/Image Bank



Printed in the United States of America
10 9 8 7 6 5

ISBN 0-13-902834-X

Prentice-Hall International (UK) Limited, *London*
Prentice-Hall of Australia Pty. Limited, *Sydney*
Prentice-Hall Canada Inc., *Toronto*
Prentice-Hall Hispanoamericana, S.A., *Mexico*
Prentice-Hall of India Private Limited, *New Delhi*
Prentice-Hall of Japan, Inc., *Tokyo*
Simon & Schuster Asia Pte. Ltd., *Singapore*
Editora Prentice-Hall do Brasil, Ltda., *Rio de Janeiro*

*To Margaret Jolind*

# CONTENTS

# PREFACE

Most of the ideas for this second edition come from the classroom. Teaching from the first edition daily for seven years, I have listened to students, observed what worked and did not, and took notes. My colleagues did the same, and all comments and suggestions were carefully saved. Most of this material was completely revised four years ago for another project, and now it has been revised again for this edition. Each revision was aided by the many reviewers obtained by Prentice Hall.

My "finished" manuscript received yet another round of reviews, including a scrutiny by a developmental editor with a mathematics teaching background and was revised once more. It was then taken in hand by a team of editors and designers at Prentice Hall who made the book as attractive, useable, and error free as possible. Those who took part in this long project are listed in the acknowledgements. They have my sincere thanks.

This book is intended for students at technical schools or two-year technical colleges. Depending on the pace and the amount of material included, it may be used for either a two- or three-semester course.

We start with *Numerical Computation*. In addition to arithmetic, other topics in Chapter 1 include calculator operations, significant digits and rounding, scientific notation, and conversion of units. Some of these may be unfamiliar to high school graduates.

Many instructors will want to start their first semester with *Introduction to Algebra,* Chapter 2, and proceed through *Complex Numbers,* Chapter 20. Following these are several chapters of precalculus material from which to choose, as time allows.

A typical one-semester course in Analytic Geometry and Calculus will usually include the material from *The Straight Line,* Chapter 25, through *Applications of the Definite Integral,* Chapter 33. We provide four more chapters of calculus beyond these to allow for a second semester of calculus.

## FEATURES OF THE BOOK

A mathematics book is never easy reading, so much care has gone into making the material as clear as possible. We follow an intuitive rather than a rigorous approach and give information in small segments. Marginal notes, many illustrations, and careful page layout are designed to make the material as interesting and easy to follow as possible. The book also has the following features:

**CHAPTER OBJECTIVES:** Each chapter starts with a list of goals. They state what the student should be able to do upon completion of the chapter.

**EXAMPLES:** Many fully worked out examples form the backbone of the textbook. They have been specially chosen to help the student do the exercises. We have added dozens of examples to supplement those found in the first edition. Examples now have lines above and below to separate them clearly from the text discussion.

**EXERCISES AND CHAPTER REVIEW PROBLEMS:** Practice is essential for learning mathematics, and so we include thousands of exercises and chapter review problems. *Exercises* after each section are graded by difficulty and grouped by type to allow practice on a particular area. However, the *chapter review problems* are scrambled as to type and difficulty. Exercises are now labeled by category, as well as by exercise number. We give answers to all odd-numbered exercises and problems in the *Answer Key,* Appendix E. Complete solutions to every exercise and problem are contained in the *Solutions Manual,* available to instructors. Complete solutions for every other odd exercise and problem are given in the *Student Solutions Manual.*

**FORMULAS:** Each important formula, both mathematical and technical, is boxed and numbered in the text. We also list these formulas in Appendix A as the *Summary of Facts and Formulas*. This listing can function as a handbook for a mathematics course and for other courses as well. It provides a common thread between chapters. We hope it also will help a student to see connections that might otherwise be overlooked. The formulas are grouped logically in the *Summary of Facts and Formulas* and are numbered sequentially. Therefore, the formulas do not necessarily appear in the text in numerical order.

**COMMON ERROR BOXES:** An instructor quickly learns the pitfalls and traps that ''get'' students year after year. Many of these are boxed in the text and are labeled *Common Error.*

**APPLICATIONS:** We include discussion of many technical applications, such as motion problems and electric circuits. They are included for classes that wish to cover those topics, and to show that mathematics has real-world uses. Because space does not permit a full discussion of each application, judgment must be used in assigning applications problems. It is not intended that every student be able to solve every application. We assume that students will have sufficient background before they attempt difficult problems in their technical area and that they will get help from their instructors. The *Index to Applications* should help in finding specific applications.

**NUMERICAL METHODS AND COMPUTER:** As numerical methods and the computer grow in importance we continue to add these topics. However they can be skipped without impairing the overall study of technical mathematics. For those calculations that are best done by computer we give computer methods in the text itself. Elsewhere, computer problems are suggested at the end of exercises, as enrichment activities. The following numerical methods are included in the second edition:

- Roots of equations by the midpoint method
- Roots of equations by the false position method
- Roots of equations by simple iteration
- Roots of equations by Newton's method
- Solving systems of equations by the Gauss-Seidel method
- Evaluation of determinants by the method of Chió
- Solving sets of equations by Gauss Elimination
- Solving sets of equations by the unit matrix method
- Solving sets of equations by matrix inversion
- Integrals by the average ordinate method
- Integrals by the midpoint method
- Integrals by the prismoidal formula
- Integrals by Simpson's rule
- Integrals by the trapezoid rule
- Solving differential equations by the Runge-Kutta method
- Numerical method for finding the coefficients in a Fourier series

As with technical applications, space does not permit the teaching of programming here, and we assume that a student who tries these has some computer background. We do, however, provide a *Summary of the Basic Language* in Appendix D. Complete programs are given in the solutions manual. A diskette containing all the programs is available to instructors, who may copy it for their students.

**WRITING ACROSS THE CURRICULUM:** There is a movement among mathematics teachers to use writing to help teach mathematics. Practice questions for mathematics requiring written responses are hard to find. We have provided one such question per chapter. These questions can be assigned as homework or can be used as models by instructors who wish to design others.

**NEW MATERIAL:** The content of the book is greatly expanded from the first edition. There are new chapters on

Matrices
Radian measure and arc length
Binary, hexadecimal, octal, and BCD numbers
Sequences, series, and the binomial theorem

Statistics and probability
Differential equations
Laplace transform, and numerical solution of differential equations
Infinite series, including Fourier series

And new sections on:

Higher order determinants
Exponents
The sine wave as a function of time
Graphing parametric equations
Graphing parametric equations in polar form
Graphs on logarithmic and semilogarithmic paper
Linear programming
Applications of the derivative and integral to electric circuits
Improper integrals

There are also new topics in numerical methods and the computer, mentioned earlier.

## TEACHING RESOURCES

There are a number of supplemental teaching resources to aid both the instructor and the student.

An *Annotated Instructor's Edition* (*AIE*) of this text contains answers to every exercise and problem. The answers, provided and checked by Susan Porter and Linda Davis of Vermont Technical College, are printed in red right next to the exercise or problem. The AIE also has marginal notes to the instructor (also printed in red), giving teaching tips, applications, and practice problems, and an introductory section giving tips on how to get the most from the text. All of this material has been prepared by Pat Hirschy of Delaware Technical and Community College.

An *Instructor's Solutions Manual* contains worked-out solutions to every problem in the text and listings of all computer programs.

The *Student Solutions Manual,* by John Knox of Vermont Technical College, gives the solution to every other odd problem. They are usually worked in more detail than in the *Instructor's Solutions Manual.*

The *Supplementary Problems for Technical Mathematics,* by Michael Calter, Teaching Fellow at Harvard University, is useful for students who need more practice problems.

*How To Study Technical Mathematics,* by Paul Dudenhefer of State Technical College at Memphis, is a short booklet that provides encouragement, motivation, and strategies for the beginning technical mathematics student. Adopters of the text may request a free copy for each student.

A *Computerized Test Item File,* by Michael Calter, is a bank of test questions, with answers. Questions may be mixed, sorted, changed, or deleted. It consists of a test file disk and a test generator disk, ready to run on IBM PC's or compatibles.

*Software Tools,* by Nathan O. Niles and Al Parish, contains programs to enhance concepts covered in the text. It is self-contained, ready to run, and available in IBM PC and Apple Macintosh formats.

A *Computer Solutions Disk,* by Michael Calter, contains each program from the text, ready to run.

*Transparency Acetates,* giving additional problems and full explanations, are available.

*Transparency Masters* provide 378 enlarged drawings from the text.

## ACKNOWLEDGMENTS

As mentioned at the beginning of the preface, this text, in its final form, is the result of an intensive developmental effort, in which many people took part. There are reviewers of the manuscripts of this edition and of the first edition, teachers of technical mathematics who participated in group discussions on how to improve the manuscript, teachers of writing who helped with the writing questions, checkers who solved problems and did proofreading, and so on.

I am indebted to Roberta Lewis, developmental editor, and to Frank Juszli, editor of the Prentice Hall series in Technical Mathematics, both of whom did detailed reviews of the entire manuscript and suggested many ways to improve it. Other significant contributors to the development of the book are Linda Davis of Vermont Technical College and Susan Porter of Randolph, Vermont, both of whom checked the entire set of proofs and the solutions manual for accuracy. I also want to express my gratitude to the following:

Reviewers of the second edition and the supplements:

Jim Beam, Savannah Area VOTECH
Michael Calter, Harvard University
John Eisley, Mott Community College
Richard Hanson, Burnsville, Minnesota
Tommy Hinson, Forsythe Community College
Wendell Johnson, University of Akron Community and Technical College
Joseph Jordan, John Tyler Community College
Rob Kimball, Wake Technical Community College
John Knox, Vermont Technical College
John Luke, Indiana University-Purdue University
Fran Leach, Delaware Technical College
Paul Maini, Suffolk County Community College
Harold Oxsen, Diablo Valley College
Thomas Stark, Cincinnati Technical College
Joel Turner, Blackhawk Technical Institute
Douglas Wolansky, North Alberta Institute of Technology
Henry Zatkis, New Jersey Institute of Technology

Participants in the focus-group discussions:

Jacquelyn Briley, Guilford Technical Community College
Cheryl Cleaves, State Technical Institute at Memphis
Ray Collings, Tri-County Technical College
Margie Hobbs, State Technical Institute at Memphis
Blin Scatterday, University of Akron Community and Technical College

Teachers of writing who reviewed the writing questions:

Miriam Conlon, Vermont Technical College
Kati Dana, Norwich University
Crystal Gromer, Vermont Technical College
Lorraine Lachs, Norwich University

Reviewers of earlier editions of this material:

Bryon Angell, Vermont Technical College
David Bashaw, New Hampshire Technical Institute
Elizabeth Bliss, Trident Technical College
Franklin Blou, Essex County College
Walt Granter, Vermont Technical College
Martin Horowitz, Queensborough Community College
Glenn Jacobs, Greenville Technical College
Ellen Kowalczyk, Madison Area Technical College
Don Nevin, Vermont Technical College
Donald Reichman, Mercer County Community College
Ursula Rodin, Nashville State Technical College
Frank Scalzo, Queensborough Community College
Edward W. Seabloom, Lane Community College
Robert Seaver, Lorain Community College

The tedious job of solving many of the exercises and problems was done by my students: Brett Benner, Keith Crowe, Jim Davis, Nancy Davis, Kelly Dennehy, Mel Emerson, Ellie Germain, Robert Morel, Jeffery Sloan, Ray Wells.

Finally, let me express my gratitude to the editorial team at Prentice Hall: Mary Carnis and Ed Jones, production editors; Mariann Hutlak, supplements editor; Ray Mullaney, editor in chief for book development; and Rick Williamson, acquisitions editor. Thank you all.

*Paul Calter*
*Randolph, Vermont*

## ABOUT THE AUTHOR

Paul Calter is Professor of Mathematics at Vermont Technical College. A graduate of The Cooper Union, New York, he received his M.S. from Columbia University.

Professor Calter, a member of the American Society of Mechanical Engineers, has been teaching Technical Mathematics for over twenty years. In 1987, he was the recipient of the Vermont State College Faculty Fellow Award.

A member of the American Mathematical Association of Two Year Colleges, Professor Calter is the Chairman of the AMATYC Summer Institute in Vermont. He is also a member of the Mathematical Association of America, the Society for Technical Communication, Volunteers in Technical Assistance, and the Author's Guild.

Professor Calter is the author of several mathematics textbooks, among which is the *Schaum's Outline of Technical Mathematics*. With Prentice Hall he has published three other books: *Practical Math Handbook for the Building Trades, Mathematics for Computer Technology,* and *Technical Calculus*. He has also published research articles in engineering journals.

He resides in rural Vermont and is an accomplished sculptor. His work often reflects themes related to mathematics.

# 1

# NUMERICAL COMPUTATION

## OBJECTIVES

**When you have completed this chapter, you should be able to:**

- Perform basic arithmetic operations on signed numbers.
- Perform basic arithmetic operations on approximate numbers.
- Take powers, roots, and reciprocals of signed and approximate numbers.
- Convert units of measurement from one system to another.
- Substitute given values into equations and formulas.
- Convert numbers between decimal and scientific notation.
- Solve common percentage problems.

In this first chapter we cover ordinary arithmetic, but in a different way. The basic operations are explained here in terms of the *calculator*. Although the calculator makes arithmetic easier than before, it also introduces a complication: that of knowing how many of the digits shown in a calculator display should be kept. We will see that it is usually incorrect to keep them all.

Also in this chapter, we learn some rules that will help us when we get to algebra.

We then consider the *units* in which quantities are measured: how to convert from one unit to another, and how to substitute into equations and formulas and have the units work out right.

We also learn *scientific notation*—a convenient way to handle very large or small numbers without having to write down lots of zeros. It is also the way in which calculators and computers display such numbers.

Finally, we cover *percentage,* one of the most often used mathematical ideas in technology as well as in everyday life.

When you finish this chapter you should be able to use most of the keys on your calculator to do some fairly difficult computations. The keys for trigonometry and logarithms are discussed in the chapters on those topics.

## 1-1 THE REAL NUMBERS

Before we start our calculator practice, we must get a few definitions out of the way. In mathematics, as in many fields, you will have trouble understanding the material if you do not clearly understand the meanings of the words that are used.

### Integers

**The three dots indicate that the sequence of numbers continues indefinitely.**

The *integers*

$$\ldots, -4, -3, -2, -1, 0, 1, 2, 3, 4, \ldots$$

are the whole numbers, including zero and negative values.

### Rational and Irrational Numbers

The *rational* numbers include the integers and all other numbers that can be expressed as the quotient of two integers: for example,

$$\frac{1}{2}, \quad -\frac{3}{5}, \quad \frac{57}{23}, \quad -\frac{98}{99}, \quad \text{and} \quad 7$$

Numbers that cannot be expressed as the quotient of two integers are called *irrational*. Some irrational numbers are

$$\sqrt{2}, \quad \sqrt[3]{5}, \quad \sqrt{7}, \quad \pi, \quad \text{and} \quad e$$

where $\pi \cong 3.1416$ and $e \cong 2.7182$.

### Real Numbers

The rational and irrational numbers together make up the *real numbers*.

Numbers such as $\sqrt{-4}$ do not belong to the real number system. They are called *imaginary numbers* and are discussed in Chapter 20. Except when otherwise noted, all the numbers we work with are real numbers.

## Decimal Numbers

Most of our computations are with numbers written in the familiar *decimal* system. The names of the *places* relative to the *decimal point* are shown in Fig. 1-1. We say that the decimal system uses a *base of 10* because it takes 10 units in any place to equal to 1 unit in the next-higher place. For example, 10 units in the hundreds position equals 1 unit in the thousands position.

**Systems having bases other than 10 are used in computer science: *binary* (base 2); *octal* (base 8), and *hexadecimal* (base 16) (see Chapter 21).**

**The numbers $10^2$, $10^3$, etc. are called *powers of 10.* Don't worry if they are unfamiliar to you. We will explain them in Sec. 1-8.**

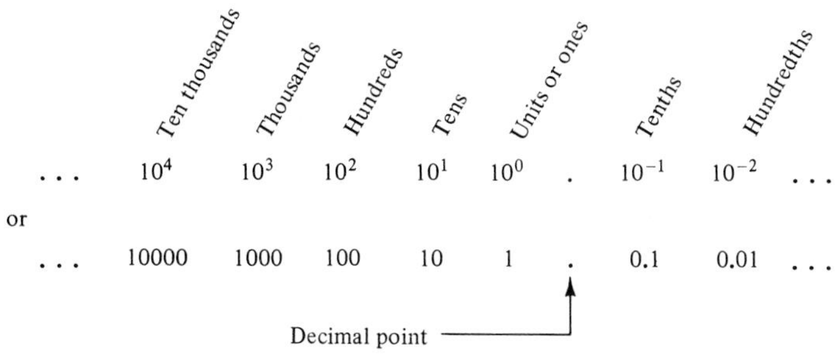


**FIGURE 1-1 Values of the positions in a decimal number.**

## Positional Number Systems

A *positional* number system is one in which the *position* of a digit determines its value. Our decimal system is positional.

---

**EXAMPLE 1:** In the number 351.3, the digit 3 on the right has the value $\frac{3}{10}$, but the digit 3 on the left has the value 300.

---

The position immediately to the left of the decimal point is called position 0. The position numbers then increase to the left and decrease to the right of the 0 position.

## Place Value

Each position in a number has a *place value* equal to the base of the number system raised to the position number. The place values in the decimal number system, as well as the place names, are shown in Fig. 1-1.

## Number Line

We can graphically represent every real number as a point on a line called the *number line* (Fig. 1-2). It is customary to show the positive numbers to the right of zero and the negative numbers to the left.

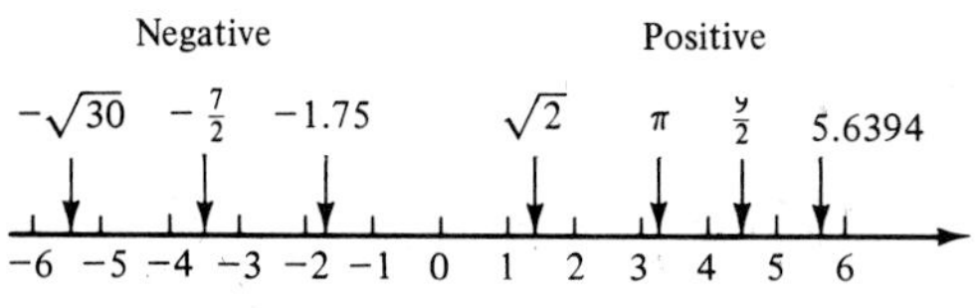


**FIGURE 1-2 The number line.**

## Signs of Equality and Inequality

Several signs are used to show the relative position of two quantities on the number line.

$a = b$ means that *a equals b* and that they occupy the same position on the number line.

$a \neq b$ means that *a* and *b* are *not equal* and have different locations on the number line.

$a > b$ means that *a* is *greater than b* and lies to the right of *b* on the number line.

$a < b$ means that *a* is *less than b* and lies to the left of *b* on the number line.

$a \cong b$ means that *a* is *approximately equal to b* and that *a* and *b* are *near* each other on the number line.

## Absolute Value

The *absolute value* of a number *n* is its *magnitude* regardless of its algebraic sign. It is written $|n|$.

**EXAMPLE 2:**

(a) $|5| = 5$

(b) $|-9| = 9$

(c) $|3 - 7| = |-4| = 4$

(d) $-|-4| = -4$

(e) $-|7 - 21| - |13 - 19| = -|-14| - |-6| = -14 - 6 = -20$

## Approximate Numbers

Most of the numbers we deal with in technology are *approximate*.

**EXAMPLE 3:**

(a) All numbers that represent *measured* quantities are approximate. A certain shaft, for example, is approximately 1.75 inch in diameter.

(b) Many *fractions* can be expressed only approximately in decimal form. Thus $\frac{2}{3}$ is approximately equal to 0.6667.

(c) *Irrational numbers* can be written only approximately in decimal form. The number $\sqrt{3}$ is approximately equal to 1.732.

## Exact Numbers

*Exact* numbers are those that *have no uncertainty*.

**EXAMPLE 4:**

(a) There are exactly 24 hours in a day, no more, no less.

(b) An automobile has exactly four wheels.

(c) Exact numbers are usually integers, but not always. For example, there are *exactly* 2.54 cm in an inch, by definition.

(d) On the other hand, not all integers are exact. For example, a certain town has a population of *approximately* 3500 people.

## Significant Digits

In a decimal number, zeros are sometimes used just to locate the decimal point. When used in that way, we say that those zeros are *not significant*. The remaining digits in the number, including zeros, are called *significant digits*.

---

**EXAMPLE 5:**

(a) The numbers 497.3, 39.05, 8003, and 2.008 each have *four* significant digits.

(b) The numbers 1570, 24,900, 0.0583, and 0.000583 each have *three* significant digits. The zeros in these numbers serve only to locate the decimal point.

(c) The numbers 18.50, 1.490, and 2.000 each have *four* significant digits. The zeros here are not needed to locate the decimal point. They are placed there to show that those digits are in fact zeros, and not something else.

---

**An overscore is sometimes placed over the last trailing zero that is significant. Thus, the numbers $32.5\bar{0}$ and $735{,}\bar{0}00$ each have four significant digits.**

## Accuracy and Precision

The *accuracy* of a decimal number is given by the number of *significant digits* in the number; the *precision* of a decimal number is given by the position of the rightmost significant digit.

---

**EXAMPLE 6:**

(a) The number 1.255 is accurate to four significant digits, and precise to three decimal places. We also say that it is precise to the nearest thousandth.

(b) The number 23,800 is accurate to three significant digits, and precise to the nearest hundred.

---

## Rounding

We will see, in the next few sections, that the numbers we get from a computation often contain *worthless digits* that must be *thrown away*. Whenever we do this, we must *round* our answer.

*Round up* (increase the last retained digit by one) when the discarded digits are greater than 5. *Round down* (do not change the last retained digit) when the discarded digits are less than 5.

---

**EXAMPLE 7:**

| *Number* | *Rounded to Three Decimal Places* |
|---|---|
| 4.3654 | 4.365 |
| 4.3656 | 4.366 |
| 4.365501 | 4.366 |
| 1.764999 | 1.765 |
| 1.927499 | 1.927 |

---

When the discarded portion is 5 *exactly*, it does not usually matter whether you round up or down. The exception is when adding or subtracting a long

This is just a convention. We could just as well round to the nearest odd number.

column of figures. If, when discarding a 5, you always rounded up, you could bias the result in that direction. To avoid that you want to round up about as many times as you round down, and a simple way to do that is to always *round to the nearest even number.*

**EXAMPLE 8:**

| *Number* | *Rounded to Two Decimal Places* |
|---|---|
| 4.365 | 4.36 |
| 4.355 | 4.36 |
| 7.76500 | 7.76 |
| 7.75500 | 7.76 |

## EXERCISE 1—THE REAL NUMBERS

### Equality and Inequality Signs

Insert the proper sign of equality or inequality (=, ≅, >, <) between each pair of numbers.

**1.** 7 and 10 **2.** 9 and −2 **3.** −3 and 4
**4.** −3 and −5 **5.** $\frac{3}{4}$ and 0.75 **6.** $\frac{2}{3}$ and 0.667

### Absolute Value

Evaluate each expression.

**7.** $-|9 - 23| - |-7 + 3|$ **8.** $|12 - 5 + 8| - |-6| + |15|$
**9.** $-|3 - 9| - |5 - 11| + |21 + 4|$

### Significant Digits

Determine the number of significant digits in each approximate number.

**10.** 78.3 **11.** 9274 **12.** 4.008 **13.** 9400
**14.** 20,000 **15.** 5000.0 **16.** 0.9972 **17.** 1.0000

Round each number to two decimal places.

**18.** 38.468 **19.** 1.996 **20.** 96.835001 **21.** 55.8650
**22.** 398.372 **23.** 2.9573 **24.** 2985.339 **25.** 278.382

Round each number to one decimal place.

**26.** 13.98 **27.** 745.62 **28.** 5.6501 **29.** 0.482
**30.** 398.36 **31.** 34.927 **32.** 9839.2857 **33.** 0.847

Round each number to the nearest hundred.

**34.** 28,583 **35.** 7550 **36.** 3,845,240 **37.** 274,837

Round each number to three significant digits.

**38.** 9.284 **39.** 2857 **40.** 0.04825 **41.** 483,982
**42.** 0.08375 **43.** 29.555 **44.** 29.45001 **45.** 8372

## 1-2 ADDITION AND SUBTRACTION

### Addition and Subtraction of Integers

Try the following addition problem on your calculator.

**EXAMPLE 9:** Evaluate 7392 + 1147.

**Solution:** On your calculator:

| *Press* | *Display* |
|---|---|
| 7392 [+] | 7392 |
| 1147 [=] | 8539 |

### AOS and RPN

Does your calculator have an [=] key, so that you were able to do the last problem as written? If so, your calculator uses algebraic notation or the *algebraic operating system* (AOS). If you have no [=] key but have an [ENTER] key instead, your calculator uses *reverse Polish notation* (RPN). Do not worry if you have one rather than the other; both are good.

Named for the Polish philosopher and mathematician Jan Lukasiewicz (1878–1956).

On an RPN calculator the computation shown above would be:

| *Press* | *Display* |
|---|---|
| 7392 [ENTER] | 7392 |
| 1147 [+] | 8539 |

Unless otherwise noted, computations in this book will be in AOS. You should have no trouble following the same computations on an RPN calculator.

### Addition and Subtraction of Negative Numbers

Let $a$ and $b$ stand for two *positive* quantities. Our first rule of signs is:

| Rule of Signs for Addition | $a + (-b) = a - b$ | **6** |
|---|---|---|

All boxed and numbered formulas are tabulated in Appendix A. There they are arranged in logical order, by type, and are numbered consecutively. Since the formulas often appear in the text in a different order than in Appendix A, they may not be in numerical order here.

*The operation of adding a negative quantity* (−b) *to the quantity* a *is equivalent to the operation of subtracting the positive quantity* b *from* a.

**EXAMPLE 10:**

(a) $7 + (-2) = 7 - 2 = 5$

(b) $9 + (-15) = 9 - 15 = -6$

(c) $-8 + (-3) = -8 - 3 = -11$

Our second rule is:

| Rule of Signs for Subtraction | $a - (-b) = a + b$ | **7** |
|---|---|---|

*The operation of subtracting a negative quantity* (−b) *from a quantity* a *is equivalent to the operation of adding the positive quantity* b *to* a.

Note that the minus sign (−) is used for *two different things:*

1. To indicate a *negative quantity*
2. For the operation of *subtraction*

This difference is clear on the calculator, which has *separate keys* for these two functions. The *change-sign* key [+/−] or [CHS] is used to enter a negative number, whereas the [−] key is used for subtraction. More on the change-sign key in Sec. 1-3.

**EXAMPLE 11:**

(a) $15 - (-3) = 15 + 3 = 18$

(b) $-5 - (-9) = -5 + 9 = 4$

(c) $-25 - (-5) = -25 + 5 = -20$

## Commutative and Associative Laws

**These laws are surely familiar to you, even if you do not recognize their names. We will run into them again when studying algebra.**

The *commutative law*

| | | |
|---|---|---|
| Commutative Law for Addition | $a + b = b + a$ | **1** |

simply says that you can add numbers in *any order*.

**EXAMPLE 12:**

$$2 + 3 = 3 + 2$$
$$= 5$$

The *associative law*

| | | |
|---|---|---|
| Associative Law for Addition | $a + (b + c) = (a + b) + c$<br>$= (a + c) + b$ | **3** |

says that you can group numbers to be added in several ways.

**EXAMPLE 13:**

$$2 + 3 + 4 = 2 + (3 + 4) = 2 + 7 = 9$$
$$= (2 + 3) + 4 = 5 + 4 = 9$$
$$= (2 + 4) + 3 = 6 + 3 = 9$$

## Addition and Subtraction of Approximate Numbers

Addition and subtraction of integers is simple enough. But now let us tackle the problem mentioned earlier: How many digits do we keep in our answer when adding or subtracting *approximate numbers?*

| Rule | When adding or subtracting approximate numbers, keep as many decimal places in your answer as contained in the number having the fewest decimal places. |
|---|---|

**EXAMPLE 14:**

$$32.4 \text{ cm} + 5.825 \text{ cm} = 38.2 \text{ cm} \quad (\textit{not } 38.225 \text{ cm})$$

We do not use the symbol ≅ when dealing with approximate numbers. We would not write, for example, 32.4 cm + 5.825 cm ≅ 38.2 cm.

**EXAMPLE 15:**

| | |
|---|---|
| 25.8 | |
| 18.3 | 125 |
| 5.4 | 07 |
| 49.5 | 195 |

discard →

Students *hate* to throw away those last digits. Remember that by keeping worthless digits you are telling whoever reads that number that it is more precise than it really is.

**EXAMPLE 16:** A certain stadium contains about 3500 people. It starts to rain and 372 people leave. How many are left in the stadium?

**Solution:** Subtracting, we obtain

$$3500 - 372 = 3128$$

which we round to 3100, because 3500 here is known to only two significant digits.

It is safer to round the answer after adding, rather than to round the original numbers before adding, although it usually does not make any difference.

## Combining Exact and Approximate Numbers

When combining an exact number and an approximate one, it is obvious that the accuracy of the result will be limited by the approximate number. Thus round the result to the number of decimal places found in the approximate number, even though the exact number may *appear* to have fewer decimal places.

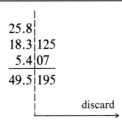

**EXAMPLE 17:** Express 2 h and 35.8 min in minutes.

**Solution:** We must add an exact number (120) and an approximate number (35.8):

$$\begin{array}{rl} 120 & \text{min} \\ + \underline{\ 35.8} & \text{min} \\ 155.8 & \text{min} \end{array}$$

Since 120 is exact, we do *not* round our answer to the nearest 10 minutes, but retain as many decimal places as in the approximate number. Our answer is thus 155.8 min.

| Common Error | Be sure to recognize which numbers in a computation are exact; otherwise, you may perform drastic rounding by mistake. |
|---|---|

## EXERCISE 2—ADDITION AND SUBTRACTION

Combine as indicated.

1. −955, +212, −347
2. 8275, −2163, − 874
3. −748, −212, −156

Add each column of figures.

4. \$99.84, 24.96, 6.24, 1.56, 12.48, .98, 3.12
5. 96256, 6016, 376, 141, 188, 1504, 752
6. 98304, 6144, 384, 24576, 3072, 144, 49152

Combine as indicated.

7. 926 + 863
8. 274 + (−412)
9. −576 + (−553)
10. −207 + (−819)
11. −575 − 275
12. −771 − (−976)
13. 1123 − (−704)
14. 818 − (−207) + 318

Combine each set of approximate numbers as indicated. Round your answer.

15. 4857 + 73.8
16. 39.75 + 27.4
17. 296.44 + 296.997
18. 385.28 − 692.8
19. 0.000583 + 0.0008372 − 0.00173
20. Mt. Blanc is 15,572 ft high, and Pike's Peak is about 14,000 ft high. What is the difference in their heights?
21. California contains 158,933 square miles ($mi^2$) and Texas 237,321 $mi^2$. How much larger is Texas than California?
22. A man willed \$125,000 to his wife and two children. To one child he gave \$44,675, to the other \$26,380 and to his wife the remainder. What was his wife's share?
23. A circular pipe has an inside radius $r$ of 10.6 cm and a wall thickness of 2.125 cm. It is surrounded by insulation having a thickness of 4.8 cm (Fig. 1-3). What is the outside diameter $D$ of the insulation?

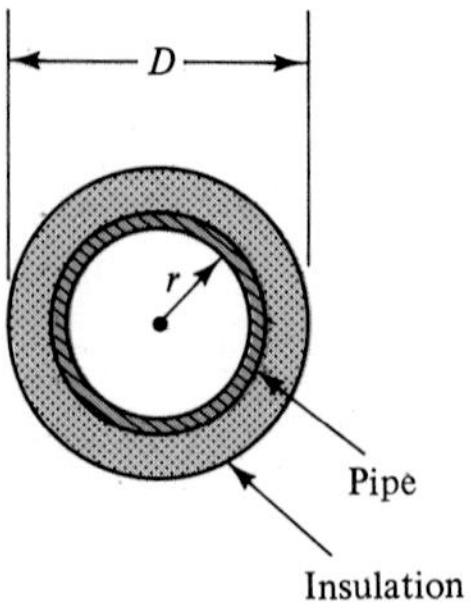


**FIGURE 1-3 An insulated pipe.**

24. A batch of concrete is made by mixing 267 kg of aggregate, 125 kg of sand, 75.5 kg of cement, and 25.25 kg of water. Find the total weight of the mixture.
25. Three resistors, having values of 27.3 ohms (Ω), 4.0155 Ω, and 9.75 Ω, are wired in series. What is the total resistance? (See Eq. A63, which says that the total series resistance is the sum of the individual resistances.)

**Formula numbers with the prefix "A" are applications. They are listed toward the end of Appendix A.**

## 1-3 MULTIPLICATION

### Factors and Product

The numbers we multiply to get a *product* are called *factors.*

$$3 \times 5 = 15$$

factors ↑ ↑ (3, 5)  ↑ product (15)

### Multiplication by Calculator

The keystrokes used to multiply the factors 23 and 57 are:

| *Press* | *Display* |
|---|---|
| 23 [×] | 23 |
| 57 [=] | 1311 |

### Commutative, Associative, and Distributive Laws

The *commutative law*

| Commutative Law for Multiplication | $ab = ba$ | **2** |
|---|---|---|

states that the *order* of multiplication is not important.

---

**EXAMPLE 18:** It is no surprise that

$$2 \times 3 = 3 \times 2$$

---

The *associative law*

| Associative Law for Multiplication | $a(bc) = (ab)c = (ac)b = abc$ | **4** |
|---|---|---|

allows us to group the numbers to be multiplied in any way.

---

**EXAMPLE 19:**

$$\begin{aligned} 2 \times 3 \times 4 &= 2(3 \times 4) = 2(12) = 24 \\ &= (2 \times 3)4 = (6)4 = 24 \\ &= (2 \times 4)3 = (8)3 = 24 \end{aligned}$$

---

The *distributive law*

| | | |
|---|---|---|
| Distributive Law | $a(b + c) = ab + ac$ | **5** |

shows how a factor may be *distributed* among several terms.

**EXAMPLE 20:**

$$2(3 + 4) = 2(7) = 14$$

But the distributive law enables us to do the same computation in a different way:

$$2(3 + 4) = 2(3) + 2(4) = 6 + 8$$
$$= 14 \quad \text{as before}$$

## Rules of Signs

The rules of signs for multiplication say that *the product of two quantities of like sign is positive:*

**As before, *a* and *b* are *positive* quantities.**

| | | |
|---|---|---|
| Rule of Signs for Multiplication | $(+a)(+b) = (-a)(-b) = +ab$ | **8** |

and that *the product of two quantities of unlike sign is negative:*

| | | |
|---|---|---|
| Rule of Signs for Multiplication | $(+a)(-b) = (-a)(+b) = -ab$ | **9** |

**EXAMPLE 21:**

(a) $2(-3) = -6$
(b) $(-2)3 = -6$
(c) $(-2)(-3) = 6$

When we multiply two negative numbers we get a positive product. So when multiplying a *string* of numbers, if an even number of them are negative, the answer will be positive, and if an odd number of them are negative, the answer will be negative.

**EXAMPLE 22:**

(a) $2(-3)(-1)(-2) = -12$
(b) $2(-3)(-1)(2) = 12$

## Multiplying Negative Numbers by Calculator

Negative numbers are entered into a calculator by first entering the positive value and then *changing its sign* by pressing a *change-sign key,* marked [+/−] or [CHS]. Check your manual to see how the change-sign key is marked on your calculator.

**EXAMPLE 23:** To multiply −96 and −83, we use the following keystrokes:

| *Press* | *Display* |
|---|---|
| 96 [+/−] [×] | −96 |
| 83 [+/−] | −83 |
| [=] | 7968 |

A simpler way to do the last problem would be to multiply +96 and +83 and determine the sign by inspection.

| Common Error | Do not try to use the [−] key to enter negative numbers into your calculator. The [−] key is only for subtraction. |
|---|---|

## Multiplication of Approximate Numbers

| Rule | When multiplying two or more approximate numbers, round the result to as many digits as in the factor having the fewest significant digits. |
|---|---|

**EXAMPLE 24:**

| 12.1 | × | 15.6 | = | 189 |
|---|---|---|---|---|
| ↑ | | ↑ | | ↑ |
| three digits | | three digits | | three digits |

When the factors have different numbers of significant digits, keep the same number of digits in your answer as is contained in the factor that has the *fewest* significant digits.

**EXAMPLE 25:**

| 123.56 | × | 2.21 | = | 273 |
|---|---|---|---|---|
| ↑ | | ↑ | | ↑ |
| five digits | | three digits | | keep three digits |

| | |
|---|---|
| Common Error | Do not confuse *significant digits* with *decimal places*. The number 274.56 has *five* significant digits and *two* decimal places.<br>Decimal places determine how we round after adding or subtracting. Significant figures determine how we round after multiplying, and, as we will soon see, after dividing, raising to a power, or taking roots. |

### Multiplication with Exact Numbers

When using *exact numbers* in a computation, treat them as if they had *more* significant figures than any of the approximate numbers in that computation.

---

**EXAMPLE 26:** If a certain car tire weighs 32.2 lb, how much will four such tires weigh?

**Solution:** Multiplying, we obtain

$$32.2(4) = 128.8 \text{ lb}$$

Since the 4 is an exact number, we retain as many significant figures as contained in 32.2, and round our answer to 129 lb.

---

## EXERCISE 3—MULTIPLICATION

Multiply each approximate number and retain the proper number of digits in your answer.

1. $3967 \times 1.84$
2. $4.900 \times 59.3$
3. $93.9 \times 0.0055908$
4. $4.97 \times 9.27 \times 5.78$
5. $69.0 \times (-258)$
6. $-385 \times (-2.2978)$
7. $2.86 \times (4.88 \times 2.97) \times 0.553$
8. $(5.93 \times 7.28) \times (8.26 \times 1.38)$

### Word Problems

9. What is the cost of 52.5 tons of cement at $63.25 a ton?
10. If 108 tons of rail is needed for 1 mi of track, how many tons will be required for 476 mi, and what will be its cost at $925 a ton?
11. Three barges carry 26 tons of gravel each and a fourth carries 35 tons. What is the value of the whole, at $12.75 per ton?
12. Two cars start from the same place and travel in opposite directions, one at the rate of 45 km/h, the other at 55 km/h. How far apart will they be at the end of 6 h?
13. What will be the cost of building a telephone line 274 km long, at $5723 per kilometer?
14. The current to a projection lamp is measured at 4.7 A when the line voltage is 115.45 V. Using Eq. A65, (power = voltage × current), find the power dissipated in the lamp.
15. A gear in a certain machine rotates at the speed of 1808 rev/min. How many revolutions will it make in 9.500 min?
16. How much will 1000 washers weigh if each weighs 2.375 g?
17. One inch equals exactly 2.54 cm. Convert 385.84 in. to centimeters.
18. If there are 360 degrees per revolution, how many degrees are there in 4.863 revolutions?

## 1-4 DIVISION

### Definitions

The *dividend,* when divided by the *divisor,* gives us the *quotient:*

$$\text{dividend} \div \text{divisor} = \text{quotient}$$

or

$$\frac{\text{dividend}}{\text{divisor}} = \text{quotient}$$

**A quantity *a/b* is also a *fraction,* and can also be referred to as the *ratio* of *a* to *b*. Fractions and ratios are treated in Chapter 8.**

### Division by Calculator

To divide 861 by 123 on an AOS calculator:

| *Press* | *Display* |
|---|---|
| 861 [÷] | 861 |
| 123 [=] | 7 |

When we multiplied two integers, we always got an integer for an answer. This is not always the case when dividing.

---

**EXAMPLE 27:** When we divide 2 by 3, we get 0.666666666. . . . We must choose how many digits we wish to retain, and round our answer. Rounding to, say, three digits, we obtain

$$2 \div 3 \cong 0.667$$

Here it is appropriate to use the ≅ symbol.

---

### Division of Approximate Numbers

The rule for rounding is almost the same as with multiplication:

| | |
|---|---|
| Rule | After dividing one approximate number by another, round the quotient to as many digits as there are in the original number having the fewest significant digits. |

---

**EXAMPLE 28:** Divide 846.2 by 4.75.

**Solution:** By calculator,

$$846.2 \div 4.75 = 178.1473684$$

Since 4.75 has three significant digits, we round our quotient to 178.

---

**EXAMPLE** Divide 846.2 into three equal parts.

**Solution:** We divide by the integer 3, and since we consider integers to be exact, we retain in our answer the same number of significant digits as in 846.2.

$$846.2 \div 3 = 282.1$$

---

### Dividing Negative Numbers

*The quotient is positive when dividend and divisor have the same sign;*

| Rule of Signs for Division | $\frac{+a}{+b} = \frac{-a}{-b} = \frac{a}{b}$ | **10** |
|---|---|---|

and *the quotient is negative when dividend and divisor have opposite signs.*

| Rule of Signs for Division | $\frac{+a}{-b} = \frac{-a}{+b} = -\frac{a}{b}$ | **11** |
|---|---|---|

---

**EXAMPLE 30:**

(a) $8 \div -4 = -2$

(b) $-8 \div 4 = -2$

(c) $-8 \div -4 = 2$

---

**EXAMPLE 31:** Divide 85.4 by −2.5386 on the calculator.

**Solution:** The keystrokes are:

| *Press* | *Display* |
|---|---|
| 85.4 [÷] | 85.4 |
| 2.5386 [+/−] [=] | −33.6405893 |

which we round to three digits, getting −33.6.

---

As with multiplication, the sign could also have been found by inspection.

### Zero

Zero divided by any quantity (except zero) is zero. But division *by* zero is not defined. It is an illegal operation in mathematics.

---

**EXAMPLE 32:** Using your calculator, divide 5 by zero.

**Solution:**

| *Press* | *Display* |
|---|---|
| 5 [÷] 0 [=] | flashing display or other indication of error |

---

### Reciprocals

The *reciprocal* of any number $n$ is $1/n$. Thus the product of a quantity and its reciprocal is equal to 1.

**EXAMPLE 33:**

(a) The reciprocal of 10 is 1/10.

(b) The reciprocal of 1/2 is 2.

### Reciprocals by Calculator

Simply enter the number and press the $\boxed{1/x}$ key. Keep as many digits in your answer as there are significant digits in the original number.

**EXAMPLE 34:**

(a) The reciprocal of 6.38 is 0.157.

(b) The reciprocal of −2.754 is −0.3631.

## EXERCISE 4—DIVISION

Divide, then round your answer to the proper number of digits.

1. 947 ÷ 5.82
2. 0.492 ÷ 0.00478
3. −99.4 ÷ 286.5
4. −4.8 ÷ −2.557
5. 5836 ÷ 8264
6. 5.284 ÷ 3.827
7. 94,840 ÷ 1.33876
8. 3.449 ÷ −6.837
9. 2,497,000 ÷ 150,000
10. 2.97 ÷ 4.828

### Word Problems Involving Division

11. A stretch of roadway 1858.54 m long is to be divided into five equal sections. Find the length of each section.
12. At the rate of 24.5 km in 8.25 h, how many kilometers would a person walk in 12.75 h? (Distance = rate × time.)
13. If 5 masons can build a wall in 8 days, how many masons are needed to build it in 4 days?
14. If 867 shares of railroad stock are valued at $84,099, what is the value of each share?

### Reciprocals

Find the reciprocal of each number, retaining the proper number of digits in your answer.

15. 693
16. 0.00630
17. −396,000
18. 39.74
19. −0.00573
20. 938.4
21. 4.992
22. −6.93
23. 11.1
24. −375
25. 1.007
26. 3.98

### Word Problems Involving Reciprocals

27. Using Eq. A64 find the equivalent resistance of a 473-Ω resistor and a 928-Ω resistor, connected in parallel.
28. When an object is placed 126 cm in front of a certain thin lens having a focal length $f$, the image will be formed 245 cm from the lens. The distances are related by

$$\frac{1}{f} = \frac{1}{126} + \frac{1}{245}$$

Find $f$.

29. The sine of an angle $\theta$ (written $\sin\theta$) is equal to the reciprocal of the cosecant of $\theta$ ($\csc\theta$). Find $\sin\theta$ if $\csc\theta = 3.58$.
30. If two straight lines are perpendicular, the slope of one line is the negative reciprocal of the slope of the other. If the slope of a line is −2.55, find the slope of a perpendicular to that line.

## 1-5 POWERS AND ROOTS

### Definitions

In the expression

$$2^4$$

the number 2 is called the *base* and the number 4 is called the *exponent*. The expression is read "two to the fourth power." Its value is

$$2^4 = 2 \cdot 2 \cdot 2 \cdot 2 = 16$$

### Powers by Calculator

To square a number on the calculator, simply enter the number and press the $\boxed{x^2}$ key. To raise a number to other powers, use the $\boxed{y^x}$ key, as in the following example. Round your result to the number of significant digits contained *in the base,* not the exponent.

---

**EXAMPLE 35:** Find $(3.85)^3$.

**Solution:**

| *Press* | *Display* |
|---|---|
| 3.85 $\boxed{y^x}$ | 3.85 |
| 3 $\boxed{=}$ | 57.066625 |

which we round to 57.1.

---

### Negative Base

A negative base raised to an *even* power gives a *positive* number. A negative base raised to an *odd* power gives a *negative* number.

---

**EXAMPLE 36:**

(a) $(-2)^2 = (-2)(-2) = 4$

(b) $(-2)^3 = (-2)(-2)(-2) = -8$

(c) $(-1)^{24} = 1$

(d) $(-1)^{25} = -1$

---

If you try to do these problems on your calculator, you will probably get an error indication. Some calculators will not work with a negative base, even though this is a valid operation.

Then how do you do it? Simply enter the base as *positive,* find the power, and determine the sign by inspection.

---

**EXAMPLE 37:** Find $(-1.45)^5$.

**Solution:** From the calculator,

$$(+1.45)^5 = 6.41$$

Since we know that a negative number raised to an odd power is negative, we write

$$(-1.45)^5 = -6.41$$

---

## Negative Exponent

A number can be raised to a negative exponent on the calculator with the $\boxed{y^x}$ key. Just remember to change the sign of the exponent before pressing the $\boxed{=}$ key.

---

**EXAMPLE 38:** Evaluate $(3.85)^{-3}$.

**Solution:** The key strokes are:

| *Press* | *Display* |
|---|---|
| 3.85 $\boxed{y^x}$ | 3.85 |
| 3 $\boxed{+/-}$ = | 0.0175233772 |

which we round to 0.0175.

---

## Fractional Exponents

We will see later that fractional exponents are another way of writing radicals. For now, we just evaluate fractional exponents on the calculator. For a calculator not having parentheses, first calculate the exponent and store it in the memory. Then enter the base, press $\boxed{y^x}$, and recall the exponent from memory.

---

**EXAMPLE 39:** Evaluate $8^{2/3}$.

**Solution:**

| *Press* | *Display* |
|---|---|
| 2 $\boxed{\div}$ 3 $\boxed{=}$ $\boxed{\text{STO}}$ | 0.6666666667 |
| 8 $\boxed{y^x}$ $\boxed{\text{RCL}}$ $\boxed{=}$ | 4 |

Some calculators use $x \rightarrow$ M instead of STO, and RM instead of RCL.

---

## Roots

If $a^n = b$, then

$$\sqrt[n]{b} = a$$

which is read "the $n$th root of $b$ equals $a$." The symbol $\sqrt{\ }$ is a *radical sign,* $b$ is the *radicand,* and $n$ is the *index* of the radical.

---

**EXAMPLE 40:**

(a) $\sqrt{4} = 2$ because $2^2 = 4$
(b) $\sqrt[3]{8} = 2$ because $2^3 = 8$
(c) $\sqrt[4]{81} = 3$ because $3^4 = 81$

---

## Principal Root

The *principal root* of a positive number is defined as the *positive* root. Thus $\sqrt{4} = +2$, not $\pm 2$.

The principal root is *negative* when we take an *odd* root of a *negative* number.

**When we speak about the root of a number we mean, unless otherwise stated, the *principal root.* The principal root of a number has the same sign as the number itself.**

**EXAMPLE 41:**

$$\sqrt[3]{-8} = -2$$

because $(-2)(-2)(-2) = -8$.

## Roots by Calculator

To find square roots, enter the number and then press the $\boxed{\sqrt{x}}$ key. To find other roots, use the $\boxed{\sqrt[x]{y}}$ key, as in the following example. (See the following section if you have no $\boxed{\sqrt[x]{y}}$ key.) Retain as many digits in your answer as there are significant digits in the original number.

**EXAMPLE 42:** Find $\sqrt[5]{28.4}$, the fifth root of 28.4.

**Solution:**

| *Press* | *Display* |
|---|---|
| 28.4 $\boxed{\sqrt[x]{y}}$ | 28.4 |
| 5 $\boxed{=}$ | 1.952826537 |

which we round to 1.95.

## Roots and Powers Related

If your calculator does not have a $\boxed{\sqrt[x]{y}}$ key, you can find roots by using the $\boxed{y^x}$ key. For this, you need to know the relationship between roots and exponents. It is

We will study this equation in detail later. For now, we just use it to find roots with the $\boxed{y^x}$ key.

$$\sqrt[n]{a} = a^{1/n} \qquad \textbf{36}$$

**EXAMPLE 43:** Find $\sqrt[4]{482}$ using the $\boxed{y^x}$ key.

**Solution:** By Eq. 36,

$$\sqrt[4]{482} = 482^{1/4}$$

On the calculator:

| *Press* | *Display* |
|---|---|
| 482 $\boxed{y^x}$ | 482 |
| 4 $\boxed{1/x}$ | 0.25 |
| $\boxed{=}$ | 4.685562762 |

which we round to 4.69.

## Odd Roots of Negative Numbers by Calculator

An *even* root of a negative number is *imaginary* (such as $\sqrt{-4}$). We will study these in Chapter 20. But an *odd* root of a negative number is *not* imaginary. It is a real, negative, number. As with powers, some calculators will not accept a negative radicand. Fortunately, we can outsmart our calculators and take odd roots of negative numbers anyway.

**EXAMPLE 44:** Find $\sqrt[5]{-875}$.

**Solution:** We know that an odd root of a negative number is real and negative. So we take the fifth root of +875, by calculator,

$$\sqrt[5]{+875} = 3.88 \quad \text{(rounded)}$$

and we only have to place a minus sign before the number.

$$\sqrt[5]{-875} = -3.88$$

## EXERCISE 5—POWERS AND ROOTS

### Powers

Evaluate each power without using a calculator.

| | | | |
|---|---|---|---|
| 1. $2^3$ | 2. $5^3$ | 3. $(-2)^3$ | 4. $9^2$ |
| 5. $1^3$ | 6. $(-1)^2$ | 7. $(-1)^{40}$ | 8. $3^2$ |
| 9. $1^8$ | 10. $(-1)^3$ | 11. $(-1)^{41}$ | 12. $(-3)^2$ |

Evaluate each power of 10.

***Powers of 10*** **will be needed for** ***scientific notation*** **later. Arrange these powers of 10 in order and try to invent a rule that will enable you to write the value of a power of 10 without doing the computation.**

| | | | |
|---|---|---|---|
| 13. $10^2$ | 14. $10^1$ | 15. $10^0$ | 16. $10^{-4}$ |
| 17. $10^3$ | 18. $10^5$ | 19. $10^{-2}$ | 20. $10^{-1}$ |
| 21. $10^4$ | 22. $10^{-3}$ | 23. $10^{-5}$ | |

Evaluate each expression, retaining the correct number of digits in your answer.

| | | | |
|---|---|---|---|
| 24. $(8.55)^3$ | 25. $(1.007)^5$ | 26. $(9.55)^3$ | 27. $(-4.82)^3$ |
| 28. $(-77.2)^2$ | 29. $(8.28)^{-2}$ | 30. $(0.0772)^{0.426}$ | 31. $(5.28)^{-2.15}$ |
| 32. $(35.2)^{1/2}$ | 33. $(462)^{2/3}$ | 34. $(88.2)^{-2}$ | 35. $(-37.3)^{-3}$ |

### Word Problems Involving Powers

36. The distance traveled by a falling body, starting from rest, is equal to $16t^2$, where $t$ is the elapsed time. In 5.448 s, the distance fallen is $16(5.448)^2$ ft. Evaluate this quantity. (Treat 16 here as an approximate number.)
37. The power dissipated in a resistance $R$ through which is flowing a current $I$ is equal to $I^2R$. Therefore, the power in a 365-Ω resistor carrying a current of 0.5855 A is $(0.5855)^2(365)$ W. Evaluate this power.
38. The volume of a cube of side 35.8 cm is $(35.8)^3$. Evaluate this volume.
39. The volume of a 59.4-cm-radius sphere is $\frac{4}{3}\pi(59.4)^3$ cm³. Find this volume.
40. An investment of $2000 at a compound interest rate of $6\frac{1}{4}\%$, left for $7\frac{1}{2}$ years, will be worth $2000(1.0625)^{7.5}$ dollars. Find this amount.

### Roots

Evaluate each radical without using your calculator.

| | | |
|---|---|---|
| 41. $\sqrt{25}$ | 42. $\sqrt[3]{27}$ | 43. $\sqrt{49}$ |
| 44. $\sqrt[3]{-27}$ | 45. $\sqrt[3]{-8}$ | 46. $\sqrt[5]{-32}$ |

Evaluate each radical by calculator, retaining the proper number of digits in your answer.

| | | |
|---|---|---|
| 47. $\sqrt{49.2}$ | 48. $\sqrt{1.863}$ | 49. $\sqrt[3]{88.3}$ |
| 50. $\sqrt{772}$ | 51. $\sqrt{3875}$ | 52. $\sqrt[3]{7295}$ |
| 53. $\sqrt[3]{-386}$ | 54. $\sqrt[5]{-18.4}$ | 55. $\sqrt[3]{-2.774}$ |

**56.** The period $T$ (time for one swing) of a simple pendulum 2.55 ft long is

$$T = 2\pi \sqrt{\frac{2.55}{32}} \qquad \text{seconds}$$

Evaluate $T$.

**57.** The magnitude $Z$ of the impedance in a circuit having a resistance of 3540 Ω and a reactance of 2750 Ω is

$$Z = \sqrt{(3540)^2 + (2750)^2} \qquad \text{ohms}$$

Find $Z$.

**58.** The geometric mean $B$ between 3.75 and 9.83 is

$$B = \sqrt{(3.75)(9.83)}$$

Evaluate $B$.

## 1-6 UNITS OF MEASUREMENT

### Units

A *unit* is a standard of measurement, such as the meter, inch, hour, or pound.

**SI stands for Le Système International d'Unités.**

The two main systems of units in use are the British system (feet, pounds, gallons, etc.) and the SI (metric) system (meters, kilograms, newtons, etc.). In addition, some special units (such as a *square* of roofing material) and some obsolete units (such as rods and chains) must occasionally be dealt with.

### Conversion Factors

Each physical quantity can be expressed in any one of a bewildering variety of units. Length, for example, can be measured in meters, inches, nautical miles, angstroms, feet, and so on.

We can *convert* from any unit of length to any other unit of length by multiplying by a suitable *conversion factor.*

---

**EXAMPLE 45:** Convert 15.4 in. to centimeters.

**Remember that the conversion factor 2.54 is an *exact number.* Thus the number of digits to be retained in the answer is determined by the other numbers in the computation.**

**Solution:** From Appendix B we find the relation between inches and centimeters:

$$2.54 \text{ cm} = 1 \text{ in.}$$

Dividing both sides by 1 in., we get the *conversion factor:*

$$\frac{2.54 \text{ cm}}{1 \text{ in.}} = 1$$

Multiplying yields

$$15.4 \text{ in.} = 15.4 \cancel{\text{in.}} \times \frac{2.54 \text{ cm}}{\cancel{\text{in.}}} = 39.1 \text{ cm} \quad \text{(rounded)}$$

---

Suppose, in Example 45, that we had divided both sides by 2.54 cm instead of by 1 in. We would have had another conversion factor:

$$\frac{1\text{ in.}}{2.54\text{ cm}} = 1$$

Thus each relation between two units of measurement gives us *two* conversion factors. Since each of these is equal to 1, we may multiply any quantity by them *without changing the value* of that quantity. We will, however, cause the units to change. But which of the two conversion factors should we use? It is simple. *Multiply by the conversion factor that will cancel the units you wish to eliminate.*

---

**EXAMPLE 46:** Convert 145 acres to square meters.

**Solution:** From Appendix B we find the equation

$$1\text{ acre} = 4047\text{ square meters}$$

We must write our conversion factor so that the unwanted units (acres) are in the denominator, so that they will cancel. So our conversion factor is

$$\frac{4047\text{ m}^2}{1\text{ acre}}$$

Multiplying, we obtain

$$\begin{aligned} 145\text{ acres} &= 145\,\cancel{\text{acres}} \times \frac{4047\text{ m}^2}{1\,\cancel{\text{acre}}} \\ &= 586{,}815\text{ m}^2 \end{aligned}$$

which we round to three significant digits, getting 587,000 m$^2$.

---

Sometimes you may not be able to find a *single* conversion factor linking the units you want to convert. You may have to use *more than one*.

---

**EXAMPLE 47:** Convert 8825 yd to nautical miles.

**Solution:** In Appendix B we find no conversion between nautical miles and yards, but we see that

$$1\text{ nautical mile} = 6076\text{ ft}$$

and also that

$$3\text{ ft} = 1\text{ yd}$$

So

$$\begin{aligned} 8825\text{ yd} &= 8825\,\cancel{\text{yd}} \times \frac{3\,\cancel{\text{ft}}}{1\,\cancel{\text{yd}}} \times \frac{1\text{ nau mi}}{6076\,\cancel{\text{ft}}} \\ &= 4.357\text{ nautical miles} \end{aligned}$$

---

### Converting Areas and Volumes

Length may be given in, say, centimeters (cm), but an *area* may be given in square centimeters (cm$^2$). Similarly, a *volume* may be in cubic centimeters (cm$^3$). So to get a conversion factor for area or volume, if not found in Appendix B, simply square or cube the conversion factor for length.

If we take the equation

$$2.54 \text{ cm} = 1 \text{ in.}$$

and square both sides, we get

$$(2.54 \text{ cm})^2 = (1 \text{ in.})^2$$

or

$$6.4516 \text{ cm}^2 = 1 \text{ in.}^2$$

the conversion between square centimeters and square inches.

---

**EXAMPLE 48:** Convert 864 yd² to acres.

**Solution:** Appendix B has no conversion for square yards. However,

$$1 \text{ yd} = 3 \text{ ft}$$

Squaring yields

$$1 \text{ yd}^2 = (3 \text{ ft})^2 = 9 \text{ ft}^2$$

Also from the table,

$$1 \text{ acre} = 43{,}560 \text{ ft}^2$$

So

$$864 \text{ yd}^2 = 864 \cancel{\text{yd}^2} \times \frac{9 \cancel{\text{ft}^2}}{1 \cancel{\text{yd}^2}} \times \frac{1 \text{ acre}}{43{,}560 \cancel{\text{ft}^2}} = 0.179 \text{ acre}$$

---

## Converting Rates to Other Units

A *rate* is the amount of one quantity expressed *per unit of some other quantity*. Some rates, with typical units, are:

| | |
|---|---|
| rate of travel (mi/h) or (km/h) | flow rate (gal/min) or (m³/s) |
| application rate (lb/acre) | unit price (dollars/lb) |

Each rate contains *two* units of measure; miles per hour, for example, has *miles* in the numerator and *hours* in the denominator. It may be necessary to convert *either or both* of those units to other units. Sometimes a single conversion factor can be found (such as 1 m/h = 1.466 ft/s), but more often you will have to convert each unit with a *separate* conversion factor.

**Be sure to write the original quantity as a built-up fraction, $\frac{a}{b}$, rather than on a single line, *a/b*. This will greatly reduce your chances of making an error.**

---

**EXAMPLE 49:** A certain chemical is to be added to a pool at the rate of 3.74 oz per gallon of water. Convert this to pounds of chemical per cubic foot of water.

**Solution:** We write the original quantity as a fraction, and multiply by the appropriate factors, themselves written as fractions.

$$3.74 \text{ oz/gal} = \frac{3.74 \text{ oz}}{\text{gal}} \times \frac{1 \text{ lb}}{16 \text{ oz}} \times \frac{7.481 \text{ gal}}{\text{ft}^3} = 1.75 \text{ lb/ft}^3$$

---

## EXERCISE 6—UNITS OF MEASUREMENT

Convert.

1. 185 ft to meters
2. 9.37 mi to kilometers
3. 993 cm to inches
4. 4935 yd to meters
5. 79.2 acres to square meters
6. 148 acres to ares
7. 88.3 $cm^2$ to square inches
8. 525 $yd^3$ to cubic meters
9. 994 kg to slugs
10. 1.05 $ft^2$ to square centimeters
11. 77.3 N to ounces
12. 96,500 lb to newtons
13. 5528 N to pounds
14. 2.5 hp to kilowatts

**Here we make a distinction between *mass* (kilograms, slugs) and *weight* (newtons, pounds). Weight has the same units as force.**

Convert units of the following time rates.

15. 63 ft/s to miles per hour
16. 555 gal/min to cubic meters per hour
17. 55.4 mi/h to kilometers per hour

# 1-7 SUBSTITUTING INTO EQUATIONS AND FORMULAS

## Substituting into Equations

We get an *equation* when two expressions are set equal to each other.

---

**EXAMPLE 50:** $x = 5a - 2b + 3c$ is an equation that enables us to find $x$ if we know $a$, $b$, and $c$.

---

We will study equations in detail later, but for now we will simply substitute into equations and use our calculators to compute the result. To *substitute into an equation* means to replace the letter quantities in an equation by their given numerical values, and to perform the computation indicated.

---

**EXAMPLE 51:** Substitute the values $a = 5$, $b = 3$, and $c = 6$ into the equation

$$x = \frac{3a + b}{c}$$

**Solution:** Substituting, we obtain

$$x = \frac{3(5) + 3}{6} = \frac{18}{6} = 3$$

---

When substituting approximate numbers, be sure to round your answer to the proper number of digits. Treat any integers in the equation as exact numbers.

## Substituting into Formulas

A *formula* is an equation expressing some general mathematical or physical fact, such as the formula for the area of a circle of radius $r$:

| | | |
|---|---|---|
| Area of a Circle | $A = \pi r^2$ | **114** |

We substitute into formulas just as we substituted into equations; except that we now carry *units* along with the numerical values. You will often need conversion factors to make the units cancel properly, so that the answer will be in the desired units.

---

**EXAMPLE 52:** A tensile load of 4500 lb is applied to a bar that is 5.2 yd long and has a cross-sectional area of 11.6 $cm^2$. The elongation is 0.38 mm. Using Eq. A54, find the modulus of elasticity $E$ in pounds per square inch.

**Solution:** Substituting the values *with units* into Eq. A54, we obtain

$$E = \frac{PL}{ae} = \frac{4500 \text{ lb} \times 5.2 \text{ yd}}{11.6 \text{ cm}^2 \times 0.38 \text{ mm}}$$

**If the units to be used in a certain formula are specified, convert all quantities to those specified units before substituting into the formula.**

Notice that we have a length (5.2 yd) in the numerator and a length (0.38 mm) in the denominator. To make these units cancel, we use the conversion factors

$$25.4 \text{ mm} = 1 \text{ in.}$$

and

$$36 \text{ in.} = 1 \text{ yd}$$

Also, our answer is to have square inches in the denominator, not square centimeters. So we use another conversion factor,

$$6.452 \text{ cm}^2 = 1 \text{ in}^2$$

$$E = \frac{4500 \text{ lb} \times 5.2 \text{ yd}}{11.6 \text{ cm}^2 \times 0.38 \text{ mm}} \times \frac{25.4 \text{ mm}}{\text{in.}} \times \frac{36 \text{ in.}}{\text{yd}} \times \frac{6.452 \text{ cm}^2}{\text{in}^2}$$

$$= 31{,}000{,}000 \text{ lb/in}^2 \quad \text{(rounded to two digits)}$$

---

| Common Error | Students often neglect to include *units* when substituting into a formula, with the result that the units often do not cancel properly. |
|---|---|

## EXERCISE 7—SUBSTITUTING INTO EQUATIONS AND FORMULAS

Substitute the given integers into each equation. Do not round your answer.

1. $y = 5x + 2$ $\quad (x = 3)$
2. $y = 2m^2 - 3m + 5$ $\quad (m = -2)$
3. $y = 2a - 3x^2$ $\quad (x = 3, a = -5)$
4. $y = 3x^3 - 2x^2 + 4x - 7$ $\quad (x = 2)$
5. $y = 2b + 3w^2 - 5z^3$ $\quad (b = 3, w = -4, z = 2)$
6. $y = \dfrac{r^2}{x} - \dfrac{x^3}{r} + \dfrac{w}{x^2}$ $\quad (x = 5, w = 3, r = -4)$

Substitute the given approximate numbers into each equation. Round your answer to the proper number of digits.

7. $y = 7x - 5$ $\quad (x = 2.73)$
8. $y = 2w^2 - 3x^2$ $\quad (x = -11.5, w = 9.83)$
9. $y = 8 - x + 3x^2$ $\quad (x = -8.49)$

10. $y = \sqrt[3]{8x + 7w}$ $(x = 1.255, w = 2.304)$
11. $y = \sqrt{x^3 - 3x}$ $(x = 4.25)$
12. $y = (w - 2x)^{1.6}$ $(x = 1.8, w = 7.2)$
13. Use Eq. A9 to find the amount to which \$3000 will accumulate in 5 years at a simple interest rate of 6.5%.
14. Using Eq. A18 find the displacement after 1.3 min of a body thrown downward with a speed of 12 ft/s.
15. Using Eq. A50 convert 128°F to degrees Celsius.
16. A bar 15.2 m long having a cross-sectional area of 12.7 cm² is subject to a tensile load of 22,500 N (Fig. 1-4). The elongation is 2.75 mm. Use Eq. A54 to find the modulus of elasticity in newtons per square centimeter.
17. Use Eq. A10 to find the amount $y$ obtained when \$9570 is allowed to accumulate for 5 years at a compound interest rate of $6\frac{3}{4}$%.
18. The resistance of a copper coil is 775 Ω at 20°C. The temperature coefficient of resistance is 0.00393 at 20°C. Use Eq. A70 to find the resistance at 80°C.

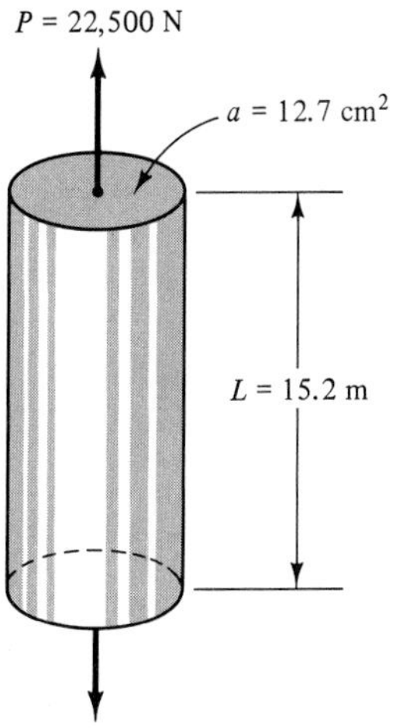


**FIGURE 1-4 A bar in tension.**

## 1-8 SCIENTIFIC NOTATION

### Definitions

Let us multiply two large numbers on the calculator: say, 500,000 and 300,000. On an algebraic calculator:

| *Press* | *Display* |
|---|---|
| 500000 [×] | 500000 |
| 300000 [=] | 1.5 11 |

We get the strange-looking display, [1.5 11]. What has happened?

Our answer (150,000,000,000) is too large to fit the calculator display, so the machine has automatically switched to *scientific notation*. Our calculator display actually contains *two* numbers: a decimal number (1.5) and an integer (11). Our answer is equal to the decimal number, multiplied by 10 raised to the value of the integer.

Calculator display: [1.5 11]

Scientific notation: $1.5 \times 10^{11}$

decimal part → 1.5; power of 10 → $10^{11}$

Ten raised to a power (such as $10^{11}$) is called a *power of 10.*

A number is said to be in *scientific notation* when it is written as a number whose absolute value is between 1 and 10, multiplied by a power of 10.

**EXAMPLE 53:**

(a) $2.74 \times 10^3$ (b) $8.84 \times 10^9$

(c) $5.4 \times 10^{-6}$ (d) $-1.2 \times 10^{-5}$

are numbers written in scientific notation.

## Evaluating Powers of 10

We did some work with powers in Sec. 1-5. We saw, for example, that $2^3$ meant

$$2^3 = 2 \cdot 2 \cdot 2 = 8$$

Here, the power 3 tells how many 2's are to be multiplied to give the product. For powers of 10, the power tells how many 10's are to be multiplied to give the product.

**EXAMPLE 54:**

(a) $10^2 = 10 \times 10 = 100$

(b) $10^3 = 10 \times 10 \times 10 = 1000$

Negative powers, as before, are calculated by means of Eq. 35, $x^{-a} = 1/x^a$.

**EXAMPLE 55:**

(a) $10^{-2} = \dfrac{1}{10^2} = \dfrac{1}{100} = 0.01$

(b) $10^{-5} = \dfrac{1}{10^5} = \dfrac{1}{100,000} = 0.00001$

Summarizing some powers of 10 in a table:

| *Positive Powers* | *Negative Powers* |
|---|---|
| $1,000,000 = 10^6$ | $0.1 = 10^{-1}$ |
| $100,000 = 10^5$ | $0.01 = 10^{-2}$ |
| $10,000 = 10^4$ | $0.001 = 10^{-3}$ |
| $1,000 = 10^3$ | $0.0001 = 10^{-4}$ |
| $100 = 10^2$ | $0.00001 = 10^{-5}$ |
| $10 = 10^1$ | $0.000001 = 10^{-6}$ |
| $1 = 10^0$ | |

## Converting Numbers to Scientific Notation

First rewrite the given number with a single digit to the left of the decimal point, discarding any nonsignificant zeros. Then multiply this number by the power of 10 that will make it equal to the original number.

**EXAMPLE 56:**

$$\begin{aligned} 346 &= 3.46 \times 100 \\ &= 3.46 \times 10^2 \end{aligned}$$

**EXAMPLE 57:**

$$\begin{aligned} 2700 &= 2.7 \times 1000 \\ &= 2.7 \times 10^3 \end{aligned}$$

Note that we have discarded the two nonsignificant zeros.

When converting a number whose absolute value is less than 1, our power of 10 will be negative, as in Example 58.

**EXAMPLE 58:**

$$0.00000950 = 9.50 \times 0.000001$$
$$= 9.50 \times 10^{-6}$$

The sign of the exponent has nothing to do with the sign of the original number. You can convert a negative number to scientific notation just as you would a positive number, and put the minus sign on afterward.

**EXAMPLE 59:** Convert $-34{,}720$ to scientific notation.

**Solution:** Converting $+34{,}720$ to scientific notation, we obtain

$$34{,}720 = 3.472 \times 10{,}000$$
$$34{,}720 = 3.472 \times 10^4$$

Then multiplying by $-1$ gives

$$-34{,}720 = -3.472 \times 10^4$$

To convert from scientific notation, simply reverse the process.

**EXAMPLE 60:**

$$4.82 \times 10^5 = 4.82 \times 100{,}000$$
$$= 482{,}000$$

**EXAMPLE 61:**

$$8.25 \times 10^{-3} = 8.25 \times 0.001$$
$$= 0.00825$$

### Addition and Subtraction

If two or more numbers to be added or subtracted have the *same power of 10,* simply combine the numbers and keep the same power of 10.

**EXAMPLE 62:**

(a) $2 \times 10^5 + 3 \times 10^5 = 5 \times 10^5$

(b) $8 \times 10^3 - 5 \times 10^3 + 3 \times 10^3 = 6 \times 10^3$

This is another application of the distributive law, $ab + ac = a(b + c)$. Thus

$$2 \times 10^5 + 3 \times 10^5$$
$$= (2 + 3)10^5$$
$$= 5 \times 10^5$$

If the powers of 10 are different, *they must be made equal* before the numbers can be combined. A shift of the decimal point of one place to the *left* will *increase* the exponent by 1. Conversely, a shift of the decimal point one place to the *right* will *decrease* the exponent by 1.

**EXAMPLE 63:**

(a) $1.5 \times 10^4 + 3 \times 10^3 = 1.5 \times 10^4 + 0.3 \times 10^4$

$$= 1.8 \times 10^4$$

(b) $1.25 \times 10^5 - 2 \times 10^4 + 4 \times 10^3$

$$= 1.25 \times 10^5 - 0.2 \times 10^5 + 0.04 \times 10^5$$
$$= 1.09 \times 10^5$$

## Multiplication

We multiply powers of 10 by *adding their exponents.*

**We are really using Eq. 29, $x^a \cdot x^b = x^{a+b}$. This equation is one of the *laws of exponents* that we will study later.**

**EXAMPLE 64:**

(a) $10^3 \cdot 10^4 = 10^{3+4} = 10^7$

(b) $10^{-2} \cdot 10^5 = 10^{-2+5} = 10^3$

To multiply two numbers in scientific notation, multiply the decimal parts and the powers of 10 *separately.*

**EXAMPLE 65:**

$$\begin{aligned}(2 \times 10^5)(3 \times 10^2) &= (2 \times 3)(10^5 \times 10^2)\\ &= 6 \times 10^{5+2} = 6 \times 10^7\end{aligned}$$

## Division

We divide powers of 10 by subtracting the exponent of the denominator from the exponent of the numerator.

**EXAMPLE 66:**

(a) $\dfrac{10^5}{10^3} = 10^{5-3} = 10^2$

(b) $\dfrac{10^{-4}}{10^{-2}} = 10^{-4-(-2)} = 10^{-2}$

As with multiplication, we divide the decimal parts and the powers of 10 separately.

**EXAMPLE 67:**

(a) $\dfrac{8 \times 10^5}{4 \times 10^2} = \dfrac{8}{4} \times \dfrac{10^5}{10^2} = 2 \times 10^{5-2} = 2 \times 10^3$

(b) $\dfrac{12 \times 10^3}{4 \times 10^5} = 3 \times 10^{3-5} = 3 \times 10^{-2}$

## Scientific Notation on Computer or Calculator

A computer or calculator will switch to scientific notation by itself when the printout or display gets too long. For example, a computer might print the number

```
2.75E-6
```

which we interpret as

$$2.75 \times 10^{-6} \quad \text{or} \quad 0.00000275$$

We may also *enter* numbers in scientific notation. In BASIC, simply type the number followed by E, followed by the exponent to which 10 is to be raised.

**EXAMPLE 68:** To enter the number 5,824,000,000, which in scientific notation is $5.824 \times 10^9$, we type

```
5.824E9
```

On a *calculator* we use the *enter exponent* key, usually marked EE or EEX.

**EXAMPLE 69:** Enter the number $3.5 \times 10^4$.

**Solution:** The keystrokes are:

| *Press* | *Display* |
|---|---|
| 3.5 [EE] | 3.5 00 |
| 4 | 3.5 04 |

**EXAMPLE 70:** Enter the number $9.5 \times 10^{-7}$.

**Solution:**

| *Press* | *Display* |
|---|---|
| 9.5 [EE] | 9.5 00 |
| 7 [+/−] | 9.5 −07 |

Simply follow the same calculator procedures as you would for decimal numbers and let the calculator worry about the powers of 10.

**EXAMPLE 71:** Multiply $5.38 \times 10^3$ and $3.973 \times 10^{-2}$.

**Solution:**

| *Press* | *Display* |
|---|---|
| 5.38 [EE] | 5.38 00 |
| 3 [×] | 5.38 03 |
| 3.973 [EE] | 3.973 00 |
| 2 [+/−] | 3.973 − 02 |
| [=] | 2.137474 02 |

We now round 2.137474 to three digits, getting $2.14 \times 10^2$, or 214.

| | |
|---|---|
| Common Error | Students often enter powers of ten (such as $10^4$) incorrectly into their calculators, forgetting that $10^4$ is really $1 \times 10^4$.<br>Thus to enter $10^4$, we press<br>1 [EE] 4<br>*and not*<br>10 [EE] 4 |

## Metric Prefixes

**A *prefix* is a group of letters placed at the beginning of a word to modify the meaning of that word.**

Multiples and submultiples of metric units are sometimes indicated by a *prefix* rather than by scientific notation. For example, the prefix *kilo* means 1000, or $10^3$. Thus a *kilo*meter is 1000 meters.

---

**EXAMPLE 72:** Express 15,485 m in kilometers.

**Solution:**

$$15{,}485 \text{ m} = 15.485 \times 10^3 \text{ m}$$

$$= 15.485 \text{ km}$$

---

Other frequently used metric prefixes are

| | | | |
|---|---|---|---|
| mega | $10^6$ | milli | $10^{-3}$ |
| centi | $10^{-2}$ | micro | $10^{-6}$ |

---

**EXAMPLE 73:** Convert 1875 microseconds ($\mu$s) to milliseconds (ms).

**Solution:**

$$1875\ \mu\text{s} = 1875 \times 10^{-6} \text{ s}$$

$$= 1.875 \times 10^{-3} \text{ s}$$

$$= 1.875 \text{ ms}$$

---

## EXERCISE 8—SCIENTIFIC NOTATION

### Powers of 10

Write each number as a power of 10.

1. 100
2. 1,000,000
3. 0.0001
4. 0.001
5. 100,000,000

Write each power of 10 as a decimal number.

6. $10^5$
7. $10^{-2}$
8. $10^{-5}$
9. $10^{-1}$
10. $10^4$

### Scientific Notation

Write each number in scientific notation.

11. 186,000
12. 0.0035
13. 25,742
14. 8000
15. $98.3 \times 10^3$
16. $0.0775 \times 10^{-2}$

Convert each number from scientific notation to decimal notation.

17. $2.85 \times 10^3$
18. $1.75 \times 10^{-5}$
19. $9 \times 10^4$
20. $9.00 \times 10^4$
21. $3.667 \times 10^{-3}$

### Multiplication and Division

Multiply the following powers of 10.

22. $10^5 \cdot 10^2$
23. $10^4 \cdot 10^{-3}$
24. $10^{-5} \cdot 10^{-4}$
25. $10^{-2} \cdot 10^5$
26. $10^{-1} \cdot 10^{-4}$

Divide the following powers of 10.

**27.** $10^8 \div 10^5$ **28.** $10^4 \div 10^6$ **29.** $10^5 \div 10^{-2}$

**30.** $10^{-3} \div 10^5$ **31.** $10^{-2} \div 10^{-4}$

Multiply without using a calculator.

**32.** $(3.0 \times 10^3)(5.0 \times 10^2)$ **33.** $(5 \times 10^4)(8 \times 10^{-3})$

**34.** $(2 \times 10^{-2})(4 \times 10^{-5})$ **35.** $(7.0 \times 10^4)(3\bar{0},000)$

Divide without using a calculator.

**36.** $(8 \times 10^4) \div (2 \times 10^2)$ **37.** $(6 \times 10^4) \div 0.03$

**38.** $(3 \times 10^3) \div (6 \times 10^5)$ **39.** $(8 \times 10^{-4}) \div 400{,}000$

**40.** $(9 \times 10^4) \div (3 \times 10^{-2})$ **41.** $49{,}000 \div (7.0 \times 10^{-2})$

### Addition and Subtraction

Combine without using a calculator.

**42.** $(3.0 \times 10^4) + (2.1 \times 10^5)$

**43.** $(75.0 \times 10^2) + 32\bar{0}0$

**44.** $(1.557 \times 10^2) + (9.000 \times 10^{-1})$

**45.** $0.037 - (6.0 \times 10^{-3})$

**46.** $(7.2 \times 10^4) + (1.1 \times 10^4)$

### Scientific Notation on the Calculator

Perform the following computations in scientific notation. Combine the powers of 10 by hand or with your calculator.

**47.** $(1.58 \times 10^2)(9.82 \times 10^3)$

**48.** $(9.83 \times 10^5) \div (2.77 \times 10^3)$

**49.** $(3.87 \times 10^{-2})(5.44 \times 10^5)$

**50.** $(2.74 \times 10^3) \div (9.13 \times 10^5)$

**51.** $(5.6 \times 10^2)(3.1 \times 10^{-1})$

**52.** $(7.72 \times 10^8) \div (3.75 \times 10^{-9})$

### Applications

**53.** Three resistors, having resistances of $4.98 \times 10^5\ \Omega$, $2.47 \times 10^4\ \Omega$, and $9.27 \times 10^6\ \Omega$, are wired in series. Find the total resistance, using Eq. A63.

**54.** Find the equivalent resistance if the three resistors of Problem 53 are wired in parallel. Use Eq. A64.

**55.** Find the power dissipated in a resistor if a current of $3.75 \times 10^{-3}$ A produces a voltage drop of $7.24 \times 10^{-4}$ V across the resistor. Use Eq. A65.

**56.** The voltage across a $8.35 \times 10^5$-$\Omega$ resistor is $2.95 \times 10^{-3}$ V. Find the power dissipated in the resistor, using Eq. A66.

**57.** Three capacitors, $8.26 \times 10^{-6}$ farad (F), $1.38 \times 10^{-7}$ F, and $5.93 \times 10^{-5}$ F, are wired in parallel. Find the equivalent capacitance by Eq. A73.

**58.** A wire $4.75 \times 10^3$ cm long when loaded is seen to stretch $9.55 \times 10^{-2}$ cm. Find the strain in the wire, using Eq. A53.

**59.** How long will it take a rocket traveling at a rate of $3.2 \times 10^6$ m/h to travel the $3.8 \times 10^8$ m from the earth to the moon? Use Eq. A17.

**60.** The oil shale reserves of the United States are estimated at $2.0 \times 10^9$ tons. How long would this last at a rate of consumption of $9.0 \times 10^7$ tons/yr?

Convert.

In these problems write your answer in scientific notation if the numerical value is greater than 1000 or less than 0.1. Be careful to retain the proper number of significant digits.

61. 174,000 m to kilometers
62. 0.00537 V to millivolts
63. 39,400 g to kilograms
64. $4.83 \times 10^{-5}$ kW to watts
65. $1.45 \times 10^{9}$ Ω to megohms
66. $396 \times 10^{4}$ N to kilonewtons
67. 2584 picofarads to microfarads
68. 35,960 nanoseconds to milliseconds

## 1-9 PERCENTAGE

### Definition of Percent

The word *percent* means *by the hundred,* or *per hundred.* A percent thus gives the number of parts in every hundred.

---

**EXAMPLE 74:** If we say that a certain concrete mix is 12% cement by weight, we mean that 12 lb out of every 100 lb of mix will be cement.

---

### Rates

The word *rate* is often used to indicate a percent, or percentage rate, as in "rejection rate," "rate of inflation," or "growth rate."

---

**EXAMPLE 75:** A failure rate of 2% means that 2 parts out of every 100 would be expected to fail.

---

### Percent as a Fraction

Percent is another way of expressing a *fraction* having 100 as the denominator.

---

**EXAMPLE 76:** If we say that a builder has finished 75% of a house, we mean that he has finished $\frac{75}{100}$ (or $\frac{3}{4}$) of the house.

---

### Converting Decimals to Percent

Before working some percentage problems, let us first get some practice in converting decimals and fractions to percents, and vice versa. To convert decimals to percent, simply move the decimal point two places to the right and affix the percent symbol (%).

---

**EXAMPLE 77:**

(a) $0.75 = \dfrac{75}{100} = 75\%$

(b) $3.65 = \dfrac{365}{100} = 365\%$

(c) $0.003 = \dfrac{0.3}{100} = 0.3\%$

---

### Converting Fractions or Mixed Numbers to Percent

First write the fraction or mixed number as a decimal, and then proceed as above.

---

**EXAMPLE 78:**

(a) $\frac{1}{4} = 0.25 = \frac{25}{100} = 25\%$

(b) $\frac{5}{2} = 2.5 = \frac{250}{100} = 250\%$

(c) $1\frac{1}{4} = 1.25 = \frac{125}{100} = 125\%$

---

### Converting Percent to Decimals

Move the decimal point two places to the left and remove the percent sign.

---

**EXAMPLE 79:**

(a) $13\% = \frac{13}{100} = 0.13$

(b) $4.5\% = \frac{4.5}{100} = 0.045$

(c) $155\% = \frac{155}{100} = 1.55$

(d) $27\frac{3}{4}\% = \frac{27.75}{100} = 0.2775$

---

### Converting Percent to a Fraction

Write a fraction with 100 in the denominator and the percent in the numerator. Remove the percent sign and reduce the fraction to lowest terms.

---

**EXAMPLE 80:**

(a) $75\% = \frac{75}{100} = \frac{3}{4}$

(b) $87.5\% = \frac{87.5}{100} = \frac{875}{1000} = \frac{7}{8}$

(c) $125\% = \frac{125}{100} = \frac{5}{4} = 1\frac{1}{4}$

---

### Amount, Base, and Rate

Percentage problems always involve three quantities:

1. The *percent rate, P*
2. The *base, B:* the quantity we are taking the percent of
3. The *amount A* we get when we take the percent of the base

In a percentage problem, you will know two of these three quantities (amount, base, or rate) and be required to find the third. This is easily done, for the rate, base, and amount are related by the equation

| | | |
|---|---|---|
| Percentage | amount = base × rate<br>$A = BP$<br>where $P$ is expressed as a decimal. | **12** |

### Finding the Amount When the Base and Rate Are Known

We substitute the given base and rate into Eq. 12 and solve for the amount:

---

**EXAMPLE 81:** What is 35 percent of 80?

**Solution:** In this problem the rate is 35%, so

$$P = 0.35$$

But is 80 the amount or the base?

| | |
|---|---|
| Tip | In a particular problem, if you have trouble telling which number is the base and which is the amount, look for the key phrase "*percent of*." The quantity following this phrase is *always the base*. |

We look for the key phrase "percent of."

$$\text{What is 35 } \underrightarrow{\text{percent of}}\ \overset{\text{base}}{\boxed{80}}\text{?}$$

Since 80 immediately follows *percent of*,

$$B = 80$$

From Eq. 12,

$$A = BP = 80(0.35) = 28$$

---

| | |
|---|---|
| Common Error | Do not forget to convert the percent rate to a *decimal* when using Eq. 12. |

---

**EXAMPLE 82:** Find 0.1% of 5600.

**Solution:** We substitute into Eq. 12 with

$$B = 5600$$

and

$$P = 0.001 \quad (\textit{not } 0.1!)$$

So

$$A = BP = 5600(0.001) = 5.6$$

---

## Finding Amounts by Calculator

We make use of the $\boxed{\%}$ key, as in the following examples.

---

**EXAMPLE 83:** Find 18% of 248 by calculator.

**Solution:** The keystrokes are:

| *Press* | *Display* |
|---|---|
| 248 $\boxed{\times}$ | 248 |
| 18 $\boxed{\%}$ $\boxed{=}$ | 44.64 |

---

## Finding the Base When a Percent of It Is Known

We see from Eq. 12 that the base equals the amount divided by the rate (expressed as a decimal), or $B = A/P$.

---

**EXAMPLE 84:** 12% of what number is 72?

**Solution:** Finding the key phrase,

$$12\ \underrightarrow{\text{percent of}}\ \underbrace{\boxed{\text{what number}}}_{\text{base}}\ \text{is } 72?$$

it is clear that we are looking for the base. So

$$A = 72 \qquad \text{and} \qquad P = 0.12$$

By Eq. 12,

$$B = \frac{A}{P} = \frac{72}{0.12} = 600$$

---

**EXAMPLE 85:** 125 is 25% of what number?

**Solution:** From Eq. 12,

$$B = \frac{A}{P} = \frac{125}{0.25} = 500$$

---

## Finding the Percent One Number Is of Another Number

From Eq. 12, the rate equals the amount divided by the base, or $P = A/B$.

---

**EXAMPLE 86:** 42 is what percent of 400?

**Solution:** By Eq. 12, with $A = 42$ and $B = 400$,

$$P = \frac{A}{B} = \frac{42}{400} = 0.105 = 10.5\%$$

---

**EXAMPLE 87:** What percent of 1.4 is 0.35?

**Solution:** From Eq. 12,

$$P = \frac{A}{B} = \frac{0.35}{1.4} = 0.25 = 25\%$$

---

## Percent Change

Percentages are often used to compare two quantities. You often hear statements like the following:

The price of steel rose 3% over last year's price.

The weights of two cars differed by 20%.

Production dropped 5% from last year.

When the two numbers being compared involve a *change* from one to the other, the *original value* is usually taken as the base.

$$\text{percent change} = \frac{\text{new value} - \text{original value}}{\text{original value}} \times 100\% \qquad \textbf{13}$$

**Be sure to show the *direction* of change with a plus or minus sign, or with words like "increase" or "decrease."**

**EXAMPLE 88:** A certain price rose from \$1.55 to \$1.75. Find the percentage change in price.

**Solution:** We use the original value, \$1.55, as the base. From Eq. 13,

$$\text{percent change} = \frac{1.75 - 1.55}{1.55} \times 100\% = 12.9\% \text{ increase}$$

A common type of problem is to *find the new value* when the original value is changed by a given percent. We see from Eq. 13 that

$$\text{new value} = \text{original value} + (\text{original value}) \times (\text{percent change})$$

**EXAMPLE 89:** Find the cost of a \$156 suit after the price increases by $2\frac{1}{2}\%$.

**Solution:** The original value is 156, and the percent change, expressed as a decimal, is 0.025. So

$$\text{new value} = 156 + 156(0.025) = \$159.90$$

The calculator is convenient for this type of problem.

**EXAMPLE 90:** What will be the cost of a \$5500 machine after an increase of 5% in price?

**Solution:**

| *Press* | *Display* |
|---|---|
| 5500 [+] 5 [%] [=] | 5775 |

## Percent Efficiency

The power output of any machine or device is always *less* than the power input, because of inevitable power losses within the device. The *efficiency* of the device is a measure of those losses.

$$\text{percent efficiency} = \frac{\text{output}}{\text{input}} \times 100\% \qquad \textbf{16}$$

**EXAMPLE 91:** A certain electric motor consumes 865 W and has an output of 1.12 hp. Find the efficiency of the motor. (1 hp = 746 W.)

**Solution:** Since output and input must be in the same units, we must convert either to horsepower or to watts. Converting the output to watts, we obtain

$$\text{output} = 1.12\ \text{hp}\left(\frac{746\ \text{W}}{\text{hp}}\right) = 836\ \text{W}$$

By Eq. 16,

$$\text{percent efficiency} = \frac{836}{865} \times 100\% = 96.6\%$$

### Percent Error

The accuracy of measurements is often specified by the *percent error*. The percent error is the difference between the measured value and the known or "true" value, expressed as a percent of the known value.

$$\text{percent error} = \frac{\text{measured value} - \text{known value}}{\text{known value}} \times 100\% \qquad \textbf{14}$$

**EXAMPLE 92:** A laboratory weight that is certified to be 500 g is placed on a scale. The scale reading is 507 g. What is the percent error in the reading?

**Solution:** From Eq. 14,

$$\text{percent error} = \frac{507 - 500}{500} \times 100\% = 1.4\%\ \text{high}$$

As with percent change, be sure to specify the *direction* of the error.

### Percent Concentration

In a mixture of two or more ingredients,

$$\begin{array}{l}\text{percent concentration}\\ \text{of ingredient } A\end{array} = \frac{\text{amount of } A}{\text{amount of mixture}} \times 100\% \qquad \textbf{15}$$

**EXAMPLE 93:** A certain fuel mixture contains 18 liters of alcohol and 84 liters of gasoline. Find the percentage of gasoline in the mixture.

**Solution:** The total amount of mixture is

$$18 + 84 = 102\ \text{liters}$$

So by Eq. 15,

$$\text{percent gasoline} = \frac{84}{102} \times 100\% = 82.4\%$$

| Common Error | The denominator in Eq. 15 must be the *total amount* of mixture, or the sum of *all* the ingredients. Do not use just one of the ingredients. |
|---|---|

## EXERCISE 9—PERCENTAGE

### Conversions

Convert each decimal to a percent.

**1.** 3.72 **2.** 0.877 **3.** 0.0055 **4.** 0.563

Convert each fraction to a percent. Round to three significant digits.

**5.** $\frac{2}{5}$ **6.** $\frac{3}{4}$ **7.** $\frac{7}{10}$ **8.** $\frac{3}{7}$

Convert each percent to a decimal.

**9.** 23% **10.** 2.97% **11.** $287\frac{1}{2}\%$ **12.** $6\frac{1}{4}\%$

Convert each percent to a fraction.

**13.** 37.5% **14.** $12\frac{1}{2}\%$ **15.** 150% **16.** 3%

### Finding the Amount

Find:

**17.** 40% of 250 tons
**18.** 15% of 300 mi
**19.** $33\frac{1}{3}\%$ of 660 kg
**20.** $12\frac{1}{2}\%$ of 72 gal
**21.** 30% of 400 liters
**22.** 50% of \$240
**23.** A resistance, now 7250 Ω, is to be increased by 15%. How much resistance should be added?
**24.** It is estimated that $\frac{1}{2}\%$ of the earth's surface receives more energy than the total projected needs for the year 2000. Assuming the earth's surface area to be $1.97 \times 10^8$ mi$^2$, find the required area in acres.
**25.** As an incentive to install solar equipment, a tax credit of 40% of the first \$1000 and 25% of the next \$6400 spent on solar equipment is proposed. How much credit would a homeowner get when installing \$5000 worth of solar equipment?
**26.** How much metal will be obtained from 375 tons of ore if the metal is $10\frac{1}{2}\%$ of the ore?

### Finding the Base

Find the number of which:

**27.** 86.5 is $16\frac{2}{3}\%$
**28.** 45 is 1.5%
**29.** $\frac{3}{8}$ is $1\frac{1}{3}\%$
**30.** 50 is 25%
**31.** 60 is 60%
**32.** 60 is 75%
**33.** A Department of Energy report on an experimental electric car gives the range of the car as 160 km and states that this is "50% better than on earlier electric vehicles." What was the range of earlier electric vehicles?
**34.** A man withdrew 25% of his bank deposits and spent $33\frac{1}{3}\%$ of the money drawn in the purchase of a radio worth \$250. How much money did he have in the bank?
**35.** Solar panels provide 55% of the heat for a certain building. If \$225 per year is now spent for heating oil, what would have been spent if the solar panels were not used?
**36.** If the United States imports 9.14 billion barrels of oil per day and if this is 48.2% of its needs, how much oil is needed per day?

### Finding the Rate

What percent of:

**37.** 24 is 12?
**38.** 36 is 12?

**39.** 40 is 8?

**40.** 840 men are 420 men?

**41.** 450 h are 150 h?

**42.** 450 tons are 300 tons?

**43.** A 50,500-liter-capacity tank contains 5840 liters of water. Express the amount of water in the tank as a percentage of the total capacity.

**44.** In a journey of 1560 km, a person traveled 195 km by car and the rest of the distance by rail. What percent of the distance was traveled by rail?

**45.** A power supply has a dc output of 50.0 V with a ripple of 0.75 V peak to peak. Express the ripple as a percentage of the dc output voltage.

**46.** The construction of a factory cost $136,000 for materials and $157,000 for labor. The labor cost was what percentage of the total?

### Percent Change

What number increased by:

**47.** 12% of itself = 560?

**48.** 17% of itself = 702?

**49.** $16\frac{2}{3}$% of itself = 700?

**50.** $2\frac{1}{2}$% of itself = 820?

What number decreased by:

**51.** 10% of itself = 90?

**52.** 20% of itself = 48?

**53.** 15% of itself = 544?

**54.** 90% of itself = 12.6?

Find the percent change when a quantity changes:

**55.** from 29.3 to 57.6

**56.** from 107 to 23.75

**57.** from 227 to 298

**58.** from 0.774 to 0.638

**59.** The temperature in a building rose from 19°C to 21°C during the day. Find the percent change in temperature.

**60.** A casting initially weighing 115 lb has 22% of its material machined off. What is its final weight?

**61.** A certain common stock rose from a value of $\$35\frac{1}{2}$ per share to $\$37\frac{5}{8}$ per share. Find the percent change in value.

**62.** A house that costs $635 per year to heat has insulation installed in the attic, causing the fuel bill to drop to $518 per year. Find the percent change in fuel cost.

### Percent Efficiency

**63.** A certain device consumes 18 hp and delivers 12 hp. Find its efficiency.

**64.** An electric motor consumes 1250 W. Find the horsepower it can deliver if it is 85% efficient. (1 hp = 746 W.)

**65.** A water pump requires an input of $\frac{1}{2}$ hp and delivers 10,000 lb of water per hour to a house 72 ft above the pump. Find its efficiency. (1 hp = 550 ft-lb/s)

**66.** A certain speed reducer delivers 1.7 hp with a power input of 2.2 hp. Find the percent efficiency of the speed reducer.

### Percent Error

**67.** A certain quantity is measured at 125 units but is known actually to be 128 units. Find the percent error in the measurement.

**68.** A shaft is known to have a diameter of 35.000 mm. You measure it and get a reading of 34.725 mm. What is the percent error of your reading?

**69.** A certain capacitor has a working voltage of 125 V dc −10%, +150%. Between what two voltages would the actual working voltage lie?

**70.** A resistor is labeled as 5500 Ω with a tolerance of ±5%. Between what two values is the actual resistance expected to lie?

### Percent Concentration

**71.** A solution is made by mixing 75 liters of alcohol with 125 liters of water. Find the percent concentration of alcohol.

72. Eight cubic feet of cement is contained in a concrete mixture that is 12% cement by volume. What is the volume of the total mixture?

73. How many liters of alcohol are contained in 500 liters of a gasohol mixture that is 5% alcohol by volume?

74. How many liters of gasoline are there in 155 gal of a methanol–gasoline blend that is 10% methanol by volume?

## CHAPTER 1 REVIEW PROBLEMS

1. Combine: $1.435 - 7.21 + 93.24 - 4.1116$
2. Give the number of significant digits in:
   **(a)** 9.886 **(b)** 1.002
   **(c)** 0.3500 **(d)** $15,\bar{0}00$
3. Multiply: $21.8(3.775 \times 1.07)$
4. Divide: $88.25 \div 9.15$
5. Find the reciprocal of 2.89.
6. Evaluate: $-|-4 + 2| - |-9 - 7| + 5$
7. Evaluate: $(9.73)^2$
8. Evaluate: $(7.75)^{-2}$
9. Evaluate: $\sqrt{29.8}$
10. Evaluate: $(123)(2.75) - (81.2)(3.24)$
11. Evaluate: $(91.2 - 88.6)^2$
12. Evaluate: $\left(\frac{77.2 - 51.4}{21.6 - 11.3}\right)^2$
13. Evaluate: $y = 3x^2 - 2x$ when $x = -2.88$
14. Evaluate: $y = 2ab - 3bc + 4ac$ when $a = 5$, $b = 2$, and $c = -6$
15. Evaluate: $y = 2x - 3w + 5z$ when $x = 7.72$, $w = 3.14$, and $z = 2.27$
16. Round to two decimal places.
    **(a)** 7.977 **(b)** 4.655
    **(c)** 11.845 **(d)** 1.004
17. Round to three significant digits.
    **(a)** 179.2 **(b)** 1.076
    **(c)** 4.8550 **(d)** 45,725
18. A news report states that a new hydroelectric generating station in Holyoke, Massachusetts, will produce 47 million kWh/yr, and that this power, for 20 years of operation, is equivalent to 2 million barrels of oil. Using these figures, how many kilowatthours is each barrel of oil equivalent to?
19. A certain generator has a power input of 2.5 hp and delivers 1310 W. Find its percent efficiency.
20. Using Eq. A52, find the stress in pounds per square inch for a force of $1.17 \times 10^3$ N distributed over an area of $3.14 \times 10^3$ mm$^2$.
21. Combine: $(8.34 \times 10^5) + (2.85 \times 10^6) - (5.29 \times 10^4)$
22. A train running at 25 mi/h increases its speed $12\frac{1}{2}\%$. How fast does it then go?
23. The average solar radiation in the continental United States is about $7.4 \times 10^5$ J/m$^2$·h. How many kilowatts would be collected by 15 acres of solar panels?
24. An item rose in price from \$29.35 to \$31.59. Find the percent increase.
25. Find the percent concentration of alcohol if 2 liters of alcohol is added to 15 gal of gasoline.
26. A bar, known to be 2.0000 in. in diameter, is measured at 2.0064 in. Find the percent error in the measurement.
27. The Department of Energy estimates that there are 700 billion barrels of oil in the oil shale deposits of Colorado, Wyoming, and Utah. Express this amount in scientific notation.
28. Multiply: $(7.23 \times 10^5) \times (1.84 \times 10^{-3})$
29. Divide: $-39.2$ by $-0.003826$
30. Convert 6930 Btu/h to foot-pounds per minute.
31. Divide: $8.24 \times 10^{-3}$ by $1.98 \times 10^7$
32. What percent of 40.8 is 11.3?
33. Evaluate: $\sqrt[5]{82.8}$
34. Multiply: $(4.92 \times 10^6) \times (9.13 \times 10^{-3})$
35. Insert the proper sign of equality or inequality between $-\frac{2}{3}$ and $-0.660$.
36. Convert 0.000426 mA to microamperes.
37. Find 49.2% of 4827.
38. Combine: $-385 - (227 - 499) - (-102) + (-284)$
39. Find the reciprocal of $-0.582$.
40. Find the percent change in a voltage that increased from 110 V to 118 V.
41. By insulating, a homeowner's yearly fuel consumption dropped from 628 gal to 405 gal. Her present oil consumption is what percent of the former?
42. Write in decimal notation: $5.28 \times 10^4$
43. Convert 49.3 pounds to newtons.
44. Evaluate: $(45.2)^{-0.45}$
45. Using Eq. A18 find the distance in feet traveled by a falling object in 5.25 s, thrown downward with an initial velocity of 284 m/min.
46. Write in scientific notation: 0.000374
47. 8463 is what percent of 38,473?

48. The U.S. energy consumption of 37 million barrels oil equivalent per day is expected to climb to 48 million in 6 years. Find the percent increase in consumption.

49. The population of a certain town is 8118, which is $12\frac{1}{2}\%$ more than it was 3 years ago. What was the population then?

50. The temperature of a room rose from 68°F to 73°F. Find the percent increase.

51. Combine: $4.928 + 2.847 - 2.836$

52. Give the number of significant digits in 2003.0.

53. Multiply: 2.84(38.4)

54. Divide: $48.3 \div 2.841$

55. Find the reciprocal of 4.82.

56. Evaluate: $|-2| - |3 - 5|$

57. Evaluate: $(3.84)^2$

58. Evaluate: $(7.62)^{-2}$

59. Evaluate: $\sqrt{38.4}$

60. Evaluate: $(49.3 - 82.4)(2.84)$

61. Evaluate: $x^2 - 3x + 2$ when $x = 3$

62. Round 45.836 to one decimal place.

63. Round 83.43 to three significant digits.

64. Multiply: $(7.23 \times 10^5) \times (1.84 \times 10^{-3})$

65. Convert 36.82 in. to centimeters.

66. What percent of 847 is 364?

67. Evaluate: $\sqrt[3]{746}$

68. Find 35.8% of 847.

69. 746 is what percent of 992?

70. Write 0.00274 in scientific notation.

71. Write $73.7 \times 10^{-3}$ in decimal notation.

72. Evaluate: $(47.3)^{-0.26}$

73. Combine: $6.128 + 8.3470 - 7.23612$

74. Give the number of significant digits in 6013.00.

75. Multiply: 7.184(16.8)

76. Divide: $78.7 \div 8.251$

77. Find the reciprocal of 0.825.

78. Evaluate: $|-5| - |2 - 7| + |-6|$

79. Evaluate: $(0.55)^2$

80. Write 23,800 in scientific notation.

*Writing*

81. Suppose you have submitted a report that contains calculations in which you have rounded the answers according to the rules given in this chapter. Jones, your company rival, has sharply attacked your work, calling it "inaccurate" because you did not keep enough digits, and your boss seems to agree.

Write a memo to your boss defending your rounding practices. Point out why it is misleading to retain too many digits. Do not write more than one page. You may use numerical examples to prove your point.

# 2

# INTRODUCTION TO ALGEBRA

## OBJECTIVES

**When you have completed this chapter, you should be able to:**

- Determine the number of terms in an expression, identify the coefficient of each term, and state the degree of the expression.
- Simplify expressions by removing symbols of grouping and by combining like terms.
- Use the laws of exponents to simplify and combine expressions containing powers.
- Add, subtract, multiply, and divide algebraic expressions.

*Algebra* is a generalization of arithmetic. For example, the statement

$$2 \cdot 2 \cdot 2 \cdot 2 = 2^4$$

can be generalized to

$$a \cdot a \cdot a \cdot a = a^4$$

where $a$ can be *any* number, not just 2. We can go further and say that

$$\underbrace{a \cdot a \cdot a \cdot a \ldots a}_{n \text{ factors}} = a^n$$

where $a$ can be any number, as before, and $n$ can be *any* positive integer, not just 4.

We learn a lot of new words in this chapter. Every field has its special terms, and algebra is no exception. But since algebra is generalized arithmetic, some of what was said in Chapter 1 (such as rules of signs) will be repeated here.

We will redo the basic operations of addition, subtraction, and so on, but now with symbols rather than numbers. Learn this material well, for it is the foundation on which later chapters rest.

In this chapter we review the algebraic operations themselves, with no attempt to show applications. But don't get the idea that algebra has no practical use, for the following chapters will show applications to many branches of technology.

## 2-1 ALGEBRAIC EXPRESSIONS

### Mathematical Expressions

A mathematical *expression* is a grouping of mathematical symbols, such as signs of operation, numbers, and letters.

---

**EXAMPLE 1:**

(a) $x^2 - 2x + 3$

(b) $4 \sin 3x$

(c) $5 \log x + e^{2x}$

are mathematical expressions.

---

### Algebraic Expressions

An *algebraic expression* is one containing only algebraic symbols and operations (addition, subtraction, multiplication, division, roots, and powers), such as in Example 1 (a). All other expressions are called *transcendental,* such as Examples 1 (b) and (c).

### Equations

None of the expressions above contains an equal sign (=). When two expressions are set equal to each other, we get an *equation*.

**We mentioned equations briefly in Chapter 1 when we substituted into equations. We treat them in detail in Chapter 3.**

**EXAMPLE 2:**

(a) $2x^2 + 3x - 5 = 0$

(b) $6x - 4 = x + 1$

(c) $y = 3x - 5$

(d) $3 \sin x = 2 \cos x$

are equations.

### Constants

A *constant* is a quantity that does not change in value in a particular problem. It is usually a number, such as 8, 4.67, or $\pi$, but can also be represented by a letter. Such letters are usually chosen from the beginning of the alphabet ($a$, $b$, $c$, etc.). The letter $k$ is often used as a constant.

**EXAMPLE 3:** The constants in the expression

$$3x^2 + 4x + 5$$

are 3, 4, and 5.

**EXAMPLE 4:** The constants in the expression

$$ax^2 + bx + c$$

are $a$, $b$, and $c$.

An expression in which the constants are represented by letters, as in the preceding example, is called a *literal* expression.

### Variables

The quantities that may change during a problem are called *variables*. They are usually represented by letters from the end of the alphabet ($x$, $y$, $z$, etc.).

**EXAMPLE 5:** In the expression

$$ax + by + cz$$

the letters $a$, $b$, and $c$ would usually represent constants, and $x$, $y$, and $z$ would represent variables.

### Symbols of Grouping

**We will show how to *remove* symbols of grouping in Sec. 2-2. We will also see in Sec. 2-4 that symbols of grouping are used to indicate *multiplication*.**

A mathematical expression will often contain parentheses ( ), brackets [ ], and braces { }. These are used to group parts of the expression together, and affect the meaning of the expression.

**EXAMPLE 6:**

$$\{[(3 + x^2) - 2] - (2x + 3)\}(x - 3)$$

shows the use of symbols of grouping.

**EXAMPLE 7:** The value of the expression

$$2 + 3(4 + 5)$$

is *different* from

$$(2 + 3)4 + 5$$

The first expression has a value of 29 and the second expression has a value of 25. The placement of symbols of grouping *is* important.

### Terms

The plus and minus signs divide an expression into *terms*.

**EXAMPLE 8:** The expression

$$4x^2 + 3x + 5$$

$$\boxed{4x^2} \quad \boxed{3x} \quad \boxed{5}$$

first term second term third term

has *three* terms.

An exception is when the plus or minus sign is *within* a symbol of grouping.

**EXAMPLE 9:** The expression

$$\underbrace{[(x + 2x^2) + 3]}_{\text{first term}} - \underbrace{(2x + 2)}_{\text{second term}}$$

has *two* terms.

### Factors

Any divisor of a term is called a *factor* of that term.

**EXAMPLE 10:** Some factors of $3axy$ are 3, $a$, $x$, and $y$.

**The number 1, as well as the entire expression $3axy$, are also factors. They are not usually stated. See Chapter 7 for a more thorough discussion of factors.**

**EXAMPLE 11:** The expression

$$2x + 3yz$$

has two *terms*, $2x$ and $3yz$. The first term has the *factors* 2 and $x$, and the second term has the factors 3, $y$, and $z$.

### Coefficient

The *coefficient* of a term is (usually) the constant part of the term.

**EXAMPLE 12:** The coefficient in the term $3axy^2$ is $3a$. We say that $3a$ is the coefficient of $xy^2$. *But* it is also possible to say that

$3ax$ is the coefficient of $y^2$
$3ay^2$ is the coefficient of $x$
$3xy^2$ is the coefficient of $a$
or even $axy^2$ is the coefficient of 3

But when we use the word "coefficient" in this book, it will almost always refer to the constant part of the term ($3a$ in the preceding example), also called the *numerical* coefficient. If there is no numerical coefficient written before a term, it is understood to be 1.

**EXAMPLE 13:** In the expression $\frac{w}{2} + x - y - 3z$

the coefficient of $w$ is $\frac{1}{2}$
the coefficient of $x$ is 1
the coefficient of $y$ is $-1$
the coefficient of $z$ is $-3$

**Do not forget to include the minus sign with the coefficient.**

## Degree

The *degree* of a term refers to the power to which the variable is raised.

**EXAMPLE 14:**

(a) $2x$ is a first-degree term.
(b) $3x^2$ is a second-degree term.
(c) $5y^9$ is a ninth-degree term.

**The term *degree* is used only when the exponents are *positive integers.* You would not say that $x^{1/2}$ is a "half-degree" term.**

If there is more than one variable, their powers are added to obtain the degree of the term.

**EXAMPLE 15:**

(a) $2x^2y^3$ is of fifth degree.
(b) $3xyz^2$ is of fourth degree.

The degree of an *expression* is the same as that of the term having the highest degree.

**EXAMPLE 16:** $3x^2 - 2x + 4$ is a second-degree expression.

A *multinomial* is an algebraic expression having *more than one term.*

**EXAMPLE 17:** Some multinomials are

(a) $3x + 5$ (b) $2x^3 - 3x^2 + 7$ (c) $\frac{1}{x} + \sqrt{2x}$

A *polynomial* is a monomial or multinomial in which the powers to which the unknown is raised are all *nonnegative integers*. The first two expressions in the preceding example are polynomials, but the third is not.

A *binomial* is a polynomial with *two* terms, and a *trinomial* is a polynomial having *three* terms. In the preceding example, the first expression is a binomial and the second is a trinomial.

## EXERCISE 1—ALGEBRAIC EXPRESSIONS

How many terms are there in each expression?

1. $x^2 - 3x$
2. $7y - (y^2 + 5)$
3. $(x + 2)(x - 1)$
4. $3(x - 5) + 2(x + 1)$

Write the coefficients of each term. Assume that letters from the beginning of the alphabet ($a$, $b$, $c$, . . .) are constants.

5. $5x^3$
6. $2ay^2$
7. $\frac{bx^3}{4}$
8. $\frac{1}{4}(bx)$
9. $\frac{3y^2}{2a}$
10. $\frac{2c}{a}(4x^2)$

## 2-2 ADDITION AND SUBTRACTION OF ALGEBRAIC EXPRESSIONS

### Like Terms

*Like terms* are those that differ only in their coefficients.

---

**EXAMPLE 18:**

(a) $2wx$ and $-3wx$ are like terms.

(b) $2wx$ and $-3wx^2$ are *not* like terms.

---

Algebraic expressions are added and subtracted by *combining like terms*. Like terms are added by adding their coefficients.

**This process is also referred to as *collecting terms*.**

---

**EXAMPLE 19:**

(a) $7x + 5x = 12x$

(b) $8w + 2w - 4w - w = 5w$

(c) $9y - 3y = 6y$

(d) $2x - 3y - 5x + 2y = -3x - y$

---

**EXAMPLE 20:** From the sum of $3x^3 + 2x - 5$ and $x^3 - 3x^2 + 7$, subtract $2x^3 + 3x^2 - 4x - 7$.

**Solution:** Combine the first two expressions:

$$\begin{array}{r} 3x^3 \phantom{{}-3x^2} + 2x - 5 \\ x^3 - 3x^2 \phantom{{}+2x} + 7 \\ \hline 4x^3 - 3x^2 + 2x + 2 \end{array}$$

and subtract the third expression:

$$\begin{array}{rr} & 4x^3 - 3x^2 + 2x + 2 \\ - & 2x^3 + 3x^2 - 4x - 7 \\ \hline & 2x^3 - 6x^2 + 6x + 9 \end{array}$$

---

### Commutative Law of Addition

We mentioned the commutative law (Eq. 1) when adding numbers in Chapter 1. All it means is that the order of addition does not affect the sum ($2 + 3 = 3 + 2$). The same law applies, of course, when we have letters instead of numbers.

---

**EXAMPLE 21:**

(a) $a + b = b + a$

(b) $2x + 3x = 3x + 2x$
$$= 5x$$

---

This law enables us to *rearrange the terms* of an expression for our own convenience.

---

**EXAMPLE 22:** Simplify the expression

$$4y - 2x + 3z - 7z + 4x - 6y$$

**Solution:** Using the commutative law, we rearrange the expression so as to get like terms together.

$$-2x + 4x + 4y - 6y + 3z - 7z$$

Collecting terms, we obtain

$$2x - 2y - 4z$$

With practice, you will soon be able to omit the first step, and collect terms by inspecting the original expression.

---

### Removal of Parentheses

To combine quantities within parentheses with other quantities not within those parentheses, we must first remove the parentheses. If the parentheses (or other symbol of grouping) are preceded only by a + sign (or no sign), the parentheses may be removed.

---

**EXAMPLE 23:**

(a) $(a + b - c) = a + b - c$

(b) $(x + y) + (w - z) = x + y + w - z$

---

When the parentheses are preceded only by a ($-$) sign, change the sign of every term within the parentheses upon removing the parentheses.

---

**EXAMPLE 24:**

(a) $-(a + b - c) = -a - b + c$

(b) $-(x + y) - (x - y) = -x - y - x + y = -2x$

(c) $(2w - 3x - 2y) - (w - 4x + 5y) - (3w - 2x - y)$
$$= 2w - 3x - 2y - w + 4x - 5y - 3w + 2x + y$$
$$= 2w - w - 3w - 3x + 4x + 2x - 2y - 5y + y$$
$$= -2w + 3x - 6y$$

---

When there are groups within groups, start simplifying the innermost groups and work outward.

**EXAMPLE 25:**

We'll do more complicated problems of this type later.

$$\{3x - [2y + 5z - (y + 2)] + 3\} - 7z$$
$$= \{3x - [2y + 5z - y - 2] + 3\} - 7z$$
$$= \{3x - [y + 5z - 2] + 3\} - 7z$$
$$= \{3x - y - 5z + 2 + 3\} - 7z$$
$$= 3x - y - 5z + 5 - 7z$$
$$= 3x - y - 12z + 5$$

## Good Working Habits

Start now to develop careful working habits. Why make things even harder for yourself by scribbling? Form the symbols with care; work in sequence, from top of page to bottom; do not crowd; use a sharp pencil, not a pen; erase mistakes instead of crossing them out. What chance do you have if you cannot read your own work?

| Common Error | Do not switch between capitals and lowercase letters without reason. Although $b$ and $B$ are the same alphabetic letter, in a math problem they could stand for entirely different quantities. |
|---|---|

## EXERCISE 2—ADDITION AND SUBTRACTION OF ALGEBRAIC EXPRESSIONS

Combine as indicated, and simplify.

1. $7x + 5x$
2. $2x + 5x - 4x + x$
3. $6ab - 7ab - 9ab$
4. $9.4x - 3.7x + 1.4x$
5. Add $7a - 3b + m$ and $3b - 7a - c + m$.
6. What is the sum of $6ab + 12bc - 8cd$, $3cd - 7cd - 9bc$, and $12cd - 2ab - 5bc$?
7. What is the sum of $3a + b - 10$, $c - d - a$, and $-4c + 2a - 3b - 7$?
8. Add $7m + 3n - 11p$, $3a - 9n - 11m$, $8n - 4m + 5p$, and $6n - m + 3p$.
9. Add $8ax + 2(x + a) + 3b$, $9ax + 6(x + a) - 9b$, and $11x + 6b - 7ax - 8(x + a)$.
10. Add $a - 9 - 8a^2 + 16a^3$, $5 + 15a^3 - 12a - 2a^2$, and $6a^2 - 10a^3 + 11a - 13$.
11. Subtract $41x^3 - 2x^2 + 13$ from $15x^3 + x - 18$.
12. Subtract $3b - 6d - 10c + 7a$ from $4d + 12a - 13c - 9b$.
13. Add $x - y - z$ and $y - x + z$.
14. What is the sum of $a + 2b - 3c - 10$, $3b - 4a + 5c + 10$, and $5b - c$?
15. Add $72ax^4 - 8ay^3$, $-38ax^4 - 3ay^4 + 7ay^3$, $8 + 12ay^4$, $-6ay^3 + 12$, and $-34ax^4 + 5ay^3 - 9ay^4$.
16. Add $2a(x - y^2) - 3mz^2$, $4a(x - y^2) - 5mz^2$, and $5a(x - y^2) + 7mz^2$.
17. What is the sum of $9b^2 - 3ac + d$, $4b^2 + 7d - 4ac$, $3d - 4b^2 + 6ac$, $5b^2 - 2ac - 12d$, and $4b^2 - d$?
18. From the sum of $2a + 3b - 4c$ and $3b + 4c - 5d$, subtract the sum of $5c - 6d - 7a$ and $-7d + 8a + 9b$.
19. What is the sum of $7ab - m^2 + q$, $-4ab - 5m^3 - 3q$, $12ab + 14m^2 - z$, and $-6m^2 - 2q$?
20. Add $14(x + y) - 17(y + z)$, $4(y + z) - 19(z + x)$, and $-7(z + x) - 3(x + y)$.
21. Add $3.52(a + b)$, $4.15(a + b)$, and $-1.84(a + b)$.

### Symbols of Grouping

Remove symbols of grouping and collect terms.

22. $(3x + 5) + (2x - 3)$
23. $3a^2 - (3a - x + b)$
24. $40xy - (30xy - 2b^2 + 3c - 4d)$
25. $(6.4 - 1.8x) - (7.5 + 2.6x)$
26. $7m^2 + 2bc - (3m^2 - bc - x)$
27. $a^2 - a - (4a - y - 3a^2 - 1)$
28. $(2x^4 + 3x^3 - 4) + (3x^4 - 2x^3 - 8)$
29. $a + b - m - (m - a - b)$
30. $(3xy - 2y + 3z) + (-3y + 8z) - (-3xy - z)$
31. $3m - z - y - (2z - y - 3m)$
32. $(2x - 3y) - (5x - 3y - 2z)$
33. $(w + 2z) - (-3w - 5x) + (x + z) - (z - w)$
34. $2a + \{-6b - [3c + (-4b - 6c + a)]\}$
35. $4a - \{a - [-7a - [8a - (5a + 3)] - (-6a - 2a - 9)]\}$
36. $9m - \{3n + [4m - (n - 6m)] - [m + 7n]\}$

## 2-3 INTEGRAL EXPONENTS

### Definitions

In this section we deal only with expressions that have integers (positive or negative, and zero) as exponents.

**EXAMPLE 26:** We study such expressions as

$$x^3 \qquad a^{-2} \qquad y^0 \qquad x^2y^3w^{-1} \qquad (m + n)^5$$

A positive *exponent* shows how many times the *base* is to be multiplied by itself.

**EXAMPLE 27:** In the expression $3^4$, 3 is the base and 4 is the exponent.

$$\text{base} \longrightarrow 3^4 \longleftarrow \text{exponent}$$

Its meaning is

$$3^4 = 3(3)(3)(3) = 81$$

In general,

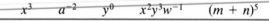

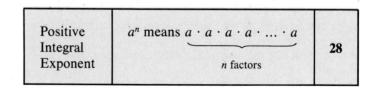

| | | |
|---|---|---|
| Positive Integral Exponent | $a^n$ means $\underbrace{a \cdot a \cdot a \cdot a \cdot \ldots \cdot a}_{n \text{ factors}}$ | **28** |

| | |
|---|---|
| Common Errors | An exponent applies *only* to the symbol directly before it. Thus $$2x^2 \neq 2^2x^2$$ but $$(2x)^2 = (2x)(2x) = 4x^2$$ |
| | Also, $$-2^2 \neq 4$$ but $$(-2)^2 = 4 \quad \text{and} \quad -2^2 = -4$$ |

### Multiplying Powers

Let us multiply the quantity $x^2$ by $x^3$. From Eq. 28, we know that

$$x^2 = x \cdot x$$

and that

$$x^3 = x \cdot x \cdot x$$

Multiplying, we obtain

$$\begin{aligned} x^2 \cdot x^3 &= (x \cdot x)(x \cdot x \cdot x) \\ &= x \cdot x \cdot x \cdot x \cdot x \\ &= x^5 \end{aligned}$$

Notice that the exponent in the result is the *sum of the two original exponents.*

$$x^2 \cdot x^3 = x^{2+3} = x^5$$

This will always be so. We summarize this rule as our first *law of exponents.*

| | | |
|---|---|---|
| Products | $x^a \cdot x^b = x^{a+b}$ | **29** |

*When multiplying powers of the same base, we keep the same base and* ***add the exponents.***

---

**EXAMPLE 28:**
(a) $x^4(x^3) = x^{4+3} = x^7$
(b) $x^5(x^a) = x^{5+a}$
(c) $y^2(y^n)(y^3) = y^{2+n+3} = y^{5+n}$

---

Every quantity has an exponent of 1, even though it is not usually written.

The "Invisible 1" appears again. We saw it before as the unwritten coefficient of every term, and now as the unwritten exponent. It is also in the denominator.

$$x = \frac{1x^1}{1}$$

Do not forget about those invisible 1's. *We use them all the time.*

---

**EXAMPLE 29:**

(a) $x(x^3) = x^{1+3} = x^4$

(b) $a^2(a^4)(a) = a^{2+4+1} = a^7$

(c) $x^a(x)(x^b) = x^{a+1+b}$

---

## Quotients

Let us divide $x^5$ by $x^3$. By Eq. 29,

$$\frac{x^5}{x^3} = \frac{x \cdot x \cdot x \cdot x \cdot x}{x \cdot x \cdot x} = \frac{x \cdot x \cdot x}{x \cdot x \cdot x} \cdot x \cdot x$$

$$= x \cdot x = x^2$$

The same result could have been obtained by *subtracting exponents,*

$$\frac{x^5}{x^3} = x^{5-3} = x^2$$

This always works, and we state it as another law of exponents:

| Quotients | $\dfrac{x^a}{x^b} = x^{a-b} \qquad (x \neq 0)$ | **30** |
|---|---|---|

*When dividing powers of the same base, we keep the same base and* ***subtract the exponents.***

---

**EXAMPLE 30:**

(a) $\dfrac{a^4b^3}{ab^2} = a^{4-1}b^{3-2} = a^3b$

(b) $\dfrac{y^{n-m}}{y^{n+m}} = y^{n-m-(n+m)} = y^{-2m}$

---

| Common Error | Remember when using rules 29 or 30 that the quantities to be multiplied or divided **must have the same base.** |
|---|---|

## Power Raised to a Power

Let us take a quantity raised to a power, say, $x^2$, and raise the entire expression to *another* power, say, 3.

$$(x^2)^3$$

By Eq. 29,

$$(x^2)^3 = (x^2)(x^2)(x^2)$$

$$= (x \cdot x)(x \cdot x)(x \cdot x)$$

$$= x \cdot x \cdot x \cdot x \cdot x \cdot x = x^6$$

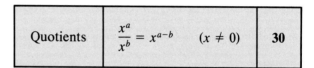

a result that could have been obtained by *multiplying the exponents.*

$$(x^2)^3 = x^{2(3)} = x^6$$

In general,

| Powers | $(x^a)^b = x^{ab}$ | **31** |
|---|---|---|

*When raising a power to a power, we keep the same base and **multiply the exponents.***

**EXAMPLE 31:**

(a) $(w^5)^2 = w^{5(2)} = w^{10}$

(b) $(a^{-3})^2 = a^{(-3)(2)} = a^{-6}$

(c) $(10^4)^3 = 10^{4(3)} = 10^{12}$

(d) $(b^{x+2})^3 = b^{3x+6}$

## Product Raised to a Power

We now raise a product, such as $xy$, to some power, say 3.

$$(xy)^3$$

By Eq. 29,

$$\begin{aligned}(xy)^3 &= (xy)(xy)(xy)\\ &= x \cdot y \cdot x \cdot y \cdot x \cdot y\\ &= x \cdot x \cdot x \cdot y \cdot y \cdot y\\ &= x^3y^3\end{aligned}$$

In general,

| Product Raised to a Power | $(xy)^n = x^ny^n$ | **32** |
|---|---|---|

*When a product is raised to a power, each factor may be **separately** raised to the power.*

**EXAMPLE 32:**

(a) $(xyz)^5 = x^5y^5z^5$

(b) $(2x)^3 = 2^3x^3 = 8x^3$

(c) $(3.75 \times 10^3)^2 = (3.75)^2 \times (10^3)^2$
$= 14.1 \times 10^6$

(d) $(3x^2y^n)^3 = 3^3(x^2)^3(y^n)^3$
$= 27x^6y^{3n}$

(e) $(-x^2y)^3 = (-1)^3(x^2)^3y^3 = -x^6y^3$

A good way to test a "rule" that you are not sure of is to *try it with numbers.* In this case, does $(2 + 3)^2$ equal $2^2 + 3^2$? Evaluating each expression, we obtain

$$(5)^2 \stackrel{?}{=} 4 + 9$$

$$25 \neq 13$$

| | |
|---|---|
| Common Error | There is *no* similar rule for the *sum* of two quantities raised to a power. $$(x + y)^n \neq x^n + y^n$$ |

### Quotient Raised to a Power

Using the same steps as in the preceding section, see if you can show that

$$\left(\frac{x}{y}\right)^3 = \frac{x^3}{y^3}$$

Or, in general,

| | | | |
|---|---|---|---|
| Quotient Raised to a Power | $\left(\frac{x}{y}\right)^n = \frac{x^n}{y^n}$ | $(y \neq 0)$ | **33** |

*When a quotient is raised to a power, the numerator and denominator may be* ***separately*** *raised to the power.*

---

**EXAMPLE 33:**

(a) $\left(\frac{x}{5}\right)^2 = \frac{x^2}{5^2} = \frac{x^2}{25}$

(b) $\left(\frac{3a}{2b}\right)^3 = \frac{3^3a^3}{2^3b^3} = \frac{27a^3}{8b^3}$

(c) $\left(\frac{2x^2}{5y^3}\right)^3 = \frac{2^3(x^2)^3}{5^3(y^3)^3} = \frac{8x^6}{125y^9}$

(d) $\left(-\frac{a}{b^2}\right)^3 = (-1)^3\frac{a^3}{(b^2)^3} = -\frac{a^3}{b^6}$

---

### Zero Exponent

If we divide $x^n$ by itself we get, by Eq. 30,

$$\frac{x^n}{x^n} = x^{n-n} = x^0$$

But any expression divided by itself equals 1, so

| | | | |
|---|---|---|---|
| Zero Exponent | $x^0 = 1$ | $(x \neq 0)$ | **34** |

*Any quantity (except 0) raised to the zero power equals 1.*

---

**EXAMPLE 34:**

(a) $(xyz)^0 = 1$

(b) $3862^0 = 1$

(c) $(x^2 - 2x + 3)^0 = 1$

(d) $5x^0 = 5(1) = 5$

---

### Negative Exponent

We now divide $x^0$ by $x^a$. By Eq. 30,

$$\frac{x^0}{x^a} = x^{0-a} = x^{-a}$$

Since $x^0 = 1$, we get

| Negative Exponent | $x^{-a} = \dfrac{1}{x^a} \qquad (x \neq 0)$ | **35** |
|---|---|---|

*When taking the reciprocal of a base raised to a power,* ***change the sign of the exponent.***

We can, of course, also give Eq. 35 in the form

$$x^a = \frac{1}{x^{-a}} \qquad (x \neq 0)$$

---

**EXAMPLE 35:**

(a) $5^{-1} = \dfrac{1}{5}$

(b) $x^{-4} = \dfrac{1}{x^4}$

(c) $\dfrac{1}{xy^{-2}} = \dfrac{y^2}{x}$

(d) $\dfrac{w^{-3}}{z^{-2}} = \dfrac{z^2}{w^3}$

(e) $\left(\dfrac{3x}{2y^2}\right)^{-b} = \left(\dfrac{2y^2}{3x}\right)^b$

---

We can also use negative exponents to rewrite an expression *without fractions*.

---

**EXAMPLE 36:**

(a) $\dfrac{1}{y} = y^{-1}$

(b) $\dfrac{a^2}{b^3} = a^2b^{-3}$

(c) $\dfrac{x^2y^{-3}}{w^4z^{-2}} = x^2y^{-3}w^{-4}z^2$

---

### EXERCISE 3—INTEGRAL EXPONENTS

Evaluate each expression.

1. $2^4$
2. $(-3)^2$
3. $(-3)^3$
4. $(|-3|)^3$
5. $(0.1)^4$
6. $(-2)^5$

Multiply.

7. $a^3 \cdot a^5$
8. $x^a \cdot x^2$
9. $y^{a+1} \cdot y^{a-3}$
10. $10^4 \cdot 10^3$
11. $10^a \cdot 10^b$
12. $10^{n+2} \cdot 10^{2n-1}$

Divide. Write your answers without negative exponents.

13. $\frac{y^5}{y^2}$
14. $\frac{5^5}{5^3}$
15. $\frac{x^{n+2}}{x^{n+1}}$
16. $\frac{10^5}{10}$
17. $\frac{10^{x+5}}{10^{x+3}}$
18. $\frac{10^2}{10^{-3}}$
19. $\frac{x^{-2}}{x^{-3}}$
20. $\frac{a^{-5}}{a}$

Simplify.

21. $(x^3)^4$
22. $(9^2)^3$
23. $(a^x)^y$
24. $(x^{-2})^{-2}$
25. $(x^{a+1})^2$

Raise to the power indicated.

26. $(xy)^2$
27. $(2x)^3$
28. $(3x^2y^3)^2$
29. $(3abc)^3$
30. $\left(\frac{3}{5}\right)^2$
31. $\left(-\frac{1}{3}\right)^3$
32. $\left(\frac{x}{y}\right)^5$
33. $\left(\frac{2a}{3b^2}\right)^3$
34. $\left(\frac{3x^2y}{4wz^3}\right)^2$

Write each expression with positive exponents only.

35. $a^{-2}$
36. $(-x)^{-3}$
37. $\left(\frac{3}{y}\right)^{-3}$
38. $a^{-2}bc^{-3}$
39. $\left(\frac{2a}{3b^3}\right)^{-2}$
40. $xy^{-4}$
41. $2x^{-2} + 3y^{-3}$
42. $\left(\frac{x}{y}\right)^{-1}$

Express without fractions, using negative exponents where needed.

43. $\frac{1}{x}$
44. $\frac{3}{y^2}$
45. $\frac{x^2}{y^2}$
46. $\frac{x^2y^{-3}}{z^{-2}}$
47. $\frac{a^{-3}}{b^2}$
48. $\frac{x^{-2}y^{-3}}{w^{-1}z^{-4}}$

Evaluate.

49. $(a + b + c)^0$
50. $8x^0y^2$
51. $\frac{a^0}{9}$
52. $\frac{y}{x^0}$
53. $\frac{x^{2n} \cdot x^3}{x^{3+2n}}$
54. $5\left(\frac{x}{y}\right)^0$

## 2-4 MULTIPLICATION OF ALGEBRAIC EXPRESSIONS

### Symbols and Definitions

Multiplication is indicated in several ways: by the usual × symbol, by a dot, or by parentheses, brackets, or braces. Thus the product of $b$ and $d$ could be written

$$b \cdot d \qquad b \times d \qquad b(d) \qquad (b)d \qquad (b)(d)$$

Most common of all is to use no symbol at all. The product of $b$ and $d$ would usually be written $bd$.

Avoid using the × symbol because it could get confused with the letter x.

We get a *product* when we multiply two or more *factors*.

$$(\text{factor})(\text{factor})(\text{factor}) = \text{product}$$

## Rules of Signs

The product of two factors having the *same* sign is *positive;* the product of two factors having *different* signs is *negative*. If $a$ and $b$ are positive quantities, we have

| | | |
|---|---|---|
| Rules of Signs | $(+a)(+b) = (-a)(-b) = +ab$ | **8** |
| | $(+a)(-b) = (-a)(+b) = -(+a)(+b) = -ab$ | **9** |

When multiplying more than two factors, the product of every *pair* of negative terms will be positive. Thus, if there is an even number of negative factors, the final result will be positive; if there is an odd number of negative factors, the final result will be negative.

**EXAMPLE 37:**

(a) $x(-y)(-z) = xyz$

(b) $(-a)(-b)(-c) = -abc$

(c) $(-p)(-q)(-r)(-s) = pqrs$

## Commutative Law

As we saw when multiplying numbers, the *order* of multiplication does not make any difference. In symbols,

| | | |
|---|---|---|
| Commutative Law | $ab = ba$ | **2** |

**EXAMPLE 38:** Multiply $3y$ by $-2x$.

**Solution:**

$$(3y)(-2x) = 3(y)(-2)(x)$$

By Eq. 2,

$$= 3(-2)(y)(x)$$

$$= -6yx$$

or

$$= -6xy$$

since it is common practice to write the letters in alphabetical order.

## Multiplying Monomials

A *monomial* is an algebraic expression having *one term*.

**EXAMPLE 39:** Some monomials are

$$\text{(a) } 2x^3 \qquad \text{(b) } \frac{3wxy}{4} \qquad \text{(c) } (2b)^3$$

To multiply monomials, we make use of the commutative law (Eq. 2), the rules of signs (Eqs. 8 and 9), and the law of exponents for products.

| | | |
|---|---|---|
| Products | $x^a \cdot x^b = x^{a+b}$ | **29** |

**EXAMPLE 40:**

$$\begin{aligned}(4a^3b)(-3ab^2) &= (4)(-3)(a^3)(a)(b)(b^2)\\ &= -12a^{3+1}b^{1+2}\\ &= -12a^4b^3\end{aligned}$$

**With some practice you should be able to omit some steps in many of these solutions. But do not rush it. Use as many steps as you need to get it right.**

**EXAMPLE 41:** Multiply $5x^2$, $-3xy$, and $-xy^3z$.

**Solution:** By Eq. 2,

$$5x^2(-3xy)(-xy^3z) = 5(-3)(-1)(x^2)(x)(x)(y)(y^3)(z)$$

By Eq. 29,

$$\begin{aligned}&= 15x^{2+1+1}y^{1+3}z\\ &= 15x^4y^4z\end{aligned}$$

**EXAMPLE 42:** Multiply $3x^2$ by $2ax^n$.

**Solution:** By Eq. 2,

$$3x^2(2ax^n) = 3(2a)(x^2)(x^n)$$

By Eq. 29,

$$= 6ax^{2+n}$$

## Multiplying a Multinomial by a Monomial

To multiply a monomial and a multinomial, we use the *distributive law*

**We will later use this law to remove common factors.**

| | | |
|---|---|---|
| Distributive Law | $a(b + c) = ab + ac$ | **5** |

*The product of a monomial and a polynomial is equal to the sum of the products of the monomial and each term of the polynomial.*

**EXAMPLE 43:**

$$\begin{aligned}2x(x - 3x^2) &= 2x(x) + 2x(-3x^2)\\ &= 2x^2 - 6x^3\end{aligned}$$

The distributive law, although written for a monomial times a binomial, can be extended for a multinomial having any number of terms. Simply *multiply every term in the multinomial by the monomial.*

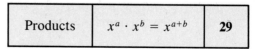

**EXAMPLE 44:**

(a) $-3x^3(3x^2 - 2x + 4) = -3x^3(3x^2) + (-3x^3)(-2x) + (-3x^3)(4)$

$= -9x^5 + 6x^4 - 12x^3$

(b) $4xy(x^2 - 3xy + 2y^2) = 4x^3y - 12x^2y^2 + 8xy^3$

## Removing Symbols of Grouping

Symbols of grouping, in addition to indicating that the enclosed expression is to be taken as a whole, also indicate multiplication.

**EXAMPLE 45:**

(a) $x(y)$ is the product of $x$ and $y$.

(b) $x(y + 2)$ is the product of $x$ and $(y + 2)$.

(c) $(x - 1)(y + 2)$ is the product of $(x - 1)$ and $(y + 2)$.

(d) $-(y + 2)$ is the product of $-1$ and $(y + 2)$.

The parentheses may be removed after the multiplication has been performed.

**EXAMPLE 46:**

(a) $x(y + 2) = xy + 2x$

(b) $5 + 3(x - 2) = 5 + 3x - 6 = 3x - 1$

| | |
|---|---|
| Common Errors | Do not forget to multiply *every* term within the grouping by the preceding factor. $-(2x + 5) \neq -2x + 5$ |
| | Multiply the terms within the parentheses *only* by the factor directly preceding it. $x + 2(a + b) \neq (x + 2)(a + b)$ |

If there are groupings within groupings, start simplifying with the *innermost* groupings.

**EXAMPLE 47:**

(a) $x - 3[2 - 4(y + 1)] = x - 3[2 - 4y - 4]$

$= x - 3[-2 - 4y]$

$= x + 6 + 12y$

(b) $3\{2[4(w - 4) - (x + 3)] - 2\} - 6x$

$= 3\{2[4w - 16 - x - 3] - 2\} - 6x$

$= 3\{2[4w - x - 19] - 2\} - 6x$

$= 3\{8w - 2x - 38 - 2\} - 6x$

$= 3\{8w - 2x - 40\} - 6x$

$= 24w - 6x - 120 - 6x$

$= 24w - 12x - 120$

### Multiplying a Multinomial by a Multinomial

Multiply every term in one multinomial by every term in the other, and combine like terms.

---

**EXAMPLE 48:**

$$(x+4)(x-3) = x(x) + x(-3) + 4(x) + 4(-3)$$
$$= x^2 - 3x + 4x - 12 = x^2 + x - 12$$

---

Some prefer a vertical arrangement for problems like the last one.

---

**EXAMPLE 49:**

$$\begin{array}{r} x + 4 \\ \underline{x - 3} \\ -3x - 12 \\ \underline{x^2 + 4x \quad\;\;} \\ x^2 + \;\; x - 12 \end{array} \qquad \text{as before}$$

---

**EXAMPLE 50:**

(a) $(x-3)(x^2-2x+1)$

$$= x(x^2) + x(-2x) + x(1) - 3(x^2) - 3(-2x) - 3(1)$$
$$= x^3 - 2x^2 + x - 3x^2 + 6x - 3$$
$$= x^3 - 5x^2 + 7x - 3$$

(b) $(2x+3y-4z)(x-2y-3z)$

$$= 2x^2 - 4xy - 6xz + 3xy - 6y^2 - 9yz - 4xz + 8yz + 12z^2$$
$$= 2x^2 - 6y^2 + 12z^2 - xy - 10xz - yz$$

---

When multiplying *more than two* expressions, first multiply any pair, and then multiply that product by the third expression, and so on.

---

**EXAMPLE 51:** Multiply $(x-2)(x+1)(2x-3)$.

**Solution:** Multiplying the last two binomials yields

$$(x-2)(x+1)(2x-3) = (x-2)(2x^2-x-3)$$

Then multiplying that product by $(x-2)$ gives

$$(x-2)(2x^2-x-3) = 2x^3 - x^2 - 3x - 4x^2 + 2x + 6$$
$$= 2x^3 - 5x^2 - x + 6$$

---

To raise an expression to a power, simply multiply the expression by itself the proper number of times.

---

**EXAMPLE 52:**

$$(x-1)^3 = (x-1)(x-1)(x-1)$$
$$= (x-1)(x^2-2x+1)$$
$$= x^3 - 2x^2 + x - x^2 + 2x - 1$$
$$= x^3 - 3x^2 + 3x - 1$$

---

### Product of the Sum and Difference of Two Terms

Let us multiply the binomials $(a - b)$ and $(a + b)$. We get

$$(a - b)(a + b) = a^2 + ab - ab - b^2$$

The two middle terms cancel, leaving

When we multiply two binomials, we get expressions that occur frequently enough in our later work for us to take special note of here. These are called *special products.* We will need them especially when we study factoring.

| | | |
|---|---|---|
| Difference of Two Squares | $(a - b)(a + b) = a^2 - b^2$ | **41** |

Thus the product is the *difference of two squares:* the squares of the original numbers.

---

**EXAMPLE 53:**

$$(x + 2)(x - 2) = x^2 - 2^2 = x^2 - 4$$

---

### Product of Two Binomials Having the Same First Term

When we multiply $(x + a)$ and $(x + b)$, we get

$$(x + a)(x + b) = x^2 + ax + bx + ab$$

Combining like terms, we obtain

| | | |
|---|---|---|
| Trinomial, Leading Coefficient = 1 | $(x + a)(x + b) = x^2 + (a + b)x + ab$ | **45** |

---

**EXAMPLE 54:**

$$(w + 3)(w - 4) = w^2 + (3 - 4)w + 3(-4)$$
$$= w^2 - w - 12$$

---

### General Quadratic Trinomial

When we multiply two binomials of the form $ax + b$ and $cx + d$, we get

$$(ax + b)(cx + d) = acx^2 + adx + bcx + bd$$

Combining like terms, we obtain

Some people like the ***FOIL*** rule. Multiply the ***F****irst* terms, then ***O****uter*, ***I****nner*, and ***L****ast* terms.

| | | |
|---|---|---|
| General Quadratic Trinomial | $(ax + b)(cx + d) = acx^2 + (ad + bc)x + bd$ | **46** |

**EXAMPLE 55:**

$$(3x - 2)(2x + 5) = 3(2)x^2 + [(3)(5) + (-2)(2)]x + (-2)(5)$$
$$= 6x^2 + 11x - 10$$

### Powers of Multinomials

As with a monomial raised to a power that is a positive integer, we raise a *multinomial* to a power by multiplying that multinomial by itself the proper number of times.

**EXAMPLE 56:**

$$(x - 3)^2 = (x - 3)(x - 3)$$
$$= x^2 - 3x - 3x + 9$$
$$= x^2 - 6x + 9$$

**EXAMPLE 57:**

$$(x - 3)^3 = (x - 3)(x - 3)(x - 3)$$
$$= (x - 3)(x^2 - 6x + 9) \quad \text{(from Example 56)}$$
$$= x^3 - 6x^2 + 9x - 3x^2 + 18x - 27$$

Combining like terms gives us

$$(x - 3)^3 = x^3 - 9x^2 + 27x - 27$$

### Perfect Square Trinomial

When we *square a binomial* such as $(a + b)$, we get

$$(a + b)^2 = (a + b)(a + b) = a^2 + ab + ab + b^2$$

Collecting terms, we obtain

| | | |
|---|---|---|
| Perfect Square Trinomial | $(a + b)^2 = a^2 + 2ab + b^2$ | **47** |

This trinomial is called a *perfect square trinomial,* because it is the square of a binomial. Its first and last terms are the squares of the two terms of the binomial; its middle term is twice the product of the terms of the binomial. Similarly, when we square $(a - b)$, we get

| | | |
|---|---|---|
| Perfect Square Trinomial | $(a - b)^2 = a^2 - 2ab + b^2$ | **48** |

**EXAMPLE 58:**

$$(3x - 2)^2 = (3x)^2 - 2(3x)(2) + 2^2$$
$$= 9x^2 - 12x + 4$$

## EXERCISE 4—MULTIPLICATION OF ALGEBRAIC EXPRESSIONS

Multiply the following monomials, and simplify.

1. $x^3(x^2)$
2. $x^2(3axy)$
3. $(-5ab)(-2a^2b^3)$
4. $(3.52xy)(-4.14xyz)$
5. $(6ab)(2a^2b)(3a^3b^3)$
6. $(-2a)(-3abc)(ac^2)$
7. $2p(-4p^2q)(pq^3)$
8. $(-3xy)(-2w^2x)(-xy^3)(-wy)$

Multiply the polynomial by the monomial, and simplify.

9. $2a(a - 5)$
10. $(x^2 + 2x)x^3$
11. $xy(x - y + xy)$
12. $3a^2b^3(ab - 2ab^2 + b)$
13. $-4pq(3p^2 + 2pq - 3q^2)$
14. $2.8x(1.5x^3 - 3.2x^2 + 5.7x - 4.6)$
15. $3ab(2a^2b^2 - 4a^2b + 3ab^2 + ab)$
16. $-3xy(2xy^3 - 3x^2y + xy^2 - 4xy)$

Multiply the following binomials, and simplify.

17. $(a + b)(a + c)$
18. $(2x + 2)(5x + 3)$
19. $(x + 3y)(x^2 - y)$
20. $(a - 5)(a + 5)$
21. $(m + n)(9m - 9n)$
22. $(m^2 + 2c)(m^2 - 5c)$
23. $(7cd^2 + 4y^3z)(7cd^2 - 4y^3z)$
24. $(3a + 4c)(2a - 5c)$
25. $(x^2 + y^2)(x^2 - y^2)$
26. $(3.5a^2 + 1.2b)(3.8a^2 - 2.4b)$

Multiply the following binomials and trinomials, and simplify.

27. $(a - c - 1)(a + 1)$
28. $(x - y)(2x + 3y - 5)$
29. $(m^2 - 3m - 7)(m - 2)$
30. $(-3x^2 + 3x - 9)(x + 3)$
31. $(x^6 + x^4 + x^2)(x^2 - 1)$
32. $(m - 1)(2m - m^2 + 1)$
33. $(1 - z)(z^2 - z + 2)$
34. $(3a^3 - 2ab - b^2)(2a - 4b)$
35. $(a^2 - ay + y^2)(a + y)$
36. $(y^2 - 7.5y + 2.8)(y + 1.6)$
37. $(a^2 + ay + y^2)(a - y)$
38. $(a^2 + ay - y^2)(a - y)$

Multiply the following binomials and polynomials, and simplify.

39. $[x^2 - (m - n)x - mn](x - p)$
40. $(x^2 + ax - bx - ab)(x + b)$
41. $(m^4 + m^3 + m^2 + m + 1)(m - 1)$

Multiply the following monomials and binomials, and simplify.

42. $(2a + 3x)(2a + 3x)9$
43. $(3a - b)(3a - b)x$
44. $c(m - n)(m + n)$
45. $(x + 1)(x + 1)(x - 2)$
46. $(m - 2.5)(m - 1.3)(m + 3.2)$

47. $(x + a)(x + b)(x - c)$
48. $(x - 4)(x - 5)(x + 4)(x + 5)$
49. $(1 + c)(1 + c)(1 - c)(1 + c^2)$

Multiply the following trinomials, and simplify.

50. $(a + b - c)(a - b + c)$
51. $(2a^3 + 5a)(2a^3 - 5ac^2 + 2a)$
52. $[3(a + b)^2 - (a + b) + 2][4(a + b)^2 - (a + b) - 5]$
53. $(a + b + c)(a - b + c)(a + b - c)$

Multiply the following polynomials, and simplify.

54. $(x^3 - 6x^2 + 12x - 8)(x^2 + 4x + 4)$
55. $(n - 5n^2 + 2 + n^3)(5n + n^2 - 10)$
56. $(mx + my - nx - ny)(mx - my + nx - ny)$

### Powers of Algebraic Expressions

Square each binomial.

57. $(a + c)$
58. $(p + q)$
59. $(m - n)$
60. $(x - y)$
61. $(A + B)$
62. $(A - C)$
63. $(3.8a - 2.2x)$
64. $(m + c)$
65. $(2c - 3d)$
66. $(x^2 - x)$
67. $(a - 1)$
68. $(a^2x - ax^3)$
69. $(y^2 - 20)$
70. $(x^n - y^2)$

Square each trinomial.

71. $(a + b + c)$
72. $(a - b - c)$
73. $(1 + x - y)$
74. $(x^2 + 2x - 3)$
75. $(x^2 + xy + y^2)$
76. $(x^3 - 2x - 1)$

Cube each binomial.

77. $(a + b)$
78. $(2x + 1)$
79. $(p - 3q)$
80. $(x^2 - 1)$
81. $(a - b)$
82. $(3m - 2n)$

### Symbols of Grouping

Remove symbols of grouping, and simplify.

83. $-3[-(a + b) - c]$
84. $-[x + (y - 2) - 4]$
85. $(p - q) - [p - (p - q) - q]$
86. $-(x - y) - [y - (y - x) - x]$
87. $-5[-2(3y - z) - z] + y$
88. $[(m + 2n) - (2m - n)][(2m + n) - (m - 2n)]$
89. $(x - 1) - \{[x - (x - 3)] - x\}$
90. $(a - b)(a^3 + b^3)[a(a + b) + b^2]$
91. $y - \{4x - [y - (2y - 9) - x] + 2\}$
92. $[2x^2 + (3x - 1)(4x + 5)][5x^2 - (4x + 3)(x - 2)]$

## 2-5 DIVISION OF ALGEBRAIC EXPRESSIONS

### Symbols for Division

Division may be indicated by any of the symbols

$$x \div y \qquad \frac{x}{y} \qquad x/y$$

The names of the parts are

$$\text{quotient} = \frac{\text{dividend}}{\text{divisor}} = \frac{\text{numerator}}{\text{denominator}}$$

The quantity $x/y$ is also called a *fraction,* and is also spoken of as the *ratio* of $x$ to $y$.

### Reciprocals

As we saw in Sec. 1-4, the *reciprocal* of a number is 1 divided by that number. The reciprocal of $n$ is $1/n$. We can use the idea of a reciprocal to show how division is related to multiplication. We may write the quotient of $x \div y$ as

$$\frac{x}{y} = \frac{x}{1} \cdot \frac{1}{y} = x \cdot \frac{1}{y}$$

We see that to *divide by a number* is the same thing as to *multiply by its reciprocal.* This fact will be especially useful for dividing by a fraction.

### Division by Zero

Division by zero is not a permissible operation.

If division by zero were allowed, we could, for example, divide 2 by zero and get a quotient $x$.

$$\frac{2}{0} = x$$

or

$$2 = 0 \cdot x$$

But there is no number $x$ which, when multiplied by zero, gives 2, so we cannot allow this operation.

---

**EXAMPLE 59:** In the fraction

$$\frac{x+5}{x-2}$$

$x$ cannot equal 2, or the illegal operation of division by zero will result.

---

### Rules of Signs

The quotient of two terms of *like* sign is *positive.*

$$\frac{+a}{+b} = \frac{-a}{-b} = \frac{a}{b}$$

The quotient of two terms of *unlike* sign is *negative.*

$$\frac{+a}{-b} = \frac{-a}{+b} = -\frac{a}{b}$$

The fraction itself carries a third sign, which, when negative, reverses the sign of the quotient. These three ideas are summarized in the following rules:

| | | |
|---|---|---|
| Rules of Signs for Division | $\frac{+a}{+b} = \frac{-a}{-b} = -\frac{-a}{+b} = -\frac{+a}{-b} = \frac{a}{b}$ | **10** |
| | $\frac{+a}{-b} = \frac{-a}{+b} = -\frac{-a}{-b} = -\frac{a}{b}$ | **11** |

These rules show that any *pair* of negative signs may be removed without changing the value of the fraction.

---

**EXAMPLE 60:** Simplify $-\frac{ax^2}{-y}$.

**Solution:** Removing the pair of minus signs, we obtain

$$-\frac{ax^2}{-y} = \frac{ax^2}{y}$$

---

### Dividing a Monomial by a Monomial

Any quantity (except zero) divided by itself equals *one*. So if the same factor appears in both the divisor and the dividend, it may be eliminated. This process is called *canceling*.

---

**EXAMPLE 61:** Divide $6ax$ by $3a$.

**Solution:**

$$\frac{6ax}{3a} = \frac{6}{3} \cdot \frac{a}{a} \cdot x = 2x$$

---

To divide quantities having exponents, we make use of the law of exponents for division:

| | | |
|---|---|---|
| Quotients | $\frac{x^a}{x^b} = x^{a-b}$ | **30** |

---

**EXAMPLE 62:** Divide $y^5$ by $y^3$.

**Solution:** By Eq. 30,

$$\frac{y^5}{y^3} = y^{5-3} = y^2$$

---

When there are numerical coefficients, divide them separately.

---

**EXAMPLE 63:** Divide $15x^6$ by $3x^4$.

**Solution:**

$$\frac{15x^6}{3x^4} = \frac{15}{3} \cdot \frac{x^6}{x^4} = 5x^{6-4} = 5x^2$$

---

If there is more than one unknown, treat each separately.

**EXAMPLE 64:** Divide $18x^5y^2z^4$ by $3x^2yz^3$.

**Solution:**

$$\frac{18x^5y^2z^4}{3x^2yz^3} = \frac{18}{3} \cdot \frac{x^5}{x^2} \cdot \frac{y^2}{y} \cdot \frac{z^4}{z^3}$$
$$= 6x^3yz$$

Sometimes negative exponents will be obtained.

**EXAMPLE 65:**

$$\frac{6x^2}{x^5} = 6x^{2-5} = 6x^{-3}$$

The answer may be left in this form, or, as is usually done, use Eq. 35 to eliminate the negative exponent. Thus

$$6x^{-3} = \frac{6}{x^3}$$

**Example 66:** Simplify the fraction

$$\frac{3x^2yz^5}{9xy^4z^2}$$

**Solution:** The procedure is no different than if we had been asked to divide $3x^2yz^5$ by $9xy^4z^2$.

$$\frac{3x^2yz^5}{9xy^4z^2} = \frac{3}{9}x^{2-1}y^{1-4}z^{5-2}$$
$$= \frac{1}{3}xy^{-3}z^3$$

or

$$= \frac{xz^3}{3y^3}$$

**The process of dividing a monomial by a monomial is also referred to as *simplifying a fraction,* or *reducing a fraction to lowest terms.* We discuss this in Chapter 8.**

Do not be dismayed if the expressions to be divided have negative exponents. Apply Eq. 30, as before.

**EXAMPLE 67:** Divide $21x^2y^{-3}z^{-1}$ by $7x^{-4}y^2z^{-3}$.

**Solution:** Proceeding as before, we obtain

$$\frac{21x^2y^{-3}z^{-1}}{7x^{-4}y^2z^{-3}} = \frac{21}{7}x^{2-(-4)}y^{-3-2}z^{-1-(-3)}$$
$$= 3x^6y^{-5}z^2$$
$$= \frac{3x^6z^2}{y^5}$$

### Dividing a Multinomial by a Monomial

Divide each term of the multinomial by the monomial.

**EXAMPLE 68:** Divide $9x^2 + 3x - 2$ by $3x$.

**Solution:**

$$\frac{9x^2 + 3x - 2}{3x} = \frac{9x^2}{3x} + \frac{3x}{3x} - \frac{2}{3x}$$

This is really a consequence of the distributive law (Eq. 5).

$$\frac{9x^2 + 3x - 2}{3x} = \frac{1}{3x}(9x^2 + 3x - 2) = \left(\frac{1}{3x}\right)(9x^2) + \left(\frac{1}{3x}\right)(3x) - \left(\frac{1}{3x}\right)(2)$$

Each of these terms is now simplified as in the preceding section.

$$\frac{9x^2}{3x} + \frac{3x}{3x} - \frac{2}{3x} = 3x + 1 - \frac{2}{3x}$$

| | |
|---|---|
| Common Error | Do not forget to divide *every* term of the multinomial by the monomial. $\frac{x^2 + 2x + 3}{x} \neq x + 2 + 3$ |

**EXAMPLE 69:**

(a) $$\frac{3ab + 2a^2b - 5ab^2 + 3a^2b^2}{15ab} = \frac{3ab}{15ab} + \frac{2a^2b}{15ab} - \frac{5ab^2}{15ab} + \frac{3a^2b^2}{15ab}$$

$$= \frac{1}{5} + \frac{2a}{15} - \frac{b}{3} + \frac{ab}{5}$$

(b) $$\frac{6x^3y - 8xy^2 - 2x^4y^4}{-2xy} = -3x^2 + 4y + x^3y^3$$

| | |
|---|---|
| Common Error | There is no similar rule for dividing a monomial by a multinomial. $\frac{a}{b + c} \neq \frac{a}{b} + \frac{a}{c}$ |

## Dividing a Polynomial by a Polynomial

Write the divisor and the dividend in the order of descending powers of the variable. Supply any missing terms, using a coefficient of zero. Set up as a long-division problem, as in the following example.

**EXAMPLE 70:** Divide $x^2 + 4x^4 - 2$ by $1 + x$.

This method is used only for *polynomials:* expressions in which the powers are all *positive integers.*

**Solution:**

1. Write the dividend in descending order of the powers.

$$4x^4 + x^2 - 2$$

2. Supply the missing terms with coefficients of zero.

$$4x^4 + 0x^3 + x^2 + 0x - 2$$

3. We write the divisor in descending order of the powers, and set up in long-division format.

$$(x + 1)\overline{)4x^4 + 0x^3 + x^2 + 0x - 2}$$

4. Divide the first term in the dividend ($4x^4$) by the first term in the divisor ($x$). The result ($4x^3$) is written above the dividend, in line with the term having the same power. It is the first term of the quotient.

$$\begin{array}{r} 4x^3 \phantom{+ x^2 + 0x - 2} \\ (x + 1)\overline{)4x^4 + 0x^3 + x^2 + 0x - 2} \end{array}$$

5. Multiply the divisor by the first term of the quotient. Write the result below the dividend. Subtract it from the dividend.

$$\begin{array}{r} 4x^3 \phantom{+ x^2 + 0x - 2} \\ (x + 1)\overline{)4x^4 + 0x^3 + x^2 + 0x - 2} \\ \underline{4x^4 + 4x^3 \phantom{+ x^2 + 0x - 2}} \\ -4x^3 + x^2 + 0x - 2 \end{array}$$

6. Repeat steps 4 and 5, each time using the new dividend obtained, until the degree of the remainder is less than the degree of the divisor.

$$\begin{array}{r} 4x^3 - 4x^2 + 5x - 5 \\ (x + 1)\overline{)4x^4 + 0x^3 + x^2 + 0x - 2} \\ \underline{4x^4 + 4x^3 \phantom{+ x^2 + 0x - 2}} \\ -4x^3 + x^2 + 0x - 2 \\ \underline{-4x^3 - 4x^2 \phantom{+ 0x - 2}} \\ 5x^2 + 0x - 2 \\ \underline{5x^2 + 5x \phantom{- 2}} \\ -5x - 2 \\ \underline{-5x - 5} \\ 3 \end{array}$$

The result is written

$$\frac{4x^4 + x^2 - 2}{x + 1} = 4x^3 - 4x^2 + 5x - 5 + \frac{3}{x + 1}$$

or $$4x^3 - 4x^2 + 5x - 5 \quad (R3)$$

---

**Common Error**

Errors are often made during the subtraction step (step 5 in the preceding example).

$$\begin{array}{r} 4x^3 \phantom{+ x^2 + 0x - 2} \\ (x + 1)\overline{)4x^4 + 0x^3 + x^2 + 0x - 2} \\ \underline{4x^4 + 4x^3 \phantom{+ x^2 + 0x - 2}} \\ 4x^3 + x^2 + 0x - 2 \end{array}$$

No! Should be $-4x^3$.

**EXAMPLE 71:** Divide $x^4 - 5x^3 + 9x^2 - 7x + 2$ by $x^2 - 3x + 2$.

**Solution:** Using long division, we get

$$
\begin{array}{r}
x^2 - 2x + 1 \\
x^2 - 3x + 2 \overline{\big)\, x^4 - 5x^3 + 9x^2 - 7x + 2} \\
\underline{x^4 - 3x^3 + 2x^2} \qquad\quad \\
-2x^3 + 7x^2 - 7x \quad \\
\underline{-2x^3 + 6x^2 - 4x} \quad \\
x^2 - 3x + 2 \\
\underline{x^2 - 3x + 2}
\end{array}
$$

## Synthetic Division

The division in Example 70 can be done more quickly and written more compactly by a method known as *synthetic division*. Using the same expressions as in Example 70, we first remove all the $x$'s:

**We'll use synthetic division later when solving third- and fourth-degree equations.**

$$
\begin{array}{r}
4 - 4 + 5 - 5 \\
1 \overline{\big)\, 4 + 0 + 1 + 0 - 2} \\
\underline{\mathbf{4} + 4} \qquad\qquad\quad \\
-4 + \mathbf{1} + \mathbf{0} - \mathbf{2} \\
\underline{-\mathbf{4} - 4} \qquad\quad \\
5 + \mathbf{0} - \mathbf{2} \\
\underline{\mathbf{5} + 5} \quad \\
-5 - \mathbf{2} \\
\underline{-\mathbf{5} - 5} \\
3
\end{array}
$$

The same information can be conveyed if we now eliminate those numbers (shown in boldface type) which are merely repeats of the numbers directly above them:

**Synthetic division works only if the divisor is of the form $x - r$, where $r$ is a positive or negative constant. Don't try to use it to divide by, say, $3x - 4$. Further, as with long division, we use it only for polynomials.**

$$
\begin{array}{r}
4 - 4 + 5 - 5 \\
1 \overline{\big)\, 4 + 0 + 1 + 0 - 2} \\
\underline{+4} \qquad\qquad\qquad \\
-4 \qquad\qquad\qquad \\
\underline{-4} \qquad\qquad \\
5 \qquad\qquad \\
\underline{+5} \qquad \\
-5 \qquad \\
\underline{-5} \\
3
\end{array}
$$

which can now be written compactly as

$$
\begin{array}{r|rrrrr}
1 & 4 & 0 & 1 & 0 & -2 \\
 & & 4 & -4 & 5 & -5 \\
\hline
 & 4 & -4 & 5 & -5 & 3
\end{array}
$$

Finally, since errors are less likely to occur when adding than when subtracting, we *change the sign of the divisor,* and add rather than subtract the numbers in the second line.

divisor

$$\begin{array}{r|rrrrr} -1 & 4 & 0 & 1 & 0 & -2 \\ & & -4 & 4 & -5 & 5 \\ \hline & 4 & -4 & 5 & -5 & 3 \end{array}$$

quotient remainder

---

**EXAMPLE 72:** Using synthetic division, divide $2x^4 - 5x^2 - 7x + 3$ by $x - 2$.

**Solution:** We set up the division, including a zero coefficient for the missing $x^3$ term:

$$\begin{array}{r|rrrrr} 2 & 2 & 0 & -5 & -7 & 3 \\ & \downarrow & & & & \\ \hline & 2 & & & & \end{array}$$

and bring down the first term. We then multiply that number by the divisor and place the product under the second number in the dividend. Then add.

$$\begin{array}{r|rrrrr} 2 & 2 & 0 & -5 & -7 & 3 \\ & & 4 & & & \\ \hline & 2 & 4 & & & \end{array}$$

Repeat this procedure until the last column is reached.

$$\begin{array}{r|rrrrr} 2 & 2 & 0 & -5 & -7 & 3 \\ & & 4 & 8 & 6 & -2 \\ \hline & 2 & 4 & 3 & -1 & 1 \end{array}$$

The quotient is thus $2x^3 + 4x^2 + 3x - 1$ with a remainder of 1.

---

## EXERCISE 5—DIVISION OF ALGEBRAIC EXPRESSIONS

Divide the following monomials.

1. $z^5$ by $z^3$
2. $45y^3$ by $15y^2$
3. $-25x^2y^2z^3 \div 5xyz^2$
4. $20a^5b^5c \div 10abc$
5. $30cd^2f \div 15cd^2$
6. $36ax^2y \div 18ay$
7. $-18x^2yz \div 9xy$
8. $(x - y)^5$ by $(x - y)^3$
9. $15axy^3$ by $-3ay$
10. $-18a^3x$ by $-3ay$
11. $6acdxy^2$ by $2adxy^2$
12. $12a^2x^2$ by $-3a^2x$
13. $15ay^2$ by $-3ay$
14. $45(a - x)^3$ by $15(a - x)^2$
15. $-21vwz^2 \div 7vz^2$
16. $-33r^2s \div 11rs^2$
17. $35m^2nx \div 5m^2x$
18. $20x^3y^3z^3 \div 10x^3yz^3$
19. $16ab$ by $4a$
20. $21acd$ by $7c$
21. $ab^2c$ by $ac$
22. $(a + b)^4$ by $(a + b)$
23. $x^m$ by $x^n$
24. $-14a^2x^3y^4 \div 7axy^2$
25. $32r^2s^2q \div 8r^2sq$
26. $-18v^2x^2y \div v^2xy$
27. $24a^{2n}bc^n \div (-a^{2n}bc^n)$
28. $36a^{2n}xy^{3n} \div (-4a^nxy^{2n})$
29. $25xyz^2 \div (-5x^2y^2z)$
30. $-28y^2z^{2m} \div 4y^2z^3x$
31. $-30n^2x^2 \div 6m^2x^3$
32. $28x^2y^2z^2 \div 7xyz$
33. $6abc$ by $2c$
34. $ax^3$ by $ax^2$
35. $3mx^6$ by $mx$
36. $210c^3b$ by $7cd$

37. $42xy$ by $xy^2$
38. $-21ac$ by $-7a$
39. $-12xy$ by $3y$
40. $72abc$ by $-8c$

Divide the polynomial by the monomial.

41. $2a^3 - a^2$ by $a$
42. $42a^5 - 6a^2$ by $6a$
43. $21x^4 - 3x^2$ by $3x^2$
44. $35m^4 - 7p^2$ by $7$
45. $27x^6 - 45x^4$ by $9x^2$
46. $24x^6 - 8x^3$ by $-8x^3$
47. $34x^3 - 51x^2$ by $17x$
48. $5x^5 - 10x^3$ by $-5x^3$
49. $-3a^2 - 6ac$ by $-3a$
50. $-5x^3 + x^2y$ by $-x^2$
51. $ax^2y - 2xy^2$ by $xy$
52. $3xy^2 - 3x^2y$ by $xy$
53. $4x^3y^2 + 2x^2y^3$ by $2x^2y^2$
54. $3a^2b^2 - 6ab^3$ by $3ab$
55. $abc^2 - a^2b^2c$ by $-abc$
56. $9x^2y^2z + 3xyz^2$ by $3xyz$
57. $x^2y^2 - x^3y - xy^3$ by $xy$
58. $a^3 - a^2b - ab^2$ by $-a$
59. $a^2b - ab + ab^2$ by $-ab$
60. $xy - x^2y^2 + x^3y^3$ by $-xy$
61. $a^2 - 3ab + ac^2$ by $a$
62. $x^2y - xy^2 + x^2y^3$ by $xy$
63. $x^2 - 2xy + y^2$ by $x$
64. $z^2 - 3xz + 3z^3$ by $z$
65. $m^2n + 2mn - 3m^2$ by $mn$
66. $c^2d - 3cd^2 + 4d^3$ by $cd$
67. $a^2x - abx - acx$ by $ax$
68. $3x^5y^3 - 3x^4y^3 - 3x^2y^4$ by $3x^3y^2$
69. $a^2b^2 - 2ab - 3ab^3$ by $ab$
70. $3a^3c^3 + 3a^2c - 3ac^2$ by $3ac$
71. $6a^3x^2 - 15a^4x^2 + 30a^3x^3$ by $-3a^3x^2$
72. $20x^2y^4 - 14xy^3 + 8x^2y^2$ by $2x^2y^2$

Divide the polynomial by the binomial.

**Use synthetic division for some of these, as directed by your instructor.**

73. $x^2 + 15x + 56$ by $x + 7$
74. $x^2 - 15x + 56$ by $x - 7$
75. $x^2 - 4x + 3$ by $x + 2$
76. $2 + 4x - x^2$ by $4 - x$
77. $2a^2 + 11a + 5$ by $2a + 1$
78. $6a^2 - 7a - 3$ by $2a - 3$
79. $27a^3 - 8b^3$ by $3a - 2b$
80. $2x^2 - 5x + 4$ by $x + 1$
81. $4 + 2x - 5x^2$ by $3 - x$
82. $4a^2 + 23a + 15$ by $4a + 3$
83. $3a^2 - 4a - 4$ by $2 - a$
84. $x^4 + x^2 + 1$ by $x^2 + 1$
85. $x^8 + x^4 + 1$ by $x^4 - x$
86. $1 - a^3b^3$ by $1 - ab$
87. $x^3 - 8x - 3$ by $x - 3$

Divide the polynomial by the trinomial.

88. $a^2 - 2ab + b^2 - c^2$ by $a - b - c$
89. $a^2 + 2ab + b^2 - c^2$ by $a + b + c$
90. $x^2 - y^2 + 2yz - z^2$ by $x - y + z$
91. $c^4 + 2c^2 - c + 2$ by $c^2 - c + 1$
92. $x^2 - 4y^2 - 4yz - z^2$ by $x + 2y + z$
93. $x^5 + 37x^2 - 70x + 50$ by $x^2 - 2x + 10$
94. $6(x - y)^2 - 7(x - y) - 20$ by $3(x - y) + 4$
95. $23x^2 - 5x^4 - 12 + 12x^5 + 8x - 14x^3$ by $x - 2 + 3x^2$
96. $16(a + b)^4 - 81$ by $2(a + b) - 3$
97. $8x^6 - 4x^5 - 2x^4 + 15x^3 + 3x^2 - 5x - 15$ by $4x^3 - x - 3$
98. $a^2 - 4b^2 - 9c^2 + 12bc$ by $a - 3c + 2b$

## CHAPTER 2 REVIEW PROBLEMS

1. Multiply: $(b^4 + b^2x^3 + x^4)(b^2 - x^2)$
2. Square: $(x + y - 2)$
3. Evaluate: $(7.28 \times 10^4)^3$
4. Square: $(xy + 5)$
5. Multiply: $(3x - m)(x^2 + m^2)(3x - m)$
6. Divide: $-x^6 - 2x^5 - x^4$ by $-x^4$
7. Cube: $(2x + 1)$
8. Simplify: $7x - \{-6x - [-5x - (-4x - 3x) - 2]\}$
9. Multiply: $3ax^2$ by $2ax^3$
10. Simplify: $\left(\frac{3a^2}{2b^3}\right)^3$
11. Square: $(4a - 3b)$
12. Multiply: $(x^3 - xy + y^2)(x + y)$
13. Multiply: $(xy - 2)(xy - 4)$
14. Divide: $x^{m+1} + x^{m+2} + x^{m+3} + x^{m+4}$ by $x^4$
15. Square: $(2a - 3b)$
16. Multiply: $(a^2 - 3a + 8)(a + 3)$
17. Divide: $a^3b^2 - a^2b^5 - a^4b^2$ by $a^2b$
18. Multiply: $(2x - 5)(x + 2)$
19. Multiply: $(2m - c)(2m + c)(4m^2 + c^2)$
20. Simplify: $\left(\frac{8x^5y^{-2}}{4x^3y^{-3}}\right)^3$
21. Divide: $2a^6$ by $a^4$
22. Multiply: $(a^2 + a^2y + ay^2 + y^3)(a - y)$
23. Simplify: $y - 3[y - 2(4 - y)]$
24. Divide: $-a^7$ by $a^5$
25. Multiply: $(2x^2 + xy - 2y^2)(3x + 3y)$
26. Divide: $x^4 - \frac{1}{2}x^3 - \frac{1}{3}x^2 - 2x - 1$ by $2x$
27. Simplify: $-2[w - 3(2w - 1)] + 3w$
28. Multiply: $(a^2 + b)(a + b^2)$
29. Multiply: $(a^4 - 2a^3c + 4a^2c^2 - 8ac^3 + 16c^4)(a + 2c)$
30. Divide: $16x^3$ by $4x$
31. Divide: $(a - c)^m$ by $(a - c)^2$
32. Divide: $7 - 8c^2 + 5c^3 + 8c$ by $5c - 3$
33. Divide: $-x^2y - xy^2$ by $-xy$
34. Combine: $(-2x^2 - x + 6) - (7x^2 - 2x + 4) + (x^2 - 3)$
35. Cube: $(b - 3)$
36. Simplify: $(2x^2y^3z^{-1})^3$
37. Divide: $2x^{-2}y^3$ by $4x^{-4}y^6$
38. Simplify: $-\{-[-(a - 3) - a]\} + 2a$
39. Multiply: $(x - 2)(x + 4)$
40. Square: $(x^2 + 2)$
41. Evaluate: $(2.83 \times 10^3)^2$
42. Simplify: $(x - 3) - [x - (2x + 3) + 4]$
43. Multiply: $2xy^3$ by $5x^2y$
44. Divide: $27xy^5z^2$ by $3x^2yz^2$
45. Multiply: $(x - 1)(x^2 + 4x)$
46. Square: $(z^2 - 3)$
47. Evaluate: $(1.33 \times 10^4)^2$
48. Simplify: $(y + 1) - [y(y + 1) + (3y - 1) + 5]$
49. Multiply: $2ab^2$ by $3a^2b$
50. Divide: $64ab^4c^3$ by $8a^2bc^3$

*Writing*

51. In the following chapter we solve simple equations, something you must have done before. Without peeking ahead, write down in your own words whatever you remember about solving equations. You may give a list of steps, or give a description in paragraph form.

# 3

# SIMPLE EQUATIONS AND WORD PROBLEMS

## OBJECTIVES

**When you have completed this chapter, you should be able to:**

- Solve simple linear equations.
- Check an apparent solution to a linear equation.
- Solve simple fractional equations.
- Write an algebraic expression to describe a given verbal statement.
- Set up and solve word problems that lead to linear equations.
- Solve number problems, financial problems, mixture problems, and statics problems.

When two mathematical expressions are set equal to each other, we get an *equation*. Much of our work in technical mathematics is devoted to *solving equations*. We start in this chapter with the simplest types, and in later chapters cover more difficult ones (quadratic equations, exponential equations, trigonometric equations, and so forth).

And why is it so important to solve equations? The main reason is that equations describe the way certain things happen in the world. Solving an equation often tells us something important we would not otherwise know. For example, the note produced by a guitar string is not a whim of nature, but can be predicted (solved for) when you know the length, the mass, and the tension in the string (and you know the *equation* relating the pitch, length, mass, and tension).

Thousands of equations exist that link together the various quantities in the physical world, in chemistry, finance, and so on, and their number is still increasing. To be able to solve and to manipulate such equations is essential for anyone who has to deal with these quantities on the job.

## 3-1 SOLVING FIRST-DEGREE EQUATIONS

### Equations

An equation has two *sides* or members, and an *equal sign*.

$$\underbrace{3x^2 - 4x}_{\text{left side}} \underset{\text{equal sign}}{\overset{}{=}} \underbrace{2x + 5}_{\text{right side}}$$

### Conditional Equation

A *conditional equation* is one whose sides are equal only for certain values of the unknown.

---

**EXAMPLE 1:** The equation

$$x - 5 = 0$$

is a conditional equation because the sides are equal only when $x = 5$. When we say "equation," we will mean "conditional equation." Other equations that are true for any value of the unknown, such as $x(x + 2) = x^2 + 2x$, are called *identities*.

**The symbol $\equiv$ is used for identities. We would write $x(x + 2) \equiv x^2 + 2x$.**

---

### The Solution

The *solution* to an equation is that value (or values) of the unknown that make the sides equal.

---

**EXAMPLE 2:** The solution to the equation

$$x - 5 = 0$$

is

$$x = 5$$

The value $x = 5$ is also called the *root* of the equation. We also say that it *satisfies* the equation.

---

## First-Degree Equations

A *first-degree* equation (also called a *linear* equation) is one in which the terms containing the unknown are all of first degree.

---

**EXAMPLE 3:** The equations $2x + 3 = 9 - 4x$, $3x/2 = 6x + 3$, and $3x + 5y - 6z = 0$ are all of first degree. The first and second equations have one variable, but the third equation has more than one variable.

## Solving an Equation

To solve an equation, we *perform the same mathematical operation on each side of the equation*. The object is to get the unknown standing alone on one side of the equation.

---

**EXAMPLE 4:** Solve $3x = 8 + 2x$.

**Solution:** Subtracting $2x$ from both sides, we obtain

$$3x - 2x = 8 + 2x - 2x$$

Combining like terms yields

$$x = 8$$

---

Check an apparent solution by substituting it back into the original equation.

---

**EXAMPLE 5:** Checking the apparent solution to Example 4, we substitute 8 for $x$ in the original equation:

**Get into the habit of checking your work, for errors creep in everywhere.**

$$3x = 8 + 2x$$

$$3(8) \stackrel{?}{=} 8 + 2(8)$$

$$24 = 8 + 16 \quad \text{checks}$$

---

| Common Error | Check your solution only in the *original* equation. Later versions may already contain errors. |
|---|---|

---

**EXAMPLE 6:** Solve the equation $3x - 5 = x + 1$.

**Solution:** We first subtract $x$ from both sides, and add 5 to both sides.

$$\begin{array}{rcr} 3x - 5 & = & x + 1 \\ -x + 5 & & -x + 5 \\ \hline 2x & = & 6 \end{array}$$

Dividing both sides by 2, we obtain

$$x = 3$$

Check: Substituting into the original equation yields

$$3(3) - 5 \stackrel{?}{=} 3 + 1$$

$$9 - 5 = 4 \quad \text{checks}$$

---

## Symbols of Grouping

When the equation contains symbols of grouping, remove them early in the solution.

---

**EXAMPLE 7:** Solve the equation $3(3x + 1) - 6 = 5(x - 2) + 15$.

**Solution:** First, remove the parentheses:

$$9x + 3 - 6 = 5x - 10 + 15$$

Combine like terms:

$$9x - 3 = 5x + 5$$

Add $-5x + 3$ to both sides:

$$\frac{\begin{array}{rcl} -5x + 3 & & -5x + 3 \end{array}}{4x = 8}$$

Divide by 4:

$$x = 2$$

Check:

$$3(6 + 1) - 6 \stackrel{?}{=} 5(2 - 2) + 15$$

$$21 - 6 \stackrel{?}{=} 0 + 15$$

$$15 = 15 \quad \text{checks}$$

---

> The mathematical operations you perform must be done *to both sides* of the equation in order to preserve the equality.

> The mathematical operations you perform on both sides of an equation must be done to each side *as a whole*—not term by term.

## Fractional Equations

A *fractional equation* is one that contains one or more fractions. When an equation contains a single fraction, the fraction can be eliminated by *multiplying both sides by the denominator of the fraction.*

---

**EXAMPLE 8:** Solve:

$$\frac{x}{3} - 2 = 5$$

**Solution:** Multiplying both sides by 3 gives us

$$3\left(\frac{x}{3} - 2\right) = 3(5)$$

$$3\left(\frac{x}{3}\right) - 3(2) = 3(5)$$

$$x - 6 = 15$$

Adding 6 to both sides yields

$$x = 15 + 6 = 21$$

Check:

$$\frac{21}{3} - 2 \stackrel{?}{=} 5$$

$$7 - 2 = 5 \quad \text{checks}$$

---

When there are two or more fractions, multiplying both sides by the product of the denominators (called a *common* denominator) will clear the fractions.

**In Chapter 8 we treat fractional equations in detail. There we will see how to find the *least* common denominator.**

**EXAMPLE 9:** Solve:

$$\frac{x}{2} + 3 = \frac{x}{3}$$

**Solution:** Multiplying by 2(3) or 6, we have

$$6\left(\frac{x}{2} + 3\right) = 6\left(\frac{x}{3}\right)$$

$$6\left(\frac{x}{2}\right) + 6(3) = 6\left(\frac{x}{3}\right)$$

$$3x + 18 = 2x$$

Rearranging gives us

$$3x - 2x = -18$$

$$x = -18$$

Check:

$$\frac{-18}{2} + 3 \stackrel{?}{=} \frac{-18}{3}$$

$$-9 + 3 \stackrel{?}{=} -6$$

$$-6 = -6 \quad \text{checks}$$

## Strategy

While solving an equation, you should keep in mind the objective of *getting x by itself on one side of the equal sign, with no x on the other side.* Use any valid operation toward this end.

Equations can be of so many different forms that we cannot give a step-by-step procedure for their solution, but the following tips should help.

**There are computer programs now that can simplify algebraic expressions and solve equations. They use a step-by-step procedure (an algorithm) something like the one given here.**

1. Eliminate fractions by multiplying both sides by the common denominator of both sides.
2. Remove any parentheses by performing the indicated multiplication.
3. All terms containing $x$ should be moved to one side of the equation, and all other terms moved to the other side.
4. Like terms on the same side of the equation should be combined at any stage of the solution.
5. Any coefficient of $x$ may be removed by dividing both sides by that coefficient.
6. Check the answer.

**EXAMPLE 10:** Solve:

$$\frac{3x-5}{2} = \frac{2(x-1)}{3} + 4$$

**Solution:**

1. We eliminate the fractions by multiplying by 6:

$$6\left(\frac{3x-5}{2}\right) = 6\left[\frac{2(x-1)}{3}\right] + 6(4)$$

$$3(3x-5) = 4(x-1) + 6(4)$$

2. Then we remove parentheses:

$$9x - 15 = 4x - 4 + 24$$

$$9x - 15 = 4x + 20$$

3. We get all $x$ terms on one side by adding 15 and subtracting $4x$ from both sides:

$$9x - 4x = 20 + 15$$

4. Combining like terms gives us

$$5x = 35$$

5. Dividing by the coefficient of $x$ yields

$$x = 7$$

6. Check:

$$\frac{3(7)-5}{2} \stackrel{?}{=} \frac{2(7-1)}{3} + 4$$

$$\frac{21-5}{2} \stackrel{?}{=} \frac{12}{3} + 4$$

$$\frac{16}{2} \stackrel{?}{=} 4 + 4$$

$$8 = 8 \qquad \text{checks}$$

**EXAMPLE 11:** Solve:

$$3(x+2)(2-x) = 3(x-2)(x-3) + 2x(4-3x)$$

**Solution:** We remove parentheses in two steps. First,

$$3(2x - x^2 + 4 - 2x) = 3(x^2 - 5x + 6) + 8x - 6x^2$$

Then

$$-3x^2 + 12 = 3x^2 - 15x + 18 + 8x - 6x^2$$

Next we combine like terms:

$$\cancel{-3x^2} + 12 = \cancel{-3x^2} - 7x + 18$$

Subtracting 18 from both sides gives

$$12 - 18 = -7x$$

$$-6 = -7x$$

Dividing by $-7$ and switching sides, we obtain

$$x = \frac{6}{7}$$

---

## EXERCISE 1—SOLVING FIRST-DEGREE EQUATIONS

Solve and check each equation.

**Treat the integers in these equations as exact numbers. Leave your answers in fractional, rather than decimal, form. When the equation has decimal coefficients, keep as many significant digits in your answer as contained in those coefficients.**

1. $x - 5 = 28$
2. $4x + 2 = 18$
3. $8 - 3x = 7x$
4. $9x + 7 = 25$
5. $9x - 2x = 3x + 2$
6. $x + 8 = 12 - 3x$
7. $x + 4 = 5x + 2$
8. $3y - 2 = 2y$
9. $3x = x + 8$
10. $3x = 2x + 5$
11. $3x + 4 = x + 10$
12. $y - 4 = 7 + 23y$
13. $2.8 - 1.3y = 4.6$
14. $3.75x + 7.24 = 5.82x$
15. $4x + 6 = x + 9$
16. $7x - 19 = 5x + 7$
17. $p - 7 = -3 + 9p$
18. $-3x + 7 = -15 + 2x$
19. $4x - 11 = 2x - 7$
20. $3x - 8 = 12x - 7$
21. $8x + 7 = 4x + 27$
22. $3x + 10 = x + 20$
23. $\frac{1}{2}x = 14$
24. $\frac{x}{5} = 3x - 25$
25. $5x + 22 - 2x = 31$
26. $4x + 20 - 6 = 34$
27. $\frac{x}{3} + 5 = 2x$
28. $5x - \frac{2x}{5} = 3$
29. $2x + 5(x - 4) = 6 + 3(2x + 3)$
30. $6x - 5(x - 2) = 12 - 3(x - 4)$
31. $4(3x + 2) - 4x = 5(2 - 3x) - 7$
32. $6 - 3(2x + 4) - 2x = 7x + 4(5 - 2x) - 8$
33. $3(6 - x) + 2(x - 3) = 5 + 2(3x + 1) - x$
34. $3x - 2 + x(3 - x) = (x - 3)(x + 2) - x(2x + 1)$
35. $(2x - 5)(x + 3) - 3x = 8 - x + (3 - x)(5 - 2x)$
36. $x + (1 + x)(3x + 4) = (2x + 3)(2x - 1) - x(x - 2)$
37. $3 - 6(x - 1) = 9 - 2(1 + 3x) + 2x$
38. $7x + 6(2 - x) + 3 = 4 + 3(6 + x)$
39. $3 - (4 + 3x)(2x + 1) = 6x - (3x - 2)(x + 1) - (6 - 3x)(1 - x)$
40. $6.1 - (x + 2.3)(3.2x - 4.3) - 5.2x = 3.3x - x(3.1 + 3.2x) - 4.7$
41. $7x - 2x(2x - 3) - 2 = 2x^2 - (x - 2)(3 + 6x) - 8$
42. $2x - 3(x - 6)(2x - 2) + 6 = 3x - (2x - 1)(3x + 2) + 6$
43. $8x - 2(2 + x)(3x - 6) = 3 - 3(2x + 5)(x - 1) - 2(2x - 3)$
44. $1.3x - (x - 2.3)(3.1x - 5.4) - 2.5 = 4.6 - (3.1x - 1.2)(x - 5.3) - 4.3x$

## 3-2 SOLVING WORD PROBLEMS

### Problems in Verbal Form

**Math skills are important, but so are reading skills. Most students find word problems difficult, but if you have an *unusual* amount of trouble with them you should seek out a reading teacher and get help.**

We have already done some simple word problems in earlier chapters, and we will have more of them with each new topic, but their main treatment will be here. After looking at these word problems you may ask: "When will I ever have to solve problems like that on the job?"

Probably never. But you will most likely have to deal with technical material in written form: instruction manuals, specifications, contracts, building codes, handbooks, and so on. This chapter will aid you in coping with precise technical documents, where a mistaken meaning of a word could cost dollars, lives, or your job.

Some tips on how to approach any word problem are given in the following paragraphs.

### Picture the Problem

Try to *visualize* the situation described in the problem. Form a picture in your mind. *Draw a diagram* showing as much of the given information as possible, including the unknown.

### Understand the Words

Look up the meanings of unfamiliar words in a dictionary, handbook, or textbook.

---

**EXAMPLE 12:** A certain problem contains the following statement:

> A simply supported beam carries a concentrated load midway between its center of gravity and one end.

To understand this statement, you must know what a "simply supported beam" is, and the meanings of "concentrated load" and "center of gravity."

---

Locate the words and expressions standing for mathematical operations, such as

. . . the difference of . . .
. . . is equal to . . .
. . . the quotient of . . .

---

**EXAMPLE 13:**

(a) The difference of $a$ and $b$ means $a - b$ (or $b - a$)
(b) Four less than $x$ means $x - 4$
(c) Three greater than $c$ means $c + 3$
(d) The sum of $x$ and 4 is equal to three times $x$, in symbols, is

$$x + 4 = 3x$$

(e) Three consecutive integers, starting with $x$, are

$$x, \quad x + 1, \quad x + 2$$

---

Many problems will contain a *rate*, either as the unknown or as one of the given quantities. The word *per* is your tipoff. Look for "miles *per* hour," "dollars *per* pound," "pounds *per* gallon," "dollars *per* mile," and the like. The word *per* also suggests *division*. Miles per gallon, for example, means miles traveled *divided* by gallons consumed.

### Identify the Unknown

Be sure that you know exactly *what is to be found* in a particular problem. Find the sentence that ends with a question mark, or the sentence that starts with "Find . . ." or "Calculate . . ." or "What is . . ." or "How much . . ." or similar phrases.

*Label the unknown* with a statement such as: "Let $x$ = rate of the train, mi/h." Be sure to include *units of measure* when defining the unknown.

---

**EXAMPLE 14:** Fifty-five liters of a mixture that is 24% alcohol by volume is mixed with 74 liters of a mixture that is 76% alcohol. Find the amount of alcohol in the final mixture.

You may be tempted to label the unknown in the following ways:

**This step of labeling the unknown annoys many students; they see it as a nitpicking thing that teachers make them do just for show, when they would rather start solving the problem. But realize that this step *is* the start of the problem. It is the crucial first step and *should never be omitted.***

| | |
|---|---|
| Let $x$ = alcohol | This is too vague. Are we speaking of volume of alcohol, or weight, or some other property? |
| Let $x$ = volume of alcohol | The units are missing. |
| Let $x$ = liters of alcohol | Do we mean liters of alcohol in one of the *initial* mixtures or in the *final* mixture? |
| Let $x$ = number of liters of alcohol in final mixture | Good. |

We will do complete solutions of this type of problem in Sec. 3-4.

---

### Define Other Unknowns

If there is more than one unknown, you will often be able to define the additional unknowns *in terms of the original unknown,* rather than introducing another symbol.

---

**EXAMPLE 15:** A word problem contains the following statement: "Find two numbers whose sum is 80 and, . . ."

Here we have two unknowns. We can label them with separate symbols:

$$\text{let } x = \text{first number}$$

$$\text{let } y = \text{second number}$$

but another way would be to label the second unknown in terms of the first:

$$\text{let } x = \text{first number}$$

$$\text{let } 80 - x = \text{second number}$$

---

Sometimes a careful reading of the word problem is needed to *identify* the unknowns before you can define them in terms of one of the variables.

**EXAMPLE 16:** A word problem contains the following statement: "A person invests part of his \$1000 savings in a bank, at 6%, and the remainder in a certificate of deposit, at 8%. . . ." We are then given more information and are asked to find the amount invested at each rate.

Since the investment is in two parts, both of which are unknown, we can say

let $x$ = amount invested at 6%

We'll finish this problem in Sec. 3-3.

and $1000 - x$ = amount invested at 8%

But isn't it equally correct to say

let $x$ = amount invested at 8%

and $1000 - x$ = amount invested at 6%?

Yes. You can define the variables either way, as long as you are consistent and *do not change the definition partway through the problem.*

## Write and Solve the Equation

The unknown quantity must now be related to the given quantities by means of an *equation*. Where does one get an equation? Either the equation is given verbally, right in the problem statement, or else the relationship between the quantities in the problem is one you are expected to know or to be able to find. These could be mathematical relationships or they could be from any branch of technology. For example:

How is the area of a circle related to its radius?
How is the current in a circuit related to the circuit resistance?
How is the deflection of a beam related to its depth?
How is the distance traveled by a car related to its speed?

and so on. The relationships you will need for the problems in this book can be found in the Summary of Facts and Formulas (Appendix A). Of course, each branch of technology has many more formulas than can be shown there. Problems that need some of these formulas are treated later in this chapter.

In the following number problems, the equation is given verbally in the statement of the problem.

## Number Problems

Number problems, although of little practical value, will give you practice at picking out the relationship between the quantities in a problem.

**EXAMPLE 17:** If twice a certain number is increased by 9, the result is equal to 2 less than triple the number. Find the number.

**Solution:** Let $x$ = the number. From the problem statement,

$$2x + 9 = 3x - 2$$

Solving for $x$, we obtain

$$9 + 2 = 3x - 2x$$

$$x = 11$$

Often the solution of a problem involves finding *more than one quantity*.

**EXAMPLE 18:** Find three consecutive odd integers whose sum is 39.

**Solution:** We first define our unknowns. Since any odd integer is two greater than the preceding odd integer, we can say

Let $x$ = the first odd integer

Then $x + 2$ = the second consecutive odd integer

and $x + 4$ = the third consecutive odd integer

Since the problem states that the sum of the three integers is 39, we write

$$x + (x + 2) + (x + 4) = 39$$

Solving for $x$ gives

$$3x + 6 = 39$$

$$3x = 33$$

$$x = 11$$

But don't stop here, we are not finished. We're looking for *three* numbers. So

$$x + 2 = 13 \qquad \text{and} \qquad x + 4 = 15$$

Our three consecutive odd integers are then 11, 13, and 15.

Sometimes there may seem to be "too many" unknowns to solve a problem. If so, carefully choose one variable to call $x$, so that the other unknowns can be expressed in terms of $x$. Then use the remaining information to write your equation.

**EXAMPLE 19:** In a group of 102 employees, there are three times as many on the day shift as on the night shift, and two more on the swing shift than on the night shift. How many are on each shift?

**Solution:** Since the number of employees on the day and swing shifts are given in terms of the number on the night shift, we write

Let $x$ = the number of employees on the night shift

Then $3x$ = the number on the day shift

and $x + 2$ = the number on the swing shift

Since we are told that there are 102 employees in all,

$$x + 3x + (x + 2) = 102$$

from which

$$5x = 100$$

$$x = 20 \text{ on the night shift}$$

$$3x = 60 \text{ on the day shift}$$

$$x + 2 = \underline{22} \text{ on the swing shift}$$

Check: Adding gives $\quad 102 \quad$ total

---

## Check Your Answer

You should check your answer in the original problem statement. For example, let us check the answer (11) to Example 17.

> Twice 11 increased by 9 is 22 plus 9, or 31.
> Triple 11 is 33, less 2 is 31. checks

| Common Error | Checking an answer by substituting in your *equation* is not good enough. The equation itself may be wrong. Check your answer with the problem statement. |
|---|---|

## EXERCISE 2—SOLVING WORD PROBLEMS

### Verbal Statements

Rewrite each expression as an algebraic expression.

1. Eight more than 5 times a certain number.
2. Four consecutive integers.
3. Two numbers whose sum is 83.
4. Two numbers whose difference is 17.
5. The amounts of antifreeze and of water in 6 gal of an antifreeze–water solution.
6. A fraction whose denominator is 7 more than 5 times its numerator.
7. The angles in a triangle, if one angle is half another.
8. The number of gallons of antifreeze in a radiator containing $x$ gallons of a mixture that is 9% antifreeze.
9. The distance traveled in $x$ hours by a car going 82 km/h.
10. If $x$ equals the length of a rectangle, write an expression for the width of the rectangle if **(a)** the perimeter is 58; **(b)** the area is 200.

### Number Problems

11. Nine more than 4 times a certain number is 29. Find the number.
12. Find a number such that the sum of 15 and that number is 4 times that number.
13. Find three consecutive integers whose sum is 174.
14. The sum of 56 and some number equals 9 times that number. Find it.
15. Twelve less than 3 times a certain number is 30. Find the number.
16. When 9 is added to 7 times a certain number, the result is equal to subtracting 3 from 10 times that number. Find the number.

17. The denominator of a fraction is 2 greater than 3 times its numerator, and the reduced value of the fraction is $\frac{3}{10}$. Find the numerator.
18. Find three consecutive odd integers whose sum is 63.
19. Seven less than 4 times a certain number is 9. Find the number.
20. Find the number whose double exceeds 70 by as much as the number itself is less than 80.
21. If $2x - 3$ stands for 29, for what number does $4 + x$ stand?
22. Separate 197 into two parts such that the smaller part divides into the greater 5 times, with a remainder of 23.
23. After a certain number is diminished by 8, the amount remaining is multiplied by 8; the result is the same as if the number were diminished by 6 and the amount remaining multiplied by 6. What is the number?
24. Thirty-one times a certain number is as much above 40 as 9 times the number is below 40. Find the number.
25. Three times the amount by which a certain number exceeds 6 is equal to the number plus 144. Find the number.
26. In a group of 90 persons, composed of men, women, and children, there are 3 times as many children as men, and twice as many women as men. How many are there of each?

## 3-3 FINANCIAL PROBLEMS

We turn now from number problems to types that have more application to technology; financial, mixture, and statics problems. Uniform motion problems, which often require the solution of a fractional equation, are left for the chapter on fractions.

**EXAMPLE 20:** A consultant had to pay income taxes of \$1720 plus 22% of the amount by which her taxable income exceeded \$18,000. Her tax bill was \$1842. What was her taxable income?

**Solution:** Let $x$ = taxable income (dollars). The amount by which her income exceeded \$18,000 is then

$$x - 18{,}000$$

Her tax is 22% of that amount, plus \$1720. So

$$\text{tax} = 1720 + 0.22(x - 18{,}000) = 1842$$

Solving for $x$, we obtain

$$0.22(x - 18{,}000) = 1842 - 1720 = 122$$

$$x - 18{,}000 = \frac{122}{0.22} = 554.55$$

$$x = 554.55 + 18{,}000 = \$18{,}554.55$$

**We usually work financial problems to the nearest penny or nearest dollar, regardless of the number of significant digits in the problem.**

**EXAMPLE 21:** A person invests part of his \$1000 savings in a bank, at 6%, and part in a certificate of deposit, at 8% simple interest. He gets a total of \$75 per year in interest from the two investments. How much is invested at each rate?

**Solution:** Does this look familiar? It is the problem we started in Example 16. There we defined our variables as

$$\text{let } x = \text{amount invested at 6\%}$$

and

$$1000 - x = \text{amount invested at 8\%}$$

If $x$ dollars are invested at 6%, the interest on that investment is $0.06x$ dollars. Similarly, $0.08(1000 - x)$ dollars are earned on the other investment. Since the total earnings are \$75, we write

$$0.06x + 0.08(1000 - x) = 75$$

We multiply by 100 to eliminate the decimals, and solve for $x$.

$$6x + 8(1000 - x) = 7500$$

$$6x + 8000 - 8x = 7500$$

$$-2x = -500$$

$$x = \$250 \text{ at } 6\%$$

and

$$1000 - x = \$750 \text{ at } 8\%$$

## EXERCISE 3—FINANCIAL PROBLEMS

Financial problems make heavy use of percentage, so you may want to review Sec. 1-9.

1. The labor costs for a certain project were \$1501 per day for 17 technicians and helpers. If the technicians earned \$92/day and the helpers \$85/day, how many technicians were employed on the project?
2. A company has \$86,500 invested in bonds, and earns \$6751 in interest annually. Part of the money is invested at 7.4% and the remainder at 8.1%, simple interest. How much is invested at each rate?
3. How much must a company earn in order to have \$95,000 left after paying 27% in taxes?
4. Three equal batches of fiberglass insulation were bought for \$408: the first for \$17 per ton, the second for \$16 per ton, and the third for \$18 per ton. How many tons of each were bought?
5. A water company changed its rates from \$1.95 per 1000 gal to \$1.16 per 1000 gal plus a service charge of \$45 per month. How much water can you purchase before your new monthly bill would equal the bill under the former rate structure?
6. What salary should a person receive in order to take home \$20,000 after deducting 23% for taxes?
7. A person was \$450 in debt, owing a certain sum to a brother, twice that amount to an uncle, and twice as much to a bank as to the uncle. How much was owed to each?
8. A student sold used skis and boots for \$210, getting four times as much for the boots as for the skis. What was the price of each?
9. A person used part of a \$100,000 inheritance to build a house and invested the remainder for 1 year, one-third of it at 6% and two-thirds at 5%, simple interest. The income from both investments was \$320. Find the cost of the house.
10. A used Jeep and a snowplow attachment are worth \$1200, the Jeep being worth seven times as much as the plow. Find the value of each.

11. A person spends one-fourth of her annual income for board, one-twelfth for clothes, one-half for other expenses, and saves \$2000. What is her income?
12. An estate is to be divided among four children in the following manner: to the first, \$2000 more than one-fourth of the whole; to the second, \$3400 more than one-fifth of the whole; to the third, \$3000 more than one-sixth of the whole; and to the fourth, \$4000 more than one-eighth of the whole. What is the value of the estate?

## 3-4 MIXTURE PROBLEMS

### Basic Relationships

The total amount of mixture is, obviously, equal to the sum of the amounts of the ingredients:

| | |
|---|---|
| amount of mixture = amount of $A$ + amount of $B$ + · · · | **A1** |

These two ideas are so obvious that it may seem unnecessary to write them down. But because they are obvious, they are often overlooked. They state that the whole is equal to the sum of its parts.

and for each ingredient,

| | |
|---|---|
| final amount of each ingredient = initial amount + amount added − amount removed | **A2** |

**EXAMPLE 22:** From 100 lb of solder, half lead and half zinc, 20 lb is removed. Then 30 lb of lead is added. How much lead is contained in the final mixture?

**Solution:**

$$\text{initial weight of lead} = 0.5(100) = 50 \text{ lb}$$

$$\text{amount of lead removed} = 0.5(20) = 10 \text{ lb}$$

$$\text{amount of lead added} = 30 \text{ lb}$$

By Eq. A2,

$$\text{final amount of lead} = 50 - 10 + 30 = 70 \text{ lb}$$

### Percent Concentration

The percent concentration of each ingredient is

| | |
|---|---|
| $\text{percent concentration of ingredient} = \dfrac{\text{amount of ingredient in mixture}}{\text{total amount of mixture}} \times 100\%$ | **15** |

**EXAMPLE 23:** The percent concentration of lead in Example 22 is, by Eq. 15,

$$\text{percent lead} = \frac{70}{110} \times 100\% = 63.6\%$$

where 110 is the total weight (lb) of the final mixture.

### Two Mixtures

When *two mixtures* are combined to make a *third* mixture, the amount of any ingredient $A$ in the final mixture is

| | |
|---|---|
| final amount of $A$ = amount of $A$ in first mixture + amount of $A$ in second mixture | **A3** |

---

**EXAMPLE 24:** One hundred liters of gasohol containing 12% alcohol is mixed with 200 liters of gasohol containing 8% alcohol. The volume of alcohol in the final mixture is

$$\text{final amount of alcohol} = 0.12(100) + 0.08(200)$$
$$= 12 + 16 \quad = 28 \text{ liters}$$

---

### A Typical Mixture Problem

We now make use of these ideas about mixtures to solve a typical mixture problem.

---

**EXAMPLE 25:** A batch of steel containing 5% nickel and 95% iron is combined with another batch having 2% nickel and 98% iron, to make 3 tons of steel having 4% nickel and 96% iron. How many tons of each original batch is needed?

**Solution:**

Let $x$ = the number of tons of 5% steel needed

Then $3 - x$ = the number of tons of 2% steel needed

since 3 tons will be made, in all. Figure 3-1 shows the three alloys, with the

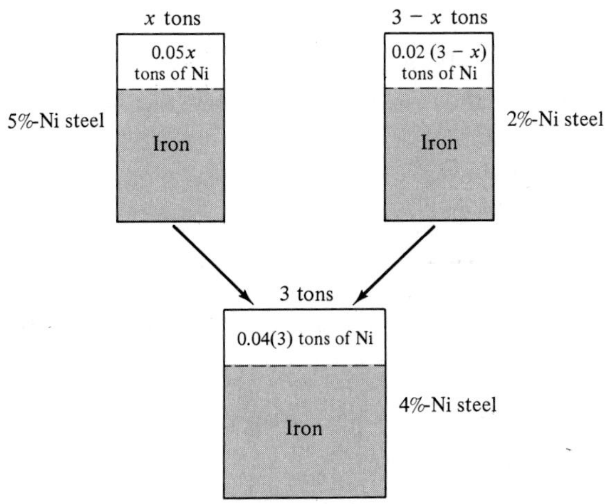


**FIGURE 3-1**

weight of the steel and the weight of the nickel in each. The amount of nickel in the 5% steel is 5% of its total weight, so

$$0.05x = \text{weight of the nickel in the 5\% steel}$$

Similarly,

$$0.02(3 - x) = \text{weight of the nickel in the 2\% steel}$$

**Notice that we are ignoring the *iron* in the alloy and writing our equations only for the nickel. Of course, we could have written our equations for the amount of iron and not for the nickel. The point is that you need to write Eq. A3 for *only one of the ingredients.***

The sum of these must give the weight of nickel in the final mixture:

$$0.05x + 0.02(3 - x) = 0.04(3)$$

Multiplying by 100 and clearing parentheses, we obtain

$$5x + 6 - 2x = 12$$

$$3x = 6$$

$$x = 2 \text{ tons of 5\% steel}$$

$$3 - x = 1 \text{ ton of 2\% steel}$$

Check:

$$\text{final tons of nickel} = 0.05(2) + 0.02(1) = 0.1 + 0.02 = 0.12 \text{ ton}$$

$$\text{percent nickel} = \frac{0.12}{3} \times 100 = 4\% \qquad \text{checks}$$

---

| | |
|---|---|
| Common Error | If you wind up with an equation that looks like this:<br>( ) lb nickel + ( ) lb iron = ( ) lb nickel<br>you know that something is wrong. When using Eq. A3 all the terms must be *for the same ingredient.* |

## EXERCISE 4—MIXTURE PROBLEMS

1. Two different mixtures of gasohol are available, one with 5% alcohol and the other containing 12% alcohol. How many gallons of the 12% mixture must be added to 250 gal of the 5% mixture to produce a mixture containing 9% alcohol?
2. How many metric tons of chromium must be added to 2.5 metric tons of stainless steel to raise the percent of chromium from 11% to 18%?
3. How many kilograms of nickel silver containing 18% zinc and how many kilograms of nickel silver containing 31% zinc must be melted together to produce 700 kg of a new nickel silver containing 22% zinc?
4. A certain bronze alloy containing 4% tin is to be added to 350 lb of bronze containing 18% tin to produce a new bronze containing 15% tin. How many pounds of the 4% bronze are required?
5. How many kilograms of brass containing 63% copper must be melted with 1100 kg of brass containing 72% copper to produce a new brass containing 67% copper?
6. A certain chain saw requires a fuel mixture of 5.5% oil and the remainder gasoline. How many liters of 2.5% mixture and how many of 9% mixture must be combined to produce 40 liters of 5.5% mixture?

7. A certain automobile cooling system contains 11 liters of coolant that is 15% antifreeze. How many liters of mixture must be removed so that, when replaced with pure antifreeze, a mixture of 25% antifreeze will result?
8. A vat contains 4000 liters of wine with an alcohol content of 10%. How much of this wine must be removed so that, when replaced with wine with a 17% alcohol content, the alcohol content in the final mixture will be 12%?
9. A certain paint mixture weighing 300 lb contains 20% solids suspended in water. How many pounds of water must be allowed to evaporate to raise the concentration of solids to 25%?
10. Fifteen liters of fuel containing 3.2% oil is available for a certain two-cycle engine. This fuel is to be used for another engine requiring a 5.5% oil mixture. How many liters of oil must be added?
11. A concrete mixture is to be made which contains 35% sand by weight, and 640 lb of mixture containing 29% sand is already on hand. How many pounds of sand must be added to this mixture to arrive at the required 35%?
12. How many liters of a solution containing 18% sulfuric acid and how many liters of another solution containing 25% sulfuric acid must be mixed together to make 550 liters of solution containing 23% sulfuric acid? (All percentages are by volume.)

## 3-5 STATICS PROBLEMS

### Moments

The *moment of a force* about some point *a* is the product of the force *F* and the perpendicular distance *d* from the force to the point.

| | | |
|---|---|---|
| Moment of a Force | $M_a = Fd$ | A12 |

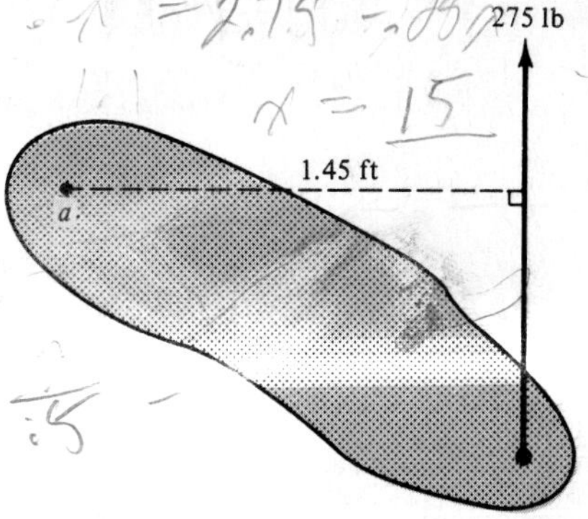


**FIGURE 3-2 Moment of a force.**

**EXAMPLE 26:** The moment of the force in Fig. 3-2 about point *a* is

$$M_a = 275 \text{ lb } (1.45 \text{ ft}) = 399 \text{ ft lb}$$

### Equations of Equilibrium

If a body is in *equilibrium* (is not moving, or moves with a constant velocity):

| | | |
|---|---|---|
| Equations of Equilibrium | The sum of all horizontal forces acting on the body = 0 | A13 |
| | The sum of all vertical forces acting on the body = 0 | A14 |
| | The sum of the moments about any point on the body = 0 | A15 |

This is Newton's first law of motion.

**EXAMPLE 27:** Find the reactions $R_1$ and $R_2$ in Fig. 3-3.

**Solution:** In a statics problem, we may consider all the weight of an object to be concentrated at a single point (called the *center of gravity*) on that object. For a uniform bar, the center of gravity is just where you would expect it to be, at the midpoint. Replacing the weights by forces gives the simplified diagram, Fig. 3-4.

The moment of the 125-lb force about $p$ is, by Eq. A12,

$$125(1.72) \text{ ft-lb clockwise}$$

Similarly, the other moments about $p$ are

$$25.0(2.07) \text{ ft-lb clockwise}$$

and

$$4.14\, R_2 \text{ ft-lb counterclockwise}$$

But Eq. A15 says that the sum of the clockwise moments must equal the sum of the counterclockwise moments, so

$$\begin{aligned} 4.14R_2 &= 125(1.72) + 25.0(2.07) \\ &= 266.8 \\ R_2 &= 64.4 \text{ lb} \end{aligned}$$

Also, Eq. A14 says that the sum of the upwards forces ($R_1$ and $R_2$) must equal the sum of the downward forces (125 lb and 25.0 lb), so

$$\begin{aligned} R_1 + R_2 &= 125 + 25.0 \\ R_1 &= 125 + 25.0 - 64.4 \\ &= 85.6 \text{ lb} \end{aligned}$$

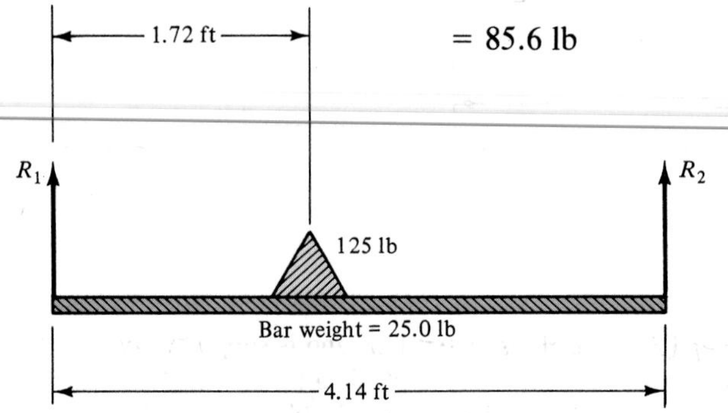


FIGURE 3-3

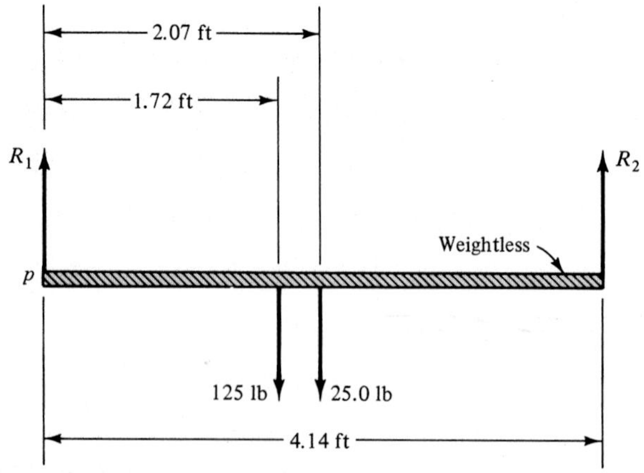


FIGURE 3-4

**EXAMPLE 28:** A horizontal uniform beam of negligible weight is 6.35 m long and is supported by columns at either end. A concentrated load of 525 N is applied to the beam. At what distance from one end must this load be located so that the vertical reaction at that same end is 315 N, and what will be the reaction at the other end?

**Solution:** We draw a diagram (Fig. 3-5) and label the required distance as $x$. By Eq. A14,

$$R + 315 = 525$$

$$R = 210 \text{ N}$$

Taking moments about $p$ we set the moments which tend to turn the bar in a clockwise (CW) direction equal to the moments which tend to turn the bar in the counterclockwise (CCW) direction. By Eq. A15,

$$525x = 210(6.35)$$

$$x = \frac{210(6.35)}{525} = 2.54 \text{ m}$$

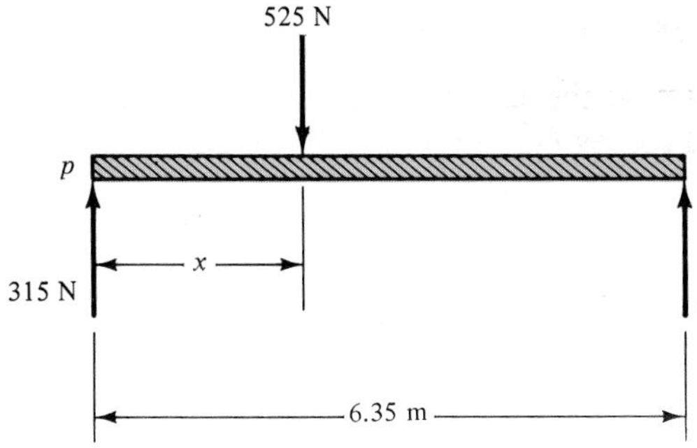


## EXERCISE 5—STATICS PROBLEMS

1. A horizontal beam of negligible weight is 18.0 ft long and is supported by columns at either end. A vertical load of 14,500 lb is applied to the beam at a distance $x$ from the left end. Find $x$ so that the reaction at the left column is 10,500 lb.
2. For the beam of Problem 1, find the reaction at the right column.
3. A certain cantilever beam of negligible weight has an additional support 13.1 ft from the built-in end, as shown in Fig. 3-6. The beam is 17.3 ft long and has a concentrated load of 2350 lb at the free end. Find the vertical reactions at the built-in end and at the support.

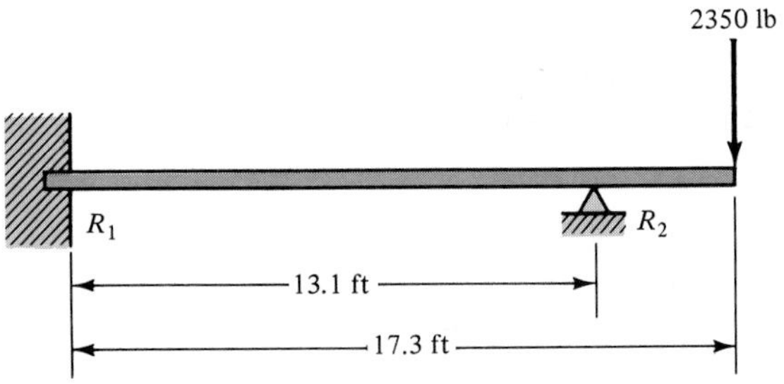


**FIGURE 3-6 Cantilever beam.**

4. A horizontal bar of negligible weight has a 55.1-lb weight hanging from the left end and a 72.0-lb weight hanging from the right end. The bar is seen to balance 97.5 in. from the left end. Find the length of the bar.
5. A horizontal bar of negligible weight hangs from two cables, one at each end. When a 624-N force is applied vertically downward from a point 185 cm from the left end of the bar, the left cable is seen to have a tension of 283 N, and the right cable a tension of 341 N. Find the length of the bar.
6. A uniform horizontal beam is 9.74 ft long and weighs 386 lb. It is supported by columns at either end. A vertical load of 3814 lb is applied to the beam at a distance $x$ from the left end. Find $x$ so that the reaction at the right column is 2000 lb.
7. A uniform horizontal beam is 19.8 ft long and weighs 1360 lb. It is supported at either end. A vertical load of 13,510 lb is applied to the beam 8.45 ft from the left end. Find the reaction at each end of the beam.
8. A bar of uniform cross section is 82.3 in. long and weighs 10.5 lb. A weight of 27.2 lb is suspended from one end. The bar and weight combination is to be suspended from a cable attached at the balance point. How far from the weight should the cable be attached, and what is the tension in the cable?

## CHAPTER 3 REVIEW PROBLEMS

*Solve for x.*

1. $2x - (3 + 4x - 3x + 5) = 4$
2. $5(2 - x) + 7x - 21 = x + 3$
3. $3(x - 2) + 2(x - 3) + (x - 4) = 3x - 1$
4. $x + 1 + x + 2 + x + 4 = 2x + 12$
5. $(2x - 5) - (x - 4) + (x - 3) = x - 4$
6. $4 - 5x - (1 - 8x) = 63 - x$
7. $3x - (x + 10) - (x - 3) = 14 - x$
8. $(2x - 9) - (x - 3) = 0$
9. $3x + 4(3x - 5) = 12 - x$
10. $6(x - 5) = 15 + 5(7 - 2x)$
11. $x^2 - 2x - 3 = x^2 - 3x + 1$
12. $(x^2 - 9) - (x^2 - 16) + x = 10$
13. $x^2 + 8x - (x^2 - x - 2) = 5(x + 3) + 3$
14. $x^2 + x - 2 + x^2 + 2x - 3 = 2x^2 - 7x - 1$
15. $10x - (x - 5) = 2x + 47$
16. $7x - 5 - (6 - 8x) + 2 = 3x - 7 + 106$
17. $3x + 2 = \dfrac{x}{5}$
18. $\dfrac{4}{x} = 3$
19. $\dfrac{2x}{3} - 5 = \dfrac{3x}{2}$
20. $5.9x - 2.8 = 2.4x + 3.4$
21. $4.5(x - 1.2) = 2.8(x + 3.7)$
22. $\dfrac{x - 4.8}{1.5} = 6.2x$
23. $6x + 3 - (3x + 2) = (2x - 1) + 9$
24. $3(x + 10) + 4(x + 20) + 5x - 170 = 15$
25. $20 - x + 4(x - 1) - (x - 2) = 30$
26. $5x + 3 - (2x - 2) + (1 - x) = 6(9 - x)$
27. Subdivide a meter of tape into two parts so that one part will be 6 cm longer than the other part.
28. A certain mine yields low-grade oil shale containing 18 gal of oil per ton of rock, and another mine has shale yielding 30 gal/ton. How many tons of each must be sent each day to a processing plant that processes 25,000 tons of rock per day, so that the overall yield will be 23 gal/ton?
29. A certain automatic soldering machine requires a solder containing half tin and half lead. How much pure tin must be added to 55 kg of a solder containing 61% lead and 39% tin to raise the tin content to 50%?
30. A person owed to $A$ a certain sum, to $B$ four times as much, to $C$ eight times as much, and to $D$ six times as much. A total of \$570 would pay all the debts. What was the debt to $A$?
31. A technician spends $\frac{2}{5}$ of his salary for board, $\frac{2}{3}$ of the remainder for clothing, and saves \$1500 per year. What is his salary?
32. Find four consecutive odd numbers such that the product of the first and third will be 64 less than the product of the second and fourth.
33. The front and rear wheels of a tractor are 10 ft and 12 ft, respectively, in circumference. How many feet will the tractor have traveled when the front wheel has made 250 revolutions more than the rear wheel?

*Solve for x.*

34. $3x - (x - 4) + (x + 1) = x - 7$
35. $4(x - 3) = 21 + 7(2x - 1)$
36. $8.2(x - 2.2) = 1.3(x + 3.3)$

37. $5(x - 1) - 3(x + 3) = 12$

38. $(x - 2)(x + 3) = x(x + 7) - 8$

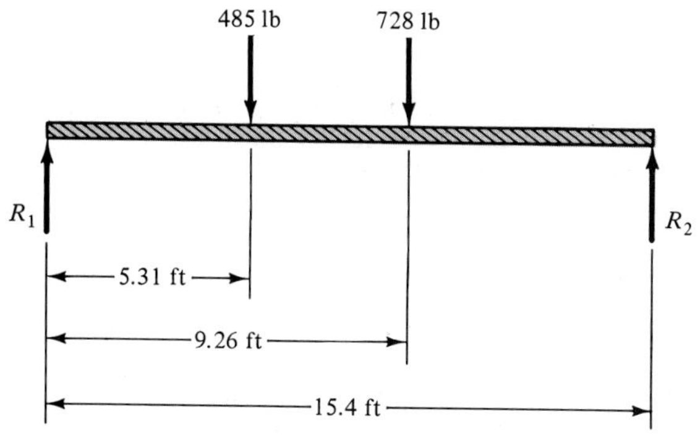


**FIGURE 3-7**

39. $5x + (2x - 3) + (x + 9) = x + 1$

40. $1.2(x - 5.1) = 7.3(x - 1.3)$

41. $2(x - 7) = 11 + 2(4x - 5)$

42. $4.4(x - 1.9) = 8.3(x + 1.1)$

43. $8x - (x - 5) + (3x + 2) = x - 14$

44. $7(x + 1) = 13 + 4(3x - 2)$

45. Find the reactions $R_1$ and $R_2$ in Fig. 3-7.

*Writing*

46. Make up a word problem. It can be very similar to one given in this chapter, or, better, very different. Swap with a classmate and solve each other's problem. Note down where the problem may be unclear, unrealistic, or ambiguous. Finally, each of you should rewrite your problem if needed.

# 4

# FUNCTIONS AND GRAPHS

## OBJECTIVES

**When you have completed this chapter, you should be able to:**

- Distinguish between relations and functions.
- Identify types of relations and functions.
- Find the domain and range of a function.
- Use functional notation to manipulate, combine, and evaluate functions.
- Graph points, relations, and functions.
- Solve equations graphically.

The equations we have been solving in the last few chapters have contained only *one variable*. For example,

$$x(x - 3) = x^2 + 7$$

contains the single variable $x$.

But many situations involve *two* (or more) variables that are somehow related to each other. Look, for example, at Fig. 4-1, and suppose that you leave your campsite and walk a path up the hill. As you walk, both the horizontal distance $x$ from your camp, and the vertical distance $y$ above your camp, will change. You cannot change $x$ without changing $y$, and vice versa (unless you jump into the air or dig a hole). The variables $x$ and $y$ are *related*. In this chapter we study the relation between two variables and introduce the concept of a *function*. The idea of a function provides us with a different way of speaking about mathematical relationships. We could say, for example, that the formula for the area of a circle *as a function of* its radius is $A = \pi r^2$.

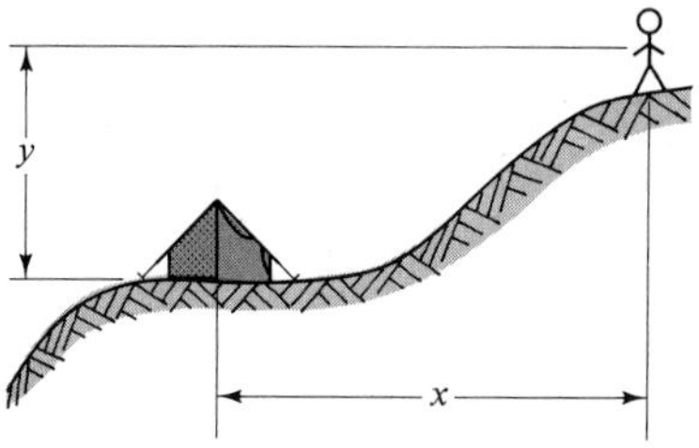

**FIGURE 4-1**

It may become apparent as you study this chapter that these same problems could be solved without ever introducing the idea of a function. Does the function concept, then, merely give us new jargon for the same old ideas? Not really.

A new way of *speaking* about something can lead to a new way of *thinking* about that thing, and so it is with functions. It also will lead to the powerful and convenient *functional notation,* which will be especially useful when you study calculus and computer programming. In addition, this introduction to functions will prepare us for the study of *functional variation* later.

Finally, we learn about *graphing,* a skill that is indispensable in all branches of technology and in many nontechnical fields, such as business. Also, when studying graphing, we learn how to find an *approximate solution* to any type of equation.

## 4-1 FUNCTIONS

### Functions and Relations

If we have an equation that enables us to find exactly one $y$ for any given $x$, then *$y$ is said to be a function of $x$.*

---

**EXAMPLE 1:** The equation

$$y = 3x - 5$$

is a function. For any $x$, say $x = 2$, there corresponds exactly one $y$ (in this case, $y = 1$).

---

Thus a function may be in the form of an equation. But *not every equation is a function.*

---

**EXAMPLE 2:** The equation $y = \pm\sqrt{x}$ is not a function. Each value of $x$ yields *two* values of $y$. This equation is called a *relation.*

---

We see from Examples 1 and 2 that a relation need not have *one* value of $y$ for each value of $x$, as is required for a function. Thus only some equations that are relations can be called functions.

**EXAMPLE 3:** The equation

$$y^2 = 2x + 4$$

is a relation but is not a function, because for some values of $x$ there are two values of $y$. When $x = 0$, for example, $y$ has the values 2 and $-2$.

Some older textbooks do not distinguish between relations and functions, but between *single-valued* functions and *two-valued* or *multiple-valued* functions (what we now call relations).

The fact that a relation is not a function does not imply second-class status. Relations are no less useful when they are not functions. In fact for practical work we often do not care whether our relation is a function or not.

The word *function* is commonly used to describe certain types of equations.

**EXAMPLE 4:**

| | |
|---|---|
| $y = 5x + 3$ | is a linear function |
| $y = 3x^3$ | is a power function |
| $y = 7x^2 + 2x - 5$ | is a quadratic function |
| $y = 3 \sin 2x$ | is a trigonometric function |
| $y = 5^{2x}$ | is an exponential function |
| $y = \log (x + 2)$ | is a logarithmic function |

## A More General Definition

When we think about a function or use one, it will most often be in the form of an equation. However, we do not want to limit our definition of a function to equations only, but would like to include other kinds of associations between numbers, such as tables of data, graphs, and verbal statements.

A more general definition of a function that includes all of these is usually given in terms of sets. A *set* is a collection of particular things, such as the set of all automobiles. The objects or members of a set are called *elements*. The only sets we will consider here are those that contain as elements all or some of the real numbers.

We do not intend to study sets here, or to introduce set notation, but will use the idea of a set to define a function, as follows:

> Let $x$ be an element in set $A$, and $y$ an element in set $B$. We say that $y$ is a function of $x$ if there is a rule that associates exactly one $y$ in set $B$ with each $x$ in set $A$.

Here, set $A$ is called the *domain* and set $B$ is called the *range*.

We can picture a function as in Fig. 4-2. Each set is represented by a shaded area, and each element by a point within the shaded area. The function $f$ associates exactly one $y$ in the range $B$ with each $x$ in the domain $A$.

Some find it helpful to think of a function as a machine (Fig. 4-3). An input of $x$ (within the domain of $x$) results in the output of a single corresponding value of $y$.

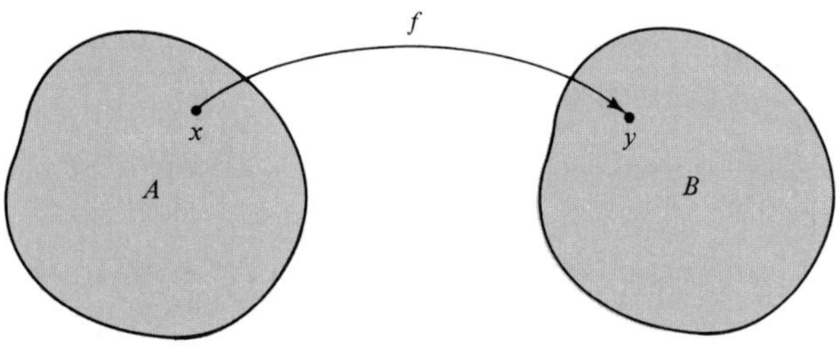


**FIGURE 4-2**

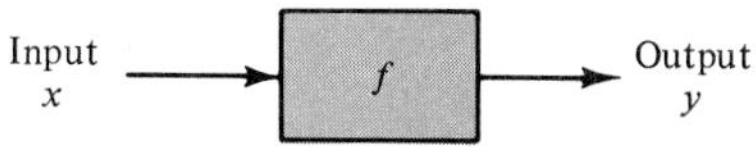


**FIGURE 4-3** **A "function machine."**

---

**EXAMPLE 5:** The equation from Example 1,

$$y = 3x - 5$$

can be thought of as a rule that associates exactly one $y$ with any given $x$. The domain of $x$ is the set of all real numbers, and the range of $y$ is the set of all real numbers.

---

To be strictly correct, the domain should be stated whenever an equation is written. However, this is often not done, and the domain should then be assumed to be the largest set of real numbers that will yield real values of $y$.

---

**EXAMPLE 6:** In the function

$$y = \sqrt{x - 4}$$

a value of $x$ less than 4 will make the quantity under the radical sign negative, causing $y$ to be nonreal. We thus assume the domain of $x$ to be

$$x \geq 4$$

With $x$ so limited, the only possible values for $y$ are the positive numbers and zero. Thus the range of $y$ is

$$y \geq 0$$

---

**We'll get more practice finding domain and range of functions a little later.**

### Other Forms of a Function

We have seen that certain equations are functions. But a function may also be (a) a set of ordered pairs, (b) a verbal statement, or (c) a graph. In technical work we must often deal with *pairs* of numbers rather than with single values, and these pairs often make up a function.

---

**EXAMPLE 7:** In the experiment shown in Fig. 4-4, we change the load on the spring, and for each load, measure and record the distance that the spring has stretched from its unloaded position. If a load of 6 kg causes a stretch of 3 cm, then 6 kg and 3 cm are called *corresponding values*. The value 3 cm has no meaning *by itself,* but is meaningful only when it is *paired* with the load that produced it (6 kg). If we always write the pair of numbers in the *same order* (the load first, and the distance second, in this example), it is then called an *ordered pair* of numbers. It is written (6, 3).

---

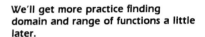

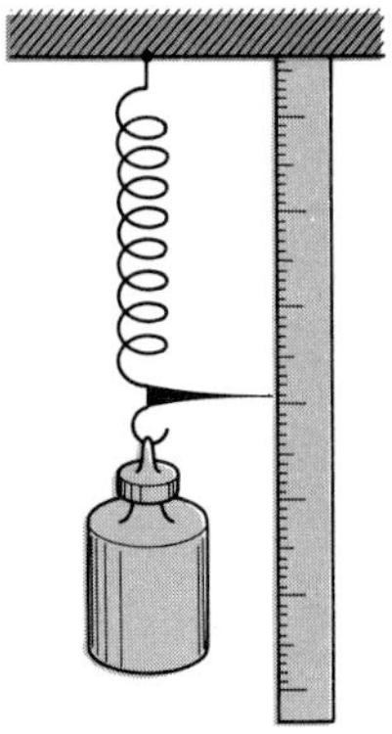

**FIGURE 4-4**

A set or table of ordered pairs is a function if for each $x$ value there is only one $y$ value given. The domain of $x$ is the set of all $x$ values in the table, and the range is the set of all $y$ values.

**EXAMPLE 8:** Suppose that the spring experiment of Example 7 gave the following results:

| *Load* (*kg*) | 0 | 1 | 2 | 3 | 4 | 5 | 6 | 7 | 8 |
|---|---|---|---|---|---|---|---|---|---|
| *Stretch* (*cm*) | 0 | 0.5 | 0.9 | 1.4 | 2.1 | 2.4 | 3.0 | 3.6 | 4.0 |

This table of ordered pairs is a function. The domain is the set of numbers (0, 1, 2, 3, 4, 5, 6, 7, 8) and the range is the set of numbers (0, 0.5, 0.9, 1.4, 2.1, 2.4, 3.0, 3.6, 4.0). The table associates exactly one number in the range with each number in the domain.

A set of ordered pairs is not always given in table form.

**EXAMPLE 9:** The function of Example 8 can also be written

(0, 0), (1, 0.5), (2, 0.9), (3, 1.4), (4, 2.1), (5, 2.4), (6, 3.0), (7, 3.6), (8, 4.0)

Functions may also be given in verbal form.

**EXAMPLE 10:** "The shipping charges are 55 cents/lb for the first 50 lb and 45 cents/lb thereafter." This verbal statement is a function relating the shipping costs to the weight of the item.

We sometimes want to switch from verbal form to an equation, or vice versa.

**EXAMPLE 11:** Write $y$ as a function of $x$ if $y$ equals twice the cube of $x$ diminished by half of $x$.

**Solution:** We replace the verbal statement by the equation

$$y = 2x^3 - \frac{x}{2}$$

**There are, of course, many different ways to express this same relationship.**

**EXAMPLE 12:** The equation $y = 5x^2 + 9$ can be stated verbally as: "$y$ equals the sum of 9 and 5 times the square of $x$."

**EXAMPLE 13:** Express the volume $V$ of a cone having a base area of 75 units *as a function of* its altitude $H$.

**Solution:** This is another way of saying: "Write an equation for *the volume of a cone in terms of its base area and altitude*."

The formula for the volume of a cone is given in the Summary of Facts and Formulas (Appendix A).

$$V = \tfrac{1}{3}(\text{base area})(\text{altitude}) = \tfrac{1}{3}(75)H$$

So

$$V = 25H$$

is the required expression.

We see, then, that a function or relation can be expressed in several different ways (Fig. 4-5): as an equation, as a table or set of ordered pairs, or as a verbal statement. A function or relation can also be expressed as a *graph,* as we'll see in Sec. 4-4.

**Different Forms of a Relation or Function**

Equation:

$$y = x^2 - 3$$

Table of values:

| $x$ | 0 | 1 | 2 | 3 |
|---|---|---|---|---|
| $y$ | −3 | −2 | 1 | 6 |

Verbal statement:

"$y$ is equal to the square of $x$ diminished by 3."

Set of ordered pairs:

(1, −2), (3, 6),
(0, −3), (2, 1)

Graph:

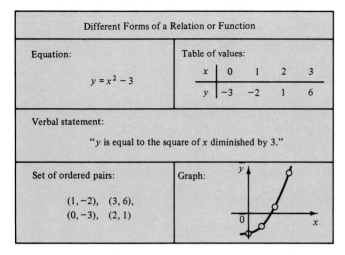

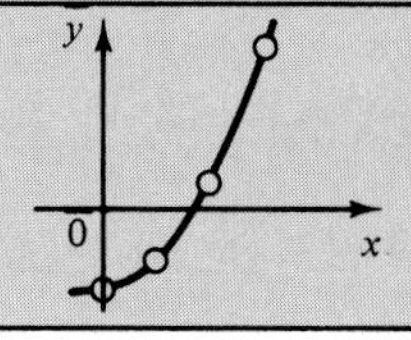


FIGURE 4-5

## Finding Domain and Range

Sometimes the domain of a function is given with the function, as in the equation for exponential growth,

$$y = ae^{nx} \qquad (x \geqslant 0)$$

That is, the domain is the set of all numbers greater than or equal to zero.

If the domain is *not* given, it is understood to be all the real numbers, except those that will make $y$ nonreal or that will result in an illegal mathematical operation such as division by zero.

---

**EXAMPLE 14:** Find the domain and range of the function

$$y = \sqrt{x - 2}$$

**Solution:** Our method is to see what values of $x$ and $y$ "do not work" (give a nonreal result or an illegal operation); then the domain and range will be those values that "do work."

Any value of $x$ less than 2 will make the quantity under the radical sign negative, resulting in an imaginary $y$. Thus the domain of $x$ is all the positive numbers equal to or greater than 2. The range of $y$ will be all the positive numbers and zero. Can you explain why?

---

**EXAMPLE 15:** Find the domain and range of the function

$$y = \frac{9}{\sqrt{4 - x}}$$

**Solution:** Since the denominator cannot be zero, $x$ can not be 4. Also, any $x$ greater than 4 will result in a negative quantity under the radical sign. The domain of $x$ is then

$$x < 4$$

Restricted to these values, the quantity $4 - x$ is positive, and ranges from very small (when $x$ is nearly 4) to very large (when $x$ is large and negative). Thus the denominator is positive, since we allow only the principal (positive) root, and can vary from near zero to infinity. The range of $y$, then, includes all the values greater than zero.

---

## EXERCISE 1—FUNCTIONS

Which of the following relations are also functions?

1. $y = 3x^2 - 5$
2. $y = \sqrt{2x}$
3. $y = \pm\sqrt{2x}$
4. $y^2 = 3x - 5$
5. $2x^2 = 3y^2 - 4$
6. $x^2 - 2y^2 - 3 = 0$
7. Is the set of ordered pairs (1, 3), (2, 5), (3, 8), (4, 12) a function? Explain.
8. Is the set of ordered pairs (0, 0), (1, 2), (1, −2), (2, 3), (2, −3), a function? Explain.

### Functions in Verbal Form

For each expression, write $y$ as a function of $x$, where the value of $y$ is equal to:

9. The cube of $x$
10. The square root of $x$, diminished by 5
11. $x$ increased by twice the square of $x$
12. The reciprocal of the cube of $x$
13. Two-thirds of the amount by which $x$ exceeds 4

Replace each function by a verbal statement of the type given in Problems 9 to 13.

14. $y = 3x^2$
15. $y = 5 - x$
16. $y = \dfrac{1}{x} + x$
17. $y = 2\sqrt[3]{x}$
18. $y = 5(4 - x)$

Write the equation called for in each following statement. Refer to the Summary of Facts and Formulas (Appendix A) if necessary.

19. Express the area $A$ of a triangle as a function of its base $b$ and altitude $h$.
20. Express the hypotenuse $c$ of a right triangle as a function of its legs, $a$ and $b$.
21. Express the volume $V$ of a sphere as a function of the radius $r$.
22. Express the power $P$ dissipated in a resistor as a function of its resistance $R$ and the current $I$ through the resistor.

Write the equations called for in each statement.

23. A car is traveling at a speed of 55 mi/h. Write the distance $d$ travelled by the car as a function of time $t$.
24. To ship their merchandise, a mail-order company charges 65 cents/lb plus \$2.25 for handling and insurance. Express the total shipping charge $s$ as a function of the item weight $w$.
25. A projectile is shot upward with an initial velocity of 125 m/s. Express the height $H$ of the projectile as a function of time $t$. (See Eq. A18.)

### Domain and Range

State the domain and range of each function.

26. (0, 2), (1, 4), (2, 8), (3, 16), (4, 32)
27. (−10, 20), (5, 7), (−7, 10), (10, 20), (0, 3)
28.

| $x$ | 2 | 4 | 6 | 8 | 10 |
|---|---|---|---|---|---|
| $y$ | 0 | −2 | −5 | −9 | −15 |

Find the domain and range for each function.

29. $y = \sqrt{x - 7}$
30. $y = \dfrac{3}{\sqrt{x - 2}}$
31. $y = x - \dfrac{1}{x}$
32. $y = \sqrt{x^2 - 25}$

33. $y = \dfrac{8 - x}{9 - x}$

34. $y = \dfrac{11}{(x + 4)(x + 2)}$

35. $y = \dfrac{4}{\sqrt{1 - x}}$

36. $y = \dfrac{x + 1}{x - 1}$

37. $y = \sqrt{x - 1}$

38. $y = \dfrac{5}{(x + 2)(x - 4)}$

39. Find the range of the function $y = 3x^2 - 5$ whose domain is $0 \leqslant x \leqslant 5$.

### Calculator and Computer

40. Write a step-by-step procedure (algorithm) for your calculator which will give you the value of $y$ for any $x$ for the function

$$y = 2x^2 - 3x + 5$$

You should have to enter the value of $x$ only once if your calculator has STORE and RECALL. Use your algorithm to compute a set of ordered pairs, taking values of $x$ from 0 to 5.

41. If you have a programmable calculator, program it to execute the computation of Problem 40 *automatically*.

42. A program to compute and print in BASIC a table of ordered pairs for the function in Problem 40 is

```
10    FOR X = 0 to 5
20    LET Y = 2*X↑2 - 3*X + 5
30    PRINT X, Y
40    NEXT X
```

Try this program on your computer, if you have BASIC, or modify it for the language you have available. Then change it for different values of $x$, different step sizes (say, FOR X = −10 TO 10 STEP .5), and for a different function.

**Your calculator or computer can be your own "function machine."**

**We don't intend to teach computer programming in this text, but we include an occasional problem for those who know simple programming.**

## 4-2 FUNCTIONAL NOTATION

### Implicit and Explicit Form

When one variable in an equation is isolated on one side of the equal sign, the equation is said to be in *explicit form*.

---

**EXAMPLE 16:** The equations $y = 2x^3 + 5$, $z = ay + b$, and $x = 3z^2 + 2z - z$ are all in explicit form.

---

When a variable is *not* isolated, the equation is in *implicit form*.

---

**EXAMPLE 17:** The equations $y = x^2 + 4y$, $x^2 + y^2 = 25$, and $w + x = y + z$ are all in implicit form.

---

### Dependent and Independent Variables

In the equation

$$y = x + 5$$

$y$ is called the *dependent variable*, because its value depends on the value of $x$, and $x$ is called the *independent* variable. Of course, the same equation can be written $x = y - 5$, so that $x$ becomes the dependent variable and $y$ the independent variable.

The terms "dependent" and "independent" are used only for an equation in explicit form.

**EXAMPLE 18:** In the implicit equation $y - x = 5$, neither $x$ nor $y$ is called dependent or independent.

### Functional Notation

Just as we use the symbol $x$ to represent a *number,* without saying which number we are specifying, we use the notation

$$f(x)$$

to represent a *function* without having to specify which particular function we are talking about.

We can also use functional notation to designate a *particular* function, such as

$$f(x) = x^3 - 2x^2 - 3x + 4$$

A functional relation between two variables $x$ and $y$, in explicit form, such as

$$y = 5x^2 - 6$$

could be written

$$y = f(x)$$

or

$$f(x) = 5x^2 - 6$$

**This is read "*y* is a function of *x*" or "*y* equals *f* of *x*." It does not mean "*y* equals *f* times *x*." The independent variable *x* is sometimes referred to as the *argument*.**

**EXAMPLE 19:** We may know that the horsepower $P$ of an engine depends (somehow) on the engine displacement $d$. We can express this fact by

$$P = f(d)$$

We are saying that $P$ is a function of $d$, even though we do not know (or perhaps even care) what the relationship is.

The letter $f$ is usually used to represent a function, but *other letters* can, of course, be used ($g$ and $h$ being common). Subscripts are also used to distinguish one function from another.

**EXAMPLE 20:** You may see functions written as

$$y = f_1(x) \qquad y = g(x)$$
$$y = f_2(x) \qquad y = h(x)$$
$$y = f_3(x)$$

*Implicit functions* can also be represented in functional notation.

**EXAMPLE 21:** The equation $x - 3xy + 2y = 0$ can be represented in functional notation by $f(x, y) = 0$.

Functions relating *more than two variables* can be represented in functional notation, as in the following examples.

**EXAMPLE 22:**

| | | |
|---|---|---|
| $y = 2x + 3z$ | can be written | $y = f(x, z)$ |
| $z = x^2 - 2y + w^2$ | can be written | $z = f(w, x, y)$ |
| $x^2 + y^2 + z^2 = 0$ | can be written | $f(x, y, z) = 0$ |

## Manipulating Functions

Functional notation provides us with a convenient way of indicating what is to be done with a function or functions. This includes solving an equation for a different variable, changing from implicit to explicit form, or vice versa, combining two or more functions to make another function, or substituting numerical or literal values into an equation.

**EXAMPLE 23:** Write the equation $y = 2x - 3$ in the form $x = f(y)$.

**Solution:** We are being asked to write the given equation with $x$, instead of $y$, as the dependent variable. Solving for $x$, we obtain

$$2x = y + 3$$

$$x = \frac{y + 3}{2}$$

**EXAMPLE 24:** Write the equation $x = \dfrac{3}{2y - 7}$ in the form $y = f(x)$.

**Solution:** Here we are asked to rewrite the equation with $y$ as the dependent variable, so we solve for $y$. Multiplying both sides by $2y - 7$ and dividing by $x$ gives

$$2y - 7 = \frac{3}{x}$$

or

$$2y = \frac{3}{x} + 7$$

Dividing by 2 gives

$$y = \frac{3}{2x} + \frac{7}{2}$$

**EXAMPLE 25:** Write the equation $y = 3x^2 - 2x$ in the form $f(x, y) = 0$.

**Solution:** We are asked here to go from explicit to implicit form. Rearranging gives us

$$3x^2 - 2x - y = 0$$

## Substituting into Functions

If we have a function, say $f(x)$, then the notation $f(a)$ means to replace $x$ by $a$ in the same function.

**EXAMPLE 26:** Given $f(x) = x^3 - 5x$, find $f(2)$.

**Solution:** The notation $f(2)$ means that 2 is to be *substituted for* $x$ in the given function. Wherever an $x$ appears, we replace it with 2.

$$\begin{array}{ccccc} f(x) & = & x^3 & - & 5x \\ \updownarrow & & \updownarrow & & \updownarrow \\ f(2) & = & (2)^3 & - & 5(2) = -2 \end{array}$$

Often, we must substitute *several values* into a function and combine them as indicated.

**EXAMPLE 27:** If $f(x) = x^2 - 3x + 4$, find

$$\frac{f(5) - 3f(2)}{2f(3)}$$

**Solution:**

$$f(2) = 2^2 - 3(2) + 4 = 2$$

and

$$f(3) = 3^2 - 3(3) + 4 = 4$$

$$f(5) = 5^2 - 3(5) + 4 = 14$$

so

$$\frac{f(5) - 3f(2)}{2f(3)} = \frac{14 - 3(2)}{2(4)}$$

$$= \frac{8}{8} = 1$$

The substitution might involve *literal* quantities instead of numerical values.

**EXAMPLE 28:** Given $f(x) = 3x^2 - 2x + 3$, find $f(5a)$.

**Solution:** We substitute $5a$ for $x$,

$$\begin{aligned} f(5a) &= 3(5a)^2 - 2(5a) + 3 \\ &= 3(25a^2) - 10a + 3 \\ &= 75a^2 - 10a + 3 \end{aligned}$$

**EXAMPLE 29:** If $f(x) = 5x - 2$, find $f(2w)$.

**Solution:**

$$f(2w) = 5(2w) - 2 = 10w - 2$$

**EXAMPLE 30:** If $f(x) = x^2 - 2x$, find $f(w^2)$.

**Solution:**

$$f(w^2) = (w^2)^2 - 2(w^2) = w^4 - 2w^2$$

It can be confusing when the expression to be substituted contains the same variable as in the original function.

**EXAMPLE 31:** If $f(x) = 5x - 2$, then

$$f(x + a) = 5(x + a) - 2 = 5x + 5a - 2$$

**EXAMPLE 32:** If $f(x) = x^2 - 2x$, then

$$\begin{aligned} f(x - 1) &= (x - 1)^2 - 2(x - 1) \\ &= x^2 - 2x + 1 - 2x + 2 \\ &= x^2 - 4x + 3 \end{aligned}$$

There may be *more than one function* in a single problem.

**EXAMPLE 33:** Given three different functions,

$$f(x) = 3x \qquad g(x) = x^2 \qquad h(x) = \sqrt{x}$$

evaluate

$$\frac{2g(3) + 4h(9)}{f(5)}$$

**Solution:** First substitute into each function:

$$f(5) = 3(5) = 15 \qquad g(3) = 3^2 = 9 \qquad h(9) = \sqrt{9} = 3$$

Then combine these as indicated.

$$\frac{2g(3) + 4h(9)}{f(5)} = \frac{2(9) + 4(3)}{15} = \frac{18 + 12}{15} = \frac{30}{15} = 2$$

Sometimes we must substitute into a function containing *more than one variable*.

**EXAMPLE 34:** Given $f(x, y, z) = 2y - 3z + x$, find $f(3, 1, 2)$.

**Solution:** We substitute the given numerical values for the variables. Be sure that the numerical values are taken in the *same order* as the variable names in the functional notation.

$$\begin{matrix} f(x, y, z) \\ \updownarrow \ \updownarrow \ \updownarrow \\ f(3, 1, 2) \end{matrix}$$

Substituting, we obtain

$$\begin{aligned} f(3, 1, 2) &= 2(1) - 3(2) + 3 = 2 - 6 + 3 \\ &= -1 \end{aligned}$$

## Composition of Functions

Just as we can substitute a constant or a variable into a given function, so we can substitute a *function* into a function.

**EXAMPLE 35:** If $g(x) = x + 1$, find

(a) $g(2)$ (b) $g(z^2)$ (c) $g[f(x)]$

**Solution:**

(a) $g(2) = 2 + 1 = 3$ (b) $g(z^2) = z^2 + 1$ (c) $g[f(x)] = f(x) + 1$

In calculus, we will take derivatives of composite functions by means of the *chain rule.*

The function $g[f(x)]$ (which we read "$g$ of $f$ of $x$"), being made up of the two functions $g(x)$ and $f(x)$, is called a *composite function*. If we think of a function as a machine, it is as if we are using the output $f(x)$ of the function machine $f$ as the input of a second function machine $g$.

$$x \rightarrow \boxed{f} \rightarrow f(x) \rightarrow \boxed{g} \rightarrow g[f(x)]$$

We thus obtain $g[f(x)]$ by replacing $x$ in $g(x)$ by the function $f(x)$.

---

**EXAMPLE 36:** Given the functions $g(x) = x + 1$, and $f(x) = x^3$, write the composite function $g[f(x)]$.

**Solution:** In the function $g(x)$ we replace $x$ by $f(x)$:

$$\begin{array}{rcl} g(x) & = & x \quad + 1 \\ \downarrow & & \downarrow \\ g[f(x)] & = & f(x) + 1 \\ & & \downarrow \\ & = & x^3 \quad + 1 \end{array}$$

since $f(x) = x^3$.

---

As we have said, the notation $g[f(x)]$ means to substitute $f(x)$ into the function $g(x)$. On the other hand, the notation $f[g(x)]$ means to substitute $g(x)$ into $f(x)$.

$$x \rightarrow \boxed{g} \rightarrow g(x) \rightarrow \boxed{f} \rightarrow f[g(x)]$$

In general, $f[g(x)]$ *will not be the same as* $g[f(x)]$.

---

**EXAMPLE 37:** Given $g(x) = x^2$ and $f(x) = x + 1$, find

(a) $f[g(x)]$ (b) $g[f(x)]$ (c) $f[g(2)]$ (d) $g[f(2)]$

**Solution:**

(a) $f[g(x)] = g(x) + 1 = x^2 + 1$

(b) $g[f(x)] = [f(x)]^2 = (x + 1)^2$

(c) $f[g(2)] = 2^2 + 1 = 5$

(d) $g[f(2)] = (2 + 1)^2 = 9$

Notice that here $f[g(x)]$ is not equal to $g[f(x)]$.

---

## Inverse of a Function

Consider a function $f$ that, given a value of $x$, returns some value of $y$.

$$x \rightarrow \boxed{f} \rightarrow y$$

If that $y$ is now put into a function $g$ that *reverses the operations* in $f$ so that its output is the original $x$, then $g$ is called the *inverse* of $f$.

$$x \rightarrow \boxed{f} \rightarrow y \rightarrow \boxed{g} \rightarrow x$$

---

**EXAMPLE 38:** Two such inverse operations are "cube" and "cube root."

$$x \rightarrow \boxed{\text{cube } x} \rightarrow x^3 \rightarrow \boxed{\text{take cube root}} \rightarrow x$$

---

Thus if a function $f(x)$ has an inverse $g(x)$ that reverses the operations in $f(x)$, then the *composite* of $f(x)$ and $g(x)$ should have no overall effect. If the input is $x$, then the output must also be $x$. In symbols, if $f(x)$ and $g(x)$ are inverse functions, then

$$g[f(x)] = x$$

and
$$f[g(x)] = x$$

Conversely, if $g[f(x)] = x$ and $f[g(x)] = x$, then $f(x)$ and $g(x)$ are inverse functions.

---

**EXAMPLE 39:** Using the example of the cube and cube root, if

$$f(x) = x^3 \qquad \text{and} \qquad g(x) = \sqrt[3]{x}$$

then

$$g[f(x)] = \sqrt[3]{f(x)} = \sqrt[3]{x^3} = x$$

and
$$f[g(x)] = [g(x)]^3 = (\sqrt[3]{x})^3 = x$$

This shows that $f(x)$ and $g(x)$ are indeed inverse functions.

---

To find the inverse of a function $y = f(x)$:

1. Solve the given equation for $x$.
2. Interchange $x$ and $y$.

---

**EXAMPLE 40:** We use the cube, cube root example one more time. Find the inverse $g(x)$ of the function

$$y = f(x) = x^3$$

**Solution:** We solve for $x$ and get

$$x = \sqrt[3]{y}$$

It is then customary to interchange variables so that the dependent variable is (as usual) $y$. This gives

$$y = g(x) = \sqrt[3]{x}$$

Thus $g(x) = \sqrt[3]{x}$ is the inverse of $f(x) = x^3$, as verified earlier.

---

**EXAMPLE 41:** Find the inverse $g(x)$ of the function

$$y = f(x) = 2x + 5$$

**Solution:** Solving for $x$ gives

$$x = \frac{y - 5}{2}$$

Interchanging $x$ and $y$, we obtain

$$y = g(x) = \frac{x - 5}{2}$$

---

Sometimes the inverse of a function gets a special name. The inverse of the sine function, for example, is the *arcsin,* and the inverse of an exponential function is a *logarithmic* function.

## Functional Notation on the Computer

Programming languages usually have a command for defining a function. BASIC, for example, uses the DEF FN statement, as in

```
50  DEF FND(X) = 3*X + 5
```

Here the *name* of the function is $D$, and the *independent variable or argument* is $x$.

If we have the line

```
80  PRINT FND(4)
```

later in the same program, the value 4 will be substituted for $x$ in the function $D$ and the value

$$3(4) + 5 = 17$$

will be printed.

Functions of more than one variable can also be defined, as in

```
100  DEF FNB(X,Y,Z) = 3*X + Y/2 - Z
```

## EXERCISE 2—FUNCTIONAL NOTATION

### Implicit and Explicit Form

Which equations are in explicit form and which in implicit form?

1. $y = 5x - 8$
2. $x = 2xy + y^2$
3. $3x^2 + 2y^2 = 0$
4. $y = wx + wz + xz$

### Dependent and Independent Variables

Label the variables in each equation as dependent or independent.

5. $y = 3x^2 + 2x$
6. $x = 3y - 8$
7. $w = 3x + 2y$
8. $xy = x + y$
9. $x^2 + y^2 = z$

### Manipulating Functions

10. If $y = 5x + 3$, write $x = f(y)$.
11. If $x = \dfrac{2}{y - 3}$, write $y = f(x)$.
12. If $y = \dfrac{1}{x} - \dfrac{1}{5}$, write $x = f(y)$.
13. If $x^2 + y = x - 2y + 3x^2$, write $y = f(x)$.
14. The power $P$ dissipated in a resistor is given by $P = I^2R$. Write $R = f(P, I)$.
15. Young's modulus $E$ is given by

$$E = \frac{PL}{ae}$$

Write $e = f(P, L, a, E)$.

### Composite Functions

16. Given the functions $g(x) = 2x + 3$, and $f(x) = x^2$, write the composite function $g[f(x)]$.
17. Given the functions $g(x) = x^2 - 1$ and $f(x) = 3 + x$, write the composite function $f[g(x)]$.

Given $g(x) = x^3$ and $f(x) = 4 - 3x$, find:

18. $f[g(x)]$
19. $g[f(x)]$
20. $f[g(3)]$
21. $g[f(3)]$

**Substituting Into Functions**

Substitute the given numerical value(s) into each function.

22. If $f(x) = 2x^2 + 4$, find $f(3)$.
23. If $f(x) = 5x + 1$, find $f(1)$.
24. If $f(x) = 15x + 9$, find $f(3)$.
25. If $f(x) = 5 - 13x$, find $f(2)$.
26. If $g(x) = 9 - 3x^2$, find $g(-2)$.
27. If $h(x) = x^3 - 2x + 1$, find $h(2.55)$.
28. If $f(x) = 7 + 2x$, find $f(3)$.
29. If $f(x) = x^2 - 9$, find $f(-2)$.
30. If $f(x) = 2x + 7$, find $f(-1)$.

Substitute and combine as indicated.

31. Given $f(x) = 2x - x^2$, find $f(5) + 3f(2)$.
32. Given $f(x) = x^2 - 5x + 2$, find $f(2) + f(3) - f(1)$.
33. Given $f(x) = x^2 - 4$, find $\dfrac{f(2) + f(3)}{4}$.
34. Given **(a)** $f(x) = 2x$ and **(b)** $f(x) = x^2$ evaluate $\dfrac{f(x + d) - f(x)}{d}$ for each given function.

**We deal with expressions of this form in calculus, when studying the derivative.**

Substitute the literal values into each function.

35. If $f(x) = 2x^2 + 4$, find $f(a)$.
36. If $f(x) = 2x - \dfrac{1}{x} + 4$, find $f(2a)$.
37. If $f(x) = 5x + 1$, find $f(a + b)$.
38. If $f(x) = 5 - 13x$, find $f(-2c)$.

Substitute in each function and combine.

39. If $f(x) = x^2$ and $g(x) = \dfrac{1}{x}$, find $f(3) + g(2)$.
40. If $f(x) = \dfrac{2}{x}$ and $g(x) = 3x$, find $f(4) - g\left(\dfrac{1}{3}\right)$.
41. If $f(x) = 8x^2 - 2x + 1$ and $g(x) = x - 5$, find $2f(3) - 4g(3)$.
42. If $h(z) = 2z$ and $g(w) = w^2$, find

    **(a)** $\dfrac{h(4) + g(1)}{5}$ **(b)** $h[g(3)]$ **(c)** $g[(h(2)]$

**Functions of More Than One Variable**

Substitute the values into each function.

43. If $f(x, y) = 3x + 2y^2 - 4$, find $f(2, 3)$.
44. If $f(x, y, z) = 3z - 2x + y^2$, find $f(3, 1, 5)$.
45. If $g(a, b) = 2b - 3a^2$, find $g(4, -2)$.
46. If $f(x, y) = y - 3x$, find $3f(2, 1) + 2f(3, 2)$.
47. If $g(a, b, c) = 2b - a^2 + c$, find $\dfrac{3g(1, 1, 1) + 2g(1, 2, 3)}{g(2, 1, 3)}$.

**Inverse Functions**

Find the inverse of:

48. $y = 8 - 3x$
49. $y = 5(2x - 3) + 4x$
50. $y = 4x + 2(5 - x)$
51. $y = 3(x - 2) - 4(x + 3)$

## Applications

52. The distance traveled by a freely falling body is a function of the elapsed time $t$,

$$f(t) = V_0 t + \tfrac{1}{2}gt^2$$

where $V_0$ is the initial velocity and $g$ is the acceleration due to gravity (about 32 ft/s$^2$). If $V_0$ is 55 ft/s, find $f(10)$, $f(15)$, and $f(20)$.

53. The resistance $R$ of a conductor is a function of temperature,

$$f(t) = R_0(1 + \alpha t)$$

where $R_0$ is the resistance at 0°C and $\alpha$ is the temperature coefficient of resistance (0.00427 for copper). If the resistance of a copper coil is 9800 Ω at 0°C, find $f(20)$, $f(25)$, and $f(30)$.

54. The maximum deflection in inches of a certain cantilever beam, with a concentrated load applied $r$ feet from the fixed end, is a function of $r$,

$$f(r) = 0.00003r^2(80 - r)$$

Find the deflections $f(10)$ and $f(15)$.

## Functional Notation on the Computer

55. Using the DEFINE instructions on your machine, compute and print

$$y = \frac{4f_1(x)}{f_2(x) + 3}$$

for integer values of $x$ from 1 to 10, where

$$f_1(x) = 2x - 5 \quad \text{and} \quad f_2(x) = 3x^2 + 2x - 3$$

# 4-3 RECTANGULAR COORDINATES

## The Rectangular Coordinate System

**Rectangular coordinates are also called *Cartesian* coordinates, after the French mathematician René Descartes (1596–1650). Another type of coordinate system we will use is called the polar coordinate system.**

In Chapter 1 we plotted numbers on the number line. Suppose, now, that we take a second number line and place it at right angles to the first one, so that each intersects the other at the zero mark, as in Fig. 4-6. We call this a *rectangular coordinate system.*

The horizontal number line is called the *x axis* and the vertical is called the *y axis*. They intersect at the *origin*. These two axes divide the plane into four *quadrants*, numbered counterclockwise, as in the figure.

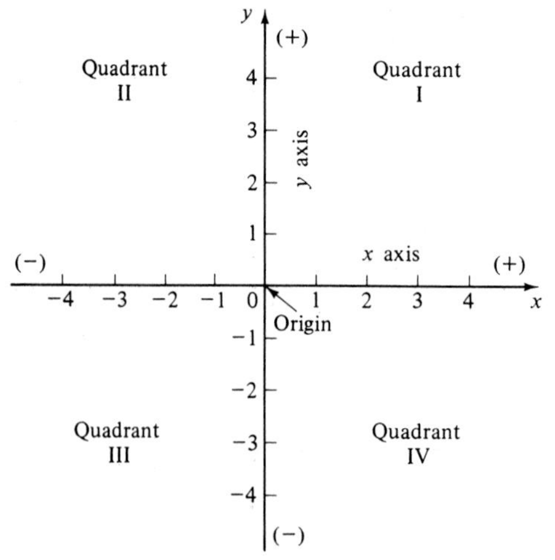


**FIGURE 4-6 The rectangular coordinate system.**

## Graphing Ordered Pairs

Figure 4-7 shows a point $P$ in the first quadrant. Its horizontal distance from the origin, called the *x coordinate* or *abscissa* of the point, is 3 units. Its vertical distance from the origin, called the *y coordinate* or *ordinate* of the point, is 2 units. The numbers in the *ordered pair* (3, 2) are called the *rectangular coordinates* (or simply, *coordinates*) of the point. They are always written in the same order, with the $x$ coordinate first. The letter identifying the point is sometimes written before the coordinates, as $P(3, 2)$.

To plot any ordered pair $(h, k)$, simply place a point at a distance $h$ units from the $y$ axis and $k$ units from the $x$ axis. Remember that negative values of $x$ are located to the left of the origin and negative $y$'s are below the origin.

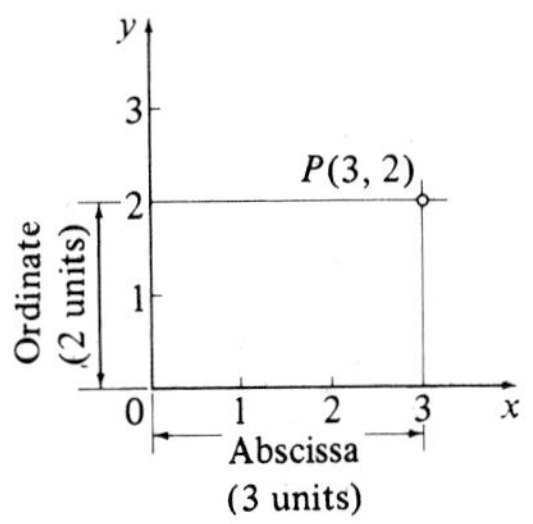


**FIGURE 4-7 Is it clear from this figure why we call these rectangular coordinates?**

---

**EXAMPLE 42:** The points

$$P(4, 1) \quad Q(-2, 3) \quad R(-1, -2) \quad S(2, -3) \quad T(1.3, 2.7)$$

are shown plotted in Fig. 4-8.

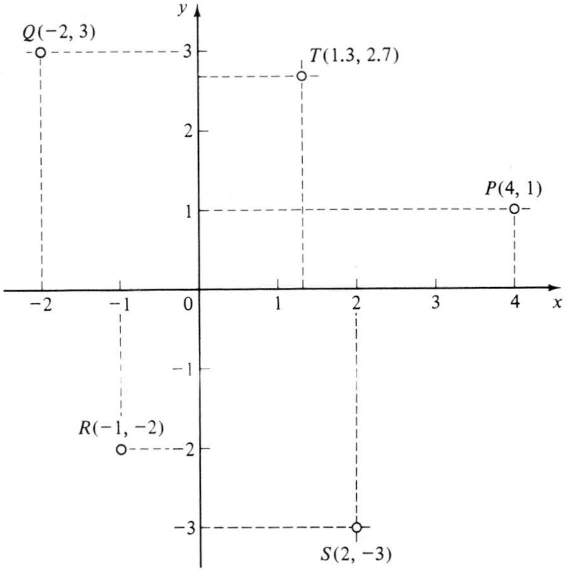


**FIGURE 4-8**

---

Notice that the abscissa is negative in the second and third quadrants, and that the ordinate is negative in the third and fourth quadrants. Thus the signs of the coordinates of a point tell us the quadrant in which the point lies.

---

**EXAMPLE 43:** The point $(-3, -5)$ lies in the third quadrant, for that is the only quadrant in which the abscissa and the ordinate are both negative.

---

## EXERCISE 3—RECTANGULAR COORDINATES

### Rectangular Coordinates

If $h$ and $k$ are positive quantities, in which quadrants would the following points lie?

1. $(h, -k)$
2. $(h, k)$
3. $(-h, k)$
4. $(-h, -k)$

5. Which quadrant contains points having a positive abscissa and a negative ordinate?
6. In which quadrants is the ordinate negative?
7. In which quadrants is the abscissa positive?
8. The ordinate of any point on a certain straight line is $-5$. Give the coordinates of the point of intersection of that line and the $y$ axis.
9. Find the abscissa of any point on a vertical straight line that passes through the point (7, 5).

### Graphing Ordered Pairs

10. Write the coordinates of points $A$, $B$, $C$, and $D$ in Fig. 4-9.

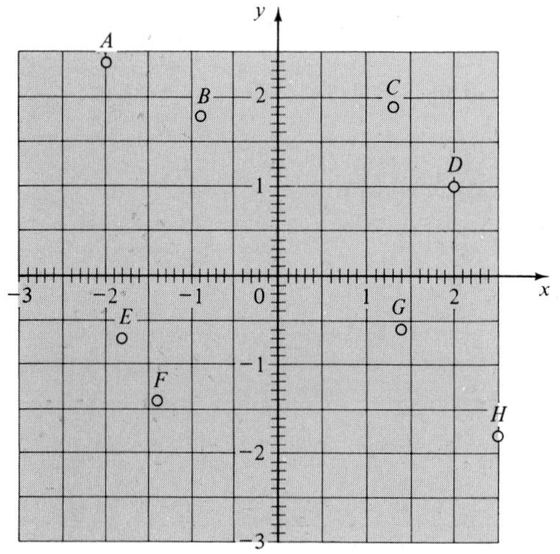


**FIGURE 4-9**

11. Write the coordinates of points $E$, $F$, $G$, and $H$ in Fig. 4-9.
12. Graph each point.

| | | |
|---|---|---|
| (a) (3, 5) | (b) (4, −2) | (c) (−2.4, −3.8) |
| (d) (−3.75, 1.42) | (e) (−4, 3) | (f) (−1, −3) |

Graph each set of points, connect them, and identify the geometric figure formed.

13. (0.7, 2.1), (2.3, 2.1), (2.3, 0.5), and (0.7, 0.5)

14. $(2, -\frac{1}{2})$, $(3, -1\frac{1}{2})$, $(1\frac{1}{2}, -3)$, and $(\frac{1}{2}, -2)$

15. $(-1\frac{1}{2}, 3)$, $(-2\frac{1}{2}, \frac{1}{2})$, and $(-\frac{1}{2}, \frac{1}{2})$

16. $(-3, -1)$, $(-1, -\frac{1}{2})$, $(-2, -3)$, and $(-4, -3\frac{1}{2})$

17. Three corners of a rectangle have the coordinates (−4, 9), (8, 3), and (−8, 1). Graphically find the coordinate of the fourth corner.

## 4-4 THE GRAPH OF A RELATION OR FUNCTION

### Graphing Relations or Functions

If the relation or function is given as a set of ordered pairs, simply plot each ordered pair. We usually connect the points with a smooth curve unless we have reason to believe that there are sharp corners, breaks, or gaps in the graph.

If the relation or function is in the form of an *equation,* we obtain a table of ordered pairs by selecting values of $x$ over the required domain, and then computing corresponding values of $y$. Since we are usually free to select any $x$ values we like, we pick "easy" integer values. We then plot the set of ordered pairs.

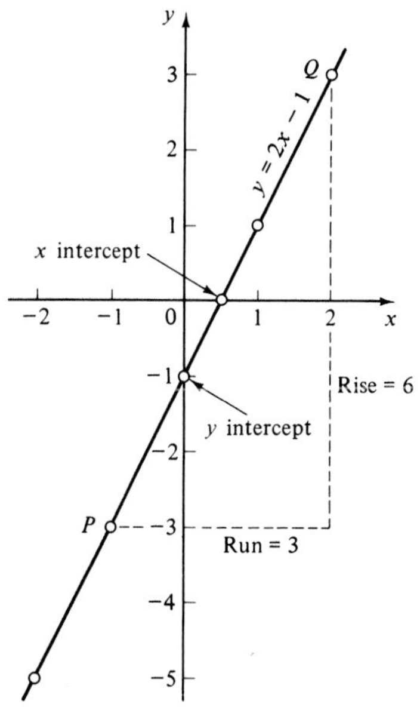


**FIGURE 4-10** **A first-degree equation will always plot as a straight line—hence the name *linear* equation. The points where the curve crosses the coordinate axes are called *intercepts*.**

---

**EXAMPLE 44:** Graph the function $y = f(x) = 2x - 1$ for values of $x$ from $-2$ to 2.

**Solution:** Substituting into the equation, we obtain

$$f(-2) = 2(-2) - 1 = -5$$
$$f(-1) = 2(-1) - 1 = -3$$
$$f(0) = 2(0) - 1 = -1$$
$$f(1) = 2(1) - 1 = 1$$
$$f(2) = 2(2) - 1 = 3$$

Thus our points are $(-2, -5)$, $(-1, -3)$, $(0, -1)$, $(1, 1)$, and $(2, 3)$. Note that each of these pairs of numbers satisfies the given equation.

These points plot as a straight line (Fig. 4-10). Had we known in advance that the graph would be a straight line, we could have saved time by plotting just two points, with perhaps a third as a check.

The horizontal distance between any two points on a straight line is called the *run* and the vertical distance between the same points is called the *rise*. For points $P$ and $Q$ the rise is 6 in a run of 3. The rise divided by the run is called the *slope* of the line, and is given the symbol $m$:

$$\text{slope} = m = \frac{\text{rise}}{\text{run}}$$

For the line in Fig. 4-10,

$$m = \frac{6}{3} = 2$$

Further, notice that in the equation of the line, the coefficient of $x$ is equal to the slope and that the constant term is equal to the $y$ intercept. For this reason, the equation is said to be in *slope-intercept* form.

$$y = 2x \underline{-1}$$

slope ↑ (coefficient 2) ↑ $y$ intercept ($-1$)

---

**The straight line and the parabola are covered in much greater detail in analytic geometry.**

Our next graph is an interesting and useful curve called the *parabola*.

**EXAMPLE 45:** Graph the function $y = f(x) = x^2 - 4x - 3$ for values of $x$ from $-1$ to 5.

**Solution:** Substituting into the equation, we obtain

$$f(-1) = (-1)^2 - 4(-1) - 3 = 1 + 4 - 3 = 2$$
$$f(0) = 0 - 0 - 3 = -3$$
$$f(1) = 1 - 4(1) - 3 = 1 - 4 - 3 = -6$$
$$f(2) = 2^2 - 4(2) - 3 = 4 - 8 - 3 = -7$$
$$f(3) = 3^2 - 4(3) - 3 = 9 - 12 - 3 = -6$$
$$f(4) = 4^2 - 4(4) - 3 = 16 - 16 - 3 = -3$$
$$f(5) = 5^2 - 4(5) - 3 = 25 - 20 - 3 = 2$$

The points obtained are plotted in Fig. 4-11.

**If we had plotted many more points than these—say billions of them, they would be crowded so close together that they would seem to form a continuous line. The curve can be thought of as a *collection* of all points that satisfy the equation. Such a curve (or the set of points) is called a *locus* of the equation.**

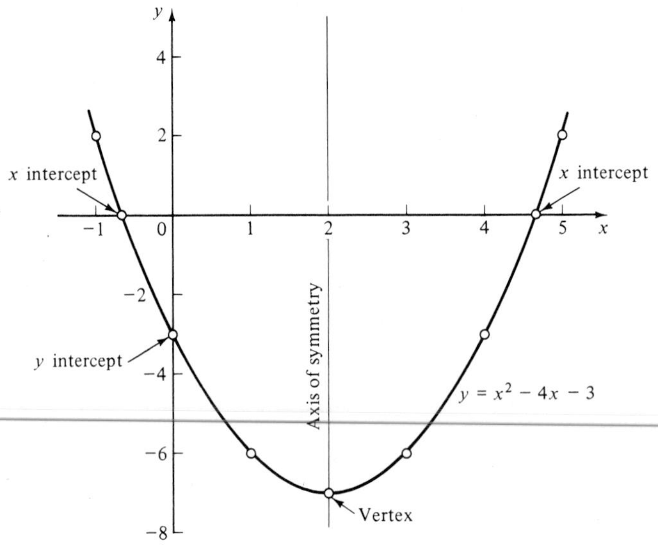


**FIGURE 4-11 This curve is called a *parabola*, which we will meet again later. Note the *vertex* and the *axis of symmetry*.**

| Common Error | Be especially careful when substituting negative values into an equation. It is easy to make an error. |
|---|---|

## Graph of a Relation versus the Graph of a Function

We can tell if a relation is also a function simply by looking at its graph. We can also determine the domain and range.

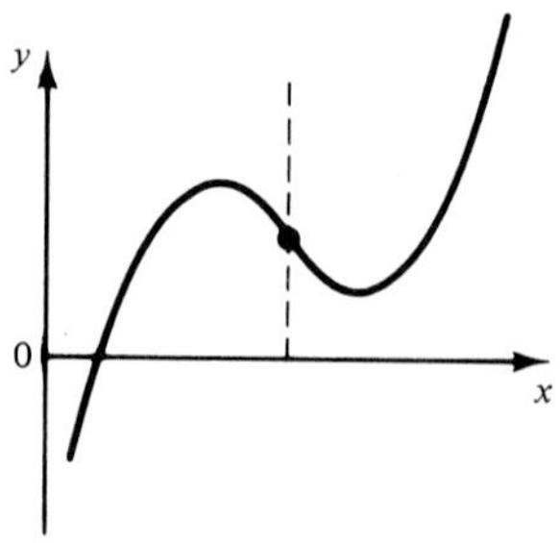


**FIGURE 4-12**

**EXAMPLE 46:** Any vertical line drawn in Fig. 4-12 crosses the curve in no more than one point. This means that for any $x$ there is not more than one $y$. Recalling our distinction between a function and a relation, we see that this is the graph of a function. The domain includes all values of $x$ and the range includes all values of $y$.

**EXAMPLE 47:** In Fig. 4-13 a vertical line can be drawn that crosses the curve in more than one point. This graph shows a relation, not a function. The domain of $x$ is

$$x \leq 5$$

and the range includes all values of $y$.

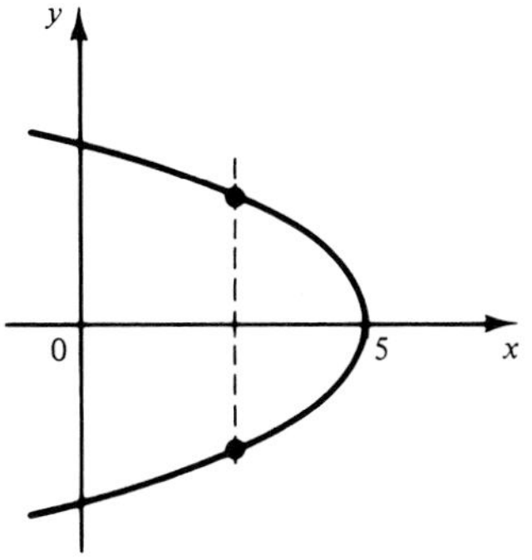


FIGURE 4-13

### Graphing Formulas

A formula is graphed in the same way that we graph a function; we substitute suitable values of the independent variable into the formula and compute the corresponding values of the dependent variable. We then plot the ordered pairs and connect the points obtained with a smooth curve. But with formulas we must be careful to handle the units properly and to label the graph more fully.

**EXAMPLE 48:** The formula for the power $P$ dissipated in a resistor carrying a current of $i$ amperes (A) is

$$P = i^2R \qquad \text{watts (W)}$$

where $R$ is the resistance in ohms. Graph $P$ versus $i$, for a resistance of 10,000 Ω. Take $i$ from 0 to 10 A.

Let us choose values of $i$ of 0, 1, 2, 3, . . . , 10, and make a table of ordered pairs by substituting these into the formula.

| $i$ (A) | 0 | 1 | 2 | 3 | . . . |
|---|---|---|---|---|---|
| $P$ (W) | 0 | 10,000 | 40,000 | 90,000 | . . . |

At this point we notice that the figures for wattage are so high that it will be more convenient to work in kilowatts (kW), where 1 kW = 1000 W.

| $i$ (A) | 0 | 1 | 2 | 3 | 4 | 5 | 6 | 7 | 8 | 9 | 10 |
|---|---|---|---|---|---|---|---|---|---|---|---|
| $P$ (kW) | 0 | 10 | 40 | 90 | 160 | 250 | 360 | 490 | 640 | 810 | 1000 |

These points are plotted in Fig. 4-14. Note the labeling of the graph and of the axes.

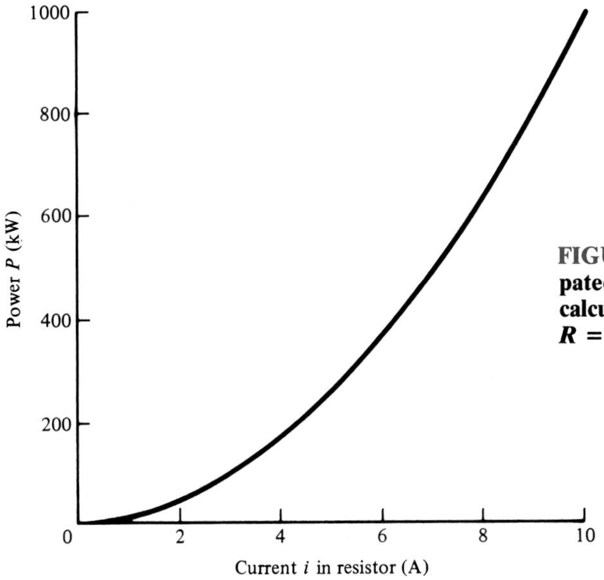


**FIGURE 4-14 Power dissipated in a 10,000-Ω resistor calculated using $P = i^2R$ with $R$ = 10,000 Ω.**

## Graphing by Addition of Ordinates

When the function to be graphed has two or more terms, it is sometimes easier to graph each term *separately* and then to combine them graphically right on the paper.

**EXAMPLE 49:** Graph the function

$$y = \frac{x^2}{2} + \frac{1}{x}$$

by addition of ordinates, for $x = 0$ to 3.

**Solution:** Let us split the function into two parts:

$$y_1 = \frac{x^2}{2} \quad \text{and} \quad y_2 = \frac{1}{x}$$

and plot $y_1$ and $y_2$ separately. Substituting into these equations, we obtain

| $x$ | 0 | $\frac{1}{2}$ | 1 | $1\frac{1}{2}$ | 2 | $2\frac{1}{2}$ | 3 |
|---|---|---|---|---|---|---|---|
| $y_1$ | 0 | $\frac{1}{8}$ | $\frac{1}{2}$ | $\frac{9}{8}$ | 2 | $\frac{25}{8}$ | $\frac{9}{2}$ |
| $y_2$ | not defined | 2 | 1 | $\frac{2}{3}$ | $\frac{1}{2}$ | $\frac{2}{5}$ | $\frac{1}{3}$ |

These points are plotted in Fig. 4-15 and connected with dashed curves. Then, at each $x$ the ordinates are added graphically, to obtain a third set of points. This can be done by measuring the ordinates of the two dashed curves and adding, or by using dividers to add the ordinate of one dashed curve to the other. The new set of points is then connected with the solid curve shown.

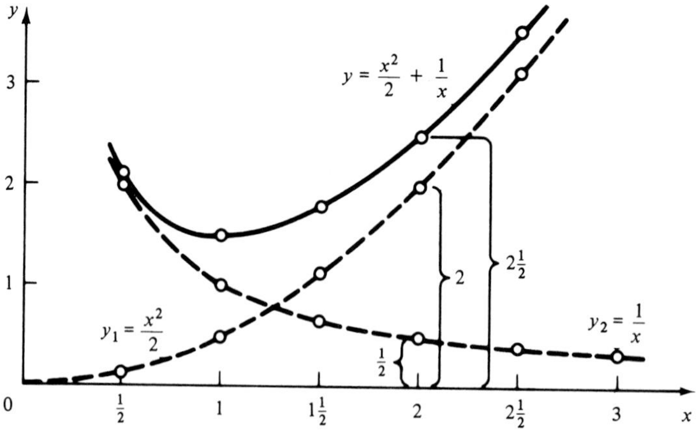


**FIGURE 4-15 Graphing by addition of ordinates.**

## Graphing Parametric Equations

For the equations we have graphed so far, $y$ has been expressed as a function of $x$.

$$y = f(x)$$

But $x$ and $y$ can also be related to each other by means of a third variable, say $t$, if both $x$ and $y$ are given as functions of $t$.

$$x = h(t)$$

and $$y = g(t)$$

Such equations are called *parametric equations*. The third variable $t$ is called the *parameter*.

To graph parametric equations we assign values to the parameter $t$ and compute $x$ and $y$ for each $t$. The table of $(x, y)$ pairs is then plotted.

**In later chapters we'll plot parametric equations in *polar* coordinates, and also use them to find *trajectories* of projectiles.**

---

**EXAMPLE 50:** Graph the parametric equations

$$x = 2t \qquad \text{and} \qquad y = t^2 - 2$$

for $t = -3$ to 3.

**Solution:** We make a table with rows for $t$, $x$, and $y$. We take values of $t$ from $-3$ to 3, and for each compute $x$ and $y$.

| $t$ | −3 | −2 | −1 | 0 | 1 | 2 | 3 |
|---|---|---|---|---|---|---|---|
| $x$ | −6 | −4 | −2 | 0 | 2 | 4 | 6 |
| $y$ | 7 | 2 | −1 | −2 | −1 | 2 | 7 |

We now plot the $(x, y)$ pairs, $(-6, 7)$, $(-4, 2)$, . . . , $(6, 7)$ and connect them with a smooth curve (Fig. 4-16). The curve obtained is a parabola, as in Example 45, but here obtained with parametric equations.

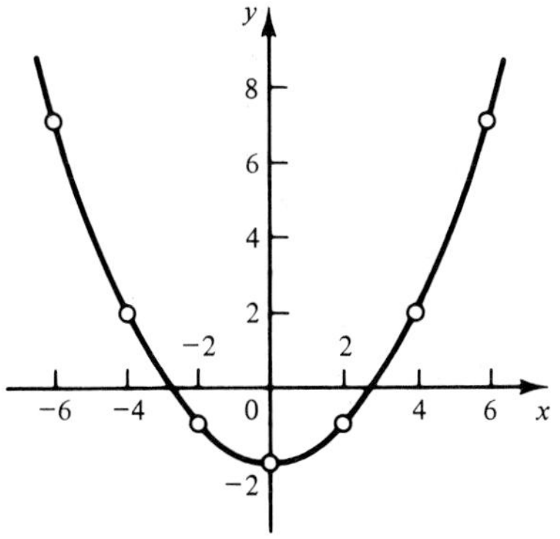


FIGURE 4-16

---

## EXERCISE 4—THE GRAPH OF A RELATION OR FUNCTION

### Graphing Sets of Ordered Pairs

Graph each set of ordered pairs. Connect them with a curve that seems to you to fit the data best.

1. (−3, −2), (9, 6), (3, 2), (−6, −4)

2. (−7, 3), (0, 3), (4, 10), (−6, 1), (2, 6), (−4, 0)

3. (−10, 9), (−8, 7), (−6, 5), (−4, 3), (−2, 4), (0, 5), (2, 6), (4, 7)

4. (0, 4), (3, 3.2), (5, 2), (6, 0), (5, −2), (3, −3.2), (0, −4)

## Graphing Empirical Data

**Data obtained by experiment or observation are called *empirical* data.**

Graph the following experimental data. Label the graph completely. Take the first quantity in each table as the abscissa, and the second quantity as the ordinate. Connect the points with a smooth curve.

5. The current $I$ (mA) through a tungsten lamp and the voltage $V$ (V) that is applied to the lamp, related as shown in the following table.

| $V$ | 10 | 20 | 30 | 40 | 50 | 60 | 70 | 80 | 90 | 100 | 110 | 120 |
|---|---|---|---|---|---|---|---|---|---|---|---|---|
| $I$ | 158 | 243 | 306 | 367 | 420 | 470 | 517 | 559 | 598 | 639 | 676 | 710 |

6. The melting point $T$ (°C) of a certain alloy and the percent of lead $P$ in the alloy, related as shown in the following table.

| $P$ | 40 | 50 | 60 | 70 | 80 | 90 |
|---|---|---|---|---|---|---|
| $T$ | 186 | 205 | 226 | 250 | 276 | 304 |

7. A steel wire in tension, with the stress $\sigma$ (lb/in$^2$) and the strain $\varepsilon$ (in./in.), related as follows.

| $\sigma$ | 5000 | 10,000 | 20,000 | 30,000 | 40,000 | 50,000 | 60,000 | 70,000 |
|---|---|---|---|---|---|---|---|---|
| $\epsilon$ | 0 | 0.00019 | 0.00057 | 0.00094 | 0.00134 | 0.00173 | 0.00216 | 0.00256 |

## Graphing Equations

8. Which of the graphs in Fig. 4-17 are functions? Explain.

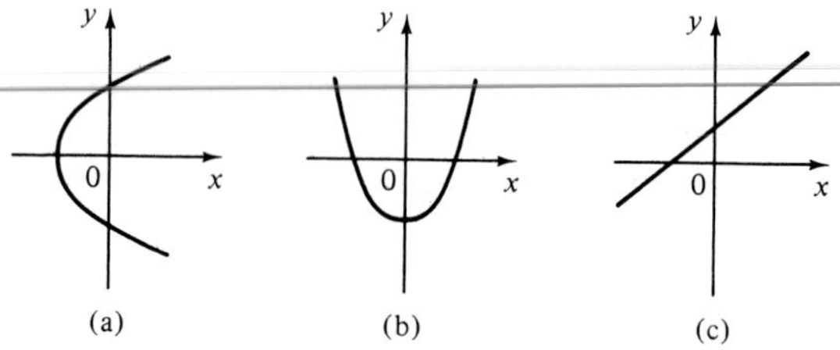


FIGURE 4-17

9. Which of the graphs in Fig. 4-18 are functions? Explain.

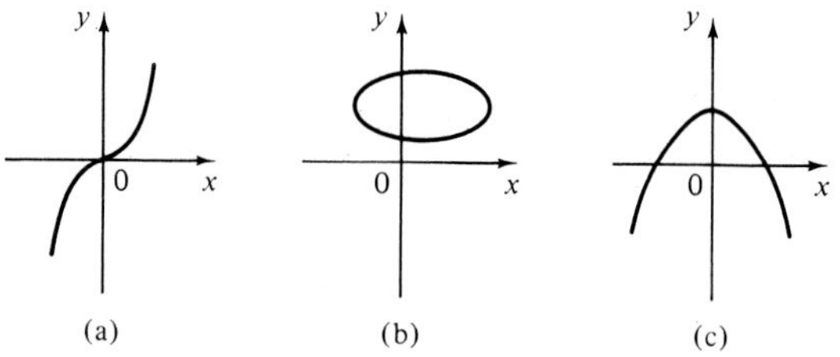


FIGURE 4-18

Fill in the missing values and graph the functions. Label any intercepts.

10. $y = x^2 - 1$

| $x$ | 0 | 1 | 2 | 3 | 4 | 5 |
|---|---|---|---|---|---|---|
| $y$ | | | | | | |

11. $y = 5x - x^2$

| $x$ | $-2$ | $-1$ | $0$ | $1$ | $2$ |
|---|---|---|---|---|---|
| $y$ | | | | | |

For each equation, make a table of ordered pairs, taking integer values of $x$ from $-3$ to 3, and graph the function. Label any intercepts.

12. $y = 3x$

13. $y = -x + 2$

14. $y = x^2$

15. $y = 4 - 2x^2$

16. $y = \dfrac{x^2}{x + 3}$

17. $y = 5$

18. $y = 0$

19. $y = x$

20. $y = x^2 - 7x + 10$

21. State the domain and range of the relations in Fig. 4-19a and b.

22. State the domain and range of the relations in Fig. 4-19c and d.

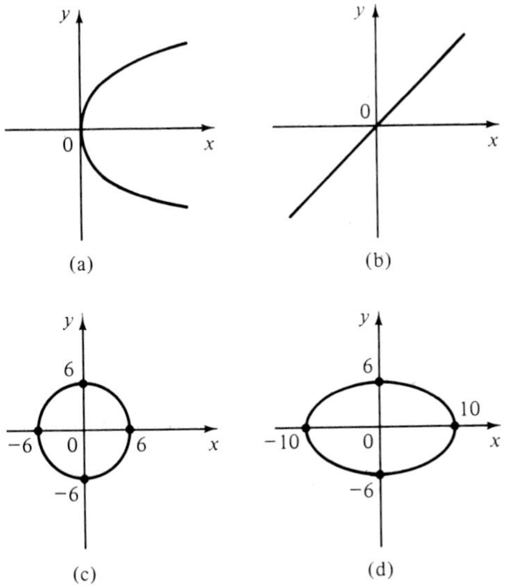


FIGURE 4-19

### Graphing Formulas

23. An amount $a$ of \$10,000 is invested at a compound interest rate $n$ of 7.5%. Graph the total amount $y$ at the end of each year $x$ for a period of 10 years, using $y = a(1 + n)^x$.

24. A milling machine having a purchase price $P$ of \$15,600 has an annual depreciation $A$ of \$1600. Graph the book value $y$ at the end of each year, for $t = 10$ years, using the equation $y = P - At$.

25. The force $f$ required to pull a block along a rough surface is given by $f = \mu N$, where $N$ is the normal force and $\mu$ is the coefficient of friction. Plot $f$ for values of $N$ from zero to 1000 N, taking $\mu$ as 0.45.

26. A 2580-Ω resistor is wired in parallel with a resistor $R_1$. Graph the equivalent resistance $R$ for values of $R_1$ from 0 to 5000 Ω, in steps of 500 Ω. Use Eq. A64.

27. Use Eq. A67 to graph the power $P$ dissipated in a 2500-Ω resistor for values of the current $I$ from 0 to 10 A.

28. A resistance of 5280 Ω is placed in series with a device that has a reactance of $X$ ohms (Fig. 4-20). Using Eq. A97, plot the absolute value of the impedance $Z$ for values of $X$ from 0 to 10,000 Ω.

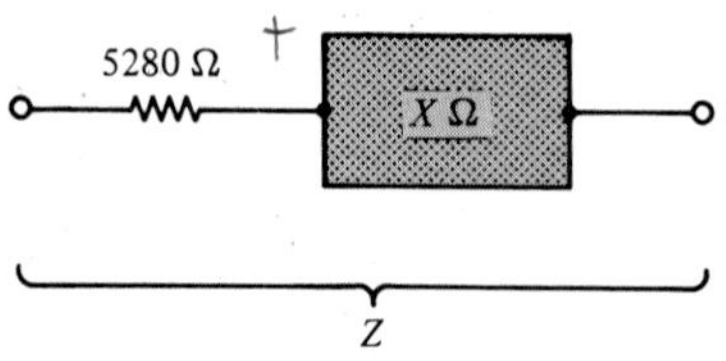


FIGURE 4-20

### Addition of Ordinates

Graph each equation by addition of ordinates for $x = -3$ to 3.

29. $y = x - \dfrac{1}{x}$

30. $y = 3x - 2$

31. $y = x^2 - \dfrac{1}{x}$

32. $y = 3x + \dfrac{1}{x}$

33. $y = \dfrac{1}{x - 2} + x$

34. $y = \dfrac{1}{x} + x^3$

### Parametric Equations

Plot the following parametric equations for values of $t$ from $-3$ to 3.

35. $x = t,\ y = t$

36. $x = 3t,\ y = t^2$

37. $x = -t,\ y = 2t^2$

38. $x = -2t,\ y = t^2 + 1$

**Computer**

39. Write a program to generate a table of $x$, $y$ point pairs for any function you enter. You should also be able to enter the domain of $x$, and the interval between your points. Use your program to obtain plotting points for any of the equations in this exercise, and plot the points by hand.
40. Some spreadsheet programs, such as LOTUS 123, are able to create graphs of data in the spreadsheet. If you have such a program available, learn how to use it. Practice with any of the graphs in this exercise.

## 4-5 GRAPHICAL SOLUTION OF EQUATIONS

### Solving Equations Graphically

We can use our knowledge of graphing functions to solve equations of the form $f(x) = 0$.

We have mentioned earlier that a point at which a graph of a function $y = f(x)$ crosses or touches the $x$ axis is called an $x$ intercept. Such an $x$ intercept is also called a *zero* of that function.

In Fig. 4-21 there are two zeros, since there are two $x$ values for which $y = 0$, and hence $f(x) = 0$. Those $x$ values for which $f(x) = 0$ are called *roots* or *solutions* to the equation $f(x) = 0$.

Thus if we were to graph the function $y = f(x)$, any value of $x$ at which $y$ is equal to zero would be a solution to $f(x) = 0$. So to solve an equation graphically, we simply put it into the form $f(x) = 0$ and then graph the function $y = f(x)$. Each $x$ intercept is then an approximate solution to the equation.

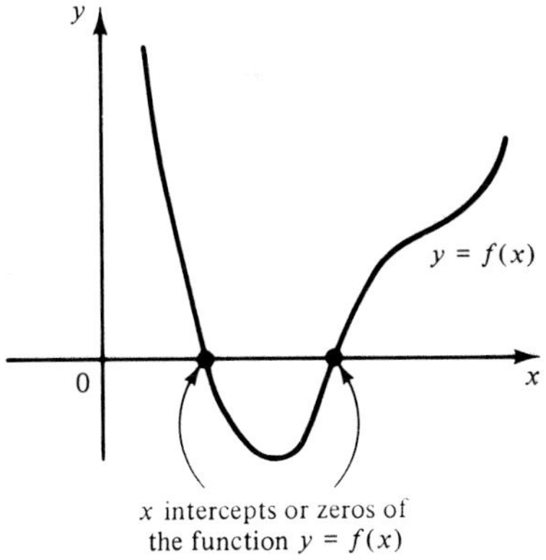


**FIGURE 4-21**

---

**EXAMPLE 51:** Graphically find the approximate root(s) of the equation

$$4.1x^3 - 5.9x^2 - 3.8x + 7.5 = 0 \qquad (1)$$

**Solution:** Let us represent the left side of the given equation by $f(x)$.

$$f(x) = 4.1x^3 - 5.9x^2 - 3.8x + 7.5$$

Any value of $x$ for which $f(x) = 0$ will clearly be a solution to (1), so we simply plot $f(x)$ and look for the $x$ intercepts. Not knowing the shape of the curve, we compute $f(x)$ at various values of $x$ until we are satisfied that we have located each region in which an $x$ intercept is located. We then make a table of point pairs for this region.

| $x$ | −2 | −1 | 0 | 1 | 2 | 3 |
|---|---|---|---|---|---|---|
| $f(x)$ | −41.3 | 1.3 | 7.5 | 1.9 | 9.1 | 53.7 |

We graph the function (Fig. 4-22) and read the approximate value of the $x$ intercept,

$$x \cong -1.1$$

This, then, is an approximate solution to the given equation. Have we found *every* root? If we compute $y$ for values of $x$ outside the region we have graphed, we would see that the curve moves even farther from the $x$ axis, so we are reasonably sure that we have found the only root. With other functions it may not be so easy to tell if the curve will reverse direction somewhere and cross the $x$ axis again.

---

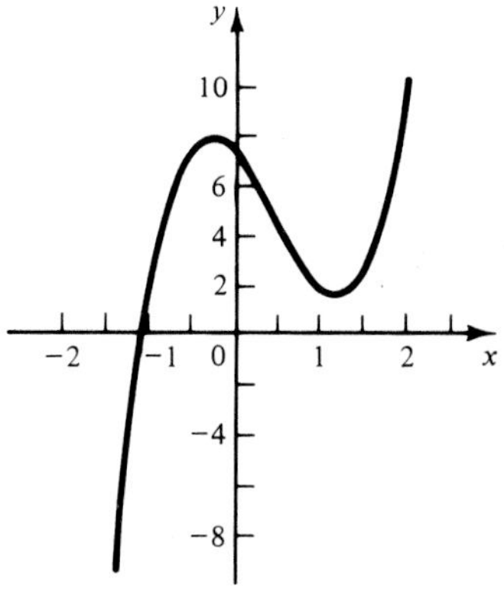


**FIGURE 4-22 The shape of this curve is typical of a third-degree (cubic) function.**

## Computer Solution of Equations by the Midpoint Method

**This method is also called the *half-interval method* or the *bisection method*.**

There are several methods for finding the roots of equations by computer. In Chapter 14 we show the method of *simple iteration,* and when you study calculus you will learn the popular *Newton's method.* Here we cover the simple *midpoint* method.

Figure 4-23 shows a graph of some function $y = f(x)$. It is a smooth, continuous curve which crosses the $x$ axis at $R$, which is a root of the equation $f(x) = 0$. We determine two values $x_1$ and $x_2$ which lie to the left and right of $R$, a fact that we verify by testing if $y_1$ and $y_2$ have opposite signs. We then locate $x_3$ midway between $x_1$ and $x_2$, and determine whether it is to the left or right of $R$ by comparing the sign of $y_3$ with that of $y_1$. Thus $x_3$ becomes a new endpoint (either the left or the right as determined by the test), and we repeat the computation, halving the interval again and again until we get as close to $R$ as we wish.

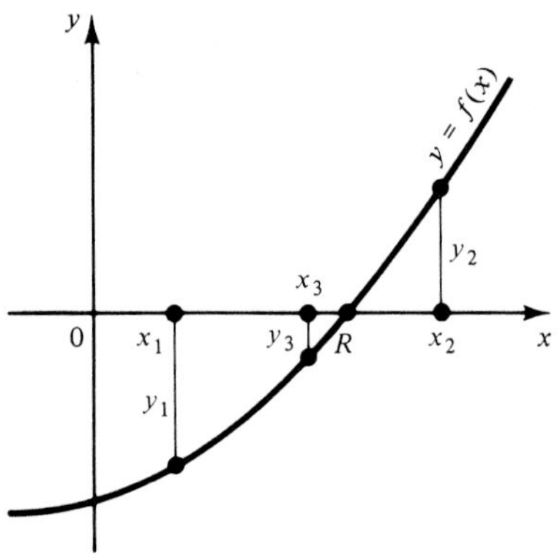


**FIGURE 4-23 The midpoint method for finding roots.**

Following is an algorithm for the midpoint method.

1. Make two initial guesses, $x_1$ and $x_2$, which must lie on either side of the root $R$.
2. Go to step 3 if $y_1$ and $y_2$ have opposite signs. Otherwise, go to step 1.
3. Compute the abscissa $x_3$ of the point midway between the two original points, with

$$x_3 = \frac{x_1 + x_2}{2}$$

4. Compute $y_1$ and $y_3$ by substituting $x_1$ and $x_3$ into the equation of the curve.
5. If the absolute value of $y_3$ is "small," then print the value of $x_3$ and end.
6. Determine if $x_3$ is to the right or left of the root by comparing the signs of $y_1$ and $y_3$.
7. If $y_1$ and $y_3$ have the same signs, replace $x_1$ with $x_3$. Otherwise, replace $x_2$ with $x_3$.
8. Go to step 3 and repeat the computation.

**A technique such as this one, in which we repeat a computation many times, each time getting an approximate result that is (with luck) closer to the true value, is called *iteration.* Iteration methods are used often on the computer, and we give several in this text.**

Using this algorithm as a guide, try to program the midpoint method on your computer. Save your program carefully because you can use it to check your manual solutions to the various equations in this text.

## EXERCISE 5—GRAPHICAL SOLUTION OF EQUATIONS

The following equations all have at least one root between $x = -10$ and $x = 10$. Put each equation into the form $y = f(x)$ and plot just enough of each to find the approximate value of the root(s).

1. $2.4x^3 - 7.2x^2 - 3.3 = 0$
2. $9.4x = 4.8x^3 - 7.2$
3. $25x^2 - 19 = 48x + x^3$
4. $1.2x + 3.4x^3 = 2.8$

5. $6.4x^4 - 3.8x = 5.5$

6. $621x^4 - 284x^3 - 25 = 0$

In each of Problems 7 to 10, two equations are given. Graph each equation as if the other were not there, but on the same coordinate axes. Each equation will graph as a straight line. Give the approximate coordinates of the point of intersection of the two lines.

7. $y = 2x + 1$
   $y = -x + 2$

8. $x - y = 2$
   $x + y = 6$

**Can you guess the significance of the point of intersection? Peek ahead at Sec. 9-1.**

9. $y = -3x + 5$
   $y = x - 4$

10. $2x + 3y = 9$
    $3x - y = -4$

**Computer**

11. Write a program for solving equations by the midpoint method. Use it to solve any of the equations in Problems 1 to 6.

## CHAPTER 4 REVIEW PROBLEMS

1. Which of the following relations are also functions?
   (a) $y = 5x^3 - 2x^2$
   (b) $x^2 + y^2 = 25$
   (c) $y = \pm\sqrt{2x}$
2. Write $y$ as a function of $x$ if $y$ is equal to half the cube of $x$, diminished by twice $x$.
3. Write an equation to express the surface area $S$ of a sphere as a function of its radius $r$.
4. Find the largest possible domain and range for each function.
   (a) $y = \dfrac{5}{\sqrt{3 - x}}$
   (b) $y = \dfrac{1 + x}{1 - x}$
5. Label each function as implicit or explicit. If explicit, name the dependent and independent variables.
   (a) $w = 3y - 7$
   (b) $x - 2y = 8$
6. Given $y = 3x - 5$, write $x = f(y)$.
7. Given $x^2 + y^2 + 2w = 3$, write $w = f(x, y)$.
8. Graph the following points and connect them with a smooth curve.

| $x$ | 0 | 1 | 2 | 3 | 4 | 5 | 6 |
|---|---|---|---|---|---|---|---|
| $y$ | 2 | $2\frac{1}{4}$ | 3 | 4 | 6 | 9 | 13 |

9. If $y = 3x^2 + 2z$, and $z = 2x^2$, write $y = f(x)$.
10. If $f(x) = 5x^2 - 7x + 2$, find $f(3)$.
11. If $f(x) = 9 - 3x$, find $2f(3) + 3f(1) - 4f(2)$.
12. Plot the function $y = x^3 - 2x$ in the domain $-3$ to $+3$. Label any roots or intercepts.
13. Graph the function $y = 3x - 2x^2$ by addition of ordinates from $x = -3$ to $x = 3$. Label any roots or intercepts.
14. Graphically find the approximate value of the roots of the equation

$$(x + 3)^2 = x - 2x^2 + 4$$

15. If $f(x) = 5x$, $g(x) = 1/x$, and $h(x) = x^3$, find

$$\frac{5h(1) - 2g(3)}{3f(2)}$$

16. Is the relation $x^2 + 3xy + y^2 = 1$ a function? Why?
17. Replace the function $y = 5x^3 - 7$ by a verbal statement.
18. What are the domain and range of the function $(10, -8)$, $(20, -5)$, $(30, 0)$, $(40, 3)$, and $(50, 7)$?
19. If $y = 6 - 3x^2$, write $x = f(y)$.

20. If $x = 6w - 5y$ and $y = 3z + 2w$, write $x = f(w, z)$.

21. If $f(x) = 8x + 3$ and $g(u) = u^2 - 4$, write $f[g(u)]$.

22. If $f(x) = 7x + 5$ and $g(x) = x^2$, find $3f(2) - 5g(3)$.

23. If $f(x, y, z) = x^2 + 3xy + z^3$, find $f(3, 2, 1)$.

24. Graphically find the point of intersection.

$$2.5x - 3.7y = 5.2$$

$$6.3x + 4.2y = 2.6$$

25. Write the inverse of the function $y = 9x - 5$.

26. Make a graph of the pressure $p$ (lb/in$^2$) in an engine cylinder, where $v$ is the volume (in$^3$) above the piston.

| $p$ | 39.6 | 44.7 | 53.8 | 73.5 | 85.8 | 113.2 | 135.8 | 178.2 |
|---|---|---|---|---|---|---|---|---|
| $v$ | 10.61 | 9.73 | 8.55 | 7.00 | 6.23 | 5.18 | 4.59 | 3.87 |

Plot the parametric equations, for $x = -4$ to 4.

27. $x = 3t$
$y = t^3$

28. $x = 2t^2$
$y = 4t$

*Writing*

29. State, in your own words, what is meant by "function" and "relation," and how a function differs from a relation. You may use examples to help describe these terms.

# 5

# GEOMETRY

## OBJECTIVES

**When you have completed this chapter, you should be able to:**

- Find the angles formed by intersecting straight lines.
- Solve practical problems which require finding the sides and angles of right triangles.
- Solve practical problems in which the area of a triangle or quadrilateral must be found.
- Solve problems involving the circumference, diameter, area, or tangent to a circle.
- Compute surface areas and volumes of spheres, cylinders, cones, and other solid figures.

We present in this chapter some of the more useful facts from geometry, together with examples of their use in solving practical problems. Geometrical theorems—provable statements—are treated as facts and are presented almost "cookbook" style, with no proofs given and with the emphasis on the applications of these facts. Our main concern will be the computation of lengths, areas, and volumes of various geometric figures.

Much of the material on angles, triangles, and radian measure will be good preparation for the later chapters on trigonometry.

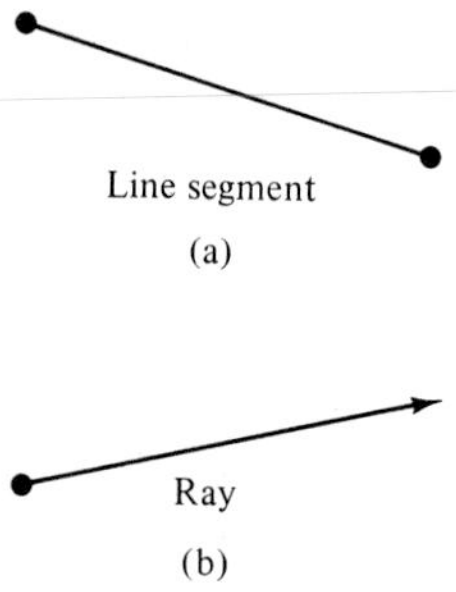


FIGURE 5-1

## 5-1 STRAIGHT LINES AND ANGLES

### Straight Line

A *line segment* is the portion of a straight line lying between two endpoints (Fig. 5-1a). A *ray,* or *half-line,* is the portion of a line lying to one side of a point (endpoint) on the line (Fig. 5-1b).

### Angles

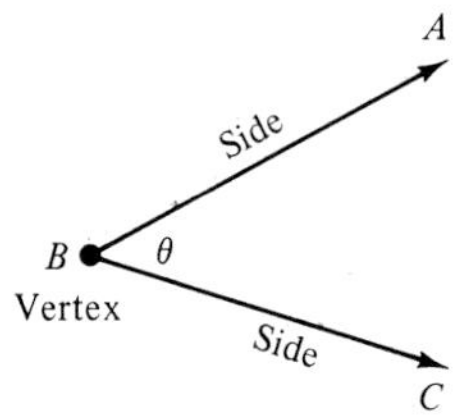


**FIGURE 5-2 An angle.**

An *angle* is formed when two rays intersect at their endpoints (Fig. 5-2). The point of intersection is called the *vertex* of the angle, and the two rays are called the *sides* of the angle.

The angle shown in Fig. 5-2 can be *designated* in any of the following ways

angle $ABC$ angle $CBA$ angle $B$ angle $\theta$

The symbol $\angle$ means *angle*. So $\angle B$ means angle $B$.

An angle can also be thought of as having been *generated* by a ray turning from some *initial position* to a *terminal position* (Fig. 5-3).

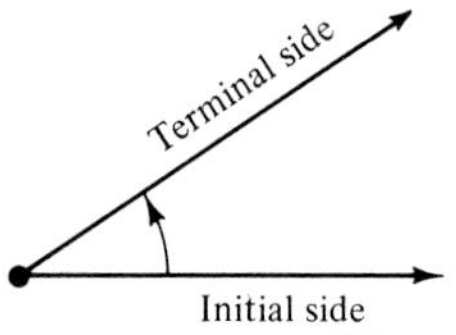


**FIGURE 5-3 An angle formed by rotation.**

One *revolution* is the amount a ray would turn to return to its original position.

The *units of angular measure* in common use are the *degree* and the *radian* (Sec. 5-4). The *measure* of an angle is the number of units of measure it contains. Two angles are *equal* if they have the same measure.

**For brevity, we say "angle *A*" instead of the cumbersome but more correct "the measure of angle *A*."**

The *degree* is a unit of angular measure equal to 1/360 of a revolution. Thus, there are 360 degrees in one complete revolution.

Figure 5-4 shows a *right* angle ($\frac{1}{4}$ revolution, or 90°), usually marked with a small square at the vertex; an *acute* angle (less than $\frac{1}{4}$ revolution); an

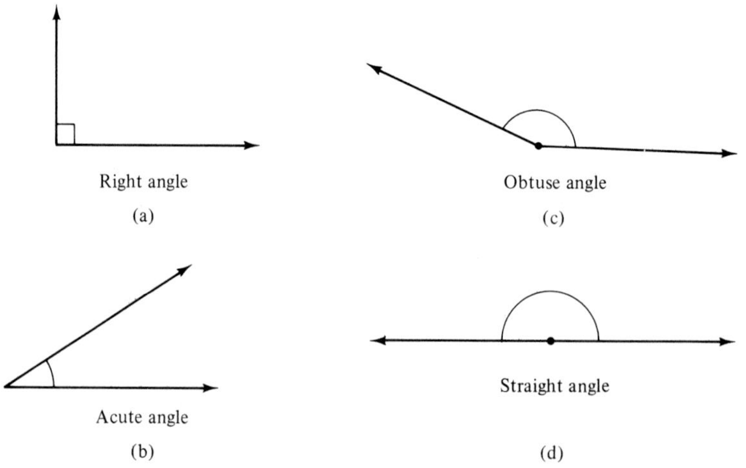


**FIGURE 5-4 Types of angles.**

*obtuse* angle (greater than $\frac{1}{4}$ revolution but less than $\frac{1}{2}$ revolution); and a *straight* angle ($\frac{1}{2}$ revolution, or 180°).

Two lines at right angles to each other are said to be *perpendicular*. Two angles are called *complementary* if their sum is a right angle, and *supplementary* if their sum is a straight angle.

Two more words we use in reference to angles are *intercept* and *subtend*. In Fig. 5-5 we say that the angle $\theta$ *intercepts* the section $PQ$ of the curve. Conversely, we say that angle $\theta$ is *subtended by PQ*.

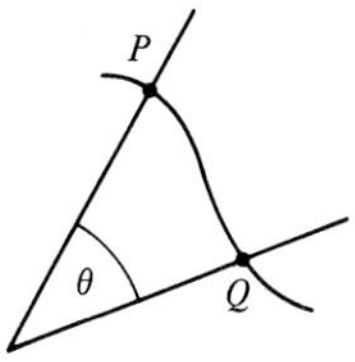

**FIGURE 5-5**

### Angles between Intersecting Lines

Angles $A$ and $B$ of Fig. 5-6 are called *opposite* or *vertical* angles. It should be apparent that $A$ and $B$ are equal.

| Opposite angles of two intersecting straight lines are equal. | **104** |
|---|---|

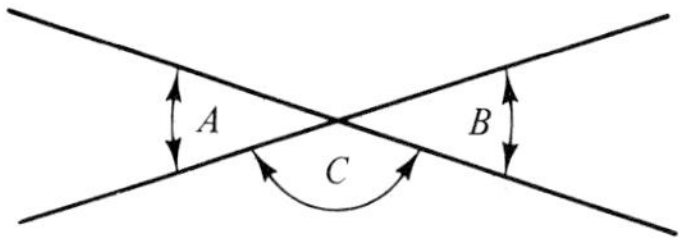

**FIGURE 5-6**

Two angles are called *adjacent* when they have a common side and a common vertex, such as angles $A$ and $C$ of Fig. 5-6. When two lines intersect, the adjacent angles are *supplementary*.

---

**EXAMPLE 1:** Find the measure (in degrees) of angles $A$ and $B$ in the structure shown in Fig. 5-7.

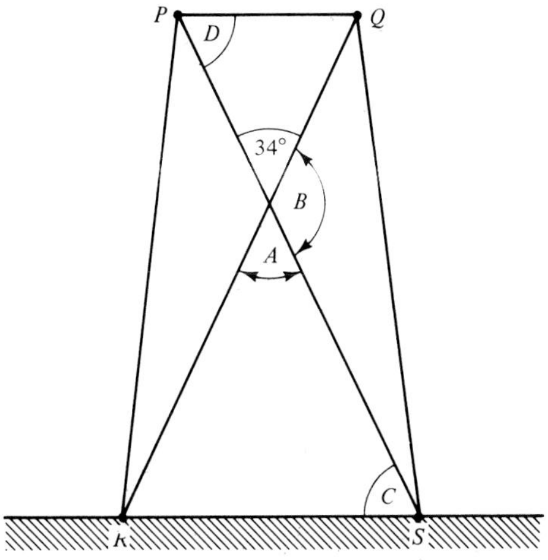

**FIGURE 5-7**

**Solution:** We see from the diagram that $RQ$ and $PS$ are straight intersecting lines (we will often assume relationships from the diagram rather than proving them). Since angle $A$ is opposite the 34° angle, angle $A = 34°$. Angle $B$ and the 34° angle are supplementary, so

$$B = 180° - 34° = 146°$$

---

### Transversals

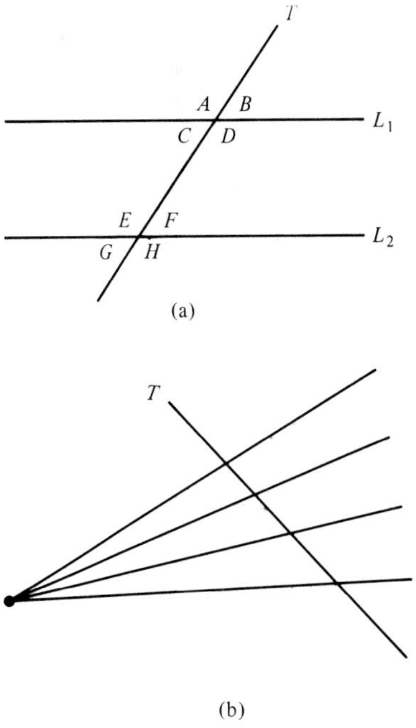

**FIGURE 5-8 Transversals.**

A *transversal* is a line that intersects a system of lines. In Fig. 5-8, the lines labeled $T$ are transversals. Lines that do not intersect are called *parallel*. In Fig. 5-8a, two parallel lines, $L_1$ and $L_2$, are cut by transversal $T$. Angles $A$, $B$,

*G*, and *H* are called *exterior* angles, and *C, D, E,* and *F* are called *interior* angles. Angles *A* and *E* are called *corresponding* angles. Other corresponding angles in Fig. 5-8a are *C* and *G*, *B* and *F*, and *D* and *H*. Angles *C* and *F* are called *alternate interior angles,* as are *D* and *E*. We have a theorem:

| | |
|---|---|
| If two parallel straight lines are cut by a transversal, corresponding angles are equal, and alternate interior angles are equal. | **105** |

Thus, in Fig. 5-8a, $\angle A = \angle E$, $\angle C = \angle G$, $\angle B = \angle F$, and $\angle D = \angle H$. Also,

$$\angle D = \angle E \quad \text{and} \quad \angle C = \angle F$$

**EXAMPLE 2:** The top girder *PQ* in the structure of Fig. 5-7 is parallel to the ground and angle *C* is 73°. Find angle *D*.

**Solution:** We have two parallel lines, *PQ* and *RS,* cut by transversal *PS*. By statement 105

$$\angle D = \angle C = 73°$$

Another useful theorem applies when a number of parallel lines are cut by *two* transversals, such as in Fig. 5-9. The portions of the transversals lying between the same parallels are called *corresponding segments*. (In Fig. 5-9, *a* and *b* are corresponding segments, *c* and *d* are corresponding segments, and *e* and *f* are corresponding segments.)

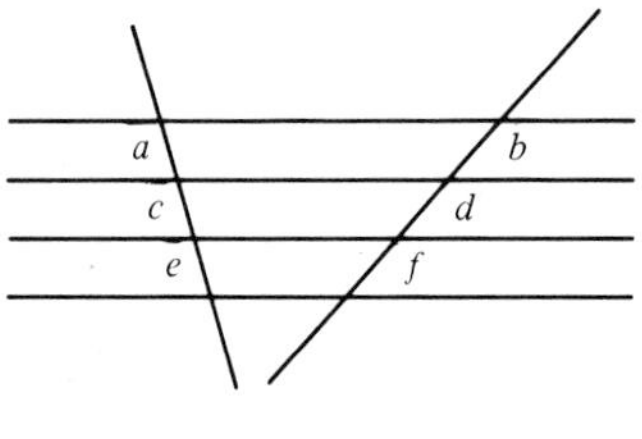


$$\frac{a}{b} = \frac{c}{d} = \frac{e}{f}$$

**FIGURE 5-9**

| | |
|---|---|
| When a number of parallel lines are cut by two transversals, the ratios of corresponding segments on the transversals are equal. | **106** |

In Fig. 5-9,

$$\frac{a}{b} = \frac{c}{d} = \frac{e}{f}$$

**EXAMPLE 3:** A portion of a street map is shown in Fig. 5-10. Find the distances $PQ$ and $QR$.

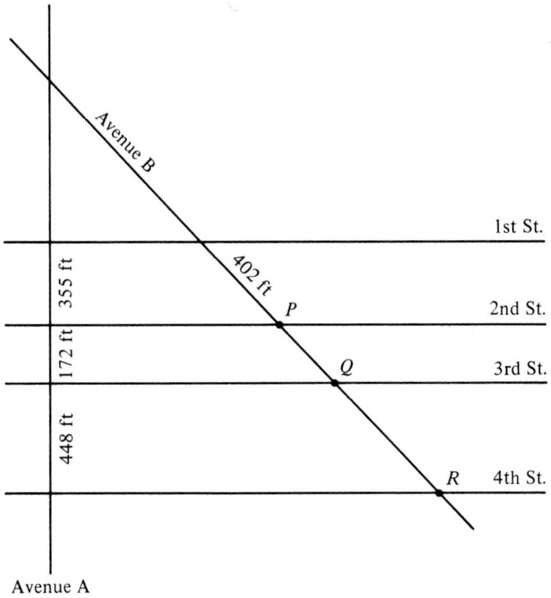


FIGURE 5-10

**Solution:** From statement 106,

$$\frac{PQ}{172} = \frac{402}{355}$$

$$PQ = \frac{402}{355}(172) = 195 \text{ ft}$$

Similarly,

$$\frac{QR}{448} = \frac{402}{355}$$

$$QR = \frac{402}{355}(448) = 507 \text{ ft}$$

## EXERCISE 1—STRAIGHT LINES AND ANGLES

1. Find angle $\theta$ in Fig. 5-11a, b, and c.

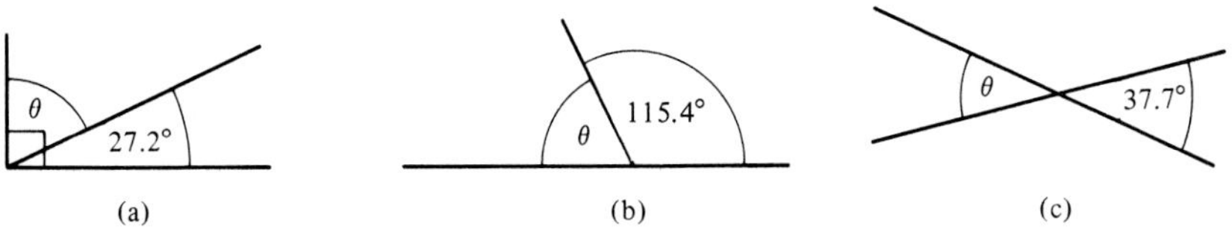


FIGURE 5-11

2. Find angles $A$, $B$, $C$, $D$, $E$, $F$, and $G$ in Fig. 5-12.

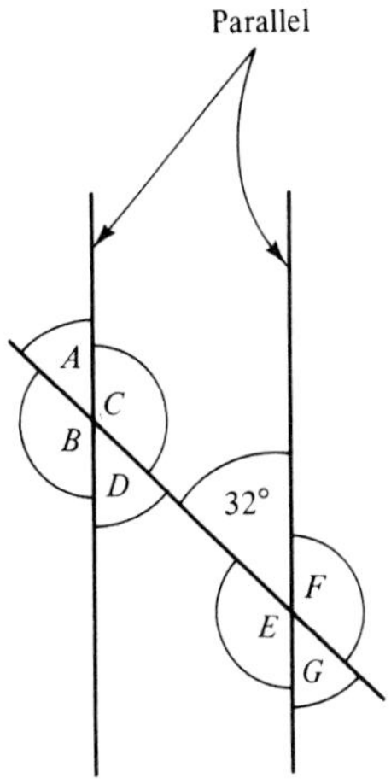


FIGURE 5-12

3. Find distance $x$ in Fig. 5-13.
4. Find angles $A$ and $B$ in Fig. 5-14.

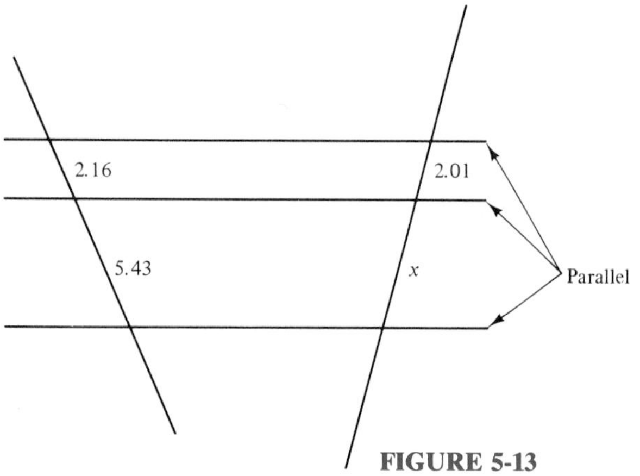


FIGURE 5-13

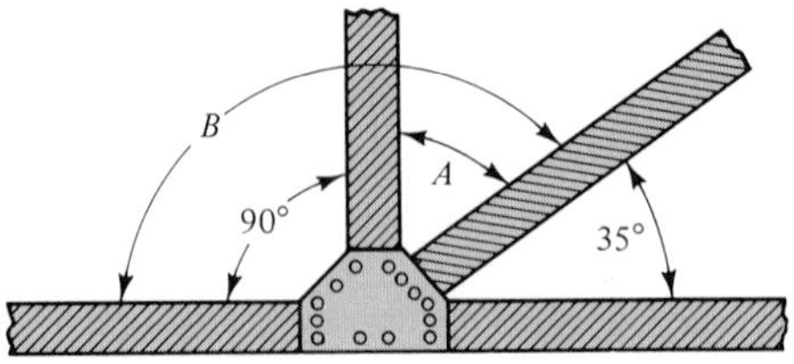


FIGURE 5-14

**The *angle of elevation* of some object above the observer is the angle between the line of sight to that object and the horizontal, as in Fig. 6-12.**

5. On a certain day, the angle of elevation of the sun is 46.3° (Fig. 5-15). Find the angles $A$, $B$, $C$, and $D$ that a ray of sunlight makes with a horizontal sheet of glass.
6. Three parallel steel cables hang from a girder to the deck of a bridge (Fig. 5-16). Find distance $x$.

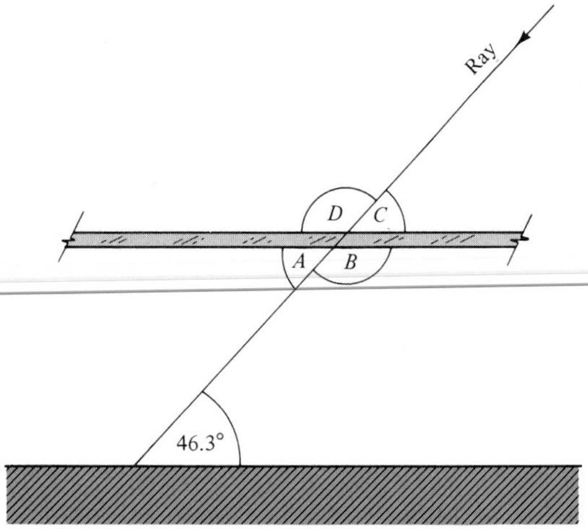


FIGURE 5-15

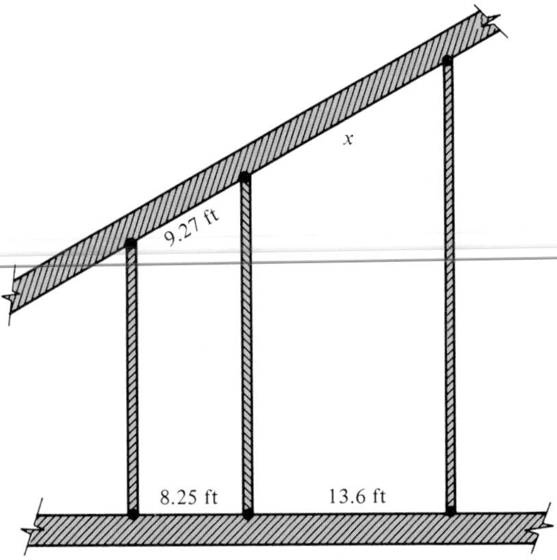


FIGURE 5-16

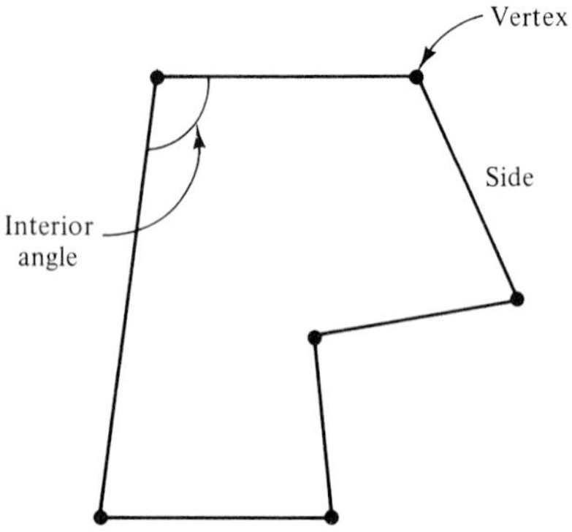


FIGURE 5-17 **A polygon.**

## 5-2 TRIANGLES

### Polygons

A *polygon* is a plane figure formed by three or more line segments, called the *sides* of the polygon, joined at their endpoints, as in Fig. 5-17. The points where the sides meet are called vertices. We say that the sides of a polygon are equal if their measures (lengths) are equal. If all the sides and angles of a polygon are equal, it is called a *regular* polygon, as in Fig. 5-18. The *perimeter* of a polygon is simply the sum of its sides.

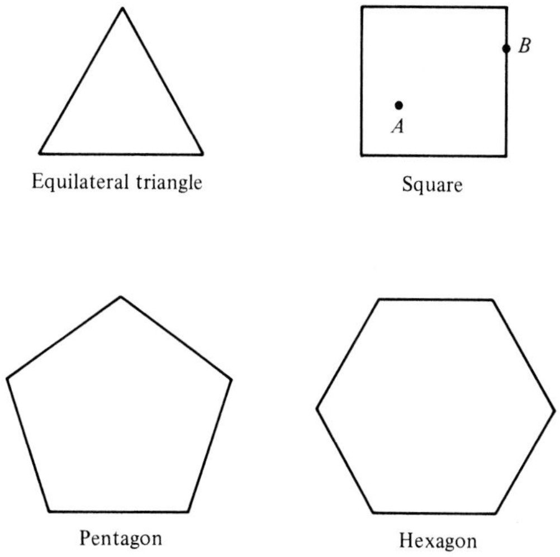


**FIGURE 5-18 Some regular polygons.**

**Modern definitions of plane figures exclude the *interior* as part of the figure. Thus, in Fig. 5-18 point *A* is *not* on the square, while point *B is* on the square. The interior is referred to as a *region*.**

## Sum of Interior Angles

A polygon of $n$ sides has $n$ interior angles, such as those shown in Fig. 5-19. Their sum is equal to:

| | | |
|---|---|---|
| Interior Angles of a Polygon | Sum of Angles $= (n - 2)180°$ | **112** |

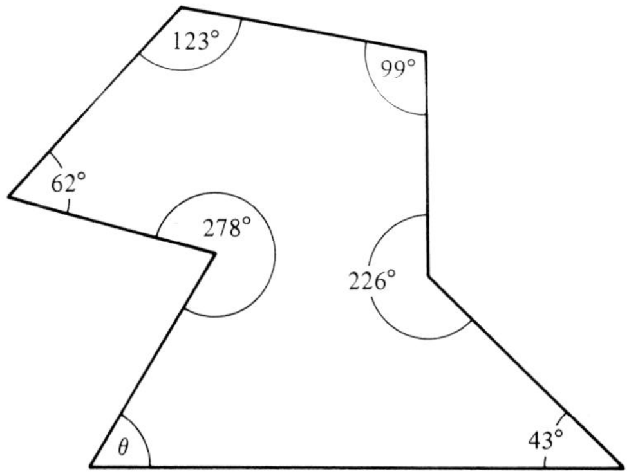


**FIGURE 5-19**

**EXAMPLE 4:** Find angle $\theta$ in Fig. 5-19.

**Solution:** The polygon shown has seven sides, so $n = 7$. By Eq. 112,

$$\text{sum of angles} = (7 - 2)(180°) = 900°$$

Adding the six given angles gives us

$$278° + 62° + 123° + 99° + 226° + 43° = 831°$$

So

$$\theta = 900° - 831° = 69°$$

## Triangles

A *triangle* is a polygon having three *sides*. The angles between the sides are the *interior angles* of the triangle, usually referred to as simply the angles of the triangle.

A *scalene* triangle (Fig. 5-20) has no equal sides, an *isosceles* triangle has two equal sides, and an *equilateral* triangle has three equal sides.

An *acute* triangle has three acute angles, an *obtuse* triangle has one obtuse angle, and a *right* triangle has one right angle.

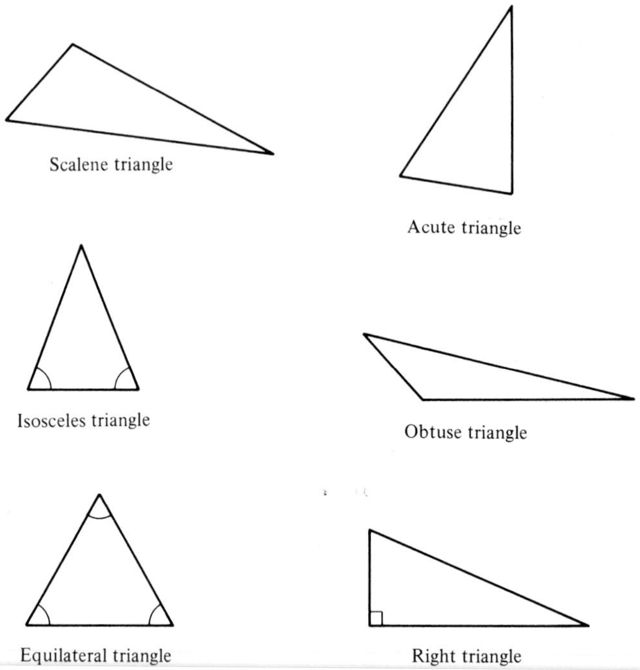


**FIGURE 5-20 Types of triangles.**

## Altitude and Base

The *altitude* of a triangle is the perpendicular distance from a vertex to the opposite side called the *base,* or an extension of that side (Fig. 5-21).

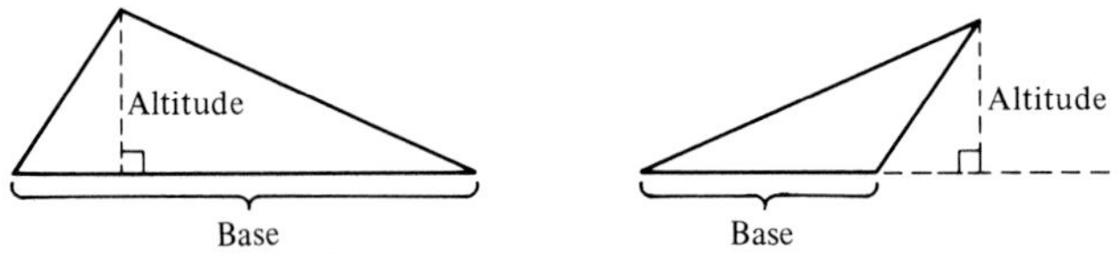


**FIGURE 5-21**

## Area of a Triangle

We make use of a familiar formula:

| | | |
|---|---|---|
| Area of a Triangle | Area equals one-half the product of the base and the altitude to that base: $$A = \frac{bh}{2}$$ | **137** |

**The measuring of areas, volumes, and lengths is sometimes referred to as *mensuration.***

**EXAMPLE 5:** Find the area of the triangle in Fig. 5-21 if the base is 52.0 and the altitude is 48.0.

**Solution:** By Eq. 137,

$$\text{area} = \frac{52.0(48.0)}{2} = 1248$$

If the altitude is not known but we have instead the lengths of the three sides, we may use *Hero's formula*. If $a$, $b$, and $c$ are the lengths of the sides,

**Named for Hero (or Heron) of Alexandria, a Greek mathematician and physicist of the 1st century A.D.**

| | | |
|---|---|---|
| Hero's Formula | area of triangle $= \sqrt{s(s-a)(s-b)(s-c)}$ where $s$ is half the perimeter $s = \frac{a+b+c}{2}$ | **138** |

**EXAMPLE 6:** Find the area of a triangle having sides of lengths 3.25, 2.16, and 5.09.

**Solution:** We first find $s$, which is half the perimeter,

$$s = \frac{3.25 + 2.16 + 5.09}{2} = 5.25$$

$$\text{area} = \sqrt{5.25(5.25 - 3.25)(5.25 - 2.16)(5.25 - 5.09)} = 2.28$$

## Sum of the Angles

An extremely useful relationship exists between the interior angles of any triangle.

| | | |
|---|---|---|
| Sum of the Angles | The sum of the three interior angles $A$, $B$, and $C$ of any triangle is 180 degrees. $A + B + C = 180°$ | **139** |

**EXAMPLE 7:** Find angle $A$ in a triangle if the other two interior angles are 38° and 121°.

**Solution:** By Eq. 139,

$$A = 180 - 121 - 38 = 21°$$

## Exterior Angles

An exterior angle is the angle between the side of a triangle and an extension of the adjacent side, such as angle $\theta$ in Fig. 5-22. The following theorem applies to exterior angles:

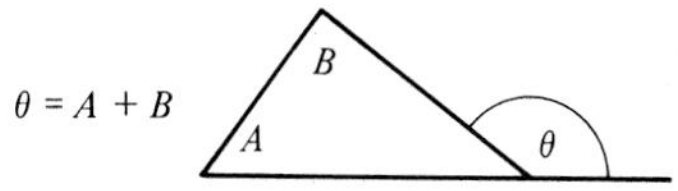


**FIGURE 5-22 An exterior angle.**

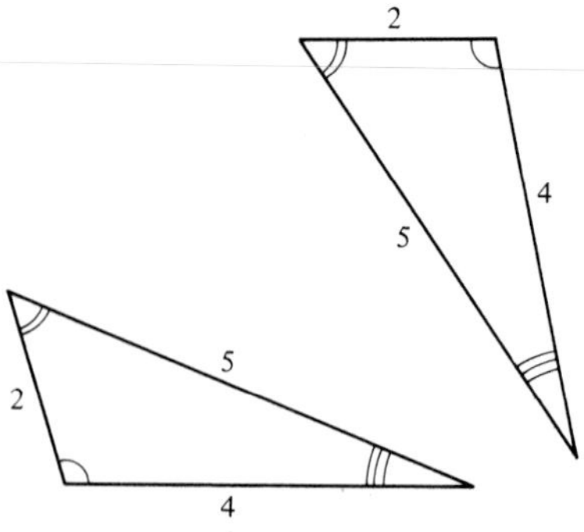


**FIGURE 5-23 Congruent triangles.**

| | | |
|---|---|---|
| Exterior Angle of a Triangle | An exterior angle equals the sum of the two opposite interior angles. $\theta = A + B$ | **142** |

## Congruent Triangles

Two triangles (or any other polygons, for that matter) are said to be *congruent* if the angles and sides of one are equal to the angles and sides of the other, as in Fig. 5-23.

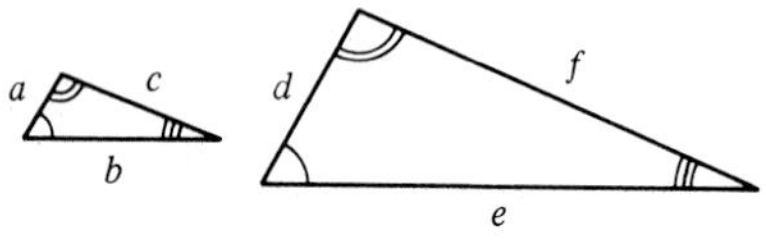


$$\frac{a}{d} = \frac{b}{e} = \frac{c}{f}$$

**FIGURE 5-24 Similar triangles.**

**We will see later in the chapter that these relationships also hold for similar figures *other than* triangles, and for similar *solids* as well.**

## Similar Triangles

Two triangles are said to be *similar* if they have the *same shape,* even if one triangle is larger than the other. This means that the angles of one of the triangles must equal the angles of the other triangle, as in Fig. 5-24. Sides that lie between the same pair of equal angles are called *corresponding sides,* such as sides $a$ and $d$. Sides $b$ and $e$, as well as sides $c$ and $f$, are also corresponding sides. We have two theorems:

| | | |
|---|---|---|
| Similar Triangles | If two angles of a triangle equal two angles of another triangle, their third angles must be equal, and the triangles are similar. | **143** |
| | If two triangles are similar, their corresponding sides are in proportion. | **144** |

**EXAMPLE 8:** Two beams, $AB$ and $CD$, in the framework of Fig. 5-25 are parallel. Find distance $AE$.

**Solution:** By statement 104, we know that angle $AEB$ equals angle $DEC$. Also, by statement 105 angle $ABE$ equals $ECD$. Thus, triangle $ABE$ is similar to triangle $CDE$. Since $AE$ and $ED$ are corresponding sides,

$$\frac{AE}{5.87} = \frac{5.14}{7.25}$$

$$AE = 5.87\left(\frac{5.14}{7.25}\right) = 4.16 \text{ m}$$

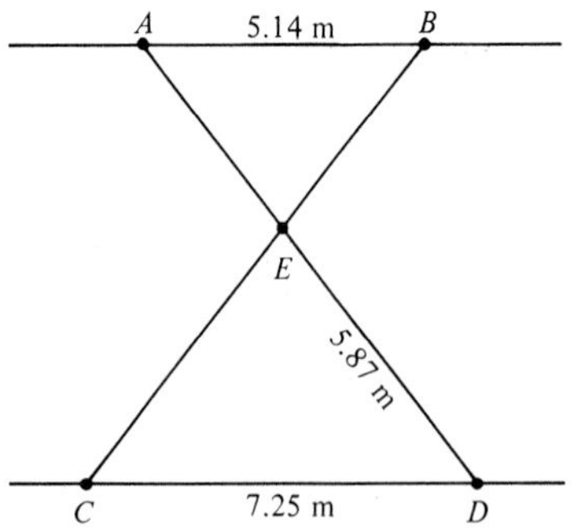


**FIGURE 5-25**

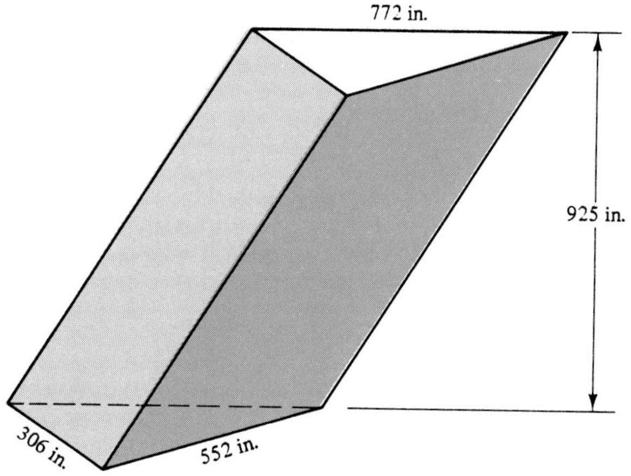


FIGURE 5-59

4. How many loads of gravel will be needed to cover 2.0 mi of roadbed, 35 ft wide, to a depth of 3.0 in. if one truckload contains 8.0 $yd^3$ of gravel?

**Cylinder**

5. Find the volume of the cylinder in Fig. 5-60.

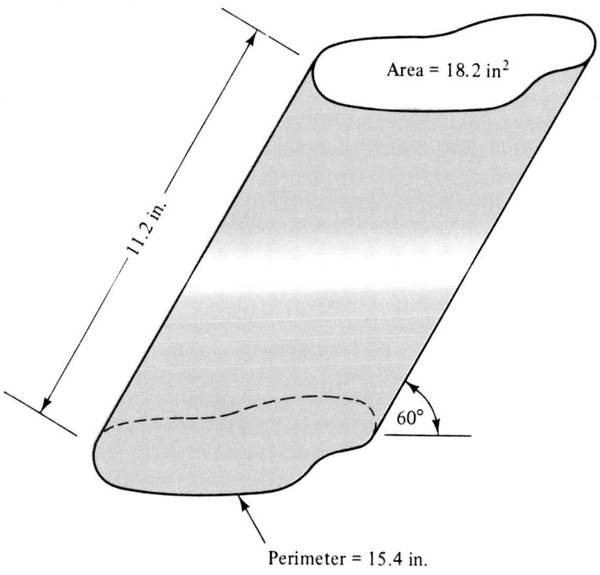


FIGURE 5-60

6. Find the volume and lateral area of a right circular cylinder having a base radius of 128 and a height of 285.
7. A 7.00-ft-long piece of iron pipe has an outside diameter of 3.50 in. and weighs 126 lb. Find the wall thickness.
8. A certain bushing is in the shape of a hollow cylinder 18.0 mm in diameter and 25.0 mm long, with an axial hole 12.0 mm in diameter. If the density of the material from which they are made is 2.70 $g/cm^3$, find the mass of 1000 bushings.
9. A steel gear is to be lightened by drilling holes through the gear. The gear is 3.50 in. thick. Find the diameter $d$ of the holes if each is to remove 12 oz.
10. A certain gasoline engine has four cylinders, each with a bore of 82.0 mm and a piston stroke of 95.0 mm. Find the engine displacement in liters.

**The engine *displacement* is the total volume swept out by all the pistons.**

## Cone and Pyramid

For some of these problems we need Eq. A43; the weight (or mass) equals the volume of the solid times the density of the material. For iron or steel take the density as 450 lb/ft³.

11. The circumference of the base of a right circular cone is 40.0 in. and the slant height is 38.0 in. What is the area of the lateral surface?
12. Find the volume of a circular cone whose altitude is 24.0 cm and whose base diameter is 30.0 cm.
13. Find the weight of the mast of a ship, its height being 50 ft, the circumference at one end 5.0 ft and at the other 3.0 ft, if the density of the wood is 58.5 lb/ft³.
14. The slant height of a right pyramid is 11 in., and the base is a 4.0-in. square. Find the area of the entire surface.
15. How many cubic feet are in a piece of timber 30.0 ft long, one end being a 15.0 in. square, and the other a 12.0 in. square?
16. Find the total surface area and volume of a tapered steel roller 12.0 ft long and having end diameters of 12.0 in. and 15.0 in.

## Sphere

A *great circle* of a sphere is one that has the same diameter as the sphere.

17. Find the volume and surface area of a sphere having a radius of 744.
18. Find the volume and radius of a sphere having a surface area of 462.
19. Find the surface area and radius of a sphere that has a volume of 5.88.
20. How many great circles of a sphere would have the same area as that of the surface of the sphere?
21. Find the weight in pounds of 100 steel balls each 2.50 in. in diameter.
22. A spherical radome encloses a volume of 9000 m³.
    (a) Find the radome radius, $r$.
    (b) If constructed of a material weighing 2.00 kg/m², find the weight of the radome.

## Computer

23. A conical tank (Fig. 5-61) which has a base diameter equal to its height is being filled with water at the rate of 2.25 m³/h. Write a program or use a spreadsheet that will compute and display the volume of water in the tank (m³) and the depth $D$ (m), for every 5 min, up to 1 h.

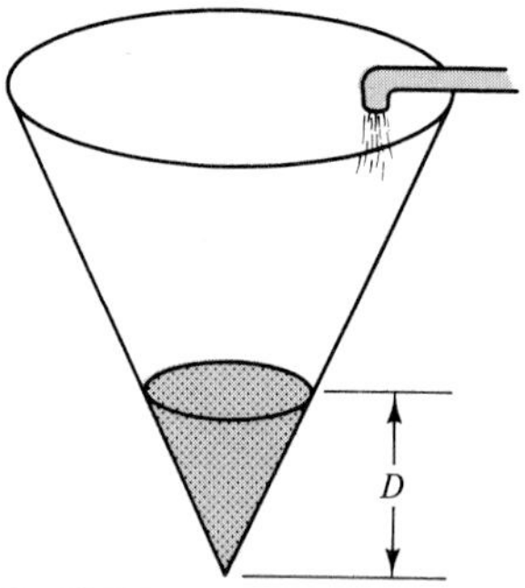


FIGURE 5-61

## CHAPTER 5 REVIEW PROBLEMS

1. A rocket ascends at an angle of 60° with the horizontal. After 1 min it is directly over a point that is a horizontal distance of 12.0 mi from the launch point. Find the speed of the rocket.
2. A rectangular beam 16 in. thick is cut from a log 20 in. in diameter. Find the greatest depth beam that can be obtained.
3. A cylindrical tank 4.00 m in diameter is placed with its axis vertical and is partially filled with water. A spherical diving bell is completely immersed in the tank, causing the water level to rise 1.00 m. Find the diameter of the diving bell.
4. Two vertical piers are 240 ft apart and support a circular bridge arch. The highest point of the arch is 30.0 ft higher than the piers. Find the radius of the arch.
5. Two antenna masts are 10 m and 15 m high and are 12 m apart. How long a wire is needed to connect the tops of the two masts?
6. Find the area and side of a rhombus whose diagonals are 100 and 140.
7. Find the area of a triangle that has sides of length 573, 638, and 972.
8. Two concentric circles have radii of 5 and 12. Find the length of a chord of the larger circle, which is tangent to the smaller circle.
9. A belt that does not cross goes around two pulleys, each with a radius of 4.00 in. and whose centers are 9.00 in. apart. Find the length of the belt.
10. A regular triangular pyramid has an altitude of 12 m, and the base is 4 m on a side. Find the area of a section made by a plane parallel to the base and 4 m from the vertex.
11. A fence parallel to one side of a triangular field cuts a second side into segments of 15.0 m and 21.0 m long. The length of the third side is 42.0 m. Find the length of the shorter segment of the third side.

12. When Fig. 5-62 is rotated about axis $AB$, we generate a cone inscribed in a hemisphere which is itself inscribed in a cylinder. Show that the volumes of these three solids are in the ratio 1:2:3. (Archimedes was so pleased with this discovery, it is said, that he ordered this figure to be engraved on his tomb.)

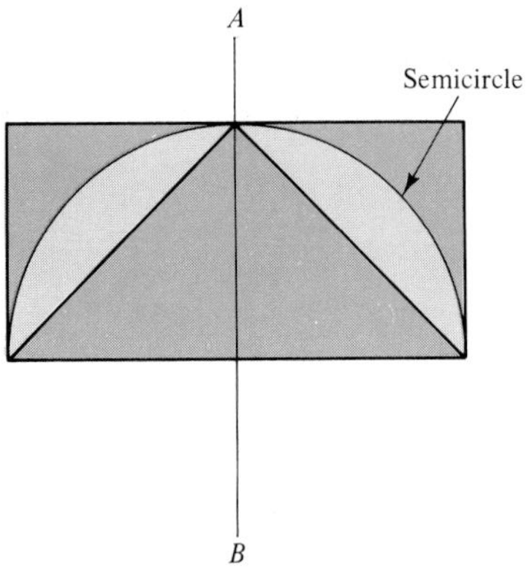


FIGURE 5-62

13. Four interior angles of a certain irregular pentagon are 38°, 96°, 112°, and 133°. Find the fifth interior angle.
14. Find the area of a trapezoid whose bases have lengths of 837 m and 583 m and are separated by a distance of 746 m.
15. A plastic drinking cup has a base diameter of 49 mm, is 63 mm wide at the top, and is 86 mm tall. Find the volume of the cup.
16. A spherical balloon is 17.4 ft in diameter and is made of a material that weighs 2.85 lb per 100 $ft^2$ of surface area. Find the volume and weight of the balloon.
17. Find the area of a triangle whose base is 38.4 in and whose altitude is 53.8 in.
18. Find the volume of a sphere of radius 33.8 cm.
19. Find the area of a parallelogram of base 39.2 m and height 29.3 m.
20. In "The Musgrave Ritual" Sherlock Holmes calculates the length of the shadow of an elm tree that is no longer standing. He does know that the elm was 64 ft high and that the shadow was cast at the instant that the sun was grazing the top of a certain oak tree.

    Holmes held a 6-ft-long fishing rod vertical and measured the length of its shadow at the proper instant. It was 9 ft long. He then said "Of course the calculation now was a simple one. If a rod of six feet threw a shadow of nine, a tree of sixty-four feet would throw one of ______." How long was the shadow of the elm?
21. Find the volume of a cylinder with base radius 22.3 cm and height 56.2 cm.
22. Find the surface area of a sphere having a diameter of 39.2 in.
23. Find the volume of a right circular cone with base diameter 2.84 ft and height 5.22 ft.
24. Find the volume of a box measuring 35.8 in. × 37.4 in. × 73.4 in.
25. Find the volume of a triangular prism having a length of 4.65 cm if the area of one end is 24.6 $cm^2$.

*Writing*

26. Write a section in an instruction manual for machinists on how to find the root diameter of a screw thread (Figure 5-34) by measuring "over wires." Your entry should have two parts. First a "how to" section giving step-by-step instructions, and then a "theory" section explaining why this method works. Keep the entry to under one page.

# 6

# RIGHT TRIANGLES AND VECTORS

## OBJECTIVES

**When you have completed this chapter, you should be able to:**

- Convert angles between decimal degrees, radians, and degrees, minutes, and seconds.
- Find the trigonometric functions of an angle.
- Find the acute angle that has a given trignometric function.
- Find the missing sides and angles of a right triangle.
- Solve practical problems involving the right triangle.
- Resolve a vector into components, and conversely, combine components into a resultant vector.
- Solve practical problems using vectors.

With this chapter we begin our study of trigonometry, the branch of mathematics that enables us to solve triangles. The trigonometric functions are introduced here and are used to solve right triangles. Other kinds of triangles (oblique triangles) are discussed in Chapter 14, and other applications of trigonometry are given in Chapters 15, 16, and 17.

We build on what we learned about triangles in Chapter 5, mainly the Pythagorean theorem and the fact that the sum of the angles of a triangle is 180°. Here we make use of coordinate axes, described in Sec. 4-3.

Also introduced in this chapter are vectors, the study of which will be continued in Chapter 14.

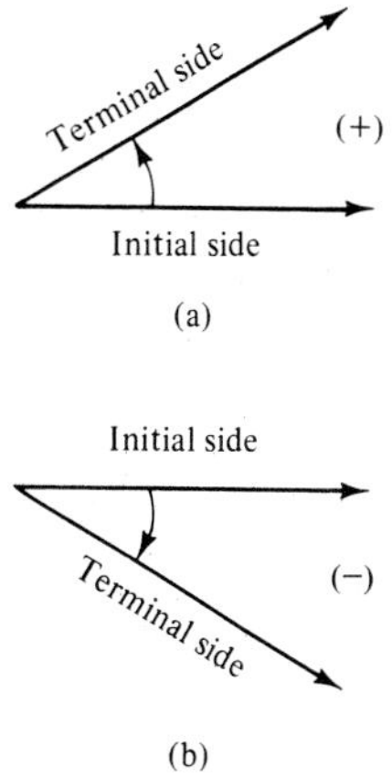


**FIGURE 6-1 Angles formed by rotation.**

## 6-1 ANGLES AND THEIR MEASURES

### Angles Formed by Rotation

We can think of an angle as the figure generated by a ray rotating from some initial position (Fig. 6-1) to some terminal position. We usually consider angles formed by rotation in the *counterclockwise* direction as *positive,* and angles formed by *clockwise* rotation as *negative*. A rotation of one *revolution* brings the ray back to its initial position.

### Degrees, Minutes, and Seconds

The degree (°) is a unit of angular measure equal to 1/360 of a revolution; thus, 360° = one revolution. A fractional part of a degree may be expressed as a common fraction (like $36\frac{1}{2}°$), as a decimal (28.74°), or as *minutes* and *seconds*.

A *minute* (′) is equal to 1/60 of a degree; a *second* (″) is equal to 1/60 of a minute, or 1/3600 of a degree.

**EXAMPLE 1:** Some angles written in degrees, minutes, and seconds are

$$85°18'42'' \qquad 62°12' \qquad 69°55'25.4'' \qquad 75°06'03''$$

**Note that minutes or seconds less than ten are written with an initial zero. Thus 6′ is written 06′.**

### Radian Measure

A *central angle* in a circle is one whose vertex is at the center of the circle. An *arc* is a portion of the circle. If an arc is laid off along a circle with a length equal to the radius of the circle, the central angle subtended by this arc is defined as one *radian*. A radian (rad) is a unit of angular measure commonly used in trigonometry. We will study radian measure in detail in Chapter 15.

Thus, an arc having a length of twice the radius will subtend a central angle of 2 radians. Similarly, an arc with a length of $2\pi$ times the radius (the circumference) will subtend a central angle of $2\pi$ radians. Thus our conversion factor is

$$2\pi \text{ radians} = 1 \text{ revolution} = 360°$$

## Conversion of Angles

We convert angles in the same way as we convert any other units. A summary of the conversions needed is:

You may prefer to use $\pi$ rad = 180°. Some calculators have special keys for angle conversions. Check your manual.

| | | |
|---|---|---|
| Angle Conversions | $1 \text{ rev} = 360° = 2\pi \text{ rad}$<br>$1° = 60'$<br>$1' = 60''$ | **117** |

---

**EXAMPLE 2:** Convert 47.6° to radians and revolutions.

**Solution:** By Eq. 117,

$$47.6°\left(\frac{2\pi \text{ rad}}{360°}\right) = 0.831 \text{ rad}$$

Note that we keep the *same number of significant digits* as we had in the original angle. To convert to revolutions,

$$47.6°\left(\frac{1 \text{ rev}}{360°}\right) = 0.132 \text{ rev}$$

---

**EXAMPLE 3:** Convert 1.8473 rad to degrees and revolutions.

**Solution:** By Eq. 117,

$$1.8473 \text{ rad}\left(\frac{360°}{2\pi \text{ rad}}\right) = 105.84°$$

and

$$1.8473 \text{ rad}\left(\frac{1 \text{ rev}}{2\pi \text{ rad}}\right) = 0.29401 \text{ rev}$$

---

Conversions involving degrees, minutes, and seconds require several steps.

---

**EXAMPLE 4:** Convert 28°17′37″ to decimal degrees and radians.

**Solution:** We separately convert the minutes and the seconds to degrees, and add them.

When the angle is known to the nearest second, work to four decimal places, and vice versa.

$$37 \text{ sec}\left(\frac{1 \text{ deg}}{3600 \text{ sec}}\right) = 0.0103°$$

$$17 \text{ min}\left(\frac{1 \text{ deg}}{60 \text{ min}}\right) = 0.2833°$$

$$28° = \underline{28.0000°}$$

Adding gives $28.2936°$

Now, by Eq. 117,

$$28.2936°\left(\frac{2\pi \text{ rad}}{360 \text{ deg}}\right) = 0.493817 \text{ rad}$$

---

**EXAMPLE 5:** Convert 1.837520 rad to degrees, minutes, and seconds.

**Solution:** We first convert to decimal degrees.

$$1.837520 \text{ rad}\left(\frac{360°}{2\pi \text{ rad}}\right) = 105.2821°$$

Then we convert the decimal part (0.2821°) to minutes,

$$0.2821°\left(\frac{60'}{1 \text{ deg}}\right) = 16.93'$$

and the decimal part of 16.93′ to seconds,

$$0.93'\left(\frac{60''}{1 \text{ min}}\right) = 56''$$

So

$$1.837520 \text{ rad} = 105°16'56''$$

**Another unit of angular measure is the *grad*. There are 400 grads in one revolution.**

| | |
|---|---|
| Common Error | When reading an angle quickly it is easy to mistake decimal degrees with degrees and minutes. Don't mistake<br>28°50′ for 28.50° |

## EXERCISE 1—ANGLES AND THEIR MEASURES

### Angle Conversions

Convert to radians.

1. 27.8°
2. 38.7°
3. 35°15′
4. 270°27′25″
5. 0.55 rev
6. 0.276 rev

Convert to revolutions.

7. 4.772 rad
8. 2.38 rad
9. 68.8°
10. 3.72°
11. 77°18′
12. 135°27′42″

Convert to degrees (decimal).

13. 2.83 rad
14. 4.275 rad
15. 0.475 rev
16. 0.236 rev
17. 29°27′
18. 275°18′35″

Convert to degrees, minutes, and seconds.

19. 4.2754 rad
20. 1.773 rad
21. 0.44975 rev
22. 0.78426 rev
23. 185.972°
24. 128.259°

### Computer

25. Write a program that will accept an angle in decimal degrees and compute and print the angle in degrees, minutes, and seconds.

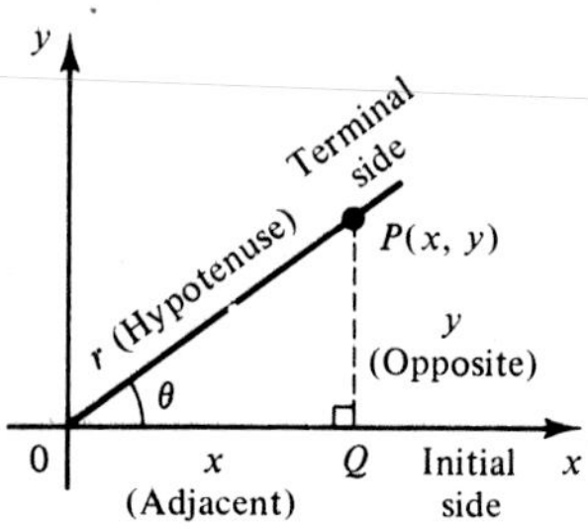


**FIGURE 6-2 An angle in standard position.**

## 6-2 THE TRIGONOMETRIC FUNCTIONS

### Trigonometric Ratios

We start with coordinate axes (Fig. 6-2) upon which we draw an angle. We always draw such angles the same way, with the vertex at the origin and with one side along the $x$ axis. Such an angle is said to be in *standard position*. We can think of the angle as being generated by a line rotating counterclockwise, whose initial position is on the $x$ axis. We now select any point $P$ on the terminal side of the angle $\theta$, and we let the coordinates of $P$ be $x$ and $y$. We form a *right triangle OPQ* by dropping a perpendicular from $P$ to the $x$ axis. The side $OP$ is the *hypotenuse* of the right triangle; we label its length $r$. Note that by the Pythagorean theorem, $r^2 = x^2 + y^2$. Also, side $OQ$ is *adjacent* to angle $\theta$ and has a length $x$, and side $PQ$ is *opposite* to angle $\theta$ and has a length $y$.

**We do only *acute* angles here and discuss larger angles in Chapter 14.**

If we had chosen to place point $P$ farther out on the terminal side of the angle, the lengths $x$, $y$, and $r$ would all be greater than they are now. *But* the *ratio* of any two sides will be the same regardless of where we place $P$, because all the triangles thus formed are similar to each other. Such a ratio will change only if we change the angle $\theta$. A ratio between any two sides of a right triangle is called a *trigonometric ratio*.

The six trigonometric ratios are defined as follows:

**The cotangent is sometimes abbreviated *ctn*.**

| | | |
|---|---|---|
| Trigonometric Ratios | $\text{sine } \theta = \sin \theta = \dfrac{y}{r} = \dfrac{\text{opposite side}}{\text{hypotenuse}}$ | **146** |
| | $\text{cosine } \theta = \cos \theta = \dfrac{x}{r} = \dfrac{\text{adjacent side}}{\text{hypotenuse}}$ | **147** |
| | $\text{tangent } \theta = \tan \theta = \dfrac{y}{x} = \dfrac{\text{opposite side}}{\text{adjacent side}}$ | **148** |
| | $\text{cotangent } \theta = \cot \theta = \dfrac{x}{y} = \dfrac{\text{adjacent side}}{\text{opposite side}}$ | **149** |
| | $\text{secant } \theta = \sec \theta = \dfrac{r}{x} = \dfrac{\text{hypotenuse}}{\text{adjacent side}}$ | **150** |
| | $\text{cosecant } \theta = \csc \theta = \dfrac{r}{y} = \dfrac{\text{hypotenuse}}{\text{opposite side}}$ | **151** |

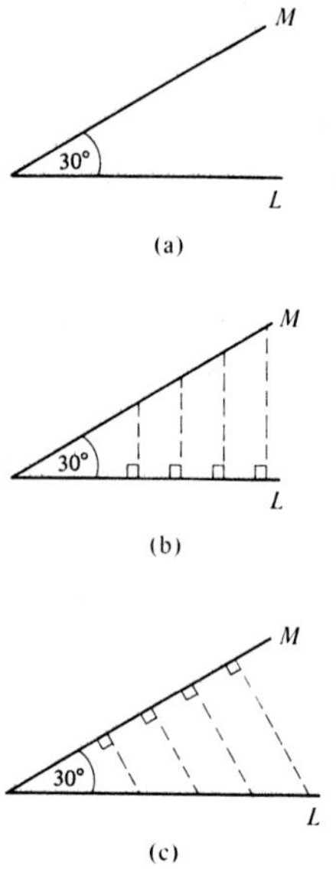


**FIGURE 6-3**

### The Trigonometric Functions

Note that each of the trigonometric ratios is *related to* the angle $\theta$. Take the ratio $y/r$ for example. The values of $y/r$ depend only on the measure of $\theta$, not on the lengths of $y$ or $r$. To convince yourself of this, draw an angle of, say, 30°, as in Fig. 6-3a. Then form a right triangle by drawing a perpendicular from any point on side $L$ *or from any point on side* $M$ (Fig. 6-3b and c). No matter how you draw the triangle, the side opposite the 30° angle *will always be half the hypotenuse* (sin 30° = $\frac{1}{2}$). You would get a similar result for any other angle and for any other trigonometric ratio. *The value of the trigonometric ratio depends only on the measure of the angle.* We have a *relation,* such as the relations we studied in Chapter 4.

Furthermore, for any angle $\theta$, there exists *only one* value for each trigonometric ratio. Thus, *each ratio is a function of* $\theta$. We call them the *trigonometric functions* of $\theta$. In functional notation, we could write

$$\sin\theta = f(\theta)$$

$$\cos\theta = g(\theta)$$

$$\tan\theta = h(\theta)$$

and so forth. The angle $\theta$ is called the *argument* of the function.

Note that while there is one and only one value of each trigonometric function for a given angle $\theta$, that for any given value of a trigonometric function (say, $\sin\theta = \frac{1}{2}$), there are *indefinitely many* values of $\theta$. We treat this idea more fully in Sec. 14-1.

The trigonometric functions are sometimes called *circular* functions.

### Reciprocal Relationships

From the trigonometric functions we see that

$$\sin\theta = \frac{y}{r}$$

and that

$$\csc\theta = \frac{r}{y}$$

Obviously, $\sin\theta$ and $\csc\theta$ are reciprocals.

$$\sin\theta = \frac{1}{\csc\theta}$$

Inspection of the trigonometric functions shows two more sets of reciprocals (see also Fig. 6-4).

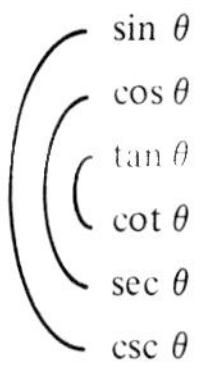


**FIGURE 6-4 This diagram shows which functions are reciprocals of each other.**

| | | |
|---|---|---|
| Reciprocal Relationships | $\sin\theta = \dfrac{1}{\csc\theta}$ | **152a** |
| | $\cos\theta = \dfrac{1}{\sec\theta}$ | **152b** |
| | $\tan\theta = \dfrac{1}{\cot\theta}$ | **152c** |

**This is a good place to use the 1/x key on your calculator.**

**EXAMPLE 6:** Find $\tan\theta$ if $\cot\theta = 0.638$.

**Solution:** From the reciprocal relationships,

$$\tan\theta = \frac{1}{\cot\theta} = \frac{1}{0.638} = 1.57$$

## Finding the Trigonometric Functions of an Angle

The six trigonometric functions of an angle can be written if we know the angle or if we know two of the sides of a right triangle containing the angle.

**EXAMPLE 7:** A right triangle (Fig. 6-5) has sides of 4.37 and 2.83. Write the six trigonometric functions of angle $\theta$.

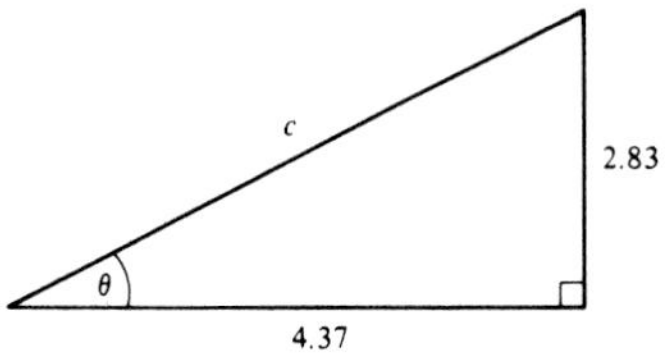


**FIGURE 6-5**

**Solution:** We first find the hypotenuse by the Pythagorean theorem,

$$c = \sqrt{(2.83)^2 + (4.37)^2} = \sqrt{27.1}$$
$$= 5.21$$

We also note that the opposite side is 2.83 and the adjacent side is 4.37. Then by Eqs. 146 to 151,

$$\sin\theta = \frac{\text{opposite side}}{\text{hypotenuse}} = \frac{2.83}{5.21} = 0.543$$

$$\cos\theta = \frac{\text{adjacent side}}{\text{hypotenuse}} = \frac{4.37}{5.21} = 0.839$$

$$\tan\theta = \frac{\text{opposite side}}{\text{adjacent side}} = \frac{2.83}{4.37} = 0.648$$

$$\cot\theta = \frac{\text{adjacent side}}{\text{opposite side}} = \frac{4.37}{2.83} = 1.54$$

$$\sec\theta = \frac{\text{hypotenuse}}{\text{adjacent side}} = \frac{5.21}{4.37} = 1.19$$

$$\csc\theta = \frac{\text{hypotenuse}}{\text{opposite side}} = \frac{5.21}{2.83} = 1.84$$

| | |
|---|---|
| Common Error | Do not omit the angle when writing a trigonometric function. To write "sin = 0.543," for example, has no meaning. |

**EXAMPLE 8:** A point $P$ on the terminal side of an angle $B$ (in standard position) has the coordinates (8.84, 3.15). Write the sine, cosine, and tangent of angle $B$.

**Solution:** We are given $x = 8.84$ and $y = 3.15$. We find $r$ from the Pythagorean theorem,

$$r = \sqrt{(8.84)^2 + (3.15)^2} = 9.38$$

Then by Eqs. 146 to 151,

$$\sin B = \frac{y}{r} = \frac{3.15}{9.38} = 0.336$$

$$\cos B = \frac{x}{r} = \frac{8.84}{9.38} = 0.942$$

$$\tan B = \frac{y}{x} = \frac{3.15}{8.84} = 0.356$$

## Trigonometric Functions by Calculator

Simply enter the angle and press the proper key, sin, cos, or tan. Make sure that your calculator is in the "degree" mode if the entered angle is in degrees, or in the "radian" mode if the angle is in radians.

**EXAMPLE 9:** Find the sine of 38.6° by calculator to four significant digits.

**Solution:** We set the mode switch to "Degrees." Then

**The keystrokes are the same for algebraic and reverse Polish calculators.**

| *Press* | *Display* |
|---|---|
| 38.6 sin | 0.6238795967 |

which we round to 0.6239.

**EXAMPLE 10:** Find tan 1.174 rad by calculator to five significant digits.

**Solution:** With the mode switch in the "radian" position: Then

| *Press* | *Display* |
|---|---|
| 1.174 tan | 2.386509704 |

which we round to 2.3865.

To find the cotangent, secant, or cosecant, first find the reciprocal function, (tangent, cosine, or sine), and then press the $1/x$ key to get the reciprocal.

**EXAMPLE 11:** Find sec 72.8° to four significant digits.

**Solution:** Set the mode switch to "Degrees." Then since $\sec \theta = 1/\cos \theta$,

| *Press* | *Display* |
|---|---|
| 72.8 cos | 0.29570805 |
| $1/x$ | 3.381713822 |

which we round to 3.382.

**EXAMPLE 12:** Find the cosecant of 0.748 rad to four significant digits.

**Solution:** We first put the calculator into "radian" mode. Then, since $\csc\theta = 1/\sin\theta$,

| *Press* | *Display* |
|---|---|
| .748 [sin] [1/x] | 1.470211991 |

which we round to 1.470.

| Common Error | It is easy to forget to set your calculator in the proper Degree/Radian mode. Be sure to check it each time. |
|---|---|

## Trigonometric Functions on the Computer

Many computer languages such as BASIC have built-in commands for some, but not all, of the trigonometric functions.

**EXAMPLE 13:** The instruction

```
50 PRINT SIN(X)
```

will cause the sine of the number X to be printed, *where X is assumed to be in radians.*

If X is in degrees, it must be converted to radians before taking the sine.

**EXAMPLE 14:** The instruction

```
50 PRINT SIN(3.1416*X/180)
```

will print the sine of X degrees.

BASIC usually includes the functions

COS ( ) and TAN ( )

in addition to SIN( ). As on the calculator, the cotangent, secant, and cosecant are obtained from the reciprocal relations (Eq. 152).

**EXAMPLE 15:** The instruction

```
80 PRINT 1/TAN(X)
```

will print the cotangent of X radians.

## Finding the Angle When the Trigonometric Function Is Given

**Instead of an [arc] key, some calculators have an [inv] key or keys marked [$\sin^{-1}$], [$\cos^{-1}$], or [$\tan^{-1}$].**

Here we are given the value of the sine (or other trigonometric function); and from this we determine the angle that has that given trigonometric function. So we reverse the previous operation of finding the trigonometric function. For this we use the [arc] key on the calculator. Thus, by pressing [arc] [sin], we find the angle whose sine is given.

**EXAMPLE 16:** If $\sin \theta = 0.7337$, find $\theta$ in degrees to three significant digits.

**Solution:** Switch the calculator into the "degree" mode. Then

| *Press* | *Display* |
|---|---|
| 0.7337 [arc] [sin] | 47.19748165 |

which we round to 47.2°. Note that a calculator gives us acute angles only.

**In this chapter we limit our work with inverse functions to *first-quadrant* angles only and consider any angle in Chapter 14.**

**EXAMPLE 17:** If $\tan x = 2.846$, find $x$ in radians to four significant digits.

**Solution:** With the calculator in the "radian" mode:

| *Press* | *Display* |
|---|---|
| 2.846 [arc] [tan] | 1.232901233 |

which we round to 1.233.

If the cotangent, secant, or cosecant is given, we first take the reciprocal of that value and then use the appropriate reciprocal relationship,

$$\sin \theta = \frac{1}{\csc \theta} \qquad \cos \theta = \frac{1}{\sec \theta} \qquad \tan \theta = \frac{1}{\cot \theta}$$

**EXAMPLE 18:** If $\sec \theta = 1.573$, find $\theta$ in degrees to four significant digits.

**Solution:** Put the calculator into the "degree" mode. Then since

$$\cos \theta = \frac{1}{\sec \theta} = \frac{1}{1.573}$$

we first find the reciprocal of 1.573 and then take the inverse cosine.

| *Press* | *Display* |
|---|---|
| 1.573 [1/x] | 0.6357279085 |
| [arc] [cos] | 50.53 (rounded) |

## Inverse Trigonometric Functions

The operation of finding the angle when the trigonometric function is given is the *inverse* of finding the function when the angle is given. There is special notation to indicate the inverse trigonometric function. If

$$A = \sin \theta$$

we write

$$\theta = \arcsin A$$

or

$$\theta = \sin^{-1} A$$

which is read "$\theta$ is the angle whose sine is $A$." Similarly, we use the symbols $\arccos A$, $\cos^{-1} A$, $\arctan A$, and so on.

| | |
|---|---|
| Common Error | Do not confuse the **inverse** with the **reciprocal.** $$\frac{1}{\sin\theta} = (\sin\theta)^{-1} \neq \sin^{-1}\theta$$ |

**EXAMPLE 19:** Evaluate arctan 1.547 to four significant digits.

**Solution:**

| *Press* | *Display* |
|---|---|
| (degree mode) 1.547 [arc] [tan] | 57.12 (degrees) |

## Inverse Trigonometric Functions on the Computer

BASIC has *only one* inverse trigonometric function, the arctangent.

**EXAMPLE 20:** The instruction

```
60 LET A = ATN(B)
```

will set $A$ equal to that angle, *in radians,* that has a tangent of $B$.

Suppose that we know the sine of some angle $\theta$ is $C$, and we want to find $\theta$. We know that

$$\sin\theta = C$$

and so

$$\theta = \arcsin C$$

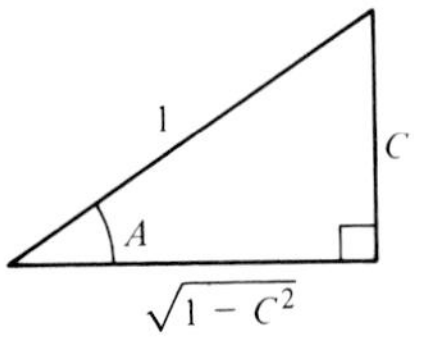


**FIGURE 6-6**

But BASIC has no arcsine function. We can, however, use the arctangent to find the arcsine. We draw a right triangle (Fig. 6-6) with a hypotenuse of 1 and a side $C$ opposite angle $\theta$. The sine of $\theta$ is thus equal to $C$. By the Pythagorean theorem, the third side is equal to

$$\sqrt{1 - C^2}$$

so

$$\tan\theta = \frac{C}{\sqrt{1 - C^2}}$$

and hence

| | | |
|---|---|---|
| Inverse Trigonometric Function | $\theta = \arcsin C = \arctan \dfrac{C}{\sqrt{1 - C^2}}$ | **183** |

In BASIC,

```
THETA = ATN(C/SQR(1 - C ∧ 2))
```

Similarly, if the cosine of some angle is equal to $D$, then the angle can be found from

| Inverse Trigonometric Function | $\theta = \arccos D = \arctan \dfrac{\sqrt{1 - D^2}}{D}$ | **184** |
|---|---|---|

Try to derive this.

## EXERCISE 2—THE TRIGONOMETRIC FUNCTIONS _____

Using a protractor, lay off the given angle on coordinate axes. Drop a perpendicular from any point on the terminal side and measure the distances $x$, $y$, and $r$, as in Fig. 6-2. Use these measured distances to write the six trigonometric ratios of the angle. Check your answer by calculator.

  **1.** 30°          **2.** 55°          **3.** 14°          **4.** 37°

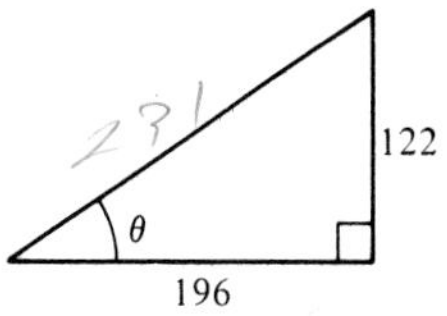

**FIGURE 6-7**

Plot the given point on coordinate axes and connect it to the origin. Measure the angle formed with a protractor. Check by taking the tangent of your measured angle. It should equal the ratio $y/x$.

  **5.** (3,4)          **6.** (5, 3)          **7.** (1, 2)          **8.** (5, 4)
  **9.** Write the six trigonometric ratios for angle $\theta$ in Fig. 6-7.

Each of the given points is on the terminal side of an angle. Compute the distance $r$ from the origin to the point and write the sin, cos, and tan of the angle.

| | |
|---|---|
| **10.** (3, 5) | **11.** (4, 2) |
| **12.** (5, 3) | **13.** (2, 3) |
| **14.** (4, 5) | **15.** (4.75, 2.68) |

**Work to three significant digits on these.**

Evaluate. Work to four decimal places.

| | | |
|---|---|---|
| **16.** sin 47.3° | **17.** tan 0.775 rad | **18.** sec 29.5° |
| **19.** cot 18.2° | **20.** csc 87°45′12″ | **21.** cos 21.27° |
| **22.** cos 86.75° | **23.** sec 12°18′ | **24.** csc 12.67° |

Write the sin, cos, and tan for each angle. Keep three decimal places.

| | |
|---|---|
| **25.** 72.8° | **26.** 19.2° |
| **27.** 33.1° | **28.** 0.722 rad |
| **29.** 42.6° | **30.** 78°15′23″ |

.722 rad $\cdot$ $\dfrac{180 \, deg}{\pi \, rad}$

41.369

### Inverse Trigonometric Functions

Find the acute angle (in decimal degrees) whose trigonometric function is given. Keep three significant digits.

| | | |
|---|---|---|
| **31.** sin $A = 0.5$ | **32.** tan $D = 1.53$ | **33.** sin $G = 0.528$   31.9 |
| **34.** cot $K = 1.774$ | **35.** sin $B = 0.483$ | **36.** cot $E = 0.847$ |

28.9

Without using tables or calculator, write the sin, cos, and tan of angle $A$. Leave your answer in fractional form.

  **37.** sin $A = \dfrac{3}{5}$        **38.** cot $A = \dfrac{12}{5}$        **39.** cos $A = \dfrac{12}{13}$

Evaluate the following, giving your answer in decimal degrees to three significant digits.

39.4  39.420

| | | |
|---|---|---|
| **40.** arcsin 0.635   39.420 | **41.** arcsec 3.86   74.985   74.985 | **42.** tan⁻¹ 2.85   70.665   71.0 |
| **43.** cot⁻¹ 1.17   40.5?? | **44.** cos⁻¹ 0.229   76.762   76.762 | **45.** arccsc 4.26   13.576   13.6 |

46. Tables of trigonometric functions are usually computed using series approximations, such as

| | | |
|---|---|---|
| Series Approximation | $\sin x = x - \dfrac{x^3}{3!} + \dfrac{x^5}{5!} - \dfrac{x^7}{7!} + \cdots$ | **210** |

where $x$ is the angle *in radians*, and 3! (read 3 factorial) is

$$3! = 3 \cdot 2 \cdot 1 = 6$$

and

$$5! = 5 \cdot 4 \cdot 3 \cdot 2 \cdot 1 = 120$$

and so on. Write a program or use a spreadsheet to compute and print the sines of the angles from 0 to 90°, taking values every 5° and using only the first three terms to the series. Compare your results with those obtained by calculator.

47. Write a program or use a spreadsheet to evaluate the cosine of an angle $x$ using the series expansion

| | | |
|---|---|---|
| Series Approximation | $\cos x = 1 - \dfrac{x^2}{2!} + \dfrac{x^4}{4!} - \dfrac{x^6}{6!} + \cdots$ | **211** |

## 6-3 SOLUTION OF RIGHT TRIANGLES

### Right Triangles

The right triangle (Fig. 6-8) was introduced in Sec. 5-2 where we also had the Pythagorean theorem

| | | |
|---|---|---|
| Pythagorean Theorem | $c^2 = a^2 + b^2$ | **145** |

Furthermore, since angle $C$ is always 90°, Eq. 139 becomes

| | | |
|---|---|---|
| Sum of the Angles | $A + B = 90°$ | **139** |

To these, we now add

| | | |
|---|---|---|
| Trigonometric Functions | $\sin \theta = \dfrac{\text{opposite side}}{\text{hypotenuse}}$ | **146** |
| | $\cos \theta = \dfrac{\text{adjacent side}}{\text{hypotenuse}}$ | **147** |
| | $\tan \theta = \dfrac{\text{opposite side}}{\text{adjacent side}}$ | **148** |

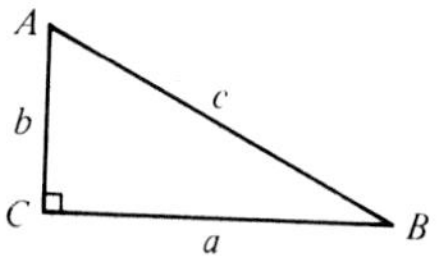


**FIGURE 6-8 A right triangle. We will usually *label* a right triangle as shown here. We label the angles with capital letters *A*, *B*, and *C*, with *C* always the right angle. We label the sides with lowercase letters *a*, *b*, and *c*, with side *a* opposite angle *A*, side *b* opposite angle *B*, and side *c* (the hypotenuse) opposite angle *C* (the right angle).**

These five equations will be our tools for solving any right triangle.

## Solving Right Triangles When One Side and One Angle Are Known

To *solve* a triangle means to find all missing sides and angles (although in most practical problems we need find only one missing side or angle). We can solve any right triangle if we know one side and either another side or one angle.

To solve a right triangle when one side and one angle are known:

1. Make a sketch.
2. Find the missing angle by using Eq. 139.
3. Relate the known side to one of the missing sides by one of the trigonometric ratios. Solve for the missing side.
4. Repeat step 3 to find the second missing side.
5. Check your work with the Pythagorean theorem.

---

**EXAMPLE 21:** Solve right triangle *ABC* if $A = 18.6°$ and $c = 135$.

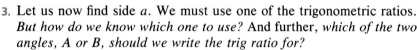

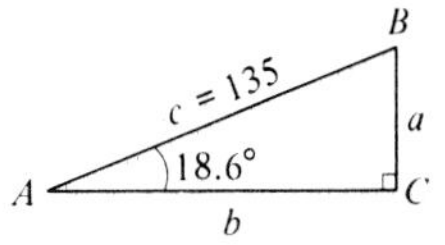


**FIGURE 6-9**

**Solution:**

1. We make a sketch (Fig. 6-9).
2. Then, by Eq. 139,

$$B = 90° - A = 90° - 18.6° = 71.4°$$

3. Let us now find side $a$. We must use one of the trigonometric ratios. *But how do we know which one to use?* And further, *which of the two angles, A or B, should we write the trig ratio for?*

   It is simple. First, always work with the *given angle,* because if you made a mistake in finding angle $B$ and then used it to find the sides, they would be wrong also. Then, to decide which trigonometric ratio to use, we note that side $a$ is *opposite* to angle $A$ and that the given side is the *hypotenuse*. Thus, our trig function must be one that relates the *opposite side* to the *hypotenuse*. Our obvious choice is Eq. 146.

**Realize that either of the two legs can be called *opposite* or *adjacent,* depending on which angle we are referring them to. Here *b* is adjacent to angle *A* but opposite to angle *B*.**

$$\sin A = \frac{\text{opposite side}}{\text{hypotenuse}}$$

Substituting the given values, we obtain

$$\sin 18.6° = \frac{a}{135}$$

Solving for $a$,

$$a = 135 \sin 18.6° = 43.1$$

4. We now find side $b$. Note that side $b$ is *adjacent* to angle $A$. We therefore use Eq. 147,

$$\cos 18.6° = \frac{b}{135}$$

So

$$b = 135 \cos 18.6° = 128$$

We have thus found all the missing parts of the triangle.

**As a rough check of any triangle, see if the longest side is opposite the largest angle and if the shortest side is opposite the smallest angle. Also check that the hypotenuse is greater than either leg but less than their sum.**

5. For a check, we see if the three sides will satisfy the Pythagorean theorem.

   Check:

$$(43.1)^2 + (128)^2 \stackrel{?}{=} (135)^2$$

$$18{,}242 \stackrel{?}{=} 18{,}225$$

Since we are working to three significant digits, this is close enough for a check.

---

**EXAMPLE 22:** In right triangle *ABC*, angle $B = 55.2°$ and $a = 207$. Solve the triangle.

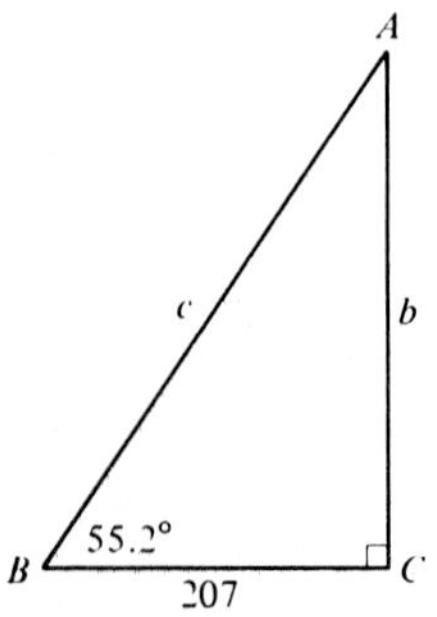

FIGURE 6-10

**Solution:**

1. We make a sketch (Fig. 6-10).
2. By Eq. 139,

$$A = 90° - 55.2° = 34.8°$$

3. By Eq. 147,

$$\cos 55.2° = \frac{207}{c}$$

$$c = \frac{207}{\cos 55.2°} = \frac{207}{0.5707} = 363$$

4. Then by Eq. 148,

$$\tan 55.2° = \frac{b}{207}$$

$$b = 207 \tan 55.2° = 207(1.439) = 298$$

5. Checking with the Pythagorean theorem,

$$(363)^2 \stackrel{?}{=} (207)^2 + (298)^2$$

$$131{,}769 \stackrel{?}{=} 131{,}653 \quad \text{checks to within three significant digits}$$

---

| Tip | Whenever possible, use the *given information* for each computation, rather than some quantity previously calculated. This way, any errors in the early computation will not be carried along. |
|---|---|

## Solving Right Triangles When Two Sides Are Known

1. Draw a diagram of the triangle.
2. Write the trigonometric ratio that relates one of the angles to the two given sides. Solve for the angle.
3. Subtract the angle just found from 90° to get the second angle.
4. Find the missing side by the Pythagorean theorem.
5. Check the computed side and angles with trigonometric ratios, as shown in Example 23.

**EXAMPLE 23:** Solve right triangle $ABC$ if $a = 1.48$ and $b = 2.25$.

**Solution:**

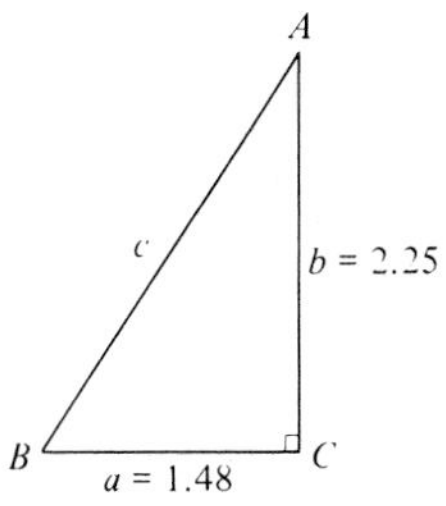


**FIGURE 6-11**

1. We sketch the triangle (Fig. 6-11).

2. To find angle $A$, we note that the 1.48 side is *opposite* angle $A$ and that the 2.25 side is *adjacent* to angle $A$. The trig ratio relating *opposite* and *adjacent* is the *tangent* (Eq. 148).

$$\tan A = \frac{1.48}{2.25} = 0.6578$$

Now using the $\boxed{\text{arc}}$ or $\boxed{\text{inv}}$ key on the calculator,

$$A = 33.3°$$

3. Solving for angle $B$,

$$B = 90° - A = 90 - 33.3 = 56.7°$$

4. We find side $c$ by the Pythagorean theorem,

$$c^2 = (1.48)^2 + (2.25)^2 = 7.253$$

$$c = 2.69$$

5. Check:

$$\sin 33.3° \stackrel{?}{=} \frac{1.48}{2.69}$$

$$0.549 \simeq 0.550 \qquad \text{checks}$$

and

$$\sin 56.7° \stackrel{?}{=} \frac{2.25}{2.69}$$

$$0.836 = 0.836 \qquad \text{checks}$$

(Note that we could just as well have used another trigonometric function, such as the cosine, for our check.)

## Cofunctions

The sine of angle $A$ in Fig. 6-8 is

$$\sin A = \frac{a}{c}$$

But $a/c$ is *also* the cosine of the complementary angle $B$. Thus we can write

$$\sin A = \cos B$$

Here $\sin A$ and $\cos B$ are called *cofunctions*. Similarly, $\cos A = \sin B$. These cofunctions and others in the same right triangle are given in the following table.

| | | |
|---|---|---|
| Cofunctions where $A + B = 90°$ | $\sin A = \cos B$ | **154a** |
| | $\cos A = \sin B$ | **154b** |
| | $\tan A = \cot B$ | **154c** |
| | $\cot A = \tan B$ | **154d** |
| | $\sec A = \csc B$ | **154e** |
| | $\csc A = \sec B$ | **154f** |

In general, *a trigonometric function of an acute angle is equal to the corresponding cofunction of the complementary angle.*

---

**EXAMPLE 24:** If the cosine of the acute angle $A$ in a right triangle $ABC$ is equal to 0.725, find the sine of the other acute angle, $B$.

**Solution:** By Eq. 154b,

$$\sin B = \cos A = 0.725$$

---

## EXERCISE 3—SOLUTION OF RIGHT TRIANGLES

### Right Triangles with One Side and One Angle Given

Sketch each right triangle and find all the missing parts. Assume the triangles to be labeled as in Fig. 6-8. Work to three significant digits.

**1.** $a = 155$  $A = 42.9°$
**2.** $b = 82.6$  $B = 61.4°$
**3.** $a = 1.74$  $B = 31.9°$
**4.** $b = 7.74$  $A = 22.5°$
**5.** $a = 284$  $A = 1.13$ rad
**6.** $b = 73.2$  $B = 0.655$ rad

### Right Triangles with Two Sides Given

Sketch each right triangle and find all missing parts. Work to three significant digits and express the angles in decimal degrees.

**7.** $a = 382$  $b = 274$
**8.** $a = 3.88$  $c = 5.37$
**9.** $b = 3.97$  $c = 4.86$
**10.** $a = 63.9$  $b = 84.3$
**11.** $a = 27.4$  $c = 37.5$
**12.** $b = 746$  $c = 957$

### Cofunctions

Express as a function of the complementary angle.

**13.** $\sin 38°$
**14.** $\cos 73°$
**15.** $\tan 19°$
**16.** $\sec 85.6°$
**17.** $\cot 63.2°$
**18.** $\csc 82.7°$
**19.** $\tan 35°14'$
**20.** $\cos \frac{\pi}{8}$ rad
**21.** $\sin 0.875$ rad

## Computer

22. Write a program or use spreadsheet to compute the perimeter of a regular polygon of $n$ sides inscribed in a circle having a diameter of one unit. Let $n$ vary from 3 to 20, and for each $n$ print **(a)** the number of sides, $n$; **(b)** the perimeter of the polygon; and **(c)** the difference between the circumference of the circle and the perimeter of the polygon.
23. Write a program to solve any right triangle. Have the program accept as input a known side and angle, or two known sides, and compute and print all the sides and angles.

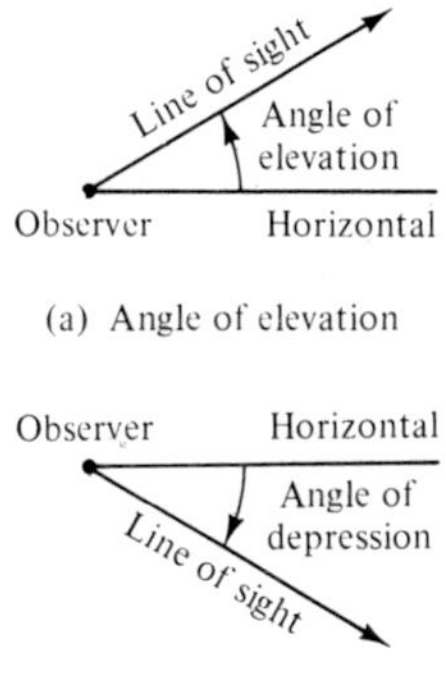

(b) Angle of depression

**FIGURE 6-12 Angles of elevation and depression.**

# 6-4 APPLICATIONS OF THE RIGHT TRIANGLE

There are, of course, a huge number of applications for the right triangle, a few of which are given in the following exercise. A typical application is that of finding a distance that cannot be measured directly.

---

**EXAMPLE 25:** From a plane at an altitude of 2750 ft, the pilot observes the angle of depression of a lake to be 18.6°. How far is the lake from a point on the ground directly beneath the plane?

**Solution:** We first note that an *angle of depression,* as well as an *angle of elevation,* is defined as the angle between a line of sight and the *horizontal,* as shown in Fig. 6-12. We then make a sketch for our problem (Fig. 6-13). Since the ground and the horizontal line drawn through the plane are parallel, angle $A$ and the angle of depression (18.6°) are alternate interior angles. Therefore, $A = 18.6°$. Then

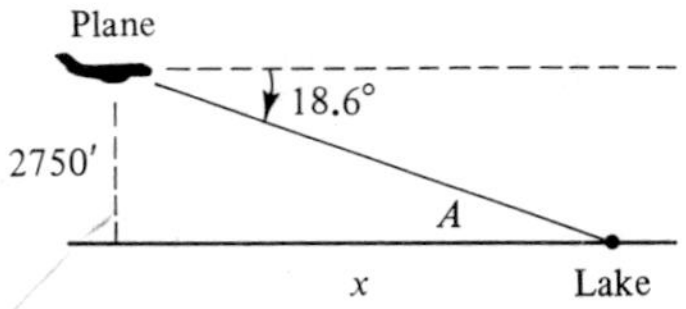


**FIGURE 6-13**

$$\tan 18.6° = \frac{2750}{x}$$

$$x = \frac{2750}{\tan 18.6°} = 8171 \text{ ft}$$

---

## EXERCISE 4—APPLICATIONS OF THE RIGHT TRIANGLE

### Measuring Inaccessible Distances

1. From a point on the ground 255 m from the base of a tower, the angle of elevation to the top of the tower is 57.6°. Find the height of the tower.
2. A pilot 4220 m directly above the front of a straight train observes that the angle of depression of the end of the train is 68.2°. Find the length of the train.
3. From the top of a lighthouse 156 ft above the surface of the water, the angle of depression of a boat is observed to be 28.7°. Find the horizontal distance from the boat to the lighthouse.
4. An observer in an airplane 1520 ft above the surface of the ocean observes that the angle of depression of a ship is 28.8°. Find the straight-line distance from the plane to the ship.
5. The distance $PQ$ across a swamp is desired. A line $PR$ of length 59.3 m is laid off at right angles to $PQ$. Angle $PRQ$ is measured at 47.6°. Find the distance $PQ$.
6. The angle of elevation of the top of a building from a point on the ground 275 ft from its base is 51.3°. Find the height of the building.
7. The angle of elevation of the top of a building from a point on the ground 75.0 yd from its base is 28.0°. How high is the building?

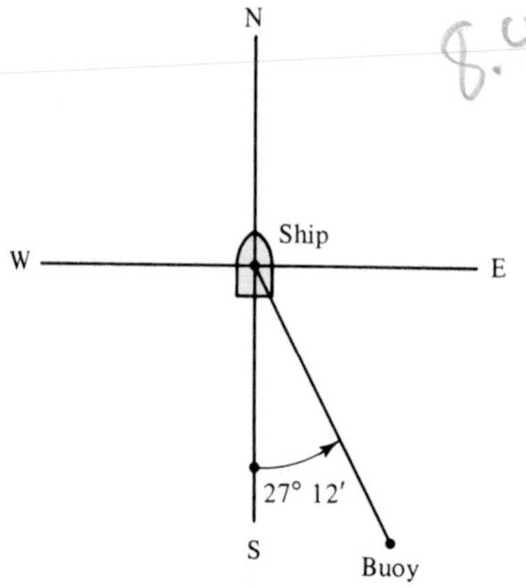


**FIGURE 6-14 A compass direction of S27°12′E means an angle of 27°12′ measured from due south to the east.**

8. From the top of a hill 125 ft above a river, the angles of depression of a point on the near shore and of a point on the opposite shore are 42.3° and 40.6°. Find the width of the river between these two points.
9. From the top of a tree 15.0 m high on the shore of a pond, the angle of depression of a point on the other shore is 6.70°. What is the width of the pond?

## Navigation

10. A passenger on a ship sailing due north at 8.20 km/h noticed that at 1 P.M. a buoy was due east of the ship (Fig. 6-14). At 1:45 P.M., the bearing of the buoy from the ship was S27°12′E. How far was the ship from the buoy at 1 P.M.?
11. An observer at a point $P$ on a coast sights a ship in a direction N43°15′E. The ship is at the same time directly east of a point $Q$, 15.6 km due north of $P$. Find the distance of the ship from point $P$ and from $Q$.
12. A ship sailing parallel to a straight coast is directly opposite one of two lights on the shore. The angle between the lines of sight from the ship to these lights is 27°50′, and it is known that the lights are 355 m apart. Find the perpendicular distance of the ship from the shore. Hint: Use parallel lines and alternate interior angles.
13. An airplane left an airport and flew 315 mi in the direction N15°18′E, then turned and flew 296 mi in the direction S74°42′E. How far, and in what direction, should it fly to return to the airport?
14. A ship is sailing south at a speed of 7.50 km/h when a light is sighted with a bearing S35°28′E. One hour later the ship is due west of the light. Find the distance from the ship to the light at that instant.
15. After leaving port a ship holds a course N46°12′E for 225 mi. Find how far north and how far east of the port the ship is now located.

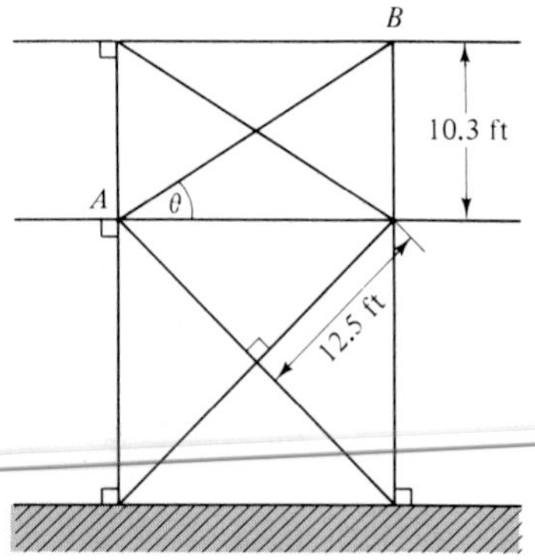


**FIGURE 6-15**

## Structures

16. A guy wire from the top of an antenna is anchored 53.5 ft from the base of the antenna and makes an angle of 85.2° with the ground. Find (a) the height $h$ of the antenna; (b) the length $L$ of the wire.
17. Find the angle $\theta$ and length $AB$ in the truss shown in Fig. 6-15.
18. A house (Fig. 6-16) is 8.75 m wide and its roof has an angle of inclination of 34.0°. Find the length $L$ of the rafters.

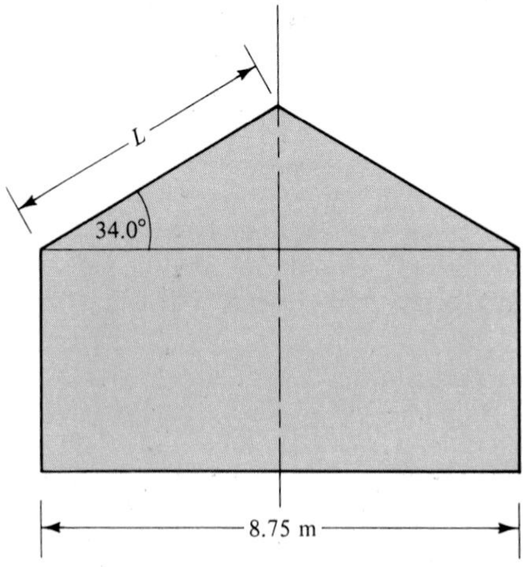


**FIGURE 6-16**

19. A guy wire 82.0 ft long is stretched from the ground to the top of a telephone pole 65.0 ft high. Find the angle between the wire and the pole.

## Geometry

20. What is the angle between a diagonal *AB* of a cube and a diagonal *AC* of a face of that cube (Fig. 6-17)?
21. Two of the sides of an isosceles triangle have a length of 150 units, and each of the base angles is 68°. Find the altitude and the base of the triangle.
22. The diagonal of a rectangle is 3 times the length of the shorter side. Find the angle between the diagonal and the longer side.
23. Find the angles between the diagonals of a rectangle whose dimensions are 580 units × 940 units.
24. Find the area of a parallelogram if the lengths of the sides are 255 units and 482 units and if one angle is 83.2°.
25. Find the length of a side of a regular hexagon inscribed in a 125-cm-radius circle. Hint: Draw lines from the center of the hexagon to each vertex and to the midpoint of each side, to form right triangles.
26. Find the length of the side of a regular pentagon circumscribed about a circle of radius 244 in.

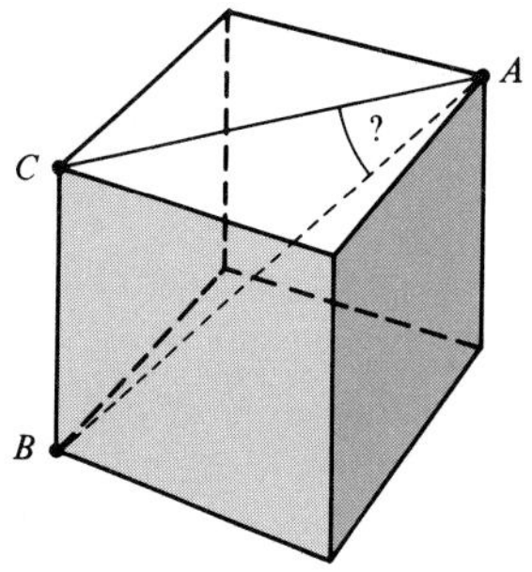


FIGURE 6-17

## Shop Trigonometry

**Trigonometry finds many uses in the machine shop. Given here are some standard shop calculations.**

27. Find the dimension *x* in Fig. 6-18.

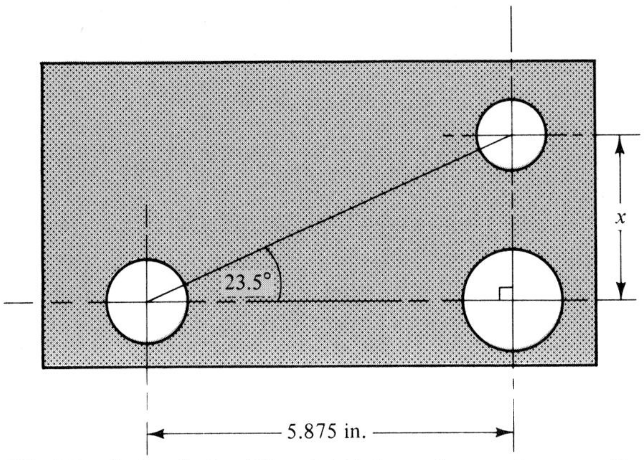


FIGURE 6-18

28. The bolt circle (Fig. 6-19) is to be made on a jig borer. Find the dimensions $x_1$, $y_1$, $x_2$, and $y_2$.
29. A bolt circle with a radius of 36.000 cm contains 24 holes equally spaced. Find the straight-line distance between the holes so that they may be stepped off with dividers.
30. A 9.000-in.-long shaft (Fig. 6-20) is to taper 3.500° and be 1.000 in. in diameter at the narrow end. Find the diameter *d* of the larger end. Hint: Draw a line (shown dashed in the figure) from the small end to the large, to form a right triangle. Solve that triangle.
31. To find a circle diameter when the center is inaccessible, a scale can be placed across the curve to be measured, as in Fig. 6-21, and the chord *c* and the perpendicular distance *h* measured. Find the radius of the curve if the chord length is 8.25 cm and *h* is 1.16 cm.

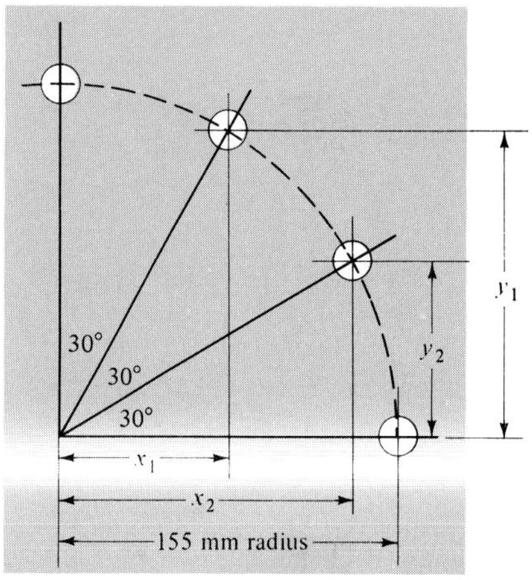


FIGURE 6-19
**A bolt circle.**

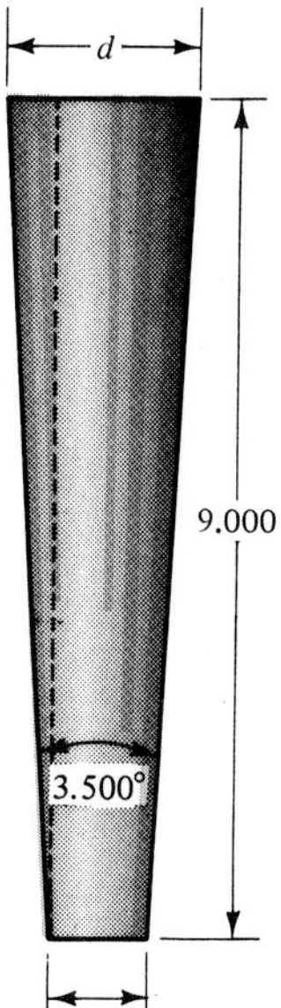


FIGURE 6-20
**Tapered shaft.**

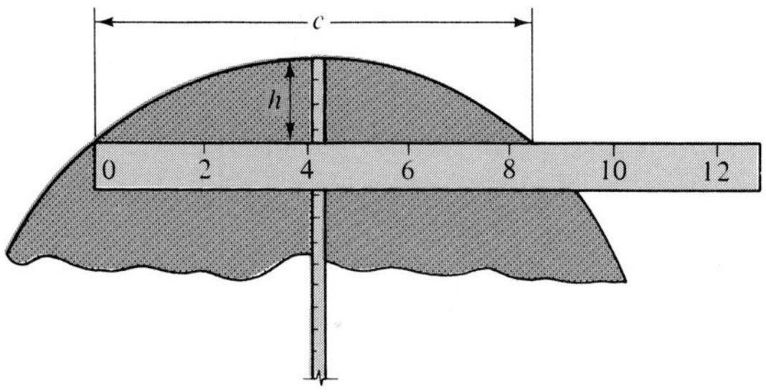


FIGURE 6-21

32. A measuring square having an included angle of 60° is placed over the pulley fragment of Fig. 6-22. The distance $d$ from the corner of the square to the pulley rim is measured at 5.53 in. Find the pulley radius $r$.
33. A common way of measuring dovetails is with the aid of round plugs, as in Fig. 6-23. Find the distance $x$ over the 0.500-in.-diameter plugs. Hint: Find the length of side $AB$ in the 30–60–90 triangle $ABC$, and then use $AB$ to find $x$.
34. A bolt head (Fig. 6-24) measures 0.750 cm across the flats. Find the distance $p$ across the corners and the width $r$ of each flat.

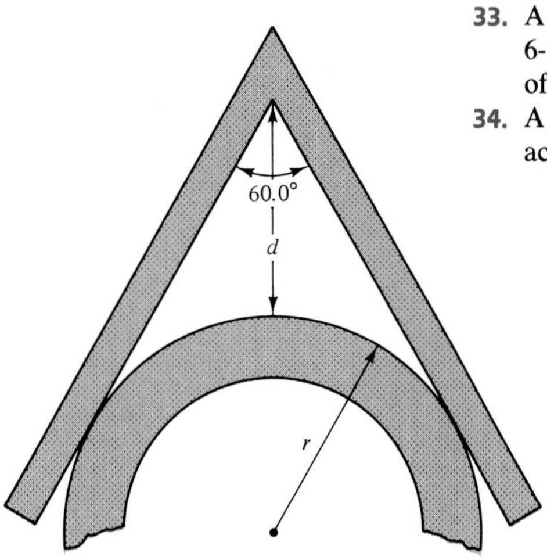


**FIGURE 6-22 Measuring the radius of a broken pulley.**

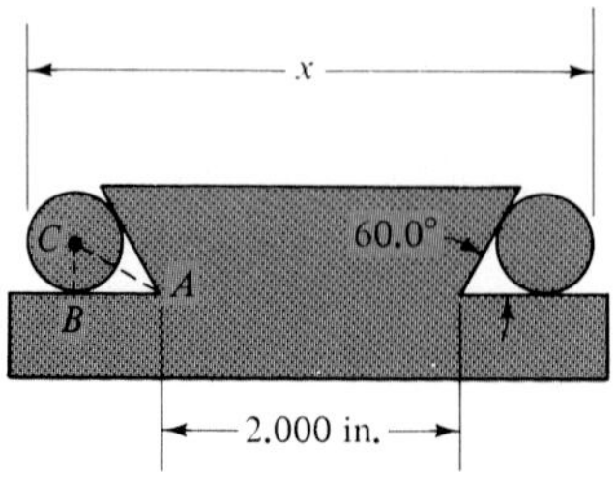


**FIGURE 6-23 Measuring a dovetail.**

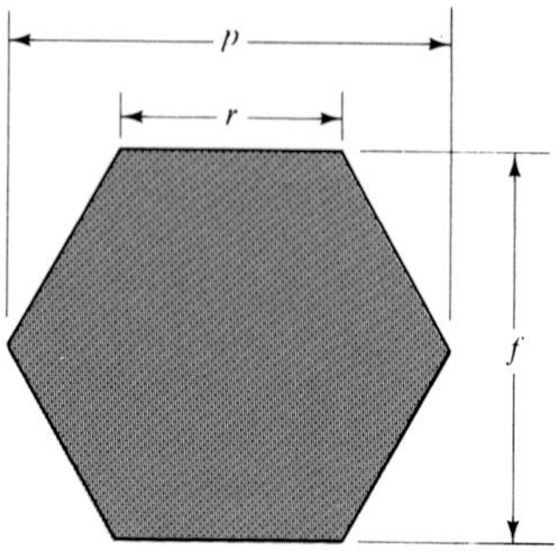


**FIGURE 6-24**

## 6-5 VECTORS

### Definition

A *vector quantity* is one that has *direction* as well as magnitude. For example, a velocity cannot be completely described without giving the direction of the motion, as well as the magnitude (the speed). Other vector quantities are force and acceleration.

Quantities that have magnitude but no direction are called *scalar* quantities. These include time, volume, mass, and so on.

### Representation of Vectors

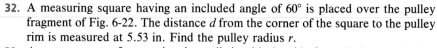

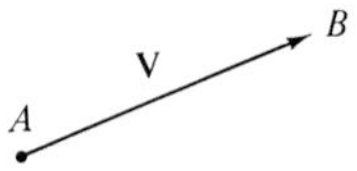


**FIGURE 6-25 Representation of a vector.**

A vector is represented by a line segment (Fig. 6-25) whose length is proportional to the magnitude of the vector, and whose direction is the same as the direction of the vector quantity.

Vectors are represented by boldface type. Thus **B** is a vector quantity, whereas $B$ is a scalar quantity. A vector can be designated by a single letter, or by two letters representing the endpoints, with the starting point given first. Thus, the vector in Fig. 6-25 can be labeled

$$\mathbf{V} \quad \text{or} \quad \mathbf{AB}$$

Boldface letters are not practical in handwritten work. Instead, it is customary to place an *arrow* over vector quantities.

or, when handwritten,

$$\vec{V} \quad \text{or} \quad \overrightarrow{AB}$$

The *magnitude* of a vector can be designated with absolute value symbols, or by ordinary (nonboldface) type. Thus the magnitude of vector **V** in Fig. 6-25 is

$$|\mathbf{V}| \quad \text{or} \quad V \quad \text{or} \quad |\mathbf{AB}| \quad \text{or} \quad AB$$

A vector drawn from the origin can also be designated by the *coordinates of its endpoint*.

**EXAMPLE 26:**

(a) The ordered pair (3, 5) is a vector. It designates a line segment drawn from the origin to the point (3, 5).

(b) The *ordered triple* of numbers (4, 6, 3) is a vector in three dimensions. It designates a point drawn from the origin to the point (4, 6, 3).

(c) The set of numbers (7, 3, 6, 1, 5) is a vector in five dimensions. We cannot draw it but it is handled mathematically in the same way as the others.

We cover vectors such as these in Chapter 11.

## Rectangular Components of a Vector

Any vector can be replaced by two (or more) vectors which, acting together, exactly duplicate the effect of the original vector. They are called the *components* of the vector. The components are usually chosen perpendicular to each other and are then called *rectangular components*. To *resolve* a vector means to replace it by its components.

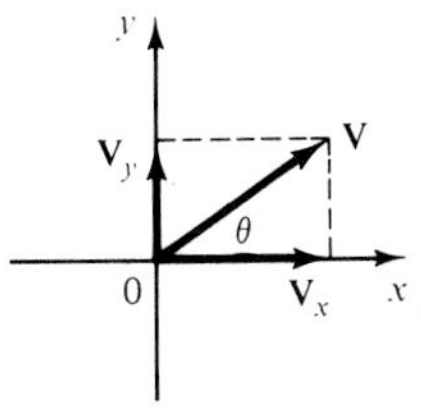


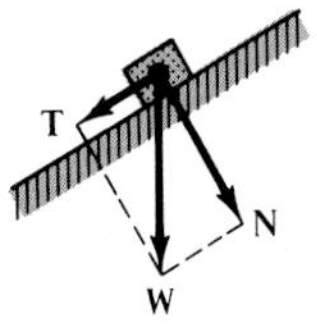


**FIGURE 6-26 Rectangular components of a vector.**

**EXAMPLE 27:** A ball thrown from one person to another moves both in the horizontal direction and in the vertical direction at the same time. The ball thus has a horizontal and a vertical component of velocity.

**EXAMPLE 28:** Figure 6-26a shows a vector **V** resolved into its $x$ component $\mathbf{V}_x$ and its $y$ component $\mathbf{V}_y$.

**EXAMPLE 29:** Figure 6-26b shows a block on an inclined plane with its weight **W** resolved into a component **N** normal (perpendicular) to the plane and a component **T** tangential (parallel) to the plane.

## Resolution of Vectors

We see in Fig. 6-26 that a vector and its rectangular components form a rectangle which has two *right triangles*. Thus we can resolve a vector into its rectangular components with the right-triangle trigonometry of this chapter.

The component of a vector along an axis is also referred to as the *projection* of the vector onto that axis.

**EXAMPLE 30:** The vector **V** in Fig. 6-26a has a magnitude of 248 units and makes an angle $\theta$ of 38.2° with the $x$ axis. Find the $x$ and $y$ components.

**Solution:**

$$\sin 38.2° = \frac{V_y}{248}$$

$$V_y = 248 \sin 38.2° = 153$$

Similarly,

$$V_x = 248 \cos 38.2° = 195$$

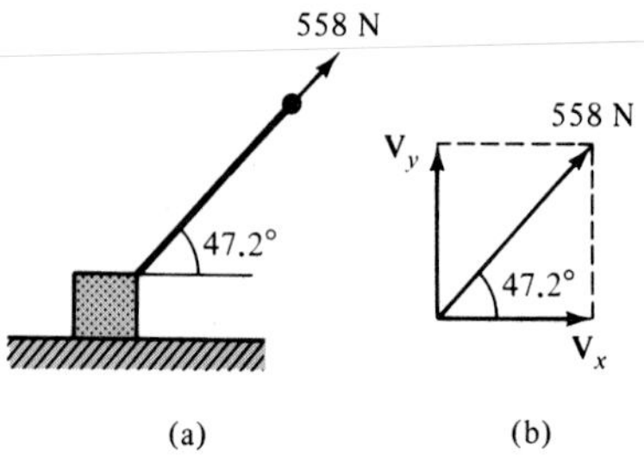


FIGURE 6-27

**EXAMPLE 31:** A cable exerts a force of 558 N at an angle of 47.2° with the horizontal (Fig. 6-27a). Resolve this force into vertical and horizontal components.

**Solution:** We draw a vector diagram (Fig. 6-27b). Then

$$\sin 47.2° = \frac{V_y}{558}$$

$$V_y = 558 \sin 47.2° = 409 \text{ N}$$

Similarly,

$$V_x = 558 \cos 47.2° = 379 \text{ N}$$

## Resultant

**Some calculators have a key for converting between rectangular and polar coordinates, which can also be used to find resultants and components of vectors. Check your manual.**

Just as any vector can be *resolved* into components, so can several vectors be *combined* into a single vector called the *resultant,* or *vector sum.* The process of combining vectors into a resultant is called *vector addition.*

When combining or adding two perpendicular vectors, we can use right-triangle trigonometry, just as we did for resolving a vector into rectangular components.

**EXAMPLE 32:** Find the resultant of two perpendicular vectors whose magnitudes are 485 and 627, and the angle it makes with the 627 magnitude vector.

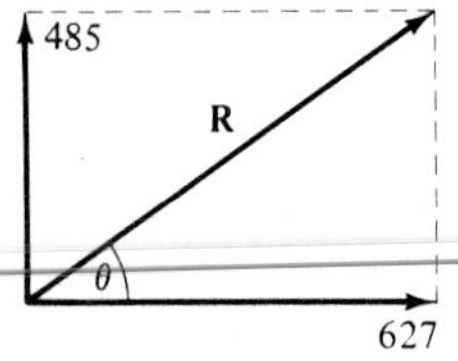


FIGURE 6-28 **Resultant of two vectors.**

**Solution:** We draw a vector diagram (Fig. 6-28). Then by the Pythagorean theorem,

$$R = \sqrt{(485)^2 + (627)^2} = 793$$

and by Eq. 148,

$$\tan \theta = \frac{485}{627} = 0.774$$

$$\theta = 37.7°$$

**EXAMPLE 33:** If the components of vector **A** are

$$A_x = 735 \quad \text{and} \quad A_y = 593$$

find the magnitude of **A** and the angle $\theta$ it makes with the $x$ axis.

**Solution:** By the Pythagorean theorem,

$$A^2 = A_x^2 + A_y^2$$

$$= (735)^2 + (593)^2 = 891{,}900$$

$$A = 944$$

and

$$\tan \theta = \frac{593}{735}$$

$$\theta = 38.9°$$

## EXERCISE 5—VECTORS

### Resolution of Vectors

Given the magnitude of each vector and the angle $\theta$ that it makes with the $x$ axis, find the $x$ and $y$ components.

1. magnitude = 4.93 $\theta = 48.3°$
2. magnitude = 835 $\theta = 25°45'$
3. magnitude = 1.884 $\theta = 58.24°$
4. magnitude = 362 $\theta = 13.8°$

### Vector Addition

In the following problems, the magnitudes of perpendicular vectors **A** and **B** are given. Find the resultant and the angle it makes with vector **B**.

5. $A = 483$ $B = 382$
6. $A = 2.85$ $B = 4.82$
7. $A = 7364$ $B = 4837$
8. $A = 46.8$ $B = 38.6$
9. $A = 1.25$ $B = 2.07$

In the following problems, vector **A** has $x$ and $y$ components $A_x$ and $A_y$. Find the magnitude of **A** and the angle it makes with the $x$ axis.

10. $A_x = 483$ $A_y = 382$
11. $A_x = 6.82$ $A_y = 4.83$
12. $A_x = 58.3$ $A_y = 37.2$
13. $A_x = 2.27$ $A_y = 3.97$

## 6-6 APPLICATIONS OF VECTORS

Any vector quantities such as force, velocity, and impedance can be combined or resolved by the methods in the preceding section.

### Force Vectors

**EXAMPLE 34:** A cable running from the top of a telephone pole (Fig. 6-29) creates a horizontal pull of 875 N. A support cable running to the ground is inclined 71.5° from the horizontal. Find the tension in the support cable.

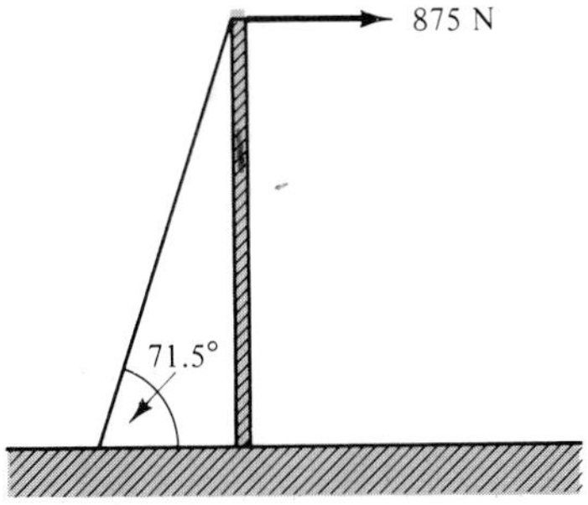


**FIGURE 6-29 A telephone pole.**

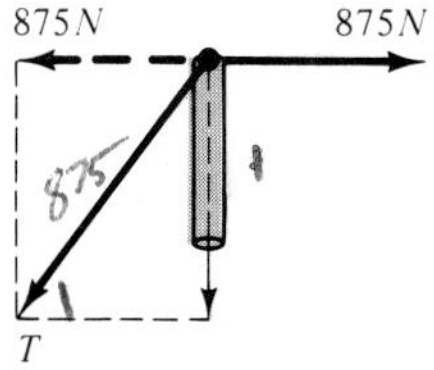


**FIGURE 6-30**

**Solution:** We draw the forces acting at the top of the pole (Fig. 6-30). We see that the horizontal component of the tension $T$ in the support cable must equal the horizontal pull of 875 N. So

$$\cos 71.5° = \frac{875}{T}$$

$$T = \frac{875}{\cos 71.5°} = 2760 \text{ N}$$

## Velocity Vectors

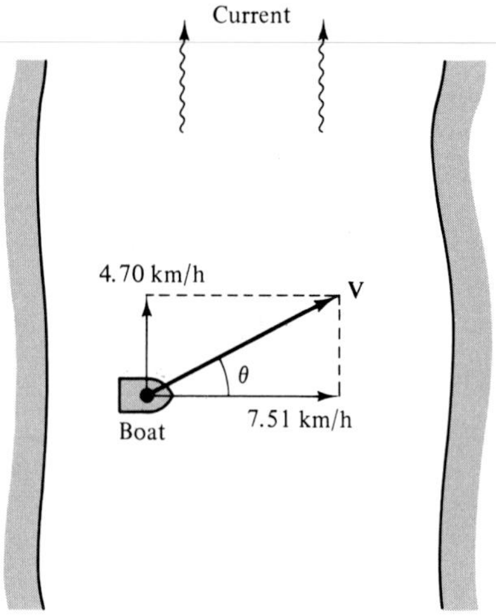


FIGURE 6-31

---

**EXAMPLE 35:** A river flows at the rate of 4.70 km/h. A rower, who can travel 7.51 km/h in still water, heads directly across the current. Find the rate and direction of travel of the boat.

**Solution:** The boat (Fig. 6-31) is crossing the current and at the same time is being carried downstream. Thus the velocity **V** has one component of 7.51 km/h across the current and another component of 4.70 km/h downstream. We find the resultant of these two components by the Pythagorean theorem,

$$V^2 = (4.70)^2 + (7.51)^2$$

from which

$$V = 8.86 \text{ km/h}$$

Now finding the angle $\theta$ yields

$$\tan \theta = \frac{4.70}{7.51}$$

from which

$$\theta = 32.0°$$

---

## Impedance Vectors

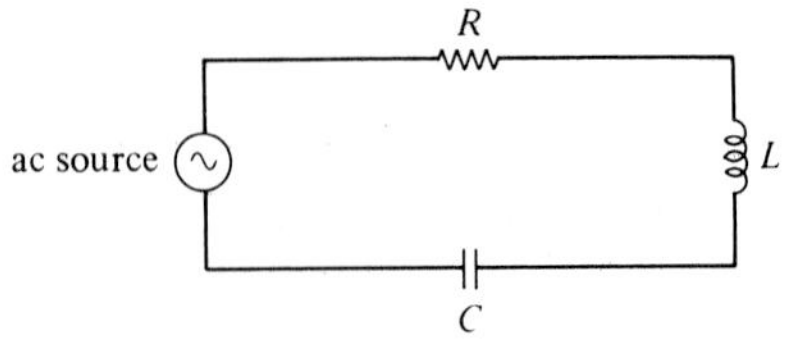


FIGURE 6-32 **Series *RLC* circuit.**

Vectors find extensive application in electrical technology, and one of the most common applications is in the calculation of impedances. Figure 6-32 shows a resistor, an inductor, and a capacitor, connected in series with an *ac* source. The *reactance X* is a measure of how much the capacitance and inductance retard the flow of current in such a circuit. It is the difference between the *capacitive reactance* $X_C$ and the *inductive reactance* $X_L$.

| | | |
|---|---|---|
| Reactance | $X = X_L - X_C$ | **A96** |

**We certainly are not attempting to teach ac circuits in a few paragraphs, but only hope to reinforce concepts learned in your other courses.**

The *impedance Z* is a measure of how much the flow of current in an *ac* circuit is retarded by all circuit elements, including the resistance. The magnitude of the impedance is related to the total resistance $R$ and reactance $X$ by

| | | |
|---|---|---|
| Magnitude of Impedance | $\lvert Z \rvert = \sqrt{R^2 + X^2}$ | **A97** |

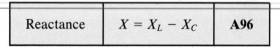

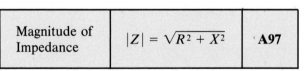

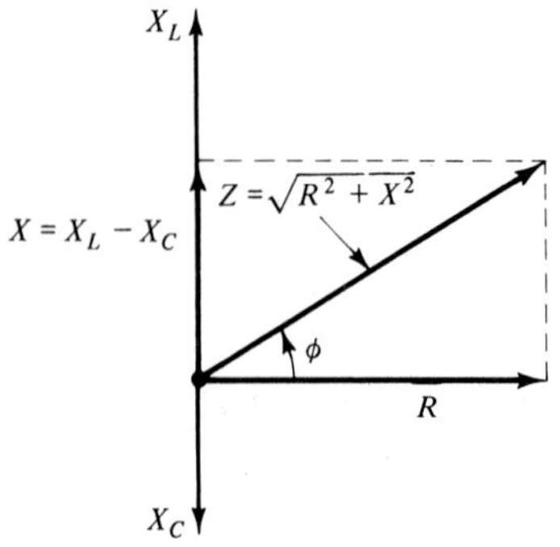


FIGURE 6-33 **Vector impedance diagram.**

The impedance, resistance, and reactance form the three sides of a right triangle, the *vector impedance diagram* (Fig. 6-33). The angle $\phi$ between $Z$ and $R$ is called the *phase angle*.

| | | |
|---|---|---|
| Phase Angle | $\phi = \arctan \dfrac{X}{R}$ | **A98** |

**EXAMPLE 36:** The capacitive reactance of a certain circuit is 2720 Ω, the inductive reactance is 3260 Ω, and the resistance is 1150 Ω. Find the reactance and magnitude of the impedance of the circuit, and the phase angle.

**Solution:** By Eq. A96,

$$X = 3260 - 2720 = 540\ \Omega$$

By Eq. A97,

$$|Z| = \sqrt{(1150)^2 + (540)^2} = 1270\ \Omega$$

and by Eq. A98,

$$\phi = \arctan \frac{540}{1150} = 25.2°$$

## EXERCISE 6—APPLICATIONS OF VECTORS

### Force Vectors

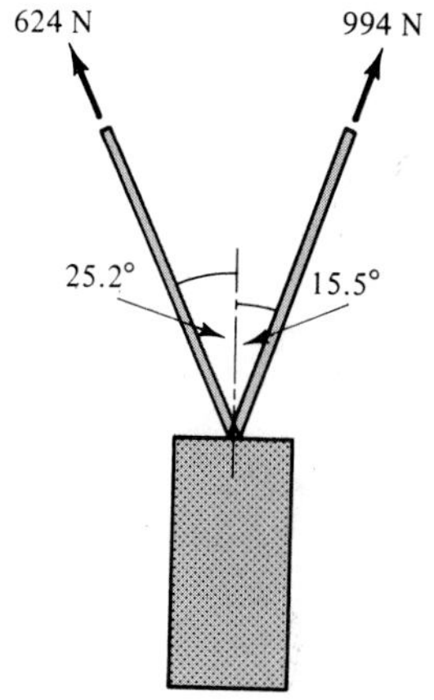


FIGURE 6-34

You will need the equations of equilibrium, Eqs. A13 and A14, for some of these problems.

1. What force, neglecting friction, must be exerted to drag a 56.5-N weight up a slope inclined 12.6° from the horizontal?
2. What is the largest weight that a tractor can drag up a slope which is inclined 21.2° from the horizontal if it is able to pull along the incline with a force of 2750 lb? Neglect the force due to friction.
3. Two ropes (Fig. 6-34) hold a crate when one is pulling with a force of 994 N in a direction 15.5° with the vertical and the other with a force of 624 N in a direction 25.2° with the vertical. How much does the crate weigh?
4. A truck weighing 7285 lb moves up a bridge inclined 4.8° from the horizontal. Find the force of the truck normal (perpendicular) to the bridge.
5. A person has just enough strength to pull a 1270-N weight up a certain slope. Neglecting friction, find the angle at which the slope is inclined to the horizontal if the person is able to exert a pull of 550 N.
6. A person wishes to pull a 255-lb weight up an incline to the top of a wall 14.5 ft high. Neglecting friction, what is the length of the shortest incline (measured along the incline) that can be used if the person's pulling strength is 145 lb?
7. A truck weighing 18.6 tons stands on a hill inclined 15.4° from the horizontal. How large a force must be counteracted by brakes to prevent the truck from rolling downhill?

### Velocity Vectors

8. A plane is headed due west with an air speed of 212 km/h. It is driven from its course by a wind from due north blowing at 23.6 km/h. Find the ground speed of the plane and the actual direction of travel. Refer to Fig. 6-35 for definitions of flight terminology.

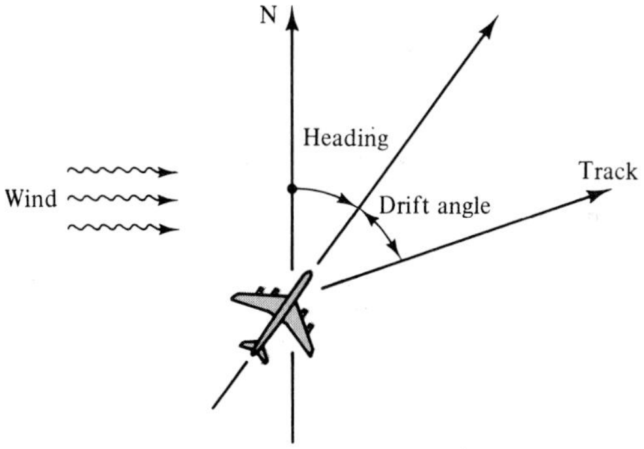


**FIGURE 6-35 Flight terminology. The *heading* of an aircraft is the direction in which the craft is *pointed*. Due to air current, it will not usually travel in that direction but in an actual path called the *track*. The angle between the heading and the track is the *drift angle*. The *air speed* is the speed relative to the surrounding air, and the *ground speed* is the craft's speed relative to the earth.**

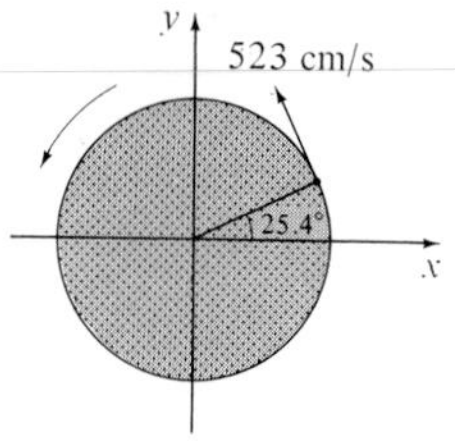

**FIGURE 6-36**

9. At what air speed should an airplane head west in order to follow a course S80°15′W if a 20.5 km/h wind is blowing from due north?

10. A pilot heading his plane due north finds the actual direction of travel to be N5°12′E. The plane's air speed is 315 mi/h and the wind is from the west. Find the plane's ground speed and the velocity of the wind.

11. A point on a rotating wheel has a tangential velocity of 523 cm/s. Find the $x$ and $y$ components of the velocity when the point is in the position shown in Fig. 6-36.

12. A certain escalator travels at a rate of 10.6 m/min and its angle of inclination is 32.5°. What is the vertical component of the velocity? How long will it take a passenger to travel 10.0 m vertically?

13. A projectile is launched at an angle of 55.6° to the horizontal with a speed of 7550 m/min. Find the vertical and horizontal components of this velocity.

14. At what speed with respect to the water should a ship head north in order to follow a course N5°15′E if a current is flowing east at the rate of 10.6 mi/h?

### Impedance Vectors

15. A circuit has a reactance of 2650 Ω and a phase angle of 44.6°. Find the resistance and the magnitude of the impedance.

16. A circuit has a resistance of 115 Ω and a phase angle of 72°. Find the reactance and the magnitude of the impedance.

17. A circuit has an impedance of 975 Ω and a phase angle of 28°. Find the resistance and the reactance.

18. A circuit has a reactance of 5.75 Ω and a resistance of 4.22 Ω. Find the magnitude of the impedance and the phase angle.

19. A circuit has a capacitive reactance of 1776 Ω, an inductive reactance of 5140 Ω, and a total impedance of 5560 Ω. Find the resistance and the phase angle.

## CHAPTER 6 REVIEW PROBLEMS

*Perform the angle conversions, and fill in the missing quantities.*

| | Decimal degrees | Degrees– minutes– seconds | Radians | Revolutions |
|---|---|---|---|---|
| 1. | 38.2 | 38,12.00 | .6667 | .106 |
| 2. | | 25°28′45″ | | |
| 3. | 157.56 | 157°33′44″ | 2.745 | .4376 |
| 4. | 99 | 99,00,00 | 1.7278 | 0.275 |

*The following points are on the terminal side of an angle θ. Write the six trigonometric functions of the angle to three significant digits, and find the angle in decimal degrees.*

5. (3, 7)
6. (2, 5)
7. (4, 3)
8. (2.3, 3.1)

sin = .919   1.09
cos = .394   2.54
tan = 2.33   .429
66.8

*Write the six trigonometric functions of each angle. Keep four decimal places.*

9. 72.9°   .9554   1.0463   3.2940   2.4009   3.2506   3.2506   .3076
10. 35°13′33″
11. 1.05 rad

*Find acute angle θ in decimal degrees.*

12. sin θ = 0.574
13. cos θ = 0.824    34.51
14. tan θ = 1.345

*Evaluate each expression. Give your answer as an acute angle in degrees.*

15. arctan 2.86   70.7
16. cos⁻¹ 0.385
17. arcsec 2.447   24.1

*Solve right triangle ABC.*

18. $a = 746$ and $A = 37.2°$
19. $b = 3.72$ and $A = 28.5°$
20. $c = 45.9$ and $A = 61.4°$

21. Find the $x$ and $y$ components of a vector that has a magnitude of 885 and makes an angle of 66.3° with the $x$ axis.

22. Find the magnitude of the resultant of two perpendicular vectors that have magnitudes of 54.8 and 39.4 and find the angle the resultant makes with the 54.8 vector.

23. A vector has $x$ and $y$ components of 385 and 275. Find the magnitude and direction of that vector.

24. A telephone pole casts a shadow 13.5 m long when the angle of elevation of the sun is 15.4°. Find the height of the pole.

25. From a point 125 ft in front of a church, the angles of elevation of the top and base of its steeple are 22.5° and 19.6°, respectively. Find the height of the steeple.

26. A circuit has a resistance of 125 Ω, an impedance of 256 Ω, an inductive reactance of 312 Ω, and a positive phase angle. Find the capacitive reactance and the phase angle.

*Evaluate to four decimal places.*

27. cos 59.2°
28. csc 19.3°
29. tan 52.8°
30. sec 37.2°
31. cot 24.7°
32. sin 42.9°

*Find the acute angle whose trigonometric function is given, to the nearest tenth of a degree.*

33. $\tan \theta = 1.7362$
34. $\sec \theta = 2.9914$
35. $\sin \theta = 0.7253$
36. $\csc \theta = 3.2746$
37. $\cot \theta = 0.9475$
38. $\cos \theta = 0.3645$

*Writing*

39. Write a short paragraph, with examples, on how you might possibly use an idea from this chapter in a real-life situation: on the job, in a hobby, or around the home.

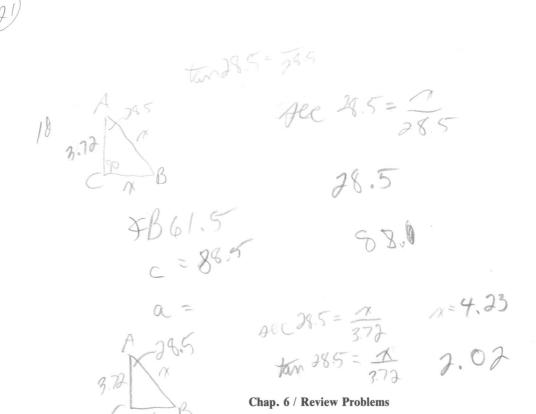

# 7

# FACTORS AND FACTORING

## OBJECTIVES

**When you have completed this chapter, you should be able to:**

- Factor expressions by removing common factors.
- Factor binomials which are the difference of two squares.
- Factor the sum or difference of two cubes.
- Factor trinomials.
- Factor the perfect square trinomial.
- Factor expressions by grouping.
- Use factoring techniques in applications.

In Chapter 2 we learned how to multiply expressions together—here we do the reverse. We find those quantities (factors) which, when multiplied together, give the original product.

Why bother factoring? What is the point of changing an expression from one form to another? Because factoring, which may seem complicated at first, is actually essential to simplifying expressions and solving equations, especially the quadratic equations in Chapter 13. The value of factoring will become clear as we use it toward the end of this chapter to solve literal equations and to manipulate the formulas that are so important in technology. We also use factoring in Chapter 8 to simplify fractions.

**The idea of a factor is not entirely new to us. You might glance back at Secs. 1-3 and 2-1, before starting this chapter.**

## 7-1 COMMON FACTORS

### Factors of an Expression

The *factors of an expression* are those quantities whose product is the original expression.

---

**EXAMPLE 1:** The factors of $x^2 - 9$ are $x + 3$ and $x - 3$, because

$$\begin{aligned}(x + 3)(x - 3) &= x^2 + 3x - 3x - 9 \\ &= x^2 - 9\end{aligned}$$

---

### Prime Factors

Many expressions have no factors other than 1 and themselves. Such expressions are called *prime*.

---

**EXAMPLE 2:** The expressions $x$, $x - 3$, and $x^2 - 3x + 7$ are all prime.

---

**EXAMPLE 3:** The expressions $x^2 - 9$ and $x^2 + 5x + 6$ are not prime, because they can be factored. They are said to be *factorable*.

---

### Factoring

*Factoring* is the process of finding the factors of an expression. It is *the reverse of finding the product* of two or more quantities.

| Multiplication (Finding the Product) |
|---|
| $x(x + 4) = x^2 + 4x$ |
| **Factoring (Finding the Factors)** |
| $x^2 + 4x = x(x + 4)$ |

We factor an expression by *recognizing the form* of that expression and then applying standard rules for factoring. In the first type of factoring, we look for *common factors*.

## Common Factors

If each term of an expression contains the same quantity (called the *common factor*), that quantity may be *factored out*.

This is nothing but the ***distributive law*** that we studied earlier.

| Common Factor | $ab + ac = a(b + c)$ | 5 |
|---|---|---|

---

**EXAMPLE 4:** In the expression

$$2x + x^2$$

each term contains an $x$ as a common factor. So we write

$$2x + x^2 = x(2 + x)$$

Most of the factoring we will do will be of this type.

---

**EXAMPLE 5:**

(a) $3x^3 - 2x + 5x^4 = x(3x^2 - 2 + 5x^3)$
(b) $3xy^2 - 9x^3y + 6x^2y^2 = 3xy(y - 3x^2 + 2xy)$
(c) $3x^3 - 6x^2y + 9x^4y^2 = 3x^2(x - 2y + 3x^2y^2)$
(d) $2x^2 + x = x(2x + 1)$

---

When factoring, don't confuse superscripts (powers) with subscripts.

---

**EXAMPLE 6:** The expression $x^3 + x^2$ is factorable

$$x^3 + x^2 = x^2(x + 1)$$

but the expression

$$x_3 + x_2$$

is not factorable.

---

**Common Error**

Students are sometimes puzzled over the 1 in Example 6. Why should it be there? After all, when you remove a chair from a room, it is *gone*; there is nothing (zero) remaining where the chair used to be. If you remove an $x$ by factoring, you might assume that nothing (zero) remains where the $x$ used to be.

$$2x^2 + x = x(2x + 0)?$$

Prove to yourself that this is not correct by multiplying the factors to see if you get back the original expression.

### Factors in the Denominator

Common factors may appear in the denominators of the terms as well as in the numerators.

---

**EXAMPLE 7:**

(a) $\dfrac{1}{x} + \dfrac{2}{x^2} = \dfrac{1}{x}\left(1 + \dfrac{2}{x}\right)$

(b) $\dfrac{x}{y^2} + \dfrac{x^2}{y} + \dfrac{2x}{3y} = \dfrac{x}{y}\left(\dfrac{1}{y} + x + \dfrac{2}{3}\right)$

---

### Checking

To check if factoring has been done correctly, simply multiply your factors together and see if you get back the original expression.

---

**EXAMPLE 8:** Are $2xy$ and $x + 3 - y^2$ the factors of

$$2x^2y + 6xy - 2xy^3?$$

**Solution:** Multiplying the factors, we obtain

$$2xy(x + 3 - y^2) = 2x^2y + 6xy - 2xy^3 \quad \text{checks}$$

---

This check will tell us if we have factored correctly but, of course, not whether we have factored *completely*, that is, found the prime factors.

### EXERCISE 1—COMMON FACTORS

Factor each expression and check your result.

1. $3y^2 + y^3$
2. $6x - 3y$
3. $x^5 - 2x^4 + 3x^3$
4. $9y - 27xy$
5. $3a + a^2 - 3a^3$
6. $8xy^3 - 6x^2y^2 + 2x^3y$
7. $5(x + y) + 15(x + y)^2$
8. $\dfrac{a}{3} - \dfrac{a^2}{4} + \dfrac{a^3}{5}$
9. $\dfrac{3}{x} + \dfrac{2}{x^2} - \dfrac{5}{x^3}$
10. $\dfrac{3ab^2}{y^3} - \dfrac{6a^2b}{y^2} + \dfrac{12ab}{y}$
11. $5a^2b + 6a^2c$
12. $a^2c + b^2c + c^2d$
13. $4x^2y + cxy^2 + 3xy^3$
14. $4abx + 6a^2x^2 + 8ax$
15. $3a^3y - 6a^2y^2 + 9ay^3$
16. $2a^2c - 2a^2c^2 + 3ac$
17. $5acd - 2c^2d^2 + bcd$
18. $4b^2c^2 - 12abc - 9c^2$
19. $8x^2y^2 + 12x^2z^2$
20. $6xyz + 12x^2y^2z$
21. $3a^2b + abc - abd$
22. $5a^3x^2 - 5a^2x^3 + 10a^2x^2z$

### Applications

23. When a bar of length $L_0$ is changed in temperature by an amount $t$, its new length $L$ will be $L = L_0 + L_0\alpha t$, where $\alpha$ is the coefficient of thermal expansion. Factor the right side of this equation.
24. A sum of money $a$ when invested for $t$ years at an interest rate $n$ will accumulate to an amount $y$, where $y = a + ant$. Factor the right side of this equation.

Compare your result with Eq. A70.

25. When a resistance $R_1$ is heated from a temperature $t_1$ to a new temperature $t$, it will increase in resistance by an amount $\alpha(t - t_1)R_1$, where $\alpha$ is the temperature coefficient of resistance. The final resistance will then be $R = R_1 + \alpha(t - t_1)R_1$. Factor the right side of this equation.
26. An item costing $P$ dollars is reduced in price by 15%. The resulting price $C$ is then $C = P - 0.15P$. Factor the right side of this equation.
27. The displacement of a uniformly accelerated body is given by

$$s = v_0 t + \frac{a}{2}t^2$$

Factor the right side of this equation.

28. The sum of the voltage drops across the resistors in Fig. 7-1 must equal the battery voltage $E$.

$$E = iR_1 + iR_2 + iR_3$$

Factor the right side of this equation.

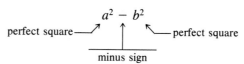

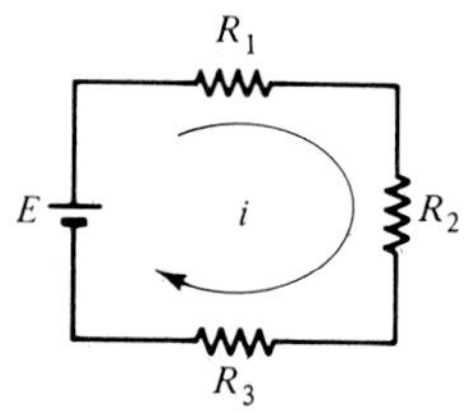


**FIGURE 7-1**

29. The mass of a spherical shell having an outside radius of $r_2$ and an inside radius $r_1$ is

$$\text{mass} = \tfrac{4}{3}\pi r_2^3 D - \tfrac{4}{3}\pi r_1^3 D$$

where $D$ is the mass density of the material. Factor the right side of this equation.

## 7-2 DIFFERENCE OF TWO SQUARES

### Form

An expression of the form

$$a^2 - b^2$$

perfect square ↗ ↑ ↖ perfect square

minus sign

where one perfect square is subtracted from another, is called a *difference of two squares*. It arises when $(a - b)$ and $(a + b)$ are multiplied together.

Remember this from Chapter 2?

| | | |
|---|---|---|
| Difference of Two Squares | $a^2 - b^2 = (a + b)(a - b)$ | **41** |

### Factoring the Difference of Two Squares

Once we recognize its form, the difference of two squares is easily factored.

**EXAMPLE 9:**

$$4x^2 - 9 = (2x)^2 - (3)^2$$

$$(2x)^2 - (3)^2 = (2x \quad\;)(2x \quad\;)$$

square root of the first term

$$(2x)^2 - (3)^2 = (2x \quad 3)(2x \quad 3)$$

square root of the last term

$$4x^2 - 9 = (2x + 3)(2x - 3)$$

opposite signs

**EXAMPLE 10:**

(a) $y^2 - 1 = (y + 1)(y - 1)$
(b) $9a^2 - 16b^2 = (3a + 4b)(3a - 4b)$
(c) $49x^2 - 9a^2y^2 = (7x - 3ay)(7x + 3ay)$
(d) $1 - a^2b^2c^2 = (1 - abc)(1 + abc)$

The difference of any two quantities that have *even powers* can be factored as the difference of two squares. *Express each quantity as a square* by means of

| | | |
|---|---|---|
| Powers | $x^{ab} = (x^a)^b$ | **31** |

**EXAMPLE 11:**

(a) $a^4 - b^6 = (a^2)^2 - (b^3)^2$
$\qquad = (a^2 + b^3)(a^2 - b^3)$
(b) $x^{2a} - y^{6b} = (x^a)^2 - (y^{3b})^2$
$\qquad = (x^a + y^{3b})(x^a - y^{3b})$

Sometimes one or both terms in the given expression may be a fraction.

**EXAMPLE 12:**

$$\frac{1}{x^2} - \frac{1}{y^2} = \left(\frac{1}{x}\right)^2 - \left(\frac{1}{y}\right)^2$$

$$= \left(\frac{1}{x} + \frac{1}{y}\right)\left(\frac{1}{x} - \frac{1}{y}\right)$$

Sometimes one (or both) of the terms in an expression is itself a binomial.

**EXAMPLE 13:**

$$(a + b)^2 - x^2 = [(a + b) + x][(a + b) - x]$$

| | |
|---|---|
| Common Error | There is no similar rule for factoring the *sum* of two squares, $$a^2 + b^2$$ |

## Factoring Completely

After factoring an expression, see if any of the factors themselves can be factored *again*.

---

**EXAMPLE 14:** Factor $a - ab^2$.

**Solution:** Always remove any common factors first. Factoring, we obtain

$$a - ab^2 = a(1 - b^2)$$

Now factoring the difference of two squares, we get

$$a - ab^2 = a(1 + b)(1 - b)$$

---

**EXAMPLE 15:**

$$x^4 - y^4 = (x^2 + y^2)(x^2 - y^2)$$

Now factoring the difference of two squares again, we obtain

$$x^4 - y^4 = (x^2 + y^2)(x + y)(x - y)$$

---

## EXERCISE 2—DIFFERENCE OF TWO SQUARES

Factor completely.

1. $4 - x^2$
2. $x^2 - 9$
3. $9a^2 - x^2$
4. $25 - x^2$
5. $4x^2 - 4y^2$
6. $9x^2 - y^2$
7. $x^2 - 9y^2$
8. $16x^2 - 16y^2$
9. $9c^2 - 16d^2$
10. $25a^2 - 9b^2$
11. $9y^2 - 1$
12. $4x^2 - 9y^2$

### Expressions with Higher Powers or Literal Powers

13. $m^4 - n^4$
14. $a^8 - b^8$
15. $m^{2n} - n^{2m}$
16. $9^{2n} - 4b^{4n}$
17. $a^{16} - b^8$
18. $9a^2b^2 - 4c^4$
19. $25x^4 - 16y^6$
20. $36y^2 - 49z^6$
21. $16a^4 - 121$
22. $121a^4 - 16$
23. $25a^4b^4 - 9$
24. $121a^2 - 36b^4$

### Applications

25. The thrust washer in Fig. 7-2 has a surface area of

$$\text{area} = \pi r_2^2 - \pi r_1^2$$

Factor this expression.

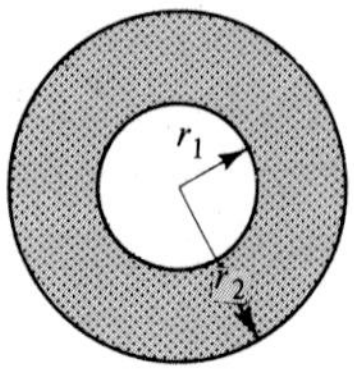


**FIGURE 7-2 Circular washer.**

26. A flywheel of diameter $d_1$ (Fig. 7-3) has a balancing hole of diameter $d_2$ drilled through it. The mass $M$ of the flywheel is then

$$\text{mass} = \frac{\pi d_1^2}{4} Dt - \frac{\pi d_2^2}{4} Dt$$

Factor this expression.

27. A spherical balloon shrinks from radius $r_1$ to radius $r_2$. The change in surface area is, from Eq. 129, $4\pi r_1^2 - 4\pi r_2^2$. Factor this expression.

28. When an object is released from rest, the distance fallen between time $t_1$ and time $t_2$ is $\frac{1}{2}gt_2^2 - \frac{1}{2}gt_1^2$, where $g$ is the acceleration due to gravity. Factor this expression.

29. When a body of mass $m$ slows down from velocity $v_1$ to velocity $v_2$, the decrease in kinetic energy is $\frac{1}{2}mv_1^2 - \frac{1}{2}mv_2^2$. Factor this expression.

30. The work required to stretch a spring from a length $x_1$ (measured from the unstretched position) to a new length $x_2$ is $\frac{1}{2}kx_2^2 - \frac{1}{2}kx_1^2$. Factor this expression.

31. The formula for the volume of a cylinder of radius $r$ and height $h$ is $v = \pi r^2 h$. Write an expression for the volume of a hollow cylinder that has an inside radius $r$, an outside radius $R$, and a height $h$. Then factor the expression completely.

32. A body at temperature $T$ will radiate an amount of heat $kT^4$ to its surroundings and will absorb from the surroundings an amount of heat $kT_s^4$, where $T_s$ is the temperature of the surroundings. Write an expression for the net heat transfer by radiation (amount radiated minus amount absorbed) and factor this expression completely.

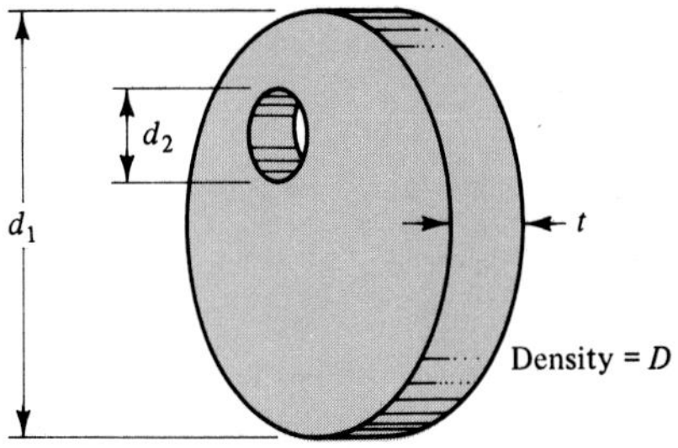


FIGURE 7-3

## 7-3 FACTORING TRINOMIALS

### Trinomials

A *trinomial,* you recall, is a polynomial having *three terms.*

**EXAMPLE 16:** $2x^3 - 3x + 4$ is a trinomial.

A *quadratic trinomial* has an $x^2$ term, an $x$ term, and a constant term.

**EXAMPLE 17:** $4x^2 + 3x - 5$ is a quadratic trinomial.

In this section we factor only quadratic trinomials.

### Test for Factorability

Not all quadratic trinomials can be factored. We test for factorability as follows.

| | | |
|---|---|---|
| Test for Factorability | The trinomial $ax^2 + bx + c$ (where $a$, $b$, and $c$ are constants) is factorable if $b^2 - 4ac$ is a perfect square. | 44 |

**$b^2 - 4ac$ is called the *discriminant,* Eq. 101. We use it later to predict the nature of the roots of a quadratic equation.**

**EXAMPLE 18:** Can the trinomial $6x^2 + x - 12$ be factored?

**Solution:** Using the test for factorability, Eq. 44 with $a = 6$, $b = 1$, and $c = -12$,

$$\begin{aligned} b^2 - 4ac &= 1^2 - 4(6)(-12) \\ &= 1 + 288 = 289 \end{aligned}$$

By taking the square root of 289 on the calculator, we see that it is a perfect square ($289 = 17^2$), so the given trinomial is factorable. We will see later that its factors are $(2x + 3)$ and $(3x - 4)$.

The coefficient of the $x^2$ term in a quadratic trinomial is called the *leading coefficient*. (It is the constant $a$ in Eq. 44). We will first factor the easier type of trinomial, in which the leading coefficient is 1.

## Trinomials with a Leading Coefficient of 1

We get such a trinomial when we multiply two binomials if the coefficients of the $x$ terms in the binomials are also 1.

$$(x + a)(x + b) = x^2 + (a + b)x + ab$$

Note that the coefficient of the middle term of the trinomial equals the *sum* of $a$ and $b$, and that the last term equals the *product* of $a$ and $b$. Knowing this, we can reverse the process and find the factors when the trinomial is given.

| | | |
|---|---|---|
| Trinomials, Leading Coefficient = 1 | sum<br>$x^2 + (a + b)x + ab = (x + a)(x + b)$<br>product | **45** |

**EXAMPLE 19:** Factor the trinomial $x^2 + 8x + 15$.

**Solution:** From Eq. 45 this trinomial, if factorable, will factor as $(x + a)(x + b)$, where $a$ and $b$ have a sum of 8 and a product of 15. The integers 5 and 3 have a sum of 8 and a product of 15, so,

$$x^2 + 8x + 15 = (x + 5)(x + 3)$$

## Using the Signs to Aid Factoring

The signs of the terms of the trinomial can tell you the signs of the factors.

1. If the sign of the last term is *positive*, both signs in the factors will be the same (both positive or both negative). The sign of the middle term

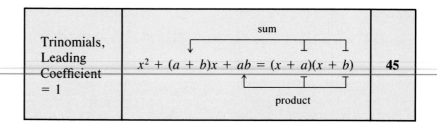

of the trinomial will tell whether both signs in the factors are positive or negative.

---

**EXAMPLE 20:**

positive, so signs in the factors are the same

(a) $x^2 + 6x + 8 = (x + 4)(x + 2)$

positive, so both signs in the factors are positive

positive, so signs in the factors are the same

(b) $x^2 - 6x + 8 = (x - 4)(x - 2)$

negative, so both signs in the factors are negative

---

2. If the sign of the last term of the trinomial is *negative,* the signs of $a$ and $b$ in the factors will differ (one positive, one negative). The sign of the middle term of the trinomial will tell which quantity, $a$ or $b$, is larger (in absolute value), the positive one or the negative one.

---

**EXAMPLE 21:**

negative, so the signs in the factors differ, one positive and one negative

$x^2 - 2x - 8 = (x - 4)(x + 2)$

negative, so the larger number (4) has a negative sign

---

**EXAMPLE 22:** Factor $x^2 + 2x - 15$.

Solution: We first look at the sign of the last term ($-15$). The negative sign tells us that the signs of the factors will differ.

$$x^2 + 2x - 15 = (x - \quad)(x + \quad)$$

We then find two numbers whose product is $-15$ and whose sum is 2. These two numbers must have opposite signs in order to have a negative product. Also, since the middle coefficient is $+2$, the positive number must be 2 greater than the negative number. Two numbers that meet these conditions are $+5$ and $-3$. So

$$x^2 + 2x - 15 = (x + 5)(x - 3)$$

---

Remember to remove any *common factors* before factoring a trinomial.

---

**EXAMPLE 23:**
$$2x^3 + 4x^2 - 30x = 2x(x^2 + 2x - 15)$$
$$= 2x(x - 3)(x + 5)$$

---

**As with the difference of two squares, many trinomials can be factored that at first glance do not seem to be in the proper form. Here are some common types.**

**EXAMPLE 24:** Factor $x^2 + 5xy + 6y^2$.

**Solution:** One way to factor such a trinomial is, first, to temporarily drop the second variable ($y$ in this example).

$$x^2 + 5x + 6$$

Then factor the resulting trinomial.

$$(x + 2)(x + 3)$$

Finally, note that if we tack $y$ onto the second term of each factor, this will cause $y$ to appear in the middle term of the trinomial and $y^2$ to appear in the last term of the trinomial. The required factors are then

$$(x + 2y)(x + 3y)$$

**EXAMPLE 25:** Factor $w^2x^2 + 3wxy - 4y^2$.

**Solution:** Temporarily dropping $w$ and $y$ yields

$$x^2 + 3x - 4$$

which factors into

$$(x - 1)(x + 4)$$

We then note that if a $w$ were appended to each $x$ term and a $y$ to each second term in each binomial

$$(wx - y)(wx + 4y)$$

that the product of these factors will give the original expression.

## Trinomials with Higher Powers

Trinomials with powers greater than *two* may be factored if one exponent can be expressed as a *multiple of two,* and if that exponent is *twice the other.*

**EXAMPLE 26:**

$$\begin{aligned} x^6 + x^3 - 2 &= (x^3)^2 + (x^3) - 2 \\ &= (x^3 - 1)(x^3 + 2) \end{aligned}$$

## Trinomials with Variables in the Exponents

If one exponent is *twice the other,* you may be able to factor such a trinomial.

**EXAMPLE 27:**

$$\begin{aligned} x^{2n} + 5x^n + 6 &= (x^n)^2 + 5(x^n) + 6 \\ &= (x^n + 3)(x^n + 2) \end{aligned}$$

## Substitution

You may prefer to handle some expressions by making a *substitution,* as in the following examples.

**This technique will be valuable when we solve *equations of quadratic type* in Chapter 13.**

---

**EXAMPLE 28:** Factor $x^{6n} + 2x^{3n} - 3$.

**Solution:** We see that one exponent is twice the other. Let us substitute a new letter for $x^{3n}$. Let

$$w = x^{3n}$$

Then

$$x^{6n} = (x^{3n})^2 = w^2$$

and our expression becomes

$$w^2 + 2w - 3$$

which factors into

$$(w + 3)(w - 1)$$

or, substituting back, we have

$$(x^{3n} + 3)(x^{3n} - 1)$$

---

**EXAMPLE 29:** Factor $(x + y)^2 + 7(x + y) - 30$.

**Solution:** We might be tempted to expand this expression, but notice that if we make the substitution

$$\text{let } w = x + y$$

we get

$$w^2 + 7w - 30$$

which factors into $(w + 10)(w - 3)$. Substituting back, we get the factors

$$(x + y + 10)(x + y - 3)$$

---

## EXERCISE 3—FACTORING TRINOMIALS

### Test for Factorability

Test each trinomial for factorability.

1. $3x^2 - 5x - 9$
2. $3a^2 + 7a + 2$
3. $2x^2 - 12x + 18$
4. $5z^2 - 2z - 1$
5. $x^2 - 30x - 64$
6. $y^2 + 2y - 7$

Factor completely.

### Trinomials with Leading Coefficient of 1

7. $x^2 - 10x + 21$
8. $x^2 - 15x + 56$
9. $x^2 - 10x + 9$
10. $x^2 + 13x + 30$

11. $x^2 + 7x - 30$
12. $x^2 + 3x + 2$
13. $x^2 + 7x + 12$
14. $c^2 + 9c + 18$
15. $x^2 - 4x - 21$
16. $x^2 - x - 56$
17. $x^2 + 6x + 8$
18. $x^2 + 12x + 32$
19. $b^2 - 8b + 15$
20. $b^2 + b - 12$
21. $b^2 - b - 12$

### Trinomials Having More Than One Variable

22. $x^2 - 13xy + 36y^2$
23. $x^2 + 19xy + 84y^2$
24. $a^2 - 9ab + 20b^2$
25. $a^2 + ab - 6b^2$
26. $a^2 + 3ab - 4b^2$
27. $a^2 + 2ax - 63x^2$
28. $x^2y^2 - 19xyz + 48z^2$
29. $a^2 - 20abc - 96b^2c^2$
30. $x^2 - 10xyz - 96y^2z^2$
31. $a^2 + 49abc + 48b^2c^2$

### Trinomials with Higher Powers or with Variables in the Exponents

32. $a^{6p} + 13a^{3p} + 12$
33. $(a + b)^2 - 7(a + b) - 8$
34. $w(x + y)^2 + 5w(x + y) + 6w$
35. $x^4 - 9x^2y^2 + 20y^4$
36. $a^6b^6 - 23a^3b^3c^2 + 132c^4$

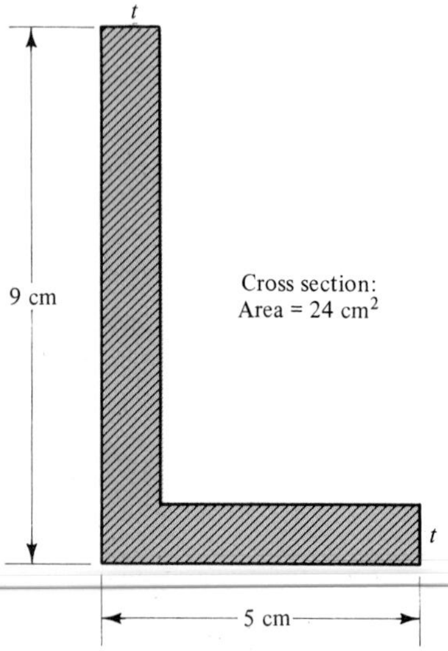


FIGURE 7-4

### Applications

37. To find the thickness $t$ of the angle iron in Fig. 7-4, it is necessary to solve the quadratic equation

$$t^2 - 14t + 24 = 0$$

Factor the left side of this equation. (*Extra:* Can you see where this equation comes from?)

38. To find the dimensions of a rectangular field having a perimeter of 70 ft and an area of 300 ft$^2$, we must solve the equation

$$x^2 - 35x + 300 = 0$$

Factor the left side of this equation.

39. To find the width $2m$ of a road that will give a sight distance of 1000 ft on a curve of radius 500 ft, we must solve the equation

$$m^2 - 1000m + 250{,}000 = 0$$

Factor the left side of this equation.

40. To find the width of the frame in Fig. 7-5, we must solve the equation

$$x^2 - 11x + 10 = 0$$

Factor the left side of this equation.

41. To find two resistors that will give an equivalent resistance of 400 Ω when wired in series and 75 Ω when wired in parallel, we must solve the equation

$$R^2 - 400R + 30{,}000 = 0$$

Factor the left side of this equation.

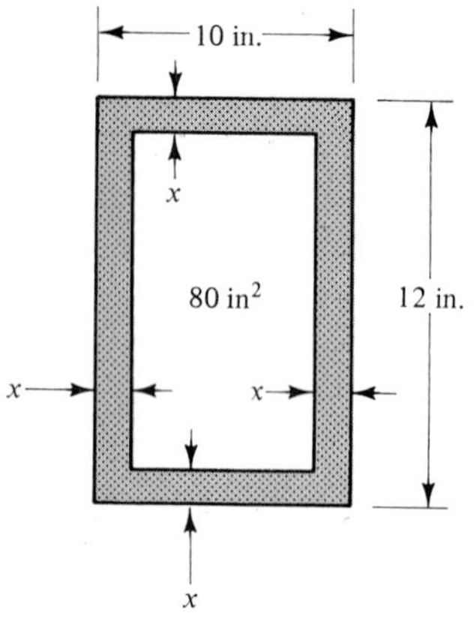


FIGURE 7-5

## 7-4 FACTORING BY GROUPING

When the expression to be factored has four or more terms, these terms can sometimes be arranged in smaller groups which can be factored separately.

---

**EXAMPLE 30:** Factor $ab + 4a + 3b + 12$.

Group the two terms containing the factor $a$, and the two containing the factor 3. Remove the common factor from each pair of terms.

**We'll use this same grouping idea later to factor certain trinomials.**

$$(ab + 4a) + (3b + 12) = a(b + 4) + 3(b + 4)$$

Both terms now have the common factor $(b + 4)$, which we factor out.

$$(b + 4)(a + 3)$$

---

**EXAMPLE 31:** Factor $x^2 - y^2 + 2x + 1$.

**Solution:** Taking our cue from Example 30, we try grouping the $x$ terms together,

$$(x^2 + 2x) + (-y^2 + 1)$$

and are no better off than before. We then notice that if the $+1$ term were grouped with the $x$ terms, we would get a trinomial that could be factored. Thus

$$(x^2 + 2x + 1) - y^2$$

$$(x + 1)(x + 1) - y^2$$

or

$$(x + 1)^2 - y^2$$

We now have the difference of two squares. Factoring again gives

$$(x + 1 + y)(x + 1 - y)$$

---

### EXERCISE 4—FACTORING BY GROUPING

Factor completely.

1. $a^3 + 3a^2 + 4a + 12$
2. $x^3 + x^2 + x + 1$
3. $x^3 - x^2 + x - 1$
4. $2x^3 - x^2 + 4x - 2$
5. $x^2 - bx + 3x - 3b$
6. $ab + a - b - 1$
7. $3x - 2y - 6 + xy$
8. $x^2y^2 - 3x^2 - 4y^2 + 12$
9. $x^2 + y^2 + 2xy - 4$
10. $x^2 - 6xy + 9y^2 - a^2$
11. $m^2 - n^2 - 4 + 4n$
12. $p^2 - r^2 - 6pq + 9q^2$

**Expressions with Three or More Letters**

13. $ax + bx + 3a + 3b$
14. $x^2 + xy + xz + yz$
15. $a^2 - ac + ab - bc$

16. $ay - by + ab - y^2$
17. $2a + bx^2 + 2b + ax^2$
18. $2xy + wy - wz - 2xz$
19. $6a^2 + 2ab - 3ac - bc$
20. $x^2 + bx + cx + bc$
21. $b^2 - bc + ab - ac$
22. $bx - cx + bc - x^2$
23. $x^2 - a^2 - 2ab - b^2$
24. $p^2 - y^2 - x^2 - 2xy$

## 7-5 GENERAL QUADRATIC TRINOMIAL

When we multiply the two binomials $(ax + b)$ and $(cx + d)$, we get a trinomial with a leading coefficient of $ac$, a middle coefficient of $(ad + bc)$, and a constant term of $bd$.

| | | |
|---|---|---|
| General Quadratic Trinomial | $(ax + b)(cx + d) = acx^2 + (ad + bc)x + bd$ | **46** |

The general quadratic trinomial may be factored by trial and error or by the grouping method.

### Trial and Error

To factor the general quadratic trinomial by trial and error, we look for four numbers, $a$, $b$, $c$, and $d$, such that

$$ac = \text{the leading coefficient}$$

$$ad + bc = \text{the middle coefficient}$$

$$bd = \text{the constant term}$$

Also, the *signs* in the factors are found in the same way as for trinomials with a leading coefficient of 1.

---

**EXAMPLE 32:** Factor $2x^2 + 5x + 3$.

**Solution:** The leading coefficient $ac = 2$ and the constant term $bd = 3$. Try

$$a = 1, \quad c = 2 \qquad \text{and} \qquad b = 3, \quad d = 1$$

Then $ad + bc = 1(1) + 3(2) = 7$. No good. It is supposed to be 5. We next try

$$a = 1, \quad c = 2, \quad b = 1, \quad \text{and} \quad d = 3$$

Then $ad + bc = 1(3) + 1(2) = 5$. This works. So

$$2x^2 + 5x + 3 = (x + 1)(2x + 3)$$

---

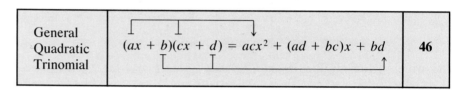

## Grouping Method

The grouping method eliminates the need for trial and error.

Some people have a knack for factoring and can quickly factor a trinomial by trial and error. Others rely on the longer but surer grouping method.

---

**EXAMPLE 33:** Factor $3x^2 - 16x - 12$.

**Solution:**

1. Multiply the leading coefficient and the constant term.

$$3(-12) = -36$$

2. Find two numbers whose product equals −36 and whose sum equals the middle coefficient, −16. Two such numbers are 2 and −18.
3. Rewrite the trinomial, splitting the middle term according to the selected factors ($-16x = 2x - 18x$)

$$3x^2 + 2x - 18x - 12$$

and group the first two terms together and the last two terms together.

$$(3x^2 + 2x) + (-18x - 12)$$

4. Remove common factors from each grouping.

$$x(3x + 2) - 6(3x + 2)$$

5. Remove the common factor $(3x + 2)$ from the entire expression:

$$(3x + 2)(x - 6)$$

which are the required factors.

| | |
|---|---|
| Common Error | It is easy to make a mistake when factoring out a negative quantity. Thus when we factored out a −6 in going from step 3 to step 4 above, we got $(-18x - 12) = -6(3x + 2)$ but not $-6(3x - 2)$ ← incorrect! |

Notice that if we had grouped the terms differently in step 3, we would have arrived at the same factors:

$$3x^2 - 18x + 2x - 12$$

$$3x(x - 6) + 2(x - 6)$$

$$(3x + 2)(x - 6)$$

as before.

---

Sometimes an expression may be simplified before factoring, as we did in Sec. 7-3.

**EXAMPLE 34:** Factor $12(x + y)^{6n} - (x + y)^{3n}z - 6z^2$.

**Solution:** If we substitute $a = (x + y)^{3n}$ our expression becomes

$$12a^2 - az - 6z^2$$

Temporarily dropping the $z$'s gives

$$12a^2 - a - 6$$

which factors into

$$(4a - 3)(3a + 2)$$

Replacing the $z$'s, we get

$$12a^2 - az - 6z^2 = (4a - 3z)(3a + 2z)$$

Finally, substituting back $(x + y)^{3n}$ for $a$, we have

$$12(x + y)^{6n} - (x + y)^{3n}z - 6z^2 = [4(x + y)^{3n} - 3z][3(x + y)^{3n} + 2z]$$

## EXERCISE 5—GENERAL QUADRATIC TRINOMIAL

Factor completely.

1. $4x^2 - 13x + 3$
2. $5a^2 - 8a + 3$
3. $5x^2 + 11x + 2$
4. $7x^2 + 23x + 6$
5. $12b^2 - b - 6$
6. $6x^2 - 7x + 2$
7. $2a^2 + a - 6$
8. $2x^2 + 3x - 2$
9. $5x^2 - 38x + 21$
10. $4x^2 + 7x - 15$
11. $3x^2 + 6x + 3$
12. $2x^2 + 11x + 12$
13. $3x^2 - x - 2$
14. $7x^2 + 123x - 54$
15. $4x^2 - 10x + 6$
16. $3x^2 + 11x - 20$
17. $4a^2 + 4a - 3$
18. $9x^2 - 27x + 18$
19. $9a^2 - 15a - 14$
20. $16c^2 - 48c + 35$

### Expressions Reducible to Quadratic Trinomials

Factor completely.

21. $49x^6 + 14x^3y - 15y^2$
22. $-18y^2 + 42x^2 - 24xy$
23. $4x^6 + 13x^3 + 3$
24. $5a^{2n} - 8a^n + 3$
25. $5x^{4n} + 11x^{2n} + 2$
26. $12(a + b)^2 - (a + b) - 6$
27. $3(a + x)^{2n} - 3(a + x)^n - 6$

### Applications

28. An object is thrown upward with an initial velocity of 32 ft/s from a building 128 ft above the ground. The height, $s$, of the object above the ground at any time $t$ is given by

$$s = 128 + 32t - 16t^2$$

Factor the right side of this equation.

29. An object is thrown into the air with an initial velocity of 82 ft/s. To find the time it takes for the object to reach a height of 45 ft, we must solve the quadratic equation

$$16t^2 - 82t + 45 = 0$$

Factor the left side of this equation.

30. To find the depth of cut $h$ needed to produce a flat of a certain width on a 1-in.-radius bar (Fig. 7-6) we must solve the equation

$$4h^2 - 8h + 3 = 0$$

Factor the left side of this equation.

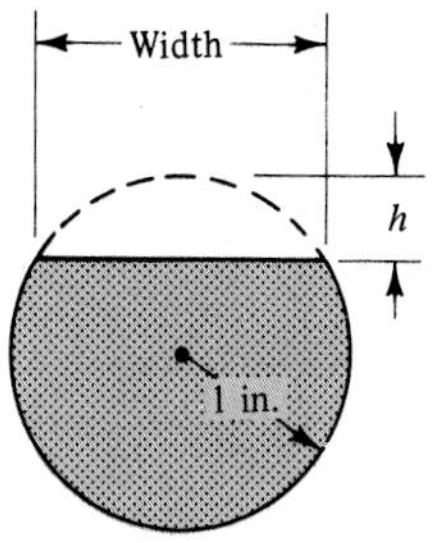


FIGURE 7-6

## 7-6 THE PERFECT SQUARE TRINOMIAL

### The Square of a Binomial

In Chapter 2 we saw that the expression obtained when a binomial is squared is called a *perfect square trinomial.*

---

**EXAMPLE 35:** Squaring the binomial $(2x + 3)$, we obtain

$$(2x + 3)^2 = 4x^2 + 6x + 6x + 9$$
$$= 4x^2 + 12x + 9$$

---

Note that in the perfect square trinomial obtained in Example 35, the first and last terms are the squares of the first and last terms of the binomial

$$(2x + 3)^2$$
square ↙ ↘ square
$$4x^2 + 12x + 9$$

and that the middle term is twice the product of the terms of the binomial.

$$(2x + 3)^2$$
product ↓
$$6x$$
twice the product
↓
$$4x^2 + 12x + 9$$

and that the constant term is always positive. In general,

| | | |
|---|---|---|
| Perfect Square Trinomials | $(a + b)^2 = a^2 + 2ab + b^2$ | **47** |
| | $(a - b)^2 = a^2 - 2ab + b^2$ | **48** |

**Any quadratic trinomial can be manipulated into the form of a perfect square trinomial by a procedure called *completing the square.* We will use that method in Sec. 13-2 to solve quadratic equations.**

### Factoring a Perfect Square Trinomial

We can factor a perfect square trinomial in the same way we factored the general quadratic trinomial in Sec. 7-5. However, the work is faster if we recognize that a trinomial is a perfect square. If it is, its factors will be the square of a binomial. The terms of that binomial are the square roots of the first and last terms of the trinomial. The sign in the binomial will be the same as the sign of the middle term of the trinomial.

**EXAMPLE 36:** Factor $a^2 - 4a + 4$.

**Solution:** The first and last terms are both perfect squares, and the middle term is twice the product of the square roots of the first and last terms. Thus the trinomial is a perfect square. Factoring, we obtain

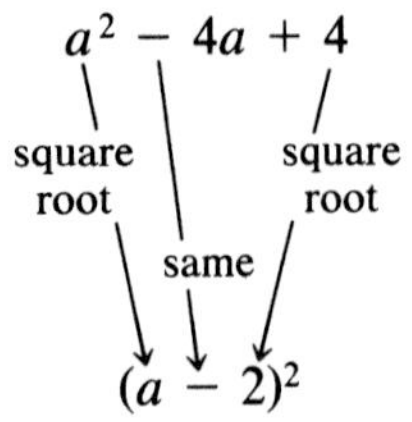


**EXAMPLE 37:**

$$9x^2y^2 - 6xy + 1 = (3xy - 1)^2$$

| | |
|---|---|
| Common Error | $(a + b)^2 \neq a^2 + b^2$ |

## EXERCISE 6—THE PERFECT SQUARE TRINOMIAL

Factor completely.

1. $x^2 + 4x + 4$
2. $x^2 - 30x + 225$
3. $y^2 - 2y + 1$
4. $x^2 + 2x + 1$
5. $2y^2 - 12y + 18$
6. $9 - 12a + 4a^2$
7. $9 + 6x + x^2$
8. $4y^2 - 4y + 1$
9. $9x^2 + 6x + 1$
10. $16x^2 + 16x + 4$
11. $9y^2 - 18y + 9$
12. $16n^2 - 8n + 1$
13. $16 + 16a + 4a^2$
14. $1 + 20a + 100a^2$
15. $49a^2 - 28a + 4$

### Perfect Square Trinomials with Two Variables

16. $a^2 - 2ab + b^2$
17. $x^2 + 2xy + y^2$
18. $a^2 - 14ab + 49b^2$
19. $a^2w^2 + 2abw + b^2$
20. $a^2 - 10ab + 25b^2$
21. $x^2 + 10ax + 25a^2$
22. $c^2 - 6cd + 9d^2$
23. $x^2 + 8xy + 16y^2$

### Expressions That Can Be Reduced to Perfect Square Trinomials

24. $1 + 2x^2 + x^4$
25. $z^6 + 16z^3 + 64$
26. $36 + 12a^2 + a^4$
27. $49 - 14x^3 + x^6$
28. $a^2b^2 - 8ab^3 + 16b^4$
29. $a^4 - 2a^2b^2 + b^4$
30. $16a^2b^2 - 8ab^2c^2 + b^2c^4$
31. $4a^{2n} + 12a^nb^n + 9b^{2n}$

## 7-7 SUM OR DIFFERENCE OF TWO CUBES

### Definition

An expression such as

$$x^3 + 27$$

is called the sum of two cubes ($x^3$ and $3^3$). In general, when we multiply the binomial $(a + b)$ and the trinomial $(a^2 - ab + b^2)$, we obtain

$$(a + b)(a^2 - ab + b^2) = a^3 - a^2b + ab^2 + a^2b - ab^2 + b^3$$
$$= a^3 + b^3$$

All but the cubed terms drop out, leaving the sum of two cubes. In general,

| | | |
|---|---|---|
| Sum of Two Cubes | $a^3 + b^3 = (a + b)(a^2 - ab + b^2)$ | **42** |

and similarly,

| | | |
|---|---|---|
| Difference of Two Cubes | $a^3 - b^3 = (a - b)(a^2 + ab + b^2)$ | **43** |

When we recognize that an expression is the sum (or difference) of two cubes, we can write the factors immediately.

---

**EXAMPLE 38:** Factor $x^3 + 27$.

**Solution:** This is the sum of two cubes, $x^3 + 3^3$. Substituting into Eq. 42, with $a = x$ and $b = 3$, yields

$$x^3 + 27 = x^3 + 3^3 = (x + 3)(x^2 - 3x + 9)$$

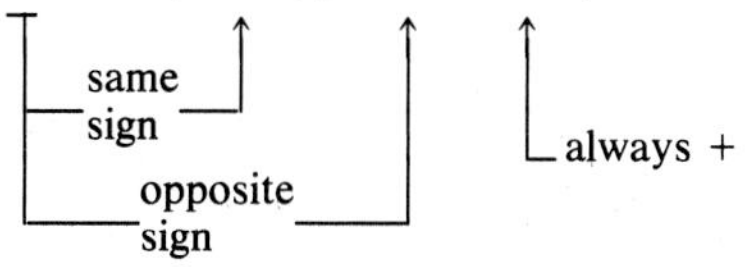


---

**EXAMPLE 39:** Factor $27x^3 - 8y^3$.

**Solution:** This is the difference of two cubes, $(3x)^3 - (2y)^3$. Factoring gives us

$$27x^3 - 8y^3 = (3x)^3 - (2y)^3 = (3x - 2y)(9x^2 + 6xy + 4y^2)$$

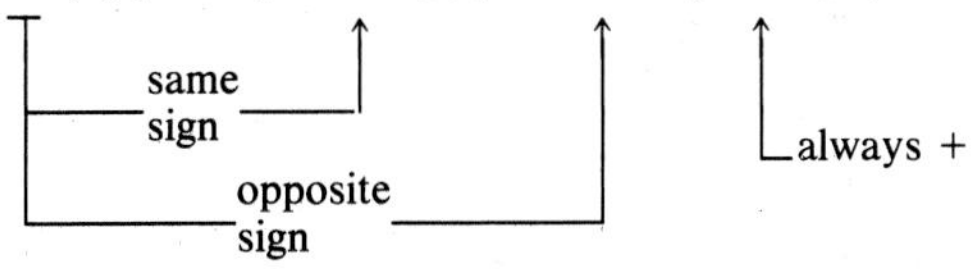


---

| | |
|---|---|
| Common Error | The middle term of the trinomials in Eqs. 42 and 43 is often mistaken as $2ab$. $$a^3 + b^3 \neq (a + b)(a^2 - 2ab + b^2)$$ ↑ no! |

When the powers are multiples of 3, we may be able to factor the expression as the sum or difference of two cubes.

**EXAMPLE 40:**

$$a^6 - b^9 = (a^2)^3 - (b^3)^3$$

Factoring, we obtain

$$= (a^2 - b^3)(a^4 + a^2b^3 + b^6)$$

## EXERCISE 7—SUM OR DIFFERENCE OF TWO CUBES

Factor completely.

1. $64 + x^3$
2. $1 - 64y^3$
3. $2a^3 - 16$
4. $a^3 - 27$
5. $x^3 - 1$
6. $x^3 - 64$
7. $x^3 + 1$
8. $a^3 - 343$
9. $a^3 + 64$
10. $x^3 + 343$
11. $x^3 + 125$
12. $64a^3 + 27$
13. $216 - 8a^3$
14. $343 - 27y^3$
15. $343 + 64x^3$

### More Difficult Types

16. $27x^9 + 512$
17. $y^9 + 64x^3$
18. $64a^{12} + x^{15}$
19. $27x^{15} + 8a^6$
20. $8x^{6p} + 8a^6$
21. $8a^{6x} - 125b^{3x}$
22. $64x^{12a} - 27y^{15a}$
23. $64x^{3n} - y^{9n}$
24. $x^3y^3 - z^3$
25. $a^3b^3 + 27x^3$
26. $a^3b^3 - 27c^3$
27. $x^3y^3z^3 - 8$
28. $(a + b)^3 - 8$
29. $27x^3 + (y - z)^3$
30. $(a - b)^3 - (c + d)^3$

### Applications

31. The volume of a hollow spherical shell having an inside radius of $r_1$ and an outside radius $r_2$ is $\frac{4}{3}\pi r_2^3 - \frac{4}{3}\pi r_1^3$. Factor this expression completely.
32. A cistern is in the shape of a hollow cube whose inside dimension is $s$ and whose outside dimension is $S$. If made of concrete of density $d$, its weight is $dS^3 - ds^3$. Factor completely.

## CHAPTER 7 REVIEW PROBLEMS

*Factor* completely.

1. $x^2 - 2x - 15$
2. $2a^2 + 3a - 2$
3. $x^6 - y^4$
4. $2ax^2 + 8ax + 8a$
5. $2x^2 + 3x - 2$
6. $a^2 + ab - 6b^2$
7. $(a - b)^3 + (c + d)^3$
8. $8x^3 - \frac{y^3}{27}$
9. $\frac{x^2}{y} - \frac{x}{y}$
10. $2ax^2y^2 - 18a$
11. $xy - 2y + 5x - 10$
12. $3a^2 - 2a - 8$
13. $x - bx - y + by$
14. $\frac{2a^2}{12} - \frac{8b^2}{27}$
15. $(y + 2)^2 - z^2$
16. $2x^2 - 20ax + 50a^2$
17. $x^2 - 7x + 12$
18. $4a^2 - (3a - 1)^2$
19. $16x^{4n} - 81y^{8n}$
20. $xy - y^2 + xz - yz$
21. $9a^2 + 12az^2 + 4z^4$
22. $4a^6 - 4b^6$
23. $x^{2m} + 2x^m + 1$
24. $4x^{6m} + 4x^{3m}y^m + y^{2m}$
25. $(x - y)^2 - z^2$
26. $x^2 - 2x - 3$
27. $1 - 16x^2$
28. $a^2 - 2a - 8$
29. $9x^4 - x^2$
30. $x^2 - 21x + 110$
31. $(p - q)^3 - 27$
32. $5a^2 - 20x^2$
33. $27a^3 - 8w^3$
34. $3x^2 - 6x - 45$
35. $16x^2 - 16xy + 4y^2$
36. $64m^3 - 27n^3$
37. $6ab + 2ay + 3bx + xy$
38. $15a^2 - 11a - 12$
39. $2y^4 - 18$
40. $ax - bx + ay - by$
41. The reduction in power in a resistance $R$ by lowering the voltage across the resistor from $V_2$ to $V_1$ is

$$\frac{V_2^2}{R} - \frac{V_1^2}{R}$$

Factor this expression.

42. A conical tank of height $h$ is filled to a depth $d$. The volume of liquid that can still be put into the tank is

$$V = \frac{\pi}{3} r_1^2 h - \frac{\pi}{3} r_2^2 d$$

where $r_1$ is the base radius of the tank and $r_2$ is the base radius of the liquid. Factor the right side of this equation.

*Writing*

43. We have studied the factoring of seven different types of expressions in this chapter. List them and give an example of each. State in words how to recognize each and to tell one from the other. Also list at least four other expressions that are *not* one of the given seven, and state why each is different from those we have studied.

# 8

# FRACTIONS AND FRACTIONAL EQUATIONS

## OBJECTIVES

**When you have completed this chapter, you should be able to:**

- Determine values of the variables that result in undefined fractions.
- Convert common fractions to decimals, and vice versa.
- Reduce algebraic fractions.
- Add, subtract, and divide algebraic fractions.
- Simplify complex fractions.
- Solve fractional equations.
- Solve applied problems requiring fractional equations.
- Solve literal equations and formulas.

Although the calculator, the computer, and the metric system (all of which use decimal notation) have somewhat reduced the importance of common fractions, they are still widely used. Algebraic fractions are, of course, as important as ever. We must be able to handle them in order to solve fractional equations and the word problems from which they arise.

This material is not all new to us. Some was covered in Chapter 2 and simple fractional equations were solved in Chapter 3. We also make heavy use of factoring in order to simplify algebraic fractions.

## 8-1 SIMPLIFICATION OF FRACTIONS

### Parts of a Fraction

A fraction has a *numerator, denominator,* and a *fraction line.*

$$\text{fraction line} \rightarrow \frac{a}{b} \begin{array}{l} \longleftarrow \text{numerator} \\ \longleftarrow \text{denominator} \end{array}$$

### Quotient

A fraction is a way of indicating a *quotient* of two quantities. The fraction $a/b$ can be read "$a$ divided by $b$."

### Ratio

The quotient of two numbers or quantities is also spoken of as the *ratio* of those quantities. Thus the ratio of $x$ to $y$ is $x/y$.

### Division by Zero

Since division by zero is not permitted, it should be understood in our work with fractions that *the denominator cannot be zero.*

---

**EXAMPLE 1:** What values of $x$ are not permitted in the fraction

$$\frac{3x}{x^2 + x - 6} \quad ?$$

**Solution:** Factoring the denominator, we get

$$\frac{3x}{x^2 + x - 6} = \frac{3x}{(x - 2)(x + 3)}$$

We see that an $x$ equal to 2 or to $-3$ will make $(x - 2)$ or $(x + 3)$ equal to zero. This will result in division by zero, so is not permitted.

---

### Common Fraction

A fraction whose numerator and denominator are both integers is called a *common fraction.*

---

**EXAMPLE 2:**

$$\frac{2}{3}, \quad -\frac{9}{5}, \quad \frac{-124}{125}, \quad \text{and} \quad \frac{18}{-11}$$

are common fractions.

---

## Algebraic Fractions

An *algebraic fraction* is one whose numerator and/or denominator contain *literal* quantities.

**EXAMPLE 3:**

$$\frac{x}{y}, \quad \frac{\sqrt{x+2}}{x}, \quad \frac{3}{y}, \quad \text{and} \quad \frac{x^2}{x-3}$$

are algebraic fractions.

## Rational Algebraic Fractions

Recall that a polynomial is an expression in which the exponents are nonnegative integers.

An algebraic fraction is called *rational* if the numerator and denominator are both *polynomials*.

**EXAMPLE 4:**

$$\frac{x}{y}, \quad \frac{3}{w^3}, \quad \text{and} \quad \frac{x^2}{x-3}$$

are rational fractions but $\sqrt{x+2}/x$ is not.

## Proper and Improper Fractions

A *proper* common fraction is one whose numerator is smaller than its denominator.

**EXAMPLE 5:** $\frac{3}{5}$, $\frac{1}{3}$, and $\frac{9}{11}$ are proper fractions, whereas $\frac{8}{5}$, $\frac{3}{2}$, and $\frac{7}{4}$ are *improper* fractions.

A proper *algebraic* fraction is a rational fraction whose numerator is of *lower degree* than the denominator.

**EXAMPLE 6:**

$$\frac{x}{x^2+2} \quad \text{and} \quad \frac{x^2+2x-3}{x^3+9}$$

are proper fractions, while

$$\frac{x^3-2}{x^2+x-3} \quad \text{and} \quad \frac{x^2}{y}$$

are improper fractions.

## Mixed Form

A *mixed number* is the sum of an integer and a fraction.

**EXAMPLE 7:**

$$2\frac{1}{2}, \quad 5\frac{3}{4}, \quad \text{and} \quad 3\frac{1}{3}$$

are mixed numbers.

A *mixed expression* is the sum or difference of a polynomial and a rational algebraic fraction.

**EXAMPLE 8:**

$$3x - 2 + \frac{1}{x} \quad \text{and} \quad y - \frac{y}{y^2 + 1}$$

are mixed expressions.

### Decimals and Fractions

To change a fraction to an equivalent decimal, simply divide the numerator by the denominator.

**EXAMPLE 9:** To write $\frac{9}{11}$ as a decimal, we divide 9 by 11:

$$\frac{9}{11} = 0.8181818\overline{181}\ldots$$

Repeating decimals are often written with a bar over the repeating part: $0.\overline{81}$.

We get a *repeating decimal;* the dots following the number indicate that the digits continue indefinitely.

To change a decimal number to a fraction, write a fraction with the decimal number in the numerator and 1 in the denominator. Multiply numerator and denominator by a multiple of 10 that will make the numerator a whole number. Finally, reduce to lowest terms.

**EXAMPLE 10:** Express 0.875 as a fraction.

**Solution:**

$$0.875 = \frac{0.875}{1} = \frac{875}{1000} = \frac{7}{8}$$

To express a *repeating* decimal as a fraction, follow the steps in the next example.

**EXAMPLE 11:** Change the repeating decimal $0.81\overline{81}$ to a fraction.

**Solution:** Let $x = 0.81\overline{81}$. Multiplying by 100, we have

$$100x = 81.\overline{81}$$

Subtracting the first equation from the second gives us

$$99x = 81 \quad \text{(exactly)}$$

Dividing by 99 yields

$$x = \frac{81}{99} = \frac{9}{11} \quad \text{(reduced)}$$

### Reducing a Fraction to Lowest Terms

Divide both numerator and denominator by any factor that is contained in both.

| | |
|---|---|
| $\frac{ad}{bd} = \frac{a}{b}$ | **50** |

**EXAMPLE 12:** Reduce to lowest terms. Write the answer without negative exponents.

(a) $\dfrac{9}{12} = \dfrac{3(3)}{4(3)} = \dfrac{3}{4}$

(b) $\dfrac{3x^2yz}{9xy^2z^3} = \dfrac{3}{9} \cdot \dfrac{x^2}{x} \cdot \dfrac{y}{y^2} \cdot \dfrac{z}{z^3} = \dfrac{x}{3yz^2}$

When possible, factor the numerator and denominator. Then divide both numerator and denominator by any factors common to both.

**EXAMPLE 13:**

(a) $\dfrac{2x^2 + x}{3x} = \dfrac{(2x + 1)x}{3(x)} = \dfrac{2x + 1}{3}$

(b) $\dfrac{ab + bc}{bc + bd} = \dfrac{b(a + c)}{b(c + d)} = \dfrac{a + c}{c + d}$

(c) $\dfrac{2x^2 - 5x - 3}{4x^2 - 1} = \dfrac{(2x + 1)(x - 3)}{(2x + 1)(2x - 1)} = \dfrac{x - 3}{2x - 1}$

(d) $\dfrac{x^2 - ax + 2bx - 2ab}{2x^2 + ax - 3a^2} = \dfrac{x(x - a) + 2b(x - a)}{(x - a)(2x + 3a)}$

$= \dfrac{(x - a)(x + 2b)}{(x - a)(2x + 3a)} = \dfrac{x + 2b}{2x + 3a}$

The process of striking out the same factors from numerator and denominator is called *canceling*.

**Most students love canceling because they think they can cross out any term that stands in their way. If you use canceling, use it carefully!**

| | |
|---|---|
| Common Errors | If a factor is missing from *even one term* in the numerator or denominator, that factor *cannot* be canceled. $$\frac{xy - z}{wx} \neq \frac{y - z}{w}$$ |
| | We may divide (or multiply) the numerator and denominator by the same quantity (Eq. 50), but we *may not add or subtract* the same quantity in the numerator and denominator, as this will change the value of the fraction. For example, $$\frac{3}{5} \neq \frac{3 + 1}{5 + 1} = \frac{4}{6} = \frac{2}{3}$$ |

## Simplifying Fractions by Changing Signs

Recall from Chapter 2 that any two of the three signs of a fraction may be changed without changing the value of a fraction.

| | | |
|---|---|---|
| Rules of Signs | $\frac{+a}{+b} = \frac{-a}{-b} = -\frac{-a}{+b} = -\frac{+a}{-b} = \frac{a}{b}$ | **10** |
| | $\frac{+a}{-b} = \frac{-a}{+b} = -\frac{-a}{-b} = -\frac{a}{b}$ | **11** |

We can sometimes use this idea to simplify a fraction.

**EXAMPLE 14:** Simplify the fraction

$$-\frac{3x-2}{2-3x}$$

**Solution:** We change the sign of the denominator *and* the sign of the entire fraction:

$$-\frac{3x-2}{2-3x} = +\frac{3x-2}{-(2-3x)} = \frac{3x-2}{-2+3x} = \frac{3x-2}{3x-2} = 1$$

changed

## EXERCISE 1—SIMPLIFICATION OF FRACTIONS

In each fraction, what values of $x$, if any, are not permitted?

1. $\frac{12}{x}$
2. $\frac{x}{12}$
3. $\frac{18}{x-5}$
4. $\frac{5x}{x^2-49}$
5. $\frac{7}{x^2-3x+2}$
6. $\frac{3x}{8x^2-14x+3}$

*Hint:* **Factor the denominators in Problems 4, 5, and 6.**

Change each fraction to a decimal. Work to four decimal places.

7. $\frac{7}{12}$
8. $\frac{5}{9}$
9. $\frac{15}{16}$
10. $\frac{125}{155}$
11. $\frac{11}{3}$
12. $\frac{25}{9}$

Change each decimal to a fraction.

13. 0.4375
14. 0.390625
15. 0.6875
16. 0.28125
17. 0.7777 . . .
18. 0.636363 . . .

Simplify each fraction by manipulating the algebraic signs.

19. $\frac{a-b}{b-a}$
20. $-\frac{2x-y}{y-2x}$
21. $\frac{(a-b)(c-d)}{b-a}$
22. $\frac{w(x-y-z)}{y-x+z}$

Reduce to lowest terms. Write your answers without negative exponents.

23. $\frac{2ab}{6b}$
24. $\frac{12m^2n}{15mn^2}$

25. $\dfrac{21m^2p^2}{28mp^4}$

26. $\dfrac{abx - bx^2}{acx - cx^2}$

27. $\dfrac{4a^2 - 9b^2}{4a^2 + 6ab}$

28. $\dfrac{3a^2 + 6a}{a^2 + 4a + 4}$

29. $\dfrac{x^2 + 5x}{x^2 + 4x - 5}$

30. $\dfrac{xy - 3y^2}{x^3 - 27y^3}$

31. $\dfrac{x^2 - 4}{x^3 - 8}$

32. $\dfrac{2a^3 + 6a^2 - 8a}{2a^3 + 2a^2 - 4a}$

33. $\dfrac{2m^3n - 2m^2n - 24mn}{6m^3 + 6m^2 - 36m}$

34. $\dfrac{9x^3 - 30x^2 + 25x}{3x^4 - 11x^3 + 10x^2}$

35. $\dfrac{2a^2 - 2}{a^2 - 2a + 1}$

36. $\dfrac{3a^2 - 4ab + b^2}{a^2 - ab}$

37. $\dfrac{x^2 - z^2}{x^3 - z^3}$

38. $\dfrac{2x^2}{6x - 4x^2}$

39. $\dfrac{2a^2 - 8}{2a^2 - 2a - 12}$

40. $\dfrac{2a^2 + ab - 3b^2}{a^2 - ab}$

41. $\dfrac{x^2 - 1}{2xy + 2y}$

42. $\dfrac{x^3 - a^2x}{x^2 - 2ax + a^2}$

43. $\dfrac{2x^4y^4 + 2}{3x^8y^8 - 3}$

44. $\dfrac{18a^2c - 6bc}{42a^2d - 14bd}$

45. $\dfrac{(x + y)^2}{x^2 - y^2}$

46. $\dfrac{mw + 3w - mz - 3z}{m^2 - m - 12}$

47. $\dfrac{b^2 - a^2 - 6b + 9}{5b - 15 - 5a}$

48. $\dfrac{3w - 3y - 3}{w^2 + y^2 - 2wy - 1}$

## 8-2 MULTIPLICATION AND DIVISION OF FRACTIONS

### Multiplication

We multiply a fraction $a/b$ by another fraction $c/d$ as follows:

| | | |
|---|---|---|
| Multiplying Fractions | $\dfrac{a}{b} \cdot \dfrac{c}{d} = \dfrac{ac}{bd}$ | **51** |

*The product of two or more fractions is a fraction whose numerator is the product of the numerators of the original fractions, and whose denominator is the product of the denominators of the original fractions.*

**EXAMPLE 15:**

(a) $\dfrac{2}{3} \cdot \dfrac{5}{7} = \dfrac{2(5)}{3(7)} = \dfrac{10}{21}$

(b) $\dfrac{1}{2} \cdot \dfrac{2}{3} \cdot \dfrac{3}{5} = \dfrac{1(2)(3)}{2(3)(5)} = \dfrac{1}{5}$

(c) $5\dfrac{2}{3} \cdot 3\dfrac{1}{2} = \dfrac{17}{3} \cdot \dfrac{7}{2} = \dfrac{119}{6}$

| Common Error | When multiplying mixed numbers, students sometimes try to multiply the whole parts and the fractional parts separately. $$2\frac{2}{3} \times 4\frac{3}{5} \neq 8\frac{6}{15}$$ The correct way is to write each mixed number as an improper fraction and to multiply as shown. $$2\frac{2}{3} \times 4\frac{3}{5} = \frac{8}{3} \times \frac{23}{5} = \frac{184}{15} = 12\frac{4}{15}$$ |
|---|---|

**EXAMPLE 16:**

(a) $\dfrac{2a}{3b} \cdot \dfrac{5c}{4a} = \dfrac{10ac}{12ab} = \dfrac{5c}{6b}$

(b) $$\frac{x}{x+2} \cdot \frac{x^2-4}{x^3} = \frac{x(x^2-4)}{(x+2)x^3}$$
$$= \frac{x(x+2)(x-2)}{(x+2)x^3} = \frac{x-2}{x^2}$$

Leave your product in factored form until *after* you simplify.

(c) $$\frac{x^2+x-2}{x^2-4x+3} \cdot \frac{2x^2-3x-9}{2x^2+7x+6}$$
$$= \frac{(x-1)(x+2)(x-3)(2x+3)}{(x-1)(x-3)(x+2)(2x+3)} = 1$$

## Division

To divide one fraction, $a/b$, by another fraction, $c/d$:

$$\frac{\frac{a}{b}}{\frac{c}{d}}$$

we multiply numerator and denominator by $d/c$:

Can you say *why* it is permissible to multiply numerator and denominator by $d/c$?

$$\frac{\frac{a}{b} \cdot \frac{d}{c}}{\frac{c}{d} \cdot \frac{d}{c}} = \frac{\frac{a}{b} \cdot \frac{d}{c}}{\frac{cd}{cd}} = \frac{\frac{a}{b} \cdot \frac{d}{c}}{1} = \frac{a}{b} \cdot \frac{d}{c} = \frac{ad}{bc}$$

We see that dividing by a fraction is the same as *multiplying by the reciprocal* of that fraction.

| Division of Fractions | $\dfrac{a}{b} \div \dfrac{c}{d} = \dfrac{a}{b} \cdot \dfrac{d}{c} = \dfrac{ad}{bc}$ | **52** |
|---|---|---|

*When dividing fractions, invert the divisor and multiply.*

**EXAMPLE 17:**

(a) $\frac{2}{3} \div \frac{5}{7} = \frac{2}{3} \cdot \frac{7}{5} = \frac{14}{15}$

(b) $3\frac{2}{5} \div 2\frac{4}{15} = \frac{17}{5} \div \frac{34}{15}$

$$= \frac{17}{5} \times \frac{15}{34} = \frac{3}{2}$$

(c) $\frac{x}{y} \div \frac{x+2}{y-1} = \frac{x}{y} \cdot \frac{y-1}{x+2}$

$$= \frac{x(y-1)}{y(x+2)}$$

Remember that all literal fractions have nonzero denominators, since division by zero is not permitted.

(d) $\frac{x^2+x-2}{x} \div \frac{x+2}{x^2} = \frac{x^2+x-2}{x} \cdot \frac{x^2}{x+2}$

$$= \frac{(x+2)(x-1)x^2}{x(x+2)} = x(x-1)$$

(e) $x \div \frac{\pi r^2 x}{4} = \frac{x}{1} \div \frac{\pi r^2 x}{4} = \frac{x}{1} \cdot \frac{4}{\pi r^2 x} = \frac{4}{\pi r^2}$

| Common Error | We invert the *divisor* and multiply. Be sure not to invert the dividend. |
|---|---|

## Changing Improper Fractions to Mixed Form

To write an improper fraction in mixed form, divide the numerator by the denominator and express the remainder as the numerator of a fraction whose denominator is the original denominator.

**EXAMPLE 18:** Write $\frac{45}{7}$ as a mixed number.

**Solution:** Dividing 45 by 7, we get 6 with a remainder of 3, so

$$\frac{45}{7} = 6\frac{3}{7}$$

The procedure is the same for changing an *algebraic* fraction to mixed form. Divide the numerator by the denominator.

**EXAMPLE 19:**

$$\frac{x^2+3}{x} = \frac{x^2}{x} + \frac{3}{x}$$

$$= x + \frac{3}{x}$$

## EXERCISE 2—MULTIPLICATION AND DIVISION OF FRACTIONS

Multiply and reduce.

1. $\frac{1}{3} \times \frac{2}{5}$
2. $\frac{3}{7} \times \frac{21}{24}$
3. $\frac{2}{3} \times \frac{9}{7}$
4. $\frac{11}{3} \times 7$
5. $\frac{2}{3} \times 3\frac{1}{5}$
6. $3 \times 7\frac{2}{5}$
7. $3\frac{3}{4} \times 2\frac{1}{2}$
8. $\frac{3}{5} \times \frac{2}{7} \times \frac{5}{9}$
9. $3 \times \frac{5}{8} \times \frac{4}{5}$
10. $\frac{15a^2}{7b^2} \cdot \frac{28ab}{9a^3c}$
11. $\frac{a^4b^4}{2a^2y^n} \cdot \frac{a^2x}{xy^n}$
12. $\frac{3x^2y^2z^3}{4a^2b^2c^2} \cdot \frac{8a^3b^2c^2}{9x^2yz^3}$
13. $\frac{5m^2n^2p^4}{3x^2yz^3} \cdot \frac{21xyz^2}{20m^2n^2p^2}$
14. $\frac{x + y}{x - y} \cdot \frac{x^2 - y^2}{(x + y)^2}$
15. $\frac{x^2 - a^2}{xy} \cdot \frac{xy}{x + a}$
16. $\frac{x^4 - y^4}{a^2x^2} \cdot \frac{ax^3}{x^2 + y^2}$
17. $\frac{a}{x - y} \cdot \frac{b}{x + y}$
18. $\frac{x + y}{10} \cdot \frac{ax}{3(x + y)}$
19. $\frac{c}{x^2 - y^2} \cdot \frac{d}{x^2 - y^2}$
20. $\frac{x - 2}{2x + 3} \cdot \frac{2x^2 + x - 3}{x^2 - 4}$
21. $\frac{x^2 - 1}{x^2 + x - 6} \cdot \frac{x^2 + 2x - 8}{x^2 - 4x - 5}$
22. $\frac{5x - 1}{x^2 - x - 2} \cdot \frac{x - 2}{10x^2 + 13x - 3}$
23. $\frac{2x^2 + x - 1}{3x^2 - 11x - 4} \cdot \frac{3x^2 + 7x + 2}{2x^2 - 7x + 3}$
24. $\frac{x^3 + 1}{x + 1} \cdot \frac{x + 3}{x^2 - x + 1}$
25. $\frac{2x^2 + x - 6}{4x^2 - 3x - 1} \cdot \frac{4x^2 + 5x + 1}{x^2 - 4x - 12}$

Divide and reduce.

26. $\frac{7}{9} \div \frac{5}{3}$
27. $3\frac{7}{8} \div 2$
28. $\frac{7}{8} \div 4$
29. $\frac{9}{16} \div 8$
30. $24 \div \frac{5}{8}$
31. $\frac{5}{8} \div 2\frac{1}{4}$
32. $2\frac{7}{8} \div 1\frac{1}{2}$
33. $2\frac{5}{8} \div \frac{1}{2}$
34. $50 \div 2\frac{3}{5}$
35. $\frac{5abc^3}{3x^2} \div \frac{10ac^3}{6bx^2}$
36. $\frac{7x^2y}{3ad} \div \frac{2xy^2}{3a^2d}$
37. $\frac{4a^3x}{6dy^2} \div \frac{2a^2x^2}{8a^2y}$
38. $\frac{5x^2y^3z}{6a^2b^2c} \div \frac{10xy^3z^2}{8ab^2c^2}$
39. $\frac{3an + cm}{x^2 - y^2} \div (x^2 + y^2)$
40. $\frac{a^2 + 4a + 4}{d + c} \div (a + 2)$
41. $\frac{ac + ad + bc + bd}{c^2 - d^2} \div (a + b)$
42. $\frac{5(x + y)^2}{x - y} \div (x + y)$
43. $\frac{1}{x^2 + 17x + 30} \div \frac{1}{x + 15}$
44. $\frac{5xy}{a - x} \div \frac{10xy}{a^2 - x^2}$

45. $\dfrac{3x - 6}{2x + 3} \div \dfrac{x^2 - 4}{4x^2 + 2x - 6}$

46. $\dfrac{a^2 - a - 2}{5a - 1} \div \dfrac{a - 2}{10a^2 + 13a - 3}$

47. $\dfrac{2p^2 + p - 6}{4p^2 - 3p - 1} \div \dfrac{2p^2 - 5p + 3}{4p^2 + 9p + 2}$

48. $\dfrac{x^3 + 1}{x + 1} \div \dfrac{x^2 - x + 1}{3x + 9}$

49. $\dfrac{z^2 - 1}{z^2 + z - 6} \div \dfrac{z^2 - 4z - 5}{z^2 + 2z - 8}$

50. $\dfrac{x^2 + 5x + 6}{2x^2 - 3x + 1} \div \dfrac{x^2 - 9}{2x^2 - 7x + 3}$

### Improper Fractions and Mixed Form

Change each improper fraction to a mixed number.

51. $\dfrac{5}{3}$ 52. $\dfrac{11}{5}$ 53. $\dfrac{29}{12}$

54. $\dfrac{17}{3}$ 55. $\dfrac{47}{5}$ 56. $\dfrac{125}{12}$

Change each improper algebraic fraction to a mixed expression.

57. $\dfrac{x^2 + 1}{x}$ 58. $\dfrac{1 - 4x}{2x}$ 59. $\dfrac{4x - x^2}{2x^2}$

### Applications

60. For a thin lens, the relationship between the focal length $f$, the object distance $p$, and the image distance $q$ is

$$f = \frac{pq}{p + q}$$

A second lens of focal length $f_1$ has the same object distance $p$ but a different image distance $q_1$:

$$f_1 = \frac{pq_1}{p + q_1}$$

Find the ratio $f/f_1$ and simplify.

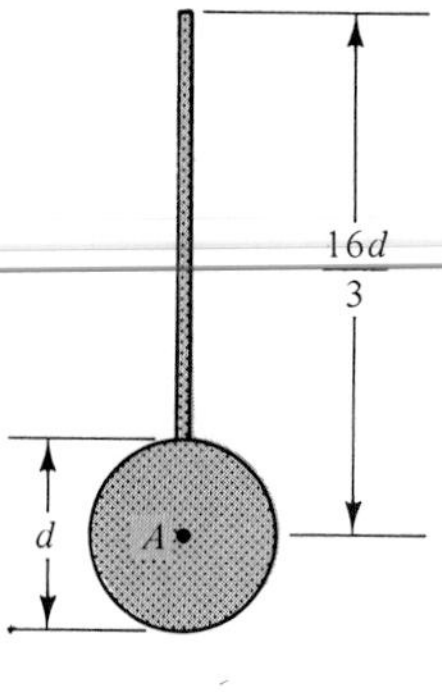

FIGURE 8-1

61. The mass density of an object is its mass divided by its volume. Write and simplify an expression for the density of a sphere having a mass $m$ and a volume equal to $4\pi r^3/3$.
62. The pressure on a surface is equal to the total force divided by the area. Write and simplify an expression for the pressure on a circular surface of area $\pi d^2/4$ subjected to a distributed load $F$.
63. The stress on a bar in tension is equal to the load divided by the cross-sectional area. Write and simplify an expression for the stress in a bar having a trapezoidal cross section of area $(a + b)h/2$, subject to a load $P$.
64. The acceleration on a body is equal to the force on the body divided by its mass, and the mass equals the volume of the object times the density. Write and simplify an expression for the acceleration of a sphere having a volume $4\pi r^3/3$ and a density $D$, subjected to a force $F$.
65. To find the moment of the area $A$ in Fig. 8-1, we must multiply the area $\pi d^2/4$ by the distance to the pivot, $16d/3$. Multiply and simplify.
66. The circle in Fig. 8-2 has an area $\pi d^2/4$, and the sector has an area equal to $ds/4$. Find the ratio of the area of the circle to the area of the sector.

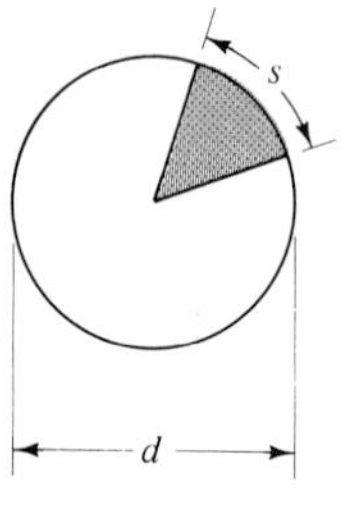

FIGURE 8-2

## 8-3 ADDITION AND SUBTRACTION OF FRACTIONS

### Similar Fractions

*Similar fractions* (also called *like* fractions) are those having the same (common) denominator. To add or subtract similar fractions, *combine the numerators and place them over the common denominator.*

| Addition and Subtraction of Fractions | $\frac{a}{b} \pm \frac{c}{b} = \frac{a \pm c}{b}$ | **53** |
|---|---|---|

**EXAMPLE 20:**

(a) $\frac{2}{3} + \frac{5}{3} = \frac{2+5}{3} = \frac{7}{3}$

(b) $\frac{1}{x} + \frac{3}{x} = \frac{1+3}{x} = \frac{4}{x}$

(c) $\frac{3x}{x+1} - \frac{5}{x+1} + \frac{x^2}{x+1} = \frac{3x - 5 + x^2}{x+1}$

### Least Common Denominator

The *least common denominator*, LCD (also called the *lowest* common denominator) is the smallest expression that is exactly divisible by each of the denominators. Thus the LCD must contain all the prime factors of each of the denominators. The common denominator of two or more fractions is simply the product of the denominators of those fractions. To find the *least* common denominator, drop any prime factor from one denominator that also appears in another denominator before multiplying.

**The least common denominator is also called the *least common multiple* (LCM) of the denominators.**

**EXAMPLE 21:** Find the LCD for the two fractions $\frac{3}{8}$ and $\frac{1}{18}$.

**Solution:** Factoring each denominator, we obtain

$$\begin{array}{cc} 8 & 18 \\ (2)(2)(2) & (2)(3)(3) \end{array}$$

↖ duplicates; include only once in LCD ↗

Our LCD is then the product of these factors, dropping one of the 2's that appears in both sets of factors.

$$\text{LCD} = (2)(2)(2)(3)(3) = 8(9) = 72$$

For this simple problem you probably found the LCD by inspection in less time than it took to read this example, and are probably wondering what all the fuss is about. What we are really doing is developing a *method* that we can use on algebraic fractions, where the LCD will *not* be obvious.

**EXAMPLE 22:** Find the LCD for the fractions

$$\frac{5}{x^2 + x}, \quad \frac{x}{x^2 - 1}, \quad \text{and} \quad \frac{9}{x^3 - x^2}$$

**Solution:** The denominator $x^2 + x$ has the factors $x$ and $x + 1$; the denominator $x^2 - 1$ has the factors $x + 1$ and $x - 1$; and the denominator $x^3 - x^2$ has the factors $x$, $x$, and $x - 1$. Our LCD is thus

$$(x)(x)(x + 1)(x - 1) \quad \text{or} \quad x^2(x^2 - 1)$$

### Combining Fractions with Different Denominators

Find the LCD. Then multiply numerator and denominator of each fraction by that quantity that will make the denominator equal to the LCD. Finally, combine as shown above, and simplify.

**EXAMPLE 23:** Add $\frac{1}{2}$ and $\frac{2}{3}$.

**Solution:** The LCD is 6, so

$$\frac{1}{2} + \frac{2}{3} = \frac{1}{2}\left(\frac{3}{3}\right) + \frac{2}{3}\left(\frac{2}{2}\right)$$

$$= \frac{3}{6} + \frac{4}{6}$$

$$= \frac{7}{6}$$

**EXAMPLE 24:** Add $\frac{3}{8}$ and $\frac{1}{18}$.

**Solution:** The LCD, from Example 21, is 72. So

$$\frac{3}{8} + \frac{1}{18} = \frac{3}{8} \cdot \frac{9}{9} + \frac{1}{18} \cdot \frac{4}{4}$$

$$= \frac{27}{72} + \frac{4}{72}$$

$$= \frac{31}{72}$$

**It is not necessary to write the fractions with the *least* common denominator; any common denominator will work as well. But your final result will then have to be reduced to lowest terms.**

The method for adding and subtracting unlike fractions can be summarized by the formula that follows.

| Combining Unlike Fractions | $\frac{a}{b} \pm \frac{c}{d} = \frac{ad}{bd} \pm \frac{bc}{bd} = \frac{ad \pm bc}{bd}$ | **54** |
|---|---|---|

**EXAMPLE 25:** Combine: $x/2y - 5/x$.

**Solution:** The LCD will be the product of the two denominators, or $2xy$. So

$$\frac{x}{2y} - \frac{5}{x} = \frac{x}{2y}\left(\frac{x}{x}\right) - \frac{5}{x}\left(\frac{2y}{2y}\right)$$

$$= \frac{x^2}{2xy} - \frac{10y}{2xy}$$

$$= \frac{x^2 - 10y}{2xy}$$

The procedure is the same, of course, even when the denominators are more complicated.

**EXAMPLE 26:** Combine the fractions

$$\frac{x+2}{x-3} + \frac{2x+1}{3x-2}$$

**Solution:** Our LCD is $(x - 3)(3x - 2)$, so

$$\frac{x+2}{x-3} + \frac{2x+1}{3x-2} = \frac{(x+2)(3x-2)}{(x-3)(3x-2)} + \frac{(2x+1)(x-3)}{(3x-2)(x-3)}$$

$$= \frac{(3x^2+4x-4)+(2x^2-5x-3)}{(x-3)(3x-2)}$$

$$= \frac{5x^2 - x - 7}{3x^2 - 11x + 6}$$

Always factor the denominators completely. This will show if the same factor appears in more than one denominator.

**EXAMPLE 27:** Combine and simplify:

$$\frac{1}{x-3} + \frac{x-1}{x^2+3x+9} + \frac{x^2+x-3}{x^3-27}$$

**Solution:** Factoring gives

$$\frac{1}{x-3} + \frac{x-1}{x^2+3x+9} + \frac{x^2+x-3}{(x-3)(x^2+3x+9)}$$

Our LCD is then $x^3 - 27$ or $(x - 3)(x^2 + 3x + 9)$. Multiplying and adding, we have

$$\frac{(x^2+3x+9)+(x-1)(x-3)+(x^2+x-3)}{(x-3)(x^2+3x+9)}$$

$$= \frac{x^2+3x+9+x^2-4x+3+x^2+x-3}{(x-3)(x^2+3x+9)}$$

$$= \frac{3x^2+9}{x^3-27}$$

### Combining Integers and Fractions

Treat the integer as a fraction having 1 as a denominator, and combine as shown above.

---

**EXAMPLE 28:**

$$3 + \frac{2}{9} = \frac{3}{1} + \frac{2}{9} = \frac{3}{1} \cdot \frac{9}{9} + \frac{2}{9}$$

$$= \frac{27}{9} + \frac{2}{9} = \frac{29}{9}$$

---

The same procedure may be used to change a *mixed number* to an *improper fraction*.

---

**EXAMPLE 29:**

$$2\frac{1}{3} = 2 + \frac{1}{3} = \frac{6}{3} + \frac{1}{3}$$

$$= \frac{7}{3}$$

---

### Changing a Mixed Algebraic Expression to an Improper Fraction

The procedure is no different from that in the preceding section. Write the nonfractional expression as a fraction with 1 as the denominator, find the LCD, and combine, as shown in the following example.

---

**EXAMPLE 30:**

$$x^2 + \frac{5}{2x} = \frac{x^2}{1}\left(\frac{2x}{2x}\right) + \frac{5}{2x}$$

Add:

$$= \frac{2x^3 + 5}{2x}$$

---

### EXERCISE 3—ADDITION AND SUBTRACTION OF FRACTIONS

#### Common Fractions and Mixed Numbers

**Don't use your calculator for these numerical problems. The practice you get working with common fractions will help you when doing algebraic fractions.**

Combine and simplify.

1. $\frac{3}{5} + \frac{2}{5}$
2. $\frac{1}{8} - \frac{3}{8}$
3. $\frac{2}{7} + \frac{5}{7} - \frac{6}{7}$
4. $\frac{5}{9} + \frac{7}{9} - \frac{1}{9}$
5. $\frac{1}{3} - \frac{7}{3} + \frac{11}{3}$
6. $\frac{1}{5} - \frac{9}{5} + \frac{12}{5} - \frac{2}{5}$
7. $\frac{1}{2} + \frac{2}{3}$
8. $\frac{3}{5} - \frac{1}{3}$
9. $\frac{3}{4} + \frac{7}{16}$
10. $\frac{2}{3} + \frac{3}{7}$
11. $\frac{5}{9} - \frac{1}{3} + \frac{3}{18}$
12. $\frac{1}{2} + \frac{1}{3} + \frac{1}{5}$

13. $2 + \frac{3}{5}$

14. $3 - \frac{2}{3} + \frac{1}{6}$

15. $\frac{1}{5} - 7 + \frac{2}{3}$

16. $\frac{1}{5} + 2 - \frac{1}{3} + 5$

17. $\frac{1}{7} + 2 - \frac{3}{7} + \frac{1}{5}$

18. $5 - \frac{3}{5} + \frac{2}{15} - 2$

Rewrite each mixed number as an improper fraction.

19. $2\frac{2}{3}$

20. $1\frac{5}{8}$

21. $3\frac{1}{4}$

22. $2\frac{7}{8}$

23. $5\frac{11}{16}$

24. $9\frac{3}{4}$

### Algebraic Fractions

Combine and simplify.

25. $\frac{1}{a} + \frac{5}{a}$

26. $\frac{3}{x} + \frac{2}{x} - \frac{1}{x}$

27. $\frac{2a}{y} + \frac{3}{y} - \frac{a}{y}$

28. $\frac{x}{3a} - \frac{y}{3a} + \frac{z}{3a}$

29. $\frac{5x}{2} - \frac{3x}{2}$

30. $\frac{7}{x + 2} - \frac{5}{x + 2}$

31. $\frac{3x}{a - b} + \frac{2x}{b - a}$

32. $\frac{a}{x^2} - \frac{b}{x^2} + \frac{c}{x^2}$

33. $\frac{7}{a + 1} + \frac{9}{1 + a} - \frac{3}{-a - 1}$

34. $\frac{2a}{x(x - 1)} - \frac{3a}{x^2 - x} + \frac{a}{x - x^2}$

35. $\frac{3a}{2x} + \frac{2a}{5x}$

36. $\frac{1}{x} + \frac{1}{y} + \frac{1}{z}$

37. $\frac{a}{3} + \frac{2}{x} - \frac{1}{2}$

38. $\frac{1}{x + 3} + \frac{1}{x - 2}$

39. $\frac{a + b}{3} - \frac{a - b}{2}$

40. $\frac{3}{a + b} + \frac{2}{a - b}$

41. $\frac{4}{x - 1} - \frac{5}{x + 1}$

42. $\frac{x + 1}{x - 1} - \frac{x - 1}{x + 1}$

43. $\frac{b}{a + b} - \frac{a}{b - a} - \frac{a - b}{a + b}$

44. $\frac{y^2 - 2xy + x^2}{x^2 - xy} + \frac{x}{x - y}$

45. $\frac{1}{2(x - 1)} - \frac{1}{2(x + 1)} + \frac{1}{x^2}$

46. $\frac{1 + x}{1 + x + x^2} + \frac{1 - x}{1 - x + x^2}$

47. $\frac{1}{x + 1} + \frac{2}{x + 2} + \frac{3}{x - 3}$

48. $\frac{a - b}{ab} + \frac{b - c}{bc} + \frac{c - a}{ac}$

49. $\frac{x}{x^2 - 9} + \frac{x}{x + 3}$

50. $\frac{x + 3}{x^2 - 2x - 8} - \frac{2x}{x - 4}$

51. $\frac{5}{x - 2} - \frac{3x + 4}{x^3 - 8}$

52. $\frac{3x}{x - 1} + \frac{5}{x^3 + x^2 - 2x} + \frac{2}{x} + \frac{x - 1}{x + 2}$

Expressions of this form arise in calculus, when taking the derivative of a fraction by the delta method.

53. $\dfrac{1}{x + d + 1} - \dfrac{1}{x + 1}$

54. $\dfrac{1}{(x + d)^2} - \dfrac{1}{x^2}$

55. $x + d + \dfrac{1}{x + d} - \left(x + \dfrac{1}{x}\right)$

56. $\dfrac{x + d}{(x + d - 1)^2} - \dfrac{x}{(x - 1)^2}$

### Mixed Form

Rewrite each mixed expression as an improper fraction.

57. $x + \dfrac{1}{x}$

58. $\dfrac{1}{2} - x$

59. $\dfrac{x}{x - 1} - 1 - \dfrac{1}{x + 1}$

60. $a + \dfrac{1}{a^2 - b^2}$

61. $3 - a - \dfrac{2}{a}$

62. $\dfrac{5}{x} - 2x + \dfrac{3}{x^2}$

63. $\left(3b + \dfrac{1}{x}\right) - \left(2b - \dfrac{b}{ax}\right)$

64. $\left(5x + \dfrac{x + 2}{3}\right) - \left(2x - \dfrac{2x - 3}{4}\right)$

### Applications

*Hint:* In Problem 65, first find the amount that each crew can do in one day (e.g., the first crew can grade 7/3 mi per day). Then add the separate amounts to get the daily total. You can use a similar approach to the other work problems in this group.

Do this group of problems without using your calculator, and leave your answers in fractional form.

65. A certain work crew can grade 7 mi of roadbed in 3 days, and another crew can do 9 mi in 4 days. How much can both crews together grade in 1 day?
66. Liquid is running into a tank from a pipe that can fill four tanks in 3 days. Meanwhile, liquid is running out from a drain that can empty two tanks in 4 days. What will be the net change in volume in 1 day?
67. A planer makes a 1-m cutting stroke at a rate of 15 m/min and returns at 75 m/min. How long does it take for the cutting stroke and return? (See Eq. A17.)
68. Two resistors, 5 Ω and 15 Ω, are wired in parallel. What is the resistance of the parallel combination? (See Eq. A64.)
69. One crew can put together five machines in 8 days. Another crew can assemble three of these machines in 4 days. How much can both crews together assemble in 1 day?
70. One crew can assemble $M$ machines in $p$ days. Another crew can assemble $N$ of these machines in $q$ days. Write an expression for the number of machines that both crews together can assemble in 1 day, combine into a single term, and simplify.
71. A steel plate in the shape of a trapezoid (Fig. 8-3) has a hole of diameter $d$. The area of the plate, less the hole, is

$$\frac{(a + b)h}{2} - \frac{\pi d^2}{4}$$

Combine these two terms and simplify.

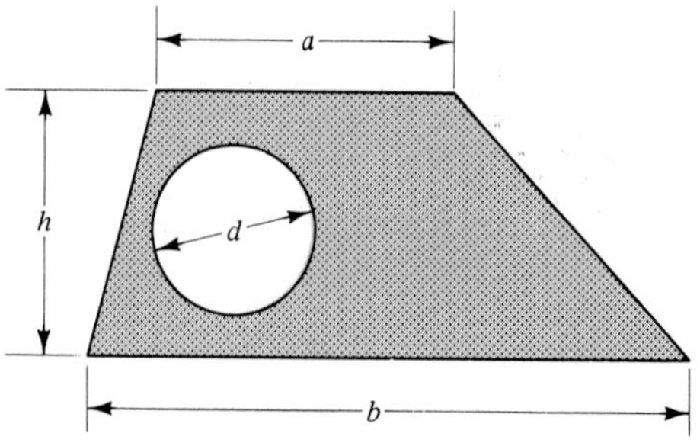


FIGURE 8-3

**72.** The resistance $R$ (Fig. 8-4) is

$$\frac{R_1R_2}{R_1 + R_2} + R_3$$

Combine into a single term and simplify.

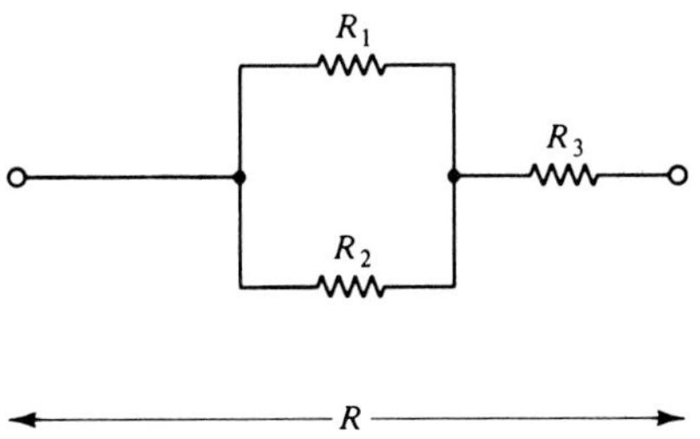

FIGURE 8-4

**73.** If a car travels a distance $d$ at a rate $V$, the time required will be $d/V$. The car then continues for a distance $d_1$ at a rate $V_1$, and a third distance $d_2$ at rate $V_2$. Write an expression for the total travel time, then combine the three terms into a single term and simplify.

## 8-4 COMPLEX FRACTIONS

Fractions that have only *one* fraction line are called *simple* fractions. Fractions with *more than one* fraction line are called *complex* fractions.

---

**EXAMPLE 31:**

$$\frac{\frac{a}{b}}{c} \quad \text{and} \quad \frac{x - \frac{1}{x}}{\frac{x}{y} - x}$$

are complex fractions.

---

We show how to simplify complex fractions in the following examples.

---

**EXAMPLE 32:** Simplify the complex fraction,

$$\frac{\frac{1}{2} + \frac{2}{3}}{3 + \frac{1}{4}}$$

**Solution:** We can simplify this fraction by multiplying numerator and denominator by the least common denominator for all the individual fractions. The denominators are 2, 3, and 4, so the LCD is 12. Multiplying, we obtain

$$\frac{\left(\frac{1}{2} + \frac{2}{3}\right) 12}{\left(3 + \frac{1}{4}\right) 12} = \frac{6 + 8}{36 + 3} = \frac{14}{39}$$

---

**EXAMPLE 33:** Simplify the complex fraction

$$\frac{1 + \dfrac{a}{b}}{1 - \dfrac{b}{a}}$$

**Solution:** The LCD for the two small fractions $a/b$ and $b/a$ is $ab$. Multiplying, we obtain

$$\frac{1 + \dfrac{a}{b}}{1 - \dfrac{b}{a}} \cdot \frac{ab}{ab} = \frac{ab + a^2}{ab - b^2}$$

or, in factored form,

$$\frac{a(b + a)}{b(a - b)}$$

In the following example, we simplify a small section at a time and work outward. Try to follow the steps.

**EXAMPLE 34:**

$$\frac{a + \dfrac{1}{b}}{a + \dfrac{1}{b + \dfrac{1}{a}}} \cdot \frac{ab}{ab + 1} = \frac{\dfrac{ab + 1}{b}}{a + \dfrac{1}{\dfrac{ab + 1}{a}}} \cdot \frac{ab}{ab + 1}$$

$$= \frac{\dfrac{ab + 1}{b}}{a + \dfrac{a}{ab + 1}} \cdot \frac{ab}{ab + 1}$$

$$= \frac{\dfrac{ab + 1}{b}}{\dfrac{a(ab + 1) + a}{ab + 1}} \cdot \frac{ab}{ab + 1}$$

$$= \frac{ab + 1}{b} \cdot \frac{ab + 1}{a(ab + 1) + a} \cdot \frac{ab}{ab + 1}$$

$$= \frac{a(ab + 1)}{a(ab + 1) + a} = \frac{ab + 1}{ab + 2}$$

## EXERCISE 4—COMPLEX FRACTIONS

Simplify.

1. $\dfrac{\dfrac{2}{3} + \dfrac{3}{4}}{\dfrac{1}{5}}$

2. $\dfrac{\dfrac{3}{4} - \dfrac{1}{3}}{\dfrac{1}{2} + \dfrac{1}{6}}$

3. $\dfrac{\frac{1}{2}+\frac{1}{3}+\frac{1}{4}}{3-\frac{4}{5}}$

4. $\dfrac{\frac{4}{5}}{\frac{1}{5}+\frac{2}{3}}$

5. $\dfrac{5-\frac{2}{5}}{6+\frac{1}{3}}$

6. $\dfrac{1}{2}+\dfrac{3}{\frac{2}{5}+\frac{1}{3}}$

7. $\dfrac{x+\frac{y}{4}}{x-\frac{y}{3}}$

8. $\dfrac{\frac{a}{b}+\frac{x}{y}}{\frac{a}{z}-\frac{x}{c}}$

9. $\dfrac{1+\frac{x}{y}}{1-\frac{x^2}{y^2}}$

10. $\dfrac{x+\frac{a}{c}}{x+\frac{b}{d}}$

11. $\dfrac{a^2+\frac{x}{3}}{4+\frac{x}{5}}$

12. $\dfrac{3a^2-3y^2}{\frac{a+y}{3}}$

13. $\dfrac{x+\frac{2d}{3ac}}{x+\frac{3d}{2ac}}$

14. $\dfrac{4a^2-4x^2}{\frac{a+x}{a-x}}$

15. $\dfrac{x^2-\frac{y^2}{2}}{\frac{x-3y}{2}}$

16. $\dfrac{\frac{ab}{7}-3d}{3c-\frac{ab}{d}}$

17. $\dfrac{1+\frac{1}{x+1}}{1-\frac{1}{x-1}}$

18. $\dfrac{xy-\frac{3x}{ac}}{\frac{ac}{x}+2c}$

19. $\dfrac{\frac{2m+n}{m+n}-1}{1-\frac{n}{m+n}}$

20. $\dfrac{\frac{x^3+y^3}{x^2-y^2}}{\frac{x^2-xy+y^2}{x-y}}$

21. $\dfrac{1}{a+\dfrac{1}{1+\frac{a+1}{3-a}}}$

22. $\dfrac{x+y}{x+y+\dfrac{1}{x-y+\frac{1}{x+y}}}$

23. $\dfrac{3abc}{bc+ac+ab}-\dfrac{\frac{a-1}{a}+\frac{b-1}{b}+\frac{c-1}{c}}{\frac{1}{a}+\frac{1}{b}+\frac{1}{c}}$

24. $\dfrac{\frac{1}{a}+\frac{1}{b+c}}{\frac{1}{a}-\frac{1}{b+c}}\left(1+\dfrac{b^2+c^2-a^2}{2bc}\right)$

25. $\dfrac{\frac{x^2-y^2-z^2-2yz}{x^2-y^2-z^2+2yz}}{\frac{x-y-z}{x+y-z}}$

## Applications

26. A car travels a distance $d_1$ at a rate $V_1$, then another distance $d_2$ at a rate $V_2$. The average speed for the entire trip is

$$\text{average speed} = \frac{d_1 + d_2}{\dfrac{d_1}{V_1} + \dfrac{d_2}{V_2}}$$

Simplify this complex fraction.

27. The equivalent resistance of two resistors in parallel is

$$\frac{R_1R_2}{R_1 + R_2}$$

If each resistor is made of wire of resistivity $\rho$, with $R_1$ using a wire of length $L_1$ and cross-sectional area $A_1$, and $R_2$ having a length $L_2$ and area $A_2$, our expression becomes

See Eq. A71, $R = \rho L/A$.

$$\frac{\dfrac{\rho L_1}{A_1} \cdot \dfrac{\rho L_2}{A_2}}{\dfrac{\rho L_1}{A_1} + \dfrac{\rho L_2}{A_2}}$$

Simplify this complex fraction.

# 8-5 FRACTIONAL EQUATIONS

## Solving Fractional Equations

An equation in which one or more terms is a fraction is called a *fractional equation*. To solve a fractional equation, first eliminate the fractions by multiplying both sides of the equation by the least common denominator (LCD) of every term. We can do this because multiplying both sides of an equation by the same quantity (the LCD in this case) does not unbalance the equation. With the fractions thus eliminated, the equation is then solved like any nonfractional equation.

---

**EXAMPLE 35:** Solve for $x$:

$$\frac{3x}{5} - \frac{x}{3} = \frac{2}{15}$$

**Solution:** Multiplying both sides of the equation by the LCD (15), we obtain

$$15\left(\frac{3x}{5} - \frac{x}{3}\right) = 15\left(\frac{2}{15}\right)$$

$$9x - 5x = 2$$

$$x = \frac{1}{2}$$

---

## Equations with the Unknown in the Denominator

The procedure is the same when the unknown appears in the denominator of one or more terms. However, the LCD will now contain the unknown. Here it is understood that $x$ cannot have a value that will make any of the denominators in the problem equal to zero. Such forbidden values are sometimes stated with the problem, but often are not.

**EXAMPLE 36:** Solve for $x$:

$$\frac{2}{3x} = \frac{5}{x} + \frac{1}{2} \qquad (x \neq 0)$$

**Solution:** The LCD is $6x$. Multiplying both sides of the equation yields

$$6x\left(\frac{2}{3x}\right) = 6x\left(\frac{5}{x} + \frac{1}{2}\right)$$

$$6x\left(\frac{2}{3x}\right) = 6x\left(\frac{5}{x}\right) + 6x\left(\frac{1}{2}\right)$$

$$4 = 30 + 3x$$

$$3x = -26$$

$$x = -\frac{26}{3}$$

We assume the integers in these equations to be exact numbers. Thus we leave our answers in fractional form rather than converting to (approximate) decimals.

Check:

$$\frac{2}{3\left(-\frac{26}{3}\right)} \stackrel{?}{=} \frac{5}{-\frac{26}{3}} + \frac{1}{2}$$

$$-\frac{2}{26} \stackrel{?}{=} -\frac{15}{26} + \frac{13}{26}$$

$$-\frac{2}{26} = -\frac{2}{26} \qquad \text{checks}$$

**EXAMPLE 37:** Solve for $x$:

$$\frac{8x+7}{5x+4} = 2 - \frac{2x}{5x+1}$$

**Solution:** Multiplying by the LCD $(5x + 4)(5x + 1)$ yields

$$(8x + 7)(5x + 1) = 2(5x + 4)(5x + 1) - 2x(5x + 4)$$

$$40x^2 + 35x + 8x + 7 = 2(25x^2 + 5x + 20x + 4) - 10x^2 - 8x$$

$$40x^2 + 43x + 7 = 50x^2 + 50x + 8 - 10x^2 - 8x$$

$$43x + 7 = 42x + 8$$

$$x = 1$$

Check:

$$\frac{8(1)+7}{5(1)+4} \stackrel{?}{=} 2 - \frac{2(1)}{5(1)+1}$$

$$\frac{15}{9} \stackrel{?}{=} 2 - \frac{2}{6}$$

$$\frac{5}{3} = \frac{6}{3} - \frac{1}{3} \qquad \text{checks}$$

| | |
|---|---|
| Common Error | The technique of multiplying by the LCD in order to eliminate the denominators is valid *only when we have an equation*. Do not multiply through by the LCD when there is no equation! |

**EXAMPLE 38:** Solve for $x$:

$$\frac{3}{x-3} = \frac{2}{x^2 - 2x - 3} + \frac{4}{x+1}$$

**Solution:** The denominator of the second fraction factors into

$$x^2 - 2x - 3 = (x - 3)(x + 1)$$

The LCD is therefore $(x - 3)(x + 1)$. Multiplying both sides of the given equation by the LCD gives

$$3(x + 1) = 2 + 4(x - 3)$$

$$3x + 3 = 4x - 10$$

$$x = 13$$

## EXERCISE 5—FRACTIONAL EQUATIONS

Solve for $x$.

1. $2x + \frac{x}{3} = 28$
2. $4x + \frac{x}{5} = 42$
3. $x + \frac{x}{5} = 24$
4. $\frac{x}{6} + x = 21$
5. $3x - \frac{x}{7} = 40$
6. $x - \frac{x}{6} = 25$
7. $\frac{2x}{3} - x = -24$
8. $\frac{3x}{5} + 7x = 38$
9. $\frac{x}{2} + \frac{x}{3} + \frac{x}{4} = 26$
10. $\frac{x-1}{2} = \frac{x+1}{3}$
11. $\frac{3x-1}{4} = \frac{2x+1}{3}$
12. $x + \frac{2x}{3} + \frac{3x}{4} = 29$
13. $2x + \frac{x}{3} - \frac{x}{4} = 50$
14. $3x - \frac{2x}{3} - \frac{5x}{6} = 18$
15. $3x - \frac{x}{6} + \frac{x}{12} = 70$
16. $\frac{x}{4} + \frac{x}{6} + \frac{x}{8} = 26$
17. $\frac{6x-19}{2} = \frac{2x-11}{3}$
18. $\frac{7x-40}{8} = \frac{9x-80}{10}$
19. $\frac{3x-116}{4} + \frac{180-5x}{6} = 0$
20. $\frac{3x-4}{2} - \frac{3x-1}{16} = \frac{6x-5}{8}$
21. $\frac{x-1}{8} - \frac{x+1}{18} = 1$
22. $\frac{3x}{4} + \frac{180-5x}{6} = 29$
23. $\frac{15x}{4} = \frac{9}{4} - \frac{3-x}{2}$
24. $\frac{x}{4} + \frac{x}{10} + \frac{x}{8} = 19$

### Equations with Unknown in Denominator

25. $\frac{2}{3x} + 6 = 5$
26. $9 - \frac{4}{5x} = 7$

27. $4 + \frac{1}{x+3} = 8$

28. $\frac{x+5}{x-2} = 5$

29. $\frac{x-3}{x+2} = \frac{x+4}{x-5}$

30. $\frac{3}{x+2} = \frac{5}{x} + \frac{4}{x^2+2x}$

31. $\frac{9}{x^2+x-2} = \frac{7}{x-1} - \frac{3}{x+2}$

32. $\frac{3x+5}{2x-3} = \frac{3x-3}{2x-1}$

33. $\frac{4}{x^2-1} + \frac{1}{x-1} + \frac{1}{x+1} = 0$

34. $\frac{x}{3} - \frac{x^2-5x}{3x-7} = \frac{2}{3}$

35. $\frac{3}{x+1} - \frac{x+1}{x-1} = \frac{x^2}{1-x^2}$

36. $\frac{x+1}{2(x-1)} - \frac{x-1}{x+1} = \frac{17-x^2}{2(x^2-1)}$

37. $\frac{9x+20}{36} - \frac{x}{4} = \frac{4x-12}{5x-4}$

38. $\frac{10x+13}{18} - \frac{x+2}{x-3} = \frac{5x-4}{9}$

39. $\frac{6x+7}{10} - \frac{3x+1}{5} = \frac{x-1}{3x-4}$

40. $\frac{5x}{1-x} - \frac{7x}{3+x} = \frac{12(1-x^2)}{x^2+2x-3}$

## 8-6 WORD PROBLEMS LEADING TO FRACTIONAL EQUATIONS

### Solving Word Problems

The method for solving the word problems in this chapter (and in any other chapter) is no different from that given in Chapter 3. The resulting equations, however, will now contain fractions.

---

**EXAMPLE 39:** One-fifth of a certain number is 3 greater than one-sixth of that same number. Find the number.

**Solution:** Let $x$ = the number. Then from the problem statement,

$$\frac{1}{5}x = \frac{1}{6}x + 3$$

Multiplying by the LCD (30), we obtain

$$6x = 5x + 90$$

$$x = 90$$

**Check:** Is one-fifth of 90 (which is 18) three greater than one-sixth of 90 (which is 15)? Yes.

---

### Rate Problems

Fractional equations often arise when we solve problems involving *rates*. Problems of this type include:

| | |
|---|---|
| Uniform motion: | amount traveled = rate of travel × time traveled |
| Work: | amount of work done = rate of work × time worked |
| Fluid flow: | amount of flow = flow rate × duration of flow |
| Energy flow: | amount of energy = rate of energy flow × time |

Notice that these equations are mathematically identical. Word problems involving motion, flow of fluids or solids, flow of heat, electricity, solar energy, mechanical energy, and so on, can be handled in the same way.

To convince yourself of the similarity of all these types of problems, consider the following work problem: *If worker M can do a certain job in 5 h, and N can do the same amount in 8 h, how long will it take M and N together to do that same amount?* Are the following problems really any different?

| | | | | | | |
|---|---|---|---|---|---|---|
| If | worker<br>machine<br>vehicle<br>pipe | $M$ can | do<br>generate<br>consume<br>travel<br>deliver | a certain | job<br>amount of energy<br>distance<br>amount of liquid | in 5 h |

| | | |
|---|---|---|
| and $N$ can | do<br>generate<br>consume<br>travel<br>deliver | the same amount in 8 h, |

| | |
|---|---|
| how long will it take $M$ and $N$ together to | do<br>generate<br>consume<br>travel<br>deliver |

that same amount?

We will first do a uniform motion problem and then proceed to work and flow problems.

### Uniform Motion

Motion is called *uniform* when the speed does not change. The distance traveled at constant speed is related to the rate of travel and the elapsed time by the following equation.

**Be careful not to use this formula for anything but *uniform* motion.**

| | | |
|---|---|---|
| Uniform Motion | distance = rate × time | **A17** |

**EXAMPLE 40:** A train departs at noon traveling at a speed of 64 km/h. A car leaves the same station $\frac{1}{2}$ h later to overtake the train, traveling on a road parallel to the track. If the car's speed is 96 km/h, at what time and at what distance from the station will it overtake the train?

**Solution:** Let $x$ = time traveled by car (hours). Then $x + \frac{1}{2}$ = time traveled by train (hours). The distance traveled by the car is

$$\text{distance} = \text{rate} \times \text{time} = 96x$$

and for the train,

$$\text{distance} = 64\left(x + \frac{1}{2}\right)$$

But the distance is the same for car and train, so we can write the equation

$$96x = 64\left(x + \frac{1}{2}\right)$$

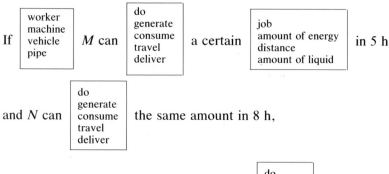

Removing parentheses yields

$$96x = 64x + 32$$

$$96x - 64x = 32$$

$$32x = 32$$

$$x = 1 \text{ h} = \text{time for car}$$

$$x + \frac{1}{2} = 1\tfrac{1}{2} \text{ h} = \text{time for train}$$

So the car overtakes the train at 1:30 P.M. at a distance of 96 km from the station.

---

### Work Problems

To tackle work problems, we need one simple idea:

| Work | (rate of work) × (time) = amount done | **A5** |
|---|---|---|

We often use this equation to find the rate of work for a person or a machine.

In a typical work problem, there are two or more persons or machines doing work, each at a different rate. For each worker the work rate is the amount done by that worker divided by the time taken to do the work. That is, if a person can stamp 9 parts in 13 min, that person's work rate is $\frac{9}{13}$ part per minute.

---

**EXAMPLE 41:** Crew *A* can assemble 2 cars in 5 days, and crew *B* can assemble 3 cars in 7 days. If both crews together assemble 100 cars, with crew *B* working 10 days longer than crew *A*, how many days must each crew work?

**Solution:** Let $x$ = days worked by crew $A$

and $x + 10$ = days worked by crew $B$

The work rate of each crew is

Crew *A*: rate $= \frac{2}{5}$ car per day  Crew *B*: rate $= \frac{3}{7}$ car per day

The amount done by each crew equals their work rate times the number of days worked. The sum of the amounts done by the two crews must equal 100 cars.

$$\frac{2}{5}x + \frac{3}{7}(x + 10) = 100$$

We clear fractions by multiplying by the LCD, 35:

$$14x + 15(x + 10) = 3500$$

$$14x + 15x + 150 = 3500$$

$$29x = 3350$$

$$x = 115.5 \text{ days for crew } A$$

$$x + 10 = 125.5 \text{ days for crew } B$$

---

## Fluid Flow and Energy Flow

For flow problems we make use of the simple equation

$$\text{amount of flow} = \text{flow rate} \times \text{time}$$

Here we assume, of course, that the flow rate is constant.

**EXAMPLE 42:** A certain small hydroelectric generating station can produce 17 megawatthours (MWh) of energy per year. After 4 months of operation, another generator is added which, by itself, can produce 11 MWh in 5 months. How many additional months are needed for a total of 25 MWh to be produced?

**Solution:** Let $x$ = additional months. The original generating station can produce 17/12 MWh per month for $4 + x$ months. The new generator can produce 11/5 MWh per month for $x$ months. The problem states that the total amount produced (in MWh) is 25, so

$$\frac{17}{12}(4 + x) + \frac{11}{5}x = 25$$

Multiplying by the LCD, 60, we obtain

$$5(17)(4 + x) + 12(11)x = 25(60)$$

$$340 + 85x + 132x = 1500$$

$$217x = 1160$$

$$x = \frac{1160}{217} = 5.3 \text{ months}$$

## EXERCISE 6—WORD PROBLEMS

### Number Problems

1. There is a number such that if $\frac{1}{5}$ of it is subtracted from 50 and this difference is multiplied by 4, the result will be 70 less than the number. What is the number?
2. Separate 100 into two parts such that if $\frac{1}{3}$ of one part is subtracted from $\frac{1}{4}$ of the other, the difference is 11.
3. The second of two numbers is 1 greater than the first, and $\frac{1}{2}$ of the first plus $\frac{1}{5}$ of the first is equal to the sum of $\frac{1}{3}$ of the second and $\frac{1}{4}$ of the second. What are the numbers?
4. Find two consecutive numbers such that one-half the larger exceeds one-third of the smaller by 9.
5. The difference between two positive numbers is 20, and $\frac{1}{7}$ of one is equal to $\frac{1}{3}$ of the other. What are the numbers?

### Uniform Motion Problems

**Depending on how you set them up, some of these problems may not result in fractional equations.**

6. A person sets out from Boston and walks toward Portland at the rate of 3 mi/h. Three hours afterward, a second person sets out from the same place and walks in the same direction at the rate of 4 mi/h. How far from Boston will the second overtake the first?
7. A courier who goes at the rate of $6\frac{1}{2}$ mi/h is followed, after 4 h, by another who goes at the rate of $7\frac{1}{2}$ mi/h. In how many hours will the second overtake the first?

8. A person walks to the top of a mountain at the rate of 2 mi/h and down the same way at the rate of 4 mi/h. If he walks for 6 h, how far is it to the top of the mountain?
9. In going a certain distance, a train traveling at the rate of 40 mi/h takes 2 h less than a train traveling 30 mi/h. Find the distance.
10. Spacecraft $A$ is over Houston at noon on a certain day and traveling at a rate of 300 km/h. Spacecraft $B$, attempting to overtake and dock with $A$, is over Houston at 1:15 P.M. and is traveling in the same direction as $A$, at 410 km/h. At what time will $B$ overtake $A$? At what distance from Houston?

### Work Problems

11. A laborer can do a certain job in 5 days, a second can do the job in 6 days, and a third in 8 days. In what time can the three together do the job?
12. Three masons build 318 m of wall. Mason $A$ builds 7 m/day, $B$ builds 6 m/day, and $C$ builds 5 m/day. Mason $B$ works twice as many days as $A$, and $C$ works half as many days as $A$ and $B$ combined. How many days did each work?
13. If a carpenter can roof a house in 10 days and another can do the same in 14 days, how many days will it take if they work together?
14. A technician can assemble an instrument in 9.5 h. After working for 2 h she is joined by another technician who, alone, could do the job in 7.5 h. How many additional hours are needed to finish the job?
15. A certain screw machine can produce a box of parts in 3.3 h. A new machine is to be ordered having a speed such that both machines working together would produce a box of parts in 1.4 h. How long would it take the new machine alone to produce a box of parts?

### Fluid Flow Problems

16. A tank can be filled by a pipe in 3 h and emptied by another pipe in 4 h. How much time will be required to fill an empty tank if both are running?
17. Two pipes empty into a tank. One pipe can fill the tank in 8 h and the other in 9 h. How long will it take both pipes together to fill the tank?
18. A tank has three pipes connected. The first, by itself, could fill the tank in $2\frac{1}{2}$ h, the second in 2 h, and the third in 1 h 40 min. In how many minutes will the tank be filled?
19. A tank can be filled by a certain pipe in 18 h. Five hours after this pipe is opened, it is supplemented by a smaller pipe which, by itself, could fill the tank in 24 h. Find the total time, measured from the opening of the larger pipe, to fill the tank.
20. At what rate must liquid be drained from a tank in order to empty it in 1.5 h if the tank takes 4.7 h to fill at the rate of 3.5 $m^3$/min?

### Energy Flow Problems

**In this group of problems, as in the previous ones, we assume that the rates are constant.**

21. A certain power plant consumes 1500 tons of coal in 4 weeks. There is a stockpile of 10,000 tons of coal available when the plant starts operating. After 3 weeks in operation, an additional boiler, capable of using 2300 tons in 3 weeks, is put on line with the first boiler. In how many more weeks will the stockpile of coal be consumed?
22. A certain array of solar cells can generate 2 megawatthours (MWh) in 5 months (under standard conditions). After operating for 3 months, another array of cells is added which alone can generate 5 MWh in 7 months. How many additional months, after the new array was added, is needed for the total energy generated from both arrays to be 10 MWh?
23. A landlord owns a house that consumes 2100 gal of heating oil in three winters. He buys another (insulated) house, and the two houses together use 1850 gal of oil in two winters. How many winters would it take the insulated house alone to use 1250 gal of oil?

24. A wind generator can charge 20 storage batteries in 24 h. After charging for 6 h, another generator, which can charge the batteries in 36 h, is also connected to the batteries. How many additional hour are needed to charge the batteries?
25. A certain solar collector can absorb 9000 Btu in 7 h. Another panel is added and together they collect 35,000 Btu in 5 h. How long would it take the new panel alone to collect 35,000 Btu?

### Financial Problems

26. After spending $\frac{1}{4}$ of my money, then $\frac{1}{5}$ of the remainder, I had \$66 left. How many dollars did I start with?
27. A person who had inherited some money spent $\frac{3}{8}$ of it the first year, $\frac{4}{5}$ of the remainder the next year, and had \$1420 left. What was the inheritance?
28. Three children were left an inheritance of which the eldest received $\frac{2}{3}$, the second $\frac{1}{5}$, and the third the rest, which was \$20,000. How much did each receive?
29. $A$'s capital was $\frac{3}{4}$ of $B$'s. If $A$'s had been \$500 less, it would have been only $\frac{1}{2}$ of $B$'s. What was the capital of each?

## 8-7 LITERAL EQUATIONS AND FORMULAS

### Literal Equations

A *literal equation* is one in which some or all of the constants are represented by letters.

---

**EXAMPLE 43:**

$$a(x + b) = b(x + c)$$

is a literal equation, whereas

$$2(x + 5) = 3(x + 1)$$

is a numerical equation.

---

### Formulas

A *formula* is a literal equation that relates two or more mathematical or physical quantities. These are the equations, mentioned in the introduction of this chapter, that describe the workings of the physical world. A listing of some of the common formulas used in technology is given in the Summary of Facts and Formulas (Appendix A).

---

**EXAMPLE 44:**

| | | |
|---|---|---|
| Newton's second law | $F = ma$ | **A20** |

is a formula relating the force acting on a body with its mass and acceleration.

---

### Solving Literal Equations and Formulas

When we solve a literal equation or formula, we cannot, of course, get a *numerical* answer, as we could with equations that had only one unknown. Our object here is to *isolate* one of the letters on one side of the equal sign. We "solve for" one of the literal quantities.

**EXAMPLE 45:** Solve for $x$:

$$a(x + b) = b(x + c)$$

**Solution:** Our goal is to isolate $x$ on one side of the equation. Removing parentheses, we obtain

$$ax + ab = bx + bc$$

Subtracting $bx$ and then $ab$ will place all the $x$ terms on one side of the equation.

$$ax - bx = bc - ab$$

Factoring to isolate $x$ yields

$$x(a - b) = b(c - a)$$

Dividing by $(a - b)$, where $a \neq b$, gives us

$$x = \frac{b(c - a)}{a - b}$$

---

**EXAMPLE 46:** Solve for $x$:

$$b\left(b + \frac{x}{a}\right) = d$$

**Solution:** Dividing both sides by $b$, we have

$$b + \frac{x}{a} = \frac{d}{b}$$

Subtracting $b$ yields

$$\frac{x}{a} = \frac{d}{b} - b$$

Multiplying both sides by $a$ gives us

$$x = a\left(\frac{d}{b} - b\right)$$

---

**EXAMPLE 47:** The formula for the amount of heat $q$ flowing by conduction through a wall of thickness $L$, conductivity $k$, and cross-sectional area $A$ is

$$q = \frac{kA(t_1 - t_2)}{L}$$

where $t_1$ and $t_2$ are the temperatures of the warmer and cooler sides, respectively. Solve this equation for $t_1$.

**Solution:** Multiplying both sides by $L/kA$ gives

$$\frac{qL}{kA} = t_1 - t_2$$

Then adding $t_2$ to both sides, we get

$$t_1 = \frac{qL}{kA} + t_2$$

## Checking Literal Equations

As with numerical equations, we can substitute our expression for $x$ back into the original equation and see if it checks.

**EXAMPLE 48:** We check the solution of Example 46 by substituting $a(d/b - b)$ for $x$ in the original equation.

$$b\left(b + \frac{x}{a}\right) \stackrel{?}{=} d$$

$$b\left[b + \frac{a(d/b - b)}{a}\right] \stackrel{?}{=} d$$

$$b\left(b + \frac{d}{b} - b\right) \stackrel{?}{=} d$$

$$b\left(\frac{d}{b}\right) = d \qquad \text{checks}$$

Another simple "check" is to substitute numerical values. We avoid simple values such as 0 and 1 that may make an incorrect solution appear to check, and we avoid any values that will make a denominator zero.

**EXAMPLE 49:** Check the results from Example 46.

**Solution:** Let us *choose* values for $a$, $b$, and $d$, say,

$$a = 2 \qquad b = 3 \quad \text{and} \quad d = 6$$

With these values,

$$x = a\left(\frac{d}{b} - b\right) = 2\left(\frac{6}{3} - 3\right) = -2$$

We now see if these values check in the original equation.

$$b\left(b + \frac{x}{a}\right) = d$$

$$3\left(3 + \frac{-2}{2}\right) \stackrel{?}{=} 6$$

$$3(3 - 1) \stackrel{?}{=} 6$$

$$3(2) = 6 \qquad \text{checks}$$

Realize that this is not a rigorous check, because we are testing the solution only for the specific values chosen. However, if it checks for those values, we can be *reasonably* sure that our solution is correct.

## Literal Fractional Equations

To solve literal fractional equations, the procedure is the same as for other fractional equations; multiply by the LCD to eliminate the fractions.

**EXAMPLE 50:** Solve for $x$ in terms of $a$, $b$, and $c$:

$$\frac{x}{b} - \frac{a}{c} = \frac{x}{a}$$

**Solution:** Multiplying by the LCD ($abc$) yields

$$acx - a^2b = bcx$$

Rearranging so that all $x$ terms are together on one side of the equation, we obtain

$$acx - bcx = a^2b$$

Factoring gives us

$$x(ac - bc) = a^2b$$

$$x = \frac{a^2b}{ac - bc}$$

---

**EXAMPLE 51:** Solve for $x$:

$$\frac{x}{x - a} - \frac{x + 2b}{x + a} = \frac{a^2 + b^2}{x^2 - a^2}$$

**Solution:** The LCD is $x^2 - a^2$, that is, $(x - a)(x + a)$. Multiplying each term by the LCD gives

$$x(x + a) - (x + 2b)(x - a) = a^2 + b^2$$

Removing parentheses and rearranging so that all the $x$ terms are together, we obtain

$$x^2 + ax - (x^2 + 2bx - ax - 2ab) = a^2 + b^2$$

$$x^2 + ax - x^2 - 2bx + ax + 2ab = a^2 + b^2$$

$$2ax - 2bx = a^2 - 2ab + b^2$$

Then we factor:

$$2x(a - b) = (a - b)^2$$

and divide by $2(a - b)$:

$$x = \frac{(a - b)^2}{2(a - b)} = \frac{a - b}{2}$$

---

Sometimes a fractional equation will have terms that themselves are not fractions, but the method of solution is the same.

---

**EXAMPLE 52:** Solve for $x$:

$$\frac{x}{b} + c = \frac{2x}{a + b} + b$$

**Solution:** We multiply both sides by the LCD, $b(a + b)$:

$$x(a + b) + c(b)(a + b) = 2x(b) + b(b)(a + b)$$

Removing parentheses and moving all the $x$ terms to one side gives

$$ax + bx + abc + b^2c = 2bx + ab^2 + b^3$$

$$ax + bx - 2bx = ab^2 + b^3 - abc - b^2c$$

Factoring yields

$$x(a - b) = ab^2 + b^3 - abc - b^2c$$

$$x = \frac{b(ab + b^2 - ac - bc)}{a - b}$$

---

## EXERCISE 7—LITERAL EQUATIONS AND FORMULAS

Solve for $x$.

1. $2ax = bc$
2. $ax + dx = a - c$
3. $a(x + y) = b(x + z)$
4. $4x = 2x + ab$
5. $4acx - 3d^2 = a^2d - d^2x$
6. $a(2x - c) = a + c$
7. $a^2x - cd = b - ax + dx$
8. $3(x - r) = 2(x + p)$
9. $\frac{a}{2}(x - 3w) = z$
10. $cx - x = bc - b$
11. $3x + m = b$
12. $ax + m = cx + n$
13. $ax - bx = c + dx - m$
14. $3m + 2x - c = x + d$
15. $ax - ab = cx - bc$
16. $p(x - b) = qx + d$
17. $3(x - b) = 2(bx + c) - c(x - b)$
18. $a(bx + d) - cdx = c(dx + d)$

### Literal Fractional Equations

19. $\frac{w + x}{x} = w(w + y)$
20. $\frac{ax + b}{c} = bx - a$
21. $\frac{p - q}{x} = 3p$
22. $\frac{bx - c}{ax - c} = 5$
23. $\frac{a - x}{5} = \frac{b - x}{2}$
24. $\frac{x}{a} - b = \frac{c}{d} - x$
25. $\frac{x}{a} - a = \frac{a}{c} - \frac{x}{c - a}$
26. $\frac{x}{a - 1} - \frac{x}{a + 1} = b$
27. $\frac{x - a}{x - b} = \left(\frac{2x - a}{2x - b}\right)^2$
28. $\frac{a - b}{bx + c} + \frac{a + b}{ax - c} = 0$
29. $\frac{x + a}{x - a} - \frac{x - b}{x + b} = \frac{2(a + b)}{x}$
30. $\frac{ax}{b} - \frac{b - x}{2c} + \frac{a(b - x)}{3d} = a$

### Word Problems with Literal Solutions

31. Two persons are $d$ miles apart. They set out at the same time and travel toward each other; one travels at the rate of $m$ miles an hour and the other at the rate of $n$ miles an hour. How far will each have traveled when they meet?
32. If $A$ can do a piece of work in $p$ hours, $B$ in $q$ hours, $C$ in $r$ hours, and $D$ in $s$ hours, how many hours will it take to do the work if all work together?
33. After spending $1/n$ and $1/m$ of my money, I had $a$ dollars left. How many dollars had I at first?

### Formulas

34. The correction for the sag in a surveyor's tape weighing $w$ lb/ft and pulled with a force of $P$ lb is

$$C = \frac{w^2L^3}{24P^2} \quad \text{feet}$$

Solve this equation for the distance measured, $L$.

35. When a bar of length $L_0$ having a coefficient of linear thermal expansion $\alpha$ is increased in temperature by an amount $\Delta t$, it will expand to a new length $L$, where

$$L = L_0(1 + \alpha\Delta t)$$

Solve this equation for $\Delta t$.

36. The formula for the equivalent resistance $R$ for the parallel combination of two resistors, $R_1$ and $R_2$, is

$$\frac{1}{R} = \frac{1}{R_1} + \frac{1}{R_2}$$

Solve this formula for $R_2$.

**Be careful to distinguish between capitals and lowercase letters—they are not the same. A given equation might have, for example, both *T* and *t*, each with a different meaning.**

37. A rod of cross-sectional area $a$ and length $L$ will stretch by an amount $e$ when subject to a tensile load of $P$. The modulus of elasticity is given by

$$E = \frac{PL}{ae}$$

Solve this equation for $a$.

38. The formula for the displacement $s$ of a freely falling body having an initial velocity $v_0$ and acceleration $a$ is

$$s = v_0t + \tfrac{1}{2}at^2$$

Solve this equation for $a$.

39. The formulas for the amount of heat flowing through a wall by conduction is

$$q = \frac{kA(t_1 - t_2)}{L}$$

Solve this equation for $t_2$.

40. If the resistance of a conductor is $R_1$ at temperature $t_1$, the resistance will change to a value $R$ when the temperature changes to $t$, where

$$R = R_1[1 + \alpha(t - t_1)]$$

and $\alpha$ is the temperature coefficient of resistance at temperature $t_1$. Solve this equation for $t_1$.

41. An amount $a$ invested at a simple interest rate $n$ for $t$ years will accumulate to an amount $y$, where $y = a + ant$. Solve for $a$.

42. Applying Kirchhoff's law (Eq. A68) to loop 1 in Fig. 8-5, we get $E = I_1R_1 + I_1R_2 - I_2R_2$. Solve for $I_1$.

43. When a bar of length $L_0$ is changed in temperature by an amount $\Delta t$, it changes to a new length $L$, where

$$L = L_0 + L_0\alpha\Delta t$$

where $\alpha$ is the coefficient of thermal expansion. Solve for $L_0$.

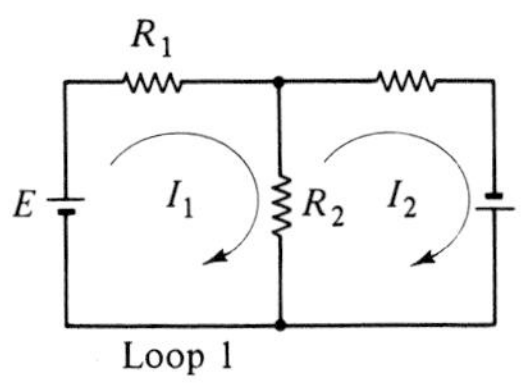


**FIGURE 8-5**

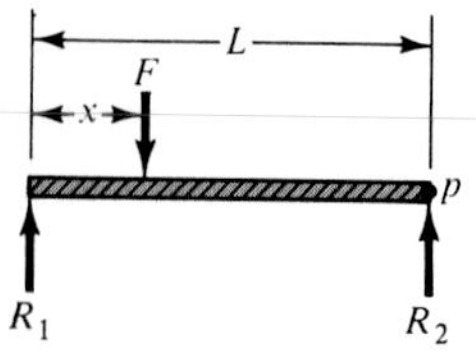


**FIGURE 8-6**

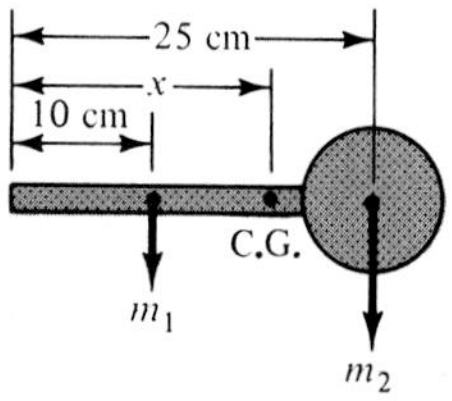


**FIGURE 8-7**

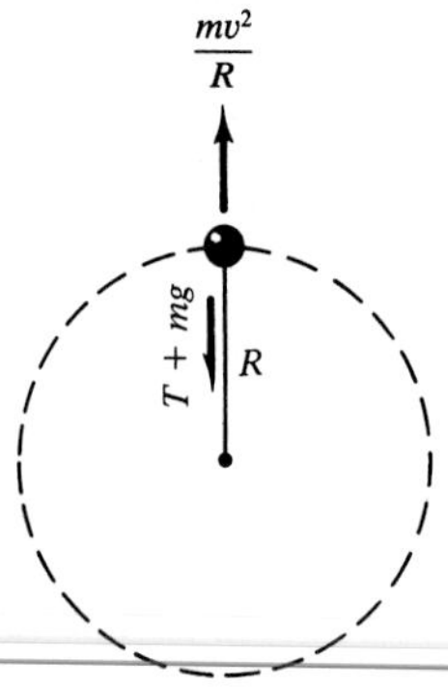


**FIGURE 8-8**

44. Taking the moment $M$ about point $p$ in Fig. 8-6, we get $M = R_1L - F(L - x)$. Solve for $L$.

45. Three masses, $m_1$, $m_2$, and $m_3$, are attached together and accelerated by means of a force $F$, where

$$F = m_1a + m_2a + m_3a$$

Solve for the acceleration $a$.

46. A resistance $R_1$, when changed in temperature from $t_1$ to $t$, will change in resistance to $R$, where

$$R = R_1 + R_1\alpha(t - t_1)$$

where $\alpha$ is the temperature coefficient of resistance. Solve for $R_1$.

47. A bar of mass $m_1$ is attached to a sphere of mass $m_2$ (Fig. 8-7). The distance $x$ to the center of gravity is

$$x = \frac{10m_1 + 25m_2}{m_1 + m_2}$$

Solve for $m_1$.

48. A ball of mass $m$ is swung in a vertical circle (Fig. 8-8). At the top of its swing, the tension $T$ in the cord plus the ball's weight $mg$ is just balanced by the centrifugal force $mv^2/R$:

$$T + mg = \frac{mv^2}{R}$$

Solve for $m$.

49. The total energy of a body of mass $m$, moving with velocity $v$ and located at a height $y$ above some datum, is the sum of the potential energy $mgy$ and the kinetic energy $\frac{1}{2}mv^2$:

$$E = mgy + \tfrac{1}{2}mv^2$$

Solve for $m$.

## CHAPTER 8 REVIEW PROBLEMS

*Perform the indicated operations and simplify.*

1. $\dfrac{a^2 + b^2}{a - b} - a + b$

2. $\dfrac{3}{5x^2} - \dfrac{2}{15xy} + \dfrac{1}{6y^2}$

3. $\dfrac{3m + 1}{3m - 1} + \dfrac{m - 4}{5 - 2m} - \dfrac{3m^2 - 2m - 4}{6m^2 - 17m + 5}$

4. $\dfrac{\dfrac{1}{1 - x} - \dfrac{1}{1 + x}}{\dfrac{1}{1 - x^2} - \dfrac{1}{1 + x^2}}$

5. $(a^2 + 1 + a)\left(1 - \dfrac{1}{a} + \dfrac{1}{a^2}\right)$

6. $\dfrac{\dfrac{a - 1}{6} - \dfrac{2a - 7}{2}}{\dfrac{3a}{4} - 3}$

7. $\dfrac{1 + \dfrac{a - c}{a + c}}{1 - \dfrac{a - c}{a + c}}$

8. $\left(1 + \dfrac{x + y}{x - y}\right)\left(1 - \dfrac{x - y}{x + y}\right)$

9. $\dfrac{x^2 + 8x + 15}{x^2 + 7x + 10} - \dfrac{x - 1}{x - 2}$

10. $\dfrac{x^2 - 5ax + 6a^2}{x^2 - 8ax + 15a^2} - \dfrac{x - 7a}{x - 5a}$

11. $\dfrac{1}{a^2 - 7a + 12} + \dfrac{2}{a^2 - 4a + 3} - \dfrac{3}{a^2 - 5a + 4}$

12. $\dfrac{x^4 - y^4}{(x - y)^2} \times \dfrac{x - y}{x^2 + xy}$

13. $\dfrac{a^3 - b^3}{a^3 + b^3} \times \dfrac{a^2 - ab + b^2}{a - b}$

14. $\dfrac{3wx^2y^3}{7axyz} \times \dfrac{4a^3xz}{6aw^2y}$

15. $\dfrac{4a^2 - 9c^2}{4a^2 + 6ac}$

16. $\dfrac{b^2 - 5b}{b^2 - 4b - 5}$

17. $\dfrac{3a^2 + 6a}{a^2 + 4a + 4}$

18. $\dfrac{20(a^3 - c^3)}{4(a^2 + ac + c^2)}$

19. $\dfrac{x^2 - y^2 - 2yz - z^2}{x^2 + 2xy + y^2 - z^2}$

20. $\dfrac{2a^2 + 17a + 21}{3a^2 + 26a + 35}$

21. $\dfrac{(a + b)^2 - (c + d)^2}{(a + c)^2 - (b + d)^2}$

22. $\dfrac{1}{x - 2a} + \dfrac{a^2}{x^3 - 8a^3} - \dfrac{x + a}{x^2 + 2ax + 4a^2}$

*Solve for x.*

23. $cx - 5 = ax + b$

24. $a(x - 3) - b(x + 2) = c$

25. $\dfrac{5 - 3x}{4} + \dfrac{3 - 5x}{3} = \dfrac{3}{2} - \dfrac{5x}{3}$

26. $\dfrac{3x - 1}{11} - \dfrac{2 - x}{10} = \dfrac{6}{5}$

27. $\dfrac{x + 3}{2} + \dfrac{x + 4}{3} + \dfrac{x + 5}{4} = 16$

28. $\dfrac{2x + 1}{4} - \dfrac{4x - 1}{10} + \dfrac{5}{4} = 0$

29. $x^2 - (x - p)(x + q) = r$

30. $mx - n = \dfrac{nx - m}{p}$

31. $p = \dfrac{q - rx}{px - q}$

32. $m(x - a) + n(x - b) + p(x - c) = 0$

33. $\dfrac{x - 3}{4} - \dfrac{x - 1}{9} = \dfrac{x - 5}{6}$

34. $\dfrac{1}{x - 5} - \dfrac{1}{4} = \dfrac{1}{3}$

35. $\dfrac{3}{y - 4} + \dfrac{5}{2(y - 4)} + \dfrac{9}{2(y - 4)} = \dfrac{1}{2}$

36. $\dfrac{2}{x - 2} = \dfrac{5}{2(x - 1)}$

37. $\dfrac{10 - 7x}{6 - 7x} = \dfrac{5x - 4}{5x}$

38. $\dfrac{9x + 5}{6(x - 1)} + \dfrac{3x^2 - 51x - 71}{18(x^2 - 1)} = \dfrac{15x - 7}{9(x + 1)}$

39. It is estimated that a bulldozer can prepare a certain site in 15 days and that a larger bulldozer can prepare the same site in 11 days. If we assume that they do not get in each other's way, how long will it take the two machines, working together, to prepare the site?

40. The acceleration $a_1$ of a body of mass $m_1$ subjected to a force $F_1$ is, by Eq. A20, $a_1 = F_1/m_1$. Write an expression for the difference in acceleration between that body and another having a mass $m_2$ and force $F_2$. Combine the terms of that expression and simplify.

41. Multiply: $2\frac{3}{5} \times 7\frac{5}{9}$

42. Divide: $\dfrac{8x^3y^2z}{5a^4b^3c}$ by $\dfrac{4xy^5z^3}{15abc^3}$

43. Multiply: $\dfrac{3wx^2y^3}{7axyz}$ by $\dfrac{4a^3xz}{6aw^2y}$

44. Divide: $9\frac{5}{7} \div 3\frac{4}{9}$

*Writing*

45. Suppose that a vocal member of your local school board says that the study of fractions is no longer important now that we have computers and insists that it be cut from the curriculum to save money.

Write a short letter to the editor of your local paper in which you agree or disagree. Give your reasons for retaining or eliminating the study of fractions.

# 9

# SYSTEMS OF LINEAR EQUATIONS

## OBJECTIVES

**When you have completed this chapter, you should be able to:**

- Find an approximate graphical solution to a system of two equations.
- Solve a system of two equations in two unknowns by the addition–subtraction method.
- Solve a system of two equations by substitution, or by a computer technique.
- Solve a system of two equations having fractional coefficients or having the unknowns in the denominators.
- Solve a system of three or more equations by addition–subtraction or by substitution.
- Write a system of equations to describe a given word problem and to solve those equations.

Usually, a physical situation can be described by a single equation. For example, the displacement of a freely falling body is described by Eq. A18, $s = v_0t + \frac{1}{2}at^2$. Often, however, a situation can only be described by *more than one* equation. For example, to find the two currents $I_1$ and $I_2$ in the circuit of Fig. 9-1, we must solve the two equations

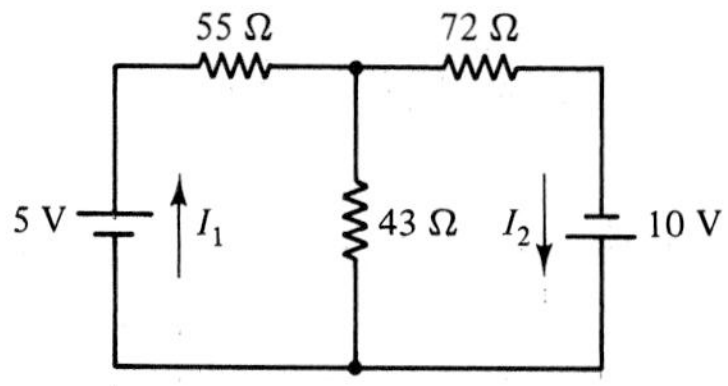


**FIGURE 9-1**

$$98I_1 - 43I_2 = 5$$
$$43I_1 - 115I_2 = 10$$

We must find values for $I_1$ and $I_2$ that satisfy *both* equations at the same time. In this chapter we learn how to solve such sets of two or three linear equations, and we also get practice in writing systems of equations to describe a variety of technical problems. In Chapter 10 we learn how to solve systems of equations by means of determinants, and in Chapter 11 we solve sets of equations using matrices.

## 9-1 SYSTEMS OF TWO LINEAR EQUATIONS

### Linear Equations

We have previously defined a *linear equation* as one of *first degree.*

---

**EXAMPLE 1:** The equation

$$3x + 5 = 20$$

is a linear equation in *one unknown*. We learned how to solve these in Chapter 3.

---

**EXAMPLE 2:** The equation

$$2x - 5y = 8$$

is a linear equation in *two* unknowns.

---

A linear equation can have *any number* of unknowns.

---

**EXAMPLE 3:** $x - 3y + 2z = 5$ is a linear equation in three unknowns.

---

**EXAMPLE 4:** $3x + 2y - 5z - w = 6$ is a linear equation in four unknowns.

---

### Systems of Equations

Systems of equations are also called *simultaneous* equations.

A set of two or more equations that simultaneously impose conditions on all the variables is called a *system of equations.*

**EXAMPLE 5:**

(a) $3x - 2y = 5$

$x + 4y = 1$

is a system of two linear equations in two unknowns.

(b) $x - 2y + 3z = 4$

$3x + y - 2z = 1$

$2x + 3y - z = 3$

is a system of three linear equations in three unknowns.

**Note that some variables may have zero coefficients and may not appear in every equation.**

(c) $2x - y = 5$

$x + 2z = 3$

$3y - z = 1$

is also a system of three linear equations in three unknowns.

(d) $y = 2x^2 - 3$

$y^2 = x + 5$

is a system of two quadratic equations in two unknowns.

## Solution to a System of Equations

The solution to a *system* of equations is a set of values of the unknowns that will *satisfy every equation* in the system.

**EXAMPLE 6:** The system of equations

$$x + y = 5$$

$$x - y = 3$$

is satisfied *only* by the values $x = 4$, $y = 1$, and by *no other* set of values. Thus, the pair (4, 1) is the solution to the system, and the equations are said to be *independent*.

To get a numerical solution for all the unknowns in a system of linear equations, if one exists, *there must be as many independent equations as there are unknowns*. We first solve two equations in two unknowns, then later, three equations in three unknowns, and then larger systems, but for a solution to be possible, the number of equations must always equal the number of unknowns.

## Approximate Graphical Solution to a Pair of Equations

Since any point on a curve satisfies the equation of that curve, the coordinates of the *points of intersection* of two curves will satisfy the equations of *both* curves. Thus we merely have to plot the two curves and find their point or points of intersection; these will be the solution to the pair of equations.

**EXAMPLE 7:** Graphically find the approximate solution to the pair of linear equations

$$3x - y = 1$$
$$x + y = 3$$

**Solution:** We plot the lines as in Chapter 4, as shown in Fig. 9-2. The lines are seen to intersect at the point (1, 2). Our solution is then $x = 1$, $y = 2$.

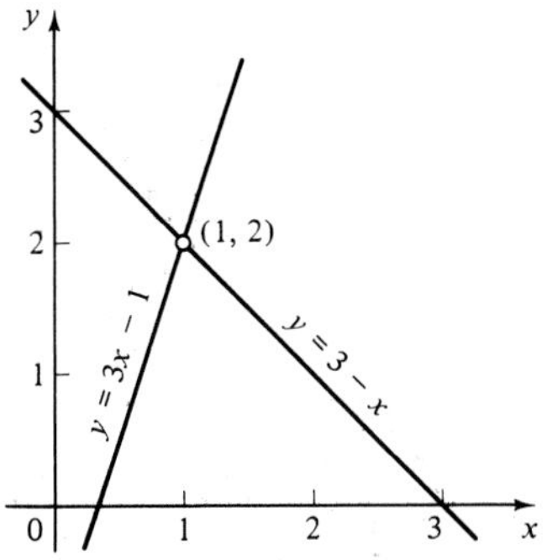


**FIGURE 9-2.** **Graphical solution to a pair of equations. This is also a good way to get an approximate solution to a pair of *nonlinear* equations.**

A *linear* equation in two unknowns plots as a *straight line*. Thus two linear equations plot as two straight lines. If the lines intersect in a single point, the coordinates of that point will satisfy both equations, and hence is a solution to the set of equations. If the lines are *parallel*, there is no point whose coordinates satisfy both equations, and the set of equations is said to be *inconsistent*. If the two lines *coincide* at every point, the coordinates of any point on one line will satisfy both equations. Such a set of equations is called *dependent*.

## Solving a Pair of Linear Equations by the Addition–Subtraction Method

The method of *addition–subtraction*, and the method of *substitution* that follows, have the object of *eliminating* one of the unknowns.

In the addition–subtraction method, this is accomplished by first (if necessary) multiplying each equation by such numbers that will make the coefficients of one unknown in both equations equal in absolute value. The two equations are then added or subtracted so as to eliminate that variable.

Let us first use this method for the same system that we solved graphically in Example 7.

**EXAMPLE 8:** Solve by the addition–subtraction method:

$$3x - y = 1$$
$$x + y = 3$$

**Solution:** Simply adding the two equations causes $y$ to drop out.

$$\begin{array}{r} 3x - y = 1 \\ \underline{x + y = 3} \\ 4x \quad\;\; = 4 \\ x = 1 \end{array}$$

Substituting into the second given equation yields

$$1 + y = 3$$
$$y = 2$$

Our solution is then $x = 1$, $y = 2$, as found graphically in Example 7.

In the next example we must multiply one equation by a constant before adding.

**EXAMPLE 9:** Solve by the addition–subtraction method:

$$2x - 3y = -4$$

$$x + y = 3$$

**Solution:** Multiply the second equation by 3:

$$\begin{aligned} 2x - 3y &= -4 \\ 3x + 3y &= 9 \\ \hline 5x \quad &= 5 \end{aligned}$$

Add:

We have thus reduced our two original equations to a single equation in one unknown. Solving for $x$ gives

$$x = 1$$

Substituting into the second original equation, we have

$$1 + y = 3$$

$$y = 2$$

So the solution is $x = 1$, $y = 2$.

*Check*: Substitute into the first original equation:

$$2(1) - 3(2) \stackrel{?}{=} -4$$

$$2 - 6 = -4 \qquad \text{checks}$$

and into the second original equation:

$$1 + 2 = 3 \qquad \text{checks}$$

Often it is necessary to multiply *both* given equations by suitable factors, as shown in the following example.

**EXAMPLE 10:** Solve by addition or subtraction:

$$5x - 3y = 19$$

$$7x + 4y = 2$$

**Solution:** Multiply the first equation by 4 and the second by 3:

$$\begin{aligned} 20x - 12y &= 76 \\ 21x + 12y &= 6 \\ \hline 41x \quad &= 82 \\ x &= 2 \end{aligned}$$

Add:

Substituting $x = 2$ into the first equation gives

$$5(2) - 3y = 19$$

$$3y = -9$$

$$y = -3$$

So the solution is $x = 2$, $y = -3$.

These values check when substituted into each of the original equations (work not shown). Notice that we could have eliminated the $x$ terms by multiplying the first equation by 7 and the second by $-5$, and adding. The results, of course, would have been the same: $x = 2$, $y = -3$.

The coefficients in the preceding examples were integers, but in applications will usually be approximate numbers. If so, we must retain the proper number of significant digits, as in the following example.

---

**EXAMPLE 11:** Solve for $x$ and $y$.

$$2.64x + 8.47y = 3.72$$
$$1.93x + 2.61y = 8.25$$

**Solution:** Let us eliminate $y$. We multiply the first equation by 2.61 and the second equation by $-8.47$. Let us carry at least three digits and/or two decimal places in our calculation, and round our answer to three digits at the end.

We are less likely to make a mistake adding rather than subtracting, so it is safer to multiply by a negative number and add the resulting equations, as we do here.

$$\begin{array}{rl} 6.89x + 22.11y &= 9.71 \\ -16.35x - 22.11y &= -69.88 \\ \hline -9.46x \quad\quad\quad &= -60.17 \\ x &= 6.36 \end{array}$$

Substituting into the first equation yields

$$2.64(6.36) + 8.47y = 3.72$$
$$8.47y = 3.72 - 16.79 = -13.07$$
$$y = -1.54$$

---

### Substitution Method

To use the *substitution method* to solve a pair of linear equations, first solve either original equation for one unknown in terms of the other unknown. Then substitute this expression into the other equation, thereby eliminating one unknown.

---

**EXAMPLE 12:** Solve by substitution:

$$7x - 9y = 15 \quad (1)$$
$$8y - 5x = -17 \quad (2)$$

**Solution:** We choose to solve the second equation for $y$.

$$8y = 5x - 17$$
$$y = \frac{5x - 17}{8} \quad (3)$$

Substituting this expression for $y$ into the first equation yields

$$7x - 9\left(\frac{5x - 17}{8}\right) = 15$$

If one or both of the given equations is already in explicit form, then the substitution method is probably the easier. Otherwise, many students find the addition–subtraction method easier.

We now have one equation in one unknown. Solving for $x$ gives

$$56x - 9(5x - 17) = 120$$
$$56x - 45x + 153 = 120$$
$$11x = -33$$
$$x = -3$$

Substituting $x = -3$ into (3), we obtain

$$y = \frac{5(-3) - 17}{8} = -4$$

So the solution is $x = -3$, $y = -4$.

## Systems Having No Solution

Certain systems of equations have no unique solution. If you try to solve either of these types, *both variables will vanish.*

**EXAMPLE 13:** Solve the system

$$2x + 3y = 5$$
$$6x + 9y = 2$$

**Solution:** Multiply the first equation by 3:

$$6x + 9y = 15$$
$$6x + 9y = 2$$

Subtract: $\quad 0 = 13 \quad$ no solution

It does not matter much whether a system is inconsistent or dependent; in either case we get no useful solution. But the practical problems we solve here will always have numerical solutions, so if your variables vanish, go back and check your work.

If both variables vanish and an *inequality* results, as in Example 13, the system is called *inconsistent*. The equations would plot as two *parallel lines*. There is no point of intersection and hence no solution.

If both variables vanish and an *equality* results (such as 4 = 4), the system is called *dependent*. The two equations would plot as a *single line*. This indicates that there are infinitely many solutions.

## A Computer Technique: The Gauss–Seidel Method

The *Gauss–Seidel method* is a simple computer technique for solving systems of equations. Suppose that we wish to solve the set of equations

$$x - 2y + 2 = 0 \qquad (1)$$
$$3x - 2y - 6 = 0 \qquad (2)$$

We first solve the first equation for $x$ in terms of $y$ and the second equation for $y$ in terms of $x$.

$$x = 2y - 2 \qquad (3)$$
$$y = \frac{3x - 6}{2} \qquad (4)$$

We then *guess* at the value of $y$, substitute this value into (3), and obtain a value for $x$. This $x$ is substituted into (4) and a value of $y$ is obtained. These values for $x$ and $y$ will not be the correct values; they are only our *first approximation* to the true values.

The latest value for $y$ is then put into (3), producing a new $x$; which is then put into (4), producing a new $y$; which is then put into (3); and so on. We repeat the computation until the values no longer change (we say that they *converge* on the true values). If the values of $x$ and $y$ get very large (*diverge* instead of converge), we solve the first equation for $y$ and the second equation for $x$, and the computation will converge.

**This type of repetitive process is called *iteration*, or the *method of successive approximations*.**

---

**EXAMPLE 14:** Calculate $x$ and $y$ for the equations above using the Gauss–Seidel method. Take $y = 0$ for the first guess.

**Solution:** From equation (3) we get

$$x = 2(0) - 2 = -2$$

substituting $x = -2$ into (4),

$$y = \frac{3(-2) - 6}{2} = -6$$

With $y$ equal to $-6$, we then use (3) to find a new $x$, and so on. We get the values

| $x$ | $y$ |
|---:|---:|
| −2 | −6 |
| −14 | −24 |
| −50 | −78 |
| −158 | −240 |
| −482 | −726 |
| −1,454 | −2,184 |
| −4,370 | −6,558 |
| −13,118 | −19,680 |
| −39,362 | −59,046 |
| ⋮ | ⋮ |

Note that the values are *diverging*. We try again, this time solving equation (1) for $y$,

$$y = \frac{x + 2}{2}$$

and solving (2) for $x$,

$$x = \frac{2y + 6}{3}$$

Repeating the computation, starting with $x = 0$, gives the values

| $x$ | $y$ |
|---|---|
| 2.666667 | 1 |
| 3.555556 | 2.333334 |
| 3.851852 | 2.777778 |
| 3.950617 | 2.925926 |
| 3.983539 | 2.975309 |
| 3.994513 | 2.991769 |
| 3.998171 | 2.997256 |
| 3.99939 | 2.999086 |
| 3.999797 | 2.999695 |
| 3.999932 | 2.999899 |
| 3.999977 | 2.999966 |
| 3.999992 | 2.999989 |
| 3.999998 | 2.999996 |
| 3.999999 | 2.999999 |
| 4 | 3 |
| 4 | 3 |
| ⋮ | ⋮ |

which converge on $x = 4$ and $y = 3$.

How do we know what value to take for our first guess? Choose whatever values seem "reasonable." If you have no idea, then choose any values, such as the zeros in this example. Actually, it does not matter much what values we start with. If the computation is going to converge, it will converge regardless of the initial values chosen. However, the closer your guess, the *faster* the convergence.

## EXERCISE 1—SYSTEMS OF TWO LINEAR EQUATIONS

### Graphical Solution

Graphically find the approximate solution to each system of equations.

There are no applications given here. They are located in Exercise 3.

1. $2x - y = 5$
   $x - 3y = 5$
2. $x + 2y = -7$
   $5x - y = 9$
3. $x - 2y = -3$
   $3x + y = 5$
4. $4x + y = 8$
   $2x - y = 7$
5. $2x + 5y = 4$
   $5x - 2y = -3$
6. $x - 2y + 2 = 0$
   $3x - 6y + 2 = 0$

### Algebraic Solution

Solve each system of equations by addition–subtraction, by substitution, or by computer, as directed by your instructor.

7. $2x + y = 11$
   $3x - y = 4$
8. $5x + 7y = 101$
   $y = 7x - 55$
9. $3x - 2y = -15$
   $5x + 6y = 3$
10. $7x + 6y = 20$
    $2x + 5y = 9$
11. $x + 5y = 11$
    $3x + 2y = 7$
12. $4x - 5y = -34$
    $2x - 3y = -22$

13. $x = 11 - 4y$
$5x - 2y = 11$

14. $2x - 3y = 3$
$4x + 5y = 39$

15. $7x - 4y = 81$
$5x - 3y = 57$

16. $3x + 4y = 85$
$5x + 4y = 107$

17. $3x - 2y = 1$
$2x + y = 10$

18. $5x - 2y = 3$
$2x + 3y = 5$

19. $y = 9 - 3x$
$x = 8 - 2y$

20. $y = 2x - 3$
$x = 19 - 3y$

21. $29.1x - 47.6y = 42.8$
$11.5x + 72.7y = 25.8$

22. $4.92x - 8.27y = 2.58$
$6.93x + 2.84y = 8.36$

23. $4n = 18 - 3m$
$m = 8 - 2n$

24. $5p + 4q - 14 = 0$
$17p = 31 + 3q$

25. $3w = 13 + 5z$
$4w - 7z - 17 = 0$

26. $3u = 5 + 2v$
$5v + 2u = 16$

**Computer**

27. Write a program or use a spreadsheet for the Gauss-Seidel method described in Section 9-1. Use it to solve any of the sets of equations in this exercise.

## 9-2 OTHER SYSTEMS OF EQUATIONS

### Systems with Fractional Coefficients

When one or more of the equations in our system has fractional coefficients, simply multiply the entire equation by the LCD, and proceed as before.

**The techniques in this section apply not only to two equations in two unknowns, but to larger systems as well.**

---

**EXAMPLE 15:** Solve for $x$ and $y$:

$$\frac{x}{2} + \frac{y}{3} = \frac{5}{6} \qquad (1)$$

$$\frac{x}{4} - \frac{y}{2} = \frac{7}{4} \qquad (2)$$

**Solution:** Multiply (1) by 6 and (2) by 4:

$$3x + 2y = 5 \qquad (3)$$
$$x - 2y = 7 \qquad (4)$$

Add (3) and (4):

$$4x = 12$$
$$x = 3$$

Having $x$, we can now substitute back to get $y$. It is not necessary to substitute back into one of the original equations. We choose instead the easiest place to substitute, such as Equation (4). (However, when *checking,* be sure to substitute your answers into both of the *original* equations.) Substituting $x = 3$ into (4) gives

$$3 - 2y = 7$$
$$2y = 3 - 7 = -4$$
$$y = -2$$

**To save space we will not show the check in many of these examples.**

---

## Fractional Equations with Unknowns in the Denominator

The same method (multiplying by the LCD) can be used to clear fractions when the unknowns appear in the denominators. Note that such equations are not linear (that is, not of first degree) as were the equations we have solved so far, but we are able to solve them by the same methods.

---

**EXAMPLE 16:** Solve the system

Of course, neither $x$ nor $y$ can equal zero in this system.

$$\frac{10}{x} - \frac{9}{y} = 8 \qquad (1)$$

$$\frac{8}{x} + \frac{15}{y} = -1 \qquad (2)$$

**Solution:** Multiply (1) by $xy$ and (2) by $xy$:

$$10y - 9x = 8xy \qquad (3)$$

$$8y + 15x = -xy \qquad (4)$$

Multiply (3) by 5, and (4) by 3:

$$50y - 45x = 40xy$$
$$24y + 45x = -3xy$$

Add: $\quad 74y = 37xy$

Divide by $37y$: $\quad 2 = x$

Substitute $x = 2$ into (1):

$$\frac{10}{2} - \frac{9}{y} = 8$$

$$\frac{9}{y} = 5 - 8 = -3$$

$$y = -3$$

---

A convenient way to solve nonlinear systems such as these is to *substitute new variables* such as to make the equations linear. Solve in the usual way and then substitute back.

---

**EXAMPLE 17:** Solve the nonlinear system of equations

$$\frac{2}{3x} + \frac{3}{5y} = 17 \qquad (1)$$

$$\frac{3}{4x} + \frac{2}{3y} = 19 \qquad (2)$$

**Solution:** We substitute $m = 1/x$ and $n = 1/y$ and get the linear system

$$\frac{2m}{3} + \frac{3n}{5} = 17 \qquad (3)$$

$$\frac{3m}{4} + \frac{2n}{3} = 19 \qquad (4)$$

This technique of substitution is very useful, and we will use it again later to reduce certain equations to quadratic form. Do not confuse this kind of substitution with the method of substitution we studied in Sec. 9-1.

Again, we clear fractions by multiplying each equation by its LCD. Multiplying (3) by 15 gives

$$10m + 9n = 255 \qquad (5)$$

and multiplying (4) by 12 gives

$$9m + 8n = 228 \qquad (6)$$

Using the addition–subtraction method, we multiply (5) by 8 and multiply (6) by −9.

$$\begin{aligned} 80m + 72n &= 2040 \\ -81m - 72n &= -2052 \\ \hline \text{Add:} \quad -m \qquad &= -12 \\ m &= 12 \end{aligned}$$

Substituting $m = 12$ into (5) yields

$$120 + 9n = 255$$

$$n = 15$$

Finally, we substitute back to get $x$ and $y$:

$$x = \frac{1}{m} = \frac{1}{12} \quad \text{and} \quad y = \frac{1}{n} = \frac{1}{15}$$

| Common Error | Students often forget that last step. We are not solving for $m$ and $n$, but for $x$ and $y$. |
|---|---|

### Literal Equations

We use the method of addition–subtraction or the method of substitution for solving systems of equations with literal coefficients, treating the literals as if they were numbers.

**EXAMPLE 18:** Solve for $x$ and $y$ in terms of $m$ and $n$.

$$2mx + ny = 3$$

$$mx + 3ny = 2$$

**Solution:** We will use the addition–subtraction method. Multiply the second equation by −2:

$$\begin{aligned} 2mx + ny &= 3 \\ -2mx - 6ny &= -4 \\ \hline \text{Add:} \quad -5ny &= -1 \\ y &= \frac{1}{5n} \end{aligned}$$

Substituting back into the second original equation, we obtain

$$mx + 3n\left(\frac{1}{5n}\right) = 2$$

$$mx + \frac{3}{5} = 2$$

$$mx = 2 - \frac{3}{5} = \frac{7}{5}$$

$$x = \frac{7}{5m}$$

**EXAMPLE 19:** Solve for $x$ and $y$ by the addition–subtraction method:

$$a_1x + b_1y = c_1$$
$$a_2x + b_2y = c_2$$

**Solution:** Multiplying the first equation by $b_2$ and the second equation by $b_1$, we obtain

$$\begin{aligned} a_1b_2x + b_1b_2y &= b_2c_1 \\ a_2b_1x + b_1b_2y &= b_1c_2 \end{aligned}$$

Subtract: $$(a_1b_2 - a_2b_1)x = b_2c_1 - b_1c_2$$

Dividing by $a_1b_2 - a_2b_1$ gives

$$x = \frac{b_2c_1 - b_1c_2}{a_1b_2 - a_2b_1}$$

Now solving for $y$, we multiply the first equation by $a_2$ and the second equation by $a_1$. Writing the second equation above the first, we get

$$\begin{aligned} a_1a_2x + a_1b_2y &= a_1c_2 \\ a_1a_2x + a_2b_1y &= a_2c_1 \end{aligned}$$

Subtract: $$(a_1b_2 - a_2b_1)y = a_1c_2 - a_2c_1$$

Dividing by $a_1b_2 - a_2b_1$ yields

$$y = \frac{a_1c_2 - a_2c_1}{a_1b_2 - a_2b_1}$$

---

This result may be summarized as follows:

We are going to need this formula again when we discuss determinants in Chapter 10.

| | | |
|---|---|---|
| Two Linear Equations in Two Unknowns | The solution to the set of equations $a_1x + b_1y = c_1$, $a_2x + b_2y = c_2$ is $x = \frac{b_2c_1 - b_1c_2}{a_1b_2 - a_2b_1}$ $y = \frac{a_1c_2 - a_2c_1}{a_1b_2 - a_2b_1}$ $(a_1b_2 - a_2b_1 \neq 0)$ | **63** |

Thus Eq. 63 is a formula for solving a pair of linear equations. We simply have to identify the six numbers $a_1$, $b_1$, $c_1$, $a_2$, $b_2$, and $c_2$, and substitute them into the equations. But be sure to put the equations into standard form first.

**EXAMPLE 20:** Solve using Eq. 63:

$$9y = 7x - 15$$
$$-17 + 5x = 8y$$

**Solution:** Rewrite our equations in the form given in Eq. 63:

$$7x - 9y = 15$$
$$5x - 8y = 17$$

Then substitute, with

$$a_1 = 7 \qquad b_1 = -9 \qquad c_1 = 15$$
$$a_2 = 5 \qquad b_2 = -8 \qquad c_2 = 17$$

$$x = \frac{b_2c_1 - b_1c_2}{a_1b_2 - a_2b_1} = \frac{-8(15) - (-9)(17)}{7(-8) - 5(-9)}$$
$$= \frac{-120 + 153}{-56 + 45} = \frac{33}{-11} = -3$$

$$y = \frac{a_1c_2 - a_2c_1}{a_1b_2 - a_2b_1} = \frac{7(17) - (5)(15)}{-11}$$
$$= \frac{119 - 75}{-11} = \frac{44}{-11} = -4$$

## EXERCISE 2—OTHER SYSTEMS OF EQUATIONS

### Fractional Coefficients

Solve simultaneously.

**1.** $\frac{x}{5} + \frac{y}{6} = 18$

$\frac{x}{2} - \frac{y}{4} = 21$

**2.** $\frac{x}{2} + \frac{y}{3} = 7$

$\frac{x}{3} + \frac{y}{4} = 5$

**3.** $\frac{x}{3} + \frac{y}{4} = 8$

$x - y = -3$

**4.** $\frac{x}{2} + \frac{y}{3} = 5$

$\frac{x}{3} + \frac{y}{2} = 5$

**5.** $\frac{3x}{5} + \frac{2y}{3} = 17$

$\frac{2x}{3} + \frac{3y}{4} = 19$

**6.** $\frac{x}{7} + 7y = 251$

$\frac{y}{7} + 7x = 299$

**7.** $\frac{m}{2} + \frac{n}{3} - 3 = 0$

$\frac{n}{2} + \frac{m}{5} = \frac{23}{10}$

**8.** $\frac{p}{6} - \frac{q}{3} + \frac{1}{3} = 0$

$\frac{2p}{3} - \frac{3q}{4} - 1 = 0$

**9.** $\frac{r}{6.2} - \frac{s}{4.3} = \frac{1}{3.1}$

$\frac{r}{4.6} - \frac{s}{2.3} = \frac{1}{3.5}$

### Unknowns in the Denominator

Solve simultaneously.

10. $\frac{8}{x} + \frac{6}{y} = 3$

    $\frac{6}{x} + \frac{15}{y} = 4$

11. $\frac{1}{x} + \frac{3}{y} = 11$

    $\frac{5}{x} + \frac{4}{y} = 22$

12. $\frac{5}{x} + \frac{6}{y} = 7$

    $\frac{7}{x} + \frac{9}{y} = 10$

13. $\frac{2}{x} + \frac{4}{y} = 14$

    $\frac{6}{x} - \frac{2}{y} = 14$

14. $\frac{6}{x} + \frac{8}{y} = 1$

    $\frac{7}{x} - \frac{11}{y} = -9$

15. $\frac{2}{5x} + \frac{5}{6y} = 14$

    $\frac{2}{5x} - \frac{3}{4y} = -5$

16. $\frac{5}{3a} + \frac{2}{5b} = 7$

    $\frac{7}{6a} - 3 = \frac{1}{10b}$

17. $\frac{1}{5z} + \frac{1}{6w} = 18$

    $\frac{1}{4w} - \frac{1}{2z} + 21 = 0$

18. $\frac{8.1}{5.1t} + \frac{1.4}{3.6s} = 1.8$

    $\frac{2.1}{1.4s} - \frac{1.3}{5.2t} + 3.1 = 0$

### Literal Equations

Solve for $x$ and $y$ in terms of the other literal quantities.

19. $3x - 2y = a$

    $2x + y = b$

20. $ax + by = r$

    $ax + cy = s$

21. $ax - dy = c$

    $mx - ny = c$

22. $px - qy + pq = 0$

    $2px - 3qy = 0$

### Computer

23. Program Eqs. 63 into your computer so that $x$ and $y$ will be computed and printed whenever the six coefficients

$$a_1, \quad b_1, \quad c_1 \qquad \text{and} \qquad a_2, \quad b_2, \quad c_2$$

are entered into the program. Test the program on a system of equations that you have already solved algebraically.

## 9-3 WORD PROBLEMS WITH TWO UNKNOWNS

In many problems there are two or more unknowns that must be found. To solve such problems, we must write *as many independent equations as there are unknowns*. Otherwise, it is not possible to obtain numerical answers, although one unknown could be expressed in terms of another.

Set up these problems as we did in Chapter 3 and solve the resulting system of equations by any of the methods of this chapter.

---

**EXAMPLE 21:** A certain investment in bonds had a value of \$248,000 after 4 years, and of \$260,000 after 5 years, at simple interest. Find the amount invested and the interest rate. (The formula $y = P(1 + nt)$ gives the amount $y$ obtained by investing an amount $P$ for $t$ years at an interest rate $n$.)

**Solution:** For the 4-year investment, $t = 4$ years and $y = \$248{,}000$. Substituting into the given formula yields:

$$\$248{,}000 = P(1 + 4n)$$

Then substitute again, with $t = 5$ years and $y = \$260{,}000$:

$$\$260{,}000 = P(1 + 5n)$$

Thus we get two equations in two unknowns, $P$ (the amount invested) and $n$ (the interest rate). Next we remove parentheses:

$$248{,}000 = P + 4Pn \qquad (1)$$

$$260{,}000 = P + 5Pn \qquad (2)$$

At this point we may be tempted to subtract one of these equations from the other to eliminate $P$. But this will not work because $P$ remains in the term containing $Pn$. Instead, we multiply (1) by 5 and (2) by $-4$ to eliminate the $Pn$ term:

$$1{,}240{,}000 = 5P + 20Pn$$

$$-1{,}040{,}000 = -4P - 20Pn$$

Add: $$\$200{,}000 = P$$

Substituting back into (1) gives us

$$248{,}000 = 200{,}000 + 4(200{,}000)n$$

from which

$$n = 0.06$$

Thus a sum of \$200,000 was invested at 6%.
Check: For the 4-year investment

$$y = 200{,}000[1 + 4(0.06)] = \$248{,}000 \qquad \text{checks}$$

and for the 5-year investment,

$$y = 200{,}000[1 + 5(0.06)] = \$260{,}000 \qquad \text{checks}$$

---

**EXAMPLE 22:** During a certain day, two computer printers are observed to process 1705 form letters, with the slower printer in use for 5.5 h and the faster for 4.0 h. On another day the slower printer works for 6.0 h and the faster for 6.5 h, and together they print 2330 form letters. How many letters can each print in an hour, working alone?

**Solution:** We let

$$x = \text{rate of slow printer, letters/h}$$

$$y = \text{rate of fast printer, letters/h}$$

We write two equations to express, for each day, the total amount of work produced, remembering from Chapter 8 that

$$\text{amount of work} = \text{work rate} \times \text{time}$$

On the first day, the slow printer produces $5.5x$ letters while the fast printer produces $4.0y$ letters. Together they produce

$$5.5x + 4.0y = 1705$$

Similarly, for the second day,

$$6.0x + 6.5y = 2330$$

Using the addition–subtraction method, we multiply the first equation by 6.5 and the second by $-4.0$.

$$\begin{aligned} 35.75x + 26y &= 11{,}083 \\ -24x - 26y &= \phantom{0}9{,}320 \end{aligned}$$

Add:

$$\begin{aligned} 11.75x &= 1{,}763 \\ x &= 150 \text{ letters/h} \end{aligned}$$

Substituting back yields

$$5.5(150) + 4y = 1705$$

$$y = 220 \text{ letters/h}$$

## EXERCISE 3—WORD PROBLEMS WITH TWO UNKNOWNS

### Number Problems

**Some of these problems can be set up using just one unknown, but for now use two unknowns for the practice.**

1. The sum of two numbers is 24 and their difference is 8. What are the numbers?
2. The sum of two numbers is 29 and their difference is 5. What are the numbers?
3. There is a fraction such that if 3 is added to the numerator its value will be $\frac{1}{3}$, and if 1 is subtracted from the denominator its value will be $\frac{1}{5}$. What is the fraction?
4. The sum of the two digits of a certain number is 9. If 9 is added to the original number, the new number will have the original digits reversed. Find the number. (*Hint:* Let $x$ = the ten's digit and $y$ = the one's digit, so that the value of the number is $10x + y$.)
5. A number consists of two digits. The number is 2 more than 8 times the sum of the digits, and if 54 is subtracted from the number, the digits will be reversed. Find the number.
6. If the larger of two numbers is divided by the smaller, the quotient is 7 and the remainder is 4. But if three times the greater is divided by twice the smaller, the quotient is 11 and the remainder is 4. What are the numbers?

7. The sum of two numbers divided by 2 gives a quotient of 24, and their difference divided by 2 gives a quotient of 17. What are the numbers?

### Geometry Problems

8. If the width of a certain rectangle is increased by 3 and the length decreased by 3, the area is seen to increase by 6. But if the width is reduced by 5 and the length increased by 3, the area decreases by 90. Find the original dimensions.
9. If the width of a certain rectangle is increased by 3 and the length reduced by 4, we get a square with the same area as the original rectangle. Find the length and width of the original rectangle.

### Uniform Motion Problems

10. In a first race, $A$ gives $B$ a headstart of 100 ft and overtakes him in 4 min. In a second race he gains 750 ft on $B$ when $B$ runs 9000 ft. Find the rate at which each runs.
11. $A$ and $B$ run two races of 280 ft. In the first race, $A$ gives $B$ a start of 70 ft and neither wins the race. In the second race, $A$ gives $B$ a start of 35 ft and beats him by $6\frac{2}{3}$ s. How many feet can each run in a second?
12. $A$ and $B$ run two races from $P$ to $Q$ and back, the distance from $P$ to $Q$ being 108 m. In the first race, $A$ reaches $Q$ first and meets $B$ on his return at a point 12 m from $Q$. In the second race, $A$ increases his speed by 2 m/s and $B$ by 1 m/s. Now $A$ meets $B$ 18 m from $Q$. How fast did each run during the first race?
13. A certain river has a speed of 2.5 mi/h. A rower travels downstream for 1.5 h and returns in 4.5 h. Find his rate in still water, and the one-way distance traveled.
14. A canoist can paddle 20.0 mi down a certain river and back in 8.0 h (40.0 mi round trip). She can also paddle 5.0 mi down-river in the same time as she rows 3.0 mi up-river. Find her rate in still water, and the rate of the current.

### Financial Problems

15. A certain investment, at simple interest, amounted in 5 years to \$3000 and in 6 years to \$3100. Find the amount invested and the rate of interest (see Eq. A9).
16. If 4 yd of velvet and 3 yd of silk are sold for \$53, and 5 yd of velvet and 6 yd of silk for \$72, what is the price per yard of each?
17. \$10,000 is invested, part at 4% and the remainder at 5%. The annual income from the 4% investment is \$40 more than from the 5% investment. Find the amounts invested at 4% and at 5%.
18. A person invested \$4400, part of it in railroad bonds bearing 6.2% interest and the remainder in state bonds bearing 9.7% interest, and received the same income from each. How much was invested in each?
19. A farmer bought 100 acres of land, part at \$370 and part at \$450 an acre, paying for the whole \$42,200. How much land was there in each part?
20. If I loan my money at 6% simple interest for a given time, I shall receive \$720 interest; but if I loan it for 3 years longer, I shall receive \$1800. Find the amount of money and the time.

### Mixture Problems

21. A certain brass alloy contains 35% zinc and 3.0% lead. Then $x$ kg of zinc and $y$ kg of lead are added to 200 kg of the original alloy to make a new alloy that is 40% zinc and 4.0% lead. **(a)** Show that the amount of zinc in the new alloy is given by

$$0.35(200) + x = 0.40(200 + x + y)$$

and that the amount of lead is given by

$$0.03(200) + y = 0.04(200 + x + y)$$

**(b)** Solve for $x$ and $y$.

**22.** A certain concrete mixture contains 5.0% cement and 8.0% sand. How many pounds of this mixture and how many pounds of sand should be combined with 255 lb of cement to make a batch that is 12% cement and 15% sand?

**23.** A distributor has two gasohol blends: one that contains 5% alcohol and another with 11% alcohol. How many gallons of each must be mixed to make 500 gal of gasohol containing 9.5% alcohol?

**24.** A potting mixture contains 12% peat moss and 6.0% vermiculite. How much peat and how much vermiculite must be added to 100 lb of this mixture to produce a new mixture having 15% peat and 15% vermiculite?

### Statics

**25.** Find the forces $F_1$ and $F_2$ in Fig. 9-3.

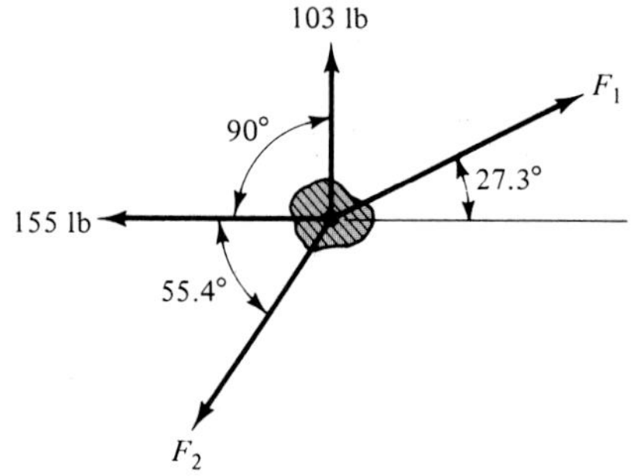


**FIGURE 9-3**

**26.** Find the tensions in the ropes in Fig. 9-4.

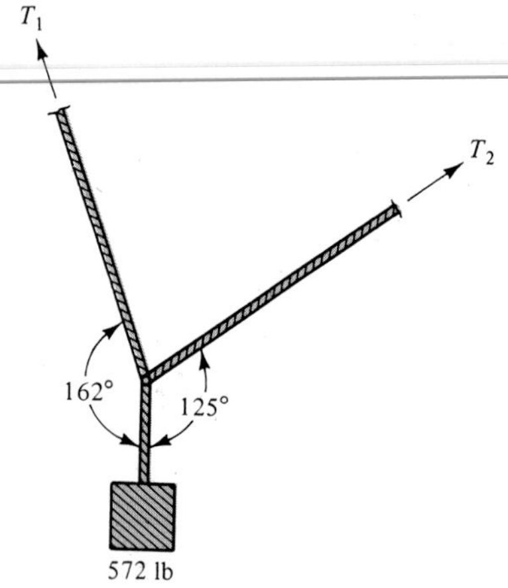


**FIGURE 9-4**

**27.** When the 45.3-kg mass in Fig. 9-5 is increased to 100 kg, the balance point shifts 15.4 cm. Find the length of the bar and the original distance from the balance point to the 45.3-kg mass.

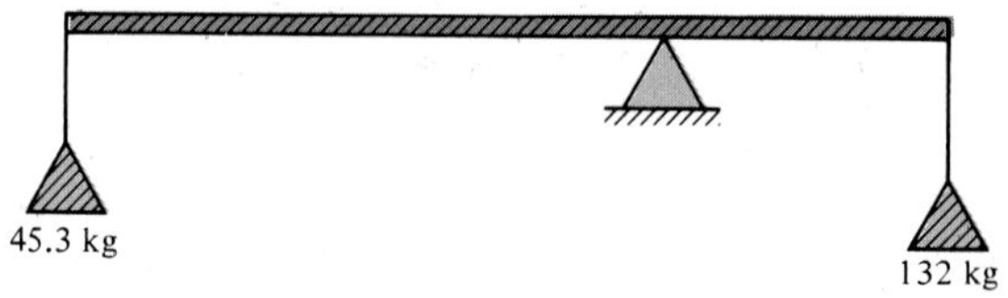


**FIGURE 9-5**

### Work Problems

**28.** A carpenter and a helper can do a certain job in 15 days. If the carpenter works 1.5 times as fast as the helper, how long would it take each to do the job, working alone?

**29.** During 1 week two machines produce a total of 27,210 parts, with the faster machine working for 37.5 h and the slower for 28.2 h. During another week, they produce 59,830 parts, with the faster machine working 66.5 h and the slower machine working for 88.6 h. How many parts can each produce in an hour working alone?

### Flow Problems

**30.** Two different-sized pipes lead from a dockside to a group of oil storage tanks. On one day the two pipes are seen to deliver 117,000 gal, with pipe $A$ in use for 3.5 h and pipe $B$ for 4.5 h. On another day the two pipes deliver 151,200 gal, with pipe $A$ operating for 5.2 h and pipe $B$ for 4.8 h. If pipe $A$ can deliver $x$ gal/h and pipe $B$ can deliver $y$ gal/h, **(a)** show that the total gallons of oil delivered for the two days are given by the equations

$$3.5x + 4.5y = 117{,}000$$

$$5.2x + 4.8y = 151{,}200$$

**(b)** Solve for $x$ and $y$.

**31.** Two conveyors can fill a certain bin in 6 h, working together. If one conveyor works 1.8 times as fast as the other, how long would it take each to fill the bin working alone?

### Energy Flow

**32.** A hydroelectric generating plant and a coal-fired generating plant together supply a city of 255,000 people, with the hydro plant producing 1.75 times the power of the coal plant. How many people could each service alone?

**33.** During a certain week a small wind generator and a small hydro unit together produce 5880 kWh, with the wind generator operating only 85% of the time. During another week the two units produce 6240 kWh, with the wind generator working 95% of the time and the hydro unit down 7.5 h for repairs. Assuming that each unit has a constant output when operating, find the number of kilowatts produced by each in 1 h.

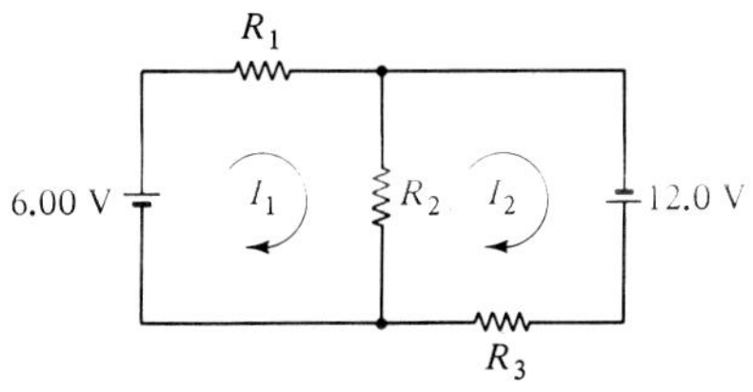

FIGURE 9-6

### Electrical

**34.** To find the currents $I_1$ and $I_2$ in Fig. 9-6 we use Eq. A68 in each loop and get the pair of equations

$$6.00 - R_1I_1 - R_2I_1 + R_2I_2 = 0$$

$$12.0 - R_3I_2 - R_2I_2 + R_2I_1 = 0$$

Solve for $I_1$ and $I_2$ if $R_1 = 736\ \Omega$, $R_2 = 386\ \Omega$, and $R_3 = 375\ \Omega$.

**35.** Use Eq. A68 to write a pair of equations for the circuit of Fig. 9-7 as in Problem 34. Solve these equations for $I_1$ and $I_2$.

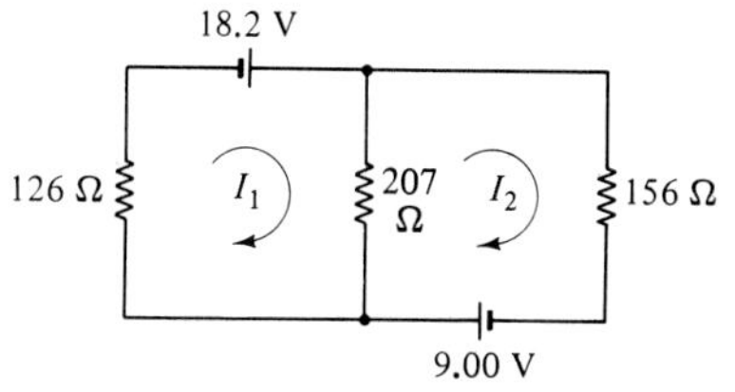

FIGURE 9-7

## 9-4 SYSTEMS OF THREE EQUATIONS

Our strategy here is to reduce a given system of three equations in three unknowns to a system of two equations in two unknowns, which we already know how to solve.

## Addition–Subtraction Method

In the following chapter we will use *determinants* to solve sets of three or more equations.

We take any two of the given equations and, by addition–subtraction or substitution, eliminate one variable, obtaining a single equation in two unknowns. We then take another pair of equations (which must include the one not yet used, as well as one of those already used), and similarly obtain a second equation in the *same two* unknowns. This pair of equations can then be solved simultaneously, and the values obtained are substituted back to obtain the third variable.

---

**EXAMPLE 23:** Solve

$$6x - 4y - 7z = 17 \quad (1)$$

$$9x - 7y - 16z = 29 \quad (2)$$

$$10x - 5y - 3z = 23 \quad (3)$$

**Solution:** Let us start by eliminating $x$ from equations (1) and (2).

It is a good idea to number your equations, as in this example, to help keep track of your work.

Multiply (1) by 3: $18x - 12y - 21z = 51$ (4)
and multiply (2) by $-2$: $-18x + 14y + 32z = -58$ (5)
Add: $2y + 11z = -7$ (6)

We now eliminate the same variable, $x$, from equations (1) and (3).

Multiply (1) by $-5$: $-30x + 20y + 35z = -85$ (7)
and multiply (3) by 3: $30x - 15y - 9z = 69$ (8)
Add: $5y + 26z = -16$ (9)

Now we solve (6) and (9) simultaneously.

Multiply (6) by 5: $10y + 55z = -35$
and (9) by $-2$: $-10y - 52z = 32$
Add: $3z = -3$

$$z = -1$$

Substituting $z = -1$ into (6) gives us

$$2y + 11(-1) = -7$$

$$y = 2$$

Substituting $y = 2$ and $z = -1$ into (1) yields

$$6x - 4(2) - 7(-1) = 17$$

$$x = 3$$

Our solution is then $x = 3$, $y = 2$, and $z = -1$.

---

### Substitution Method

A *sparse* system (one in which many terms are missing) is often best solved by substitution, as in the following example.

---

**EXAMPLE 24:** Solve by substituting:

$$x - y = 4$$

$$x + z = 8$$

$$x - y + z = 10$$

**Solution:** From the first two equations we can write both $y$ and $z$ in terms of $x$:

$$y = x - 4$$

and

$$z = 8 - x$$

Substituting these back into the third equation yields

$$x - (x - 4) + (8 - x) = 10$$

from which

$$x = 2$$

Substituting back gives

$$y = x - 4 = 2 - 4 = -2$$

and

$$z = 8 - x = 8 - 2 = 6$$

Our solution is then $x = 2$, $y = -2$, and $z = 6$.

---

### Fractional Equations

We use the same techniques for solving a set of three fractional equations that we did for two equations: (1) multiply each equation by its LCD to eliminate fractions; and (2) if the unknowns are in the denominators, substitute new variables which are the reciprocals of the originals.

**EXAMPLE 25:** Solve for $x$, $y$, and $z$.

$$\frac{4}{x} + \frac{9}{y} - \frac{8}{z} = 3 \tag{1}$$

$$\frac{8}{x} - \frac{6}{y} + \frac{4}{z} = 3 \tag{2}$$

$$\frac{5}{3x} + \frac{7}{2y} - \frac{2}{z} = \frac{3}{2} \tag{3}$$

**Solution:** We make the substitution

$$p = \frac{1}{x} \qquad q = \frac{1}{y} \qquad \text{and} \qquad r = \frac{1}{z}$$

and also multiply (3) by its LCD, 6, to clear fractions.

$$4p + 9q - 8r = 3 \tag{4}$$

$$8p - 6q + 4r = 3 \tag{5}$$

$$10p + 21q - 12r = 9 \tag{6}$$

We multiply (5) by 2 and add it to (4):

$$20p - 3q = 9 \tag{7}$$

Then we multiply (5) by 3 and add it to (6):

$$34p + 3q = 18 \tag{8}$$

Adding (7) and (8) gives

$$54p = 27$$

$$p = \frac{1}{2}$$

Then from (8),

$$3q = 18 - 17 = 1$$

$$q = \frac{1}{3}$$

and from (5),

$$4r = 3 - 4 + 2 = 1$$

$$r = \frac{1}{4}$$

Returning to our original variables, $x = 2$, $y = 3$, and $z = 4$.

## EXERCISE 4—SYSTEMS OF THREE EQUATIONS

Solve each system of equations.

**1.** $$\begin{aligned} x + y &= 35 \\ x + z &= 40 \\ y + z &= 45 \end{aligned}$$

**2.** $$\begin{aligned} x + y + z &= 12 \\ x - y &= 2 \\ x - z &= 4 \end{aligned}$$

**3.** $$\begin{aligned} 3x + y &= 5 \\ 2y - 3z &= -5 \\ x + 2z &= 7 \end{aligned}$$

**4.** $$\begin{aligned} x - y &= 5 \\ y - z &= -6 \\ 2x - z &= 2 \end{aligned}$$

**5.** $$\begin{aligned} x + y + z &= 18 \\ x - y + z &= 6 \\ x + y - z &= 4 \end{aligned}$$

**6.** $$\begin{aligned} x + y + z &= 90 \\ 2x - 3y &= -20 \\ 2x + 3z &= 145 \end{aligned}$$

**7.** $$\begin{aligned} x + 2y + 3z &= 14 \\ 2x + y + 2z &= 10 \\ 3x + 4y - 3z &= 2 \end{aligned}$$

**8.** $$\begin{aligned} x + y + z &= 35 \\ x - 2y + 3z &= 15 \\ y - x + z &= -5 \end{aligned}$$

**9.** $$\begin{aligned} x - 2y + 2z &= 5 \\ 5x + 3y + 6z &= 57 \\ x + 2y + 2z &= 21 \end{aligned}$$

**10.** $$\begin{aligned} 1.21x + 1.48y + 1.63z &= 6.83 \\ 4.94x + 4.27y + 3.63z &= 21.7 \\ 2.88x + 4.15y - 2.79z &= 2.76 \end{aligned}$$

**11.** $$\begin{aligned} 2.51x - 4.48y + 3.13z &= 10.8 \\ 2.84x + 1.37y - 1.66z &= 6.27 \\ 1.58x + 2.85y - 1.19z &= 27.3 \end{aligned}$$

**12.** $$\begin{aligned} 5a + b - 4c &= -5 \\ 3a - 5b - 6c &= -20 \\ a - 3b + 8c &= -27 \end{aligned}$$

**13.** $$\begin{aligned} p + 3q - r &= 10 \\ 5p - 2q + 2r &= 6 \\ 3p + 2q + r &= 13 \end{aligned}$$

### Fractional Equations

**14.** $$\begin{aligned} x + \frac{y}{3} &= 5 \\ x + \frac{z}{3} &= 6 \\ y + \frac{z}{3} &= 9 \end{aligned}$$

**15.** $$\begin{aligned} \frac{1}{x} + \frac{1}{y} &= 5 \\ \frac{1}{y} + \frac{1}{z} &= 7 \\ \frac{1}{x} + \frac{1}{z} &= 6 \end{aligned}$$

**16.** $$\begin{aligned} \frac{x}{10} + \frac{y}{5} + \frac{z}{20} &= \frac{1}{4} \\ x + y + z &= 6 \\ \frac{x}{3} + \frac{y}{2} + \frac{z}{6} &= 1 \end{aligned}$$

**17.** $$\begin{aligned} \frac{1}{x} + \frac{2}{y} - \frac{1}{z} &= -3 \\ \frac{3}{x} + \frac{1}{y} + \frac{1}{z} &= 4 \\ \frac{1}{x} - \frac{1}{y} + \frac{2}{z} &= 6 \end{aligned}$$

### Literal Equations

Solve for $x$, $y$, and $z$.

**18.** $$\begin{aligned} x - y &= a \\ y + z &= 3a \\ 5z - x &= 2a \end{aligned}$$

**19.** $$\begin{aligned} x + a &= y + z \\ y + a &= 2x + 2z \\ z + a &= 3x + 3y \end{aligned}$$

**20.** $$\begin{aligned} ax + by &= (a + b)c \\ by + cz &= (c + a)b \\ ax + cz &= (b + c)a \end{aligned}$$

**21.** $$\begin{aligned} x + y + 2z &= 2(b + c) \\ x + 2y + z &= 2(a + c) \\ 2x + y + z &= 2(a + b) \end{aligned}$$

**22.** $$\begin{aligned} a_1x + b_1y + c_1z &= k_1 \\ a_2x + b_2y + c_2z &= k_2 \\ a_3x + b_3y + c_3z &= k_3 \end{aligned}$$

**Solving Problem 22 gives us formulas for solving a set of three equations. We'll find them useful in the next chapter.**

### Number Problems

**23.** The sum of the digits in a certain 3-digit number is 21. The sum of the first and third digits is twice the middle digit. If the hundreds and tens digits are interchanged, the number is reduced by 90. Find the number.

24. There are two fractions which have the same denominator. If 1 is subtracted from the numerator of the smaller fraction, its value will be $\frac{1}{3}$ of the larger fraction; but if 1 is subtracted from the numerator of the larger, its value will be twice that of the smaller fraction. The difference between the fractions is $\frac{1}{3}$. What are the fractions?
25. A certain number is expressed by three digits whose sum is 10. The sum of the first and last digits is $\frac{2}{3}$ of the second digit; and if 198 is subtracted from the number, the digits will be reversed. What is the number?

### Financial Problems

26. Three persons, $A$, $B$, and $C$, jointly purchased some lumber for \$900. One-half of what $A$ paid added to $\frac{1}{4}$ of what $B$ paid and $\frac{1}{5}$ of what $C$ paid equals \$279; and the sum that $A$ paid increased by $\frac{2}{3}$ of what $B$ paid and diminished by $\frac{1}{2}$ of what $C$ paid is \$320. How much did each pay?
27. A person wants to invest \$13,000, part in a checking account, part in a money market account (MMA), and the rest in a certificate of deposit (CD). She wants twice as much in the MMA as in the CD, and \$1000 more in the checking account than in the CD. How much should be placed in each account?
28. A merchant found, on counting her cash, that she had 84 coins and bills, in dollars, half-dollars, and quarters, worth \$42. She also found that $\frac{1}{3}$ of her half-dollars and $\frac{1}{4}$ of her quarters were worth \$6.50. How many pieces of each kind did she have? (*Hint:* If $q$ = the number of quarters, then $0.25q$ is the *value* of those quarters, in dollars.)

### Electrical Problems

29. When writing Kirchhoff's law (Eq. A68) for a certain three-loop network, we get the set of equations

$$3I_1 + 2I_2 - 4I_3 = 4$$
$$I_1 - 3I_2 + 2I_3 = -5$$
$$2I_1 + I_2 - I_3 = 3$$

where $I_1$, $I_2$, and $I_3$ are the loop currents in amperes. Solve for these currents.




FIGURE 9-8

30. For the three-loop network of Fig. 9-8, **(a)** use Kirchhoff's law to show that the currents may be found from

$$159I_1 - 50.9I_2 = 1$$
$$407I_1 - 2400I_2 + 370I_3 = 1$$
$$142I_2 - 380I_3 = 1$$

**(b)** Solve this set of equations for the three currents.

### Computer

31. The Gauss–Seidel method described in Sec. 9-1 works for any number of equations. For three equations in three unknowns, for example, write the equations in the form

$$x = f(y, z)$$
$$y = g(x, z)$$
$$z = h(x, y)$$

and proceed as before. Try it on one of the systems given in this exercise. If the computation diverges, then solve each equation for a different variable than before and try again. It may take two or three tries to find a combination that works.

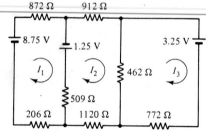

## CHAPTER 9 REVIEW PROBLEMS

*Solve each system of equations by any method.*

1. $$\begin{aligned} 4x + 3y &= 27 \\ 2x - 5y &= -19 \end{aligned}$$

2. $$\begin{aligned} \frac{x}{3} + \frac{y}{2} &= \frac{4}{3} \\ \frac{x}{2} + \frac{y}{3} &= \frac{7}{6} \end{aligned}$$

3. $$\begin{aligned} \frac{15}{x} + \frac{4}{y} &= 1 \\ \frac{5}{x} - \frac{12}{y} &= 7 \end{aligned}$$

4. $$\begin{aligned} 2x + 4y - 3z &= 22 \\ 4x - 2y + 5z &= 18 \\ 6x + 7y - z &= 63 \end{aligned}$$

5. $$\begin{aligned} 5x + 3y - 2z &= 5 \\ 3x - 4y + 3z &= 13 \\ x + 6y - 4z &= -8 \end{aligned}$$

6. $$\begin{aligned} 3x - 5y - 2z &= 14 \\ 5x - 8y - z &= 12 \\ x - 3y - 3z &= 1 \end{aligned}$$

7. $$\begin{aligned} \frac{3}{x+y} + \frac{4}{x-z} &= 2 \\ \frac{6}{x+y} - \frac{5}{y-z} &= 1 \\ \frac{4}{x-z} + \frac{5}{y-z} &= 2 \end{aligned}$$

8. $$\begin{aligned} 4x + 2y - 26 &= 0 \\ 3x + 4y &= 39 \end{aligned}$$

9. $$\begin{aligned} 2x - 3y + 14 &= 0 \\ 3x + 2y &= 44 \end{aligned}$$

10. $$\begin{aligned} \frac{2}{x} + \frac{1}{y} &= \frac{4}{3} \\ \frac{3}{x} + \frac{5}{y} &= \frac{19}{6} \end{aligned}$$

11. $$\begin{aligned} \frac{9}{x} + \frac{8}{y} &= \frac{43}{6} \\ \frac{3}{x} + \frac{10}{y} &= \frac{29}{6} \end{aligned}$$

12. $$\begin{aligned} \frac{x}{a} + \frac{y}{b} &= p \\ \frac{x}{b} + \frac{y}{a} &= q \end{aligned}$$

13. $$\begin{aligned} x + y &= a \\ x + z &= b \\ y + z &= c \end{aligned}$$

14. $$\begin{aligned} \frac{5x}{6} + \frac{2y}{5} &= 14 \\ \frac{3x}{4} - \frac{2y}{5} &= 5 \end{aligned}$$

15. $$\begin{aligned} \frac{2x}{7} + \frac{2y}{3} &= \frac{16}{3} \\ x + y &= 12 \end{aligned}$$

16. $$\begin{aligned} x + y + z &= 35 \\ 5x + 4y + 3z &= 22 \\ 3x + 4y - 3z &= 2 \end{aligned}$$

17. $$\begin{aligned} 2x - 4y + 3z &= 10 \\ 3x + y - 2z &= 6 \\ x + 3y - z &= 20 \end{aligned}$$

18. $$\begin{aligned} 5x + y - 4z &= -5 \\ 3x - 5y - 6z &= -20 \\ x - 3y + 8z &= -27 \end{aligned}$$

19. $$\begin{aligned} x + 3y - z &= 10 \\ 5x - 2y + 2z &= 6 \\ 3x + 2y + z &= 13 \end{aligned}$$

20. $$\begin{aligned} x + 21y &= 2 \\ 2x + 27y &= 19 \end{aligned}$$

21. $$\begin{aligned} 2x - y &= 9 \\ 5x - 3y &= 14 \end{aligned}$$

22. $$\begin{aligned} 6x - 2y + 5z &= 53 \\ 5x + 3y + 7z &= 33 \\ x + y + z &= 5 \end{aligned}$$

23. $$\begin{aligned} 9x + 4y &= 54 \\ 4x + 9y &= 89 \end{aligned}$$

24. A sum of money was divided between $A$ and $B$ so that $A$'s share was to $B$'s share as 5 is to 3. Also, $A$'s share exceeded $\frac{5}{9}$ of the whole sum by \$50. What was each share?

25. $A$ and $B$ can do a job in 12 days, and $B$ and $C$ can do the same job in 16 days. How long would it take them all working together to do the job if $A$ does $1\frac{1}{2}$ times as much as $C$?

26. If the numerator of a certain fraction is increased by 2 and its denominator diminished by 2, its value will be 1. If the numerator is increased by the denominator and the denominator is diminished by 5, its value will be 5. Find the fraction.

*Writing*

27. Suppose that you have a number of pairs of equations to solve in order to find the loop currents in a circuit you are designing. Each pair of equations has different values of the variables. You must leave for a week but want your assistant to solve the equations in your absence. Write step-by-step instructions to be followed, using the addition/subtraction method, including instructions as to the number of digits to be retained.

# 10

# DETERMINANTS

## OBJECTIVES

**When you have completed this chapter, you should be able to:**

- Evaluate second-order determinants.
- Evaluate third-order determinants by minors.
- Use the properties of determinants to reduce the order of a determinant.
- Evaluate determinants by a systematic method suitable for the computer.
- Solve a system of any number of linear equations by determinants.

In Chapter 9 we learned how to solve sets of two or three equations by graphing, addition–subtraction, substitution, and by computer using the Gauss–Seidel method. These methods, however, become unwieldy when the number of equations to be solved is greater than three.

In this chapter we introduce the idea of a determinant. We learn how to evaluate determinants and then how to use them to solve a set of equations. We start with sets of two equations and then expand our use of determinants so that we can solve systems with any number of equations.

## 10-1 SECOND-ORDER DETERMINANTS

### Definitions

In Chapter 9 we solved the set of equations

$$a_1x + b_1y = c_1$$
$$a_2x + b_2y = c_2$$

and got the solution

$$x = \frac{b_2c_1 - b_1c_2}{a_1b_2 - a_2b_1} \quad \text{and} \quad y = \frac{a_1c_2 - a_2c_1}{a_1b_2 - a_2b_1}$$

The denominator $(a_1b_2 - a_2b_1)$ of each of these fractions is called the *determinant* of the coefficients $a_1$, $b_1$, $a_2$, and $b_2$. It is commonly expressed by the symbol

$$\begin{vmatrix} a_1 & b_1 \\ a_2 & b_2 \end{vmatrix}$$

**It is common practice to use the same word *determinant* to refer to the *symbol***

$$\begin{vmatrix} a_1 & b_1 \\ a_2 & b_2 \end{vmatrix}$$

**as well as to the *value* $(a_1b_2 - a_2b_1)$. We'll follow this practice.**

with it being understood that this signifies the product of the upper left and lower right numbers, minus the product of the upper right and lower left numbers. Thus

$$\begin{vmatrix} a_1 & b_1 \\ a_2 & b_2 \end{vmatrix} = a_1b_2 - a_2b_1$$

---

**EXAMPLE 1:** In the determinant

$$\begin{vmatrix} 2 & 5 \\ 6 & 1 \end{vmatrix}$$

each of the numbers 2, 5, 6, and 1 is called an *element*. There are two *rows*, the first row containing the elements 2 and 5, and the second row having the elements 6 and 1. There are also two *columns*, the first with elements 2 and 6, and the second with elements 5 and 1. The *value* of this determinant is

$$2(1) - 5(6) = 2 - 30 = -28$$

---

Thus a determinant is written as a *square array*, enclosed between vertical bars. The number of rows equals the number of columns. The *order* of a determinant is equal to the number of rows or columns it contains. Thus the determinants above are of second order.

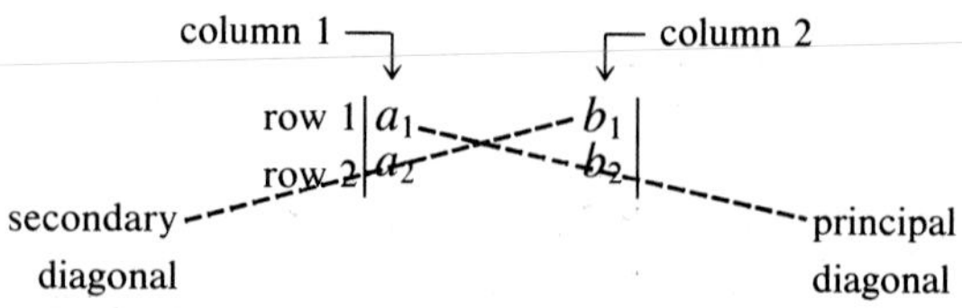


The diagonal running from upper left to lower right is called the *principal diagonal.* The other is called the *secondary diagonal.* Thus

| | | |
|---|---|---|
| Second-Order Determinant | $\begin{vmatrix} a_1 & b_1 \\ a_2 & b_2 \end{vmatrix} = a_1b_2 - a_2b_1$ | **65** |

*The value of a second-order determinant is equal to the product of the elements on the principal diagonal minus the product of the elements on the secondary diagonal.*

---

**EXAMPLE 2:**

(a) $\begin{vmatrix} 2 & -1 \\ 3 & 5 \end{vmatrix} = 2(5) - 3(-1) = 13$ (b) $\begin{vmatrix} 4 & a \\ x & 3b \end{vmatrix} = 4(3b) - x(a) = 12b - ax$

---

| | |
|---|---|
| Common Errors | Students sometimes add the numbers on a diagonal instead of *multiplying*. Be careful not to do that. |
| | Don't just ignore a zero element. It causes the product along its diagonal to be zero. |

---

**EXAMPLE 3:**

$$\begin{vmatrix} 0 & 3 \\ 5 & 2 \end{vmatrix} = 0(2) - 3(5) = -15$$

---

## Solving a System of Equations by Determinants

We now return to the set of equations

$$\begin{aligned} a_1x + b_1y &= c_1 \\ a_2x + b_2y &= c_2 \end{aligned} \tag{1}$$

If, in the solution for $x$,

$$x = \frac{b_2c_1 - b_1c_2}{a_1b_2 - a_2b_1} \tag{2}$$

This denominator is called the ***determinant of the coefficients,*** or sometimes the ***determinant of the system.*** It is usually given the Greek capital letter delta ($\Delta$). If this determinant equals zero, there is no solution to the set of equations.

we replace the denominator by the square array,

$$\begin{vmatrix} a_1 & b_1 \\ a_2 & b_2 \end{vmatrix} = a_1b_2 - a_2b_1$$

we get

$$x = \frac{b_2c_1 - b_1c_2}{\begin{vmatrix} a_1 & b_1 \\ a_2 & b_2 \end{vmatrix}} \tag{3}$$

Now look at the numerator of (3), which is $b_2c_1 - b_1c_2$. If this expression were the value of some determinant, it is clear that the elements on the principal diagonal must be $b_2$ and $c_1$ and that the elements on the secondary diagonal must be $b_1$ and $c_2$. One such determinant is

column of constants

$$\begin{vmatrix} c_1 & b_1 \\ c_2 & b_2 \end{vmatrix} = b_2c_1 - b_1c_2$$

Our solution for $x$ can then be expressed

$$x = \frac{\begin{vmatrix} c_1 & b_1 \\ c_2 & b_2 \end{vmatrix}}{\begin{vmatrix} a_1 & b_1 \\ a_2 & b_2 \end{vmatrix}}$$

An expression for $y$ can be developed in a similar way. Thus the solution to the system of equations (1), in terms of determinants, is

| | | |
|---|---|---|
| Cramer's Rule | $x = \frac{\begin{vmatrix} c_1 & b_1 \\ c_2 & b_2 \end{vmatrix}}{\begin{vmatrix} a_1 & b_1 \\ a_2 & b_2 \end{vmatrix}}$ and $y = \frac{\begin{vmatrix} a_1 & c_1 \\ a_2 & c_2 \end{vmatrix}}{\begin{vmatrix} a_1 & b_1 \\ a_2 & b_2 \end{vmatrix}}$ | **70** |

After the Swiss mathematician Gabriel Cramer, 1704–1752. Cramer's rule, as we will see, is valid for systems with any number of linear equations.

*The solution for any variable is a fraction whose denominator is the determinant of the coefficients, and whose numerator is that same determinant, except that the column of coefficients for the variable for which we are solving is replaced by the column of constants.*

Of course, the denominator $\Delta$ cannot be zero or we get division by zero. This would indicate that the set of equations has no unique solution.

---

**EXAMPLE 4:** Solve by determinants:

$$2x - 3y = 1$$
$$x + 4y = -5$$

**Solution:** We first evaluate the determinant of the coefficients, because, if $\Delta = 0$ there is no unique solution, and we have no need to proceed further.

$$\Delta = \begin{vmatrix} 2 & -3 \\ 1 & 4 \end{vmatrix} = 2(4) - 1(-3) = 8 + 3 = 11$$

Solving for $x$, we have

column of constants

$$x = \frac{\begin{vmatrix} 1 & -3 \\ -5 & 4 \end{vmatrix}}{\Delta} = \frac{1(4) - (-5)(-3)}{11} = \frac{4 - 15}{11} = -1$$

It's easier to find $y$ by substituting $x$ back into one of the previous equations, but we use determinants instead, to show how it is done.

Solving for $y$ yields

column of constants

$$y = \frac{\begin{vmatrix} 2 & 1 \\ 1 & -5 \end{vmatrix}}{\Delta} = \frac{2(-5) - 1(1)}{11} = \frac{-10 - 1}{11} = -1$$

---

| Common Error | Don't forget to first arrange the given equations into the form $a_1x + b_1y = c_1$, $a_2x + b_2y = c_2$ before writing the determinants. |
|---|---|

**EXAMPLE 5:** Solve for $x$ and $y$ by determinants:

$$2ax - by + 3a = 0$$
$$4y = 5x - a$$

**Solution:** We first rearrange our equations.

$$2ax - by = -3a$$
$$5x - 4y = a$$

The determinant of the coefficients is

$$\Delta = \begin{vmatrix} 2a & -b \\ 5 & -4 \end{vmatrix} = 2a(-4) - 5(-b) = -8a + 5b$$

So

$$x = \frac{\begin{vmatrix} -3a & -b \\ a & -4 \end{vmatrix}}{\Delta} = \frac{(-3a)(-4) - a(-b)}{5b - 8a}$$

$$= \frac{12a + ab}{5b - 8a}$$

and

$$y = \frac{\begin{vmatrix} 2a & -3a \\ 5 & a \end{vmatrix}}{\Delta} = \frac{2a(a) - 5(-3a)}{5b - 8a} = \frac{2a^2 + 15a}{5b - 8a}$$

## EXERCISE 1—SECOND-ORDER DETERMINANTS

Find the value of each second-order determinant.

1. $\begin{vmatrix} 3 & 2 \\ 1 & -4 \end{vmatrix}$
2. $\begin{vmatrix} -2 & 5 \\ 3 & -3 \end{vmatrix}$
3. $\begin{vmatrix} 0 & 5 \\ -3 & 4 \end{vmatrix}$
4. $\begin{vmatrix} 1 & 1 \\ 1 & 1 \end{vmatrix}$
5. $\begin{vmatrix} 5 & 5 \\ 5 & 5 \end{vmatrix}$
6. $\begin{vmatrix} 3 & 0 \\ 7 & 0 \end{vmatrix}$
7. $\begin{vmatrix} 4.82 & 2.73 \\ 2.97 & 5.28 \end{vmatrix}$
8. $\begin{vmatrix} 48.7 & -53.6 \\ 4.93 & 9.27 \end{vmatrix}$
9. $\begin{vmatrix} -2/3 & 2/5 \\ -1/3 & 4/5 \end{vmatrix}$
10. $\begin{vmatrix} 1/2 & 1/4 \\ -1/2 & -1/4 \end{vmatrix}$
11. $\begin{vmatrix} a & b \\ c & d \end{vmatrix}$
12. $\begin{vmatrix} 2m & 3n \\ -m & 4n \end{vmatrix}$
13. $\begin{vmatrix} \sin\theta & 3 \\ \tan\theta & \sin\theta \end{vmatrix}$
14. $\begin{vmatrix} 2i_1 & i_2 \\ -3i_1 & 4i_2 \end{vmatrix}$
15. $\begin{vmatrix} 3i_1 & i_2 \\ \sin\theta & \cos\theta \end{vmatrix}$
16. $\begin{vmatrix} 3x & 2 \\ x & 5 \end{vmatrix}$

Solve each pair of equations by determinants.

17. $2x + y = 11$
$3x - y = 4$

18. $5x + 7y = 101$
$y = 7x - 55$

19. $3x - 2y = -15$
$5x + 6y = 3$

20. $7x + 6y = 20$
$2x + 5y = 9$

21. $x + 5y = 11$
$3x + 2y = 7$

22. $4x - 5y = -34$
$2x - 3y = -22$

23. $x = 11 - 4y$
$5x - 2y = 11$

24. $2x - 3y = 3$
$4x + 5y = 39$

25. $7x - 4y = 81$
$5x - 3y = 57$

26. $3x + 4y = 85$
$5x + 4y = 107$

27. $3x - 2y = 1$
$2x + y = 10$

28. $5x - 2y = 3$
$2x + 3y = 5$

29. $y = 9 - 3x$
$x = 8 - 2y$

30. $y = 2x - 3$
$x = 19 - 3y$

31. $29.1x - 47.6y = 42.8$
$11.5x + 72.7y = 25.8$

32. $4.92x - 8.27y = 2.58$
$6.93x + 2.84y = 8.36$

33. $4n = 18 - 3m$
$m = 8 - 2n$

34. $5p + 4q - 14 = 0$
$17p = 31 + 3q$

35. $3w = 13 + 5z$
$4w - 7z - 17 = 0$

36. $3u = 5 + 2v$
$5v + 2u = 16$

**Fractional and Literal Equations**

Solve for $x$ and $y$ by determinants.

37. $\frac{x}{3} + \frac{y}{2} = 1$
$\frac{x}{2} - \frac{y}{3} = -1$

38. $px - qy + pq = 0$
$2px - 3qy = 0$

39. $ax + by = p$
$cx + dy = q$

40. $\frac{x}{2} - \frac{3y}{4} = 0$
$\frac{x}{3} + \frac{y}{4} = 0$

41. $\frac{3}{x} + \frac{8}{y} = 3$
$\frac{15}{x} - \frac{4}{y} = 4$

42. $4mx - 3ny = 6$
$3mx + 2ny = -7$

## 10-2 THIRD-ORDER DETERMINANTS

### Minors

There is a method for evaluating a third-order determinant by multiplying along diagonals. We omit it here because it works *only* for third-order determinants, and also is not suitable for computer solution.

The technique we used for finding the value of a second-order determinant *cannot* be used for higher-order determinants. Instead, we use a technique called *development by minors* to reduce the order of a determinant to second order so that it may be evaluated by the method described in Sec. 10-1.

The *minor of an element in a determinant* is a determinant of next lower order, obtained by deleting the row and the column in which that element lies.

**EXAMPLE 6:** The minor of element $c$ in the determinant

$$\begin{vmatrix} a & b & c \\ d & e & f \\ g & h & i \end{vmatrix}$$

is found by striking out the first row and third column

$$\begin{vmatrix} \cancel{a} & \cancel{b} & \textcircled{c} \\ d & e & \cancel{f} \\ g & h & \cancel{i} \end{vmatrix} \quad \text{or} \quad \begin{vmatrix} d & e \\ g & h \end{vmatrix}$$

## Sign Factor

The *sign factor* for an element depends upon the position of the element in the determinant. To find the sign factor, add the row number and the column number of the element. If the sum is even, the sign factor is (+1), if the sum is odd, the sign factor is (−1).

**A minor, with the sign factor attached, is called a *signed minor*. It is also called a *cofactor* of the given element.**

**EXAMPLE 7:**

(a) For the determinant in Example 6, element $c$ is in the first row and third column. The sum of the row and column numbers

$$1 + 3 = 4$$

is even, so the sign factor for the element $c$ is (+1).

(b) The sign factor for the element $b$ in Example 6 is (−1).

The sign factor for the element in the upper left-hand corner of the determinant is (+1) and the signs alternate according to the pattern

$$\begin{vmatrix} + & - & + & - & + & - & \cdot \\ - & + & - & + & - & \cdot & \cdot \\ + & - & + & - & \cdot & \cdot & \cdot \\ - & + & - & \cdot & \cdot & \cdot & \cdot \\ + & - & \cdot & \cdot & \cdot & \cdot & \cdot \\ - & \cdot & \cdot & \cdot & \cdot & \cdot & \cdot \\ \cdot & \cdot & \cdot & \cdot & \cdot & \cdot & \cdot \end{vmatrix}$$

so that the sign factor may be found simply by *counting off* from the upper left corner.

Thus for any element in a determinant we can write three quantities:

1. The element itself
2. The sign factor of the element
3. The minor of the element

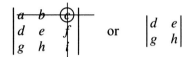

**EXAMPLE 8:** For the element in the second row and first column in the determinant

$$\begin{vmatrix} -2 & 5 & 1 \\ 3 & -2 & 4 \\ 1 & -4 & 2 \end{vmatrix}$$

1. The element itself is 3.
2. The sign factor of that element is $(-1)$.
3. The minor of that element is

$$\begin{vmatrix} 5 & 1 \\ -4 & 2 \end{vmatrix}$$

### Evaluating a Determinant by Minors

We now define the value of a determinant as follows:

| | |
|---|---|
| Value of a Determinant | A determinant is equal to the sum of the products of the elements in any row (or column), the sign factors of those elements, and the minors of those elements. |

**EXAMPLE 9:** Evaluate by minors:

$$\begin{vmatrix} 3 & 2 & 4 \\ 1 & 6 & -2 \\ -1 & 5 & -3 \end{vmatrix}$$

*Solution:* We first choose a row or a column for development. The work of expansion is greatly reduced if that row or column contains zeros. Our given determinant has no zeros, so let us choose column 1, which at least contains some 1's.

The elements of column 1

$$\begin{vmatrix} 3 & 2 & 4 \\ 1 & 6 & -2 \\ -1 & 5 & -3 \end{vmatrix} = 3(1)\begin{vmatrix} 6 & -2 \\ 5 & -3 \end{vmatrix} + 1(-1)\begin{vmatrix} 2 & 4 \\ 5 & -3 \end{vmatrix} + (-1)(1)\begin{vmatrix} 2 & 4 \\ 6 & -2 \end{vmatrix}$$

The sign factor of each element

The minor of each element

$$= 3[(6)(-3) - (-2)(5)] - 1[(2)(-3) - (4)(5)] - 1[(2)(-2) - 4(6)]$$

$$= -24 + 26 + 28$$

$$= 30$$

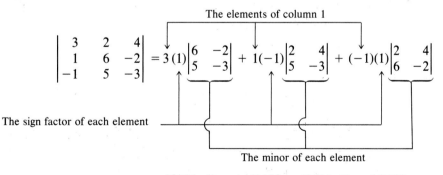

**EXAMPLE 10:** Evaluate by minors:

$$\begin{vmatrix} a_1 & b_1 & c_1 \\ a_2 & b_2 & c_2 \\ a_3 & b_3 & c_3 \end{vmatrix}$$

**Solution:** Choosing, say, the first row for development, we get

$$a_1 \begin{vmatrix} b_2 & c_2 \\ b_3 & c_3 \end{vmatrix} - b_1 \begin{vmatrix} a_2 & c_2 \\ a_3 & c_3 \end{vmatrix} + c_1 \begin{vmatrix} a_2 & b_2 \\ a_3 & b_3 \end{vmatrix}$$

$$= a_1(b_2c_3 - b_3c_2) - b_1(a_2c_3 - a_3c_2) + c_1(a_2b_3 - a_3b_2)$$

$$= a_1b_2c_3 - a_1b_3c_2 - a_2b_1c_3 + a_3b_1c_2 + a_2b_3c_1 - a_3b_2c_1$$

$$= a_1b_2c_3 + a_3b_1c_2 + a_2b_3c_1 - a_3b_2c_1 - a_1b_3c_2 - a_2b_1c_3$$

The value of the determinant is thus given by the following formula:

| | | |
|---|---|---|
| Third-Order Determinant | $\begin{vmatrix} a_1 & b_1 & c_1 \\ a_2 & b_2 & c_2 \\ a_3 & b_3 & c_3 \end{vmatrix} = a_1b_2c_3 + a_3b_1c_2 + a_2b_3c_1 - a_3b_2c_1 - a_1b_3c_2 - a_2b_1c_3$ | **66** |

## Solving a System of Three Equations by Determinants

If we were to *algebraically* solve the system of equations

$$a_1x + b_1y + c_1z = k_1$$

$$a_2x + b_2y + c_2z = k_2$$

$$a_3x + b_3y + c_3z = k_3$$

**The solution of this system was given as an exercise in Chapter 9, Exercise 4.**

we would get the solution

| | | |
|---|---|---|
| Three Equations in Three Unknowns | $x = \dfrac{b_2c_3k_1 + b_1c_2k_3 + b_3c_1k_2 - b_2c_1k_3 - b_3c_2k_1 - b_1c_3k_2}{a_1b_2c_3 + a_3b_1c_2 + a_2b_3c_1 - a_3b_2c_1 - a_1b_3c_2 - a_2b_1c_3}$ <br> $y = \dfrac{a_1c_3k_2 + a_3c_2k_1 + a_2c_1k_3 - a_3c_1k_2 - a_1c_2k_3 - a_2c_3k_1}{a_1b_2c_3 + a_3b_1c_2 + a_2b_3c_1 - a_3b_2c_1 - a_1b_3c_2 - a_2b_1c_3}$ <br> $z = \dfrac{a_1b_2k_3 + a_3b_1k_2 + a_2b_3k_1 - a_3b_2k_1 - a_1b_3k_2 - a_2b_1k_3}{a_1b_2c_3 + a_3b_1c_2 + a_2b_3c_1 - a_3b_2c_1 - a_1b_3c_2 - a_2b_1c_3}$ | **64** |

Notice that the three denominators are identical and are equal to the value of the determinant formed from the coefficients of the unknowns (Eq. 66). As before, we call it the *determinant of the coefficients*, Δ.

Furthermore, the numerator of each can be obtained from the determinant of the coefficients by replacing the coefficients of the variable in question with the constants $k_1$, $k_2$, and $k_3$. The solution to our set of equations

$$a_1x + b_1y + c_1z = k_1$$
$$a_2x + b_2y + c_2z = k_2$$
$$a_3x + b_3y + c_3z = k_3$$

is then

| Cramer's Rule | $x = \dfrac{\begin{vmatrix} k_1 & b_1 & c_1 \\ k_2 & b_2 & c_2 \\ k_3 & b_3 & c_3 \end{vmatrix}}{\Delta} \quad y = \dfrac{\begin{vmatrix} a_1 & k_1 & c_1 \\ a_2 & k_2 & c_2 \\ a_3 & k_3 & c_3 \end{vmatrix}}{\Delta} \quad z = \dfrac{\begin{vmatrix} a_1 & b_1 & k_1 \\ a_2 & b_2 & k_2 \\ a_3 & b_3 & k_3 \end{vmatrix}}{\Delta}$ where $\Delta = \begin{vmatrix} a_1 & b_1 & c_1 \\ a_2 & b_2 & c_2 \\ a_3 & b_3 & c_3 \end{vmatrix} \neq 0$ | **71** |
|---|---|---|

Cramer's rule, although we do not prove it, works for higher-order systems as well. We restate it now in words.

| Cramer's Rule | The solution for any variable is a fraction whose denominator is the determinant of the coefficients, and whose numerator is the same determinant, except that the column of coefficients for the variable for which we are solving is replaced by the column of constants. | **69** |
|---|---|---|

**EXAMPLE 11:** Solve by determinants:

$$x + 2y + 3z = 14$$
$$2x + y + 2z = 10$$
$$3x + 4y - 3z = 2$$

**Solution:** We first write the determinant of the coefficients

$$\Delta = \begin{vmatrix} 1 & 2 & 3 \\ 2 & 1 & 2 \\ 3 & 4 & -3 \end{vmatrix}$$

and expand it by minors. Choosing the first row for expansion gives

$$\Delta = \begin{vmatrix} 1 & 2 & 3 \\ 2 & 1 & 2 \\ 3 & 4 & -3 \end{vmatrix} = 1\begin{vmatrix} 1 & 2 \\ 4 & -3 \end{vmatrix} - 2\begin{vmatrix} 2 & 2 \\ 3 & -3 \end{vmatrix} + 3\begin{vmatrix} 2 & 1 \\ 3 & 4 \end{vmatrix}$$
$$= 1(-3 - 8) - 2(-6 - 6) + 3(8 - 3)$$
$$= -11 + 24 + 15 = 28$$

As when solving a set of two equations, a $\Delta$ of zero would mean that the system had no unique solution, and we would stop here.

We now make a new determinant by replacing the coefficients of $x$, $\begin{vmatrix} 1 \\ 2 \\ 3 \end{vmatrix}$, by the column of constants $\begin{vmatrix} 14 \\ 10 \\ 2 \end{vmatrix}$, and expand it by, say, the second column.

$$\begin{vmatrix} 14 & 2 & 3 \\ 10 & 1 & 2 \\ 2 & 4 & -3 \end{vmatrix} = -2\begin{vmatrix} 10 & 2 \\ 2 & -3 \end{vmatrix} + 1\begin{vmatrix} 14 & 3 \\ 2 & -3 \end{vmatrix} - 4\begin{vmatrix} 14 & 3 \\ 10 & 2 \end{vmatrix}$$

$$= -2(-30 - 4) + 1(-42 - 6) - 4(28 - 30)$$

$$= 68 - 48 + 8 = 28$$

Dividing this value by $\Delta$ gives $x$.

$$x = \frac{28}{\Delta} = \frac{28}{28} = 1$$

Next we replace the coefficients of $y$ with the column of constants and expand the first column by minors.

$$\begin{vmatrix} 1 & 14 & 3 \\ 2 & 10 & 2 \\ 3 & 2 & -3 \end{vmatrix} = 1\begin{vmatrix} 10 & 2 \\ 2 & -3 \end{vmatrix} - 2\begin{vmatrix} 14 & 3 \\ 2 & -3 \end{vmatrix} + 3\begin{vmatrix} 14 & 3 \\ 10 & 2 \end{vmatrix}$$

$$= 1(-30 - 4) - 2(-42 - 6) + 3(28 - 30)$$

$$= -34 + 96 - 6 = 56$$

Dividing by $\Delta$ gives the value of $y$.

$$y = \frac{56}{\Delta} = \frac{56}{28} = 2$$

We can get $z$ also by determinants or, more easily, by substituting back. Substituting $x = 1$ and $y = 2$ into the first equation, we obtain

$$1 + 2(2) + 3z = 14$$

$$3z = 9$$

$$z = 3$$

The solution is thus (1, 2, 3).

---

Often, some terms will have zero coefficients or decimal coefficients. Further, the terms may be out of order, as in the following example.

---

**EXAMPLE 12:** Solve for $x$, $y$, and $z$.

$$23.7y + 72.4x = 82.4 - 11.3x$$

$$25.5x - 28.4z + 19.3 = 48.2y$$

$$13.4 + 66.3z = 39.2x - 10.5$$

**Solution:** We rewrite each equation in the form $ax + by + cz = k$, combining like terms as we go, and putting in the missing terms with zero coefficients. We do this *before* writing the determinant.

$$83.7x + 23.7y + 0z = 82.4 \qquad (1)$$

$$25.5x - 48.2y - 28.4z = -19.3$$

$$-39.2x + 0y + 66.3z = -23.9 \qquad (2)$$

The determinant of the system is then

$$\Delta = \begin{vmatrix} 83.7 & 23.7 & 0 \\ 25.5 & -48.2 & -28.4 \\ -39.2 & 0 & 66.3 \end{vmatrix}$$

Let us develop the first row by minors.

$$\Delta = 83.7\begin{vmatrix} -48.2 & -28.4 \\ 0 & 66.3 \end{vmatrix} - 23.7\begin{vmatrix} 25.5 & -28.4 \\ -39.2 & 66.3 \end{vmatrix}$$

$$= 83.7(-48.2)(66.3) - 23.7[25.5(66.3) - (-28.4)(-39.2)]$$

$$= -281{,}000 \qquad \text{(to three significant digits)}$$

Now solving for $x$ gives us

$$x = \frac{\begin{vmatrix} 82.4 & 23.7 & 0 \\ -19.3 & -48.2 & -28.4 \\ -23.9 & 0 & 66.3 \end{vmatrix}}{\Delta}$$

Let us develop the first row of the determinant by minors. We get

$$82.4\begin{vmatrix} -48.2 & -28.4 \\ 0 & 66.3 \end{vmatrix} - 23.7\begin{vmatrix} -19.3 & -28.4 \\ -23.7 & 66.3 \end{vmatrix}$$

$$= 82.4(-48.2)(66.3) - 23.7[(-19.3)(66.3) - (-28.4)(-23.7)]$$

$$= -217{,}000$$

Dividing by $\Delta$ yields

$$x = \frac{-217{,}000}{-281{,}000} = 0.772$$

We solve for $y$ and $z$ by substituting back. From (1) we get

$$23.7y = 82.4 - 83.7(0.772) = 17.8$$

$$y = 0.751$$

and from (2)

$$66.3z = -23.9 + 39.2(0.772) = 6.34$$

$$z = 0.0960$$

The solution is then $x = 0.772$, $y = 0.751$, and $z = 0.0960$.

## EXERCISE 2—THIRD-ORDER DETERMINANTS

Evaluate each determinant.

1. $\begin{vmatrix} 1 & 0 & 2 \\ 3 & 1 & 0 \\ 1 & 2 & 1 \end{vmatrix}$

2. $\begin{vmatrix} 2 & -1 & 3 \\ 0 & 2 & 1 \\ 3 & -2 & 4 \end{vmatrix}$

3. $\begin{vmatrix} -3 & 1 & 2 \\ 0 & -1 & 5 \\ 6 & 0 & 1 \end{vmatrix}$

4. $\begin{vmatrix} -1 & 0 & 3 \\ 2 & 0 & -2 \\ 1 & -3 & 4 \end{vmatrix}$

5. $\begin{vmatrix} 5 & 1 & 2 \\ -3 & 2 & -1 \\ 4 & -3 & 5 \end{vmatrix}$

6. $\begin{vmatrix} 1.0 & 2.4 & -1.5 \\ -2.6 & 0 & 3.2 \\ -2.9 & 1.0 & 4.1 \end{vmatrix}$

7. $\begin{vmatrix} 2 & 1 & 3 \\ 0 & -2 & 4 \\ 0 & 1 & 5 \end{vmatrix}$

8. $\begin{vmatrix} 1 & 5 & 4 \\ -3 & 6 & -2 \\ -1 & 5 & 3 \end{vmatrix}$

Solve by determinants.

**Some of these are identical to those given in Chapter 9, Exercise 4.**

9. $x + y + z = 18$
   $x - y + z = 6$
   $x + y - z = 4$

10. $x + y + z = 12$
    $x - y = 2$
    $x - z = 4$

11. $x + y = 35$
    $x + z = 40$
    $y + z = 45$

12. $x + y + z = 35$
    $x - 2y + 3z = 15$
    $y - x + z = -5$

13. $x + 2y + 3z = 14$
    $2x + y + 2z = 10$
    $3x + 4y - 3z = 2$

14. $x + y + z = 90$
    $2x - 3y = -20$
    $2x + 3z = 145$

15. $x - 2y + 2z = 5$
    $5x + 3y + 6z = 57$
    $x + 2y + 2z = 21$

16. $3x + y = 5$
    $2y - 3z = -5$
    $x + 2z = 7$

17. $2x - 4y + 3z = 10$
    $3x + y - 2z = 6$
    $x + 3y - z = 20$

18. $x - y = 5$
    $y - z = -6$
    $2x - z = 2$

### Fractional Equations

Solve by determinants.

19. $x + \dfrac{y}{3} = 5$
    $x + \dfrac{z}{3} = 6$
    $y + \dfrac{z}{3} = 9$

20. $\dfrac{1}{x} + \dfrac{1}{y} = 5$
    $\dfrac{1}{y} + \dfrac{1}{z} = 7$
    $\dfrac{1}{x} + \dfrac{1}{z} = 6$

21. $\dfrac{x}{10} + \dfrac{y}{5} + \dfrac{z}{20} = \dfrac{1}{4}$
    $x + y + z = 6$
    $\dfrac{x}{3} + \dfrac{y}{2} + \dfrac{z}{6} = 1$

22. $\dfrac{1}{x} + \dfrac{2}{y} - \dfrac{1}{z} = -3$
    $\dfrac{3}{x} + \dfrac{1}{y} + \dfrac{1}{z} = 4$
    $\dfrac{1}{x} - \dfrac{1}{y} + \dfrac{2}{z} = 6$

## 10-3 HIGHER-ORDER DETERMINANTS

We can evaluate a determinant of any order by repeated use of the method of minors. Thus any row or column of a fifth-order determinant can be developed, thus reducing the determinant to five fourth-order determinants. Each of these can then be developed into four third-order determinants, and so on, until only second-order determinants remain.

Obviously, this is lots of work. To avoid this, we first *simplify* the determinant, using the properties of determinants in the following section, and then apply a systematic method for reducing the determinants by minors shown after that.

### Properties of Determinants

A determinant larger than third order can usually be evaluated faster if we first reduce its order before expanding by minors. We do this by applying the various *properties of determinants* as shown in the following examples.

| | | |
|---|---|---|
| Zero Row or Column | If all elements in a row (or column) are zero, the value of the determinant is zero. | **72** |

**EXAMPLE 13:**

$$\begin{vmatrix} 4 & 8 & 0 \\ 5 & 2 & 0 \\ 3 & 9 & 0 \end{vmatrix} = 0$$

| | | |
|---|---|---|
| Identical Rows or Columns | The value of a determinant is zero if two rows (or columns) are identical. | **73** |

**EXAMPLE 14:** The value of the determinant

$$\begin{vmatrix} 2 & 1 & 3 \\ 4 & 2 & 5 \\ 4 & 2 & 5 \end{vmatrix}$$

is found by expanding by minors using the first row,

$$2\begin{vmatrix} 2 & 5 \\ 2 & 5 \end{vmatrix} - 1\begin{vmatrix} 4 & 5 \\ 4 & 5 \end{vmatrix} + 3\begin{vmatrix} 4 & 2 \\ 4 & 2 \end{vmatrix}$$

$$= 2(10 - 10) - 1(20 - 20) + 3(8 - 8)$$

$$= 0$$

| | | |
|---|---|---|
| Zeros below the Principal Diagonal | If all elements below the principal diagonal are zeros, then the value of the determinant is the product of the elements along the principal diagonal. | **74** |

**EXAMPLE 15:** The determinant

$$\begin{vmatrix} 2 & 9 & 7 & 8 \\ 0 & 1 & 5 & 9 \\ 0 & 0 & 3 & 4 \\ 0 & 0 & 0 & 1 \end{vmatrix}$$

**Expand this determinant by minors and see for yourself that its value is indeed 6.**

has the value

$$(2)(1)(3)(1) = 6$$

| | | |
|---|---|---|
| Interchanging Rows with Columns | The value of a determinant is unchanged if we change the rows to columns and the columns to rows. | **75** |

**EXAMPLE 16:** The value of the determinant

$$\begin{vmatrix} 5 & 2 \\ 4 & 3 \end{vmatrix}$$

is

$$(5)(3) - (2)(4) = 7$$

If the first column now becomes the first row, and the second column becomes the second row, we get the determinant

$$\begin{vmatrix} 5 & 4 \\ 2 & 3 \end{vmatrix}$$

which has the value

$$(5)(3) - (4)(2) = 7$$

as before.

| | | |
|---|---|---|
| Interchange of Two Rows (or Columns) | A determinant will change sign when we interchange two rows (or columns). | **76** |

**EXAMPLE 17:** The value of the determinant

$$\begin{vmatrix} 3 & 0 & 2 \\ 1 & 4 & 3 \\ 2 & 1 & 2 \end{vmatrix}$$

is found by expanding by minors using the first row.

$$3\begin{vmatrix} 4 & 3 \\ 1 & 2 \end{vmatrix} - 0 + 2\begin{vmatrix} 1 & 4 \\ 2 & 1 \end{vmatrix} = 3(8 - 3) + 2(1 - 8) = 1$$

Let us now interchange, say, the first and second columns.

$$\begin{vmatrix} 0 & 3 & 2 \\ 4 & 1 & 3 \\ 1 & 2 & 2 \end{vmatrix}$$

Its value is again found by expanding using the first row,

$$0 - 3\begin{vmatrix} 4 & 3 \\ 1 & 2 \end{vmatrix} + 2\begin{vmatrix} 4 & 1 \\ 1 & 2 \end{vmatrix} = -3(8 - 3) + 2(8 - 1) = -1$$

So interchanging two columns has reversed the sign of the determinant.

| | | |
|---|---|---|
| Multiplying by a Constant | If each element in a row (or column) is multiplied by some constant, the value of the determinant is multiplied by that constant. | 77 |

**EXAMPLE 18:** We saw that the value of the determinant in Example 17 was equal to 1.

$$\begin{vmatrix} 3 & 0 & 2 \\ 1 & 4 & 3 \\ 2 & 1 & 2 \end{vmatrix} = 1$$

Let us now multiply the elements in the third row by some constant, say, 3.

$$\begin{vmatrix} 3 & 0 & 2 \\ 1 & 4 & 3 \\ 6 & 3 & 6 \end{vmatrix}$$

This new determinant has the value

$$3(24 - 9) + 2(3 - 24) = 45 - 42 = 3$$

or three times its previous value.

We can also use this rule *in reverse,* to *remove a factor* from a row or column.

**EXAMPLE 19:** The determinant

$$\begin{vmatrix} 3 & 1 & 0 \\ 30 & 10 & 15 \\ 3 & 2 & 1 \end{vmatrix}$$

could be evaluated as it is, but let us first factor a 3 from the first column,

$$\begin{vmatrix} 3 & 1 & 0 \\ 30 & 10 & 15 \\ 3 & 2 & 1 \end{vmatrix} = 3\begin{vmatrix} 1 & 1 & 0 \\ 10 & 10 & 15 \\ 1 & 2 & 1 \end{vmatrix}$$

and then factor a 5 from the second row,

$$3\begin{vmatrix} 1 & 1 & 0 \\ 10 & 10 & 15 \\ 1 & 2 & 1 \end{vmatrix} = 3(5)\begin{vmatrix} 1 & 1 & 0 \\ 2 & 2 & 3 \\ 1 & 2 & 1 \end{vmatrix}$$

$$= 15[(2 - 6) - (2 - 3)] = -45$$

| Multiples of One Row (or Column) Added to Another | The value of a determinant is unchanged when the elements of a row (or column) are multiplied by some factor, and then added to the corresponding elements of another row or column. | **78** |
|---|---|---|

**EXAMPLE 20:** Again using the determinant from Example 17,

$$\begin{vmatrix} 3 & 0 & 2 \\ 1 & 4 & 3 \\ 2 & 1 & 2 \end{vmatrix} = 1$$

let us get a new second row by multiplying the third row by $-4$ and adding those elements to the second row. The new elements in the second row are then

$$1 + 2(-4) = -7, \quad 4 + 1(-4) = 0, \quad \text{and} \quad 3 + 2(-4) = -5$$

giving a new determinant

$$\begin{vmatrix} 3 & 0 & 2 \\ -7 & 0 & -5 \\ 2 & 1 & 2 \end{vmatrix}$$

whose value we find by developing the second column by minors,

$$-1\begin{vmatrix} 3 & 2 \\ -7 & -5 \end{vmatrix} = -1[3(-5) - 2(-7)] = -[-15 + 14] = 1$$

as before. Notice, however, that we have *introduced another zero* into the determinant, making it easier to evaluate. This, of course, is the point of the whole operation.

## A Systematic Method for Evaluating a Determinant

In one method for evaluating a determinant, you first select the row or column having the most zeros, and then use the properties of determinants to introduce zeros into all but one element of that row or column, and then develop that row or column by minors. That method is fine for manual computation, but not as good for a computer. Although we could program a computer to select the best row for development, the logic gets complicated.

The following algorithm, based on the properties of determinants in Eqs. 77 and 78, gives a systematic method for manual or computer evaluation of a determinant of any size. It is sometimes called the *method of Chió*.

1. Factor out the element in row 1, column 1 from each element in row 1.
2. Multiply each element in row 1 by the first element in row 2. Subtract these products from the corresponding elements in row 2.
3. Repeat step 2 for each row until all but the first element in the first column are zeros.
4. Delete the first row and first column, thus reducing the order of the determinant by 1.
5. Repeat steps 1 through 4 until the determinant is reduced to a single element. The value of the determinant is then the product of that element and all the factors removed in step 1.

---

**EXAMPLE 21:** Evaluate the following determinant by the method of Chió. Work to three significant digits.

$$\begin{vmatrix} 5 & 3 & 8 \\ 2 & 6 & 1 \\ 4 & 7 & 9 \end{vmatrix}$$

**Solution:**

---------------------- **First Pass** ----------------------

Step 1: We factor out the 5.

$$5\begin{vmatrix} 1 & 0.6 & 1.6 \\ 2 & 6 & 1 \\ 4 & 7 & 9 \end{vmatrix}$$

Step 2: Multiply row 1 by 2 and subtract it from row 2.

$$5\begin{vmatrix} 1 & 0.6 & 1.6 \\ 0 & 4.8 & -2.2 \\ 4 & 7 & 9 \end{vmatrix}$$

Step 3: Multiply row 1 by 4 and subtract it from row 3.

$$5\begin{vmatrix} 1 & 0.6 & 1.6 \\ 0 & 4.8 & -2.2 \\ 0 & 4.6 & 2.6 \end{vmatrix}$$

Step 4: We develop the first column by minors. Since the first element is a 1 and the others are zeros, we simply delete the first row and first column.

$$5(1)\begin{vmatrix} 4.8 & -2.2 \\ 4.6 & 2.6 \end{vmatrix}$$

Thus we have reduced the order of the determinant from three to two. We then repeat the whole process.

If, during step 1, the element in row 1, column 1 happens to be zero, we will get division by zero. If this happens, interchange two rows or two columns and try again. Remember to change the sign of the determinant when interchanging rows or columns.

---------------------- **Second Pass** ----------------------

Step 1: Factor out 4.8.

$$5(4.8)\begin{vmatrix} 1 & -0.458 \\ 4.6 & 2.6 \end{vmatrix}$$

Step 2:

$$5(4.8)\begin{vmatrix} 1 & -0.458 \\ 0 & 4.707 \end{vmatrix}$$

Step 3: Not needed.

Step 4: Delete the first row and first column. The value of the determinant is then

$$5(4.8)[4.707] = 113$$

---

**EXAMPLE 22:** Evaluate the following fourth-order determinant by the method of Chió. Work to three significant digits.

$$\begin{vmatrix} 5 & 2 & 6 & 1 \\ 4 & 2 & 9 & 4 \\ 3 & 6 & 1 & 0 \\ 7 & 1 & 3 & 4 \end{vmatrix}$$

**Solution:**

---------------------- **First Pass** ----------------------

Step 1: We factor out the 5.

$$5\begin{vmatrix} 1 & 0.4 & 1.2 & 0.2 \\ 4 & 2 & 9 & 4 \\ 3 & 6 & 1 & 0 \\ 7 & 1 & 3 & 4 \end{vmatrix}$$

Step 2: Multiply row 1 by 4 and subtract it from row 2.

$$5\begin{vmatrix} 1 & 0.4 & 1.2 & 0.2 \\ 0 & 0.4 & 4.2 & 3.2 \\ 3 & 6 & 1 & 0 \\ 7 & 1 & 3 & 4 \end{vmatrix}$$

Step 3: Multiply row 1 by 3 and subtract it from row 3. Multiply row 1 by 7 and subtract it from row 4.

$$5\begin{vmatrix} 1 & 0.4 & 1.2 & 0.2 \\ 0 & 0.4 & 4.2 & 3.2 \\ 0 & 4.8 & -2.6 & -0.6 \\ 0 & -1.8 & -5.4 & 2.6 \end{vmatrix}$$

Step 4: Delete the first row and first column.

$$5\begin{vmatrix} 0.4 & 4.2 & 3.2 \\ 4.8 & -2.6 & -0.6 \\ -1.8 & -5.4 & 2.6 \end{vmatrix}$$

**---------------------- Second Pass ----------------------**

Step 1: Factor out 0.4.

$$5(0.4)\begin{vmatrix} 1 & 10.5 & 8 \\ 4.8 & -2.6 & -0.6 \\ -1.8 & -5.4 & 2.6 \end{vmatrix}$$

Steps 2 and 3:

$$5(0.4)\begin{vmatrix} 1 & 10.5 & 8 \\ 0 & -53 & -39 \\ 0 & 13.5 & 17 \end{vmatrix}$$

Step 4: Delete the first row and first column.

$$5(0.4)\begin{vmatrix} -53 & -39 \\ 13.5 & 17 \end{vmatrix}$$

**---------------------- Third Pass ----------------------**

Step 1: Factor out −53.

$$5(0.4)(-53)\begin{vmatrix} 1 & 0.736 \\ 13.5 & 17 \end{vmatrix}$$

Step 2: Multiply row 1 by 13.5 and subtract it from row 2.

$$5(0.4)(-53)\begin{vmatrix} 1 & 0.736 \\ 0 & 7.06 \end{vmatrix}$$

Step 3: Not needed.

Step 4: Delete the first row and first column. The value of the determinant is, then,

$$5(0.4)(-53)(7.06) = -749$$

---

## EXERCISE 3—HIGHER-ORDER DETERMINANTS

Evaluate each determinant.

1. $\begin{vmatrix} 4 & 3 & 1 & 0 \\ -1 & 2 & -3 & 5 \\ 0 & 1 & -1 & 2 \\ 0 & 2 & -3 & 5 \end{vmatrix}$

2. $\begin{vmatrix} -1 & 3 & 0 & 2 \\ 2 & -1 & 1 & 0 \\ 5 & 2 & -2 & 0 \\ 1 & -1 & 3 & 1 \end{vmatrix}$

3. $\begin{vmatrix} 2 & 0 & -1 & 0 \\ 0 & 0 & 2 & -1 \\ 1 & 3 & 2 & 1 \\ 3 & 1 & 1 & -2 \end{vmatrix}$

4. $\begin{vmatrix} 1 & 2 & -1 & 1 \\ -1 & 1 & 2 & 3 \\ 3 & -1 & 1 & 2 \\ 1 & 2 & -1 & 1 \end{vmatrix}$

5. $\begin{vmatrix} 3 & 1 & 0 & 2 & 4 \\ 1 & 2 & 4 & 0 & 1 \\ 2 & 3 & 1 & 4 & 2 \\ 1 & 2 & 0 & 2 & 1 \\ 3 & 4 & 1 & 3 & 1 \end{vmatrix}$

6. $\begin{vmatrix} 2 & 1 & 5 & 3 & 6 \\ 1 & 4 & 2 & 4 & 3 \\ 3 & 1 & 2 & 4 & 1 \\ 5 & 2 & 3 & 1 & 4 \\ 4 & 5 & 2 & 3 & 1 \end{vmatrix}$

### Four Equations in Four Unknowns

Solve by determinants.

7. $$\begin{aligned} x + y + 2z + w &= 18 \\ x + 2y + z + w &= 17 \\ x + y + z + 2w &= 19 \\ 2x + y + z + w &= 16 \end{aligned}$$

8. $$\begin{aligned} 2x - y - z - w &= 0 \\ x - 3y + z + w &= 0 \\ x + y - 4z + w &= 0 \\ x + y + w &= 36 \end{aligned}$$

9. $$\begin{aligned} 3x - 2y - z + w &= -3 \\ -x - y + 3z + 2w &= 23 \\ x + 3y - 2z + w &= -12 \\ 2x - y - z - 3w &= -22 \end{aligned}$$

10. $$\begin{aligned} x + 2y &= 5 \\ y + 2z &= 8 \\ z + 2u &= 11 \\ u + 2x &= 6 \end{aligned}$$

11. $$\begin{aligned} x + y &= a + b \\ y + z &= b + c \\ z + w &= a - b \\ w - x &= c - b \end{aligned}$$

12. $$\begin{aligned} 2x - 3y + z - w &= -6 \\ x + 2y - z &= 8 \\ 3y + z + 3w &= 0 \\ 3x - y + w &= 0 \end{aligned}$$

### Five Equations in Five Unknowns

Solve by determinants.

13. $$\begin{aligned} x + y &= 9 \\ y + z &= 11 \\ z + w &= 13 \\ w + u &= 15 \\ u + x &= 12 \end{aligned}$$

14. $$\begin{aligned} 3x + 4y + z &= 35 \\ 3z + 2y - 3w &= 4 \\ 2x - y + 2w &= 17 \\ 3z - 2w + v &= 9 \\ w + y &= 13 \end{aligned}$$

15. $$\begin{aligned} w + v + x + y &= 14 \\ w + v + x + z &= 15 \\ w + v + y + z &= 16 \\ w + x + y + z &= 17 \\ v + x + y + z &= 18 \end{aligned}$$

16. Separate 125 into four parts such that if the first is increased by 4, the second diminished by 4, the third multiplied by 4, and the fourth divided by 4, these four quantities will all be equal.

17. A person divided a sum of money among four children so that the share of the eldest was $\frac{1}{2}$ of the sum of the shares of the other three, the share of the second was $\frac{1}{3}$ of the sum of the shares of the other three, and the share of the third was $\frac{1}{4}$ of the sum of the shares of the other three. The eldest has \$14,000 more than the youngest. What was the share of each?

18. Applying Kirchhoff's law to a certain four-loop network gives the equations

$$\begin{aligned} 57.2I_1 + 92.5I_2 - 23.0I_3 - 11.4I_4 &= 38.2 \\ 95.3I_1 - 14.9I_2 + 39.0I_3 + 59.9I_4 &= 29.3 \\ 66.3I_1 + 81.4I_2 - 91.5I_3 + 33.4I_4 &= -73.6 \\ 38.2I_1 - 46.6I_2 + 30.1I_3 + 93.2I_4 &= 55.7 \end{aligned}$$

Solve for the four loop currents, by any method.

19. We have available four bronze alloys containing the following percentages of copper, zinc, and lead.

| | Alloy 1 | Alloy 2 | Alloy 3 | Alloy 4 |
|---|---|---|---|---|
| Copper | 52 | 48 | 61 | 55 |
| Zinc | 30 | 38 | 20 | 38 |
| Lead | 3 | 2 | 4 | 3 |

How many pounds of each alloy should be taken to produce 600 kg of a new alloy that is 53.8% copper, 30.1% zinc, and 3.2% lead?

### Computer

20. Write a program to evaluate a determinant of any order. Use the algorithm given for the method of Chió. Use DATA statements to input the order of the determinant, and then the elements, row by row. Use it to evaluate any of the determinants in this chapter.

## CHAPTER 10 REVIEW PROBLEMS

*Evaluate.*

1. $\begin{vmatrix} 6 & 1/2 & -2 \\ 3 & 1/4 & 4 \\ 2 & -1/2 & 3 \end{vmatrix}$

2. $\begin{vmatrix} 2 & 7 & -2 & 8 \\ 4 & 1 & 1 & -3 \\ 0 & 3 & -1 & 4 \\ 6 & 4 & 2 & -8 \end{vmatrix}$

3. $\begin{vmatrix} 0 & n & m \\ -n & 0 & l \\ -m & -l & 0 \end{vmatrix}$

4. $\begin{vmatrix} 8 & 2 & 0 & 1 & 4 \\ 0 & 1 & 4 & 2 & 7 \\ 2 & 6 & 3 & 8 & 0 \\ 1 & 4 & 2 & 6 & 5 \\ 4 & 6 & 8 & 3 & 5 \end{vmatrix}$

5. $\begin{vmatrix} 25 & 23 & 19 \\ 14 & 11 & 9 \\ 21 & 17 & 14 \end{vmatrix}$

6. $\begin{vmatrix} x + y & x - y \\ x - y & x + y \end{vmatrix}$

7. $\begin{vmatrix} 9 & 13 & 17 \\ 11 & 15 & 19 \\ 17 & 21 & 25 \end{vmatrix}$

8. $\begin{vmatrix} 5 & -3 & -2 & 0 \\ 4 & 1 & -6 & 2 \\ -1 & 4 & 3 & -5 \\ 0 & 6 & -4 & 2 \end{vmatrix}$

9. $\begin{vmatrix} 1 & 2 & 3 \\ 3 & 1 & 2 \\ 2 & 3 & 1 \end{vmatrix}$

10. $\begin{vmatrix} 2 & -3 & 1 \\ -2 & 4 & 5 \\ 3 & -1 & -4 \end{vmatrix}$

11. $\begin{vmatrix} 1 & 2 & 2 & 4 \\ 1 & 4 & 4 & 1 \\ 1 & 1 & 2 & 2 \\ 4 & 8 & 11 & 13 \end{vmatrix}$

12. $\begin{vmatrix} a - b & -2a \\ 2b & a - b \end{vmatrix}$

13. $\begin{vmatrix} 3 & 1 & 5 & 2 \\ 4 & 10 & 14 & 6 \\ 8 & 9 & 1 & 4 \\ 6 & 15 & 21 & 9 \end{vmatrix}$

14. $\begin{vmatrix} 7 & 8 & 9 \\ 28 & 35 & 40 \\ 21 & 26 & 30 \end{vmatrix}$

15. $\begin{vmatrix} 1 & 5 & 2 \\ 4 & 7 & 3 \\ 9 & 8 & 6 \end{vmatrix}$

16. $\begin{vmatrix} 3 & -6 \\ -5 & 4 \end{vmatrix}$

17. $\begin{vmatrix} 6 & 4 & 7 \\ 9 & 0 & 8 \\ 5 & 3 & 2 \end{vmatrix}$

18. $\begin{vmatrix} 3 & 2 & 1 & 3 \\ 4 & 2 & 2 & 4 \\ 2 & 3 & 1 & 6 \\ 10 & 4 & 5 & 8 \end{vmatrix}$

*Solve by determinants.*

19. $$\begin{aligned} x + y + z + w &= -4 \\ x + 2y + 3z + 4w &= 0 \\ x + 3y + 6z + 10w &= 9 \\ x + 4y + 10z + 20w &= 24 \end{aligned}$$

20. $$\begin{aligned} x + 2y + 3z &= 4 \\ 3x + 5y + z &= 18 \\ 4x + y + 2z &= 12 \end{aligned}$$

21. $$\begin{aligned} 4x + 3y &= 27 \\ 2x - 5y &= -19 \end{aligned}$$

22. $$\begin{aligned} 8x - 3y - 7z &= 85 \\ x + 6y - 4z &= -12 \\ 2x - 5y + z &= 33 \end{aligned}$$

23. $$\begin{aligned} x + 2y &= 10 \\ 2x - 3y &= -1 \end{aligned}$$

24. $$\begin{aligned} x + y + z + u &= 1 \\ 2x + 3y - 4z + 5u &= -31 \\ 3x - 4y + 5z + 6u &= -22 \\ 4x + 5y - 6z - u &= -13 \end{aligned}$$

25. $$\begin{aligned} 2x - 3y &= 7 \\ 5x + 2y &= 27 \end{aligned}$$

26. $$\begin{aligned} 6x + y &= 60 \\ 3x + 2y &= 39 \end{aligned}$$

27. $$\begin{aligned} 2x + 5y &= 29 \\ 2x - 5y &= -21 \end{aligned}$$

28. $$\begin{aligned} 4x + 3y &= 7 \\ 2x - 3y &= -1 \end{aligned}$$

29. $$\begin{aligned} 4x - 5y &= 3 \\ 3x + 5y &= 11 \end{aligned}$$

30. $$\begin{aligned} x + 5y &= 41 \\ 3x - 2y &= 21 \end{aligned}$$

31. $$\begin{aligned} x + 2y &= 7 \\ x + y &= 5 \end{aligned}$$

32. $$\begin{aligned} 8x - 3y &= 22 \\ 4x + 5y &= 18 \end{aligned}$$

33. $$\begin{aligned} 3x + 4y &= 25 \\ 4x + 3y &= 21 \end{aligned}$$

*Writing*

34. Suppose that you often solve sets of three or more equations on the job. Write a memo to your boss asking for a desktop computer to do the work. It should include a description of the method of Chió which can be understood by an engineer who does not have extensive computer experience.

# 11

# MATRICES

## OBJECTIVES

**When you have completed this chapter, you should be able to:**

- Identify various types of arrays, matrices, and vectors, and give their dimensions.
- Write the transpose of a matrix.
- Add, subtract, and multiply matrices.
- Solve systems of equations using Gauss elimination.
- Solve systems of equations using matrix inversion.

The method of determinants is fine for solving sets of equations by hand, but it is not easy to program for the computer. Two other methods that are frequently used on the computer are called *Gauss elimination* and *matrix inversion*. We cover these methods after first learning how to add, subtract, and multiply matrices.

## 11-1 DEFINITIONS

### Arrays

A set of numbers, called *elements*, arranged in a pattern, is called an *array*. Arrays are named for the shape of the pattern made by the elements.

---

**EXAMPLE 1:** The array

$$\begin{pmatrix} 7 & 3 & 9 & 1 \\ & 8 & 3 & 2 \\ & & 9 & 1 \\ & & & 5 \end{pmatrix}$$

is called a *triangular array*.

---

### Matrices

A *matrix* is a *rectangular* array.

---

**EXAMPLE 2:** The rectangular arrays

(a) $\begin{pmatrix} 2 & 5 & 7 \\ 6 & 3 & 1 \end{pmatrix}$

(b) $\begin{pmatrix} 7 & 3 & 5 \\ 1 & 9 & 3 \\ 8 & 3 & 2 \end{pmatrix}$

(c) $\begin{pmatrix} 5 \\ 0 \\ 4 \\ 1 \end{pmatrix}$

(d) $(3 \quad 2 \quad 9 \quad 6)$

are matrices. Further, (b) is a *square* matrix, (c) is called a *column vector*, and (d) is a *row vector*.

---

In everyday language, an array is a *table* and a vector is a *list*.

### Subscripts

Each element in an array is located in a horizontal *row* and a vertical *column*. We indicate the row and column by means of *subscripts*.

---

**EXAMPLE 3:** The element $a_{25}$ is located in row 2 and column 5.

---

Thus, an element in an array needs *double subscripts* to give its location. An element in a list, such as $b_7$, needs only a *single subscript*.

## Dimensions

A matrix will have, in general, $m$ rows and $n$ columns. The numbers $m$ and $n$ are the *dimensions* of the matrix, as, for example, a $4 \times 5$ matrix. Thus, the matrices in Example 2 have the dimensions (a) $2 \times 3$ (b) $3 \times 3$ (c) $4 \times 1$ (d) $1 \times 4$.

## Scalars

A single number, as opposed to an array of numbers, is called a *scalar*.

**EXAMPLE 4:** Some scalars are 5, 693.6, or −24.3.

A scalar can also be thought of as an array having just one row and one column.

## Naming a Matrix

We will often denote or *name* a matrix with a single letter.

**EXAMPLE 5:** We can let

$$\mathbf{A} = \begin{pmatrix} 2 & 5 & 1 \\ 0 & 4 & 3 \end{pmatrix}$$

Thus we can represent an entire array by a single symbol.

## The Null or Zero Matrix

A matrix in which all elements are zero is called the *null* or *zero* matrix.

**EXAMPLE 6:** The matrix

$$\mathbf{O} = \begin{pmatrix} 0 & 0 & 0 \\ 0 & 0 & 0 \end{pmatrix}$$

is a $2 \times 3$ zero matrix. We will denote the zero matrix by the boldface letter **O**.

## The Unit Matrix

If all the elements *not* on the main diagonal of a square matrix are 0, we have a *diagonal matrix*. If, in addition, the diagonal elements are equal, we have a *scalar matrix*. If the diagonal elements of a scalar matrix are 1's, we have a *unit matrix* or *identity matrix*.

**EXAMPLE 7:** The matrices

$$\mathbf{A} = \begin{pmatrix} 3 & 0 & 0 \\ 0 & 5 & 0 \\ 0 & 0 & 7 \end{pmatrix} \qquad \mathbf{B} = \begin{pmatrix} 6 & 0 & 0 \\ 0 & 6 & 0 \\ 0 & 0 & 6 \end{pmatrix} \qquad \mathbf{I} = \begin{pmatrix} 1 & 0 & 0 \\ 0 & 1 & 0 \\ 0 & 0 & 1 \end{pmatrix}$$

are all diagonal matrices. **B** and **I** are, in addition, scalar matrices, and **I** is a unit matrix. We usually denote the unit matrix by the letter **I.**

## Equality of Matrices

When comparing two matrices *of the same dimension,* elements in each matrix having the same row and column subscripts are called *corresponding elements*. Two matrices of the same dimension are equal if their corresponding elements are equal.

**EXAMPLE 8:** The matrices

$$\begin{pmatrix} x & 2 & 5 \\ 3 & 1 & 7 \end{pmatrix} \quad \text{and} \quad \begin{pmatrix} 4 & 2 & 5 \\ 3 & 1 & y \end{pmatrix}$$

are equal only if $x = 4$ and $y = 7$.

## Transpose of a Matrix

The *transpose* $\mathbf{A}'$ of a matrix **A** is obtained by changing the rows to columns and changing the columns to rows.

**EXAMPLE 9:** The transpose of the matrix

$$\begin{pmatrix} 1 & 2 & 3 & 4 \\ 5 & 6 & 7 & 8 \end{pmatrix}$$

is

$$\begin{pmatrix} 1 & 5 \\ 2 & 6 \\ 3 & 7 \\ 4 & 8 \end{pmatrix}$$

## EXERCISE 1—DEFINITIONS

Given

$$\mathbf{A} = \begin{pmatrix} 2 & 5 & 1 \\ 6 & 3 & 7 \\ 1 & 6 & 9 \\ 7 & 4 & 2 \end{pmatrix} \qquad \mathbf{B} = \begin{pmatrix} 7 \\ 3 \\ 9 \\ 2 \end{pmatrix} \qquad \mathbf{C} = \begin{pmatrix} f & i & q & w \\ & g & w & k \\ & & c & z \\ & & & b \end{pmatrix}$$

$$\mathbf{D} = \begin{pmatrix} 6 & 2 & 0 & 1 \\ 2 & 8 & 3 & 9 \end{pmatrix} \qquad \mathbf{E} = \begin{pmatrix} x & y \\ z & w \end{pmatrix} \qquad \mathbf{F} = \begin{pmatrix} 0 & 0 & 0 & 0 \\ 0 & 0 & 0 & 0 \end{pmatrix}$$

$$\mathbf{G} = 7 \qquad \mathbf{H} = \begin{pmatrix} 3 & 8 \\ & 5 \end{pmatrix} \qquad \mathbf{I} = (3 \quad 8 \quad 4 \quad 6)$$

Which of the nine arrays shown above is:

1. a rectangular array?
2. a square array?
3. a triangular array?
4. a column vector?
5. a row vector?
6. a table?
7. a list?
8. a scalar?
9. a null matrix?
10. a matrix?

For the matrix **A** above, find the element:

11. $a_{32}$
12. $a_{41}$

Give the dimensions of:

13. matrix **A**
14. matrix **B**
15. matrix **D**
16. matrix **E**

Write the transpose of:

17. matrix **F**
18. matrix **D**

19. Under what conditions will the matrices

$$\begin{pmatrix} 6 & x \\ w & 8 \end{pmatrix} \quad \text{and} \quad \begin{pmatrix} y & 2 \\ 7 & z \end{pmatrix}$$

be equal?

**Computer**

20. Using subscripted variables, write a computer program to find and print the transpose of any matrix. Use DATA statements to enter the number of rows and of columns, and then the elements of the original matrix, row by row, into your program.

## 11-2 OPERATIONS WITH MATRICES

In earlier chapters we have learned how to add, subtract, and multiply numbers and algebraic quantities. We now perform these operations on matrices.

### Sum of Two Matrices

**Addition and subtraction are *not defined* for two matrices of *different* dimensions.**

To add two matrices *of the same dimensions,* simply combine corresponding elements.

**EXAMPLE 10:**

$$\begin{pmatrix} 2 & -3 & 4 & 1 \\ 0 & 5 & 6 & -8 \end{pmatrix} + \begin{pmatrix} 5 & 8 & 1 & 2 \\ 9 & -3 & 0 & 1 \end{pmatrix} = \begin{pmatrix} 7 & 5 & 5 & 3 \\ 9 & 2 & 6 & -7 \end{pmatrix}$$

Subtraction is done in a similar way.

**EXAMPLE 11:**

$$\begin{pmatrix} 2 & -3 & 4 & 1 \\ 0 & 5 & 6 & -8 \end{pmatrix} - \begin{pmatrix} 5 & 8 & 1 & 2 \\ 9 & -3 & 0 & 1 \end{pmatrix} = \begin{pmatrix} -3 & -11 & 3 & -1 \\ -9 & 8 & 6 & -9 \end{pmatrix}$$

As with ordinary addition, the *commutative law* (Eq. 1) and the *associative law* (Eq. 3) apply.

| | | |
|---|---|---|
| Commutative Law | $\mathbf{A} + \mathbf{B} = \mathbf{B} + \mathbf{A}$ | **79** |

| | | |
|---|---|---|
| Associative Law | $\mathbf{A} + (\mathbf{B} + \mathbf{C}) = (\mathbf{A} + \mathbf{B}) + \mathbf{C}$<br>$= (\mathbf{A} + \mathbf{C}) + \mathbf{B}$ | **80** |

### Product of a Scalar and a Matrix

To multiply a matrix by a scalar, multiply each element in the matrix by the scalar.

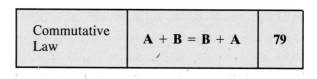

**EXAMPLE 12:**

$$3\begin{pmatrix} 2 & 4 & 1 \\ 5 & 7 & 3 \end{pmatrix} = \begin{pmatrix} 6 & 12 & 3 \\ 15 & 21 & 9 \end{pmatrix}$$

Worked in reverse, we can *factor* a scalar from a matrix. Thus, in the example shown, we can think of the number 3 as being *factored out.*

In symbols

| | | |
|---|---|---|
| Product of a Scalar and a Matrix | $k\begin{pmatrix} a & b \\ c & d \end{pmatrix} = \begin{pmatrix} ka & kb \\ kc & kd \end{pmatrix}$ | **87** |

## Multiplication of Matrices

Not all matrices can be multiplied. Further, for matrices **A** and **B,** it may be possible to find the product **AB** but *not* the product **BA.**

| | | |
|---|---|---|
| Conformable Matrices | The product **AB** of two matrices **A** and **B** is defined only when the number of columns in **A** equals the number of rows in **B.** | **82** |

**EXAMPLE 13:** If

$$\mathbf{A} = \begin{pmatrix} 3 & 5 \\ 2 & 1 \\ 7 & 4 \\ 9 & 0 \end{pmatrix} \quad \text{and} \quad \mathbf{B} = \begin{pmatrix} 7 & 3 & 8 \\ 4 & 7 & 1 \end{pmatrix}$$

Two matrices for which multiplication is defined are called *conformable matrices.*

we can find the product **AB,** because the number of columns in **A** (2) equals the number of rows in **B** (2). However, we *cannot* find the product **BA** because the number of columns in **B** (3) does not equal the number of rows in **A** (4).

For two or more conformable matrices, the *associative* law and the *distributive* law for multiplication applies.

| | | |
|---|---|---|
| Associative Law | **A(BC) = (AB)C = ABC** | **84** |

| | | |
|---|---|---|
| Distributive Law | **A(B + C) = AB + AC** | **85** |

These are similar to Eqs. 4 and 5 for ordinary multiplication.

The *dimensions* of the matrix obtained by multiplying an $m \times p$ matrix with a $p \times n$ matrix is equal to $m \times n$. In other words, the product matrix **AB** will have the same number of *rows* as **A** and the same number of *columns* as **B**.

| | | |
|---|---|---|
| Dimensions of the Product | $(m \times p)(p \times n) = (m \times n)$ | **86** |

**EXAMPLE 14:** The product **AB** of a $3 \times 5$ matrix **A** and a $5 \times 7$ matrix **B** is a $3 \times 7$ matrix.

Whatever we said about matrices applies, of course, to vectors as well.

## Scalar Product of Two Vectors

We will show matrix multiplication first with vectors, and then extend it to other matrices.

The *scalar product* (also called the *dot product*) of a row vector **A** and a column vector **B** is defined only when the number of columns in **A** is equal to the number of rows in **B**.

If each vector contains $m$ elements, the dimensions of the product will be, by Eq. 86,

$$(1 \times m)(m \times 1) = 1 \times 1$$

Thus, the product is a scalar, hence the name *scalar product*.

To multiply a row vector by a column vector, multiply the first elements of each, then the second, and so on, and add the products obtained.

**EXAMPLE 15:**

$$(2 \quad 6 \quad 4 \quad 8)\begin{pmatrix}3\\5\\1\\7\end{pmatrix} = [(2)(3) + (6)(5) + (4)(1) + (8)(7)]$$

$$= (6 + 30 + 4 + 56) = 96$$

**EXAMPLE 16:** A certain store has four types of radios, priced at \$71, \$62, \$83, and \$49. On a particular day, they sell quantities of 8, 3, 7, and 6, respectively. If we represent the prices by a row vector and the quantities by a column vector, and multiply, we get

$$(71 \quad 62 \quad 83 \quad 49)\begin{pmatrix}8\\3\\7\\6\end{pmatrix} = [71(8) + 62(3) + 83(7) + 49(6)]$$

$$= 1629$$

The interpretation of this number is obviously the total dollar income from the sale of the four types of radios on that day.

## Product of a Row Vector and a Matrix

By Eq. 86, the product of a row vector and a matrix having $n$ columns will be a row vector having $n$ elements.

$$(1 \times p)(p \times n) = (1 \times n)$$

We saw earlier that the product of a row vector and a column vector is a single number, or scalar. Thus, the product of the given row vector and the first column of the matrix will give a scalar. This scalar will be *the first element in the product vector.*

The *second* element in the product vector is found by multiplying the second column in the matrix by the row vector, and so on.

---

**EXAMPLE 17:** Multiply

$$(1 \quad 3 \quad 0)\begin{pmatrix} 7 & 5 \\ 1 & 0 \\ 4 & 9 \end{pmatrix}$$

**Solution:** The dimensions of the product will be

$$(1 \times 3)(3 \times 2) = (1 \times 2)$$

Multiplying the vector and the first column of the matrix gives

$$(1 \quad 3 \quad 0)\begin{pmatrix} \mathbf{7} & 5 \\ \mathbf{1} & 0 \\ \mathbf{4} & 9 \end{pmatrix} = [\mathbf{1(7) + 3(1) + 0(4)} \qquad * \qquad ]$$

$$= (\mathbf{10} \qquad *)$$

Here we use the asterisk (*) as a placeholder to denote the other element still to be computed.

Multiplying the vector and the second column of the matrix yields

$$(1 \quad 3 \quad 0)\begin{pmatrix} 7 & \mathbf{5} \\ 1 & \mathbf{0} \\ 4 & \mathbf{9} \end{pmatrix} = [10 \qquad \mathbf{1(5) + 3(0) + 0(9)}]$$

$$= (10 \quad \mathbf{5})$$

---

## Product of Two Matrices

We now multiply two matrices **A** and **B.** We already know how to multiply a row vector and a matrix. So let us think of the matrix **A** as several row vectors. Each of these row vectors, multiplied by the matrix **B,** will produce a row vector in the product **AB.**

**EXAMPLE 18:** Find the product **AB** of the matrices

$$\mathbf{A} = \begin{pmatrix} 1 & 0 \\ 3 & 2 \end{pmatrix} \qquad \mathbf{B} = \begin{pmatrix} 5 & 4 & 8 \\ 6 & 7 & 9 \end{pmatrix}$$

**Solution:** The dimensions of the product will be

$$(2 \times 2)(2 \times 3) = (2 \times 3)$$

We think of the matrix **A** as being two row vectors, each of which, when multiplied by matrix **B,** will produce a row vector in the product matrix.

$$\underset{\mathbf{A}}{\begin{matrix} (1 & 0) \\ (3 & 2) \end{matrix}} \underset{\mathbf{B}}{\begin{pmatrix} 5 & 4 & 8 \\ 6 & 7 & 9 \end{pmatrix}} = \underset{\mathbf{AB}}{\begin{matrix} ( & * & * & * & ) \\ ( & * & * & * & ) \end{matrix}}$$

Multiplying by the first row vector in **A,**

$$\underset{\mathbf{A}}{\begin{matrix} \mathbf{(1} & \mathbf{0)} \\ (3 & 2) \end{matrix}} \underset{\mathbf{B}}{\begin{pmatrix} 5 & 4 & 8 \\ 6 & 7 & 9 \end{pmatrix}} = \underset{\mathbf{AB}}{\begin{matrix} \mathbf{[1(5) + 0(6)} & \mathbf{1(4) + 0(7)} & \mathbf{1(8) + 0(9)]} \\ [\quad * & * & * \quad] \end{matrix}}$$

$$= \begin{matrix} \mathbf{[5} & \mathbf{4} & \mathbf{8]} \\ [* & * & *] \end{matrix}$$

Then multiplying by the second row vector in **A,**

$$\underset{\mathbf{A}}{\begin{matrix} (1 & 0) \\ \mathbf{(3} & \mathbf{2)} \end{matrix}} \underset{\mathbf{B}}{\begin{pmatrix} 5 & 4 & 8 \\ 6 & 7 & 9 \end{pmatrix}} = \underset{\mathbf{AB}}{\begin{matrix} [\quad 5 & 4 & 8 \quad] \\ \mathbf{[3(5) + 2(6)} & \mathbf{3(4) + 2(7)} & \mathbf{3(8) + 2(9)]} \end{matrix}}$$

giving the final product

$$\underset{\mathbf{A}}{\begin{pmatrix} 1 & 0 \\ 3 & 2 \end{pmatrix}} \underset{\mathbf{B}}{\begin{pmatrix} 5 & 4 & 8 \\ 6 & 7 & 9 \end{pmatrix}} = \underset{\mathbf{AB}}{\begin{pmatrix} 5 & 4 & 8 \\ 27 & 26 & 42 \end{pmatrix}}$$

**EXAMPLE 19:** Find the product **AB** of the matrices

$$\underset{\mathbf{A}}{\begin{pmatrix} 1 & 0 & 2 \\ 0 & 1 & 1 \\ 3 & 0 & 2 \\ 1 & 1 & 0 \\ 2 & 1 & 1 \end{pmatrix}} \underset{\mathbf{B}}{\begin{pmatrix} 3 & 0 & 1 & 1 \\ 1 & 2 & 3 & 0 \\ 0 & 1 & 2 & 1 \end{pmatrix}}$$

**Solution:** We think of the matrix **A** as five row vectors. Each of these five row vectors, multiplied by the matrix **B,** will produce a row vector in the product **AB.**

$$\underset{\substack{\mathbf{A} \\ (5 \times 3)}}{\begin{matrix} (1 & 0 & 2) \\ (0 & 1 & 1) \\ (3 & 0 & 2) \\ (1 & 1 & 0) \\ (2 & 1 & 1) \end{matrix}} \underset{\substack{\mathbf{B} \\ (3 \times 4)}}{\begin{pmatrix} 3 & 0 & 1 & 1 \\ 1 & 2 & 3 & 0 \\ 0 & 1 & 2 & 1 \end{pmatrix}} = \underset{\substack{\mathbf{AB} \\ (5 \times 4)}}{\begin{matrix} ( & & ) \\ ( & & ) \\ ( & & ) \\ ( & & ) \\ ( & & ) \end{matrix}}$$

Multiplying the first row vector (1  0  2) in **A** by the matrix **B** gives the first row vector in **AB.**

$$\begin{array}{c} \mathbf{(1\quad 0\quad 2)} \\ (0\quad 1\quad 1) \\ (3\quad 0\quad 2) \\ (1\quad 1\quad 0) \\ (2\quad 1\quad 1) \end{array} \begin{pmatrix} 3 & 0 & 1 & 1 \\ 1 & 2 & 3 & 0 \\ 0 & 1 & 2 & 1 \end{pmatrix} = \begin{array}{c} \mathbf{(3\quad 2\quad 5\quad 3)} \\ (\qquad\qquad\quad) \\ (\qquad\qquad\quad) \\ (\qquad\qquad\quad) \\ (\qquad\qquad\quad) \end{array}$$

**A** **B** **AB**

Multiplying the *second* row vector in **A** and the matrix **B** gives a second row vector in **AB.**

$$\begin{array}{c} (1\quad 0\quad 2) \\ \mathbf{(0\quad 1\quad 1)} \\ (3\quad 0\quad 2) \\ (1\quad 1\quad 0) \\ (2\quad 1\quad 1) \end{array} \begin{pmatrix} 3 & 0 & 1 & 1 \\ 1 & 2 & 3 & 0 \\ 0 & 1 & 2 & 1 \end{pmatrix} = \begin{array}{c} (3\quad 2\quad 5\quad 3) \\ \mathbf{(1\quad 3\quad 5\quad 1)} \\ (\qquad\qquad\quad) \\ (\qquad\qquad\quad) \\ (\qquad\qquad\quad) \end{array}$$

**A** **B** **AB**

The *third* row vector in **A** produces a third row vector in **AB.**

$$\begin{array}{c} (1\quad 0\quad 2) \\ (0\quad 1\quad 1) \\ \mathbf{(3\quad 0\quad 2)} \\ (1\quad 1\quad 0) \\ (2\quad 1\quad 1) \end{array} \begin{pmatrix} 3 & 0 & 1 & 1 \\ 1 & 2 & 3 & 0 \\ 0 & 1 & 2 & 1 \end{pmatrix} = \begin{array}{c} (3\quad 2\quad 5\quad 3) \\ (1\quad 3\quad 5\quad 1) \\ \mathbf{(9\quad 2\quad 7\quad 5)} \\ (\qquad\qquad\quad) \\ (\qquad\qquad\quad) \end{array}$$

**A** **B** **AB**

And so on, until we get the final product

$$\begin{pmatrix} 1 & 0 & 2 \\ 0 & 1 & 1 \\ 3 & 0 & 2 \\ 1 & 1 & 0 \\ 2 & 1 & 1 \end{pmatrix} \begin{pmatrix} 3 & 0 & 1 & 1 \\ 1 & 2 & 3 & 0 \\ 0 & 1 & 2 & 1 \end{pmatrix} = \begin{pmatrix} 3 & 2 & 5 & 3 \\ 1 & 3 & 5 & 1 \\ 9 & 2 & 7 & 5 \\ 4 & 2 & 4 & 1 \\ 7 & 3 & 7 & 3 \end{pmatrix}$$

**A** **B** **AB**

---

We can state the product of two matrices as a formula as follows:

| | | |
|---|---|---|
| Product of Two Matrices | $\begin{pmatrix} a & b & c \\ d & e & f \end{pmatrix} \begin{pmatrix} u & x \\ v & y \\ w & z \end{pmatrix} = \begin{pmatrix} au + bv + cw & ax + by + cz \\ du + ev + fw & dx + ey + fz \end{pmatrix}$ | **92** |

### Product of Square Matrices

We multiply square matrices **A** and **B** just as we do other matrices. But unlike for rectangular matrices, both the products **AB** and **BA** will be defined, although they will not usually be equal.

**EXAMPLE 20:** Find the products **AB** and **BA** for the square matrices

$$\mathbf{A} = \begin{pmatrix} 0 & 1 & 2 \\ 1 & 1 & 0 \\ 2 & 3 & 1 \end{pmatrix} \qquad \mathbf{B} = \begin{pmatrix} 2 & 1 & 0 \\ 1 & 0 & 3 \\ 3 & 1 & 1 \end{pmatrix}$$

and show that the products are not equal.

**Solution:** Multiplying as shown earlier,

$$\mathbf{AB} = \underset{(3 \times 3)}{\begin{pmatrix} 0 & 1 & 2 \\ 1 & 1 & 0 \\ 2 & 3 & 1 \end{pmatrix}} \underset{(3 \times 3)}{\begin{pmatrix} 2 & 1 & 0 \\ 1 & 0 & 3 \\ 3 & 1 & 1 \end{pmatrix}} = \underset{(3 \times 3)}{\begin{pmatrix} 7 & 2 & 5 \\ 3 & 1 & 3 \\ 10 & 3 & 10 \end{pmatrix}}$$

and

$$\mathbf{BA} = \begin{pmatrix} 2 & 1 & 0 \\ 1 & 0 & 3 \\ 3 & 1 & 1 \end{pmatrix} \begin{pmatrix} 0 & 1 & 2 \\ 1 & 1 & 0 \\ 2 & 3 & 1 \end{pmatrix} = \begin{pmatrix} 1 & 3 & 4 \\ 6 & 10 & 5 \\ 3 & 7 & 7 \end{pmatrix}$$

We see that multiplying the matrices in the reverse order does *not* give the same result.

| | | |
|---|---|---|
| Commutative Law | **AB ≠ BA**<br>Matrix multiplication is *not* commutative. | **83** |

### Product of a Matrix and a Column Vector

We have already multiplied a row vector **A** by a matrix **B.** Now we multiply a matrix **A** by a column vector **B.**

**EXAMPLE 21:** Multiply

$$\begin{pmatrix} 3 & 1 & 2 & 0 \\ 1 & 3 & 5 & 2 \end{pmatrix} \begin{pmatrix} 2 \\ 1 \\ 3 \\ 4 \end{pmatrix}$$

**Solution:** The dimensions of the product will be

$$(2 \times 4)(4 \times 1) = (2 \times 1)$$

Multiplying the column vector by the first row in the matrix,

$$\begin{pmatrix} \mathbf{3} & \mathbf{1} & \mathbf{2} & \mathbf{0} \\ 1 & 3 & 5 & 2 \end{pmatrix} \begin{pmatrix} 2 \\ 1 \\ 3 \\ 4 \end{pmatrix} = \begin{pmatrix} \mathbf{3(2) + 1(1) + 2(3) + 0(4)} \\ \end{pmatrix} = \begin{pmatrix} \mathbf{13} \\ \end{pmatrix}$$

and then by the second row,

$$\begin{pmatrix} 3 & 1 & 2 & 0 \\ \mathbf{1} & \mathbf{3} & \mathbf{5} & \mathbf{2} \end{pmatrix} \begin{pmatrix} 2 \\ 1 \\ 3 \\ 4 \end{pmatrix} = \begin{pmatrix} 13 \\ \mathbf{1(2) + 3(1) + 5(3) + 2(4)} \end{pmatrix} = \begin{pmatrix} 13 \\ \mathbf{28} \end{pmatrix}$$

**Most of the matrix multiplication we do later will be of a matrix and a column vector.**

Note that the product of a matrix and a column vector is a column vector.

**EXAMPLE 22:** Five students decided to raise money by selling T-shirts for three weeks. The following table shows the number of shirts sold by each student during each week.

| | | *Week* 1 | 2 | 3 |
|---|---|---|---|---|
| | *1* | 8 | 2 | 6 |
| | *2* | 3 | 5 | 2 |
| *Student* | *3* | 6 | 5 | 4 |
| | *4* | 7 | 3 | 1 |
| | *5* | 2 | 3 | 4 |

They charged \$7.50 per shirt the first week, and then lowered the price to \$6.00 the second week, and then \$5.50 for the third week. Find the total amount raised by each student for the three weeks.

**Solution:** To get the total amounts, we know that we must multiply the matrix **B** representing the number of T-shirts sold,

$$\mathbf{B} = \begin{pmatrix} 8 & 2 & 6 \\ 3 & 5 & 2 \\ 6 & 5 & 4 \\ 7 & 3 & 1 \\ 2 & 3 & 4 \end{pmatrix}$$

by a vector **C** representing the price per shirt. But should **C** be a row vector or a column vector? And should we find the product **BC** or the product **CB**?

If we let **C** be a $1 \times 3$ row vector and try to multiply it by the $5 \times 3$ matrix **B**, we see that we cannot, because these matrices are not conformable for either product **BC** or **CB**.

On the other hand, if we let **C** be a $3 \times 1$ column vector,

$$\mathbf{C} = \begin{pmatrix} 0.75 \\ 0.60 \\ 0.55 \end{pmatrix}$$

the product **CB** is not defined, but the product **BC** gives a column vector

$$\mathbf{BC} = \begin{pmatrix} 8 & 2 & 6 \\ 3 & 5 & 2 \\ 6 & 5 & 4 \\ 7 & 3 & 1 \\ 2 & 3 & 4 \end{pmatrix} \begin{pmatrix} 0.75 \\ 0.60 \\ 0.55 \end{pmatrix} = \begin{pmatrix} 10.50 \\ 6.35 \\ 9.70 \\ 7.60 \\ 5.50 \end{pmatrix}$$

which represents the total earned by each student.

### Tensor Product of Two Vectors

We have already found the product **AB** of a row vector **A** and a column vector **B**. We found it to be a scalar, and we called it the scalar or dot product. We now find the product **BA**, and find that the product is a matrix, rather than a scalar. It is called the *tensor product*.

**EXAMPLE 23:**

$$\begin{pmatrix}2\\0\\1\end{pmatrix}(1 \quad 2 \quad 0 \quad 3)=\begin{pmatrix}2(1) & 2(2) & 2(0) & 2(3)\\0(1) & 0(2) & 0(0) & 0(3)\\1(1) & 1(2) & 1(0) & 1(3)\end{pmatrix}$$

$$=\begin{pmatrix}2 & 4 & 0 & 6\\0 & 0 & 0 & 0\\1 & 2 & 0 & 3\end{pmatrix}$$

$(3 \times 1)\ (1 \times 4) \qquad (3 \times 4)$

## EXERCISE 2—OPERATIONS WITH MATRICES

### Addition and Subtraction of Matrices

Combine the following vectors and matrices.

1. $(3 \quad 8 \quad 2 \quad 1) + (9 \quad 5 \quad 3 \quad 7)$
2. $\begin{pmatrix}6 & 3 & 9\\2 & 1 & 3\end{pmatrix} + \begin{pmatrix}8 & 2 & 7\\1 & 6 & 3\end{pmatrix}$
3. $\begin{pmatrix}5\\-3\\5\\8\end{pmatrix} + \begin{pmatrix}7\\0\\-4\\8\end{pmatrix} - \begin{pmatrix}-2\\7\\-5\\2\end{pmatrix}$
4. $\begin{pmatrix}6 & 2 & -7\\-3 & 5 & 9\\2 & 5 & 3\end{pmatrix} - \begin{pmatrix}5 & 2 & 9\\1 & 4 & 8\\7 & -3 & 4\end{pmatrix} + \begin{pmatrix}1 & -6 & 7\\2 & 4 & -9\\-6 & 2 & 7\end{pmatrix}$
5. $\begin{pmatrix}1 & 5 & -2 & 1\\0 & -2 & 3 & 5\\-2 & 9 & 3 & -2\\5 & -1 & 2 & 6\end{pmatrix} - \begin{pmatrix}-6 & 4 & 2 & 8\\9 & 2 & -6 & 3\\-5 & 2 & 6 & 7\\6 & 3 & 7 & 0\end{pmatrix}$

### Product of a Scalar and a Matrix

Multiply each scalar and vector.

6. $6\begin{pmatrix}9\\-2\\6\\-3\end{pmatrix}$
7. $-3(5 \quad 9 \quad -2 \quad 7 \quad 3)$

Multiply each scalar and matrix.

8. $7\begin{pmatrix}7 & -3 & 7\\-2 & 9 & 5\\9 & 1 & 0\end{pmatrix}$
9. $-3\begin{pmatrix}4 & 2 & -3 & 0\\-1 & 6 & 3 & -4\end{pmatrix}$

Remove any factors from the vector or matrix.

10. $\begin{pmatrix}25\\-15\\30\\-5\end{pmatrix}$
11. $\begin{pmatrix}21 & 7 & -14\\49 & 63 & 28\\14 & -21 & 56\\-42 & 70 & 84\end{pmatrix}$

If $\mathbf{A} = \begin{pmatrix} 2 & 5 & -1 \\ -3 & 2 & 6 \\ 9 & -4 & 5 \end{pmatrix}$ and $\mathbf{B} = \begin{pmatrix} 0 & 1 & 6 \\ -7 & 2 & 1 \\ 4 & -2 & 0 \end{pmatrix}$

find:

12. $\mathbf{A} - 4\mathbf{B}$
13. $3\mathbf{A} + \mathbf{B}$
14. $2\mathbf{A} + 2\mathbf{B}$
15. Show that $2(\mathbf{A} + \mathbf{B}) = 2\mathbf{A} + 2\mathbf{B}$, thus demonstrating the distributive law (Eq. 85).

### Scalar Product of Two Vectors

Multiply each pair of vectors.

16. $(6 \quad -2)\begin{pmatrix} -1 \\ 7 \end{pmatrix}$
17. $(5 \quad 3 \quad 7)\begin{pmatrix} 2 \\ 6 \\ 1 \end{pmatrix}$
18. $(-3 \quad 6 \quad -9)\begin{pmatrix} 8 \\ -3 \\ 4 \end{pmatrix}$
19. $(-4 \quad 3 \quad 8 \quad -5)\begin{pmatrix} 7 \\ -6 \\ 0 \\ 3 \end{pmatrix}$

### Product of a Row Vector and a Matrix

Multiply each row vector and matrix.

20. $(4 \quad -2)\begin{pmatrix} 3 & 6 \\ -1 & 5 \end{pmatrix}$
21. $(3 \quad -1)\begin{pmatrix} 4 & 7 & -2 & 4 & 0 \\ 2 & -6 & 8 & -3 & 7 \end{pmatrix}$
22. $(2 \quad 8 \quad 1)\begin{pmatrix} 5 & 6 \\ 3 & 1 \\ 1 & 0 \end{pmatrix}$
23. $(-3 \quad 6 \quad -1)\begin{pmatrix} 5 & -2 & 7 \\ -6 & 2 & 5 \\ 9 & 0 & -1 \end{pmatrix}$
24. $(9 \quad 1 \quad 4)\begin{pmatrix} 6 & 2 & 1 & 5 & -8 \\ 7 & 1 & -3 & 0 & 3 \\ 1 & -5 & 0 & 3 & 2 \end{pmatrix}$
25. $(2 \quad 4 \quad 3 \quad -1 \quad 0)\begin{pmatrix} 4 & 3 & 5 & -1 & 0 \\ 1 & 3 & 5 & -2 & 6 \\ 2 & -4 & 3 & 6 & 5 \\ 3 & 4 & -1 & 6 & 5 \\ 2 & 3 & 1 & 4 & -6 \end{pmatrix}$

### Product of Two Matrices

Multiply each pair of matrices.

26. $\begin{pmatrix} 7 & 2 \\ 1 & 3 \end{pmatrix}\begin{pmatrix} 4 & 1 & 2 \\ 3 & 2 & 5 \end{pmatrix}$
27. $\begin{pmatrix} 3 & 1 & 4 \\ 1 & -3 & 5 \end{pmatrix}\begin{pmatrix} 4 & 2 \\ -1 & 5 \\ 2 & -6 \end{pmatrix}$
28. $\begin{pmatrix} 4 & 2 & -1 \\ -2 & 1 & 6 \end{pmatrix}\begin{pmatrix} 5 & -1 & 2 & 3 \\ 1 & 3 & 5 & 2 \\ 0 & 1 & -2 & 4 \end{pmatrix}$

29. $\begin{pmatrix} 5 & 1 \\ -2 & 0 \\ 1 & 3 \end{pmatrix}\begin{pmatrix} -4 & 1 \\ 0 & 5 \end{pmatrix}$

30. $\begin{pmatrix} 2 & 4 \\ 0 & -1 \\ 2 & 5 \end{pmatrix}\begin{pmatrix} 0 & -2 & 3 & -2 \\ 1 & 3 & -4 & 1 \end{pmatrix}$

31. $\begin{pmatrix} 4 & 1 & -2 & 3 \\ 0 & 2 & 1 & 5 \end{pmatrix}\begin{pmatrix} 3 & -1 \\ 0 & 2 \\ 3 & 1 \\ 0 & -2 \end{pmatrix}$

32. $\begin{pmatrix} -1 & 0 & 3 & 1 \\ 2 & 1 & 0 & -3 \end{pmatrix}\begin{pmatrix} 1 & 2 & 3 & 0 \\ 3 & 0 & -1 & 2 \\ -1 & 4 & 3 & 0 \\ 2 & 0 & -2 & 1 \end{pmatrix}$

33. $\begin{pmatrix} 3 & 1 \\ -2 & 0 \\ 1 & -3 \\ 5 & 0 \end{pmatrix}\begin{pmatrix} -1 & 0 \\ 3 & 2 \end{pmatrix}$

34. $\begin{pmatrix} 1 & 0 & -2 \\ 2 & 4 & 1 \\ 0 & -1 & 3 \end{pmatrix}\begin{pmatrix} 4 & 0 & 1 & -2 \\ 0 & 2 & 1 & 3 \\ 1 & 0 & -3 & 2 \end{pmatrix}$

35. $\begin{pmatrix} 2 & 1 & 0 & -1 \\ 1 & 2 & -1 & 1 \\ 0 & -2 & 4 & 0 \end{pmatrix}\begin{pmatrix} 1 & 2 & 0 & -1 \\ 3 & 1 & -1 & 3 \\ 2 & 1 & 0 & 3 \\ 3 & 5 & 1 & 0 \end{pmatrix}$

**Product of Square Matrices**

Find the product **AB** and the product **BA** for each pair of square matrices.

36. $\mathbf{A} = \begin{pmatrix} 2 & 3 \\ 5 & 1 \end{pmatrix}$ $\quad \mathbf{B} = \begin{pmatrix} 4 & 6 \\ 7 & 9 \end{pmatrix}$

37. $\mathbf{A} = \begin{pmatrix} 2 & -1 & 0 \\ 1 & 0 & 2 \\ -2 & 1 & 0 \end{pmatrix}$ $\quad \mathbf{B} = \begin{pmatrix} 2 & 1 & 0 \\ 3 & 0 & 1 \\ -2 & 4 & 1 \end{pmatrix}$

Using the matrices in Problem 36, find

38. $\mathbf{A}^2$

39. $\mathbf{B}^3$

**Product of a Matrix and a Column Vector**

Multiply each matrix and column vector.

40. $\begin{pmatrix} 2 & 1 \\ 0 & 3 \end{pmatrix}\begin{pmatrix} 3 \\ 1 \end{pmatrix}$

41. $\begin{pmatrix} 2 & 0 & 1 \\ 1 & 5 & 2 \end{pmatrix}\begin{pmatrix} 3 \\ 0 \\ -1 \end{pmatrix}$

42. $\begin{pmatrix} 2 & 1 & 0 \\ 1 & 0 & -2 \\ 3 & 5 & 0 \\ 0 & -1 & 3 \end{pmatrix}\begin{pmatrix} 3 \\ 0 \\ -1 \end{pmatrix}$

43. $\begin{pmatrix} 0 & 2 & -1 & 4 \\ 1 & 5 & 3 & 0 \\ -2 & 0 & 1 & 5 \end{pmatrix}\begin{pmatrix} 2 \\ 0 \\ -1 \\ 3 \end{pmatrix}$

### Tensor Product of a Column Vector and a Row Vector

Find the tensor product.

44. $\begin{pmatrix} 3 \\ 1 \end{pmatrix}(4 \quad 2 \quad 1)$

45. $\begin{pmatrix} 2 \\ 0 \\ 3 \\ -1 \end{pmatrix}(3 \quad 2 \quad -4)$

46. $\begin{pmatrix} 1 \\ 0 \\ 3 \end{pmatrix}(5 \quad -1 \quad 2 \quad 3)$

47. The unit matrix **I** is a square matrix that has 1's on the main diagonal and 0's elsewhere. Show that the product of **I** and another square matrix **A** is simply **A**.

| | | |
|---|---|---|
| Multiplying by the Unit Matrix | $\mathbf{AI} = \mathbf{IA} = \mathbf{A}$ | 94 |

### Computer

Write computer programs or use spreadsheets to:

48. Add two matrices
49. Multiply a matrix by a scalar
50. Multiply two matrices

Test your programs on any of the problems in this exercise.

## 11-3 SOLVING SYSTEMS OF LINEAR EQUATIONS BY MATRICES

### Representing a System of Equations by Matrices

Suppose that we have the set of equations

$$\begin{aligned} 2x + y + 2z &= 10 \\ x + 2y + 3z &= 14 \\ 3x + 4y - 3z &= 2 \end{aligned}$$

The left sides of these equations can be represented by the product of a square matrix **A** made up of the coefficients, called the *coefficient matrix*, and a column vector **X** made up of the unknowns. If

$$\mathbf{A} = \begin{pmatrix} 2 & 1 & 2 \\ 1 & 2 & 3 \\ 3 & 4 & -3 \end{pmatrix} \quad \text{and} \quad \mathbf{X} = \begin{pmatrix} x \\ y \\ z \end{pmatrix}$$

then

$$\mathbf{AX} = \begin{pmatrix} 2 & 1 & 2 \\ 1 & 2 & 3 \\ 3 & 4 & -3 \end{pmatrix}\begin{pmatrix} x \\ y \\ z \end{pmatrix} = \begin{pmatrix} 2x + y + 2z \\ x + 2y + 3z \\ 3x + 4y - 3z \end{pmatrix}$$

Note that our product is a column vector having *three* elements. The expression $(2x + y + 2z)$ is a *single element* in that vector.

If we now represent the constants in our set of equations by the column vector

$$\mathbf{B} = \begin{pmatrix} 10 \\ 14 \\ 2 \end{pmatrix}$$

our system of equations can be written, using matrices, as

$$\begin{pmatrix} 2 & 1 & 2 \\ 1 & 2 & 3 \\ 3 & 4 & -3 \end{pmatrix} \begin{pmatrix} x \\ y \\ z \end{pmatrix} = \begin{pmatrix} 10 \\ 14 \\ 2 \end{pmatrix}$$

or in the more compact form

| | | |
|---|---|---|
| Matrix Form for a System of Equations | $\mathbf{AX} = \mathbf{B}$ | **95** |

We now show two methods for solving a system of equations.

## Gauss Elimination

When we solved sets of linear equations in Chapter 9, we were able to:

1. Interchange two equations.
2. Multiply an equation by a nonzero constant.
3. Add a constant multiple of one equation to another equation.

Thus for a matrix that represents such a system of equations, we may perform the following transformations without altering the meaning of a matrix.

| | | |
|---|---|---|
| Elementary Transformations of a Matrix | 1. Interchange any rows. 2. Multiply a row by a nonzero constant. 3. Add a constant multiple of one row to another row. | **96** |

In the *Gauss elimination* method our strategy is to apply the elementary transforms to put our system of equations into the form

**This is just one of many mathematical ideas named for Carl Friedrich Gauss (1777–1855). It is also called the *Gauss–Jordan* method.**

$$\begin{pmatrix} 1 & k_1 & k_2 \\ 0 & 1 & k_3 \\ 0 & 0 & 1 \end{pmatrix} \begin{pmatrix} x \\ y \\ z \end{pmatrix} = \begin{pmatrix} r_1 \\ r_2 \\ r_3 \end{pmatrix}$$

where the $k$'s and $r$'s are constants. The coefficient matrix is now said to be in *triangular form* (all 1's on the main diagonal, and 0's below it). The advantage to this form becomes clear when we multiply the coefficient matrix by the column of unknowns. We get

$$x + k_1y + k_2z = r_1$$

$$y + k_3z = r_2$$

$$z = r_3$$

Thus, we immediately get a solution for $z$, and the other variables can be found by back substitution.

## Algorithm for Gauss Elimination

We use the following algorithm for applying the elementary transformations to the coefficient matrix *and* to the column of constants.

This algorithm can be programmed for the computer (see Exercise 3).

In this algorithm, a *pivot element* for any row is the one that lies on the main diagonal of the coefficient matrix.

1. Divide a row $p$ by its pivot element. This is called *normalizing*.
2. Multiply the row $p$ just normalized by the first element in another row $j$. Subtract these products from row $j$. This will make the first element in row $j$ equal to 0.
3. Repeat step 2 for all remaining rows. When finished, the first column should consist of a 1, with all 0's below it.
4. Repeat steps 1 through 3 until the last row has been finished.
5. One unknown is now equal to the last element in the column of constants. The others are found by back substitution.

## Solving Linear Equations by Gauss Elimination

**EXAMPLE 24:** Solve the set of equations

$$\begin{aligned} 2x + y + 2z &= 10 \\ x + 2y + 3z &= 14 \\ 3x + 4y - 3z &= 2 \end{aligned}$$

by Gauss elimination.

**Solution:** We first represent our system by the matrix

$$\begin{pmatrix} 2 & 1 & 2 & 10 \\ 1 & 2 & 3 & 14 \\ 3 & 4 & -3 & 2 \end{pmatrix}$$

Note that this matrix consists of the coefficient matrix as well as the column of constants. It is called an *augmented matrix*.

---------------------- **First Pass** ----------------------

Step 1: We divide the first row by its pivot element, 2.

$$\begin{pmatrix} 1 & 1/2 & 1 & 5 \\ 1 & 2 & 3 & 14 \\ 3 & 4 & -3 & 2 \end{pmatrix}$$

Step 2: We multiply the first row by the first element (1) of the second row, and subtract the resulting elements from the second row.

$$\begin{pmatrix} 1 & 1/2 & 1 & 5 \\ 0 & 3/2 & 2 & 9 \\ 3 & 4 & -3 & 2 \end{pmatrix}$$

Step 3: We multiply the first row by 3 and subtract it from the third row.

$$\begin{pmatrix} 1 & 1/2 & 1 & 5 \\ 0 & 3/2 & 2 & 9 \\ 0 & 5/2 & -6 & -13 \end{pmatrix}$$

Thus we have introduced zeros into the first column.

**Second Pass**

When the pivot element happens to be 0, we get division by zero when trying to normalize. Should this happen, *write the rows in a different order* and try again.

Step 1: We normalize the second row by dividing it by $\frac{3}{2}$.

$$\begin{pmatrix} 1 & 1/2 & 1 & 5 \\ 0 & 1 & 4/3 & 6 \\ 0 & 5/2 & -6 & -13 \end{pmatrix}$$

thus obtaining another 1 on the main diagonal.

Step 2: Multiply the second row by $\frac{5}{2}$ and subtract it from the third row.

$$\begin{pmatrix} 1 & 1/2 & 1 & 5 \\ 0 & 1 & 4/3 & 6 \\ 0 & 0 & -28/3 & -28 \end{pmatrix}$$

**Third Pass**

Step 1: Normalize the third row.

$$\begin{pmatrix} 1 & 1/2 & 1 & 5 \\ 0 & 1 & 4/3 & 6 \\ 0 & 0 & 1 & 3 \end{pmatrix}$$

The coefficient matrix is now in triangular form. The third row gives

$$z = 3$$

Back-substituting, we get from the second row,

$$y + \frac{4}{3}z = 6$$

$$y = 6 - 4 = 2$$

and from the first row,

$$x + \frac{y}{2} + z = 5$$

$$x = 5 - 1 - 3 = 1$$

Our solution is then (1, 2, 3).

## Solving Linear Equations by Transforming to the Unit Matrix

Let us return to our original system of equations, in matrix form,

This method is *also* called Gauss Elimination. However, to avoid confusion we'll refer to this as the *unit matrix method.*

| | | |
|---|---|---|
| Matrix Form for a System of Equations | **AX = B** | **95** |

Instead of putting the coefficient matrix **A** into triangular form, we use another method. We use Gauss elimination to transform **A** into the unit matrix **I** (1's on the main diagonal, 0's elsewhere), and in so doing transform the column of unknowns **B** into a new column vector which we call **C**.

$$\begin{array}{ccc} \mathbf{AX} & = & \mathbf{B} \\ \downarrow & & \downarrow \\ \mathbf{IX} & = & \mathbf{C} \end{array}$$

But by Eq. 94, **IX** = **X**, so

$$\mathbf{X} = \mathbf{C}$$

Thus the column of unknowns **X** is equal to the transformed column of constants **C**, and is our solution.

| | | |
|---|---|---|
| Unit Matrix Method | When we transform the coefficient matrix **A** into the unit matrix **I**, the column vector **B** gets transformed into the solution set. | **97** |

We use this method now without proof to find the general solution to a system of two linear equations in two unknowns.

---

**EXAMPLE 25:** Solve the system of equations

$$a_1x + b_1y = c_1$$

$$a_2x + b_2y = c_2$$

by the unit matrix method.

**Solution:** Our augmented matrix is

$$\begin{pmatrix} a_1 & b_1 & c_1 \\ a_2 & b_2 & c_2 \end{pmatrix}$$

Let us normalize both rows by dividing the first by $a_1$ and the second by $a_2$.

$$\begin{pmatrix} 1 & \dfrac{b_1}{a_1} & \dfrac{c_1}{a_1} \\ 1 & \dfrac{b_2}{a_2} & \dfrac{c_2}{a_2} \end{pmatrix}$$

Subtracting the second row from the first gives, after combining fractions,

$$\begin{pmatrix} 1 & \dfrac{b_1}{a_1} & \dfrac{c_1}{a_1} \\ 0 & \dfrac{a_1b_2 - a_2b_1}{-a_1a_2} & \dfrac{a_1c_2 - a_2c_1}{-a_1a_2} \end{pmatrix}$$

We normalize again by dividing the second row by its second element.

$$\begin{pmatrix} 1 & \dfrac{b_1}{a_1} & \dfrac{c_1}{a_1} \\ 0 & 1 & \dfrac{a_1c_2 - a_2c_1}{a_1b_2 - a_2b_1} \end{pmatrix}$$

Next we multiply the second row by $b_1/a_1$ and subtract it from the first row. After combining fractions, we get

$$\begin{pmatrix} 1 & 0 & \dfrac{b_2c_1 - b_1c_2}{a_1b_2 - a_2b_1} \\ 0 & 1 & \dfrac{a_1c_2 - a_2c_1}{a_1b_2 - a_2b_1} \end{pmatrix}$$

which is equivalent to

$$\begin{pmatrix} 1 & 0 \\ 0 & 1 \end{pmatrix} \begin{pmatrix} x \\ y \end{pmatrix} \begin{pmatrix} \dfrac{b_2c_1 - b_1c_2}{a_1b_2 - a_2b_1} \\ \dfrac{a_1c_2 - a_2c_1}{a_1b_2 - a_2b_1} \end{pmatrix}$$

Our original matrix has been transformed into the unit matrix, so our column of constants has become the solution set. Our solution is then

| | |
|---|---|
| $x = \dfrac{b_2c_1 - b_1c_2}{a_1b_2 - a_2b_1} \qquad y = \dfrac{a_1c_2 - a_2c_1}{a_1b_2 - a_2b_1}$ | **63** |

These, of course, are the same equations we got back in Sec. 9-2 when we solved the same set of equations by the addition–subtraction method.

## EXERCISE 3—SOLVING LINEAR EQUATIONS BY MATRICES

1. Solve any of the systems of equations in Chapter 9 or 10 by Gauss elimination, or by the unit matrix method.

**Computer**

2. Write a program to solve a system of equations by Gauss elimination, using the algorithm given. Use DATA statements to enter the number of equations, and the coefficients. Test your program by solving any of the systems of equations in this chapter.

# 11-4 THE INVERSE OF A MATRIX

## Inverse of a Square Matrix

Recall from Sec. 11-1 that the unit matrix **I** is a square matrix which has 1's along its main diagonal and zeros elsewhere, such as

$$\mathbf{I} = \begin{pmatrix} 1 & 0 & 0 & 0 \\ 0 & 1 & 0 & 0 \\ 0 & 0 & 1 & 0 \\ 0 & 0 & 0 & 1 \end{pmatrix}$$

We now define the *inverse* of any given matrix **A** as another matrix $\mathbf{A}^{-1}$ such that the product of the matrix and its inverse is equal to the unit matrix.

| | | |
|---|---|---|
| Product of a Matrix and Its Inverse | $\mathbf{AA}^{-1} = \mathbf{A}^{-1}\mathbf{A} = \mathbf{I}$ | **93** |

*The product of a matrix and its inverse is the unit matrix.*

In *ordinary* algebra, if $b$ is some nonzero quantity, then

$$b\left(\frac{1}{b}\right) = bb^{-1} = 1$$

In *matrix* algebra, the unit matrix **I** has properties similar to the number 1, and the inverse of a matrix has properties similar to the *reciprocal* in ordinary algebra.

**EXAMPLE 26:** The inverse of the matrix

$$\mathbf{A} = \begin{pmatrix} 1 & 0 \\ 2 & 0.5 \end{pmatrix} \quad \text{is} \quad \mathbf{A}^{-1} = \begin{pmatrix} 1 & 0 \\ -4 & 2 \end{pmatrix}$$

because

$$\begin{pmatrix} 1 & 0 \\ 2 & 0.5 \end{pmatrix}\begin{pmatrix} 1 & 0 \\ -4 & 2 \end{pmatrix} = \begin{bmatrix} 1(1) + 0(-4) & 1(0) + 0(2) \\ 2(1) + 0.5(-4) & 2(0) + 0.5(2) \end{bmatrix}$$

$$= \begin{pmatrix} 1 & 0 \\ 0 & 1 \end{pmatrix}$$

and

$$\begin{pmatrix} 1 & 0 \\ -4 & 2 \end{pmatrix}\begin{pmatrix} 1 & 0 \\ 2 & 0.5 \end{pmatrix} = \begin{bmatrix} 1(1) + 0(2) & 1(0) + 0(0.5) \\ -4(1) + 2(2) & -4(0) + 2(0.5) \end{bmatrix}$$

$$= \begin{pmatrix} 1 & 0 \\ 0 & 1 \end{pmatrix}$$

## Finding the Inverse of a Matrix

One method to find the inverse of a matrix is to apply the rules of matrix transformation to transform the given matrix into a unit matrix, while at the same time performing the same operations on a unit matrix. The unit matrix, after the transformations, becomes the inverse of the original matrix.

**Not every matrix has an inverse. If the inverse exists, the matrix is said to be *invertible*.**

$$\begin{matrix} \mathbf{A} & \mathbf{I} \\ \downarrow & \downarrow \\ \mathbf{I} & \mathbf{A}^{-1} \end{matrix}$$

We will not try to prove the method here, but only show its use.

**EXAMPLE 27:** Find the inverse of the matrix of Example 26,

$$\mathbf{A} = \begin{pmatrix} 1 & 0 \\ 2 & 0.5 \end{pmatrix}$$

**Solution:** To the given matrix **A**, we append a unit matrix,

$$\left(\begin{array}{cc|cc} 1 & 0 & 1 & 0 \\ 2 & 0.5 & 0 & 1 \end{array}\right)$$

obtaining a *double-square* matrix. We then double the first row and subtract it from the second.

$$\left(\begin{array}{cc|cc} 1 & 0 & 1 & 0 \\ 0 & 0.5 & -2 & 1 \end{array}\right)$$

We now double the second row.

$$\left(\begin{array}{cc|cc} 1 & 0 & 1 & 0 \\ 0 & 1 & -4 & 2 \end{array}\right)$$

*Our original matrix **A** is now the unit matrix, and the original unit matrix is now the inverse of **A**.*

$$\mathbf{A}^{-1} = \begin{pmatrix} 1 & 0 \\ -4 & 2 \end{pmatrix}$$

**EXAMPLE 28:** Find the inverse of the matrix

$$\mathbf{A} = \begin{pmatrix} a_1 & b_1 \\ a_2 & b_2 \end{pmatrix}$$

**Solution:** We append the unit matrix

$$\left(\begin{array}{cc|cc} a_1 & b_1 & 1 & 0 \\ a_2 & b_2 & 0 & 1 \end{array}\right)$$

Then we multiply the first row by $b_2$ and the second by $b_1$, and subtract the second from the first.

$$\left(\begin{array}{cc|cc} a_1b_2 - a_2b_1 & 0 & b_2 & -b_1 \\ a_2b_1 & b_1b_2 & 0 & b_1 \end{array}\right)$$

We normalize the first row by dividing it by $a_1b_2 - a_2b_1$.

$$\left(\begin{array}{cc|cc} 1 & 0 & \dfrac{b_2}{a_1b_2 - a_2b_1} & \dfrac{-b_1}{a_1b_2 - a_2b_1} \\ a_2b_1 & b_1b_2 & 0 & b_1 \end{array}\right)$$

Then we multiply the first row by $a_2b_1$ and subtract it from the second. After combining fractions, we get

$$\left(\begin{array}{cc|cc} 1 & 0 & \dfrac{b_2}{a_1b_2 - a_2b_1} & \dfrac{-b_1}{a_1b_2 - a_2b_1} \\ 0 & b_1b_2 & \dfrac{-a_2b_1b_2}{a_1b_2 - a_2b_1} & \dfrac{a_1b_1b_2}{a_1b_2 - a_2b_1} \end{array}\right)$$

We then divide the second row by $b_1b_2$.

$$\left(\begin{array}{cc|cc} 1 & 0 & \dfrac{b_2}{a_1b_2 - a_2b_1} & \dfrac{-b_1}{a_1b_2 - a_2b_1} \\ 0 & 1 & \dfrac{-a_2}{a_1b_2 - a_2b_1} & \dfrac{a_1}{a_1b_2 - a_2b_1} \end{array}\right)$$

The inverse of our matrix is then

$$\mathbf{A}^{-1} = \frac{1}{a_1b_2 - a_2b_1} \begin{pmatrix} b_2 & -b_1 \\ -a_2 & a_1 \end{pmatrix}$$

after factoring out $1/(a_1b_2 - a_2b_1)$.

Comparing this result with the original matrix, we see that:

1. The elements on the main diagonal are interchanged.
2. The elements on the secondary diagonal have reversed sign.
3. It is divided by the determinant of the original matrix.

## Shortcut for Finding the Inverse of a 2 × 2 Matrix

The results of Example 28 provide us with a way to quickly write the inverse of a 2 × 2 matrix. We obtain the inverse of a 2 × 2 matrix by:

1. Interchanging the elements on the main diagonal
2. Reversing the signs of the elements on the secondary diagonal
3. Dividing by the determinant of the original matrix

**EXAMPLE 29:** Find the inverse of

$$\mathbf{A} = \begin{pmatrix} 5 & -2 \\ 1 & 3 \end{pmatrix}$$

**Solution:** We interchange the 5 and the 3, and reverse the signs of the −2 and 1,

$$\begin{pmatrix} 3 & 2 \\ -1 & 5 \end{pmatrix}$$

and divide by the determinant of the original matrix

$$5(3) - (-2)(1) = 17$$

So

$$\mathbf{A}^{-1} = \frac{1}{17}\begin{pmatrix} 3 & 2 \\ -1 & 5 \end{pmatrix}$$

## Solving Systems of Equations by Matrix Inversion

Suppose that we have a system of equations, which we represent in matrix form as

$$\mathbf{AX} = \mathbf{B}$$

where **A** is the coefficient matrix, **X** is the column vector of the unknowns, and **B** is the column of constants. Multiplying both sides by $\mathbf{A}^{-1}$, the inverse of **A**,

$$\mathbf{A}^{-1}\mathbf{AX} = \mathbf{A}^{-1}\mathbf{B}$$

or

$$\mathbf{IX} = \mathbf{A}^{-1}\mathbf{B}$$

since $\mathbf{A}^{-1}\mathbf{A} = \mathbf{I}$. But, by Eq. 94, $\mathbf{IX} = \mathbf{X}$, so

| | | |
|---|---|---|
| Solving a Set of Equations Using the Inverse | $\mathbf{X} = \mathbf{A}^{-1}\mathbf{B}$ | **98** |

*The solutions to a set of equations can be obtained by multiplying the column of constants by the inverse of the coefficient matrix.*

This is called the *matrix inversion* method. To use this method, set up the equations in standard form. Write **A**, the matrix of the coefficients. Then find $\mathbf{A}^{-1}$ by matrix inversion or by the shortcut method (for a 2 × 2 matrix). Finally, multiply $\mathbf{A}^{-1}$ by the column of constants **B** to get the solution.

**EXAMPLE 30:** Solve the system of equations

$$x - 2y = -3$$
$$3x + y = 5$$

by the matrix inversion method.

**Solution:** The coefficient matrix **A** is

$$\mathbf{A} = \begin{pmatrix} 1 & -2 \\ 3 & 1 \end{pmatrix}$$

We now find the inverse, $\mathbf{A}^{-1}$, of **A**. We append the unit matrix

We could have used the shortcut method here, since this is a 2 × 2 matrix. We have, however, used the longer general method to give another example of its use.

$$\left(\begin{array}{rr|rr} 1 & -2 & 1 & 0 \\ 3 & 1 & 0 & 1 \end{array}\right)$$

and then multiply the first row by $-3$ and add it to the second row.

$$\left(\begin{array}{rr|rr} 1 & -2 & 1 & 0 \\ 0 & 7 & -3 & 1 \end{array}\right)$$

Then we multiply the second row by 2/7 and add it to the first row.

$$\left(\begin{array}{rr|rr} 1 & 0 & 1/7 & 2/7 \\ 0 & 7 & -3 & 1 \end{array}\right)$$

Finally, we divide the second row by 7.

$$\left(\begin{array}{rr|rr} 1 & 0 & 1/7 & 2/7 \\ 0 & 1 & -3/7 & 1/7 \end{array}\right)$$

Thus the inverse of our coefficient matrix **A** is

$$\mathbf{A}^{-1} = \begin{pmatrix} 1/7 & 2/7 \\ -3/7 & 1/7 \end{pmatrix}$$

We then get our solution by multiplying the inverse $\mathbf{A}^{-1}$ by the column of constants, **B**.

$$\begin{array}{ccc} \mathbf{A}^{-1} & \mathbf{B} = & \mathbf{X} \end{array}$$

$$\begin{pmatrix} 1/7 & 2/7 \\ -3/7 & 1/7 \end{pmatrix}\begin{pmatrix} -3 \\ 5 \end{pmatrix} = \begin{pmatrix} -3/7 + 10/7 \\ 9/7 + 5/7 \end{pmatrix} = \begin{pmatrix} 1 \\ 2 \end{pmatrix}$$

so $x = 1$ and $y = 2$.

**EXAMPLE 31:** Solve the general system of two equations

$$a_1x + b_1y = c_1$$
$$a_2x + b_2y = c_2$$

by matrix inversion.

**Solution:** We write the coefficient matrix

$$\mathbf{A} = \begin{pmatrix} a_1 & b_1 \\ a_2 & b_2 \end{pmatrix}$$

The inverse of our coefficient matrix has already been found in Example 28. It is

$$\mathbf{A}^{-1} = \frac{1}{a_1b_2 - a_2b_1}\begin{pmatrix} b_2 & -b_1 \\ -a_2 & a_1 \end{pmatrix}$$

By Eq. 98 we now get the column of unknowns by multiplying the inverse by the column of coefficients.

$$\mathbf{X} = \mathbf{A}^{-1}\mathbf{B}$$

$$= \frac{1}{a_1b_2 - a_2b_1}\begin{pmatrix} b_2 & -b_1 \\ -a_2 & a_1 \end{pmatrix}\begin{pmatrix} c_2 \\ c_2 \end{pmatrix}$$

$$= \begin{pmatrix} \dfrac{b_2c_1 - b_1c_2}{a_1b_2 - a_2b_1} \\ \dfrac{a_1c_2 - a_2c_1}{a_1b_2 - a_2b_1} \end{pmatrix}$$

Thus we again get the now familiar equations,

$$x = \frac{b_2c_1 - b_1c_2}{a_1b_2 - a_2b_1} \qquad y = \frac{a_1c_2 - a_2c_1}{a_1b_2 - a_2b_1} \qquad \mathbf{63}$$

---

**EXAMPLE 32:** Solve the system

$$\begin{aligned} 3x - y + z + 4w &= 15 \\ -x + 3y + 5z - w &= -9 \\ 2x + y - z + 3w &= 18 \\ x - 2y - 3z + w &= 7 \end{aligned}$$

**Solution:** We write the coefficient matrix, with the unit matrix appended.

$$\left(\begin{array}{rrrr|rrrr} 3 & -1 & 1 & 4 & 1 & 0 & 0 & 0 \\ -1 & 3 & 5 & -1 & 0 & 1 & 0 & 0 \\ 2 & 1 & -1 & 3 & 0 & 0 & 1 & 0 \\ 1 & -2 & -3 & 1 & 0 & 0 & 0 & 1 \end{array}\right)$$

We first use Gauss elimination to put the coefficient matrix into triangular form in exactly the same way as we did in Sec. 11-3. We normalize one row at a time and then introduce zeros below the pivot element. Then we create zeros *above* the main diagonal by the same method. The various steps in the computation are shown in the following computer-generated table.

Even though this printout shows only two significant digits, the work was actually done using the full accuracy of the computer.

| | | | | | | | | |
|---|---|---|---|---|---|---|---|---|
| Original matrix | 3.0 | −1.0 | 1.0 | 4.0 | 1.0 | 0.0 | 0.0 | 0.0 |
| | −1.0 | 3.0 | 5.0 | −1.0 | 0.0 | 1.0 | 0.0 | 0.0 |
| | 2.0 | 1.0 | −1.0 | 3.0 | 0.0 | 0.0 | 1.0 | 0.0 |
| | 1.0 | −2.0 | −3.0 | 1.0 | 0.0 | 0.0 | 0.0 | 1.0 |
| Normalize row 1 | **1.0** | **−0.3** | **0.3** | **1.3** | **0.3** | **0.0** | **0.0** | **0.0** |
| | −1.0 | 3.0 | 5.0 | −1.0 | 0.0 | 1.0 | 0.0 | 0.0 |
| | 2.0 | 1.0 | −1.0 | 3.0 | 0.0 | 0.0 | 1.0 | 0.0 |
| | 1.0 | −2.0 | −3.0 | 1.0 | 0.0 | 0.0 | 0.0 | 1.0 |
| Create zeros below pivot element | 1.0 | −0.3 | 0.3 | 1.3 | 0.3 | 0.0 | 0.0 | 0.0 |
| | **0.0** | 2.7 | 5.3 | 0.3 | 0.3 | 1.0 | 0.0 | 0.0 |
| | **0.0** | 1.7 | −1.7 | 0.3 | −0.7 | 0.0 | 1.0 | 0.0 |
| | **0.0** | −1.7 | −3.3 | −0.3 | −0.3 | 0.0 | 0.0 | 1.0 |
| | | Unit matrix | | | | Inverse matrix | | |

Normalize row 2

$$\left(\begin{array}{rrrr|rrrr} 1.0 & -0.3 & 0.3 & 1.3 & 0.3 & 0.0 & 0.0 & 0.0 \\ 0.0 & \mathbf{1.0} & \mathbf{2.0} & \mathbf{0.1} & \mathbf{0.1} & \mathbf{0.4} & \mathbf{0.0} & \mathbf{0.0} \\ 0.0 & 1.7 & -1.7 & 0.3 & -0.7 & 0.0 & 1.0 & 0.0 \\ 0.0 & -1.7 & -3.3 & -0.3 & -0.3 & 0.0 & 0.0 & 1.0 \end{array}\right)$$

Create zeros below pivot element

$$\left(\begin{array}{rrrr|rrrr} 1.0 & -0.3 & 0.3 & 1.3 & 0.3 & 0.0 & 0.0 & 0.0 \\ 0.0 & 1.0 & 2.0 & 0.1 & 0.1 & 0.4 & 0.0 & 0.0 \\ 0.0 & \mathbf{0.0} & -5.0 & 0.1 & -0.9 & -0.6 & 1.0 & 0.0 \\ 0.0 & \mathbf{0.0} & 0.0 & -0.1 & -0.1 & 0.6 & 0.0 & 1.0 \end{array}\right)$$

Normalize row 3

$$\left(\begin{array}{rrrr|rrrr} 1.0 & -0.3 & 0.3 & 1.3 & 0.3 & 0.0 & 0.0 & 0.0 \\ 0.0 & 1.0 & 2.0 & 0.1 & 0.1 & 0.4 & 0.0 & 0.0 \\ 0.0 & 0.0 & \mathbf{1.0} & \mathbf{-0.0} & \mathbf{0.2} & \mathbf{0.1} & \mathbf{-0.2} & \mathbf{0.0} \\ 0.0 & 0.0 & 0.0 & -0.1 & -0.1 & 0.6 & 0.0 & 1.0 \end{array}\right)$$

Create zeros below pivot element

$$\left(\begin{array}{rrrr|rrrr} 1.0 & -0.3 & 0.3 & 1.3 & 0.3 & 0.0 & 0.0 & 0.0 \\ 0.0 & 1.0 & 2.0 & 0.1 & 0.1 & 0.4 & 0.0 & 0.0 \\ 0.0 & 0.0 & 1.0 & 0.0 & 0.2 & 0.1 & -0.2 & 0.0 \\ 0.0 & 0.0 & \mathbf{0.0} & -0.1 & -0.1 & 0.6 & 0.0 & 1.0 \end{array}\right)$$

Normalize row 4

$$\left(\begin{array}{rrrr|rrrr} 1.0 & -0.3 & 0.3 & 1.3 & 0.3 & 0.0 & 0.0 & 0.0 \\ 0.0 & 1.0 & 2.0 & 0.1 & 0.1 & 0.4 & 0.0 & 0.0 \\ 0.0 & 0.0 & 1.0 & 0.0 & 0.2 & 0.1 & -0.2 & 0.0 \\ 0.0 & 0.0 & 0.0 & \mathbf{1.0} & \mathbf{1.0} & \mathbf{-5.0} & \mathbf{0.0} & \mathbf{-8.0} \end{array}\right)$$

Create zeros above pivot element

$$\left(\begin{array}{rrrr|rrrr} 1.0 & -0.3 & 0.3 & \mathbf{0.0} & -1.0 & 6.7 & 0.0 & 10.7 \\ 0.0 & 1.0 & 2.0 & \mathbf{0.0} & 0.0 & 1.0 & 0.0 & 1.0 \\ 0.0 & 0.0 & 1.0 & \mathbf{0.0} & 0.2 & 0.0 & -0.2 & -0.2 \\ 0.0 & 0.0 & 0.0 & 1.0 & 1.0 & -5.0 & 0.0 & -8.0 \end{array}\right)$$

$$\left(\begin{array}{rrrr|rrrr} 1.0 & -0.3 & \mathbf{0.0} & 0.0 & -1.1 & 6.7 & 0.1 & 10.7 \\ 0.0 & 1.0 & \mathbf{0.0} & 0.0 & -0.4 & 1.0 & 0.4 & 1.4 \\ 0.0 & 0.0 & 1.0 & 0.0 & 0.2 & 0.0 & -0.2 & -0.2 \\ 0.0 & 0.0 & 0.0 & 1.0 & 1.0 & -5.0 & 0.0 & -8.0 \end{array}\right)$$

Inversion complete

$$\left(\begin{array}{rrrr|rrrr} 1.0 & \mathbf{0.0} & 0.0 & 0.0 & -1.2 & 7.0 & 0.2 & 11.2 \\ 0.0 & 1.0 & 0.0 & 0.0 & -0.4 & 1.0 & 0.4 & 1.4 \\ 0.0 & 0.0 & 1.0 & 0.0 & 0.2 & 0.0 & -0.2 & -0.2 \\ 0.0 & 0.0 & 0.0 & 1.0 & 1.0 & -5.0 & 0.0 & -8.0 \end{array}\right)$$

Unit matrix　　　　Inverse matrix

We get our solution by multiplying the inverse matrix by the column of constants.

$$\begin{pmatrix} -1.2 & 7 & 0.2 & 11.2 \\ -0.4 & 1 & 0.4 & 1.4 \\ 0.2 & 0 & -0.2 & -0.2 \\ 1 & -5 & 0 & -8 \end{pmatrix} \begin{pmatrix} 15 \\ -9 \\ 18 \\ 7 \end{pmatrix} = \begin{pmatrix} 1 \\ 2 \\ -2 \\ 4 \end{pmatrix}$$

so $x = 1$, $y = 2$, $z = -2$, and $w = 4$.

## EXERCISE 4—THE INVERSE OF A MATRIX

Find the inverse of each matrix.

1. $\begin{pmatrix} 4 & 8 \\ -5 & 0 \end{pmatrix}$

2. $\begin{pmatrix} -2 & 4 \\ 3 & 1 \end{pmatrix}$

3. $\begin{pmatrix} 8 & -3 \\ 1 & 6 \end{pmatrix}$

4. $\begin{pmatrix} 1 & 5 & 3 \\ 2 & 3 & 0 \\ 6 & 8 & 1 \end{pmatrix}$

5. $\begin{pmatrix} -1 & 3 & 7 \\ 4 & -2 & 0 \\ 7 & 2 & -9 \end{pmatrix}$

6. $\begin{pmatrix} 18 & 23 & 71 \\ 82 & 49 & 28 \\ 36 & 19 & 84 \end{pmatrix}$

7. $\begin{pmatrix} 9 & 1 & 0 & 3 \\ 4 & 3 & 5 & 2 \\ 7 & 1 & 0 & 3 \\ 1 & 4 & 2 & 7 \end{pmatrix}$

8. $\begin{pmatrix} -2 & 1 & 5 & -6 \\ 7 & -3 & 2 & -6 \\ -3 & 2 & 6 & 1 \\ 5 & 2 & 7 & 5 \end{pmatrix}$

9. Solve any of the equations of Chapter 9 or 10 by the matrix inversion method.

### Computer

10. Write a computer program to find and print the inverse of a matrix. Such a program will have many similarities to the one for Gauss elimination. Test your program by finding the inverse of some of the matrices given in Problems 1 to 8.

## CHAPTER 11 REVIEW PROBLEMS

1. Combine.

$$(4 \quad 9 \quad -2 \quad 6) + (2 \quad 4 \quad 9 \quad -1) - (4 \quad 1 \quad -3 \quad 7)$$

2. Multiply.

$$\begin{pmatrix} 8 & 4 \\ 9 & -5 \end{pmatrix}\begin{pmatrix} 3 & 7 \\ -3 & 6 \end{pmatrix}$$

3. Multiply.

$$(7 \quad -2 \quad 5)\begin{pmatrix} 6 \\ -2 \\ 8 \end{pmatrix}$$

4. Find the inverse.

$$\begin{pmatrix} 5 & -2 & 5 & 1 \\ 2 & 8 & -3 & -1 \\ -4 & 2 & 7 & 5 \\ 9 & 2 & 5 & 3 \end{pmatrix}$$

5. Find the inverse.

$$\begin{pmatrix} 4 & -4 \\ 3 & 1 \end{pmatrix}$$

6. Combine.

$$\begin{pmatrix} 5 & 9 \\ -4 & 7 \end{pmatrix} + \begin{pmatrix} -2 & 6 \\ 3 & 8 \end{pmatrix}$$

7. Multiply.

$$\begin{pmatrix} 1 & 2 \\ 0 & 3 \end{pmatrix}\begin{pmatrix} 1 & 2 & 4 \\ 3 & 0 & 1 \end{pmatrix}$$

8. Multiply.

$$\begin{pmatrix} 8 & 2 & 6 & 1 \\ 9 & 0 & 1 & 4 \\ 1 & 6 & 9 & 3 \\ 6 & 9 & 3 & 5 \end{pmatrix}\begin{pmatrix} 5 \\ 3 \\ 2 \\ 8 \end{pmatrix}$$

9. Solve by Gauss elimination.

$$\begin{aligned} x + y + z + w &= -4 \\ x + 2y + 3z + 4w &= 0 \\ x + 3y + 6z + 10w &= 9 \\ x + 4y + 10z + 20w &= 24 \end{aligned}$$

10. Multiply.

$$7\begin{pmatrix} 5 & -3 \\ -4 & 8 \end{pmatrix}$$

11. Solve by Gauss elimination.

$$\begin{aligned} 5x + 3y - 2z &= 5 \\ 3x - 4y + 3z &= 13 \\ x + 6y - 4z &= -8 \end{aligned}$$

12. Find the inverse.

$$\begin{pmatrix} 5 & 1 & 4 \\ -2 & 6 & 9 \\ 4 & -1 & 0 \end{pmatrix}$$

**13.** Solve by matrix inversion.

$$\begin{aligned} 4x + 3y &= 27 \\ 2x - 5y &= -19 \end{aligned}$$

**14.** Multiply.

$$9(6 \quad 3 \quad -2 \quad 8)$$

**15.** The four loop currents (mA) in a certain circuit satisfy the equations

$$\begin{aligned} 376I_1 + 374I_2 - 364I_3 - 255I_4 &= 284 \\ 898I_1 - 263I_2 + 229I_3 + 274I_4 &= -847 \\ 364I_1 + 285I_2 - 857I_3 - 226I_4 &= 746 \\ 937I_1 + 645I_2 - 387I_3 + 847I_4 &= -254 \end{aligned}$$

Solve for the four currents by matrix inversion.

*Writing*

**16.** Suppose you have submitted a lab report in which you stated that you have used Gauss elimination to solve some sets of equations. The report was rejected because you failed to say anything about that method.

Write one paragraph explaining Gauss elimination to be inserted into the next draft of your report. Start by telling what Gauss elimination is, and then go on to say how it works.

# 12

# Exponents and Radicals

## OBJECTIVES

**When you have completed this chapter, you should be able to:**

- Use the laws of exponents to simplify and combine expressions having integral exponents.
- Simplify radicals by removing perfect powers, by rationalizing the denominator, and by reducing the index.
- Add, subtract, multiply, and divide radicals.
- Solve radical equations.

In our introduction to algebra, we studied expressions with integral exponents and derived some laws of exponents. We expand on that work in this chapter, using the laws of exponents and our additional skills of factoring and manipulation of fractions to simplify more complex expressions.

We also give meaning to a fractional exponent, which we'll see is another way of writing a radical. We had a brief look at radicals in Chapter 1, where we computed roots by calculator. Here, we'll simplify, add, subtract, multiply, and divide radicals and solve some simple radical equations. We must postpone the harder radical equations until we can solve quadratic equations (Chapter 13).

## 12-1 INTEGRAL EXPONENTS

Here we continue the study of exponents that we started in Sec. 2-3. We repeat the laws of exponents derived there, and use them to simplify harder expressions than before. You should glance back at that section before starting here.

### Negative Exponents and Zero Exponents

The two laws that apply are given again here.

| | | |
|---|---|---|
| Negative Exponent | $x^{-a} = \dfrac{1}{x^a} \quad (x \neq 0)$ | **35** |

*When taking the reciprocal of a base raised to a power,* ***change the sign of the exponent.***

and

| | | |
|---|---|---|
| Zero Exponent | $x^0 = 1 \quad (x \neq 0)$ | **34** |

*Any quantity (except 0) raised to the zero power equals 1.*

We use these laws to rewrite expressions so that they contain no negative or zero exponents.

---

**EXAMPLE 1:**

(a) $xy^{-1} = \dfrac{x}{y}$

(b) $\dfrac{ab^{-2}}{c^{-3}d} = \dfrac{ac^3}{b^2d}$

(c) $x^2y^0z = x^2z$

(d) $(abc)^0 = 1$

(e) $\left(\dfrac{a}{b}\right)^{-1} = \dfrac{b}{a}$

---

We'll use these laws frequently in the examples to come.

### Power Raised to a Power

| | | |
|---|---|---|
| Powers | $(x^a)^b = x^{ab}$ | **31** |

*When raising a power to a power, keep the same base and* ***multiply the exponents.***

---

**EXAMPLE 2:**

(a) $(a^2)^3 = a^6$ (b) $(z^{-3})^{-2} = z^6$

(c) $(x^5)^{-2} = x^{-10} = \dfrac{1}{x^{10}}$

---

### Products

The two laws of exponents that apply to products are given here again.

| | | |
|---|---|---|
| Products | $x^a \cdot x^b = x^{a+b}$ | **29** |

*When multiplying powers of the same base, keep the same base and* ***add the exponents.***

and

| | | |
|---|---|---|
| Product Raised to a Power | $(xy)^n = x^n y^n$ | **32** |

*When a product is raised to a power, each factor may be* ***separately*** *raised to the power.*

---

**EXAMPLE 3:**

(a) $2axy^3z(3a^2xyz^3) = 6a^3x^2y^4z^4$ (b) $(3bx^3y^2)^3 = 27b^3x^9y^6$

(c) $\left(\dfrac{xy^2}{3}\right)^2 = \dfrac{x^2y^4}{9}$

---

| | |
|---|---|
| Common Error | Parentheses are important when raising products to a power.<br>$(2x)^0 = 1$ but $2x^0 = 2(1) = 2$<br>$(4x)^{-1} = \dfrac{1}{4x}$ but $4x^{-1} = \dfrac{4}{x}$ |

We try to leave our expressions without zero or negative exponents.

**EXAMPLE 4:**

(a) $3p^3q^2r^{-4}(4p^{-5}qr^4) = 12p^{-2}q^3r^0 = \dfrac{12q^3}{p^2}$

(b) $(2m^2n^3x^{-2})^{-3} = 2^{-3}m^{-6}n^{-9}x^6 = \dfrac{x^6}{8m^6n^9}$

Of course, our old rules for multiplying multinomials still apply.

**EXAMPLE 5:**

(a) $3x^{-2}y^3(2ax^3 - x^2y + ax^4y^{-3})$

$$= 6axy^3 - 3x^0y^4 + 3ax^2y^0 = 6axy^3 - 3y^4 + 3ax^2$$

(b) $(p^2x - q^2y^{-2})(p^{-2} - q^{-2}y^2)$

$$= p^0x - p^2q^{-2}xy^2 - p^{-2}q^2y^{-2} + q^0y^0$$

$$= x - \frac{p^2xy^2}{q^2} - \frac{q^2}{p^2y^2} + 1$$

Letters in the exponents are handled just as if they were numbers.

**EXAMPLE 6:**

$$x^{n+1}y^{1-n}(x^{2n-3}y^{n-1}) = x^{n+1+2n-3}y^{1-n+n-1}$$
$$= x^{3n-2}y^0 = x^{3n-2}$$

## Quotients

Our two laws of exponents that apply to quotients are repeated here.

| | | | |
|---|---|---|---|
| Quotients | $\dfrac{x^a}{x^b} = x^{a-b}$ | $(x \neq 0)$ | **30** |

*When dividing powers of the same base, keep the same base and* ***subtract the exponents.***

and

| | | | |
|---|---|---|---|
| Quotient Raised to a Power | $\left(\dfrac{x}{y}\right)^n = \dfrac{x^n}{y^n}$ | $(y \neq 0)$ | **33** |

*When a quotient is raised to a power, the numerator and denominator may be* ***separately*** *raised to the power.*

**EXAMPLE 7:**

(a) $\dfrac{x^5}{x^3} = x^2$

(b) $\dfrac{16p^5q^{-3}}{8p^2q^{-5}} = 2p^3q^2$

(c) $\dfrac{x^{2p-1}y^2}{x^{p+4}y^{1-p}} = x^{2p-1-p-4}y^{2-1+p} = x^{p-5}y^{1+p}$

(d) $\left(\dfrac{x^3y^2}{p^2q}\right)^3 = \dfrac{x^9y^6}{p^6q^3}$

As before, we try to leave our expressions without zero or negative exponents.

**EXAMPLE 8:**

(a) $\dfrac{8a^3b^{-5}c^3}{4a^5b^{-1}c^3} = 2a^{-2}b^{-4}c^0 = \dfrac{2}{a^2b^4}$

(b) $\left(\dfrac{a}{b}\right)^{-2} = \dfrac{a^{-2}}{b^{-2}} = \dfrac{b^2}{a^2} = \left(\dfrac{b}{a}\right)^2$

Note in Example 8(b) how the negative exponent had the effect of inverting the fraction.

Don't forget our hard-won skills of factoring, combining of fractions over a common denominator, and of simplifying complex fractions. Use them where needed.

**EXAMPLE 9:**

(a) $\dfrac{1}{(5a)^2} + \dfrac{1}{(2a^2b^{-4})^2} = \dfrac{1}{25a^2} + \dfrac{1}{\dfrac{4a^4}{b^8}}$

We simplify the compound fraction and combine the two fractions over a common denominator, getting

$$\frac{1}{25a^2} + \frac{b^8}{4a^4} = \frac{4a^2}{100a^4} + \frac{25b^8}{100a^4}$$

$$= \frac{4a^2 + 25b^8}{100a^4}$$

(b) $(x^{-2} - 3y)^{-2} = \dfrac{1}{(x^{-2} - 3y)^2} = \dfrac{1}{\left(\dfrac{1}{x^2} - 3y\right)^2}$

$$= \frac{1}{\left(\dfrac{1 - 3x^2y}{x^2}\right)^2}$$

$$= \frac{x^4}{(1 - 3x^2y)^2}$$

(c) $\dfrac{p^{4m} - q^{2n}}{p^{2m} + q^n} = \dfrac{(p^{2m})^2 - (q^n)^2}{p^{2m} + q^n}$

Factoring the difference of two squares in the numerator and canceling gives us

$$\frac{(p^{2m} + q^n)(p^{2m} - q^n)}{p^{2m} + q^n} = p^{2m} - q^n$$

## EXERCISE 1—INTEGRAL EXPONENTS

Simplify, and write without negative exponents.

1. $3x^{-1}$
2. $4a^{-2}$
3. $2p^0$
4. $(3m)^{-1}$
5. $a(2b)^{-2}$
6. $x^2y^{-1}$
7. $a^3b^{-2}$
8. $m^2n^{-3}$
9. $p^3q^{-1}$
10. $(2x^2y^3z^4)^{-1}$
11. $(3m^3n^2p)^0$
12. $(5p^2q^2r^2)^3$
13. $(4a^3b^2c^6)^{-2}$
14. $(a^p - b^q)^2$
15. $(x + y)^{-1}$
16. $(2a + 3b)^{-1}$
17. $(m^{-2} - 6n)^{-2}$
18. $(4p^2 + 5q^{-4})^{-2}$
19. $2x^{-1} + y^{-2}$
20. $p^{-1} - 3q^{-2}$
21. $(3m)^{-3} - 2n^{-2}$
22. $(5a)^{-2} + (2a^2b^{-4})^{-2}$
23. $(x^n + y^m)^2$
24. $(x^{a-1} + y^{a-2})(x^a + y^{a-1})$
25. $(16x^6y^0 \div 8x^4y) \div 4xy^6$
26. $(a^{n-1} + b^{n-2})(a^n + b^{n-1})$
27. $(72p^6q^7 \div 9p^4q) \div 8pq^6$
28. $\dfrac{(p^2 - pq)^8}{(p - q)^4}$
29. $\left(\dfrac{x^{m+n}}{x^n}\right)^m$
30. $\dfrac{27x^{3n} + 8y^{3n}}{3x^n + 2y^n}$
31. $\left(\dfrac{3}{b}\right)^{-1}$
32. $\left(\dfrac{x}{4}\right)^{-1}$
33. $\left(\dfrac{2x}{3y}\right)^{-2}$
34. $\left(\dfrac{3p}{4y}\right)^{-1}$
35. $\left(\dfrac{2px}{3qy}\right)^{-3}$
36. $\left(\dfrac{p^2}{q^0}\right)^{-1}$
37. $\left(\dfrac{3a^4b^3}{5x^2y}\right)^2$
38. $\left(\dfrac{x^2}{y}\right)^{-3}$
39. $\left(\dfrac{-2a^3x^3}{3b^2y}\right)^{2n}$
40. $\left(\dfrac{3p^2y^3}{4qx^4}\right)^{-2}$
41. $\left(\dfrac{2p^{-2}z^3}{3q^{-2}x^{-4}}\right)^{-1}$
42. $\left(\dfrac{9m^4n^3}{3p^3q}\right)^3$
43. $\left(\dfrac{5w^2}{2z}\right)^p$
44. $\left(\dfrac{5a^5b^3}{2x^2y}\right)^{3n}$
45. $\left(\dfrac{3n^{-2}y^3}{5m^{-3}x^4}\right)^{-2}$
46. $\left(\dfrac{z^0z^{2n-2}}{z^n}\right)^3$
47. $\left(\dfrac{1 + a}{1 + b}\right)^3 \div \dfrac{1 + a^3}{b^2 - 1}$
48. $\left(\dfrac{p^2 - q^2}{p + q}\right)^2 \cdot \dfrac{(p^3 + q^3)^2}{(p + q)^2}$
49. $\dfrac{(x^2 - xy)^7}{(x - y)^5}$
50. $\dfrac{(a^n)^{2n} - (b^{2n})^n}{(a^n)^n + (b^n)^n}$
51. $\dfrac{27a^{3n} - 8b^{3n}}{3a^n - 2b^n}$
52. $\left(\dfrac{a^2a^{2n-2}}{a^n}\right)^2$
53. $\left(\dfrac{x + 1}{y + 1}\right)^3 \div \dfrac{x^3 + 1}{y^2 - 1}$
54. $\left(\dfrac{a^2 - b^2}{x + y}\right)^2 \cdot \dfrac{(x^3 + y^3)^2}{(a + b)^2}$
55. $\left(\dfrac{7x^3}{15y^3a^{m-1}} \div \dfrac{x^{n+1}}{a^{m+n}}\right) \dfrac{y^n}{a^{m-1}}$
56. $\dfrac{a^7b^8c^9 + a^6b^7c^8 - 3a^5b^6c^7}{a^5b^6c^7}$

57. $\dfrac{128a^4b^3 - 48a^5b^2 - 40a^6b + 15a^7}{3a^2 - 8ab}$

58. $\left(\dfrac{x^{m+n}}{x^n}\right)^m \div \left(\dfrac{x^n}{x^{m+n}}\right)^{n+m}$

59. $\left(\dfrac{5a^3}{25b^3a^{m-1}} \div \dfrac{a^{n+1}}{a^{m+n}}\right)\dfrac{b^n}{a^{m-1}}$

60. $\dfrac{p^4q^8r^9 - p^6q^4r^8 + 3p^5q^6r^4}{p^3q^2r^4}$

61. $\dfrac{3x^4y^3 + 4x^5y^2 + 3x^6y + 4x^7}{4x^2 + 3xy}$

62. $\left(\dfrac{p^n}{p^{m+n}}\right)^{n-m} \div \left(\dfrac{p^{m+n}}{p^n}\right)^m$

63. $\dfrac{a^{-1} + b^{-1}}{a^{-1} - b^{-1}} \cdot \dfrac{a^{-2} - b^{-2}}{a^{-2} + b^{-2}} \cdot \dfrac{1}{\left(1 + \dfrac{b}{a}\right)^{-2} + \left(1 + \dfrac{a}{b}\right)^{-2}}$

## 12-2 SIMPLIFICATION OF RADICALS

### Relation between Fractional Exponents and Radicals

A quantity raised to a *fractional exponent* can also be written as a *radical*. A radical consists of a *radical sign*, a quantity under the radical sign called the *radicand*, and the *index* of the radical. A radical is written:

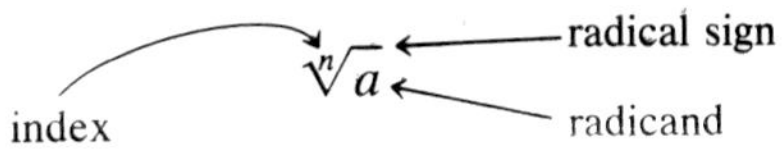


where $n$ is an integer. An index of 2, for a square root, is usually not written.

**Unless otherwise stated, we'll assume that when the index is even, that the radicand is *positive*. We'll consider square roots and fourth roots of** negative numbers when we take up **imaginary** numbers in Chapter 20.

---

**EXAMPLE 10:**

$$x^{1/2} = \sqrt{x} \qquad y^{1/4} = \sqrt[4]{y} \qquad w^{-1/2} = \frac{1}{\sqrt{w}}$$

---

In general,

| | | |
|---|---|---|
| Relation between Exponents and Radicals | $a^{1/n} = \sqrt[n]{a}$ | **36** |

If a quantity is raised to a fractional exponent $m/n$ (where $m$ and $n$ are integers), we have

$$a^{m/n} = (a^m)^{1/n} = \sqrt[n]{a^m}$$

Further,

$$a^{m/n} = (a^{1/n})^m = (\sqrt[n]{a})^m$$

So,

| | |
|---|---|
| $a^{m/n} = \sqrt[n]{a^m} = (\sqrt[n]{a})^m$ | **37** |

We use these definitions to switch between *exponential form* and *radical form.*

---

**EXAMPLE 11:** Express $x^{1/2}y^{-1/3}$ in radical form.

**Solution:**

$$x^{1/2}y^{-1/3} = \frac{x^{1/2}}{y^{1/3}} = \frac{\sqrt{x}}{\sqrt[3]{y}}$$

---

**EXAMPLE 12:** Express $\sqrt[3]{y^2}$ in exponential form.

**Solution:**

$$\sqrt[3]{y^2} = (y^2)^{1/3} = y^{2/3}$$

---

**EXAMPLE 13:** Find the value of $\sqrt[4]{(81)^3}$ without using a calculator.

**Solution:** Instead of cubing 81 and then taking the fourth root, let us take the fourth root first.

$$\sqrt[4]{(81)^3} = (\sqrt[4]{81})^3 = (3)^3 = 27$$

---

**EXAMPLE 14:** Since

$$a^{1/3} \cdot a^{1/3} \cdot a^{1/3} = a$$

then

$$\sqrt[3]{a} \cdot \sqrt[3]{a} \cdot \sqrt[3]{a} = a$$

---

| | |
|---|---|
| Common Error | Don't confuse the **coefficient** of a radical with the **index** of a radical. $$3\sqrt{x} \neq \sqrt[3]{x}$$ |

### Root of a Product

We have several *rules of radicals,* which are similar to the laws of exponents, and, in fact, are derived from them. The first rule is for products. By our definition of a radical,

$$\sqrt[n]{ab} = (ab)^{1/n}$$

Using the law of exponents for a product and then returning to radical form,

$$(ab)^{1/n} = a^{1/n}b^{1/n} = \sqrt[n]{a}\,\sqrt[n]{b}$$

So our first rule of radicals is

| | | |
|---|---|---|
| Root of a Product | $\sqrt[n]{ab} = \sqrt[n]{a}\,\sqrt[n]{b}$ | **38** |

*The root of a product equals the product of the roots of the factors.*

---

**EXAMPLE 15:** We may split the radical $\sqrt{9x}$ into two radicals,

$$\sqrt{9x} = \sqrt{9}\,\sqrt{x} = 3\sqrt{x}$$

since $\sqrt{9} = 3$.

---

**EXAMPLE 16:** Write as a single radical $\sqrt{7}\,\sqrt{2}\,\sqrt{x}$.

**Solution:** By Eq. 38,

$$\sqrt{7}\,\sqrt{2}\,\sqrt{x} = \sqrt{7(2)x} = \sqrt{14x}$$

---

| | |
|---|---|
| Common Error | There is no similar rule for the root of a sum. $\sqrt[n]{a+b} \neq \sqrt[n]{a} + \sqrt[n]{b}$ |

---

**EXAMPLE 17:** $\sqrt{9} + \sqrt{16}$ does *not* equal $\sqrt{25}$.

---

| | |
|---|---|
| Common Error | Equation 38 does not hold when $a$ and $b$ are both *negative* and the index is even. |

---

**EXAMPLE 18:**

$$(\sqrt{-4})^2 = \sqrt{-4}\,\sqrt{-4} \neq \sqrt{(-4)(-4)}$$
$$\neq \sqrt{16} = +4$$

Instead, we convert to *imaginary numbers,* as we will show in Sec. 20-1.

---

### Root of a Quotient

Just as the root of a product can be split up into the roots of the factors, the root of a quotient can be expressed as the root of the numerator divided by the root of the denominator. We first write the quotient in exponential form.

$$\sqrt[n]{\frac{a}{b}} = \left(\frac{a}{b}\right)^{1/n} = \frac{a^{1/n}}{b^{1/n}}$$

by the law of exponents for quotients. Returning to radical form, we have

| | | |
|---|---|---|
| Root of a Quotient | $\sqrt[n]{\frac{a}{b}} = \frac{\sqrt[n]{a}}{\sqrt[n]{b}}$ | **39** |

*The root of a quotient equals the quotient of the roots of numerator and denominator.*

---

**EXAMPLE 19:** The radical $\sqrt{\frac{w}{25}}$ can be written $\frac{\sqrt{w}}{\sqrt{25}}$ or $\frac{\sqrt{w}}{5}$.

---

### Simplest Form for a Radical

Rather than "simplifying," we are actually putting radicals into a *standard form,* so that they may be combined or compared.

A radical is said to be in *simplest form* when:

1. The radicand has been reduced as much as is possible.
2. There are no radicals in the denominator and no fractional radicands.
3. The index has been made as small as is possible.

### Removing Factors from the Radicand

Using the fact that

$$\sqrt{x^2} = x, \qquad \sqrt[3]{x^3} = x, \ldots, \qquad \text{and } \sqrt[n]{x^n} = x$$

try to factor the radicand so that one or more of the factors is a perfect $n$th power (where $n$ is the index of the radical). Then use Eq. 38 to split the radical into two or more radicals, some of which can then be reduced.

---

**EXAMPLE 20:**
(a) $\sqrt{50} = \sqrt{(25)(2)} = \sqrt{25}\,\sqrt{2} = 5\sqrt{2}$
(b) $\sqrt{x^3} = \sqrt{x^2 x} = \sqrt{x^2}\,\sqrt{x} = x\sqrt{x}$

---

**EXAMPLE 21:** Simplify $\sqrt{50x^3}$.

**Solution:** We factor the radicand so that some factors are perfect squares,

$$\sqrt{50x^3} = \sqrt{(25)(2)x^2x}$$

Then, by Eq. 38:

$$= \sqrt{25}\,\sqrt{x^2}\,\sqrt{2x} = 5x\sqrt{2x}$$

---

**EXAMPLE 22:** Simplify $\sqrt{24y^5}$.

**Solution:** We look for factors of the radicand that are perfect squares.

$$\sqrt{24y^5} = \sqrt{4(6)y^4y} = 2y^2\sqrt{6y}$$

---

**EXAMPLE 23:**

$$\sqrt[4]{24y^5} = \sqrt[4]{24y^4y} = y\sqrt[4]{24y}$$

**EXAMPLE 24:** Simplify $\sqrt[3]{24y^5}$.

**Solution:** We look for factors of the radicand that are perfect cubes.

$$\sqrt[3]{24y^5} = \sqrt[3]{8(3)y^3y^2}$$
$$= \sqrt[3]{8y^3}\,\sqrt[3]{3y^2} = 2y\sqrt[3]{3y^2}$$

When the radicand contains more than one term, try to *factor out* a perfect $n^{\text{th}}$ power (where $n$ is the index).

**EXAMPLE 25:** Simplify $\sqrt{4x^2y + 12x^4z}$.

**Solution:** We factor $4x^2$ from the radicand, and then remove it from under the radical sign.

$$\sqrt{4x^2y + 12x^4z} = \sqrt{4x^2(y + 3x^2z)}$$
$$= 2x\sqrt{y + 3x^2z}$$

## Rationalizing the Denominator

An expression is considered in simpler form when its denominators contain no radicals. To put it into this form is called *rationalizing* the denominator. We will show how to rationalize the denominator when it is a square root, a cube root, or a root with any index, and where it has more than one term.

If the denominator is a square root, multiply numerator and denominator of the fraction by a quantity that will make the radicand in the denominator a perfect square. Note that we are eliminating radicals from the *denominator,* and that the numerator may still contain radicals. Further, even though we call this process *simplifying,* the resulting radical may look more complicated than the original.

**EXAMPLE 26:**

$$\frac{5}{\sqrt{2}} = \frac{5}{\sqrt{2}} \cdot \frac{\sqrt{2}}{\sqrt{2}} = \frac{5\sqrt{2}}{\sqrt{4}} = \frac{5\sqrt{2}}{2}$$

When the *entire* fraction is under the radical sign, we make the denominator of that fraction a perfect square and remove it from the radical sign.

**EXAMPLE 27:**

$$\sqrt{\frac{3x}{2y}} = \sqrt{\frac{3x(2y)}{2y(2y)}} = \sqrt{\frac{6xy}{4y^2}} = \frac{\sqrt{6xy}}{2y}$$

If the denominator is a *cube* root, we must multiply numerator and denominator by a quantity that will make the quantity under the radical sign a perfect cube.

**EXAMPLE 28:** Simplify

$$\frac{7}{\sqrt[3]{4}}$$

**Solution:** To rationalize the denominator $\sqrt[3]{4}$, we multiply numerator and denominator of the fraction by $\sqrt[3]{2}$. We chose this as the factor because it results in a perfect cube (8) under the radical sign.

$$\frac{7}{\sqrt[3]{4}} = \frac{7}{\sqrt[3]{4}} \cdot \frac{\sqrt[3]{2}}{\sqrt[3]{2}} = \frac{7\sqrt[3]{2}}{\sqrt[3]{8}} = \frac{7\sqrt[3]{2}}{2}$$

The same principle applies regardless of the index. In general, if the index is $n$, we must make the quantity under the radical sign (in the denominator) a perfect $n$th power.

**EXAMPLE 29:**

$$\frac{2y}{3\sqrt[5]{x}} = \frac{2y}{3\sqrt[5]{x}} \cdot \frac{\sqrt[5]{x^4}}{\sqrt[5]{x^4}} = \frac{2y\sqrt[5]{x^4}}{3\sqrt[5]{x^5}} = \frac{2y\sqrt[5]{x^4}}{3x}$$

Sometimes the denominator will have more than one term.

**EXAMPLE 30:**

(a) $\sqrt{\dfrac{a}{a^2 + b^2}} = \sqrt{\dfrac{a}{a^2 + b^2} \cdot \dfrac{a^2 + b^2}{a^2 + b^2}} = \dfrac{\sqrt{a(a^2 + b^2)}}{a^2 + b^2}$

(b) $\sqrt{\dfrac{ab}{a + b}} = \sqrt{\dfrac{ab(a + b)}{(a + b)(a + b)}} = \sqrt{\dfrac{ab(a + b)}{(a + b)^2}}$

$= \dfrac{\sqrt{ab(a + b)}}{a + b}$

## Reducing the Index

We can sometimes reduce the index by writing the radical in exponential form and then reducing the fractional exponent, as in the next example.

**EXAMPLE 31:**

(a) $\sqrt[6]{x^3} = x^{3/6} = x^{1/2} = \sqrt{x}$

(b) $\sqrt[4]{4x^2y^2} = \sqrt[4]{(2xy)^2}$

$= (2xy)^{2/4} = (2xy)^{1/2}$

$= \sqrt{2xy}$

## EXERCISE 2—SIMPLIFICATION OF RADICALS

### Exponential and Radical Form

Express in radical form.

1. $a^{1/4}$
2. $x^{1/2}$
3. $z^{3/4}$
4. $a^{1/2}b^{1/4}$
5. $(m - n)^{1/2}$
6. $(x^2y)^{-1/2}$
7. $\left(\dfrac{x}{y}\right)^{-1/3}$
8. $a^0b^{-3/4}$

Express in exponential form.

9. $\sqrt{b}$
10. $\sqrt[3]{x}$
11. $\sqrt{y^2}$

12. $4\sqrt[3]{xy}$ 13. $\sqrt[n]{a+b}$ 14. $\sqrt[n]{x^m}$

15. $\sqrt{x^2y^2}$ 16. $\sqrt[n]{a^nb^{3n}}$

**Simplifying Radicals**

Write in simplest form.

**Do not use your calculator for any of these numerical problems. Leave the answers in radical form.**

17. $\sqrt{18}$ 18. $\sqrt{75}$ 19. $\sqrt{63}$

20. $\sqrt[3]{16}$ 21. $\sqrt[3]{-56}$ 22. $\sqrt[4]{48}$

23. $\sqrt{a^3}$ 24. $3\sqrt{50x^5}$ 25. $\sqrt{36x^2y}$

26. $\sqrt[3]{x^2y^5}$ 27. $x\sqrt[3]{16x^3y}$ 28. $\sqrt[4]{64m^2n^4}$

29. $3\sqrt[5]{32xy^{11}}$ 30. $6\sqrt[3]{16x^4}$

31. $\sqrt{a^3 - a^2b}$ 32. $x\sqrt{x^4 - x^3y^2}$

33. $\sqrt{9m^3 + 18n}$ 34. $\sqrt{2x^3 + x^4y}$

35. $\sqrt[3]{x^4 - a^2x^3}$ 36. $x\sqrt{a^3 + 2a^2b + ab^2}$

37. $(a+b)\sqrt{a^3 - 2a^2b + ab^2}$ 38. $\sqrt{\frac{2}{5}}$

39. $\sqrt{\frac{3}{7}}$ 40. $\sqrt{\frac{2}{3}}$

41. $\sqrt[3]{\frac{1}{4}}$ 42. $\sqrt{\frac{5}{8}}$

43. $\sqrt[3]{\frac{2}{9}}$ 44. $\sqrt[4]{\frac{7}{8}}$

45. $\sqrt{\frac{1}{2x}}$ 46. $\sqrt{\frac{5m}{7n}}$

47. $\sqrt{\frac{3a^3}{5b}}$ 48. $\sqrt{\frac{5ab}{6xy}}$

49. $\sqrt[3]{\frac{1}{x^2}}$ 50. $\sqrt[3]{\frac{81x^4}{16yz^2}}$

51. $\sqrt[6]{\frac{4x^6}{9}}$ 52. $\sqrt{x^2 - \left(\frac{x}{2}\right)^2}$

**In these more complicated types, you should start by simplifying the expression under the radical sign.**

53. $(x^2 - y^2)\sqrt{\frac{x}{x+y}}$ 54. $\sqrt{a^3 + \left(\frac{a^3}{2}\right)^3}$

55. $(m+n)\sqrt{\frac{m}{m-n}}$ 56. $\sqrt{\left(\frac{x+1}{2}\right)^2 - x}$

57. $\sqrt{\frac{8a^2 - 48a + 72}{3a}}$

58. A stone is thrown upward with a horizontal velocity of 40 ft/s and an upward velocity of 60 ft/s. At $t$ seconds it will have a horizontal displacement $H$ equal to $40t$ and a vertical displacement $V$ equal to $60t - 16t^2$. The straight-line distance $S$ from the stone to the launch point is found by the Pythagorean theorem. Write an equation for $S$ in terms of $t$, and simplify.

## 12-3 OPERATIONS WITH RADICALS

### Adding and Subtracting Radicals

Radicals are called *similar* if they have the same index and the same radicand, such as $5\sqrt[3]{2x}$ and $3\sqrt[3]{2x}$. We add and subtract radicals by *combining similar radicals*.

**One reason we learned to put radicals into a standard form is to combine them.**

**EXAMPLE 32:**

$$5\sqrt{y} + 2\sqrt{y} - 4\sqrt{y} = 3\sqrt{y}$$

Radicals that may look similar at first glance may not actually be similar.

**EXAMPLE 33:** The radicals

$$\sqrt{2x} \quad \text{and} \quad \sqrt{3x}$$

are *not* similar.

| Common Error | Do not try to combine radicals that are not similar. $\sqrt{2x} + \sqrt{3x} \neq \sqrt{5x}$ |
|---|---|

Radicals that do not appear to be similar at first may turn out to be so after simplification.

**EXAMPLE 34:**

(a) $\sqrt{18x} - \sqrt{8x} = 3\sqrt{2x} - 2\sqrt{2x} = \sqrt{2x}$

(b) $\sqrt[3]{24y^4} + \sqrt[3]{81x^3y} = 2y\sqrt[3]{3y} + 3x\sqrt[3]{3y}$

Factoring, $\quad = (2y + 3x)\sqrt[3]{3y}$

(c) $5y\sqrt{\frac{x}{y}} - 2\sqrt{xy} + x\sqrt{\frac{y}{x}}$

$= 5y\sqrt{\frac{xy}{y^2}} - 2\sqrt{xy} + x\sqrt{\frac{yx}{x^2}}$

$= 5\sqrt{xy} - 2\sqrt{xy} + \sqrt{xy} = 4\sqrt{xy}$

### Multiplying Radicals

Radicals having the *same index* can be multiplied by Eq. 38:

$$\sqrt[n]{a}\sqrt[n]{b} = \sqrt[n]{ab}$$

**EXAMPLE 35:**

(a) $\sqrt{x}\sqrt{2y} = \sqrt{2xy}$

(b) $(5\sqrt[3]{3a})(2\sqrt[3]{4b}) = 10\sqrt[3]{12ab}$

(c) $(2\sqrt{m})(5\sqrt{n})(3\sqrt{mn}) = 30\sqrt{m^2n^2} = 30\ mn$

(d) $\sqrt{2x}\sqrt{3x} = \sqrt{6x^2} = x\sqrt{6}$

(e) $\sqrt{3x}\sqrt{3x} = 3x$

(f) $\sqrt[3]{2x}\sqrt[3]{4x^2} = \sqrt[3]{8x^3} = 2x$

Radicals having *different indices* are multiplied by first going to exponential form, then multiplying using Eq. 29 ($x^ax^b = x^{a+b}$) and finally returning to radical form.

**EXAMPLE 36:**

$$\sqrt{a}\sqrt[4]{b} = a^{1/2}b^{1/4}$$

$$= a^{2/4}b^{1/4} = (a^2b)^{1/4}$$

Or, in radical form,

$$= \sqrt[4]{a^2b}$$

When the radicands are the same, the work is even easier.

**EXAMPLE 37:** Multiply $\sqrt[3]{x}$ by $\sqrt[6]{x}$.

**Solution:**

$$\sqrt[3]{x}\sqrt[6]{x} = x^{1/3}x^{1/6} = x^{2/6}x^{1/6}$$

By Eq. 29,

$$= x^{3/6} = x^{1/2}$$

$$= \sqrt{x}$$

*Multinomials* containing radicals are multiplied in the way we learned in Sec. 2-4.

**EXAMPLE 38:**

(a) $(3 + \sqrt{x})(2\sqrt{x} - 4\sqrt{y}) = 3(2\sqrt{x}) - 3(4\sqrt{y}) + 2\sqrt{x}\sqrt{x} - 4\sqrt{x}\sqrt{y}$

$$= 6\sqrt{x} - 12\sqrt{y} + 2x - 4\sqrt{xy}$$

(b) $\sqrt{50x}(\sqrt{2x} - \sqrt{xy}) = \sqrt{100x^2} - \sqrt{50x^2y}$

$$= 10x - 5x\sqrt{2y}$$

(c) $(\sqrt{x} - \sqrt{y})(\sqrt{x} + \sqrt{y}) = \sqrt{x^2} + \sqrt{xy} - \sqrt{xy} - \sqrt{y^2}$

$$= \sqrt{x^2} - \sqrt{y^2}$$

$$= x - y$$

To raise a radical to a *power,* we simply multiply the radical by itself the proper number of times.

**EXAMPLE 39:** Cube the expression

$$2x\sqrt{y}$$

**Solution:**

$$(2x\sqrt{y})^3 = (2x\sqrt{y})(2x\sqrt{y})(2x\sqrt{y})$$

$$= 2^3x^3(\sqrt{y})^3$$

$$= 8x^3\sqrt{y^3} = 8x^3y\sqrt{y}$$

## Dividing Radicals

Radicals having the same indices can be divided using Eq. 39,

$$\frac{\sqrt[n]{a}}{\sqrt[n]{b}} = \sqrt[n]{\frac{a}{b}}$$

**EXAMPLE 40:**

(a) $$\frac{\sqrt{4x^5}}{\sqrt{2x}} = \sqrt{\frac{4x^5}{2x}} = \sqrt{2x^4} = \sqrt{2}x^2$$

(b) $$\frac{\sqrt{3a^7} + \sqrt{12a^5} - \sqrt{6a^3}}{\sqrt{3a}} = \sqrt{\frac{3a^7}{3a}} + \sqrt{\frac{12a^5}{3a}} - \sqrt{\frac{6a^3}{3a}}$$
$$= \sqrt{a^6} + \sqrt{4a^4} - \sqrt{2a^2}$$
$$= a^3 + 2a^2 - \sqrt{2}a$$

It is a common practice to rationalize the denominator after division.

**EXAMPLE 41:** $$\frac{\sqrt{10x}}{\sqrt{5y}} = \sqrt{\frac{10x}{5y}} = \sqrt{\frac{2x}{y}}$$

Rationalizing the denominator,

$$\sqrt{\frac{2x}{y}} = \sqrt{\frac{2x}{y} \cdot \frac{y}{y}} = \frac{\sqrt{2xy}}{y}$$

If the indices are different, we go to exponential form, as we did for multiplication. Divide using Eq. 30 and then return to radical form.

**EXAMPLE 42:** $$\frac{\sqrt[3]{a}}{\sqrt[4]{b}} = \frac{a^{1/3}}{b^{1/4}} = \frac{a^{4/12}}{b^{3/12}} = \left(\frac{a^4}{b^3}\right)^{1/12}$$

Returning to radical form, we obtain

$$\left(\frac{a^4}{b^3}\right)^{1/12} = \sqrt[12]{\frac{a^4}{b^3}}$$
$$= \sqrt[12]{\frac{a^4}{b^3} \cdot \frac{b^9}{b^9}}$$
$$= \sqrt[12]{\frac{a^4b^9}{b^{12}}}$$
$$= \frac{1}{b}\sqrt[12]{a^4b^9}$$

Thus the conjugate of the binomial $a + b$ is $a - b$. The conjugate of $x - y$ is $x + y$.

When dividing by a binomial containing *square* roots, multiply the divisor and the dividend by the *conjugate* of that binomial (a binomial that differs from the original dividend only in the sign of one term). This operation will remove all square roots from the denominator.

**EXAMPLE 43:** Divide $(3 + \sqrt{x})$ by $(2 - \sqrt{x})$.

**Solution:** The conjugate of the divisor $2 - \sqrt{x}$ is $2 + \sqrt{x}$. Multiplying divisor and dividend by $2 + \sqrt{x}$, we get

$$\frac{3 + \sqrt{x}}{2 - \sqrt{x}} = \frac{3 + \sqrt{x}}{2 - \sqrt{x}} \cdot \frac{2 + \sqrt{x}}{2 + \sqrt{x}}$$
$$= \frac{6 + 3\sqrt{x} + 2\sqrt{x} + \sqrt{x}\sqrt{x}}{4 + 2\sqrt{x} - 2\sqrt{x} - \sqrt{x}\sqrt{x}}$$
$$= \frac{6 + 5\sqrt{x} + x}{4 - x}$$

## EXERCISE 3—OPERATIONS WITH RADICALS

### Addition and Subtraction of Radicals

Combine as indicated.

**As in Exercise 2, do not use your calculator for any of these numerical problems. Leave the answers in radical form.**

1. $2\sqrt{24} - \sqrt{54}$
2. $\sqrt{300} + \sqrt{108} - \sqrt{243}$
3. $\sqrt{24} - \sqrt{96} + \sqrt{54}$
4. $\sqrt{128} - \sqrt{18} + \sqrt{32}$
5. $2\sqrt{50} + \sqrt{72} + 3\sqrt{18}$
6. $\sqrt[3]{384} - \sqrt[3]{162} + \sqrt[3]{750}$
7. $2\sqrt[3]{2} - 3\sqrt[3]{16} + \sqrt[3]{-54}$
8. $3\sqrt[3]{108} + 2\sqrt[3]{32} - \sqrt[3]{256}$
9. $\sqrt[3]{625} - 2\sqrt[3]{135} - \sqrt[3]{320}$
10. $5\sqrt[3]{320} + 2\sqrt[3]{40} - 4\sqrt[3]{135}$
11. $\sqrt[4]{768} - \sqrt[4]{48} - \sqrt[4]{243}$
12. $3\sqrt{\frac{5}{4}} + 2\sqrt{45}$
13. $\sqrt{128x^2y} - \sqrt{98x^2y} + \sqrt{162x^2y}$
14. $3\sqrt{\frac{1}{3}} - 2\sqrt{\frac{3}{4}} + 4\sqrt{3}$
15. $7\sqrt{\frac{27}{50}} - 3\sqrt{\frac{2}{3}}$
16. $4\sqrt{50} - 2\sqrt{72} + \frac{3}{\sqrt{2}}$
17. $\sqrt{a^2x} + \sqrt{b^2x}$
18. $\sqrt{2b^2xy} - \sqrt{2a^2xy}$
19. $\sqrt{x^2y} - \sqrt{4a^2y}$
20. $\sqrt{80a^3} - 3\sqrt{20a^3} - 2\sqrt{45a^3}$
21. $4\sqrt{3a^2x} - 2a\sqrt{48x}$
22. $\sqrt[3]{125x^2} - 2\sqrt[3]{8x^2}$
23. $\sqrt[3]{1250a^3b} + \sqrt[3]{270c^3b}$
24. $2\sqrt[3]{ab^7} + 3\sqrt[3]{a^7b} + 2\sqrt[3]{8a^4b^4}$
25. $\sqrt[5]{a^{13}b^{11}c^{12}} - 2\sqrt[5]{a^8bc^2} + \sqrt[5]{a^3b^6c^7}$
26. $\sqrt{\frac{9x}{16}} - \frac{3\sqrt{x}}{4}$
27. $\sqrt[6]{8x^3} - 2\sqrt[4]{\frac{x^2}{4}} + 3\sqrt{8x}$
28. $\sqrt{\frac{x}{a^2}} + 2\sqrt{\frac{x}{b^2}} - 3\sqrt{\frac{x}{c^2}}$
29. $3\sqrt{\frac{3x}{4y^2}} + 2\sqrt{\frac{x}{27y^2}} - \sqrt{\frac{x}{3y^2}}$
30. $\sqrt{(a+b)^2x} + \sqrt{(a-b)^2x}$

### Multiplication of Radicals

Multiply.

31. $3\sqrt{3}$ by $5\sqrt{3}$
32. $2\sqrt{3}$ by $3\sqrt{8}$
33. $\sqrt{8}$ by $\sqrt{160}$
34. $\sqrt{\frac{5}{8}}$ by $\sqrt{\frac{3}{4}}$
35. $4\sqrt[3]{45}$ by $2\sqrt[3]{3}$
36. $3\sqrt{3}$ by $2\sqrt[3]{2}$
37. $3\sqrt{2}$ by $2\sqrt[3]{3}$
38. $2\sqrt{3}$ by $\sqrt[4]{5}$
39. $2\sqrt[3]{3}$ by $5\sqrt[4]{4}$
40. $2\sqrt[3]{24}$ by $\sqrt[9]{\frac{8}{27}}$
41. $2x\sqrt{3a}$ by $3\sqrt{y}$
42. $3\sqrt[3]{9a^2}$ by $\sqrt[3]{3abc}$
43. $\sqrt[5]{4xy^2}$ by $\sqrt[5]{8x^2y}$
44. $\sqrt{\frac{a}{b}}$ by $\sqrt{\frac{c}{d}}$
45. $\sqrt{a}$ by $\sqrt[4]{b}$
46. $\sqrt[3]{x}$ by $\sqrt{y}$
47. $\sqrt{xy}$ by $2\sqrt{xz}$ and $\sqrt[3]{x^2y^2}$
48. $\sqrt[3]{a^2b}$ by $\sqrt[3]{2a^2b^2}$ and $\sqrt{3a^3b^2}$
49. $(\sqrt{5} - \sqrt{3})$ by $2\sqrt{3}$
50. $(x + \sqrt{y})$ by $\sqrt{y}$
51. $\sqrt{x^3 - x^4y}$ by $\sqrt{x}$
52. $(\sqrt{x} - \sqrt{2})$ by $(2\sqrt{x} + \sqrt{2})$
53. $(a + \sqrt{b})$ by $(a - \sqrt{b})$
54. $(\sqrt{x} + \sqrt{y})$ by $(\sqrt{x} + \sqrt{y})$
55. $(4\sqrt{x} + 2\sqrt{y})$ by $(4\sqrt{x} - 5\sqrt{y})$
56. $(x + \sqrt{xy})$ by $(\sqrt{x} - \sqrt{y})$

### Powers

Square the following expressions.

57. $3\sqrt{y}$

58. $4\sqrt[3]{4x^2}$

59. $3x\sqrt[3]{2x^2}$

60. $5 + 4\sqrt{x}$

61. $3 - 5\sqrt{a}$

62. $\sqrt{a} + 5a\sqrt{b}$

Cube the following radicals.

63. $5\sqrt{2x}$

64. $2x\sqrt{3x}$

65. $5\sqrt[3]{2ax}$

### Division of Radicals

Divide.

66. $8 \div 3\sqrt{2}$

67. $6\sqrt{72} \div 12\sqrt{32}$

68. $(4\sqrt[3]{3} - 3\sqrt[3]{2}) \div \sqrt[3]{2}$

69. $5\sqrt[3]{12} \div 10\sqrt{8}$

70. $9\sqrt[3]{18} \div 3\sqrt{6}$

71. $12\sqrt[3]{4a^3} \div 4\sqrt[3]{2a^5}$

72. $12\sqrt{256} \div 18\sqrt{2}$

73. $(4\sqrt[3]{4} + 2\sqrt[3]{3} + 3\sqrt[3]{6}) \div \sqrt[3]{6}$

74. $2 \div \sqrt[3]{6}$

75. $\sqrt{72} \div 2\sqrt[4]{64}$

76. $8 \div 2\sqrt[3]{4}$

77. $8\sqrt[3]{ab} \div 4\sqrt{ac}$

78. $\sqrt[3]{4ab} \div \sqrt[4]{2ab}$

79. $2\sqrt[4]{a^2b^3c^2} \div 4\sqrt[3]{ab^2c^2}$

80. $6\sqrt[3]{x^4y^7z} \div \sqrt[4]{xy}$

81. $(3 + \sqrt{2}) \div (2 - \sqrt{2})$

82. $5 \div \sqrt{3x}$

83. $4\sqrt{x} \div \sqrt{a}$

84. $10 \div \sqrt[3]{9x^2}$

85. $12 \div \sqrt[3]{4x^2}$

86. $5\sqrt{a^2b} \div 3\sqrt[3]{a^2b^2}$

87. $(10\sqrt{6} - 3\sqrt{5}) \div (2\sqrt{6} - 4\sqrt{5})$

88. $\sqrt{x} \div (\sqrt{x} + \sqrt{y})$

89. $a \div (a + \sqrt{b})$

90. $(a + \sqrt{b}) \div (a - \sqrt{b})$

91. $(\sqrt{x} - \sqrt{y}) \div (\sqrt{x} + \sqrt{y})$

92. $(3\sqrt{m} - \sqrt{2n}) \div (\sqrt{3n} + \sqrt{m})$

93. $(5\sqrt{x} + \sqrt{2y}) \div (2\sqrt{3x} + 2\sqrt{y})$

94. $(5\sqrt{2p} - 3\sqrt{3q}) \div (3\sqrt{3p} + 2\sqrt{5q})$

95. $(2\sqrt{3a} + 4\sqrt{5b}) \div (3\sqrt{2a} + 4\sqrt{b})$

## 12-4 RADICAL EQUATIONS

An equation in which the unknown is under a radical sign is called a *radical equation*. To solve a radical equation it is necessary to isolate the radical term on one side of the equal sign and then to raise both sides to whatever power will eliminate the radical.

**EXAMPLE 44:** Solve for $x$:

$$\sqrt{x-5} - 4 = 0$$

**Solution:** Rearranging yields

$$\sqrt{x-5} = 4$$

Squaring both sides:

$$x - 5 = 16$$

$$x = 21$$

We will do more difficult radical equations in Chapter 13.

Check:

$$\sqrt{21-5} - 4 \stackrel{?}{=} 0$$

$$\sqrt{16} - 4 \stackrel{?}{=} 0$$

$$4 - 4 = 0 \quad \text{checks}$$

---

**EXAMPLE 45:** Solve for $x$:

$$\sqrt{x^2 - 6x} = x - 9$$

**Solution:** Squaring, we obtain

$$x^2 - 6x = x^2 - 18x + 81$$

$$12x = 81$$

$$x = \frac{81}{12} = \frac{27}{4}$$

Remember that we take only the principal (positive) root.

Check:

$$\sqrt{\left(\frac{27}{4}\right)^2 - 6\left(\frac{27}{4}\right)} \stackrel{?}{=} \frac{27}{4} - 9$$

$$\sqrt{\frac{729}{16} - \frac{162}{4}} \stackrel{?}{=} \frac{27}{4} - \frac{36}{4}$$

$$\sqrt{\frac{81}{16}} = \frac{9}{4} \neq -\frac{9}{4} \quad \text{does not check}$$

The given equation has no solution.

---

| Common Error | The squaring process often introduces *extraneous roots*. These are discarded because they do not satisfy the original equation. Check you answer in the original equation. |
|---|---|

If the equation has more than one radical, isolate one at a time and square both sides, as in the following example.

**EXAMPLE 46:** Solve for $x$:

$$\sqrt{x - 32} + \sqrt{x} = 16$$

*Tip:* **It is usually better to isolate and square the *most complicated* radical first.**

**Solution:** Rearranging gives

$$\sqrt{x - 32} = 16 - \sqrt{x}$$

Squaring yields

$$x - 32 = (16)^2 - 32\sqrt{x} + x$$

We rearrange again to isolate the radical,

$$32\sqrt{x} = 256 + 32 = 288$$
$$\sqrt{x} = 9$$

Squaring again, we obtain

$$x = 81$$

Check:

$$\sqrt{81 - 32} + \sqrt{81} \stackrel{?}{=} 16$$
$$\sqrt{49} + 9 \stackrel{?}{=} 16$$
$$7 + 9 = 16 \qquad \text{checks}$$

If a radical equation contains a fraction, we proceed as we did with other fractional equations and multiply both sides by the least common denominator.

**EXAMPLE 47:** Solve

$$\sqrt{x - 3} + \frac{1}{\sqrt{x - 3}} = \sqrt{x}$$

**Solution:** Multiplying both sides by $\sqrt{x - 3}$ gives

$$x - 3 + 1 = \sqrt{x}\sqrt{x - 3}$$
$$x - 2 = \sqrt{x^2 - 3x}$$

Squaring both sides yields

$$x^2 - 4x + 4 = x^2 - 3x$$
$$x = 4$$

Check:

$$\sqrt{4 - 3} + \frac{1}{\sqrt{4 - 3}} \stackrel{?}{=} \sqrt{4}$$

$$1 + \frac{1}{1} = 2 \qquad \text{checks}$$

Radical equations having indices other than 2 are solved in a similar way.

**EXAMPLE 48:** Solve for $x$:

$$\sqrt[3]{x - 5} = 2$$

**Solution:** Cubing both sides, we obtain

$$x - 5 = 8$$

$$x = 13$$

Check:

$$\sqrt[3]{13 - 5} \stackrel{?}{=} 2$$

$$\sqrt[3]{8} = 2 \qquad \text{checks}$$

## EXERCISE 4—RADICAL EQUATIONS

### Radical Equations

Solve for $x$ and check.

1. $\sqrt{x} = 6$
2. $\sqrt{x} + 5 = 9$
3. $\sqrt{7x + 8} = 6$
4. $\sqrt{3x - 2} = 5$
5. $\sqrt{x + 1} = \sqrt{2x - 7}$
6. $2 = \sqrt[4]{1 + 3x}$
7. $\sqrt{3x + 1} = 5$
8. $\sqrt[3]{2x} = 4$
9. $\sqrt{x - 3} = \dfrac{4}{\sqrt{x - 3}}$
10. $x + 2 = \sqrt{x^2 + 6}$
11. $\sqrt{x^2 - 7.25} = 8.75 - x$
12. $\sqrt{x - 15.5} = 5.85 - \sqrt{x}$
13. $\dfrac{6}{\sqrt{3 + x}} = \sqrt{x + 3}$
14. $\sqrt{12 + x} = 2 + \sqrt{x}$
15. $\sqrt[5]{x - 7} = 1$
16. $\sqrt{x^2 - 2x - \sqrt{1 - x}} = 1 - x$
17. $\sqrt{\dfrac{1}{x}} - \sqrt{x} = \sqrt{1 + x}$
18. $\sqrt{x - 2} - \sqrt{x} = \dfrac{1}{\sqrt{x - 2}}$
19. $\sqrt{5x - 19} - \sqrt{5x + 14} = -3$
20. $\sqrt[3]{x - 3} = 2$
21. Solve for $C$.

$$Z = \sqrt{R^2 + \left(\omega L - \frac{1}{\omega C}\right)^2}$$

### Computer

22. If you have written a program to solve equations by the *midpoint method,* as suggested in Sec. 4-5, use your program to solve any of the radical equations above. If not, you may want to write that program now, for it will be very useful in later chapters as well.

## CHAPTER 12 REVIEW PROBLEMS

*Simplify.*

1. $\sqrt{52}$
2. $\sqrt{108}$
3. $\sqrt[3]{162}$
4. $\sqrt[4]{9}$
5. $\sqrt[6]{4}$
6. $\sqrt{81a^2x^3y}$
7. $3\sqrt[4]{81x^5}$
8. $\sqrt{ab^2 - b^3}$

9. $\sqrt[3]{(a-b)^5x^4}$

10. $5\sqrt{\dfrac{7x}{12y}}$

11. $\sqrt{\dfrac{a-2}{a+2}}$

*Perform the indicated operations and simplify. Don't use a calculator.*

12. $4b\sqrt{3y} \cdot 5\sqrt{x}$

13. $\sqrt{x} \cdot \sqrt{x^3 - x^4y}$

14. $(\sqrt{x} + \sqrt{y})(\sqrt{x} - \sqrt{y})$

15. $\dfrac{3 - 2\sqrt{3}}{2 - 5\sqrt{2}}$

16. $\dfrac{\sqrt{x} - \sqrt{y}}{\sqrt{x} + \sqrt{y}}$

17. $\dfrac{x + \sqrt{x^2 - y^2}}{x - \sqrt{x^2 - y^2}}$

18. $\sqrt{98x^2y^2} - \sqrt{128x^2y^2}$

19. $4\sqrt[3]{125x^2} + 3\sqrt[3]{8x^2}$

20. $3\sqrt{x^2 - y^2} - 2\sqrt{\dfrac{x+y}{x-y}}$

21. $\sqrt{a}\sqrt[3]{b}$

22. $\sqrt[3]{2abc^2} \cdot \sqrt{abc}$

23. $(3 + 2\sqrt{x})^2$

24. $(4x\sqrt{2x})^3$

25. $3\sqrt{50} - 2\sqrt{32}$

26. $3\sqrt{72} - 4\sqrt{8} + \sqrt{128}$

27. $5\sqrt{\dfrac{9}{8}} - 2\sqrt{\dfrac{25}{18}}$

28. $2\sqrt{2} \div \sqrt[3]{2}$

29. $\sqrt{2ab} \div \sqrt{4ab^2}$

30. $9 \div \sqrt[3]{7x^2}$

31. $3\sqrt{9} \cdot 4\sqrt{8}$

*Solve for x and check.*

32. $\dfrac{\sqrt{x} - 8}{\sqrt{x} - 6} = \dfrac{\sqrt{x} - 4}{\sqrt{x} + 2}$

33. $\sqrt{x + 6} = 4$

34. $\sqrt{2x - 7} = \sqrt{x - 3}$

35. $\sqrt[5]{2x + 4} = 2$

36. $\sqrt{4x^2 - 3} = 2x - 1$

37. $\sqrt{x} + \sqrt{x - 9.75} = 6.23$

38. $\sqrt[3]{21.5x} = 2.33$

*Simplify, and write without negative exponents.*

39. $(x^{n-1} + y^{n-2})(x^n + y^{n-1})$

40. $(72a^6b^7 \div 9a^4b) \div 8ab^6$

41. $\left(\dfrac{9x^4y^3}{6x^3y}\right)^3$

42. $\left(\dfrac{5w^2}{2x}\right)^n$

43. $\left(\dfrac{2x^5y^3}{x^2y}\right)^3$

44. $5a^{-1}$

45. $3w^{-2}$

46. $2r^0$

47. $(3x)^{-1}$

48. $2a^{-1} + b^{-2}$

49. $x^{-1} - 2y^{-2}$

50. $(5a)^{-3} - 3b^{-2}$

51. $(3x)^{-2} + (2x^2y^{-4})^{-2}$

52. $(a^n + b^m)^2$

53. $(p^{a-1} + q^{a-2})(p^a + q^{a-1})$

54. $(16a^6b^0 \div 8a^4b) \div 4ab^6$

55. $p^2q^{-1}$

56. $x^3y^{-2}$

57. $r^2s^{-3}$

58. $a^3b^{-1}$

59. $(3x^3y^2z)^0$

60. $(5a^2b^2c^2)^3$

*Writing*

61. Explain how exponents and radicals are really two different ways of writing the same expression. Also explain why, if they are the same, we need both.

# 13

# QUADRATIC EQUATIONS

## OBJECTIVES

**When you have completed this chapter, you should be able to:**

- Solve quadratic equations by factoring, by completing the square, and by quadratic formula.
- Solve literal and fractional quadratic equations.
- Solve radical equations that lead to quadratics.
- Solve word problems that result in quadratic equations.
- Graph quadratic equations, and solve them graphically.
- Solve polynomial equations of higher degree by factoring.
- Solve equations by the numerical method of simple iteration.
- Solve systems of two quadratic equations.

So far we have learned how to solve linear (first-degree) equations, and sets of first-degree equations. But many technical problems require us to solve more complicated equations than that. In this chapter we study second-degree equations (quadratics), and later we cover logarithmic, exponential, and trigonometric equations. That does not include every equation type that exists, but many of the common types.

Quadratic equations arise in many technical problems, such as the simple falling-body problem: "An object is thrown downward with a speed of 15 ft/s. How long will it take to fall 100 ft?"

The displacement $s$ of a freely falling body is given by $s = v_0t + \frac{1}{2}at^2$, and if we substitute $s = 100$, $v_0 = 15$, and $a = 32$, we get

$$100 = 15t + 16t^2$$

Now how do we solve that equation? How shall we find the values of $t$ that will make the equation balance? Even this simple problem requires us to solve a quadratic equation.

Let us leave this problem and return to it after we have learned how to solve quadratics. Before we start, you might want to leaf through Chapter 7 if you have gotten rusty on factoring. Look especially at the sections on common factors, trinomials, and the perfect square trinomial.

## 13-1 SOLVING QUADRATICS BY FACTORING

### Terminology

**Recall that in a polynomial, all the powers of $x$ are *positive integers*.**

A polynomial equation of second degree is called a *quadratic equation*. It is common practice to refer to it simply as a *quadratic*.

---

**EXAMPLE 1:** The equations

(a) $4x^2 - 5x + 2 = 0$

(b) $x^2 = 58$

(c) $9x^2 - 5x = 0$

(d) $2x^2 - 7 = 0$

are quadratic equations. Equation (a) is called a complete quadratic, (b) and (d), which have no $x$ terms, are called pure quadratics, and (c), which has no constant term, is called an incomplete quadratic.

---

A quadratic is in *general form* when it is written in the form

| | | |
|---|---|---|
| General Form of a Quadratic | $ax^2 + bx + c = 0$ | **99** |

where $a$, $b$, and $c$ are constants.

**EXAMPLE 2:** Write the quadratic equation

$$7 - 4x = \frac{5x^2}{3}$$

in general form, and identify $a$, $b$, and $c$.

**Solution:** Subtracting $5x^2/3$ from both sides and writing the terms in descending order of the exponents, we obtain

$$-\frac{5x^2}{3} - 4x + 7 = 0$$

Quadratics in general form are usually written without fractions and with the first term positive. Multiplying by $-3$, we get

$$5x^2 + 12x - 21 = 0$$

The equation is now in general form, with $a = 5$, $b = 12$, and $c = -21$.

### Number of Roots

We'll see later in this chapter that the maximum number of roots that certain types of equations may have is equal to the degree of the equation. Thus, a quadratic equation, being of degree 2, has *two solutions* or roots. The two roots are sometimes equal, or they may be imaginary or complex numbers. However, in applications, we'll see that one of the two roots must sometimes be discarded.

**EXAMPLE 3:** The quadratic equation

$$x^2 = 4$$

has the two roots, $x = 2$ and $x = -2$.

### Solving Pure Quadratics

We simply isolate the $x^2$ term and then take the square root of both sides, as in the following example.

**EXAMPLE 4:** Solve $3x^2 - 75 = 0$.

**Solution:** Adding 75 to both sides and dividing by 3, we obtain

$$3x^2 = 75$$

$$x^2 = 25$$

Taking the square root yields

$$x = \pm\sqrt{25} = \pm 5$$

Check: We check our solution the same way as for other equations, by substituting back into the original equation. Now, however, we must check two solutions.

Substitute $+5$: $\quad 3(5)^2 - 75 = 0$

$\quad 75 - 75 = 0 \quad$ checks

Substitute $-5$: $\quad 3(-5)^2 - 75 = 0$

$\quad 75 - 75 = 0 \quad$ checks

When taking the square root, be sure to keep both the plus *and* the minus values. *Both* will satisfy the equation.

At this point students usually grumble: "First we're told that $\sqrt{4} = +2$ only, not $\pm 2$ (Sec. 1-5). *Now* we're told to keep both plus and minus. What's going on here?" Here's the difference. When we solve a quadratic, we know that it must have two roots, so we keep both values. When we evaluate a square root, such as $\sqrt{4}$, which is not the solution of a quadratic, we agree to take only the positive value to avoid ambiguity.

The roots might sometimes be irrational, as in the following example.

---

**EXAMPLE 5:** Solve $4x^2 - 15 = 0$.

**Solution:** Following the same steps as before, we get

$$4x^2 = 15$$

$$x^2 = \frac{15}{4}$$

$$x = \pm\sqrt{\frac{15}{4}} = \pm\frac{\sqrt{15}}{2}$$

---

## Solving Incomplete Quadratics

To solve an incomplete quadratic, remove the common factor $x$ from each term, and set each factor equal to zero.

---

**EXAMPLE 6:** Solve

$$x^2 + 5x = 0$$

**Solution:** Factoring yields

$$x(x + 5) = 0$$

**We use this idea often in this chapter. If we have the product of two quantities *a* and *b* set equal to zero, *ab* = 0, this equation will be true if *a* = 0 (0·*b* = 0) or if *b* = 0 (*a*·0 = 0), or if both are zero (0·0 = 0).**

Note that this expression will be true if either or both of the two factors equals zero. We therefore set each factor in turn equal to zero.

$$x = 0, \qquad x + 5 = 0$$

$$x = -5$$

The two solutions are thus $x = 0$ and $x = -5$.

---

| Common Error | Do not cancel an $x$ from the terms of an incomplete quadratic. That will cause a root to be lost. If, in the last example, we had said $$x^2 = -5x$$ and had divided by $x$, $$x = -5$$ we would have obtained the correct root $x = -5$ but would have lost the root $x = 0$. |
|---|---|

## Solving the Complete Quadratic

We now consider a quadratic that has all its terms in place, the complete quadratic.

| General Form | $ax^2 + bx + c = 0$ | **99** |
|---|---|---|

First write the quadratic in general form, Eq. 99. Factor the trinomial (if possible) by the methods of Chapter 7 and set each factor equal to zero.

---

**EXAMPLE 7:** Solve by factoring:

$$x^2 - x - 6 = 0$$

**Solution:** Factoring gives

$$(x - 3)(x + 2) = 0$$

This equation will be satisfied if either or both of the two factors $(x - 3)$ and $(x + 2)$ are zero. We therefore set each factor in turn to equal zero:

$$x - 3 = 0 \qquad x + 2 = 0$$

so the roots are

$$x = 3 \qquad \text{and} \qquad x = -2$$

---

| Common Error | When the product of two quantities is zero, as in $$(x - 3)(x + 2) = 0$$ we can set each factor equal to zero, getting $x - 3 = 0$ and $x + 2 = 0$. <br> But *this is valid only when the product is zero.* Thus if $$(x - 3)(x + 2) = 5$$ we *cannot* say that $x - 3 = 5$ and $x + 2 = 5$. |
|---|---|

Often an equation must first be simplified before factoring.

**EXAMPLE 8:** Solve for $x$,

$$x(x - 8) = 2x(x - 1) + 9$$

**Solution:** Removing parentheses gives

$$x^2 - 8x = 2x^2 - 2x + 9$$

Collecting terms, we get

$$x^2 + 6x + 9 = 0$$

Factoring yields

$$(x + 3)(x + 3) = 0$$

which gives the double root,

$$x = -3, -3$$

Sometimes an equation will not look like a quadratic, at first glance. The following example shows a fractional equation which, after simplification, turns out to be a quadratic.

**EXAMPLE 9:** Solve for $x$,

$$\frac{3x - 1}{4x + 7} = \frac{x + 1}{x + 7}$$

**Solution:** We start by multiplying both sides by the LCD, $(4x + 7)(x + 7)$. We get

$$(3x - 1)(x + 7) = (x + 1)(4x + 7)$$

or

$$3x^2 + 20x - 7 = 4x^2 + 11x + 7$$

Collecting terms gives

$$x^2 - 9x + 14 = 0$$

Factoring yields

$$(x - 7)(x - 2) = 0$$

so $x = 7$ and $x = 2$.

## Writing the Equation When the Roots Are Known

Given the roots, we simply reverse the process to find the equation.

**EXAMPLE 10:** Write the quadratic equation that has the roots $x = 2$ and $x = -5$.

**Solution:** If the roots are 2 and $-5$, we know that the factors of the equation must be $(x - 2)$ and $(x + 5)$. So

$$(x - 2)(x + 5) = 0$$

Multiplying gives us

$$x^2 + 5x - 2x - 10 = 0$$

So

$$x^2 + 3x - 10 = 0$$

is the original equation.

## Solving Radical Equations

In Chapter 12 we solved simple radical equations. We isolated a radical on one side of the equation and then squared both sides. Here we solve equations in which this squaring operation results in a quadratic equation.

---

**EXAMPLE 11:** Solve for $x$:

$$3\sqrt{x-1} - \frac{4}{\sqrt{x-1}} = 4$$

**Solution:** We clear fractions by multiplying through by $\sqrt{x-1}$.

$$3(x-1) - 4 = 4\sqrt{x-1}$$

$$3x - 7 = 4\sqrt{x-1}$$

Squaring both sides yields

$$9x^2 - 42x + 49 = 16(x-1)$$

Removing parentheses and collecting terms gives

$$9x^2 - 58x + 65 = 0$$

Factoring gives

$$(x-5)(9x-13) = 0$$

$$x = 5 \quad \text{and} \quad x = \frac{13}{9}$$

Check: When $x = 5$:

$$3\sqrt{5-1} - \frac{4}{\sqrt{5-1}} \stackrel{?}{=} 4$$

$$3(2) - \frac{4}{2} = 4 \qquad \text{checks}$$

When $x = \frac{13}{9}$:

$$\sqrt{\frac{13}{9} - 1} = \sqrt{\frac{4}{9}} = \frac{2}{3}$$

So

$$3 \cdot \frac{2}{3} - \frac{4}{\frac{2}{3}} \stackrel{?}{=} 4$$

$$2 - 6 \neq 4 \qquad \text{does not check}$$

Our solution is then $x = 5$.

Remember that the squaring operation sometimes gives an extraneous root that will not check.

---

## EXERCISE 1—SOLVING QUADRATICS BY FACTORING

### Pure and Incomplete Quadratics

Solve for $x$.

1. $2x = 5x^2$
2. $2x - 40x^2 = 0$
3. $3x(x-2) = x(x+3)$
4. $2x^2 - 6 = 66$
5. $5x^2 - 3 = 2x^2 + 24$
6. $7x^2 + 4 = 3x^2 + 40$
7. $(x+2)^2 = 4x + 5$
8. $5x^2 - 2 = 3x^2 + 6$
9. $8.25x^2 - 2.93x = 0$
10. $284x = 827x^2$

### Complete Quadratics

Solve for $x$.

11. $x^2 + 2x - 15 = 0$
12. $x^2 + 6x - 16 = 0$
13. $x^2 - x - 20 = 0$
14. $x^2 + 13x + 42 = 0$
15. $x^2 - x - 2 = 0$
16. $x^2 + 7x + 12 = 0$
17. $x^2 + 3x + 2 = 0$
18. $x^2 - 4x - 21 = 0$
19. $x^2 - 7x - 18 = 0$
20. $x^2 + 6x + 8 = 0$
21. $x^2 + 12x + 32 = 0$
22. $x^2 - 10x - 39 = 0$
23. $x^2 - 12x - 64 = 0$
24. $x^2 + 14x + 33 = 0$
25. $2x^2 - 3x - 5 = 0$
26. $4x^2 - 10x + 6 = 0$
27. $2x^2 + 5x - 12 = 0$
28. $3x^2 - x - 2 = 0$
29. $5x^2 + 14x - 3 = 0$
30. $5x^2 + 3x - 2 = 0$
31. $2 + x - x^2 = 0$
32. $20 + 19x - 6x^2 = 0$
33. $-26x + 5 - 24x^2 = 0$
34. $15x^2 = 8 + 14x$
35. $2x^3 + x^2 - 18 = x(2x^2 + 3)$
36. $\frac{x}{2} + \frac{2}{x} - \frac{5}{2} = 0$
37. $\frac{2 - x}{3} = 2x^2$
38. $(x - 6)(x + 6) = 5x$
39. $x(x - 5) = 36$
40. $(2x - 3)^2 = 2x - x^2$
41. $x^2 - \frac{5}{6}x = \frac{1}{6}$
42. $\frac{7}{x + 4} - \frac{1}{4 - x} = \frac{2}{3}$
43. $\frac{2x - 1}{x + 3} = \frac{x + 3}{2x - 1}$
44. $1 - \frac{x}{2} = 5 - \frac{36}{x + 2}$
45. $3 - x^2 = \frac{2x - 76}{3}$
46. $\frac{x}{x - 1} - \frac{x - 1}{x} = \frac{3}{2}$
47. $\frac{2x}{3} - \frac{5}{4x} = \frac{7x}{9} - \frac{21}{4x}$
48. $\frac{2x + 5}{2x - 5} = \frac{7x - 5}{2x}$
49. $x^2 + \frac{x}{2} = \frac{2x^2}{5} - \frac{x}{5} + \frac{13}{10}$
50. $\frac{2x - 1}{x - 1} + \frac{1}{6} = \frac{2x - 3}{x - 2}$

### Literal Equations

Solve for $x$.

51. $4x^2 + 16ax + 12a^2 = 0$
52. $14x^2 - 23ax + 3a^2 = 0$
53. $9x^2 + 30bx + 24b^2 = 0$
54. $24x^2 - 17xy + 3y^2 = 0$
55. $21x^2 + 23xy^2 + 6y^4 = 0$
56. $28x^2 - mx - 2m^2 = 0$

### Writing the Equation from the Roots

Write the quadratic equations that have the following roots.

57. $x = 4$ and $x = 7$
58. $x = -3$ and $x = -5$
59. $x = \frac{2}{3}$ and $x = \frac{3}{5}$
60. $x = p$ and $x = q$
61. $x = -\frac{5}{3}$ and $x = -\frac{7}{2}$
62. $x = 3 + \sqrt{2}$ and $x = 3 - \sqrt{2}$
63. $x = 1 + \sqrt{5}$ and $x = 1 - \sqrt{5}$
64. $x = a$ and $x = a - b$

### Radical Equations

Solve each radical equation for $x$ and check.

65. $\sqrt{5x^2 - 3x - 41} = 3x - 7$
66. $3\sqrt{x - 1} - \frac{4}{\sqrt{x - 1}} = 4$

67. $\sqrt{x + 5} = \dfrac{12}{\sqrt{x + 12}}$

68. $\sqrt{7x + 8} - \sqrt{5x - 4} = 2$

69. $2x + \sqrt{4x^2 - 7} = \dfrac{21}{\sqrt{4x^2 - 7}}$

**Computer**

70. Use your program for solving equations by the midpoint method (Sec. 4-5) to find the roots of any of the quadratics in this exercise.

## 13-2 SOLVING QUADRATICS BY COMPLETING THE SQUARE

If a quadratic equation is not factorable, it is possible to manipulate it into factorable form by a procedure called *completing the square*. The form into which we shall put our expression is the *perfect square trinomial*, Eqs. 47 and 48, that we studied in Sec. 7-6.

If you recall the perfect square trinomial,

1. The first and last terms were perfect squares.
2. The middle term was twice the product of the square roots of the outer terms.

**The method of completing the square is really too cumbersome to be a practical tool for solving quadratics. The main reason we learn it is to derive the quadratic formula.**

To complete the square, we manipulate our given expression so that these two conditions are met. This is best shown by an example.

---

**EXAMPLE 12:** Solve the quadratic $x^2 - 8x + 6 = 0$ by completing the square.

**Solution:** Subtracting 6 from both sides, we obtain

$$x^2 - 8x = -6$$

We complete the square by *adding the square of half the coefficient of the x term* to both sides. The coefficient of $x$ is $-8$. We take half of $-8$ and square it, getting $(-4)^2$ or 16. Adding 16 to both sides,

$$x^2 - 8x + 16 = -6 + 16 = 10$$

Factoring,

$$(x - 4)^2 = 10$$

Taking the square root of both sides,

$$x - 4 = \pm\sqrt{10}$$

Finally, adding 4 to both sides,

$$x = 4 \pm \sqrt{10}$$

$$\cong 7.16 \quad \text{or} \quad 0.838$$

---

| Common Error | When adding the quantity needed to complete the square to the left-hand side, it is easy to forget to add the same quantity to the right-hand side. $$x^2 - 8x \boxed{+16} = -6 \boxed{+16}$$ (↑ don't forget) |
|---|---|

If the $x^2$ term has a coefficient other than 1, divide through by this coefficient *before* completing the square.

---

**EXAMPLE 13:** Solve

$$2x^2 + 4x - 3 = 0$$

**Solution:** Rearranging and dividing by 2 gives

$$x^2 + 2x = \frac{3}{2}$$

Completing the square by adding 1 (half the coefficient of the $x$ term, squared) to both sides of the equation,

$$x^2 + 2x + 1 = \frac{3}{2} + 1$$

$$(x + 1)^2 = \frac{5}{2}$$

$$x + 1 = \pm\sqrt{\frac{5}{2}} = \frac{\pm\sqrt{10}}{2}$$

$$x = -1 \pm \frac{1}{2}\sqrt{10}$$

---

| Common Error | Don't forget to divide through by the coefficient of the $x^2$ term *before* completing the square. |
|---|---|

## EXERCISE 2—COMPLETING THE SQUARE

Solve each quadratic by completing the square.

1. $x^2 - 8x + 2 = 0$
2. $x^2 + 4x - 9 = 0$
3. $x^2 + 7x - 3 = 0$
4. $x^2 - 3x - 5 = 0$
5. $4x^2 - 3x - 5 = 0$
6. $2x^2 + 3x - 3 = 0$
7. $3x^2 + 6x - 4 = 0$
8. $5x^2 - 2x - 1 = 0$
9. $4x^2 + 7x - 5 = 0$
10. $8x^2 - 4x - 3 = 0$

## 13-3 SOLVING QUADRATICS BY FORMULA

Of the several methods we have for solving quadratics, the most useful is the quadratic formula. It will work for any quadratic, regardless of the type of roots, and can easily be programmed for the computer.

### Derivation of the Quadratic Formula

We wish to find the roots of the equation

$$ax^2 + bx + c = 0$$

by completing the square. We start by subtracting $c$ from both sides and dividing by $a$:

$$x^2 + \frac{b}{a}x = -\frac{c}{a}$$

Completing the square gives

$$x^2 + \frac{b}{a}x + \left(\frac{b}{2a}\right)^2 = \frac{b^2}{4a^2} - \frac{c}{a}$$

Factoring, by Eq. 47, we obtain

$$\left(x + \frac{b}{2a}\right)^2 = \frac{b^2 - 4ac}{4a^2}$$

Taking the square root yields

$$x + \frac{b}{2a} = \pm\frac{\sqrt{b^2 - 4ac}}{2a}$$

Rearranging, we get the well-known formula for finding the roots of the quadratic equation, $ax^2 + bx + c = 0$.

| | | |
|---|---|---|
| Quadratic Formula | $x = \dfrac{-b \pm \sqrt{b^2 - 4ac}}{2a}$ | **100** |

### Using the Quadratic Formula

Simply put the given equation into general form (Eq. 99); list $a$, $b$, and $c$; and substitute them into the formula.

---

**EXAMPLE 14:** Solve $2x^2 - 5x - 3 = 0$ by the quadratic formula.

**Solution:** The equation is already in general form, with

$$a = 2 \qquad b = -5 \qquad c = -3$$

Substituting into Eq. 100, we obtain

$$x = \frac{-(-5) \pm \sqrt{(-5)^2 - 4(2)(-3)}}{2(2)}$$

$$= \frac{5 \pm \sqrt{25 + 24}}{4}$$

$$= \frac{5 \pm \sqrt{49}}{4} = \frac{5 \pm 7}{4}$$

$$= 3 \quad \text{and} \quad -\frac{1}{2}$$

---

| Common Error | Always rewrite a quadratic in general form before trying to use the quadratic formula. |
|---|---|

**EXAMPLE 15:** Solve the equation $5.25x - 2.94x^2 + 6.13 = 0$ by the quadratic formula.

**Solution:** The constants $a$, $b$, and $c$ are *not* 5.25, 2.94, and 6.13, as you might think at first glance. We must rearrange the terms.

$$-2.94x^2 + 5.25x + 6.13 = 0$$

Although not a necessary step, dividing through by the coefficient of $x^2$ simplifies the work a bit.

$$x^2 - 1.79x - 2.09 = 0$$

Substituting into the quadratic formula with $a = 1$, $b = -1.79$, and $c = -2.09$,

$$x = \frac{1.79 \pm \sqrt{(1.79)^2 - 4(1)(-2.09)}}{2(1)}$$

$$= 2.59 \quad \text{and} \quad -0.805$$

We now have the tools to finish the falling-body problem started in the introduction to this chapter.

**EXAMPLE 16:** Solve the equation

$$100 = 15t + 16t^2$$

(where $t$ is in seconds), to three significant digits.

**Solution:** We first go to general form,

$$16t^2 + 15t - 100 = 0$$

Substituting into the quadratic formula gives us

$$t = \frac{-15 \pm \sqrt{225 - 4(16)(-100)}}{2(16)}$$

$$= \frac{-15 \pm \sqrt{6625}}{32} = \frac{-15 \pm 81.4}{32}$$

$$\simeq 2.08 \text{ s} \quad \text{and} \quad -3.01 \text{ s}$$

In this problem, a negative elapsed time makes no sense, so we discard the negative root. Thus it will take 2.08 s for an object to fall 100 ft when thrown downward with a speed of 15 ft/s.

### Nonreal Roots

Quadratics resulting from practical problems will usually yield real roots if the equation is properly set up. Other quadratics, however, could give nonreal roots, as shown in the next example.

**EXAMPLE 17:** Solve

$$x^2 + 2x + 5 = 0$$

**Solution:** From Eq. 100,

$$x = \frac{-2 \pm \sqrt{4 - 4(1)(5)}}{2} = \frac{-2 \pm \sqrt{-16}}{2}$$

The expression $\sqrt{-16}$ can be written $\sqrt{-1}\,\sqrt{16}$ or $j4$, where $j = \sqrt{-1}$. So

$$x = \frac{-2 \pm j4}{2} = -1 \pm j2$$

See Chapter 20 for a complete discussion of imaginary and complex numbers.

### Predicting the Nature of the Roots

As seen from the preceding example, when the quantity in Eq. 100 under the radical sign ($b^2 - 4ac$) is negative, the roots are not real. It should also be evident that if $b^2 - 4ac$ is zero, the roots will be equal. The quantity $b^2 - 4ac$, called the *discriminant,* can thus be used to predict what kind of roots an equation will give, without having to actually find the roots.

| | | |
|---|---|---|
| The Discriminant | If $a$, $b$, and $c$ are real, and<br>$b^2 - 4ac > 0$, the roots are real and unequal.<br>$b^2 - 4ac = 0$, the roots are real and equal.<br>$b^2 - 4ac < 0$, the roots are not real. | **101** |

**EXAMPLE 18:** What sort of roots can you expect from the equation

$$3x^2 - 5x + 7 = 0?$$

**Solution:** Computing the discriminant, we obtain

$$b^2 - 4ac = (-5)^2 - 4(3)(7) = 25 - 84 = -59$$

A negative discriminant tells us that the roots are not real.

Do not lose too much sleep over the discriminant. If you need to know the nature of the roots of a quadratic and cannot remember Eq. 101, all you have to do is calculate the roots themselves. This is not much more work than finding the discriminant.

### EXERCISE 3—QUADRATIC FORMULA

Solve by quadratic formula, to three significant digits.

1. $x^2 - 12x + 28 = 0$
2. $x^2 - 6x + 7 = 0$
3. $x^2 + x - 19 = 0$
4. $x^2 - x - 13 = 0$
5. $3x^2 + 12x - 35 = 0$
6. $29.4x^2 - 48.2x - 17.4 = 0$
7. $36x^2 + 3x - 7 = 0$
8. $28x^2 + 29x + 7 = 0$
9. $49x^2 + 21x - 5 = 0$
10. $16x^2 - 16x + 1 = 0$

11. $3x^2 - 10x + 4 = 0$
12. $x^2 - 34x + 22 = 0$
13. $3x^2 + 5x = 7$
14. $4x + 5 = x^2 + 2x$
15. $x^2 - 4 = 4x + 7$
16. $x^2 - 6x - 14 = 3$
17. $(4.2x - 5.8)(7.2x - 9.2) = 8.2x + 9.9$
18. $3x^2 - 25x = 5x - 73$
19. $2x^2 + 100 = 32x - 11$
20. $33 - 3x^2 - 10x = 0$
21. $x(2x - 3) = 3x(x + 4) - 2$
22. $2.95(x^2 + 8.27x) = 7.24x\,(4.82x - 2.47) + 8.73$
23. $6x - 300 = 205 - 3x^2$
24. $(2x - 1)^2 + 6 = 6(2x - 1)$

### Nature of the Roots

Determine if the roots of each quadratic are real or nonreal, and whether they are equal or unequal.

25. $x^2 - 5x - 11 = 0$
26. $x^2 - 6x + 15 = 0$
27. $2x^2 - 4x + 7 = 0$
28. $2x^2 + 3x - 2 = 0$
29. $3x^2 - 3x + 5 = 0$
30. $4x^2 + 4x + 1 = 0$

### Computer

31. Write a program or use a spreadsheet to compute and print the roots of any quadratic once the constants $a$, $b$, and $c$ are entered. Have the computer evaluate the discriminant before computing the roots. If the discriminant is negative, have the run stop. This will prevent the program from crashing when the computer tries to take the square root of a negative number.
32. Use your program for solving equations by the midpoint method (Sec. 4-5) to find the roots of any of the quadratics in this exercise.

## 13-4 APPLICATIONS AND WORD PROBLEMS

Now that we have the tools to solve any quadratic, let us go on to problems from technology that require us to solve these equations. At this point you might want to take a quick look at Chapter 3 and review some of the suggestions for setting up and solving word problems. You should set up these problems just as you did then.

When you solve the resulting quadratic you will get two roots, of course. If one of the roots does not make sense in the physical problem (like a beam having a length of $-2000$ ft), throw it away. But do not be too hasty. Often a second root will give an unexpected but equally good answer.

**EXAMPLE 19:** The angle iron in Fig. 13-1 has a cross-sectional area of 53.4 cm². Find the thickness $x$.

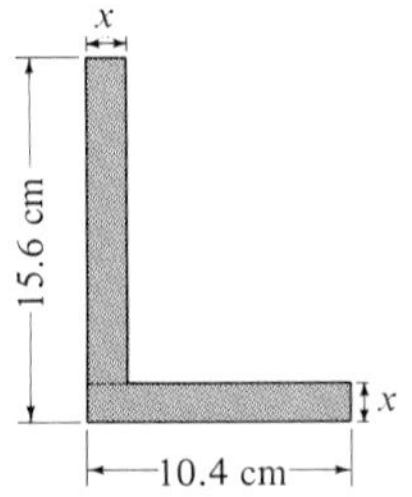


**FIGURE 13-1**

**Solution:** We divide the area into two rectangles, as shown by the dashed line. One rectangle has an area of $10.4x$ and the other has an area of $(15.6 - x)x$. Since the sum of these areas must be 53.4,

$$10.4x + (15.6 - x)x = 53.4$$

Putting this equation into standard form, we get

$$10.4x + 15.6x - x^2 = 53.4$$

$$x^2 - 26.0x + 53.4 = 0$$

By the quadratic formula,

$$x = \frac{26.0 \pm \sqrt{676 - 4(53.4)}}{2} = \frac{26.0 \pm 21.5}{2}$$

So

$$x = \frac{26.0 - 21.5}{2} = 2.25 \text{ cm}$$

and

$$x = \frac{26.0 + 21.5}{2} = 23.8 \text{ cm}$$

We discard 23.8 cm because it is an impossible solution in this problem.

Check: Does our answer meet the requirements of the original statement? Let us compute the area of the angle iron using our value of 2.25 cm for the thickness. We get

$$\text{area} = 2.25(13.35) + 2.25(10.4)$$

$$= 30.0 + 23.4 = 53.4 \text{ cm}^2$$

which is the required area.

**EXAMPLE 20:** A certain train is to be replaced with a "bullet" train that goes 40 mi/h faster than the old train and will make the 850-mi run in 3.0 h less time. Find the speed of each train.

**Solution:** Let

$$x = \text{rate of old train, mi/h}$$

Then

$$x + 40 = \text{rate of bullet train, mi/h}$$

The time it takes the old train to travel 850 mi at $x$ mi/h is, by Eq. A17,

$$\text{time} = \frac{\text{distance}}{\text{rate}} = \frac{850}{x} \quad \text{h}$$

The time for the bullet train is then $(850/x - 3)$ h. Applying Eq. A17 for the bullet train gives us

$$\text{rate} \times \text{time} = \text{distance}$$

$$(x + 40)\left(\frac{850}{x} - 3\right) = 850$$

Removing parentheses, we have

$$850 - 3x + \frac{34{,}000}{x} - 120 = 850$$

Collecting terms and multiplying through by $x$ gives

$$-3x^2 + 34{,}000 - 120x = 0$$

or

$$x^2 + 40x - 11{,}333 = 0$$

Solving for $x$ by the quadratic formula yields

$$x = \frac{-40 \pm \sqrt{1600 - 4(-11{,}333)}}{2}$$

If we drop the negative root, we get

$$x = \frac{-40 + 217}{2} = 88.3 \text{ mi/h} = \text{speed of slow train}$$

and

$$x + 40 = 128.3 \text{ mi/h} = \text{speed of bullet train}$$

## EXERCISE 4—APPLICATIONS

### Number Problems

1. What number added to its reciprocal gives $2\frac{1}{6}$?
2. Find three consecutive numbers such that the sum of their squares will be 434.
3. Find two numbers whose difference is 7 and the difference of whose cubes is 1267.
4. Find two numbers whose sum is 11 and whose product is 30.

5. Find two numbers whose difference is 10 and the sum of whose squares is 250.
6. A number increased by its square is equal to 9 times the next higher number. Find the number.

## Geometry Problems

7. A rectangle is to be 2 m longer than it is wide and have an area of 24 m$^2$. Find its dimensions.
8. One leg of a right triangle is 3 cm greater than the other leg, and the hypotenuse is 15 cm. Find the legs of the triangle.
9. A rectangular sheet of brass is twice as long as it is wide. Squares, 3 cm × 3 cm, are cut from each corner (Fig. 13-2), and the ends are turned up to form an open box having a volume of 648 cm$^3$. What are the dimensions of the original sheet of brass?
10. The length, width, and height of a cubical shipping container are all decreased by 1.0 ft, thereby decreasing the volume of the cube by 37 ft$^3$. What was the volume of the original container?
11. Find the dimensions of a rectangular field that has a perimeter of 724 m and an area of 32,400 m$^2$.
12. A flat of width $w$ is to be cut on a bar of radius $r$ (Fig. 13-3). Show that the required depth of cut $x$ is given by the formula

$$x = r \pm \sqrt{r^2 - \frac{w^2}{4}}$$

13. A rectangular field is surrounded by a fence 160 ft long. The cost of this fence, at 96 cents/ft, was one-tenth as many dollars as there are square feet in the area of the field. What are the dimensions of the field?
14. The cylinder in Fig. 13-4 has a surface area of 846 cm$^2$, including the ends. Find its radius.
15. A cylindrical tank having a diameter of 75.5 in. is placed so that it touches a wall, as in Fig. 13-5. Find the radius $x$ of the largest pipe that can fit into the space between the tank, the wall, and the floor. *Hint:* First use the Pythagorean theorem to show that $OC = 53.4$ in. and that $OP = 15.6$ in. Then in triangle $OQR$, $(OP - x)^2 = x^2 + x^2$.

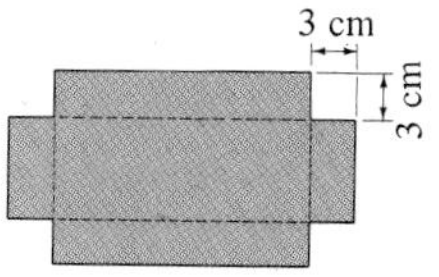

FIGURE 13-2

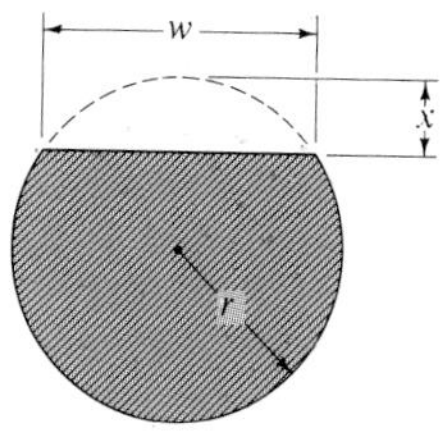

FIGURE 13-3

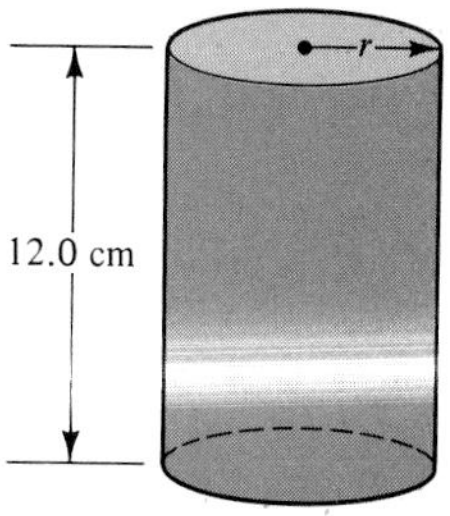

FIGURE 13-4

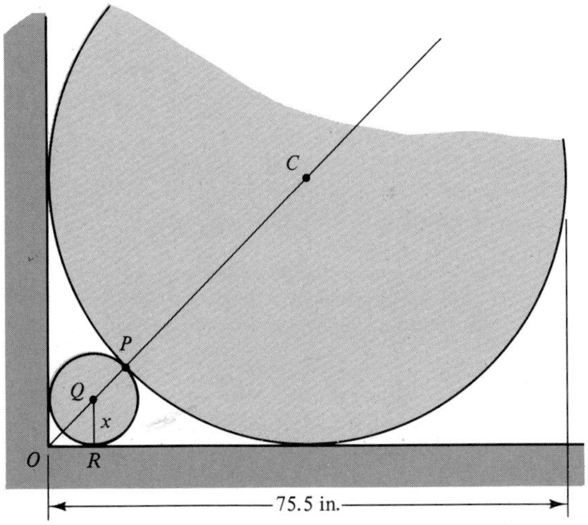

FIGURE 13-5

## Uniform Motion

16. A truck travels 350 mi to a delivery point, unloads, and, now empty, returns to the starting point at a speed 8.0 mi/h greater than on the outward trip. What was the speed of the outward trip if the total round-trip driving time was 14.4 h?

17. An airplane flies 355 mi to city $A$. Then, with better winds, it continues on to city $B$, 448 mi from $A$, at a speed 15.8 mi/h greater than on the first leg of the trip. The total flying time was 5.2 h. Find the speed at which the plane traveled to city $A$.
18. An express bus travels a certain 250-mi route in 1.0 h less time than it takes a local bus to travel a 240-mi route. Find the speed of each bus if the speed of the local is 10 mi/h less than the express.
19. A trucker calculates that if he increases his average speed by 20 km/h, he could travel his 800-km route in 2.0 h less time than usual. Find his usual speed.
20. A boat sails 30 km at a uniform rate. If the rate had been 1 km/h more, the time of the sailing would have been 1 h less. Find the rate of travel.

### Work Problems

21. A certain punch press requires 3 h longer to stamp a box of parts than does a newer-model punch press. After the older press has been punching a box of parts for 5 h, it is joined by the newer machine. Together, they finish the box of parts in 3 additional hours. How long does it take each machine, working alone, to punch a box of parts?
22. Two water pipes together can fill a certain tank in 8.4 h. The smaller pipe alone takes 2.5 h longer than the larger pipe to fill that same tank. How long would it take the larger pipe alone to fill the tank?
23. A laborer built 35 m of stone wall. If she had built 2 m less each day, it would have taken her 2 days longer. How many meters did she build each day, working at her usual rate?
24. A woman worked a certain number of days, receiving for her pay $1800. If she had received $10 per day less than she did, she would have had to work 3 days longer to earn the same sum. How many days did she work?

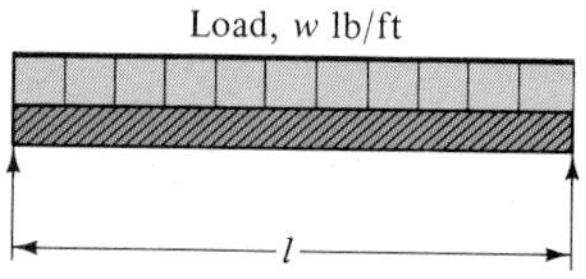


**FIGURE 13-6 Simply supported beam with a uniformly distributed load.**

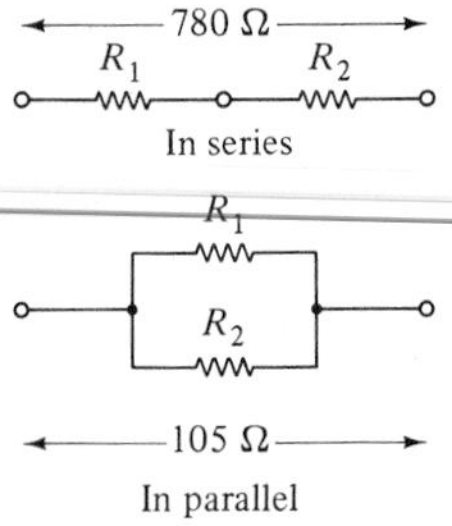


**FIGURE 13-7**

### Simply Supported Beam

25. For a simply supported beam of length $l$ having a distributed load of $w$ lb/ft (Fig. 13-6), the bending moment $M$ at any distance $x$ from one end is given by

$$M = \frac{1}{2}wlx - \frac{1}{2}wx^2$$

Find the locations on the beam where the bending moment is zero.

26. A simply supported beam, 25 ft long, carries a distributed load of 1550 lb/ft. At what distance from an end of the beam will the bending moment be 112,000 ft-lb? (Use the equation from Problem 25.)

### Freely Falling Body

**Use $s = v_0t + \frac{1}{2}gt^2$, Eq. A18, for these falling-body problems, but be careful of the signs. If you take the upward direction as positive, $g$ will be *negative*.**

27. An object is thrown upward with a velocity of 145 ft/s. When will it be 85 ft above its initial position?
28. An object is thrown upward with an initial speed of 120 m/s. Find the time for it to return to its starting point. (In the metric system, $g$ = 980 cm/s$^2$).

### Electrical Problems

29. What two resistances (Fig. 13-7) will give a total resistance of 780 Ω when wired in series and 105 Ω when wired in parallel? (See Eqs. A63 and A64.)
30. Find two resistances that will give an equivalent resistance of 9070 Ω in series and 1070 Ω in parallel.
31. In the circuit of Fig. 13-8 the power $P$ dissipated in the load resistor $R_1$ is

$$P = EI - I^2R$$

If the voltage $E$ is 115 V, and $R$ = 100 Ω, find the current $I$ needed to produce a power of 29.3 W in the load.

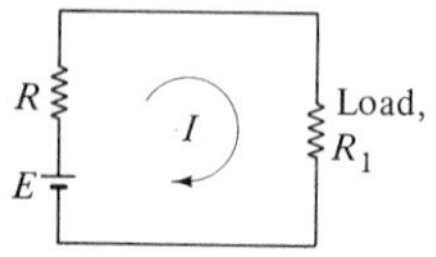


**FIGURE 13-8**

32. The reactance $X$ of a capacitance $C$ and an inductance $L$ in series (Fig. 13-9) is

$$X = \omega L - \frac{1}{\omega C}$$

where $\omega$ is the angular frequency in rad/s, $L$ is in henries, and $C$ is in farads. Find the angular frequency needed to make the reactance equal 1500 Ω, if $L = 0.5$ henry, and $C = 0.2 \times 10^{-6}$ farad (0.2 microfarad).

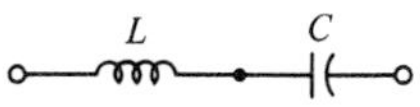


FIGURE 13-9

33. Figure 13-10 shows two currents flowing in a single resistor $R$. The total current in the resistor will be $I_1 + I_2$, so the power dissipated is

$$P = (I_1 + I_2)^2 R$$

If $R = 100$ Ω and $I_2 = 0.2$ A, find the current $I_1$ needed to produce a power of 9.0 W.

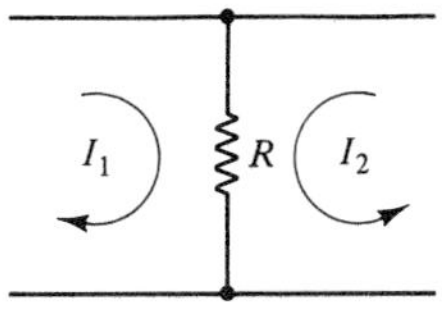


FIGURE 13-10

34. A *square-law device* is one whose output is proportional to the square of the input. A junction field-effect transistor (JFET) (Fig. 13-11) is such a device. The current $I$ that will flow through an $n$-channel JFET when a voltage $V$ is applied is

$$I = A\left(1 - \frac{V}{B}\right)^2$$

where $A$ is the drain saturation current and $B$ is the gate source pinch-off voltage. Solve this equation for $V$.

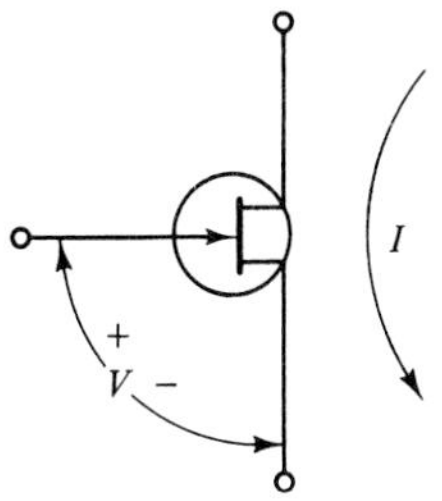


**FIGURE 13-11.** ***N*-channel JFET.**

35. A certain JFET has a drain saturation current of 4.8 mA and a gate source pinch-off voltage of −2.5 V. What input voltage is needed to produce a current of 1.5 mA?

## 13-5 GRAPHING THE QUADRATIC FUNCTION

### Quadratic Function

A *quadratic function* is one whose highest-degree term is second degree.

---

**EXAMPLE 21:** The functions

(a) $f(x) = 5x^2 - 3x + 2$ (b) $f(x) = 9 - 3x^2$

(c) $f(x) = x(x + 7)$ (d) $f(x) = x - 4 - 3x^2$

are quadratic functions.

---

### The Parabola

When we plot a quadratic function, we get the well-known and extremely useful curve called the *parabola*.

**EXAMPLE 22:** Plot the quadratic function $y = x^2 + x - 3$, for $x = -3$ to 3.

The parabola is one of the four *conic sections* (the curves obtained when a cone is intersected by a plane, at various angles), and it has many interesting properties and applications. We take only a very brief look at the parabola here, with the main treatment saved for our study of the conic sections.

**Solution:** As shown in Sec. 4-4, we compute a table of point pairs:

| $x$ | −3 | −2 | −1 | 0 | 1 | 2 |
|---|---|---|---|---|---|---|
| $y$ | 3 | −1 | −3 | −3 | −1 | 3 |

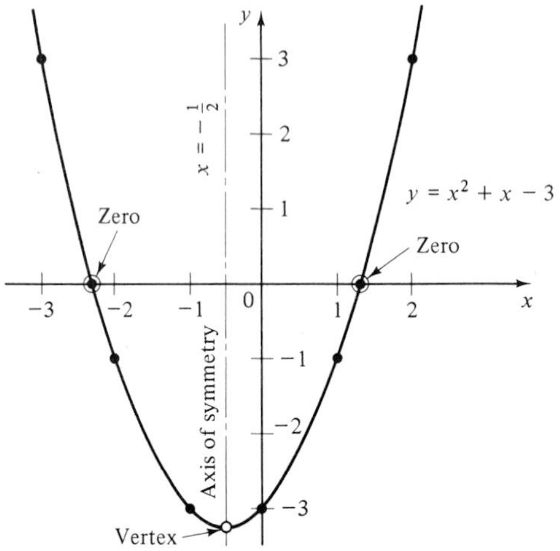


**FIGURE 13-12 A parabola.**

We plot these points (Fig. 13-12) and connect them with a smooth curve, a parabola.

Note that the curve in this example is symmetrical about the line $x = -\frac{1}{2}$. This line is called the *axis of symmetry*. The point where the parabola crosses the axis of symmetry is called the *vertex*.

## Solving Quadratics Graphically

In Sec. 4-5 we saw that a graphical solution to the equation

$$f(x) = 0$$

could be obtained by plotting the function

$$y = f(x)$$

and locating the points where the curve crosses the $x$ axis. Those points are called the zeros of the function. Those points are then the solution, or roots, of the equation. Notice that in Fig. 13-12 there are two zeros, corresponding to the two roots of the equation.

**EXAMPLE 23:** Graphically find the approximate roots of the equation

This, of course, is not a practical way to solve a quadratic. It is much faster and more accurate to use the formula. However, the same method is valuable for finding the roots of other equations that cannot be as easily solved algebraically.

$$x^2 + x - 3 = 0$$

**Solution:** The function

$$y = x^2 + x - 3$$

is already plotted in Fig. 13-12. Reading the $x$ intercepts as accurately as possible, we get for the roots

$$x \simeq -2.3 \quad \text{and} \quad x \simeq 1.3$$

## EXERCISE 5—GRAPHING THE QUADRATIC FUNCTION

Plot the following quadratic functions. Find **(a)** the coordinates of the vertex; **(b)** the equation of the axis of symmetry; and **(c)** the approximate values of the zeros.

1. $y = x^2 + 3x - 15$
2. $y = 5x^2 + 14x - 3$
3. $y = 8x^2 - 6x + 1$
4. $y = 2.74x^2 - 3.12x + 5.38$

### Computer

5. Use your program for generating a table of point pairs from Chapter 4, Exercise 4, to obtain point pairs for any of the equations in Problems 1 to 4 above. Plot these points to obtain a graph of the equation.

### Parabolic Arch

6. The parabolic arch is often used in construction because of its great strength. The equation of the bridge arch in Fig. 13-13 is $y = 0.0625x^2 - 5x + 100$. Find the distances $a$, $b$, $c$, and $d$.

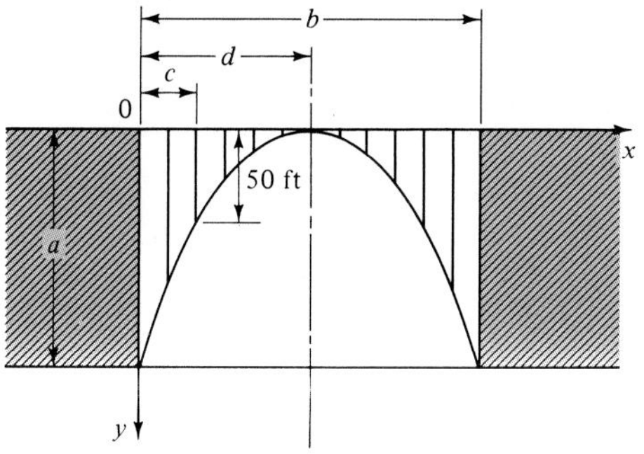


**FIGURE 13-13 A parabolic bridge arch. Note that $y$ axis is positive downward.**

7. Plot the parabolic arch of Problem 6, taking values of $x$ every 10 ft.

### Parabolic Reflector

8. The parabolic solar collector in Fig. 13-14 has the equation $x^2 = 125y$. Plot the curve, taking values of $x$ every 10 cm.

**A parabolic surface has the property that all the incoming rays that are parallel to the axis of symmetry, after being reflected from the surface of the parabola, will converge to a single point called the *focus*. This property is used in optical and radio telescopes, solar collectors, searchlights, microwave and radar antennas, and other devices.**

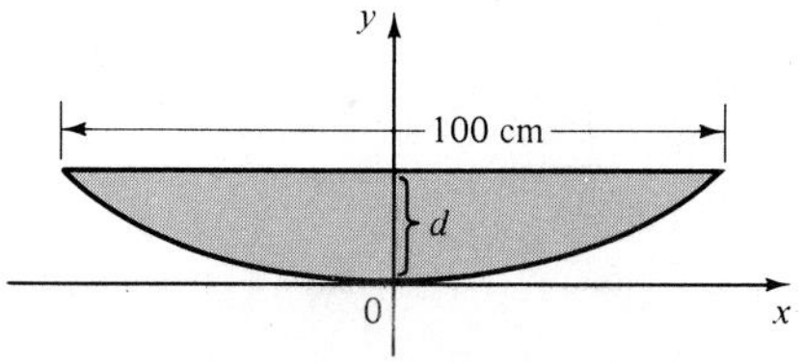


**FIGURE 13-14**

9. Find the depth $d$ of the collector of Problem 8 at its center.

### Vertical Highway Curves

Where there is a change in the slope of a highway surface, such as at the top of a hill or the bottom of a dip, the roadway is often made parabolic in shape to provide a smooth transition between the two different grades.

To construct a vertical highway curve (Fig. 13-15) we make use of the following property of the parabola:

> *The offset C from a tangent to a parabola is proportional to the square of the distance x from the point of tangency:*

or $C = kx^2$, where $k$ is a constant of proportionality, found from a known offset.

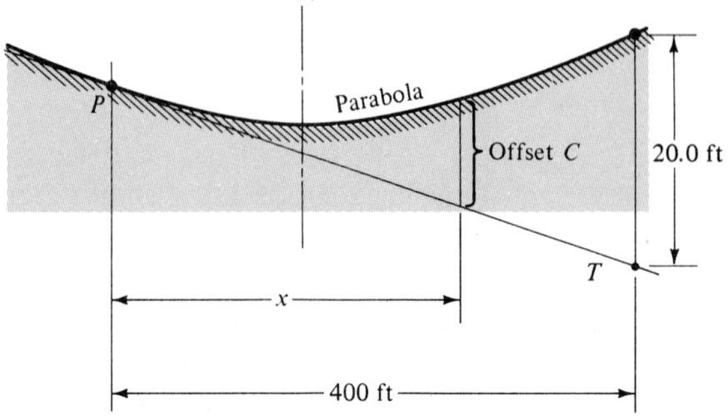


**FIGURE 13-15 A dip in a highway.**

10. If the offset is 20.0 ft at a point 400 ft from the point of tangency, find the constant of proportionality $k$.
11. For the highway curve of Problem 10, find the offset at a point 225 ft from the point of tangency.

**Simply Supported Beam**

12. The bending moment for a simply supported beam carrying a distributed load $w$ (Fig. 13-6) is $M = \frac{1}{2}wlx - \frac{1}{2}wx^2$. The bending moment curve is therefore a parabola. Plot the curve of $M$ vs. $x$ for a 10-ft-long beam carrying a load of 1000 lb/ft. Take 1-ft intervals along the beam. Graphically locate **(a)** the points of zero bending moment; **(b)** the point of maximum bending moment.

**Square-Law Devices**

13. For the $n$-channel JFET of Exercise 4, Problem 35, plot the curve of current $I$ versus applied voltage $V$, the vertical axis for current and the horizontal axis for voltage. Compute $I$ for values of $V$ from −2.5 V to 0 V. (The curve obtained is a parabola and is called the *transconductance curve*.) Graphically determine the voltage needed to produce a current of 1.0 mA.

## 13-6 EQUATIONS OF QUADRATIC TYPE

### Recognition of Form

The general quadratic, Eq. 99, can be written

$$a(x)^2 + b(x) + c = 0$$

The form of this equation obviously does not change if we put a different symbol in the parentheses, as long as we put the same one in both places. The equations

$$a(Q)^2 + b(Q) + c = 0$$

and

$$a(\$)^2 + b(\$) + c = 0$$

are still quadratics. The form still does not change if we put more complicated expressions in the parentheses, such as

$$a(x^2)^2 + b(x^2) + c = 0$$

or

$$a(\sqrt{x})^2 + b(\sqrt{x}) + c = 0$$

or

$$a(x^{-1/3})^2 + b(x^{-1/3}) + c = 0$$

The preceding three equations can no longer be called quadratics, however, because they are no longer of second degree. They are called *equations of quadratic type*. They can be easily recognized because the power of $x$ in one term is always twice the power of $x$ in another term (and there are no other terms containing powers of $x$).

## Solving Equations of Quadratic Type

The solution of these equations is best shown by examples.

---

**EXAMPLE 24:** Solve

$$x^4 - 5x^2 + 4 = 0$$

**Solution:** We see that the power of $x$ in the first term (4) is twice the power of $x$ in the second term (2). Make the substitution; let $w = x^2$. The original equation then becomes the quadratic,

$$w^2 - 5w + 4 = 0$$

Factoring yields

$$(w - 1)(w - 4) = 0$$

So

$$w = 1 \quad \text{and} \quad w = 4$$

but

$$w = x^2 = 1 \quad \text{and} \quad w = x^2 = 4$$

So

$$x = \pm 1 \quad \text{and} \quad x = \pm 2$$

Check: The four values, 1, −1, 2, and −2, each satisfy the original equation (work not shown).

---

**EXAMPLE 25:** Solve the equation

$$2x^{-1/3} + x^{-2/3} + 1 = 0$$

**Solution:** Inspecting the powers of $x$, we see that one of them, $-\frac{2}{3}$, is twice the other, $-\frac{1}{3}$. We make the substitution,

$$\text{let } w = x^{-1/3}$$

and our equation becomes

$$2w + w^2 + 1 = 0$$

or

$$w^2 + 2w + 1 = 0$$

which factors into

$$(w + 1)(w + 1) = 0$$

Setting each factor equal to zero yields

$$w + 1 = 0 \quad \text{and} \quad w + 1 = 0$$

So

$$w = -1 = x^{-1/3}$$

Cubing both sides, we get

$$x^{-1} = (-1)^3 = -1$$

$$x^{-1} = \frac{1}{x} = -1$$

So

$$x = \frac{1}{-1} = -1$$

Check:

$$2(-1)^{-1/3} + (-1)^{-2/3} + 1 \stackrel{?}{=} 0$$

$$2(-1) \quad + (1) \quad + 1 = 0$$

$$0 = 0 \quad \text{Checks}$$

---

## EXERCISE 6—EQUATIONS OF QUADRATIC TYPE

Solve for all values of $x$.

1. $x^6 - 6x^3 + 8 = 0$
2. $x^4 - 10x^2 + 9 = 0$
3. $x^{-1} - 2x^{-1/2} + 1 = 0$
4. $9x^4 - 6x^2 + 1 = 0$
5. $x^{-2/3} - x^{-1/3} = 0$
6. $x^{-6} + 19x^{-3} = 216$
7. $x^{-10/3} + 244x^{-5/3} + 243 = 0$
8. $x^6 - 7x^3 = 8$
9. $x^{2/3} + 3x^{1/3} = 4$
10. $64 + 63x^{-3/2} - x^{-3} = 0$
11. $81x^{-3/4} - 308 - 64x^{3/4} = 0$
12. $27x^6 + 46x^3 = 16$
13. $x^{1/2} - x^{1/4} = 20$
14. $3x^{-2} + 14x^{-1} = 5$
15. $2.35x^{1/2} - 1.82x^{1/4} = 43.8$
16. $1.72x^{-2} + 4.75x^{-1} = 2.24$

### Computer

17. Use your program for finding roots of equations from Sec. 4-5 to solve any of the equations in this exercise.

## 13-7 SIMPLE EQUATIONS OF HIGHER DEGREE

### Polynomial of Degree $n$

In Sec. 13-1 we used factoring to find the roots of a quadratic equation. We now expand this idea to include polynomials of higher degree. A polynomial, remember, has exponents that are all positive integers.

| | | |
|---|---|---|
| Polynomial of Degree $n$ | $a_0x^n + a_1x^{n-1} + \cdots + a_{n-1}x + a_n$ | **102** |

### The Factor Theorem

In Sec. 13-1 the expression

$$x^2 - x - 6 = 0$$

for example, factored into

$$(x - 3)(x + 2) = 0$$

and hence had the roots

$$x = 3 \qquad \text{and} \qquad x = -2$$

In a similar way, the cubic equation

$$x^3 - 7x - 6 = 0$$

has the factors

$$(x + 1)(x - 3)(x + 2) = 0$$

and hence has the roots

$$x = -1 \qquad x = 3 \qquad x = -2$$

In general,

| | | |
|---|---|---|
| Factor Theorem | If a polynomial equation $f(x) = 0$ has a root $r$, then $(x - r)$ is a factor of the polynomial $f(x)$. Conversely, if $(x - r)$ is a factor of a polynomial $f(x)$, then $r$ is a root of $f(x) = 0$. | **103** |

### Number of Roots

A *polynomial equation of degree n has n and only n roots*. Two or more of the roots may be equal, however, and some may be nonreal, giving the appearance of fewer than $n$ roots; but there can never be *more* than $n$ roots.

## Solving Polynomial Equations

A graph of the function is very helpful in finding the roots.

If the expression cannot be factored directly, we proceed largely by trial and error to obtain one root. Once a root $r$ is obtained (and verified by substituting in the equation), we obtain a *reduced* or *depressed* equation by dividing the original equation by $(x - r)$. The process is then repeated.

---

**EXAMPLE 26:** One root of the cubic equation $16x^3 - 13x + 3 = 0$ is $x = -1$. Find the other roots.

**Solution:** For a cubic, we expect a maximum of three roots. By the factor theorem, since $x = -1$ is a root, then $(x + 1)$ must be a factor of the given equation.

$$(x + 1)(\text{other factor}) = 0$$

We obtain the "other factor" by dividing the given equation by $(x + 1)$. Using synthetic division gives

Synthetic division is explained in Chapter 2. Of course, "ordinary" division works as well.

| | | | | |
|---|---|---|---|---|
| −1 | 16 | 0 | −13 | 3 |
| | | −16 | 16 | −3 |
| | 16 | −16 | 3 | 0 |

The "other factor," or reduced equation, is then

$$16x^2 - 16x + 3 = 0, \text{ a quadratic.}$$

By the quadratic formula,

$$x = \frac{16 \pm \sqrt{16^2 - 4(16)(3)}}{2(16)} = \frac{1}{4} \text{ and } \frac{3}{4}$$

Our three roots are therefore $x = -1$, $x = \frac{1}{4}$, and $x = \frac{3}{4}$.

---

## Simple Iteration: A Computer Method

This is also known as the *method of Picard,* after Charles Émile Picard, 1856–1941.

We already have the midpoint method for computer solution of equations, and now we learn the method of *simple iteration.*

With this method, we split the given equation into two or more terms, one of which is the unknown itself, and move the unknown to the left side of the equation. We then guess the value of $x$ and substitute it into the right side of the equation to obtain another value of $x$, which is then substituted into the equation, and so on, until our values converge on the correct value of $x$.

---

**EXAMPLE 27:** Find one root of the equation from Example 26

$$16x^3 - 13x + 3 = 0$$

by simple iteration. Assume that there is a root somewhere near $x = 1$.

**Solution:** The equation contains an $x$ in both the first and second terms. Either can be chosen to isolate on the left side of the equation. Let us choose the first term. Rearranging yields

$$16x^3 = 13x - 3$$

Dividing gives

$$x^3 = 0.8125x - 0.1875$$

if we work to four significant digits. Taking the cube root of both sides, we get

$$x = (0.8125x - 0.1875)^{1/3} \tag{1}$$

Note that we have not *solved* the equation, because there is still an $x$ on the right side of the equation.

For a first approximation, let $x = 1$. Substituting into the right of equation (1) gives

$$x = (0.8125 - 0.1875)^{1/3}$$
$$= 0.8550$$

This, of course, is not a root, but only a *second approximation* to a root. We now take the value 0.8550 and substitute it into the right side of equation (1).

$$x = [0.8125(0.8550) - 0.1875]^{1/3}$$
$$= 0.7975$$

This third approximation is substituted into (1), giving

$$x = [0.8125(0.7975) - 0.1875]^{1/3}$$
$$= 0.7722$$

We continue, getting the following table of values:

0.8550
0.7975
0.7722
0.7605
0.7550
0.7524
0.7512
0.7506
0.7503

Notice that we are converging on the value

$$x = 0.750$$

which is one of the roots found in Example 26. With a different first guess, we can usually make the computation converge on one of the other roots.

**If the computation diverges instead of converges, we split the equation in a different way and try again. In the example here, we would isolate the $x$ in the middle term, rather than the one in the first term.**

## EXERCISE 7—EQUATIONS OF HIGHER DEGREE

Find the remaining roots of each polynomial, given one root.

1. $x^3 + x^2 - 17x + 15 = 0, \quad x = 1$
2. $x^3 + 9x^2 + 2x - 48 = 0, \quad x = 2$
3. $x^3 - 2x^2 - x + 2 = 0, \quad x = 1$
4. $x^3 + 9x^2 + 26x + 24 = 0, \quad x = -2$
5. $2x^3 + 7x^2 - 7x - 12 = 0, \quad x = -1$
6. $3x^3 - 7x^2 + 4 = 0, \quad x = 1$
7. $9x^3 + 21x^2 - 20x - 32 = 0, \quad x = -1$
8. $16x^4 - 32x^3 - 13x^2 + 29x - 6 = 0, x = -1$

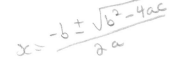

### Computer

9. Write a program or use a spreadsheet for solving equations by simple iteration. Use it to find the roots of any of the equations in this exercise.
10. Use your program for finding roots by the midpoint method for the equations in this exercise.

## 13-8 SYSTEMS OF QUADRATIC EQUATIONS

In Chapter 9 we solved systems of *linear* equations. We now use the same techniques, substitution or addition–subtraction, to solve sets of two equations in two unknowns, where one or both equations are of second degree.

### One Equation Linear and One Quadratic

These are best done by substitution.

---

**EXAMPLE 28:** Solve for $x$ and $y$:

$$2x + y = 5$$

$$2x^2 + y^2 = 17$$

**Solution:** From the first equation,

$$x = \frac{5 - y}{2}$$

Squaring gives

$$x^2 = \frac{25 - 10y + y^2}{4}$$

Substituting into the second equation yields

$$\frac{2(25 - 10y + y^2)}{4} + y^2 = 17$$

$$25 - 10y + y^2 + 2y^2 = 34$$

We get the quadratic

$$3y^2 - 10y - 9 = 0$$

which, by quadratic formula, gives

$$y = 4.07 \qquad \text{and} \qquad y = -0.74$$

Substituting back for $x$ gives

$$x = \frac{5 - 4.07}{2} = 0.47 \qquad \text{and} \qquad x = \frac{5 - (-0.74)}{2} = 2.87$$

So our two solutions are (0.47, 4.07) and (2.87, −0.74). The graphs (Fig. 13-16) of the two given equations show points of intersection at those coordinates.

---

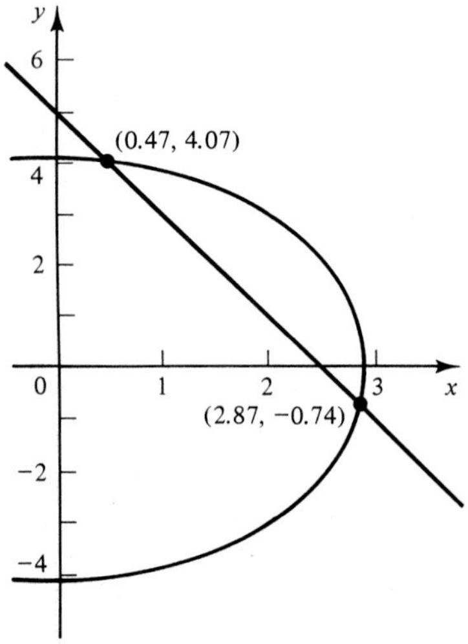


FIGURE 13-16

**EXAMPLE 29:** Solve for $x$ and $y$.

$$x + y = 2$$
$$xy + 15 = 0$$

**Solution:** From the first equation,

$$y = 2 - x$$

Substituting into the second equation, we have

$$x(2 - x) + 15 = 0$$

from which

$$x^2 - 2x - 15 = 0$$

Factoring gives

$$(x - 5)(x + 3) = 0$$

$$x = 5 \quad \text{and} \quad x = -3$$

Substituting back yields

$$y = 2 - 5 = -3$$

and

$$y = 2 - (-3) = 5$$

so the solutions are

$$x = 5,\ y = -3 \quad \text{and} \quad x = -3,\ y = 5$$

See Fig. 13-17.

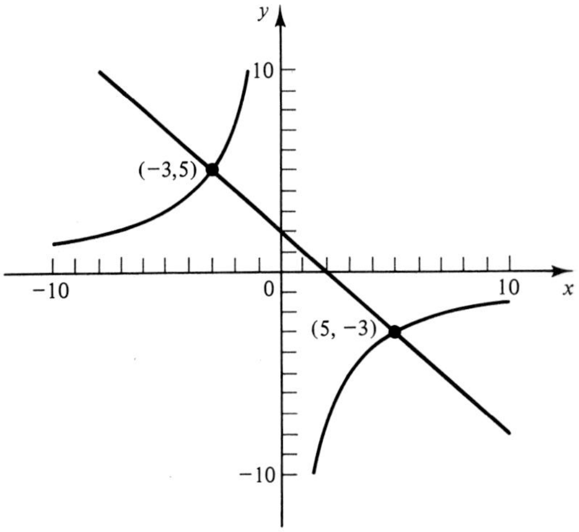


**FIGURE 13-17**

## Both Equations Quadratic

These may be solved either by substitution, or by addition–subtraction.

**EXAMPLE 30:** Solve for $x$ and $y$:

$$3x^2 - 4y = 47$$
$$7x^2 + 6y = 33$$

**Solution:** We multiply the first equation by 3 and the second by 2:

$$\begin{aligned} 9x^2 - 12y &= 141 \\ 14x^2 + 12y &= 66 \\ \hline \text{Add:}\quad 23x^2 \quad &= 207 \\ x^2 &= 9 \\ x &= \pm 3 \end{aligned}$$

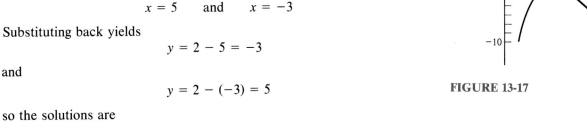

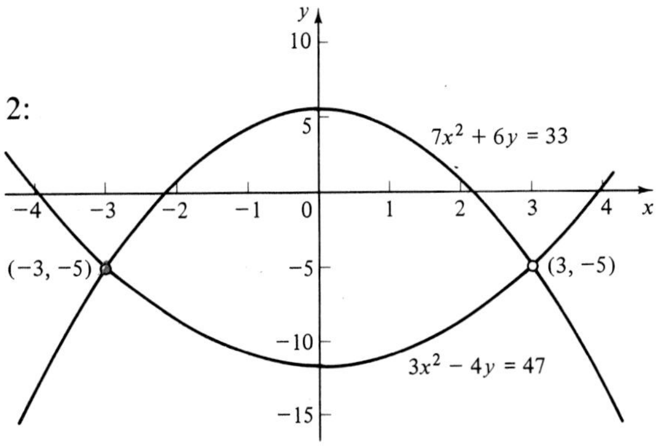


Substituting back, we get $y = -5$ for both $x = 3$ and $x = -3$. A plot of the given curves (Fig. 13-18) clearly shows the points of intersection $(3, -5)$ and $(-3, -5)$.

**FIGURE 13-18 Intersection of two parabolas. This is a good way to get an approximate solution to any pair of equations that can readily be graphed.**

The graphs of the equations in the preceding examples intersected in two points. However, the graphs of two second-degree equations can sometimes intersect in as many as four points.

**EXAMPLE 31:** Solve for $x$ and $y$:

$$3x^2 + 4y^2 = 76$$
$$-11x^2 + 3y^2 = 4$$

**Solution:** We multiply the first equation by 3 and the second by $-4$.

$$9x^2 + 12y^2 = 228$$
$$44x^2 - 12y^2 = -16$$

Add:

$$53x^2 = 212$$
$$x^2 = 4$$
$$x = \pm 2$$

From the first equation, when $x$ is either $+2$ or $-2$,

$$4y^2 = 76 - 12 = 64$$
$$y = \pm 4$$

So our solutions are

$$x = 2 \quad x = 2 \quad x = -2 \quad x = -2$$
$$y = 4 \quad y = -4 \quad y = 4 \quad y = -4$$

Sometimes we can eliminate the squared terms from a pair of equations, and obtain a linear equation. This linear equation can then be solved simultaneously with one of the original equations.

**EXAMPLE 32:** Solve for $x$ and $y$:

$$x^2 + y^2 - 3y = 4$$
$$x^2 + y^2 + 2x - 5y = 2$$

**Solution:** Subtracting the second equation from the first gives

$$-2x + 2y = 2$$

or

$$y = x + 1$$

Substituting $x + 1$ for $y$ in the first equation gives

$$x^2 + (x + 1)^2 - 3(x + 1) = 4$$
$$2x^2 - x - 6 = 0$$

Factoring yields

$$(x - 2)(2x + 3) = 0$$

From which

$$x = 2 \quad \text{and} \quad x = -3/2$$

Substituting back, we have

$$y = 3 \quad \text{and} \quad y = -1/2$$

Figure 13-19 shows the graphs of the solutions.

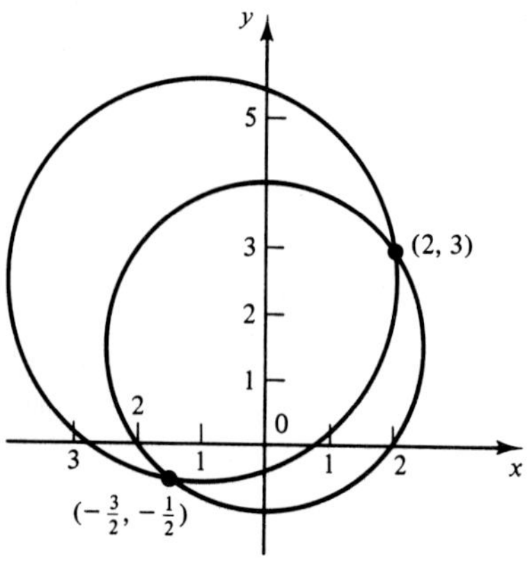


FIGURE 13-19

## EXERCISE 8—SYSTEMS OF QUADRATIC EQUATIONS

Solve for $x$ and $y$.

1. $x - y = 4$
   $x^2 - y^2 = 32$
2. $x + y = 5$
   $x^2 + y^2 = 13$
3. $x + 2y = 7$
   $2x^2 - y^2 = 14$
4. $x - y = 15$
   $x = 2y^2$
5. $x + 4y = 14$
   $y^2 - 2y + 4x = 11$
6. $x^2 + y = 3$
   $x^2 - y = 4$
7. $2x^2 + 3y = 10$
   $x^2 - y = 8$
8. $2x^2 - 3y = 15$
   $5x^2 + 2y = 12$
9. $x + 3y^2 = 7$
   $2x - y^2 = 9$
10. $x^2 - y^2 = 3$
    $x^2 + 3y^2 = 7$
11. $x - y = 11$
    $xy + 28 = 0$
12. $x^2 + y^2 - 8x - 4y + 16 = 0$
    $x^2 + y^2 - 4x - 8y + 16 = 0$

## CHAPTER 13 REVIEW PROBLEMS

*Solve each equation.*

1. $y^2 - 5y - 6 = 0$
2. $2x^2 - 5x = 2$
3. $x^4 - 13x^2 + 36 = 0$
4. $(a + b)x^2 + 2(a + b)x = -a$
5. $w^2 - 5w = 0$
6. $x(x - 2) = 2(-2 - x + x^2)$
7. $\dfrac{r}{3} = \dfrac{r}{r + 5}$
8. $6y^2 + y - 2 = 0$
9. $3t^2 - 10 = 13t$
10. $2.73x^2 + 1.47x - 5.72 = 0$
11. $\dfrac{1}{t^2} + 2 = \dfrac{3}{t}$
12. $3w^2 + 2w - 11 = 0$
13. $9 - x^2 = 0$
14. $\dfrac{2y}{3} = \dfrac{3}{5} + \dfrac{2}{y}$
15. $2x(x + 2) = x(x + 3) + 5$
16. $\dfrac{ax^2}{b} + \dfrac{bx}{c} + \dfrac{c}{a} = 0$
17. $y + 6 = 5y^{1/2}$
18. $18 + w^2 + 11w = 0$
19. $9y^2 + y = 5$
20. $\dfrac{5}{w} - \dfrac{4}{w^2} = \dfrac{3}{5}$
21. $\dfrac{z}{2} = \dfrac{5}{z}$
22. $3x - 6 = \dfrac{5x + 2}{4x}$
23. $2x^2 + 3x = 2$
24. $x^2 + Rx - R^2 = 0$
25. $27x = 3x^2$
26. $\dfrac{n}{n + 1} = \dfrac{2}{n + 3} - 1$
27. $z^{2/3} + 8 = 9z^{1/3}$
28. $6w^2 + 13w + 6 = 0$
29. $5y^2 = 125$
30. $x^2 - 12 = x$
31. $z^2 + 2z - 3 = 0$
32. $\dfrac{t + 2}{t} = \dfrac{4t}{3}$
33. A person purchased some bags of insulation for \$1000. If she had purchased 5 more bags for the same sum, they would have cost 12 cents less per bag. How many did she buy?
34. The perimeter of a rectangular field is 184 ft and its area 1920 ft². Find its dimensions.
35. A rectangular yard (Fig. 13-20) is to be enclosed by fence on three sides, and an existing wall is to form the fourth side. The area of the yard is to be 450 m², and its length is to be twice its width. Find the dimensions of the yard.

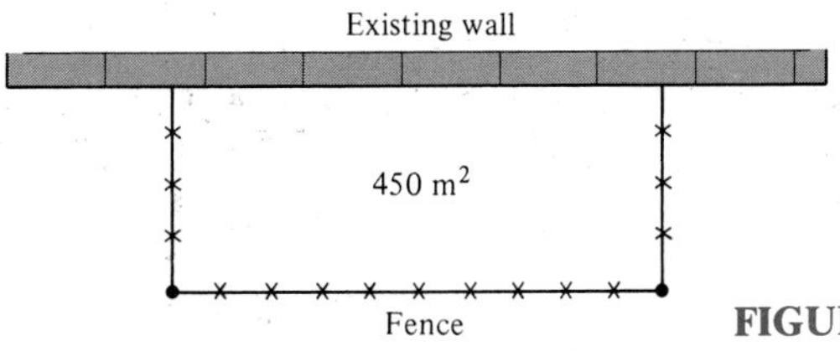


**FIGURE 13-20**

36. Solve for $x$ and $y$.

$$x - y^2 = 7$$
$$x^2 + 3y^2 = 7$$

**37.** If one root is $x = -1$, find the other roots of the equation

$$x^4 - 2x^3 - 13x^2 + 14x + 24 = 0$$

**38.** Plot the parabola $y = 3x^2 + 2x - 6$. Locate the vertex, the axis of symmetry, and any zeros.

**39.** Solve for $x$ and $y$.

$$x^2 + y^2 = 55$$
$$x + y = 7$$

**40.** A fast train runs 8 mi/h faster than a slow train, and takes 3 h less to travel 288 mi. Find the rates of the trains.

**41.** A man started to walk 3 mi, intending to arrive at a certain time. After walking 1 mi, he was detained 10 min and had to walk the rest of the way 1 mi/h faster in order to arrive at the intended time. What was his original speed?

**42.** A rectangular field is 12 m longer than it is wide and contains 448 $m^2$. What are the lengths of its sides?

**43.** A tractor wheel, 15 ft in circumference, revolves in a certain number of seconds. If it revolved in a time longer by 1 s, the tractor would travel 14,400 ft less in 1 h. In how many seconds does it revolve?

**44.** The iron counterweight in Fig. 13-21 is to have its weight increased by 50% by bolting plates of iron along the top and side (but not at the ends). The top plate and the side plate have the same thickness. Find their thickness.

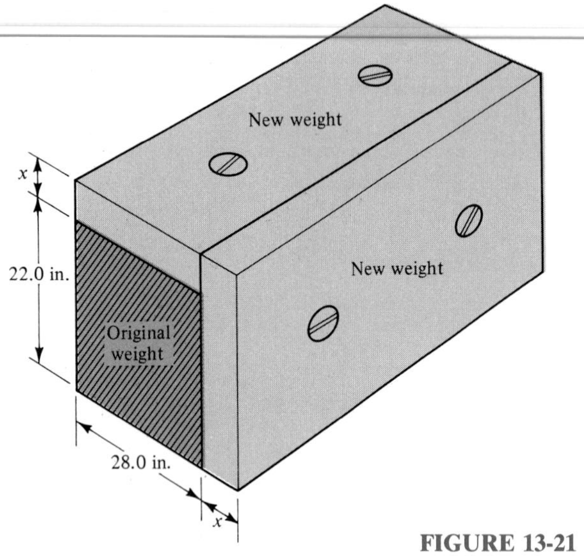


**FIGURE 13-21**

**45.** A 26-in.-wide strip of steel is to have its edges bent up at right angles to form an open trough, as in Fig. 13-22. The cross-sectional area is to be 80 $in^2$. Disregarding the thickness of the steel sheet, find the width and depth of the trough.

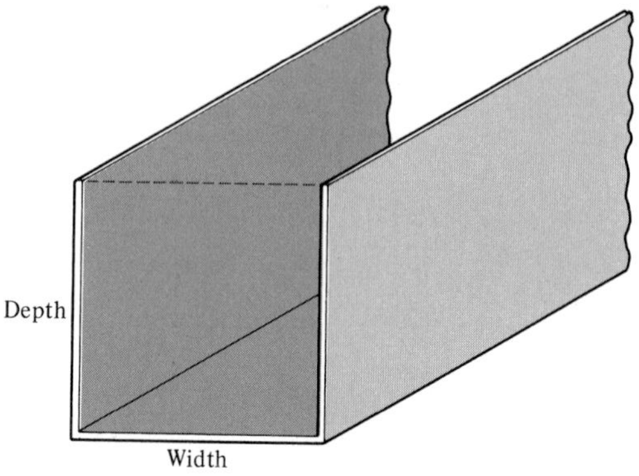


**FIGURE 13-22**

**46.** A boat sails 30 mi at a uniform rate. If the rate had been 1 mi/h less, the time of the sailing would have been 1 h more. Find the rate of travel.

**47.** In a certain number of hours a woman traveled 36 km. If she had traveled 1.5 km more per hour, it would have taken her 3 h less to make the journey. How many kilometers did she travel per hour?

**48.** The length of a rectangular court exceeds its width by 2 m. If the length and width were each increased by 3 m, the area of the court would be 80 $m^2$. Find the dimensions of the court.

**49.** The area of a certain square will double if its dimensions are increased by 6 ft and 4 ft, respectively. Find its dimensions.

**50.** A mirror 18 in. by 12 in. is to set in a frame of uniform width, and the area of the frame is to be equal to that of the glass. Find the width of the frame.

*Writing*

**51.** As in Chapter 3, our writing assignment is to make up a word problem. But now we want a word problem that leads to a quadratic equation. As before, swap with a classmate, solve each other's problem, note anything unclear, unrealistic, or ambiguous, and then rewrite your problem if needed.

# 14

# OBLIQUE TRIANGLES

## OBJECTIVES

**When you have completed this chapter, you should be able to:**

- Identify the algebraic sign of a given trig function for an angle in any quadrant.
- Write the trig function of an angle given a point on the terminal side.
- Determine the reference angle for a given angle.
- Find a trig function for any angle using a calculator.
- Determine the angle(s) given the value of a trig function.
- Solve oblique triangles using the law of sines.
- Solve oblique triangles using the law of cosines.
- Solve applied problems requiring oblique triangles.
- Determine the resultant of two or more vectors.
- Resolve a vector into its components.
- Solve applied problems requiring vectors.

In Chapter 6 we defined the trigonometric functions for angles of any size, but used them only for acute angles. Here we learn how to find the trigonometric functions of obtuse angles, negative angles, angles greater than one revolution, and angles with terminal sides on the coordinate axes. We need the trigonometric functions of obtuse angles to solve oblique triangles, and the trigonometric functions of angles larger than 180° for vectors and other applications.

An *oblique triangle* is one that does not contain a right angle. In this chapter we derive two new formulas, the *law of sines* and the *law of cosines,* to enable us to solve oblique triangles quickly. We cannot, of course, use the methods we derived for solving right triangles (the Pythagorean theorem or the six trigonometric functions) to solve oblique triangles, although we will use these relationships to derive the law of sines and law of cosines.

We also continue our study of vectors in this chapter. In Chapter 6 we dealt with vectors at right angles to each other; here we consider vectors at any angle.

## 14-1 TRIGONOMETRIC FUNCTIONS OF ANY ANGLE

### Definition of the Trigonometric Functions

We defined the trigonometric functions of any angle in Sec. 6-2 but have so far done problems only with acute angles. We turn now to larger angles.

Figure 14-1a–c shows angles in the second, third, and fourth quadrants, and Fig. 14-1d shows an angle greater than 360°. The trigonometric functions of any of these angles are defined exactly as for an acute angle in quadrant I. From any point $P$ on the terminal side of the angle we drop a perpendicular to the $x$ axis, forming a right triangle with legs $x$ and $y$ and with a hypotenuse $r$. The six trigonometric ratios are then given by Eqs. 146 to 151, just as before, except that some of the ratios might now be negative.

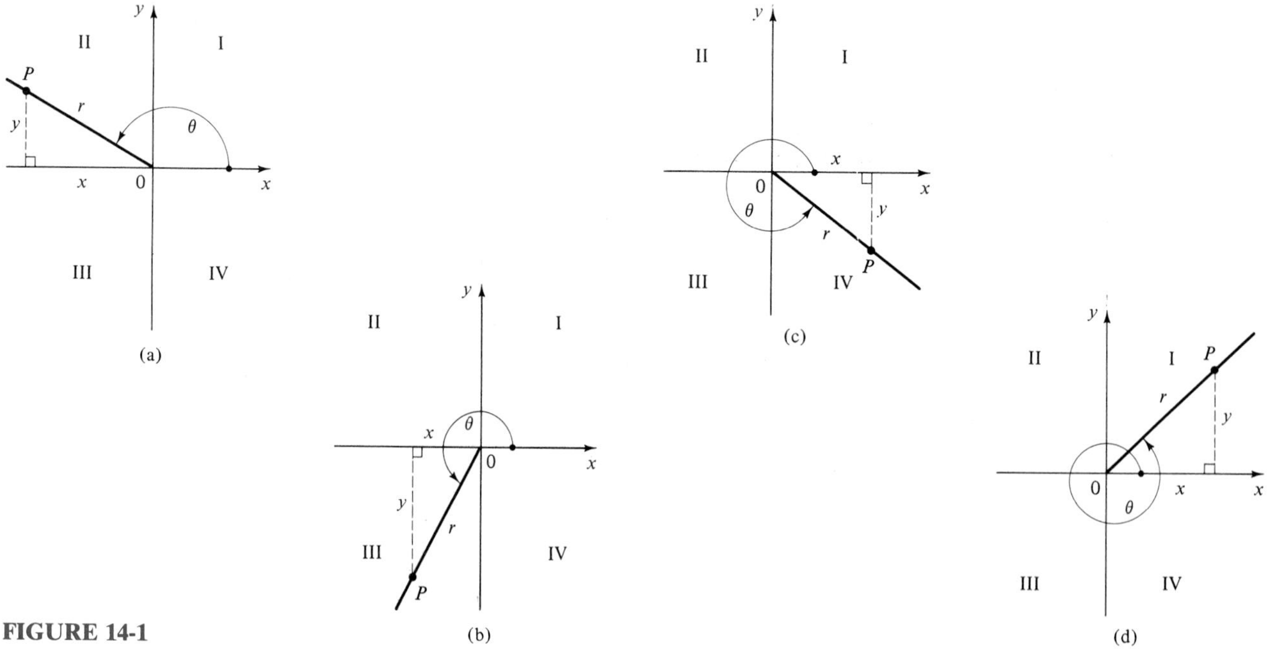


**FIGURE 14-1**

**EXAMPLE 1:** A point on the terminal side of angle $\theta$ has the coordinates $(-3, -5)$. Write the six trigonometric functions of $\theta$ to three significant digits.

**Solution:** We sketch the angle (Fig. 14-2) and see that it lies in the third quadrant. We find distance $r$ by the Pythagorean theorem:

$$r^2 = (-3)^2 + (-5)^2 = 9 + 25 = 34$$

$$r = 5.83$$

Then by Eqs. 146 to 151, with $x = -3$, $y = -5$, and $r = 5.83$,

$$\sin\theta = \frac{y}{r} = \frac{-5}{5.83} = -0.858$$

$$\cos\theta = \frac{x}{r} = \frac{-3}{5.83} = -0.515$$

$$\tan\theta = \frac{y}{x} = \frac{-5}{-3} = 1.67$$

$$\cot\theta = \frac{x}{y} = \frac{-3}{-5} = 0.600$$

$$\sec\theta = \frac{r}{x} = \frac{5.83}{-3} = -1.94$$

$$\csc\theta = \frac{r}{y} = \frac{5.83}{-5} = -1.17$$

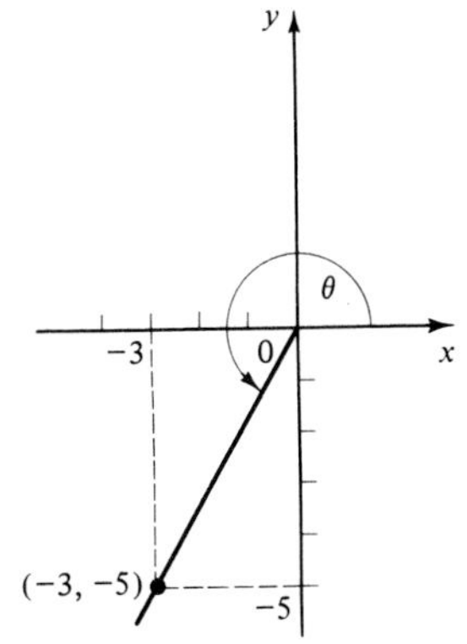

FIGURE 14-2

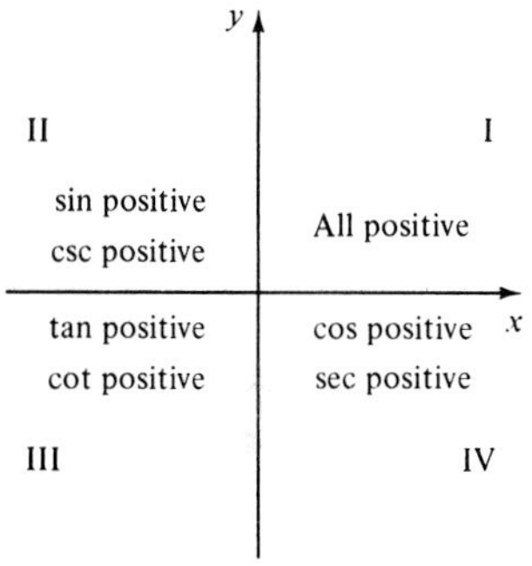

FIGURE 14-3

## Algebraic Signs of the Trigonometric Functions

We saw in Chapter 6 that the trigonometric functions of first quadrant angles were always positive. From the preceding example, it is clear that some of the trigonometric functions of angles in the second, third, and fourth quadrants are negative, because $x$ or $y$ can be negative ($r$ is always positive). Figure 14-3 shows the signs of the trigonometric functions in each quadrant.

Instead of trying to remember which trigonometric functions are negative in which quadrants, just sketch the angle and note whether $x$ or $y$ is negative. From this you can figure out whether the function you want is positive or negative.

**Why bother learning the signs when a calculator gives them to us automatically? One reason is that you will need them when using a *calculator* to find the inverse of a function, as discussed below.**

**EXAMPLE 2:** What is the algebraic sign of csc 315°?

**Solution:** We make a sketch, such as in Fig. 14-4. It is not necessary to draw the angle accurately, but it must be shown in the proper quadrant, quadrant IV in this case. We note that $y$ is negative and that $r$ is (always) positive. So

$$\csc 315° = \frac{r}{y} = \frac{(+)}{(-)} = \text{negative}$$

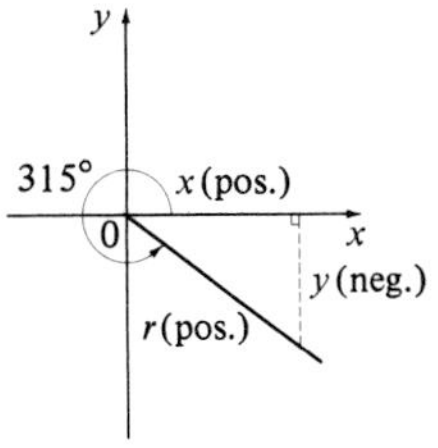

FIGURE 14-4

## Trigonometric Functions of Any Angle by Calculator

Simply enter the angle and press the proper trigonometric key. Be sure that the Degree/Radian switch is in the proper position. The calculator will give the correct algebraic sign.

**EXAMPLE 3:** Find sin 212° to four significant digits.

**Solution:**

| *Press* | *Display* |
|---|---|
| Degree mode 212 [sin] | −0.5299 (rounded) |

For angles in quadrants II, III, or IV, as with acute angles, we use the reciprocal relations,

| | | |
|---|---|---|
| Reciprocal Relationships | $\cot\theta = \dfrac{1}{\tan\theta}$ $\quad \sec\theta = \dfrac{1}{\cos\theta}$ $\quad \csc\theta = \dfrac{1}{\sin\theta}$ | **152** |

to find the cotangent, secant, and cosecant.

**EXAMPLE 4:** Find sec 124° to four significant digits.

**Solution:**

| *Press* | *Display* |
|---|---|
| Degree mode 124 [cos] [1/*x*] | −1.788 (rounded) |

For negative angles, use the change-sign key [CHS] or [+/−] after entering the angle.

**EXAMPLE 5:** Find tan (−35°) to four significant digits.

**Solution:**

| *Press* | *Display* |
|---|---|
| Degree mode 35 [+/−] | −35 |
| [tan] | −0.7002 (rounded) |

Angles greater than 360° are handled in the same manner as any other angle.

**EXAMPLE 6:** Find cos 412° to four significant digits.

**Solution:**

| *Press* | *Display* |
|---|---|
| Degree mode 412 [cos] | 0.6157 (rounded) |

**EXAMPLE 7:** Evaluate the expression

$$(\sin^2 48° + \cos 62°)^3$$

to four significant digits.

**Solution:** The notation $\sin^2 48°$ is the same as $(\sin 48°)^2$. Let us carry five digits and round to four in the last step.

$$(\sin^2 48° + \cos 62°)^3 = [(0.74314)^2 + 0.46947]^3$$

$$= (1.0217)^3 = 1.067$$

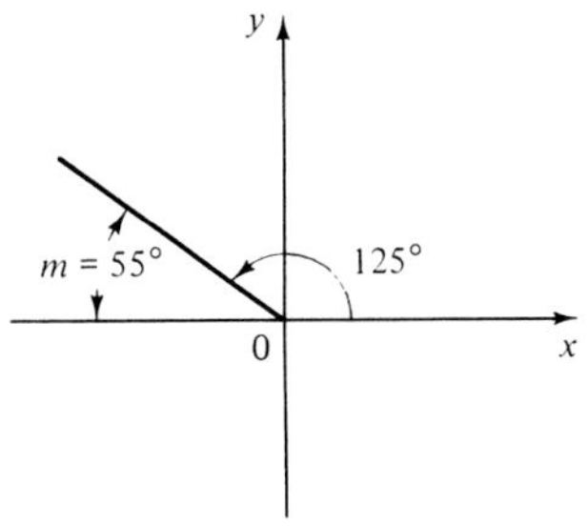


**FIGURE 14-5 Reference angle *m*. It is also called the working angle.**

| | |
|---|---|
| Common Error | Do not confuse an exponent that is on the angle with one that is on the entire function. $(\sin\theta)^2 = \sin^2\theta \neq \sin\theta^2$ |

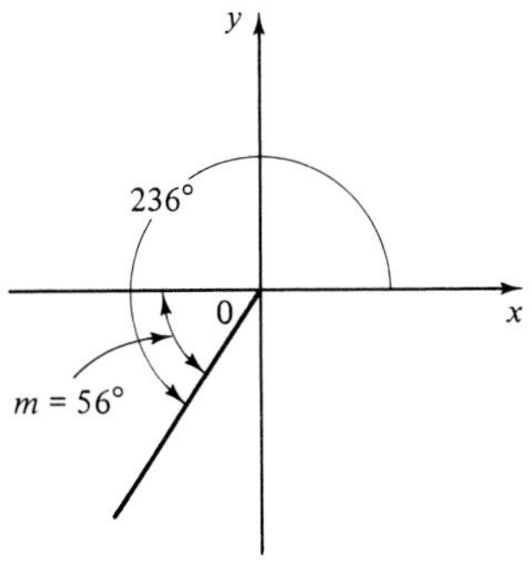


**FIGURE 14-6**

## Reference Angle

Finding the trigonometric function of any angle by calculator is no problem—the calculator does all the thinking. This is not the case, however, when we are given the function and asked to find the angle, or when we have to find trigonometric functions in a table. For both these operations we will make use of the *reference angle*.

For an angle in standard position on coordinate axes, the *acute* angle that its terminal side makes with the $x$ axis is called the *reference angle*.

**EXAMPLE 8:** The reference angle $m$ for an angle of 125° is

$$m = 180 - 125 = 55°$$

as in Fig. 14-5.

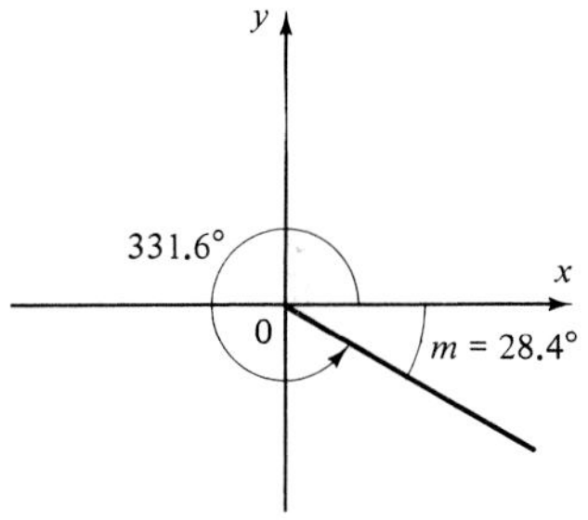


**FIGURE 14-7**

**EXAMPLE 9:** The reference angle $m$ for an angle of 236° is

$$m = 236° - 180° = 56°$$

as in Fig. 14-6.

**EXAMPLE 10:** The reference angle $m$ for an angle of 331.6° is (Fig. 14-7)

$$m = 360 - 331.6 = 28.4°$$

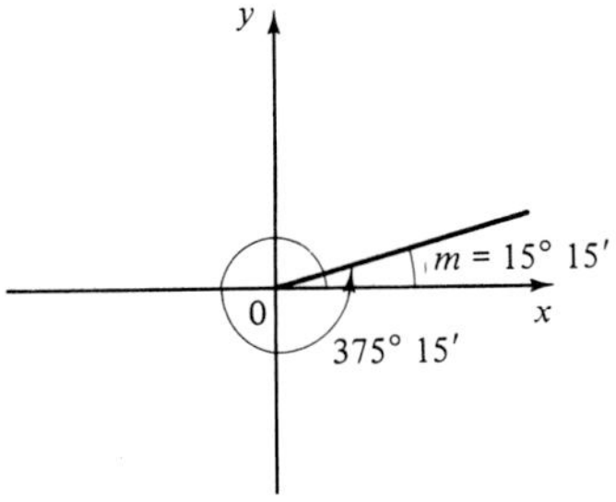


**FIGURE 14-8**

**EXAMPLE 11:** The reference angle $m$ for an angle of 375°15′ is

$$m = 375°15' - 360° = 15°15'$$

as in Fig. 14-8.

| | |
|---|---|
| Common Error | The reference angle is always measured from the *x* axis, but never measured from the *y* axis. |

It may be helpful to remember that for an acute angle *m*:

| | |
|---|---|
| $180° - m$ | is in quadrant II |
| $180° + m$ | is in quadrant III |
| $360° - m$ | is in quadrant IV |

### Finding the Angle When the Function Is Given

We use the same procedure as was given for acute angles. However, a calculator gives us just *one* angle, when we enter the value of a trigonometric function. But we know that there are infinitely many angles that have the same value of that trigonometric function. We must use the angle given by calculator to determine the reference angle, which is then used to compute as many larger angles as we wish having the same trigonometric ratio. Usually, we want only the two positive angles less than 360° that have the required trigonometric function.

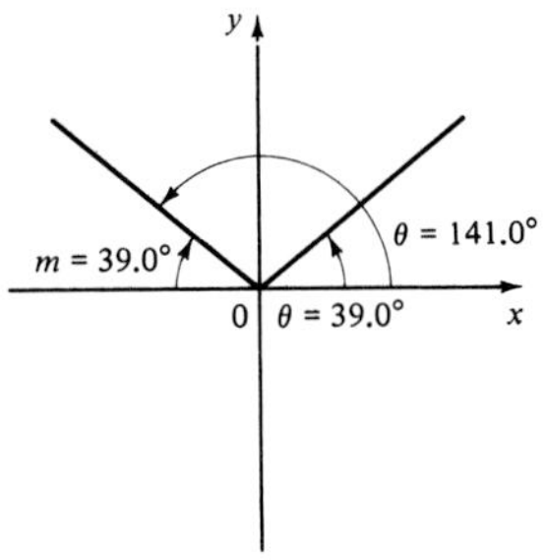


FIGURE 14-9

---

**EXAMPLE 12:** If $\sin\theta = 0.6293$, find two positive values of $\theta$ less than 360°.

**Solution:** Using the arc or inv key on the calculator, or from the tables,

$$m = 39.0°$$

The sine is positive in the first and second quadrants. Our first quadrant angle is 39.0° (Fig. 14-9), and our second quadrant angle is, taking 39.0° as the reference angle,

$$\theta = 180 - 39.0 = 141.0°$$

**Check your work by taking the sine. In this case, sin 141.0° = 0.6293. Checks.**

---

**EXAMPLE 13:** Find the two positive angles less than 360° that have a tangent of $-2.25$.

**Solution:** From the calculator, $\tan^{-1}(-2.25) = -66.0°$, so our reference angle is 66°. The tangent is negative in the second and fourth quadrants. Our second quadrant angle is

$$180 - 66.0° = 114.0°$$

and our fourth quadrant angle is (Fig. 14-10)

$$360 - 66.0° = 294.0°$$

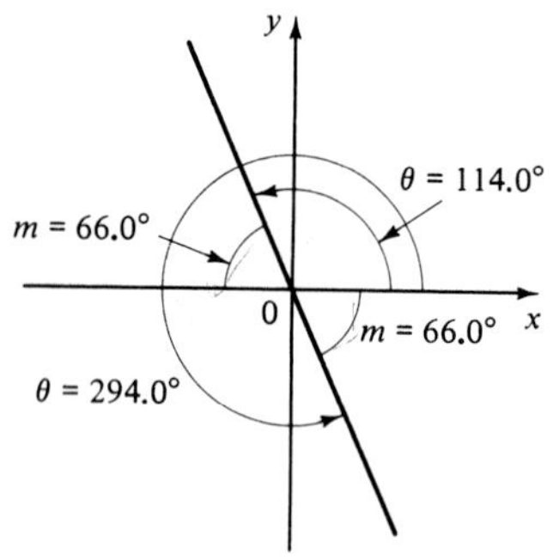


FIGURE 14-10

---

**EXAMPLE 14:** Find $\cos^{-1} 0.575$.

**Solution:** From the calculator,

$$\cos^{-1}0.575 = 54.9°$$

The cosine is positive in the first and fourth quadrants. Our fourth quadrant angle is

$$360° - 54.9° = 305.1°$$

---

| Common Error | Do not confuse the *inverse* with the *reciprocal*. $$\underbrace{\sin^{-1}\theta}_{inverse} \neq \underbrace{(\sin\theta)^{-1} = \frac{1}{\sin\theta}}_{reciprocal}$$ |
|---|---|

When the cotangent, secant, or cosecant is given, we make use of the reciprocal relationships (Eqs. 152) as in the following example.

---

**EXAMPLE 15:** Find two positive angles less than 360° that have a secant of −4.22.

**Solution:** If we let the angle be $\theta$, then, by Eq. 152b,

$$\cos\theta = \frac{1}{\sec\theta} = \frac{1}{-4.22} = -0.237$$

By calculator,

$$\theta = 103.7°$$

The secant is also negative in the third quadrant. Our reference angle (Fig. 14-11) is

$$m = 180 - 103.7 = 76.3°$$

so the third quadrant angle is

$$\theta = 180 + 76.3 = 256.3°$$

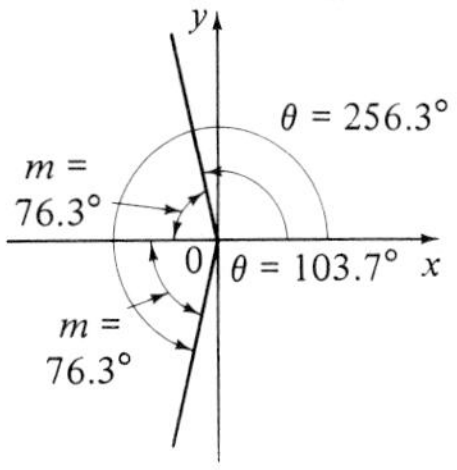


**FIGURE 14-11**

---

**EXAMPLE 16:** Evaluate arcsin (−0.528).

**Solution:** As before, we seek only two positive angles less than 360°. By calculator,

$$\arcsin(-0.528) = -31.9°$$

which is a fourth quadrant angle. As a positive angle, it is

$$\theta = 360° - 31.9° = 328.1°$$

The sine is also negative in the third quadrant. Using 31.9° as our reference angle,

$$\theta = 180° + 31.9° = 211.9°$$

---

## Special Angles

The angles in the following table appear so frequently in problems that it is convenient to be able to write their trigonometric functions from memory.

| *Angle* | *Sin* | *Cos* | *Tan* |
|---|---|---|---|
| 0 | 0 | 1 | 0 |
| 30 | $\frac{1}{2}$ | $\sqrt{3}/2$ (or 0.8660) | $\sqrt{3}/3$ (or 0.5774) |
| 45 | $\sqrt{2}/2$ (or 0.7071) | $\sqrt{2}/2$ (or 0.7071) | 1 |
| 60 | $\sqrt{3}/2$ (or 0.8660) | $\frac{1}{2}$ | $\sqrt{3}$ (or 1.732) |
| 90 | 1 | 0 | undefined |
| 180 | 0 | −1 | 0 |
| 270 | −1 | 0 | undefined |
| 360 | 0 | 1 | 0 |

The angles 0°, 90°, 180°, and 360° are called *quadrantal* angles because the terminal side of each of them lies along one of the coordinate axes. Notice that the tangent is undefined for 90° and 270°, angles whose terminal side is on the $y$ axis. This is because for any point $(x, y)$ on the $y$ axis, the value of $x$ is zero. Since the tangent is equal to $y/x$, we have division by zero, which is not defined.

## EXERCISE 1—TRIGONOMETRIC FUNCTIONS OF ANY ANGLE

### Signs of the Trigonometric Functions

**Assume all angles in this exercise to be in standard position.**

State in what quadrant or quadrants the terminal side of $\theta$ can lie if:

1. $\theta = 123°$
2. $\theta = 272°$
3. $\theta = -47°$
4. $\theta = -216°$
5. $\theta = 415°$
6. $\theta = -415°$
7. $\theta = 845°$
8. $\sin\theta$ is positive
9. $\cos\theta$ is negative
10. $\sec\theta$ is positive
11. $\cos\theta$ is positive and $\sin\theta$ is negative
12. $\sin\theta$ and $\cos\theta$ are both negative
13. $\tan\theta$ is positive and $\csc\theta$ is negative

State whether the following expressions are positive or negative. Do not use your calculator, and try not to refer to your book.

14. sin 174°
15. cos 329°
16. tan 227°
17. sec 332°
18. cot 206°
19. csc 125°
20. sin(−47°)
21. tan(−200°)
22. cos 400°

Give the algebraic signs of the sine, cosine, and tangent of:

23. 110°
24. 206°
25. 335°
26. −48°
27. 500°

### Trigonometric Functions

Sketch the angle and write the six trigonometric functions if the terminal side of the angle passes through the point:

28. (3, 5)
29. (−4, 12)
30. (24, −7)
31. (−15, −8)
32. (0, −3)
33. (5, 0)

Write the sine, cosine, and tangent of $\theta$ if:

Leave your answers for this set in fractional or radical form.

34. $\sin\theta = -\frac{3}{5}$ and $\tan\theta$ is positive.
35. $\cos\theta = -\frac{4}{5}$ and $\theta$ is in the third quadrant.
36. $\csc\theta = -\frac{25}{7}$ and $\cos\theta$ is negative.
37. $\tan\theta = 2$ and $\theta$ is not in the first quadrant.
38. $\cot\theta = -\frac{4}{3}$ and $\sin\theta$ is positive.
39. $\sin\theta = \frac{2}{3}$ and $\theta$ is not in the first quadrant.
40. $\cos\theta = -\frac{7}{9}$ and $\tan\theta$ is negative.

Write, to four significant digits, the sine, cosine, and tangent of each angle.

41. 101°
42. 216°
43. 331°
44. 125.8°
45. −62.85°
46. −227.4°
47. 486°
48. −527°
49. 114°23′
50. 264°15′45″
51. −166°55′
52. 1.15°
53. −11°18′
54. 412°
55. 238°

### Evaluating Trigonometric Expressions

Find the numerical value of each expression to four significant digits.

56. $\sin 35° + \cos 35°$
57. $\sin 125° \tan 225°$
58. $\cos 270° \cos 150° + \sin 270° \sin 150°$
59. $\dfrac{\sin^2 155°}{1 + \cos 155°}$
60. $\sin^2 75°$
61. $\tan^2 125° - \cos^2 125°$
62. $(\cos^2 206° + \sin 206°)^2$
63. $\sqrt{\sin^2 112° - \cos 112°}$

### Inverse Trigonometric Functions

Find, to the nearest tenth of a degree, all positive angles less than 360° whose trigonometric function is:

64. $\sin\theta = \frac{1}{2}$
65. $\tan\theta = -1$
66. $\cot\theta = -\sqrt{3}$
67. $\cos\theta = 0.8372$
68. $\csc\theta = -3.85$
69. $\tan\theta = 6.372$
70. $\cos\theta = -\frac{1}{2}$
71. $\cot\theta = 0$
72. $\cos\theta = -1$

Evaluate.

73. arcsin(−0.736)
74. arcsec 2.85
75. $\cos^{-1} 0.827$
76. arccot 5.22
77. arctan(−4.48)
78. $\csc^{-1} 5.02$

Evaluate from memory. Do not use a calculator or tables.

79. sin 30°
80. cos 45°
81. tan 180°
82. sin 360°
83. tan 45°
84. cos 30°
85. sin 180°
86. cos 360°
87. tan 0°
88. sin 90°
89. cos 180°
90. sin 45°
91. tan 360°
92. cos 270°
93. cos 60°
94. sin 60°
95. cos 0°
96. tan 90°
97. cos 90°

### Computer

98. Figure 14-12 shows a ray of light passing from glass to air. The angle of refraction $R$ is related to the angle of incidence $I$ by Snell's law,

$$\frac{\sin R}{\sin I} = \text{constant}$$

Using a value for the constant (called the index of refraction) of 1.5, write a program or use a spreadsheet that will compute and print angle $R$ for values of $I$ from 0 to 90°. Have the computer inspect $\sin R$ each time it is computed, and when its value exceeds 1.00, stop the computation and print "TOTAL INTERNAL REFLECTION."

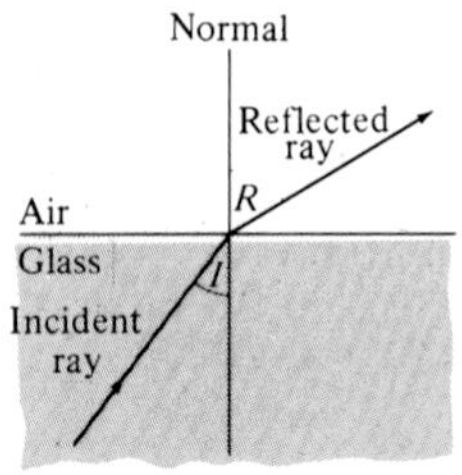


**FIGURE 14-12 Refraction of light.**

## 14-2 LAW OF SINES

### Derivation

We first derive the law of sines for an oblique triangle in which all three angles are acute, such as in Fig. 14-13. We start by breaking the given triangle into two right triangles by drawing altitude $h$ to side $AB$.

In right triangle $ACD$,

$$\sin A = \frac{h}{b} \quad \text{or} \quad h = b \sin A$$

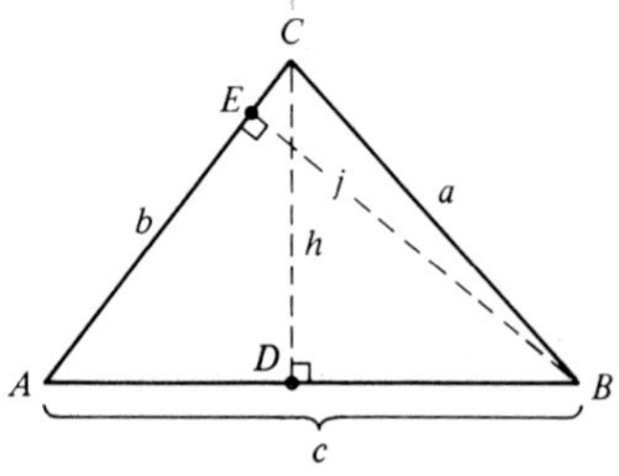


**FIGURE 14-13 Derivation of law of sines.**

In right triangle $BCD$,

$$\sin B = \frac{h}{a} \quad \text{or} \quad h = a \sin B$$

So

$$b \sin A = a \sin B$$

Dividing by $\sin A \sin B$, we have

$$\frac{a}{\sin A} = \frac{b}{\sin B}$$

Similarly, drawing altitude $j$ to side $AC$, and using triangles $BEC$ and $AEB$, we get

$$j = a \sin C = c \sin A$$

or

$$\frac{a}{\sin A} = \frac{c}{\sin C}$$

Combining this with the previous result, we obtain

| Law of Sines | $\dfrac{a}{\sin A} = \dfrac{b}{\sin B} = \dfrac{c}{\sin C}$ | **140** |
|---|---|---|

*The sides of a triangle are proportional to the sines of the opposite angles.*

When one of the angles of the triangle is obtuse, as in Fig. 14-14, the derivation is nearly the same. We draw an altitude $h$ to the extension of side $AB$. Then:

In right triangle $ACD$: $\sin A = \frac{h}{b}$, or $h = b \sin A$

In right triangle $BCD$: $\sin(180 - B) = \frac{h}{a}$

but $\sin(180 - B) = \sin B$, so

$$h = a \sin B$$

The derivation then proceeds exactly as before.

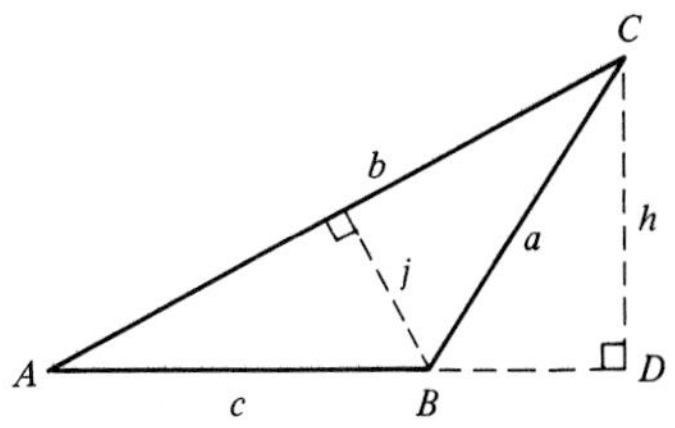


FIGURE 14-14

**A diagram drawn more or less to scale can serve as a good check of your work and reveal inconsistencies in the given data. As another rough check, see that the longest side is opposite the largest angle and that the shortest side is opposite the smallest angle.**

## Solving a Triangle When Two Angles and One Side Are Given (AAS) or (ASA)

Recall that "solving a triangle" means to find all missing sides and angles. Here we use the law of sines to solve an oblique triangle. To do this, we must have *a known side opposite to a known angle*, as well as another angle.

---

**EXAMPLE 17:** Solve triangle $ABC$ where $A = 32.5°$, $B = 49.7°$, and $a = 226$.

**Solution:** We want to find the unknown angle $C$ and sides $b$ and $c$. We sketch the triangle (Fig. 14-15). The missing angle is found by Eq. 139.

$$C = 180 - 32.5 - 49.7 = 97.8°$$

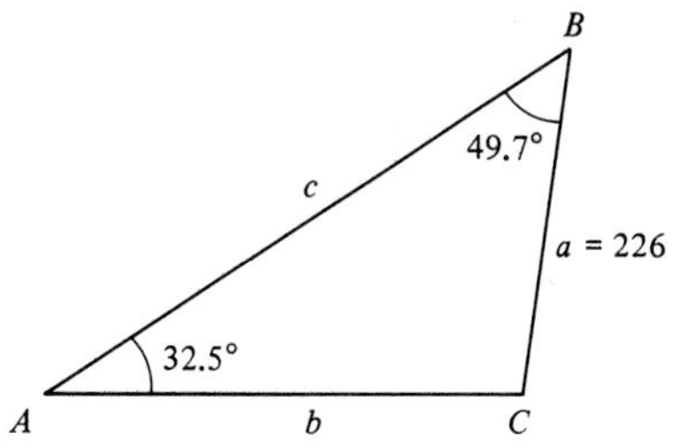


FIGURE 14-15

Then by the law of sines,

$$\frac{a}{\sin A} = \frac{b}{\sin B}$$

$$\frac{226}{\sin 32.5°} = \frac{b}{\sin 49.7°}$$

Solving for $b$, we get

$$b = \frac{226 \sin 49.7°}{\sin 32.5°} = 321$$

Again using the law of sines,

$$\frac{a}{\sin A} = \frac{c}{\sin C}$$

$$\frac{226}{\sin 32.5°} = \frac{c}{\sin 97.8°}$$

So

$$c = \frac{226 \sin 97.8°}{\sin 32.5°} = 417$$

---

Notice that if two angles are given, the third can easily be found, since all angles of a triangle must have a sum of 180°. This means that if two sides and an included angle are given (ASA), the problem can be solved as shown above for AAS.

## Solving a Triangle When Two Sides and One Angle Are Given (SSA): The Ambiguous Case

In Example 17, *two angles* and *one side* were given. We can also use the law of sines when *one angle* and *two sides* are given, provided that the given angle is opposite one of the given sides. But we must be careful here, because we can sometimes get *two solutions,* as in the following example.

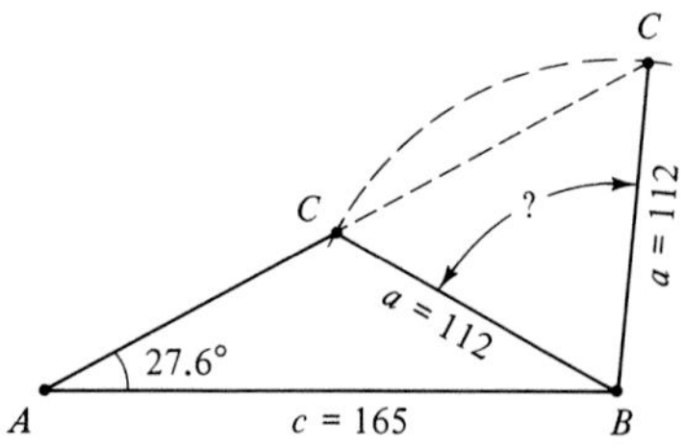


**FIGURE 14-16 The ambiguous case.**

**EXAMPLE 18:** Solve triangle $ABC$ where $A = 27.6°$, $a = 112$, and $c = 165$.

**Solution:** We want to find the unknown angles $B$ and $C$, and side $b$. When we draw the triangle (Fig. 14-16) we see that there are two possible positions for side $a$, both of which satisfy the given data. We will solve both cases. By the law of sines,

$$\frac{\sin C}{165} = \frac{\sin 27.6°}{112}$$

$$\sin C = \frac{165 \sin 27.6°}{112} = 0.6825$$

$$C = 43.0°$$

Recall from Sec. 14-1 that there are two angles less than 360° for which the sine is positive. One of them, $\theta$, is in the first quadrant, and the other, $180° - \theta$, is in the second quadrant.

This is *one* of the possible values for $C$. The other is

$$C = 180° - 43.0° = 137.0°$$

We now find the two corresponding values for side $b$ and angle $B$.

When $C = 43.0°$:

$$B = 180° - 27.6° - 43.0° = 109.4°$$

So

$$\frac{b}{\sin 109.4°} = \frac{112}{\sin 27.6°}$$

from which $b = 228$.

When $C = 137.0°$:

$$B = 180° - 27.6° - 137.0° = 15.4°$$

So

$$\frac{b}{\sin 15.4°} = \frac{112}{\sin 27.6°}$$

from which $b = 64.2$.

| | |
|---|---|
| Common Error | In a problem like the preceding one, it is easy to forget the second possible solution ($C = 137.0°$), especially since a calculator will give only the acute angle when computing the arc sine. |

Not every SSA problem has two solutions, as we'll see in the following example.

**EXAMPLE 19:** Solve triangle $ABC$ if $A = 35.2°$, $a = 525$, and $c = 412$.

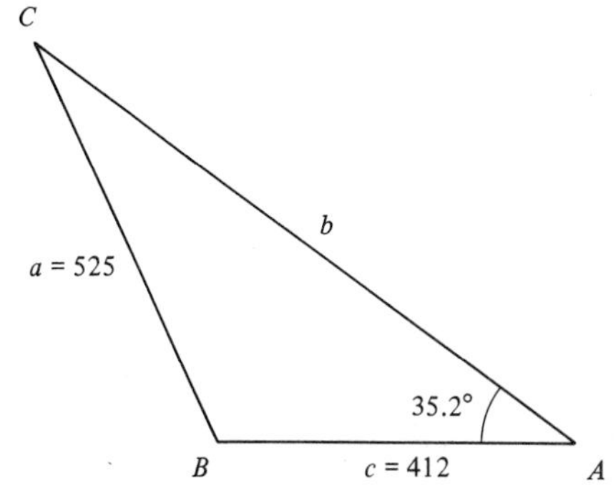


**FIGURE 14-17**

**Solution:** When we sketch the triangle (Fig. 14-17) we see that the given information allows us to draw the triangle in only one way. We will get a unique solution. By the law of sines,

$$\frac{\sin C}{412} = \frac{\sin 35.2°}{525}$$

$$\sin C = \frac{412 \sin 35.2°}{525} = 0.4524$$

Thus the two possible values for $C$ are

$$C = 26.9° \qquad \text{and} \qquad C = 180° - 26.9° = 153.1°$$

Our sketch, even if crudely drawn, shows that $C$ cannot be obtuse, so we discard the 153.1° value. Then, by Eq. 139,

$$B = 180° - 35.2° - 26.9° = 117.9°$$

Using the law of sines once again, we get

**It is sometimes convenient, as in this problem, to use the *reciprocals* of the expressions in the law of sines.**

$$\frac{b}{\sin 117.9°} = \frac{525}{\sin 35.2°}$$

$$b = \frac{525 \sin 117.9°}{\sin 35.2°} = 805$$

## EXERCISE 2—LAW OF SINES

Data for triangle $ABC$ is given in the following table. Solve for the missing parts.

| | Angles | | | Sides | | |
|---|---|---|---|---|---|---|
| | $A$ | $B$ | $C$ | $a$ | $b$ | $c$ |
| **1.** | | 46.25° | | 228 | 304 | |
| **2.** | | 65.9° | | | 1.59 | 1.46 |
| **3.** | | | 43.0° | | 15 | 21 |
| **4.** | 21.45° | 35.17° | | 276.0 | | |
| **5.** | 61.9° | | 47.0° | 7.65 | | |
| **6.** | 126° | 27.0° | | | 119 | |
| **7.** | | 31.6° | 44.8° | | 11.7 | |
| **8.** | 15.0° | | 72.0° | | | 375 |
| **9.** | | 125° | 32° | | | 58 |
| **10.** | 24.14° | 38.27° | | 5562 | | |
| **11.** | | 55.38° | 18.20° | | 77.85 | |
| **12.** | 44.47° | | 63.88° | | | 1.065 |
| **13.** | 18.0° | 12.0° | | 50.7 | | |
| **14.** | | | 45.55° | | 1137 | 1010 |

### Computer

15. Write a program to solve a triangle by the law of sines. Have it *menu-driven* with the following choices.

    1. Two angles and an opposite side given (AAS)
    2. Two sides and one angle given (SSA)
    3. Two angles and the included side given (ASA)

    These triangles are shown in Fig. 14-20. For each, have the program ask for the given sides and angles, and then compute and print the missing sides and angles.

## 14-3 LAW OF COSINES

### Derivation

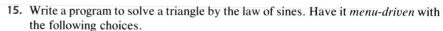

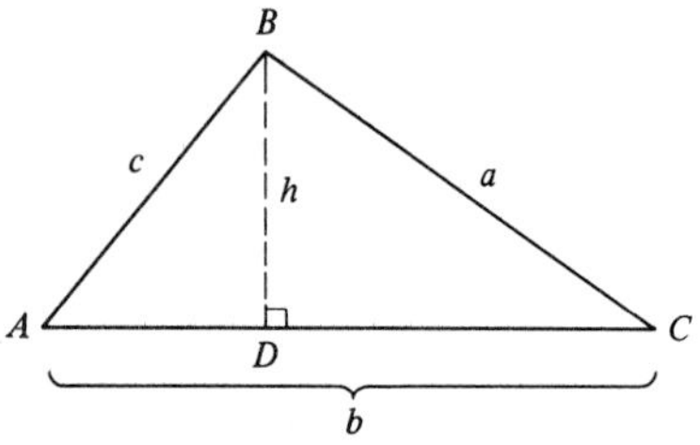


**FIGURE 14-18 Derivation of law of cosines.**

Consider an oblique triangle $ABC$ in Fig. 14-18. As we did for the law of sines, we start by dividing the triangle into two right triangles by drawing an altitude $h$ to side $AC$.

In right triangle $ABD$,

$$c^2 = h^2 + (AD)^2$$

But $AD = b - CD$. Substituting, we get

$$c^2 = h^2 + (b - CD)^2 \qquad (1)$$

Now, in right triangle $BCD$, by Eq. 147,

$$\frac{CD}{a} = \cos C$$

or

$$CD = a \cos C$$

Substituting $a \cos C$ for $CD$ in (1) yields

$$c^2 = h^2 + (b - a \cos C)^2$$

Squaring, we have

$$c^2 = h^2 + b^2 - 2ab \cos C + a^2 \cos^2 C \qquad (2)$$

Let us leave this expression for the moment, and write the Pythagorean theorem for the same triangle $BCD$:

$$h^2 = a^2 - (CD)^2$$

Again substituting $a \cos C$ for $CD$, we obtain

$$h^2 = a^2 - (a \cos C)^2$$
$$= a^2 - a^2 \cos^2 C$$

Substituting this expression for $h^2$ back into (2), we get

$$c^2 = a^2 - a^2 \cos^2 C + b^2 - 2ab \cos C + a^2 \cos^2 C$$

Collecting terms we get the law of cosines,

$$c^2 = a^2 + b^2 - 2ab \cos C$$

We can repeat the derivation, with perpendiculars drawn to side $AB$ and to side $BC$, and get two more forms of the law of cosines:

| | | |
|---|---|---|
| Law of Cosines | $a^2 = b^2 + c^2 - 2bc \cos A$<br>$b^2 = a^2 + c^2 - 2ac \cos B$<br>$c^2 = a^2 + b^2 - 2ab \cos C$ | **141** |

*The square of any side equals the sum of the squares of the other two sides minus twice the product of the other sides and the cosine of the opposite angle.*

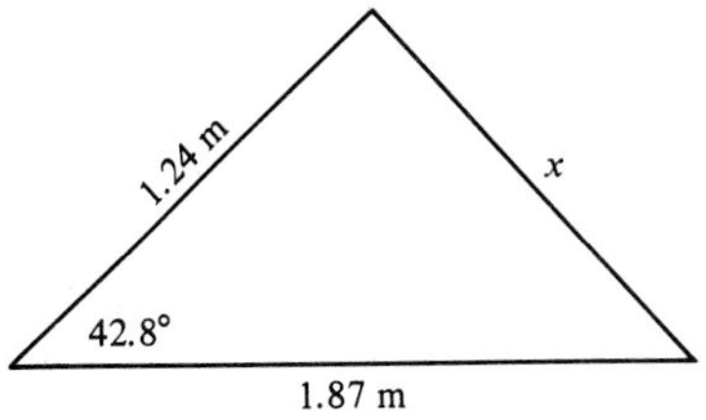


**FIGURE 14-19**

**EXAMPLE 20:** Find side $x$, Fig. 14-19.

**Solution:** By the law of cosines,

$$x^2 = (1.24)^2 + (1.87)^2 - 2(1.24)(1.87) \cos 42.8°$$

$$= 5.03 - 4.64(0.7337) = 1.63$$

$$x = 1.28 \text{ m}$$

## When to Use the Law of Sines or Law of Cosines

It is sometimes not clear whether to use the law of sines or the law of cosines to solve a triangle. We use the law of sines when we have a *known side opposite a known angle*. We use the law of cosines only when the law of sines does not work, that is, for all other cases. In Fig. 14-20, the heavy lines indicate the known information and may help in choosing the proper law.

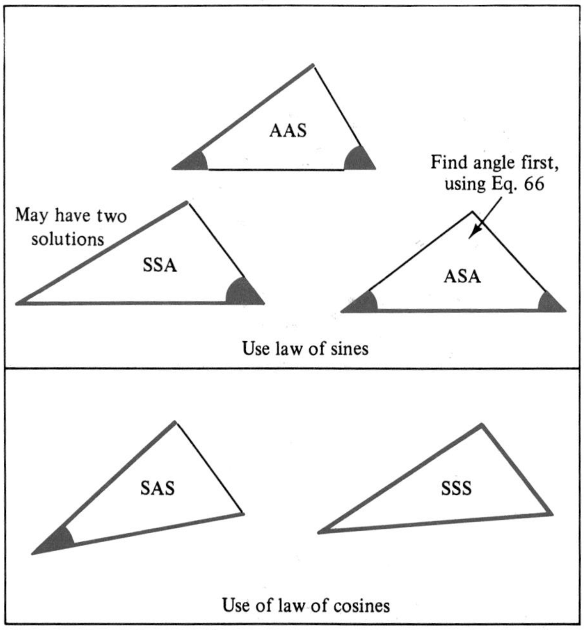


**FIGURE 14-20 When to use the law of sines or law of cosines.**

$a^2 = b^2 + c^2 - 2bc \cos A$

$b^2 = a^2 + c^2 - 2ac \cos B$

$c^2 = a^2 + b^2 - 2ab \cos C$

$-2ab \cos C = a^2 + b^2 - c^2$

$\cos C = \dfrac{a^2 + b^2 - c^2}{-2ab}$

## Using the Cosine Law When Two Sides and the Included Angle Are Known

We can solve triangles by the law of cosines if we know *two sides and the angle between them,* or if we know *three sides*. We consider the first of these in the following example.

---

**EXAMPLE 21:** Solve triangle $ABC$ where $a = 184$, $b = 125$, and $C = 27.2°$.

**Solution:** We make a sketch (Fig. 14-21). Then by the law of cosines,

**Notice that we cannot initially use the law of sines because we do not have a known side opposite a known angle.**

$$c^2 = a^2 + b^2 - 2ab \cos C$$

$$= (184)^2 + (125)^2 - 2(184)(125) \cos 27.2°$$

$$c = \sqrt{(184)^2 + (125)^2 - 2(184)(125) \cos 27.2°}$$

$$= 92.6$$

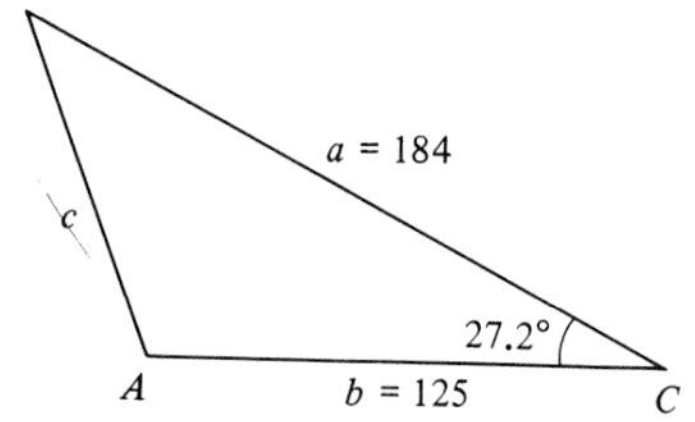

**FIGURE 14-21**

Now that we have a known side opposite a known angle, we can use the law of sines to find angle $A$ or angle $B$. Which shall we find first?

Use the law of sines to *find the acute angle first* (angle $B$ in this example). If, instead, you solve for the obtuse angle first, you may forget to subtract the angle obtained from the calculator from 180°. Further, if one of the angles is so close to 90° that you cannot tell from your sketch if it is acute or obtuse, find the other angle first and then subtract the two known angles from 180° to obtain the third angle.

By the law of sines,

$$\frac{\sin B}{125} = \frac{\sin 27.2°}{92.6}$$

$$\sin B = \frac{125 \sin 27.2°}{92.6} = 0.617$$

$$B = 38.1° \quad \text{and} \quad B = 180 - 38.1 = 141.9°$$

We drop the larger value because our sketch shows us that $B$ must be acute. Then, by Eq. 139,

$$A = 180° - 27.2° - 38.1° = 114.7°$$

---

In our next example, the given angle is *obtuse*.

---

**EXAMPLE 22:** Solve triangle $ABC$ where $b = 16.4$, $c = 10.6$, and $A = 128.5°$.

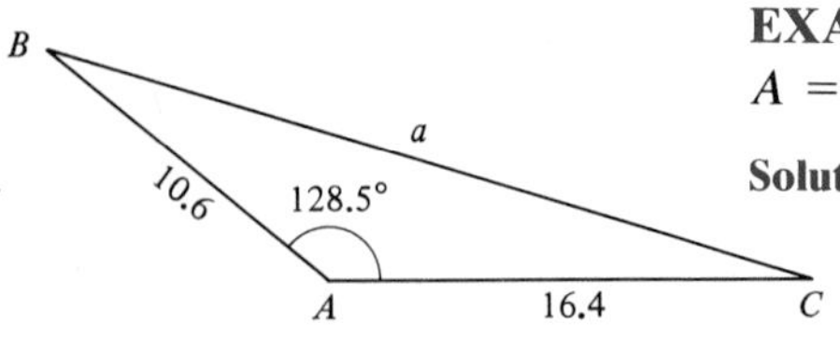

**FIGURE 14-22**

**Solution:** We make a sketch (Fig. 14-22). Then, by the law of cosines,

$$a^2 = (16.4)^2 + (10.6)^2 - 2(16.4)(10.6) \cos 128.5°$$

| Common Error | The cosine of an obtuse angle is *negative*. Be sure to use the proper algebraic sign when applying the law of cosines to an obtuse angle. |
|---|---|

In our example,

$$\cos 128.5° = -0.6225$$

So

$$a = \sqrt{(16.4)^2 + (10.6)^2 - 2(16.4)(10.6)(-0.6225)} = 24.4$$

By the law of sines,

$$\frac{\sin B}{16.4} = \frac{\sin 128.5°}{24.4}$$

$$\sin B = \frac{16.4 \sin 128.5°}{24.4} = 0.526$$

$$B = 31.7° \quad \text{and} \quad B = 180 - 31.7 = 148.3°$$

We drop the larger value because our sketch shows us that $B$ must be acute. Then, by Eq. 139,

$$C = 180° - 31.7° - 128.5° = 19.8°$$

---

## Using the Cosine Law When Three Sides Are Known

When three sides of an oblique triangle are known, we can use the law of cosines to solve for one of the angles. A second angle is found using the law of sines, and the third angle is found by subtracting the other two from 180°.

---

**EXAMPLE 23:** Solve triangle $ABC$ in Fig. 14-23, where $a = 128$, $b = 146$, and $c = 222$.

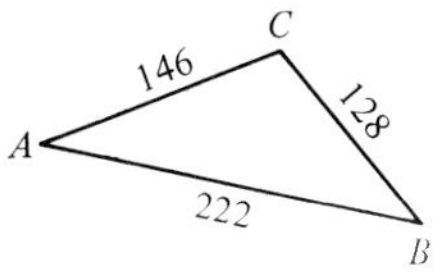


**FIGURE 14-23**

**Solution:** We start by writing the law of cosines for any of the three angles. A good way to avoid ambiguity is to find the largest angle first (the law of cosines will tell us if it is acute or obtuse) and then we are sure that the other two angles are acute.

Writing the law of cosines for angle $C$ gives

$$(222)^2 = (128)^2 + (146)^2 - 2(128)(146) \cos C$$

Solving for cos $C$ gives

$$\cos C = -0.3099$$

Since the cosine is negative, $C$ must be obtuse, so

$$C = 108.1°$$

Then by the law of sines

$$\frac{\sin A}{128} = \frac{\sin 108.1°}{222}$$

from which $\sin A = 0.5482$. Since we know that $A$ is acute, we get

$$A = 33.2°$$

Finally,

$$B = 180° - 108.1° - 33.2° = 38.7°$$

---

## EXERCISE 3—LAW OF COSINES

Data for triangle *ABC* are given in the following table. Solve for the missing parts.

| | *Angles* | | | *Sides* | | |
|---|---|---|---|---|---|---|
| | *A* | *B* | *C* | *a* | *b* | *c* |
| 1. | | | 106.0° | 15.7 | 11.2 | |
| 2. | | 51.4° | | 1.95 | | 1.46 |
| 3. | 68.3° | | | | 18.3 | 21.7 |
| 4. | | | | 728 | 906 | 663 |
| 5. | | | 27.3° | 128 | 152 | |
| 6. | | 128° | | 1.16 | | 1.95 |
| 7. | 135° | | | | 275 | 214 |
| 8. | | | 35.2° | 77.3 | 81.4 | |
| 9. | | | | 11.3 | 15.6 | 12.8 |
| 10. | | 41.73° | | 199 | | 202 |
| 11. | 115.3° | | | | 46.8 | 51.3 |
| 12. | | | | 1.475 | 1.836 | 2.017 |
| 13. | | | 67.0° | 9.08 | 6.75 | |
| 14. | | 129° | | 186 | | 179 |
| 15. | 158° | | | | 1.77 | 1.99 |
| 16. | | | | 41.8 | 57.2 | 36.7 |
| 17. | | | 41.77° | 1445 | 1502 | |
| 18. | | 108.8° | | 7.286 | | 6.187 |
| 19. | | | | 97.3 | 81.4 | 88.5 |
| 20. | 36.29° | | | | 47.28 | 51.36 |

### Computer

21. Write a program to solve a triangle by the law of cosines. Have it *menu-driven* with the following choices.

    1. Two sides and the included angle given (SAS)
    2. Three sides given (SSS)

    These triangles are shown in Fig. 14-20. For each, have the program ask for the given sides and angles, and then compute and print the missing sides and angles.

## 14-4 APPLICATIONS

As with right triangles, oblique triangles have many applications in technology, as you will see in the exercises for this section. Follow the same procedures for setting up these problems as we used for other word problems, and solve the resulting triangle by the law of sines or the law of cosines, or both.

If an *area* of an oblique triangle is needed, either compute all the sides and use Hero's formula (Eq. 138) or find an altitude with right triangle trigonometry and use Eq. 137.

**EXAMPLE 24:** Find the area of the gusset in Fig. 14-24.

**Solution:** We first find $\theta$:

$$\theta = 90° + 35.0° = 125.0°$$

We now know two sides and the angle between them, so can find side $x$ by the law of cosines:

$$x^2 = (18.0)^2 + (20.0)^2 - 2(18.0)(20.0) \cos 125° = 1137$$

$$x = 33.7 \text{ in.}$$

We find the area of the gusset by Hero's formula (Eq. 138),

$$s = \tfrac{1}{2}(18.0 + 20.0 + 33.7) = 35.9$$

$$\text{area} = \sqrt{35.9(35.9 - 18.0)(35.9 - 20.0)(35.9 - 33.7)}$$

$$= 150 \text{ in}^2.$$

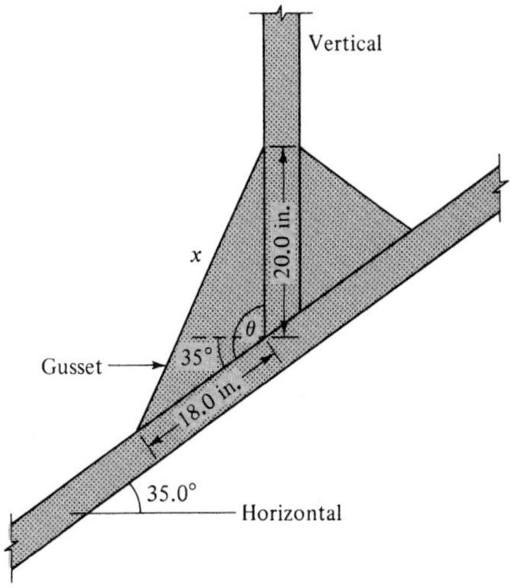


**FIGURE 14-24**

**EXAMPLE 25:** A ship takes a sighting on two buoys. At a certain instant the bearing of buoy $A$ is N 44.23° W and of buoy $B$ is N 62.17° E. The distance between the buoys is 3.60 km and bearing of $B$ from $A$ is N 87.87° E. Find the distance of the ship from each buoy.

**Solution:** We draw a diagram (Fig. 14-25). Notice how the compass directions are laid out, starting from north and turning in the indicated direction. Finding the angles of the triangle $ABS$,

$$\theta = 44.23° + 62.17° = 106.40°$$

And since angle $SAC$ is 44.23°:

$$\alpha = 180° - 87.87° - 44.23° = 47.90°$$

and

$$\beta = 180° - 106.40° - 47.90° = 25.70°$$

From the law of sines,

$$\frac{SA}{\sin 25.70°} = \frac{SB}{\sin 47.90°} = \frac{3.60}{\sin 106.40°}$$

from which

$$SA = \frac{3.60 \sin 25.70°}{\sin 106.40°} = 1.63 \text{ km}$$

and

$$SB = \frac{3.60 \sin 47.90°}{\sin 106.40°} = 2.78 \text{ km}$$

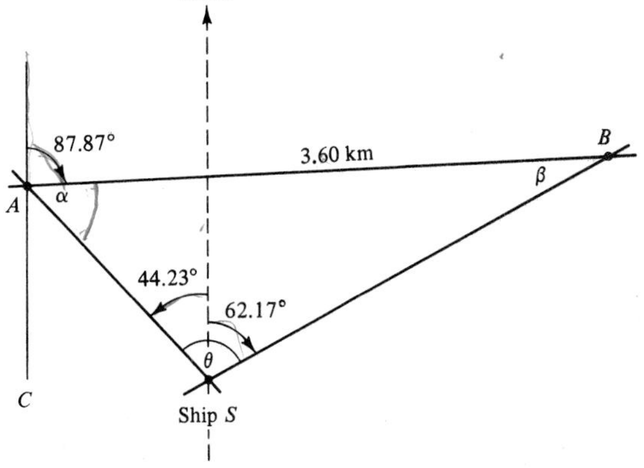


**FIGURE 14-25**

### Determining Inaccessible Distances

**This set of problems contains no vectors, which are covered in the next section.**

**Remember that angles of elevation or depression are always measured from the *horizontal.***

1. Two stakes, $A$ and $B$, are 88.6 m apart. From a third stake $C$, the angle $ACB$ is 85.4°, and from $A$, the angle $BAC$ is 74.3°. Find the distance from $C$ to each of the other stakes.
2. From a point on level ground, the angles of elevation of the top and the bottom of an antenna standing on top of a building are 32.6° and 27.8°, respectively. If the building is 125 ft high, how tall is the antenna?
3. A triangular lot measures 115 m, 187 m, and 215 m along its sides. Find the angles between the sides.
4. Two boats are 45.5 km apart. Both are traveling toward the same point, which is 87.6 km from one of them and 77.8 km from the other. Find the angle at which their paths intersect.

### Navigation

***Hint:* Draw your diagram after $t$ hours have elapsed.**

5. A ship is moving at 15 km/h in the direction N 15° W. A helicopter with a speed of 22 km/h is due east of the ship. In what direction should the helicopter travel if it is to meet the ship?
6. City $A$ is 215 miles N 12° E from city $B$. The bearing of city $C$ from $B$ is S 55° E. The bearing of $C$ from $A$ is S 15° E. How far is $C$ from $A$? From $B$?
7. A ship is moving in a direction S 24.25° W at a rate of 8.6 mi/h. If a launch that travels at 15.4 mi/h is due west of the ship, in what direction should it travel in order to meet the ship?
8. A ship is 9.5 km directly east of a port. If the ship sails southeast for 2.5 km, how far will it be from the port?
9. From a plane flying due east, the bearing of a radio station is S 31° E at 1 P.M. and S 11° E at 1:20 P.M. The ground speed of the plane is 625 km/h. Find the distance of the plane from the station at 1 P.M.

### Structures

10. A tower stands vertically on sloping ground whose inclination with the horizontal is 11.6°. From a point 42.0 m downhill from the tower (measured along the slope) the angle of elevation of the top of the tower is 18.8°. How tall is the tower?
11. A vertical antenna stands on a slope that makes an angle of 8.7° with the horizontal. From a point directly uphill from the antenna the angle of elevation of its top is 61°. From a point 16 m farther up the slope (measured along the slope) the angle of elevation of its top is 38°. How tall is the antenna?
12. A power pole on level ground is supported by two wires that run from the top of the pole to the ground (Fig. 14-26). One wire is 18.5 m long and makes an angle of 55.6° with the ground, and the other wire is 17.8 m long. Find the angle the second wire makes with the ground.
13. A 71.6-m-high antenna mast (Fig. 14-27) is to be placed on sloping ground with the cables making an angle of 42.5° with the top of the mast. Find the lengths of the two cables.

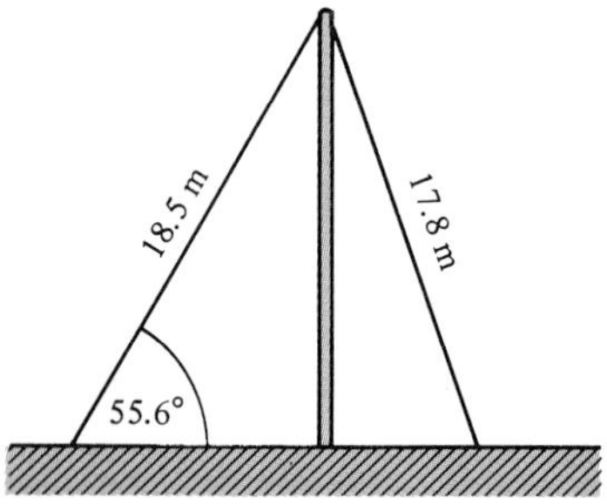


**FIGURE 14-26**

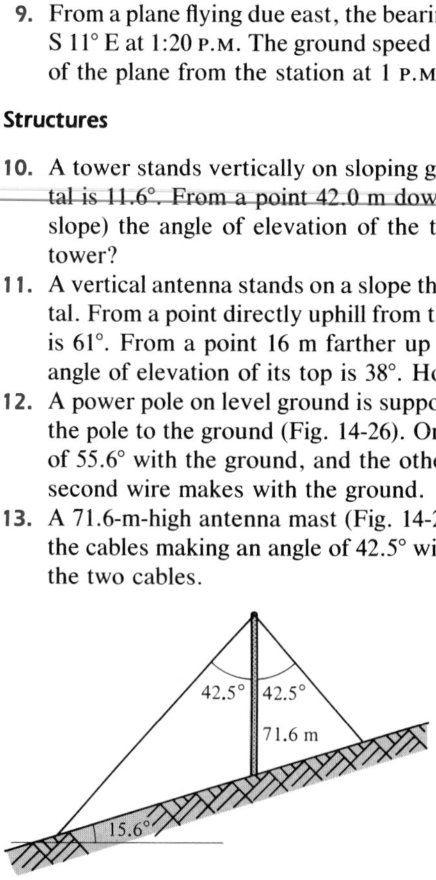


**FIGURE 14-27**

14. In the roof truss in Fig. 14-28 find the lengths of members $AB$, $BD$, $AC$, and $AD$.

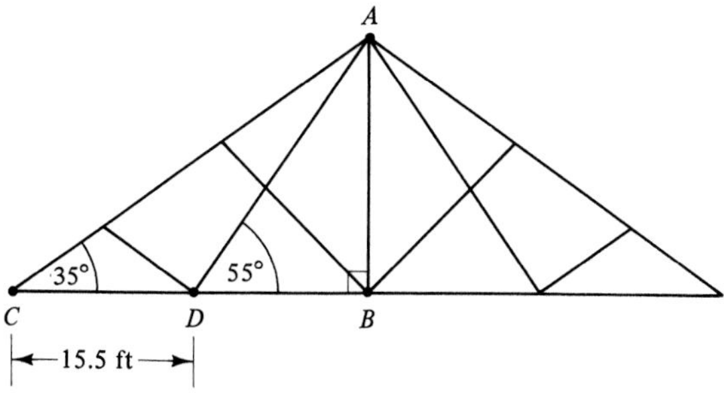


**FIGURE 14-28 Roof truss.**

15. From a point on level ground between two power poles of the same height, cables are stretched to the top of each pole. One cable is 52.6 ft long, the other is 67.5 ft long, and the angle of intersection between the two cables is 125°. Find the distance between the poles.
16. A pole standing on level ground makes an angle of 85.8° with the horizontal. The pole is supported by a 22-ft prop whose base is 12.5 ft from the base of the pole. Find the angle made by the prop with the horizontal.

### Mechanisms

17. In the slider crank mechanism of Fig. 14-29, find the distance $x$ between the wrist pin $W$ and the crank center $C$, when $\theta = 35.7°$.

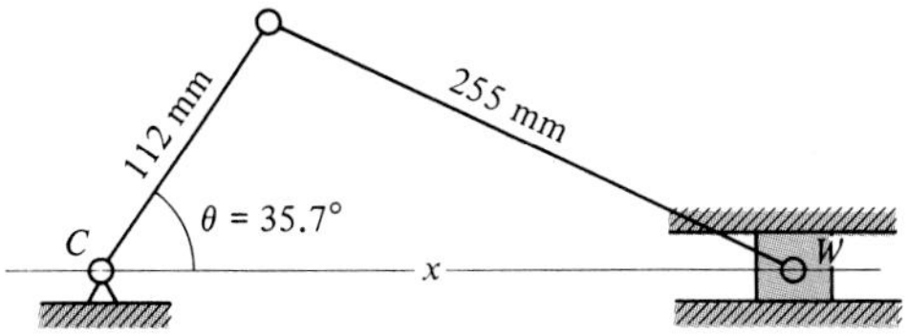


FIGURE 14-29

18. In the four-bar linkage of Fig. 14-30, find angle $\theta$ when angle $BAD$ is 41.5°.

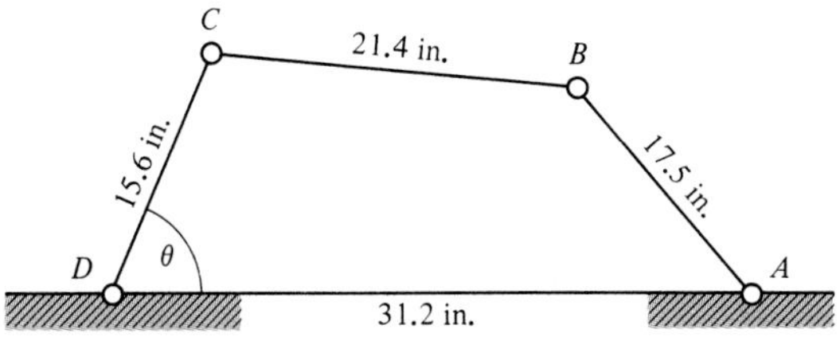


FIGURE 14-30

19. Two links, $AC$ and $BC$ (Fig. 14-31) are pivoted at $C$. How far apart are $A$ and $B$ when angle $ACB$ is 66.3°?

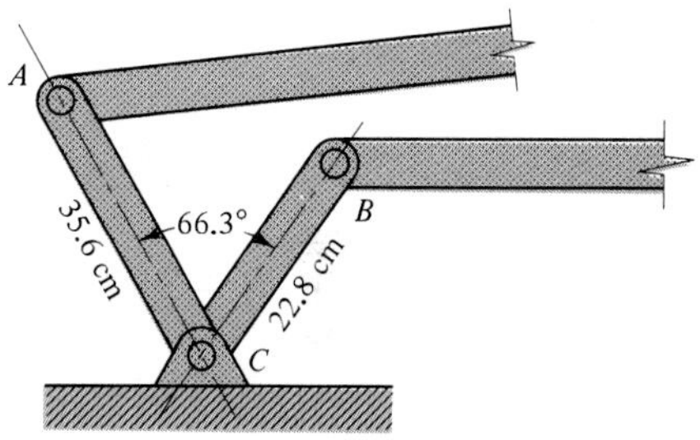


FIGURE 14-31

### Geometry

20. Find angles $A$, $B$, and $C$ in the quadrilateral in Fig. 14-32.
21. Find side $AB$ in the quadrilateral in Fig. 14-33.
22. Two sides of a parallelogram are 22.8 and 37.8 m and one of the diagonals is 42.7 m. Find the angles of the parallelogram.
23. Find the lengths of the sides of a parallelogram if its diagonal, which is 125 mm long, makes angles with the sides of 22.7° and 15.4°.
24. Find the lengths of the diagonals of a parallelogram, two of whose sides are 3.75 m and 1.26 m; their included angle is 68.4°.
25. In triangle $ABC$, $A = 62.3°$, $b = 112$, and the median from $C$ to the midpoint of $c$ is 186. Find $c$.

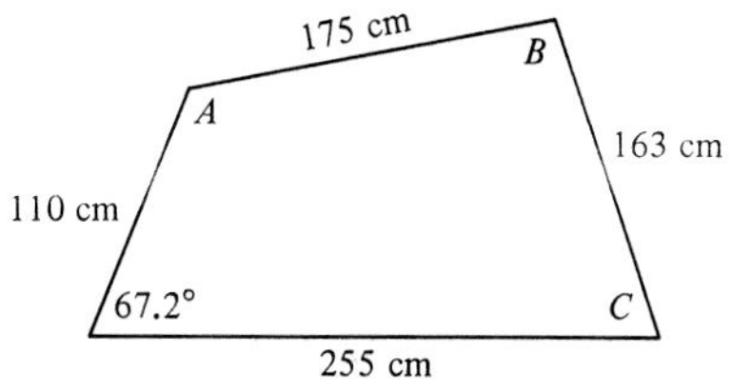


FIGURE 14-32

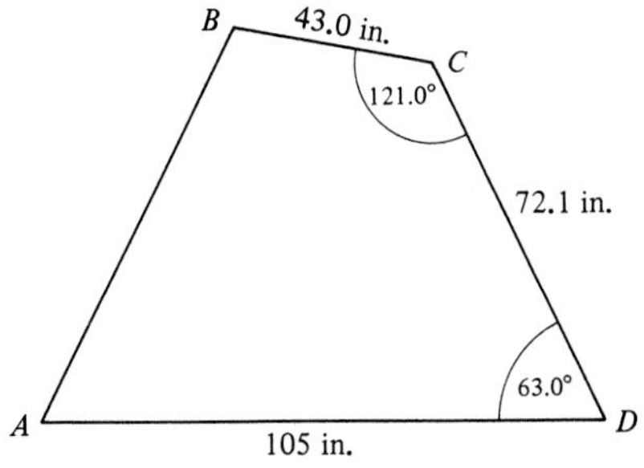


FIGURE 14-33

**A *median* of a triangle is a line joining a vertex to the midpoint of the opposite side.**

26. The sides of a triangle are 124, 175, and 208. Find the length of the median drawn to the longest side.
27. The angles of a triangle are in the ratio 3 : 4 : 5 and the shortest side is 994. Solve the triangle.
28. The sides of a triangle are in the ratio 2 : 3 : 4. Find the cosine of the largest angle.
29. Two solar panels are to be placed as in Fig. 14-34. Find the minimum distance $x$ so that the first panel will not cast a shadow on the second when the angle of elevation of the sun is 18.5°.

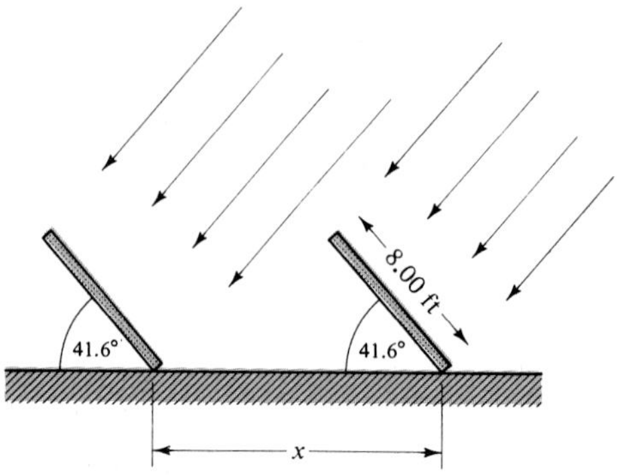

**FIGURE 14-34 Solar panels.**

30. Find the overhang $x$ so that the window in Fig. 14-35 will be in complete shade when the sun is 60° above the horizontal.

### Computer

31. Combine your triangle-solving programs from Exercises 2 and 3 into a single program to solve any triangle. The menu should contain the five options shown in Fig. 14-20. Also compute the area of the triangle by Hero's formula.
32. For the slider crank mechanism of Fig. 14-29, compute and print the position $x$ of the wrist pin for values of $\theta$ from 0 to 180°, in 15° steps.

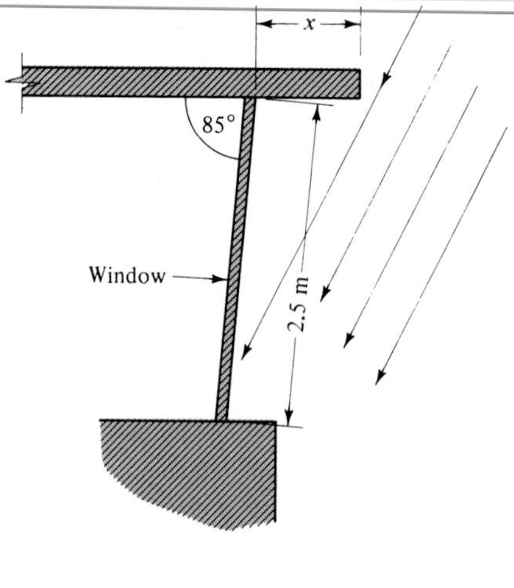

**FIGURE 14-35**

## 14-5 ADDITION OF VECTORS

We can add two nonperpendicular vectors by first resolving each into components, then adding the components, and finally resolving the components into a single resultant. But, by using the law of sines and the law of cosines, we can combine two nonperpendicular vectors directly, with much less time and effort. However, when more than two vectors must be added, it is faster to resolve each into its $x$ and $y$ components, combine the $x$ components and the $y$ components, and then find the resultant of those two perpendicular vectors.

In this section we show both methods.

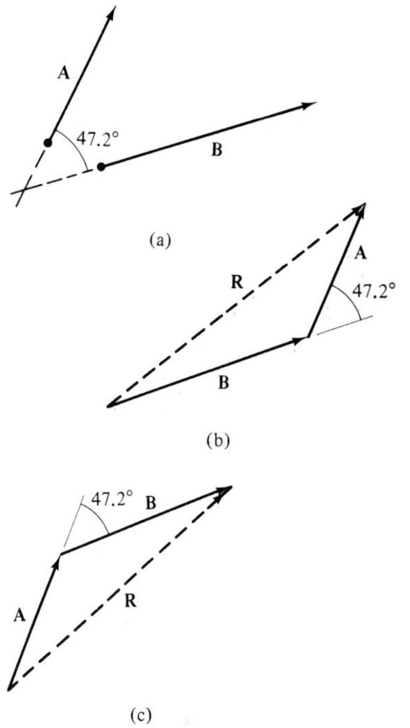


**FIGURE 14-36 Addition of vectors.**

## Vector Diagram

We can illustrate the resultant, or vector sum, of two vectors by means of a diagram. Suppose that we wish to add the vectors **A** and **B** (Fig. 14-36a). If we draw the two vectors *tip to tail,* as in Fig. 14-36b, the resultant **R** will be the vector that will complete the triangle when drawn from the tail of the first vector to the tip of the second vector. It does not matter whether vector **A** or vector **B** is drawn first; the same resultant will be obtained either way, as shown in Fig. 14-36c.

The *parallelogram method* will give the same result. To add the same two vectors **A** and **B** as before, we first draw the given vectors *tail to tail* (Fig. 14-37) and complete a parallelogram by drawing lines from the tips of the given vectors, parallel to the given vectors. The resultant **R** is then the diagonal of the parallelogram drawn from the intersections of the tails of the original vectors.

## Finding the Resultant of Two Nonperpendicular Vectors

Whichever method we use for drawing the two vectors, *the resultant is one side of an oblique triangle.* To find the length of the resultant and the angle it makes with one of the original vectors, we simply solve the oblique triangle by the methods we learned earlier in this chapter.

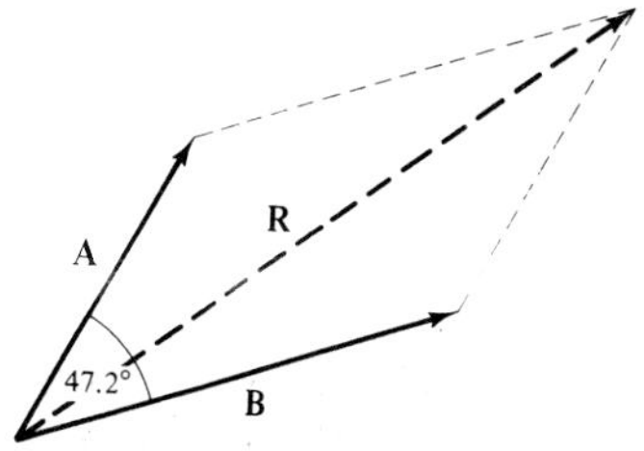


**FIGURE 14-37 Parallelogram method.**

---

**EXAMPLE 26:** Two vectors, **A** and **B**, make an angle of 47.2° with each other (Fig. 14-37). If their magnitudes are $A = 125$ and $B = 146$, find the magnitude of the resultant **R** and the angle **R** makes with vector **B**.

**Solution:** We make a vector diagram, either tip to tail or by the parallelogram method. Either way, we must solve the oblique triangle in Fig. 14-38 for $R$ and $\phi$. Finding $\theta$ yields

$$\theta = 180° - 47.2° = 132.8°$$

By the law of cosines,

$$R^2 = (125)^2 + (146)^2 - 2(125)(146)\cos 132.8° = 61{,}740$$

$$R = 248$$

Then, by the law of sines,

$$\frac{\sin\phi}{125} = \frac{\sin 132.8}{248}$$

$$\sin\phi = \frac{125 \sin 132.8}{248} = 0.3698$$

$$\phi = 21.7°$$

---

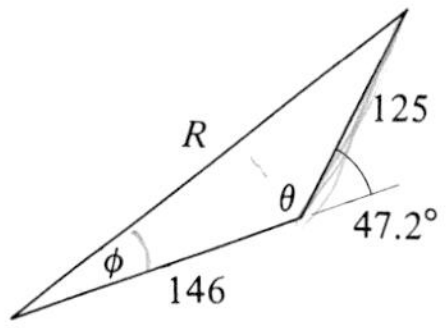


**FIGURE 14-38**

## Addition of Several Vectors

The law of sines and the law of cosines are good for adding *two* nonperpendicular vectors. However, when *several* vectors are to be added, we usually break each into its $x$ and $y$ components and combine them, as in the following example.

**EXAMPLE 27:** Find the resultant of the vectors shown in Fig. 14-39a.

**Solution:** The $x$ component of a vector of magnitude $V$ at an angle $\theta$ is

$$V \cos \theta$$

and the $y$ component is

$$V \sin \theta$$

We compute and tabulate the $x$ and $y$ components of each original vector, and find the sums of each.

| Vector | *x component* | | *y component* | |
|---|---|---|---|---|
| **A** | 42.0 cos 58.0° = | 22.3 | 42.0 sin 58.0° = | 35.6 |
| **B** | 56.1 cos 148° = | −47.6 | 56.1 sin 148° = | 29.7 |
| **C** | 52.7 cos 232° = | −32.4 | 52.7 sin 232° = | −41.5 |
| **D** | 45.3 cos 291° = | 16.2 | 45.3 sin 291° = | −42.3 |
| **R** | $R_x$ = | −41.5 | $R_y$ = | −18.5 |

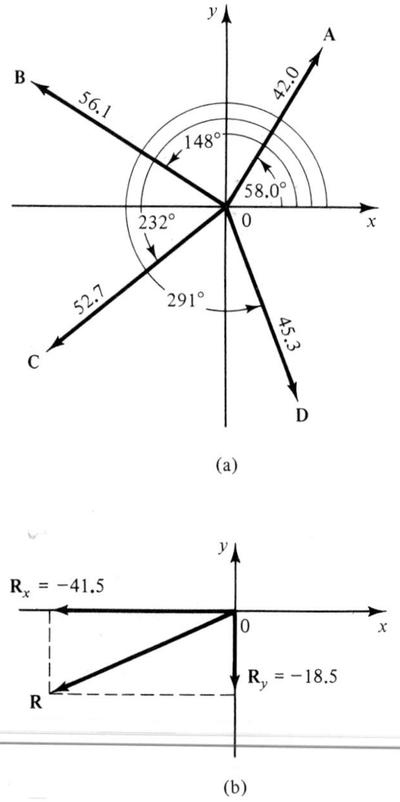


FIGURE 14-39

The two vectors $\mathbf{R}_x$ and $\mathbf{R}_y$ are shown in Fig. 14-39b. We find their resultant **R** by the Pythagorean theorem,

$$R^2 = (-41.5)^2 + (-18.5)^2 = 2065$$

$$R = 45.4$$

and the angle $\theta$ by

$$\theta = \arctan \frac{R_y}{R_x}$$

$$= \arctan \frac{-18.5}{-41.5}$$

$$= 24.0° \quad \text{or} \quad 204°$$

Since our resultant is in the third quadrant, we drop the 24.0° value. Thus, the resultant has a magnitude of 45.4 and a direction of 204°. This is often written in the form

$$\mathbf{R} = 45.4 \,\underline{/204°}$$

This is called *polar form,* which we'll cover in Chap. 16.

## EXERCISE 5—ADDITION OF VECTORS

The magnitudes of vectors **A** and **B** are given in the table, as well as the angle between the vectors. For each, find the magnitude $R$ of the resultant and the angle that the resultant makes with vector **B**.

| | *Magnitudes* | | |
|---|---|---|---|
| | *A* | *B* | *Angle* |
| **1.** | 244 | 287 | 21.8° |
| **2.** | 1.85 | 2.06 | 136° |
| **3.** | 55.9 | 42.3 | 55.5° |
| **4.** | 1.006 | 1.745 | 148.4° |
| **5.** | 4483 | 5829 | 100° |
| **6.** | 35.2 | 23.8 | 146° |

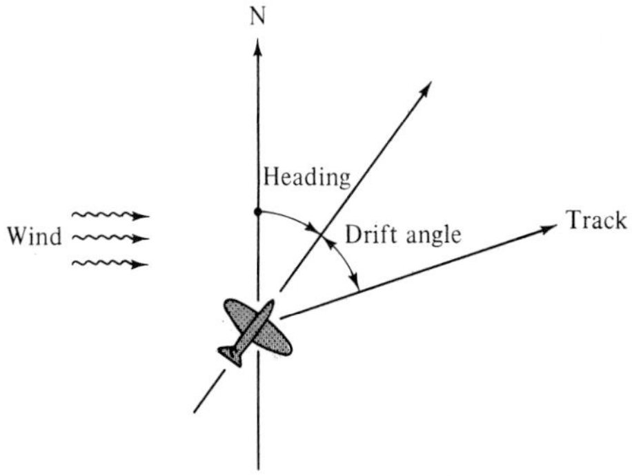


**FIGURE 14-40 Flight terminology. The *heading* of an aircraft is the direction in which the craft is pointed. Due to air currents, it usually will not travel in that direction but in an actual path called the *track*. The angle between the heading and the track is the *drift angle*. The *airspeed* is the speed relative to the surrounding air, and the *ground speed* is the craft's speed relative to the earth.**

### Force Vectors

**7.** Two forces of 18.6 N and 21.7 N are applied to a point on a body. The angle between the forces is 44.6°. Find the magnitude of the resultant and the angle it makes with the larger force.

**8.** Two forces whose magnitudes are 187 lb and 206 lb act on an object. The angle between the forces is 88.4°. Find the magnitude of the resultant force.

**9.** A force of 125 N pulls due west on a body, and a second force pulls N 28°44′ W. The resultant force is 212 N. Find the second force and the direction of the resultant.

**10.** Forces of 675 lb and 828 lb act on a body. The smaller force acts due north; the larger force acts N 52.3° E. Find the direction and magnitude of the resultant.

**11.** Two forces, of 925 N and 1130 N, act on an object. Their lines of action make an angle of 67.2° with each other. Find the magnitude and direction of their resultant.

**12.** Two forces, of 136 lb and 251 lb, act on an object, with an angle of 53.9° between their lines of action. Find the magnitude of their resultant and its direction.

**13.** The resultant of two forces of 1120 N and 2210 N is 2870 N. What angle does the resultant make with each of the two forces?

**14.** Three forces are in equilibrium: 212 N, 325 N, and 408 N. Find the angles between their lines of action.

### Velocity Vectors

See Fig. 14-40 for definitions of the terms used in these problems.

**15.** As an airplane heads west with an air speed of 325 mi/h, a wind with a speed of 35 mi/h causes the plane to travel slightly south of west with a ground speed of 305 mi/h. In what direction is the wind blowing? In what direction does the plane travel?

**16.** A boat heads S 15° E on a river that flows due west. The boat travels S 11° W with a speed of 25 km/h. Find the speed of the current and the speed of the boat in still water.

**17.** A pilot wishes to fly in the direction N 45° E. The wind is from the west at 36 km/h and the plane's speed in still air is 388 km/h. Find the heading and the ground speed.

**18.** The heading of a plane is N 27°44′ E, and its air speed is 255 mi/h. If the wind is blowing from the south with a velocity of 42 mi/h, find the actual direction of travel of the plane and its ground speed.

**19.** A plane flies with a heading of N 48° W and an air speed of 584 km/h. It is driven from its course by a wind of 58 km/h from S 12° E. Find the ground speed and the drift angle of the plane.

### Current and Voltage Vectors

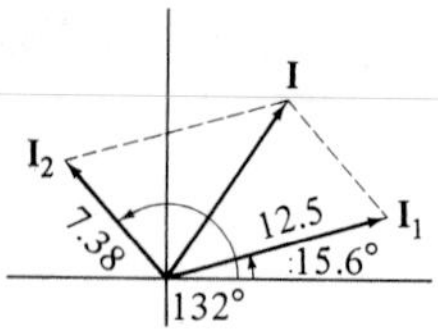


FIGURE 14-41

20. We will see later that it is possible to represent an alternating current or voltage by a vector whose length is equal to the maximum amplitude of the current or voltage, placed at an angle that we later define as the *phase angle*. Then to add two alternating currents or voltages, we *add the vectors* representing those voltages or currents in the same way that we add force or velocity vectors.
A current $\mathbf{I}_1$ is represented by a vector of magnitude 12.5 A at an angle of 15.6°, and a second current $\mathbf{I}_2$ is represented by a vector of magnitude 7.38 A at an angle of 132°, as in Fig. 14-41. Find the magnitude and direction of the sum of these currents, represented by the vector **I**.

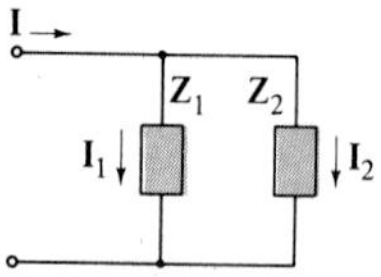


FIGURE 14-42

21. Figure 14-42 shows two impedances in parallel, with the currents in each represented by

$$\mathbf{I}_1 = 18.4 \text{ at } 51.5° \quad \text{A}$$

and

$$\mathbf{I}_2 = 11.3 \text{ at } 0° \quad \text{A}$$

The current **I** will be the sum of $\mathbf{I}_1$ and $\mathbf{I}_2$. Find the magnitude and direction of the vector representing **I**.

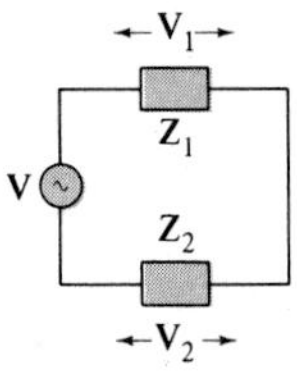


FIGURE 14-43

22. Figure 14-43 shows two impedances in series, with the voltage drop $\mathbf{V}_1$ equal to 92.4 V at 71.5° and $\mathbf{V}_2$ equal to 44.2 V at −53.8°. Find the magnitude and direction of the vector representing the total drop **V**.

### Addition of Several Vectors

Find the resultant of the following vectors.

23. 273 $\angle 34.0°$, 179 $\angle 143°$, 203 $\angle 225°$, 138 $\angle 314°$
24. 72.5 $\angle 284°$, 28.5 $\angle 331°$, 88.2 $\angle 104°$, 38.9 $\angle 146°$

### Computer

25. Write a program or use a spreadsheet that will accept as input the magnitude and direction of any number of vectors. Have the computer resolve each vector into $x$ and $y$ components, combine these components into a single $x$ component and $y$ component, and compute and print the magnitude and direction of the resultant. Try your program on Problems 23 and 24.
26. A certain airplane has an airspeed of 225 mi/h, and the wind is from the northeast at 45 mi/h. Write a program or use a spreadsheet to compute and print the groundspeed and actual direction of travel of the plane, for headings that vary from due north completely around the compass, in steps of 15°.

## CHAPTER 14 REVIEW PROBLEMS

*Solve oblique triangle ABC, if:*

1. $C = 135°$ $a = 44.9$ $b = 39.1$
2. $A = 92.4°$ $a = 129$ $c = 83.6$
3. $B = 38.4°$ $a = 1.84$ $c = 2.06$
4. $B = 22°38'$ $a = 2840$ $b = 1170$
5. $A = 132°$ $b = 38.2$ $c = 51.8$

*In what quadrant(s) will the terminal side of θ lie if:*

6. $\theta = 227°$
7. $\theta = -45°$
8. $\theta = 126°$
9. $\theta = 170°$
10. $\tan\theta$ is negative

*Without using book or calculator, state the algebraic sign of:*

11. tan 275°
12. sec (−58°)
13. cos 183°
14. cos 45°
15. sin 300°

*Write the sin, cos, and tan, to three digits, for the angle whose terminal side passes through the point*

16. $(-2, 5)$
17. $(-3, -4)$
18. $(5, -1)$

*Two vectors of magnitudes A and B are separated by an angle θ. Find the resultant and the angle the resultant makes with vector* **B**.

19. $A = 837$ $B = 527$ $\theta = 58.2°$
20. $A = 2.58$ $B = 4.82$ $\theta = 82.7°$
21. $A = 44.9$ $B = 29.4$ $\theta = 155°$
22. $A = 8374$ $B = 6926$ $\theta = 115.4°$
23. From a ship sailing north at the rate of 18 km/h, the bearing of a lighthouse is N 18°15′ E. Ten minutes later the bearing is N 75°46′ E. How far is the ship from the lighthouse at the time of the second observation?

*Write the sin, cos, and tan, to four decimal places, of:*

24. 273°
25. 175°
26. 334°36′
27. 127°22′
28. 114°

*Evaluate to four significant digits the expression*

29. $\sin 35° \cos 35°$
30. $\tan^2 68°$
31. $(\cos 14° + \sin 14°)^2$
32. What angle does the slope of a hill make with the horizontal if a vertical tower 18.5 m tall, located on the slope of the hill, is found to subtend an angle of 25.5° from a point 35.0 m directly downhill from the foot of the tower, measured along the slope?
33. Three forces are in equilibrium. One force of 457 lb acts in the direction N 28° W. The second force acts N 37° E. Find the direction of the third force of magnitude 638 lb.

*Find to the nearest tenth of a degree all values of θ less than 360°.*

34. $\cos \theta = 0.736$
35. $\tan \theta = -1.16$
36. $\sin \theta = 0.774$

*Evaluate to the nearest tenth of a degree.*

37. arcsin 0.737
38. $\tan^{-1} 4.37$
39. $\cos^{-1} 0.174$
40. A ship wishes to travel in the direction N 38° W. The current from due east at 4.2 mi/h and the speed of the ship in still water is 18.5 mi/h. Find the direction in which the ship should head, and the speed of the ship in the actual direction of travel.
41. Two forces, of 483 lb and 273 lb, act on a body. The angle between the lines of action of these forces is 48.2°. Find the magnitude of the resultant and the angle it makes with the 483-lb force.
42. Find the vector sum of two voltages, $\mathbf{V}_1 = 134 \angle 24.5°$ and $\mathbf{V}_2 = 204 \angle 85.7°$.

*Writing*

43. Suppose that you have analyzed a complex bridge truss made up of many triangles, sometimes using the law of sines and sometimes the more time-consuming law of cosines. Your client's accountant, angry over the size of your bill, has attacked your report for often using the longer law when it is clear to him that the shorter law of sines is also "good for solving triangles." Write a letter to your client explaining why you sometimes had to use one law and sometimes the other.

# 15

# RADIAN MEASURE, ARC LENGTH, AND ROTATION

## OBJECTIVES

**When you have completed this chapter, you should be able to:**

- Convert angles between radians, degrees, and revolutions.
- Write the trigonometric functions of an angle given in radians.
- Compute arc length, radius, or central angle.
- Compute the angular velocity of a rotating body.
- Compute the linear speed of a point on a rotating body.
- Solve applied problems involving arc length or rotation.

37. $\frac{7\pi}{8}$ 38. $\frac{5\pi}{9}$ 39. $\frac{2\pi}{15}$

40. $\frac{6\pi}{7}$ 41. $\frac{\pi}{12}$ 42. $\frac{8\pi}{9}$

Use a calculator to evaluate to four significant digits.

43. $\sin \frac{\pi}{3}$ 44. $\tan 0.442$ 45. $\cos 1.063$

46. $\tan \left(-\frac{2\pi}{3}\right)$ 47. $\cos \frac{3\pi}{5}$ 48. $\sin \left(-\frac{7\pi}{8}\right)$

49. $\sec 0.355$ 50. $\csc \frac{4\pi}{3}$ 51. $\cot \frac{8\pi}{9}$

52. $\tan \frac{9\pi}{11}$ 53. $\cos \left(-\frac{6\pi}{5}\right)$ 54. $\sin 1.075$

55. $\sin^2 \frac{\pi}{6} + \cos \frac{\pi}{6}$ 56. $7 \tan^2 \frac{\pi}{9}$ 57. $\cos^2 \frac{3\pi}{4}$

58. $\frac{\pi}{6} \sin \frac{\pi}{6}$ 59. $\sin \frac{\pi}{8} \tan \frac{\pi}{8}$ 60. $3 \sin \frac{\pi}{9} \cos^2 \frac{\pi}{9}$

61. The area of a *segment* of a circle (Fig. 15-3) is

$$\text{area of segment} = r^2 \arccos \frac{r-h}{r} - (r-h)\sqrt{2rh - h^2}$$

where $\arccos \frac{r-h}{r}$ is in radians. Compute the area of a segment in a circle of radius 25.0 if $h = 10.0$.

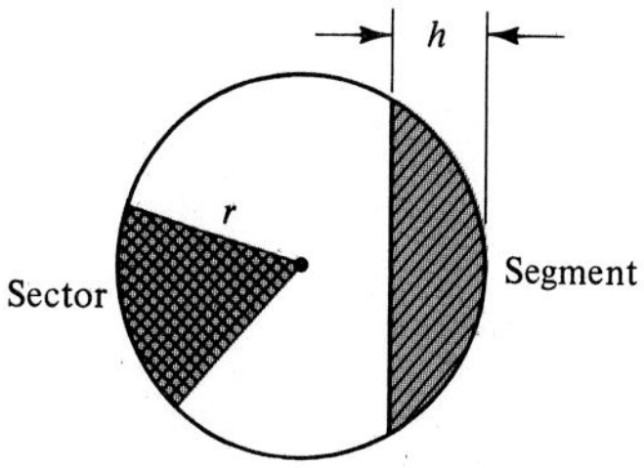


**FIGURE 15-3 Segment and sector of a circle.**

62. The area of a *sector* having a radius $r$, an arc length $s$, and a central angle $\theta$ (in radians) is

| | | |
|---|---|---|
| Area of a Sector | $A = \frac{rs}{2} = \frac{r^2\theta}{2}$ | **116** |

Find the area of a sector having a radius of 8.25 m and a central angle of 46.8°.

63. A weight bouncing on the end of a spring moves with *simple harmonic motion* according to the equation $y = 4 \cos 25t$, where $y$ is in inches. Find the displacement $y$ when $t = 2$ s. (In this equation, the angle $25t$ must be in radians.)

64. The angle $D$ (measured at the earth's center) between two points on the earth's surface is found by

$$\cos D = \sin L_1 \sin L_2 + \cos L_1 \cos L_2 \cos (M_1 - M_2)$$

where $L_1$ and $M_1$ are the latitude and longitude, respectively, of one point, and $L_2$ and $M_2$ are the latitude and longitude of the second point. Find the angle between Pittsburgh (latitude 40°N, longitude 81°W) and Houston (latitude 30°N, longitude 95°W) by substituting into this equation.

## 15-2 ARC LENGTH

We have seen that arc length, radius, and central angle are related to each other. In Fig. 15-4, if the arc length $s$ is equal to the radius $r$, we have $\theta$ equal to 1 radian. For other lengths $s$, the angle $\theta$ is equal to the *ratio* of $s$ to $r$.

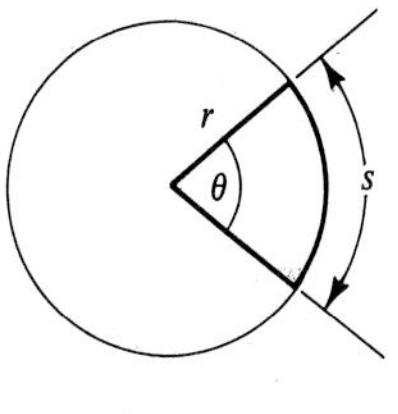


**FIGURE 15-4 Relationship between arc length, radius, and central angle.**

| | | |
|---|---|---|
| Central Angle | $\theta = \dfrac{s}{r}$ ($\theta$ must be in radians) | **115** |

*The central angle in a circle (in radians) is equal to the ratio of the intercepted arc and the radius of the circle.*

When dividing $s$ by $r$ to obtain $\theta$, $s$ and $r$ must have the same units. The units cancel, *leaving $\theta$ as a dimensionless ratio* of two lengths. Thus the radian is not a unit of measure like the degree or inch, although we usually carry it along as if it were.

We can use Eq. 115 to find any of the quantities $\theta$, $r$, or $s$, when the other two are known.

---

**EXAMPLE 11:** Find the angle that would intercept an arc of 27.0 ft in a circle of radius 21.0 ft.

**Solution:** From Eq. 115,

$$\theta = \frac{s}{r} = \frac{27.0 \text{ ft}}{21.0 \text{ ft}} = 1.29 \text{ rad}$$

---

**EXAMPLE 12:** Find the arc length intercepted by a central angle of 62.5° in a 10.4-cm-radius circle.

**Solution:** Converting the angle to radians, we get

$$62.5° \left(\frac{\pi \text{ rad}}{180°}\right) = 1.09 \text{ rad}$$

By Eq. 115,

$$s = r\theta = 10.4 \text{ cm } (1.09) = 11.3 \text{ cm}$$

---

**EXAMPLE 13:** Find the radius of a circle in which an angle of 2.06 rad intercepts an arc of 115 ft.

**Solution:** By Eq. 115,

$$r = \frac{s}{\theta} = \frac{115 \text{ ft}}{2.06} = 55.8 \text{ ft}$$

---

| | |
|---|---|
| Common Error | Be sure that $r$ and $s$ have the same units. Convert if necessary. |

---

**EXAMPLE 14:** Find the angle that intercepts a 35.8-in. arc in a circle of radius 49.2 cm.

**Solution:** We divide $s$ by $r$, and convert units at the same time.

$$\theta = \frac{s}{r} = \frac{35.8 \text{ in.}}{49.2 \text{ cm}} \cdot \frac{2.54 \text{ cm}}{1 \text{ in.}} = 1.85 \text{ rad}$$

---

**EXAMPLE 15:** How far will the rack in Fig. 15-5 move when the pinion rotates 300.0°?

**Solution:** Converting to radians gives us

$$\theta = 300.0° \left(\frac{\pi \text{ rad}}{180°}\right) = 5.236 \text{ rad}$$

Then by Eq. 115,

$$s = r\theta = 11.25 \text{ mm } (5.236) = 58.91 \text{ mm}$$

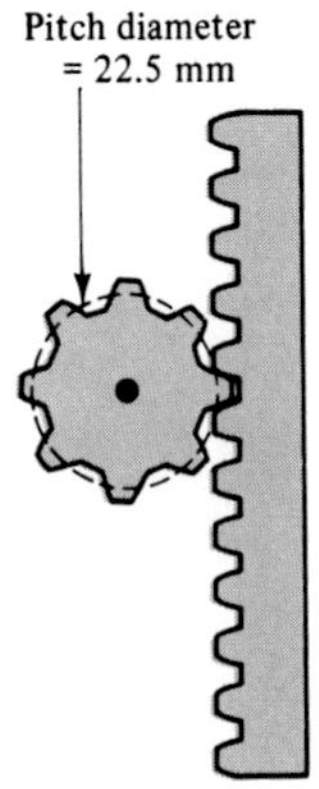


**FIGURE 15-5 Rack and pinion.**

**EXAMPLE 16:** How many miles north of the equator is a town of latitude 43.6°N? Assume that the earth is a sphere of radius 3960 mi.

**Solution:** The definitions of latitude and longitude are shown in Fig. 15-6. The latitude angle, in radians, is

$$\theta = 43.6° \left(\frac{\pi \text{ rad}}{180°}\right) = 0.761 \text{ rad}$$

Then by Eq. 115,

$$s = r\theta = 3960 \text{ mi } (0.761) = 3013 \text{ mi}$$

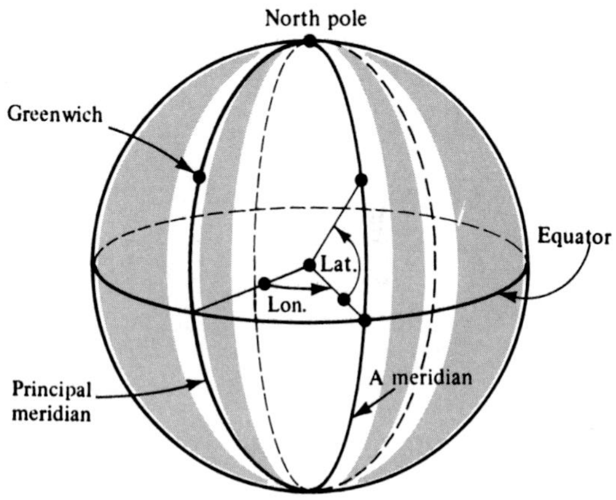


**FIGURE 15-6** ***Latitude*** **is the angle (measurement at the earth's center) between a point on the earth and the equator.** ***Longitude*** **is the angle between the meridian passing through a point on the earth, and the principal (or prime) meridian passing through Greenwich, England.**

## EXERCISE 2—ARC LENGTH

In the following table, $s$ is the length of arc subtended by a central angle $\theta$ in a circle of radius $r$. Fill in the missing values.

| | $r$ | $\theta$ | $s$ |
|---|---|---|---|
| 1. | 4.83 | $\frac{2\pi}{5}$ | |
| 2. | 11.5 | 1.36 rad | |
| 3. | 284 | 46°24′ | |
| 4. | 2.87 m | 1.55 rad | |
| 5. | 64.8 in. | 38.5° | |
| 6. | 28.3 | | 32.5 |
| 7. | 263 mm | | 582 mm |
| 8. | 21.5 ft | | 18.2 ft |
| 9. | 3.87 m | | 15.8 ft |
| 10. | | $\frac{\pi}{12}$ | 88.1 |
| 11. | | 77.2° | 1.11 |
| 12. | | 2.08 rad | 3.84 m |
| 13. | | 12°55′ | 28.2 ft |
| 14. | | $\frac{5\pi}{6}$ | 125 mm |

## Applications

**Assume the earth to be a sphere with a radius of 3960 mi. Actually, the distance from pole to pole is about 27 mi less than the diameter at the equator.**

15. A certain town is at a latitude of 35.2° N. Find the distance in miles from the town to the north pole.
16. The hour hand of a clock is 85.5 mm long. How far does the tip of the hand travel between 1:00 A.M. and 11:00 A.M.?
17. Find the radius of a circular railroad track that will cause a train to change direction by 17.5° in a distance of 180 m.
18. The pulley attached to the tuning knob of a radio (Fig. 15-7) has a radius of 35 mm. How far will the needle move if the knob is turned a quarter of a revolution?
19. Find the length of contact $ABC$ between the belt and pulley in Fig. 15-8.
20. One circular "track" on a magnetic disk used for computer data storage is located at a radius of 155 mm from the center of the disk. If 10 "bits" of data can be stored in 1 mm of track, how many bits can be stored in the length of track subtending an angle of $\pi/12$ rad?
21. If we assume the earth's orbit around the sun to be circular, with a radius of 93 million miles, how many miles does the earth travel (around the sun) in 125 days?
22. Find the latitude of a city that is 1265 mi from the equator.
23. A 1.25-m-long pendulum swings 5.75° on each side of the vertical. Find the length of arc traveled by the end of the pendulum.
24. A brake band is wrapped around a drum (Fig. 15-9). If the band has a width of 92.0 mm, find the area of contact between the band and the drum.
25. Sheet metal is to be cut from the pattern of Fig. 15-10a and bent to form the frustum of a cone (Fig. 15-10b), with top and bottom open. Find the dimensions $r$ and $R$ and the angle $\theta$ in degrees.
26. An isosceles triangle is to be inscribed in a circle of radius 1.000. Find the angles of the triangle if its base subtends an arc of length 1.437.
27. A satellite is in a circular orbit 1.25 mi above the equator of the earth. How many miles must it travel for its longitude to change by 85.0°?
28. City $B$ is due north of city $A$. City $A$ has a latitude of 14°37′N and city $B$ has a latitude of 47°12′N. Find the distance in kilometers between the cities.
29. The link $AB$ in the mechanism of Fig. 15-11 rotates through an angle of 28.3°. Find the distance traveled by point $A$.

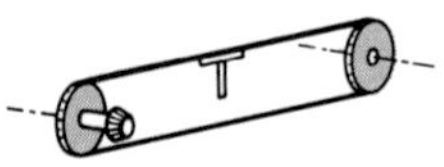

**FIGURE 15-7**

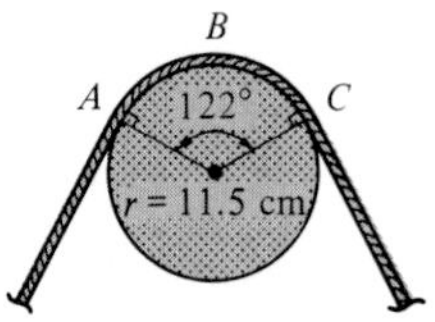


**FIGURE 15-8 Belt and pulley.**

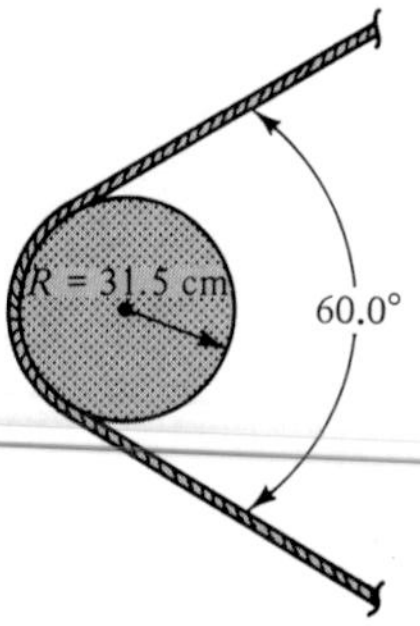


**FIGURE 15-9 Brake drum.**

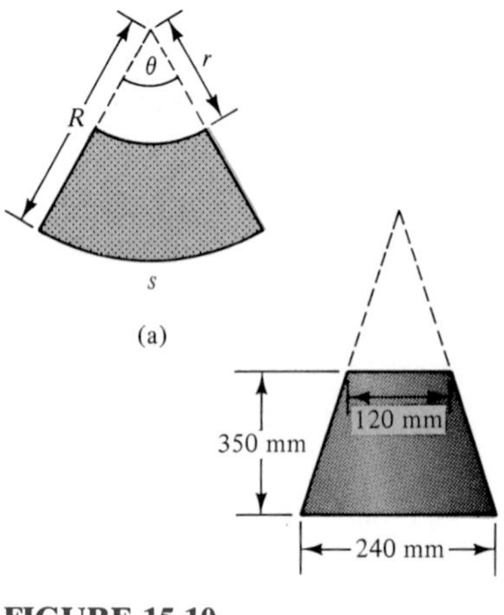


**FIGURE 15-10**

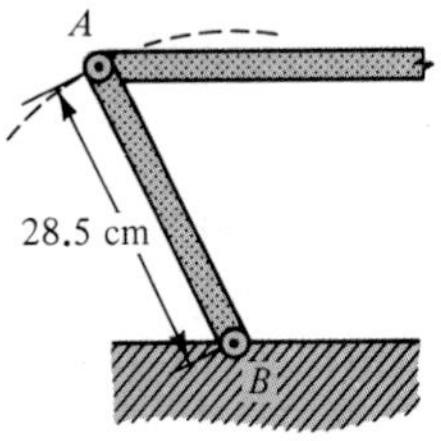


**FIGURE 15-11**

30. Find the radius $R$ of the sector gear of Fig. 15-12.

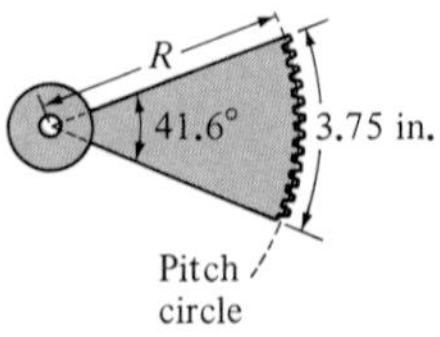


**FIGURE 15-12 Sector gear.**

31. A circular highway curve has a radius of 325.500 ft and a central angle of 15°25′15″ measured to the centerline of the road. Find the length of the curve.

32. When bending metal (Fig. 15-13), the amount $s$ that must be allowed for the bend is called the *bending allowance*. If we assume that the neutral axis (the line at which there is no stretching or compression of the metal) is at a distance from the inside of the bend equal to four-tenths of the metal thickness $t$: **(a)** show that the bending allowance is

$$s = \frac{A(r + 0.4t)\pi}{180}$$

where $A$ is the angle of bend in degrees; and **(b)** find the bending allowance for a 60° bend in $\frac{1}{4}$-in.-thick steel with a radius of 1.5 in.

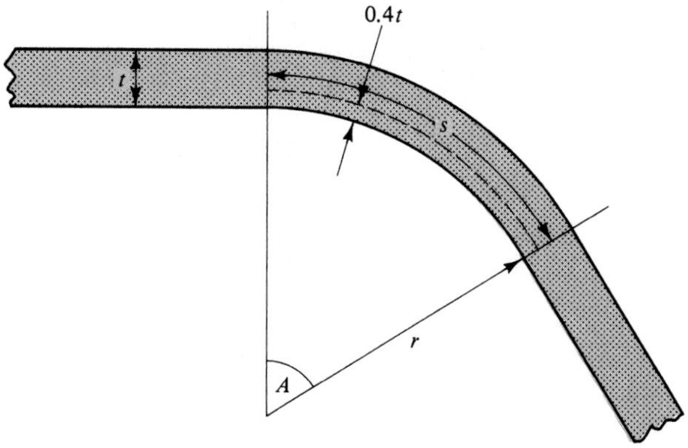


**FIGURE 15-13 Bending allowance.**

**Computer**

33. For the angles from 0 to 10°, with steps every $\frac{1}{2}$ degree, compute and print the angle in radians, the sine of the angle, and the tangent of the angle. What do you notice about these three columns of figures? What is the largest angle for which the sine and tangent are equal to the angle in radians, to three significant digits?

## 15-3 UNIFORM CIRCULAR MOTION

### Angular Velocity

Let us consider a rigid body (Fig. 15-14) that is rotating about a point $O$. The *angular velocity* $\omega$ is a measure of the *rate* at which the object rotates. The motion is called *uniform* when the angular velocity is constant. The units of angular velocity are degrees, radians, or revolutions, per unit time.

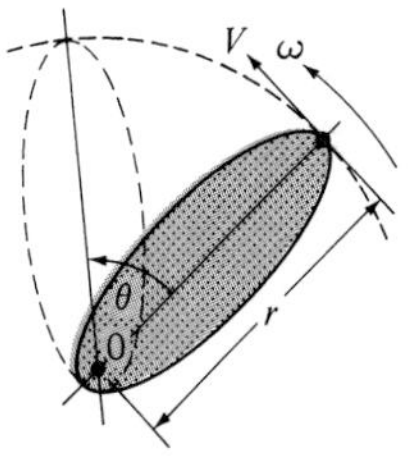


**FIGURE 15-14 Rotating body.**

### Angular Displacement

The angle $\theta$ through which the body rotates in time $t$ is called the *angular displacement*. It is related to $\omega$ and $t$ by

**This is similar to our old formula (distance = rate × time) for linear motion.**

| | | |
|---|---|---|
| Angular Displacement | $\theta = \omega t$ | **A26** |

*The angular displacement is the product of the angular velocity and the elapsed time.*

**EXAMPLE 17:** A wheel is rotating with an angular velocity of 1800 rev/min. How many revolutions does the wheel make in 1.50 s?

**Solution:** We first make the units of time consistent. Converting yields

$$\omega = \frac{1800 \text{ rev}}{\text{min}} \cdot \frac{1 \text{ min}}{60 \text{ s}} = 30 \text{ rev/s}$$

Then by Eq. A26,

$$\theta = \omega t = \frac{30 \text{ rev}}{\text{s}} (1.5 \text{ s}) = 45 \text{ rev}$$

---

**EXAMPLE 18:** Find the angular velocity in revolutions per minute of a pulley that rotates 275° in 0.750 s.

**Solution:** By Eq. A26,

$$\omega = \frac{\theta}{t} = \frac{275°}{0.750 \text{ s}} = 367 \text{ deg/s}$$

Converting to rev/min, we obtain

$$\omega = \frac{367 \text{ deg}}{\text{s}} \cdot \frac{60 \text{ s}}{\text{min}} \cdot \frac{1 \text{ rev}}{360 \text{ deg}} = 61.2 \text{ rev/min}$$

---

**EXAMPLE 19:** How long will it take a spindle rotating at 3.55 rad/s to make 1000 revolutions?

**Solution:** Converting revolutions to radians, we get

$$\theta = 1000 \text{ rev} \cdot \frac{2\pi \text{ rad}}{1 \text{ rev}} = 6280 \text{ rad}$$

Then by Eq. A26,

$$t = \frac{\theta}{\omega} = \frac{6280 \text{ rad}}{3.55 \text{ rad/s}} = 1770 \text{ s} = 29.5 \text{ min}$$

---

## Linear Speed

For any point on the rotating body, the linear displacement per unit time along the circular path is called the *linear speed.* The linear speed is zero for a point at the center of rotation and is directly proportional to the distance $r$ from the point to the center of rotation. If $\omega$ is expressed in radians per unit time, the linear speed $v$ is

| | | |
|---|---|---|
| Linear Speed | $v = \omega r$ | **A27** |

*The linear speed of a point on a rotating body is proportional to the distance from the center of rotation.*

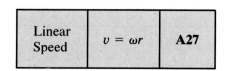

**EXAMPLE 20:** A wheel is rotating at 2450 rev/min. Find the linear speed of a point 35.0 cm from the center.

**Solution:** We first express the angular velocity in terms of radians.

$$\omega = \frac{2450 \text{ rev}}{\text{min}} \cdot \frac{2\pi \text{ rad}}{\text{rev}} = 15{,}400 \text{ rad/min}$$

Then, by Eq. A27,

$$v = \omega r = \frac{15{,}400 \text{ rad}}{\text{min}} (35.0 \text{ cm}) = 539{,}000 \text{ cm/min}$$

$$= 89.8 \text{ m/s}$$

**What became of "radians" in our answer? Shouldn't the final units be rad cm/min? No. Remember that radians is a dimensionless ratio; it is the ratio of two lengths (arc length and radius) whose units cancel.**

| Common Error | Remember when using Eq. A27 that the angular velocity must be expressed in *radians* per unit time. |
|---|---|

**EXAMPLE 21:** A belt having a speed of 885 in./min turns a 12.5-in.-radius pulley. Find the angular velocity of the pulley in rev/min.

**Solution:** By Eq. A27,

$$\omega = \frac{v}{r} = \frac{885 \text{ in.}}{\text{min}} \div 12.5 \text{ in.} = 70.8 \text{ (rad)/min}$$

Converting yields

$$\omega = \frac{70.8 \text{ rad}}{\text{min}} \cdot \frac{1 \text{ rev}}{2\pi \text{ rad}} = 11.3 \text{ rev/min}$$

## EXERCISE 3—UNIFORM CIRCULAR MOTION

### Angular Velocity

Fill in the missing values.

| | *rev/min* | *rad/s* | *deg/s* |
|---|---|---|---|
| **1.** | 1850 | | |
| **2.** | | 5.85 | |
| **3.** | | | 77.2 |
| **4.** | | $3\pi/5$ | |
| **5.** | | | 48.1 |
| **6.** | 22,600 | | |

7. A flywheel makes 725 revolutions in a minute. How many degrees does it rotate in 1.00 s?
8. A propeller on a wind generator rotates 60° in 1.00 s. Find the angular velocity of the propeller in revolutions per minute.
9. A gear is rotating at 2550 rev/min. How many seconds will it take to rotate through an angle of 2.00 rad?

10. A milling machine cutter has a diameter of 75.0 mm and is rotating at 5650 rev/min. What is the linear speed at the edge of the cutter?
11. A sprocket 3.00 in. in diameter is driven by a chain that moves at a speed of 55.5 in./s. Find the angular velocity of the sprocket in rev/min.
12. A capstan on a magnetic tape drive rotates at 3600 rad/min and drives the tape at a speed of 45.0 m/min. Find the diameter of the capstan in millimeters.
13. A blade on a water turbine turns 155° in 1.25 s. Find the linear speed of a point on the tip of the blade 0.750 m from the axis of rotation.
14. A steel bar 6.50 in. in diameter is being ground in a lathe. The surface speed of the bar is 55.0 ft/min. How many revolutions will the bar make in 10.0 s?
15. Assuming the earth to be a sphere 7920 mi in diameter, calculate the linear speed in miles per hour of a point on the equator due to the rotation of the earth about its axis.
16. Assuming the earth's orbit about the sun to be a circle with a radius of $93 \times 10^6$ mi, calculate the linear speed of the earth around the sun.
17. A car is traveling at a rate of 65.5 km/h and has tires that have a radius of 31.6 cm. Find the angular velocity of the wheels of the car.
18. A wind generator has a propeller 21.7 ft in diameter and the gearbox between the propeller and the generator has a gear ratio of 1 : 44 (with the generator shaft rotating faster than the propeller). Find the tip speed of the propeller when the generator is rotating at 1800 rev/min.

## CHAPTER 15 REVIEW PROBLEMS

*Convert to degrees.*

1. $\frac{3\pi}{7}$
2. $\frac{9\pi}{4}$
3. $\frac{\pi}{9}$
4. $\frac{2\pi}{5}$
5. $\frac{11\pi}{12}$
6. Find the central angle in radians that would intercept an arc of 5.83 m in a circle of radius 7.29 m.
7. Find the angular velocity in rad/s of a wheel that rotates 33.5 revolutions in 1.45 min.
8. A wind generator has blades 3.50 m long. Find the tip speed when the blades are rotating at 35 rev/min.

*Convert to radians in terms of $\pi$.*

9. 300°
10. 150°
11. 230°
12. 145°

*Evaluate to four significant digits.*

13. $\sin \frac{\pi}{9}$
14. $\cos^2 \frac{\pi}{8}$
15. $\sin \left(\frac{2\pi}{5}\right)^2$
16. $3 \tan \frac{3\pi}{7} - 2 \sin \frac{3\pi}{7}$
17. $4 \sin^2 \frac{\pi}{9} - 4 \sin \left(\frac{\pi}{9}\right)^2$
18. What arc length is subtended by a central angle of 63.4° in a circle of radius 4.85 in.?
19. A winch has a drum 30.0 cm in diameter. The steel cable wrapped around the drum is to be pulled in at a rate of 8.0 ft/s. Ignoring the thickness of the cable, find the angular speed of the drum, in revolutions per minute.
20. Find the linear velocity of the tip of a 5.50-in.-long minute hand of a clock.
21. A satellite has a circular orbit around the earth with a radius of 4250 mi. The satellite makes a complete orbit of the earth in 3 days, 7 h, and 35 min. Find the linear speed of the satellite.

*Evaluate to four decimal places. (Angles are in radians.)*

22. cos 1.83
23. tan 0.837
24. csc 2.94
25. sin 4.22
26. cot 0.371
27. sec 3.38
28. $\cos^2 1.74$
29. $\sin(2.84)^2$
30. $(\tan 0.475)^2$
31. $\sin^2(2.24)^2$
32. Find the area of a sector with a central angle of 29.3° and a radius of 37.2 in.
33. Find the radius of a circle in which an arc of length 384 mm intercepts a central angle of 1.73 rad.
34. Find the arc length intercepted in a circle of radius 3.85 m by a central angle of 1.84 rad.

*Writing*

35. When we multiply an angular velocity in radians per second by a length in feet, we get a linear speed in feet per second. Explain why radians do not appear in the units for linear speed.

# 16

# GRAPHS OF THE TRIGONOMETRIC FUNCTIONS

## OBJECTIVES

**When you have completed this chapter, you should be able to:**

- Find the amplitude, period, frequency, and phase shift for a sine wave or a cosine wave.
- Graph the sine function, cosine function, or tangent function.
- Write a sine function or cosine function, given the amplitude, period, and phase shift.
- Graph composite functions.
- Graph the sine or cosine wave as a function of time.
- Convert between polar and rectangular coordinates.
- Graph points and equations in polar coordinates.

Here we apply the graphing techniques of Chapter 4 to graph some *periodic* functions—the sine, cosine, and tangent—and we'll see how the constants in the equation for each affect the shape and position of the curves. We define some terms that will be applicable to the study of periodic functions in general, such as *cycle, period,* and *frequency*. This material has applications to such fields as alternating current, sound waves, mechanical vibrations, reciprocating machinery such as a gasoline or steam engine, and so on.

Finally, we introduce a second coordinate system (we already have rectangular coordinates) called the *polar coordinate system,* and learn how to make transformations from one system to the other.

## 16-1 THE SINE CURVE

### Periodic Functions

A curve that *repeats* its shape over and over is called a *periodic* curve or *periodic waveform,* as in Fig. 16-1. The function that has this graph is thus called a *periodic function*. Each repeated portion of the curve is called a *cycle*.

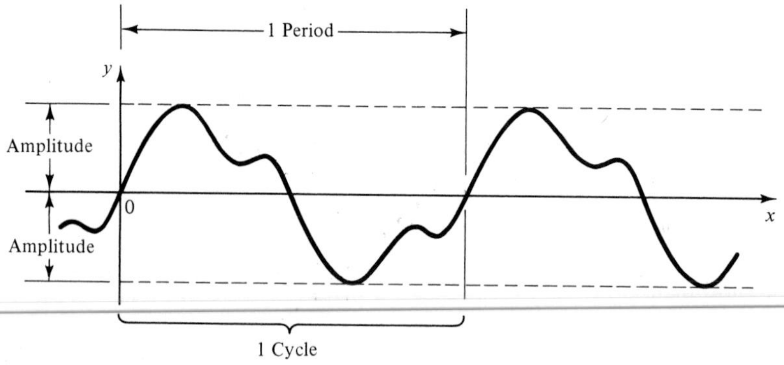


**FIGURE 16-1 A periodic waveform.**

The $x$ axis represents either *an angle* (in degrees or radians) or *time* (usually seconds or milliseconds).

The *period* is the $x$ distance taken for one cycle. Depending on the units on the $x$ axis, the period is expressed in degrees per cycle, radians per cycle, or seconds per cycle.

The *frequency* of a periodic waveform is the number of cycles that will fit into one unit (degree, radian, or second) along the $x$ axis. It is the reciprocal of the period:

$$\text{frequency} = \frac{1}{\text{period}}$$

Units of frequency are cycles/degree, cycles/radian, or cycles/second (called *hertz*), depending on the units of the period.

The *amplitude* is half the difference between the greatest and least value of a waveform (Fig. 16-1). The peak-to-peak value of a curve is equal to twice the amplitude.

The *instantaneous value* of a waveform is its $y$ value at any given $x$.

**EXAMPLE 1:** For the voltage curve in Fig. 16-2, find the period, frequency, amplitude, peak-to-peak value, and the instantaneous value of $y$ when $x$ is 50 ms.

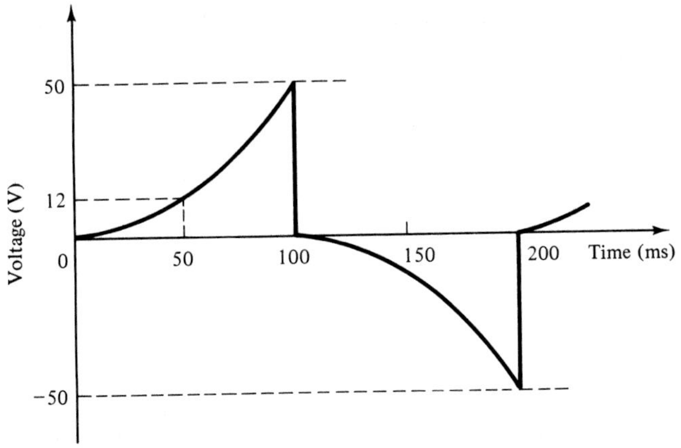


**FIGURE 16-2**

**Solution:** By observation of Fig. 16-2,

$$\text{period} = 200 \text{ ms/cycle}$$

$$\text{frequency} = \frac{1 \text{ cycle}}{200 \text{ ms}}$$

$$= 5 \text{ cycles/s} = 5 \text{ hertz (Hz)}$$

$$\text{amplitude} = 50 \text{ V}$$

$$\text{peak-to-peak value} = 100 \text{ V}$$

$$\text{instantaneous value when } x \text{ is } 50 \text{ ms} = 12 \text{ V}$$

## Graph of the Sine Function, $y = \sin x$

The general equation of the sine function contains three constants, $a$, $b$, and $c$. Graphing the equation produces a *sine curve*.

| | | |
|---|---|---|
| General Sine Curve | $y = a \sin (bx + c)$ | **206** |

The sine curve or sine wave is also called a *sinusoid*.

We will show, one at a time, the effect that each constant has on the sine curve. We'll see as we go along that the constant $a$ (the amplitude) determines how squeezed or stretched the sine wave is in the $y$ direction, that $b$ determines how squeezed or stretched the curve is in the $x$ direction, and that $c$ determines how far the entire curve is shifted to the right or left.

We start with the simplest case, where $a = 1$, $b = 1$, and $c = 0$, or $y = \sin x$. For now we will graph the sine curve just as we graphed other functions in Sec. 4-4. We choose values of $x$ and substitute into the equation to get corresponding values of $y$, and plot the resulting table of point pairs. A little later we will learn a faster method.

The units of angular measure used for the $x$ axis are sometimes degrees and sometimes radians. We will show examples of each.

**EXAMPLE 2:** Plot the sine curve $y = \sin x$, for values of $x$ from 0 to 360°.

**Solution:** Let us choose 30° intervals and make a table of point pairs.

| $x$ | 0 | 30° | 60° | 90° | 120° | 150° | 180° | 210° | 240° | 270° | 300° | 330° | 360° |
|---|---|---|---|---|---|---|---|---|---|---|---|---|---|
| $y$ | 0 | 0.5 | 0.866 | 1 | 0.866 | 0.5 | 0 | −0.5 | −0.866 | −1 | −0.866 | −0.5 | 0 |

Plotting these points gives the periodic curve shown in Fig. 16-3. If we had plotted points from $x = 360°$ to 720°, we would have gotten another cycle of the sine curve, identical to the one just plotted. We see then that the sine curve is periodic, requiring 360° for each cycle. Thus the period is equal to 360°.

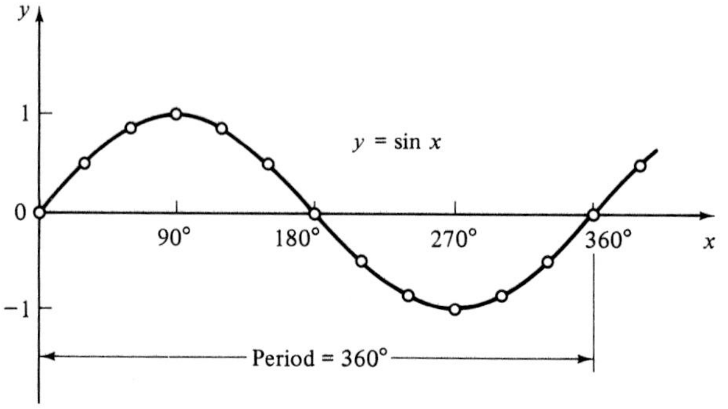


**FIGURE 16-3** $y = \sin x$ ($x$ is in degrees).

We'll now do an example in radians.

It is not easy to work in radian measure when we are used to thinking of angles in degrees. You might want to glance back at Sec. 15-1.

**EXAMPLE 3:** Plot $y = \sin x$ for values of $x$ from 0 rad to $2\pi$ rad.

**Solution:** We could choose integer values for $x$, such as 1, 2, 3, . . . and so on, or follow the usual practice of using multiples and fractions of $\pi$. This will give us points at the peaks and valleys of the curve, and where the curve crosses the $x$ axis. We take intervals of $\pi/6$ and make the following table of point pairs.

| $x$ (rad) | 0 | $\frac{\pi}{6}$ | $\frac{\pi}{3}$ | $\frac{\pi}{2}$ | $\frac{2\pi}{3}$ | $\frac{5\pi}{6}$ | $\pi$ | $\frac{7\pi}{6}$ | $\frac{4\pi}{3}$ | $\frac{3\pi}{2}$ | $\frac{5\pi}{3}$ | $\frac{11\pi}{6}$ | $2\pi$ |
|---|---|---|---|---|---|---|---|---|---|---|---|---|---|
| $y$ | 0 | 0.5 | 0.866 | 1 | 0.866 | 0.5 | 0 | −0.5 | −0.866 | −1 | −0.866 | −0.5 | 0 |

Our curve (Fig. 16-4) is the same as before, except for the scale on the $x$ axis. There, we show both integer values of $x$, and multiples of $\pi$.

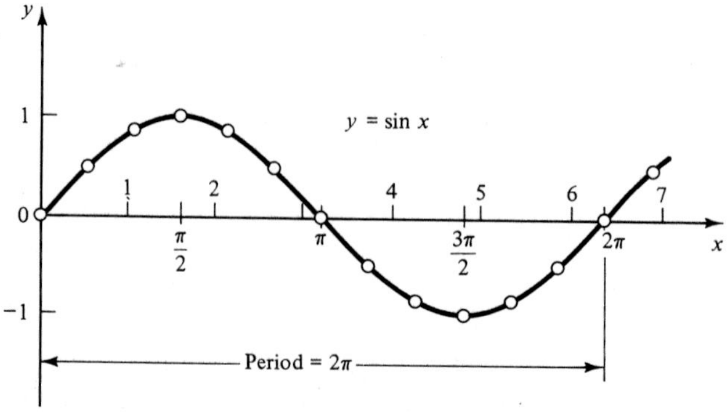


**FIGURE 16-4** $y = \sin x$ ($x$ is in radians).

Figure 16-5 shows the same curve with both radian and degree scales on the $x$ axis. Notice that one cycle of the curve fits into a rectangle whose height is twice the amplitude of the sine curve and whose width is equal to the period. We use this later for rapid sketching of periodic curves.

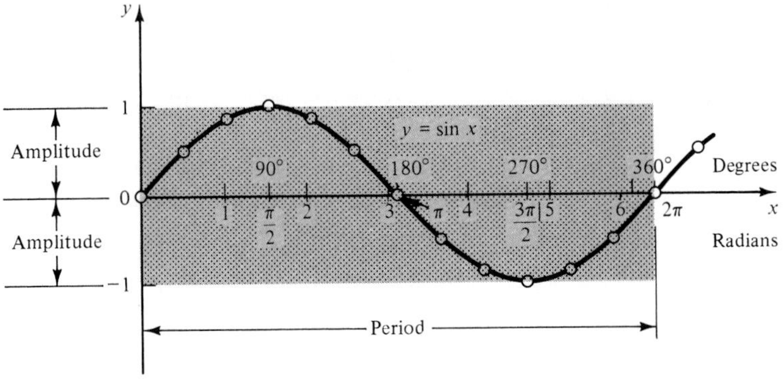


**FIGURE 16-5 The sine curve ($x$ is in degrees and radians).**

## Graph of $y = a \sin x$

We'll now see that the constant $a$ in our general sine wave

$$y = \underset{\text{amplitude}}{a} \sin (bx + c)$$

is the *amplitude* of the sine wave. In the preceding examples, the constant $a$ was equal to 1, and we got sine waves with amplitudes of 1. In examples that follow, $a$ will have other values. For now, we still let $b = 1$ and $c = 0$.

**EXAMPLE 4:** Plot the sine curve $y = 3 \sin x$ for $x = 0$ to $2\pi$.

**Solution:** When we make a table of point pairs, it becomes apparent that all the $y$ values will be three times as large as the ones in the preceding example. These points are plotted in Fig. 16-6.

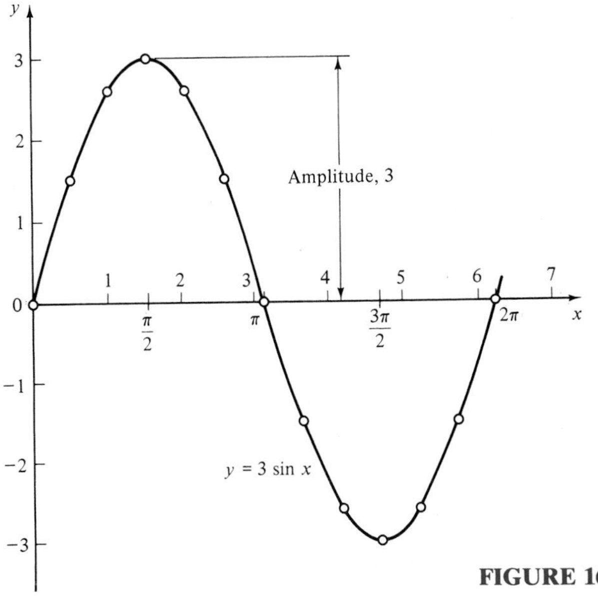


**FIGURE 16-6 Graph of $y = 3 \sin x$.**

## Graph of $y = a \sin bx$

For the sine function $y = a \sin x$, we saw that the curve repeated in an interval of 360° (or $2\pi$ radians). We called this interval the *period*. Now we will see that the constant $b$ in the equation affects the period of the graph.

That is, the curve

$$y = a \sin bx$$

will have a period of $360°/b$, if $x$ is in degrees, or $2\pi/b$ radians, if $x$ is in radians.

| | | |
|---|---|---|
| Period | $P = \dfrac{360}{b}$ degrees/cycle<br>$= \dfrac{2\pi}{b}$ radians/cycle | **207** |

**EXAMPLE 5:** Find the period for the sine curve $y = \sin 4x$ in degrees and also in radians per cycle.

**Solution:**

$$P = \frac{360°}{4} = 90°/\text{cycle}$$

or

$$P = \frac{2\pi}{4} = \frac{\pi}{2} \quad \text{radians/cycle}$$

The *frequency* $f$ is the reciprocal of the period.

| | | |
|---|---|---|
| Frequency | $f = \dfrac{b}{360}$ cycles/degree<br>$= \dfrac{b}{2\pi}$ cycles/radian | **208** |

**EXAMPLE 6:** Determine the amplitude, period, and frequency of the sine curve $y = 2 \sin(\pi x/2)$, where $x$ is in radians, and graph one cycle of the curve.

**Solution:** Here, $a = 2$ and $b = \pi/2$, so the amplitude is 2, and the period is

$$P = \frac{2\pi}{\pi/2} = 4 \text{ rad/cycle}$$

or, in degrees,

$$P = \frac{360°}{\pi/2} = 229.2°/\text{cycle}$$

We make a table of point pairs, for $x = 0$ to 4 rad.

| x (rad) | 0 | $\frac{1}{2}$ | 1 | $\frac{3}{2}$ | 2 | $\frac{5}{2}$ | 3 | $\frac{7}{2}$ | 4 |
|---|---|---|---|---|---|---|---|---|---|
| y | 0 | 1.41 | 2 | 1.41 | 0 | −1.41 | −2 | −1.41 | 0 |

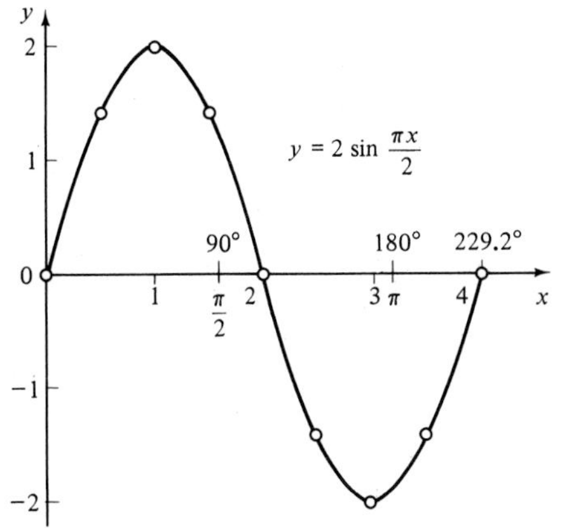


FIGURE 16-7 **Graph of $y = 2 \sin(\pi x/2)$.**

These points are plotted in Fig. 16-7. The frequency is

$$f = \frac{1}{P} = \frac{1}{4} \text{ cycle/rad}$$

In degrees,

$$f = \frac{1}{229.2^\circ} = 0.00436 \text{ cycle/deg}$$

## Graph of the General Sine Function, $y = a \sin(bx + c)$

The graph of $y = a \sin bx$, we have seen, passes through the origin. However, the graph of $y = a \sin(bx + c)$ will be shifted in the $x$ direction by an amount called the *phase displacement,* or *phase shift.*

**EXAMPLE 7:** Graph $y = \sin(3x - 90^\circ)$.

**Solution:** Since $a = 1$ and $b = 3$, we expect an amplitude of 1 and a period of $360/b = 120^\circ$. The effect that the constant $c$ has on the curve will be clearer if we put the given expression into the form

$$y = a \sin b(x + c/b)$$

by factoring. Thus our equation becomes

$$y = \sin 3(x - 30^\circ)$$

We make a table of point pairs and plot each point (Fig. 16-8).

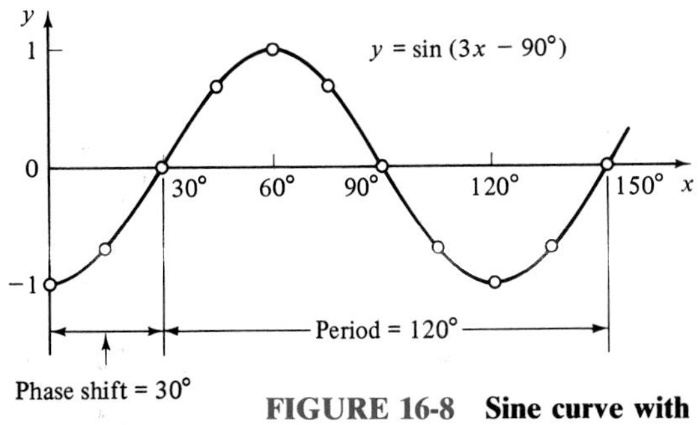


FIGURE 16-8 **Sine curve with phase shift.**

| x | 0 | 15° | 30° | 45° | 60° | 75° | 90° | 105° | 120° | 135° | 150° |
|---|---|---|---|---|---|---|---|---|---|---|---|
| y | −1 | −0.707 | 0 | 0.707 | 1 | 0.707 | 0 | −0.707 | −1 | −0.707 | 0 |

Our curve, unlike those plotted earlier, does not pass through the origin.

We see from Fig. 16-8 that the phase shift is that $x$ distance from the origin at which the positive half-cycle of the sine wave starts. To find this value for the general sine function $y = a \sin(bx + c)$ we set $y = 0$ and solve for $x$.

$$y = a \sin(bx + c) = 0$$

$$\sin(bx + c) = 0$$

$$bx + c = 0^\circ, 180^\circ, \ldots$$

We discard 180° because that is the start of the negative half-cycle. So $bx + c = 0$ and

$$x = -\frac{c}{b}$$

or

| | | |
|---|---|---|
| Phase Shift | phase shift $= -\dfrac{c}{b}$ | **209** |

The phase shift will have the same units, radians or degrees, as the constant $c$. Where no units are indicated for $c$, assume it to be in radians. If the phase shift is *negative*, it means that the positive half-cycle starts at a point to the left of the origin, so *the curve is shifted to the left*. Conversely, a *positive* phase shift means that the curve is shifted to the *right*.

**You will see, when you study *translation of axes* in analytic geometry, that replacing $x$ in a function by $(x - h)$, where $h$ is a constant, causes the graph of the function to be shifted $h$ units in the $x$ direction.**

**EXAMPLE 8:** In the equation $y = \sin(x + 45°)$

$$a = 1 \qquad b = 1 \qquad \text{and} \qquad c = 45°$$

$$\text{phase shift} = -\frac{c}{b} = -45°$$

so the curve is shifted 45° to the left (Fig. 16-9b). For comparison, Fig. 16-9 also shows (a) the unshifted sine wave and (c) the sine wave shifted 45° to the right.

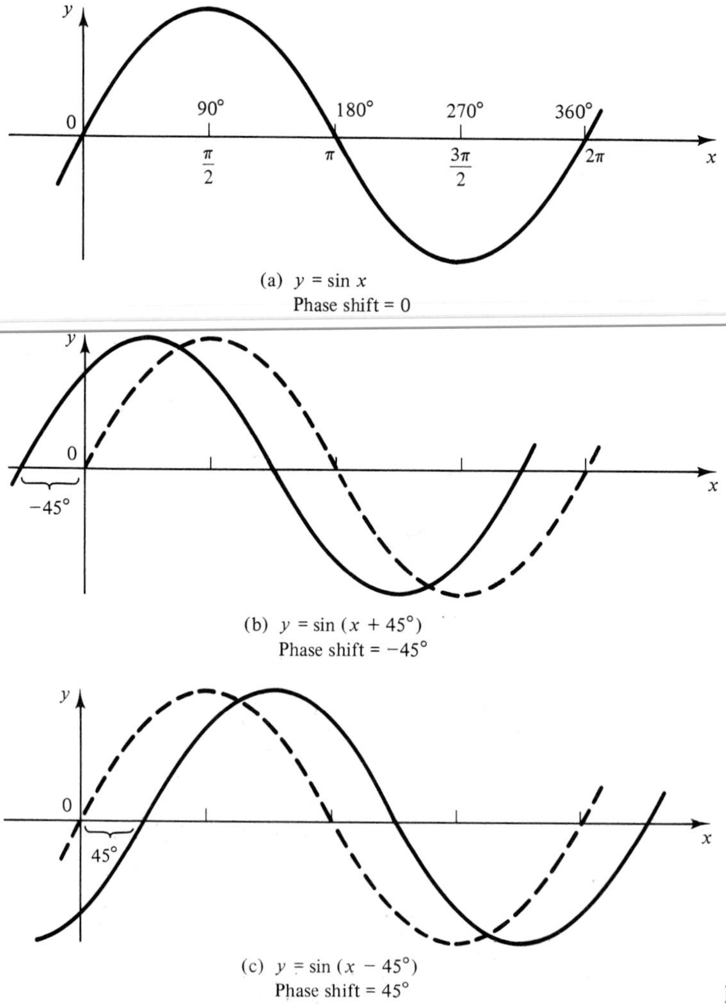


(a) $y = \sin x$
Phase shift = 0

(b) $y = \sin(x + 45°)$
Phase shift = −45°

(c) $y = \sin(x - 45°)$
Phase shift = 45°

**FIGURE 16-9**

**EXAMPLE 9:** For the sine function $y = 3\sin(4x - 2)$, we have

$$a = 3 \qquad b = 4 \qquad \text{and} \qquad c = -2$$

so

$$\text{phase shift} = -\frac{c}{b} = \frac{1}{2}\text{rad}$$

The curve is thus shifted $\frac{1}{2}$ rad to the right (Fig. 16-10).

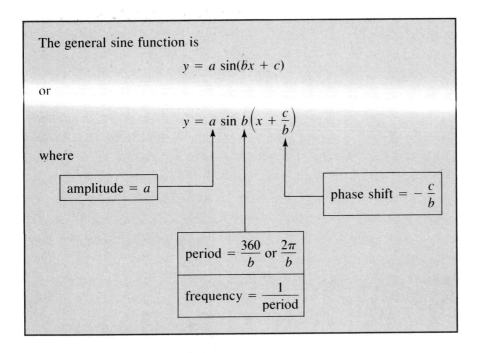

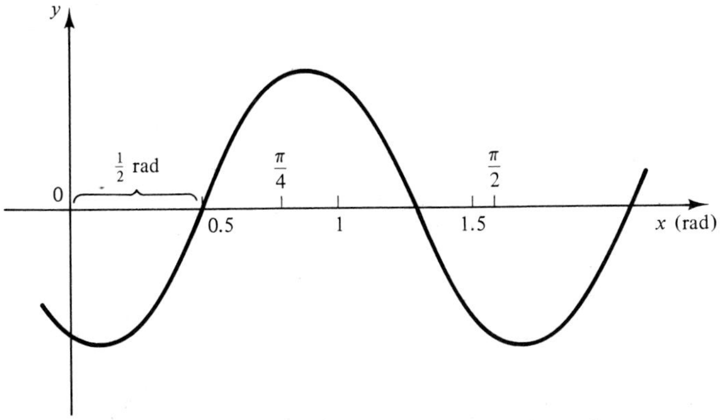


**FIGURE 16-10**

To summarize,

The general sine function is

$$y = a\sin(bx + c)$$

or

$$y = a \sin b\left(x + \frac{c}{b}\right)$$

where

amplitude $= a$

$$\text{phase shift} = -\frac{c}{b}$$

$$\text{period} = \frac{360}{b} \text{ or } \frac{2\pi}{b}$$

$$\text{frequency} = \frac{1}{\text{period}}$$

### Quick Sketching of the Sine Curve

We can plot any curve as in the preceding examples, by computing and plotting a set of point pairs. But a faster way to get a sketch is to draw a rectangle of width $P$ and height $2a$, and then sketch the sine curve (whose *shape* does not vary) within that box. First determine the amplitude, period, and phase shift. Then:

1. Draw two horizontal lines each at a distance $a$ from the $x$ axis.
2. Draw a vertical line at a distance $P$ from the origin.
3. Subdivide the period $P$ into four equal parts. Label the $x$ axis at these points and draw vertical lines through them.
4. Lightly sketch in the sine curve.
5. Shift the curve by the amount of the phase shift.

---

**EXAMPLE 10:** Make a quick sketch of $y = 2 \sin (3x + 60°)$.

**Solution:** We have $a = 2$, $b = 3$, and $c = 60°$. From Eq. 207, the period is

$$P = \frac{360°}{3} = 120°$$

and by Eq. 209,

$$\text{phase shift} = -\frac{60°}{3} = -20°$$

The steps for sketching the curve are shown in Fig. 16-11.

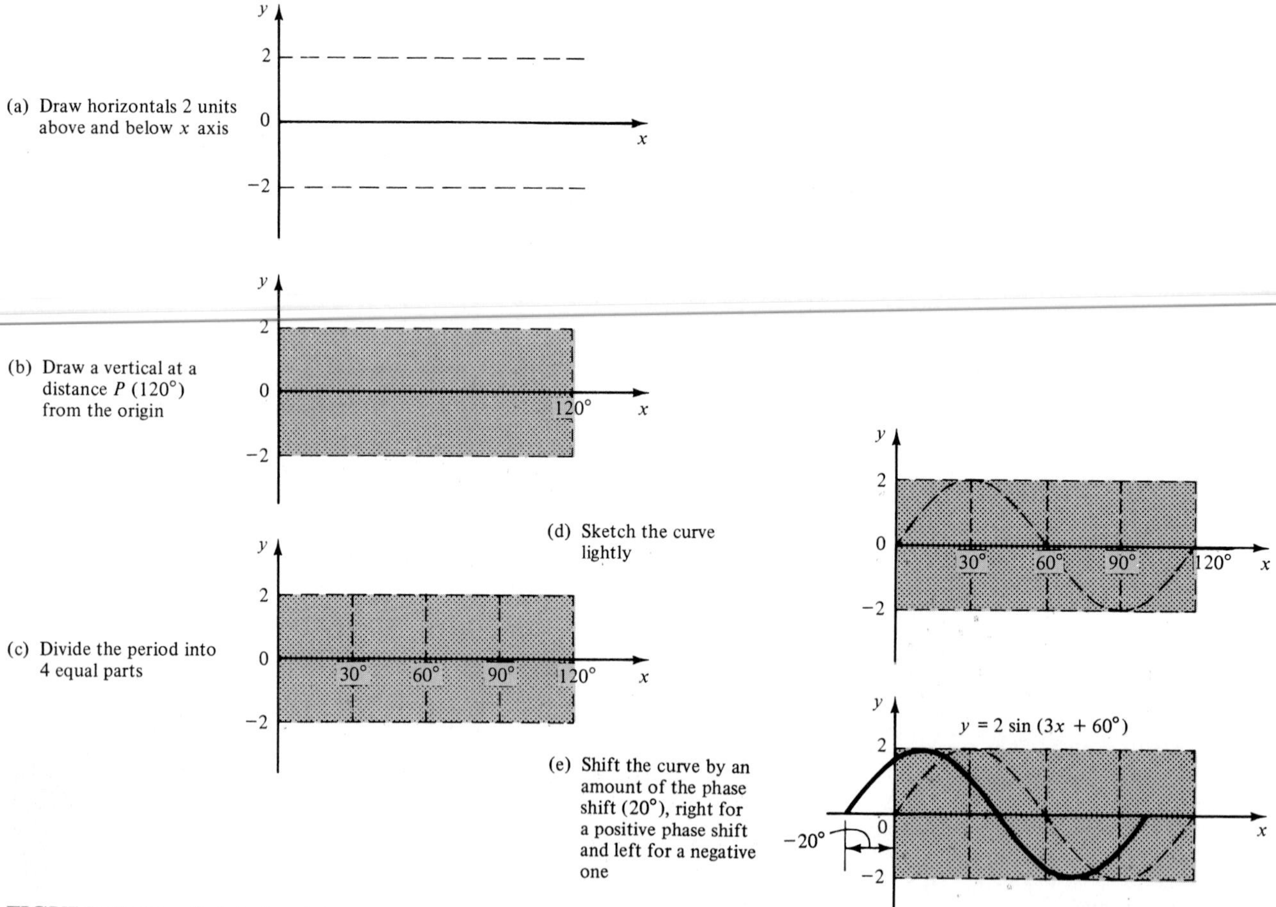

(a) Draw horizontals 2 units above and below $x$ axis
(b) Draw a vertical at a distance $P$ (120°) from the origin
(c) Divide the period into 4 equal parts
(d) Sketch the curve lightly
(e) Shift the curve by an amount of the phase shift (20°), right for a positive phase shift and left for a negative one
$y = 2 \sin (3x + 60°)$
30°
60°
90°
120°
−20°


**FIGURE 16-11 Quick sketching of the sine curve.**

---

We now do an example where the units are radians.

**EXAMPLE 11:** Make a quick sketch of

$$y = 1.5 \sin (5x - 2)$$

**Solution:** From the given equation

$$a = 1.5, \qquad b = 5, \qquad \text{and} \qquad c = -2 \text{ rad}$$

so

$$P = \frac{2\pi}{5} = 1.26 \text{ rad}$$

and

$$\text{phase shift} = -\left(\frac{-2}{5}\right) = 0.4 \text{ rad}$$

We draw a rectangle (Fig. 16-12a) whose height is 2 × 1.5 = 3 units, and whose width is 1.26 rad, and then draw three verticals spaced apart by 1.26/4 = 0.315 rad. We sketch a sine wave (shown dashed) into this rectangle, and then shift it 0.4 rad to the right to get the final curve (Fig. 16-12b).

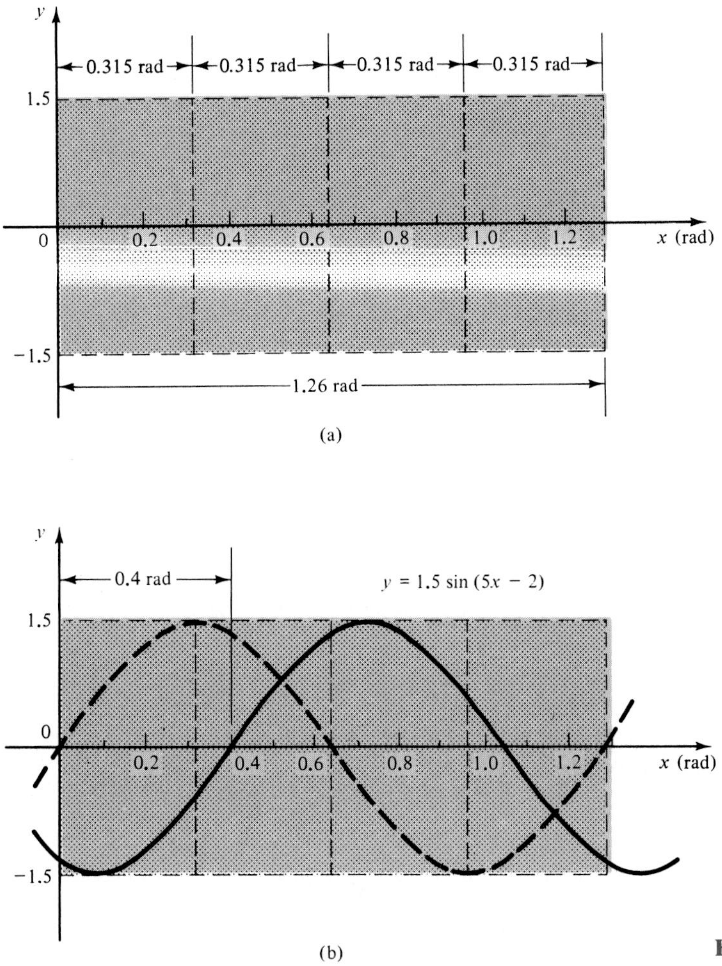


**FIGURE 16-12**

**EXAMPLE 12:** Write the equation of a sine curve having an amplitude of 5, a period of $\pi$, and a phase shift of $-0.5$ rad.

**Solution:** The period is given as $\pi$, so

$$P = \pi = \frac{2\pi}{b}$$

so

$$b = \frac{2\pi}{\pi} = 2$$

Then

$$\text{phase shift} = -0.5 = -\frac{c}{b}$$

Since $b = 2$, we get

$$c = -2(-0.5) = 1$$

Substituting into the general equation of the sine function

$$y = a \sin(bx + c)$$

with $a = 5$, $b = 2$, and $c = 1$ gives

$$y = 5 \sin(2x + 1)$$

## EXERCISE 1—THE SINE CURVE

### Periodic Waveforms

Find the period, frequency, and amplitude for each periodic waveform.

1. Figure 16-13a.
2. Figure 16-13b.

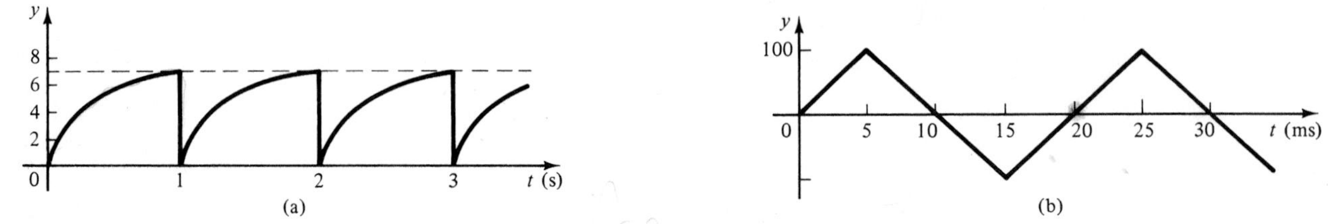


FIGURE 16-13

### The Sine Function

Graph each function. Find the amplitude, period, frequency, and phase shift. Do a point-by-point graph or a quick plot, as directed by your instructor.

3. $y = 2 \sin x$
4. $y = 1.5 \sin x$
5. $y = -3 \sin x$

6. $y = 4.25 \sin x$

7. $y = \sin 2x$

8. $y = \sin \frac{x}{3}$

9. $y = 4 \sin(x - 180°)$

10. $y = -2 \sin(x + 90°)$

11. $y = \sin(x - 90°)$

12. $y = 20 \sin(2x - 45°)$

13. $y = 0.5 \sin 2x$

14. $y = 100 \sin \frac{3x}{2}$

15. $y = 2 \sin 3x$

16. $y = \sin \pi x$

17. $y = \sin 3\pi x$

18. $y = -3 \sin \pi x$

19. $y = \sin(x - 1)$

20. $y = 0.5 \sin \left(\frac{x}{2} - 1\right)$

21. $y = \sin\left(x + \frac{\pi}{2}\right)$

22. $y = \sin\left(x + \frac{\pi}{8}\right)$

23. Write the equations of the sine curves in Fig. 16-14.

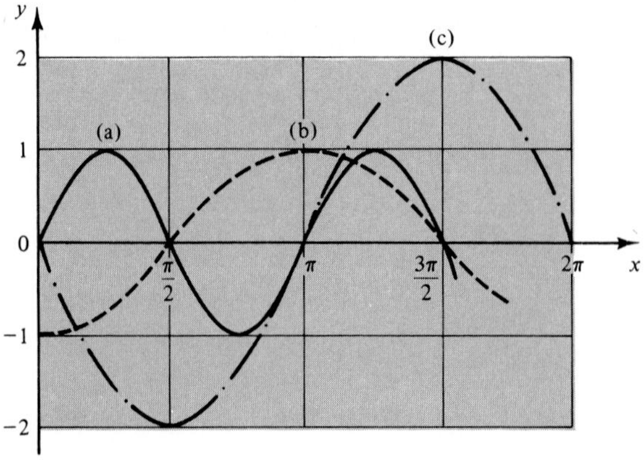


**FIGURE 16-14**

24. Write the equation of a sine curve that has an amplitude of 3, a period of $4\pi$, and a phase shift of zero.
25. Write the equation of a sine curve having an amplitude of $-2$, a period of $6\pi$, and a phase shift of $-\pi/4$.

### Computer

26. Use your program for generating a table of point pairs from Sec. 4-4 with some of the equations in this exercise, and plot the resulting sine waves.

## 16-2 GRAPHS OF MORE TRIGONOMETRIC FUNCTIONS

### Graph of the Cosine Function, $y = \cos x$

Let us graph one cycle of the function $y = \cos x$. We make a table of point pairs, and plot each point. The graph of the cosine function is shown in Fig. 16-15.

| $x$ | 0 | $\frac{\pi}{6}$ | $\frac{\pi}{3}$ | $\frac{\pi}{2}$ | $\frac{2\pi}{3}$ | $\frac{5\pi}{6}$ | $\pi$ | $\frac{7\pi}{6}$ | $\frac{4\pi}{3}$ | $\frac{3\pi}{2}$ | $\frac{5\pi}{3}$ | $\frac{11\pi}{6}$ | $2\pi$ |
|---|---|---|---|---|---|---|---|---|---|---|---|---|---|
| $y$ | 1 | 0.866 | 0.5 | 0 | −0.5 | −0.866 | −1 | −0.866 | −0.5 | 0 | 0.5 | 0.866 | 1 |

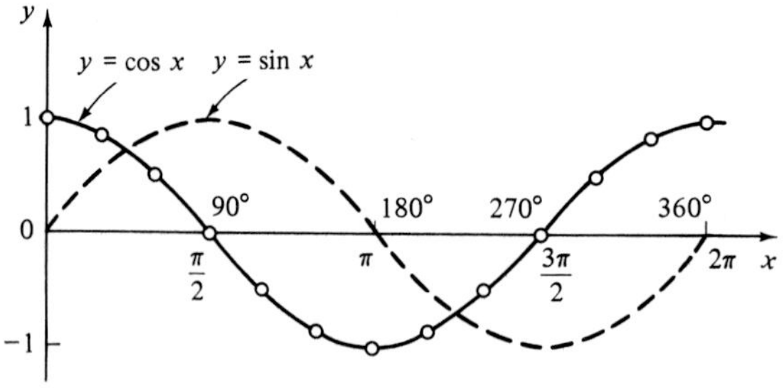


**FIGURE 16-15** $y = \cos x$

### Cosine and Sine Curves Related

Note that the cosine curve (Fig. 16-15) and the sine curve, shown dashed, have the same shape. In fact, the cosine curve appears to be identical to a sine curve shifted 90° to the left, or that

$$\cos\theta = \sin(\theta + 90°) \tag{1}$$

We can show that (1) is true. We lay out the two angles $\theta$ and $\theta + 90°$ (Fig. 16-16), choose points $P$ and $Q$ so that $OP = OQ$, and drop perpendiculars $PR$ and $QS$ to the $x$ axis. Since triangles $OPR$ and $OQS$ are congruent, we have $OR = QS$. The cosine of $\theta$ is then

$$\cos\theta = \frac{OR}{OP} = \frac{QS}{OQ} = \sin(\theta + 90°)$$

which verifies (1).

**Since the sine and cosine curves are identical except for phase shift, we can use either one to describe periodically varying quantities.**

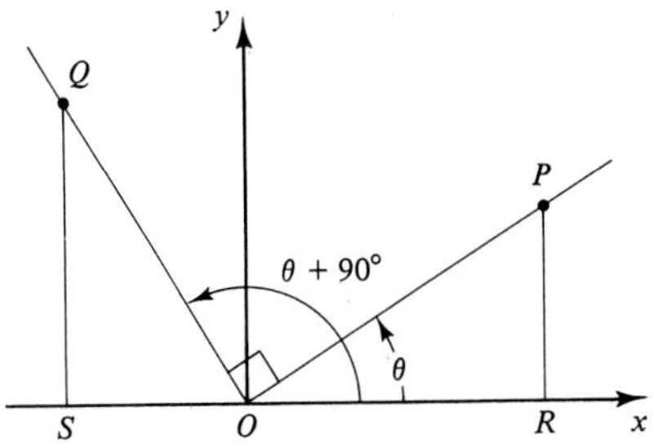


FIGURE 16-16

## Graph of the General Cosine Function, $y = a\cos(bx + c)$

The equation for the general cosine function is

$$y = a\cos(bx + c)$$

The amplitude, period, frequency, and phase shift are found the same way as for the sine curve, and, of course, the quick plotting method works here, too.

---

**EXAMPLE 13:** Find the amplitude, period, frequency, and phase shift for the curve $y = 3\cos(4x - 120°)$ and make a graph.

**Solution:** From the equation, $a$ = amplitude = 3. Since $b = 4$, the period is

$$\text{period} = \frac{360°}{4} = 90°/\text{cycle}$$

and

$$\text{frequency} = \frac{1}{P} = 0.0111 \text{ cycle/degree}$$

Since $c = -120°$,

$$\text{phase shift} = -\frac{c}{b} = -\frac{-120°}{4} = 30°$$

so we expect the curve to be shifted 30° to the right. For a quick plot we draw a rectangle with height twice the amplitude and a length of one period (Fig. 16-17). We sketch in the cosine curve, shown dashed, and then shift it 30° to the right to get the final curve, shown solid.

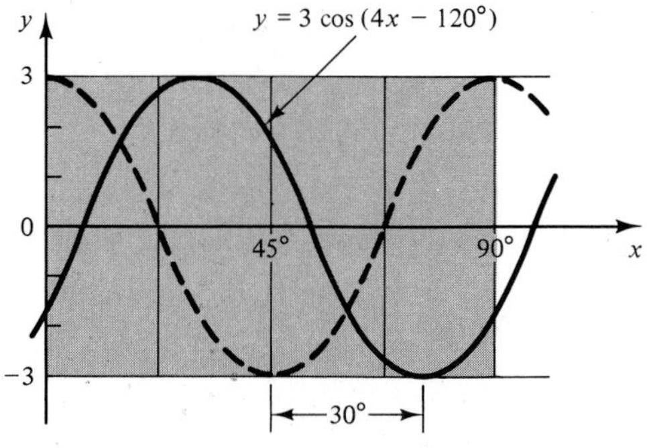


FIGURE 16-17

---

## Graph of the Tangent Function, $y = \tan x$

We now plot the function $y = \tan x$. As usual, we make a table of point pairs.

| $x$ | 0 | 30° | 60° | 90° | 120° | 150° | 180° | 210° | 240° | 270° | 300° | 330° | 360° |
|---|---|---|---|---|---|---|---|---|---|---|---|---|---|
| y | 0 | 0.577 | 1.73 | — | −1.73 | −0.577 | 0 | 0.577 | 1.73 | — | −1.73 | −0.577 | 0 |

The tangent is not defined at 90° (or 270°), as we saw in Sec. 14-1. But for angles slightly less than 90° (or 270°) the tangent is large and grows larger the closer the angle gets to 90° (for example, tan 89.9° = 573, tan 89.99° = 5730, etc.). The same thing happens for angles slightly larger than 90° (or 270°) except that the tangent is now negative (tan 90.1° = −573, tan 90.01° = −5730, etc.). With this information, we graph the tangent curve in Fig. 16-18.

The terms "period" and "frequency" have the same meaning as before, but *amplitude* is not used in connection with the tangent function, as the curve extends to infinity in the positive and negative directions.

The vertical lines at $x = 90°$ and 270° that the tangent graph approaches but never touches are called *asymptotes*.

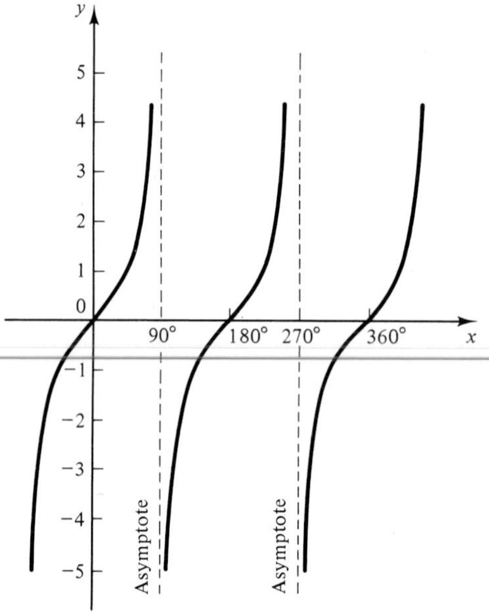


**FIGURE 16-18 The tangent curve.**

## Graphs of Cotangent, Secant, and Cosecant Functions

For completeness, the graphs of all six trigonometric functions are shown in Fig. 16-19. To make them easier to compare, the same horizontal scale is used for each curve. Only the sine and cosine curves find much use in technology. Notice that of the six, they are the only curves that have no "gaps" or "breaks." They are called *continuous* curves, while the others are called *discontinuous*.

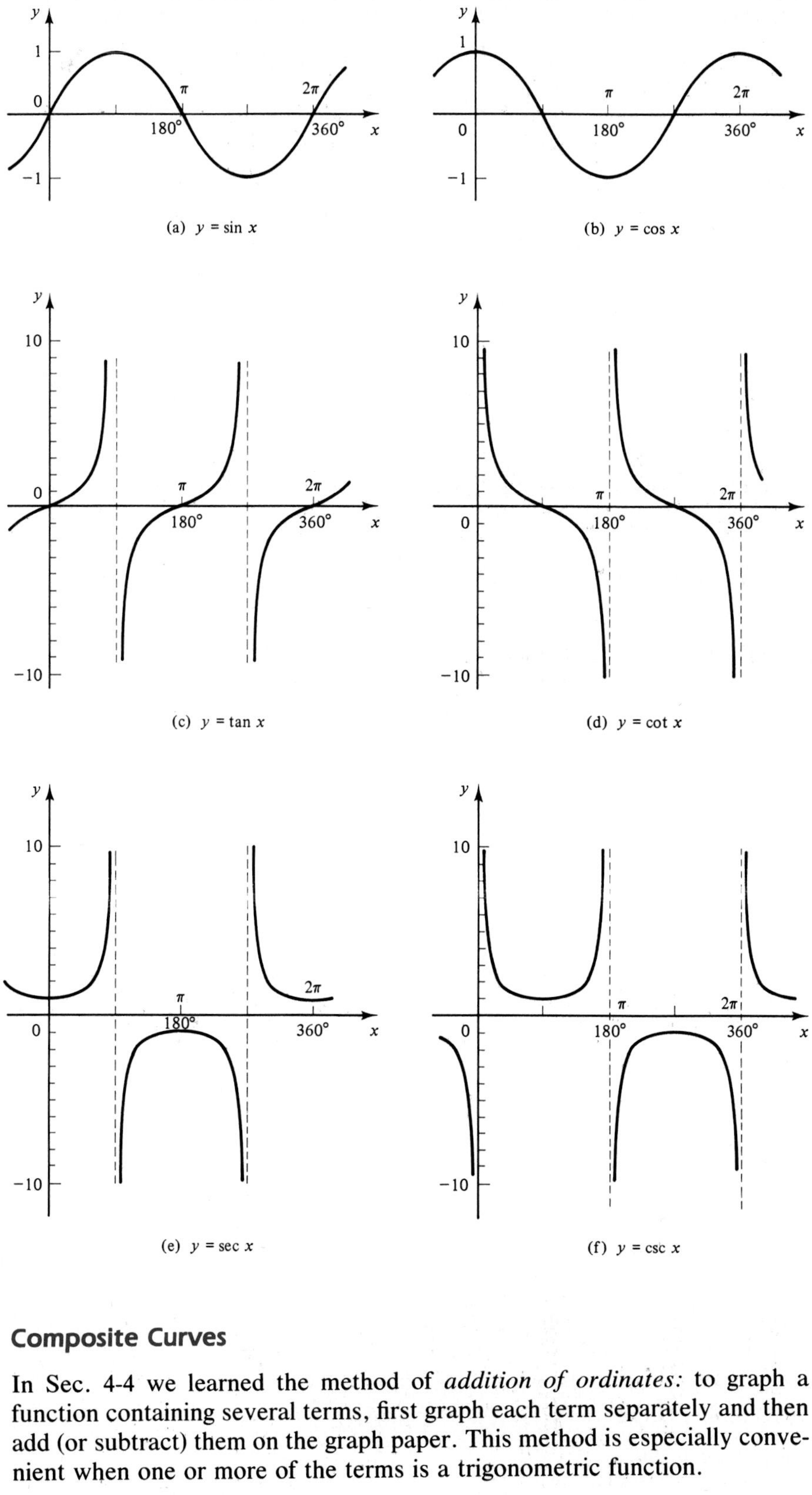


**FIGURE 16-19 Graphs of the six trigonometric functions.**

## Composite Curves

In Sec. 4-4 we learned the method of *addition of ordinates:* to graph a function containing several terms, first graph each term separately and then add (or subtract) them on the graph paper. This method is especially convenient when one or more of the terms is a trigonometric function.

**EXAMPLE 14:** Graph the function $y = 8/x + \cos x$, where $x$ is in radians, from $x = 0$ to 12 rad.

**Solution:** We separately graph $y_1 = 8/x$ and $y_2 = \cos x$, shown dashed in Fig. 16-20. We add the ordinates graphically to get the composite curve, shown as a solid line.

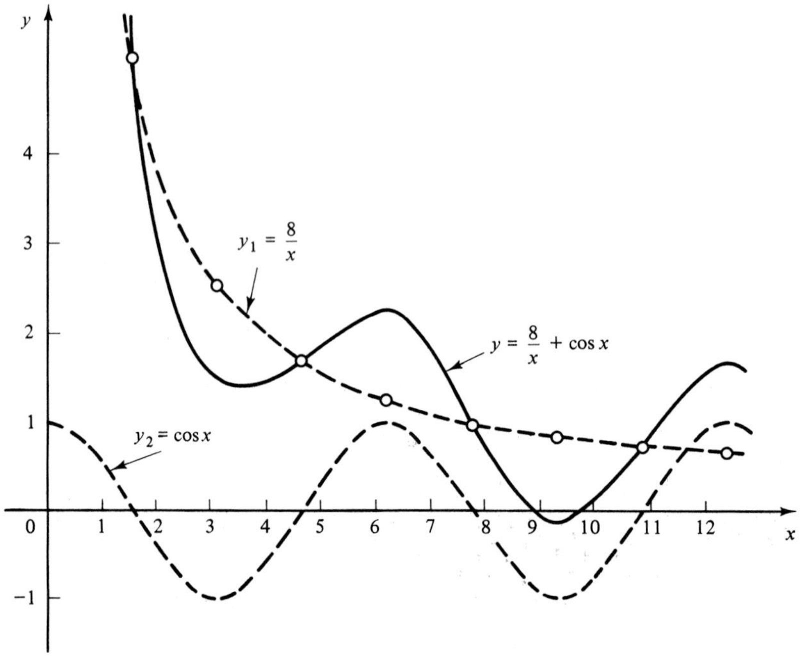


**FIGURE 16-20 Graphing by addition of ordinates.**

## EXERCISE 2—MORE TRIGONOMETRIC FUNCTIONS

### The Cosine Curve

Graph each function, and find the amplitude, period, and phase shift.

1. $y = 3 \cos x$
2. $y = \cos 2x$
3. $y = 2 \cos 3x$
4. $y = \cos(x + 2)$
5. $y = 3 \cos\left(x - \frac{\pi}{4}\right)$
6. $y = 2 \cos(3x + 1)$

### The Tangent Curve

Graph each function.

7. $y = 2 \tan x$
8. $y = \tan 4x$
9. $y = 3 \tan 2x$

10. $y = \tan\left(x - \frac{\pi}{2}\right)$ 11. $y = 2\tan(3x - 2)$ 12. $y = 4\tan\left(x + \frac{\pi}{6}\right)$

### Composite Curves

Graph by addition of ordinates. ($x$ is in radians.)

13. $y = x + \sin x$

14. $y = \cos x - x$

15. $y = \frac{2}{x} - \sin x$

16. $y = 2x + 5\cos x$

17. $y = \sin x + \cos 2x$

18. $y = \sin x + \sqrt{x}$

## 16-3 THE SINE WAVE AS A FUNCTION OF TIME

The sine curve can be generated in a simple geometric way. Figure 16-21 shows a vector $OP$ rotating with an angular velocity $\omega$. A rotating vector is called a *phasor*. Its angular velocity $\omega$ is almost always given in radians per second (rad/s).

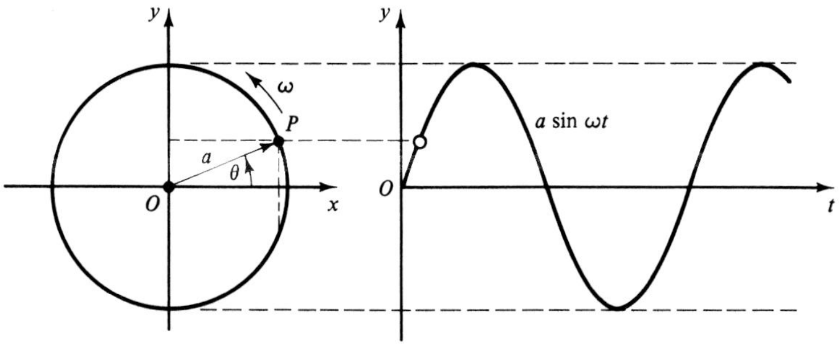


**FIGURE 16-21 The sine curve generated by a rotating vector.**

If the length of the phasor is $a$, then its projection on the $y$ axis is

$$y = a \sin \theta$$

But since by Eq. A26 the angle $\theta$ at any instant $t$ is equal to $\omega t$,

$$y = a \sin \omega t$$

Further, if the phasor does not start from the $x$ axis but has some *phase angle* $\phi$, we get

$$y = a \sin(\omega t + \phi)$$

Notice in this equation, that *y is a function of time,* rather than of an angle. Thus it is more appropriate for representing alternating current, which we will discuss shortly.

**EXAMPLE 15:** Write the equation for a sine wave generated by a phasor of length 8 rotating with an angular velocity of 300 rad/s, and with a phase angle of 0.

**Solution:** Substituting, with $a = 8$, $\omega = 300$ rad/s, and $\phi = 0$,

$$y = 8 \sin 300t$$

## Period

Recall that the period was defined as the distance along the $x$ axis taken by one cycle of the waveform. When the units on the $x$ axis represented an angle, the period was in radians or degrees. Now that the $x$ axis shows time, the period will be in seconds. From Eq. 207,

$$P = \frac{2\pi}{b}$$

But in the equation $y = a \sin \omega t$, $b = \omega$, so

| Period of a Sine Wave | $P = \dfrac{2\pi}{\omega}$ | **A77** |
|---|---|---|

where $P$ is in seconds and $\omega$ is in rad/s.

**EXAMPLE 16:** Find the period and amplitude of the sine wave $y = 6 \sin 10t$ and make a graph with time $t$ shown on the horizontal axis.

**Solution:** The period is, from Eq. A77,

$$P = \frac{2\pi}{\omega} = \frac{2(3.142)}{10} = 0.628 \text{ s}$$

Thus, it takes 628 ms for a full cycle, 314 ms for a half cycle, 157 ms for a quarter cycle, and so forth. This sine wave, with amplitude 6, is plotted in Fig. 16-22.

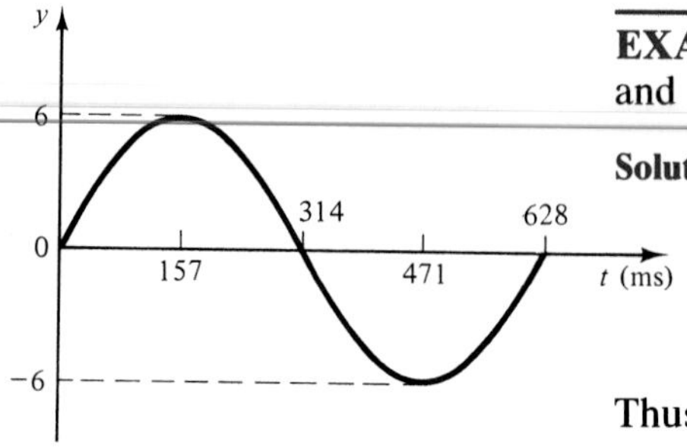

**FIGURE 16-22**

| Common Error | Don't confuse the function $y = \sin bx$ with the function $y = \sin \omega t$. In the first case, $y$ is a function of an angle $x$, and $b$ is a coefficient that has no units. In the second case, $y$ is a function of time, $t$, and $\omega$ is an angular velocity in radians per second. |
|---|---|

## Frequency

From Eq. 208 we see that the frequency $f$ of a periodic waveform is equal to the reciprocal of the period $P$. So, for the sine wave,

| | | |
|---|---|---|
| Frequency of a Sine Wave | $f = \frac{1}{P} = \frac{\omega}{2\pi}$ | **A78** |

where $P$ is in seconds and $\omega$ is in rad/s.

The unit of frequency is therefore cycles/s, or *hertz* (Hz):

$$1 \text{ Hz} = 1 \text{ cycle/s}$$

Higher frequencies are often expressed in kilohertz (kHz), where

$$1 \text{ kHz} = 10^3 \text{ Hz}$$

or in megahertz (MHz), where

$$1 \text{ MHz} = 10^6 \text{ Hz}$$

---

**EXAMPLE 17:** The frequency of the sine wave of Example 16 is

$$f = \frac{1}{P} = \frac{1}{0.628} = 1.59 \text{ Hz}$$

---

When the period $P$ is not wanted, the angular velocity can be obtained directly from Eq. A78 by noting that $\omega = 2\pi f$.

---

**EXAMPLE 18:** Find the angular velocity of a 1000-Hz source.

**Solution:** From Eq. A78,

$$\omega = 2\pi f = 2(3.14)(1000) = 6280 \text{ rad/s}$$

---

A sine wave as a function of time can also have a phase shift, expressed either in degrees or radians, as in the following example.

---

**EXAMPLE 19:** Given the sine wave $y = 5.83 \sin(114t + 15°)$, find (a) the amplitude, (b) the angular velocity, (c) the period, (d) the frequency, (e) the phase angle, and (f) make a graph.

**Solution:** (a) The amplitude is 5.83 units.

(b) The angular velocity is $\omega = 114$ rad/s.

(c) From Eq. A77, the period is

$$P = \frac{2\pi}{\omega} = \frac{2\pi}{114} = 0.0551 \text{ s} = 55.1 \text{ ms}$$

(d) From Eq. A78, the frequency is

$$f = \frac{1}{P} = \frac{1}{0.0551} = 18.1 \text{ Hz}$$

(e) The phase angle is 15°. It is not unusual to see a sine wave given with $\omega t$ in radians and the phase angle in degrees. We get the phase shift, in units of time, the same way we got the phase shift for the general sine function in Sec. 16-1. We find the value of $t$ at which the positive half-cycle starts. As before, we set $y$ equal to zero and solve for $t$.

$$y = \sin(\omega t + \phi) = 0$$

$$\omega t = -\phi$$

$$t = -\frac{\phi}{\omega} = \text{phase shift}$$

Substituting our values for $\omega$ and $\phi$ (which we first convert to 0.262 radians) gives

$$\text{phase shift} = -\frac{0.262 \text{ rad}}{114 \text{ rad/s}} = -0.00230 \text{ s}$$

so our curve is shifted 2.30 milliseconds to the left.

(f) This sine wave is graphed in Fig. 16-23. We draw a rectangle of height 2(5.83) and width 55.1 ms. We subdivide the rectangle into four rectangles of equal width and sketch in the sine wave, shown dashed. We then shift the sine wave 2.30 ms to the left to get the final (solid) curve.

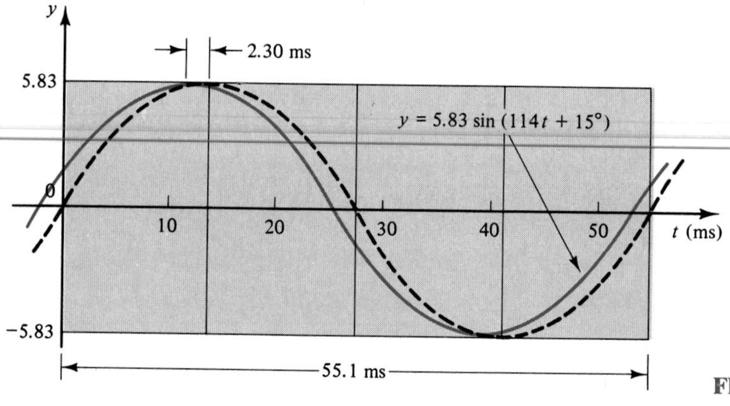


FIGURE 16-23

## Adding Sine Waves of the Same Frequency

We have seen that $A \sin \omega t$ is the $y$ component of a phasor of magnitude $A$ rotating at angular velocity $\omega$. Similarly, $B \sin(\omega t + \theta)$ is the $y$ component of a phasor of magnitude $B$ rotating at the same angular velocity $\omega$, but with a phase angle $\phi$ between $A$ and $B$. Since each is the $y$ component of a phasor, their sum is equal to the sum of the $y$ components of the two phasors, in other words, simply the $y$ component of the resultant of those phasors.

Thus to add two sine waves of the same frequency, we simply *find the resultant of the phasors representing those sine waves.*

**EXAMPLE 20:** Express $y = 2.00 \sin \omega t + 3.00 \sin(\omega t + 60°)$ as a single sine wave.

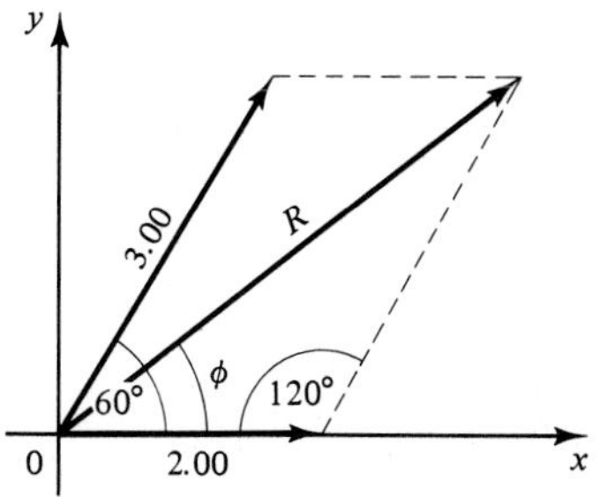


FIGURE 16-24

**Solution:** We represent $2.00 \sin \omega t$ by a phasor of length 2.00, and we represent $3.00 \sin(\omega t + 60°)$ as a phasor of length 3.00, at an angle of 60°. These phasors are shown in Fig. 16-24. We wish to add $y = 2.00 \sin \omega t$ and $y = 3.00 \sin(\omega t + 60°)$. By the law of cosines,

$$R = \sqrt{2.00^2 + 3.00^2 - 2(2.00)(3.00) \cos 120°} = 4.36$$

Then by the law of sines,

$$\frac{\sin \phi}{3.00} = \frac{\sin 120°}{4.36}$$

$$\sin \phi = 0.596$$

$$\phi = 36.6°$$

Thus

$$y = 2.00 \sin \omega t + 3.00 \sin(\omega t + 60°)$$

$$= 4.36 \sin(\omega t + 36.6°)$$

Figure 16-25 shows the two original sine waves as well as a graph of $4.36 \sin(\omega t + 36.6°)$. Note that the sum of the ordinates of the original waves, at any value of $\omega t$, is equal to the ordinate of the final wave.

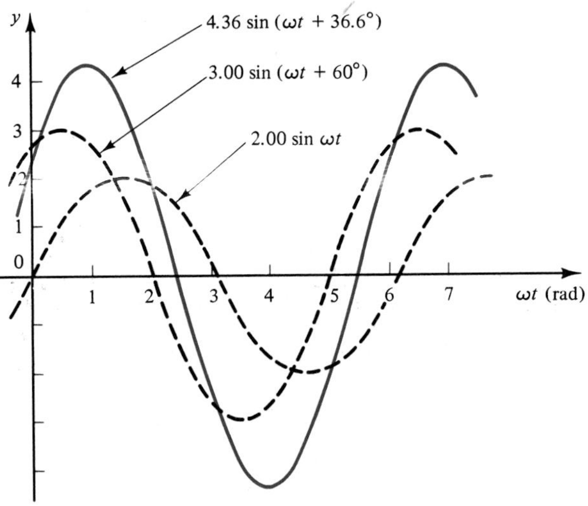


FIGURE 16-25

## Adding Sine and Cosine Waves of the Same Frequency

In Sec. 16-2 we verified that

$$\cos \omega t = \sin(\omega t + 90°)$$

and saw that a cosine wave $A \cos \omega t$ is identical to the sine wave $A \sin \omega t$, except for a phase difference of 90° between the two curves (Fig. 16-15). Thus we can find the sum $(A \sin \omega t + B \cos \omega t)$ of a sine and a cosine wave the same way that we added two sine waves in the preceding section, by

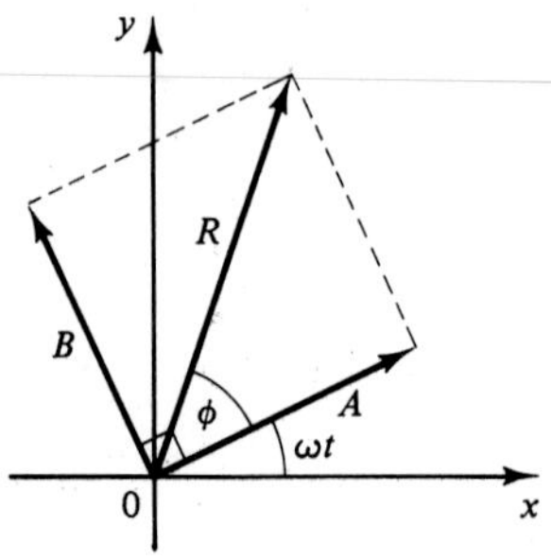


FIGURE 16-26

finding the resultant of the phasors representing each sine wave. Since the angle between the two phasors is 90° (Fig. 16-26), the resultant has a magnitude

$$R = \sqrt{A^2 + B^2}$$

and is at an angle

$$\phi = \arctan \frac{B}{A}$$

So

| Addition of a Sine Wave and a Cosine Wave | $A \sin \omega t + B \cos \omega t = R \sin(\omega t + \phi)$<br>where<br>$R = \sqrt{A^2 + B^2}$ and $\phi = \arctan \frac{B}{A}$ |
|---|---|

Thus we can represent the sum of a sine and a cosine function by a single sine function.

---

**EXAMPLE 21:** Express $y = 274 \sin \omega t + 371 \cos \omega t$ as a single sine function.

**Solution:** The magnitude of the resultant is

$$R = \sqrt{(274)^2 + (371)^2} = 461$$

Note that you can use the rectangular to polar conversion keys on your calculator to do this calculation.

and the phase angle is

$$\phi = \arctan \frac{371}{274} = \arctan 1.354 = 53.6°$$

so

$$y = 274 \sin \omega t + 371 \cos \omega t$$
$$= 461 \sin(\omega t + 53.6°)$$

Figure 16-27 shows the original sine and cosine functions, and a graph of $y = 461 \sin(\omega t + 53.6°)$. Note that the sum of the ordinates of the original waves equals the ordinate of the computed wave at any abscissa.

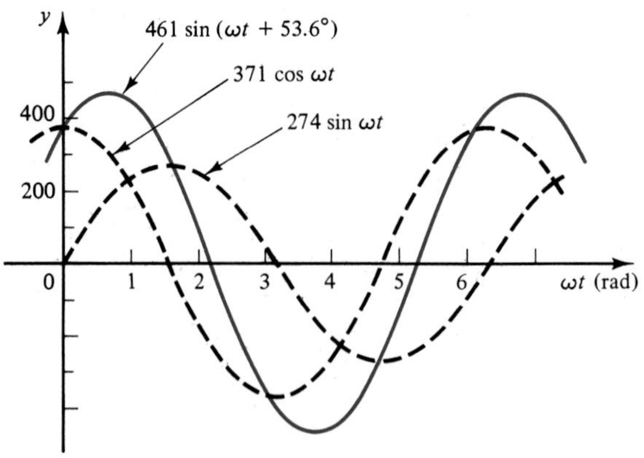


FIGURE 16-27

---

### Alternating Current

When a loop of wire rotates in a magnetic field, any portion of the wire cuts the field while traveling first in one direction and then in the other direction. Since the polarity of the voltage induced in the wire depends on the direction in which the field is cut, the induced current will travel first in one direction and then in the other. It *alternates*.

The same is true with an armature more complex than a simple loop. We get an *alternating current*. Just as the rotating point $P$ in Fig. 16-21 generated a sine wave, the current induced in the rotating armature will be sinusoidal.

The sinusoidal voltage or current can be described by the equation

$$y = a \sin(\omega t + \phi)$$

or, in electrical terms,

$$i = I_m \sin(\omega t + \phi_2)$$

where $i$ is the current at time $t$, $I_m$ the maximum current (the amplitude), and $\phi_2$ the phase angle. And if we let $V_m$ stand for the amplitude of the voltage wave, the instantaneous voltage $v$ becomes

$$v = V_m \sin(\omega t + \phi_1)$$

**We have given the phase angle $\phi$ subscripts because, in a given circuit, the current and voltage waves will usually have different phase angles.**

Equations A77 and A78 still apply here.

$$\text{period} = \frac{2\pi}{\omega} \quad \text{s}$$

$$\text{frequency} = \frac{\omega}{2\pi} \quad \text{Hz}$$

---

**EXAMPLE 22:** Utilities in the United States supply alternating current at a frequency of 60 Hz. Find the angular velocity and the period $P$.

**Solution:** By Eq. A78,

$$P = \frac{1}{f} = \frac{1}{60} = 0.0167 \text{ s}$$

and by Eq. A77,

$$\omega = \frac{2\pi}{P} = \frac{2\pi}{0.0167} = 377 \text{ rad/s}$$

These are good numbers to remember.

---

**EXAMPLE 23:** A certain alternating current has an amplitude of 1.5 A and a frequency of 60 Hz (cycles/s). Taking the phase angle as zero, write the equation for the current as a function of time, find the period, and find the current at $t = 0.01$ s.

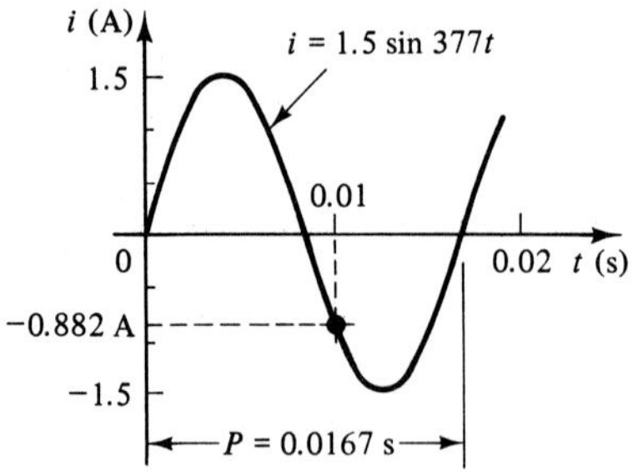


FIGURE 16-28

**Solution:** From Example 22,

$$P = 0.0167 \text{ s} \quad \text{and} \quad \omega = 377 \text{ rad/s}$$

so the equation is

$$i = 1.5 \sin(377t)$$

When $t = 0.01$ s,

$$i = 1.5 \sin(377)(0.01) = -0.882 \text{ A}$$

as shown in Fig. 16-28.

## Phase Shift

When writing the equation of a single alternating voltage or current we are usually free to choose the origin anywhere along the time axis, and so can place it to make the phase angle equal to zero. However, when there are *two* curves on the same graph that are *out of phase,* we usually locate the origin so that the phase angle of one curve is zero.

**EXAMPLE 24:** Write the equations for the voltage and current waves in Fig. 16-29.

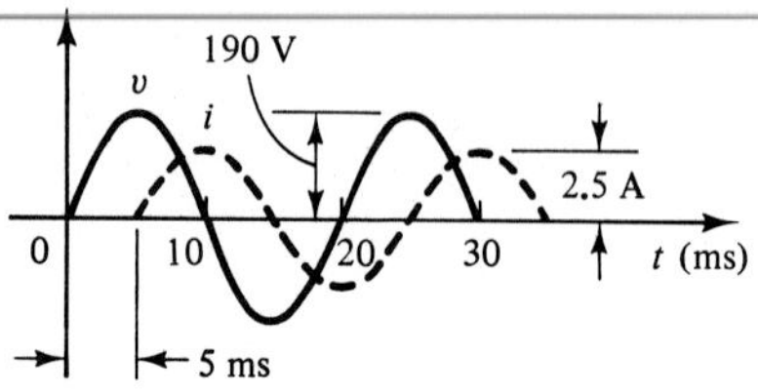


FIGURE 16-29

**Solution:** For the voltage wave,

$$V_m = 190 \quad \phi = 0 \quad \text{and} \quad P = 0.02 \text{ s}$$

By Eq. A77,

$$\omega = \frac{2\pi}{0.02} = 100\pi = 314 \text{ rad/s}$$

So

$$v = 190 \sin 314t$$

where $v$ is expressed in volts. For the current wave, $I_m = 2.5$ A. Since the curve is shifted to the right by 5 ms,

$$\text{phase shift} = 0.005 \text{ s} = -\frac{\phi}{\omega}$$

so the phase angle $\phi$ is

$$\phi = -(0.005 \text{ s})(100\pi \text{ rad/s}) = -\frac{\pi}{2} \text{ rad} = -90°$$

The angular frequency is the same as before, so

$$i = 2.5 \sin(314t - 90°)$$

where $i$ is expressed in amperes. With $\phi$ in radians,

$$i = 2.5 \sin\left(314t - \frac{\pi}{2}\right)$$

where $i$ is expressed in amperes.

## EXERCISE 3—THE SINE WAVE AS A FUNCTION OF TIME

Find the period and angular velocity of a repeating waveform that has a frequency of:

1. 68 Hz
2. 10 Hz
3. 5000 Hz

Find the frequency (in hertz) and angular velocity of a repeating waveform whose period is:

4. 1 s
5. $\frac{1}{8}$ s
6. 95 ms
7. If a periodic waveform has a frequency of 60 Hz, how many seconds will it take to complete 200 cycles?
8. Find the frequency in Hz for a wave that completes 150 cycles in 10 s.

Find the period and frequency of a sine wave that has an angular velocity of:

9. 455 rad/s
10. 2.58 rad/s
11. 500 rad/s

Find the period, amplitude, and phase angle for:

12. The sine wave shown in Figure 16-30a.
13. The sine wave shown in Figure 16-30b.

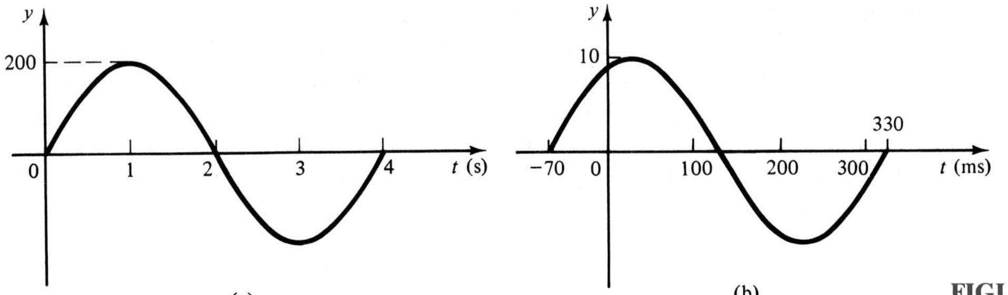


(a) (b) **FIGURE 16-30**

14. Write an equation for a sine wave generated by a phasor of length 5 rotating with an angular velocity of 750 rad/s, with a phase angle of 0°.
15. Repeat Problem 14, with a phase angle of 15°.

Plot each sine wave.

16. $y = \sin t$
17. $y = 3 \sin 377t$
18. $y = 54 \sin(83t - 20°)$
19. $y = 375 \sin\left(55t + \frac{\pi}{4}\right)$

### Adding Sine Waves and Cosine Waves

Express as a single sine wave.

20. $y = 3.00 \sin \omega t + 5.50 \sin(\omega t + 30°)$
21. $y = 18.3 \sin \omega t + 26.3 \sin(\omega t + 75°)$
22. $y = 2.47 \sin \omega t + 1.83 \sin(\omega t - 24°)$
23. $y = 384 \sin \omega t + 536 \sin(\omega t - 48°)$
24. $y = 8370 \sin \omega t + 7570 \cos \omega t$
25. $y = 7.37 \cos \omega t + 5.83 \sin \omega t$
26. $y = 74.2 \sin \omega t + 69.3 \cos \omega t$
27. $y = 364 \sin \omega t + 550 \cos \omega t$

### Alternating Current

28. An alternating current has the equation

$$i = 25 \sin(635t - 18°)$$

where $i$ is given in amperes, A. Find the maximum current, the period, the frequency, the phase angle, and the instantaneous current at $t = 0.01$ s.

29. Given an alternating voltage $v = 4.27 \sin(463t + 27°)$, find the maximum voltage, period, frequency, phase angle, and the instantaneous voltage at $t = 0.12$ s.
30. Write an equation of an alternating voltage that has a peak value of 155 V, a frequency of 60 Hz, and a phase angle of 22°.
31. Write an equation for an alternating current that has a peak amplitude of 49.2 mA, a frequency of 35 Hz, and a phase angle of 63.2°.

## 16-4 POLAR COORDINATES

### The Polar Coordinate System

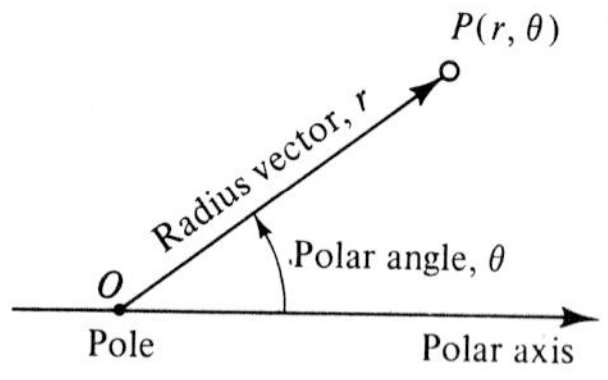


**FIGURE 16-31 Polar coordinates.**

The *polar coordinate system* (Fig. 16-31) consists of a *polar axis,* passing through point $O$, which is called the *pole*. The location of a point $P$ is given by its distance $r$ from the origin, called the *radius vector,* and by the angle $\theta$, called the *polar angle* (sometimes called the vectorial angle or reference angle). The polar angle is called positive when measured counterclockwise from the polar axis, and negative when measured clockwise.

The *polar coordinates* of a point $P$ are thus $r$ and $\theta$, usually written in the form $P(r, \theta)$, or as $r\underline{/\theta}$ (read $r$ at an angle of $\theta$).

**Most of our graphing will continue to be in rectangular coordinates, but in some cases polar coordinates will be more convenient.**

---

**EXAMPLE 25:** A point at a distance 5 from the origin with a polar angle of 28° can be written

$$P(5, 28°) \qquad \text{or} \qquad 5\underline{/28°}$$

---

### Plotting Points in Polar Coordinates

For plotting in polar coordinates it is convenient, although not essential, to have *polar coordinate paper* (Fig. 16-32). This paper has concentric circles, equally spaced, and an angular scale in degrees or radians.

---

**EXAMPLE 26:** The points $P(3, 45°)$, $Q(-2, 320°)$, $R(1.5, 7\pi/6)$, and $S(2.5, -5\pi/3)$ are plotted in Fig. 16-32.

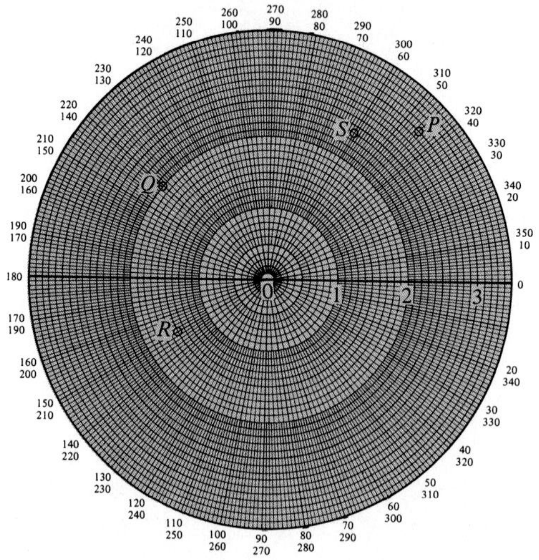


FIGURE 16-32

Notice that to plot $Q(-2, 320°)$, which has a negative value for $r$, we first locate (2, 320°) and then find the corresponding point along the diameter and in the opposite quadrant.

---

## Graphing Equations in Polar Coordinates

To graph a function $r = f(\theta)$, simply assign convenient values to $\theta$ and compute the corresponding value for $r$. Then plot the resulting table of point pairs.

**EXAMPLE 27:** Graph the function $r = \cos\theta$.

**Solution:** Let us take values for $\theta$ every 30° and make a table.

| $\theta$ | 0 | 30° | 60° | 90° | 120° | 150° | 180° | 210° | 240° | 270° | 300° | 330° | 360° |
|---|---|---|---|---|---|---|---|---|---|---|---|---|---|
| $r$ | 1 | 0.87 | 0.5 | 0 | −0.5 | −0.87 | −1 | −0.87 | −0.5 | 0 | 0.5 | 0.87 | 1 |

Plotting these points, we get a circle (Fig. 16-33).

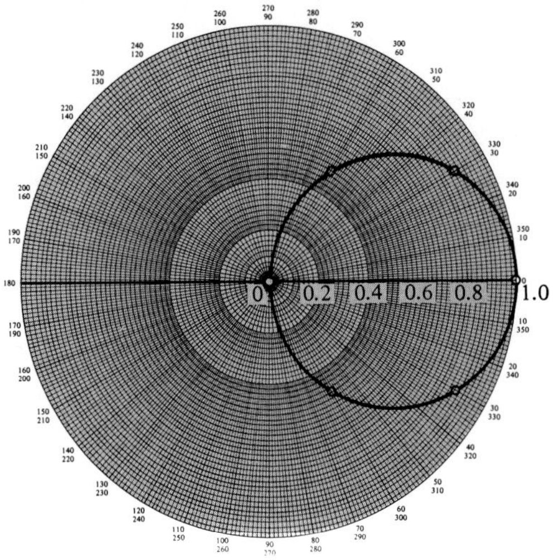


**FIGURE 16-33 Graph of $r = \cos\theta$. The polar equation of a circle that passes through the pole and has its center at $(c, \alpha)$ is $r = 2c\cos(\theta - \alpha)$.**

In Chapter 4 we graphed a parabola in rectangular coordinates. We now graph an equation in polar coordinates that also results in a parabola.

**EXAMPLE 28:** Graph the parabola

$$r = \frac{p}{1 - \cos\theta}$$

with $p = 1$.

**Solution:** As before, we compute $r$ for selected values of $\theta$.

| $\theta$ | 0° | 30° | 60° | 90° | 120° | 150° | 180° | 210° | 240° | 270° | 300° | 330° | 360° |
|---|---|---|---|---|---|---|---|---|---|---|---|---|---|
| $r$ | — | 7.46 | 2.00 | 1.00 | 0.667 | 0.536 | 0.500 | 0.536 | 0.667 | 1.00 | 2.00 | 7.46 | — |

Note that we get division by zero at $\theta = 0°$ and 360°, so that the curve does not exist there. Plotting these points (except for 7.46, which is off the graph), we get the parabola in Fig. 16-34.

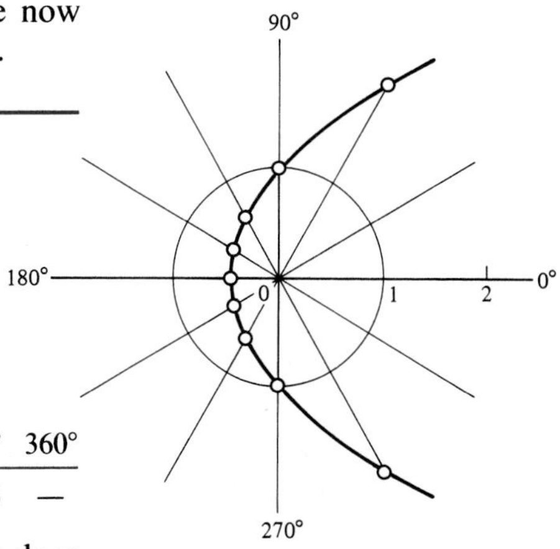


**FIGURE 16-34**

Some of the more interesting curves in polar coordinates are shown in Fig. 16-35.

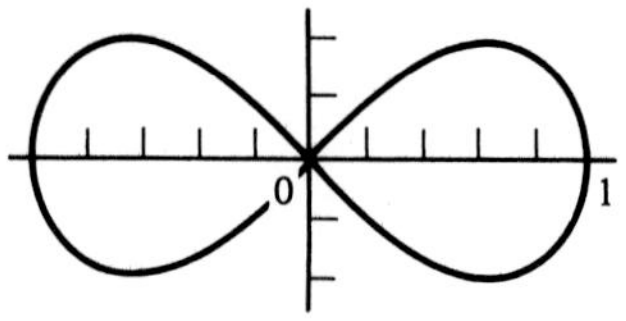


Lemniscate of Bernoulli

$r = a\sqrt{\cos 2\theta}$

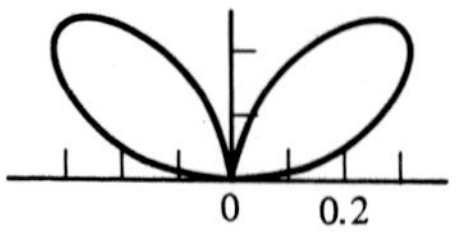


Bifolium

$r = a \sin \theta \cos^2 \theta$

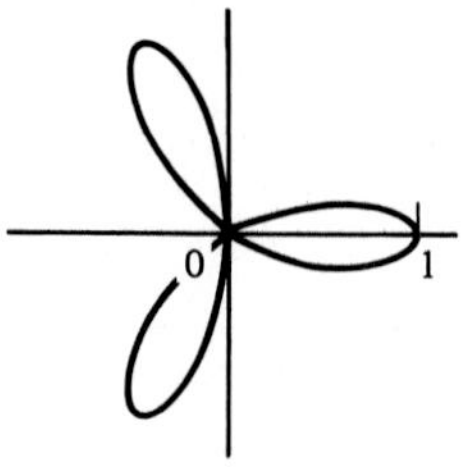


Three-leaved rose

$r = a \cos 3\theta$

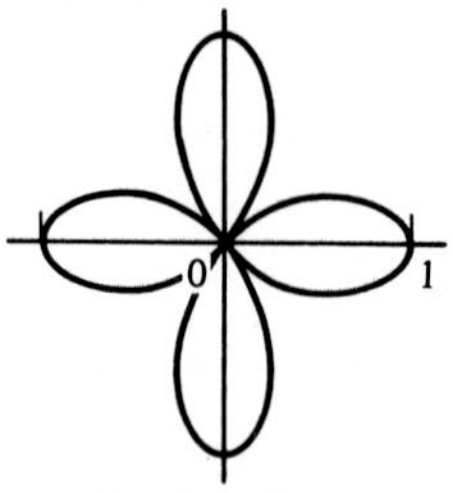


Four-leaved rose

$r = a \cos 2\theta$

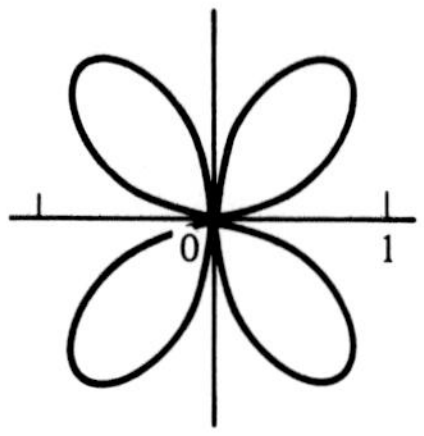


Four-leaved rose

$r = a \sin 2\theta$

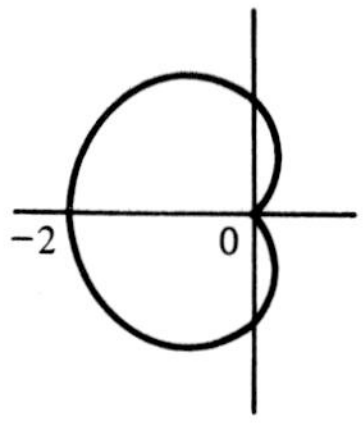


Cardioid

$r = a(1 - \cos \theta)$

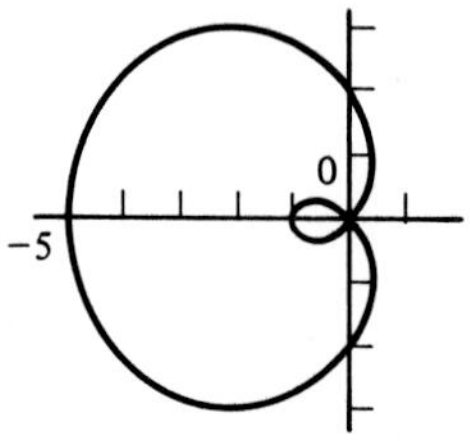


Limaçon of Pascal

$r = b - a \cos \theta$

(graphed with $a = 3$, $b = 2$)

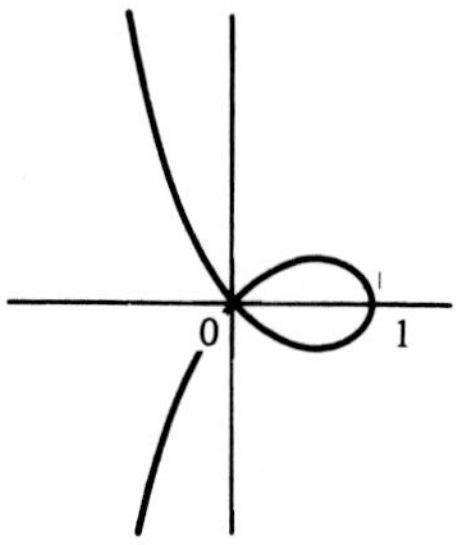


Strophoid

$r = a \cos 2\theta \sec \theta$

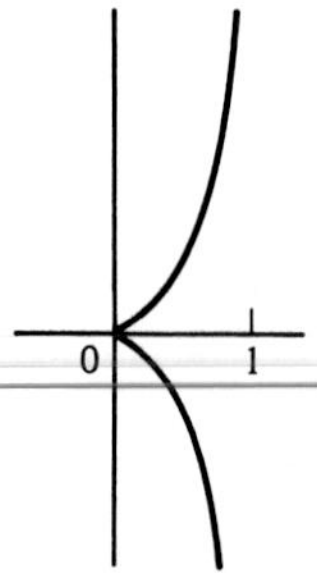


Cissoid of Diocles

$r = a(\sec \theta - \cos \theta)$

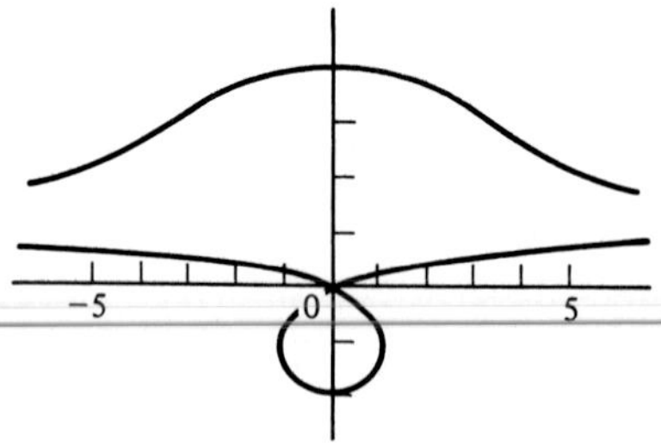


Conchoid of Nicodemus

$r = a \csc \theta + b$

(graphed with $a = 1$, $b = 3$)

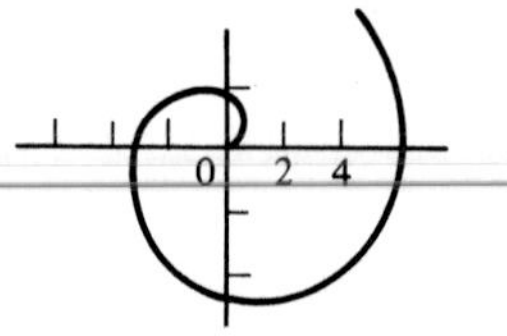


Spiral of Archimedes

$r = a\theta$

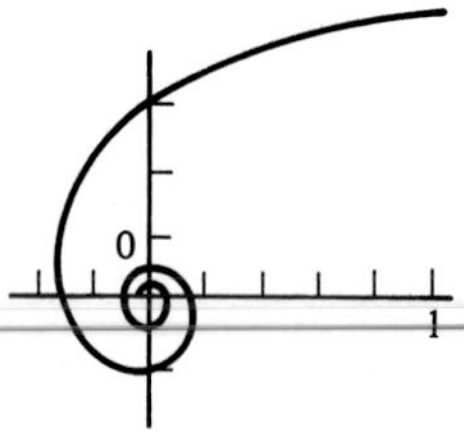


Hyperbolic spiral

$r = a/\theta$

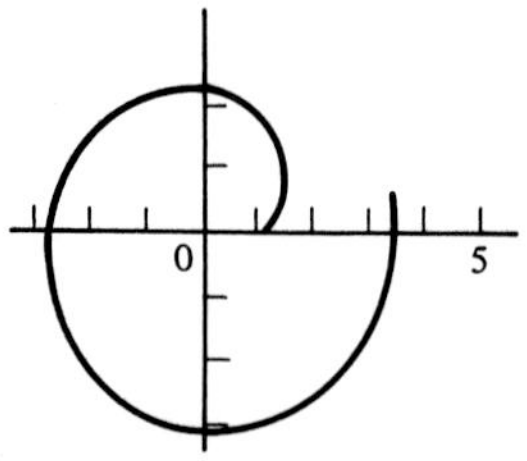


Parabolic spiral

$r = a\sqrt{\theta} + b$

(graphed with $a = b = 1$)

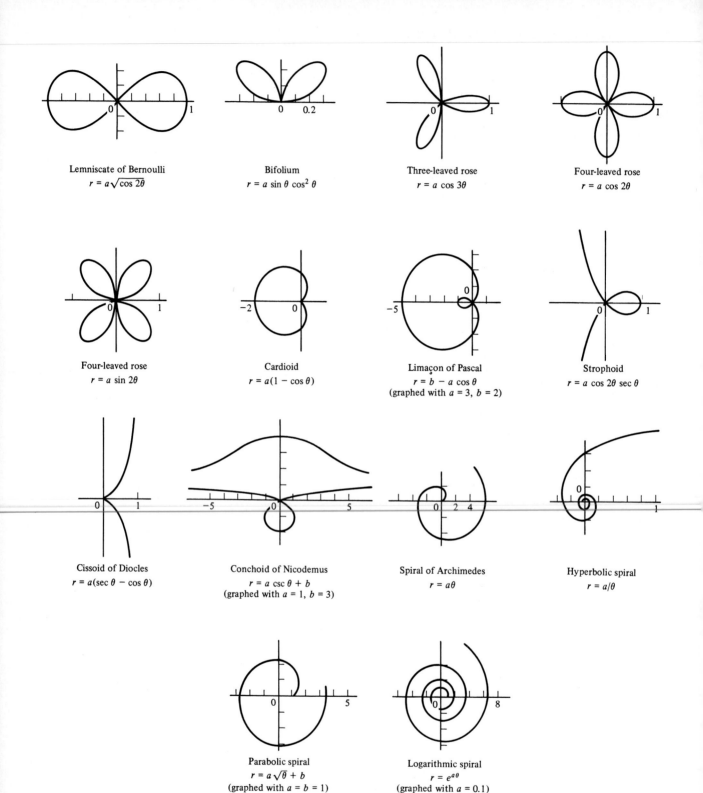

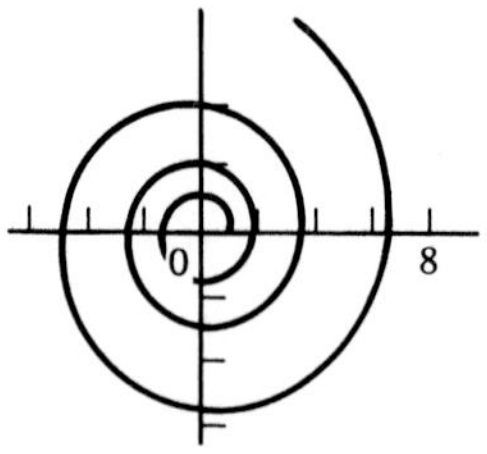


Logarithmic spiral

$r = e^{a\theta}$

(graphed with $a = 0.1$)

**FIGURE 16-35 Some curves in polar coordinates. Unless otherwise noted, the curves were graphed with $a = 1$.**

## Transforming between Polar and Rectangular Coordinates

The relationships between rectangular coordinates and polar coordinates are easily seen when we put both systems on a single diagram (Fig. 16-36). Using the trigonometric functions and the Pythagorean theorem, we get

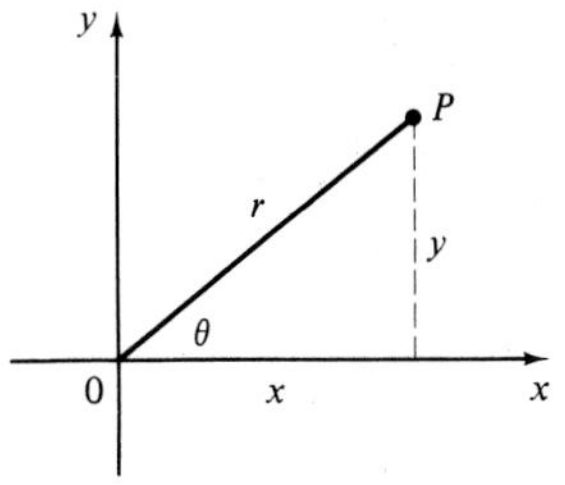


FIGURE 16-36 **Rectangular and polar coordinates of a point.**

| | | |
|---|---|---|
| Rectangular | $x = r \cos \theta$ | **158** |
| | $y = r \sin \theta$ | **159** |
| Polar | $r = \sqrt{x^2 + y^2}$ | **160** |
| | $\theta = \arctan \dfrac{y}{x}$ | **161** |

---

**EXAMPLE 29:** What are the polar coordinates of $P(3, 4)$?

**Solution:**

$$r = \sqrt{9 + 16} = 5$$

$$\theta = \arctan \frac{4}{3} = 53.1°$$

So the polar coordinates of $P$ are $(5, 53.1°)$.

**Some calculators have special keys for converting between polar and rectangular coordinates. Check your manual.**

---

**EXAMPLE 30:** The polar coordinates of a point are $(8, 125°)$. What are the rectangular coordinates?

**Solution:**

$$x = 8 \cos 125° = -4.59$$

$$y = 8 \sin 125° = 6.55$$

So the rectangular coordinates are $(-4.59, 6.55)$.

---

We may also use Eqs. 158 to 161 to transform *equations* from one system of coordinates to the other.

---

**EXAMPLE 31:** Transform the polar equation $r = \cos \theta$ to rectangular coordinates.

**Solution:** Multiplying both sides by $r$ yields

$$r^2 = r \cos \theta$$

Then by Eqs. 158 and 160,

$$x^2 + y^2 = x$$

---

**EXAMPLE 32:** Transform the rectangular equation $2x + 3y = 5$ into polar form.

**Solution:** By Eqs. 158 and 160,

$$2r \cos \theta + 3r \sin \theta = 5$$

or

$$r(2 \cos \theta + 3 \sin \theta) = 5$$

$$r = \frac{5}{2 \cos \theta + 3 \sin \theta}$$

## EXERCISE 4—POLAR COORDINATES

Plot each point in polar coordinates.

1. $(4, 35°)$
2. $(3, 120°)$
3. $(2.5, 215°)$
4. $(3.8, 345°)$
5. $\left(2.7, \frac{\pi}{6}\right)$
6. $\left(3.9, \frac{7\pi}{8}\right)$
7. $\left(-3, \frac{\pi}{2}\right)$
8. $\left(4.2, \frac{2\pi}{5}\right)$
9. $(3.6, -20°)$
10. $(-2.5, -35°)$
11. $\left(-1.8, -\frac{\pi}{6}\right)$
12. $\left(3.7, -\frac{3\pi}{5}\right)$

Graph the following curves from Fig. 16-35.

13. The lemniscate of Bernoulli
14. The bifolium
15. The three-leaved rose
16. The four-leaved rose, $r = a \cos 2\theta$
17. The cardioid
18. The limaçon of Pascal
19. The strophoid
20. The cissoid of Diocles
21. The conchoid of Nicodemus
22. The spiral of Archimedes
23. The hyperbolic spiral
24. The parabolic spiral
25. The logarithmic spiral

Write the polar coordinates of each point.

26. $(2, 5)$
27. $(3, 6)$
28. $(1, 4)$
29. $(-4, 3)$
30. $(2.7, -1.8)$
31. $(-4.8, -5.9)$
32. $(207, 186)$
33. $(-312, -509)$
34. $(1.08, -2.15)$

Write the rectangular coordinates of each point.

35. $(5, 47°)$
36. $(6.3, 227°)$
37. $(445, 312°)$
38. $\left(3.6, \frac{\pi}{5}\right)$
39. $\left(-4, \frac{3\pi}{4}\right)$
40. $\left(18.3, \frac{2\pi}{3}\right)$
41. $(15, -35°)$
42. $(-12, -48°)$
43. $\left(-9.8, -\frac{\pi}{5}\right)$

Write each polar equation in rectangular form.

44. $r = 6$
45. $r = 2 \sin \theta$
46. $r = \sec \theta$
47. $r^2 = 1 - \tan \theta$
48. $r(1 - \cos \theta) = 1$
49. $r^2 = 4 - r \cos \theta$

Write each rectangular equation in polar form.

50. $x = 2$
51. $y = -3$
52. $x = 3 - 4y$
53. $x^2 + y^2 = 1$
54. $3x - 2y = 1$
55. $y = x^2$

## 16-5 GRAPHING PARAMETRIC EQUATIONS IN POLAR COORDINATES

In Chapter 4 we discussed parametric equations of the form

$$x = h(t)$$

and

$$y = g(t)$$

where $x$ and $y$ are each expressed in terms of a third variable $t$, called the parameter. There we graphed parametric equations in rectangular coordinates; here we graph some in polar coordinates.

The procedure is the same. We assign values to the parameter (which will now be $\theta$) and compute $x$ and $y$ for each $\theta$. The table of $(x, y)$ pairs is then plotted.

---

**EXAMPLE 33:** Plot the parametric equations

$$x = 3 \cos \theta$$

$$y = 2 \sin 2\theta$$

**Solution:** We select values for $\theta$ and compute $x$ and $y$.

| $\theta$ | 0 | $\frac{\pi}{4}$ | $\frac{\pi}{2}$ | $\frac{3\pi}{4}$ | $\pi$ | $\frac{5\pi}{4}$ | $\frac{3\pi}{2}$ | $\frac{7\pi}{4}$ | $2\pi$ |
|---|---|---|---|---|---|---|---|---|---|
| $x$ | 3 | 2.12 | 0 | −2.12 | −3 | −2.12 | 0 | 2.12 | 3 |
| $y$ | 0 | 2 | 0 | −2 | 0 | 2 | 0 | −2 | 0 |

Each $(x, y)$ pair is plotted in Fig. 16-37.

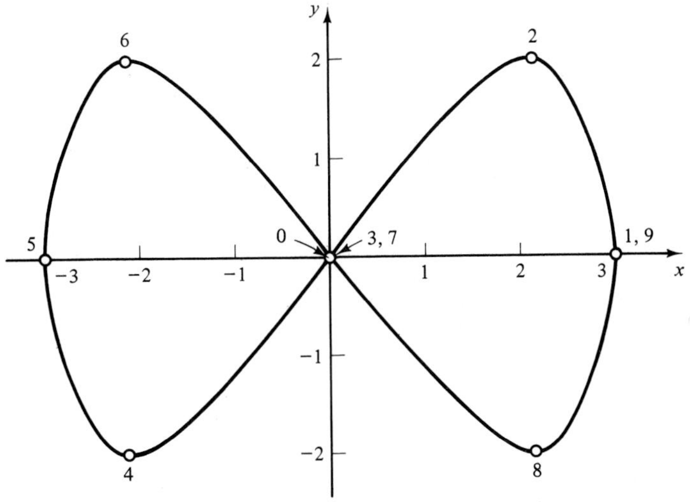


**FIGURE 16-37** Patterns of this sort can be obtained by applying ac voltages to both the horizontal and vertical deflection plates of an oscilloscope. Called *Lissajous figures,* they can indicate the relative amplitudes, frequencies, and phase angles of the two applied voltages.

---

The graphs of some parametric equations in polar coordinates are given in Fig. 16-38.

Cycloid

$x = a(\theta - \sin\theta)$
$y = a(1 - \cos\theta)$

Prolate cycloid

$x = a\theta - b\sin\theta$
$y = a - b\cos\theta$
(graphed with $a = 1$, $b = 2$)

Hypocycloid of four Cusps
or Astroid

$x = a\cos^3\theta$
$y = a\sin^3\theta$

Evolute of ellipse

$x = a\cos^3\theta$
$y = b\sin^3\theta$
(graphed with $a = 1$, $b = 2$)

Witch of Agnesi

$x = a\cot\theta$
$y = a\sin^2\theta$
(graphed with $a = 2$)

Serpentine

$x = a\cot\theta$
$y = b\sin\theta\cos\theta$
(graphed with $a = 1$, $b = 5$)

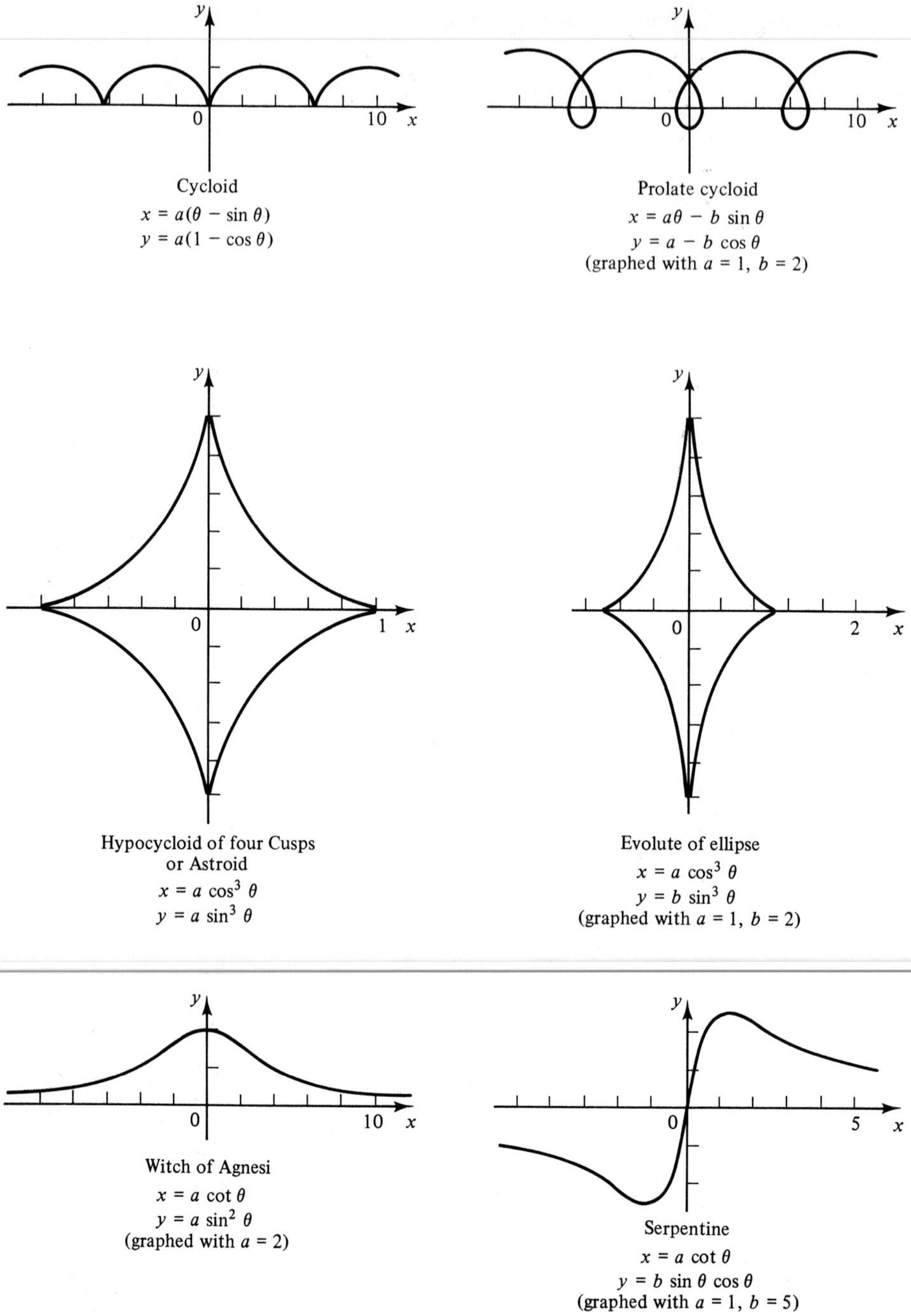

**FIGURE 16-38** **Graphs of some parametric equations in polar coordinates. Unless otherwise noted, the curves were graphed with $a = 1$.**

## EXERCISE 5—PARAMETRIC EQUATIONS IN POLAR COORDINATES __

Graph each curve from Fig. 16-38. Use $a = 1$ unless otherwise noted.

1. The cycloid
2. The prolate cycloid
3. The hypocycloid of four cusps
4. The evolute of ellipse
5. The witch of Agnesi
6. The serpentine

## CHAPTER 16 REVIEW PROBLEMS

*Graph one cycle of each curve and find the period, amplitude, and phase shift.*

1. $y = 3 \sin 2x$
2. $y = 5 \cos 3x$
3. $y = 1.5 \sin \left(3x + \frac{\pi}{2}\right)$
4. $y = -5 \cos \left(x - \frac{\pi}{6}\right)$
5. $y = 2.5 \sin \left(4x + \frac{2\pi}{9}\right)$
6. $y = 4 \tan x$
7. Write the equation of a sine curve that has an amplitude of 5, a period of $3\pi$, and a phase shift of $-\pi/6$.

*Graph by addition of ordinates.*

8. $y = \sin x + \cos 2x$
9. $y = x + \cos x$
10. $y = 2 \sin x - \frac{1}{x}$

*Plot each point in polar coordinates.*

11. (3.4, 125°)
12. (−2.5, 228°)
13. $\left(1.7, -\frac{\pi}{6}\right)$

*Graph each equation in polar coordinates.*

14. $r^2 = \cos 2\theta$
15. $r + 2 \cos \theta = 1$

*Write the polar coordinates of each point.*

16. (7, 3)
17. (−5.3, 3.8)
18. (−24, −52)

*Write the rectangular coordinates of each point.*

19. (3.8, 48°)
20. $\left(-65, \frac{\pi}{9}\right)$
21. (3.8, −44°)

*Transform into rectangular form.*

22. $r(1 - \cos \theta) = 2$
23. $r = 2 \cos \theta$

*Transform into polar form.*

24. $x - 3y = 2$
25. $5x + 2y = 1$
26. Find the frequency and angular velocity of a sine wave that has a period of 2.5 s.
27. Find the period and angular velocity of a cosine wave that has a frequency of 120 Hz.
28. Find the period and frequency of a sine wave that has an angular velocity of 44.8 rad/s.
29. If a sine wave has a frequency of 30 Hz, how many seconds will it take to complete 100 cycles?
30. What frequency must a sine wave have in order to complete 500 cycles in 2 s?
31. Graph the parametric equations $x = 2 \sin \theta$, $y = 3 \sin 4\theta$.
32. Write an equation for an alternating current that has a peak amplitude of 92.6 mA, a frequency of 82 Hz, and a phase angle of 28.3°.

**33.** Write as a single sine wave:

**(a)** $y = 63.7 \sin \omega t + 42.9 \sin(\omega t - 38°)$

**(b)** $y = 1.73 \sin \omega t + 2.64 \cos \omega t$

**34.** Given an alternating voltage $v = 27.4 \sin(736t + 37°)$, find the maximum voltage, period, frequency, phase angle, and the instantaneous voltage at $t = 0.25$ s.

*Writing*

**35.** Given the general sine wave, $y = a \sin(bx + c)$, explain in your own words how a change in one of the constants ($a$, $b$, or $c$) will affect the appearance of the sine wave.

# 17

# TRIGONOMETRIC IDENTITIES AND EQUATIONS

## OBJECTIVES

**When you have completed this chapter, you should be able to:**

- Write a trigonometric expression in terms of the sine and cosine.
- Simplify a trigonometric expression using the fundamental identities.
- Prove trigonometric identities using the fundamental identities.
- Simplify expressions or prove identities using the sum or difference formulas, the double-angle formulas, or the half-angle formulas.
- Solve trigonometric equations.
- Add a sine wave and a cosine wave of the same frequency.
- Evaluate inverse trigonometric functions.

In Sec. 3-1 we said that an *identity* is an equation that is true for *all* meaningful values of the variables, as contrasted with a conditional equation, which is true only for certain values of the variable. Thus $(x + 2)(x - 2) = x^2 - 4$ is true for every value of $x$. A *trigonometric identity* is one that contains one or more trigonometric functions. We have already seen that $\sin\theta = 1/\csc\theta$ for any angle $\theta$ (except, of course, for values of $\theta$ for which $\csc\theta = 0$).

We derive some new identities in this chapter, which we can then use to simplify trigonometric expressions, to verify other identities, and to aid in the solution of trigonometric equations.

## 17-1 FUNDAMENTAL IDENTITIES

### Reciprocal Relations

We have already encountered the reciprocal relations in Sec. 6-2, and we repeat them here.

| | | |
|---|---|---|
| Reciprocal Relations | $\sin\theta = \dfrac{1}{\csc\theta}$, or $\csc\theta = \dfrac{1}{\sin\theta}$, or $\sin\theta\csc\theta = 1$ | **152a** |
| | $\cos\theta = \dfrac{1}{\sec\theta}$, or $\sec\theta = \dfrac{1}{\cos\theta}$, or $\cos\theta\sec\theta = 1$ | **152b** |
| | $\tan\theta = \dfrac{1}{\cot\theta}$, or $\cot\theta = \dfrac{1}{\tan\theta}$, or $\tan\theta\cot\theta = 1$ | **152c** |

---

**EXAMPLE 1:** Simplify $\dfrac{\cos\theta}{\sec^2\theta}$.

**Solution:** Using Eq. 152b gives us

$$\frac{\cos\theta}{\sec^2\theta} = \cos\theta(\cos^2\theta) = \cos^3\theta$$

---

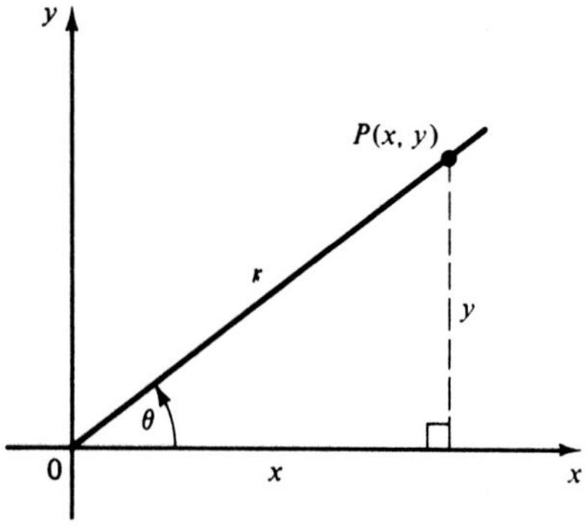


**FIGURE 17-1 Angle in standard position.**

### Quotient Relations

Figure 17-1 shows an angle $\theta$ in standard position, as when we first defined the trigonometric functions in Sec. 6-2. We see that

$$\sin\theta = \frac{y}{r} \quad \text{and} \quad \cos\theta = \frac{x}{r}$$

Dividing yields

$$\frac{\sin\theta}{\cos\theta} = \frac{\frac{y}{r}}{\frac{x}{r}} = \frac{y}{x}$$

But $y/x = \tan\theta$, so

| | | |
|---|---|---|
| Quotient Relations | $\tan \theta = \dfrac{\sin \theta}{\cos \theta}$ | **162** |
| | $\cot \theta = \dfrac{\cos \theta}{\sin \theta}$ | **163** |

where $\cot \theta$ is found by taking the reciprocal of $\tan \theta$.

---

**EXAMPLE 2:** Rewrite the expression

$$\frac{\cot \theta}{\csc \theta} - \frac{\tan \theta}{\sec \theta}$$

so that it contains only the sine and cosine, and simplify.

**Solution:**

$$\frac{\cot \theta}{\csc \theta} - \frac{\tan \theta}{\sec \theta} = \frac{\cos \theta}{\sin \theta} \cdot \frac{\sin \theta}{1} - \frac{\sin \theta}{\cos \theta} \cdot \frac{\cos \theta}{1}$$

$$= \cos \theta - \sin \theta$$

---

## Pythagorean Relations

We can get three more relations by applying the Pythagorean theorem to the triangle in Fig. 17-1.

$$x^2 + y^2 = r^2$$

Dividing through by $r^2$, we get

$$\frac{x^2}{r^2} + \frac{y^2}{r^2} = 1$$

or

$$\left(\frac{x}{r}\right)^2 + \left(\frac{y}{r}\right)^2 = 1$$

but $x/r = \cos \theta$ and $y/r = \sin \theta$, so

| | | |
|---|---|---|
| Pythagorean Relation | $\sin^2 \theta + \cos^2 \theta = 1$ | **164** |

Recall that $\sin^2 \theta$ is a short way of writing $(\sin \theta)^2$.

Returning to equation $x^2 + y^2 = r^2$, we now divide through by $x^2$,

$$1 + \frac{y^2}{x^2} = \frac{r^2}{x^2}$$

But $y/x = \tan \theta$ and $r/x = \sec \theta$, so

| | | |
|---|---|---|
| Pythagorean Relation | $1 + \tan^2 \theta = \sec^2 \theta$ | **165** |

Now dividing $x^2 + y^2 = r^2$ by $y^2$,

$$\frac{x^2}{y^2} + 1 = \frac{r^2}{y^2}$$

But $x/y = \cot\theta$ and $r/y = \csc\theta$, so

| | | |
|---|---|---|
| Pythagorean Relation | $1 + \cot^2\theta = \csc^2\theta$ | **166** |

**EXAMPLE 3:** Simplify

$$\sin^2\theta - \csc^2\theta - \tan^2\theta + \cot^2\theta + \cos^2\theta + \sec^2\theta$$

**Solution:** By the Pythagorean relations,

$$\sin^2\theta - \csc^2\theta - \tan^2\theta + \cot^2\theta + \cos^2\theta + \sec^2\theta$$

$$= (\sin^2\theta + \cos^2\theta) - (\tan^2\theta - \sec^2\theta) + (\cot^2\theta - \csc^2\theta)$$

$$= 1 - (-1) + (-1) = 1$$

## Simplification of Trigonometric Expressions

One use of the trigonometric identities is the simplification of expressions, as in the preceding example. We now give a few more examples.

**EXAMPLE 4:** Simplify $(\cot^2\theta + 1)(\sec^2\theta - 1)$.

**Solution:** By Eqs. 165 and 166,

$$(\cot^2\theta + 1)(\sec^2\theta - 1) = \csc^2\theta\tan^2\theta$$

and by Eqs. 152a and 162,

$$= \frac{1}{\sin^2\theta} \cdot \frac{\sin^2\theta}{\cos^2\theta}$$

$$= \frac{1}{\cos^2\theta}$$

and by Eq. 152b,

$$= \sec^2\theta$$

**EXAMPLE 5:** Simplify $\dfrac{1 - \sin^2\theta}{\sin\theta + 1}$.

**Solution:** Factoring the difference of two squares in the numerator gives

$$\frac{1 - \sin^2\theta}{\sin\theta + 1} = \frac{(1 - \sin\theta)(1 + \sin\theta)}{\sin\theta + 1} = 1 - \sin\theta$$

**EXAMPLE 6:** Simplify $\dfrac{\cos\theta}{1-\sin\theta}+\dfrac{\sin\theta-1}{\cos\theta}$.

**Solution:** Combining the two fractions over a common denominator, we have

$$\frac{\cos\theta}{1-\sin\theta}+\frac{\sin\theta-1}{\cos\theta}=\frac{\cos^2\theta+(\sin\theta-1)(1-\sin\theta)}{(1-\sin\theta)\cos\theta}$$

$$=\frac{\cos^2\theta+\sin\theta-\sin^2\theta-1+\sin\theta}{(1-\sin\theta)\cos\theta}$$

From Eq. 164:

$$=\frac{-\sin^2\theta-\sin^2\theta+2\sin\theta}{(1-\sin\theta)\cos\theta}$$

$$=\frac{-2\sin^2\theta+2\sin\theta}{(1-\sin\theta)\cos\theta}$$

Factor the numerator:

$$=\frac{2\sin\theta(-\sin\theta+1)}{(1-\sin\theta)\cos\theta}$$

$$=\frac{2\sin\theta}{\cos\theta}$$

By Eq. 162:

$$=2\tan\theta$$

Students tend to forget the rules of *algebraic operations* when working trigonometric expressions. We still need to factor, to combine fractions over a common denominator, and so on.

## Trigonometric Identities

We know that an *identity* is an equation that is true for any values of the unknowns, and we have presented eight trigonometric identities so far in this chapter. We now use these eight identities to *verify* or *prove* whether a given identity is true. We do this by trying to transform one side of the identity (usually the more complicated side) until it is identical to the other side.

A good way to start is to rewrite each side so that it contains only sines and cosines.

**EXAMPLE 7:** Prove the identity $\sec\theta\csc\theta=\tan\theta+\cot\theta$.

**Solution:** Expressing the given identity in terms of sines and cosines gives

$$\sec\theta\csc\theta=\tan\theta+\cot\theta$$

$$\frac{1}{\cos\theta}\cdot\frac{1}{\sin\theta}=\frac{\sin\theta}{\cos\theta}+\frac{\cos\theta}{\sin\theta}$$

If we are to make one side identical to the other, each must finally have the same number of terms. But the left side now has one term and the right has two, so we combine the two fractions on the right over a common denominator.

$$\frac{1}{\sin\theta\cos\theta}=\frac{\sin^2\theta+\cos^2\theta}{\sin\theta\cos\theta}$$

By Eq. 164:

$$\frac{1}{\sin\theta\cos\theta}=\frac{1}{\sin\theta\cos\theta}$$

Proving identities is *not easy*. It takes a good knowledge of the fundamental identities, lots of practice, and often several false starts. Do not be discouraged if you fail to get them right away.

| | |
|---|---|
| Common Error | Each side of an identity must be worked *separately;* we cannot transpose, or multiply both sides by the same thing, the way we do with equations. The reason for this is that we do not yet know if the two sides are in fact equal; that is what we are trying to prove. |

---

**EXAMPLE 8:** Prove the identity $\csc x + \tan x = \dfrac{\cot x + \sin x}{\cos x}$.

**Solution:** As in Example 7, we have a single term on one side of the equals sign, and two terms on the other. Now, however, it may be easier to split the single term into two, rather than combine the two terms into one, as we did before,

$$\csc x + \tan x = \frac{\cot x}{\cos x} + \frac{\sin x}{\cos x}$$

Switching to sines and cosines yields

$$\frac{1}{\sin x} + \frac{\sin x}{\cos x} = \frac{\frac{\cos x}{\sin x}}{\cos x} + \frac{\sin x}{\cos x}$$

$$= \frac{1}{\sin x} + \frac{\sin x}{\cos x}$$

---

## EXERCISE 1—FUNDAMENTAL IDENTITIES

Change to an expression containing only sin and cos.

1. $\tan x - \sec x$
2. $\cot x + \csc x$
3. $\tan \theta \csc \theta$
4. $\sec \theta - \tan \theta \sin \theta$
5. $\dfrac{\tan \theta}{\csc \theta} + \dfrac{\sin \theta}{\tan \theta}$
6. $\cot x + \tan x$

Simplify.

7. $1 - \sec^2 x$
8. $\dfrac{\csc \theta}{\sin \theta}$
9. $\dfrac{\cos \theta}{\cot \theta}$
10. $\sin \theta \csc \theta$
11. $\tan \theta \csc \theta$
12. $\dfrac{\sin \theta}{\csc \theta}$
13. $\sec x \sin x$
14. $\sec x \sin x \cot x$
15. $\csc \theta \tan \theta - \tan \theta \sin \theta$
16. $\dfrac{\cos x}{\cot x \sin x}$
17. $\cot \theta \tan^2 \theta \cos \theta$
18. $\dfrac{\tan x(\csc^2 x - 1)}{\sin x + \cot x \cos x}$
19. $\dfrac{\sin \theta}{\cos \theta \tan \theta}$
20. $\dfrac{\sin^2 x + \cos^2 x}{1 - \cos^2 x}$
21. $\dfrac{1}{\sec^2 x} + \dfrac{1}{\csc^2 x}$
22. $\sin \theta(\csc \theta + \cot \theta)$
23. $\csc x - \cot x \cos x$
24. $1 + \dfrac{\tan^2 \theta}{1 + \sec \theta}$

25. $\dfrac{\sec x - \csc x}{1 - \cot x}$

26. $\dfrac{1}{1 + \sin x} + \dfrac{1}{1 - \sin x}$

27. $\sec^2 x(1 - \cos^2 x)$

28. $\tan x + \dfrac{\cos x}{\sin x + 1}$

29. $\cos\theta \sec\theta - \dfrac{\sec\theta}{\cos\theta}$

30. $\cot^2 x \sin^2 x + \tan^2 x \cos^2 x$

Prove each identity.

**All identities in this chapter *can* be proven.**

31. $\tan x \cos x = \sin x$

32. $\tan x = \dfrac{\sec x}{\csc x}$

33. $\dfrac{\sin x}{\csc x} + \dfrac{\cos x}{\sec x} = 1$

34. $\sin\theta = \dfrac{1}{\cot\theta \sec\theta}$

35. $(\cos^2\theta + \sin^2\theta)^2 = 1$

36. $\tan x = \dfrac{\tan x - 1}{1 - \cot x}$

37. $\dfrac{\csc\theta}{\sec\theta} = \cot\theta$

38. $\cot^2 x = \dfrac{\cos x}{\tan x \sin x}$

39. $\cos x + 1 = \dfrac{\sin^2 x}{1 - \cos x}$

40. $\csc x - \sin x = \cot x \cos x$

41. $\cot^2 x - \cos^2 x = \cos^2 x \cot^2 x$

42. $\csc x = \cos x \cot x + \sin x$

43. $1 = (\csc x - \cot x)(\csc x + \cot x)$

44. $\tan x = \dfrac{\tan x + \sin x}{1 + \cos x}$

45. $\dfrac{\tan x + 1}{1 - \tan x} = \dfrac{\sin x + \cos x}{\cos x - \sin x}$

46. $\cot x = \cot x \sec^2 x - \tan x$

47. $\dfrac{\sin\theta + 1}{1 - \sin\theta} = (\tan\theta + \sec\theta)^2$

48. $\dfrac{1 + \sin\theta}{1 - \sin\theta} = \dfrac{1 + \csc\theta}{\csc\theta - 1}$

49. $(\sec\theta - \tan\theta)(\tan\theta + \sec\theta) = 1$

50. $\dfrac{1 + \cot\theta}{\csc\theta} = \dfrac{\tan\theta + 1}{\sec\theta}$

## 17-2 SUM OR DIFFERENCE OF TWO ANGLES

### The Sine of the Sum of Two Angles

We wish now to derive a formula for the sine of the sum of two angles, say $\alpha$ and $\beta$. For example, is it true that

$$\sin 20° + \sin 30° = \sin(50°)?$$

Try it on your calculator—you will see that it is not true.

We start by drawing two positive acute angles $\alpha$ and $\beta$ (Fig. 17-2), small enough so that their sum $(\alpha + \beta)$ is also acute. From any point $P$ on the

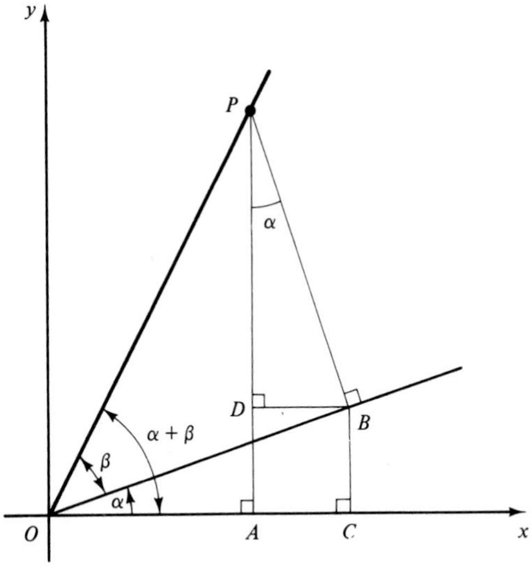


FIGURE 17-2

terminal side of $\beta$ we draw perpendicular $AP$ to the $x$ axis, and draw perpendicular $BP$ to line $OB$. Since the angle between two lines equals the angle between the perpendiculars to those lines, we note that angle $APB$ is equal to $\alpha$. Then

$$\sin(\alpha + \beta) = \frac{AP}{OP} = \frac{AD + PD}{OP} = \frac{BC + PD}{OP}$$

$$= \frac{BC}{OP} + \frac{PD}{OP}$$

But in triangle $OBC$,

$$BC = OB \sin \alpha$$

and in triangle $PBD$,

$$PD = PB \cos \alpha$$

Substituting, we obtain

$$\sin(\alpha + \beta) = \frac{OB \sin \alpha}{OP} + \frac{PB \cos \alpha}{OP}$$

But in triangle $OPB$,

$$\frac{OB}{OP} = \cos \beta$$

and

$$\frac{PB}{OP} = \sin \beta$$

So

We have proven this true for acute angles whose sum is also acute. These identities are, in fact, true for any size angles, positive or negative, although we will not take the space to prove this.

$$\sin(\alpha + \beta) = \sin \alpha \cos \beta + \cos \alpha \sin \beta$$

| Common Error | The sine of the sum of two angles *is not* the sum of the sines of each angle. $\sin(\alpha + \beta) \neq \sin \alpha + \sin \beta$ |
|---|---|

## Cosine of the Sum of Two Angles

Using the same figure, we can derive an expression for $\cos(\alpha + \beta)$. From Eq. 147,

$$\cos(\alpha + \beta) = \frac{OA}{OP} = \frac{OC - AC}{OP} = \frac{OC - BD}{OP}$$

$$= \frac{OC}{OP} - \frac{BD}{OP}$$

Now, in triangle $OBC$,

$$OC = OB \cos \alpha$$

and in triangle $PDB$,

$$BD = PB \sin \alpha$$

Substituting, we obtain

$$\cos(\alpha + \beta) = \frac{OB \cos \alpha}{OP} - \frac{PB \sin \alpha}{OP}$$

As before,

$$\frac{OB}{OP} = \cos \beta \quad \text{and} \quad \frac{PB}{OP} = \sin \beta$$

So

$$\cos(\alpha + \beta) = \cos \alpha \cos \beta - \sin \alpha \sin \beta$$

### The Difference of Two Angles

We can obtain a formula for the sine of the difference of two angles merely by substituting $-\beta$ for $\beta$ in the equation previously derived for $\sin(\alpha + \beta)$.

$$\sin[\alpha + (-\beta)] = \sin \alpha \cos(-\beta) + \cos \alpha \sin(-\beta)$$

But for $\beta$ in the first quadrant, $(-\beta)$ is in the fourth, so

$$\cos(-\beta) = \cos \beta \quad \text{and} \quad \sin(-\beta) = -\sin \beta$$

Therefore,

$$\sin(\alpha - \beta) = \sin \alpha \cos \beta + \cos \alpha(-\sin \beta)$$

or

$$\sin(\alpha - \beta) = \sin \alpha \cos \beta - \cos \alpha \sin \beta$$

We see that the result is identical to the formula for $\sin(\alpha + \beta)$ except for a change in sign. This enables us to write the two identities in a single expression using the $\pm$ sign:

**When the double signs are used, it is understood that the upper signs correspond and the lower signs correspond.**

| | | |
|---|---|---|
| Sine of Sum or Difference of Two Angles | $\sin(\alpha \pm \beta) = \sin \alpha \cos \beta \pm \cos \alpha \sin \beta$ | **167** |

Similarly, finding $\cos(\alpha - \beta)$, we have

$$\cos[\alpha + (-\beta)] = \cos \alpha \cos(-\beta) - \sin \alpha \sin(-\beta)$$

So

$$\cos(\alpha - \beta) = \cos \alpha \cos \beta + \sin \alpha \sin \beta$$

or

| | | |
|---|---|---|
| Cosine of Sum or Difference of Two Angles | $\cos(\alpha \pm \beta) = \cos \alpha \cos \beta \mp \sin \alpha \sin \beta$ | **168** |

---

**EXAMPLE 9:** Expand the expression $\sin(x + 3y)$.

**Solution:** By Eq. 167,

$$\sin(x + 3y) = \sin x \cos 3y + \cos x \sin 3y$$

---

**EXAMPLE 10:** Simplify

$$\cos 5x \cos 3x - \sin 5x \sin 3x$$

**Solution:** We see that this has a similar form to Eq. 168, with $\alpha = 5x$ and $\beta = 3x$, so

$$\begin{aligned}\cos 5x \cos 3x - \sin 5x \sin 3x &= \cos(5x + 3x) \\ &= \cos 8x\end{aligned}$$

---

**EXAMPLE 11:** Prove that

$$\cos(180° - \theta) = -\cos \theta$$

**Solution:** Expanding the left side by means of Eq. 168, we get

$$\cos(180° - \theta) = \cos 180° \cos \theta + \sin 180° \sin \theta$$

But $\cos 180° = -1$ and $\sin 180° = 0$, so

$$\cos(180° - \theta) = (-1) \cos \theta + (0) \sin \theta = -\cos \theta$$

---

**EXAMPLE 12:** Prove $\dfrac{\tan x - \tan y}{\tan x + \tan y} = \dfrac{\sin (x - y)}{\sin (x + y)}$.

**Solution:** The right side contains functions of the sum of two angles, but the left side contains functions of single angles. We thus start by expanding the right side by using Eq. 167,

$$\frac{\sin (x - y)}{\sin (x + y)} = \frac{\sin x \cos y - \cos x \sin y}{\sin x \cos y + \cos x \sin y}$$

Our expression now contains $\sin x$, $\sin y$, $\cos x$, and $\cos y$, but we want an expression containing only $\tan x$ and $\tan y$.

To have $\tan x$ instead of $\sin x$, we can divide numerator and denominator by $\cos x$. Similarly, to obtain $\tan y$ instead of $\sin y$, we can divide by $\cos y$. We thus divide numerator and denominator by $\cos x \cos y$.

$$\frac{\sin x \cos y - \cos x \sin y}{\sin x \cos y + \cos x \sin y} = \frac{\dfrac{\sin x \cancel{\cos y}}{\cos x \cancel{\cos y}} - \dfrac{\cancel{\cos x} \sin y}{\cancel{\cos x} \cos y}}{\dfrac{\sin x \cancel{\cos y}}{\cos x \cancel{\cos y}} + \dfrac{\cancel{\cos x} \sin y}{\cancel{\cos x} \cos y}}$$

Then by Eq. 162:

$$= \frac{\tan x - \tan y}{\tan x + \tan y}$$

---

### Tangent of the Sum or Difference of Two Angles

Since, by Eq. 162,

$$\tan\theta = \frac{\sin\theta}{\cos\theta}$$

we simply divide Eq. 167 by Eq. 168,

$$\tan(\alpha + \beta) = \frac{\sin(\alpha + \beta)}{\cos(\alpha + \beta)} = \frac{\sin\alpha\cos\beta + \cos\alpha\sin\beta}{\cos\alpha\cos\beta - \sin\alpha\sin\beta}$$

Dividing numerator and denominator by $\cos\alpha\cos\beta$ yields

$$\tan(\alpha + \beta) = \frac{\dfrac{\sin\alpha}{\cos\alpha} + \dfrac{\sin\beta}{\cos\beta}}{1 - \dfrac{\sin\alpha}{\cos\alpha}\cdot\dfrac{\sin\beta}{\cos\beta}}$$

Applying Eq. 162 again, we get

$$\tan(\alpha + \beta) = \frac{\tan\alpha + \tan\beta}{1 - \tan\alpha\tan\beta}$$

A similar derivation (which we will not do) will show that $\tan(\alpha - \beta)$ is identical to the expression just derived, except, as we might expect, for a reversal of signs. We combine the two expressions using double signs:

| | | |
|---|---|---|
| Tangent of Sum or Difference of Two Angles | $\tan(\alpha \pm \beta) = \dfrac{\tan\alpha \pm \tan\beta}{1 \mp \tan\alpha\tan\beta}$ | **169** |

---

**EXAMPLE 13:** Simplify

$$\frac{\tan 3x + \tan 2x}{\tan 2x\tan 3x - 1}$$

**Solution:** This can be put into the form of Eq. 169 if we factor $(-1)$ from the denominator,

$$\frac{\tan 3x + \tan 2x}{\tan 2x\tan 3x - 1} = \frac{\tan 3x + \tan 2x}{-(-\tan 2x\tan 3x + 1)}$$

$$= -\frac{\tan 3x + \tan 2x}{1 - \tan 3x\tan 2x}$$

$$= -\tan(3x + 2x)$$

$$= -\tan 5x$$

---

**EXAMPLE 14:** Prove that

$$\tan(45° + x) = \frac{1 + \tan x}{1 - \tan x}$$

**Solution:** Expanding the left side by Eq. 169, we get

$$\tan(45° + x) = \frac{\tan 45° + \tan x}{1 - \tan 45°\tan x}$$

But $\tan 45° = 1$, so

$$\tan(45° + x) = \frac{1 + \tan x}{1 - \tan x}$$

---

**EXAMPLE 15:** Prove

$$\frac{\cot y - \cot x}{\cot x \cot y + 1} = \tan(x - y)$$

**Solution:** Using Eq. 152c on the left side yields

$$\frac{\dfrac{1}{\tan y} - \dfrac{1}{\tan x}}{\dfrac{1}{\tan x} \cdot \dfrac{1}{\tan y} + 1} = \tan(x - y)$$

Multiply numerator and denominator by tan $x$ tan $y$:

$$\frac{\tan x - \tan y}{1 + \tan x \tan y} = \tan(x - y)$$

Then by Eq. 169:

$$\tan(x - y) = \tan(x - y)$$

## Adding a Sine Wave and a Cosine Wave of the Same Frequency

In Sec. 16-3, we used vectors to show that the sum of a sine wave and a cosine wave of the same frequency could be written as a single sine wave, at the original frequency, but with some phase angle. The resulting equation is useful in electrical applications. Here we use the formula for the sum or difference of two angles to derive that equation.

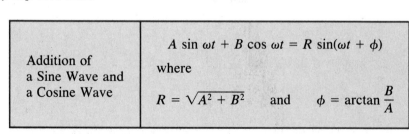

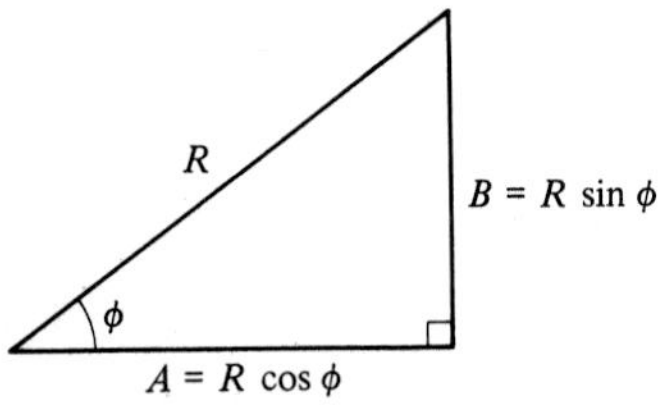

**FIGURE 17-3**

Let $A \sin \omega t$ be a sine wave of amplitude $A$, and $B \cos \omega t$ be a cosine wave of amplitude $B$, each of frequency $\omega/2\pi$. If we draw a right triangle (Fig. 17-3) with sides $A$ and $B$ and hypotenuse $R$, then

$$A = R \cos \phi \quad \text{and} \quad B = R \sin \phi$$

The sum of the sine wave and the cosine wave is then

$$\begin{aligned} A \sin \omega t + B \cos \omega t &= R \sin \omega t \cos \phi + R \cos \omega t \sin \phi \\ &= R(\sin \omega t \cos \phi + \cos \omega t \sin \phi) \end{aligned}$$

The expression on the right will be familiar. It is the relationship of the sine of the *sum* of two quantities, $\omega t$ and $\phi$. Continuing, we have

$$\begin{aligned} A \sin \omega t + B \cos \omega t &= R(\sin \omega t \cos \phi + \cos \omega t \sin \phi) \\ &= R \sin(\omega t + \phi) \end{aligned}$$

by Eq. 167. Thus

| | |
|---|---|
| Addition of a Sine Wave and a Cosine Wave | $A \sin \omega t + B \cos \omega t = R \sin(\omega t + \phi)$ where $R = \sqrt{A^2 + B^2}$ and $\phi = \arctan \dfrac{B}{A}$ |

**EXAMPLE 16:** Express as a single sine function,

$$y = 3.46 \sin \omega t + 2.28 \cos \omega t.$$

**Solution:** The magnitude of the resultant is

$$R = \sqrt{(3.46)^2 + (2.28)^2} = 4.14$$

and the phase angle is

$$\phi = \arctan \frac{2.28}{3.46} = \arctan 0.659 = 33.4°$$

So

$$y = 3.46 \sin \omega t + 2.28 \cos \omega t$$
$$= 4.14 \sin(\omega t + 33.4°)$$

**This is sometimes referred to as the *magnitude and phase* form.**

## EXERCISE 2—SUM OR DIFFERENCE OF TWO ANGLES

Expand by means of the addition and subtraction formulas and simplify.

1. $\sin(\theta + 30°)$
2. $\cos(45° - x)$
3. $\sin(x + 60°)$
4. $\tan(\pi + \theta)$
5. $\cos\left(x + \frac{\pi}{2}\right)$
6. $\tan(2x + y)$
7. $\sin(\theta + 2\phi)$
8. $\cos[(\alpha + \beta) + \gamma]$
9. $\tan(2\theta - 3\alpha)$

Simplify.

10. $\cos 2x \cos 9x + \sin 2x \sin 9x$
11. $\cos(\pi + \theta) + \sin(\pi + \theta)$
12. $\sin 3\theta \cos 2\theta - \cos 3\theta \sin 2\theta$
13. $\sin\left(\frac{\pi}{3} - x\right) - \cos\left(\frac{\pi}{6} - x\right)$

Prove each identity.

14. $\cos x = \sin(x + 90°)$
15. $\sin(30° - x) + \cos(60° - x) = \cos x$
16. $2 \sin \alpha \cos \beta = \sin(\alpha + \beta) + \sin(\alpha - \beta)$
17. $\sin(\theta + 60°) + \cos(\theta + 30°) = \sqrt{3} \cos \theta$
18. $\cos(2\pi - x) = \cos x$
19. $\sin \alpha = \cos (30° - \alpha) - \sin(60° - \alpha)$
20. $\cos(x + y) + \cos(x - y) = 2 \cos x \cos y$
21. $\cos x = \sin\left(x + \frac{\pi}{6}\right) - \sin\left(x - \frac{\pi}{6}\right)$
22. $\tan(360° - \beta) = -\tan \beta$
23. $\cos(60° + \alpha) + \sin (330° + \alpha) = 0$
24. $\sin 5x \sec x \csc x = \frac{\sin 4x}{\sin x} + \frac{\cos 4x}{\cos x}$
25. $\frac{\sin(x - y)}{\sin(x + y)} = \frac{\tan x - \tan y}{\tan x + \tan y}$
26. $\cos(x + 60°) + \cos(60° - x) = \cos x$

27. $\dfrac{\cos(x - y)}{\sin x \cos y} = \tan y + \cot x$

28. $\cot\left(x + \dfrac{\pi}{4}\right) + \tan\left(x - \dfrac{\pi}{4}\right) = 0$

29. $\tan(\alpha + 45^\circ) = \dfrac{\cos \alpha + \sin \alpha}{\cos \alpha - \sin \alpha}$

30. $\dfrac{\cot \alpha \cot \beta - 1}{\cot \alpha + \cot \beta} = \cot(\alpha + \beta)$

31. $\dfrac{1 + \tan x}{1 - \tan x} = \tan\left(\dfrac{\pi}{4} + x\right)$

Express as a single sine function.

32. $y = 47.2 \sin \omega t + 64.9 \cos \omega t$
33. $y = 8470 \sin \omega t + 7360 \cos \omega t$
34. $y = 1.83 \sin \omega t + 2.74 \cos \omega t$
35. $y = 84.2 \sin \omega t + 74.2 \cos \omega t$

## 17-3 FUNCTIONS OF DOUBLE ANGLES

### Sine of $2\alpha$

An equation for the sine of $2\alpha$ is easily derived by setting $\beta = \alpha$ in Eq. 167,

$$\sin(\alpha + \alpha) = \sin \alpha \cos \alpha + \cos \alpha \sin \alpha$$

or

| | | |
|---|---|---|
| Sine of Twice an Angle | $\sin 2\alpha = 2 \sin \alpha \cos \alpha$ | **170** |

### Cosine of $2\alpha$

Similarly, setting $\beta = \alpha$ in Eq. 168, we have

$$\cos(\alpha + \alpha) = \cos \alpha \cos \alpha - \sin \alpha \sin \alpha$$

or

| | | |
|---|---|---|
| Cosine of Twice an Angle | $\cos 2\alpha = \cos^2\alpha - \sin^2\alpha$ | **171a** |

There are two alternative forms for Eq. 171a. From Eq. 164,

$$\cos^2\alpha = 1 - \sin^2\alpha$$

Substituting into Eq. 171a yields

$$\cos 2\alpha = 1 - \sin^2\alpha - \sin^2\alpha$$

| | | |
|---|---|---|
| Cosine of Twice an Angle | $\cos 2\alpha = 1 - 2 \sin^2\alpha$ | **171b** |

We can similarly use Eq. 164 to eliminate the $\sin^2\alpha$ term,

$$\cos 2\alpha = \cos^2\alpha - (1 - \cos^2\alpha)$$

So

| Cosine of Twice an Angle | $\cos 2\alpha = 2\cos^2\alpha - 1$ | **171c** |
|---|---|---|

---

**EXAMPLE 17:** Express $\sin 3x$ in terms of the single angle $x$.

**Solution:** We can consider $3x$ to be $x + 2x$, and, using Eq. 167, we have

$$\sin 3x = \sin(x + 2x) = \sin x \cos 2x + \cos x \sin 2x$$

We now replace $\cos 2x$ by $\cos^2 x - \sin^2 x$ and $\sin 2x$ by $2 \sin x \cos x$.

$$\begin{aligned}\sin 3x &= \sin x(\cos^2 x - \sin^2 x) + \cos x(2 \sin x \cos x)\\ &= \sin x \cos^2 x - \sin^3 x + 2 \sin x \cos^2 x\\ &= 3 \sin x \cos^2 x - \sin^3 x\end{aligned}$$

---

**EXAMPLE 18:** Prove that

$$\cos 2A + \sin(A - B) = 0$$

where $A$ and $B$ are the two acute angles of a right triangle.

**Solution:** By Eqs. 171a and 167,

$$\cos 2A + \sin(A - B) = \cos^2 A - \sin^2 A + \sin A \cos B - \cos A \sin B$$

But by the cofunctions, Eq. 154,

$$\cos B = \sin A \qquad \text{and} \qquad \sin B = \cos A$$

So

$$\begin{aligned}\cos 2A + \sin(A - B) &= \cos^2 A - \sin^2 A + \sin A \sin A - \cos A \cos A\\ &= \cos^2 A - \sin^2 A + \sin^2 A - \cos^2 A = 0\end{aligned}$$

---

**EXAMPLE 19:** Simplify the expression

$$\frac{\sin 2x}{1 + \cos 2x}$$

**Solution:** By Eqs. 170 and 171c,

$$\begin{aligned}\frac{\sin 2x}{1 + \cos 2x} &= \frac{2 \sin x \cos x}{1 + 2\cos^2 x - 1}\\ &= \frac{2 \sin x \cos x}{2\cos^2 x}\\ &= \frac{\sin x}{\cos x} = \tan x\end{aligned}$$

---

### Tangent of $2\alpha$

Setting $\beta = \alpha$ in Eq. 169 gives

$$\tan(\alpha + \alpha) = \frac{\tan \alpha + \tan \alpha}{1 - \tan \alpha \tan \alpha}$$

So

| | | |
|---|---|---|
| Tangent of Twice an Angle | $\tan 2\alpha = \dfrac{2 \tan \alpha}{1 - \tan^2\alpha}$ | **172** |

**EXAMPLE 20:** Prove

$$\frac{2 \cot x}{\csc^2 x - 2} = \tan 2x$$

**Solution:** By Eq. 172,

$$\tan 2x = \frac{2 \tan x}{1 - \tan^2 x}$$

and by Eq. 152c:

$$= \frac{\dfrac{2}{\cot x}}{1 - \dfrac{1}{\cot^2 x}}$$

Multiply numerator and denominator by $\cot^2 x$:

$$= \frac{2 \cot x}{\cot^2 x - 1}$$

Then by Eq. 166:

$$= \frac{2 \cot x}{(\csc^2 x - 1) - 1} = \frac{2 \cot x}{\csc^2 x - 2}$$

| | |
|---|---|
| Common Error | The sine of twice an angle is not twice the sine of the angle. Nor is the cosine (or tangent) of twice an angle equal to twice the cosine (or tangent) of that angle. Remember to use the formulas from this section for all the trig. functions of double angles. |

## EXERCISE 3—FUNCTIONS OF DOUBLE ANGLES

Simplify.

1. $2 \sin^2 x + \cos 2x$
2. $2 \sin 2\theta \cos 2\theta$
3. $\dfrac{2 \tan x}{1 + \tan^2 x}$
4. $\dfrac{2 - \sec^2 x}{\sec^2 x}$
5. $\dfrac{\sin 6x}{\sin 2x} - \dfrac{\cos 6x}{\cos 2x}$

Prove.

6. $\dfrac{2 \tan \theta}{1 - \tan^2\theta} = \tan 2\theta$
7. $\dfrac{1 - \tan^2 x}{1 + \tan^2 x} = \cos 2x$
8. $2 \csc 2\theta = \tan \theta + \cot \theta$
9. $2 \cot 2x = \cot x - \tan x$

10. $\sec 2x = \dfrac{1 + \cot^2 x}{\cot^2 x - 1}$

11. $\dfrac{1 - \tan x}{\tan x + 1} = \dfrac{\cos 2x}{\sin 2x + 1}$

12. $\dfrac{\sin 2\theta + \sin \theta}{1 + \cos \theta + \cos 2\theta} = \tan \theta$

13. $1 + \cot x = \dfrac{2 \cos 2x}{\sin 2x - 2 \sin^2 x}$

14. $4 \cos^3 x - 3 \cos x = \cos 3x$

15. $\sin \alpha + \cos \alpha = \dfrac{\sin 2\alpha + 1}{\cos \alpha + \sin \alpha}$

16. $\dfrac{\cot^2 x - 1}{2 \cot x} = \cot 2x$

If $A$ and $B$ are the two acute angles in a right triangle, show that:

17. $\sin 2A = \sin 2B$

18. $\tan(A - B) = -\cot 2A$

19. $\sin 2A = \cos(A - B)$

## 17-4 FUNCTIONS OF HALF-ANGLES

### Sine of $\alpha/2$

The double-angle formulas derived in Sec. 17-3 can also be regarded as half-angle formulas, because if one angle is double another, the second angle must be half the first.

Starting with Eq. 171b, we obtain

$$\cos 2\theta = 1 - 2 \sin^2\theta$$

We solve for $\sin \theta$.

$$2 \sin^2\theta = 1 - \cos 2\theta$$

$$\sin \theta = \pm \sqrt{\frac{1 - \cos 2\theta}{2}}$$

For emphasis, we replace $\theta$ by $\alpha/2$.

| | | |
|---|---|---|
| Sine of Half an Angle | $\sin\dfrac{\alpha}{2} = \pm \sqrt{\dfrac{1 - \cos \alpha}{2}}$ | **173** |

**In Equations 173, 174, and 175c, we choose the plus or minus sign depending on which quadrant the angle $\alpha/2$ is in. Thus if $\alpha/2$ is in the third quadrant, then $\sin \alpha/2$ is negative, $\cos \alpha/2$ is negative, and $\tan \alpha/2$ is positive.**

### Cosine of $\alpha/2$

Similarly, starting with Eq. 171c, we have

$$\cos 2\theta = 2 \cos^2\theta - 1$$

Then we solve for $\cos \theta$,

$$2 \cos^2\theta = 1 + \cos 2\theta$$

$$\cos \theta = \pm \sqrt{\frac{1 + \cos 2\theta}{2}}$$

and replace $\theta$ by $\alpha/2$.

| | | |
|---|---|---|
| Cosine of Half an Angle | $\cos\dfrac{\alpha}{2} = \pm \sqrt{\dfrac{1 + \cos \alpha}{2}}$ | **174** |

### Tangent of $\alpha/2$

There are three formulas for the tangent of a half-angle; we will show the derivation of each in turn. From Eq. 162,

$$\tan\frac{\alpha}{2} = \frac{\sin\frac{\alpha}{2}}{\cos\frac{\alpha}{2}}$$

Multiply numerator and denominator by $2 \sin(\alpha/2)$:

$$= \frac{2\sin^2\frac{\alpha}{2}}{2\sin\frac{\alpha}{2}\cos\frac{\alpha}{2}}$$

Then by Eqs. 171b and 170,

| | | |
|---|---|---|
| Tangent of Half an Angle | $\tan\frac{\alpha}{2} = \frac{1-\cos\alpha}{\sin\alpha}$ | **175a** |

Another form of this identity is obtained by multiplying numerator and denominator by $1 + \cos\alpha$:

$$\tan\frac{\alpha}{2} = \frac{1-\cos\alpha}{\sin\alpha}\cdot\frac{1+\cos\alpha}{1+\cos\alpha}$$

$$= \frac{1-\cos^2\alpha}{\sin\alpha(1+\cos\alpha)} = \frac{\sin^2\alpha}{\sin\alpha(1+\cos\alpha)}$$

or

| | | |
|---|---|---|
| Tangent of Half an Angle | $\tan\frac{\alpha}{2} = \frac{\sin\alpha}{1+\cos\alpha}$ | **175b** |

We can obtain a third formula for the tangent by dividing Eq. 173 by Eq. 174:

$$\tan\frac{\alpha}{2} = \frac{\sin\frac{\alpha}{2}}{\cos\frac{\alpha}{2}} = \frac{\pm\sqrt{\frac{1-\cos\alpha}{2}}}{\pm\sqrt{\frac{1+\cos\alpha}{2}}}$$

| | | |
|---|---|---|
| Tangent of Half an Angle | $\tan\frac{\alpha}{2} = \pm\sqrt{\frac{1-\cos\alpha}{1+\cos\alpha}}$ | **175c** |

**EXAMPLE 21:** Prove

$$\frac{1 + \sin^2\dfrac{\theta}{2}}{1 + \cos^2\dfrac{\theta}{2}} = \frac{3 - \cos\theta}{3 + \cos\theta}$$

**Solution:** By Eqs. 173 and 174,

$$\frac{1 + \sin^2\dfrac{\theta}{2}}{1 + \cos^2\dfrac{\theta}{2}} = \frac{1 + \dfrac{1 - \cos\theta}{2}}{1 + \dfrac{1 + \cos\theta}{2}}$$

Multiply numerator and denominator by 2:

$$= \frac{2 + 1 - \cos\theta}{2 + 1 + \cos\theta}$$

$$= \frac{3 - \cos\theta}{3 + \cos\theta}$$

## EXERCISE 4—Functions of Half-Angles

Prove each identity.

1. $2\sin^2\dfrac{\theta}{2} + \cos\theta = 1$
2. $4\cos^2\dfrac{x}{2}\sin^2\dfrac{x}{2} = 1 - \cos^2 x$
3. $2\cos^2\dfrac{x}{2} = \dfrac{\sec x + 1}{\sec x}$
4. $\cot\dfrac{\theta}{2} = \csc\theta + \cot\theta$
5. $\dfrac{\sin\alpha}{\cos\alpha + 1} = \sin\dfrac{\alpha}{2}\sec\dfrac{\alpha}{2}$
6. $\dfrac{\cos^2\dfrac{\theta}{2} - \cos\theta}{\sin^2\dfrac{\theta}{2}} = 1$
7. $\sec\theta = \tan\dfrac{\theta}{2}\tan\theta + 1$
8. $\sin x + 1 = \left(\cos\dfrac{x}{2} + \sin\dfrac{x}{2}\right)^2$
9. $2\sin\alpha + \sin 2\alpha = 4\sin\alpha\cos^2\dfrac{\alpha}{2}$
10. $\dfrac{1 - \cos x}{\sin x} = \tan\dfrac{x}{2}$
11. $\dfrac{1 - \tan^2\dfrac{\theta}{2}}{1 + \tan^2\dfrac{\theta}{2}} = \cos\theta$

In right triangle *ABC*, show that:

12. $\tan\dfrac{A}{2} = \dfrac{a}{b + c}$
13. $\sin\dfrac{A}{2} = \sqrt{\dfrac{c - b}{2c}}$

# 17-5 TRIGONOMETRIC EQUATIONS

## Solving Trigonometric Equations

One use for the trigonometric identities we have just studied is in the solution of trigonometric equations.

A trigonometric equation will usually have an infinite number of roots. However, it is customary to list only *nonnegative values less than 360°* that satisfy the equation.

**EXAMPLE 22:** Solve the equation $\sin\theta = \frac{1}{2}$.

**Solution:** Even though there are infinitely many values of $\theta$,

$$\theta = 30°, 150°, 390°, 510°, \ldots$$

whose sine is $\frac{1}{2}$, we will follow the common practice of listing only the nonnegative values less than 360°, and give as our solution

$$\theta = 30°, 150°$$

The solution could also be expressed in radian measure: $\theta = \pi/6, \frac{5}{6}\pi$ radians.

It is difficult to give general rules for solving the great variety of trigonometric equations possible, but a study of the following examples should be helpful.

## Equations Containing a Single Trigonometric Function and a Single Angle

We start with the simplest type. These equations contain, after simplification, only one trigonometric function (only sine, or only cosine, for example) and have only one angle, as in Example 22. If an equation does not *appear* to be of this type, simplify it first. Then isolate the trigonometric function on one side of the equation and find the angles.

**EXAMPLE 23:** Solve the equation $\dfrac{2.73\sec\theta}{\csc\theta} + \dfrac{1.57}{\cos\theta} = 0$.

**Solution:** As with identities, it is helpful to rewrite the given expression with only sines and cosines. Thus

$$\frac{2.73\sin\theta}{\cos\theta} + \frac{1.57}{\cos\theta} = 0$$

Multiplying by $\cos\theta$ and rearranging gives an expression with only one function, sine.

$$\sin\theta = -\frac{1.57}{2.73} = -0.575$$

Our reference angle is arcsin(0.575) or 35.1°. The sine is negative in the third and fourth quadrants, so we give the third and fourth quadrant values for $\theta$:

$$\theta = 180° + 35.1° = 215.1°$$

and

$$\theta = 360° - 35.1° = 324.9°$$

Our next example contains a double angle.

**EXAMPLE 24:** Solve the equation $2 \cos 2\theta - 1 = 0$.

**Solution:** Rearranging and dividing, we have

$$\cos 2\theta = \tfrac{1}{2}$$

$$2\theta = 60°, 300°, 420°, 660°, \text{etc.}$$

$$\theta = 30°, 150°, 210°, \text{and } 330°$$

if we limit our solution to angles less than 360°.

**Note that although this problem contained a double angle, *we did not need the double-angle formula.* It would have been needed, however, if the same problem contained both a double angle *and* a single angle.**

| Common Error | It is easy to forget to find *all the angles* less than 360°, especially when the given equation contains a double angle, such as in Example 24. |
|---|---|

If one of the functions is *squared,* we may have an equation in *quadratic form,* which can be solved by the methods of Chapter 13.

**EXAMPLE 25:** Solve the equation $\sec^2\theta = 4$.

**Solution:** Taking the square root of both sides gives us

$$\sec \theta = \pm\sqrt{4} = \pm 2$$

We thus have two solutions. By the reciprocal relations, we get

$$\frac{1}{\cos \theta} = 2 \quad \text{and} \quad \frac{1}{\cos \theta} = -2$$

$$\cos \theta = \tfrac{1}{2} \quad \text{and} \quad \cos \theta = -\tfrac{1}{2}$$

$$\theta = 60°, 300° \quad \text{and} \quad \theta = 120°, 240°$$

Our solution is then

$$\theta = 60°, 120°, 240°, 300°$$

**EXAMPLE 26:** Solve the equation $2 \sin^2\theta - \sin \theta = 0$.

**Solution:** This is an *incomplete quadratic* in $\sin \theta$. Factoring, we obtain

$$\sin \theta(2 \sin \theta - 1) = 0$$

Setting each factor equal to zero yields

| | |
|---|---|
| $\sin \theta = 0$ | $2 \sin \theta - 1 = 0$ |
| $\theta = 0°, 180°$ | $\sin \theta = \tfrac{1}{2}$ |
| | Since sine is positive in the first and second quadrants, |
| | $\theta = 30°, 150°$ |

If the equation is in the form of a quadratic that cannot be factored, use the quadratic formula.

**EXAMPLE 27:** Solve the equation $\cos^2\theta = 3 + 5\cos\theta$.

**Solution:** Rearranging into standard quadratic form, we have

$$\cos^2\theta - 5\cos\theta - 3 = 0$$

This cannot be factored, so we use the quadratic formula,

$$\cos\theta = \frac{5 \pm \sqrt{25 - 4(-3)}}{2}$$

There are two values for $\cos\theta$:

| | |
|---|---|
| $\cos\theta = 5.54$ | $\cos\theta = -0.541$ |
| (not possible) | Since cosine is negative in the second and third quadrants, |
| | $\theta = 123°, 237°$ |

## Equations with One Angle but More Than One Function

If the equation contains two or more trigonometric functions of the same angle, first transpose all the terms to one side and try to factor that side into factors, *each containing only a single function,* and proceed as above.

**EXAMPLE 28:** Solve $\sin\theta\sec\theta - 2\sin\theta = 0$.

**Solution:** Factoring,

$$\sin\theta(\sec\theta - 2) = 0$$

| | |
|---|---|
| $\sin\theta = 0$ | $\sec\theta = 2$ |
| | $\cos\theta = \frac{1}{2}$ |
| | Since cosine is positive in the first and fourth quadrants, |
| $\theta = 0°, 180°$ | $\theta = 60°, 300°$ |

If the expression is *not factorable* at first, use the fundamental identities to *express everything in terms of a single trigonometric function* and proceed as above.

**EXAMPLE 29:** Solve $\sin^2\theta + \cos\theta = 1$.

**Solution:** By Eq. 164, $\sin^2\theta = 1 - \cos^2\theta$. Substituting gives

$$1 - \cos^2\theta + \cos\theta = 1$$

$$\cos\theta - \cos^2\theta = 0$$

Factoring, we obtain

$$\cos\theta(1 - \cos\theta) = 0$$

| | |
|---|---|
| $\cos\theta = 0$ | $\cos\theta = 1$ |
| $\theta = 90°, 270°$ | $\theta = 0°, 360°$ |

In order to simplify an equation using the Pythagorean relations, it is sometimes necessary to *square both sides*.

---

**EXAMPLE 30:** Solve $\sec\theta + \tan\theta = 1$.

**Solution:** We have no identity that enables us to write $\sec\theta$ in terms of $\tan\theta$, but we do have an identity for $\sec^2\theta$. We rearrange and square both sides, getting

$$\sec\theta = 1 - \tan\theta$$

$$\sec^2\theta = 1 - 2\tan\theta + \tan^2\theta$$

Replacing $\sec^2\theta$ by $1 + \tan^2\theta$ yields

$$1 + \tan^2\theta = 1 - 2\tan\theta + \tan^2\theta$$

$$\tan\theta = 0$$

$$\theta = 0, 180°$$

Since squaring can introduce extraneous roots that do not satisfy the original equation, we substitute back to check our answers. We find that the only angle that satisfies the given equation is $\theta = 0°$.

---

### Equations Containing More Than One Angle

If the equation contains two angles and one angle is a multiple of the other, use the double-angle or half-angle identities to *express everything in terms of a single angle* and proceed as above.

---

**EXAMPLE 31:** Solve the equation

$$2\sin 2\theta + 2\sin\theta + \cos 2\theta - 1 = 0$$

for nonnegative values of $\theta$ less than 360°.

**Solution:** Since the equation contains both $\theta$ and $2\theta$, we use the double-angle identities to get every expression in terms of $\theta$ only. By Eq. 170,

$$\sin 2\theta = 2\sin\theta\cos\theta$$

and by Eq. 171b,

$$\cos 2\theta = 1 - 2\sin^2\theta$$

Substituting gives us

$$2(2\sin\theta\cos\theta) + 2\sin\theta + 1 - 2\sin^2\theta - 1 = 0$$

$$4\sin\theta\cos\theta + 2\sin\theta - 2\sin^2\theta = 0$$

$$2\sin\theta\cos\theta + \sin\theta - \sin^2\theta = 0$$

Factoring yields

$$\sin\theta(2\cos\theta + 1 - \sin\theta) = 0$$

Setting each factor equal to zero, we get

$$\sin \theta = 0$$

$$\theta = 0, \theta = 180°$$

$$2 \cos \theta + 1 - \sin \theta = 0$$

$$2 \cos \theta + 1 = \sin \theta$$

But, by Eq. 164, $\sin^2\theta = 1 - \cos^2\theta$ and $\sin \theta = \pm\sqrt{1 - \cos^2\theta}$

So, substituting,

$$2 \cos \theta + 1 = \pm\sqrt{1 - \cos^2\theta}$$

Squaring gives us

$$4 \cos^2\theta + 4 \cos \theta + 1 = 1 - \cos^2\theta$$

$$5 \cos^2\theta + 4 \cos \theta = 0$$

$$\cos \theta(5 \cos \theta + 4) = 0$$

| | |
|---|---|
| $\cos \theta = 0$ | $\cos \theta = -\frac{4}{5}$ |
| $\theta = 90°$ | $\theta \simeq 143°$ |
| $\theta = 270°$ | $\theta \simeq 217°$ |

These four values should be checked because the squaring process used in the solution often introduces extraneous roots.

Check:

$\theta = 90°$: $\quad 2 \sin 180° + 2 \sin 90° + \cos 180° - 1 \stackrel{?}{=} 0$

$$0 + 2 - 1 - 1 = 0 \qquad \text{checks}$$

$\theta = 143°$: $\quad 2 \sin 286° + 2 \sin 143° + \cos 286° - 1 \stackrel{?}{=} 0$

$$-1.92 + (-1.60) + (0.276) - 1 \stackrel{?}{=} 0$$

does not check

$\theta = 217°$: $\quad 2 \sin(434°) + 2 \sin(217°) + \cos(434°) - 1 \stackrel{?}{=} 0$

$$1.92 - 1.20 + 0.276 - 1 = 0 \text{ checks}$$

$\theta = 270°$: $\quad 2 \sin(540°) + 2 \sin(270°) + \cos(540°) - 1 \stackrel{?}{=} 0$

$$0 + (-2) + (-1) - 1 \neq 0$$

does not check

The roots $\theta = 0$ and $180°$ check, so our solutions are

$$\theta = 0 \qquad \theta = 90° \qquad \theta = 180° \qquad \theta = 217°$$

## EXERCISE 5—TRIGONOMETRIC EQUATIONS

Solve each equation for all nonnegative values of $\theta$ less than 360°.

1. $4 \sin^2\theta = 3$
2. $2 \sin 3\theta = \frac{1}{2}$
3. $2 \sin(\theta + 15°) = 1$
4. $\csc^2\theta = 4$
5. $2 \cos^2\theta = 1 + 2 \sin^2\theta$
6. $2 \sec \theta = -3 - \cos \theta$
7. $4 \sin^4\theta = 1$
8. $2 \csc \theta - \cot \theta = \tan \theta$
9. $1 + \tan \theta = \sec^2\theta$
10. $1 + \cot^2\theta = \sec^2\theta$

11. $3 \cot \theta = \tan \theta$

12. $\sin^2 \theta = 1 - 6 \sin \theta$

13. $3 \sin\frac{\theta}{2} - 1 = 2 \sin^2\frac{\theta}{2}$

14. $\sin \theta = \cos \theta$

15. $4 \cos^2 \theta + 4 \cos \theta = -1$

16. $\cos \theta \sin 2\theta = 0$

17. $3 \tan \theta = 4 \sin^2 \theta \tan \theta$

18. $3 \sin \theta \tan \theta + 2 \tan \theta = 0$

19. $\sec \theta = -\csc \theta$

20. $\sin \theta = 2 \cos\frac{\theta}{2}$

21. $\sin \theta = 2 \sin \theta \cos \theta$

22. $1 + \sin \theta = \sin \theta \cos \theta + \cos \theta$

**Computer**

23. Use your program for solving equations by the midpoint method (Sec. 4-5) to find the roots of any of the trigonometric equations in this section.

## 17-6 INVERSE TRIGONOMETRIC FUNCTIONS

### Inverse of the Sine Function: Arc Sine

The inverse of the sine function

$$y = \sin x$$

is written

$$x = \sin^{-1}y \qquad \text{or} \qquad x = \arcsin y$$

and is read "$x$ is the angle whose sine is $y$."

**Glance back at what we have already said about inverse trigonometric functions in Chaps. 6 and 14.**

We now graph the inverse curve, but first let us interchange $x$ and $y$ so that $y$ is, as usual, the dependent variable. The graph of $y = \arcsin x$ is shown in Fig. 17-4. Note that it is identical to the graph of the sine function but with the $x$ and $y$ axes interchanged.

Recalling the distinction between a relation and a function from Sec. 4-1, we see that $y = \arcsin x$ is a relation, but not a function, because for a single $x$ there may be more than one $y$. For this relation to also be a function, we must limit the range so that repeated values of $y$ are not encountered.

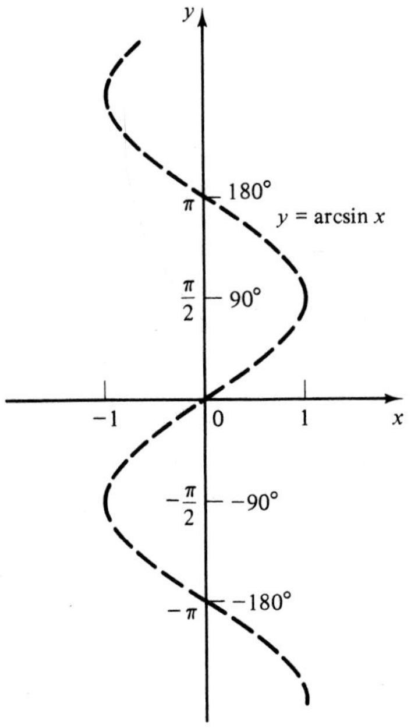


**FIGURE 17-4 Graph of arcsin $x$.**

The first letter of the function is sometimes capitalized. Thus, arcsin $x$ would be the relation and Arcsin $x$ would be the function.

The arc sine, for example, is customarily limited to the interval between $-\pi/2$ and $\pi/2$ (or $-90°$ to $90°$), as shown in Fig. 17-5. The numbers within this interval are called the *principal values* of the arc sine. It is also, of course, the *range* of arcsin $x$.

Given these limits, our relation $y = \text{Arcsin } x$ now becomes a *function*.

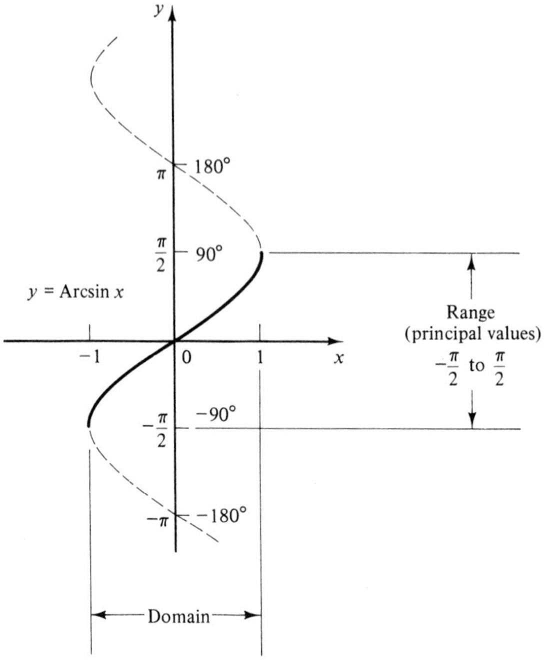


**FIGURE 17-5 Principal values.**

## Arc Cosine and Arc Tangent

In a similar way, the inverse of the cosine and tangent functions are limited to principal values, as shown by the solid lines in Figs. 17-6 and 17-7.

The numerical values of any of the arc functions are found by calculator, as was shown in Secs. 6-2 and 14-1.

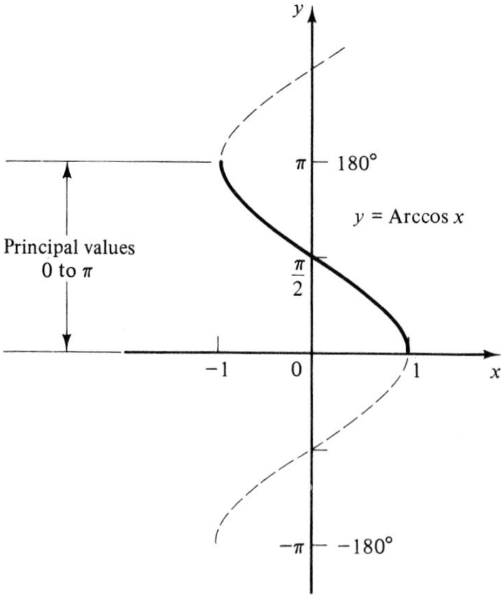


**FIGURE 17-6**

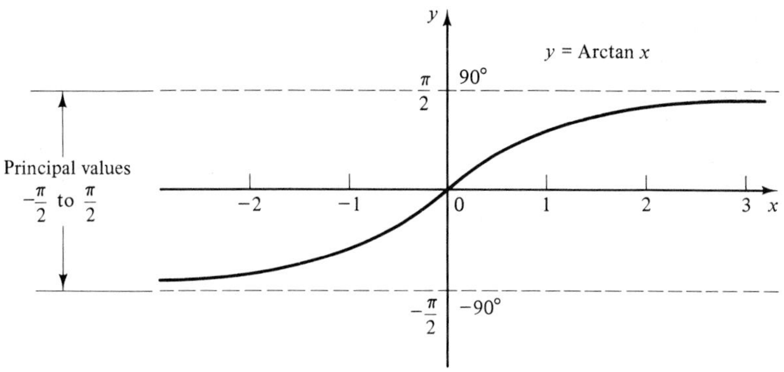


**FIGURE 17-7**

To summarize, if $x$ *is positive,* then Arcsin $x$ is an angle between 0 and 90°, and hence lies in quadrant I. The same is true of Arccos $x$ and Arctan $x$. If $x$ *is negative,* then Arcsin $x$ is an angle between −90° and 0° and hence is in quadrant IV. The same is true of Arctan $x$. However, Arccos $x$ is in quadrant II when $x$ is negative. These facts are summarized in the following table:

| $x$ | *Arcsin x* | *Arccos x* | *Arctan x* |
|---|---|---|---|
| Positive | Quadrant I | Quadrant I | Quadrant I |
| Negative | Quadrant IV | Quadrant II | Quadrant IV |

**EXAMPLE 32:** Find (a) Arcsin 0.638, (b) $\text{Tan}^{-1}(-1.63)$, and (c) Arcsec(−2.34). Give your answer in degrees, to one decimal place.

*Solution:* (a) Since $x$ is positive, we seek a first-quadrant angle. By calculator,

$$\text{Arcsin } 0.638 = 39.6°$$

(b) By calculator

$$\text{Tan}^{-1}(-1.63) = -58.5°$$

Most calculators will give the principal value directly, for first- and second-quadrant angles, and as a negative angle for fourth-quadrant angles. Try your calculator to see what it does.

(c) Using the reciprocal relations, we get

$$\text{Arcsec}(-2.34) = \text{Arccos } \frac{1}{-2.34} = \text{Arccos}(-0.427)$$

$$= 115.3°$$

## EXERCISE 6—INVERSE TRIGONOMETRIC FUNCTIONS

Evaluate each expression. Give your answer in degrees to one decimal place.

1. Arcsin 0.374
2. Arccos 0.826
3. Arctan(−4.92)
4. $\text{Cos}^{-1}0.449$
5. $\text{Tan}^{-1}6.92$
6. $\text{Sin}^{-1}(-0.822)$
7. Arcsec 3.96
8. $\text{Cot}^{-1}4.97$
9. $\text{Csc}^{-1}1.824$
10. $\text{Sec}^{-1}2.89$

## CHAPTER 17 REVIEW PROBLEMS

*Prove.*

1. $\sec^2\theta + \tan^2\theta = \sec^4\theta - \tan^4\theta$
2. $\dfrac{1 + \csc\theta}{\cot\theta} = \dfrac{\cot\theta}{\csc\theta - 1}$
3. $\tan^2\theta\sin^2\theta = \tan^2\theta - \sin^2\theta$
4. $\cot\theta + \csc\theta = \dfrac{1}{\csc\theta - \cot\theta}$
5. $\sin(45° + \theta) - \sin(45° - \theta) = \sqrt{2}\sin\theta$
6. $\dfrac{\cot\alpha\cot\beta + 1}{\cot\beta - \cot\alpha} = \cot(\alpha - \beta)$
7. $\dfrac{\sin^3\theta + \cos^3\theta}{\sin\theta + \cos\theta} = 1 - \dfrac{\sin 2\theta}{2}$
8. $\dfrac{1 + \tan\theta}{1 - \tan\theta} = \sec 2\theta + \tan 2\theta$
9. $\sin\theta\cot\dfrac{\theta}{2} = \cos\theta + 1$
10. $\sin^4\theta + 2\sin^2\theta\cos^2\theta + \cos^4\theta = 1$
11. $\cot\alpha\cot\beta = \dfrac{\csc\beta + \cot\alpha}{\tan\alpha\sec\beta + \tan\beta}$
12. $\cos\theta\cot\theta = \dfrac{\cos\theta + \cot\theta}{\tan\theta + \sec\theta}$

*Simplify.*

13. $\dfrac{\sin\theta\sec\theta}{\tan\theta}$
14. $\sec^2\theta - \sin^2\theta\sec^2\theta$
15. $\dfrac{\sec\theta}{\sec\theta - 1} + \dfrac{\sec\theta}{\sec\theta + 1}$
16. $\dfrac{\sin^3\theta + \cos^3\theta}{\sin\theta + \cos\theta}$
17. $\cos\theta\tan\theta\csc\theta$
18. $(\sec\theta - 1)(\sec\theta + 1)$
19. $\dfrac{\cot\theta}{1 + \cot^2\theta} \cdot \dfrac{\tan^2\theta + 1}{\tan\theta}$
20. $\dfrac{\cos\theta}{1 - \tan\theta} + \dfrac{\sin\theta}{1 - \cot\theta}$
21. $(1 - \sin^2\theta)\sec^2\theta$
22. $\tan^2\theta(1 + \cot^2\theta)$
23. $\dfrac{\sin\theta - 2\sin\theta\cos\theta}{2 - 2\sin^2\theta - \cos\theta}$

*Solve for all positive values of $\theta$ less than 360°.*

24. $1 + 2\sin^2\theta = 3\sin\theta$
25. $3 + 5\cos\theta = 2\cos^2\theta$
26. $\cos\theta - 2\cos^3\theta = 0$
27. $\tan\theta = 2 - \cot\theta$
28. $\sin 3\theta - \sin 6\theta = 0$
29. $\cot 2\theta - 2\cos 2\theta = 0$
30. $\sin\theta + \cos 2\theta = 1$
31. $\sin\theta = 1 - 3\cos\theta$
32. $5\tan\theta = 6 + \tan^2\theta$
33. $\sin^2\theta = 1 + 6\sin\theta$
34. $\tan^2 2\theta = 1 + \sec 2\theta$
35. $16\cos^4\dfrac{\theta}{2} = 9$
36. $\tan^2\theta + 1 = \sec\theta$

*Evaluate.*

37. Arcsin 0.825
38. $\text{Cos}^{-1}0.232$
39. Arctan 3.55

*Express as a single sine function.*

40. $y = 1.84\sin\omega t + 2.18\cos\omega t$
41. $y = 47.6\sin\omega t + 62.1\cos\omega t$

*Writing*

42. Leaf through this chapter and locate any points that are not clear. Write a letter to the author in which you give specific examples of what you think can be explained better. Point out where you would like more examples, and be specific about the type. Instead of putting this letter into your journal, you are urged to send it to the author at Vermont Technical College, Randolph Center, VT 05061. We promise an answer. (You may, of course, do the same for anything in this text.)

# 18

# RATIO, PROPORTION, AND VARIATION

## OBJECTIVES

**When you have completed this chapter, you should be able to:**

- Set up and solve a proportion for a missing quantity.
- Solve applied problems using proportions.
- Set up and solve problems involving direct variation, inverse variation, joint variation, and combined variation.
- Solve applied problems involving variation.
- Set up and solve power function problems.
- Find dimensions, areas, and volumes of similar geometric figures.

In Chapter 8 we introduced the idea of a *ratio* as the quotient of two quantities. Here we expand upon that idea, and also do problems where two ratios are set equal to each other, that is, problems involving *proportions*. The computations will be no different, but we will learn another way of looking at problems involving fractions.

When two quantities, say $x$ and $y$, are connected by some functional relation, $y = f(x)$, some variation in the independent variable $x$ will imply a variation in the dependent variable $y$. Under the heading of *functional variation*, we study the changes in $y$ brought about by changes in $x$, for several simple functions. We often want to know the amount by which $y$ will change when we make a certain change in $x$. The quantities $x$ and $y$ will not be abstract quantities but quantities we care about, such as: By how much will the deflection $y$ of a beam increase when we decrease the beam thickness $x$ by 50%?

We have, in fact, already studied some aspects of functional variation—first when we substituted numerical values for $x$ into a function and computed corresponding values of $y$ and later when we graphed functions in Chapter 4. We do some graphing here, too, and also learn some better techniques for solving numerical problems.

## 18-1 RATIO AND PROPORTION

### Ratio

In our work so far we have often dealt with expressions of the form

$$\frac{a}{b}$$

There are several different ways of looking at such an expression. We can say that $a/b$ is

- A *fraction,* with a numerator $a$ and a denominator $b$
- A *quotient,* where $a$ is *divided* by $b$
- The *ratio* of $a$ to $b$

Thus a ratio can be thought of as a fraction, or as the quotient of two quantities. For a ratio of two physical quantities, it is usual to express the numerator and denominator *in the same units,* so that they cancel and leave the ratio *dimensionless*. The quantities $a$ and $b$ are called the *terms* of the ratio.

**EXAMPLE 1:** A corridor is 8 ft wide and 12 yd long. Find the ratio of length to width.

**Solution:** We first express the length and width in the same units, say, feet.

$$12 \text{ yd} = 36 \text{ ft}$$

So the ratio of length to width is

$$\frac{36}{8} = \frac{9}{2}$$

Another way of writing a ratio is with the colon (:). The ratio of $a$ to $b$ is

$$\frac{a}{b} = a:b$$

In Example 1 we could write that the ratio of length to width is 9 : 2.

Dimensionless ratios are handy because you do not have to worry about units, and are often used in technology. Some examples are:

| | | |
|---|---|---|
| Poisson's ratio | fuel–air ratio | turn ratio |
| endurance ratio | load ratio | gear ratio |
| radian measure of angles | pi | trigonometric ratio |

The word *specific* is often used to denote a ratio when there is a standard unit to which a given quantity is being compared under standard conditions. A few such ratios are:

| | | |
|---|---|---|
| specific heat | specific volume | specific conductivity |
| specific weight | specific gravity | specific speed |

For example, the specific gravity of a solid or liquid is the ratio of the density of the substance to the density of water, at a standard temperature.

## Proportion

A *proportion* is an equation obtained when two ratios are set equal to each other. If the ratio $a:b$ is equal to the ratio $c:d$, we have the proportion

$$a:b = c:d$$

**This can also be written $a:b::c:d$.**

which reads "the ratio of $a$ to $b$ equals the ratio of $c$ to $d$" or "$a$ is to $b$ as $c$ is to $d$."

The two inside terms of a proportion are called the *means*, and the two outside terms are the *extremes*:

extremes ($a$ and $d$)

$$a:b = c:d$$

means ($b$ and $c$)

## Finding a Missing Term

Solve a proportion just as you would any other fractional equation.

---

**EXAMPLE 2:** Find $x$ if $\frac{3}{x} = \frac{7}{9}$.

**Solution:** Multiplying both sides by the LCD, $9x$, we obtain

$$27 = 7x$$

$$x = \frac{27}{7}$$

---

**EXAMPLE 3:** Find $x$ if $\frac{x+2}{3} = \frac{x-1}{5}$.

**Solution:** Multiplying through by 15 gives

$$5(x + 2) = 3(x - 1)$$

$$5x + 10 = 3x - 3$$

$$2x = -13$$

$$x = -\frac{13}{2}$$

**EXAMPLE 4:** Insert the missing quantity,

$$\frac{x-3}{2x} = \frac{?}{4x^2}$$

**Solution:** Replacing the question mark with a variable, say $z$, and solving for $z$ gives us

$$z = \frac{4x^2(x-3)}{2x}$$

$$= 2x(x - 3)$$

## Mean Proportional

When the means of a proportion are equal, as in

$$\frac{a}{b} = \frac{b}{c}$$

the term $b$ is called the *mean proportional* between $a$ and $c$. Solving for $b$, we get

$$b^2 = ac$$

The mean proportional $b$ is also called the *geometric mean* between $a$ and $c$, because $a$, $b$, and $c$ form a *geometric progression* (a series of numbers in which each term is obtained by multiplying the previous term by the same quantity).

| | | |
|---|---|---|
| Mean Proportional | $b = \pm\sqrt{ac}$ | **59** |

**EXAMPLE 5:** Find the mean proportional between 3 and 12.

**Solution:** From Eq. 59,

$$b = \pm\sqrt{3(12)} = \pm 6$$

Some books define $b$ as positive if both $a$ and $c$ are positive. Here, we'll show both values.

### Word Problems Involving Proportions

Use the same guidelines for setting up these word problems as were given in Sec. 3-2 for any word problem. Just be careful not to reverse the terms in the proportion. Match the first item in the verbal statement with the first number in the ratio, and the second with the second.

---

**EXAMPLE 6:** A man is 25 years old and his brother is 15. In how many years will their ages be in the ratio 5 : 4?

**Solution:** Let $x$ = required number of years. After $x$ years, the man's age will be

$$25 + x$$

and the brother's age will be

$$15 + x$$

Forming a proportion, we have

$$\frac{25 + x}{15 + x} = \frac{5}{4}$$

Multiplying by $4(15 + x)$, we obtain

$$4(25 + x) = 5(15 + x)$$

$$100 + 4x = 75 + 5x$$

$$x = 25 \text{ years}$$

---

## EXERCISE 1—RATIO AND PROPORTION

Find the value of $x$.

1. $3:x = 4:6$
2. $x:5 = 3:10$
3. $4:6 = x:4$
4. $3:x = x:12$
5. $x:(14 - x) = 4:3$
6. $x:12 = (x - 12):3$
7. $x:6 = (x + 6):10\frac{1}{2}$
8. $(x - 7):(x + 7) = 2:9$

Insert the missing quantity.

9. $\frac{x}{3} = \frac{?}{9}$
10. $\frac{?}{4x} = \frac{7}{16x}$
11. $\frac{5a}{7b} = \frac{?}{-7b}$
12. $\frac{a - b}{c - d} = \frac{b - a}{?}$
13. $\frac{x + 2}{5x} = \frac{?}{5}$

Find the mean proportional between

14. 2 and 50
15. 3 and 48
16. 6 and 150
17. 5 and 45
18. 4 and 36

#### Word Problems Involving Proportions

19. Find two numbers that are to each other as 5 : 7 and whose sum is 72.
20. Separate $150 into two parts so that the smaller may be to the greater as 7 is to 8.

21. Into what two parts may the number 56 be separated so that one may be to the other as 3 is to 4?
22. The difference between two numbers is 12 and the larger is to the smaller as 11 is to 7. What are the numbers?
23. An estate of \$75,000 was divided between two heirs so that the elder's share was to the younger's as 8 is to 7. What was the share of each?
24. Two partners jointly bought some stock. The first put in \$400 more than the second, and the stock of the first was to that of the second as 5 is to 4. How much money did each invest?
25. The sum of two numbers is 20 and their sum is to their difference as 10 is to 1. What are the numbers?
26. The sum of two numbers is $a$ and their sum is to their difference as $m$ is to $n$. What are the numbers?
27. The *turn ratio* of a transformer (Fig. 18-1) is

$$a = \frac{\text{number of turns in secondary winding}}{\text{number of turns in primary winding}} = \frac{N_2}{N_1}$$

Find the number of turns in the primary for a transformer having 4500 turns in the secondary and a turn ratio of 15.

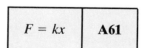

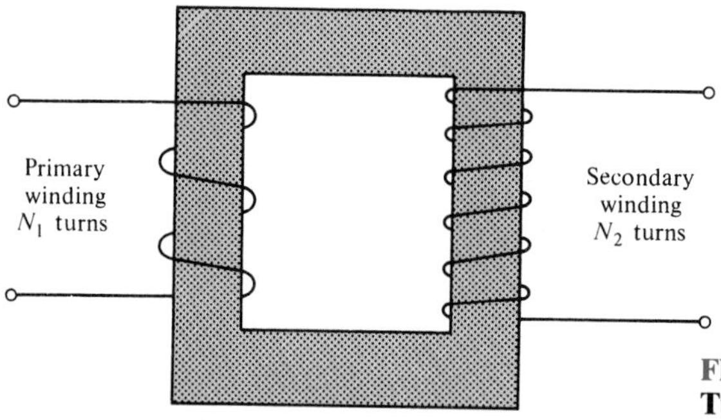


**FIGURE 18-1 A transformer. Turn ratio $a = N_2/N_1$.**

## 18-2 DIRECT VARIATION

### Constant of Proportionality

If two variables are related by an equation of the form

| | | |
|---|---|---|
| Direct Variation | $y = kx$ | **60** |

where $k$ is a constant, we say that *$y$ varies directly* as $x$, or that $y$ is *directly proportional* to $x$. The constant $k$ is called the *constant* of *proportionality*.

---

**EXAMPLE 7:** The force $F$ needed to stretch or compress a spring a distance $x$ is

| | |
|---|---|
| $F = kx$ | **A61** |

Comparing this formula with Eq. 60, we see that the force varies directly as the distance and that the constant of proportionality is $k$ (the spring constant).

---

Direct variation may also be written using the special symbol $\propto$:

$$y \propto x$$

It is read "$y$ varies directly as $x$" or "$y$ is directly proportional to $x$."

## Solving Variation Problems

Variation problems can be solved with or without evaluating the constant of proportionality. We first show a solution in which the constant *is* found by substituting the given values into Eq. 60.

---

**EXAMPLE 8:** If $y$ is directly proportional to $x$, and $y$ is 27 when $x$ is 3, find $y$ when $x$ is 6.

**Solution:** Since $y$ varies directly as $x$, we use Eq. 60:

$$y = kx$$

To find the constant of proportionality, we substitute the given values for $x$ and $y$, 3 and 27:

$$27 = k(3)$$

So $k = 9$. Our equation is then

$$y = 9x$$

When $x = 6$,

$$y = 9(6) = 54$$

---

We now show how to solve such a problem without finding the constant of proportionality.

---

**EXAMPLE 9:** Solve Example 8 *without* finding the constant of proportionality.

**Solution:** When quantities vary directly with each other, we can set up a proportion and solve it. Let us represent the initial values of $x$ and $y$ by $x_1$ and $y_1$, and the second set of values by $x_2$ and $y_2$. Substituting each set of values into Eq. 60 gives us

$$y_1 = kx_1$$

and

$$y_2 = kx_2$$

We divide the second equation by the first, and $k$ cancels.

$$\frac{y_2}{y_1} = \frac{x_2}{x_1}$$

The proportion says: The new $y$ is to the old $y$ as the new $x$ is to the old $x$. We now substitute the old $x$ and $y$ (3 and 27), as well as the new $x$, (6).

**Of course, the same proportion can also be written in the form $y_1/x_1 = y_2/x_2$. Two other forms are also possible:**

$$\frac{x_1}{y_1} = \frac{x_2}{y_2}$$

**and**

$$\frac{x_1}{x_2} = \frac{y_1}{y_2}.$$

$$\frac{y_2}{27} = \frac{6}{3}$$

Solving yields

$$y_2 = 27\left(\frac{6}{3}\right) = 54 \qquad \text{as before.}$$

---

**EXAMPLE 10:** If $y$ varies directly as $x$, fill in the missing numbers in the table of values.

| $x$ | 1 | 2 | | 5 | |
|---|---|---|---|---|---|
| $y$ | | | 16 | 20 | 28 |

**Solution:** We find the constant of proportionality from the given pair of values (5, 20). Starting with Eq. 60, we have

$$y = kx$$

and substituting gives

$$20 = k(5)$$

$$k = 4$$

So

$$y = 4x$$

With this equation we find the missing values.

When $x = 1$: $\quad y = 4$

When $x = 2$: $\quad y = 8$

When $y = 16$: $\quad x = \frac{16}{4} = 4$

When $y = 28$: $\quad x = \frac{28}{4} = 7$

So the completed table is

| $x$ | 1 | 2 | 4 | 5 | 7 |
|---|---|---|---|---|---|
| $y$ | 4 | 8 | 16 | 20 | 28 |

## Word Problems

Many practical problems can be solved using the idea of direct variation. Once you know that two quantities are directly proportional, you may assume an equation of the form of Eq. 60. Substitute the two given values to obtain the constant of proportionality, which you then put back into Eq. 60, to obtain the complete equation. From it you may find any other corresponding values.

Alternatively, you may decide not to find $k$, but to form a proportion in which three values will be known, enabling you to find the fourth.

**EXAMPLE 11:** The force $F$ needed to stretch a spring is directly proportional to the distance $x$ stretched. If it takes 15 N to stretch a certain spring 28 cm, how much force is needed to stretch it 34 cm?

Solution: Assuming an equation of the form

$$F = kx$$

and substituting the first set of values, we have

$$15 = k(28)$$

$$k = \frac{15}{28}$$

So the equation is $F = \frac{15}{28} x$. When $x = 34$,

$$F = \left(\frac{15}{28}\right)34 = 18 \text{ N} \quad \text{(rounded)}$$

---

## EXERCISE 2—DIRECT VARIATION

1. If $y$ varies directly as $x$, and $y$ is 56 when $x$ is 21, find $y$ when $x$ is 74.
2. If $w$ is directly proportional to $z$, and $w$ has a value of 136 when $z$ is 10.8, find $w$ when $z$ is 37.3.
3. If $p$ varies directly as $q$, and $p$ is 846 when $q$ is 135, find $q$ when $p$ is 448.
4. If $y$ is directly proportional to $x$, and $y$ has a value of 88.4 when $x$ is 23.8: **(a)** Find the constant of proportionality. **(b)** Write the equation $y = f(x)$. **(c)** Find $y$ when $x = 68.3$. **(d)** Find $x$ when $y = 164$.

   Assuming that $y$ varies directly as $x$, fill in the missing values in each table of ordered pairs.

5.

| $x$ | 9 | 11 | 15 |
|---|---|---|---|
| $y$ | 45 | 55 | 75 |

6.

| $x$ | 3.40 | 7.20 | | 12.3 |
|---|---|---|---|---|
| $y$ | | 50.4 | 68.6 | |

7.

| $x$ | 115 | 125 | 137.5 | 154 |
|---|---|---|---|---|
| $y$ | 139.6 | 151.8 | 167 | 187 |

8. Graph the linear function $y = 2x$ for values of $x$ from $-5$ to $5$

### Applications

9. The distance between two cities is 828 km and they are 29.5 cm apart on a map. Find the distance between two points 15.6 cm apart on the same map.
10. If the weight of 2500 steel balls is 3.65 kg, find the number of balls in 10.0 kg.
11. If 80 transformer laminations make a stack 1.75 cm thick, how many laminations are contained in a stack 3 cm thick?
12. If your car now gets 21 mi/gal of gas and you can go 250 mi on a tank of gasoline, how far could you drive with the same amount of gasoline with a car that gets 35 mi/gal?
13. A certain automobile engine delivers 53 hp and has a displacement (the total volume swept out by the pistons) of 3.0 liters. If the power is directly proportional to the displacement, what horsepower would you expect from a similar engine that has a displacement of 3.8 liters?

14. The resistance of a conductor is directly proportional to its length. If the resistance of 2.6 mi of a certain transmission line is 155 Ω, find the resistance of 75 mi of that line.
15. The resistance of a certain spool of wire is 1120 Ω. A piece 10 m long is found to have a resistance of 12.3 Ω. Find the length of wire on the spool.
16. If a certain machine can make 1850 parts in 55 min, how many parts can it make in 7.5 h?

For problem 17, use Charles' law: ***The volume of a gas at constant pressure is directly proportional to its absolute temperature.*** **K is the abbreviation for *kelvin*, the SI absolute temperature scale. Add 273.15 to Celsius temperatures to obtain temperatures on the kelvin scale.**

17. In Fig. 18-2, the constant force on the plunger keeps the pressure of the gas in the cylinder constant. The piston raises when the gas is heated and falls when the gas is cooled. If the volume of the gas is 1520 $cm^3$ when the temperature is 302 K, find the volume when the temperature is 358 K.
18. If the power generated by a hydroelectric plant is directly proportional to the flow rate through the turbines, and a flow rate of 5625 gallons of water per minute produces 41.2 MW, how much power would you expect when a drought reduces the flow to 5000 gal/min?

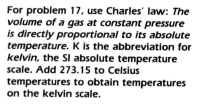

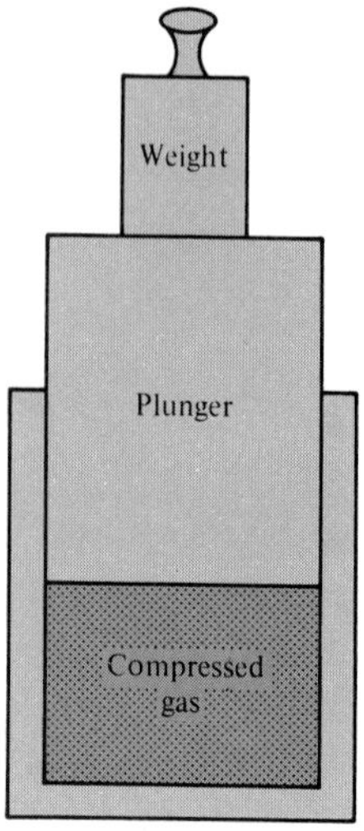

FIGURE 18-2

## 18-3 THE POWER FUNCTION

### Power Function

In Sec. 18-2 we saw that we could represent the statement "$y$ varies directly as $x$" by Eq. 60,

$$y = kx$$

Similarly, if $y$ varies directly as the *square* of $x$, we have

$$y = kx^2$$

or if $y$ varies directly as the *square root* of $x$,

$$y = k\sqrt{x} = kx^{1/2}$$

These are all examples of the *power function:*

| | | |
|---|---|---|
| Power Function | $y = kx^n$ | **194** |

where the constants $k$ and $n$ can be any positive or negative number. This simple function gives us a great variety of relations useful in technology whose forms depend on the value of the exponent, $n$. A few of these are shown in the following table.

**The constants can, of course, be represented by any letter. Appendix A shows *a* instead of *k*.**

**For exponents of 4 and 5, we have the *quartic* and *quintic* functions, respectively.**

| *When:* | *We Get the:* | | *Whose Graph Is a:* |
|---|---|---|---|
| $n = 1$ | Linear function | $y = kx$ (direct variation) | Straight line |
| $n = 2$ | Quadratic function | $y = kx^2$ | Parabola |
| $n = 3$ | Cubic function | $y = kx^3$ | Cubical parabola |
| $n = -1$ | | $y = \frac{k}{x}$ (inverse variation) | Hyperbola |

The first three of these are special cases of the *general polynomial function of degree n,*

| $y = a_0x^n + a_1x^{n-1} + \cdots + a_{n-3}x^3 + a_{n-2}x^2 + a_{n-1}x + a_n$ | **102** |
|---|---|

where $n$ is a positive integer and the $a$'s are constants. For example, the quadratic function $y = kx^2$ is a polynomial function of second degree, where $a_{n-2} = k$ and all the other $a$'s are zeros. It is called *incomplete* because it is lacking an $x$ term and a constant term.

We also consider power functions that are not polynomials; those with noninteger positive powers will be covered in this section and those with negative exponents in the following section on inverse variation.

## Graph of the Power Function

The graph of a power function varies greatly, depending on the exponent. We first show some power functions with positive, integral exponents.

**EXAMPLE 12:** Graph the power functions $y = x^2$ and $y = x^3$ for a range of $x$ from $-3$ to $+3$.

**Solution:** Make a table of values.

| $x$ | $-3$ | $-2$ | $-1$ | 0 | 1 | 2 | 3 |
|---|---|---|---|---|---|---|---|
| $y = x^2$ | 9 | 4 | 1 | 0 | 1 | 4 | 9 |
| $y = x^3$ | $-27$ | $-8$ | $-1$ | 0 | 1 | 8 | 27 |

The curves $y = x^2$ and $y = x^3$ (shown dashed) are plotted in Fig. 18-3. Notice that the plot of $y = x^2$ has no negative values of $y$. This is typical of power functions that have *even* positive integers for exponents. The plot of $y = x^3$ does have negative $y$'s for negative values of $x$. The shape of this curve is typical of a power function that has an *odd* positive integer as exponent.

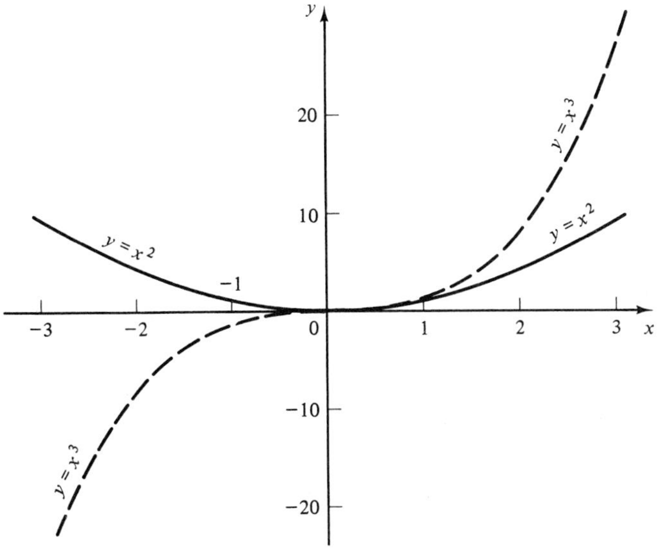


FIGURE 18-3

The graph of $y = kx^3$ is sometimes called a *cubical parabola.*

When graphing a power function with a *fractional* exponent, the curve may not exist for negative values of $x$.

**EXAMPLE 13:** Graph the function $y = x^{1/2}$ for $x = -9$ to 9.

**Solution:** We see that for negative values of $x$ there are no real number values of $y$. For the nonnegative values, we get

| $x$ | 0 | 1 | 2 | 3 | 4 | 5 | 6 | 7 | 8 | 9 |
|---|---|---|---|---|---|---|---|---|---|---|
| $y$ | 0 | 1 | 1.41 | 1.73 | 2.00 | 2.24 | 2.45 | 2.65 | 2.83 | 3.00 |

which are plotted in Fig. 18-4.

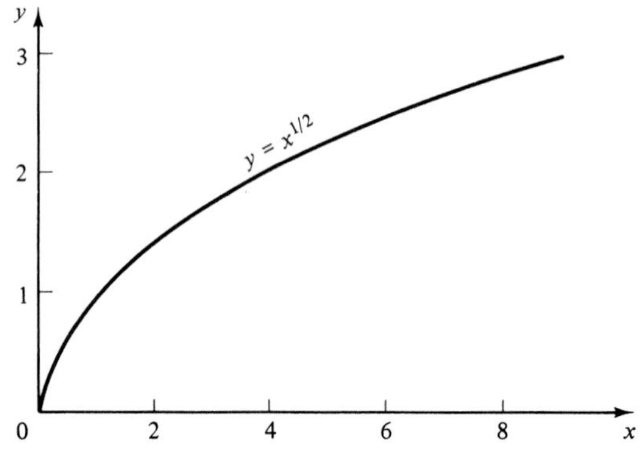


**FIGURE 18-4**

## Solving Power Function Problems

As with direct variation (and with inverse variation and combined variation treated later), we can solve problems involving the power function with or without finding the constant of proportionality. The following example will illustrate both methods.

**EXAMPLE 14:** If $y$ varies directly as the $\frac{3}{2}$ power of $x$, and $y$ is 54 when $x$ is 9, find $y$ when $x$ is 25.

**Solution by Solving for the Constant of Proportionality:** Here we set up an equation with $k$, and solve for the constant of proportionality. We let the exponent $n$ in Eq. 194 be equal to $\frac{3}{2}$,

$$y = kx^{3/2}$$

As before, we evaluate the constant of proportionality by substituting the known values into the equation. Using $x = 9$ and $y = 54$, we obtain

$$54 = k(9)^{3/2} = k(27)$$

so $k = 2$. Our power function is then

$$y = 2x^{3/2}$$

This equation can then be used to find other pairs of corresponding values. For example, when $x = 25$,

$$y = 2(25)^{3/2}$$
$$= 2(\sqrt{25})^3$$
$$= 2(5)^3 = 250$$

**Solution by Setting Up a Proportion:** Here we set up a proportion. Three values are known, and we then solve for the fourth. If

$$y_1 = kx_1^{3/2}$$

and

$$y_2 = kx_2^{3/2}$$

then $y_2$ is to $y_1$ as $kx_2^{3/2}$ is to $kx_1^{3/2}$:

$$\frac{y_2}{y_1} = \left(\frac{x_2}{x_1}\right)^{3/2}$$

Substituting $y_1 = 54$, $x_1 = 9$, and $x_2 = 25$ yields

$$\frac{y_2}{54} = \left(\frac{25}{9}\right)^{3/2}$$

from which

$$y_2 = 54\left(\frac{25}{9}\right)^{3/2} = 250$$

---

**EXAMPLE 15:** The horizontal distance $S$ traveled by a projectile is directly proportional to the square of its initial velocity $V$. If the distance traveled is 1250 m when the initial velocity is 190 m/s, find the distance traveled when the initial velocity is 250 m/s.

**Solution:** The problem statement implies a power function of the form $S = kV^2$. Substituting the initial set of values gives us

$$1250 = k(190)^2$$

$$k = \frac{1250}{(190)^2} = 0.03463$$

So our relationship is

$$S = 0.03463\ V^2$$

When $V = 250\text{m/s}$,

$$S = 0.03463(250)^2 = 2160\text{ m}$$

---

Some problems may contain no numerical values at all, as in the following example.

---

**EXAMPLE 16:** If $y$ varies directly as the cube of $x$, by what factor will $y$ change if $x$ is increased by 25%?

**Solution:** The relationship between $x$ and $y$ is

$$y = kx^3$$

If we give subscripts 1 to the initial values and subscripts 2 to the final values, we may write the proportion

$$\frac{y_2}{y_1} = \frac{kx_2^3}{kx_1^3} = \left(\frac{x_2}{x_1}\right)^3$$

If the new $x$ is 25% greater than the old $x$, then $x_2 = x_1 + 0.25x_1$ or $1.25x_1$. We thus substitute

$$x_2 = 1.25x_1$$

so

$$\frac{y_2}{y_1} = \left(\frac{1.25x_1}{x_1}\right)^3 = \left(\frac{1.25}{1}\right)^3 = 1.95$$

or

$$y_2 = 1.95y_1$$

So $y$ has increased by a factor of 1.95.

---

### Similar Figures

We considered similar triangles in Chapter 5, and said that corresponding sides were in proportion. We now expand the idea to cover similar plane figures of *any* shape, and also similar *solids*. Similar figures (plane or solid) are those in which the distance between any two points on one of the figures is a *constant multiple* of the distance between two corresponding points on the other figure. In other words, if two corresponding dimensions are in the ratio of, say, 2 : 1, all other corresponding dimensions must be in the ratio of 2 : 1. In other words,

| | | |
|---|---|---|
| Dimensions of Similar Figures | Corresponding dimensions of plane or solid similar figures are in proportion. | **134** |

---

**EXAMPLE 17:** Two similar solids are shown in Fig. 18-5. Find the hole diameter $D$.

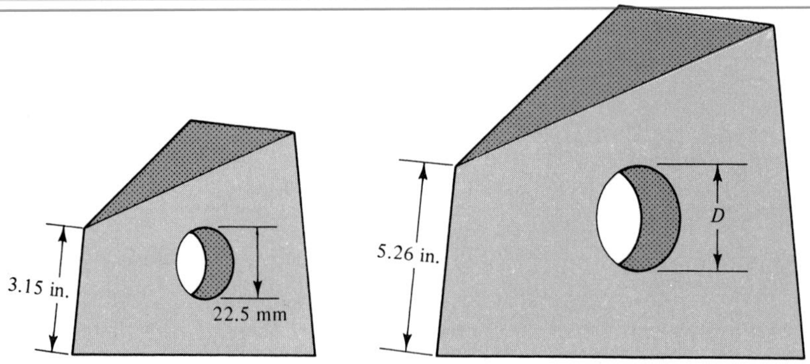


**FIGURE 18-5 Similar solids.**

**Solution:** By statement 134,

$$\frac{D}{22.5} = \frac{5.26}{3.15}$$

$$D = \frac{22.5(5.26)}{3.15} = 37.6 \text{ mm}$$

---

Note that since we are dealing with *ratios* of corresponding dimensions, it was not necessary to convert all dimensions to the same units.

## Areas of Similar Figures

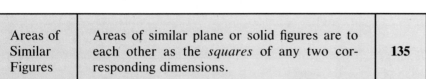

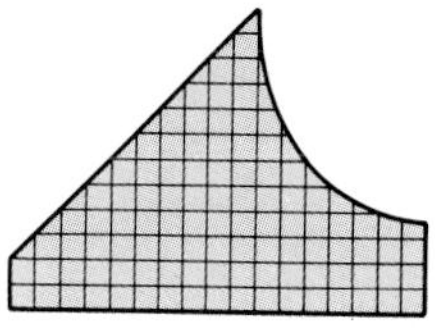

FIGURE 18-6 **We can, in our minds, make the squares so small that they completely fill any irregular area. We use a similar idea in calculus when finding areas by integration.**

The area of a square of side $s$ is equal to the square of one side, that is, $s \times s$ or $s^2$. Thus if a side is multiplied by a factor $k$, the area of the larger square is $(ks) \times (ks)$ or $k^2s^2$. We see that the area has increased by a factor of $k^2$.

An area (Fig. 18-6) more complicated than a square can be thought of as made up of many small squares. Then if a dimension of that area is multiplied by $k$, the area of each small square increases by a factor of $k^2$, and hence the entire area of the figure increases by a factor of $k^2$. Thus

| | | |
|---|---|---|
| Areas of Similar Figures | Areas of similar plane or solid figures are to each other as the *squares* of any two corresponding dimensions. | **135** |

Thus if a figure has its sides doubled, the new area would be *four* times the original area. This relationship is valid not only for plane areas and for surface areas of solids, but also for cross-sectional areas of solids.

The units of measure for area are square inches ($in^2$), square centimeters ($cm^2$), and so on, depending on the units used for the sides of the figure.

---

**EXAMPLE 18:** The triangular top surface of the smaller solid in Fig. 18-5 has an area of 4.84 $in^2$. Find the area $A$ of the corresponding surface on the larger solid.

**Solution:** By statement 135, the area $A$ is to 4.84 as the *square* of the ratio of 5.26 to 3.15.

$$\frac{A}{4.84} = \left(\frac{5.26}{3.15}\right)^2$$

$$A = 4.84\left(\frac{5.26}{3.15}\right)^2 = 13.5 \text{ in}^2$$

---

## Volumes of Similar Solids

A cube of side $s$ has volume $s^3$. We can think of any solid as being made up of many tiny cubes, each of which has a volume equal to the cube of its side. Thus if the dimensions of the solid are multiplied by a factor of $k$, the volume of each cube (and hence the entire solid) will increase by a factor of $k^3$.

| | | |
|---|---|---|
| Volumes of Similar Figures | Volumes of similar solid figures are to each other as the *cubes* of any two corresponding dimensions. | **136** |

The units of measure for volume are cubic inches ($in^3$), cubic centimeters ($cm^3$), and so on, depending on the units used for the sides of the solid figure.

**EXAMPLE 19:** If the volume of the smaller solid in Fig. 18-5 is 15.6 in³, find the volume $V$ of the larger solid.

**Solution:** By statement 136, the volume $V$ is to 15.6 as the *cube* of the ratio of 5.26 to 3.15.

$$\frac{V}{15.6} = \left(\frac{5.26}{3.15}\right)^3$$

$$V = 15.6\left(\frac{5.26}{3.15}\right)^3 = 72.6 \text{ in}^3$$

| Common Error | Students often forget to *square* corresponding dimensions when finding areas, and to *cube* corresponding dimensions when finding volumes. |
|---|---|

## Scale Drawings

An important application of similar figures is in the use of *scale drawings,* such as maps, engineering drawings, surveying layouts, and so on. The ratio of distances on the drawing, to corresponding distances on the actual object, is called the *scale* of the drawing. Use statement 135 to convert between the areas on the drawing and areas on the actual object.

**EXAMPLE 20:** A certain map has a scale of 1 : 5000. How many acres on the land are represented by 168 in² on the map?

**Solution:** If $A$ = the area on the land, then by statement 135,

$$\frac{A}{168} = \left(\frac{5000}{1}\right)^2 = 25{,}000{,}000$$

$$A = 168(25{,}000{,}000) = 42.0 \times 10^8 \text{ in}^2$$

Converting to acres, we have

$$A = 42.0 \times 10^8 \text{ in}^2\left(\frac{1 \text{ ft}^2}{144 \text{ in}^2}\right)\left(\frac{1 \text{ acre}}{43{,}560 \text{ ft}^2}\right) = 670 \text{ acres}$$

## EXERCISE 3—THE POWER FUNCTION

1. If $y$ varies directly as the square of $x$, and $y$ is 285 when $x$ is 112, find $y$ when $x$ is 315.
2. If $y$ varies directly as the square root of $x$, and $y$ is 11.8 when $x$ is 342, find $y$ when $x$ is 288.
3. If $y$ is directly proportional to the cube of $x$, and $y$ is 638 when $x$ is 145, find $y$ when $x$ is 68.3.
4. If $y$ is directly proportional to the five-halves power of $x$, and $y$ has the value 55.3 when $x$ is 17.3: **(a)** find the constant of proportionality; **(b)** write the equation $y = f(x)$; **(c)** find $y$ when $x = 27.4$; **(d)** find $x$ when $y = 83.6$.

5. If $y$ varies directly as the fourth power of $x$, fill in the missing values in the following table of ordered pairs.

| $x$ | 18.2 | 75.6 | |
|---|---|---|---|
| $y$ | 29.7 | | 154 |

6. If $y$ is directly proportional to the $\frac{3}{2}$ power of $x$, fill in the missing values in the following table of ordered pairs.

| $x$ | 1.054 | 1.135 | |
|---|---|---|---|
| $y$ | | 4.872 | 6.774 |

7. If $y$ varies directly as the cube root of $x$, fill in the missing values in the following table of ordered pairs.

| $x$ | 315 | | 782 |
|---|---|---|---|
| $y$ | | 148 | 275 |

8. Graph the power function $y = 1.04x^2$ for $x = -5$ to 5.
9. Graph the power function $y = 0.553x^5$ for $x = -3$ to 3.
10. Graph the power function $y = 1.25x^{3/2}$ for $x = -5$ to 5.

### Freely Falling Body

11. If a body falls 176 m in 6 s, how far will it fall in 9 s?
12. If a body falls 4.90 m during the first second, how far will it fall during the third second?
13. If a body falls 129 ft in 2.00 s, how many seconds will it take to fall 525 ft?
14. If a body falls 738 units in 3.00 s, how far will it fall in 6.00 s?

**From Eq. A18 we see that the distance fallen by a body (from rest) varies directly as the square of the elapsed time. Assume an initial velocity of zero for each of these problems.**

### Power in a Resistor

15. If the power dissipated in a resistor is 486 W when the current is 2.75 A, find the power when the current is 3.45 A.
16. If the current through a resistor is increased by 28%, by what percent will the power increase?
17. By what factor must the current in an electric heating coil be increased to triple the power consumed by the heater?

**The power dissipated in a resistor varies directly as the square of the current in the resistor (Eq. A67).**

### Photographic Exposures

18. A certain photograph will be correctly exposed at a shutter speed of 1/100 s with a lens opening of f5.6. What shutter speed is required if the lens opening is changed to f8?
19. For the photograph in Problem 18, what lens opening is needed for a shutter speed of 1/50 s?
20. Most cameras have the following f stops: f2.8, f4, f5.6, f8, f11, and f16. To keep the same correct exposure, by what factor must the shutter speed be increased when the lens is opened one stop?
21. A certain enlargement requires an exposure time of 24 s when the enlarger's lens is set at f22. What lens opening is needed to reduce the exposure time to 8 s?

**The exposure time for a photograph is directly proportional to the square of the f stop. (The f stop of a lens is its focal length divided by its diameter.)**

**Photographer's rule of thumb: *Double the exposure time for each increase in f stop.***

### Similar Figures

22. A container for storing propane gas is 0.755 m high and contains 20.0 liters. How high would a container of similar shape have to be to have a volume of 40.0 liters?

23. A certain wood stove is 24 in. wide and has a firebox volume of 4.25 $ft^3$. What firebox volume would be expected if all dimensions of the stove are increased by a factor of 1.25?

24. If the stove in Problem 23 weighs 327 lb, how much would the larger stove be expected to weigh?

25. A certain solar house stores heat in 155 metric tons of stones which are in a chamber beneath the house. Another solar house is to have a chamber of similar shape but with all dimensions increased by 15%. How many metric tons of stone will it hold?

26. Each side of a square is increased by 15 mm and the area is seen to increase by 2450 $mm^2$. What were the dimensions of the original square?

27. The floor plan of a certain building has a scale of $\frac{1}{4}$ in. = 1 ft and shows a room having an area of 40 $in^2$. What is the actual room area in square feet?

28. The area of a window of a car is 18.2 $in^2$ on a drawing having a scale of 1 : 4. Find the actual window area in square feet.

Assume that the amount of flow through a pipe is proportional to its cross-sectional area.

29. A pipe 3 in. in diameter discharges 500 gal of water in a certain time. What must be the diameter of a pipe that will discharge 750 gal in the same time?

30. If it cost $756 to put a fence around a circular pond, what will it cost to enclose another pond having $\frac{1}{5}$ the area?

31. A triangular field whose base is 215 m contains 12,400 $m^2$. Find the area of a field of similar form whose base is 328 m.

**Computer**

32. The power function $y = 11.9x^{2.32}$ has been proposed as an equation to fit the following data:

| $x$ | 1 | 2 | 3 | 4 | 5 | 6 | 7 | 8 | 9 | 10 |
|---|---|---|---|---|---|---|---|---|---|---|
| $y$ | 11.9 | 59.4 | 152 | 292 | 496 | 754 | 1097 | 1503 | 1901 | 2433 |

Write a program or use a spreadsheet that will compute and print for each given $x$: **(a)** the value of $y$ from the formula; **(b)** the difference between the value of $y$ from the table and that obtained from the formula. This is called a *residual*.

## 18-4 INVERSE VARIATION

### Inverse Variation

When we say that "$y$ varies inversely as $x$" or that "$y$ is inversely proportional to $x$," we mean that $x$ and $y$ are related by the following equation, where, as before, $k$ is a constant of proportionality.

We could also say that we have inverse variation when the product of $x$ and $y$ is a constant, that is, when $xy = k$.

| | | |
|---|---|---|
| Inverse Variation | $y = \frac{k}{x}$ | 61 |

### The Hyperbola

The graph of the function $y = k/x$ is called a *hyperbola*. It is one of the *conic sections,* which you will study in more detail when you take analytic geometry.

**EXAMPLE 21:** Graph the function $y = 2/x$ from $x = 0$ to $x = 5$.

**Solution:** Make a table of point pairs.

| $x$ | 0 | $\frac{1}{2}$ | 1 | 2 | 3 | 4 | 5 |
|---|---|---|---|---|---|---|---|
| $y$ | Not defined | 4 | 2 | 1 | $\frac{2}{3}$ | $\frac{1}{2}$ | $\frac{2}{5}$ |

These points are graphed in Fig. 18-7. Note that as $x$ takes on smaller and smaller values, $y$ gets larger and larger; and as $x$ gets very large, $y$ gets smaller and smaller but never reaches zero. Thus the curve gets ever closer to the $x$ and $y$ axes, but never touches them. The $x$ and $y$ axes are *asymptotes* for this curve.

**Do not confuse inverse variation with the inverse of a function.**

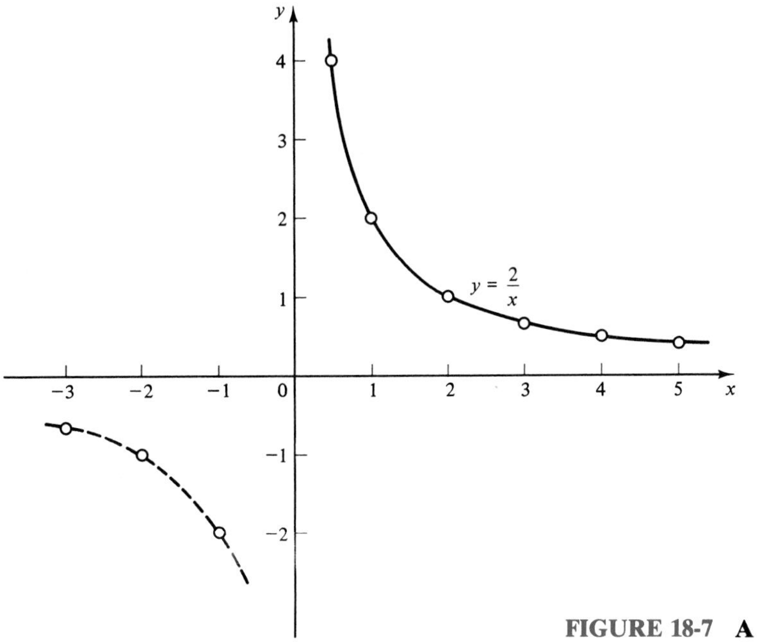


**FIGURE 18-7 A hyperbola.**

We did not plot the curve for negative values of $x$, but if we had, we would have gotten the second branch of the hyperbola, shown dashed in the figure.

## Power Function with Negative Exponent

Suppose, for example, that $y$ varies inversely as the cube of $x$. This would be written

$$y = \frac{k}{x^3}$$

This is the same as

$$y = kx^{-3}$$

The equation is seen to be a power function with a negative exponent. Such power functions are sometimes referred to as *hyperbolic*.

## Solving Inverse Variation Problems

Inverse variation problems are solved by the same methods as for any other power function. As before, we can work these problems with or without finding the constant of proportionality.

---

**EXAMPLE 22:** If $y$ is inversely proportional to $x$, and $y$ is 8 when $x$ is 3, find $y$ when $x$ is 12.

**Solution:** $x$ and $y$ are related by $y = k/x$. We find the constant of proportionality by substituting the given values of $x$ and $y$, 3 and 8.

$$8 = \frac{k}{3}$$

$$k = 3(8) = 24$$

so $y = 24/x$. When $x = 12$,

$$y = \frac{24}{12} = 2$$

---

**EXAMPLE 23:** If $y$ is inversely proportional to the square of $x$, by what factor will $y$ change if $x$ is tripled?

**Solution:** From the problem statement we know that $y = k/x^2$ or $k = x^2y$. If we use the subscript 1 for the original values and 2 for the new values, we can write

$$k = x_1^2y_1 = x_2^2y_2$$

Since the new $x$ is triple the old, we substitute $3x_1$ for $x_2$,

$$x_1^2y_1 = (3x_1)^2y_2$$

$$x_1^2y_1 = 9x_1^2y_2$$

$$y_1 = 9y_2$$

or

$$y_2 = \frac{y_1}{9}$$

So the new $y$ is one-ninth of its initial value.

---

**EXAMPLE 24:** If $y$ varies inversely as $x$, fill in the missing values in the table.

| $x$ | 1 | | 9 | 12 |
|---|---|---|---|---|
| $y$ | | 12 | 4 | |

**Solution:** Substituting the ordered pair (9, 4) into Eq. 61,

$$4 = \frac{k}{9}$$

$$k = 36$$

So

$$y = \frac{36}{x}$$

Filling in the table, we have

| $x$ | 1 | 3 | 9 | 12 |
|---|---|---|---|---|
| $y$ | 36 | 12 | 4 | 3 |

---

**EXAMPLE 25:** The number of oscillations $N$ that a pendulum makes per unit time is inversely proportional to the square root of the length $L$ of the pendulum, and $N = 2.00$ per second when $L = 85.0$ cm. (a) Write the equation $N = f(L)$; (b) find $N$ when $L = 115$ cm.

**Solution:** (a) The equation will be of the form

$$N = \frac{k}{\sqrt{L}}$$

Substituting $N = 2.00$ and $L = 85.0$ gives us

$$2.00 = \frac{k}{\sqrt{85.0}}$$

$$k = 2.00\sqrt{85.0} = 18.4$$

So the equation is

$$N = \frac{18.4}{\sqrt{L}}$$

(b) When $L = 115$,

$$N = \frac{18.4}{\sqrt{115}} = 1.72 \text{ per second}$$

## EXERCISE 4—INVERSE VARIATION

1. If $y$ varies inversely as $x$, and $y$ is 385 when $x$ is 832, find $y$ when $x$ is 226.
2. If $y$ is inversely proportional to the square of $x$, and $y$ has the value 1.55 when $x$ is 7.38, find $y$ when $x$ is 44.2.
3. If $y$ is inversely proportional to $x$, how does $y$ change when $x$ is doubled?
4. If $y$ varies inversely as $x$, and $y$ has the value 104 when $x$ is 532: **(a)** find the constant of proportionality; **(b)** write the equation $y = f(x)$; **(c)** find $y$ when $x$ is 668; and **(d)** find $x$ when $y$ is 226.
5. If $y$ is inversely proportional to $x$, fill in the missing values.

| $x$ | 306 | 622 | 91.5 |
|-----|-----|-----|-----|
| $y$ | 125 | 61.5 | 418 |

6. Fill in the missing values, assuming that $y$ is inversely proportional to the square of $x$.

| $x$ | 2.69 | | 7.83 |
|-----|------|------|------|
| $y$ | | 1.16 | 1.52 |

7. If $y$ varies inversely as the square root of $x$, fill in the missing values.

| $x$ | 226 | 3567 | 5725 |
|-----|------|------|------|
| $y$ | 1136 | 1828 | 1442.9 |

$$y = \frac{k}{\sqrt{x}} \qquad k = y\sqrt{x}$$

$$\frac{k}{\sqrt{x}} = y \qquad \sqrt{\frac{k}{y}} = \sqrt{x}$$

8. If $y$ is inversely proportional to the cube root of $x$, by what factor will $y$ change when $x$ is tripled?
9. If $y$ is inversely proportional to the square root of $x$, by what percentage will $y$ change when $x$ is decreased by 50%?
10. Plot the curve $y = 5/x^2$ for the domain 0 to 5.
11. Plot the function $y = 1/x$ for the domain 0 to 10.

## Boyle's Law

Boyle's law states that for a confined gas at a constant temperature, the product of the pressure and the volume is a constant. Another way of stating this is that the pressure is inversely proportional to the volume; or that the volume is inversely proportional to the pressure. Assume a constant temperature in the following problems.

**The *Pascal* (Pa) is the SI unit of pressure. It equals one newton per square meter.**

12. A certain quantity of gas, when compressed to a volume of 2.50 $m^3$, has a pressure of 184 Pa. Find the pressure resulting when that gas is further compressed to 1.60 $m^3$.
13. The air in the cylinder in Fig. 18-2 is at a pressure of 14.7 $lb/in^2$ and occupies a volume of 175 $in^3$. Find the pressure when it is compressed to 25.0 $in^3$.
14. A balloon contains 320 $m^3$ of gas at a pressure of 140,000 Pa. What would be the volume if the same quantity of gas were at a pressure of 250,000 Pa?

## Gravitational Attraction

**Newton's law of gravitation states that any two bodies attract each other with a force that is inversely proportional to the square of the distance between them.**

15. The force of attraction between two certain steel spheres is $3.75 \times 10^{-5}$ dyne when the spheres are placed 18 cm apart. Find the force of attraction when they are 52 cm apart.
16. How far apart must the spheres in Problem 15 be placed to cause the force of attraction between them to be $3.00 \times 10^{-5}$ dyne?
17. The force of attraction between the earth and some object is called the *weight* of that object. The law of gravitation states, then, that the weight of an object is inversely proportional to the square of its distance from the center of the earth. If a person weighs 150 lb on the surface of the earth (assume this to be 3960 mi from the center), how much will he weigh 1500 mi above the surface of the earth?
18. How much will a satellite, whose weight on earth is 675 N, weigh at an altitude of 250 km above the surface of the earth?

## Illumination

**For a surface illuminated by a light source, the *inverse square law* states that the intensity of illumination on the surface is inversely proportional to the *square* of the distance between the source and the surface.**

19. A certain light source produces an illumination of 800 lux (a lux is 1 lumen per square meter) on a surface. Find the illumination on that surface if the distance to the light source is doubled.
20. A light source located 2.75 m from a surface produces an illumination of 528 lux on that surface. Find the illumination if the distance is changed to 1.55 m.
21. A light source located 7.50 m from a surface produces an illumination of 426 lux on that surface. At what distance must that light source be placed to give an illumination of 850 lux?
22. When photographing a document on a certain copy stand, an exposure time of $\frac{1}{25}$ s is needed with the light source 0.75 m from the document. At what distance must the light be located to reduce the exposure time to $\frac{1}{100}$ s?

## Electrical

23. The current in a resistor is inversely proportional to the resistance. By what factor will the current change if a resistor increases 10% due to heating?
24. The resistance of a wire is inversely proportional to the square of its diameter. If an AWG size 12 conductor (0.0808 in. diameter) has a resistance of 14.8 Ω, what will be the resistance of an AWG size 10 conductor (0.1019 in. diameter) of the same length and material?
25. The capacitive reactance $X_c$ of a circuit varies inversely as the capacitance $C$ of the circuit. If the capacitance of a certain circuit is decreased by 25%, by what percentage will $X_c$ change?

### Computer

26. Write a program that will compute and print the values of the function $y = 1/x$ for values of $x$ from $-10$ to $+10$, taking values of $x$ every tenth of a unit. Be sure to include instructions so that the program will not "crash" when $x$ becomes zero. Make a graph of the resulting hyperbola and compare it with Fig. 18-7.

## 18-5 FUNCTIONS OF MORE THAN ONE VARIABLE

### Joint Variation

So far in this chapter we have considered only cases where $y$ was a function of a *single* variable $x$. In functional notation, this is represented by $y = f(x)$. In this section we cover functions of two or more variables, such as

$$y = f(x, w)$$

$$y = f(x, w, z)$$

and so forth. When $y$ varies directly as $x$ *and* $w$, we say that $y$ varies *jointly* as $x$ and $w$. The three variables are related by the following equation, where, as before, $k$ is a constant of proportionality.

| | | |
|---|---|---|
| Joint Variation | $y = kxw$ | **62** |

**EXAMPLE 26:** If $y$ varies jointly as $x$ and $w$, how will $y$ change when $x$ is doubled and $w$ is one-fourth of its original value?

**Solution:** Let $y'$ be the new value of $y$ obtained when $x$ is replaced by $2x$ and $w$ is replaced by $w/4$, while the constant of proportionality $k$, of course, does not change. Substituting in Eq. 62, we obtain

$$y' = k(2x)\left(\frac{w}{4}\right)$$

$$= \frac{kxw}{2}$$

but, since $kxw = y$, then

$$y' = \frac{y}{2}$$

So the new $y$ is half as large as its former value.

### Combined Variation

When one variable varies with two or more variables in ways that are more complex than in Eq. 62, it is referred to as *combined* variation. This term is applied to many different kinds of relationships, so a formula cannot be given that will cover all types. You must carefully read the problem statement in order to write the equation. Once you have the equation, the solution of combined variation problems is no different than for other types of variation.

**EXAMPLE 27:** If $y$ varies directly as the cube of $x$ and inversely as the square of $z$, we write

$$y = \frac{kx^3}{z^2}$$

**EXAMPLE 28:** If $y$ is directly proportional to $x$ and the square root of $w$, and inversely proportional to the square of $z$, we write

$$y = \frac{kx\sqrt{w}}{z^2}$$

$y = \frac{kx\sqrt{w}}{z^2}$

**EXAMPLE 29:** If $y$ varies directly as the square root of $x$ and inversely as the square of $w$, by what factor will $y$ change when $x$ is made four times larger and $w$ is tripled?

**Solution:** First write the equation linking $y$ to $x$ and $w$, including a constant of proportionality $k$:

$$y = \frac{k\sqrt{x}}{w^2}$$

We get a new value for $y$, let's call it $y'$, when $x$ is replaced by $4x$ and $w$ is replaced by $3w$.

$$y' = \frac{k\sqrt{4x}}{(3w)^2} = k\frac{2\sqrt{x}}{9w^2}$$

$$= \frac{2}{9}\left(\frac{k\sqrt{x}}{w^2}\right) = \frac{2}{9}y$$

We see that $y'$ is $\frac{2}{9}$ as large as the original $y$.

**EXAMPLE 30:** If $y$ varies directly as the square of $w$ and inversely as the cube root of $x$, and $y = 225$ when $w = 103$ and $x = 157$, find $y$ when $w = 126$ and $x = 212$.

**Solution:** From the problem statement,

$y = \frac{kw^2}{\sqrt[3]{x}}$

$$y = k\frac{w^2}{\sqrt[3]{x}}$$

Solving for $k$, we have

$$225 = k\frac{(103)^2}{\sqrt[3]{157}} = 1967k$$

or $k = 0.1144$. So

$$y = \frac{0.1144w^2}{\sqrt[3]{x}}$$

When $w = 126$ and $x = 212$,

$$y = \frac{0.1144(126)^2}{\sqrt[3]{212}} = 305$$

**EXAMPLE 31:** $y$ is directly proportional to the square of $x$ and inversely proportional to the cube of $w$. By what factor will $y$ change if $x$ is increased by 15% and $w$ is decreased by 20%?

**Solution:** The relationship between $x$, $y$, and $w$ is

$$y = k\frac{x^2}{w^3}$$

so we can write the proportion

$$\frac{y_2}{y_1} = \frac{k\dfrac{x_2^2}{w_2^3}}{k\dfrac{x_1^2}{w_1^3}} = \frac{x_2^2}{w_2^3} \cdot \frac{w_1^3}{x_1^2} = \left(\frac{x_2}{x_1}\right)^2 \left(\frac{w_1}{w_2}\right)^3$$

We now replace $x_2$ with $1.15x_1$ and replace $w_2$ with $0.8w_1$.

$$\frac{y_2}{y_1} = \left(\frac{1.15x_1}{x_1}\right)^2 \left(\frac{w_1}{0.8w_1}\right)^3 = \frac{(1.15)^2}{(0.8)^3} = 2.58$$

So $y$ will increase by a factor of 2.58.

## EXERCISE 5—FUNCTIONS OF MORE THAN ONE VARIABLE

### Joint Variation

1. If $y$ varies jointly as $w$ and $x$, and $y$ is 483 when $x$ is 742 and $w$ is 383, find $y$ when $x$ is 274 and $w$ is 756.
2. If $y$ varies jointly as $x$ and $w$, by what factor will $y$ change if $x$ is tripled and $w$ is halved?
3. If $y$ varies jointly as $w$ and $x$, by what percent will $y$ change if $w$ is increased by 12% and $x$ is decreased by 7%?
4. If $y$ varies jointly as $w$ and $x$, and $y$ is 3.85 when $w$ is 8.36 and $x$ is 11.6, evaluate the constant of proportionality and write the complete expression for $y$ in terms of $w$ and $x$.
5. If $y$ varies jointly as $w$ and $x$, fill in the missing values.

| $w$ | $x$ | $y$ |
|---|---|---|
| 46.2 | 18.3 | 127 |
| 19.5 | 41.2 | |
| | 8.86 | 155 |
| 12.2 | | 79.8 |

### Combined Variation

6. If $y$ is directly proportional to the square of $x$ and inversely proportional to the cube of $w$, and $y$ is 11.6 when $x$ is 84.2 and $w$ is 28.4, find $y$ when $x$ is 5.38 and $w$ is 2.28.
7. If $y$ varies directly as the square root of $w$ and inversely as the cube of $x$, by what factor will $y$ change if $w$ is tripled and $x$ is halved?
8. If $y$ is directly proportional to the cube root of $x$ and to the square root of $w$, by what percent will $y$ change if $x$ and $w$ are both increased by 7%?
9. If $y$ is directly proportional to the $\frac{3}{2}$ power of $x$ and inversely proportional to $w$, and $y$ is 284 when $x$ is 858 and $w$ is 361, evaluate the constant of proportionality and write the complete equation for $y$ in terms of $x$ and $w$.

10. If $y$ varies directly as the cube of $x$ and inversely as the square root of $w$, fill in the missing values in the table.

| $w$ | $x$ | $y$ |
|---|---|---|
| 1.27 | | 3.05 |
| | 5.66 | 1.93 |
| 4.66 | 2.75 | 3.87 |
| 7.07 | 1.56 | |

### Geometry

11. The area of a triangle varies jointly as its base and altitude. By what percent will the area change if the base is increased by 15% and the altitude decreased by 25%?
12. If the base and altitude of a triangle are both halved, by what factor will the area change?

### Electrical Applications

13. When an electric current flows through a wire, the resistance to the flow varies directly as the length and inversely as the cross-sectional area of the wire. If the length and the diameter are both tripled, by what factor will the resistance change?
14. If 750 m of 3-mm-diameter wire has a resistance of 27.6 Ω, what length of similar wire 5 mm in diameter will have the same resistance?

### Gravitation

15. Newton's law of gravitation states that every body in the universe attracts every other body with a force that varies directly as the product of their masses and inversely as the square of the distance between them. By what factor will the force change when the distance is doubled and each mass is tripled?
16. If both masses are increased by 60% and the distance between them is halved, by what percent will the force of attraction increase?

### Illumination

17. The intensity of illumination at a given point is directly proportional to the intensity of the light source and inversely proportional to the square of the distance from the light source. If a desk is properly illuminated by a 75-W lamp 8.0 ft from the desk, what size lamp will be needed to provide the same lighting at a distance of 12 ft?
18. How far from a 150-candela light source would a picture have to be placed so as to receive the same illumination as when it is placed 12 m from an 85-candela source?

### Gas Laws

19. The volume of a given weight of gas varies directly as the absolute temperature $t$ and inversely as the pressure $p$. If the volume is 4.45 $m^3$ when $p = 225$ kilopascals (kPa) and $t = 305$ K, find the volume when $p = 325$ kPa and $t = 354$ K.
20. If the volume of a gas is 125 $ft^3$, find its volume when the absolute temperature is increased 10% and the pressure is doubled.

### Work

21. The amount paid to a work crew varies jointly as the number of persons working and the length of time worked. If 5 workers earn \$5123.73 in 3 weeks, in how many weeks will 6 workers earn a total of \$6148.48?

22. If 5 bricklayers take 6 days to finish a certain job, how long would it take 7 bricklayers to finish a similar job requiring 4 times the number of bricks?

### Strength of Materials

23. The maximum safe load of a rectangular beam varies jointly as the width and the square of the depth, and inversely as the length of the beam. If a beam 8.0 in. wide, 11.5 in. deep, and 16 ft long can safely support 15,000 lb, find the safe load for a beam 6.5 in. wide, 13.4 in. deep, and 21 ft long made of the same material.
24. If the width of a rectangular beam is increased by 11%, the depth decreased by 8%, and the length increased 6%, by what percent will the safe load change?

### Mechanics

25. The number of vibrations per second made when a stretched wire is plucked varies directly as the square root of the tension in the wire, and inversely as the length. If a 1.00-m-long wire will vibrate 325 times a second when the tension is 115 N, find the frequency of vibration if the wire is shortened to 0.750 m and the tension is decreased to 95.0 N.
26. The kinetic energy of a moving body is directly proportional to its mass and the square of its speed. If the mass of a bullet is halved, by what factor must its speed be increased to have the same kinetic energy as before?

### Fluid Flow

27. The time needed to empty a vertical cylindrical tank varies directly as the square root of the height of the tank and the square of the radius. By what factor will the emptying time change if the height is doubled and the radius increased by 25%?
28. The power available in a jet of liquid is directly proportional to the cross-sectional area of the jet and to the cube of the velocity. By what factor will the power increase if the area and velocity are both increased 50%?

## CHAPTER 18 REVIEW PROBLEMS

1. If $y$ varies inversely as $x$, and $y$ is 736 when $x$ is 822, find $y$ when $x$ is 583.
2. If $y$ is directly proportional to the $\frac{5}{2}$ power of $x$, by what factor will $y$ change when $x$ is tripled?
3. If $y$ varies jointly as $x$ and $z$, by what percent will $y$ change when $x$ is increased by 15% and $z$ is decreased by 4%?
4. The braking distance of an automobile varies directly as the square of the speed. If the braking distance of a certain automobile is 34 ft at 25 mi/h, find the braking distance at 55 mi/h.
5. The rate of flow of liquid from a hole in the bottom of a tank is directly proportional to the square root of the liquid depth. If the flow rate is 225 liters/min when the depth is 3.46 m, find the flow rate when the depth is 1.00 m.
6. If the power needed to drive a ship varies directly as the cube of the speed of the ship and a 77.4-hp engine will drive a certain ship at 11.2 knots, find the horsepower needed to propel that ship at 18.0 knots.
7. If the tensile strength of a cylindrical steel bar varies as the square of its diameter, by what factor must the diameter be increased to triple the strength of the bar?
8. If the life of an incandescent lamp varies inversely as the 12th power of the applied voltage and the light output varies directly as the 3.5th power of the applied voltage, by what factor will the life increase if the voltage is lowered by an amount that will decrease the light output by 10%?
9. One of Kepler's laws states that the time for a planet to orbit the sun varies directly as the $\frac{3}{2}$ power of its distance from the sun. How many years will it take for Saturn, which is about 9.5 times as far from the sun as is the earth, to orbit the sun?
10. The volume of a cone varies directly as the square of the base radius and as the altitude. By what factor will the volume change if the altitude is doubled and the base radius is halved?
11. The number of oscillations made by a pendulum in a given time is inversely proportional to the length of the pendulum. A certain clock with a 30.00-in.-long pendulum is losing 15 min/day. Should the pendulum be lengthened or shortened, and by how much?

12. A trucker usually makes a trip in 18 h at an average speed of 60 mi/h. Find the traveling time if the speed were reduced to 50 mi/h.

13. The force on the vane of a wind generator varies directly as the area of the vane and the square of the wind velocity. By what factor must the area of a vane be increased so that the wind force on it will be the same in a 12-mi/h wind as it was in a 35-mi/h wind?

14. The maximum deflection of a rectangular beam varies inversely as the product of the width and the cube of the depth. If the deflection of a beam having a width of 15 cm and a depth of 35 cm is 7.5 mm, find the deflection if the width is made 20 cm and the depth 45 cm.

15. When an object moves at a constant speed, the travel time is inversely proportional to the speed. If a satellite now circles the earth in 26 h 18 min, how long will it take if booster rockets increased the speed of the satellite by 15%?

16. The life of an incandescent lamp is inversely proportional to the 12th power of the applied voltage. If a lamp has a life of 2500 h when run at 115 V, what would be its life expectancy when run at 110 V?

17. How far from the earth will a spacecraft be equally attracted by the earth and the moon? The distance from the earth to the moon is approximately 239,000 mi, and the mass of the earth is about 82 times that of the moon.

18. Ohm's law states that the electric current flowing in a circuit varies directly as the applied voltage and inversely as the resistance. By what percent will the current change when the voltage is increased by 25% and the resistance increased by 15%?

19. The allowable strength $F$ of a column varies directly as its length $L$ and inversely as its radius of gyration $r$. If $F = 35$ kg when $L$ is 12 m and $r$ is 3.6 cm, find $F$ when $L$ is 18 m and $r$ is 4.5 cm.

20. The time it takes a pendulum to go through one complete oscillation (the *period*) is directly proportional to the square root of its length. If the period of a 1-m pendulum is 1.25 s, how long must the pendulum be to have a period of 2.5 s?

21. If the maximum safe current in a wire is directly proportional to the $\frac{3}{2}$ power of the wire diameter, by what factor will the safe current increase when the wire diameter is doubled?

22. Four workers take ten 6-h days to finish a job. How many workers are needed to finish a similar job which is 3 times as large, in five 8-h days?

23. When a jet of water strikes the vane of a water turbine and is deflected through an angle $\theta$, the force on the vane varies directly as the square of the jet velocity and the sine of $\theta/2$. If $\theta$ is decreased from 55° to 40° and the jet velocity increases by 40%, by what percentage will the force change?

24. By what factor will the kinetic energy change if the speed of a projectile is doubled and its mass is halved?

25. The fuel consumption for a ship is directly proportional to the cube of the ship's speed. If a certain tanker uses 1584 gal of diesel fuel on a certain run at 15 knots, how much fuel would it use for the same run when the speed is reduced to 10 knots? A *knot* is equal to one nautical mile per hour. (1 nautical mile = 1.151 statute miles.)

26. The diagonal of a certain square is doubled. By what factor does the area change?

27. A certain parachute is made from 52 $m^2$ of fabric. If all its dimensions are increased by a factor of 1.4, how many square meters of fabric will be needed?

28. The rudder of a certain airplane has an area of 7.5 $ft^2$. By what factor must the dimensions be scaled up to triple the area?

29. A 1.5-in.-diameter pipe fills a cistern in 5 h. Find the diameter of a pipe that will fill the same cistern in 9 h.

30. If 315 m of fence will enclose a circular field containing 2 acres, what length will enclose 8 acres?

31. The surface area of a one-quarter model of an automobile measures 1.1 $m^2$. Find the surface area of the full-sized car.

32. A certain water tank is 25 ft high and holds 5000 gal. How much would a 35-ft-high tank of similar shape hold?

33. A certain tractor weighs 5.8 tons. If all its dimensions were scaled up by a factor of 1.3, what would the larger tractor be expected to weigh?

34. If a trough 5 ft long holds 12 pailfuls of water, how many pailfuls will a similar trough hold that is 8 ft long?

35. A ball 4.5 in. in diameter weighs 18 oz. What is the weight of another ball of the same density that is 9 in. in diameter?

36. A ball 4 in. in diameter weighs 9 lb. What is the weight of a ball 25 cm in diameter made of the same material?

37. A worker's contract has a cost-of-living clause that requires that his salary be proportional to the cost-of-living index. If he earned $450 per week when the index was 9.4, how much should he earn when the index is 12.7?

38. A woodsman pays a landowner $8.00 for each cord of wood he cuts on the landowner's property, and sells the wood for $75 per cord. When the price of firewood rose to $95, the landowner asked for a proportional share of the price. How much should he get per cord?

39. The series of numbers 3.5, 5.25, 7.875, . . . form a *geometric progression,* where the ratio of any term to the one preceding it is a constant. This is called the *common ratio*. Find this ratio.

40. The *specific gravity* of a solid or liquid is the ratio of the density of the substance to the density of water, at a standard temperature.

| | | |
|---|---|---|
| Specific Gravity | specific gravity $= \dfrac{\text{density of substance}}{\text{density of water}}$ | **A45** |

Taking the density of water as 62.4 lb/ft$^3$, find the density of copper having a specific gravity of 8.89.

41. An iron casting having a surface area of 7463 cm$^2$ is heated so that its length increases by 1.0%. Find the new area.

42. A certain boiler pipe will permit a flow of 35.5 gal/min. Buildup of scale inside the pipe eventually reduces its diameter to three-fourths of its previous value. Assuming that the flow rate is proportional to the cross-sectional area, find the new flow rate for the pipe.

### Writing

43. Suppose your company makes plastic trays and is planning new ones with dimensions double those now being made. Your company president is convinced that they will need only twice as much plastic as the older version. "Twice the size, twice the plastic," he proclaims, and no one is willing to challenge him. Your job is to make a presentation to the president where you tactfully point out that he is wrong and where you explain that the new trays will require eight times as much plastic. Write your presentation.

# 19

# Exponential and Logarithmic Functions

## OBJECTIVES

**When you have completed this chapter, you should be able to:**

- Graph the exponential function.
- Solve exponential growth and decay problems by formula or by the universal growth and decay curves.
- Convert expressions between exponential and logarithmic form.
- Evaluate common and natural logarithms and antilogarithms.
- Evaluate, manipulate, and simplify logarithmic expressions.
- Solve exponential and logarithmic equations.
- Solve applied problems involving exponential or logarithmic equations.
- Make graphs on logarithmic or semilogarithmic paper.
- Graph empirical data and determine an approximate equation describing the data.

You may have heard the story about the inventor of chess, who asked that the reward for his invention be a grain of wheat on the first square of the chessboard, two grains on the second square, four on the third, and so on, each square getting double the amount as on the previous square. It turns out that he would receive more wheat than has been grown by man during all history; an illustration of the remarkable properties of the exponential function, which we study in this chapter.

We also study the logarithmic function and show how it is related to the exponential function, and do a few logarithmic computations. We then solve exponential and logarithmic equations and use them in some interesting applications, and do some graphing on logarithmic and semilogarithmic paper.

## 19-1 THE EXPONENTIAL FUNCTION

### Definition

An *exponential function* is one in which the independent variable appears in the exponent. The quantity that is raised to the power is called the *base*.

**EXAMPLE 1:**

$$y = 5^x \qquad y = b^x \qquad y = e^x \qquad y = 10^x$$

$$y = 3a^x \qquad y = 7^{x-3} \qquad y = 5e^{-2x}$$

are exponential functions if $a$, $b$, and $e$ represent positive constants.

**Realize that these are different from the *power function*, $y = ax^n$. Here the unknown is *in the exponent*.**

We shall see that $e$, like $\pi$, has a specific value that we will discuss later in this chapter.

### Graph of the Exponential Function

We may make a graph of an exponential function in the same way as we graphed other functions in Chapter 4. Choose values of the independent variable $x$ throughout the region of interest, and for each $x$ compute the corresponding value of the dependent variable $y$. Plot the resulting table of ordered pairs and connect with a smooth curve.

**EXAMPLE 2:** Graph the exponential function $y = 2^x$ from $x = -2$ to $x = +4$.

**Solution:** We compute values of $y$ and obtain the following table of point pairs:

| $x$ | $-2$ | $-1$ | $0$ | $1$ | $2$ | $3$ | $4$ |
|---|---|---|---|---|---|---|---|
| $y$ | $\frac{1}{4}$ | $\frac{1}{2}$ | $1$ | $2$ | $4$ | $8$ | $16$ |

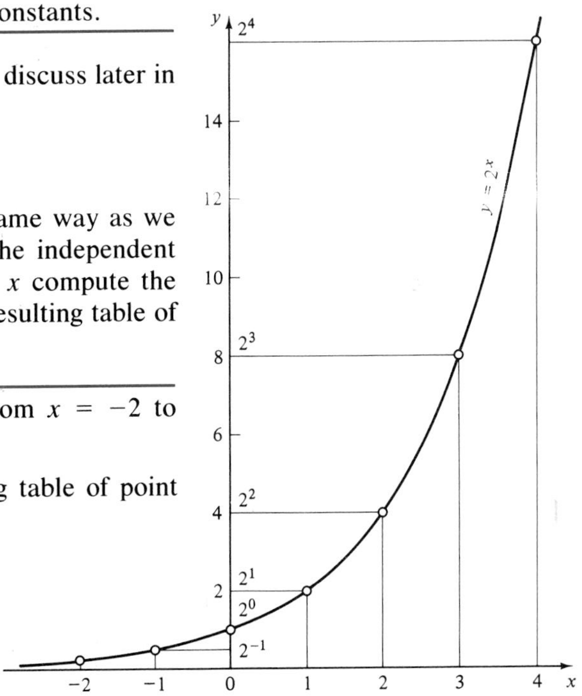

**FIGURE 19-1** **Graph of the exponential function, $y = 2^x$.**

These points pairs are plotted in Fig. 19-1.

Note that as $x$ increases, $y$ increases ever more greatly, and that the curve gets steeper and steeper. This is a characteristic of *exponential growth,* which we cover more fully below. Also note that the $y$ intercept is 1 and that there is no $x$ intercept, the $x$ axis being an asymptote. Finally, note that each successive $y$ value is two times the previous one. We say that $y$ *increases geometrically*.

**EXAMPLE 3:** Graph the exponential functions $y = e^x$ and $y = e^{-x}$, for $x = -3$ to 3, if $e$ equals 2.718.

**Solution:** We make a table of point pairs, which we then graph in Fig. 19-2.

| $x$ | −3 | −2 | −1 | 0 | 1 | 2 | 3 |
|---|---|---|---|---|---|---|---|
| $e^x$ | 0.0498 | 0.135 | 0.368 | 1 | 2.72 | 7.39 | 20.1 |
| $e^{-x}$ | 20.1 | 7.39 | 2.72 | 1 | 0.368 | 0.135 | 0.0498 |

Notice that when the exponent is *positive*, the graph *increases exponentially*, and that when the exponent is *negative*, the graph *decreases exponentially*.

Also notice that neither curve reaches the $x$ axis. Thus for the exponential functions $y = e^x$ and $y = e^{-x}$, $y$ cannot equal zero.

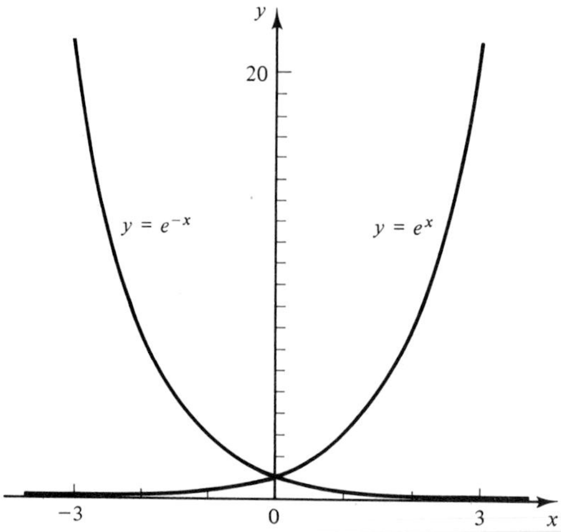


**FIGURE 19-2**

**EXAMPLE 4:** Graph the exponential function $y = 1.85e^{-x/2}$, for $x = -4$ to 4, if $e$ equals 2.718.

**Solution:** We make a table to obtain the $(x, y)$ point pairs, which we then graph in Fig. 19-3.

| $x$ | $-x/2$ | $e^{-x/2}$ | $y$ |
|---|---|---|---|
| −4 | 2.000 | 7.388 | 13.667 |
| −3 | 1.500 | 4.481 | 8.290 |
| −2 | 1.000 | 2.718 | 5.028 |
| −1 | 0.500 | 1.649 | 3.050 |
| 0 | 0.000 | 1.000 | 1.850 |
| 1 | −0.500 | 0.607 | 1.122 |
| 2 | −1.000 | 0.368 | 0.681 |
| 3 | −1.500 | 0.223 | 0.413 |
| 4 | −2.000 | 0.135 | 0.250 |

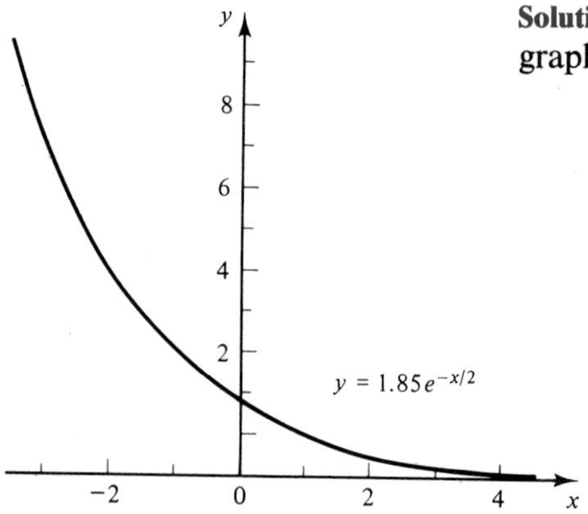


**FIGURE 19-3**

Notice that the negative sign in the exponent has made the graph *decrease* exponentially as $x$ increases, and that the curve approaches the $x$ axis as an asymptote. The shape of the curve, however, is that of an exponential function.

## Compound Interest

In this and the next few sections, we will see some problems that require exponential equations for their solution.

When an amount of money $a$ (called the *principal*) is invested at an interest rate $n$ (expressed as a decimal), it will earn an amount of interest $an$ at the end of the first interest period. If this interest is then added to the principal and both continue to earn interest, we have what is called *compound interest,* the type of interest commonly given on bank accounts.

Suppose that the interest rate is 5% (0.05). As the balance increases, the previous balance (100% of $a$ or $1.00a$) is added to the amount received for the current period (5% of $a$ or $0.05a$) to give a new total of $1.05a$. So we could say we multiply $a$ by $(1 + 0.05)$ or 1.05.

Similarly, for an interest rate of $n$, the balance at the end of any period is obtained by multiplying the preceding balance by $(1 + n)$. Thus

| *Period* | *Balance y at End of Period* |
|---|---|
| 1 | $a(1 + n)$ |
| 2 | $a(1 + n)^2$ |
| 3 | $a(1 + n)^3$ |
| 4 | $a(1 + n)^4$ |
| . | . |
| . | . |
| $t$ | $a(1 + n)^t$ |

We thus get the following exponential function, with $a$ equal to the principal, $n$ equal to the rate of interest, and with $t$ the number of interest periods.

| | | |
|---|---|---|
| Compound Interest Computed Once per Period | $y = a(1 + n)^t$ | **A10** |

**EXAMPLE 5:** To what amount will an initial deposit of $500 accumulate if invested at a compound interest rate of $6\frac{1}{2}$% per year for 8 years?

**Solution:** From the compound interest formula,

$$\begin{aligned} y &= 500(1 + 0.065)^8 \\ &= \$827.50 \end{aligned}$$

## Compound Interest Computed *m* Times per Period

The formula for compound interest is valuable not only in itself, but also because it leads us to an equation for the type of growth (and decline) that occurs in nature. In nature, growth and decline do not occur in steps, once a year, or once a minute, but occur *continuously*.

In Sec. 19-2 we will use this equation to derive an equation for exponential (continuous) growth.

Let us modify the compound interest formula. Instead of computing interest *once* per period, let us compute it *m times* per period. The exponent in Eq. A10 thus becomes $mt$. The interest rate $n$, however, has to be reduced to $(1/m)$th of its old value, or to $n/m$.

| | | |
|---|---|---|
| Compound Interest Computed $m$ Times per Period | $y = a\left(1 + \frac{n}{m}\right)^{mt}$ | **A11** |

**EXAMPLE 6:** Repeat the computation of Example 5 with the interest computed monthly rather than annually.

**Solution:** Notice that the rate $n$ is a yearly rate. If the interest is to be compounded more frequently (here it is compounded monthly), we use Eq. A11. From Eq. A11, with $m = 12$,

$$y = 500\left(1 + \frac{0.065}{12}\right)^{12(8)} = 500(1.00542)^{96}$$

$$= \$839.83$$

or $12.33 more than in Example 5.

## EXERCISE 1—THE EXPONENTIAL FUNCTION

### Exponential Function

Graph each exponential function over the given domains of $x$. Take $e = 2.718$.

1. $y = 0.2(3.2)^x \quad (x = -4 \text{ to } +4)$
2. $y = 3(1.5)^{-2x} \quad (x = 0 \text{ to } 5)$
3. $y = 5(1 - e^{-x}) \quad (x = 0 \text{ to } 10)$
4. $y = 4e^{x/2} \quad (x = 0 \text{ to } 4)$
5. A rope or chain between two points hangs in the shape of a curve called a *catenary* (Fig. 19-4). Its equation is

$$y = \frac{a(e^{x/a} + e^{-x/a})}{2}$$

Using values of $a = 2.5$ and $e = 2.718$, plot $y$ for $x = -5$ to 5.

**FIGURE 19-4 A rope or chain hangs in the shape of a catenary.**

Mount your finished graph for Problem 5 vertically and hang a fine chain from the endpoints. Does its shape (after you have adjusted its length) match that of your graph?

### Compound Interest

6. Find the amount to which $500 will accumulate in 6 years at a compound interest rate of 6% per year, compounded annually.
7. If our compound interest formula is solved for $a$, we get

$$a = \frac{y}{(1 + n)^t}$$

where $a$ is the amount that must be deposited now at interest rate $n$ to produce an amount $y$ in $t$ years. How much money would have to be invested at $7\frac{1}{2}$% per year to accumulate to $10,000 in 10 years?

The amount $a$ is called the ***present worth*** of the future amount $y$. This formula, and the three that follow, are standard formulas used in business and finance.

8. What annual compound interest rate (compounded annually) is needed to enable an investment of $5000 to accumulate to $10,000 in 12 years?

9. Using Eq. A11, find the amount to which $1 will accumulate in 20 years at a compound interest rate of 10% per year (a) compounded annually; (b) compounded monthly; (c) compounded daily.

10. If an amount of money $R$ is deposited *every year* for $t$ years at a compound interest rate $n$, it will accumulate to an amount $y$, where

$$y = R\left[\frac{(1 + n)^t - 1}{n}\right]$$

*A series of annual payments such as these is called an* annuity.

If a worker pays $2000 per year into a retirement plan yielding 6% annual interest, what will be the value of this annuity after 25 years?

11. If the equation for an annuity in Problem 10 is solved for $R$, we get

$$R = \frac{ny}{(1 + n)^t - 1}$$

*This is called a* sinking fund.

where $R$ is the annual deposit required to produce a total amount $y$ at the end of $t$ years. How much would the worker in Problem 10 have to deposit each year (at 6% per year) to have a total of $100,000 after 25 years?

12. The yearly payment $R$ that can be obtained for $t$ years from a present investment $a$ is

$$R = a\left[\frac{n}{(1 + n)^t - 1} + n\right]$$

*This is called* capital recovery.

If a person has $85,000 in a retirement fund, how much can be withdrawn each year so that the fund will be exhausted in 20 years if the amount remaining in the fund is earning $6\frac{1}{4}$% per year?

$$R = 85000\left[\frac{6.25}{\phantom{x}}\right.$$

## 19-2 EXPONENTIAL GROWTH AND DECAY

### Exponential Growth

Equation A11, $y = a(1 + n/m)^{mt}$, allows us to compute the amount obtained, with interest compounded any number of times per period that we choose. We could use this formula to compute the interest if it were compounded every week, or every day, or every second. But what about *continuous* compounding, or continuous growth? What would happen to Eq. A11 if $m$ got very, very large, in fact, infinite? Let us make $m$ increase to large values, but first, to simplify the work, we make the substitution; let

$$k = \frac{m}{n}$$

Equation A11 then becomes

$$y = a\left(1 + \frac{1}{k}\right)^{knt}$$

which can be written

$$y = a\left[\left(1 + \frac{1}{k}\right)^k\right]^{nt}$$

Then as $m$ grows large, so will $k$. Let us try some values on the calculator, and evaluate only the quantity in the square brackets, to four decimal places.

| $k$ | $\left(1 + \frac{1}{k}\right)^k$ |
|---:|---|
| 1 | 2 |
| 10 | 2.5937 ... |
| 1000 | 2.7048 ... |
| 1,000 | 2.7169 ... |
| 10,000 | 2.7181 ... |
| 100,000 | 2.7183 ... |
| 1,000,000 | 2.7183 ... |

We get the surprising result that as $m$ (and $k$) continue to grow infinitely large, the value of $(1 + 1/k)^k$ does *not* grow without limit, but approaches a specific, well-known value, the value 2.7183. . . . This important number is given the special symbol $e$. We can express the same idea in *limit notation,* by writing

We will not study limits here. They are covered in detail in calculus.

$$\lim_{k \to \infty}\left(1 + \frac{1}{k}\right)^k = e$$

*As $k$ grows without bound, the value of $(1 + 1/k)^k$ approaches $e$ ($\approx 2.7183. . .$).*

The number $e$ is named after a Swiss mathematician, Leonhard Euler (1707–1783). Its value has been computed to thousands of decimal places. In Exercise 2 we show how to compute $e$ using series.

Thus when $m$ gets infinitely large, the quantity $(1 + 1/k)^k$ approaches $e$, and the formula for continuous growth becomes

| Exponential Growth | $y = ae^{nt}$ | **197** |
|---|---|---|

To raise $e$ to a power on a calculator, simply enter the power and press the $\boxed{e^x}$ key.

---

**EXAMPLE 7:** Repeat the computation of Example 5 if the interest were compounded *continuously* rather than monthly.

Banks that offer "continuous compounding" of your account do not have a clerk constantly computing your interest. They use Eq. 197.

**Solution:** From Eq. 197,

$$y = 500e^{0.065(8)}$$
$$= 500e^{0.52}$$
$$= \$841.01$$

or $1.18 more than when compounded monthly.

---

| Common Error | When using Eq. 197 the time units of $n$ and $t$ *must agree.* |
|---|---|

**EXAMPLE 8:** A quantity grows exponentially at the rate of 2% per minute. By how many times will it have increased after 4 h?

**Solution:** The units of $n$ and $t$ do *not* agree (yet).

$$n = 2\% \text{ per } \textit{minute} \qquad \text{and} \qquad t = 4 \textit{ hours}$$

So we convert one of them to agree with the other.

$$t = 4 \text{ h} = 240 \text{ min}$$

Using Eq. 197,

$$y = ae^{nt} = ae^{0.02(240)}$$

so we can find the ratio of $y$ to $a$ by dividing both sides by $a$:

$$\frac{y}{a} = e^{0.02(240)} = e^{4.8} = 122$$

So the final amount is 122 times as great as the initial amount.

## Exponential Decay

When a quantity *decreases* so that the amount of decrease is proportional to the present amount, we have what is called *exponential decay*. Take, for example, a piece of steel just removed from a furnace. The amount of heat leaving the steel depends upon its temperature; the hotter it is, the faster heat will leave, and the faster it will cool. So as the steel cools, its lower temperature causes slower heat loss. Slower heat loss causes slower rate of temperature drop, and so on. The result is the typical exponential decay curve shown in Fig. 19-5. Finally, the temperature is said to *asymptotically approach* the outside (room) temperature.

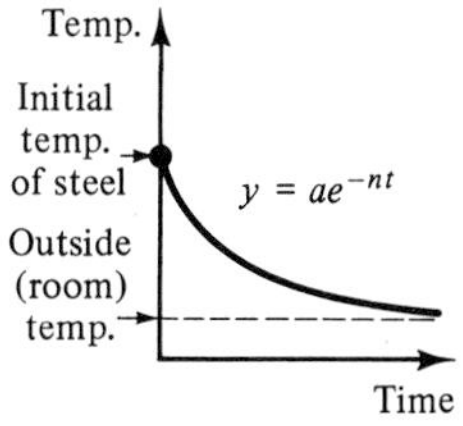


**FIGURE 19-5 Exponential decrease in temperature. This is called *Newton's law of cooling*.**

The equation for exponential decay is the same as for exponential growth, except that the exponent is negative.

| Exponential Decay | $y = ae^{-nt}$ | **199** |
|---|---|---|

**EXAMPLE 9:** A room initially 80°F above the outside temperature cools exponentially at the rate of 25% per hour. Find the temperature of the room (above the outside temperature) at the end of 135 min.

**Solution:** We first make the units consistent; 135 min = 2.25 h. Then by Eq. 199,

$$y = 80e^{-0.25(2.25)} = 45.6°\text{F}$$

## Exponential Growth to an Upper Limit

Suppose that the bucket in Fig. 19-6 initially contains $a$ gallons of water. Then assume that $y_1$, the amount remaining in the bucket, decreases exponentially. By Eq. 199,

$$y_1 = ae^{-nt}$$

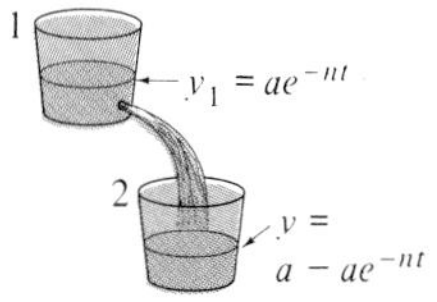


**FIGURE 19-6 The amount in bucket 2 grows exponentially to an upper limit *a*.**

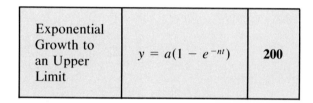

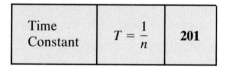

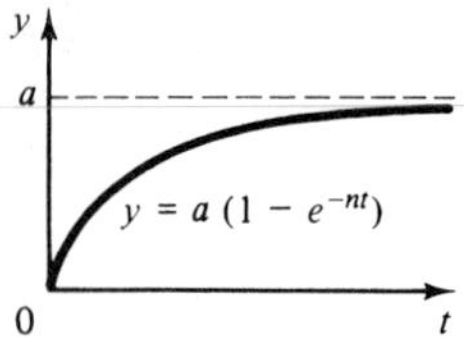


**FIGURE 19-7 Exponential growth to an upper limit.**

The amount of water in bucket 2 thus increases from zero to a final amount $a$. The amount $y$ in bucket 2 at any instant is equal to the amount that has left bucket 1, or

$$y = a - y_1$$
$$= a - ae^{-nt}$$

or,

| | | |
|---|---|---|
| Exponential Growth to an Upper Limit | $y = a(1 - e^{-nt})$ | **200** |

So the equation for this type of growth to an upper limit is exponential, where $a$ equals the upper limit quantity, $n$ is equal to the rate of growth, and $t$ represents time (see Fig. 19-7).

**EXAMPLE 10:** The voltage across a certain capacitor grows exponentially from 0 V to a maximum of 75 V, at a rate of 7% per second. Write the equation for the voltage at any instant, and find the voltage at 2.0 s.

**Solution:** From Eq. 200, with $a = 75$ and $n = 0.07$,

$$v = 75(1 - e^{-0.07t})$$

where $v$ represents the instantaneous voltage. When $t = 2.0$ s,

$$v = 75(1 - e^{-0.14})$$
$$= 9.80 \text{ V}$$

## Time Constant

Our equations for exponential growth and decay all contained a rate of growth $n$. Often it is more convenient to work with the *reciprocal* of $n$, rather than $n$ itself. This reciprocal is called the time *constant, T.*

| | | |
|---|---|---|
| Time Constant | $T = \dfrac{1}{n}$ | **201** |

**EXAMPLE 11:** If a quantity decays exponentially at a rate of 25% per second, what is the time constant?

**Solution:**

$$T = \frac{1}{0.25} = 4 \text{ s}$$

Notice that the time constant has the units of *time*.

## Universal Growth and Decay Curves

If we replace $n$ by $1/T$ in Eq. 199, we get

$$y = ae^{-t/T}$$

and then divide both sides by $a$,

$$\frac{y}{a} = e^{-t/T}$$

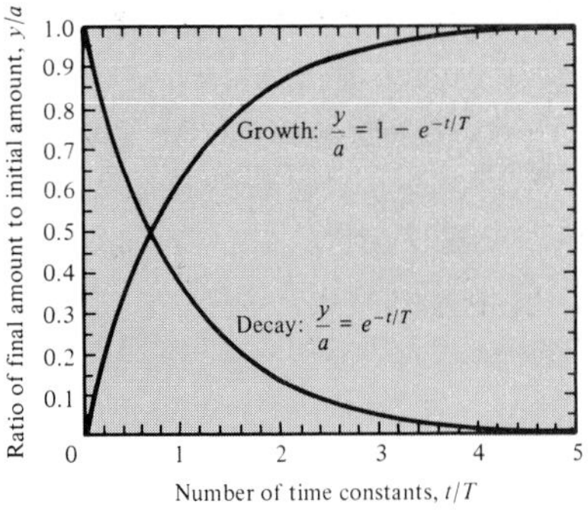


**FIGURE 19-8 Universal growth and decay curves.**

We plot this curve (Fig. 19-8) with the dimensionless ratio $t/T$ for our horizontal axis and the dimensionless ratio $y/a$ for the vertical axis. Thus the horizontal axis is *the number of time constants* that have elapsed, and the vertical axis is *the ratio of the final amount to the initial amount*. It is called the *universal decay curve*.

We can use the universal decay curve to obtain *graphical solutions* to exponential decay problems.

---

**EXAMPLE 12:** A current decays from an initial value of 300 mA, at an exponential rate of 20% per second. Find the current after 7 s graphically, using the universal decay curve.

**Solution:** We have $n = 0.20$, so the time constant is

$$T = \frac{1}{n} = \frac{1}{0.20} = 5 \text{ s}$$

Our given time $t = 7$ s is then equivalent to

$$\frac{t}{T} = \frac{7}{5} = 1.4 \text{ time constants}$$

From the universal decay curve, at $t/T = 1.4$, we read

$$\frac{y}{a} = 0.25$$

Since $a = 300$ mA,

$$y = 300(0.25) = 75 \text{ mA}$$

---

Figure 19-7 also shows the *universal growth curve*

$$\frac{y}{a} = 1 - e^{-t/T}$$

It is used in the same way as the universal decay curve.

## EXERCISE 2—EXPONENTIAL GROWTH AND DECAY

### Exponential Growth

1. A quantity grows exponentially at the rate of 5% per year for 7 years. Find the final amount if the initial amount is 200 units.
2. A yeast manufacturer finds that the yeast will grow exponentially at a rate of 15% per hour. How many pounds of yeast must they start with to obtain 500 lb at the end of 8 h?
3. If the world population were 4 billion in 1975 and grew at an annual rate of 1.7%, what would be the population in 1990?

**Use either the formulas or the universal growth and decay curves, as directed by your instructor.**

4. If the rate of inflation is 12% per year, how much could a camera that now costs \$225 be expected to cost 5 years from now?
5. The U.S. energy consumption in 1979 was estimated at 37 million barrels (bbl) per day oil equivalent. If the rate of consumption increases at an annual rate of 7%, what will be the energy need in 1990?
6. A pharmaceutical company makes a vaccine that grows at a rate of 2.5% per hour. How many units of this organism must they have initially to have 1000 units after 5 days?

### Exponential Growth to an Upper Limit

7. A steel casting initially at 0°C is placed in a furnace at 800°C. If it increases in temperature at the rate of 5% per minute, find its temperature after 20 min.
8. Equation 200 approximately describes the hardening of concrete, where $y$ is the compressive strength (lb/in$^2$) $t$ days after pouring. Using values of $a = 4000$ and $n = 0.0696$ in the equation, find the compressive strength after 14 days.
9. When the switch in Fig. 19-9 is closed, the current $i$ will grow exponentially according to the equation

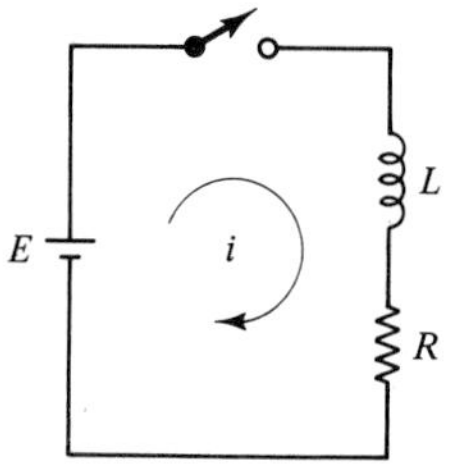

FIGURE 19-9

| | | | |
|---|---|---|---|
| Inductor Current When Charging | $i = \frac{E}{R}(1 - e^{-Rt/L})$ | amperes | **A87** |

where $L$ is the inductance in henries and $R$ is the resistance in ohms. Find the current at 0.075 s if $R = 6.25\ \Omega$, $L = 186$ henries, and $E = 250$ V.

### Exponential Decay

10. A flywheel is rotating at a speed of 1800 rev/min. When the power is disconnected, the speed decreases exponentially at the rate of 32% per minute. Find the speed after 10 min.
11. A steel forging is 1500°F above room temperature. If it cools exponentially at the rate of 2% per minute, how much will its temperature drop in 1 h?
12. As light passes through glass or water, its intensity decreases exponentially according to the equation

$$I = I_0 e^{-kx}$$

where $I$ is the intensity at a depth $x$ and $I_0$ is the intensity before entering the glass or water. If, for a certain filter glass, $k = 0.5$/cm (which means that each centimeter of filter thickness removes half the light reaching it), find the fraction of the original intensity that will pass through a filter glass 2.0 cm thick.
13. The atmospheric pressure $p$ decreases exponentially with the height $h$ (miles) above the earth according to the function $p = 29.92e^{-h/5}$ inches of mercury. Find the pressure at a height of 30,000 ft.
14. A certain radioactive material loses its radioactivity at the rate of $2\frac{1}{2}$% per year. What fraction of its initial radioactivity will remain after 10 years?
15. When a capacitor $C$, charged to a voltage $E$, is discharged through a resistor $R$ (Fig. 19-10) the current $i$ will decay exponentially according to the equation

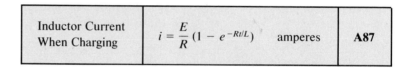

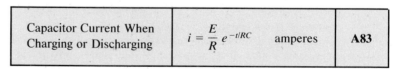

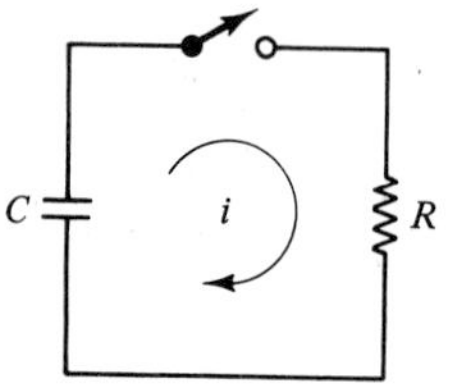

FIGURE 19-10

| | | | |
|---|---|---|---|
| Capacitor Current When Charging or Discharging | $i = \frac{E}{R} e^{-t/RC}$ | amperes | **A83** |

Find the current after 45 ms (milliseconds) in a circuit where $E = 220$ V, $C = 130$ microfarads, and $R = 2700\ \Omega$.

16. When a fully discharged capacitor $C$ (Fig. 19-11) is connected across a battery, the current $i$ flowing into the capacitor will decay exponentially according to Equation A83.

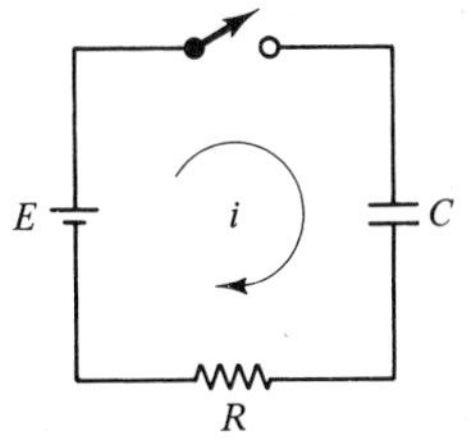

**FIGURE 19-11**

If $E = 115$ V, $R = 350 \, \Omega$, and $C = 0.00075$ F, find the current after 75 ms.

17. In a certain fabric mill, cloth is removed from a dye bath and is then observed to dry exponentially at the rate of 24% per hour. What percent of the original moisture will still be present after 5 h?

## Computer

18. Program the computer to print a two-column table, the first column giving the value of $k$ and the second column the value of $(1 + 1/k)^k$. Take a starting value for $k$ equal to 1 and increase it each time by a factor of 10. Stop the run when $1/k$ is less than 0.00001. Does the value of $(1 + 1/k)^k$ seem to be converging on some value? If not, why?

19. We can get an approximate value for $e$ from the infinite series

| Series Approximation for $e$ | $e = 2 + \dfrac{1}{2!} + \dfrac{1}{3!} + \dfrac{1}{4!} + \cdots$ | 202 |
|---|---|---|

where 4! (read 4 *factorial*) is 4(3)(2)(1) = 24. Write a program to compute $e$ using the first five terms of the series.

20. The infinite series often used to calculate $e^x$ is

| Series Approximation for $e^x$ | $e^x = 1 + x + \dfrac{x^2}{2!} + \dfrac{x^3}{3!} + \dfrac{x^4}{4!} + \cdots$ | 203 |
|---|---|---|

Write a program to compute $e^x$ by using the first 15 terms of this series, and use it to find $e^5$.

## 19-3 LOGARITHMS

In Sec. 19-1 we studied the exponential function

$$y = b^x$$

Given $b$ or $x$, we know how to find $y$. Thus if $b = 2.48$ and $x = 3.50$, then, by calculator,

$$y = 2.48^{3.50} = 24.0$$

We are also able, given $y$ and $x$, to find $b$. For example, if $y$ were 24.0 and $x$ were 3.50,

$$24.0 = b^{3.50}$$

$$b = 24.0^{1/3.50} = 2.48$$

$24 = b^{3.50}$

$b = 24^{1/3.50}$

$a = b^x$

$a^{1/x} = b$

Suppose, now, that we have $y$ and $b$ and want to find $x$. Say that $y = 24.0$ when $b = 2.48$:

$$24.0 = 2.48^x$$

We will finish this problem in Sec. 19-5.

Try it. You will find that none of the math we have learned so far will enable us to find $x$. To solve this and similar problems, we need *logarithms,* which are explained in this section.

### Definition of a Logarithm

We have the same relationship between the same quantities as we had before but a different way of expressing it. We have a similar situation in ordinary language. Take "Mary is John's *sister*." If we had only the word *sister* (and not *brother*) to describe the relationship between children, we would have to say something like "John is the person whom Mary is the sister of." That is as awkward as "the exponent to which the base must be raised to obtain the number." So we introduce new words, such as *brother* and *logarithm.*

The logarithm of some positive number $y$ *is the exponent* to which a base $b$ must be raised to obtain $y$.

---

**EXAMPLE 13:** Since $10^2 = 100$, we say that "2 is the logarithm of 100, to the base 10" because 2 is the exponent to which the base 10 must be raised to obtain 100. The logarithm is written

$$\log_{10} 100 = 2$$

Notice that the base is written in small numerals below the word "log." The statement is read: "log of 100 (to the base 10) is 2." This means that 2 is an exponent because *a logarithm is an exponent.* 100 is the result of raising the base to that power. So these two expressions are equivalent:

$$10^2 = 100$$

$$\log_{10} 100 = 2$$

---

Given the exponential function $y = b^x$, we say that "$x$ is the logarithm of $y$, to the base $b$" because $x$ is the exponent to which the base $b$ must be raised to obtain $y$.

So, in general,

| | | |
|---|---|---|
| Definition of a Logarithm | If $y = b^x$ $(y > 0,\ b > 0,\ b \neq 1)$<br>Then $x = \log_b y$<br>($x$ is the logarithm of $y$ to the base $b$.) | **185** |

In practice, the base $b$ is taken as either 10 for common logarithms, or $e$ for natural logarithms. If "log" is written without a base, the base is assumed to be 10.

---

**EXAMPLE 14:**

$$\log 5 \equiv \log_{10} 5$$

---

### Converting between Logarithmic and Exponential Form

In working with logarithmic and exponential expressions, we often have to switch between exponential and logarithmic form. While converting from one form to the other, remember that (1) the base in exponential form is also the base in logarithmic form, and (2) "the *log* is the *exponent*."

---

**EXAMPLE 15:** Change the exponential expression $3^2 = 9$ to logarithmic form.

**Solution:** The base (3) in the exponential expression remains the base in the logarithmic expression

$$3^2 = 9$$
$$\downarrow$$
$$\log_3(\quad) = (\quad)$$

and the log is the exponent

$$3^2 = 9$$
$$\log_3(\quad) = 2$$

So our expression is

$$\log_3 9 = 2$$

This is read "the log *of* 9 (to the base 3) *is* 2."

We could also write the same expression in *radical form*, $\sqrt{9} = 3$.

---

**EXAMPLE 16:** Change $\log_e x = 3$ to exponential form.

**Solution:**

$$\log_e x = 3$$
$$e^3 = x$$

---

**EXAMPLE 17:** Change $10^3 = 1000$ to logarithmic form.

**Solution:**

$$\log_{10} 1000 = 3$$

---

## Common Logarithms

Although we may write logarithms to any base, there are only two bases in regular use, the base 10 and the base $e$. Logarithms to the base $e$, called *natural* logarithms are discussed later. In this section we study logarithms to the base 10, called *common* logarithms.

Also called *Briggs' logarithms*, after Henry Briggs (1561–1630).

## Common Logarithms by Calculator

Enter the number you wish to take the log of (the *argument*) and press the [log] key.

Remember that when a logarithm is written without a base (such as $\log x$), we assume the base to be 10.

---

**EXAMPLE 18:** Find log 74.82 to five significant digits.

**Solution:** Enter 74.82. Press [log]. Read 1.8740 (rounded).

---

**EXAMPLE 19:** Find log 0.0375 to four significant digits.

**Solution:** Enter .0375. Press [log]. Read −1.426.

---

**EXAMPLE 20:** Find log (−6.73).

**Solution:** Enter 6.73. Press [+/−]. Press [log]. Read *error indication*. We cannot take the logarithm of a negative number.

### Antilogarithms by Calculator

Suppose that we have the logarithm of a number, and want to find the number itself: say, $\log x = 1.852$. This process is known as finding the antilogarithm (antilog) of 1.852. First we switch to exponential form,

$$x = 10^{1.852}$$

**Some calculators use [INV] [LOG] to find antilogarithms.**

This quantity can be found using the [$10^x$] key on the calculator (or the [$y^x$] key if you have no [$10^x$] key).

Enter 1.852. Press [$10^x$]. Read 71.12 (to four digits).

*or*

Enter 10. Press [$y^x$]. Enter 1.852. Press [=]. Read 71.12.

**EXAMPLE 21:** Find $x$ to four digits if $\log x = -1.738$.

**Solution:** Enter 1.738. Press [+/−]. Press [$10^x$]. Read .01828.

### The Logarithmic Function

We wish to graph the logarithmic function $x = \log_b y$, but first let us follow the usual convention of calling the dependent variable $y$ and the independent variable $x$. Our function then becomes

| Logarithmic Function | $y = \log_b x$ $(x > 0,\ b > 0,\ b \neq 1)$ | **204** |
|---|---|---|

**EXAMPLE 22:** Graph the logarithmic function $y = \log x$, for $x = -1$ to 14.

**Solution:** We make a table of point pairs, getting the values of $y$ by using the common log key on the calculator.

| $x$ | −1 | 0 | 1 | 2 | 3 | 4 | 5 | 6 | 7 | 8 | 9 | 10 | 11 | 12 | 13 | 14 |
|---|---|---|---|---|---|---|---|---|---|---|---|---|---|---|---|---|
| $y$ | — | — | 0 | 0.30 | 0.48 | 0.60 | 0.70 | 0.78 | 0.85 | 0.90 | 0.95 | 1.00 | 1.04 | 1.08 | 1.11 | 1.15 |

First we notice that when $x$ is negative or zero, our calculator gives an error indication. In other words, the domain of $x$ is the positive numbers only. *We cannot take the common logarithm of a negative number or zero.* We'll see that this is true for logarithms to other bases as well.

If we try a value of $x$ between zero and 1, say, $x = 0.5$, we get a negative $y$ (−0.3 in this case). Plotting our point pairs gives the curve in Fig. 19-12.

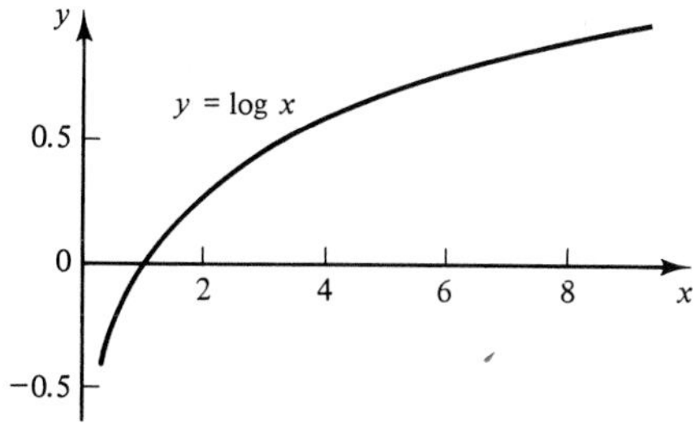


FIGURE 19-12

### Natural Logarithms

You may want to refer to Sec. 19-2 for our earlier discussion of the number $e$ (2.718 approximately). After we have defined $e$, we can use it as a base; logarithms using $e$ as a base are called *natural* logarithms. They are written $\ln x$. Thus $\ln x$ is understood to mean $\log_e x$.

**Also called *Napierian logarithms,* after John Napier (1550–1617), who first wrote on logarithms.**

### Natural Logarithms by Calculator

To find the natural logarithm of a (positive) number by calculator, enter the number and press the [ln] key.

---

**EXAMPLE 23:** Find ln 19.3 to four significant digits.

**Solution:** Enter 19.3. Press [ln]. Read 2.960 (rounded).

---

### Antilogarithms by Calculator

If the natural logarithm of a number is known, we can find the number itself by using the [$e^x$] key.

**Some calculators do not have an [$e^x$] key. Instead, use [inv] [ln] or [2nd] [ln].**

---

**EXAMPLE 24:** Find $x$ to four digits if $\ln x = 3.825$.

**Solution:** Rewriting the expression $\ln x = 3.825$ in exponential form, we have

$$x = e^{3.825}$$

On the calculator: Enter 3.825. Press [$e^x$]. Read 45.83 (rounded).

---

## EXERCISE 3—LOGARITHMICS

### Converting between Logarithmic and Exponential Form

Convert to logarithmic form.

1. $3^4 = 81$
2. $5^3 = 125$
3. $4^6 = 4096$
4. $7^3 = 343$
5. $x^5 = 995$
6. $a^3 = 6.83$

Convert to exponential form.

7. $\log_{10} 100 = 2$
8. $\log_2 16 = 4$
9. $\log_5 125 = 3$
10. $\log_4 1024 = 5$
11. $\log_3 x = 57$
12. $\log_x 54 = 285$
13. $\log_5 x = y$
14. $\log_w 3 = z$

## Logarithmic Function

Graph each logarithmic function from $x = \frac{1}{2}$ to $x = 4$.

**15.** $y = \log_e x$ **16.** $y = \log_{10} 2x$

## Common Logarithms

Find the common logarithm of each number to four decimal places.

**17.** 27.6 **18.** 4.83 **19.** 5.93
**20.** 9.26 **21.** 48.3 **22.** 385
**23.** 836 **24.** 1.03 **25.** 27.4
**26.** 0.0573 **27.** 0.00737 **28.** 9738
**29.** 34,970 **30.** $5.28 \times 10^6$

Find $x$ to three significant digits if the common logarithm of $x$ is equal to:

**31.** 1.584 **32.** 2.957 **33.** 5.273
**34.** 0.886 **35.** −2.857 **36.** 0.366
**37.** −2.227 **38.** 4.973 **39.** 3.979
**40.** 1.295 **41.** −3.972 **42.** −0.8835

## Natural Logarithms

Find the natural logarithm of each number to four decimal places.

**43.** 48.3 **44.** 846 **45.** 2365
**46.** 1.005 **47.** 0.845 **48.** 4.97
**49.** 0.00836 **50.** 45,900 **51.** 0.0000462
**52.** 82,900 **53.** $3.84 \times 10^4$ **54.** $8.24 \times 10^{-3}$

Find to four significant digits the number whose natural logarithm is:

**55.** 2.846 **56.** 4.263 **57.** 0.879
**58.** −2.846 **59.** −0.365 **60.** 5.937
**61.** 0.936 **62.** −4.97 **63.** 15.84
**64.** −9.47 **65.** −18.36 **66.** 21.83

## Computer

**67.** The series approximation for the natural logarithm of a positive number $x$ is

| | | |
|---|---|---|
| Series Approximation | $\ln x = 2a + \dfrac{2a^3}{3} + \dfrac{2a^5}{5} + \dfrac{2a^7}{7} + \cdots$ where $a = \dfrac{x-1}{x+1}$ | **205** |

Write a program to compute $\ln x$ using the first 10 terms of the series.

**68.** The series expansion for $b^x$ is

| | | |
|---|---|---|
| Series Approximation for $b^x$ | $b^x = 1 + x \ln b + \dfrac{(x \ln b)^2}{2!} + \dfrac{(x \ln b)^3}{3!} + \cdots$ $(b > 0)$ | **196** |

Write a program to compute $b^x$ using the first $x$ terms of the series.

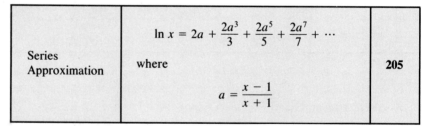

## 19-4 PROPERTIES OF LOGARITHMS

In this section we use the properties of logarithms along with our ability to convert back and forth between exponential and logarithmic forms. These together will enable us to later solve equations that we could not previously handle, such as exponential equations.

### Products

We wish to write an expression for the log of the product of two positive numbers, say, $M$ and $N$.

$$\log_b MN = ?$$

Let us substitute, $M = b^c$ and $N = b^d$. Then

$$MN = b^c b^d$$

Here, $b > 0$, $b \neq 1$.

But by the laws of exponents,

$$MN = b^{c+d}$$

Writing this expression in logarithmic form,

$$\log_b MN = c + d$$

But, since $b^c = M$, $c = \log_b M$. Similarly, $d = \log_b N$. Substituting, we get

| | | |
|---|---|---|
| Log of a Product | $\log_b MN = \log_b M + \log_b N$ | **186** |

*The log of a product equals the sum of the logs of the factors.*

**EXAMPLE 25:**

$$\log 7x = \log 7 + \log x$$

**EXAMPLE 26:**

$$\log 3 + \log x + \log y = \log 3xy$$

| | |
|---|---|
| Common Errors | The log of a sum is *not* equal to the sum of the logs. $$\log_b(M + N) \neq \log_b M + \log_b N$$ |
| | The product of two logs is *not* equal to the sum of the logs. $$(\log_b M)(\log_b N) \neq \log_b M + \log_b N$$ |

**EXAMPLE 27:** Multiply (2.947)(9.362) by logarithms.

We don't pretend that this is a practical problem, now that we have calculators. But we recommend doing a few logarithmic computations just to get practice in using the properties of logarithms, which we'll need later to solve logarithmic and exponential equations.

**Solution:** We take the log of the product

$$\log(2.947 \times 9.362) = \log 2.947 + \log 9.362$$
$$= 0.46938 + 0.97137 = 1.44075$$

This gives us the logarithm of the product. To get the product itself, we take the antilog,

$$(2.947)(9.362) = 10^{1.44075} = 27.59$$

### Quotients

Let us now consider the quotient $M$ divided by $N$. Using the same substitution as above,

$$\frac{M}{N} = \frac{b^c}{b^d} = b^{c-d}$$

Going to logarithmic form,

$$\log_b \frac{M}{N} = c - d$$

Finally, substituting back,

| | | |
|---|---|---|
| Log of a Quotient | $\log_b \frac{M}{N} = \log_b M - \log_b N$ | **187** |

*The log of a quotient equals the log of the dividend minus the log of the divisor.*

**EXAMPLE 28:**

$$\log \frac{3}{4} = \log 3 - \log 4$$

**EXAMPLE 29:** Express

$$\log \frac{ab}{xy}$$

as the sum or difference of two or more logarithms.

**Solution:** By Eq. 187,

$$\log \frac{ab}{xy} = \log ab - \log xy$$

and by Eq. 186,

$$\log ab - \log xy = \log a + \log b - (\log x + \log y)$$
$$= \log a + \log b - \log x - \log y$$

**EXAMPLE 30:** The expression $\log 10x^2 - \log 5x$ can be expressed as a single logarithm,

$$\log \frac{10x^2}{5x} \quad \text{or} \quad \log 2x$$

$$log \frac{10x^2}{5x} \qquad log\,2x$$

| | |
|---|---|
| Common Error | Similar errors are made with quotients as with products. $$\log_b(M - N) \neq \log_b M - \log_b N$$ $$\frac{\log_b M}{\mathrm{lob}_b N} \neq \log_b M - \log_b N$$ |

**EXAMPLE 31:** Divide using logarithms:

$$28.47 \div 1.883$$

**Solution:** By Eq. 187,

$$\log \frac{28.47}{1.883} = \log 28.47 - \log 1.883$$

$$= 1.4544 - 0.2749 = 1.1795$$

This gives us the logarithm of the quotient. To get the quotient itself, we take the antilog,

$$28.47 \div 1.883 = 10^{1.1795} = 15.12$$

## Powers

Consider now the quantity $M$ raised to a power $p$. Substituting $b^c$ for $M$, as before,

$$M^p = (b^c)^p = b^{cp}$$

In logarithmic form,

$$\log_b M^p = cp$$

Substituting back, we get

| Log of a Power | $\log_b M^p = p \log_b M$ | **188** |
|---|---|---|

*The log of a number raised to a power equals the power times the log of the number.*

**EXAMPLE 32:**

$$\log 3.85^{1.4} = 1.4 \log 3.85$$

**EXAMPLE 33:** The expression 7 log $x$ can be expressed as a single logarithm with a coefficient of 1, as

$$\log x^7$$

**EXAMPLE 34:** Express 2 log 5 + 3 log 2 − 4 log 1 as a single logarithm.

**Solution:**

$$2 \log 5 + 3 \log 2 - 4 \log 1 = \log 5^2 + \log 2^3 - \log 1^4$$

$$= \log \frac{25(8)}{1} = \log 200$$

**EXAMPLE 35:** Express as a single logarithm with a coefficient of 1.

$$2 \log \frac{2}{3} + 4 \log \frac{a}{b} - 3 \log \frac{x}{y}$$

**Solution:**

$$2 \log \frac{2}{3} + 4 \log \frac{a}{b} - 3 \log \frac{x}{y} = \log \frac{2^2}{3^2} + \log \frac{a^4}{b^4} - \log \frac{x^3}{y^3}$$

$$= \log \frac{4a^4y^3}{9b^4x^3}$$

**EXAMPLE 36:** Rewrite without logarithms, the equation

$$\log(a^2 - b^2) + 2 \log a = \log(a - b) + 3 \log b$$

**Solution:** We first try to combine all the logarithms into a single logarithm. Rearranging gives

$$\log(a^2 - b^2) - \log(a - b) + 2 \log a - 3 \log b = 0$$

By the properties of logarithms,

$$\log \frac{a^2 - b^2}{a - b} + \log \frac{a^2}{b^3} = 0$$

or

$$\log \frac{a^2(a - b)(a + b)}{b^3(a - b)} = 0$$

Simplifying yields

$$\log \frac{a^2(a + b)}{b^3} = 0$$

We then rewrite without logarithms by going from logarithmic to exponential form,

$$\frac{a^2(a + b)}{b^3} = 10^0 = 1$$

or

$$a^2(a + b) = b^3$$

**EXAMPLE 37:** Find $35.82^{1.4}$ by logarithms.

**Solution:** By Eq. 188,

$$\log 35.82^{1.4} = 1.4 \log 35.82$$
$$= 1.4(1.5541) = 2.1758$$

Taking the antilog yields

$$35.82^{1.4} = 10^{2.1758} = 149.9$$

## Roots

We can write a given radical expression in exponential form, and then use the rule for powers. Thus

| | |
|---|---|
| $\sqrt[q]{M} = M^{1/q}$ | **36** |

Taking the logarithm of both sides, we obtain

$$\log_b \sqrt[q]{M} = \log_b M^{1/q}$$

Then by Eq. 188,

| | | |
|---|---|---|
| Log of a Root | $\log_b \sqrt[q]{M} = \frac{1}{q} \log_b M$ | **189** |

*The log of the root of a number equals the log of the number divided by the index of the root.*

We may take the logarithm of both sides of an equation, just as we took the square root of both sides of an equation, or took the sine of both sides of an equation.

**EXAMPLE 38:**

(a) $\log \sqrt[5]{8} = \frac{1}{5} \log 8 \simeq \frac{1}{5}(0.9031) \simeq 0.1806$

(b) $\frac{\log x}{2} + \frac{\log z}{4} = \log \sqrt{x} + \log \sqrt[4]{z}$

**EXAMPLE 39:** Find $\sqrt[5]{3824}$ by logarithms.

**Solution:** By Eq. 189,

$$\log \sqrt[5]{3824} = \frac{1}{5} \log 3824 = \frac{3.5825}{5}$$
$$= 0.7165$$

Taking the antilog gives us

$$\sqrt[5]{3824} = 10^{0.7165} = 5.206$$

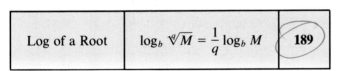

### Log of 1

Let us take the log to the base $b$ of 1, and call it $x$,

$$x = \log_b 1$$

Switching to exponential form, we have

$$b^x = 1$$

This equation is satisfied, for any positive $b$, by $x = 0$. Therefore,

| | | |
|---|---|---|
| Log of 1 | $\log_b 1 = 0$ | **190** |

*The log of 1 is zero.*

### Log of the Base

We now take the $\log_b$ of its own base $b$. Let us call this quantity $x$.

$$\log_b b = x$$

Going to exponential form gives us

$$b^x = b$$

So $x$ must equal 1.

| | | |
|---|---|---|
| Log of the Base | $\log_b b = 1$ | **191** |

*The log (to the base b) of b is equal to 1.*

---

**EXAMPLE 40:**

(a) $\log_5 5 = 1$ (b) $\log 10 = 1$ (c) $\ln e = 1$

---

### Log of the Base Raised to a Power

Consider the expression

$$\log_b b^n$$

We set this expression equal to $x$, and as before, go to exponential form,

$$\log_b b^n = x$$

$$b^x = b^n$$

$$x = n$$

So

| | | |
|---|---|---|
| Log of the Base Raised to a Power | $\log_b b^n = n$ | **192** |

*The log (to the base b) of b raised to a power is equal to the power.*

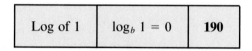

**EXAMPLE 41:**

(a) $\log_2 2^{4.83} = 4.83$

(b) $\log_e e^{2x} = 2x$

(c) $\log_{10} 0.0001 = \log_{10} 10^{-4} = -4$

## Base Raised to a Logarithm of the Same Base

We want to evaluate an expression of the form

$$b^{\log_b x}$$

Setting this expression equal to $y$ and taking the logarithms of both sides gives us

$$y = b^{\log_b x}$$

$$\log_b y = \log_b b^{\log_b x} = \log_b x \log_b b$$

Since $\log_b b = 1$, we have $\log_b y = \log_b x$. Taking the antilog of both sides,

$$x = y = b^{\log_b x}$$

so,

| Base Raised to a Logarithm of the Same Base | $b^{\log_b x} = x$ |
|---|---|

**EXAMPLE 42:**

(a) $10^{\log w} = w$

(b) $e^{\log_e 3x} = 3x$

## Change of Base

We can convert between natural logarithms and common logarithms (or between logarithms to any base) by the following procedure. Suppose that log $N$ is the common logarithm of some number $N$. We set it equal to $x$,

$$\log N = x$$

Going to exponential form, we obtain

$$10^x = N$$

Now we take the natural log of both sides,

$$\ln 10^x = \ln N$$

By Eq. 188,

$$x \ln 10 = \ln N$$

$$x = \frac{\ln N}{\ln 10}$$

But $x = \log N$, so

Some computer languages enable us to find the natural log but not the common log. This equation lets us convert from one to the other.

| | | |
|---|---|---|
| Change of Base | $\log N = \dfrac{\ln N}{\ln 10}$ where $\ln 10 \simeq 2.3026$. | **193** |

*The common logarithm of a number is equal to the natural log of that number divided by the natural log of 10.*

---

**EXAMPLE 43:** Find $\log N$ if $\ln N = 5.824$.

**Solution:** By Eq. 193,

$$\log N = \frac{5.824}{2.3026} = 2.529$$

---

**EXAMPLE 44:** What is $\ln N$ if $\log N = 3.825$?

**Solution:** By Eq. 193,

$$\ln N = \ln 10 (\log N)$$

$$= 2.3026(3.825) = 8.807$$

---

## EXERCISE 4—PROPERTIES OF LOGARITHMS

Write as the sum or difference of two or more logarithms.

1. $\log \frac{2}{3}$
2. $\log 4x$
3. $\log ab$
4. $\log \dfrac{x}{2}$
5. $\log xyz$
6. $\log 2ax$
7. $\log \dfrac{3x}{4}$
8. $\log \dfrac{5}{xy}$
9. $\log \dfrac{1}{2x}$
10. $\log \dfrac{2x}{3y}$
11. $\log \dfrac{abc}{d}$
12. $\log \dfrac{x}{2ab}$

Express as a single logarithm with a coefficient of 1. Assume that the logarithms in each problem have the same base.

13. $\log 3 + \log 4$
14. $\log 7 - \log 5$
15. $\log 2 + \log 3 - \log 4$
16. $3 \log 2$
17. $4 \log 2 + 3 \log 3 - 2 \log 4$
18. $\log x + \log y + \log z$
19. $3 \log a - 2 \log b + 4 \log c$
20. $\dfrac{\log x}{3} + \dfrac{\log y}{2}$
21. $\log \dfrac{x}{a} + 2 \log \dfrac{y}{b} + 3 \log \dfrac{z}{c}$
22. $2 \log x + \log(a + b) - \frac{1}{2} \log(a + bx)$
23. $\frac{1}{2} \log(x + 2) + \log(x - 2)$

Rewrite each equation so that it contains no logarithms.

24. $\log x + 3 \log y = 0$
25. $\log_2 x + 2 \log_2 y = x$
26. $2 \log(x - 1) = 5 \log(y + 2)$
27. $\log_2(a^2 + b^2) + 1 = 2 \log_2 a$
28. $\log(x + y) - \log 2 = \log x + \log(x^2 - y^2)$
29. $\log(p^2 - q^2) - \log(p + q) = 2$

Simplify.

30. $\log_2 2$
31. $\log_e e$
32. $\log_{10} 10$
33. $\log_3 3^2$
34. $\log_e e^x$
35. $\log_{10} 10^x$
36. $e^{\log_e x}$
37. $2^{\log_2 3y}$
38. $10^{\log x^2}$

### Logarithmic Computation

Evaluate using logarithms.

39. $5.937 \times 92.47$
40. $6923 \times 0.003846$
41. $3.97 \times 8.25 \times 9.82$
42. $88.25 \div 42.94$
43. $\sqrt[9]{8563}$
44. $4.836^{3.97}$
45. $83.62^{0.572}$
46. $\sqrt[3]{587}$
47. $\sqrt[7]{8364}$
48. $\sqrt[4]{62.4}$

**This may not seem like hot stuff in the age of calculators and computers, but the use of logarithms revolutionized computation in its time. The French mathematician Laplace wrote: "The method of logarithms, by reducing to a few days the labor of many months, doubles, as it were, the life of the astronomer, besides freeing him from the errors and disgust inseparable from long calculation."**

### Change of Base

Find the common logarithm of the number whose natural logarithm is:

49. 8.36
50. −3.846
51. 3.775
52. 15.36
53. 5.26
54. −0.638

Find the natural logarithm of the number whose common logarithm is:

55. 84.9
56. 2.476
57. −3.82
58. 73.9
59. 2.37
60. −2.63

## 19-5 SOLVING EXPONENTIAL EQUATIONS USING LOGARITHMS

### Solving Exponential Equations

We return now to the problem we started in Sec. 19-3 but could not finish, that of finding the *exponent* when the other quantities in an exponential equation were known. We tried to solve the equation

$$24.0 = 2.48^x$$

The key to solving exponential equations is to *take the logarithm of both sides*. This enables us to use Eq. 188 to extract the unknown from the exponent. Taking the logarithm of both sides gives

$$\log 24.0 = \log 2.48^x$$

By Eq. 188,

$$\log 24.0 = x \log 2.48$$

$$x = \frac{\log 24.0}{\log 2.48} = \frac{1.380}{0.3945} = 3.50$$

In this example we took *common* logarithms, but *natural* logarithms would have worked just as well.

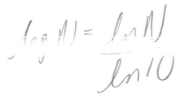

**EXAMPLE 45:** Solve for $x$ to three significant digits.

$$3.25^{x+2} = 1.44^{3x-1}$$

**Solution:** We take the log of both sides. This time choosing to use natural logs, we get

$$(x + 2) \ln 3.25 = (3x - 1) \ln 1.44$$

By calculator,

$$(x + 2)(1.179) = (3x - 1)(0.3646)$$

or

$$3.233(x + 2) = 3x - 1$$

Removing parentheses and collecting terms yields

$$0.233x = -7.467$$

$$x = -32.0$$

### Solving Exponential Equations with Base *e*

If the exponential equation contains the base $e$, taking *natural* logs rather than common logarithms will simplify the work.

**EXAMPLE 46:** Solve for $x$:

$$157 = 112e^{3x+2}$$

**Solution:** Dividing by 112, we have

$$1.402 = e^{3x+2}$$

Taking natural logs gives us

$$\ln 1.402 = \ln e^{3x+2}$$

By Eq. 188,

$$\ln 1.402 = (3x + 2) \ln e$$

But $\ln e = 1$, so

$$\ln 1.402 = 3x + 2$$

$$x = \frac{\ln 1.402 - 2}{3} = -0.554$$

**EXAMPLE 47:** Solve for $x$ to three significant digits:

$$3e^x + 2e^{-x} = 4(e^x - e^{-x})$$

**Solution:** Removing parentheses gives

$$3e^x + 2e^{-x} = 4e^x - 4e^{-x}$$

Combining like terms yields

$$e^x - 6e^{-x} = 0$$

Factoring gives

$$e^{-x}(e^{2x} - 6) = 0$$

We set each factor equal to zero (just as when solving a quadratic by factoring) and get

$$e^{-x} = 0 \qquad e^{2x} = 6$$

There is no value of $x$ that will make $e^{-x}$ equal to zero, so we get no root from that equation. We solve the other equation by taking the natural log of both sides,

$$2x \ln e = \ln 6 = 1.792$$

$$x = 0.896$$

Our exponential equation might be of *quadratic type,* and can then be solved by substitution as we did in Chapter 13.

**EXAMPLE 48:** Solve for $x$ to three significant digits:

$$3e^{2x} - 2e^x - 5 = 0$$

**Solution:** If we substitute

$$w = e^x$$

our equation becomes the quadratic

$$3w^2 - 2w - 5 = 0$$

By the quadratic formula,

$$w = \frac{2 \pm \sqrt{4 - 4(3)(-5)}}{6} = -1 \quad \text{and} \quad \frac{5}{3}$$

Substituting back, we have

$$e^x = -1 \quad \text{and} \quad e^x = \frac{5}{3}$$

So, $x = \ln(-1)$ which we discard and

$$x = \ln \frac{5}{3} = 0.511$$

## Solving Exponential Equations with Base 10

If the equation contains 10 as a base, the work will be simplified by taking *common* logarithms.

**EXAMPLE 49:** Solve $10^{5x} = 2(10^{2x})$.

**Solution:** Taking common logs of both sides gives us

$$\log 10^{5x} = \log[2(10^{2x})]$$

By Eqs. 186 and 188,

$$5x \log 10 = \log 2 + 2x \log 10$$

By Eq. 191, log 10 = 1, so

$$5x = \log 2 + 2x$$

$$3x = \log 2$$

$$x = \frac{\log 2}{3} = 0.100$$

## Doubling Time

Being able to solve an exponential equation for the exponent allows us to return to the formulas for exponential growth and decay (Sec. 19-2) and derive two interesting quantities: doubling time and half-life.

If a quantity grows exponentially according to the function

| | | |
|---|---|---|
| Exponential Growth | $y = ae^{nt}$ | **197** |

it will eventually double ($y$ would be twice $a$). Setting $y = 2a$, we get

$$2a = ae^{nt}$$

or $2 = e^{nt}$. Taking natural logs, we have

$$\ln 2 = \ln e^{nt} = nt \ln e = nt$$

or

| | | |
|---|---|---|
| Doubling Time | $t = \frac{\ln 2}{n}$ | **198** |

Since ln 2 = 0.693, if we let $P$ be the rate of growth expressed as a percent (*not* as a decimal, $P = 100n$), we get a convenient *rule of thumb:*

$$\text{doubling time} \approx \frac{70}{P}$$

**EXAMPLE 50:** In how many years will a quantity double if it grows at the rate of 5% per year?

**Solution:**

$$\text{Doubling time} = \frac{\ln 2}{0.05}$$

$$= \frac{0.693}{0.05}$$

$$= 13.9 \text{ years}$$

## Half-Life

When a material decays exponentially according to the function

| | | |
|---|---|---|
| Exponential Decay | $y = ae^{-nt}$ | **199** |

the time it takes for the material to be half gone is called the *half-life*. If we let $y = a/2$,

$$\frac{1}{2} = e^{-nt} = \frac{1}{e^{nt}}$$

$$2 = e^{nt}$$

Taking natural logarithms gives us

$$\ln 2 = \ln e^{nt} = nt$$

**Notice that the equation for half-life is the same as for doubling time. The rule of thumb is, of course, also the same.**

| | | |
|---|---|---|
| Half-Life | $t = \dfrac{\ln 2}{n}$ | **198** |

**EXAMPLE 51:** Find the half-life of a radioactive material that decays exponentially at the rate of 2% per year.

**Solution:**

$$\text{Half-life} = \frac{\ln 2}{0.02} = 34.7 \text{ years}$$

## Change of Base

If we have an exponential expression, such as $5^x$, we can convert it to another exponential expression with base $e$, or any other base.

**EXAMPLE 52:** Convert $5^x$ into an exponential expression with base $e$.

**Solution:** We write 5 as $e$ raised to some power, say $n$.

$$5 = e^n$$

Taking the natural logarithm, we obtain

$$\ln 5 = \ln e^n = n$$

So $n = \ln 5 \simeq 1.61$, and $5 = e^{1.61}$. Then

$$5^x = (e^{1.61})^x = e^{1.61x}$$

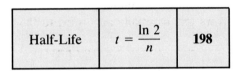

**EXAMPLE 53:** Convert $2^{4x}$ to an exponential expression with base $e$.

**Solution:** We first express 2 as $e$ raised to some power.

$$2 = e^n$$

$$\ln 2 = \ln e^n = n$$

$$n = \ln 2 = 0.693$$

Then

$$2^{4x} = (e^{0.693})^{4x} = e^{2.773x}$$

## EXERCISE 5—SOLVING EXPONENTIAL EQUATIONS USING LOGARITHMS

Solve for $x$ to three significant digits.

*Hint:* Problems 18 to 20 are *equations of quadratic type.*

1. $2^x = 7$
2. $(7.26)^x = 86.8$
3. $(1.15)^{x+2} = 12.5$
4. $(2.75)^x = (0.725)^{x^2}$
5. $(15.4)^{\sqrt{x}} = 72.8$
6. $e^{5x} = 125$
7. $5.62e^{3x} = 188$
8. $1.05e^{4x+1} = 5.96$
9. $e^{2x-1} = 3e^{x+3}$
10. $14.8e^{3x^2} = 144$
11. $5^{2x} = 7^{3x-2}$
12. $3^{x^2} = 175^{x-1}$
13. $10^{3x} = 3(10^x)$
14. $e^x + e^{-x} = 2(e^x - e^{-x})$
15. $2^{3x+1} = 3^{2x+1}$
16. $5^{2x} = 3^{3x+1}$
17. $7e^{1.5x} = 2e^{2.4x}$
18. $e^{4x} - 2e^{2x} - 3 = 0$
19. $e^x + e^{-x} = 2$
20. $e^{6x} - e^{3x} - 2 = 0$

Convert to an exponential expression with base $e$.

21. $3(4^{3x})$
22. $4(2.2^{4x})$

### Applications

23. The current $i$ in a certain circuit is given by

$$i = 6.25e^{-125t} \quad \text{amperes}$$

where $t$ is the time in seconds. At what time will the current be 1.00 A?

24. The temperature above its surroundings of an iron casting initially at 2000°F will be

$$T = 2000e^{-0.062t}$$

after $t$ seconds. Find the time for the casting to be at a temperature of 500°F above its surroundings.

25. A certain long pendulum, released from a height $y_0$ above its rest position, will be at a height

$$y = y_0e^{-0.75t}$$

at $t$ seconds. If the pendulum is released at a height of 15 cm, at what time will the height be 5 cm?

26. A population growing at a rate of 2% per year from an initial population of 9000 will grow in $t$ years to an amount

$$P = 9000e^{0.02t}$$

How many years will it take for the population to triple?

27. The barometric pressure in inches of mercury at a height of $h$ feet above sea level is

$$p = 30e^{-kh}$$

where $k = 3.83 \times 10^{-5}$. At what height will the pressure be 10 inches of mercury?

28. The approximate density of seawater at a depth of $h$ miles is

$$d = 64.0e^{0.00676h} \qquad \text{lb/ft}^3$$

Find the depth at which the density will be 64.5 $\text{lb/ft}^3$.

29. A rope passing over a rough cylindrical beam (Fig. 19-13) supports a weight $W$. The force $F$ needed to hold the weight is

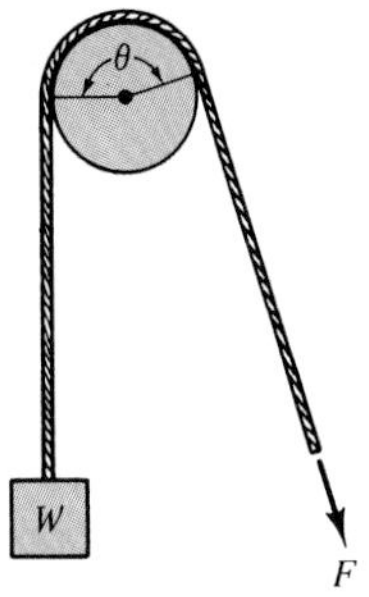

**FIGURE 19-13**

$$F = We^{-\mu\theta}$$

where $\mu$ is the coefficient of friction and $\theta$ is the angle of wrap in radians. If $\mu = 0.15$, what angle of wrap is needed for a force of 100 lb to hold a weight of 200 lb?

30. Using the formula for compound interest, Eq. A10, calculate the number of years it will take for a sum of money to triple when invested at a rate of 12% per year.
31. Using the formula for present worth, Exercise 1, Problem 7, in how many years will \$50,000 accumulate to \$70,000 at 15% interest?
32. Using the annuity formula, Exercise 1, Problem 10, find the number of years it will take a worker to accumulate \$100,000 with an annual payment of \$3000 if the interest rate is 8%.
33. Using the capital recovery formula from Exercise 1, Problem 12, calculate the number of years a person can withdraw \$10,000/yr from a retirement fund containing \$60,000 if the rate of interest is $6\frac{3}{4}$%.
34. Find the half-life of a material that decays exponentially at the rate of 3.5% per year.
35. How long will it take the U.S. annual oil consumption to double if it is increasing exponentially at a rate of 7% per year?
36. How long will it take the world population to double at an exponential growth rate of 1.64% per year?
37. What is the maximum annual growth in energy consumption permissible if the consumption is not to double in the next 20 years?

### Computer

38. An iron ball with a mass of 1 kg is cut into two equal pieces. It is cut again, the second cut producing four equal pieces, the third cut making eight pieces, and so on. Write a program to compute the number of cuts needed to make each piece smaller than one atom (mass = $9.4 \times 10^{-26}$ kg). Assume that no material is removed during cutting.
39. Assuming that the present annual world oil consumption is $17 \times 10^9$ barrels/yr, that this rate of consumption is increasing at a rate of 5% per year, and that the total world oil reserves are $1700 \times 10^9$ barrels, compute and print the following table.

| *Year* | *Annual Consumption* | *Oil Remaining* |
|---|---|---|
| 0 | 17 | 1700 |
| 1 | 17.85 | 1682.15 |
| ⋮ | ⋮ | ⋮ |

Have the computation stop when the oil reserves are gone.

40. Use your program for finding roots of equations by simple iteration or the midpoint method to solve any of Problems 1 to 20.

## 19-6 LOGARITHMIC EQUATIONS

Often a logarithmic expression can be evaluated, or an equation containing logarithms can be solved, simply by rewriting it in exponential form.

---

**EXAMPLE 54:** Evaluate $x = \log_5 25$.

**Solution:** Changing to exponential form, we have

$$5^x = 25 = 5^2$$

$$x = 2$$

Since $5^2 = 25$, the answer checks.

---

**EXAMPLE 55:** Solve for $x$:

$$\log_x 4 = \frac{1}{2}$$

**Solution:** Going to exponential form gives us

$$x^{1/2} = 4$$

Squaring both sides yields

$$x = 16$$

Since $16^{1/2} = 4$, the answer checks.

---

**EXAMPLE 56:** Solve for $x$:

$$\log_{25} x = -\frac{3}{2}$$

**Solution:** Writing the equation in exponential form, we obtain

$$25^{-3/2} = x$$

$$x = \frac{1}{25^{3/2}} = \frac{1}{5^3} = \frac{1}{125}$$

We check that $25^{-3/2} = \frac{1}{125}$. Checks.

---

**EXAMPLE 57:** Evaluate $x = \log_{81} 27$.

**Solution:** Changing to exponential form yields

$$81^x = 27$$

But 81 and 27 are both powers of 3.

$$(3^4)^x = 3^3$$

$$3^{4x} = 3^3$$

$$4x = 3$$

$$x = 3/4$$

To check, we verify that $81^{3/4} = 3^3 = 27$.

---

If only one term in our equation contains a logarithm, we isolate that term on one side of the equation and then go to exponential form.

**EXAMPLE 58:** Solve for $x$ to three significant digits:

$$3 \log(x^2 + 2) - 6 = 0$$

**Solution:** Rearranging and dividing by 3 gives

$$\log(x^2 + 2) = 2$$

Going to exponential form, we obtain

$$x^2 + 2 = 10^2 = 100$$
$$x^2 = 98$$
$$x = \pm 9.90$$

If every term contains "log," we use the properties of logarithms to combine them into a single logarithm on each side of the equation, and then take the antilog of both sides.

**EXAMPLE 59:** Solve for $x$ to three significant digits.

$$3 \log x - 2 \log 2x = 2 \log 5$$

**Solution:** Using the properties of logarithms gives

$$\log \frac{x^3}{(2x)^2} = \log 5^2$$

$$\log \frac{x}{4} = \log 25$$

Taking the antilog of both sides, we have

$$\frac{x}{4} = 25$$

$$x = 100$$

If one or more terms does not contain a log, combine all the terms that do contain a log on one side of the equation. Then go to exponential form.

**EXAMPLE 60:** Solve for $x$:

$$\log(5x + 2) - 1 = \log(2x - 1)$$

**Solution:** Rearranging yields

$$\log(5x + 2) - \log(2x - 1) = 1$$

By Eq. 187,

$$\log \frac{5x + 2}{2x - 1} = 1$$

Expressing in exponential form gives

$$\frac{5x + 2}{2x - 1} = 10^1 = 10$$

Solving for $x$, we have

$$5x + 2 = 20x - 10$$
$$12 = 15x$$
$$x = \frac{4}{5}$$

## EXERCISE 6—LOGARITHMIC EQUATIONS

Find the value of $x$ in each expression.

| | | |
|---|---|---|
| 1. $x = \log_3 9$ | 2. $x = \log_2 8$ | 3. $x = \log_8 2$ |
| 4. $x = \log_9 27$ | 5. $x = \log_{27} 9$ | 6. $x = \log_4 8$ |
| 7. $x = \log_8 4$ | 8. $x = \log_{27} 81$ | 9. $\log_x 8 = 3$ |
| 10. $\log_3 x = 4$ | 11. $\log_x 27 = 3$ | 12. $\log_x 16 = 4$ |
| 13. $\log_5 x = 2$ | 14. $\log_x 2 = \frac{1}{2}$ | 15. $\log_{36} x = \frac{1}{2}$ |
| 16. $\log_2 x = 3$ | 17. $x = \log_{25} 125$ | 18. $x = \log_5 125$ |

Solve for $x$ to three significant digits. Check your answers.

| | |
|---|---|
| 19. $\log(2x + 5) = 2$ | 20. $2 \log(x + 1) = 3$ |
| 21. $\log(2x + x^2) = 2$ | 22. $\ln x - 2 \ln x = \ln 64$ |
| 23. $\ln 6 + \ln(x - 2) = \ln(3x - 2)$ | 24. $\ln(x + 2) - \ln 36 = \ln x$ |
| 25. $\ln(5x + 2) - \ln(x + 6) = \ln 4$ | 26. $\log x + \log 4x = 2$ |
| 27. $\ln x + \ln(x + 2) = 1$ | 28. $\log 8x^2 - \log 4x = 2.54$ |
| 29. $2 \log x - \log(1 - x) = 1$ | 30. $3 \log x - 1 = 3 \log(x - 1)$ |
| 31. $\log(x^2 - 4) - 1 = \log(x + 2)$ | 32. $2 \log x - 1 = \log(20 - 2x)$ |
| 33. $\log(x^2 - 1) - 2 = \log(x + 1)$ | 34. $\ln 2x - \ln 4 + \ln(x - 2) = 1$ |
| 35. $\log(4x - 3) + \log 5 = 6$ | |

### Applications

**This equation is derived from the compound interest formula. The equations in the next two problems are obtained from the equations for an annuity and for capital recovery from Exercise 1. Can you see how they were derived?**

36. An amount of money $a$ invested at a compound interest rate of $n$ per year will take $t$ years to accumulate to an amount $y$, where $t$ is

$$t = \frac{\log y - \log a}{\log(1 + n)} \quad \text{years}$$

How many years will it take for an investment to triple in value when deposited at 8% per year?

37. If an amount $R$ is deposited once every year at a compound interest rate $n$, the number of years it will take to accumulate to an amount $y$ is

$$t = \frac{\log\left[\dfrac{ny}{R} + 1\right]}{\log(1 + n)} \quad \text{years}$$

How many years will it take an annual payment of \$1500 to accumulate to \$13,800 at 9% per year?

38. If an amount $a$ is invested at a compound interest rate $n$, it will be possible to withdraw a sum $R$ at the end of every year for $t$ years until the deposit is exhausted. The number of years is given by

$$t = \frac{\log\left[\dfrac{an}{R - an} + 1\right]}{\log(1 + n)} \quad \text{years}$$

If \$200,000 is invested at 12% interest, for how many years can an annual withdrawal of \$30,000 be made before the money is used up?

39. The *magnitude* $M$ of a star of intensity $I$ is

$$M = 2.5 \log \frac{I_1}{I} + 1$$

where $I_1$ is the intensity of a first-magnitude star. What is the magnitude of a star whose intensity is one-tenth that of a first-magnitude star?

40. The difference in elevation $h$ (ft) between two locations having barometer readings of $B_1$ and $B_2$ inches of mercury is given by the logarithmic equation $h = 60{,}470 \log(B_2/B_1)$, where $B_1$ is the pressure at the *upper* station. Find the difference in elevation between two stations having barometer readings of 29.14 in. at the lower station and 26.22 in. at the upper.
41. What will be the barometer reading 815.0 ft above a station having a reading of 28.22 in.?

Use the following information for Problems 42 to 45. If the power input to a network or device is $P_1$ and the power output is $P_2$, the amount of decibels gained or lost in the device is given by the logarithmic equation

| | | | |
|---|---|---|---|
| Decibels Gained or Lost | $G = 10 \log_{10} \dfrac{P_2}{P_1}$ | dB | **A102** |

42. A certain amplifier gives a power output of 1000 W for an input of 50 W. Find the dB gain.
43. A transmission line has a loss of 3.25 dB. Find the power transmitted for an input of 2750 kW.
44. What power input is needed to produce a 250-W output with an amplifier having a 50-dB gain?
45. The output of a certain device is half the input. This is a loss of how many decibels?
46. The heat loss $q$ per foot of cylindrical pipe insulation (Fig. 19-14) having an inside radius $r_1$ and outside radius $r_2$ is given by the logarithmic equation

$$q = \frac{2\pi k(t_1 - t_2)}{\ln (r_2/r_1)} \quad \text{Btu/h}$$

where $t_1$ and $t_2$ are the inside and outside temperatures (°F) and $k$ is the conductivity of the insulation. Find $q$ for a 4-in.-thick insulation having a conductivity of 0.036 wrapped around a 9-in.-diameter pipe at 550°F if the surroundings are at 90°F.

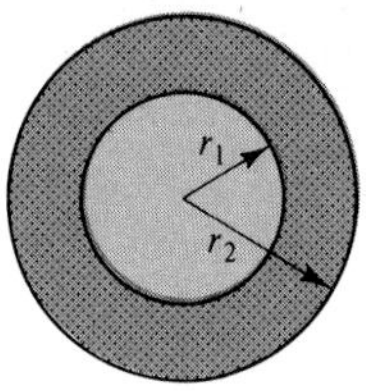


**FIGURE 19-14 An insulated pipe.**

47. If a resource is being used up at a rate that increases exponentially, the time it takes to exhaust the resource (called the exponential expiration time, EET) is

$$\text{EET} = \frac{1}{n} \ln\left(\frac{nR}{r} + 1\right)$$

where $n$ is the rate of increase in consumption, $R$ the total amount of the resource, and $r$ the initial rate of consumption. If we assume that the United States has oil reserves of $207 \times 10^9$ barrels and that our present rate of consumption is $6 \times 10^9$ barrels/yr, how long will it take to exhaust these reserves if our consumption increases by 7% per year?

Use the following information for Problems 48 to 50. The pH value of a solution having a concentration $C$ of hydrogen ions is

| | | |
|---|---|---|
| pH | $\text{pH} = -\log_{10} C$ | **A49** |

48. Find the concentration of hydrogen ions in a solution having a pH of 4.65.
49. A pH of 7 is considered *neutral*, while a lower pH is *acid* and a higher pH is *alkaline*. What is the hydrogen ion concentration at a pH of 7?

50. The acid rain during a certain storm had a pH of 3.5. Find the hydrogen ion concentration. How does it compare with that for a pH of 7? Show that **(a)** when the pH doubles, the hydrogen ion concentration is *squared*, and **(b)** when the pH increases by a factor of $n$, the hydrogen ion concentration is raised *to the nth power*.

### Computer

51. Use your program for solving equations by simple iteration or by the midpoint method to solve any one of Problems 19 to 35.

## 19-7 GRAPHS ON LOGARITHMIC AND SEMILOGARITHMIC PAPER

### Logarithmic and Semilogarithmic Paper

Our graphing so far has all been done on ordinary graph paper, on which the lines are equally spaced. For some purposes, though, it is better to use *logarithmic* paper (Fig. 19-15), also called *log-log* paper, or *semilogarithmic* paper (Fig. 19-16), also called *semilog* paper. Looking at the logarithmic scales of these graphs, we note the following:

1. The lines are not equally spaced. The distance in inches from, say, 1 to 2, is equal to the distance from 2 to 4, which, in turn, is equal to the distance from 4 to 8.
2. Each tenfold increase in the scale, say from 1 to 10 or from 10 to 100, is called a *cycle*. Each cycle requires the same distance in inches along the scale.
3. The log scales do not include zero.

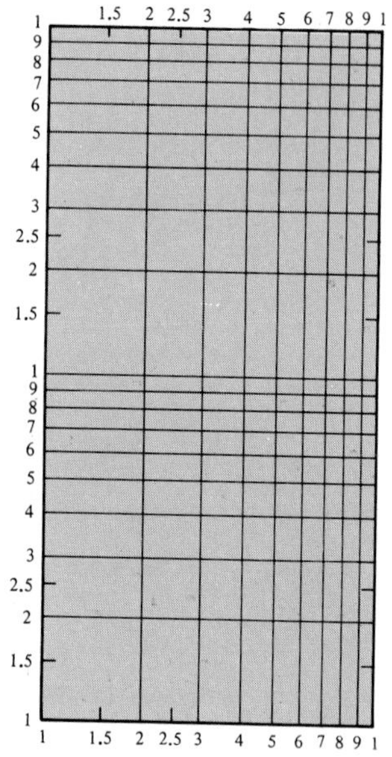


**FIGURE 19-15 Logarithmic graph paper.**

Looking at Fig. 19-16, notice that although the numbers on the vertical scale are in equal increments (1, 2, 3, . . . , 10), the spacing on that scale is proportional to the logarithms of those numbers. So the numeral "2" is placed at a position corresponding to log 2 (which is 0.3010, or about one-third of the distance along the vertical), "4" is placed at log 4 (about 0.6 of the way), and "10" is at log 10 (which equals 1, at the top of the scale).

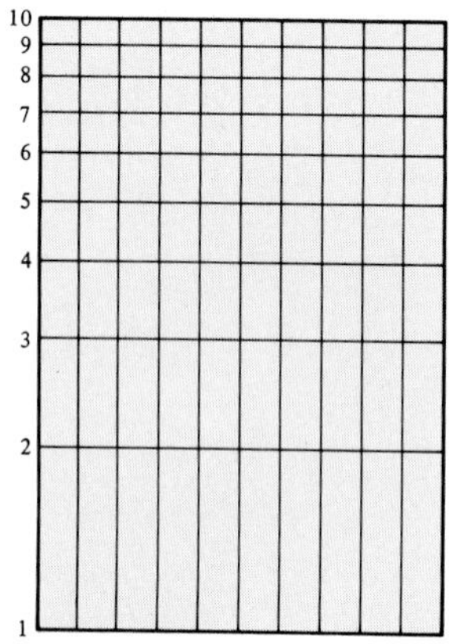


**FIGURE 19-16 Semilogarithmic graph paper.**

### When to Use Logarithmic or Semilog Paper

We use these special papers for the following reasons:

1. When the range of the variables is too large for ordinary paper.
2. When we want to graph a power function or an exponential function. Each of these will plot as a straight line on the appropriate paper, as shown in Fig. 19-17.
3. When we want to find an equation that will approximately represent a set of empirical data.

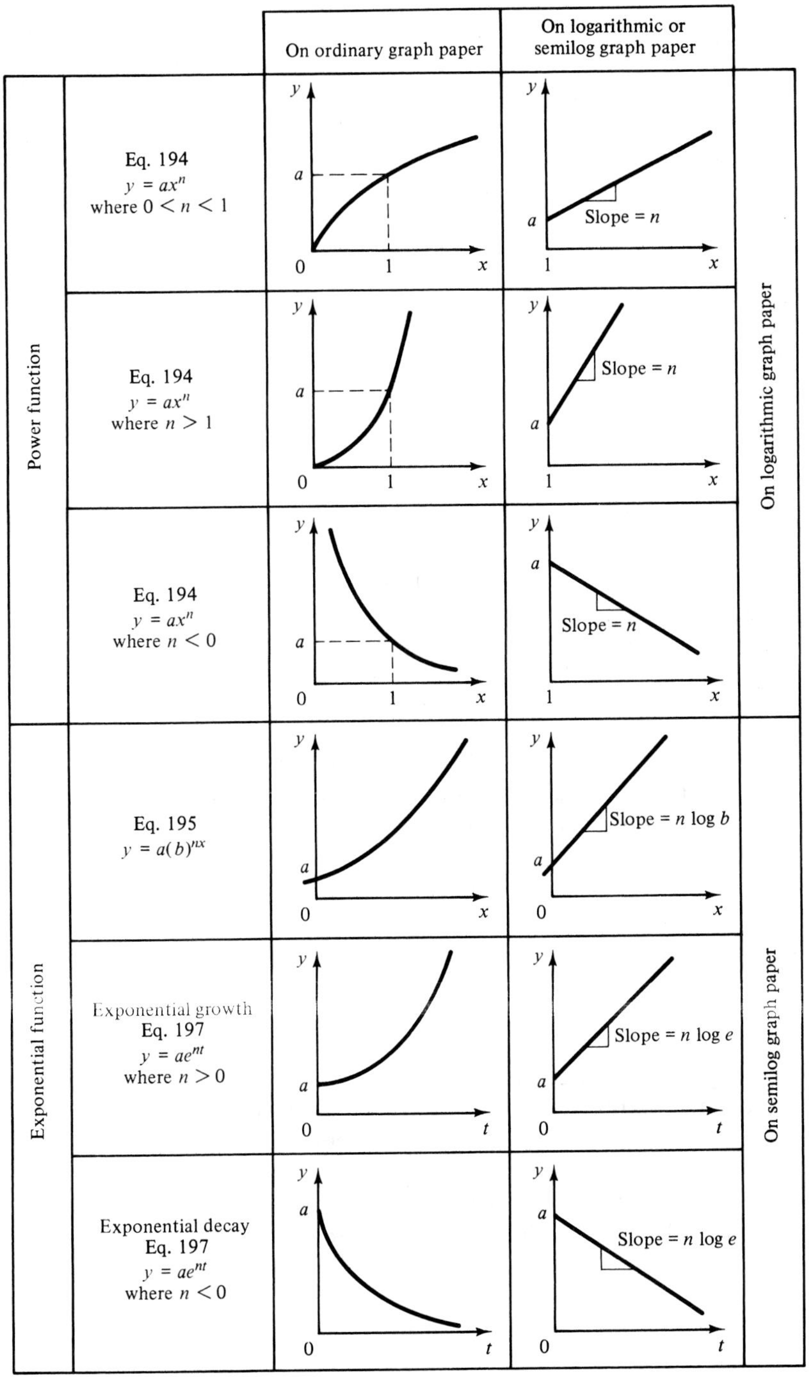


**FIGURE 19-17 The power function and the exponential function, graphed on ordinary paper and log-log or semilog paper.**

## Graphing the Power Function

A *power function* is one whose equation is of the form

| | | |
|---|---|---|
| Power Function | $y = ax^n$ | **194** |

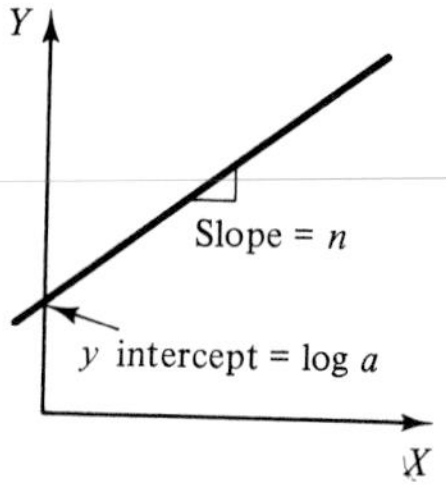


**FIGURE 19-18 Graph of $y = ax^n$ on rectangular graph paper, where $X = \log x$ and $Y = \log y$.**

where $a$ and $n$ are nonzero constants. This equation is nonlinear (except when $n = 1$), and the shape of its graph depends upon whether $n$ is positive or negative, and whether $n$ is greater than or less than 1. Figure 19-17 shows the shapes that this curve can have for various ranges of $n$. If we take the logarithm of both sides of Eq. 194, we get

$$\log y = \log(ax^n)$$
$$= \log a + n \log x$$

If we now make the substitution $X = \log x$ and $Y = \log y$, our equation becomes

$$Y = nX + \log a$$

This equation *is* linear, and, on rectangular graph paper, graphs as a straight line with a slope of $n$ and a $y$ intercept of $\log a$ (Fig. 19-18). However, we do not have to make the substitutions shown above *if we use logarithmic paper,* where the scales are proportional to the logarithms of the variables $x$ and $y$. We simply have to plot the original equation on log-log paper, and it will be a straight line with a slope $n$ which cuts the $y$ axis at $a$.

---

**EXAMPLE 61:** Plot the equation $y = 2.5x^{1.4}$, for values of $x$ from 1 to 10. Choose graph paper so that the equation plots as a straight line.

**Solution:** We make a table of point pairs. Since the graph will be a straight line we need only two points, with a third as a check. Here, we will plot four points to show that all points do lie on a straight line. We choose values of $x$ and for each compute the value of $y$.

| $x$ | 1 | 4 | 7 | 10 |
|---|---|---|---|---|
| $y$ | 2.5 | 17.4 | 38.1 | 62.8 |

We choose log-log rather than semilog paper, because we are graphing a power function, which plots as a straight line on this paper (see Fig. 19-18). We choose the number of cycles for each scale by looking at the range of values for $x$ and $y$. Thus on the $x$ axis we need one cycle. On the $y$ axis, we must go from 2.5 to 62.8. With two cycles we can span a range of 1 to 100. Thus we need log-log paper, one cycle by two cycles.

We mark the scales and plot the points as shown in Fig. 19-19, and get a straight line as expected. We note that the value of $y$ at $x = 0$ is equal to 2.5, and is the same as the coefficient of $x^{1.4}$ in the given equation.

We can get the slope of the straight line by measuring the rise and run with a scale, and dividing rise by run. Or we can use the values from the graph. But since the spacing on the graph really tells the logarithmic value of the position of the pictured points, we must remember to take the logarithm of those values. (Either common or natural will give the same result.) Thus

$$\text{slope} = \frac{\text{rise}}{\text{run}} = \frac{\ln 62.8 - \ln 2.5}{\ln 10 - \ln 1} = \frac{3.22}{2.30} = 1.40$$

The slope of the line is thus equal to the power of $x$, as expected from Fig. 19-18. We will use these ideas later when we try to write an equation to fit a set of data.

---

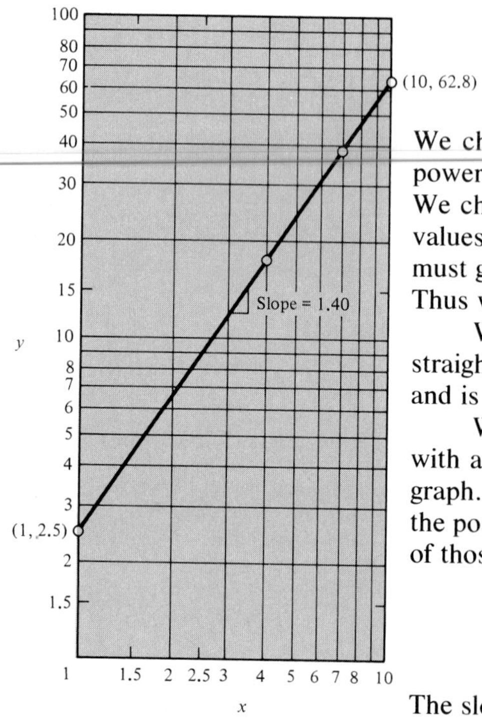


**FIGURE 19-19 Graph of $y = 2.5x^{1.4}$.**

| Common Error | Be sure to take the logs of the values on the $x$ and $y$ axes, when computing the slope. |
|---|---|

## Graphing the Exponential Function

Consider the exponential function given by

| Exponential Function | $y = a(b)^{nx}$ | **195** |
|---|---|---|

If we take the logarithm of both sides, we get

$$\log y = \log a + \log b^{nx}$$

$$= \log a + nx \log b$$

If we replace log $y$ with $Y$, we get the linear equation

$$Y = (n \log b)x + \log a$$

If we graph the given equation on semilog paper with the logarithmic scale along the $y$ axis, we get a straight line with a slope of $n \log b$ which cuts the $y$ axis at $a$, Fig. 19-17. Also shown are the special cases where the base $b$ is equal to $e$ (2.718 . . .). Here, the independent variable is shown as $t$, because exponential growth and decay are usually functions of time.

**EXAMPLE 62:** Plot the exponential function

$$y = 100\, e^{-0.2x}$$

for values of $x$ from 0 to 10.

**Solution:** We make a table of point pairs

| $x$ | 0 | 2 | 4 | 6 | 8 | 10 |
|---|---|---|---|---|---|---|
| $y$ | 100 | 67.0 | 44.9 | 30.1 | 20.2 | 13.5 |

We choose semilog paper for graphing the exponential function, and use the linear scale for $x$. The range of $y$ is from 13.5 to 100, thus we need *one cycle* of the logarithmic scale. The graph is shown in Fig. 19-20. Note that the line obtained has a $y$ intercept of 100. Also, the slope is equal to $n \log e$, or, if we use natural logs, is equal to $n$.

$$\text{slope} = n = \frac{\ln 13.5 - \ln 100}{10} = -0.2$$

This is the coefficient of $x$ in our given equation.

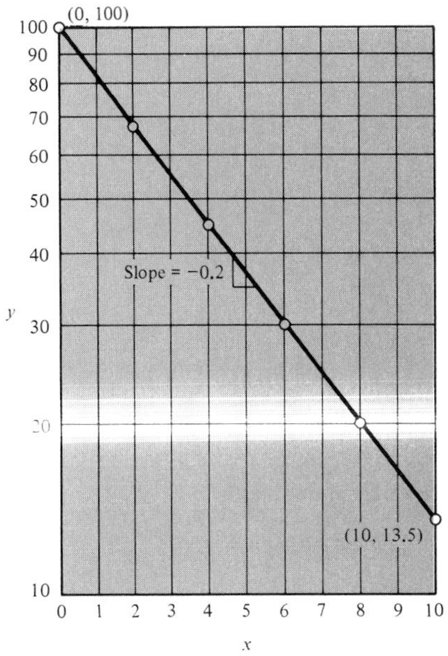


**FIGURE 19-20 Graph of $y = 100e^{-0.2x}$.**

When computing the slope on semilog paper, we take the logarithms of the values on the log scale, but not on the linear scale. Computing the slope using common logs, we get

$$\text{slope} = n \log e$$

$$= \frac{\log 13.5 - \log 100}{10}$$

$$= -0.0870$$

$$n = \frac{-0.0870}{\log e} = -0.2$$

as we got using natural logs.

## Empirical Functions

We choose logarithmic or semilog paper to plot a set of empirical data when:

1. The range of values is too large for ordinary paper.
2. We suspect that the relation between our variables may be a power function or an exponential function, and we want to find that function.

The process of finding an approximate equation to fit a set of data points is called *curve fitting.* Here we are able to do only some very simple cases.

We show the second case by means of an example.

---

**EXAMPLE 63:** A test of a certain electronic device shows it to have an output current $i$ versus input voltage $v$ as in Table 19-1. Plot the given empirical data and try to find an approximate formula for $y$ in terms of $x$.

**TABLE 19-1**

| $v$ (V) | 1 | 2 | 3 | 4 | 5 |
|---|---|---|---|---|---|
| $i$ (A) | 5.61 | 15.4 | 27.7 | 42.1 | 58.1 |

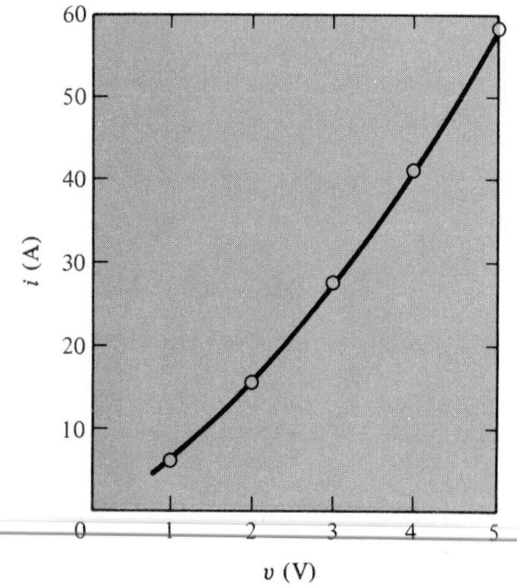


**FIGURE 19-21 Plot of Table 19-1 on linear graph paper.**

**Solution:** We first make a graph on linear graph paper (Fig. 19-21) and get a curve that is concave upward. Comparing its shape with the curves in Fig. 19-17, we suspect that the equation of the curve (if we can find one at all) may be either a power function or an exponential function.

Next, we make a plot on semilog paper (Fig. 19-22) and do *not* get a straight line, but a plot on log-log paper (Fig. 19-23) *is* linear. We thus assume that our equation has the form

$$i = av^n$$

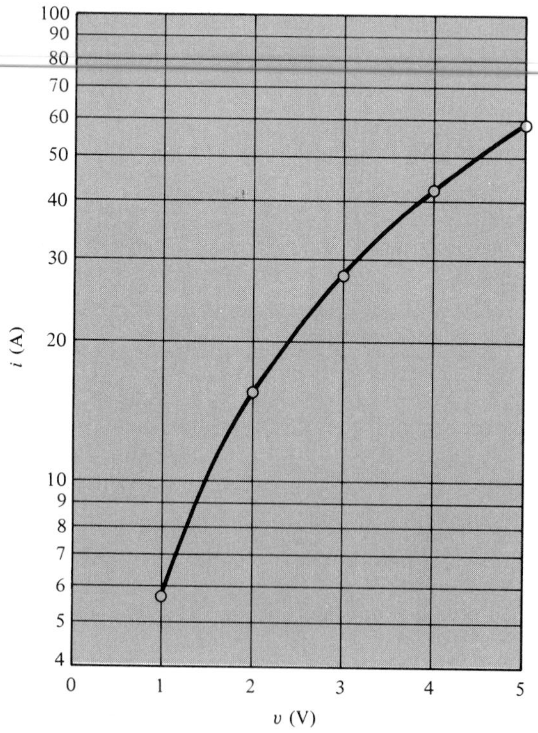


**FIGURE 19-22 Plot of Table 19-1 on semilog paper.**

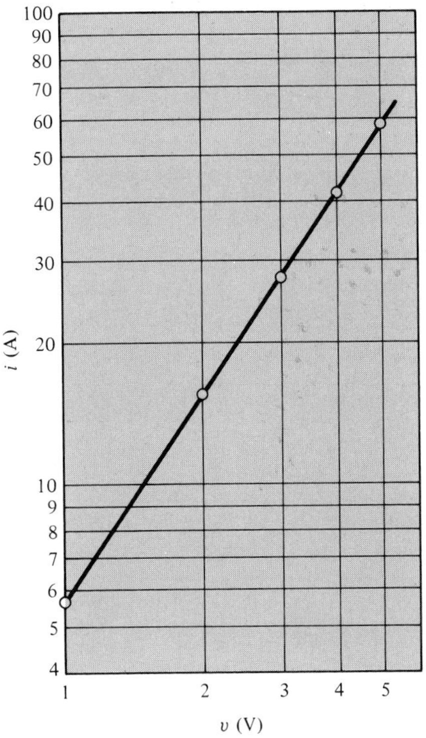


**FIGURE 19-23 Plot of Table 19-1 on log-log paper.**

Here our data plotted as a nice straight line on logarithmic paper. But with real data we are often unable to draw a straight line that passes through every point. The *method of least squares,* not shown here, is often used to draw a line that is considered the "best fit" for a scattering of data points.

In our earlier plot of the power function, we saw that the coefficient $a$ was equal to the value of the function when $x$ was 1, and that the exponent $n$ was equal to the slope of the straight line. In our present example,

$$a = 5.61$$

and the slope of the line is, using natural logs,

$$n = \frac{\ln 58.1 - \ln 5.61}{\ln 5 - \ln 1} = 1.45$$

Common logs would give the same result. Our equation is then

$$i = 5.61v^{1.45}$$

Finally, we test this formula by computing values of $i$ and comparing them with the original data.

| $v$ | *1* | *2* | *3* | *4* | *5* |
|---|---|---|---|---|---|
| Original $i$ | 5.61 | 15.4 | 27.7 | 42.1 | 58.1 |
| Calculated $i$ | 5.61 | 15.3 | 27.6 | 41.9 | 57.9 |

We get values very close to the original.

## EXERCISE 7—GRAPHS ON LOGARITHMIC AND SEMILOGARITHMIC PAPER

### Graphing the Power Function

Graph each power function on log-log paper, for $x = 1$ to 10.

**1.** $y = 2x^3$ **2.** $y = 3x^4$ **3.** $y = x^5$

4. $y = x^{3/2}$

5. $y = 5\sqrt{x}$

6. $y = 2\sqrt[3]{x}$

7. $y = 1/x$

8. $y = 3/x^2$

9. $y = 2x^{2/3}$

Graph each set of data on log-log paper, determine the coefficients graphically, and write an approximate equation to fit the given data.

10.

| $x$ | 2 | 4 | 6 | 8 | 10 |
|---|---|---|---|---|---|
| $y$ | 3.48 | 12.1 | 25.2 | 42.2 | 63.1 |

11.

| $x$ | 2 | 4 | 6 | 8 | 10 |
|---|---|---|---|---|---|
| $y$ | 240 | 31.3 | 9.26 | 3.91 | 2.00 |

### Graphing the Exponential Function

Graph each exponential function on semilog paper.

12. $y = 3^x$

13. $y = 5^x$

14. $y = e^x$

15. $y = e^{-x}$

16. $y = 4^{x/2}$

17. $y = 3^{x/4}$

18. $y = 3e^{2x/3}$

19. $y = 5e^{-x}$

20. $y = e^{x/2}$

Graph each set of data on semilog paper, determine the coefficients graphically, and write an approximate equation to fit the given data.

21.

| $x$ | 0 | 1 | 2 | 3 | 4 | 5 |
|---|---|---|---|---|---|---|
| $y$ | 1.00 | 2.50 | 6.25 | 15.6 | 39.0 | 97.7 |

22.

| $x$ | 1 | 2 | 3 | 4 |
|---|---|---|---|---|
| $y$ | 48.5 | 29.4 | 17.6 | 10.8 |

### Graphing Empirical Functions

Graph each set of data on log-log or semilog paper, determine the coefficients graphically, and write an approximate equation to fit the given data.

23. Current in a tungsten lamp, $i$, for various voltages, $v$.

| $v$ (V) | 2 | 8 | 25 | 50 | 100 | 150 | 200 |
|---|---|---|---|---|---|---|---|
| $i$ (mA) | 24.6 | 56.9 | 113 | 172 | 261 | 330 | 387 |

24. Difference in temperature, $T$, between a cooling body and its surroundings at various times, $t$.

| $t$ (s) | 0 | 3.51 | 10.9 | 19.0 | 29.0 | 39.0 | 54.0 | 71.0 |
|---|---|---|---|---|---|---|---|---|
| $T$ (°F) | 20.0 | 19.0 | 17.0 | 15.0 | 13.0 | 11.0 | 8.50 | 6.90 |

25. Pressure, $p$, of 1 lb of saturated steam at various volumes, $v$.

| $v$ (ft$^3$) | 26.4 | 19.1 | 14.0 | 10.5 | 8.00 |
|---|---|---|---|---|---|
| $p$ (lb/in$^2$) | 14.7 | 20.8 | 28.8 | 39.3 | 52.5 |

26. Maximum height $y$ reached by a long pendulum $t$ seconds after being set in motion.

| $t$ (s) | 0 | 1 | 2 | 3 | 4 | 5 | 6 |
|---|---|---|---|---|---|---|---|
| $y$ (in.) | 10 | 4.97 | 2.47 | 1.22 | 0.61 | 0.30 | 0.14 |

## CHAPTER 19 REVIEW PROBLEMS

*Convert to logarithmic form.*

1. $x^{5.2} = 352$
2. $4.8^x = 58$
3. $2.4^{1.4} = x$

*Convert to exponential form.*

4. $\log_3 56 = x$
5. $\log_x 5.2 = 124$
6. $\log_2 x = 48.2$

*Solve for x.*

7. $\log_{81} 27 = x$
8. $\log_3 x = 4$
9. $\log_x 32 = -\frac{5}{7}$

*Write as the sum or difference of two or more logarithms.*

10. $\log xy$
11. $\log \frac{3x}{z}$
12. $\log \frac{ab}{cd}$

*Express as a single logarithm.*

13. $\log 5 + \log 2$
14. $2 \log x - 3 \log y$
15. $\frac{1}{2} \log p - \frac{1}{4} \log q$

*Find the common logarithm of each number to four decimal places.*

16. 6.83
17. 364
18. 0.00638

*Find x if* log *x is equal to:*

19. 2.846

20. 1.884

21. −0.473

*Evaluate using logarithms.*

22. $5.836 \times 88.24$

23. $\dfrac{8375}{2846}$

24. $(4.823)^{1.5}$

*Find the natural logarithm of each number to four decimal places.*

25. 84.72

26. 2846

27. 0.00873

*Find the number whose natural logarithm is:*

28. 5.273

29. 1.473

30. −4.837

*Find* $\log x$ *if:*

31. $\ln x = 4.837$

32. $\ln x = 8.352$

*Find* $\ln x$ *if:*

33. $\log x = 5.837$

34. $\log x = 7.264$

*Solve for x.*

35. $(4.88)^x = 152$

36. $e^{2x+1} = 72.4$

37. $3 \log x - 3 = 2 \log x^2$

38. $\log (3x - 6) = 3$

39. A sum of \$1500 is deposited at a compound interest rate of $6\frac{1}{2}\%$ compounded quarterly. How much will it accumulate to in 5 years?

40. A quantity grows exponentially at a rate of 2% per day for 10 weeks. Find the final amount if the initial amount is 500 units.

41. A flywheel decreases in speed exponentially at the rate of 5% per minute. Find the speed after 20 min if the initial speed is 2250 rev/min.

42. Find the half-life of a radioactive substance that decays exponentially at the rate of $3\frac{1}{2}\%$ per year.

43. Find the doubling time for a population growing at the rate of 3% per year.

44. Graph the power function $y = 3x^{2/3}$ on log-log paper, for $x = 1$ to 6.

45. Graph the exponential function $y = e^{x/3}$ on semilog paper, for $x = 0$ to 6.

*Writing*

46. In this chapter we have introduced two more kinds of equations, the exponential equation and the logarithmic equation. Earlier we have had five other types. List the seven types, give an example of each, and write one sentence for each telling how it differs from the others.

# 20

# COMPLEX NUMBERS

## OBJECTIVES

**When you have completed this chapter, you should be able to:**

- Simplify radicals having negative radicands.
- Write complex numbers in rectangular, polar, trigonometric, and exponential forms.
- Evaluate powers of $j$.
- Find the sums, differences, products, quotients, powers, and roots of complex numbers.
- Solve quadratic equations having complex roots.
- Factor polynomials that have complex factors.
- Add, subtract, multiply, and divide vectors using complex numbers.
- Solve alternating current problems using complex numbers.

So far we have dealt entirely with real numbers. Recall that these are all the numbers that correspond to points on the number line, that is, the rational and irrational numbers. These include integers, fractions, roots, and the negatives of these numbers. In this chapter we consider imaginary and complex numbers. We learn what they are, how to express them in several different forms, and how to perform the basic operations with them.

Why bother with a new type of number when, as we have seen, the real numbers can do so much? The complex number system is a natural extension of the real number system. We use complex numbers mainly to manipulate *vectors,* which are so important in many branches of technology. Complex numbers (despite their unfortunate name) really simplify computations with vectors, as we see especially with the alternating current calculations at the end of this chapter.

## 20-1 COMPLEX NUMBERS IN RECTANGULAR FORM

### The Imaginary Unit

Recall that in the real number system, the equation

$$x^2 = -1$$

had no solution. This was because there was no real number such that its square was $-1$. Now we extend our number system to allow the quantity $\sqrt{-1}$ to have a meaning. We define the *imaginary unit* as the square root of $-1$, and represent it by the symbol $j$.

**The letter *i* is often used for the imaginary unit. In technical work, however, we save *i* for electric current. Further, *j* (or *i*) is sometimes written before the *b*, and sometimes after. So the number *j*5 may also be written 5*i*, 5*j*, or *i*5.**

$$\text{Imaginary unit} \qquad j \equiv \sqrt{-1}$$

### Complex Numbers

A *complex number* is any number, real or imaginary, that can be written in the form

$$a + jb$$

where $a$ and $b$ are real numbers, and $j = \sqrt{-1}$ is the *imaginary unit.*

---

**EXAMPLE 1:** The numbers

$$4 + j2 \qquad -7 + j8 \qquad 5.92 - j2.93 \qquad 83 \qquad j27$$

are complex numbers.

---

### Real and Imaginary Numbers

When $b = 0$ in a complex number $a + jb$, we have a *real number.* When $a = 0$, the number is called a *pure imaginary number.*

---

**EXAMPLE 2:**

(a) The complex number 48 is also a real number.

(b) The complex number $j9$ is a pure imaginary number.

---

The two parts of a complex number are called the *real part* and the *imaginary part.*

| | | |
|---|---|---|
| Complex Number | $a + jb$ (real part ↑ $a$, ↑ imaginary part $b$) | **213** |

This is called the *rectangular form* of a complex number.

## Addition and Subtraction of Complex Numbers

To combine complex numbers, separately combine the real parts, and then the imaginary parts, and express the result in the form $a + jb$.

| | | |
|---|---|---|
| Addition of Complex Numbers | $(a + jb) + (c + jd) = (a + c) + j(b + d)$ | **214** |
| Subtraction of Complex Numbers | $(a + jb) - (c + jd) = (a - c) + j(b - d)$ | **215** |

---

**EXAMPLE 3:** Add or subtract, as indicated.

(a) $j3 + j5 = j8$

(b) $j2 + (6 - j5) = 6 - j3$

(c) $(2 - j5) + (-4 + j3) = (2 - 4) + j(-5 + 3) = -2 - j2$

(d) $(-6 + j2) - (4 - j) = (-6 - 4) + j[2 - (-1)] = -10 + j3$

---

## Powers of $j$

We often have to evaluate powers of the imaginary unit, especially the square of $j$. Since

$$j = \sqrt{-1}$$
$$j^2 = \sqrt{-1}\sqrt{-1} = -1$$

Higher powers are easily found.

$$j^3 = j^2 j = (-1)j = -j$$
$$j^4 = (j^2)^2 = (-1)^2 = 1$$
$$j^5 = j^4 j = (1)j = j$$
$$j^6 = j^4 j^2 = (1)j^2 = -1$$

We see that the values are starting to repeat. $j^5 = j$, $j^6 = j^2$, and so on. The first four values keep repeating.

Note that when the exponent $n$ is a multiple of 4, then $j^n = 1$.

| | | |
|---|---|---|
| Powers of $j$ | $j = \sqrt{-1}$, $j^2 = -1$, $j^3 = -j$, $j^4 = 1$, $j^5 = j$, etc. | **212** |

**EXAMPLE 4:** Evaluate $j^{17}$.

**Solution:** Using the laws of exponents, we express $j^{17}$ in terms of one of the first four powers of $j$.

$$j^{17} = j^{16}j = (j^4)^4 j = (1)^4 j = j$$

## Multiplication of Imaginary Numbers

Multiply as with ordinary numbers, but use Eq. 212 to simplify any powers of $j$.

**EXAMPLE 5:**

(a) $5 \times j3 = j15$

(b) $j2 \times j4 = j^2 8 = (-1)8 = -8$

(c) $3 \times j4 \times j5 \times j = j^3 60 = (-j)60 = -j60$

(d) $(j3)^2 = j^2 3^2 = (-1)9 = -9$

## Multiplication of Complex Numbers

Multiply complex numbers as you would any algebraic expressions, replace $j^2$ by $-1$, and put the expression into the form $a + jb$.

| | | |
|---|---|---|
| Multiplication of Complex Numbers | $(a + jb)(c + jd) = (ac - bd) + j(ad + bc)$ | **216** |

**EXAMPLE 6:**

(a) $3(5 + j2) = 15 + j6$

(b) $(j3)(2 - j4) = j6 - j^2 12 = j6 - (-1)12 = 12 + j6$

(c) $(3 - j2)(-4 + j5) = 3(-4) + 3(j5) + (-j2)(-4) + (-j2)(j5)$

$$= -12 + j15 + j8 - j^2 10$$

$$= -12 + j15 + j8 - (-1)10$$

$$= -2 + j23$$

(d) $(3 - j5)^2 = (3 - j5)(3 - j5)$

$$= 9 - j15 - j15 + j^2 25 = 9 - j30 - 25$$

$$= -16 - j30$$

To multiply radicals that contain negative quantities in the radicand, first express all quantities in terms of $j$. Then proceed to multiply. Always be sure to convert radicals to imaginary numbers *before* performing other operations, or contradictions may result.

**EXAMPLE 7:** Multiply $\sqrt{-4}$ by $\sqrt{-4}$.

**Solution:** Converting to imaginary numbers, we obtain

$$\sqrt{-4}\sqrt{-4} = (j2)(j2) = j^2 4$$

We will see in the next section that multiplication and division are easier in *polar* form.

Since $j^2 = -1$,

$$j^2 4 = -4$$

| Common Error | It is *incorrect* to write $$\sqrt{-4}\sqrt{-4} = \sqrt{(-4)(-4)} = \sqrt{16} = +4$$ Our previous rule of $\sqrt{a} \cdot \sqrt{b} = \sqrt{ab}$ applied only to positive $a$ and $b$. |
|---|---|

## The Conjugate of a Complex Number

The *conjugate* of a complex number is obtained by changing the sign of its imaginary part.

**EXAMPLE 8:**

(a) The conjugate of $2 + j3$ is $2 - j3$.

(b) The conjugate of $-5 - j4$ is $-5 + j4$.

(c) The conjugate of $a + jb$ is $a - jb$.

Multiplying any complex number by its conjugate will eliminate the $j$ term.

**EXAMPLE 9:**

$$(2 + j3)(2 - j3) = 4 - j6 + j6 - j^2 9 = 4 - (-1)(9) = 13$$

This has the same form as the difference of two squares.

## Division of Complex Numbers

Division involving single terms, real or imaginary, is shown by examples.

**EXAMPLE 10:**

(a) $j8 \div 2 = j4$

(b) $j6 \div j3 = 2$

(c) $(4 - j6) \div 2 = 2 - j3$

**EXAMPLE 11:** Divide 6 by $j3$.

**Solution:**

$$\frac{6}{j3} = \frac{6}{j3} \times \frac{j3}{j3} = \frac{j18}{j^2 9} = \frac{j18}{-9} = -j2$$

To divide by a complex number, multiply dividend and divisor by the conjugate of the divisor. This will make the divisor a real number, as in the following example.

**EXAMPLE 12:** Divide $3 - j4$ by $2 + j$.

This is very similar to *rationalizing the denominator* of a fraction containing radicals (Sec. 13-2).

**Solution:** We multiply numerator and denominator by the conjugate $(2 - j)$ of the denominator.

$$\frac{3 - j4}{2 + j} = \frac{3 - j4}{2 + j} \cdot \frac{2 - j}{2 - j}$$

$$= \frac{6 - j3 - j8 + j^2 4}{4 + j2 - j2 - j^2}$$

$$= \frac{2 - j11}{5}$$

$$= \frac{2}{5} - j\frac{11}{5}$$

In general,

| | | |
|---|---|---|
| Division of Complex Numbers | $\frac{a + jb}{c + jd} = \frac{ac + bd}{c^2 + d^2} + j\frac{bc - ad}{c^2 + d^2}$ | **217** |

## Quadratics with Complex Roots

When we solved quadratic equations in Chapter 13, the roots were real numbers. Some quadratics, however, will yield complex roots, as in the following example.

**EXAMPLE 13:** Solve for $x$ to three significant digits,

$$2x^2 - 5x + 9 = 0$$

**Solution:** By the quadratic formula (Eq. 100),

$$x = \frac{5 \pm \sqrt{25 - 4(2)(9)}}{4} = \frac{5 \pm \sqrt{-47}}{4}$$

$$= \frac{5 \pm j6.86}{4} = 1.25 \pm j1.72$$

## Complex Factors

It is sometimes necessary to factor an expression into *complex factors*, as shown in the following example.

**EXAMPLE 14:** Factor the expression $a^2 + 4$ into complex factors.

**Solution:**

$$a^2 + 4 = a^2 - j^2 4 = a^2 - (j2)^2$$

$$= (a - j2)(a + j2)$$

## EXERCISE 1—COMPLEX NUMBERS IN RECTANGULAR FORM

Write as imaginary numbers.

1. $\sqrt{-9}$
2. $\sqrt{-81}$
3. $\sqrt{-\frac{4}{16}}$
4. $\sqrt{-\frac{1}{5}}$

Write as a complex number in rectangular form.

5. $4 + \sqrt{-4}$
6. $\sqrt{-25} + 3$
7. $\sqrt{-\frac{9}{4}} - 5$
8. $\sqrt{-\frac{1}{3}} + 2$

Combine and simplify.

9. $\sqrt{-9} + \sqrt{-4}$
10. $\sqrt{-4a^2} - a\sqrt{-25}$
11. $(3 - j2) + (-4 + j3)$
12. $(-1 - j2) - (j + 6)$
13. $(a - j3) + (a + j5)$
14. $(p + jq) + (q + jp)$
15. $\left(\frac{1}{2} + \frac{j}{3}\right) + \left(\frac{1}{4} - \frac{j}{6}\right)$
16. $(-84 + j91) - (28 + j72)$
17. $(2.28 - j1.46) + (1.75 + j2.66)$

Evaluate each power of $j$.

18. $j^{11}$
19. $j^5$
20. $j^{10}$
21. $j^{21}$
22. $j^{14}$

Multiply and simplify.

23. $7 \times j2$
24. $9 \times j3$
25. $j3 \times j5$
26. $j \times j4$
27. $4 \times j2 \times j3 \times j4$
28. $j \times 5 \times j4 \times j3 \times 5$
29. $(j5)^2$
30. $(j3)^2$
31. $2(3 - j4)$
32. $-3(7 + j5)$
33. $j4(5 - j2)$
34. $j5(2 + j3)$
35. $(3 - j5)(2 + j6)$
36. $(5 + j4)(4 + j2)$
37. $(6 + j3)(3 - j8)$
38. $(5 - j3)(8 + j2)$
39. $(5 - j2)^2$
40. $(3 + j6)^2$

Write the conjugate of each complex number.

41. $2 - j3$
42. $-5 - j7$
43. $p + jq$
44. $-j5 + 6$
45. $-jm + n$
46. $5 - j8$

Divide and simplify.

47. $j8 \div 4$
48. $9 \div j3$
49. $j12 \div j6$
50. $j44 \div j2$
51. $(4 + j2) \div 2$
52. $8 \div (4 - j)$
53. $(-2 + j3) \div (1 - j)$
54. $(5 - j6) \div (-3 + j2)$
55. $(j7 + 2) \div (j3 - 5)$
56. $(-9 + j3) \div (2 - j4)$

### Quadratics with Complex Roots

Solve for $x$ to three significant digits.

57. $3x^2 - 5x + 7 = 0$
58. $2x^2 + 3x + 5 = 0$
59. $x^2 - 2x + 6 = 0$
60. $4x^2 + x + 8 = 0$

### Complex Factors

Factor each expression into complex factors.

61. $x^2 + 9$
62. $b^2 + 25$
63. $4y^2 + z^2$
64. $25a^2 + 9b^2$

### Computer

65. Modify your program for solving quadratic equations (Chapter 13) so that the run will not end when the discriminant is negative, but will instead compute and print complex roots in the form $a + jb$.

## 20-2 GRAPHING COMPLEX NUMBERS

### The Complex Plane

A complex number in rectangular form is made up of two parts: a real part $a$ and an imaginary part $jb$. To graph a complex number we can use a rectangular coordinate system (Fig. 20-1) in which the horizontal axis is the *real* axis and the vertical axis is the *imaginary* axis. Such a coordinate system defines what is called the *complex plane*. Real numbers are graphed as points on the horizontal axis; pure imaginary numbers are graphed as points on the vertical axis. Complex numbers, such as $3 + j5$ (Fig. 20-1), are graphed elsewhere within the plane.

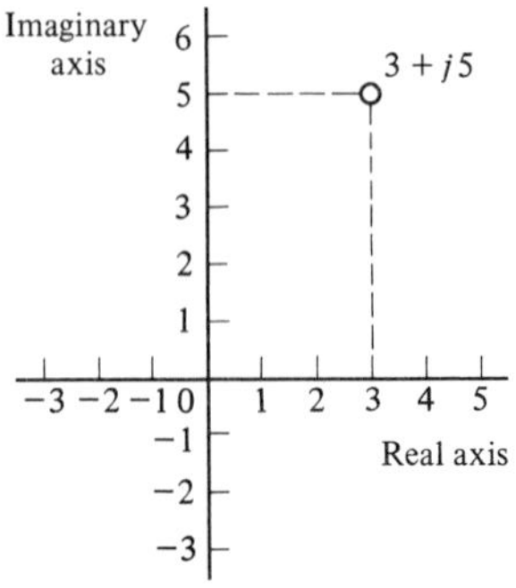


**FIGURE 20-1 The complex plane.**

### Graphing a Complex Number

**Such a plot is called an *Argand diagram,* named for Jean Robert Argand (1768–1822).**

To plot a complex number $a + jb$ in the complex plane, simply locate a point with a horizontal coordinate of $a$ and a vertical coordinate of $b$.

---

**EXAMPLE 15:** Plot the complex numbers $(2 + j3)$, $(-1 + j2)$, $(-3 - j2)$, and $(1 - j3)$.

**Solution:** The points are plotted in Fig. 20-2.

---

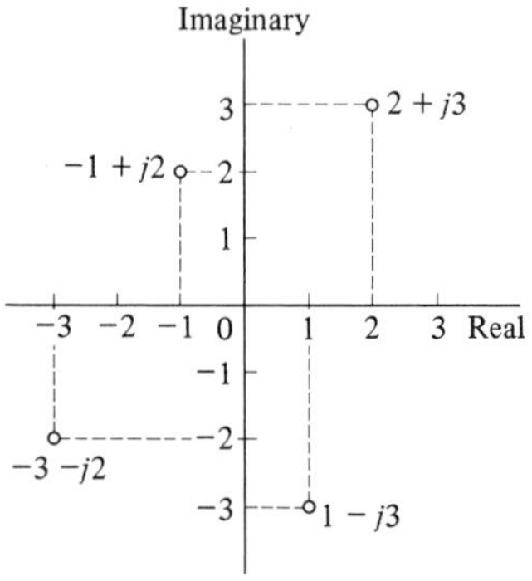


**FIGURE 20-2**

### EXERCISE 2—GRAPHING COMPLEX NUMBERS

Graph each complex number.

1. $2 + j5$
2. $-1 - j3$
3. $3 - j2$
4. $2 - j$
5. $j5$
6. $2.25 - j3.62$
7. $-2.7 - j3.4$
8. $5.02 + j$
9. $1.46 - j2.45$

## 20-3 COMPLEX NUMBERS IN TRIGONOMETRIC AND POLAR FORMS

### Polar Form

**In some books, the terms "polar form" and "trigonometric form" are used interchangeably. Here, we distinguish between them.**

In Chapter 16 we saw that a point in a plane could be located by *polar coordinates,* as well as by rectangular coordinates, and we learned how to convert from one set of coordinates to the other. We will do the same thing now with complex numbers, converting between rectangular and polar form.

In Fig. 20-3, we plot a complex number $a + jb$. We connect that point to the origin with a line of length $r$ at an angle of $\theta$ with the horizontal axis. Now in addition to expressing the complex number in terms of $a$ and $b$, we can express it in terms of $r$ and $\theta$:

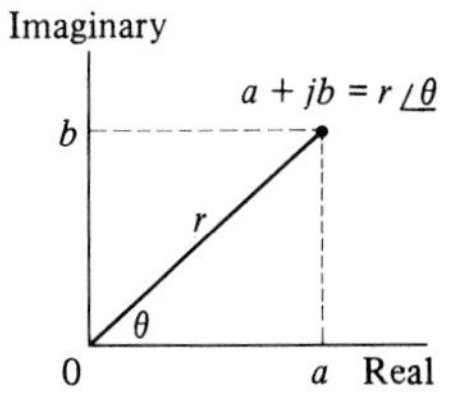


**FIGURE 20-3 Polar form of a complex number shown on a complex plane.**

| | | |
|---|---|---|
| Polar Form | $a + jb = r\underline{/\theta}$ | **223** |

Since $\cos\theta = a/r$ and $\sin\theta = b/r$, we get

| | |
|---|---|
| $a = r\cos\theta$ | **219** |
| $b = r\sin\theta$ | **220** |

The radius $r$ is called the *absolute value*. It can be found from the Pythagorean theorem:

| | |
|---|---|
| $r = \sqrt{a^2 + b^2}$ | **221** |

The angle $\theta$ is called the *argument* of the complex number. Since $\tan\theta = b/a$

| | |
|---|---|
| $\theta = \arctan\dfrac{b}{a}$ | **222** |

---

**EXAMPLE 16:** Write the complex number $2 + j3$ in polar form.

**Solution:** The absolute value is

$$r = \sqrt{2^2 + 3^2} \cong 3.61$$

And the argument is found from

$$\tan\theta = \frac{3}{2}$$

$$\theta \simeq 56.3^\circ$$

So

$$2 + j3 = 3.61\underline{/56.3^\circ}$$

---

**If your calculator has keys for converting between rectangular and polar coordinates, you can use these keys to convert complex numbers between rectangular and polar forms.**

## Trigonometric Form

If we substitute the values of $a$ and $b$ from Eqs. 219 and 220 into $a + jb$, we get

$$a + jb = (r \cos \theta) + j(r \sin \theta)$$

The angle $\theta$ is sometimes written $\theta = \arg z$, which means "$\theta$ is the argument of the complex number $z$." Also, the expression in parentheses in Eq. 218 is sometimes abbreviated as cis $\theta$: so cis $\theta = \cos \theta + j \sin \theta$.

Factoring gives

| | | |
|---|---|---|
| Trigonometric Form | $\underset{\text{rectangular form}}{a + jb} = \underset{\text{trigonometric form}}{r(\cos \theta + j \sin \theta)}$ | **218** |

---

**EXAMPLE 17:** Write the complex number $2 + j3$ in polar and trigonometric form.

**Solution:** We already have $r$ and $\theta$ from Example 16,

$$r = 3.61 \quad \text{and} \quad \theta = 56.3°$$

That is,

$$2 + j3 = 3.61 \underline{/56.3°}$$

So by Eq. 219,

$$2 + j3 = 3.61(\cos 56.3° + j \sin 56.3°)$$

---

**EXAMPLE 18:** Write the complex number $6(\cos 30° + j \sin 30°)$ in polar and rectangular form.

**Solution:** By inspection,

$$r = 6 \quad \text{and} \quad \theta = 30°$$

So

$$a = r \cos \theta = 6 \cos 30° = 5.20$$

and

$$b = r \sin \theta = 6 \sin 30° = 3.00$$

So our complex number, in rectangular form, is $5.20 + j3.00$, and in polar form, $6 \underline{/30°}$.

---

## Arithmetic Operations in Polar Form

It is common practice to switch back and forth between rectangular and polar forms during a computation, using that form in which a particular operation is easiest. We saw that addition and subtraction are fast and easy in rectangular form, and we now show that multiplication, division, and raising to a power are best done in polar form.

### Products

We will use trigonometric form to work out the formula for multiplication and then express the result in the simpler polar form.

Let us multiply $r(\cos\theta + j\sin\theta)$ by $r'(\cos\theta' + j\sin\theta')$.

$$[r(\cos\theta + j\sin\theta)][r'(\cos\theta' + j\sin\theta')]$$
$$= rr'(\cos\theta\cos\theta' + j\cos\theta\sin\theta' + j\sin\theta\cos\theta' + j^2\sin\theta\sin\theta')$$
$$= rr'[(\cos\theta\cos\theta' - \sin\theta\sin\theta') + j(\sin\theta\cos\theta' + \cos\theta\sin\theta')]$$

But, by Eq. 168,

$$\cos\theta\cos\theta' - \sin\theta\sin\theta' = \cos(\theta + \theta')$$

and by Eq. 167,

$$\sin\theta\cos\theta' + \cos\theta\sin\theta' = \sin(\theta + \theta')$$

So

$$r(\cos\theta + j\sin\theta)\cdot r'(\cos\theta' + j\sin\theta') = rr'[\cos(\theta + \theta') + j\sin(\theta + \theta')]$$

Switching now to polar form, we get

| Products | $r\underline{/\theta}\cdot r'\underline{/\theta'} = rr'\underline{/\theta + \theta'}$ | **224** |
|---|---|---|

*The absolute value of the product of two complex numbers is the product of their absolute values, and the argument is the sum of the individual arguments.*

---

**EXAMPLE 19:** Multiply $5\underline{/30^\circ}$ by $3\underline{/20^\circ}$.

**Solution:** The absolute value of the product will be 5(3) = 15 and the argument of the product will be 30° + 20° = 50°, so

$$5\underline{/30^\circ}\cdot 3\underline{/20^\circ} = 15\underline{/50^\circ}$$

---

The angle $\theta$ is not usually written greater than 360°. Subtract multiples of 360° if necessary.

---

**EXAMPLE 20:** Multiply $6.27\underline{/300^\circ}$ by $2.75\underline{/125^\circ}$.

**Solution:**

$$6.27\underline{/300^\circ}\times 2.75\underline{/125^\circ} = (6.27)(2.75)\underline{/300^\circ + 125^\circ}$$
$$= 17.2\underline{/425^\circ}$$
$$= 17.2\underline{/65^\circ}$$

---

### Quotients

The rule for division of complex numbers in trigonometric form is similar to that for multiplication:

You might try deriving this yourself.

| | | |
|---|---|---|
| Quotients | $\dfrac{r\,\underline{/\theta}}{r'\,\underline{/\theta'}} = \dfrac{r}{r'}\underline{/\theta - \theta'}$ | **225** |

*The absolute value of the quotient of two complex numbers is the quotient of their absolute values; and the argument is the difference (numerator minus denominator) of their arguments.*

---

**EXAMPLE 21:**

$$\frac{6\,\underline{/70^\circ}}{2\,\underline{/50^\circ}} = 3\,\underline{/20^\circ}$$

---

## Powers

To raise a complex number to a power, we merely have to multiply it by itself the proper number of times, using Eq. 224,

$$(r\,\underline{/\theta})^2 = r\,\underline{/\theta} \cdot r\,\underline{/\theta} = r \cdot r\,\underline{/\theta + \theta} = r^2\,\underline{/2\theta}$$

$$(r\,\underline{/\theta})^3 = r\,\underline{/\theta} \cdot r^2\,\underline{/2\theta} = r \cdot r^2\,\underline{/\theta + 2\theta} = r^3\,\underline{/3\theta}$$

Do you see a pattern developing? In general, we get

After Abraham DeMoivre (1667–1754).

| | | |
|---|---|---|
| DeMoivre's Theorem | $(r\,\underline{/\theta})^n = r^n\,\underline{/n\theta}$ | **226** |

*When a complex number is raised to the* n*th power, the new absolute value is equal to the original absolute value raised to the* n*th power, and the new argument is n times the original argument.*

---

**EXAMPLE 22:**

$$(2\,\underline{/10^\circ})^5 = 2^5\,\underline{/5(10^\circ)} = 32\,\underline{/50^\circ}$$

---

## Roots

We know that $1^4 = 1$, so, conversely, the fourth root of 1 is 1.

$$\sqrt[4]{1} = 1$$

But we have seen that $j^4 = 1$, so shouldn't it also be true that

$$\sqrt[4]{1} = j?$$

In fact, both 1 and $j$ are fourth roots of 1, and there are two other roots as well. In this section we learn how to use DeMoivre's theorem to find all the roots of a number.

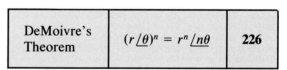

First we note that a complex number $r\underline{/\theta}$ in polar form is unchanged if we add multiples of 360° to the angle $\theta$ (in degrees). Thus

$$r\underline{/\theta} = r\underline{/\theta + 360°} = r\underline{/\theta + 720°} = r\underline{/\theta + 360k}$$

where $k$ is an integer. Rewriting DeMoivre's theorem with an exponent $1/p$ (where $p$ is an integer), we get

$$(r\underline{/\theta})^{1/p} = (r\underline{/\theta + 360k})^{1/p} = r^{1/p}\underline{/(\theta + 360k)/p}$$

so

$$(r\underline{/\theta})^{1/p} = r^{1/p}\underline{/(\theta + 360k)/p} \qquad (1)$$

We use this equation to find all the roots of a complex number by letting $k$ take on the values $k = 0, 1, 2, \ldots, (p - 1)$. We'll see in the following example that values of $k$ greater than $(p - 1)$ will give duplicate roots.

---

**EXAMPLE 23:** Find the fourth roots of 1.

**Solution:** We write the given number in polar form, thus $1 = 1\underline{/0°}$. Then using (1) with $p = 4$,

$$1^{1/4} = (1\underline{/0})^{1/4} = 1^{1/4}\underline{/(0 + 360k)/4} = 1\underline{/90k}$$

We now let $k = 0, 1, 2, \ldots$:

| | Root | |
|---|---|---|
| *k* | *Polar Form* | *Rectangular Form* |
| 0 | $1\underline{/0°}$ | 1 |
| 1 | $1\underline{/90°}$ | $j$ |
| 2 | $1\underline{/180°}$ | $-1$ |
| 3 | $1\underline{/270°}$ | $-j$ |
| 4 | $1\underline{/360°}$ | $1\underline{/0°} = 1$ |
| 5 | $1\underline{/450°}$ | $1\underline{/90°} = j$ |

Notice that the roots repeat for $k = 4$ and higher, so we look no further. Thus the fourth roots of 1 are $1, j, -1$, and $-j$. We check them by noting that each raised to the fourth power equals 1.

---

Thus the number 1 has four fourth roots. In general, any number has two square roots, three cube roots, and so on.

> There are $p$ $p$th roots of a complex number.

If a complex number is in rectangular form, we convert to polar form before taking the root.

**EXAMPLE 24:** Find $\sqrt[3]{256 + j192}$.

**Solution:** We convert the given complex number to polar form.

$$r = \sqrt{(256)^2 + (192)^2} = 320 \qquad \theta = \arctan(192/256) = 36.9°$$

Then using Eq. (1), we have

$$\sqrt[3]{256 + j192} = (320\underline{/36.9°})^{1/3} = (320)^{1/3}\underline{/(36.9° + 360k)/3}$$
$$= 6.84\underline{/12.3° + 120k}$$

We let $k = 0, 1, 2$, to obtain the three roots.

| | Root | |
|---|---|---|
| $k$ | *Polar Form* | *Rectangular Form* |
| 0 | $6.84\underline{/12.3° + 0°} = 6.84\underline{/12.3°}$ | $6.68 + j1.46$ |
| 1 | $6.84\underline{/12.3° + 120°} = 6.84\underline{/132.3°}$ | $-4.60 + j5.06$ |
| 2 | $6.84\underline{/12.3° + 240°} = 6.84\underline{/252.3°}$ | $-2.08 - j6.52$ |

## EXERCISE 3—COMPLEX NUMBERS IN TRIGONOMETRIC AND POLAR FORMS

Write each complex number in polar and trigonometric form.

1. $5 + j4$
2. $-3 - j7$
3. $4 - j3$
4. $8 + j4$
5. $-5 - j2$
6. $-4 + j7$
7. $-9 - j5$
8. $7 - j3$
9. $-4 - j7$

Write in rectangular and in polar form.

10. $4(\cos 25° + j \sin 25°)$
11. $3(\cos 57° + j \sin 57°)$
12. $2(\cos 110° + j \sin 110°)$
13. $9(\cos 150° + j \sin 150°)$
14. $7(\cos 12° + j \sin 12°)$
15. $5.46(\cos 47.3° + j \sin 47.3°)$

Write in rectangular and trigonometric form.

16. $5\underline{/28°}$
17. $9\underline{/59°}$
18. $4\underline{/63°}$
19. $7\underline{/-53°}$
20. $6\underline{/-125°}$

### Multiplication and Division

Multiply.

21. $3(\cos 12° + j \sin 12°)$ by $5(\cos 28° + j \sin 28°)$
22. $7(\cos 48° + j \sin 48°)$ by $4(\cos 72° + j \sin 72°)$

23. $5.82(\cos 44.8^\circ + j \sin 44.8^\circ)$ by $2.77(\cos 10.1^\circ + j \sin 10.1^\circ)$
24. $5\angle 30^\circ$ by $2\angle 10^\circ$
25. $8\angle 45^\circ$ by $7\angle 15^\circ$
26. $2.86\angle 38.2^\circ$ by $1.55\angle 21.1^\circ$

Divide.

27. $8(\cos 46^\circ + j \sin 46^\circ)$ by $4(\cos 21^\circ + j \sin 21^\circ)$
28. $49(\cos 27^\circ + j \sin 27^\circ)$ by $7(\cos 15^\circ + j \sin 15^\circ)$
29. $58.3(\cos 77.4^\circ + j \sin 77.4^\circ)$ by $12.4(\cos 27.2^\circ + j \sin 27.2^\circ)$
30. $24\angle 50^\circ$ by $12\angle 30^\circ$
31. $50\angle 72^\circ$ by $5\angle 12^\circ$
32. $71.4\angle 56.4^\circ$ by $27.7\angle 15.2^\circ$

**Powers and Roots**

Evaluate.

33. $[2(\cos 15^\circ + j \sin 15^\circ)]^3$
34. $[9(\cos 10^\circ + j \sin 10^\circ)]^2$
35. $(7\angle 20^\circ)^2$
36. $(1.55\angle 15^\circ)^3$
37. $\sqrt{57\angle 52^\circ}$
38. $\sqrt{22\angle 12^\circ}$
39. $\sqrt[3]{38\angle 73^\circ}$
40. $\sqrt[3]{15\angle 89^\circ}$
41. $\sqrt{135 + j204}$
42. $\sqrt[3]{56.3 + j28.5}$

## 20-4 COMPLEX NUMBERS IN EXPONENTIAL FORM

### Euler's Formula

We have already expressed a complex number in *rectangular* form

$$a + jb$$

in *polar* form

$$r\angle\theta$$

and in *trigonometric* form

$$r(\cos \theta + j \sin \theta)$$

Our next (and final) form for a complex number is *exponential* form, given by Euler's formula.

| | | |
|---|---|---|
| Euler's Formula | $re^{j\theta} = r(\cos \theta + j \sin \theta)$ | **227** |

Leonhard Euler (1707–1783). We give this formula without proof here.

Here $e$ is the base of natural logarithms (approximately 2.718), discussed in Chapter 19, and $\theta$ is the argument expressed in radians.

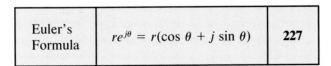

**EXAMPLE 25:** Write the complex number $5(\cos 180° + j \sin 180°)$ in exponential form.

**Solution:** We first convert the angle to radians, $180° = \pi$ rad. So

$$5(\cos \pi + j \sin \pi) = 5e^{j\pi}$$

| Common Error | Make sure that $\theta$ is in *radians* when using Eq. 227. |
|---|---|

**EXAMPLE 26:** Write the complex number $3e^{j2}$ in rectangular, trigonometric, and polar forms.

**Solution:** Changing $\theta$ to degrees gives us

$$\theta = 2 \text{ rad} = 115°$$

and from Eq. 227,

$$r = 3$$

From Eq. 219,

$$a = r \cos \theta = 3 \cos 115° = -1.27$$

and from Eq. 220,

$$b = r \sin \theta = 3 \sin 115° = 2.72$$

So, in rectangular form,

$$3e^{j2} = -1.27 + j2.72$$

in polar form,

$$= 3\underline{/115°}$$

and in trigonometric form,

$$= 3(\cos 115° + j \sin 115°)$$

## Products

We now find products, quotients, powers, and roots of complex numbers in exponential form. These operations are quite simple in this form, as we merely have to use the laws of exponents. Thus, by Eq. 29,

| Products | $r_1e^{j\theta_1} \cdot r_2e^{j\theta_2} = r_1r_2e^{j(\theta_1+\theta_2)}$ | **228** |
|---|---|---|

**EXAMPLE 27:**

(a) $2e^{j3} \cdot 5e^{j4} = 10e^{j7}$

(b) $2.83e^{j2} \cdot 3.15e^{j4} = 8.91e^{j6}$

(c) $-13.5e^{-j3} \cdot 2.75e^{j5} = -37.1e^{j2}$

### Quotients

By Eq. 30,

| | | |
|---|---|---|
| Quotients | $\dfrac{r_1e^{j\theta_1}}{r_2e^{j\theta_2}} = \dfrac{r_1}{r_2}\,e^{j(\theta_1-\theta_2)}$ | **229** |

**EXAMPLE 28:**

(a) $\dfrac{8e^{j5}}{4e^{j2}} = 2e^{j3}$

(b) $\dfrac{63.8e^{j2}}{13.7e^{j5}} = 4.66e^{-j3}$

(c) $\dfrac{5.82e^{-j4}}{9.83e^{-j7}} = 0.592e^{j3}$

### Powers and Roots

By Eq. 31,

| | | |
|---|---|---|
| Powers and Roots | $(re^{j\theta})^n = r^ne^{jn\theta}$ | **230** |

**EXAMPLE 29:**

(a) $(2e^{j3})^4 = 16e^{j12}$

(b) $(-3.85e^{-j2})^3 = -57.1e^{-j6}$

(c) $(0.223e^{-j3})^{-2} = \dfrac{e^{j6}}{(0.223)^2} = \dfrac{e^{j6}}{0.0497} = 20.1e^{j6}$

## EXERCISE 4—COMPLEX NUMBERS IN EXPONENTIAL FORM

Express each complex number in exponential form.

1. $2 + j3$
2. $-1 + j2$
3. $3(\cos 50° + j \sin 50°)$
4. $12\underline{/14°}$
5. $2.5\underline{/\pi/6}$
6. $7\left(\cos\dfrac{\pi}{3} + j\sin\dfrac{\pi}{3}\right)$
7. $5.4\underline{/\pi/12}$
8. $5 + j4$

Express in rectangular, polar, and trigonometric form.

9. $5e^{j3}$
10. $7e^{j5}$
11. $2.2e^{j1.5}$
12. $4e^{j2}$

Multiply.

13. $9e^{j2} \cdot 2e^{j4}$
14. $8e^{j} \cdot 6e^{j3}$
15. $7e^{j} \cdot 3e^{j3}$
16. $6.2e^{j1.1} \cdot 5.8e^{j2.7}$
17. $1.7e^{j5} \cdot 2.1e^{j2}$
18. $4e^{j7} \cdot 3e^{j5}$

Divide.

19. $18e^{j6}$ by $6e^{j3}$
20. $45e^{j4}$ by $9e^{j2}$
21. $55e^{j9}$ by $5e^{j6}$
22. $123e^{j6}$ by $105e^{j2}$
23. $21e^{j2}$ by $7e^{j}$
24. $7.7e^{j4}$ by $2.3e^{j2}$

Evaluate.

25. $(3e^{j5})^2$
26. $(4e^{j2})^3$
27. $(2e^{j})^3$

## 20-5 VECTOR OPERATIONS USING COMPLEX NUMBERS

### Vectors Represented by Complex Numbers

One of the major uses of complex numbers is that they can represent vectors, and, as we will soon see, enable us to manipulate them in ways that are easier than we learned when studying oblique triangles.

Take the complex number $2 + j3$, for example, which is plotted in Fig. 20-4. If we connect that point with a line to the origin, we can think of the complex number $2 + j3$ *as representing a vector* **R** *having a horizontal component of 2 units and a vertical component of 3 units.* The complex number used to represent a vector can, of course, be expressed in any of the forms of a complex number.

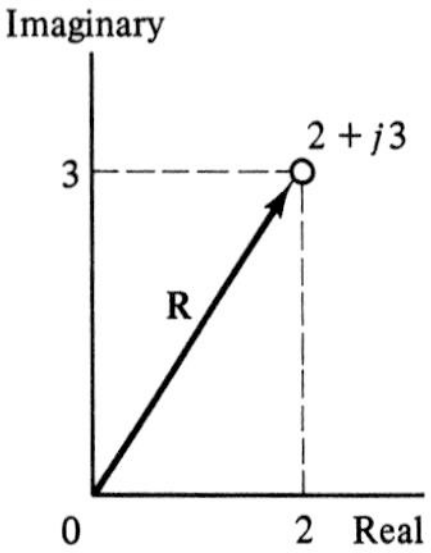


**FIGURE 20-4 A vector represented by a complex number.**

### Vector Addition and Subtraction

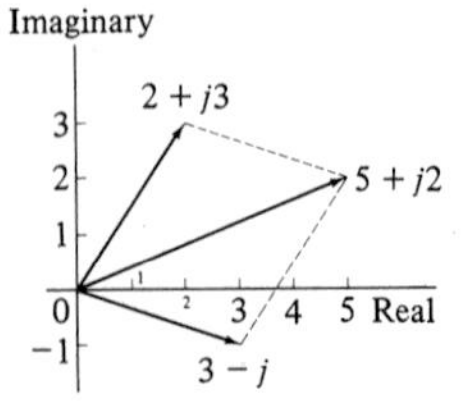


**FIGURE 20-5 Vector addition with complex numbers.**

Let us place a second vector on our diagram, represented by the complex number $3 - j$ (Fig. 20-5). We can add them graphically by the parallelogram method and we get a resultant $5 + j2$. But let us add the two original complex numbers, $2 + j3$ and $3 - j$.

$$(2 + j3) + (3 - j) = 5 + j2$$

This is the same as we obtained graphically. In others words, *the resultant of two vectors is equal to the sum of the complex numbers representing those vectors.*

**EXAMPLE 30:** Subtract the vectors $25\underline{/48°} - 18\underline{/175°}$.

**Solution:** Vector addition and subtraction is best done in rectangular form. Converting the first vector,

$$a_1 = 25 \cos 48° = 16.7$$

$$b_1 = 25 \sin 48° = 18.6$$

so

$$25\underline{/48°} = 16.7 + j18.6$$

Similarly for the second vector,

$$a_2 = 18 \cos 175° = -17.9$$

$$b_2 = 18 \sin 175° = 1.57$$

so

$$18\underline{/175°} = -17.9 + j1.57$$

Combining, we have

$$(16.7 + j18.6) - (-17.9 + j1.57) = (16.7 + 17.9) + j(18.6 - 1.57)$$
$$= 34.6 + j17.0$$

These vectors are shown in Fig. 20-6.

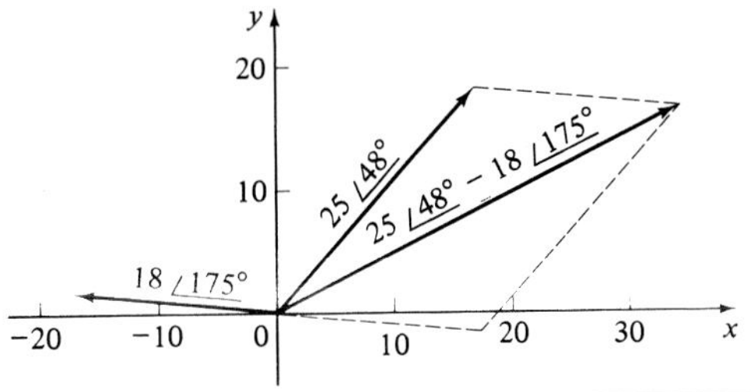


FIGURE 20-6

## Multiplication and Division of Vectors

Multiplication or division is best done if the vector is expressed as a complex number in polar form.

**EXAMPLE 31:** Multiply the vectors $2\underline{/25°}$ and $3\underline{/15°}$.

**Solution:** By Eq. 224, the product vector will have a magnitude of 3(2) = 6 and an angle of 25° + 15° = 40°. The product is thus

$$6\underline{/40°}$$

**EXAMPLE 32:** Divide the vector $8\underline{/44°}$ by $4\underline{/12°}$.

**Solution:** By Eq. 225,

$$(8\underline{/44°}) \div (4\underline{/12°}) = \frac{8}{4}\underline{/44° - 12°} = 2\underline{/32°}$$

**EXAMPLE 33:** If a certain current **I** is represented by the complex number $1.15\angle 23.5^\circ$ amperes, and a complex impedance **Z** is represented by $24.6\angle 14.8^\circ$ ohms, multiply **I** by **Z** to obtain the voltage **V**.

**Solution:**

$$\mathbf{V} = \mathbf{IZ} = (1.15\angle 23.5^\circ)(24.6\angle 14.8^\circ) = 28.3\angle 38.3^\circ \quad \text{V}$$

### The *j* Operator

Let us multiply a complex number $r\angle\theta$ by $j$. First we must express $j$ in polar form. By Eq. 219,

$$j = 0 + j1 = 1(\cos 90^\circ + j \sin 90^\circ) = 1\angle 90^\circ$$

After multiplying by $j$, the new absolute value will be

$$r \cdot 1 = r$$

and the argument will be

$$\theta + 90^\circ$$

So

$$r\angle\theta \cdot j = r\angle\theta + 90^\circ$$

Thus, the only effect of multiplying our original complex number by $j$ was to change the angle by 90°. If our complex number represents a vector, *we may think of j as an operator that causes the vector to rotate through one-quarter revolution.*

## EXERCISE 5—VECTOR OPERATIONS USING COMPLEX NUMBERS

Express each vector in rectangular and polar form.

| | *Magnitude* | *Angle* |
|---|---|---|
| 1. | 7 | 49° |
| 2. | 28 | 136° |
| 3. | 193 | 73.5° |

| | *Magnitude* | *Angle* |
|---|---|---|
| 4. | 34.2 | 1.1 rad |
| 5. | 39 | 2.5 rad |
| 6. | 59.4 | −58° |

Combine the vectors.

7. $(3 + j2) + (5 - j4)$
8. $(7 - j3) - (4 + j)$
9. $58\angle 72^\circ + 21\angle 14^\circ$
10. $8.6\angle 58^\circ - 4.2\angle 160^\circ$
11. $9(\cos 42^\circ + j \sin 42^\circ) + 2(\cos 8^\circ + j \sin 8^\circ)$
12. $8(\cos 15^\circ + j \sin 15^\circ) - 5(\cos 9^\circ - j \sin 9^\circ)$

Multiply the vectors.

13. $(7 + j3)(2 - j5)$
14. $(2.5\angle 18^\circ)(3.7\angle 48^\circ)$
15. $2(\cos 25^\circ - j \sin 25^\circ) \cdot 6(\cos 7^\circ - j \sin 7^\circ)$

Divide the vectors.

16. $(25 - j2) \div (3 + j4)$
17. $(7.7\angle 47^\circ) \div (2.5\angle 15^\circ)$
18. $5(\cos 72^\circ + j \sin 72^\circ) \div 3(\cos 31^\circ + j \sin 31^\circ)$

## 20-6 ALTERNATING-CURRENT APPLICATIONS

### Rotating Vectors in the Complex Plane

We have already shown that a vector can be represented by a complex number, $r\underline{/\theta}$. For example, the complex number $5.00\underline{/28^\circ}$ represents a vector of magnitude 5.00 at an angle of 28° with the real axis.

A phasor (a rotating vector) may also be represented by a complex number $R\underline{/\omega t}$ by replacing the angle $\theta$ by $\omega t$, where $\omega$ is the angular velocity and $t$ is time.

---

**EXAMPLE 34:** The complex number

$$11\underline{/5t}$$

represents a phasor of magnitude 11 rotating with an angular velocity of 5 rad/s.

---

In Sec. 16-3 we showed that a phasor of magnitude $R$ has a projection on the $x$ axis of $R \cos \omega t$. Similarly, a phasor $R\underline{/\omega t}$ *in the complex plane* (Fig. 20-7) will have a projection on the real axis of $R \cos \omega t$. Thus

$$R \cos \omega t = \text{real part of } R\underline{/\omega t}$$

Similarly,

$$R \sin \omega t = \text{imaginary part of } R\underline{/\omega t}$$

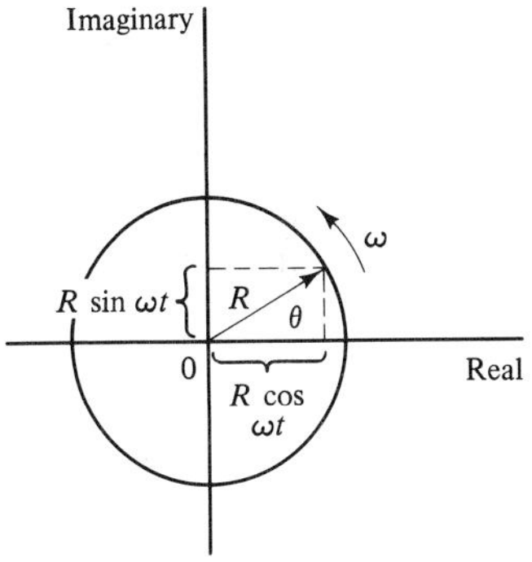


FIGURE 20-7

Thus a complex number $R\underline{/\omega t}$ can be represented in terms of sine or cosine, depending on whether we project onto the real or the imaginary axis. It does not usually matter which we choose, because the sine function and the cosine function are identical except for a phase difference of 90°. What does matter is *that we are consistent*. Here we will follow the convention of projecting the phasor onto the real axis, and drop the phrase "real part of." Thus we say that

$$R\underline{/\omega t} = R \cos \omega t$$

Similarly, if there is a phase angle $\phi$,

$$R \cos(\omega t + \phi) = R\underline{/\omega t + \phi}$$

One final simplification. It is customary to draw phasors at $t = 0$, so that $\omega t$ vanishes from the expression. Thus we write

$$R \cos(\omega t + \phi) = R\underline{/\phi}$$

---

**EXAMPLE 35:**

(a) $21.5 \cos 177t = 21.5\underline{/0^\circ}$

(b) $88 \cos(56t + 18^\circ) = 88\underline{/18^\circ}$

(c) $95\underline{/47^\circ} = 95 \cos(\omega t + 47^\circ)$

---

**Note that the "complex" expressions are the simpler. This happy situation is made use of in the ac computations to follow, which are much more cumbersome without complex numbers.**

When we write a complex number in terms of sine or cosine, we are expressing it in *sinusoidal form*. If a number is written in terms of cosine, to convert to the sine function, simply write sine in place of cosine and add 90° to the angle, since we know that $\cos \theta = \sin(\theta + 90^\circ)$.

**EXAMPLE 36:** Write a sinusoidal expression for the complex form.

(a) $21.5\underline{/177t} = 21.5 \cos 177t$

$= 21.5 \sin(177t + 90°)$

(b) $95\underline{/47°} = 95 \cos(\omega t + 47°) = 95 \sin(\omega t + 47° + 90°)$

$= 95 \sin(\omega t + 137°)$

### Alternating Current and Voltage in Complex Form

Since alternating current and voltage are sinusoidal quantities, they can be expressed in *complex form*. We call this form "complex," not because it is complicated but because it is in the form of a complex number: rectangular, polar, or trigonometric.

**EXAMPLE 37:** The alternating current

$$i = I \cos(\omega t + 33°) \quad \text{A}$$

can be written

$$i = I\underline{/33°} \quad \text{A}$$

**EXAMPLE 38:** The alternating voltage written as

$$v = V\underline{/73°} \quad \text{mV}$$

can be written in sinusoidal form

$$v = V \cos(\omega t + 73°) \quad \text{mV}$$

In general:

**In practice, it is customary to divide $V_m$ and $I_m$ by $\sqrt{2}$ in order to get the *rms* or *effective values* of voltage and current. We will not introduce that complication here, but work instead with the *peak* values $V_m$ and $I_m$.**

| | | |
|---|---|---|
| Complex Voltage and Current | the voltage $v = V_m \cos(\omega t + \phi_1)$ is represented by $\mathbf{V} = V_m\underline{/\phi_1}$ | **A75** |
| | the current $i = I_m \cos(\omega t + \phi_2)$ is represented by $\mathbf{I} = I_m\underline{/\phi_2}$ | **A76** |

Note that the complex expressions for voltage and current do not contain $t$. We say that the voltage and current have been converted from the *time domain* to the *phasor domain*.

### Complex Impedance

In Sec. 6-6 we drew a vector impedance diagram in which the impedance was the resultant of two perpendicular vectors: the resistance $R$ along the horizontal axis and the reactance $X$ along the vertical axis. If we now use the complex plane, and draw the resistance along the *real* axis and the reactance along the *imaginary* axis, we can represent impedance by a complex number.

| Complex Impedance | $\mathbf{Z} = R + jX = Z\angle\phi$ | A99 |
|---|---|---|

where

$$R = \text{circuit resistance}$$

$$X = \text{circuit reactance} = X_L - X_c \qquad \text{(A96)}$$

$$Z = \text{magnitude of impedance} = \sqrt{R^2 + X^2} \qquad \text{(A97)}$$

$$\phi = \text{phase angle} = \arctan \frac{X}{R} \qquad \text{(A98)}$$

---

**EXAMPLE 39:** A circuit has a resistance of 5 Ω in series with a reactance of 7 Ω (Fig. 20-8a). Represent the impedance by a complex number.

**Solution:** We draw the vector impedance diagram (Fig. 20-8b). The impedance, in rectangular form, is

$$Z = 5 + j7$$

The magnitude of the impedance is

$$\sqrt{5^2 + 7^2} = 8.60$$

The phase angle is

$$\arctan \frac{7}{5} = 54.5°$$

So we can write the impedance as a complex number in polar form,

$$\mathbf{Z} = 8.60\angle 54.5°$$

---

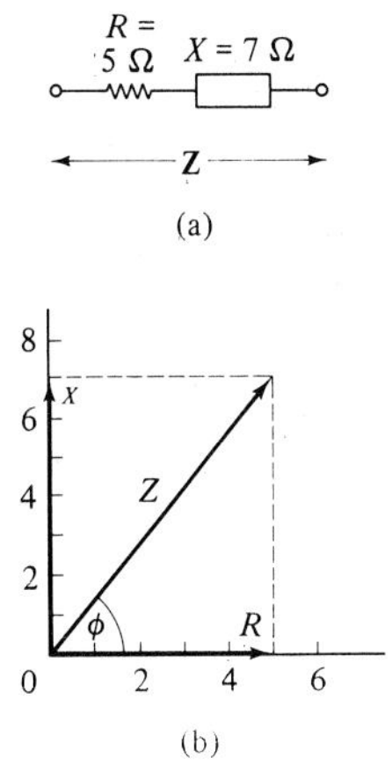


**FIGURE 20-8**

## Ohm's Law for AC

We stated at the beginning of this section that the use of complex numbers would make calculation with alternating current almost as easy as for direct current. We do this by means of the following relationship:

| Ohm's Law for AC | $\mathbf{V} = \mathbf{ZI}$ | A101 |
|---|---|---|

Note the similarity between this equation and Ohm's law, Eq. A62. It is used in the same way.

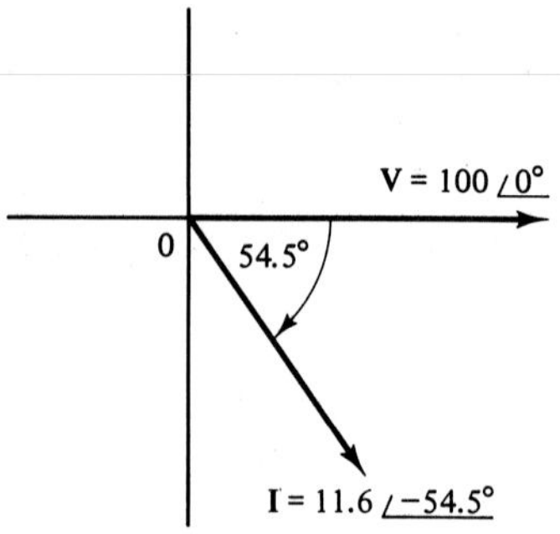


FIGURE 20-9

**EXAMPLE 40:** A voltage of $v = 100 \cos 200t$ is applied to the circuit of Fig. 20-8a. Write a sinusoidal expression for the current $i$.

**Solution:** We first write the voltage in complex form. By Eq. A75,

$$\mathbf{V} = 100\underline{/0°}$$

Next we find the complex impedance $Z$. From Example 39,

$$\mathbf{Z} = 8.60\underline{/54.5°}$$

The complex current $I$, by Ohm's law for ac, is

$$\mathbf{I} = \frac{\mathbf{V}}{\mathbf{Z}} = \frac{100\underline{/0°}}{8.60\underline{/54.5°}} = 11.6\underline{/-54.5°}$$

Then the current in sinusoidal form is (Eq. A76)

$$i = 11.6 \cos(200t - 54.5°)$$

The current and voltage phasors are plotted in Fig. 20-9, and the instantaneous current and voltage are plotted in Fig. 20-10. Note the phase difference between the voltage and current waves. This phase difference, 54.5°, converted to time measurement, is

$$54.5° = 0.951 \text{ rad} = 200t$$

$$t = \frac{0.951}{200} = 0.00475 \text{ s} = 4.75 \text{ ms}$$

We say that the voltage *leads* the current by 54.5° or 4.75 ms.

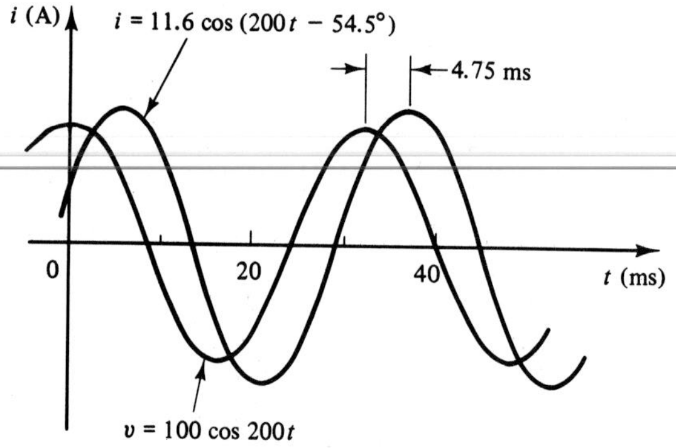


FIGURE 20-10

## EXERCISE 6—ALTERNATING-CURRENT APPLICATIONS

Express each current or voltage in complex form.

1. $i = 250 \cos(\omega t + 25°)$
2. $v = 1.5 \cos(112t - 30°)$
3. $v = 57 \cos(\omega t - 90°)$
4. $i = 2.7 \cos \omega t$
5. $v = 144 \cos 160t$
6. $i = 2.7 \cos(275t - 15°)$

Express each current or voltage in sinusoidal form.

7. $\mathbf{V} = 150\underline{/0°}$
8. $\mathbf{V} = 1.75\underline{/70°}$
9. $\mathbf{V} = 300\underline{/-90°}$
10. $\mathbf{I} = 25\underline{/30°}$
11. $\mathbf{I} = 7.5\underline{/0°}$
12. $\mathbf{I} = 15\underline{/-130°}$

Express the impedance of each circuit as a complex number in rectangular and in polar form.

13. Figure 20-11a.
14. Figure 20-11b.
15. Figure 20-11c.
16. Figure 20-11d.
17. Figure 20-11e.
18. Figure 20-11f.
19. Figure 20-11g.

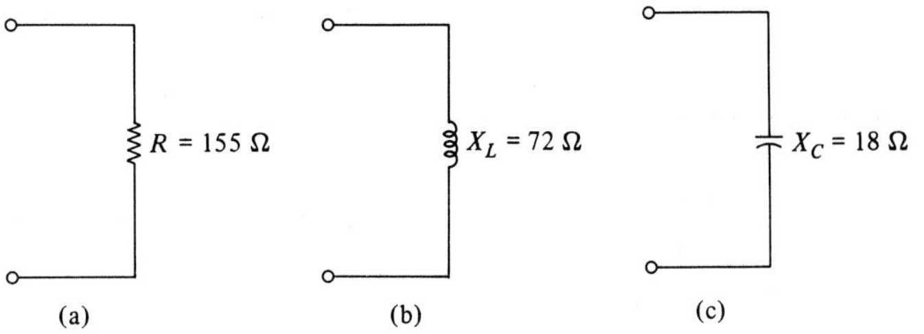


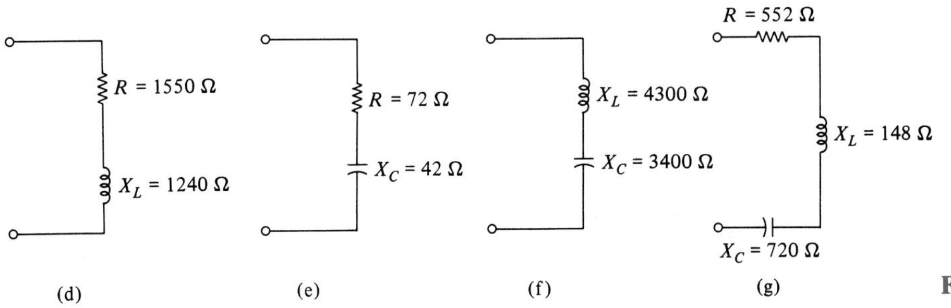


FIGURE 20-11

20. Write a sinusoidal expression for the current $i$ in each part of Fig. 20-12.

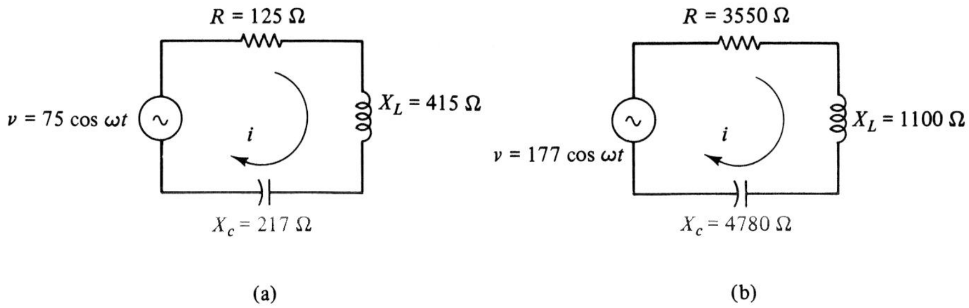


FIGURE 20-12

21. Write a sinusoidal expression for the voltage $v$ in each part of Fig. 20-13.

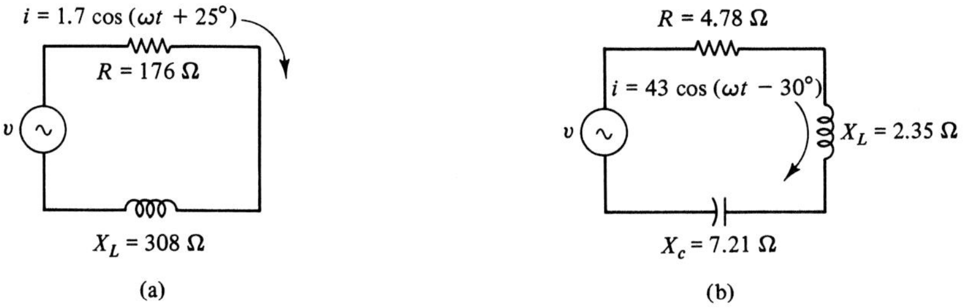


FIGURE 20-13

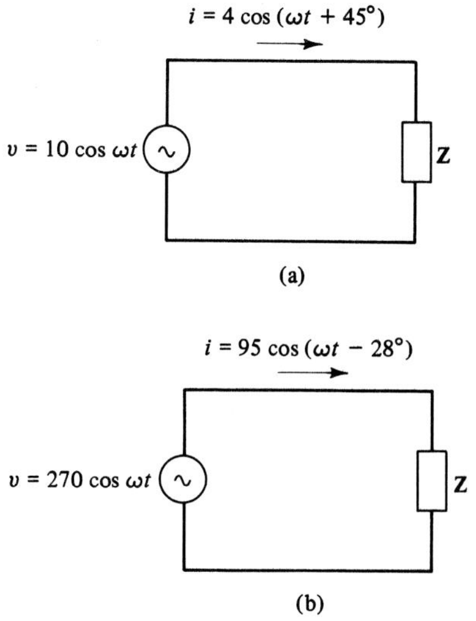


FIGURE 20-14

22. Find the complex impedance **Z** in each part of Fig. 20-14.

## CHAPTER 20 REVIEW PROBLEMS

*Express the following as complex numbers in rectangular, polar, trigonometric, and exponential forms.*

1. $3 + \sqrt{-4}$
2. $-\sqrt{-9} - 5$
3. $-2 + \sqrt{-49}$
4. $3.25 + \sqrt{-11.6}$

*Evaluate.*

5. $j^{17}$
6. $j^{25}$

*Combine the complex numbers. Leave your answer in rectangular form.*

7. $(7 - j3) + (2 + j5)$
8. $4.8\angle 28^\circ - 2.4\angle 72^\circ$
9. $52(\cos 50^\circ + j \sin 50^\circ) + 28(\cos 12^\circ + j \sin 12^\circ)$
10. $2.7e^{j7} - 4.3e^{j5}$

*Multiply and leave your answer in the same form as the complex numbers.*

11. $(2 - j)(3 + j5)$
12. $(7.3\angle 21^\circ)(2.1\angle 156^\circ)$
13. $2(\cos 20^\circ + j \sin 20^\circ) \cdot 6(\cos 18^\circ + j \sin 18^\circ)$
14. $(93e^{j2})(5e^{j7})$

*Divide and leave your answer in the same form as the complex numbers.*

15. $(9 - j3) \div (4 + j)$
16. $(18\angle 72^\circ) \div (6\angle 22^\circ)$
17. $16(\cos 85^\circ + j \sin 85^\circ) \div 8(\cos 40^\circ + j \sin 40^\circ)$
18. $127e^{j8} \div 4.75e^{j5}$

*Graph each complex number.*

19. $7 + j4$
20. $2.75\angle 44^\circ$
21. $6(\cos 135^\circ + j \sin 135^\circ)$
22. $4.75e^{j2.2}$

*Evaluate each power.*

23. $(-4 + j3)^2$
24. $(5\angle 12^\circ)^3$
25. $[5(\cos 10^\circ + j \sin 10^\circ)]^3$
26. $[2e^{j3}]^5$
27. Express in complex form $i = 45 \cos(\omega t + 32^\circ)$.
28. Express in sinusoidal form $\mathbf{V} = 283\angle -22^\circ$.
29. Write a complex expression for the current $i$ in Fig. 20-15.

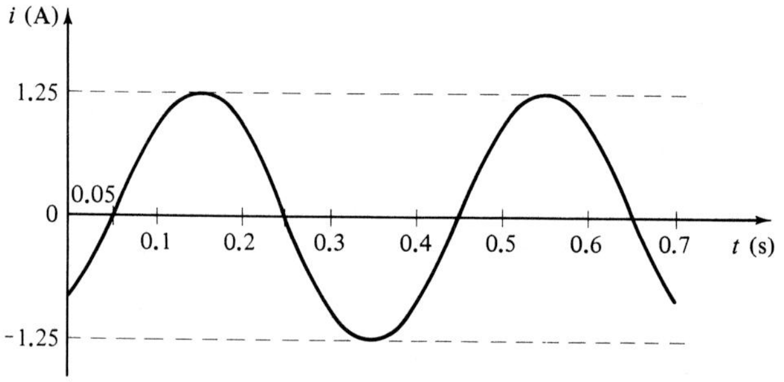


**FIGURE 20-15**

*Writing*

30. Suppose you are completing a job application to an electronics company which asks you to "Explain, in writing, how you would use complex numbers in an electrical calculation, and illustrate your explanation with an example." How would you respond?

# 21

# BINARY, HEXADECIMAL, OCTAL, AND BCD NUMBERS

## OBJECTIVES

**When you have completed this chapter, you should be able to:**

- Convert between binary and decimal numbers.
- Convert between decimal and binary fractions.
- Convert between binary and hexadecimal numbers.
- Convert between decimal and hexadecimal numbers.
- Convert between binary and octal numbers.
- Convert between binary and 8421 BCD numbers.

The decimal number system of Chapter 1 is fine for calculations done by humans, but it is not the easiest system for a computer to use. A digital computer contains elements that can be in either of two states; on or off, magnetized or not magnetized, and so on. For such devices, calculations are most conveniently done using *binary numbers*. In this chapter we learn what binary numbers are and how to convert between binary and decimal numbers.

Binary numbers are useful in a computer, where each binary digit (bit) can be represented by one state of a "binary switch" that is either on or off, but they are hard to read, partly because of their great length. To represent a nine-digit social security number, for example, requires a binary number *29 bits long*. So, in addition to binary numbers, we also study other ways in which numbers can be represented. *Hexadecimal, octal, and binary-coded decimal* systems allow us to express binary numbers more compactly, and make the transfer of data between computers and people much easier. Here we learn how to convert numbers between each of these systems, and between decimal and binary as well.

## 21-1 THE BINARY NUMBER SYSTEM

### Binary Numbers

A *binary number* is a sequence of the digits 0 and 1, such as

1101001

The number shown has no fractional part, and so is called a *binary integer*. A binary number having a fractional part contains a *binary point* (also called a *radix point*), as in the number

1001.01

### Base or Radix

The *base* of a number system (also called the *radix*) is equal to the number of digits used in the system.

---

**EXAMPLE 1:**

(a) The decimal system uses the ten digits

0 1 2 3 4 5 6 7 8 9

and has a base of 10.

(b) The binary system uses two digits

0 1

and has a base of 2.

---

### Bits, Bytes, and Words

Each of the digits is called a *bit,* from

Binary digIT

A *byte* is a group of 8 bits, and a *word* is the largest string of bits that a computer can handle in one operation. The number of bits in a word is called the *word length*. Different computers have different word lengths, with 8, 16, or 32 bits being common for desktop or personal computers. The longer words are often broken down into bytes for easier handling. Half a byte (4 bits) is called a *nibble*.

A *kilobyte* (Kbyte or KB) is 1024 ($2^{10}$) bytes, and a *megabyte* (Mbyte or MB) is 1,048,575 ($2^{20}$) bytes.

**Note that this is different from the usual meaning of these prefixes, where kilo means 1000 and mega means 1,000,000.**

### Writing Binary Numbers

A binary number is sometimes written with a subscript 2 when there is a chance that the binary number would otherwise be mistaken for a decimal number.

---

**EXAMPLE 2:** The binary number 110 could easily be mistaken for the decimal number 110, unless we write it

$$110_2$$

Similarly, a decimal number that may be mistaken for binary is often written with a subscript 10, as in

$$101_{10}$$

---

Long binary numbers are sometimes written with their bits in groups of four for easier reading.

---

**EXAMPLE 3:** The number 100100010100.001001 is easier to read when written

$$1001\ 0001\ 0100.0010\ 01$$

---

The leftmost bit in a binary number is called the *high-order* or *most significant bit (MSB)*. The bit at the extreme right of the number is the *low-order* or *least significant bit (LSB)*.

**TABLE 21-1**

| *Position* | *Place Value* |
|---|---|
| 10 | $2^{10} = 1024$ |
| 9 | $2^9 = 512$ |
| 8 | $2^8 = 256$ |
| 7 | $2^7 = 128$ |
| 6 | $2^6 = 64$ |
| 5 | $2^5 = 32$ |
| 4 | $2^4 = 16$ |
| 3 | $2^3 = 8$ |
| 2 | $2^2 = 4$ |
| 1 | $2^1 = 2$ |
| 0 | $2^0 = 1$ |
| −1 | $2^{-1} = 0.5$ |
| −2 | $2^{-2} = 0.25$ |
| −3 | $2^{-3} = 0.125$ |
| −4 | $2^{-4} = 0.0625$ |
| −5 | $2^{-5} = 0.03125$ |
| −6 | $2^{-6} = 0.015625$ |

### Place Value

We saw in Chapter 1 that a positional number system is one in which the position of a digit determines its value, and that each position in a number has a *place value* equal to the base of the number system raised to the position number.

The place values in the binary number system are

$$\ldots\ 2^4\quad 2^3\quad 2^2\quad 2^1\quad 2^0\ \ .\ \ 2^{-1}\quad 2^{-2}\ \ldots$$

or

$$\ldots\ 16\quad 8\quad 4\quad 2\quad 1\ \ .\ \ \tfrac{1}{2}\quad \tfrac{1}{4}\quad \ldots$$

binary point ↑

A more complete list of place values for a binary number is given in Table 21-1.

**Compare these place values to those in the decimal system shown in Fig. 1-1.**

### Expanded Notation

Thus the value of any digit in a number is the product of that digit and the place value. The value of the entire number is then the sum of these products.

**EXAMPLE 4:**

(a) The decimal number 526 can be expressed as

$$5 \times 10^2 + 2 \times 10^1 + 6 \times 10^0$$

or

$$5 \times 100 + 2 \times 10 + 6 \times 1$$

or

$$500 + 20 + 6$$

(b) The binary number 1011 can be expressed as

$$1 \times 2^3 + 0 \times 2^2 + 1 \times 2^1 + 1 \times 2^0$$

or

$$1 \times 8 + 0 \times 4 + 1 \times 2 + 1 \times 1$$

or

$$8 + 0 + 2 + 1$$

Numbers written in this way are said to be in *expanded notation.*

### Converting Binary Numbers to Decimal

Simply write the binary number in expanded notation (omitting those where the bit is 0), and add the resulting values.

**EXAMPLE 5:** Convert the binary number

$$1001.011$$

to decimal.

**Solution:** In expanded notation,

$$\begin{aligned} 1001.011 &= 1 \times 8 + 1 \times 1 + 1 \times \frac{1}{4} + 1 \times \frac{1}{8} \\ &= 8 + 1 + \frac{1}{4} + \frac{1}{8} \\ &= 9\frac{3}{8} \\ &= 9.375 \end{aligned}$$

### Largest Decimal Number Obtainable with *n* Bits

The largest possible 3-bit binary number is

$$1\ 1\ 1 = 7$$

or

$$2^3 - 1$$

The largest possible 4-bit number is

$$1\ 1\ 1\ 1 = 15$$

or

$$2^4 - 1$$

Similarly, the largest *n*-bit binary number is

| | | |
|---|---|---|
| Largest *n*-Bit Binary Number | $2^n - 1$ | **17** |

**EXAMPLE 6:** If a computer stores numbers with 15 bits, what is the largest decimal number that can be represented?

**Solution:** The largest decimal number is

$$2^{15} - 1 = 32{,}767$$

## Significant Digits

In Example 6 we saw that a 15-bit binary number could represent a decimal number no greater than 32,767. Thus it took 15 binary digits to represent five decimal digits. As a rule of thumb, *it takes about 3 bits for each decimal digit.*

Stated another way, if we have a computer that stores numbers with 15 bits, we should assume that the decimal numbers it prints do not contain more than five significant digits.

**EXAMPLE 7:** If we want a computer to print decimal numbers containing seven significant digits, how many bits must it use to store those numbers?

**Solution:** Using our rule of thumb, we need

$$3 \times 7 = 21 \text{ bits}$$

## Converting Decimal Numbers to Binary

To convert a decimal *integer* to binary, we first divide it by 2, obtaining a quotient and a remainder. We write down the remainder, and divide the quotient by 2, getting a new quotient and remainder. We then repeat this process until the quotient is zero.

**EXAMPLE 8:** Convert the decimal integer 59 to binary.

**Solution:** We divide 59 by 2, getting a quotient of 29 and a remainder of 1. Then dividing 29 by 2 gives a quotient of 14 and a remainder of 1. These calculations, and those that follow can be arranged in a table, as follows.

| | Remainder | | |
|---|---|---|---|
| 2 ⟌59 | | | ↑ |
| 2 ⟌29 | 1 | LSB | |
| 2 ⟌14 | 1 | | |
| 2 ⟌7 | 0 | | read up |
| 2 ⟌3 | 1 | | |
| 2 ⟌1 | 1 | | |
| 2 ⟌0 | 1 | MSB | |

Our binary number then consists of the digits in the remainders, with those at the *top* of the column appearing to the *right* in the binary number. Thus

$$59_{10} = 111011_2$$

**The conversion can, of course, be checked by converting back to decimal.**

To convert a decimal *fraction* to binary, we first multiply it by 2, remove the integral part of the product, and multiply by 2 again. The procedure is then repeated.

**EXAMPLE 9:** Convert the decimal fraction 0.546875 to binary.

**Solution:** We multiply the given number by 2, getting 1.09375. We remove the integral part, 1, leaving 0.09375, which we again multiply by 2, getting 0.1875. We repeat the computation until we get a product that has a fractional part of zero, as in the following table.

|  | 0.546875 | Integral Part | |
|---|---|---|---|
| × | 2 | | |
| | **1**.09375 | 1 | MSB |
| × | 2 | | |
| | **0**.1875 | 0 | |
| × | 2 | | |
| | **0**.375 | 0 | |
| × | 2 | | read |
| | **0**.75 | 0 | down |
| × | 2 | | |
| | **1**.50 | 1 | |
| × | 2 | | |
| | **1**.00 | 1 | LSB |

Notice that we remove the integral part, shown in boldface, before multiplying by 2.

We stop now that the fractional part of the product (1.00) is zero. The column containing the integral parts is now our binary number, with the digits at the top appearing at the *left* of the binary number. So

Note that this is the reverse of what we did when converting a decimal integer to binary.

$$0.546875_{10} = .1000\ 11_2$$

In Example 9 we were able to find an exact binary equivalent of a decimal fraction. This is not always possible, as shown in the following example.

**EXAMPLE 10:** Convert the decimal fraction 0.743 to binary.

**Solution:** We follow the same procedure as before and get the following values.

|  | 0.743 | |
|---|---|---|
| × | 2 | Integral Part |
| | **1**.486 | 1 MSB |
| × | 2 | |
| | **0**.972 | 0 |
| × | 2 | |
| | **1**.944 | 1 |
| × | 2 | |
| | **1**.888 | 1 |
| × | 2 | |
| | **1**.776 | 1 |
| × | 2 | |
| | **1**.552 | 1 |
| × | 2 | |
| | **1**.104 | 1 |
| × | 2 | |
| | **0**.208 | 0 LSB |

Chap. 21 / **Binary, Hexadecimal, Octal, and BCD Numbers**

It is becoming clear that this computation can continue indefinitely. This shows that *not all decimal fractions can be exactly converted to binary*. This inability to make an exact conversion is an unavoidable source of inaccuracy in some computations.

The result of our conversion is, then,

$$0.743_{10} \simeq .10111110_2$$

To convert a decimal number having both an integer part and a fractional part to binary, convert each part separately as shown above, and combine.

**EXAMPLE 11:** Convert the number 59.546875 to binary.

**Solution:** From the preceding examples,

$$59 = 11\ 1011$$

and

$$0.546875 = .1000\ 11$$

So

$$59.546875 = 11\ 1011.1000\ 11$$

## Converting Binary Fractions to Decimal

We use Table 21-1 to find the decimal equivalent of each binary bit located to the right of the binary point, and add.

**EXAMPLE 12:** Convert the binary fraction 0.101 to decimal.

**Solution:** From Table 21-1,

$$0.1 = 2^{-1} = 0.5$$
$$0.001 = 2^{-3} = 0.125$$

Add: $0.101 = 0.625$

The procedure is no different when converting a binary number that has both a whole and a fractional part.

**EXAMPLE 13:** Convert the binary number 10.01 to decimal.

**Solution:** From Table 21-1,

$$10 = 2$$
$$0.01 = 0.25$$

Add: $10.01 = 2.25$

## EXERCISE 1—THE BINARY NUMBER SYSTEM

### Conversion between Binary and Decimal

Convert each binary number to decimal.

| | | |
|---|---|---|
| 1. 10 | 2. 01 | 3. 11 |
| 4. 1001 | 5. 0110 | 6. 1111 |
| 7. 1101 | 8. 0111 | 9. 1100 |

10. 1011
11. 0101
12. 1101 0010
13. 0110 0111
14. 1001 0011
15. 0111 0111
16. 1001 1010 1000 0010

Convert each decimal number to binary.

17. 5
18. 9
19. 2
20. 7
21. 72
22. 28
23. 93
24. 17
25. 274
26. 937
27. 118
28. 267
29. 8375
30. 2885
31. 82740
32. 72649

Convert each decimal fraction to binary. Retain 8 bits to the right of the binary point.

33. 0.5
34. 0.25
35. 0.75
36. 0.375
37. 0.3
38. 0.8
39. 0.55
40. 0.35
41. 0.875
42. 0.4375
43. 0.3872
44. 0.8462

Convert each binary fraction to decimal.

45. 0.1
46. 0.11
47. 0.01
48. 0.011
49. 0.1001
50. 0.0111

Convert each decimal number to binary. Keep 8 bits to the right of the binary point.

51. 5.5
52. 2.75
53. 4.375
54. 29.381
55. 948.472
56. 2847.22853

Convert each binary number to decimal.

57. 1.1
58. 10.11
59. 10.01
60. 101.011
61. 1 1001.0110 1
62. 10 1001.1001 101

## 21-2 THE HEXADECIMAL NUMBER SYSTEM

### Hexadecimal Numbers

Hexadecimal numbers (or *hex* for short) are obtained by grouping the bits in a binary number into sets of four, and representing each such set by a single number or letter. A hex number one-fourth the length of the binary number is thus obtained.

### Base 16

Since a 4-bit group of binary digits can have a value between 0 and 15, we need 16 symbols to represent all these values. The base of hexadecimal numbers is thus 16. We use the digits from 0 to 9, and the capital letters A to F, as in Table 21-2.

**TABLE 21-2 Hexadecimal and Octal Number Systems**

| *Decimal* | *Binary* | *Hexadecimal* | *Octal* |
|---|---|---|---|
| 0 | 0000 | 0 | 0 |
| 1 | 0001 | 1 | 1 |
| 2 | 0010 | 2 | 2 |
| 3 | 0011 | 3 | 3 |
| 4 | 0100 | 4 | 4 |
| 5 | 0101 | 5 | 5 |
| 6 | 0110 | 6 | 6 |
| 7 | 0111 | 7 | 7 |
| 8 | 1000 | 8 | 10 |
| 9 | 1001 | 9 | 11 |
| 10 | 1010 | A | 12 |
| 11 | 1011 | B | 13 |
| 12 | 1100 | C | 14 |
| 13 | 1101 | D | 15 |
| 14 | 1110 | E | 16 |
| 15 | 1111 | F | 17 |

Table 21-2 also shows octal numbers, which we will use later.

### Converting Binary to Hexadecimal

Group the bits into sets of four starting at the binary point, adding zeros as needed to fill out the groups. Then assign to each group the appropriate letter or number from Table 21-2.

---

**EXAMPLE 14:** Convert 10110100111001 to hexadecimal.

**Solution:** Grouping, we get

10 1101 0011 1001

or 0010 1101 0011 1001

From Table 21-2: 2 D 3 9

So the hexadecimal equivalent is 2D39. Hexadecimal numbers are sometimes written with the subscript 16, as in

$$2D39_{16}$$

---

The procedure is no different for binary fractions.

---

**EXAMPLE 15:** Convert 101111.0011111 to hexadecimal.

**Solution:** Grouping from the binary point gives us

10 1111 . 0011 111

or 0010 1111 . 0011 1110

From Table 21-2: 2 F . 3 E

or 2F.3E.

---

### Converting Hexadecimal to Binary

Here we simply reverse the procedure.

**EXAMPLE 16:** Convert 3B25.E to binary.

**Solution:** We write the group of 4 bits corresponding to each hexadecimal symbol.

| 3 | B | 2 | 5 | . | E |
|---|---|---|---|---|---|
| 0011 | 1011 | 0010 | 0101 | . | 1110 |

or 11 1011 0010 0101.111.

## Convert Hexadecimal to Decimal

Instead of converting directly between decimal and hex, many find it easier to first convert to binary.

As with decimal and binary numbers, each hex digit has a *place value,* equal to the base, 16, raised to the position number. Thus the place value of the first position to the left of the decimal point is

$$16^0 = 1$$

and the next is

$$16^1 = 16$$

and so on.

To convert from hex to decimal, first replace each letter in the hex number by its decimal equivalent. Write the number in expanded notation, multiplying each hex digit by its place value. Add the resulting numbers.

**EXAMPLE 17:** Convert the hex number 3B.F to decimal.

**Solution:** We replace the hex B with 11, and the hex F with 15, and write the number in expanded form.

| 3 | B | . | F |
|---|---|---|---|
| 3 | 11 | . | 15 |

$$(3 \times 16^1) + (11 \times 16^0) + (15 \times 16^{-1})$$
$$= 48 + 11 + 0.9375$$
$$= 59.9375$$

## Converting Decimal to Hexadecimal

We repeatedly divide the given decimal number by 16, and convert each remainder to hex. The remainders form our hex number, the last remainder obtained being the most significant digit.

**EXAMPLE 18:** Convert the decimal number 83759 to hex.

**Solution:**

| | | |
|---|---|---|
| $83759 \div 16 = 5234$ | with remainder of | 15 ↑ |
| $5234 \div 16 = 327$ | with remainder of | 2 |
| $327 \div 16 = 20$ | with remainder of | 7 read up |
| $20 \div 16 = 1$ | with remainder of | 4 |
| $1 \div 16 = 0$ | with remainder of | 1 |

Changing the number 15 to hex F and reading the remainders from bottom up, our hex equivalent is 1472F.

## EXERCISE 2—THE HEXADECIMAL SYSTEM

### Binary–Hex Conversions

Convert each binary number to hexadecimal.

1. 1101
2. 1010
3. 1001
4. 1111
5. 1001 0011
6. 0110 0111
7. 1101 1000
8. 0101 1100
9. 1001 0010 1010 0110
10. 0101 1101 0111 0001
11. 1001.0011
12. 10.0011
13. 1.0011 1
14. 101.101

Convert each hex number to binary.

15. 6F
16. B2
17. 4A
18. CC
19. 2F35
20. D213
21. 47A2
22. ABCD
23. 5.F
24. A4.E
25. 9.AA
26. 6D.7C

### Decimal–Hex Conversions

Convert each hex number to decimal.

27. F2
28. 5C
29. 33
30. DF
31. 37A4
32. A3F6
33. F274
34. C721
35. 3.F
36. 22.D
37. ABC.DE
38. C.284

Convert each decimal number to hex.

39. 39
40. 13
41. 921
42. 554
43. 2741
44. 9945
45. 1736
46. 2267

## 21-3 THE OCTAL NUMBER SYSTEM

### The Octal System

The *octal* number system uses eight digits, 0 to 7, and hence has a base of eight. A comparison of the decimal, binary, hex, and octal digits is given in Table 21-2.

### Binary–Octal Conversions

To convert from binary to octal, write the bits of the binary number in groups of three, starting at the binary point. Then write the octal equivalent for each group.

**EXAMPLE 19:** Convert the binary number 1 1010 0011 0110 to octal.

**Solution:** Grouping the bits in sets of three from the binary point,

| | | | | | |
|---|---|---|---|---|---|
| | 001 | 101 | 000 | 110 | 110 |
| and the octal equivalents, | 1 | 5 | 0 | 6 | 6 |

so the octal equivalent of 1 1010 0011 0110 is 15066.

To convert from octal to binary, simply reverse the procedure shown in the preceding example.

**EXAMPLE 20:** Convert the octal number 7364 to binary.

**Solution:** We write the group of 3 bits corresponding to each octal digit.

| | | | |
|---|---|---|---|
| 7 | 3 | 6 | 4 |
| 111 | 011 | 110 | 100 |

or 1110 1111 0100.

### EXERCISE 3—THE OCTAL NUMBER SYSTEM

Convert each binary number to octal.

1. 110
2. 010
3. 111
4. 101
5. 11 0011
6. 01 1010
7. 0110 1101
8. 1 1011 0100
9. 100 1001
10. 1100 1001

Convert each octal number to binary.

11. 26
12. 35
13. 623
14. 621
15. 5243
16. 1153
17. 63150
18. 2346

## 21-4 BCD CODES

In Sec. 21-1 we converted decimal numbers into binary. Recall that each bit had a place value equal to 2 raised to the position number. Thus the binary number 10001 is equal to

$$2^4 + 2^0 = 16 + 1 = 17$$

We shall now refer to numbers such as 10001 as *straight binary.*

With a BCD or *binary-coded-decimal* code, a decimal number is not converted *as a whole* to binary, but rather *digit by digit*. For example, the 1 in the decimal number 17 is 0001 in 4-bit binary, and the 7 is equal to 0111. Thus

$$17 = 10001 \text{ in straight binary}$$

and

$$17 = 0001\ 0111 \text{ in BCD}$$

When we converted the digits in the decimal number 17 in the preceding example, we used the same 4-bit binary equivalents as in Table 21-2. The bits have the place values 8, 4, 2, and 1, and BCD numbers written in this manner are said to be in *8421 code*.

### Converting Decimal Numbers to BCD

Simply convert each decimal digit to its equivalent 4-bit code.

---

**EXAMPLE 21:** Convert the decimal number 25.3 to 8421 BCD code.

**Solution:** We find the BCD equivalent of each decimal digit from Table 21-3.

$$\begin{matrix} 2 & 5 & . & 3 \\ 0010 & 0101 & . & 0011 \end{matrix}$$

or 0010 0101.0011.

**TABLE 21-3 Binary-Coded-Decimal Numbers**

| *Decimal* | *8421* | *2421* | *5211* |
|---|---|---|---|
| 0 | 0000 | 0000 | 0000 |
| 1 | 0001 | 0001 | 0001 |
| 2 | 0010 | 0010 | 0011 |
| 3 | 0011 | 0011 | 0101 |
| 4 | 0100 | 0100 | 0111 |
| 5 | 0101 | 1011 | 1000 |
| 6 | 0110 | 1100 | 1010 |
| 7 | 0111 | 1101 | 1100 |
| 8 | 1000 | 1110 | 1110 |
| 9 | 1001 | 1111 | 1111 |

---

### Converting from BCD to Decimal

Separate the BCD number into 4-bit groups, and write the decimal equivalent of each group.

---

**EXAMPLE 22:** Convert the 8421 BCD number 1 0011.0101 to decimal.

**Solution:** We have

$$\begin{matrix} 0001 & 0011 & . & 0101 \end{matrix}$$

From Table 21-3:

$$\begin{matrix} 1 & 3 & . & 5 \end{matrix}$$

or 13.5.

---

### Other BCD Codes

There are other BCD codes in use which represent each decimal digit by a 6-bit binary number or an 8-bit binary number, and other 4-bit codes in which the place values are other than 8421. Some of these are shown in Table 21-3. Each is used in the same way as shown for the 8421 code.

**EXAMPLE 23:** Convert the decimal number 825 to 2421 BCD code.

**Solution:** We convert each decimal digit separately,

| | | |
|---|---|---|
| 8 | 2 | 5 |
| 1110 | 0010 | 1011 |

## Other Computer Codes

In addition to the binary, hexadecimal, octal, and BCD codes we have discussed, there are many other codes used in the computer industry. Numbers can be represented by the *excess-3 code* or the *Gray code*. There are error-detecting and parity-checking codes. Other codes are used to represent letters of the alphabet and special symbols, as well as numbers. Some of these are the *Morse code,* the American Standard Code for Information Interchange or *ASCII* (pronounced *as-key*) *code,* the Extended Binary-Coded-Decimal Interchange Code or *EBCDIC* (pronounced *eb-si-dik*) *code,* and the *Hollerith code* used for punched cards.

Space does not permit discussion of each code. However, if you have understood the manipulation of the codes we have described, you should have no trouble when faced with a new one.

### EXERCISE 4—BCD CODES

Convert each decimal number to 8421 BCD.

1. 62
2. 25
3. 274
4. 284
5. 42.91
6. 5.014

Convert each 8421 BCD number to decimal.

7. 1001
8. 101
9. 110 0001
10. 100 0111
11. 11 0110.1000
12. 11 1000.1000 1

## CHAPTER 21 REVIEW PROBLEMS

1. Convert the binary number 110.001 to decimal.
2. Convert the binary number 1100.0111 to hex.
3. Convert the hex number 2D to BCD.
4. Convert the binary number 1100 1010 to decimal.
5. Convert the hex number 2B4 to octal.
6. Convert the binary number 110 0111.011 to octal.
7. Convert the decimal number 26.875 to binary.
8. Convert the hexadecimal number 5E.A3 to binary.
9. Convert the octal number 534 to decimal.
10. Convert the BCD number 1001 to hex.
11. Convert the octal number 276 to hex.
12. Convert the 8421 BCD number 1001 0100 0011 to decimal.
13. Convert the decimal number 8362 to hex.
14. Convert the octal number 2453 to BCD.
15. Convert the BCD number 1001 0010 to octal.
16. Convert the decimal number 482 to octal.
17. Convert the octal number 47135 to binary.
18. Convert the hexadecimal number 3D2.A4F to decimal.
19. Convert the BCD number 0111 0100 to straight binary.
20. Convert the decimal number 382.4 to 8421 BCD.

*Writing*

21. A legislator, trying to trim the college budget, has suggested dropping the study of binary numbers. "Why do we need other numbers?" he asks. "Plain numbers were good enough when I went to school."

    Write a letter to this legislator, explaining why, in this computer age, "plain" numbers are not enough. Include specific examples of the use of binary numbers. Be polite but forceful, and limit your letter to one page.

# 22

# INEQUALITIES AND LINEAR PROGRAMMING

## OBJECTIVES

**When you have completed this chapter, you should be able to:**

- Graph linear inequalities on the number line.
- Graph linear inequalities on the coordinate axes.
- Graph a system of linear inequalities.
- Solve conditional inequalities graphically and algebraically.
- Use inequalities to determine the range of a function or relation.
- Solve absolute value inequalities.
- Solve linear programming problems.

So far we have dealt only with equations, where one expression is equal to another, and our aim was to solve them—that is, to find any value of the variables that makes the equation true. In this chapter we work with *inequalities,* in which one expression is not necessarily equal to another, but may be greater than or less than the other expression. We will graph inequalities, and also solve them. Here we seek a whole range of values of the variable to satisfy the inequality.

We then introduce *linear programming* (a term that should not be confused with *computer* programming), which plays an important part in certain business decisions. It helps us decide how to allocate resources in order to reduce costs and maximize profit—a concern of technical people as well as management. We then write a computer program to do linear programming.

## 22-1 DEFINITIONS

### Inequalities

An *inequality* is a statement that one quantity is greater than, or less than, another quantity.

---

**EXAMPLE 1:** The statement $a > b$ means "$a$ is greater than $b$."

---

### Parts of an Inequality

$$a > b$$

left side or member ↗ (a) ↑ (>) inequality sign ↖ (b) right side or member

### Inequality Symbols

In Sec. 1-1 we introduced the inequality symbols $>$ and $<$. These can be combined with the equals sign to produce the compound symbols $\geq$ and $\leq$, which mean "greater than or equal to" and "less than or equal to." Further, a slash through any of these symbols indicates the negation of the inequality.

---

**EXAMPLE 2:** The statement $a \ngtr b$ means "$a$ is not greater than $b$."

---

### Sense of an Inequality

The *sense* of an inequality refers to the *direction* in which the inequality sign points. We use this term mostly when operating on an inequality, where certain operations will *change the sense,* whereas others will not.

### Inequalities with Three Members

The inequality $3 < x$ has two members, left and right. However, many inequalities have *three* members.

---

**EXAMPLE 3:** The statement $3 < x < 9$ means "$x$ is greater than 3 and less than 9," or "$x$ is a number *between* 3 and 9."

---

## Conditional and Unconditional Inequalities

Just as we have conditional equations and identities, we also have conditional and unconditional inequalities. As with equations, a conditional inequality is one that is satisfied only by certain values of the unknown, whereas unconditional equalities are true for any values of the unknown.

---

**EXAMPLE 4:** The statement $x - 2 > 5$ is a conditional inequality because it is true only for values of $x$ greater than 7.

---

**EXAMPLE 5:** The statement $x^2 + 5 > 0$ is an unconditional inequality (also called an *absolute* inequality) because for any value of $x$, positive, negative, or zero, $x^2$ cannot be negative. Thus $x^2 + 5$ is always greater than zero.

---

**In this chapter we limit ourselves to the *real* numbers.**

## Graphical Representation

An inequality having *two members* can be represented on the number line by a *ray,* as in Fig. 22-1. The ray is called *open* or *closed,* depending on whether or not the *endpoint* is included.

An inequality having *three members* can be represented by an *interval* on the number line. The interval is called open, closed, or half-open, depending on how many of the endpoints are included, as shown in Fig. 22-2.

An inequality having *two variables* can be represented by a *region,* in the *xy* plane. To graph an inequality we first graph the corresponding equality with a dashed curve. Then, if the equality is included in the graph, we replace the dashed curve by a solid one. If the points above (or below) the curve satisfy the inequality, we shade the region above (or below) the curve.

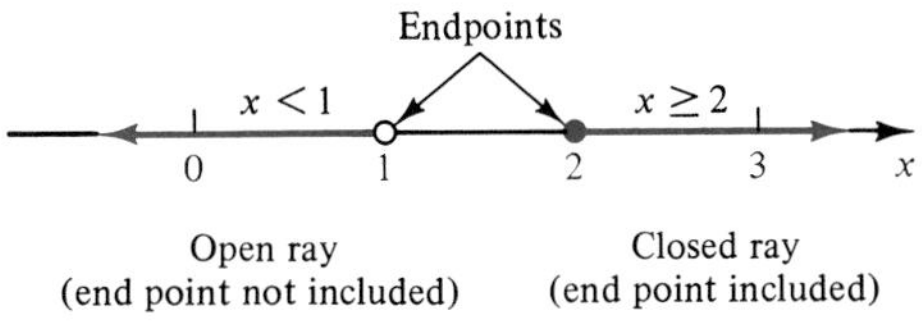


**FIGURE 22-1 Rays.**

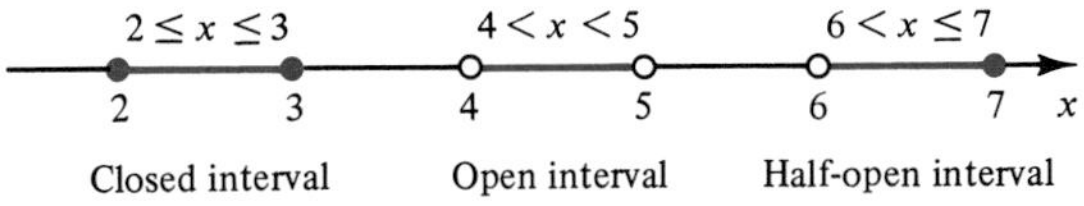


**FIGURE 22-2 Intervals.**

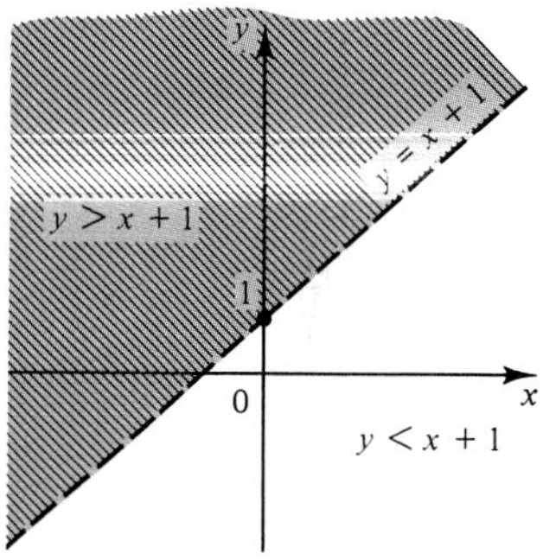


**FIGURE 22-3 Points within the shaded region satisfy the inequality $y > x + 1$.**

---

**EXAMPLE 6:** Graph the inequality $y > x + 1$.

**Solution:** We graph the line $y = x + 1$ (Fig. 22-3) using a solid line if points on the line satisfy the given inequality, or a dashed line if they do not. Here, we need a dashed line. All points in the shaded region above the dashed line satisfy the given equality. For example, if we substitute the values for the point (1, 3) into our inequality, we get

$$3 > 1 + 1$$

which is true. So (1, 3) will be in the region that satisfies the inequality $y > x + 1$. This trial point helps us choose the region to shade (here, the region above the line) and serves as a check for our work.

---

To satisfy two inequalities simultaneously we look for those points that satisfy the first, those points that satisfy the second, and then select those points that satisfy *both*. Therefore, all points in the shaded region in Fig. 22-4 satisfy the *two* inequalities

$$y < x + 1 \qquad \text{and} \qquad y \geq -2x - 2$$

If we substitute a point, say, (5, 2) into both our inequalities we get

$$2 < 5 + 1$$

which is true, and

$$2 \geq -2(5) - 2$$

or

$$2 > -12$$

which is also true. This serves to check that we have selected the correct region.

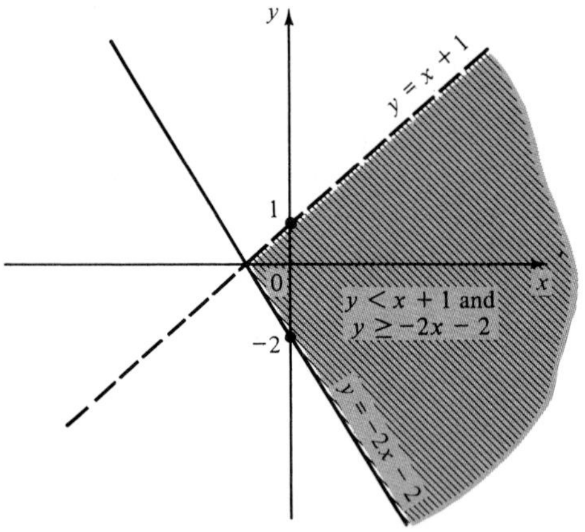


**FIGURE 22-4 Points within the shaded region satisfy two inequalities.**

---

**EXAMPLE 7:** A certain automobile company can make a sedan in 3 h and a hatchback in 2 h. How many of each type of car can be made if the working time is not to exceed 12 h?

**Solution:** We let

$$x = \text{number of sedans made in 12 h}$$

and $\qquad y = \text{number of hatchbacks made in 12 h}$

We look at the time it takes to make each vehicle. Since each sedan takes 3 h to make, it will take $3x$ hours to make the $x$ sedans. Similarly, it takes $2y$ hours for the hatchbacks. So we get the inequality which states that the sum of the times is less than or equal to 12 h.

$$3x + 2y \leq 12$$

We plot the equation $3x + 2y = 12$ in Fig. 22-5. Any point on the line or in the shaded region satisfies the inequality. For example, the company could make two sedans and three hatchbacks in the 12-h period, or it could make no cars at all, or it could make four sedans and no hatchbacks, and so on. It could not, however, make three sedans and three hatchbacks in the allotted 12 h.

---

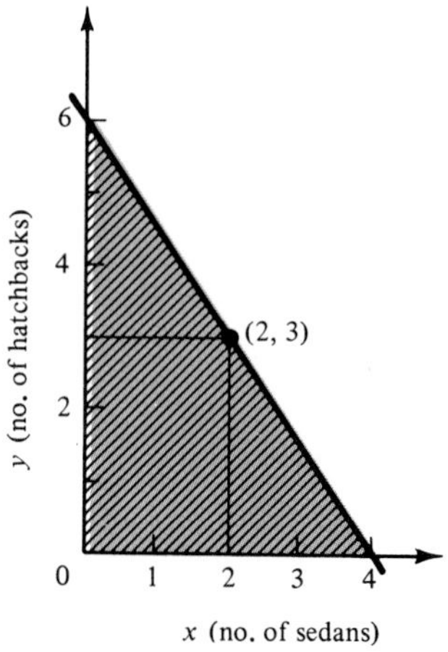


**FIGURE 22-5**

### Inequalities on the Computer

Most computer languages allow us to use the inequality symbols when comparing two quantities.

**EXAMPLE 8:** In BASIC, the instruction

```
100  IF X + Y =< 5 THEN 200
```

will send us to line 200 if the sum of $x$ and $y$ is equal to or less than 5.

**EXAMPLE 9:** For the automobile problem in Example 7 we could test if $x$ and $y$ gave us a point within the shaded region of Fig. 22-5 with the program statements

```
100  IF X < 0 THEN 200
110  IF Y < 0 THEN 200
120  IF 3 * X + 2 * Y > 12 THEN 200
130    .
 .     .
 .     .
200    .
```

Here we test whether the point $(x, y)$ is within or on the edge of the shaded region with statements which eliminate those points that are *outside* the region. If a point passes the test, control passes to line 130. Otherwise, we go to line 200.

**We will make use of statements such as these in Sec. 22-3, where we use the computer to do linear programming.**

## EXERCISE 1—DEFINITIONS

### Graphing Inequalities

Graph each inequality as a ray or interval on the number line.

1. $x > 5$
2. $x < 3$
3. $x \geq -2$
4. $x \leq 1$
5. $2 < x < 5$
6. $-1 \leq x \leq 3$
7. $4 \leq x < 7$
8. $-4 < x \leq 0$

Graph each inequality as a region in the coordinate plane.

9. $y < 3x - 2$
10. $x + y \geq 1$
11. $2x - y < 3$
12. $y > -x^2 + 4$

Graph the region in which both inequalities are satisfied.

13. $y > -x + 3$
    $y < 2x - 2$

14. $y < x^2 - 2$
$y \geq x$

15. $y > x^2 - 4$
$y < 4 - x^2$

16. The supply voltage $V$ to a certain device is required to be over 100 V, but not over 150 V. Express this as an inequality.

17. A manufacturer makes skis and snowshoes. A pair of skis takes 2.5 h to make and a pair of snowshoes takes 3.1 h. Write and graph an inequality to represent the number of pairs of skis and snowshoes that can be made in 8 h.

**Computer**

18. Write a program that will verify whether a value or values satisfy one or more inequalities. Use your program to check whether selected values of $x$ and $y$ satisfy any of the inequalities in this exercise.

## 22-2 SOLVING CONDITIONAL INEQUALITIES

Unlike the solution to a conditional equation, which consists of one or more discreet values of the variable, the solution to an inequality usually will be an *infinite set of values*. We first show a graphical method for finding these values, and then give an algebraic method.

### Graphical Solution

First transpose all terms to one side of the inequality to get it into the form

$$f(x) > 0 \qquad [\text{or } f(x) < 0]$$

Then plot $y = f(x)$ in the usual way. The solution will then be all values of $x$ for which the curve is above (or, alternatively, if $f(x) < 0$, below) the $x$ axis.

---

**EXAMPLE 10:** Solve the inequality

$$x^2 - 3 > 2x$$

**Solution:** Rearranging gives

$$x^2 - 2x - 3 > 0$$

We now plot the function

$$y = x^2 - 2x - 3$$

as shown in Fig. 22-6.

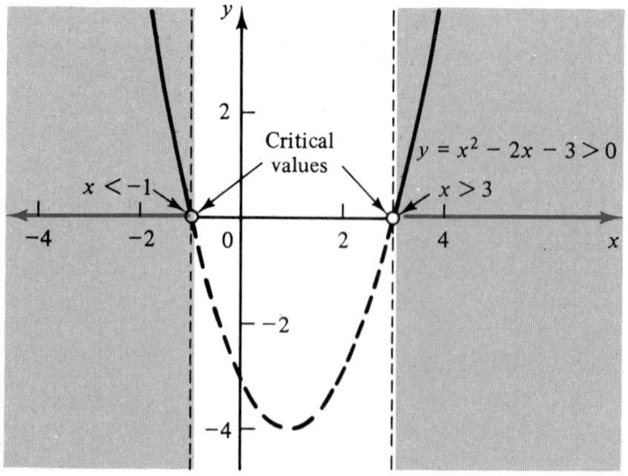


**FIGURE 22-6** **The points $x = -1$ and $x = 3$ where the curve crosses the $x$ axis are called *critical values*. $x < -1$ and $x > 3$ are solutions because those $x$ values satisfy the inequality.**

Note that our solutions to the inequality are all *values of x* that make the inequality true, that is, those values of $x$ that here make $y$ positive. The curve (a parabola) is above the $x$ axis when $x < -1$ and $x > 3$. These values of $x$ are the solutions to our inequality, because, for these values, $x^2 - 2x - 3 > 0$.

We can gain confidence in our solution by substituting values of $x$ into the inequality. Those values within the range of our solution should check, while all others should not.

**EXAMPLE 11:** For the inequality of Example 10, the values of, say, $x = 4$ and $x = -2$ should check, while $x = 1$ should not. Substituting $x = 4$ into our original inequality gives

$$4^2 - 3 > 2(4)$$

or

$$16 - 3 > 8$$

which is true. Similarly, the value $x = -2$ gives

$$(-2)^2 - 3 > 2(-2)$$

or

$$4 - 3 > -4$$

which is also true. On the other hand, a value between $-1$ and 3, such as $x = 1$, gives

$$1^2 - 3 > 2(1)$$

or

$$-2 \not> 2$$

which is not true, as expected.

## Algebraic Solution

The object of an *algebraic solution* to a conditional inequality is to locate the *critical values* (the points such as in Fig. 22-6 where the curve crossed the $x$ axis). To do this it is usually a good idea to transpose all terms to one side of the inequality and simplify the expression as much as possible. We make use of the following principles for operations with inequalities:

1. *Addition or subtraction.* The sense of an inequality is not changed if the same quantity is added to or subtracted from both sides.
2. *Multiplication or division by a positive quantity.* The sense of an inequality is not changed if both sides are multiplied or divided by the same *positive* quantity.
3. *Multiplication or division by a negative quantity.* The sense of an inequality *is reversed* if both sides are multiplied or divided by the same *negative* quantity.

**EXAMPLE 12:** Solve the inequality $21x - 23 < 2x + 15$.

**Solution:** Rearranging gives us

$$19x < 38$$

Dividing by 19, we get

$$x < 2$$

as shown in Fig. 22-7.

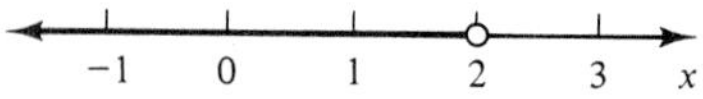


FIGURE 22-7

Sometimes we get *more than one* critical value, so that the $x$ axis is divided into more than two intervals. To decide which of these intervals contain the correct values, we test each for algebraic sign, as shown in the following example.

**EXAMPLE 13:** Solve the inequality $2x - \frac{x^2}{2} > -\frac{21}{2}$.

**Solution:** Multiplying by 2, we have

$$4x - x^2 > -21$$

Rearranging yields $\qquad 4x - x^2 + 21 > 0$

or $\qquad -x^2 + 4x + 21 > 0$

Multiplying by $-1$, which reverses the sense of the inequality, we obtain

$$x^2 - 4x - 21 < 0$$

Factoring gives us $\qquad (x - 7)(x + 3) < 0$

The critical values are those that make the left side equal zero, or $x = 7$ and $x = -3$. These divide the $x$ axis into three intervals,

$$x < -3 \qquad -3 < x < 7 \qquad x > 7$$

Which of these is correct? We can decide by computing the sign of $(x - 7)(x + 3)$ in each interval. Since this product is supposed to be less than zero, we choose the interval(s) in which the product is negative. Thus

| *Interval* | $(x - 7)$ | $(x + 3)$ | $(x - 7)(x + 3)$ |
|---|---|---|---|
| $x < -3$ | − | − | + |
| $-3 < x < 7$ | − | + | − |
| $x > 7$ | + | + | + |

The sign of the product is correct in only one interval, so our solution is

$$x > -3 \qquad \text{and} \qquad x < 7$$

as shown in Fig. 22-8. To relate this process to a graphical solution, we see that we have selected the interval $-3 < x < 7$ that corresponds to values on the parabola $y = x^2 - 4x - 21$ *below* the $x$ axis, as shown in Fig. 22-9.

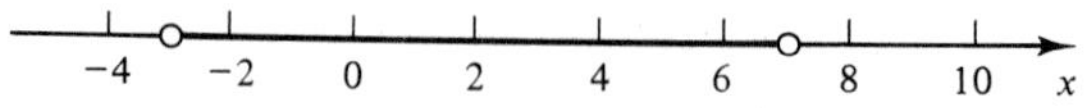


FIGURE 22-8

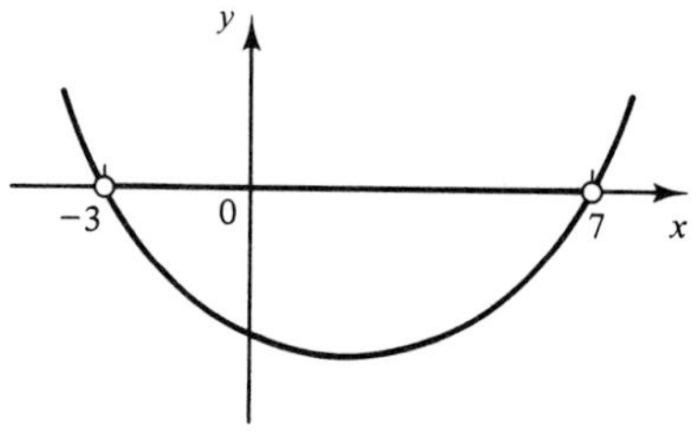


FIGURE 22-9

Inequalities containing the $\geq$ or $\leq$ signs are solved the same way as in the preceding examples, except that now the endpoint of an interval may be included in the solution.

**EXAMPLE 14:** Solve the inequality $x^3 - x^2 \geq 2x$.

**Solution:** Rearranging and factoring gives

$$x^3 - x^2 - 2x \geq 0$$

$$x(x^2 - x - 2) \geq 0$$

$$x(x + 1)(x - 2) \geq 0$$

The critical values are then

$$x = -1 \qquad x = 0 \qquad x = 2$$

We now analyze the signs, as in Example 13, looking for those intervals in which the product $x(x + 1)(x - 2)$ is positive.

| *Interval* | $x$ | $(x + 1)$ | $(x - 2)$ | $x(x + 1)(x - 2)$ |
|---|---|---|---|---|
| $x < -1$ | $-$ | $-$ | $-$ | $-$ |
| $-1 < x < 0$ | $-$ | $+$ | $-$ | $+$ |
| $0 < x < 2$ | $+$ | $+$ | $-$ | $-$ |
| $x > 2$ | $+$ | $+$ | $+$ | $+$ |

Our solution is then

$$-1 \leq x \leq 0 \qquad \text{and} \qquad x \geq 2$$

Note that the endpoints $x = -1$, 0, and 2 are included in the solution.

If we had graphed the inequality (Fig. 22-10), we would have selected the two intervals that correspond to values on the curve that are on or above the $x$ axis.

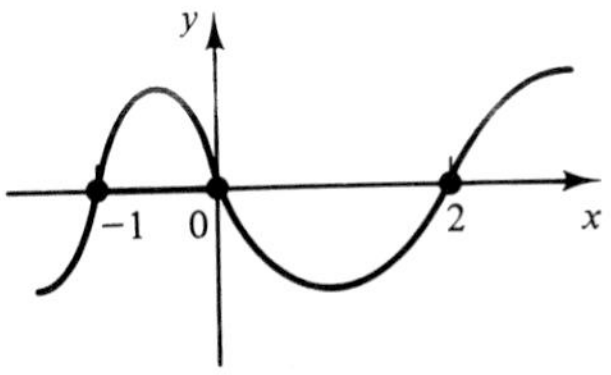


**FIGURE 22-10**

We include the graphs for comparison, but we emphasize solving equations algebraically, since that method is usually surer and quicker.

We can use the method of solving inequalities for graphing complicated functions. By finding the range of a function or relation we can locate, in advance, those regions in which the curve exists or does not exist.

**EXAMPLE 15:** Find the values of $x$ for which the relation $x^2 - y^2 - 4x + 3 = 0$ is real.

**Solution:** We solve for $y$,

$$y^2 = x^2 - 4x + 3$$

$$y = \pm\sqrt{x^2 - 4x + 3}$$

For $y$ to be real, the quantity under the radical sign must be nonnegative, so we seek the values of $x$ that satisfy the inequality

$$x^2 - 4x + 3 \geq 0$$

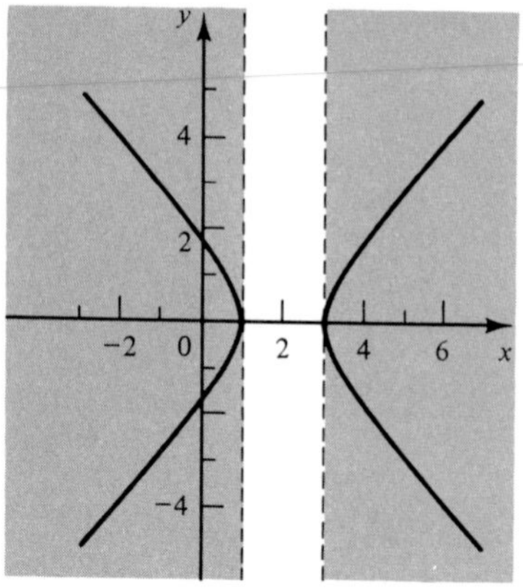


**FIGURE 22-11**

Factoring gives

$$(x-1)(x-3) \geq 0$$

The critical values are then $x = 1$ and $x = 3$. Analyzing the signs, we obtain

| *Interval* | $(x-1)$ | $(x-3)$ | $(x-1)(x-3)$ |
|---|---|---|---|
| $x < 1$ | − | − | + |
| $1 < x < 3$ | + | − | − |
| $x > 3$ | + | + | + |

We conclude that the function exists for the ranges $x \leq 1$ and $x \geq 3$, but not for $1 < x < 3$. This information helps when graphing the curve, for we now know where the curve does and does not exist. (The curve is actually a hyperbola, Fig. 22-11, which is studied in analytic geometry.)

## Inequalities Containing Absolute Values

Recall from Sec. 1-1 that the *absolute value* of a quantity is its *magnitude* and is always positive.

**EXAMPLE 16:**

(a) $|5| = 5$ (b) $|-5| = 5$

It would thus be correct to say that the quantity inside the absolute value sign could be 5 or −5. So if we see the expression $|x| > 2$, we know that either $x$ or $-x$, will be greater than 2.

$$\underbrace{|x| > 2}_{\text{expression with absolute value signs}} \quad \text{is equivalent to} \quad \underbrace{x > 2 \quad \text{and} \quad -x > 2}_{\text{two inequalities, without absolute value signs}}$$

The two inequalities can be written $x > 2$ and $x < -2$ because multiplying the right-hand inequality by −1 reverses the sense of the inequality.

Similarly, $|x| < 2$ is equivalent to $x < 2$ and $-x < 2$, or (after multiplying on the right by −1) $x < 2$ and $x > -2$. Stated as a rule,

$$|x| > a \quad \text{is equivalent to} \quad x > a \quad \text{and} \quad x < -a$$

$$|x| < a \quad \text{is equivalent to} \quad x < a \quad \text{and} \quad x > -a$$

To solve an inequality containing an absolute value sign, replace the inequality with two inequalities having no absolute value sign and proceed as before.

**EXAMPLE 17:** Solve the inequality $|7 - 3x| > 4$.

**Solution:** We replace this inequality with

$$(7 - 3x) > 4 \qquad \text{and} \qquad (7 - 3x) < -4$$

Subtracting 7, we obtain

$$-3x > -3 \qquad \text{and} \qquad -3x < -11$$

Dividing by $-3$ and reversing the sense yields

$$x < 1 \qquad \text{and} \qquad x > \frac{11}{3}$$

So $x$ may have values less than 1 or greater than $3\frac{2}{3}$.

## EXERCISE 2—SOLVING CONDITIONAL INEQUALITIES

### Solving Inequalities

Solve each inequality.

Solve some of these graphically, as directed by your instructor.

1. $2x - 5 > x + 4$
2. $x^2 + 2x < 3$
3. $x^2 + 4 \geq 2x^2 - 5$
4. $2x^2 + 15x \leq -25$
5. $2x^2 - 5x + 3 > 0$
6. $6x^2 < 19x - 10$
7. $\frac{x}{5} + \frac{x}{2} \geq 4$
8. $\frac{x+1}{2} - \frac{x-2}{3} \leq 4$
9. For what values of $x$ is $\sqrt{x^2 + 4x - 5}$ real?
10. For what values of $x$ is $\sqrt{x^3 - 4x^2 - 12x}$ real?
11. Find the values of $x$ for which the relation $y^2 = x^2 - x - 6$ is real.
12. Find the values of $x$ for which the relation $x^2 + y^2 + x - 2 = 0$ is real.

Solve each inequality.

13. $|3x| > 9$
14. $2|x + 3| < 8$
15. $5|x| > 8$
16. $|3x + 2| \leq 11$
17. $|2 - 3x| \geq 8$
18. $3|x + 2| > 21$
19. $|x + 5| > 2x$
20. $|x + 7| > 3x$
21. $|3 - 5x| \geq 4 - 3x$
22. $|x + 1| \geq 2x - 1$
23. A certain machine is rented for \$155 per day, and it costs \$25 per hour to operate. The total cost per day for this machine is not to exceed \$350. Write an inequality for this situation, and find the maximum permissible hours of operation.

## 22-3 LINEAR PROGRAMMING

Linear programming is a method for finding the maximum value of some function (usually representing *profit*) when the variables it contains are themselves restricted to within certain limits. The function that is maximized is called the *objective function,* and the limits on the variables are called *constraints*. The constraints will be in the form of inequalities.

We will do problems with only two variables here. To see how to *solve* problems with more than two variables, look up the *simplex method* in a text on linear programming.

We show the method by means of examples.

**EXAMPLE 18:** Find the values of $x$ and $y$ that will make $z$ a maximum, where the objective function is

$$z = 5x + 10y$$

In applications, ***x*** **and** ***y*** **usually represent amounts of a certain resource, or the number of items produced, and will be restricted to** ***positive*** **values.**

and $x$ and $y$ are positive and subject to the constraints

$$x + y \leq 5$$

and

$$2y - x \leq 4$$

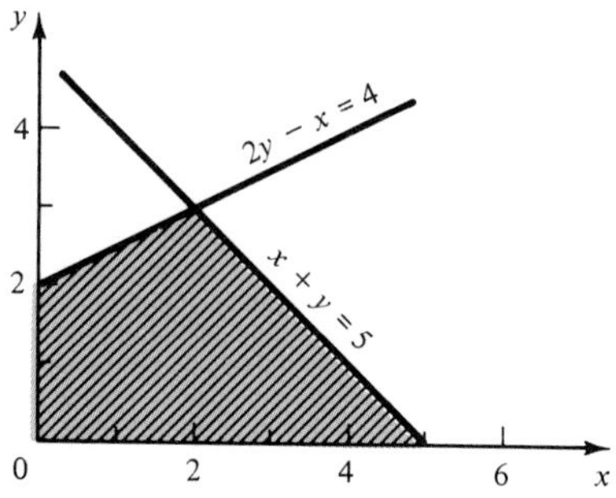


**FIGURE 22-12 Feasible solutions.**

**Solution:** We plot the two inequalities (Fig. 22-12). The only permissible values for $x$ and $y$ are the coordinates of points on the edges of or within the shaded region. These are called *feasible solutions*.

But which $(x, y)$ pair selected from all those in the shaded region will give the greatest value of $z$? In other words, which of the infinite number of feasible solutions is the *optimum* solution?

To help us answer this, let us plot the equation $z = 5x + 10y$ for different values of $z$. For example:

When $z = 0$:

$$5x + 10y = 0$$

$$y = -\frac{x}{2}$$

When $z = 10$:

$$5x + 10y = 10$$

$$y = -\frac{x}{2} + 1$$

**We'll learn more about the slopes of lines in Chapter 25.**

and so on. We get a *family* of parallel straight lines, each with a slope of $-\frac{1}{2}$, Fig. 22-13. The point $P$ at which the line having the greatest $z$ value intersects the region gives us the $x$ and $y$ values we seek. Such a point will always occur at a *corner* or *vertex* of the region. Thus, instead of plotting the family of lines, it is only necessary to compute $z$ at the vertices of the region. The coordinates of the vertices are called the *basic feasible solutions*. Of these, the one that gives the greatest $z$ is the *optimum solution*.

We will now show the process of finding the coordinates of $P$ algebraically. In our problem, the vertices are (0, 0), (0, 2), (5, 0), and $P$, whose

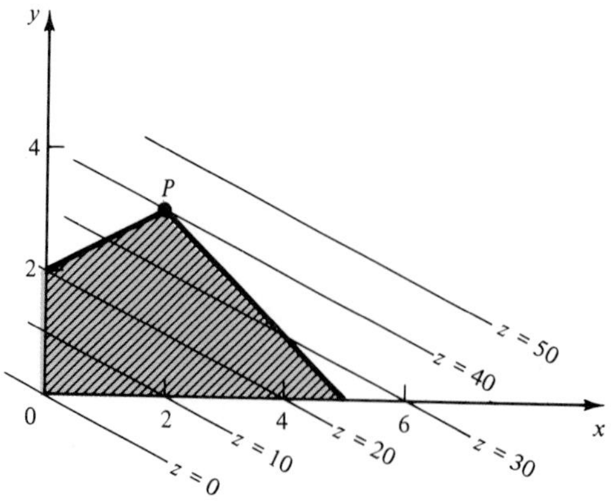


**FIGURE 22-13 Optimum solution at *P*.**

coordinates we get by simultaneous solution of the equations of the boundary lines,

$$x + y = 5$$

and $$x - 2y = -4$$

Subtract: $$3y = 9$$

$$y = 3$$

Substitute back: $$x = 2$$

So $P$ has the coordinates (2, 3). Now computing $z$ at each vertex, we get:

At (0, 0): $z = 5(0) + 10(0) = 0$

At (0, 2): $z = 5(0) + 10(2) = 20$

At (5, 0): $z = 5(5) + 10(0) = 25$

At (2, 3): $z = 5(2) + 10(3) = 40$

So $x = 2$, $y = 3$, which gives the largest $z$, is our optimum solution.

---

**EXAMPLE 19:** Find the nonnegative values of $x$ and $y$ that will maximize the function $z = 2x + y$ with the *four* constraints

$$y - x \leq 6$$

$$x + 4y \leq 40$$

$$x + y \leq 16$$

$$2x - y \leq 20$$

The points of intersection are found by simultaneous solution of the equations of the intersecting lines, and is not shown here.

**Solution:** We graph the inequalities (Fig. 22-14) and compute the points of intersection. This gives the six basic feasible solutions

(0, 0) (0, 6) (3.2, 9.2) (8, 8) (12, 4) (10, 0)

We now compute $z$ at each point of intersection.

| *Point* | $z = 2x + y$ |
|---|---|
| (0, 0) | $z = 2(0) + 0 = 0$ |
| (0, 6) | $z = 2(0) + 6 = 6$ |
| (3.2, 9.2) | $z = 2(3.2) + 9.2 = 15.6$ |
| (8, 8) | $z = 2(8) + 8 = 24$ |
| (12, 4) | $z = 2(12) + 4 = 28$ |
| (10, 0) | $z = 2(10) + 0 = 20$ |

The greatest $z$ (28) occurs at $x = 12$, $y = 4$, which is thus our optimum solution.

---

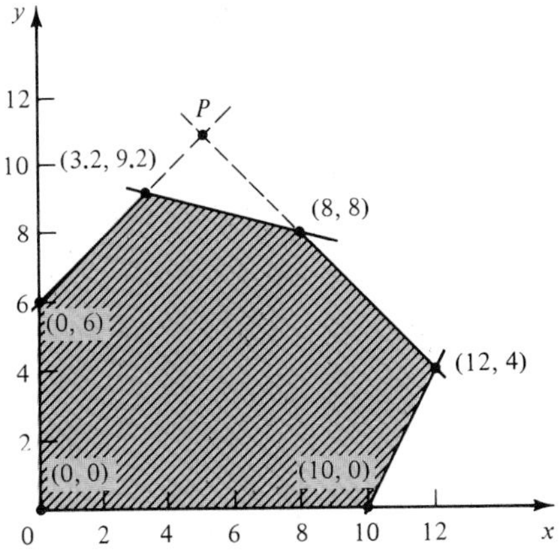


**FIGURE 22-14 The coordinates of the corners of the shaded region are the basic feasible solutions. One of these is the optimum solution.**

## Applications

Our applications will usually be business problems, where our objective function $z$ (the one we maximize) represents profit, and our variables $x$ and $y$ represent the quantities of two products to be made. A typical problem is to determine how much of each product should be manufactured so that the profit is a maximum.

**EXAMPLE 20:** Each month a company makes $x$ stereo sets and $y$ television sets. The manufacturing hours, material costs, testing time, and profit for each is as follows:

| | *Manufacturing Time (h)* | *Material Costs* | *Test Time (h)* | *Profit* |
|---|---|---|---|---|
| Stereo | 4 | \$ 95 | 1 | \$45 |
| Television | 5 | 48 | 2 | 38 |
| Available | 180 | 3600 | 60 | |

Also shown in the table is the maximum amount of manufacturing and testing time, and the cost of material available per month. How many of each item should be made for maximum profit?

**Solution:** First, why should we make any TV sets at all, if we get more profit on stereos? The answer is that we are limited by the money available. In 180 h we could make

$$180 \div 4 = 45 \text{ stereo sets}$$

But 45 stereo sets, at \$95 each, would require \$4275, but there is only \$3600 available. Then why not make

$$\$3600 \div 95 = 37 \text{ stereo sets}$$

using all our money on the most profitable item? But this would leave us with manufacturing time unused but no money left to make any TV sets.

Thus we wonder if we could do better by making fewer than 37 stereo sets, and using the extra time and money to make some TV sets. We seek the *optimum* solution. We start by writing the equations and inequalities to describe our situation.

Our total profit $z$ will be the profit per item times the number of items made.

$$z = 45x + 38y$$

The manufacturing time must not exceed 180 h, so

$$4x + 5y \leq 180$$

The materials cost must not exceed \$3600, so

$$95x + 48y \leq 3600$$

And the test time must not exceed 60 h, so

$$x + 2y \leq 60$$

This gives us our objective function and three constraints. We plot the constraints in Fig. 22-15.

We find the vertices of the region by simultaneously solving pairs of equations. To find point $A$, we take

$$x + 2y = 60$$

When $x = 0$,

$$y = 30$$

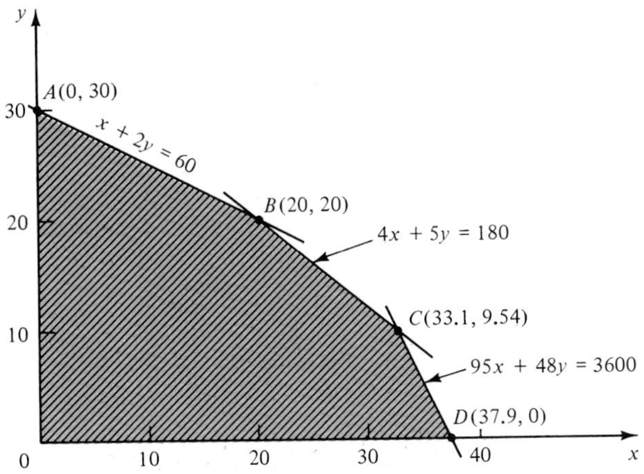


**FIGURE 22-15**

To find point $B$, we solve simultaneously

$$x + 2y = 60$$
$$4x + 5y = 180$$

and get

$$B(20, 20)$$

Again, the actual work is omitted here.

To find point $C$, we solve the equations

$$4x + 5y = 180$$
$$95x + 48y = 3600$$

and get

$$x = 33.1 \qquad y = 9.54$$

Since $x$ and $y$ must be integers, we round and get

$$C(33, 10)$$

To find point $D$, we use the equation

$$95x + 48y = 3600$$

Letting $y = 0$ gives us

$$x = 38 \quad \text{(rounded)}$$

We then evaluate the profit $z$ at each vertex (omitting the point 0, 0 at which $z = 0$).

At $A(0, 30)$: $z = 45(0) + 38(30) = \$1140$

At $B(20, 20)$: $z = 45(20) + 38(20) = \$1660$

At $C(33, 10)$: $z = 45(33) + 38(10) = \$1865$

At $D(38, 0)$: $z = 45(38) + 38(0) = \$1710$

Thus our maximum profit, \$1865, is obtained when we make 33 stereo sets and 10 television sets.

## Linear Programming on the Computer

The computer can reduce the work of linear programming, especially when the number of constraints gets large. The following algorithm is for a problem with two variables and any number of constraints.

1. Using Eq. 63, find the point of intersection of any two of the given boundary lines.
2. If the point of intersection is within the region of permissible solutions, go to step 3. If not, go to step 5. (This will eliminate points such as $P$ in Fig. 22-12.)
3. Compute $z$ at the point of intersection.
4. If $z$ is the largest so far, save it and the coordinates of the point. If not, discard it.
5. If there are combinations of equations yet unused, go to step 1.
6. Print the largest $z$ and the coordinates of the point at which it occurs.

A flowchart for this computation is shown in Fig. 22-16.

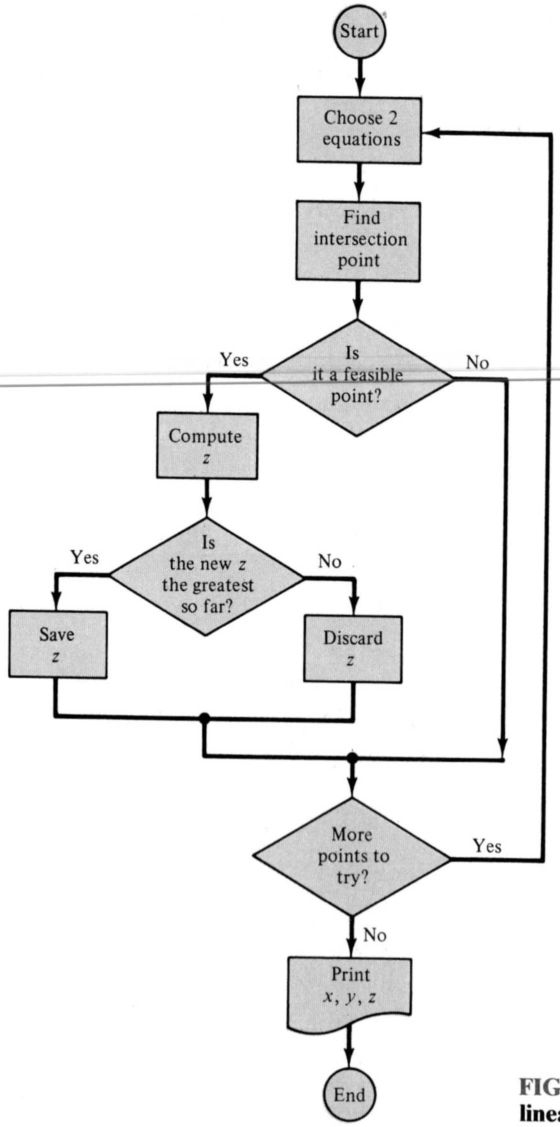


FIGURE 22-16 **Flowchart for linear programming by computer.**

## EXERCISE 3—LINEAR PROGRAMMING

1. Maximize the function $z = x + 2y$ under the constraints

$$x + y \leq 5 \quad \text{and} \quad x - y \leq -2$$

2. Maximize the function $z = 4x + 3y$ under the constraints

$$x - 2y \leq -2 \quad \text{and} \quad x + 2y \leq 4$$

3. Maximize the function $z = 5x + 4y$ under the constraints

$$7x + 4y \leq 35 \quad 2x - 3y \geq 7 \quad \text{and} \quad x - 2y \leq -5$$

4. Maximize the function $z = x + 2y$ under the constraints

$$x - y \leq -5 \quad x + 5y \leq 45 \quad \text{and} \quad x + y \leq 15$$

For Problems 5, 6, and 7, round your answer to the nearest integer.

5. A company makes pulleys and sprockets. A pulley takes 3 min to make, requires \$1.25 in materials, and gives a profit of \$2.10. A sprocket takes 4 min to make, requires \$1.30 in materials, and gives a profit of \$2.35. The time to make $x$ pulleys and $y$ sprockets is not to exceed 8 h, and the cost for materials must not exceed \$175. How many of each item should be made for maximum profit?
6. A company makes two computers, a mini and a micro, in plants in Boston and Seattle. The Boston plant operates for 180 h per month and the Seattle plant for 210 h per month. The Boston plant can make a mini in 6 h and a micro in 4 h, while the Seattle plant takes 5 h for either a mini or a micro. The profit is \$1000 for a mini and \$900 for a micro. How many of each computer should be made to give maximum profit?
7. Each week a company makes $x$ disk drives and $y$ monitors. The manufacturing hours, material costs, inspection time, and profit for each is as follows:

| | *Manufacturing Time (h)* | *Material Costs* | *Inspection Time (h)* | *Profit* |
|---|---|---|---|---|
| Disk drive | 5 | \$ 21 | 4.5 | \$34 |
| Monitor | 3 | 28 | 4 | 22 |
| Available | 50 | 320 | 50 | |

Also shown in the table is the maximum amount of manufacturing and inspection time, and the cost of material available per week. How many of each item should be made for maximum profit?

**Computer**

8. Using the flowchart (Fig. 22-16) as a guide, write a computer program to do linear programming. Try it on any of the problems in this exercise.

## CHAPTER 22 REVIEW PROBLEMS

*Represent each inequality graphically.*

1. $x < 8$
2. $y > 3x - 4$
3. $2 \leq x < 6$
4. $2x + y > 3$

5. $y > 2x^2 - 3$

6. $y < -2x - 3$ and $y > x + 2$

*Solve each inequality.*

7. $2x - 5 > x + 3$

8. $|2x| < 8$

9. $|x - 4| > 7$

10. $x^2 > x + 6$

11. Maximize the function $z = 7x + y$ under the constraints

$$2x + 3y \leq 5 \quad \text{and} \quad x - 4y \leq -4$$

12. A company makes four-cylinder engine blocks and six-cylinder engine blocks. A "four" requires 4 h of machine time and 5 h of labor, and gives a profit of $172. A "six" needs 5 h on the machine and 3 h of labor, and yields $195 in profit. If 130 h of labor and 155 h of machine time are available, how many (to the nearest integer) of each engine block should be made for maximum profit?

13. For what values of $x$ is $\sqrt{x^2 + 8x + 15}$ real?

14. For what values of $x$ is $\sqrt{x^2 - 10x + 16}$ real?

15. Find the values of $x$ for which the relation $y^2 = x^2 + 5x - 14$ is real.

16. Find the values of $x$ for which the relation $x^2 - y^2 - 10x + 9 = 0$ is real.

*Writing*

17. A job applicant says on her résumé that she knows linear programming, and the personnel director of your company insists that she fits your request for a "programmer." Write a memo to the personnel director explaining the difference between linear programming and computer programming.

# 23

# SEQUENCES, SERIES, AND THE BINOMIAL THEOREM

## OBJECTIVES

**When you have completed this chapter, you should be able to:**

- Identify various types of sequences and series.
- Write the general term or a recursion relation for many series.
- Compute any term or the sum of any number of terms of an arithmetic progression or a geometric progression.
- Compute any term of a harmonic progression.
- Insert any number of arithmetic means, harmonic means, or geometric means between two given numbers.
- Compute the sum of an infinite geometric progression.
- Solve applications problems using series.
- Raise a binomial to a power using the binomial theorem.
- Find any term in a binomial expansion.

We start this chapter with a general introduction to sequences and series. Sequences and series are of great interest, since a computer or calculator uses series internally to calculate many functions, such as the sine of an angle, the logarithm of a number, and so on. We then cover some specific sequences, such as the *arithmetic progression* and the *geometric progression*. Progressions are used to describe such things as the sequence of heights reached by a swinging pendulum on subsequent swings, or the monthly balance on a savings account that is growing with compound interest.

We finish this chapter with an introduction to the *binomial theorem*. The binomial theorem enables us to expand a binomial, such as $(3x^2 - 2y)^5$, without actually having to multiply the terms. It is also useful in deriving formulas in calculus, statistics, and probability.

## 23-1 SEQUENCES AND SERIES

### Sequences

A *sequence*

$$u_1, u_2, u_3, \ldots, u_n$$

is a set of quantities, called *terms,* arranged in the same order as the positive integers.

---

**EXAMPLE 1:**

(a) The sequence $1, \frac{1}{2}, \frac{1}{3}, \frac{1}{4}, \frac{1}{5}, \ldots, 1/n$ is called a *finite* sequence, because it has a finite number of terms, $n$.

(b) The sequence $1, \frac{1}{2}, \frac{1}{3}, \frac{1}{4}, \frac{1}{5}, \ldots, 1/n, \ldots$ is called an *infinite* sequence. The three dots at the end indicate that the sequence continues indefinitely.

(c) The sequence 3, 7, 11, 15, . . . is called an *arithmetic sequence,* or *arithmetic progression* (AP), because each term after the first is equal to the sum of the preceding term and a constant. That constant (4, in this example) is called the *common difference*.

(d) The sequence 2, 6, 18, 54, . . . is called a *geometric sequence,* or *geometric progression* (GP), because each term after the first is equal to the *product* of the preceding term and a constant. That constant (3, in this example) is called the *common ratio*.

(e) The sequence 1, 1, 2, 3, 5, 8, . . . is called a *Fibonacci* sequence. Each term after the first is the sum of the two preceding terms.

---

### Series

A *series* is the indicated sum of a sequence.

$$u_1 + u_2 + u_3 + \cdots + u_n + \cdots \qquad (428)$$

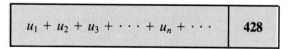

**EXAMPLE 2:**

(a) The series $1 + \frac{1}{2} + \frac{1}{3} + \frac{1}{4} + \frac{1}{5}$ is called a *finite series*. It is also called a *positive* series because all its terms are positive.

(b) The series $x - x^2 + x^3 - \cdots x^n \cdots$ is an *infinite series*. It is also called an *alternating* series because the signs of the terms alternate in sign.

(c) The series $6 + 9 + 12 + 15 + \cdots$ is an infinite arithmetic series. The terms of this series form an AP.

(d) $1 - 2 + 4 - 8 + 16 - \cdots$ is an infinite, alternating, geometric series. The terms of this series form a GP.

### General Term

The *general term* $u_n$ in a sequence or series is an expression involving $n$ (where $n = 1, 2, 3, \ldots$) by which we can obtain any term. The general term is also called the $n$th term. If we have an expression for the general term, we can then find any specific term of the sequence or series.

**EXAMPLE 3:** The general term of a certain series is

$$u_n = \frac{n}{2n + 1}$$

Write the first three terms of the series.

**Solution:** Substituting $n = 1$, $n = 2$, and $n = 3$, in turn, we get

$$u_1 = \frac{1}{2(1) + 1} \qquad u_2 = \frac{2}{2(2) + 1} \qquad u_3 = \frac{3}{2(3) + 1}$$

so our series is

$$\frac{1}{3} + \frac{2}{5} + \frac{3}{7} + \cdots + \frac{n}{2n + 1} + \cdots$$

### Recursion Relations

We have seen that we can find the terms of a series, given an expression for the $n$th term. Sometimes we can find each term from one or more immediately preceding terms. The relationship between a term and those preceding it is called a *recursion relation* or *recursion formula*.

**EXAMPLE 4:** Each term (after the first) in the series $1 + 4 + 13 + 40 + \cdots$ is found by multiplying the preceding term by 3 and adding 1. The recursion relation is then

$$u_n = 3u_{n-1} + 1$$

**EXAMPLE 5:** In a *Fibonacci sequence*

$$1, 1, 2, 3, 5, 8, 13, 21, 34, \ldots$$

each term after the first is the sum of the two preceding terms. Its recursion relation is

$$u_n = u_{n-1} + u_{n-2}$$

**Named for Leonardo Fibonacci (c. 1170–c. 1250), an Italian number theorist and algebraist who studied this sequence. He is also known as *Leonardo of Pisa*.**

## Series by Computer

Computers now make series easier to study than before. If we have either an expression for the general term of a series or the recursion formula, we can use a computer to generate large numbers of terms. By inspecting them we can get a good idea about how the series behaves.

---

**EXAMPLE 6:** Table 23-1 shows a portion of a computer printout of 80 terms of the series

$$\frac{1}{e} + \frac{2}{e^2} + \frac{3}{e^3} + \cdots + \frac{n}{e^n} + \cdots$$

Also shown for each term $u_n$ is the sum of that term and all preceding terms, called the *partial sum,* and the ratio of that term to the one preceding. We notice several things about this series.

1. The terms get smaller and smaller, and approach a value of zero.
2. The partial sums approach a limit (0.920674).
3. The ratio of two successive terms approaches a limit that is less than 1.

What do those facts tell us about this series? We'll see in the following section.

**TABLE 23-1**

| $n$ | Term $u_n$ | Partial Sum $S_n$ | Ratio |
|---|---|---|---|
| 1 | 0.367879 | 0.367879 | |
| 2 | 0.270671 | 0.638550 | 0.735759 |
| 3 | 0.149361 | 0.787911 | 0.551819 |
| 4 | 0.073263 | 0.861174 | 0.490506 |
| 5 | 0.033690 | 0.894864 | 0.459849 |
| 6 | 0.014873 | 0.909736 | 0.441455 |
| 7 | 0.006383 | 0.916119 | 0.429193 |
| 8 | 0.002684 | 0.918803 | 0.420434 |
| 9 | 0.001111 | 0.919914 | 0.413864 |
| 10 | 0.000454 | 0.920368 | 0.408755 |
| ⋮ | ⋮ | ⋮ | ⋮ |
| 75 | 0.000000 | 0.920674 | 0.372851 |
| 76 | 0.000000 | 0.920674 | 0.372785 |
| 77 | 0.000000 | 0.920674 | 0.372720 |
| 78 | 0.000000 | 0.920674 | 0.372657 |
| 79 | 0.000000 | 0.920674 | 0.372596 |
| 80 | 0.000000 | 0.920674 | 0.372536 |

## Convergence or Divergence of an Infinite Series

When we use a series to do computations, we cannot, of course, work with an infinite number of terms. In fact, for practical computation, we prefer as few terms as possible, as long as we get the accuracy that is needed. We want a series in which the terms decrease rapidly and approach zero, and for which the sum of the first several terms is not too different from the sum of all the terms of the series. Such a series is said to *converge* on some limit. A series that does not converge is said to *diverge*.

There are many tests for convergence. A particular test may tell us with certainty if the series converges or diverges. Sometimes, though, a test will fail and we must try another test. Here we consider only three tests.

1. *Magnitude of the terms.* If the terms of an infinite series do not approach zero, the series diverges. However, if the terms do approach zero, we cannot say with certainty if the series converges. Thus, having the terms approach zero is a necessary but not a sufficient condition for convergence.

2. *Sum of the terms.* If the sum of the first $n$ terms of a series reaches a limiting value as we take more and more terms, the series converges. Otherwise, the series diverges.

3. *Ratio test.* We compute the ratio of each term to the one preceding it, and see how that ratio changes as we go further out in the series. If that ratio approaches a number that is

   a. Less than 1, the series converges.

   b. Greater than 1, the series diverges.

   c. Equal to 1, the test is not conclusive.

---

**EXAMPLE 7:** Does the series of Example 6 converge or diverge?

**Solution:** We apply the three tests by inspecting the computed values.

1. The terms approach zero, so this test is not conclusive.

2. The sum of the terms appears to approach a limiting value (0.920674), indicating convergence.

3. The ratio of the terms appears to approach a value of less than 1, indicating convergence.

Note that we say *appears to* converge. The computer run, while convincing, is still not a *proof* of convergence.

Thus the series appears to converge.

---

**EXAMPLE 8:** Table 23-2 shows part of a computer run for the series

$$1 + \frac{1}{2} + \frac{1}{3} + \cdots + \frac{1}{n} + \cdots$$

Does the series converge or diverge?

TABLE 23-2

| $n$ | *Term* $u_n$ | *Partial Sum* $S_n$ | *Ratio* |
|---|---|---|---|
| 1 | 1.000000 | 1.000000 | |
| 2 | 0.500000 | 1.500000 | 0.500000 |
| 3 | 0.333333 | 1.833333 | 0.666667 |
| 4 | 0.250000 | 2.083334 | 0.750000 |
| 5 | 0.200000 | 2.283334 | 0.800000 |
| 6 | 0.166667 | 2.450000 | 0.833333 |
| 7 | 0.142857 | 2.592857 | 0.857143 |
| 8 | 0.125000 | 2.717857 | 0.875000 |
| 9 | 0.111111 | 2.828969 | 0.888889 |
| 10 | 0.100000 | 2.928969 | 0.900000 |
| ⋮ | ⋮ | ⋮ | ⋮ |
| 75 | 0.013333 | 4.901356 | 0.986667 |
| 76 | 0.013158 | 4.914514 | 0.986842 |
| 77 | 0.012987 | 4.927501 | 0.987013 |
| 78 | 0.012821 | 4.940322 | 0.987180 |
| 79 | 0.012658 | 4.952980 | 0.987342 |
| 80 | 0.012500 | 4.965480 | 0.987500 |
| ⋮ | ⋮ | ⋮ | ⋮ |
| 495 | 0.002020 | 6.782784 | 0.997980 |
| 496 | 0.002016 | 6.784800 | 0.997984 |
| 497 | 0.002012 | 6.786813 | 0.997988 |
| 498 | 0.002008 | 6.788821 | 0.997992 |
| 499 | 0.002004 | 6.790825 | 0.997996 |
| 500 | 0.002000 | 6.792825 | 0.998000 |

**Solution:** Again we apply the three tests.

1. We see that the terms are getting smaller, so this test is not conclusive.
2. The ratio of the terms appears to approach 1, so this test also is not conclusive.
3. The partial sum $S_n$ does not seem to approach a limit, but appears to keep growing even after 495 terms.

It thus appears that this series diverges.

## EXERCISE 1—SEQUENCES AND SERIES

Write the first five terms of each series, given the general term.

1. $u_n = 3n$
2. $u_n = 2n + 3$

3. $u_n = \dfrac{n+1}{n^2}$

4. $u_n = \dfrac{2^n}{n}$

Deduce the general term of each series. Use it to predict the next two terms.

5. $2 + 4 + 6 + \cdots$
6. $1 + 8 + 27 + \cdots$
7. $\dfrac{2}{4} + \dfrac{4}{5} + \dfrac{8}{6} + \dfrac{16}{7} + \cdots$
8. $2 + 5 + 10 + \cdots$

Deduce a recursion relation for each series. Use it to predict the next two terms.

9. $1 + 5 + 9 + \cdots$
10. $5 + 15 + 45 + \cdots$
11. $3 + 9 + 81 + \cdots$
12. $5 + 8 + 14 + 26 + \cdots$

**Computer**

13. Write a program or use a spreadsheet to generate the terms of a series, given the general term or a recursion relation. Have the program compute and print each term, the partial sum, and the ratio of that term to the preceding one. Use the program to determine if each of the following series converges or diverges.
   (a) $1 + \dfrac{1}{2} + \dfrac{1}{3} + \dfrac{1}{4} + \cdots + \dfrac{1}{n} + \cdots$

   (b) $\dfrac{3}{2} + \dfrac{9}{8} + \dfrac{27}{24} + \dfrac{81}{64} + \cdots + \dfrac{3^n}{n \cdot 2^n} + \cdots$

   (c) $1 + \dfrac{4}{7} + \dfrac{9}{49} + \dfrac{16}{343} + \cdots + \dfrac{n^2}{7^{n-1}} + \cdots$

## 23-2 ARITHMETIC PROGRESSIONS

### Recursion Formula

We stated earlier that an *arithmetic progression* (or AP) is a sequence of terms in which each term after the first equals the sum of the preceding term and a constant, called the common difference $d$. If $a_n$ is any term of an AP, the recursion formula for an AP is

| | | |
|---|---|---|
| AP: Recursion Formula | $a_n = a_{n-1} + d$ | **231** |

*Each term of an AP after the first equals the sum of the preceding term and the common difference.*

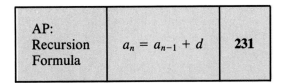

**EXAMPLE 9:** The sequences

(a) 1, 5, 9, 13, . . . $(d = 4)$

(b) 20, 30, 40, 50, . . . $(d = 10)$

(c) 75, 70, 65, 60, . . . $(d = -5)$

are arithmetic progressions. The common difference for each is given in parentheses. We see that the series is increasing when $d$ is positive and decreasing when $d$ is negative.

## General Term

For an AP whose first term is $a$ and whose common difference is $d$, the terms are

$$a,\ a + d,\ a + 2d,\ a + 3d,\ a + 4d,\ \ldots$$

We see that each term is the sum of the first term and a multiple of $d$, where the coefficient of $d$ is one less than the number $n$ of the term. So the $n$th term $a_n$ is

The general term is sometimes called the last term, but this is not an accurate name since the AP continues indefinitely.

| | | |
|---|---|---|
| AP: General Term | $a_n = a + (n - 1)d$ | **232** |

*The* $n$th *term of an AP is found by adding the first term and* $(n - 1)$ *times the common difference.*

**EXAMPLE 10:** Find the twentieth term of an AP that has a first term of 5 and common difference of 4.

**Solution:** By Eq. 232, with $a = 5$, $n = 20$, and $d = 4$,

$$a_{20} = 5 + 19(4) = 81$$

Of course, Eq. 232 can be used to find any of the four quantities ($a$, $n$, $d$, or $a_n$) given the other three.

**EXAMPLE 11:** Write the AP whose eighth term is 19 and whose fifteenth term is 33.

**Solution:** Applying Eq. 232 twice gives

$$33 = a + 14d$$

$$19 = a + 7d$$

We now have two equations in two unknowns, which we solve simultaneously. Subtracting the second from the first gives $14 = 7d$, so $d = 2$. Also, $a = 33 - 14d = 5$, so the general term of our AP is

$$a_n = 5 + (n - 1)2$$

and the AP is then

$$5, 7, 9, \ldots$$

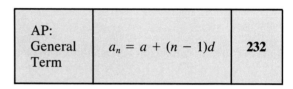

### AP: Sum of $n$ Terms

Let us derive a formula for the sum $s_n$ of the first $n$ terms of an AP. Adding term by term gives

$$s_n = a + (a + d) + (a + 2d) + \cdots + (a_n - d) + a_n \tag{1}$$

or, written in reverse order,

$$s_n = a_n + (a_n - d) + (a_n - 2d) + \cdots + (a + d) + a \tag{2}$$

Adding (1) and (2) term by term gives

$$\begin{aligned} 2s_n &= (a + a_n) + (a + a_n) + (a + a_n) + \cdots \\ &= n(a + a_n) \end{aligned}$$

Dividing both sides by 2 gives the formula

| | | |
|---|---|---|
| AP: Sum of $n$ Terms | $s_n = \frac{n}{2}(a + a_n)$ | **233** |

*The sum of n terms of an AP is half the product of n and the sum of the first and n*th *terms.*

---

**EXAMPLE 12:** Find the sum of 10 terms of the AP

$$2, 5, 8, 11, \ldots$$

**Solution:** From Eq. 232, with $a = 2$, $d = 3$, and $n = 10$,

$$a_{10} = 2 + 9(3) = 29$$

Then, from Eq. 233,

$$s_{10} = \frac{10(2 + 29)}{2} = 155$$

---

We get another form of Eq. 233 by substituting the expression for $a_n$ from Eq. 232,

| | | |
|---|---|---|
| AP: Sum of $n$ Terms | $s_n = \frac{n}{2}[2a + (n - 1)d]$ | **234** |

This form is useful for finding the sum without first computing the $n$th term.

---

**EXAMPLE 13:** We repeat Example 12 without first having to find the tenth term. From Eq. 234,

$$s_{10} = \frac{10[2(2) + 9(3)]}{2} = 155$$

---

Sometimes the sum may be given and we must find one of the other quantities in the AP.

**EXAMPLE 14:** How many terms of the AP 5, 9, 13, . . . give a sum of 275?

**Solution:** We seek $n$ so that $s_n = 275$. From Eq. 234, with $a = 5$ and $d = 4$,

$$275 = \frac{n}{2}[2(5) + (n - 1)4]$$

$$= 5n + 2n(n - 1)$$

Removing parentheses and collecting terms gives the quadratic equation

$$2n^2 + 3n - 275 = 0$$

From the quadratic formula,

$$n = \frac{-3 \pm \sqrt{9 - 4(2)(-275)}}{2(2)} = 11 \quad \text{or} \quad -12.5$$

We discard the negative root and get 11 terms as our answer.

### Arithmetic Means

The first term $a$ and last ($n$th) term $a_n$ of an AP are sometimes called the *extremes,* while the intermediate terms, $a_2, a_3, \ldots, a_{n-1}$, are called *arithmetic means.* We now show, by example, how to insert any number of arithmetic means between two extremes.

**EXAMPLE 15:** Insert five arithmetic means between 3 and −9.

**Solution:** Our AP will have seven terms, with a first term of 3 and a seventh term of −9. From Eq. 232,

$$-9 = 3 + 6d$$

from which $d = -2$. The progression is then

$$3, 1, -1, -3, -5, -7, -9$$

and the five arithmetic means are 1, −1, −3, −5, and −7.

### Average Value

Let us insert a single arithmetic mean between two numbers.

**EXAMPLE 16:** Find a single arithmetic mean $m$ between the extremes $a$ and $b$.

**Solution:** The sequence is $a, m, b$. The common difference $d$ is $m - a = b - m$. Solving for $m$ gives

$$2m = b + a$$

Dividing by 2 gives

$$m = \frac{a + b}{2}$$

which agrees with the common idea of an *average* of two numbers as the sum of those numbers divided by 2.

### Harmonic Progressions

A sequence is called a *harmonic progression* if the reciprocals of its terms form an arithmetic progression.

---

**EXAMPLE 17:** The sequence

$$1, \frac{1}{3}, \frac{1}{5}, \frac{1}{7}, \frac{1}{9}, \frac{1}{11}, \ldots$$

is a harmonic progression because the reciprocals of the terms, 1, 3, 5, 7, 9, 11, . . . form an AP.

---

It is not possible to derive an equation for the $n$th term or for the sum of a harmonic progression. However, we can solve problems involving harmonic progressions by taking the reciprocals of the terms and using the formulas for the AP.

---

**EXAMPLE 18:** Find the tenth term of the harmonic progression

$$2, \frac{2}{3}, \frac{2}{5}, \ldots$$

**Solution:** We write the reciprocals of the terms,

$$\frac{1}{2}, \frac{3}{2}, \frac{5}{2}, \ldots$$

and note that they form an AP with $a = \frac{1}{2}$ and $d = 1$. The tenth term of the AP is then

$$a_n = a + (n - 1)d$$

$$a_{10} = \frac{1}{2} + (10 - 1)(1)$$

$$= \frac{19}{2}$$

The tenth term of the harmonic progression is the reciprocal of $\frac{19}{2}$, or $\frac{2}{19}$.

---

### Harmonic Means

To insert harmonic means between two terms of a harmonic progression, we simply take the reciprocals of the given terms, insert arithmetic means between those terms, and take reciprocals again.

**EXAMPLE 19:** Insert three harmonic means between $\frac{2}{9}$ and 2.

**Solution:** Taking reciprocals, our AP is

$$\frac{9}{2}, \_\_\_\_, \_\_\_\_, \_\_\_\_, \frac{1}{2}$$

In this AP, $a = \frac{9}{2}$, $n = 5$, and $a_5 = \frac{1}{2}$. We can find the common difference $d$ from the equation

$$a_n = a + (n - 1)d$$

$$\frac{1}{2} = \frac{9}{2} + 4d$$

$$-4 = 4d$$

$$d = -1$$

Having the common difference, we can fill in the missing terms of the AP,

$$\frac{9}{2}, \underline{\frac{7}{2}}, \underline{\frac{5}{2}}, \underline{\frac{3}{2}}, \frac{1}{2}$$

Taking reciprocals again, our harmonic progression is

$$\frac{2}{9}, \frac{2}{7}, \frac{2}{5}, \frac{2}{3}, 2$$

## EXERCISE 2—ARITHMETIC PROGRESSIONS

### General Term

1. Find the fifteenth term of an AP with first term 4 and common difference 3.
2. Find the tenth term of an AP with first term 8 and common difference 2.
3. Find the twelfth term of an AP with first term $-1$ and common difference 4.
4. Find the ninth term of an AP with first term $-5$ and common difference $-2$.
5. Find the eleventh term of the AP

$$9, 13, 17, \ldots$$

6. Find the eighth term of the AP

$$-5, -8, -11, \ldots$$

7. Find the ninth term of the AP

$$x, x + 3y, x + 6y, \ldots$$

8. Find the fourteenth term of the AP

$$1, \frac{6}{7}, \frac{5}{7}, \ldots$$

Write the first five terms of each AP.

9. First term is 3 and thirteenth term is 55.
10. First term is 5 and tenth term is 32.
11. Seventh term is 41 and fifteenth term is 89.
12. Fifth term is 7 and twelfth term is 42.

13. Find the first term of an AP whose common difference is 3 and whose seventh term is 11.
14. Find the first term of an AP whose common difference is 6 and whose tenth term is 77.

### Sum of an Arithmetic Progression

15. Find the sum of the first 12 terms of the AP; 3, 6, 9, 12, . . .
16. Find the sum of the first five terms of the AP; 1, 5, 9, 13, . . .
17. Find the sum of the first nine terms of the AP; 5, 10, 15, 20, . . .
18. Find the sum of the first 20 terms of the AP; 1, 3, 5, 7, . . .
19. How many terms of the AP 4, 7, 10, . . . will give a sum of 375?
20. How many terms of the AP 2, 9, 16, . . . will give a sum of 270?

### Arithmetic Means

21. Insert two arithmetic means between 5 and 20.
22. Insert five arithmetic means between 7 and 25.
23. Insert four arithmetic means between $-6$ and $-9$.
24. Insert three arithmetic means between 20 and 56.

### Harmonic Progressions

25. Find the fourth term of the harmonic progression

$$\frac{3}{5}, \frac{3}{8}, \frac{3}{11}, \ldots$$

26. Find the fifth term of the harmonic progression

$$\frac{4}{19}, \frac{4}{15}, \frac{4}{11}, \ldots$$

### Harmonic Means

27. Insert two harmonic means between $\frac{7}{9}$ and $\frac{7}{15}$.
28. Insert three harmonic means between $\frac{6}{21}$ and $\frac{6}{5}$.

### Applications

29. *Loan Repayment:* A person agrees to repay a loan of $10,000 with an annual payment of $1000 plus 8% of the unpaid balance. **(a)** Show that the interest payments alone form the AP: $800, $720, $640, . . . . **(b)** Find the total amount of interest paid.
30. *Simple Interest:* A person deposits $50 in a bank on the first day of each month, at the same time withdrawing all interest earned on the money already in the account. **(a)** If the rate is 1% per month, computed monthly, write an AP whose terms are the amounts withdrawn each month. **(b)** How much interest will have been earned in the 36 months following the first deposit?
31. *Straight-Line Depreciation:* A certain milling machine has an initial value of $150,000 and a scrap value of $10,000 twenty years later. Assuming that the machine depreciates the same amount each year, find its value after 8 years.

**To find the amount of depreciation for each year, divide the total depreciation (initial value − scrap value) by the number of years of depreciation.**

32. *Salary or Price Increase:* A person is hired at a salary of $40,000 and receives a raise of $2500 at the end of each year. Find the total amount earned during 10 years.
33. *Freely Falling Body:* A freely falling body falls $g/2$ feet during the first second, $3g/2$ feet during the next second, $5g/2$ feet during the third second, and so on, where $g \approx 32.2$ ft/s$^2$. Find the total distance the body falls during the first 10 s.
34. Using the information of Problem 33, show that the total distance $s$ fallen in $t$ seconds is $s = \frac{1}{2}gt^2$.

## 23-3 GEOMETRIC PROGRESSIONS

### Recursion Formula

A geometric sequence or *geometric progression* (GP) is one in which each term after the first is formed by multiplying the preceding term by a factor $r$, called the *common ratio*. Thus if $a_n$ is any term of a GP, the recursion relation is

| | | |
|---|---|---|
| GP: Recursion Formula | $a_n = ra_{n-1}$ | **235** |

*Each term of a GP after the first equals the product of the preceding term and the common ratio.*

---

**EXAMPLE 20:** Some geometric progressions, with their common ratios given, are

(a) 2, 4, 8, 16, . . . $(r = 2)$

(b) 27, 9, 3, 1, $\frac{1}{3}$, . . . $(r = \frac{1}{3})$

(c) $-1, 3, -9, 27$, . . . $(r = -3)$

---

### General Term

For a GP whose first term is $a$ and whose common ratio is $r$, the terms are

$$a, ar, ar^2, ar^3, ar^4, \ldots$$

We see that each term after the first is the product of the first term and a power of $r$, where the power of $r$ is one less than the number $n$ of the term. So the $n$th term $a_n$ is

| | | |
|---|---|---|
| GP: General Term | $a_n = ar^{n-1}$ | **236** |

*The* $n$th *term of a GP is found by multiplying the first term by the* $n - 1$ *power of the common ratio.*

---

**EXAMPLE 21:** Find the sixth term of a GP with first term 5 and common ratio 4.

**Solution:** We substitute into Eq. 236 for the general term of a GP, with $a = 5$, $n = 6$, and $r = 4$,

$$a_6 = 5(4^5) = 5(1024) = 5120$$

---

### GP: Sum of $n$ Terms

We find a formula for the sum $s_n$ of the first $n$ terms of a GP (also called the sum of $n$ terms of a geometric series) by adding the terms of the GP:

$$s_n = a + ar + ar^2 + ar^3 + \cdots + ar^{n-2} + ar^{n-1} \tag{1}$$

Multiplying each term in (1) by $r$ gives

$$rs_n = ar + ar^2 + ar^3 + \cdots + ar^{n-1} + ar^n \tag{2}$$

Subtracting (1) from (2) term by term, we get

$$(1 - r)s_n = a - ar^n$$

Dividing both sides by $(1 - r)$ gives

| | | |
|---|---|---|
| GP: Sum of $n$ Terms | $s_n = \dfrac{a(1 - r^n)}{1 - r}$ | **237** |

---

**EXAMPLE 22:** Find the sum of the first six terms of the GP in Example 21.

**Solution:** We substitute into Eq. 237 using $a = 5$, $n = 6$, and $r = 4$.

$$s_n = \frac{a(1 - r^n)}{1 - r} = \frac{5(1 - 4^6)}{1 - 4} = \frac{5(-4095)}{-3} = 6825$$

---

We can get another equation for the sum of a GP in terms of the $n$th term $a_n$. We substitute into Eq. 237, using $ar^n = r(ar^{n-1}) = ra_n$,

| | | |
|---|---|---|
| GP: Sum of $n$ Terms | $s_n = \dfrac{a - ra_n}{1 - r}$ | **238** |

---

**EXAMPLE 23:** Repeat Example 22, given that the sixth term (found in Example 21) is $a_6 = 5120$.

**Solution:** Substitution, with $a = 5$, $r = 4$, and $a_6 = 5120$, yields

$$s_n = \frac{a - ra_n}{1 - r} = \frac{5 - 4(5120)}{1 - 4} = 6825$$

as before.

---

### Geometric Means

As with the AP, the intermediate terms between any two terms are called *means*. A single number inserted between two numbers is called the *geometric mean* between those numbers.

**EXAMPLE 24:** Insert a geometric mean $b$ between two numbers $a$ and $c$.

**Solution:** Our GP is $a$, $b$, $c$. The common ratio $r$ is then

$$r = \frac{b}{a} = \frac{c}{b}$$

from which $b^2 = ac$ or

This is not new. We studied the mean proportional in Chapter 18.

| | | |
|---|---|---|
| Geometric Mean | $b = \pm\sqrt{ac}$ | **59** |

*The geometric mean, or mean proportional, between two numbers is equal to the square root of their product.*

**EXAMPLE 25:** Find the geometric mean between 3 and 48.

**Solution:** Letting $a = 3$ and $c = 48$ gives us

$$b = \pm\sqrt{3(48)} = \pm 12$$

Our GP is then

$$3, \mathbf{12}, 48$$

or

$$3, \mathbf{-12}, 48$$

Note that we get *two* solutions.

To insert *several* geometric means between two numbers, we first find the common ratio.

**EXAMPLE 26:** Insert four geometric means between 2 and $15\frac{3}{16}$.

**Solution:** Here $a = 2$, $a_6 = 15\frac{3}{16}$, and $n = 6$. Then

$$a_6 = ar^5$$

$$15\frac{3}{16} = 2r^5$$

$$r^5 = \frac{243}{32}$$

$$r = \frac{3}{2}$$

Having $r$, we can write the terms of the GP. They are

$$2,\ \mathbf{3},\ \mathbf{4\frac{1}{2}},\ \mathbf{6\frac{3}{4}},\ \mathbf{10\frac{1}{8}},\ 15\frac{3}{16}$$

## EXERCISE 3—GEOMETRIC PROGRESSIONS

1. Find the fifth term of a GP with first term 5 and common ratio 2.
2. Find the fourth term of a GP with first term 7 and common ratio −4.
3. Find the sixth term of a GP with first term −3 and common ratio 5.
4. Find the fifth term of a GP with first term −4 and common ratio −2.
5. Find the sum of the first 10 terms of the GP in Problem 1.
6. Find the sum of the first nine terms of the GP in Problem 2.
7. Find the sum of the first eight terms of the GP in Problem 3.
8. Find the sum of the first five terms of the GP in Problem 4.

### Geometric Means

9. Insert a geometric mean between 5 and 45.
10. Insert a geometric mean between 7 and 112.
11. Insert a geometric mean between −10 and −90.
12. Insert a geometric mean between −21 and −84.
13. Insert two geometric means between 8 and 216.
14. Insert two geometric means between 9 and −243.
15. Insert three geometric means between 5 and 1280.
16. Insert three geometric means between 144 and 9.

### Applications

17. *Exponential Growth:* Using the equation for exponential growth,

| $y = ae^{nt}$ | **197** |
|---|---|

with $a = 1$ and $n = 0.5$, compute values of $y$ for $t = 0, 1, 2, \ldots, 10$. Show that while the values of $t$ form an AP, the values of $y$ form a GP. Find the common ratio.

18. *Exponential Decay:* Repeat Problem 17 with the formula for exponential decay

| $y = ae^{-nt}$ | **199** |
|---|---|

19. *Cooling:* A certain iron casting is at 1800°F and cools so that its temperature at each minute is 10% less than its temperature the preceding minute. Find its temperature after 1 h.
20. *Light Through an Absorbing Medium:* Sunlight passes through a glass filter. Each millimeter of glass absorbs 20% of the light passing through it. What percentage of the original sunlight will remain after passing through 5 mm of the glass?
21. *Radioactive Decay:* A certain radioactive material decays so that after each year the radioactivity is 8% less than at the start of that year. How many years will it take for its radioactivity to be 50% of its original value?
22. *Pendulum:* Each swing of a certain pendulum is 85% as long as the one before. If the first swing is 12 in., find the entire distance traveled in eight swings.
23. *Bouncing Ball:* A ball dropped from a height of 10 ft rebounds to half its height on each bounce. Find the total distance traveled when it hits the ground for the fifth time.
24. *Population Growth:* Each day the size of a certain colony of bacteria is 25% larger than on the preceding day. If the original size of the colony was 10,000 bacteria, find its size after 5 days.

**One of the most famous and controversial references to arithmetic and geometric progressions was made by Thomas Malthus in 1798. He wrote: *"Population, when unchecked, increases in a geometrical ratio, and subsistence for man in an arithmetical ratio."***

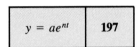

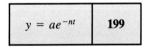

25. A person has two parents, and each parent has two parents, and so on. We can write a GP for the number of ancestors as 2, 4, 8, . . . . Find the total number of ancestors in five generations, starting with the parents' generation.

26. *Musical Scale:* The frequency of the A note above middle C is, by international agreement, equal to 440 Hz. A note one *octave* higher is at twice that frequency, or 880 Hz. The octave is subdivided into 12 *half-tone* intervals, where each half-tone is higher than the one preceding by a factor equal to the twelfth root of 2. Write a GP showing the frequency of each half-tone, from 440 to 880 Hz.

**This is called the *equally tempered scale* and is usually attributed to Johann Sebastian Bach (1685–1750).**

27. *Chemical Reactions:* Increased temperature usually causes chemicals to react faster. If a certain reaction proceeds 15% faster for each 10°C increase in temperature, by what factor is the reaction speed increased when the temperature rises by 50°C?

28. *Mixtures:* A radiator contains 30% antifreeze and 70% water. One-fourth of the mixture is removed and replaced by pure water. If this procedure is repeated three more times, find the percent antifreeze in the final mixture.

29. *Energy Consumption:* If the U.S. energy consumption is 7% higher each year, by what factor will the energy consumption have increased after 10 years?

30. *Atmospheric Pressure:* The pressures measured at 1-mi intervals above sea level form a GP, with each value smaller than the preceding by a factor of 0.819. If the pressure at sea level is 29.92 in. Hg, find the pressure at an altitude of 5 mi.

31. *Compound Interest:* A person deposits $10,000 in a bank giving 6% interest, compounded annually. Find the value of the deposit after 50 years.

32. *Inflation:* The price of a certain house, now $126,000, is expected to increase by 5% each year. Write a GP whose terms are the value of the house at the end of each year, and find the value of the house after 5 years.

33. *Depreciation:* When calculating depreciation by the *declining-balance method,* a taxpayer claims as a deduction a fixed percentage of the book value of an asset each year. The new book value is then the last book value less the amount of the depreciation. Thus for a machine having an initial book value of $100,000 and a depreciation rate of 40%, the first year's depreciation is 40% of $100,000, or $40,000, and the new book value is $100,000 − $40,000 = $60,000. Thus the book values for each year form a GP:

$100,000, $60,000, $36,000, . . .

Find the book value after 5 years.

## 23-4 INFINITE GEOMETRIC PROGRESSIONS

### Limit Notation

**We're just scratching the surface of limit notation here. It is covered more fully in a calculus course.**

We introduce notation here that will be of great importance in the study of calculus. First, we use an arrow (→) to indicate that a quantity *approaches* a given value.

---

**EXAMPLE 27:**

(a) $x \to 5$ means that $x$ gets closer and closer to the value 5.

(b) $n \to \infty$ means that $n$ grows larger and larger, without bound.

(c) $y \to 0$ means that $y$ continuously gets closer and closer to zero.

---

If some function $f(x)$ approaches some value $L$ as $x$ approaches $a$, we could write this as

$$f(x) \to L \quad \text{as} \quad x \to a$$

or, in more compact notation,

| | | |
|---|---|---|
| Limit Notation | $\lim\limits_{x \to a} f(x) = L$ | **326** |

Limit notation lets us write compactly the value to which a function is tending, as in the following example.

---

**EXAMPLE 28:**

(a) $\lim\limits_{b \to 0} b = 0$

(b) $\lim\limits_{b \to 0} (a + b) = a$

(c) $\lim\limits_{b \to 0} \dfrac{a + b}{c + d} = \dfrac{a}{c + d}$

(d) $\lim\limits_{x \to 0} \dfrac{x + 1}{x + 2} = \dfrac{1}{2}$

---

### Decreasing Geometric Progression

If the absolute value of the common ratio $r$ in a GP is less than 1, each term in the progression will be less than those preceding it. Such a progression is called a decreasing GP. As the number of terms $n$ increases without bound, the term $a_n$ will become smaller and approach zero. Using our new notation we may write

$$\lim_{n \to \infty} a_n = 0 \qquad \text{when} \quad |r| < 1$$

Thus, in the equation for the sum of $n$ terms,

$$s_n = \frac{a - ra_n}{1 - r}$$

the expression $ra_n$ will approach zero as $n$ approaches infinity, so the sum $s_\infty$ of infinitely many terms (called the sum to infinity) is

$$s_\infty = \lim_{n \to \infty} \frac{a - ra_n}{1 - r} = \frac{a}{1 - r}$$

Thus, letting $S = s_\infty$, the sum to infinity,

| | | |
|---|---|---|
| GP: Sum to Infinity | $S = \dfrac{a}{1 - r}$ when $\lvert r \rvert < 1$ | **239** |

TABLE 23-3

| | Term | Sum |
|---|---|---|
| 1 | 9.0000 | 9.0000 |
| 2 | 3.0000 | 12.0000 |
| 3 | 1.0000 | 13.0000 |
| 4 | 0.3333 | 13.3333 |
| 5 | 0.1111 | 13.4444 |
| 6 | 0.0370 | 13.4815 |
| 7 | 0.0123 | 13.4938 |
| 8 | 0.0041 | 13.4979 |
| 9 | 9.0014 | 13.4993 |
| 10 | 0.0005 | 13.4998 |
| 11 | 0.0002 | 13.4999 |
| 12 | 0.0001 | 13.5000 |
| 13 | 0.0000 | 13.5000 |
| 14 | 0.0000 | 13.5000 |
| 15 | 0.0000 | 13.5000 |
| 16 | 0.0000 | 13.5000 |
| 17 | 0.0000 | 13.5000 |
| 18 | 0.0000 | 13.5000 |

**EXAMPLE 29:** Find the sum of infinitely many terms of the GP

$$9, 3, 1, \frac{1}{3}, \ldots$$

**Solution:** Here $a = 9$ and $r = \frac{1}{3}$, so

$$S = \frac{a}{1 - r} = \frac{9}{1 - 1/3} = \frac{9}{2/3} = 13\frac{1}{2}$$

To convince ourselves that this is correct, we show the calculated values of the first 18 terms of this GP in Table 23-3, along with the sums of all the terms up to and including that term. Notice that the terms get smaller and smaller, and that the sum of the terms appears to approach a limit of 13.5, as predicted by our work in Example 29.

## Repeating Decimals

A *finite decimal* (or terminating decimal) is one that has a finite number of digits to the right of the decimal point, followed by zeros. An *infinite decimal* (or nonterminating decimal) has an infinite string of digits to the right of the decimal point. A *repeating decimal* (or periodic decimal) is one that has a block of one or more digits that repeats indefinitely. It can be proved that when one integer is divided by another, we get either a finite decimal or a repeating decimal. Irrational numbers, on the other hand, yield *nonrepeating decimals*.

**EXAMPLE 30:**

(a) $\frac{3}{4} = 0.75000000000\ldots$ (finite decimal)

(b) $\frac{3}{7} = 0.428571\ 428571\ 428571\ldots$ (infinite repeating decimal)

(c) $\sqrt{2} = 1.414213562\ldots$ (infinite nonrepeating decimal)

We can write a repeating decimal as a rational fraction by writing the repeating part as an infinite geometric series, as shown in the following example.

**EXAMPLE 31:** Write the number 4.85151 . . . as a rational fraction.

**Solution:** We rewrite the given number as

$$4.8 + 0.051 + 0.00051 + \cdots$$

The terms of this series, except for the first, form an infinite GP where $a = 0.051$ and $r = 0.01$. The sum $S$ to infinity is then

$$S = \frac{a}{1 - r} = \frac{0.051}{1 - 0.01} = \frac{0.051}{0.99} = \frac{51}{990} = \frac{17}{330}$$

Our given number is then equal to $4.8 + S$, so

$$4.85151 = 4 + \frac{8}{10} + \frac{17}{330} = 4\frac{281}{330}$$

## EXERCISE 4—INFINITE GEOMETRIC PROGRESSIONS

Evaluate each limit.

1. $\lim_{b\to 0} (b - c + 5)$
2. $\lim_{b\to 0} (a + b^2)$
3. $\lim_{b\to 0} \dfrac{3 + b}{c + 4}$
4. $\lim_{b\to 0} \dfrac{a + b + c}{b + c - 5}$

Find the sum of the infinitely many terms of each GP.

5. 144, 72, 36, 18, . . .
6. 8, 4, 1, $\frac{1}{4}$, . . .
7. 10, 2, 0.4, 0.08, . . .
8. 1, $\frac{1}{4}$, $\frac{1}{16}$, . . .

Write each decimal number as a rational fraction.

9. $0.57\overline{57}\ldots$
10. $0.69\overline{69}\ldots$
11. $7.68\overline{181}\ldots$
12. 5.86111 . . .
13. Each swing of a certain pendulum is 78% as long as the one before. If the first swing is 10 in., find the entire distance traveled by the pendulum before it comes to rest.
14. A ball dropped from a height of 20 ft rebounds to three-fourths of its height on each bounce. Find the total distance traveled by the ball before it comes to rest.
15. *Zeno's Paradox:* "A runner can never reach a finish line 1 mile away because first he would have to run half a mile, and then must run half of the remaining distance, or $\frac{1}{4}$ mile, and then half of that, and so on. Since there are an infinite number of distances that must be run, it will take an infinite length of time, and so the runner will never reach the finish line." Show that the distances to be run form an infinite series,

$$\frac{1}{2}, \frac{1}{4}, \frac{1}{8}, \ldots$$

Then disprove the paradox by actually finding the sum of that series and showing that the sum is *not* infinite.

**Named for Zeno of Elea (c. 490–c. 435 B.C.), a Greek philosopher and mathemetician. He wrote several famous paradoxes, including this one.**

## 23-5 THE BINOMIAL THEOREM

### Powers of a Binomial

Recall that a *binomial,* such as $(a + b)$, is a polynomial with two terms. By actual multiplication we can show that

$$(a + b)^1 = a + b$$

$$(a + b)^2 = a^2 + 2ab + b^2$$

$$(a + b)^3 = a^3 + 3a^2b + 3ab^2 + b^3$$

$$(a + b)^4 = a^4 + 4a^3b + 6a^2b^2 + 4ab^3 + b^4$$

We now want a formula for $(a + b)^n$ with which to expand a binomial without actually carrying out the multiplication. In the expansion of $(a + b)^n$, where $n$ is a positive integer, we note the following patterns:

1. There are $n + 1$ terms.
2. The power of $a$ is $n$ in the first term, decreases by 1 in each later term, and reaches 0 in the last term.

3. The power of $b$ is 0 in the first term, increases by 1 in each later term, and reaches $n$ in the last term.
4. Each term has a total degree of $n$. (That is, the sum of the degrees of the variables is $n$.)
5. The first coefficient is 1. Further, the product of the coefficient of any term and its power of $a$, divided by the number of the term, gives the coefficient of the next term. (This property gives a *recursion formula* for the coefficients of the binomial expansion.)

Expressed as a formula, we have

$$(a + b)^n = a^n + na^{n-1}b + \frac{n(n-1)}{2}a^{n-2}b^2 + \frac{n(n-1)(n-2)}{2(3)}a^{n-3}b^3 + \cdots + b^n$$

---

**EXAMPLE 32:** Use the binomial theorem to expand $(a + b)^5$.

**Solution:** From the binomial theorem,

$$\begin{aligned}(a + b)^5 &= a^5 + 5a^4b + \frac{5(4)}{2}a^3b^2 + \frac{5(4)(3)}{2(3)}a^2b^3 + \frac{5(4)(3)(2)}{2(3)(4)}ab^4 + b^5\\ &= a^5 + 5a^4b + 10a^3b^2 + 10a^2b^3 + 5ab^4 + b^5\end{aligned}$$

which can be verified by actual multiplication.

---

### Factorial Notation

We now introduce *factorial notation*. For a positive integer $n$, factorial $n$, written $n!$, is the product of all the positive integers less than or equal to $n$.

Many calculators have a key for evaluating factorials.

$$n! = 1 \cdot 2 \cdot 3 \cdots n$$

Of course, it does not matter in which order we multiply the numbers. We may also write

$$n! = n(n-1) \cdots 3 \cdot 2 \cdot 1$$

---

**EXAMPLE 33:**

(a) $3! = 1 \cdot 2 \cdot 3 = 6$

(b) $5! = 1 \cdot 2 \cdot 3 \cdot 4 \cdot 5 = 120$

(c) $\dfrac{6!\,3!}{7!} = \dfrac{6!\,(2)(3)}{7(6!)} = \dfrac{6}{7}$

(d) $0! = 1$ by definition

---

### Binomial Theorem Written with Factorial Notation

The use of factorial notation allows us to write the binomial theorem more compactly. Thus

| Binomial Theorem | $(a + b)^n = a^n + na^{n-1}b + \dfrac{n(n-1)}{2!}a^{n-2}b^2 + \dfrac{n(n-1)(n-2)}{3!}a^{n-3}b^3 + \cdots + b^n$ | **240** |
|---|---|---|

**EXAMPLE 34:** Using the binomial formula, expand $(a + b)^6$.

**Solution:** There will be $(n + 1)$ or seven terms. Let us first find the seven coefficients. The first coefficient is always 1, and the second coefficient is always $n$, or 6. We calculate the remaining coefficients in the following table.

| *Term* | *Coefficient* |
|---|---|
| 1 | 1 |
| 2 | $n = 6$ |
| 3 | $\dfrac{n(n-1)}{2!} = \dfrac{6(5)}{1(2)} = 15$ |
| 4 | $\dfrac{n(n-1)(n-2)}{3!} = \dfrac{6(5)(4)}{1(2)(3)} = 20$ |
| 5 | $\dfrac{n(n-1)(n-2)(n-3)}{4!} = \dfrac{6(5)(4)(3)}{1(2)(3)(4)} = 15$ |
| 6 | $\dfrac{n(n-1)(n-2)(n-3)(n-4)}{5!} = \dfrac{6(5)(4)(3)(2)}{1(2)(3)(4)(5)} = 6$ |
| 7 | $\dfrac{n(n-1)(n-2)(n-3)(n-4)(n-5)}{6!} = \dfrac{6(5)(4)(3)(2)(1)}{1(2)(3)(4)(5)(6)} = 1$ |

The seven terms thus have the coefficients

$$1 \quad 6 \quad 15 \quad 20 \quad 15 \quad 6 \quad 1$$

and our expansion is

$$(a + b)^6 = a^6 + 6a^5b + 15a^4b^2 + 20a^3b^3 + 15a^2b^4 + 6ab^5 + b^6$$

## Pascal's Triangle

We now have expansions for $(a + b)^n$, for values of $n$ from 1 to 6, to which we add $(a + b)^0 = 1$. Let us now write the coefficients only of the terms of each expansion.

| Exponent $n$ | |
|---|---|
| 0 | 1 |
| 1 | 1 1 |
| 2 | 1 2 1 |
| 3 | 1 3 3 1 |
| 4 | 1 4 6 4 1 |
| 5 | 1 5 10 10 5 1 |
| 6 | 1 6 15 20 15 6 1 |

This array is called *Pascal's triangle*. We may use it to predict the coefficients of expansions with powers higher than 6 by noting that each number in the triangle is equal to the sum of the two numbers above it (thus the 15 in the lowest row is the sum of the 5 and the 10 above it). When expanding a binomial, we may take the coefficients directly from Pascal's triangle, as shown in the following example.

**Named for Blaise Pascal (1623–1662), the French geometer, probabilist, combinatorist, physicist, and philosopher.**

**EXAMPLE 35:** Expand $(1 - x)^6$, obtaining the binomial coefficients from Pascal's triangle.

**Solution:** From Pascal's triangle, for $n = 6$, we read the coefficients

$$1 \quad 6 \quad 15 \quad 20 \quad 15 \quad 6 \quad 1$$

We now substitute into the binomial formula with $n = 6$, $a = 1$, and $b = -x$. When one of the terms of the binomial is negative, as here, we must be careful to include the minus sign whenever that term appears in the expansion. Thus

$$\begin{aligned}(1 - x)^6 &= [1 + (-x)]^6 \\ &= 1^6 + 6(1^5)(-x) + 15(1^4)(-x)^2 + 20(1^3)(-x)^3 \\ &\quad + 15(1^2)(-x)^4 + 6(1)(-x)^5 + (-x)^6 \\ &= 1 - 6x + 15x^2 - 20x^3 + 15x^4 - 6x^5 + x^6\end{aligned}$$

If either term in the binomial is itself a power, product, or a fraction, it is a good idea to enclose that entire term in parentheses before substituting.

**EXAMPLE 36:** Expand $(3/x^2 + 2y^3)^4$.

**Solution:** The binomial coefficients are 1 4 6 4 1. We substitute $3/x^2$ for $a$ and $2y^3$ for $b$.

$$\begin{aligned}\left(\frac{3}{x^2} + 2y^3\right)^4 &= \left[\left(\frac{3}{x^2}\right) + (2y^3)\right]^4 \\ &= \left(\frac{3}{x^2}\right)^4 + 4\left(\frac{3}{x^2}\right)^3(2y^3) + 6\left(\frac{3}{x^2}\right)^2(2y^3)^2 + 4\left(\frac{3}{x^2}\right)(2y^3)^3 + (2y^3)^4 \\ &= \frac{81}{x^8} + \frac{216y^3}{x^6} + \frac{216y^6}{x^4} + \frac{96y^9}{x^2} + 16y^{12}\end{aligned}$$

Sometimes we may not need the entire expansion but only the first several terms.

**EXAMPLE 37:** Find the first four terms of $(x^2 - 2y^3)^{11}$.

**Solution:** From the binomial formula, with $n = 11$, $a = x^2$, and $b = (-2y^3)$,

$$\begin{aligned}(x^2 - 2y^3)^{11} &= [(x^2) + (-2y^3)]^{11} \\ &= (x^2)^{11} + 11(x^2)^{10}(-2y^3) + \frac{11(10)}{2}(x^2)^9(-2y^3)^2 \\ &\quad + \frac{11(10)(9)}{2(3)}(x^2)^8(-2y^3)^3 + \cdots \\ &= x^{22} - 22x^{20}y^3 + 220x^{18}y^6 - 1320x^{16}y^9 + \cdots\end{aligned}$$

A *trinomial* may be expanded by the binomial formula by grouping the terms, as in the following example.

**EXAMPLE 38:** Expand $(1 + 2x - x^2)^3$ by the binomial formula.

**Solution:** Let $z = 2x - x^2$. Then

$$(1 + 2x - x^2)^3 = (1 + z)^3 = 1 + 3z + 3z^2 + z^3$$

Substituting back,

$$(1 + 2x - x^2)^3 = 1 + 3(2x - x^2) + 3(2x - x^2)^2 + (2x - x^2)^3$$

$$= 1 + 6x + 9x^2 - 4x^3 - 9x^4 + 6x^5 - x^6$$

## General Term

We can write the general or $r$th term of a binomial expansion $(a + b)^n$ by noting the following pattern.

1. The power of $b$ is one less than $r$. Thus a fifth term would contain $b^4$, and the $r$th term would contain $b^{r-1}$.
2. The power of $a$ is $n$ minus the power of $b$. Thus a fifth term would contain $a^{n-5+1}$, and the $r$th term would contain $a^{n-r+1}$.
3. The coefficient of the $r$th term is

$$\frac{n!}{(r-1)!\,(n-r+1)!}$$

Expressed as a formula,

| | | |
|---|---|---|
| General Term | In the expansion for $(a + b)^n$, $r\text{th term} = \dfrac{n!}{(r-1)!\,(n-r+1)!}\,a^{n-r+1}b^{r-1}$ | **241** |

**EXAMPLE 39:** Find the eighth term of $(3a + b^5)^{11}$.

**Solution:** Here, $n = 11$ and $r = 8$. So $n - r + 1 = 4$. Substitution yields

$$\frac{n!}{(r-1)!\,(n-r+1)!} = \frac{11!}{7!\,4!} = 330$$

Then

$$\text{eighth term} = 330(3a)^{11-8+1}(b^5)^{8-1}$$

$$= 330(81a^4)(b^{35}) = 26{,}730a^4b^{35}$$

## Proof of the Binomial Theorem

We have shown that the binomial theorem is true for $n = 1, 2, 3, 4$, or 5. But what about some other exponent, say $n = 50$? To show that the theorem works for any positive integral exponent, we give the following proof.

We first show that if the binomial theorem is true for some exponent $n = k$, it is also true for $n = k + 1$. We assume that

$$(a + b)^k = a^k + ka^{k-1}b + \cdots + Pa^{k-r}b^r + Qa^{k-r-1}b^{r+1} + Ra^{k-r-2}b^{r+2} + \cdots + b^k \tag{1}$$

where we show three intermediate terms with coefficients $P$, $Q$, and $R$. Multiplying both sides of (1) by $(a + b)$ gives

**We have not written the $b^r$ term or the $b^{r+3}$ term because we do not have expressions for their coefficients, and they are not needed for the proof.**

$$(a + b)^{k+1} = a^{k+1} + ka^kb + \cdots + Pa^{k-r+1}b^r + Qa^{k-r}b^{r+1} + Ra^{k-r-1}b^{r+2} + \cdots + ab^k + \cdots + a^kb + \cdots + Pa^{k-r}b^{r+1} + Qa^{k-r-1}b^{r+2} + Ra^{k-r-2}b^{r+3} + \cdots + b^{k+1}$$

$$(a + b)^{k+1} = a^{k+1} + (k + 1)a^kb + \cdots + (P + Q)a^{k-r}b^{r+1} + (Q + R)a^{k-r-1}b^{r+2} + \cdots + ab^k + b^{k+1} \tag{2}$$

after combining like terms. Does (2) obey the binomial theorem? First, the power of $b$ is 0 in the first term and increases by 1 in each later term, reaching $k + 1$ in the last term, so property 3 is met. Thus there are $k + 1$ terms containing $b$, plus the first term that does not, making a total of $k + 2$ terms, so property 1 is met. Next, the power of $a$ is $k + 1$ in the first term and decreases by 1 in each succeeding term until it equals 0 in the last term, making the total degree of each term equal to $k + 1$, so properties 2 and 4 are met. Further, the coefficient of the first term is 1 and the coefficient of the second term is $k + 1$, which agrees with property 5. Finally, in order to satisfy property 5 for the intermediate terms, the identity

$$(P + Q)\frac{n - r}{r + 2} = Q + R$$

must be true. But since

$$Q = \frac{P(n - r)}{r + 1} \quad \text{and} \quad R = \frac{Q(n - r - 1)}{r + 2}$$

we must show that

$$\left[P + \frac{P(n - r)}{r + 1}\right]\frac{n - r}{r + 2} = \frac{P(n - r)}{r + 1} + \frac{P(n - r)(n - r - 1)}{(r + 1)(r + 2)}$$

Working each side separately, we combine each over a common denominator.

$$\frac{P(n - r)(r + 1)}{(r + 1)(r + 2)} + \frac{P(n - r)(n - r)}{(r + 1)(r + 2)} = \frac{P(n - r)(r + 2)}{(r + 1)(r + 2)} + \frac{P(n - r)(n - r - 1)}{(r + 1)(r + 2)}$$

Since the denominators are the same on both sides, we simply have to show that the numerators are identical. Factoring gives us

$$P(n - r)(r + 1 + n - r) = P(n - r)(r + 2 + n - r - 1)$$

$$P(n - r)(n + 1) = P(n - r)(n + 1)$$

This proves that (2) satisfies property 5 as well, and completes the proof that if the binomial theorem is true for some exponent $n = k$, it is also true for the exponent $n = k + 1$. But since we have shown, by direct multiplication, that the binomial theorem *is true* for $n = 5$, it must be true for $n = 6$. And since it is true for $n = 6$, it must be true for $n = 7$, and so on, for all the positive integers.

**This type of proof is called mathematical induction.**

### Fractional and Negative Exponent

We have shown that the binomial formula is valid when the exponent $n$ is a positive integer. We shall not prove it here, but the binomial formula also holds when the exponent is negative or fractional. However, we now obtain an infinite series, called a binomial series, with no last term. Further, the binomial series is equal to $(a + b)^n$ only if the series converges.

---

**EXAMPLE 40:** Write the first four terms of the infinite binomial expansion for $\sqrt{5 + x}$.

**Solution:** We substitute into the binomial formula in the usual way,

$$(5 + x)^{1/2} = 5^{1/2} + \left(\frac{1}{2}\right)5^{-1/2}x + \frac{(1/2)(-1/2)}{2}5^{-3/2}x^2 + \frac{(1/2)(-1/2)(-3/2)}{6}5^{-5/2}x^3 + \cdots$$

$$= \sqrt{5} + \frac{x}{2\sqrt{5}} - \frac{x^2}{8(5)^{3/2}} + \frac{x^3}{16(5)^{5/2}} - \cdots$$

Switching to decimal form and working to five decimal places, we get

$$(5 + x)^{1/2} \approx 2.23607 + 0.22361x - 0.01118x^2 + 0.00112x^2 - \cdots$$

---

We now ask if this series is valid for any $x$. Let's try a few values. For $x = 0$,

$$(5 + 0)^{1/2} \approx 2.23607$$

which checks within the number of digits retained. For $x = 1$,

$$(5 + 1)^{1/2} = \sqrt{6} \approx 2.23607 + 0.22361 - 0.01118 + 0.00112 - \cdots$$

$$\approx 2.44962$$

which compares with a value for $\sqrt{6}$ of 2.44949. Not bad, considering that we are using only four terms of an infinite series.

To determine the effect of larger values of $x$, we have used a computer to generate eight terms of the binomial series. Into it we have substituted various $x$ values, from 0 to 20, and compared the value obtained by series with that from a calculator.

| $x$ | $\sqrt{5 + x}$ *By Series* | $\sqrt{5 + x}$ *By Calculator* | *Error* |
|---|---|---|---|
| 0 | 2.236068 | 2.236068 | 0 |
| 2 | 2.645766 | 2.645751 | 0.000015 |
| 4 | 3.002957 | 3 | 0.002957 |
| 6 | 3.379908 | 3.316625 | 0.063283 |
| 8 | 4.148339 | 3.605551 | 0.543788 |
| 11 | 9.723964 | 4 | 5.723964 |
| 20 | 453.9 | 5 | 448.9 |

Notice that the accuracy gets worse as $x$ gets larger, with useless values being obtained when $x$ is 6 or greater. This illustrates the fact that

> A binomial series is equal to $(a + b)^n$ only if the series converges, and that occurs when $|a| > |b|$.

Our series in Example 40 thus converges when $x < 5$.

The binomial series, then, is given by the following formula.

| | | |
|---|---|---|
| Binomial Series | $(a + b)^n = a^n + na^{n-1}b + \frac{n(n-1)}{2!}a^{n-2}b^2 + \frac{n(n-1)(n-2)}{3!}a^{n-3}b^3 + \cdots$ where $|a| > |b|$ | 242 |

The binomial series is often written with $a = 1$ and $b = x$, so we give that form here also.

| | | |
|---|---|---|
| Binomial Series | $(1 + x)^n = 1 + nx + \frac{n(n-1)}{2!}x^2 + \frac{n(n-1)(n-2)}{3!}x^3 + \cdots$ where $|x| < 1$ | 243 |

---

**EXAMPLE 41:** Expand to four terms,

$$\frac{1}{\sqrt[3]{1/a - 3b^{1/3}}}$$

**Solution:** We rewrite the given expression without radicals or fractions, and then expand.

$$\frac{1}{\sqrt[3]{1/a - 3b^{1/3}}} = [(a^{-1}) + (-3b^{1/3})]^{-1/3}$$

$$= (a^{-1})^{-1/3} - \frac{1}{3}(a^{-1})^{-4/3}(-3b^{1/3}) + \frac{2}{9}(a^{-1})^{-7/3}(-3b^{1/3})^2$$

$$- \frac{14}{81}(a^{-1})^{-10/3}(-3b^{1/3})^3 + \cdots$$

$$= a^{1/3} + a^{4/3}b^{1/3} + 2a^{7/3}b^{2/3} + \frac{14}{3}a^{10/3}b + \cdots$$

---

## EXERCISE 5—THE BINOMIAL THEOREM

### Factorial Notation

Evaluate each factorial.

1. $6!$
2. $8!$
3. $\frac{7!}{5!}$
4. $\frac{11!}{9!\,2!}$
5. $\frac{7!}{3!\,4!}$
6. $\frac{8!}{3!\,5!}$

### Binomials Raised to an Integral Power

Expand and simplify each binomial. Obtain the binomial coefficients by formula or from Pascal's triangle, as directed by your instructor.

7. $(x + y)^7$
8. $(4 + 3b)^4$
9. $(3a - 2b)^4$
10. $(x^3 + y)^7$
11. $(x^{1/2} + y^{2/3})^5$
12. $\left(\frac{1}{x} + \frac{1}{y^2}\right)^3$
13. $\left(\frac{a}{b} - \frac{b}{a}\right)^6$
14. $\left(\frac{1}{\sqrt{x}} + y^2\right)^4$
15. $(2a^2 + \sqrt{b})^5$
16. $(\sqrt{x} - c^{2/3})^4$

Write the first four terms of each binomial expansion and simplify.

17. $(x^2 + y^3)^8$
18. $(x^2 - 2y^3)^{11}$
19. $(a - b^4)^9$
20. $(a^3 + 2b)^{12}$

### Trinomials

Expand each trinomial and simplify.

21. $(2a^2 + a + 3)^3$
22. $(4x^2 - 3xy - y^2)^3$
23. $(2a^2 + a - 4)^4$

### General Term

Write the requested term of each binomial expansion, and simplify.

24. Seventh term of $(a^2 - 2b^3)^{12}$
25. Eleventh term of $(2 - x)^{16}$
26. Fourth term of $(2a - 3b)^7$
27. Eighth term of $(x + a)^{11}$
28. Fifth term of $(x - 2\sqrt{y})^{25}$
29. Ninth term of $(x^2 + 1)^{15}$

### Binomials Raised to a Fractional or Negative Power

Write the first four terms of each infinite binomial series.

30. $(1 - a)^{2/3}$
31. $\sqrt{1 + a}$
32. $(1 + 5a)^{-5}$
33. $(1 + a)^{-3}$
34. $1/\sqrt[6]{1 - a}$

## CHAPTER 23 REVIEW PROBLEMS

1. Find the sum of seven terms of the AP $-4, -1, -2, \ldots$.
2. Find the tenth term of $(1 - y)^{14}$.
3. Insert four arithmetic means between 3 and 18.
4. How many terms of the AP $-5, -2, 1, \ldots$ will give a sum of 63?
5. Insert four harmonic means between 2 and 12.
6. Find the twelfth term of an AP with first term 7 and common difference 5.
7. Find the sum of the first 10 terms of the AP $1, 2\frac{2}{3}, 4\frac{1}{3}, \ldots$.

*Expand each binomial.*

8. $(2x^2 - \sqrt{x}/2)^5$
9. $(a - 2)^5$
10. $(x - y^3)^7$
11. Find the fifth term of a GP with first term 3 and common ratio 2.
12. Find the fourth term of a GP with first term 8 and common ratio $-4$.

*Evaluate each limit.*

13. $\lim_{b\to 0}(2b - x + 9)$
14. $\lim_{b\to 0}(a^2 + b^2)$

*Find the sum of infinitely many terms of each GP.*

15. $4, 2, 1, \ldots$
16. $\frac{1}{4}, -\frac{1}{16}, \frac{1}{64}, \ldots$
17. $1, -\frac{2}{5}, \frac{4}{25}, \ldots$

*Write each decimal number as a rational fraction.*

18. $0.86\overline{86}\ldots$
19. $0.5\overline{444}\ldots$
20. Find the seventh term of the harmonic progression $3, 3\frac{3}{7}, 4, \ldots$.
21. Find the sixth term of a GP with first term 5 and common ratio 3.
22. Find the sum of the first seven terms of the GP in Problem 21.
23. Find the fifth term of a GP with first term $-4$ and common ratio 4.
24. Find the sum of the first seven terms of the GP in Problem 23.
25. Insert two geometric means between 8 and 125.
26. Insert three geometric means between 14 and 224.

*Evaluate each factorial.*

27. $\dfrac{4!\,5!}{2!}$ 28. $\dfrac{8!}{4!\,4!}$ 29. $\dfrac{7!\,3!}{5!}$ 30. $\dfrac{7!}{2!\,5!}$

*Find the first four terms of each binomial expansion.*

31. $\left(\dfrac{2x}{y^2} - y\sqrt{x}\right)^7$
32. $(1 + a)^{-2}$
33. $\dfrac{1}{\sqrt{1 + a}}$
34. Expand using the binomial theorem, $(1 - 3a + 2a^2)^4$.
35. Find the eighth term of $(3a - b)^{11}$.
36. Find the tenth term of an AP with first term 12 if the sum of the first 10 terms is 10.

*Writing*

37. State in your own words the difference between an arithmetic progression and a geometric progression. Give a real-world example of each.

# 24

# INTRODUCTION TO STATISTICS AND PROBABILITY

## OBJECTIVES

**When you have completed this chapter, you should be able to:**

- Identify data as continuous, discrete, or categorical.
- Represent data as X-Y graph, bar graph, or pie chart.
- Organize data into frequency distributions, frequency histograms, frequency polygons, and cumulative frequency distributions.
- Calculate the mean, the median, and the mode.
- Find the range, the quartiles, the deciles, the percentiles, the variance, and the standard deviation.
- Determine the probability of a single event or of several events.
- Calculate probabilities for a normal distribution.
- Estimate population means or population standard deviations within a given confidence interval.

Why study statistics? There are several reasons. First, for your work you may have to interpret statistical data or even need to collect such data and make inferences from it. You may, perhaps, be asked to determine if replacing a certain machine on a production line actually reduced the number of defective parts made. Further, a knowledge of statistics may help you to evaluate the claims made in news reports, polls, and advertisements. A toothpaste manufacturer may claim that their product "significantly reduces" tooth decay. A pollster may claim that 52% of the electorate is going to vote for candidate Jones. A newspaper may show a chart that pictures oil consumption going down. It is easy to use statistics to distort the truth, either on purpose or unintentionally. Some knowledge of the subject may help you to separate fact from distortion.

In this chapter we introduce many new terms. We then show how to display statistical data in graphical and numerical form. How reliable are those statistical data? To answer that question we include a brief introduction to *probability,* which enables us to describe the "confidence" we can place on those statistics.

## 24-1 DEFINITIONS AND TERMINOLOGY

### Populations and Samples

In statistics, an entire group of people or things is called a *population* or *universe*. A population can be *infinite* or *finite*. A small part of the entire group chosen for study is called a *sample*.

---

**EXAMPLE 1:** In a study of the heights of students at Tech College, the entire student body is our *population*. This is a *finite* population. Instead of studying every student in the population, we may choose to work with a *sample* of only 100 students.

---

**EXAMPLE 2:** Suppose that we want to determine what percentage of tosses of a coin will be heads. Then all possible tosses of the coin is an example of an *infinite* population. Theoretically, there is no limit to the number of tosses that can be made.

---

### Parameters and Statistics

We often use just a few numbers to describe an entire population. Such numbers are called *parameters*. The branch of statistics in which we use parameters to describe a population is referred to as *descriptive statistics*.

---

**EXAMPLE 3:** For the population of all students at Tech College, a computer read the height of each student and computed the average height of all students. It found this *parameter,* average height, to be 68.2 in.

---

Similarly, we use just a few numbers to describe a sample. Such numbers are called *statistics*. When we then use these sample statistics to *infer* things about the entire population, we are engaging in what is called *inductive statistics*.

**EXAMPLE 4:** From the population of all students in Tech College, a sample of 100 students was selected. The average of their heights was found to be 67.9 in. This is a sample *statistic*. From it we infer that for the entire population, the average height is 67.9 in. plus or minus some *standard error*. We show how to compute standard error in Sec. 24-5.

---

Thus parameters describe populations, while statistics describe samples. The first letters of the words help us to keep track of which goes with which.

**P**opulation **P**arameter

versus

**S**ample **S**tatistic

## Variables

In statistics we distinguish between data that are *continuous* (obtained by measurement), *discrete* (obtained by counting), or *categorical* (belonging to one of several categories).

---

**EXAMPLE 5:** A survey at Tech College asked each student for (a) height, *continuous data;* (b) number of courses taken, *discrete data;* and (c) whether they were in favor (yes or no) of eliminating final exams, *categorical data.*

---

A *variable,* as in algebra, is a quantity that can take on different values within a problem; *constants* do not change value. We call variables *continuous, discrete,* or *categorical,* depending on the type of data they represent.

---

**EXAMPLE 6:** If, in the survey of Example 5, we let

$$H = \text{student's height}$$

$$N = \text{number of courses taken}$$

$$F = \text{opinion on eliminating finals}$$

then $H$ is a continuous variable, $N$ is a discrete variable, and $F$ is a categorical variable.

---

## Ways to Display Data

Statistical data are usually presented as a table of values or displayed in a graph or chart. Common types of graphs are the $x$–$y$ graph, bar graph, or pie chart. Here we give examples of the display of discrete and categorical data. We cover display of continuous data in Sec. 24-2.

**TABLE 24-1**

| *Month* | *Pairs Sold* |
|---|---|
| Jan. | 13,946 |
| Feb. | 7,364 |
| Mar. | 5,342 |
| Apr. | 3,627 |
| May | 1,823 |
| June | 847 |
| July | 354 |
| Aug. | 425 |
| Sept. | 3,745 |
| Oct. | 5,827 |
| Nov. | 11,837 |
| Dec. | 18,475 |
| Total | 73,612 |

**EXAMPLE 7:** Table 24-1 shows the number of skis made per month by the Ace Ski Company. We show these discrete data as an $x$–$y$ graph, a bar chart, and a pie chart (Fig. 24-1).

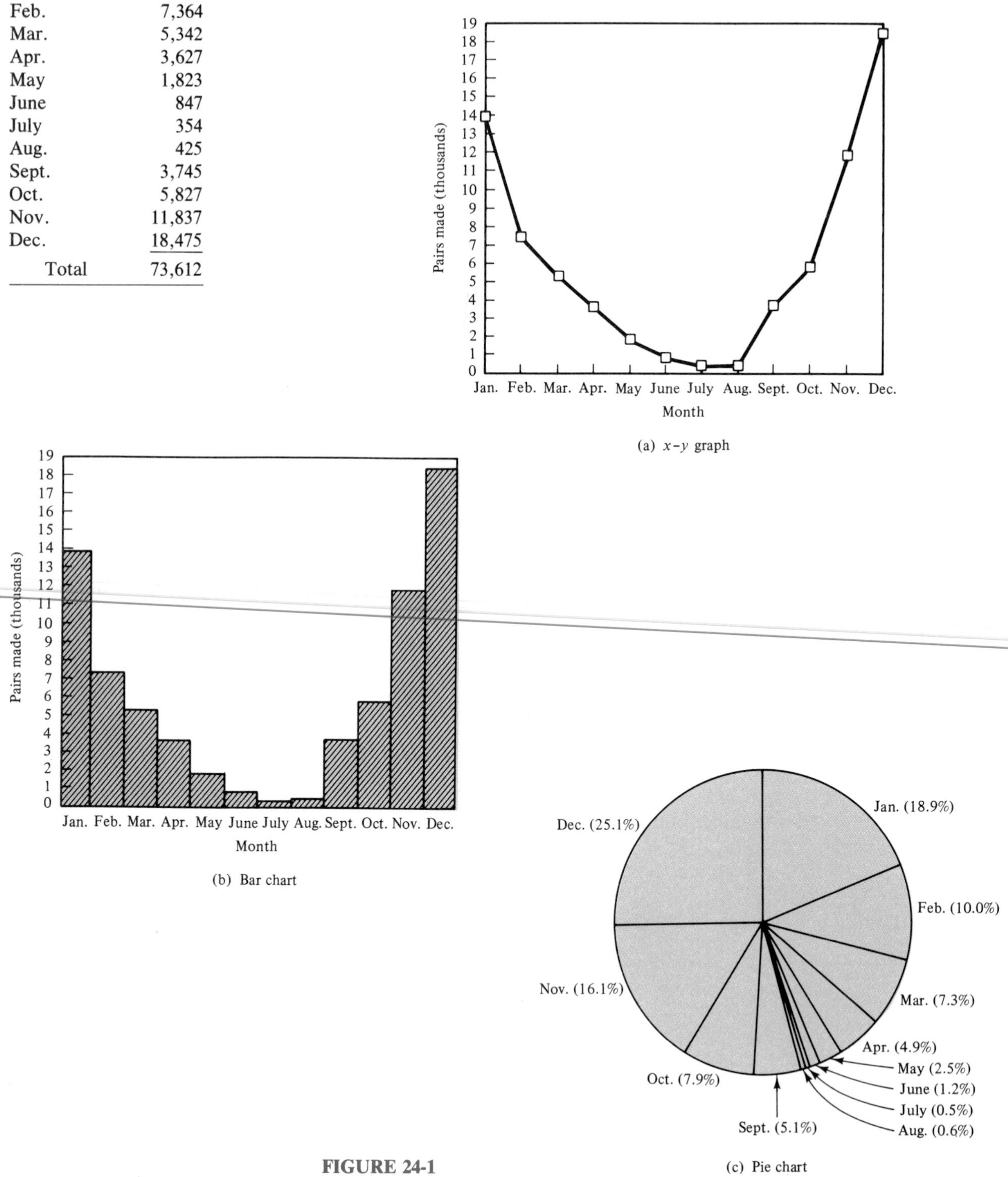


**FIGURE 24-1**

**EXAMPLE 8:** Table 24-2 shows the number of skis of various types made in a certain year by the Ace Ski Company. We show these discrete data as an $x$–$y$ graph, a bar chart, and a pie chart (Fig. 24-2).

**TABLE 24-2**

| *Type* | *Pairs Sold* |
|---|---|
| Downhill | 35,725 |
| Racing | 7,735 |
| Jumping | 2,889 |
| Cross-country | 27,263 |
| Total | 73,612 |

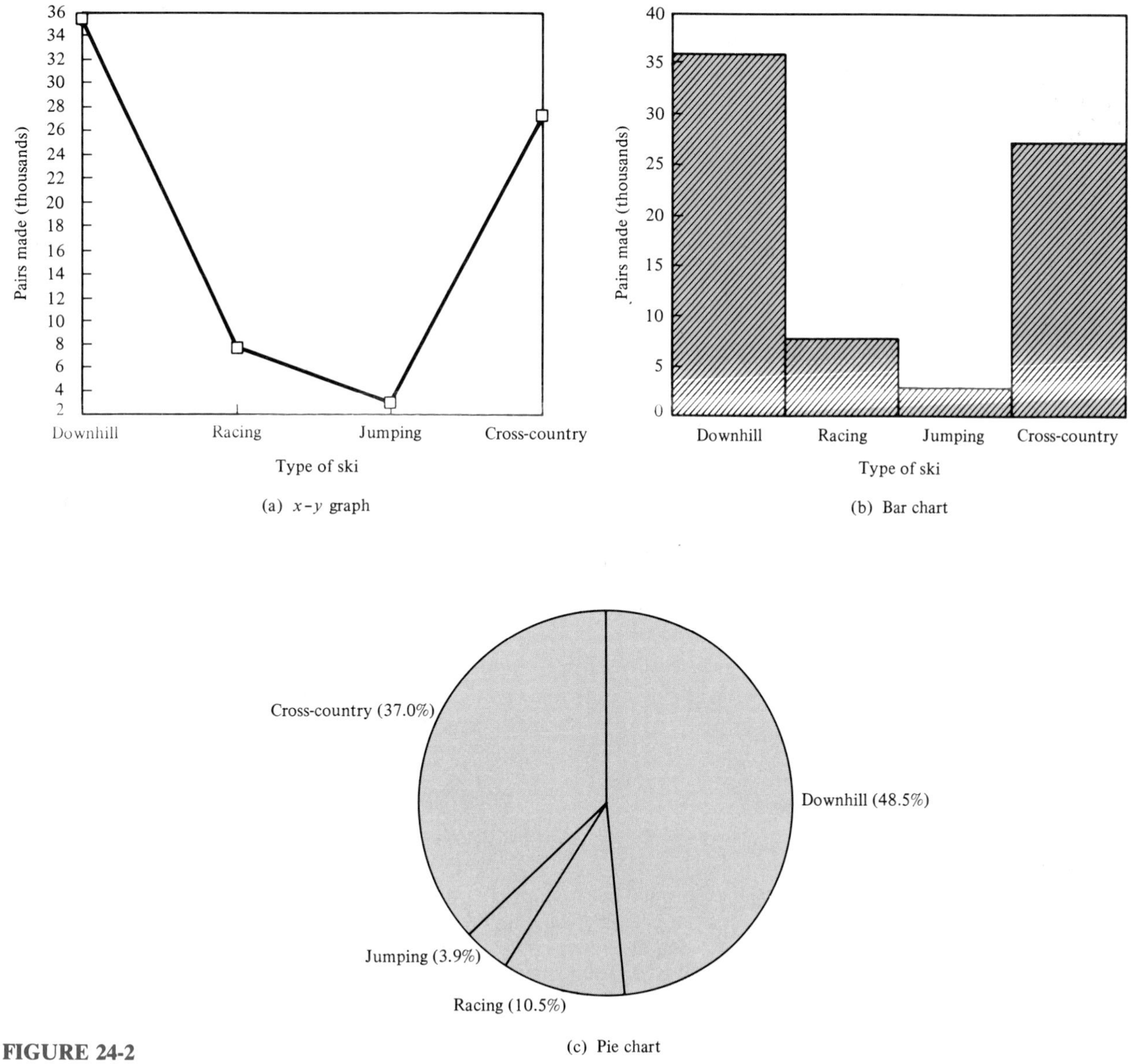


**FIGURE 24-2**

## EXERCISE 1—DEFINITIONS AND TERMINOLOGY

### Types of Data

Label each type of data as continuous, discrete, or categorical.

1. The number of people in each county that voted for Jones.
2. The life of certain 100-W light bulbs.
3. The blood types of patients at a certain hospital.
4. The number of Ford cars sold each day.
5. The models of Ford cars sold each day.
6. The weights of steers in a given herd.

### Graphical Representation of Data

7. The following table shows the population of a certain town for the years 1920–1930.

| *Year* | *Population* |
|---|---|
| 1920 | 5364 |
| 1921 | 5739 |
| 1922 | 6254 |
| 1923 | 7958 |
| 1924 | 7193 |
| 1925 | 6837 |
| 1926 | 7245 |
| 1927 | 7734 |
| 1928 | 8148 |
| 1929 | 8545 |
| 1930 | 8623 |

Make an $x$–$y$ graph of the population versus the year.

8. Make a bar graph for the data of Problem 7.

9. In a certain election, the tally was:

| *Candidate* | *Number of Votes* |
|---|---|
| Smith | 746,000 |
| Jones | 623,927 |
| Doe | 536,023 |
| Not voting | 163,745 |

Make a bar chart showing the election results.

10. Make a pie chart showing the election results of Problem 9. To make the pie chart, simply draw a circle and subdivide the area of the circle into four sectors. Make the central angle (and hence the area) of each sector proportional to the votes obtained by each candidate as a percentage of the total voting population. Label the percent in each sector.

## 24-2 FREQUENCY DISTRIBUTIONS

### Raw Data

Data that have not been organized in any way are called *raw data*. Data that have been sorted into ascending or descending order are called an *array*. The *range* of the data is the difference between the largest and the smallest number in that array.

---

**EXAMPLE 9:** Five students were chosen at random and their heights (in inches) were measured. The *raw data* obtained were:

56.2 72.8 58.3 56.9 67.5

If we sort the numbers into ascending order, we get the *array:*

56.2 56.9 58.3 67.5 72.8

The *range* of the data is

$$\text{range} = 72.8 - 56.2 = 16.6 \text{ in.}$$

---

### Grouped Data

For discrete and categorical data, we have seen that $x$–$y$ plots and bar charts are effective. These graphs can also be used for *continuous* data, but first we must collect those data into groups. Such groups are usually called *classes*. We do this by dividing the range into a convenient number of *intervals*. Each interval has an upper and a lower *limit* or *class boundary*. The *class width* is equal to the upper class boundary minus the lower class limit. The intervals are chosen so that a given data point cannot fall directly on a class boundary. We call the center of each interval the *class midpoint*.

### Frequency Distribution

Once the classes have been defined, we may tally the number of measurements falling within each class. Such a tally is called a *frequency distribution*. The number of entries in a given class is called the *absolute frequency*. That number, divided by the total number of entries, is called the *relative frequency*. The relative frequency is often expressed as a percent. To make a frequency distribution:

1. Subtract the smallest number in the data from the largest to find the range of the data.
2. Divide the range into class intervals, usually between 5 and 20. Avoid class limits that coincide with actual data by defining the limits to half a unit greater accuracy than the accuracy of the given data.
3. Get the class frequencies by tallying the number of observations that fall with each class.
4. Get the relative frequency of each class by dividing its class frequency by the total number of observations.

**EXAMPLE 10:** Table 24-3 gives the grades of 30 students. Group these raw data into convenient classes. Choose the class width, the class midpoints, and the class limits. Find the absolute frequency and relative frequency of each class.

TABLE 24-3

| | | | | | | | | | |
|---|---|---|---|---|---|---|---|---|---|
| 85.0 | 62.5 | 94.6 | 76.3 | 77.4 | 78.9 | 58.4 | 86.4 | 88.4 | 69.8 |
| 91.3 | 75.0 | 69.6 | 86.4 | 79.5 | 74.7 | 65.7 | 88.6 | 90.6 | 69.7 |
| 86.7 | 70.5 | 78.4 | 86.4 | 94.6 | 60.7 | 79.8 | 97.2 | 85.7 | 78.5 |

**Solution:**

1. The highest value in the data is 97.2 and the lowest is 58.4, so

$$\text{range} = 97.2 - 58.4 = 38.8$$

2. If we divide the range into five classes, we get a class width of 38.8/5 or 7.76. If we divide the range into 20 classes, we get a class width of 1.94. A convenient width between those two values is 5, so

$$\text{Let class width} = 5.0$$

Let us further choose class midpoints of 55, 60, . . . 95, so that our class limits are then

$$52.5, 57.5, 62.5, \ldots, 97.5$$

We must avoid ambiguity about which class to place a grade like 62.5. Our data are given to the nearest tenth, so let us make define our class limits to the nearest 0.05 unit. We then let our class limits be

$$52.45\text{–}57.45, \quad 57.45\text{–}62.45, \quad \ldots, \quad 92.45\text{–}97.45$$

3. Our next step is to tally the number of grades falling within each class and get a frequency distribution, shown in Table 24-4.
4. We then divide each frequency by 30 to get the relative frequency.

TABLE 24-4 Frequency Distribution of Student Grades

| *Class Limits* | *Class Midpoint* | *Tally* | *Absolute Frequency* | *Relative Frequency* |
|---|---|---|---|---|
| 52.45–57.45 | 55 | | 0 | 0/30 (0%) |
| 57.45–62.45 | 60 | // | 2 | 2/30 (6.7%) |
| 62.45–67.45 | 65 | // | 2 | 2/30 (6.7%) |
| 67.45–72.45 | 70 | //// | 4 | 4/30 (13.3%) |
| 72.45–77.45 | 75 | //// | 4 | 4/30 (13.3%) |
| 77.45–82.45 | 80 | ~~////~~ | 5 | 5/30 (16.7%) |
| 82.45–87.45 | 85 | ~~////~~ / | 6 | 6/30 (20%) |
| 87.45–92.45 | 90 | //// | 4 | 4/30 (13.3%) |
| 92.45–97.45 | 95 | /// | 3 | 3/30 (10%) |
| Totals | | | 30 | 30/30 (100%) |

## Frequency Histogram

A *frequency histogram* is a bar chart in which each bar is centered at its class midpoint. The width of each bar is equal to the class width, and the height of each bar is equal to the frequency. Either the absolute or the relative frequency can be used for the histogram, or the vertical axis can show both.

Since all the bars have the same width, the *areas* of the bars are proportional to the heights of the bars. The areas are thus proportional to the class frequencies. So if one bar in a relative frequency histogram has a height of 20%, that bar will contain 20% of the area of the whole graph. Thus there is a 20% chance that one grade chosen at random will fall within that class. We will make use of this connection between areas and probabilities in Sec. 24-4.

**EXAMPLE 11:** We show a frequency histogram for the data of Example 10 in Fig. 24-3. Note that the vertical scales show both absolute frequency and relative frequency.

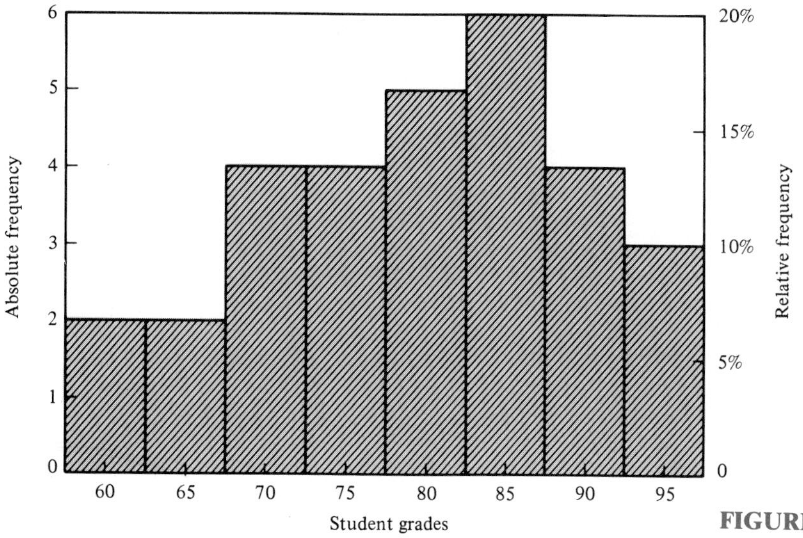


**FIGURE 24-3**

## Frequency Polygon

A *frequency polygon* is simply an $x$–$y$ graph in which frequency is plotted against class midpoint. As with the histogram, it can show either absolute or relative frequency, or both.

**EXAMPLE 12:** We show a frequency polygon for the data of Example 10 in Fig. 24-4.

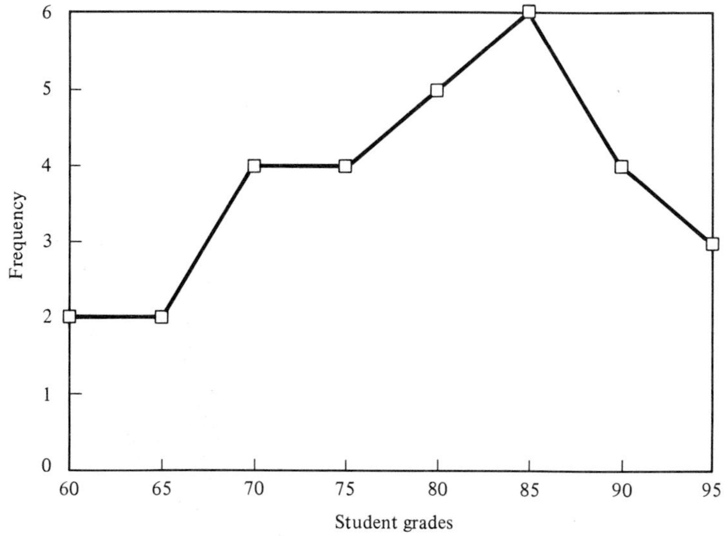


**FIGURE 24-4 A frequency polygon for the data of Example 10. Note that this polygon could have been obtained by connecting the midpoints of the tops of the bars in Fig. 24-3.**

### Cumulative Frequency Distributions

We obtain a *cumulative* frequency distribution by summing all values *less than* a given class limit. The graph of a cumulative frequency distribution is called a *cumulative frequency polygon,* or *ogive.*

**EXAMPLE 13:** Make and graph a cumulative frequency distribution for the data of Example 10.

**Solution:** We compute the cumulative frequencies by adding up the values below the given limits (Table 24-5) and plot them in Fig. 24-5.

**TABLE 24-5 Cumulative Frequency Distribution for Student Grades**

| *Grade* | *Cumulative Absolute Frequency* | *Cumulative Relative Frequency* |
|---|---|---|
| Under 62.45 | 2 | 2/30 (6.7%) |
| Under 67.45 | 4 | 4/30 (13.3%) |
| Under 72.45 | 8 | 8/30 (26.7%) |
| Under 77.45 | 12 | 12/30 (40%) |
| Under 82.45 | 17 | 17/30 (56.7%) |
| Under 87.45 | 23 | 23/30 (76.7%) |
| Under 92.45 | 27 | 27/30 (90%) |
| Under 97.45 | 30 | 30/30 (100%) |

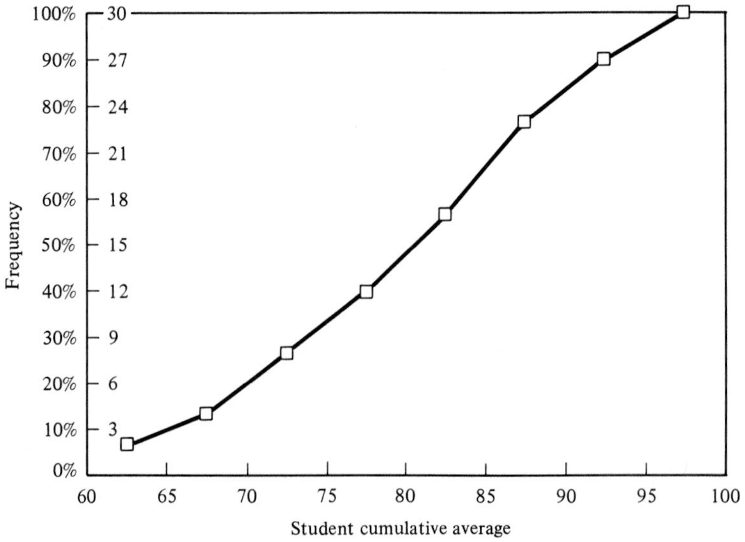


**FIGURE 24-5**

## EXERCISE 2—FREQUENCY DISTRIBUTIONS

### Frequency Distributions

1. The weights (in pounds) of 40 students at Tech College are

| | | | | | | | |
|---|---|---|---|---|---|---|---|
| 127 | 136 | 114 | 147 | 158 | 149 | 155 | 162 |
| 154 | 139 | 144 | 114 | 163 | 147 | 155 | 165 |
| 172 | 168 | 146 | 154 | 111 | 149 | 117 | 152 |
| 166 | 172 | 158 | 148 | 116 | 127 | 162 | 153 |
| 118 | 141 | 128 | 153 | 166 | 125 | 117 | 161 |

**(a)** Determine the range of the data. **(b)** Make an absolute frequency distribution using class widths of 5 lb. **(c)** Make a relative frequency distribution.

2. The times (in minutes) for 30 racers to complete a cross-country ski race are

| | | | | | | | | | |
|---|---|---|---|---|---|---|---|---|---|
| 61.4 | 72.5 | 88.2 | 71.2 | 48.5 | 48.9 | 54.8 | 71.4 | 99.2 | 74.5 |
| 84.6 | 73.6 | 69.3 | 49.6 | 59.3 | 71.4 | 89.4 | 92.4 | 48.4 | 66.3 |
| 85.7 | 59.3 | 74.9 | 59.3 | 72.7 | 49.4 | 83.8 | 50.3 | 72.8 | 69.3 |

**(a)** Determine the range of the data. **(b)** Make an absolute frequency distribution using class widths of 5 min. **(c)** Make a relative frequency distribution.

3. The prices (in dollars) of various computer printers in a distributer's catalog are

| | | | | | | | | | |
|---|---|---|---|---|---|---|---|---|---|
| 850 | 625 | 946 | 763 | 774 | 789 | 584 | 864 | 884 | 698 |
| 913 | 750 | 696 | 864 | 795 | 747 | 657 | 886 | 906 | 697 |
| 867 | 705 | 784 | 864 | 946 | 607 | 798 | 972 | 857 | 785 |

**(a)** Determine the range of the data. **(b)** Make an absolute frequency distribution using class widths of \$50. **(c)** Make a relative frequency distribution.

### Frequency Histograms and Frequency Polygons

Draw a histogram, showing both absolute and relative frequency, for the data of:

4. Problem 1 5. Problem 2 6. Problem 3

Draw a frequency polygon, showing both absolute and relative frequency, for the data of:

7. Problem 1 8. Problem 2 9. Problem 3

### Cumulative Frequency Distributions

Make a cumulative frequency distribution, showing **(a)** absolute frequency and **(b)** relative frequency, for the data of:

10. Problem 1 11. Problem 2 12. Problem 3

Draw a cumulative frequency polygon, showing both absolute and relative frequency, for the data of:

13. Problem 10 14. Problem 11 15. Problem 12

**Computer**

16. Some spreadsheets, such as Lotus, have commands for automatically making a frequency distribution, and can also draw frequency histograms and polygons. If you have such a spreadsheet available, use it for any of the problems in this exercise.

## 24-3 NUMERICAL DESCRIPTION OF DATA

We saw in Sec. 24-2 how we can describe a set of data by frequency distribution, a frequency histogram, a frequency polygon, or a cumulative frequency distribution. We can also describe a set of data with just a few numbers, and this more compact description is more convenient for some purposes. For example, if, in a report, you wanted to describe the heights of a group of students, and did not want to give the entire frequency distribution, you might simply say:

*The mean height is 58 inches, with a standard deviation of 3.5 inches.*

The *mean* is a number that shows the *center* of the data; the *standard deviation* is a measure of the *spread* of the data. As mentioned earlier, the mean and the standard deviation may be found either for an entire population (and are thus population parameters) or for a sample drawn from that population (and are thus sample statistics).

Thus to describe a population or a sample, we need numbers that give both the center of the data and the spread. Further, for sample statistics, we need to give the uncertainty of each figure. This will enable us to make inferences about the larger population.

---

**EXAMPLE 14:** A student recorded the running times for a sample of participants in a race. From that sample she inferred (by methods we'll learn later) that for the entire population of racers:

$$\text{mean time} = 23.65 \pm 0.84 \text{ minutes}$$

$$\text{standard deviation} = 5.83 \pm 0.34 \text{ minutes}$$

---

The mean time is called a *measure of central tendency*. We show how to calculate the mean, and other measures of central tendency, later in this section. The standard deviation is called a *measure of dispersion*. We also show how to compute it, and other measures of dispersion, in this section. The "plus-or-minus" values show a degree of uncertainty, called the *standard error*. The uncertainty, ±0.84 min, is called the *standard error of the mean,* and the uncertainty, ±0.34 min, is called the *standard error of the standard deviation*. We show how to calculate standard errors in Sec. 24-5.

## Measures of Central Tendency: The Mean

Some common measures of the center of a distribution are the mean, the median, and the mode. The arithmetic mean, or simply the *mean,* of a set of measurements is equal to the sum of the measurements divided by the number of measurements. It is what we commonly call the *average*. It is the most commonly used measure of central tendency. If our $n$ measurements are $x_1, x_2, \ldots, x_n$, then

$$\sum x = x_1 + x_2 + \cdots + x_n$$

where we used the Greek symbol $\Sigma$ (sigma) to represent "the sum of." The mean, which we call $\bar{x}$ ($x$ bar), is then

| | | |
|---|---|---|
| Arithmetic Mean | $\bar{x} = \dfrac{\sum x}{n}$ | **244** |

*The arithmetic mean of n measurements is the sum of those measurements divided by n.*

We can calculate the mean for a sample or for the entire population. However, we use a *different symbol* in each case.

$\bar{x}$ is the sample mean

$\mu$ is the population mean

---

**EXAMPLE 15:** Find the mean of the following sample.

746 574 645 894 736 695 635 794

**Solution:** Adding the values gives

$$\sum x = 746 + 574 + 645 + 894 + 736 + 695 + 635 + 794 = 5719$$

Then, with $n = 8$,

$$\bar{x} = \frac{5719}{8} = 715$$

rounded to three digits.

---

## Median

To find the *median,* we simply arrange the data in order of magnitude and pick the *middle value*. For an even number of measurements, we take the mean of the two middle values.

| | | |
|---|---|---|
| Median | The median of a set of numbers arranged in order of magnitude is the middle value of an odd number of measurements, or the mean of the two middle values of an even number of measurements. | **245** |

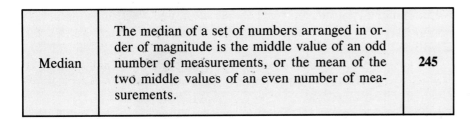

**EXAMPLE 16:** Find the median of the data in Example 15.

**Solution:** We rewrite the data in order of magnitude.

574 635 645 695 736 746 794 894

The two middle values are 695 and 736. Taking their mean gives

$$\text{median} = \frac{695 + 736}{2} = 715.5$$

### Mode

Our third measure of central tendency is the *mode*.

| | | |
|---|---|---|
| Mode | The mode of a set of numbers is the measurement(s) that occur most often in the set. | **246** |

A set of numbers may have no mode, or more than one mode.

**EXAMPLE 17:**

(a) The set 1 3 3 5 6 7 7 7 9 has the mode 7.

(b) The set 1 1 3 3 5 5 7 7 has no mode.

(c) The set 1 3 3 3 5 6 7 7 7 9 has two modes, 3 and 7. It is called *bimodal*.

### Measures of Dispersion

There is a story, probably untrue, about a statistician who drowned in a lake that had an average depth of 1 foot.

We will usually use the mean to describe a set of numbers, but it alone can be misleading.

**EXAMPLE 18:** The sets of numbers

1 1 1 1 1 1 1 1 1 91

and the set 10 10 10 10 10 10 10 10 10 10

each has a mean value of 10 but are otherwise quite different. Each set has a sum of 100, but in the first set most of this sum is concentrated in a single value, but in the second set the sum is *dispersed* among all the values.

Thus we need some *measure of dispersion* of a set of numbers. Four common ones are the range, the percentile range, the variance, and the standard deviation. We will cover each of these.

### Range

We have already introduced range in Sec. 24-2, and we define it here.

| | | |
|---|---|---|
| Range | The range of a set of numbers is the difference between the largest and the smallest number in the set. | **247** |

**EXAMPLE 19:** For the set of numbers

$$6 \quad 3 \quad 9 \quad 12 \quad 44 \quad 2 \quad 53 \quad 1 \quad 8$$

the largest value is 53 and the smallest is 1, so the range is

$$\text{range} = 53 - 1 = 52$$

We sometimes give the range by stating the end values themselves. Thus in Example 19 we might say that the range is from 1 to 53.

### Quartiles, Deciles, and Percentiles

For a set of data arranged in order of magnitude, we have seen that the median divides the data into two equal parts. There are as many numbers below the median as there are above it.

Similarly, we can determine two more values that divide each half of the data in half again. We call these values *quartiles*. Thus one-fourth of the values will fall into each quartile. The quartiles are labeled $Q_1$, $Q_2$, and $Q_3$. Thus $Q_2$ is the median.

Those values that divide the data into 10 equal parts we call *deciles* and label them $D_1, D_2, \ldots$. Those values that divide the data into 100 equal parts are *percentiles,* labeled $P_1, P_2, \ldots$. Thus the 25th percentile is the same as the first quartile ($P_{25} = Q_1$). The median is the 50th percentile, the fifth decile, and the second quartile ($P_{50} = D_5 = Q_2$). The 70th percentile is the seventh decile ($P_{70} = D_7$).

One measure of dispersion sometimes used is to give the range of values occupied by some given percentiles. Thus we have a *quartile range, decile range,* and *percentile range*. For example, the quartile range is the range from the first to the third quartile.

**EXAMPLE 20:** Find the quartile range for the set of data

$$2 \quad 4 \quad 5 \qquad 7 \quad 8 \quad 11 \qquad 15 \quad 18 \quad 19 \qquad 21 \quad 24 \quad 25$$

**Solution:** The quartiles are

$$Q_1 = \frac{5+7}{2} = 6$$

$$Q_2 = \text{the median} = \frac{11+15}{2} = 13$$

$$Q_3 = \frac{19+21}{2} = 20$$

Then $$\text{quartile range} = Q_3 - Q_1 = 20 - 6 = 14$$

We see that half the values fall within the quartile range. We may similarly compute the range of any percentiles or deciles. The range from the 10th to the 90th percentile is the one commonly used.

### Variance

We now give another measure of dispersion called the *variance,* but we must first mention deviation. We define the *deviation* of any number $x$ in a set of data as the difference between that number and the mean $\bar{x}$ of that set of data.

**EXAMPLE 21:** A certain set of measurements has a mean of 48.3. What are the deviations of the values 24.2 and 69.3 in that set?

**Solution:** The deviation of 24.2 is

$$24.2 - 48.3 = -24.1$$

and the deviation of 69.3 is

$$69.3 - 48.3 = 21.0$$

To get the *variance* of a set of $n$ numbers, we add up the squares of the deviations of each number in the set and divide by $n$. As with the mean, we use one symbol for the sample variance and a different symbol for the population variance.

$s^2$ is the sample variance

$\sigma^2$ is the population variance

**Sometimes the sample variance is defined with $n - 1$ in the denominator rather than $n$. This is done to make the sample variance $s^2$ a better estimator of the population variance $\sigma^2$. However, for large samples (over 30 or so) the variance found by either formula is practically the same. Here we will use $n$ in the denominator.**

Expressed as a formula,

| Sample Variance | $s^2 = \dfrac{\sum (x - \bar{x})^2}{n}$ | **248** |
|---|---|---|

**EXAMPLE 22:** Compute the variance for the sample

1.74 2.47 3.66 4.73 5.14 6.23 7.29 8.93 9.56

**Solution:** We first compute the mean, $\bar{x}$.

$$\bar{x} = \frac{1.74 + 2.47 + 3.66 + 4.73 + 5.14 + 6.23 + 7.29 + 8.93 + 9.56}{9}$$

$$= 5.53$$

We then subtract the mean from each of the nine values to obtain deviations. The deviations are then squared and added, as in the following table.

**Computers are ideal for long, tedious statistical computations.**

| *Measurement* $x$ | *Deviation* $x - \bar{x}$ | *Deviation Squared* $(x - \bar{x})^2$ |
|---|---|---|
| 1.74 | −3.79 | 14.35 |
| 2.47 | −3.06 | 9.35 |
| 3.66 | −1.87 | 3.49 |
| 4.73 | −0.80 | 0.64 |
| 5.14 | −0.39 | 0.15 |
| 6.23 | 0.70 | 0.49 |
| 7.29 | 1.76 | 3.11 |
| 8.93 | 3.40 | 11.58 |
| 9.56 | 4.03 | 16.26 |
| $\sum x = 49.75$ | $\sum (x - \bar{x}) = 0$ | $\sum (x - \bar{x})^2 = 59.41$ |

The variance $s^2$ is then

$$s^2 = \frac{59.41}{9} = 6.60$$

## Standard Deviation

Once we have the variance, it is a simple matter to get the *standard deviation*. It is the most common measure of dispersion.

| Standard Deviation | The standard deviation of a set of numbers is the positive square root of the variance. | **249** |
|---|---|---|

As before, we use $s$ for the sample standard deviation and $\sigma$ for the population standard deviation.

---

**EXAMPLE 23:** Find the standard deviation for the data of Example 22.

**Solution:** We have already found the variance in Example 22.

$$s^2 = 6.60$$

Taking the square root gives the standard deviation,

$$s = \sqrt{6.60} = 2.57$$

---

## EXERCISE 3—NUMERICAL DESCRIPTION OF DATA

### The Mean

1. Find the mean of the following set of grades.

   85 74 69 59 60 96 84 48 89 76 96 68 98 79 76

2. Find the mean of the following set of weights.

   173 127 142 164 163 153 116 199

3. Find the mean of the weights in Exercise 2, Problem 1.
4. Find the mean of the times in Exercise 2, Problem 2.
5. Find the mean of the prices in Exercise 2, Problem 3.

### The Median

6. Find the median of the grades in Problem 1.
7. Find the median of the weights in Problem 2.
8. Find the median of the weights in Exercise 2, Problem 1.
9. Find the median of the times in Exercise 2, Problem 2.
10. Find the median of the prices in Exercise 2, Problem 3.

### The Mode

11. Find the mode of the grades in Problem 1.
12. Find the mode of the weights in Problem 2.
13. Find the mode of the weights in Exercise 2, Problem 1.
14. Find the mode of the times in Exercise 2, Problem 2.
15. Find the mode of the prices in Exercise 2, Problem 3.

### Range

16. Find the range of the grades in Problem 1.
17. Find the range of the weights in Problem 2.

### Percentiles

18. Find the quartiles and give the quartile range of the data.

28 39 46 53 69 71 83 94 102 117 126

19. Find the quartiles and give the quartile range of the data.

1.33 2.28 3.59 4.96 5.23 6.89 7.91 8.13 9.44 10.6 11.2 12.3

### Variance and Standard Deviation

20. Find the variance and standard deviation of the grades in Problem 1.
21. Find the variance and standard deviation of the weights in Problem 2.
22. Find the variance and standard deviation of the weights in Exercise 2, Problem 1.
23. Find the variance and standard deviation of the times in Exercise 2, Problem 2.
24. Find the variance and standard deviation of the prices in Exercise 2, Problem 3.

### Calculator or Computer

25. Some calculators and some computer spreadsheet programs, such as Lotus, have built-in functions for finding the mean and standard deviation. If you have access to such a calculator or computer, use it for any of the problems in this exercise.

## 24-4 INTRODUCTION TO PROBABILITY

Why do we need probability to learn statistics? In Sec. 24-3 we learned how to compute certain sample statistics, such as the mean and the standard deviation. Knowing, for example, that the mean height $\bar{x}$ of a sample of students is 69.5 in., we might *infer* that the mean height $\mu$ of the entire population is 69.5 in. But how reliable is that number? Is $\mu$ equal to 69.5 in. *exactly?* We'll see later that we give population parameters such as $\mu$ as a *range* of values, say $69.5 \pm 0.6$ in., and we state the *probability* that the true mean lies within that range. We might say, for example, that there is a 68% chance that the true value lies within the stated range.

Further, we may report that the sample standard deviation is, say, 2.6 in. What does that mean? Of 1000 students, for example, how many can be expected to have a height falling within, say, 1 standard deviation of the mean height at that college? We use probability to help us answer questions such as that.

Further, statistics are often used to "prove" many things, and it takes a knowledge of probability to help decide whether the claims are valid or are merely a result of chance.

---

**EXAMPLE 24:** Suppose that a nurse measures the heights of 14-year-old students in a town located near a chemical plant. Of 100 students measured, she finds 75 students to be shorter than 59 in. Parents then claim that this is caused by chemicals in the air, but the company claims that such heights can occur by chance. We cannot resolve this question without a knowledge of probability.

---

We distinguish between *statistical* probability and *mathematical* probability.

**EXAMPLE 25:** We may toss a die 6000 times and count the number of times that "one" comes up, say 1006 times. We conclude that the chance of tossing a "one" is 1006 out of 6000, or about 1 in 6. This is an example of what is called *statistical* probability.

On the other hand, we may reason, without even tossing the die, that each face has an equal chance of coming up. Since there are six faces, the chance of tossing a "one" is 1 in 6. This is an example of *mathematical* probability.

We first give a brief introduction to mathematical probability and then apply it to statistics.

## Probability of a Single Event Occurring

Suppose that an event $A$ can happen in $m$ ways out of a total of $n$ equally likely ways. The probability $P(A)$ of $A$ occurring is defined as

| | | |
|---|---|---|
| Probability of a Single Event Occurring | $P(A) = \frac{m}{n}$ | **250** |

Note that the probability $P(A)$ is a number between 0 and 1. $P(A) = 0$ means that there is no chance that $A$ will occur. $P(A) = 1$ means that it is certain that $A$ will occur. Also note that if the probability of $A$ occurring is $P(A)$, the probability of $A$ *not* occurring is $1 - P(A)$.

**EXAMPLE 26:** A bag contains 15 balls, 6 of which are red. Assuming that any ball is as likely to be drawn as any other, the probability of drawing a red ball from the bag is $\frac{6}{15}$ or $\frac{2}{5}$. The probability of drawing a ball that is not red is $\frac{9}{15}$ or $\frac{3}{5}$.

**EXAMPLE 27:** A nickel and a dime are each tossed once. What is (a) the probability $P(A)$ that both coins will show heads; (b) the probability $P(B)$ that not both coins will show heads?

The possible outcomes of a single toss of each coin are

| *Nickel* | *Dime* |
|---|---|
| Heads | Heads |
| Heads | Tails |
| Tails | Heads |
| Tails | Tails |

Thus there are four equally likely outcomes. For event $A$ (both heads) there is only one favorable outcome (heads, heads). Thus

$$P(A) = \frac{1}{4} = 0.25$$

Event $B$ (that not both coins show heads) can happen in three ways (HT, TH, and TT). Thus $P(B) = \frac{3}{4} = 0.75$. Or, alternatively,

$$P(B) = 1 - P(A) = 0.75$$

### Probability of Two Events Both Occurring

If $P(A)$ is the probability that $A$ will occur, and $P(B)$ is the probability that $B$ will occur, the probability $P(A, B)$ that *both* $A$ and $B$ will occur is

| | | |
|---|---|---|
| Probability of Two Events Both Occurring | $P(A, B) = P(A)P(B)$ | **251** |

***Independent events*** **are those in which the outcome of the first in no way affects the outcome of the second.**

*The probability that two independent events will both occur is the product of their individual probabilities.*

The rule above can be extended as follows:

| | | |
|---|---|---|
| Probability of Several Events All Occurring | $P(A, B, C\ldots) = P(A)P(B)P(C)\cdots$ | **252** |

*The probability that several independent events, A, B, C, . . . , will all occur is the product of the probabilities for each.*

---

**EXAMPLE 28:** What is the probability that in tossing a coin three times, all will be heads?

**Solution:** Each toss is independent of the others, and the probability of getting a head on a single toss is $\frac{1}{2}$. Thus the probability of three heads being tossed is

$$\left(\frac{1}{2}\right)\left(\frac{1}{2}\right)\left(\frac{1}{2}\right) = \frac{1}{8}$$

or 1 in 8. If we were to make, say, 800 sets of tosses, we would expect to toss three heads about 100 times.

---

### Probability of Either of Two Events Occurring

If $P(A)$ and $P(B)$ are the probabilities that events $A$ and $B$ will occur, the probability $P(A + B)$ that *either A or B* or both will occur is

| | | |
|---|---|---|
| Probability of Either of Two Events Occurring | $P(A + B) = P(A) + P(B) - P(A, B)$ | **253** |

*The probability that either A or B or both will occur is the sum of the individual probabilities for A and B, less the probability that both will occur.*

**EXAMPLE 29:** In a certain cup production line, 6% are chipped, 8% are cracked, and 2% are both chipped and cracked. What is the probability that a cup picked at random will be either chipped or cracked, or both?

**Solution:** If $P(C)$ = probability of being chipped = 0.06
and $P(S)$ = probability of being cracked = 0.08
and $P(C, S) = 0.02$, then

$$P(C + S) = 0.06 + 0.08 - 0.02 = 0.12$$

We would thus predict that 12 out of 100 chosen would be chipped or cracked, or both chipped and cracked.

## Probability of Two Mutually Exclusive Events Occurring

Mutually exclusive events $A$ and $B$ are those that cannot both occur at the same time. Thus the probability $P(A, B)$ of both $A$ and $B$ occurring is zero.

**EXAMPLE 30:** A tossed coin lands either heads or tails. Thus the probability $P(\text{H, T})$ of a single toss being both heads and tails is

$$P(\text{H, T}) = 0$$

Setting $P(A, B) = 0$ in Eq. 253 gives

| | | |
|---|---|---|
| Probability of Two Mutually Exclusive Events Occurring | $P(A + B) = P(A) + P(B)$ | **254** |

*The probability that either of two mutually exclusive events will occur in a certain trial is the sum of the two individual probabilities.*

We can generalize this to $n$ mutually exclusive events. The probability that one of $n$ mutually exclusive events will occur in a certain trial is the sum of the $n$ individual probabilities.

**EXAMPLE 31:** In a certain school, 10% of the students have red hair and 20% have black hair. If we select one student at random, what is the probability that this student will have either red hair or black hair?

**Solution:** The probability $P(A)$ of having red hair is 0.1 and the probability $P(B)$ of having black hair is 0.2, and the two events (having red hair or having black hair) are mutually exclusive. Thus the probability $P(A + B)$ of a student having either red or black hair is

$$P(A + B) = 0.1 + 0.2 = 0.3$$

## Probability Distributions

If an experiment has $X$ different outcomes, we say that the probability of a particular outcome is $P(X)$. If we calculate $P(X)$ for each $X$, we have what is called a *probability distribution*. Such a distribution may be presented as a table, a graph, or a formula.

**EXAMPLE 32:** Make a probability distribution for a single roll of a single die.

**Solution:** Let $X$ equal the number of the face that turns up and $P(X)$ be the probability of that face turning up. That is, $P(5)$ is the probability of rolling a five. Since each number has an equal chance of being rolled, we have

| $X$ | $P(X)$ |
|---|---|
| 1 | $\frac{1}{6}$ |
| 2 | $\frac{1}{6}$ |
| 3 | $\frac{1}{6}$ |
| 4 | $\frac{1}{6}$ |
| 5 | $\frac{1}{6}$ |
| 6 | $\frac{1}{6}$ |

We graph this probability distribution in Fig. 24-6.

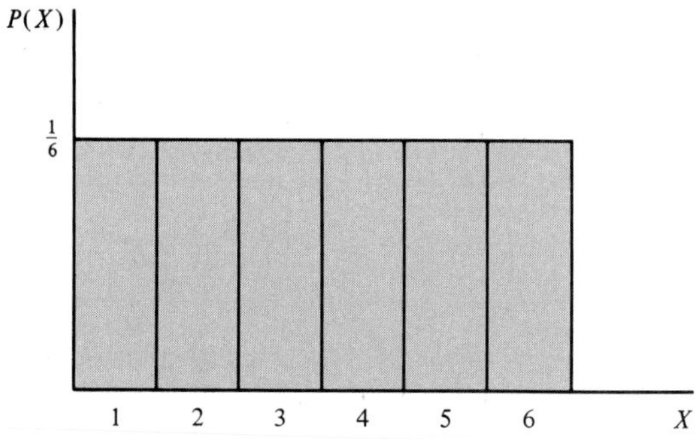


**FIGURE 24-6 Probability distribution for the toss of a single die.**

**EXAMPLE 33:** Make a probability distribution for the two-coin toss experiment of Example 27.

**Solution:** Let $X$ be the number of heads tossed and $P(X)$ be the probability of $X$ heads being tossed. The experiment has four possible outcomes (HH, HT, TH, TT). The outcome $X = 0$ (no heads) can occur in only one way (TT), so the probability $P(0)$ of not tossing a head is one in four, or $\frac{1}{4}$. The probabilities of the other outcomes are found in a similar way. They are:

| $X$ | $P(X)$ |
|---|---|
| 0 | $\frac{1}{4}$ |
| 1 | $\frac{1}{2}$ |
| 2 | $\frac{1}{4}$ |

which is shown graphically in Fig. 24-7.

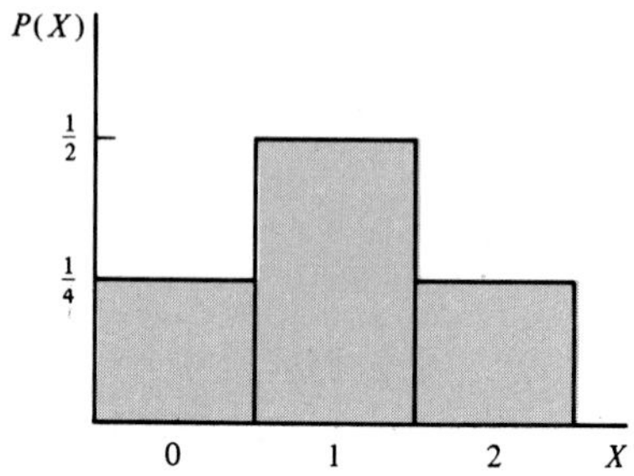


**FIGURE 24-7 Probability distribution.**

## Probabilities Obtained from Relative Frequency Histogram

We now show that the probability of a continuous variable falling between two limits is given by the area of the relative frequency histogram for that variable, between those limits. Look again at the relative frequency histogram for student's grades, Fig. 24-3. Take, say, the class with limits 92.5 to 97.5, with a class midpoint of 95. That class has a relative frequency of 10%. This means that the class occupies 10% of the entire area of the histogram. This further implies that the grade of a student chosen at random has a 10% chance of falling within that class. On the relative frequency histogram, all the bars have the same width, so the *areas* of the bars are proportional to the heights of the bars. Thus the bar for the class with midpoint 95 occupies 10% of the area of the entire histogram. Further, consider the three rightmost bars in Fig. 24-3. We find the chance of a randomly chosen grade falling within one of those three bars by adding their relative frequencies.

$$20\% + 13.3\% + 10\% = 43.3\%$$

Also note that these three bars occupy 43.3% of the total area of the histogram. Thus we see that *the area between two limits on the histogram is a measure of the probability of one measurement falling within those limits.*

---

**EXAMPLE 34:** Using the frequency histogram Fig. 24-3, find the chance of a grade chosen at random of falling between 62.5 and 77.4.

**Solution:** This range spans three classes in Fig. 24-3, having relative frequencies of 6.7%, 13.3%, and 13.3%. The chance of falling within one of the three is

$$6.7 + 13.3 + 13.3 = 33.3\%$$

so the probability is $\frac{1}{3}$.

---

### Continuous Probability Distributions

Suppose now that we collect data from a "large" population. Suppose, for example, that we record the age of every mother giving birth to a child in a given year. Since the population is large, we can use very narrow class widths and still have many observations that fall within each class. If we do so, our relative frequency histogram (Fig. 24-8a) will have very narrow bars, and our relative frequency polygon (Fig. 24-8b) will be practically *a smooth curve*.

We get probabilities from the smooth curve just as we did from the relative frequency histogram. *The probability of an observation falling between two limits a and b is proportional to the area bounded by the curve and the x axis, between those limits.* Let us designate the total area under the curve as 100% (or $P = 1$). Then the probability of an observation falling between two limits is equal to the area between those two limits.

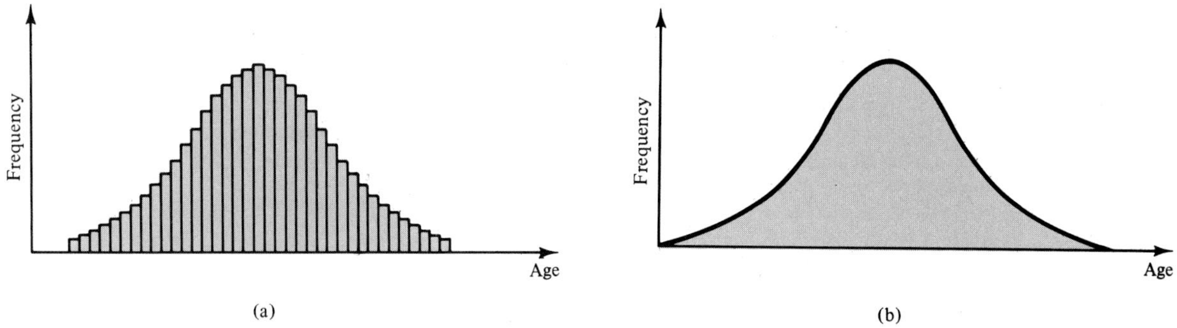


**FIGURE 24-8**

**EXAMPLE 35:** Suppose that 40% of the entire area under the curve in Fig. 24-8b lies between the limits 25 and 30. That means that there is a 40% chance ($P = 0.4$) that one mother chosen at random will be between 25 and 30 years of age.

## Normal Distribution

The frequency distribution for many real-life observations, such as a person's height, is found to have a typical "bell" shape. We have seen that the chance of one observation falling between two limits is equal to the area under the relative frequency histogram (such as Fig. 24-8a) between those limits. But how are we to get those areas for a curve such as Fig. 24-8b?

It is named for the Gauss whose *Gauss–Seidel method* we studied in Chapter 9 and whose *Gauss elimination* we studied in Chapter 11. Some consider him to be one of the greatest mathematicians of all time. He has many theorems and proofs named after him.

One way is to make a *mathematical model* of the probability distribution, and from it get the areas of any section of the distribution. One such mathematical model is called the *normal distribution, normal curve,* or *Gaussian distribution.*

The normal distribution is obtained by plotting the equation

Don't be frightened by this equation. We will not be calculating with it but give it only to show where the normal curve and Table 24-6 come from.

| | | |
|---|---|---|
| Normal Distribution | $y = \dfrac{1}{\sigma\sqrt{2\pi}} e^{-(x-\mu)^2/2\sigma^2}$ | **255** |

The normal curve (Fig. 24-9) is bell-shaped and is symmetrical about its mean, $\mu$. It closely matches many frequency distributions in the real world.

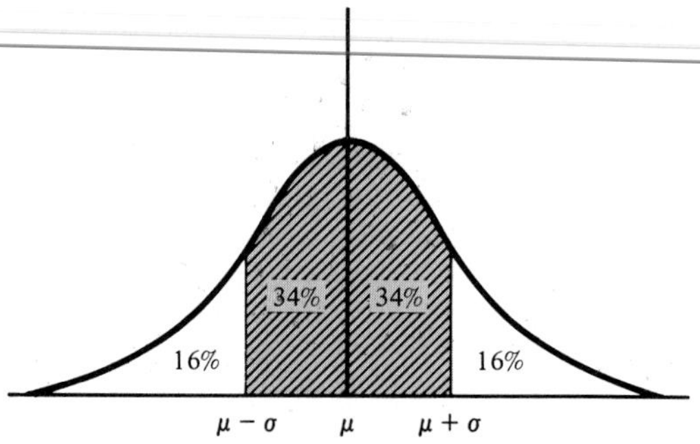


**FIGURE 24-9 The normal curve.**

To use the normal curve to determine probabilities, we must know the areas between any given limits. Since the equation of the curve is known, it is possible, by methods too advanced to show here, to compute these areas. They are given in Table 24-6. The areas are given "within $z$ standard deviations of the mean." This means that each area given is that which lies between the mean and $z$ standard deviations *on one side* of the mean, as shown in Fig. 24-10.

The number $z$ is called the *standardized variable,* or *normalized variable*. When used for educational testing it is often called the *standard score,* or *z score*. Since the normal curve is symmetrical about the mean, the given values apply to either the left or the right side.

**TABLE 24-6 Areas under the Normal Curve within *z* Standard Deviations of the Mean**

| *z* | *Area* | *z* | *Area* |
|---|---|---|---|
| 0.1 | 0.0398 | 2.1 | 0.4821 |
| 0.2 | 0.0793 | 2.2 | 0.4861 |
| 0.3 | 0.1179 | 2.3 | 0.4893 |
| 0.4 | 0.1554 | 2.4 | 0.4918 |
| 0.5 | 0.1915 | 2.5 | 0.4938 |
| 0.6 | 0.2258 | 2.6 | 0.4952 |
| 0.7 | 0.2580 | 2.7 | 0.4965 |
| 0.8 | 0.2881 | 2.8 | 0.4974 |
| 0.9 | 0.3159 | 2.9 | 0.4981 |
| 1.0 | 0.3413 | 3.0 | 0.4987 |
| 1.1 | 0.3643 | 3.1 | 0.4990 |
| 1.2 | 0.3849 | 3.2 | 0.4993 |
| 1.3 | 0.4032 | 3.3 | 0.4995 |
| 1.4 | 0.4192 | 3.4 | 0.4997 |
| 1.5 | 0.4332 | 3.5 | 0.4998 |
| 1.6 | 0.4452 | 3.6 | 0.4998 |
| 1.7 | 0.4554 | 3.7 | 0.4999 |
| 1.8 | 0.4641 | 3.8 | 0.4999 |
| 1.9 | 0.4713 | 3.9 | 0.5000 |
| 2.0 | 0.4772 | 4.0 | 0.5000 |

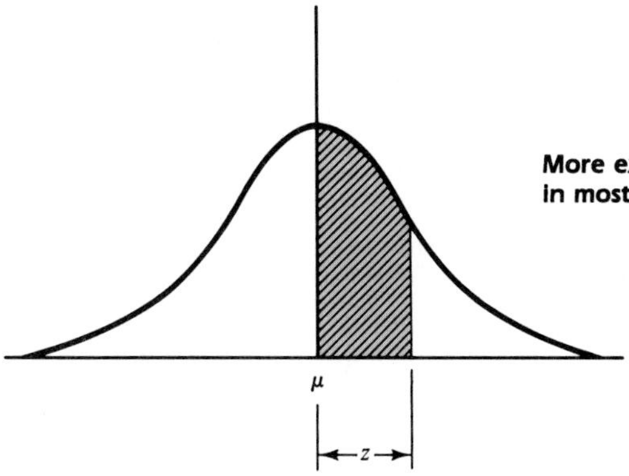

FIGURE 24-10

**More extensive tables can be found in most statistics books.**

**EXAMPLE 36:** Find the area under the normal curve between the following limits: from one standard deviation to the left of the mean, to one standard deviation to the right of the mean.

**Solution:** From Table 24-6, the area between the mean and one standard deviation is 0.3413. By symmetry, the same area lies to the other side of the mean. The total is thus 2(0.3413), or 0.6816. Thus *about 68% of the area under the normal curve lies within one standard deviation of the mean* (Fig. 24-9). Thus 100 − 68 or 32% of the area lies in the "tails," or 16% in each tail.

**EXAMPLE 37:** Find the area under the normal curve between 0.8 standard deviation to the left of the mean, to 1.1 standard deviations to the right of the mean (Fig. 24-11).

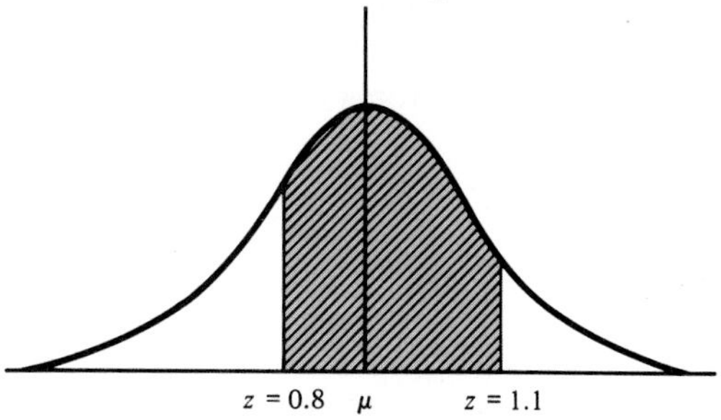

FIGURE 24-11

**Solution:** From Table 24-6, the area between the mean and 0.8 standard deviation ($z = 0.8$) is 0.2881. Also, the area between the mean and 1.1 standard deviations ($z = 1.1$) is 0.3643. The combined area is thus

$$0.2881 + 0.3643 = 0.6524$$

or 65.24% of the total area.

**EXAMPLE 38:** Find the area under the tail of the normal curve to the right of $z = 1.4$ (Fig. 24-12).

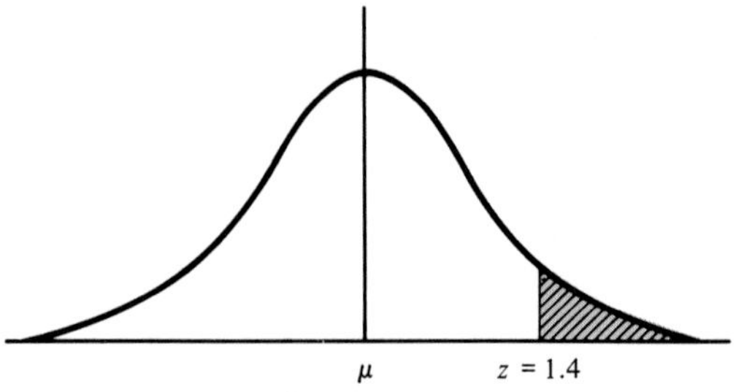


FIGURE 24-12

**Solution:** From Table 24-6, the area between the mean and $z = 1.4$ is 0.4192. But since the total area to the right of the mean is $\frac{1}{2}$, the area in the tail is

$$0.5000 - 0.4192 = 0.0808$$

or 8.08% of the total area.

Since, for a normal distribution, the probability of a measurement falling within a given interval is equal to the area under the normal curve within that interval, we can use the areas to assign probabilities. We must first convert the measurement $x$ to the standard variable $z$, which tells the number of standard deviations between $x$ and the mean $\bar{x}$. To get the value of $z$, we simply find the difference between $x$ and $\bar{x}$, and divide that difference by the standard deviation.

$$z = \frac{x - \bar{x}}{s}$$

**EXAMPLE 39:** Suppose that the heights of 2000 students have a mean of 69.3 in. with a standard deviation of 3.2 in. Assuming the heights to have a normal distribution, predict the number of students (a) shorter than 65 in.; (b) taller than 70 in.; (c) between 65 in. and 70 in.

**Solution:**

(a) Let us compute $z$ for $x = 65$ in. The mean $\bar{x}$ is 69.3 in. and the standard deviation $s$ is 3.2 in., so

$$z = \frac{65 - 69.3}{3.2} = -1.3$$

**Since the normal curve is symmetrical about the mean, either a positive or a negative value of *z* will give the same area. Thus, for compactness, Table 24-6 shows only positive values of *z*.**

From Table 24-6, with $z = 1.3$, we read an area of 0.4032. The area in the tail to the left of $z = -1.3$ is thus

$$0.5000 - 0.4032 = 0.0968$$

Thus there is a 9.68% chance that a single student will be shorter than 65 in. Since there are 2000 students, we predict that

$$9.68\% \text{ of } 2000 \text{ students} = 194 \text{ students}$$

will be within this range.

(b) For a height of 70 in., the value of $z$ is

$$z = \frac{70 - 69.3}{3.2} = 0.2$$

The area between the mean and $z = 0.2$ is 0.0793, so to the right of $z = 0.2$ is

$$0.5000 - 0.0793 = 0.4207$$

Since there is a 42.07% chance that a student will be taller than 70 in., we predict that 0.4207(2000) or 841 students will be within that range.

(c) Since 191 + 841 or 1032 students are either shorter than 65 in. or taller than 70 in., we predict that 968 students will be between these two heights.

## EXERCISE 4—PROBABILITY

### Probability of a Single Event

1. We draw a ball from a bag that contains 8 green balls and 7 blue balls. What is the probability that a ball drawn at random will be green?
2. A card is drawn from a deck containing 13 hearts, 13 diamonds, 13 clubs, and 13 spades. What chance is there that a card drawn at random will be a heart?
3. If we toss four coins, what is the probability of getting two heads and two tails? [*Hint:* This experiment has 16 possible outcomes (HHHH, HHHT, . . . , TTTT).]
4. If we toss four coins, what is the probability of getting one head and three tails?
5. If we throw two dice, what is the probability that their sum is 9? [*Hint:* List all possible outcomes, (1, 1), (1, 2), and so on, and count those that have a sum of 9.
6. If two dice are thrown, what is the probability that their sum is 7?

### Probability That Several Events Will All Occur

7. A die is rolled twice. What is the probability that both rolls will give a six?
8. We draw four cards from a deck, replacing each before the next is drawn. What chance is there that all four draws will be a red card?

### Probability That Either of Two Events Will Occur

9. At a certain school, 55% of the students have brown hair, 15% have blue eyes, and 7% have both brown hair and blue eyes. What is the probability that a student chosen at random will have either brown hair or blue eyes or both?
10. Find the probability that a card drawn from a deck will be either a "picture" card (jack, queen, or king) or a spade or a spade picture card.

### Probability of Mutually Exclusive Events

11. What chance is there of tossing a sum of 7 or 9 when two dice are tossed? You may use your results from Problems 5 and 6 here.
12. On a certain production line, 5% of the parts are underweight and 8% are overweight. What is the probability that one part selected at random is either underweight or overweight?

## The Normal Curve

Use Table 24-6 for the following problems.

13. Find the area under the normal curve between the mean and 1.5 standard deviations in Fig. 24-13.

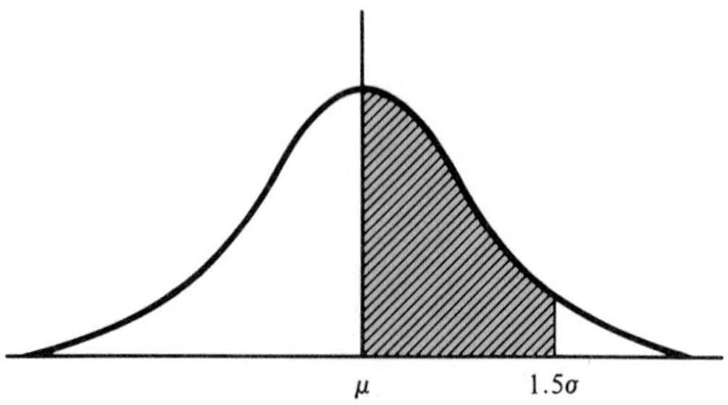


FIGURE 24-13

14. Find the area under the normal curve between 1.6 standard deviations on both sides of the mean in Fig. 24-14.

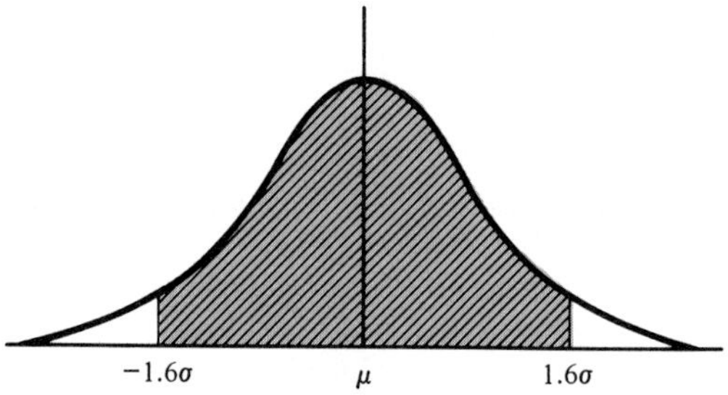


FIGURE 24-14

15. Find the area in the tail of the normal curve to the left of $z = -0.8$ in Fig. 24-15.

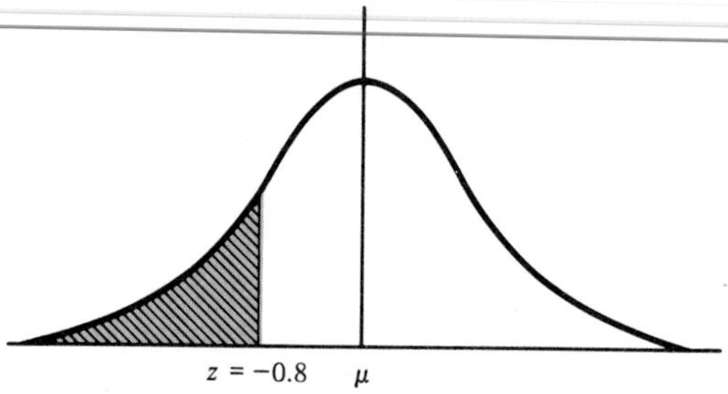


FIGURE 24-15

16. Find the area in the tail of the normal curve to the right of $z = 1.3$ in Fig. 24-16.

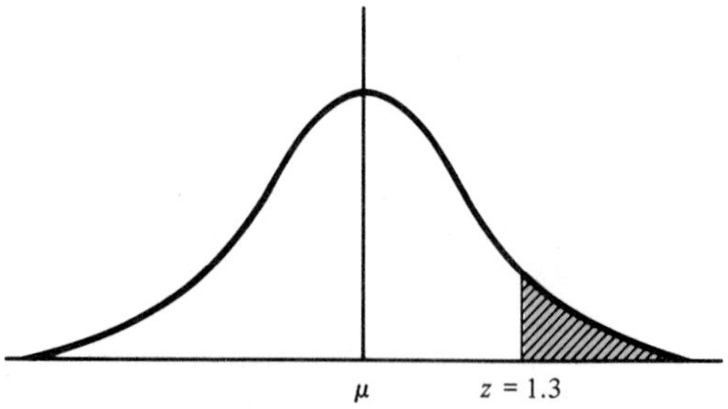


FIGURE 24-16

17. The distribution of the weights of 1000 students at Tech College has a mean of 163 lb with a standard deviation of 18 lb. Assuming that the weights are normally distributed, predict the number of students who have weights between 130 and 170 lb.

18. For the data of Problem 17, predict the number of students who will weigh less than 130 lb.
19. For the data of Problem 17, predict the number of students who will weigh more than 195 lb.
20. On a test given to 500 students, the mean grade was 82.6 with a standard deviation of 7.4. Assuming a normal distribution, predict the number of A grades (90 or over).
21. For the test of Problem 20, predict the number of failing grades (60 or less).

## 24-5 STANDARD ERRORS

When we draw a random sample from a population, it is usually to *infer* something about that population. Typically, from our sample we compute a *statistic* (such as the sample mean $\bar{x}$) and use it to infer a population *parameter* (such as the population mean $\mu$).

---

**EXAMPLE 40:** A researcher measured the heights of a randomly drawn sample of students at Tech College and calculated the mean height of that sample. It was

$$\text{mean height } \bar{x} = 67.50 \text{ in.}$$

From this he inferred that the mean height of the entire population of students at Tech College was

$$\text{mean height } \mu = 67.50 \text{ in.}$$

---

In general,

$$\text{population parameter} \approx \text{sample statistic}$$

*How accurate* is the population parameter that we find in this way? We know that all measurements are approximate, and we usually give some indication of the accuracy of a measurement. In fact, a population parameter such as the mean height is often given in the form

$$\text{mean height } \mu = 67.50 \pm 0.24 \text{ in.}$$

where $\pm 0.24$ is called the *standard error* of the mean. In this section we show how to compute the standard error of the mean and the standard error of the standard deviation.

Further, when we give a range of values for a statistic, we know it is possible that another sample can have a mean that lies outside that range. In fact, the true mean itself can lie outside the given range. That is why a statistician will *give the probability* that the true mean falls within the given range.

---

**EXAMPLE 41:** Using the same example of heights at Tech College, we might say that

$$\text{mean height } \mu = 67.50 \pm 0.24 \text{ in.}$$

*with a 68% chance* that the mean student height $\mu$ falls within the range

$$67.50 - 0.24 \quad \text{to} \quad 67.50 + 0.24$$

or from 67.26 to 67.74 in.

---

We call the interval $67.50 \pm 0.24$ a *68% confidence interval*. This means that there is a 68% chance that the true population mean falls within the given range. So the complete way to report a population parameter is

$$\text{population parameter} = \text{sample statistic} \pm \text{standard error}$$

within a given confidence interval. We will soon show how to compute this confidence interval, and others as well, but first we must lay some groundwork.

## Frequency Distribution of a Statistic

Suppose that you draw a sample of, say, 40 heights from the population of students in your school, and find the mean $\bar{x}$ of those 40 heights, say 65.6 in. Now measure another 40 heights from the same population and get another $\bar{x}$, say 69.3 in. Then repeat, drawing one sample of 40 after another, and for each compute the statistic $\bar{x}$. This collection of $\bar{x}$'s now has a frequency distribution of its own (called the $\bar{x}$ distribution) with its own mean and standard deviation.

We now ask: *How is the $\bar{x}$ distribution related to the original distribution?* To answer this we need a statistical theorem called the *central limit theorem,* which we give here without proof.

## Standard Error of the Mean

If we were to draw *all possible* samples of size $n$ from a given population, and for each sample compute the mean $\bar{x}$, and then make a frequency distribution of the $\bar{x}$'s, the central limit theorem states that:

1. The mean of the $\bar{x}$ distribution is the same as the mean $\mu$ of the population from which the samples were drawn.
2. The standard deviation of the $\bar{x}$ distribution (which we'll call $SE_{\bar{x}}$) is equal to the standard deviation $\sigma$ of the population divided by the square root of the sample size $n$.

$$SE_{\bar{x}} = \frac{\sigma}{\sqrt{n}}$$

3. For a "large" sample (over 30 or so) the $\bar{x}$ distribution is approximately a normal distribution, even if the population from which the samples were drawn is not a normal distribution.

We will illustrate parts of the central limit theorem with an example.

---

**EXAMPLE 42:** Consider the population of 256 integers from 100 to 355 (Table 24-7). The frequency distribution of these integers (Fig. 24-17a) shows that each class has a frequency of 32. We computed the population mean $\mu$ by adding all the integers and dividing by 256 and got 227. We also got a population standard deviation $\sigma$ of 76.

**TABLE 24-7**

| | | | | | | | | | | | | | | | |
|---|---|---|---|---|---|---|---|---|---|---|---|---|---|---|---|
| 174 | 230 | 237 | 173 | 178 | 242 | 113 | 177 | 238 | 175 | 303 | 172 | 114 | 179 | 307 | 176 |
| 190 | 246 | 253 | 189 | 194 | 258 | 129 | 193 | 254 | 191 | 319 | 188 | 130 | 111 | 108 | 103 |
| 206 | 262 | 269 | 205 | 210 | 274 | 145 | 209 | 270 | 207 | 335 | 204 | 146 | 127 | 124 | 119 |
| 222 | 278 | 285 | 221 | 226 | 290 | 161 | 225 | 286 | 223 | 351 | 220 | 162 | 143 | 140 | 135 |
| 300 | 229 | 302 | 228 | 296 | 233 | 298 | 232 | 301 | 239 | 236 | 231 | 304 | 241 | 306 | 240 |
| 316 | 245 | 318 | 244 | 312 | 249 | 314 | 248 | 317 | 255 | 252 | 247 | 320 | 257 | 322 | 256 |
| 332 | 261 | 334 | 260 | 328 | 265 | 330 | 264 | 333 | 271 | 268 | 263 | 336 | 273 | 338 | 272 |
| 348 | 277 | 350 | 276 | 344 | 281 | 346 | 280 | 349 | 287 | 284 | 279 | 352 | 289 | 354 | 288 |
| 166 | 102 | 109 | 165 | 170 | 234 | 105 | 169 | 110 | 167 | 295 | 164 | 106 | 171 | 299 | 168 |
| 182 | 118 | 125 | 181 | 186 | 250 | 121 | 185 | 126 | 183 | 311 | 180 | 122 | 187 | 315 | 184 |
| 198 | 134 | 141 | 197 | 202 | 266 | 137 | 201 | 142 | 199 | 327 | 196 | 138 | 203 | 331 | 200 |
| 178 | 150 | 157 | 213 | 218 | 282 | 153 | 217 | 158 | 215 | 343 | 212 | 154 | 219 | 347 | 216 |
| 292 | 101 | 294 | 100 | 297 | 107 | 104 | 235 | 293 | 195 | 323 | 192 | 305 | 115 | 112 | 243 |
| 308 | 117 | 310 | 116 | 313 | 123 | 120 | 251 | 309 | 211 | 339 | 208 | 321 | 131 | 128 | 259 |
| 324 | 133 | 326 | 132 | 329 | 139 | 136 | 267 | 325 | 227 | 355 | 224 | 337 | 147 | 144 | 275 |
| 340 | 149 | 342 | 148 | 345 | 155 | 152 | 283 | 341 | 159 | 156 | 151 | 353 | 163 | 160 | 291 |

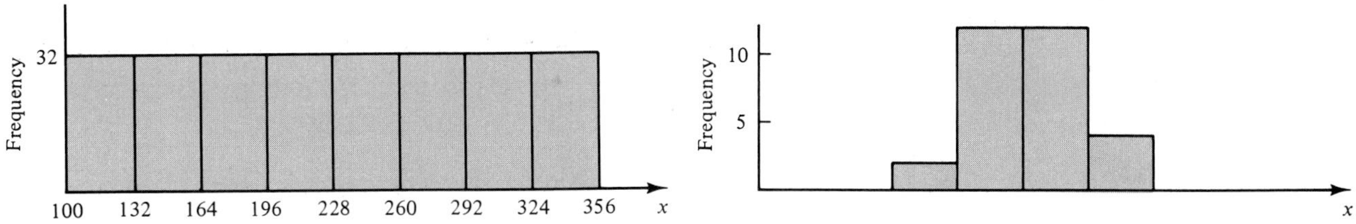


(a) Frequency distribution of the population in Table 24-7

(b) Sampling distribution of the means (frequency distribution of the means of 32 samples drawn from the population in Table 24-7)

**FIGURE 24-17**

Then 32 samples of 16 integers each was drawn from the population. For each sample we got the sample mean $\bar{x}$. The 32 values we got for $\bar{x}$ are

**The calculations in this example were done by computer, but you will not need a computer to make use of the results of this demonstration.**

234 229 236 235 227 232 229 224 266 230 218 212 234 226 237 216
170 222 220 218 232 247 262 277 265 252 238 225 191 204 216 223

We now make a frequency distribution for the sample means, which is called a *sampling distribution of the mean* (Fig. 24-17b). We further compute the mean of the sample means and get 229, nearly the same value as for the population mean $\mu$. The standard deviation of the sample means is computed to be 20. We see that:

1. The mean of the $\bar{x}$ distribution (229) is approximately equal to the mean $\mu$ of the population (227).
2. By the central limit theorem, the standard deviation $SE_{\bar{x}}$ of the $\bar{x}$ distribution is found by dividing $\sigma$ by the square root of the sample size.

$$SE_{\bar{x}} = \frac{\sigma}{\sqrt{n}} = \frac{76}{\sqrt{16}} = 19$$

This compares well with the actual standard deviation of the $\bar{x}$ distribution, which was 20.

We now take the final step and show how to predict the population mean $\mu$ from the mean $\bar{x}$ and standard deviation $s$ *of a single sample*. We also give the probability that our prediction is correct.

## Predicting $\mu$ from a Single Sample

If the standard deviation of the $\bar{x}$ distribution, $SE_{\bar{x}}$, is small, most of the $\bar{x}$'s will be near the center $\mu$ of the population. Thus a particular $\bar{x}$ has a good chance of being close to $\mu$, and will hence be a good estimator of $\mu$. Conversely, a large $SE_{\bar{x}}$ means that a given $\bar{x}$ will be a poor estimator of $\mu$.

However, since the frequency distribution of the $\bar{x}$'s is normal, the chance of a single $\bar{x}$ lying within one standard deviation of the mean $\mu$ is 68%. Conversely, the mean $\mu$ has a 68% chance of lying within one standard deviation of *a single randomly chosen* $\bar{x}$. Thus there is a 68% chance that the true population mean $\mu$ falls within the interval

$$\bar{x} \pm SE_{\bar{x}}$$

In this way we can estimate the population mean $\mu$ from a *single sample.* Not only that, but we are able to give a range of values within which $\mu$ must lie, and give *the probability* that $\mu$ will fall within that interval.

There is just one difficulty: To compute the standard error $SE_{\bar{x}}$ by the central limit theorem, we must divide the population standard deviation $\sigma$ by the square root of the sample size. But $\sigma$ *is not usually known*. Thus it is common practice to use the sample standard deviation $s$ instead. Thus the standard error of the mean is approximated by the following formula (valid if the population is large).

| | | |
|---|---|---|
| Standard Error of the Mean $\bar{x}$ | $SE_{\bar{x}} = \dfrac{\sigma}{\sqrt{n}} \approx \dfrac{s}{\sqrt{n}}$ | **256** |

---

**EXAMPLE 43:** For the population of 256 integers (Table 24-7), the central limit theorem says that 68% of the sample means should fall within one standard error of $\mu$, or within the interval $227 \pm 19$. Is this statement correct?

**Solution:** We count the number of sample means that actually do fall within this interval and get 23, which is 72% of the total number of samples.

---

**EXAMPLE 44:** One of the 32 samples in Example 42 has a mean of 234 and a standard deviation $s$ of 67. Use these figures to estimate the mean $\mu$ of the population.

**Solution:** Computing the standard error from Eq. 256, with $s = 67$ and $n = 16$, gives

$$SE_{\bar{x}} \approx \frac{67}{\sqrt{16}} = 17$$

We may then predict that there is a 68% chance that $\mu$ will fall within the range

$$\mu = 234 \pm 17$$

We see that the true population mean 227 does fall within that range.

---

**EXAMPLE 45:** The heights of 64 randomly chosen students at Tech College were measured. The mean $\bar{x}$ was found to be 68.25 in. and the standard deviation $s$ was 2.21 in. Estimate the mean $\mu$ of the entire population of students at Tech College.

**Solution:** From Eq. 256, with $s = 2.21$ and $n = 64$,

$$\text{SE}_{\bar{x}} \approx \frac{s}{\sqrt{n}} = \frac{2.21}{\sqrt{64}} = 0.28$$

We may then state that there is a 68% chance that the mean height $\mu$ of all students at Tech College is

$$\mu = 68.25 \pm 0.28 \text{ in.}$$

In other words, the true mean has a 68% chance of falling within the interval

$$67.97 \text{ to } 68.53 \text{ in.}$$

### Confidence Intervals

The interval 67.97 to 68.53 in. found in Example 45 is called the *68% confidence interval*. Similarly, there is a 95% chance that the population mean falls within *two* standard errors of the mean,

$$\mu = \bar{x} \pm 2\text{SE}_{\bar{x}}$$

so we call it the *95% confidence interval.*

**EXAMPLE 46:** For the data of Example 45, find the 95% confidence interval.

**Solution:** We had found that $\text{SE}_{\bar{x}}$ was 0.28 in. Thus we may state that $\mu$ will lie within the interval

$$68.25 \pm 2(0.28)$$

or

$$68.25 \pm 0.56 \text{ in.}$$

This is the 95% confidence interval.

We find other confidence intervals in a similar way.

### Standard Error of the Standard Deviation

We may compute standard errors for sample statistics other than the mean. The most common is for the standard deviation. The standard error of the standard deviation is called $\text{SE}_s$, and is given by

| | | |
|---|---|---|
| Standard Error of the Standard Deviation | $\text{SE}_s = \dfrac{\sigma}{\sqrt{2n}}$ | **257** |

where again, we use the sample standard deviation $s$ if we don't know $\sigma$.

**EXAMPLE 47:** A single sample of size 32 drawn from a population is found to have a standard deviation $s$ of 21.55. Give the population standard deviation $\sigma$ with a confidence level of 68%.

**Solution:** From Eq. 257, with $\sigma \approx s = 21.55$ and $n = 32$,

$$SE_s = \frac{21.55}{\sqrt{2(32)}} = 2.69$$

Thus there is a 68% chance that $\sigma$ falls within the interval

$$21.55 \pm 2.69$$

## EXERCISE 5—STANDARD ERRORS

1. The heights of 49 randomly chosen students at Tech College were measured. Their mean $\bar{x}$ was found to be 69.47 in. and their standard deviation $s$ was 2.35 in. Estimate the mean $\mu$ of the entire population of students at Tech College with a confidence level of 68%.
2. For the data of Problem 1, estimate the population mean with a 95% confidence interval.
3. For the data of Problem 1, estimate the population standard deviation with a 68% confidence interval.
4. For the data of Problem 1, estimate the population standard deviation with a 95% confidence interval.
5. A single sample of size 32 drawn from a population is found to have a mean of 164 and a standard deviation $s$ of 16.31. Give the population mean with a confidence level of 68%.
6. For the data of Problem 5, estimate the population mean with a 95% confidence interval.
7. For the data of Problem 5, estimate the population standard deviation with a 68% confidence interval.
8. For the data of Problem 5, estimate the population standard deviation with a 95% confidence interval.

## CHAPTER 24 REVIEW PROBLEMS

*Label each type of data as continuous, discrete, or categorical.*

1. The life of certain radios.
2. The number of houses sold each day.
3. The colors of the cars at a certain dealership.
4. The sales figures for a certain product for the years 1970–1980 are:

| *Year* | *Number Sold* |
|---|---|
| 1970 | 1344 |
| 1971 | 1739 |
| 1972 | 2354 |
| 1973 | 2958 |
| 1974 | 3153 |
| 1975 | 3857 |
| 1976 | 3245 |
| 1977 | 4736 |
| 1978 | 4158 |
| 1979 | 5545 |
| 1980 | 6493 |

Make an $x$–$y$ graph of the sales versus the year.

5. Make a bar graph for the data of Problem 4.

6. If we toss three coins, what is the probability of getting one head and two tails?

7. If we throw two dice, what chance is there that their sum is 8?

8. A die is rolled twice. What is the probability that both rolls will give a 2?

9. At a certain factory, 72% of the workers have brown hair, 6% are left-handed, and 3% have both brown hair and are left-handed. What is the probability that a worker chosen at random will have either brown hair or be left-handed or both?

10. What is the probability of tossing a 5 or a 9 when two dice are tossed?

11. Find the area under the normal curve between the mean and 0.5 standard deviation to one side of the mean.

12. Find the area under the normal curve between 1.1 standard deviations on both sides of the mean.

13. Find the area in the tail of the normal curve to the left of $z = 0.4$.

14. Find the area in the tail of the normal curve to the right of $z = 0.3$.

*Use the following table of data for Problems 16–29.*

| | | | | | | | |
|---|---|---|---|---|---|---|---|
| 146 | 153 | 183 | 148 | 116 | 127 | 162 | 153 |
| 168 | 161 | 117 | 153 | 116 | 125 | 173 | 131 |
| 117 | 183 | 193 | 137 | 188 | 159 | 154 | 112 |
| 174 | 182 | 144 | 144 | 133 | 167 | 192 | 145 |
| 162 | 138 | 137 | 154 | 141 | 129 | 137 | 152 |

15. Determine the range of the data.

16. Make a frequency distribution using class widths of 5. Show both absolute and relative frequency.

17. Draw a frequency histogram showing both absolute and relative frequency.

18. Draw a frequency polygon showing both absolute and relative frequency.

19. Make a cumulative frequency distribution.

20. Draw a cumulative frequency polygon.

21. Find the mean.

22. Find the median.

23. Find the mode.

24. Find the variance.

25. Find the standard deviation.

26. Predict the population mean with a 68% confidence interval.

27. Predict the population mean with a 95% confidence interval.

28. Predict the population standard deviation with a 68% confidence interval.

29. Predict the population standard deviation with a 95% confidence interval.

30. On a test given to 300 students, the mean grade was 79.3 with a standard deviation of 11.6. Assuming a normal distribution, estimate the number of A grades (90 or over).

31. For the data of Problem 30, estimate the number of failing grades (60 or less).

32. Determine the quartiles and give the quartile range of the data

118 135 133 143 164 173 179 199 212 216 256

33. Determine the quartiles and give the quartile range of the data

167 245 327 486 524 639 797 853 974 1136 1162 1183

*Writing*

34. Give an example of one statistical claim (such as an advertisement, commercial, political message) that you have heard or read lately that has made you skeptical. State your reasons for being suspicious, and point out how the claim could otherwise have been presented to make it more plausible.

# 25

# THE STRAIGHT LINE

## OBJECTIVES

**When you have completed this chapter, you should be able to:**

- Calculate the distance between two points.
- Determine the slope of a line, given two points on the line.
- Find the slope of a line given its angle of inclination, and vice versa.
- Determine the slope of a line perpendicular to a given line.
- Calculate the angle between two lines.
- Write the equation of a line using the slope-intercept form, the point-slope form, or the two-point form.
- Solve applied problems involving the straight line.

This chapter begins our study of *analytic geometry* with the straight line and continues it in Chapter 26 with the conic sections. We place geometric figures, such as points, lines, circles, and so on, on coordinate axes, where they may be studied using the methods of algebra. The basic idea of analytic geometry is that geometric figures (made up of points) and algebraic equations (representing sets of numbers) are related in a special way. If, for example, we are talking about a line, the points *on* the line can be represented by pairs of numbers $(x, y)$ that satisfy the line's equation. Conversely, if we know that numbers satisfy the equation, they must represent points that are *on* the line.

We start by computing *lengths of line segments* and then study the idea of *slope* of a line and of a curve. We then learn to write the equation of a straight line, just as we will learn to write the equations of circles, parabolas, ellipses, and hyperbolas in the following chapter. Why? Because these equations enable us to understand and to use these geometric figures in ways that are not otherwise possible. By combining geometry and algebra in analytic geometry we gain powerful tools that can be used in applications.

## 25-1 LENGTH OF A LINE SEGMENT

### Line Segments on or Parallel to a Coordinate Axis

When we speak about the length of a line segment, or about the distance between two points, we usually mean the *magnitude* of that length or distance.

To find the magnitude of the length of a line segment lying on or parallel to the $x$ (or $y$) axis, simply subtract the abscissa (or ordinate) of either endpoint from the abscissa (or ordinate) of the other endpoint, and take the absolute value of this result.

**EXAMPLE 1:** The magnitudes of the lengths of the lines in Fig. 25-1 are

(a) $AB = 5 - 2 = 3$

(b) $PQ = 5 - (-2) = 7$

(c) $RS = -1 - (-5) = 4$

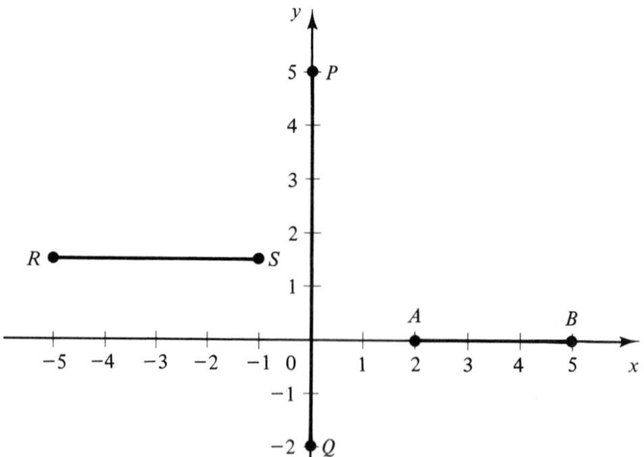


**FIGURE 25-1**

## Directed Distance

Sometimes when speaking about the length of a line segment or the distance between two points, it is necessary to specify *direction* as well as magnitude. When we specify the directed distance *AB*, for example, we mean the distance *from A to B*. This is sometimes written $\overline{AB}$.

**Note that directed distance applies only to a line parallel to a coordinate axis.**

**EXAMPLE 2:** For the two points $P(5, 2)$ and $Q(1, 2)$, the directed distance $PQ$ is $PQ = 1 - 5 = -4$ and the directed distance $QP$ is $QP = 5 - 1 = 4$.

## Increments

Let us say that a particle is moving along a curve (Fig. 25-2) from point $P$ to point $Q$. In doing so, its abscissa changes from $x_1$ to $x_2$. We call this change an *increment* in $x$, and label it $\Delta x$ (read *delta x*). Similarly, the ordinate changes from $y_1$ to $y_2$ and is labeled $\Delta y$. The increments are found simply by subtracting the coordinates at $P$ from those at $Q$.

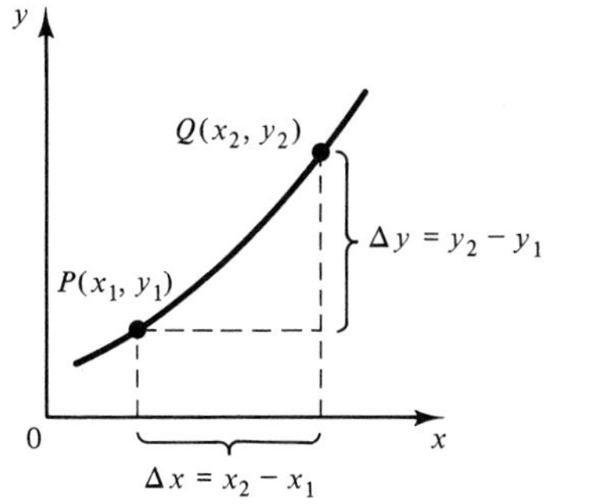


**FIGURE 25-2 Increments.**

| Increments | $\Delta x = x_2 - x_1$ and $\Delta y = y_2 - y_1$ | 327 |
|---|---|---|

**EXAMPLE 3:** A particle moves from $P_1(2, 5)$ to $P_2(7, 3)$. The increments in its coordinates are

$$\Delta x = x_2 - x_1 = 7 - 2 = 5$$

and

$$\Delta y = y_2 - y_1 = 3 - 5 = -2$$

## Distance Formula

We now find the length of a line inclined at any angle, such as $PQ$ in Fig. 25-3. We first draw a horizontal line through $P$ and drop a perpendicular from $Q$, forming a right triangle $PQR$. The sides of the triangle are $d$, $\Delta x$, and $\Delta y$. By the Pythagorean theorem,

$$d^2 = (\Delta x)^2 + (\Delta y)^2 = (x_2 - x_1)^2 + (y_2 - y_1)^2$$

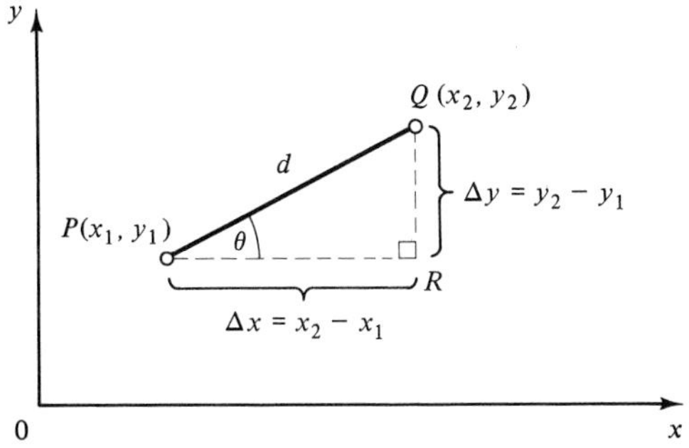


**FIGURE 25-3 Length of a line segment.**

Since we want the magnitude of the distance, we take only the positive root and get

| Distance Formula | $d = \sqrt{(\Delta x)^2 + (\Delta y)^2}$<br>$= \sqrt{(x_2 - x_1)^2 + (y_2 - y_1)^2}$ | **273** |
|---|---|---|

---

**EXAMPLE 4:** Find the length of the line segment with the endpoints (3, −5) and (−1, 6).

**Solution:** Let us give the first point (3, −5) the subscripts 1, and the other point the subscripts 2 (it does not matter which we label 1). So

$$x_1 = 3, \quad y_1 = -5, \quad x_2 = -1, \quad y_2 = 6$$

Substituting into the distance formula, we obtain

$$d = \sqrt{(-1 - 3)^2 + [6 - (-5)]^2} = \sqrt{(-4)^2 + (11)^2} = 11.7 \quad \text{(rounded)}$$

---

| Common Error | Do not take the square root of each term separately.<br>$d \neq \sqrt{(x_2 - x_1)^2} + \sqrt{(y_2 - y_1)^2}$ |
|---|---|

## EXERCISE 1—LENGTH OF A LINE SEGMENT

Find the distance between the given points to three significant digits.

1. (5, 0) and (2, 0)
2. (0, 3) and (0, −5)
3. (−2, 0) and (7, 0)
4. (−4, 0) and (−6, 0)
5. (0, −2.74) and (0, 3.86)
6. (55.34, 0) and (25.38, 0)
7. (7, 2) and (3, 2)
8. (−1, 6) and (−1, 3)
9. (−5, −5) and (−5, −6)
10. (9, −2) and (−9, −2)
11. (5.59, 3.25) and (8.93, 3.25)
12. (−2.06, −5.83) and (−2.06, −8.34)
13. (3, 9) and (−3, 5)
14. (−6, −7) and (−9, −2)
15. (−2.9, 5.3) and (5.8, −3.7)
16. (8.38, −3.95) and (2.25, −4.99)
17. (−47, 34) and (55, −48)
18. (−1.1, 4.2) and (4.2, −1.1)

Find the directed distance $AB$.

19. $A(3,0)$; $B(5, 0)$
20. $A(3.95, -2.07)$; $B(-3.95, -2.07)$
21. $B(-8, -2)$; $A(-8, -5)$
22. $A(-9, -2)$; $B(17, -2)$
23. $B(-6, -6)$; $A(-6, -7)$
24. $B(11.5, 3.68)$; $A(11.5, -5.38)$

25. Find the length of girder $AB$ in Fig. 25-4.

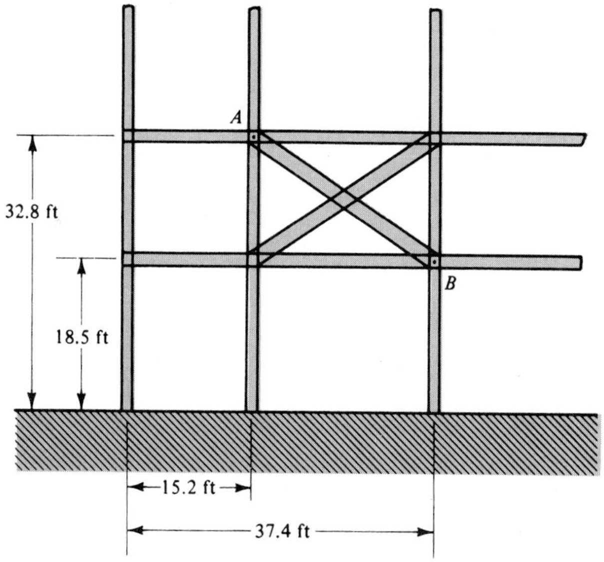


FIGURE 25-4

26. Find the distance between the holes in Fig. 25-5.

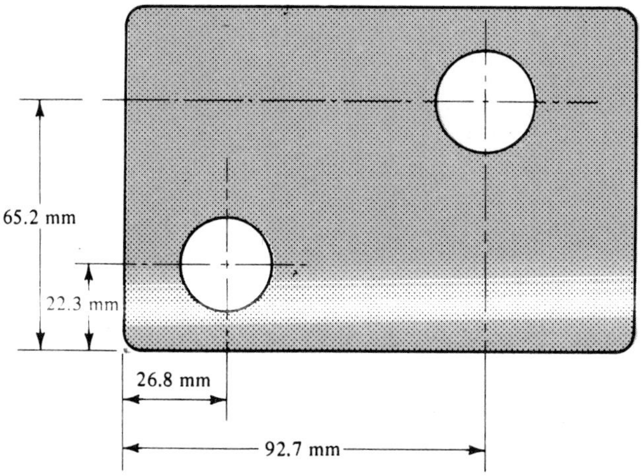


FIGURE 25-5

27. A triangle has vertices at (3, 5), (−2, 4), and (4, −3). Find the length of each side. Then compute the area using Hero's formula (Eq. 138). Work to three significant digits.

**Computer**

28. Write a program that will accept as input the coordinates of the vertices of a triangle, and then compute and print the length of each side and the area of the triangle.

## 25-2 SLOPE AND ANGLE OF INCLINATION

### Definition

Referring back to Fig. 25-3, we call the horizontal distance between the points $P$ and $Q$ the *run* from $P$ to $Q$, which is equal to $\Delta x$ for this line. The vertical distance $\Delta y$ between the same points is called the *rise*. The *slope* of the line $PQ$ is the ratio of the rise to the run. It is usually given the symbol $m$.

| | | |
|---|---|---|
| Slope | $m = \dfrac{\text{rise}}{\text{run}} = \dfrac{\Delta y}{\Delta x} = \dfrac{y_2 - y_1}{x_2 - x_1}$ | **274** |

*The slope is equal to the rise divided by the run.*

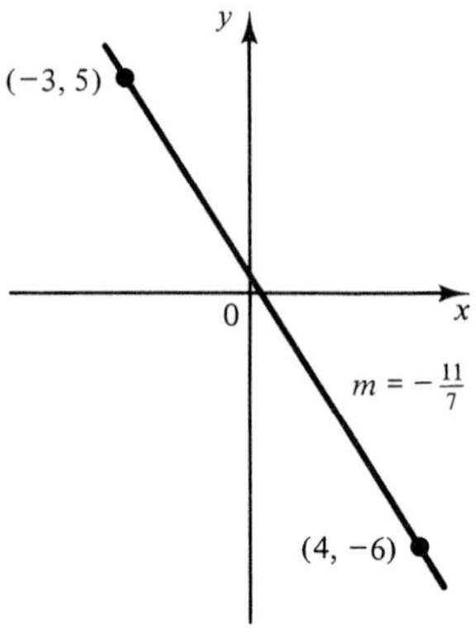

**FIGURE 25-6**

**EXAMPLE 5:** Find the slope of the line connecting the points $(-3, 5)$ and $(4, -6)$ (Fig. 25-6).

**Solution:** As when using the distance formula, it does not matter which is called point 1 and which is point 2. Let us choose

$$x_1 = -3, \qquad y_1 = 5, \qquad x_2 = 4, \qquad y_2 = -6$$

Then by Eq. 274,

$$\text{slope } m = \frac{-6 - 5}{4 - (-3)} = \frac{-11}{7} = -\frac{11}{7}$$

If we had chosen $x_1 = 4$, $y_1 = -6$ and $x_2 = -3$, $y_2 = 5$, the computation for slope,

$$\text{slope } m = \frac{5 - (-6)}{-3 - 4} = \frac{11}{-7} = -\frac{11}{7}$$

would have given the same result.

**EXAMPLE 6:** Find the slope of the line in Fig. 25-7.

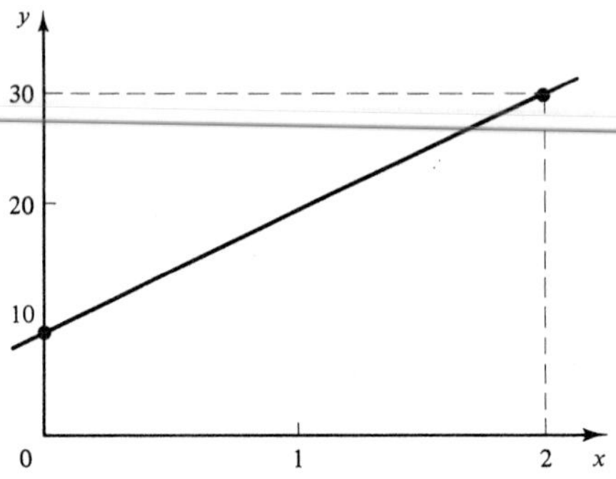

**FIGURE 25-7**

**Solution:** The rise is $30 - 10 = 20$ and the run is $2 - 0 = 2$, so

$$m = \frac{20}{2} = 10$$

(Do not be thrown off by the different scales on each axis).

| | |
|---|---|
| Common Error | Be careful not to mix up the subscripts. $m \neq \dfrac{y_2 - y_1}{x_1 - x_2}$ |

### Horizontal and Vertical Lines

For any two points on a horizontal line, the values of $y_1$ and $y_2$ in Eq. 274 are equal, making the slope equal to zero. For a vertical line, the values of $x_1$ and $x_2$ are equal, giving division by zero. Hence the slope is undefined for a vertical line. The slopes of various lines are shown in Fig. 25-8.

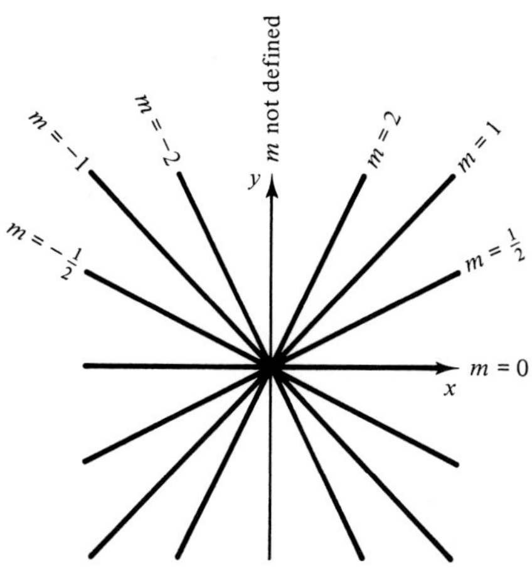


FIGURE 25-8

### Angle of Inclination

The smallest positive angle that a line makes with the positive $x$ direction is called the *angle of inclination* $\theta$ of the line (Fig. 25-9). A line parallel to the $x$ axis has an angle of inclination of zero. Thus the angle of inclination can have values from 0° up to 180°. From Fig. 25-3,

$$\tan\theta = \frac{\text{opposite side}}{\text{adjacent side}} = \frac{y_2 - y_1}{x_2 - x_1} = \frac{\text{rise}}{\text{run}} = \frac{\Delta y}{\Delta x}$$

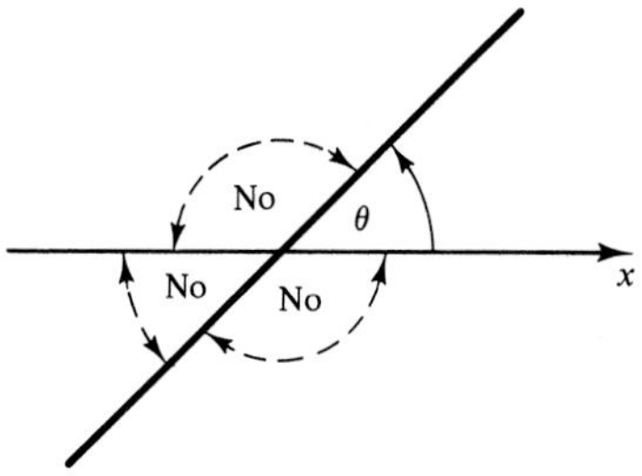


**FIGURE 25-9 Angle of inclination, θ.**

But this is the slope $m$ of the line, so

| $m = \tan\theta$ <br> $0° \leqq \theta < 180°$ | **275** |
|---|---|

*The slope of a line is equal to the tangent of the angle of inclination θ (except when θ = 90°).*

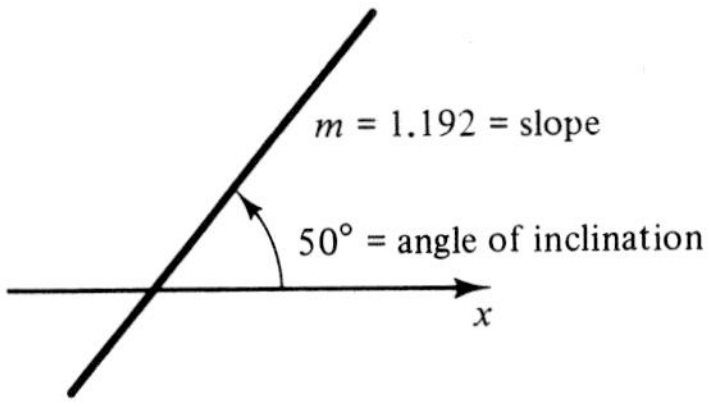


FIGURE 25-10

**EXAMPLE 7:** The slope of a line having an angle of inclination of 50° (Fig. 25-10) is, by Eq. 275,

$$m = \tan 50° = 1.192 \quad \text{(rounded)}$$

**EXAMPLE 8:** Find the angle of inclination of a line (Fig. 25-11) having a slope of 3.

**Solution:** By Eq. 275, $\tan\theta = 3$, so

$$\theta = \arctan 3 = 71.6° \quad \text{(rounded)}$$

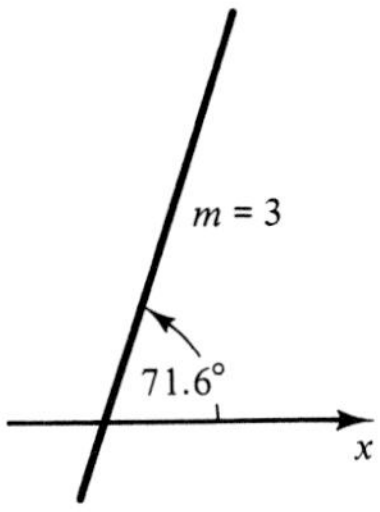


FIGURE 25-11

When the slope is negative, our calculator will give us a negative angle, which we then use to obtain a positive angle of inclination less than 180°.

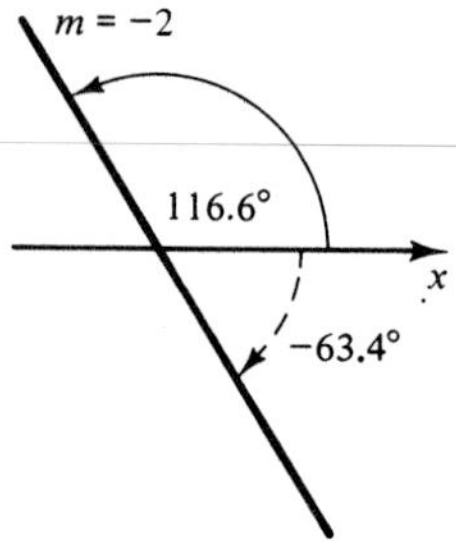


FIGURE 25-12

**EXAMPLE 9:** For a line having a slope of −2 (Fig. 25-12), arctan (−2) = −63.4° (rounded), so

$$\theta = 180° - 63.4° = 116.6°$$

**EXAMPLE 10:** Find the angle of inclination of the line passing through (−2.47, 1.74) and (3.63, −4.26), Fig. 25-13.

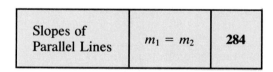

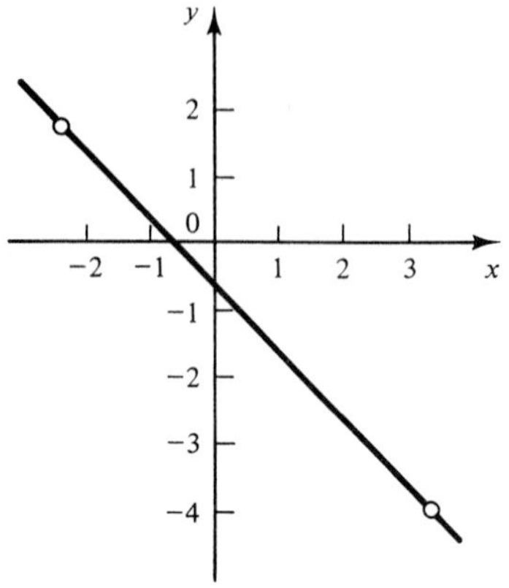


FIGURE 25-13

**Solution:** The slope, from Eq. 274, is

$$m = \frac{-4.26 - 1.74}{3.63 - (-2.47)} = -0.984$$

from which $\theta = 135.5°$.

When *different scales* are used for the $x$ and $y$ axes, the angle of inclination *will appear distorted.*

**EXAMPLE 11:** The angle of inclination of the line in Fig. 25-7 is $\theta$ = arctan 10 = 84.3°. Notice that the angle in the graph appears much smaller than 84.3°, due to the different scales on each axis.

## Slopes of Parallel and Perpendicular Lines

*Parallel* lines have, of course, *equal* slopes.

| | | |
|---|---|---|
| Slopes of Parallel Lines | $m_1 = m_2$ | **284** |

*Parallel lines have equal slopes.*

Now consider line $L_1$, with slope $m_1$ and angle of inclination $\theta_1$ (Fig. 25-14) and a *perpendicular* line $L_2$ with slope $m_2$ and angle of inclination $\theta_2 = \theta_1 + 90°$. Note that $\angle QOS = \theta_2 - 90° = \theta_1 + 90° - 90° = \theta_1$, so right

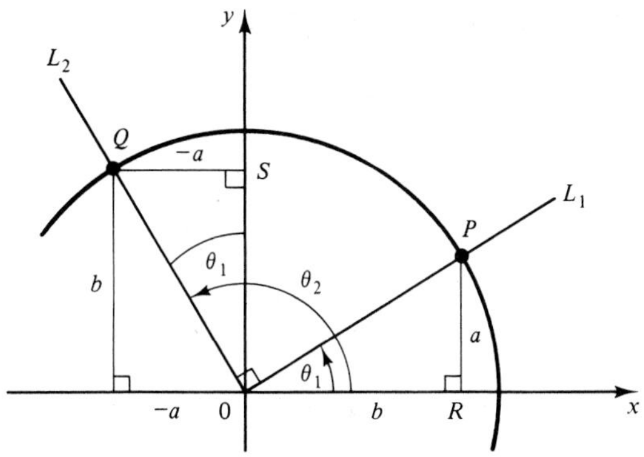


**FIGURE 25-14 Slopes of perpendicular lines.**

triangles $OPR$ and $OQS$ are congruent. Thus the magnitudes of their corresponding sides $a$ and $b$ are equal. The slope $m_1$ of $L_1$ is $a/b$, and the slope $m_2$ of $L_2$ is $-b/a$, so

| | | |
|---|---|---|
| Slopes of Perpendicular Lines | $m_1 = -\dfrac{1}{m_2}$ | **285** |

*The slope of a line is the negative reciprocal of the slope of a perpendicular to that line.*

---

**EXAMPLE 12:** Any line perpendicular to a line whose slope is 5 has a slope of $-\frac{1}{5}$.

---

| | |
|---|---|
| Common Error | The minus sign in Eq. 285 is often forgotten. $$m_1 \neq \frac{1}{m_2}$$ |

---

**EXAMPLE 13:** Is angle $ACB$ (Fig. 25-15) a right angle?

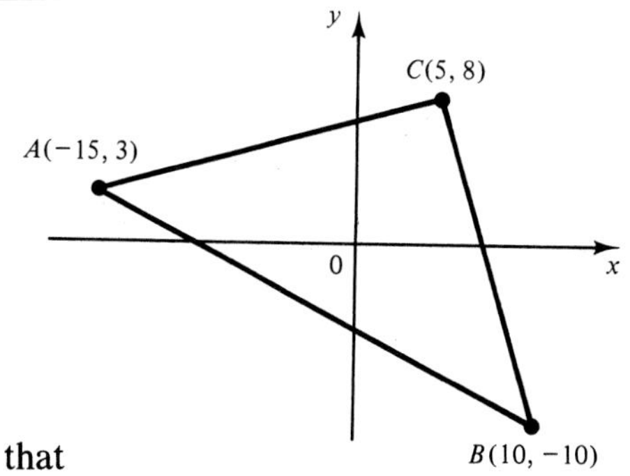


**FIGURE 25-15**

**Solution:** The slope of $AC$ is

$$\frac{8 - 3}{5 - (-15)} = \frac{5}{20} = \frac{1}{4}$$

To be perpendicular, $BC$ must have a slope of $-4$. Its actual slope is

$$\frac{-10 - 8}{10 - 5} = \frac{-18}{5} = -3.6$$

Although lines $AC$ and $BC$ *appear* to be perpendicular, we have shown that they are not. Thus $\angle ACB$ is not a right angle.

---

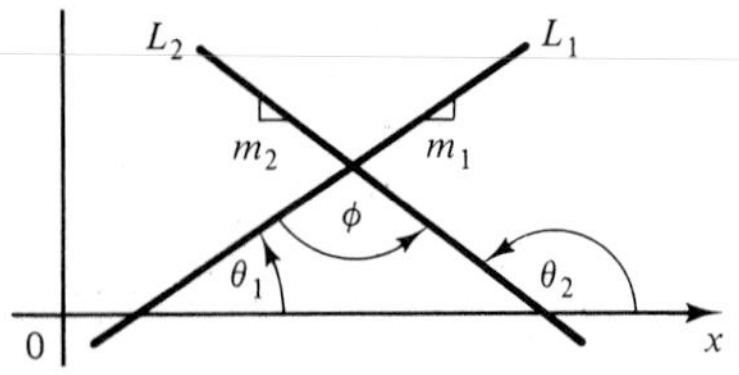


**FIGURE 25-16 Angle of intersection between two lines.**

## Angle of Intersection between Two Lines

Figure 25-16 shows two lines $L_1$ and $L_2$ intersecting at an angle $\phi$, measured counterclockwise from line 1 to line 2. We want a formula for $\phi$ in terms of the slopes $m_1$ and $m_2$ of the lines. By Eq. 142, $\theta_2$ equals the sum of the angle of inclination $\theta_1$ and the angle of intersection $\phi$. So $\phi = \theta_2 - \theta_1$. Taking the tangent of both sides gives $\tan \phi = \tan(\theta_2 - \theta_1)$, or

$$\tan \phi = \frac{\tan \theta_2 - \tan \theta_1}{1 + \tan \theta_1 \tan \theta_2}$$

by the trigonometric identity for the tangent of the difference of two angles (Eq. 169). But the tangent of an angle of inclination is the slope, so

| | | |
|---|---|---|
| Angle between Two Lines | $\tan \phi = \dfrac{m_2 - m_1}{1 + m_1 m_2}$ | **286** |

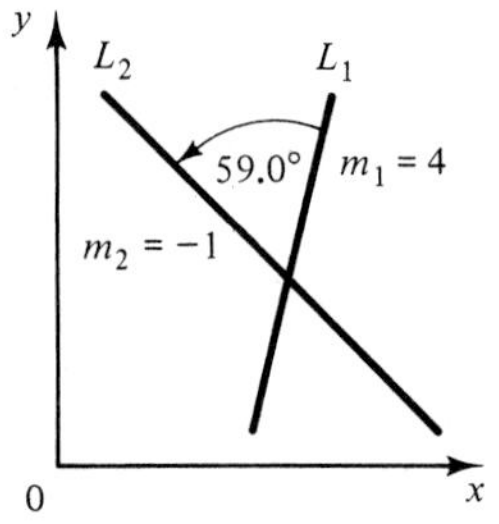


**FIGURE 25-17**

**EXAMPLE 14:** The tangent of the angle of intersection between line $L_1$ having a slope of 4, and line $L_2$ having a slope of $-1$, is

$$\tan \phi = \frac{-1 - 4}{1 + 4(-1)} = \frac{-5}{-3} = \frac{5}{3}$$

from which $\phi = \arctan \frac{5}{3} = 59.0°$ (rounded). This angle is measured counterclockwise from line 1 to line 2, as shown in Fig. 25-17.

## Tangents and Normals to Curves

We know what the tangent to a circle is, and probably have an intuitive idea of what the *tangent to a curve* is. We'll speak a lot about tangents, so let us try to get a clear idea of what one is.

To determine the tangent line at $P$ (Fig. 25-18), we start by selecting a second point $Q$, on the curve. As shown, $Q$ can be on either side of $P$. The line $PQ$ is called a *secant line*. We then let $Q$ approach $P$ (from either side). If the secant lines $PQ$ approach a single line $T$ as $Q$ approaches $P$, then line $T$ is called the tangent line at $P$.

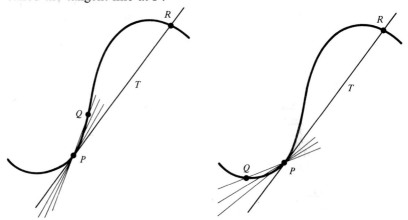

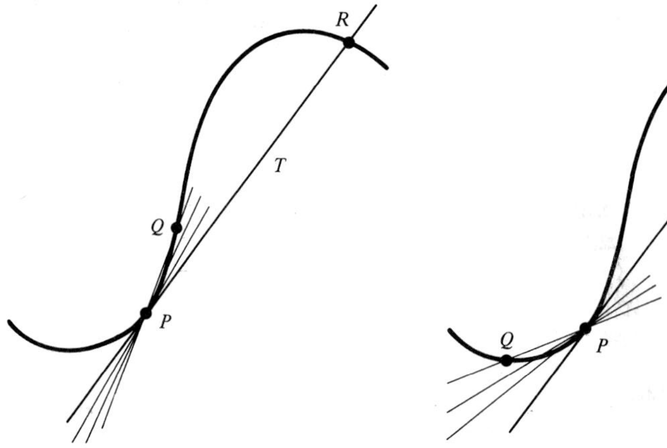


**FIGURE 25-18 A tangent line $T$ at point $P$, here defined as the limiting position of a secant line $PQ$, as $Q$ approaches $P$.**

Note that the tangent line can intersect the curve at more than one point, such as point $R$. However, at points "near" $P$, the tangent line intersects the curve just once while a secant line intersects the curve twice. The *slope of a curve* at some point $P$ is defined as the *slope of the tangent* to the curve drawn through that point.

**In calculus you will see that the slope gives the *rate of change* of the function, an extremely useful quantity to find. You will see that it is found by taking the *derivative* of the function.**

**EXAMPLE 15:** Plot the curve $y = x^2$ for integer values of $x$ from 0 to 4. By eye, draw the tangent to the curve at $x = 2$ and find the approximate value of the slope.

**Solution:** We make a table of point pairs.

| $x$ | 0 | 1 | 2 | 3 | 4 |
|---|---|---|---|---|---|
| $y$ | 0 | 1 | 4 | 9 | 16 |

The plot and the tangent line are shown in Fig. 25-19. Using the scales on the $x$ and $y$ axes, we measure a rise of 12 units for the tangent line in a run of 3 units. The slope of the tangent line is then

$$m_t \simeq \frac{12}{3} = 4$$

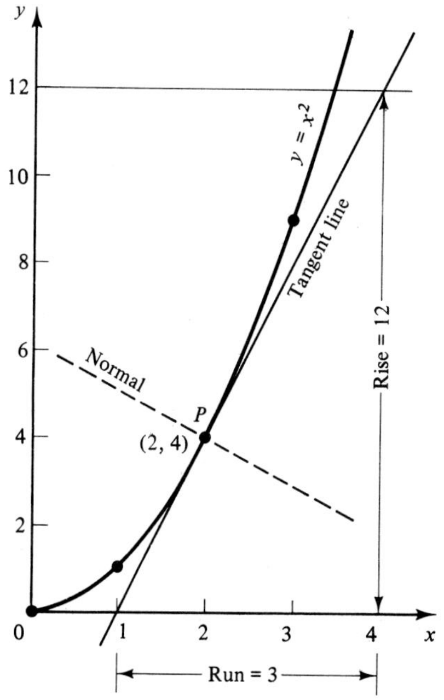


**FIGURE 25-19** **Slope of a curve.**

The *normal* to a curve at a given point is the line perpendicular to the tangent at that point, as in Fig. 25-19. Its slope is the negative reciprocal of the slope of the tangent. For Example 15, $m_n = -\frac{1}{4}$.

## EXERCISE 2—SLOPE AND ANGLE OF INCLINATION

Find the slope of the line passing through the given points.

**Work to three significant digits here.**

1. (8, 3) and (−2, 4)
2. (−3, 7) and (5, 1)
3. (−2, $a$) and ($a$, 3)
4. (5.2, 8.3) and (−2.8, −1.4)
5. (284, 184) and (338, 402)
6. (−2.14, −7.36) and (−1.37, −2.08)

Find the slope of the line having the given angle of inclination.

7. 38.2°
8. 77.9°
9. 1.83 rad
10. 58°14′
11. 156.3°
12. 132.8°

Find the angle of inclination, in decimal degrees, of a line having the given slope.

13. $m = 3$
14. $m = 1.84$
15. $m = -4$
16. $m = -2.75$
17. $m = 0$
18. $m = -15$

Find the slope of a line perpendicular to a line having the given slope.

19. $m = 5$
20. $m = 2$
21. $m = 4.8$
22. $m = -1.85$
23. $m = -2.85$
24. $m = -5.372$

Find the slope of a line perpendicular to a line having the given angle of inclination.

25. 58.2°
26. 1.84 rad
27. 136°44′

Find the angle of inclination, in decimal degrees, of a line passing through the given points.

28. (5, 2) and (−3, 4)
29. (−2.5, −3.1) and (5.8, 4.2)
30. (6, 3) and (−1, 5)
31. ($x$, 3) and ($x$ + 5, 8)

### Angle between Two Lines

32. Find the angle of intersection between line $L_1$ having a slope of 1 and line $L_2$ having a slope of 6.
33. Find the angle of intersection between line $L_1$ having a slope of 3 and line $L_2$ having a slope of $-2$.
34. Find the angle of intersection between line $L_1$ having an angle of inclination of 35° and line $L_2$ having an angle of inclination of 160°.
35. Find the angle of intersection between line $L_1$ having an angle of inclination of 22° and line $L_2$ having an angle of inclination of 86°.

### Tangents to Curves

Plot the functions from $x = 0$ to $x = 4$. Graphically find the slope of the curve at $x = 2$.

36. $y = x^3$
37. $y = -x^2 + 3$
38. $y = \dfrac{x^2}{4}$

### Applications

39. The distance between two stakes on a slope is taped at 2055 ft, and the angle of the slope with the horizontal is 12.3°. Find the horizontal distance between the stakes.
40. What is the angle of inclination with the horizontal of a roadbed that rises 15 ft in each 250 ft, measured horizontally?
41. How far apart must two stakes on a 7° slope be placed so that the horizontal distance between them is 1250 m?
42. On a 5% road grade, at what angle is the road inclined to the horizontal? How far does one rise in traveling upward 500 ft; measured along the road?
43. A straight tunnel under a river is 755 ft long and descends 12 ft in this distance. What angle does the tunnel make with the horizontal?
44. A straight driveway slopes downward from a house to a road and is 28 m in length. If the angle of inclination from the road to the house is 3.6°, find the height of the house above the road.
45. An escalator is built so as to rise 2 m for each 3 m of horizontal travel. Find its angle of inclination.
46. A straight highway makes an angle of 4.5° with the horizontal. How much does the highway rise in a distance of 2500 ft, measured along the road?

**In some fields, such as highway work, the word *grade* is used instead of slope. It is usually expressed as a percent:**

***percent grade = 100 × slope.***

**Thus, a 5% grade rises 5 units for every 100 units of run.**

## 25-3 EQUATION OF A STRAIGHT LINE

We want to be able to write the equation of a straight line, given any two pieces of information about the line. We may be given, say, the slope of the line and a point through which it passes. Instead, we may know two points through which the line passes, or some other pieces of data.

The equation of a straight line, like any equation, can be written in several different forms. We'll find it convenient to develop one form for each of the ways in which the data about the line may be presented. Thus there is a point-slope form for when we know a point on the line and the slope, a two-point form, and so on. We start with the slope-intercept form.

### Slope-Intercept Form

Suppose that $P(x, y)$ is any point on a straight line. We seek an equation that links $x$ and $y$ in a functional relationship, so that, for any $x$, a value of $y$ can be found. We can get such an equation by applying the definition of slope to our point $P$, and some *known* point on the line. We use $(0, b)$ as the coordinates for the known point. For $P$, a general point on the line, we use coordinates $(x, y)$. For a line that intersects the $y$ axis $b$ units from the origin (Fig. 25-20) the rise is $y - b$ and the run is $x - 0$, so by Eq. 274,

$$m = \frac{y - b}{x - 0}$$

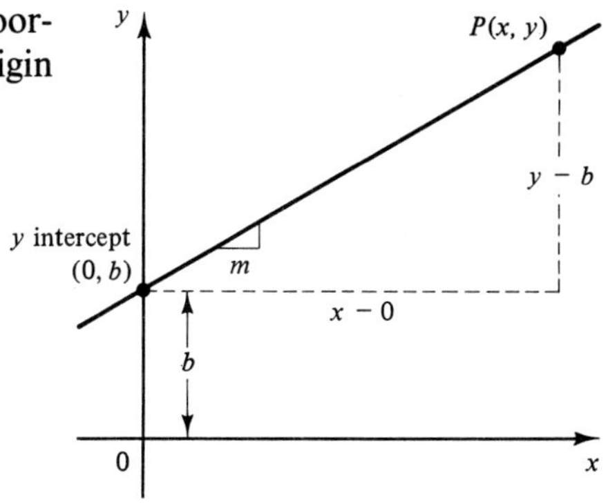

FIGURE 25-20

Simplifying, $mx = y - b$, or

| | | |
|---|---|---|
| Slope-Intercept Form | $y = mx + b$ | **279** |

This is called the slope-intercept form, because the slope $m$ and $y$ intercept $b$ are easily identified once the equation is in this form. For example, in the equation $y = 2x + 1$,

$$y = 2x + 1$$

slope ↑ (2) ↑ $y$ intercept (+1)

Here $m$, the slope, is 2 and $b$, the $y$ intercept, is 1.

---

**EXAMPLE 16:** Find the slope and $y$ intercept of the line $5x - 2y + 3 = 0$ (Fig. 25-21).

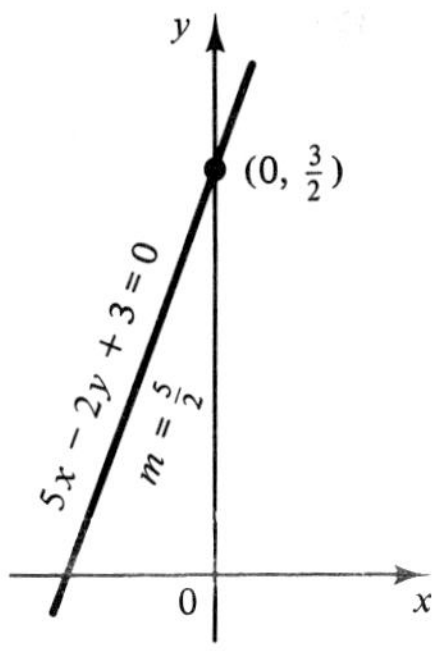

FIGURE 25-21

**Solution:** We first put the equation into slope-intercept form by solving for $y$,

$$2y = 5x + 3$$

$$y = \frac{5}{2}x + \frac{3}{2}$$

So

$$m = \frac{5}{2} \quad \text{and} \quad b = \frac{3}{2}$$

Therefore, the slope of the line is $\frac{5}{2}$ and the line intercepts the $y$ axis at $\frac{3}{2}$.

---

**EXAMPLE 17:** Write the equation of the line having a slope of 5 and a $y$ intercept of 7 (Fig. 25-22).

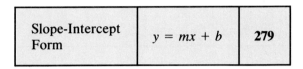

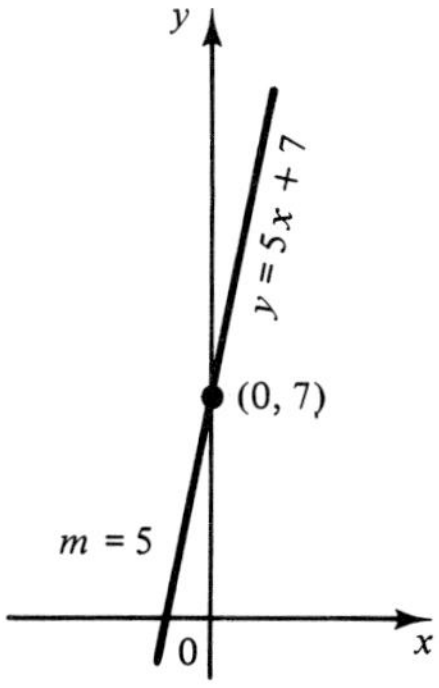

FIGURE 25-22

**Solution:** Substituting $m = 5$ and $b = 7$ into Eq. 279, we obtain

$$y = 5x + 7$$

---

## General Form

The slope-intercept form is the equation of a straight line in *explicit* form, $y = f(x)$. We can also write the equation of a straight line in *implicit* form, $f(x, y) = 0$, simply by transposing all terms to one side of the equal sign and simplifying. We usually write the $x$ term first, then the $y$ term, and finally the constant term. This form is referred to as the *general form* of the equation of a straight line.

| | | |
|---|---|---|
| General Form | $Ax + By + C = 0$ | **276** |

where $A$, $B$, and $C$ are constants.

You will see as we go along that there are several forms for the equation of a straight line, and we use the one that is most convenient in a particular problem. But to make it easy to compare answers, we usually rearrange the equation into general form.

**EXAMPLE 18:** Change the equation $y = \frac{6}{7}x - 5$ from slope-intercept form to general form.

**Solution:** Subtracting $y$ from both sides, we have

$$0 = \frac{6}{7}x - 5 - y$$

Multiplying by 7 and rearranging gives $6x - 7y - 35 = 0$.

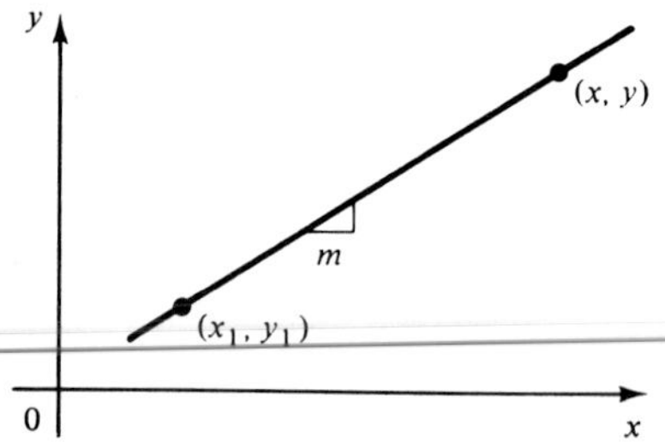


FIGURE 25-23

## Point-Slope Form

Let us write an equation for a line of slope $m$, but which passes through a given point $(x_1, y_1)$ which is *not,* in general, on either axis, as in Fig. 25-23. Again using the definition of slope (Eq. 274), with a general point $(x, y)$, we get the following form.

| | | |
|---|---|---|
| Point-Slope Form | $m = \dfrac{y - y_1}{x - x_1}$ | **281** |

This form of the equation of a straight line is most useful when we know the slope of a line and one point through which the line passes.

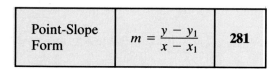

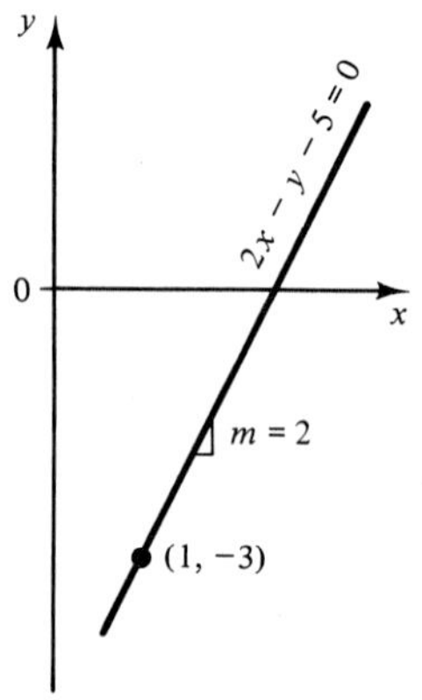


FIGURE 25-24

**EXAMPLE 19:** Write the equation in general form of the line having a slope of 2 and passing through the point $(1, -3)$ (Fig. 25-24).

**Solution:** Substituting $m = 2$, $x_1 = 1$, $y_1 = -3$ into Eq. 281 gives us

$$2 = \frac{y - (-3)}{x - 1}$$

Multiplying by $x - 1$ yields

$$2x - 2 = y + 3$$

or, in general form, $2x - y - 5 = 0$.

**EXAMPLE 20:** Write the equation in general form of the line passing through the point (3, 2) and perpendicular to the line $y = 3x - 7$ (Fig. 25-25).

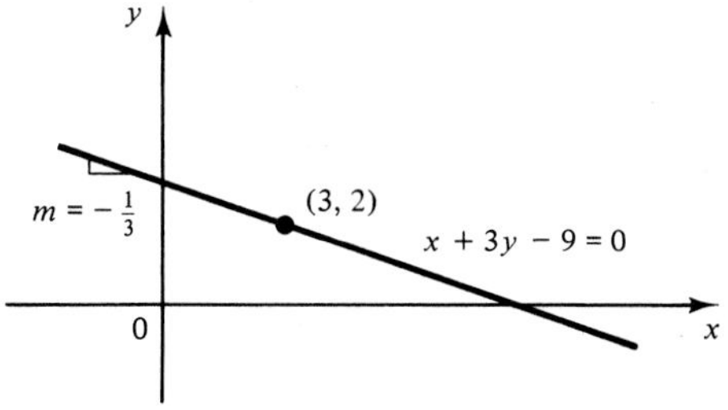


FIGURE 25-25

**Solution:** The slope of the given line is 3, so the slope of our perpendicular line is $-\frac{1}{3}$. Using the point-slope form, we obtain

$$-\frac{1}{3} = \frac{y - 2}{x - 3}$$

Going to the general form, $x - 3 = -3y + 6$, or

$$x + 3y - 9 = 0$$

**EXAMPLE 21:** Write the equation in general form of the line passing through the point (3.15, 5.88) and having an angle of inclination of 27.8° (Fig. 25-26).

**Solution:** From Eq. 275, $m = \tan 27.8° = 0.527$. From Eq. 281,

$$0.527 = \frac{y - 5.88}{x - 3.15}$$

Changing to general form, $0.527x - 1.66 = y - 5.88$, or

$$0.527x - y + 4.22 = 0$$

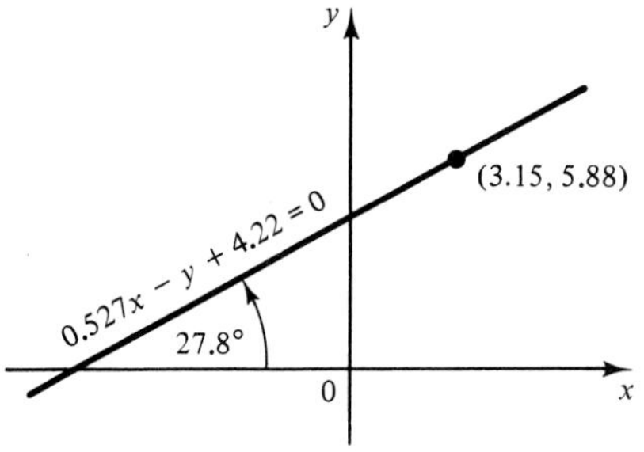


FIGURE 25-26

## Two-Point Form

If two points on a line are known, the equation of the line is easily written using the two-point form, which we now derive. If we call the points $P_1$ and $P_2$ in Fig. 25-27, the slope of the line is

$$m = \frac{y_2 - y_1}{x_2 - x_1}$$

The slope of the line segment connecting $P_1$ with any other point $P$ on the same line is

$$m = \frac{y - y_1}{x - x_1}$$

Since these slopes must be equal, we get

| | | |
|---|---|---|
| Two-Point Form | $\frac{y - y_1}{x - x_1} = \frac{y_2 - y_1}{x_2 - x_1}$ | **280** |

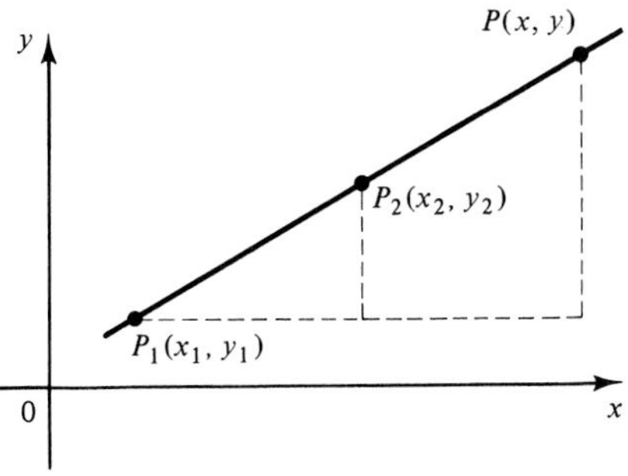


FIGURE 25-27

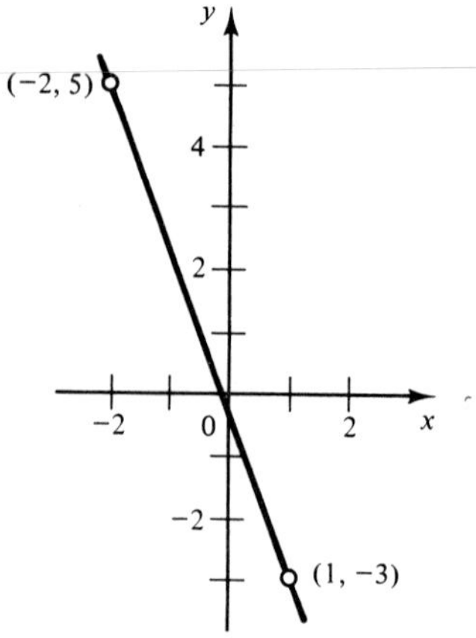


**FIGURE 25-28**

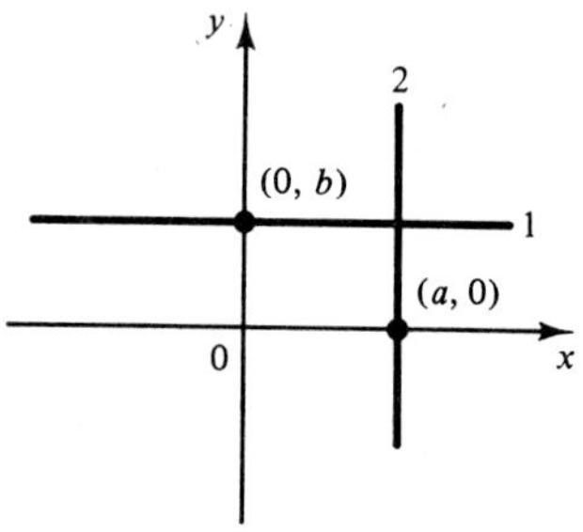


**FIGURE 25-29**

---

**EXAMPLE 22:** Write the equation in general form of the line passing through the points (1, −3) and (−2, 5), Fig. 25-28.

**Solution:** Calling the first given point $P_1$ and the second $P_2$ and substituting into Eq. 280, we have

$$\frac{y-(-3)}{x-1}=\frac{5-(-3)}{-2-1}=\frac{8}{-3}$$

$$\frac{y+3}{x-1}=-\frac{8}{3}$$

Putting the equation into general form, $3y + 9 = -8x + 8$, or

$$8x + 3y + 1 = 0$$

---

## Lines Parallel to the Coordinate Axes

Line 1 in Fig. 25-29 is parallel to the $x$ axis. Its slope is 0 and it cuts the $y$ axis at $(0, b)$. From the point-slope form, $y - y_1 = m(x - x_1)$, we get $y - b = 0(x - 0)$, or

| | | |
|---|---|---|
| Line Parallel to $x$ Axis | $y = b$ | **277** |

Line 2 has an undefined slope, but we get its equation by noting that $x = a$ at every point on the line, *regardless of the value of y*. Thus

| | | |
|---|---|---|
| Line Parallel to $y$ Axis | $x = a$ | **278** |

---

**EXAMPLE 23:**

(a) A line that passes through the point (5, −2) and is parallel to the $x$ axis has the equation $y = -2$.

(b) A line that passes through the point (5, −2) and is parallel to the $y$ axis has the equation $x = 5$.

---

## EXERCISE 3—EQUATION OF A STRAIGHT LINE

Find the slope and $y$ intercept for each equation.

1. $y = 3x - 5$
2. $y = 7x + 2$
3. $y = -\frac{1}{2}x - \frac{1}{4}$
4. $y = -x + 4$
5. $y = 6$
6. $3x - 2y = 4$
7. $x + 2y - 3 = 0$
8. $3x - \frac{1}{2}y - 5 = 0$
9. $2(x + 3) - 1 = 3(y - 2)$
10. $\frac{2x}{3} - \frac{3y}{5} = 4$

Write the equation, in general form, of each line.

11. slope = 4; $y$ intercept = $-3$
12. slope = $-1$; $y$ intercept = 2
13. slope = 2.25; $y$ intercept = $-1.48$
14. slope = $r$; $y$ intercept = $p$
15. passes through points (3, 5) and ($-1$, 2)
16. passes through points (4.24, $-1.25$) and (3.85, 4.27)
17. slope = $-4$, passes through ($-2$, 5)
18. slope = $-2$, passes through ($-2$, $-3$)
19. $x$ intercept = 5, $y$ intercept = $-3$
20. $x$ intercept = $-2$, $y$ intercept = 6
21. $y$ intercept = 3, parallel to $y = 5x - 2$
22. $y$ intercept = $-2.3$, parallel to $2x - 3y + 1 = 0$
23. $y$ intercept = $-5$, perpendicular to $y = 3x - 4$
24. $y$ intercept = 2, perpendicular to $4x - 3y = 7$
25. passes through ($-2$, 5), parallel to $y = 5x - 1$
26. passes through (4, $-1$), parallel to $4x - y = -3$
27. passes through ($-4$, 2), perpendicular to $y = 5x - 3$
28. passes through (6, 1), perpendicular to $6y - 2x = 3$
29. $y$ intercept = 4, angle of inclination = 48°
30. $y$ intercept = $-3.52$, angle of inclination = 154°44′
31. passes through (4, $-1$), angle of inclination = 22.8°
32. passes through ($-2.24$, 5.17), angle of inclination = 68°14′
33. passes through (5, 2) and is parallel to the $x$ axis.
34. passes through ($-3$, 6) and is parallel to the $y$ axis.
35. line $A$ in Fig. 25-30
36. line $B$ in Fig. 25-30
37. line $C$ in Fig. 25-30
38. line $D$ in Fig. 25-30

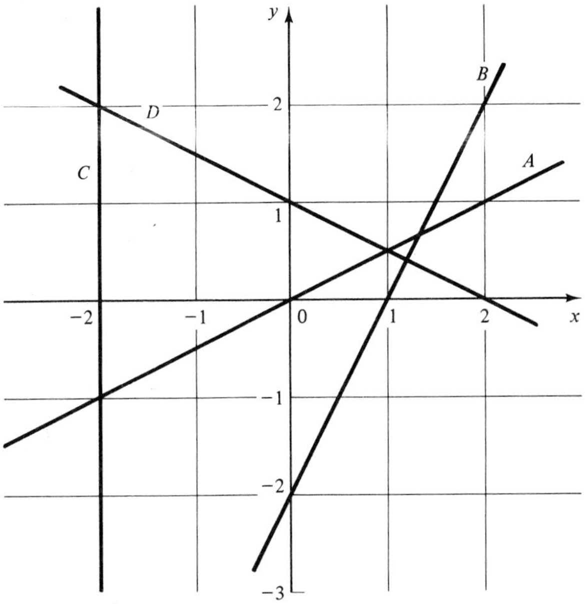


**FIGURE 25-30**

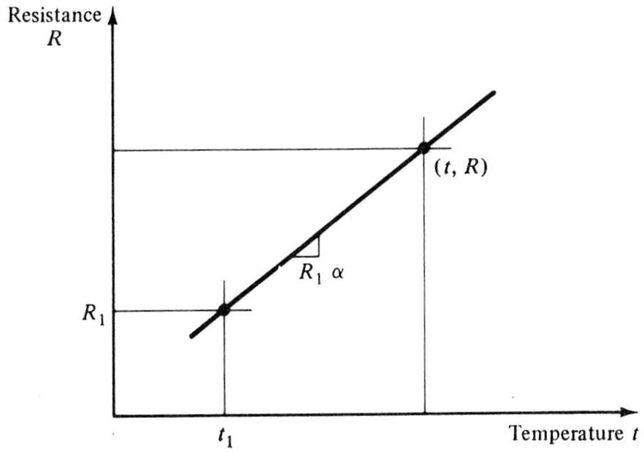


**FIGURE 25-31 Resistance change with temperature.**

**Resistance Change with Temperature**

39. The resistance of metals is a linear function of the temperature (Fig. 25-31) for certain ranges of temperature. The slope of the line is $R_1\alpha$, where $\alpha$ is the temperature coefficient of resistance at temperature $t_1$. If $R$ is the resistance of any temperature $t$, write an equation for $R$ as a function of $t$.

40. Using a value for $\alpha$ of $\alpha = 1/234.5t_1$ (for copper), find the resistance of a copper conductor at 75.0°C if its resistance at 20.0°C is 148.4 Ω.

41. If the resistance of the copper conductor in Problem 40 is 1255 Ω at 20.0°C, at what temperature will the resistance be 1265 Ω?

### Spring Constant

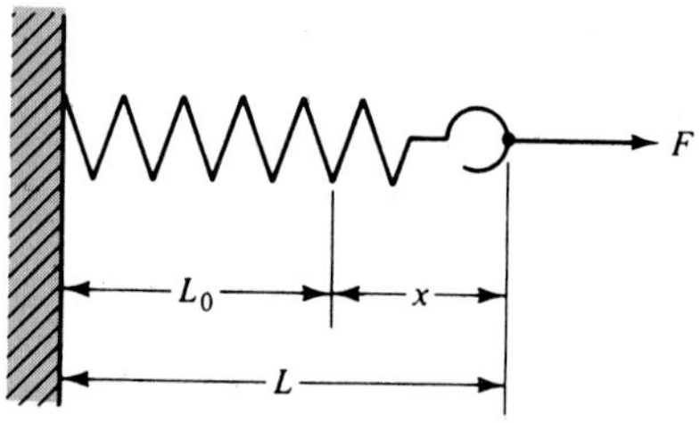


FIGURE 25-32

42. A spring whose length is $L_0$ with no force applied (Fig. 25-32) stretches an amount $x$ with an applied force of $F$, where $F = kx$. The constant $k$ is called the spring constant. Write an equation, in slope-intercept form, for $F$ in terms of $k$, $L$, and $L_0$.

43. What force would be needed to stretch a spring ($k = 14.5$ lb/in.) from an unstretched length of 8.50 in. to a length of 12.50 in.?

### Velocity of Uniformly Accelerated Body

44. When a body moves with constant acceleration $a$ (such as in free fall), its velocity $v$ at any time $t$ is given by $v = v_0 + at$, where $v_0$ is the initial velocity. Note that this is the equation of a straight line. If a body has a constant acceleration of 2.15 m/s$^2$ and has a velocity of 21.8 m/s at 5.25 s, find **(a)** the initial velocity and **(b)** the velocity at 25.0 s.

### Thermal Expansion

45. When a bar is heated, its length will increase from an initial length $L_0$ at temperature $t_0$ to a new length $L$ at temperature $t$. The plot of $L$ versus $t$ is a straight line (Fig. 25-33) with a slope of $L_0\alpha$, where $\alpha$ is the coefficient of thermal expansion. Derive the equation $L = L_0(1 + \alpha\ \Delta t)$, where $\Delta t$ is the change in temperature, $t - t_0$.

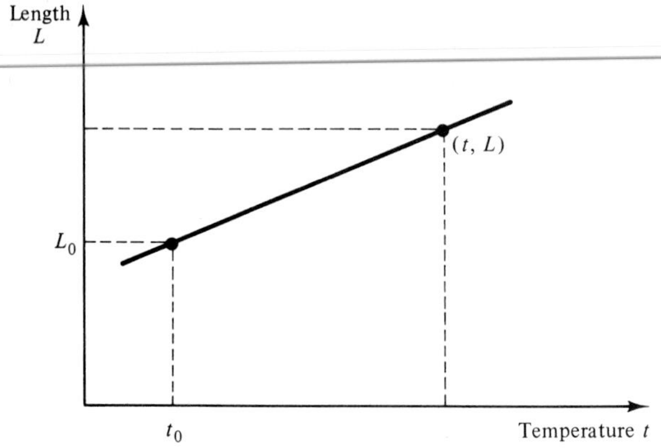


FIGURE 25-33

46. A steel pipe is 21.50 m long at 0°C. Find its length at 75.0°C if $\alpha$ for steel is $12.0 \times 10^{-6}$ per Celsius degree.

### Fluid Pressure

47. The pressure at a point located at a depth $x$ ft from the surface of the liquid varies directly as the depth. If the pressure at the surface is 20.6 lb/in$^2$ and increases by 0.432 lb/in$^2$ for every foot of depth, write an equation for $P$ as a function of the depth $x$ (in feet). At what depth will the pressure be 30.0 lb/in$^2$?

48. A straight pipe slopes downward from a reservoir to a water turbine (Fig. 25-34). The pressure head at any point in the pipe, expressed in feet, is equal to the vertical distance between the point and the surface of the reservoir. If the reservoir surface is 25 ft above the upper end of the pipe, write an expression for the head $H$ as a function of the horizontal distance $x$. At what distance $x$ will the head be 35 ft?

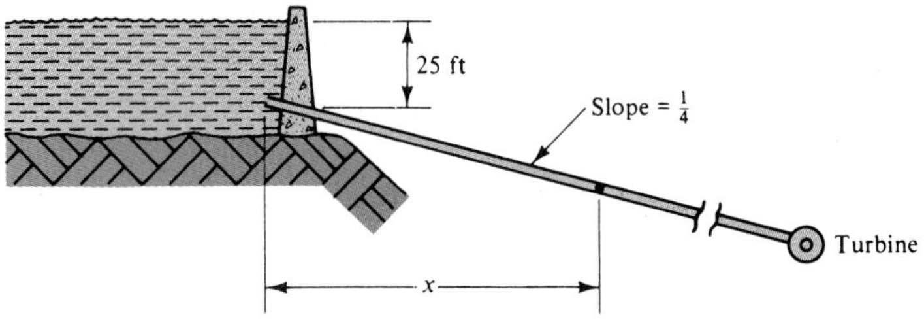


**FIGURE 25-34**

### Temperature Gradient

49. Figure 25-35 shows a uniform wall whose inside face is at temperature $t_i$ and whose outside face is at $t_o$. The temperatures within the wall plot as a straight line connecting $t_i$ and $t_o$. Write the equation $t = f(x)$ of that line, taking $x = 0$ at the inside face, if $t_i = 25°C$ and $t_o = -5°C$. At what $x$ will the temperature be 0°C? What is the slope of the line?

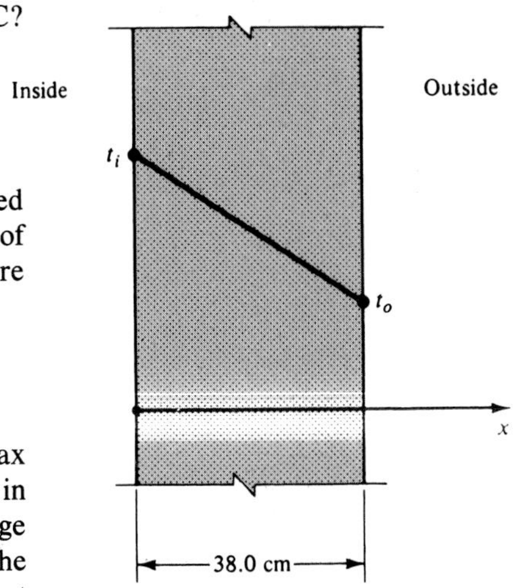


**FIGURE 25-35 Temperature drop in a wall. The slope of the line (in °C/cm) is called the *temperature gradient*. The amount of heat flowing through the wall is proportional to the temperature gradient.**

### Hooke's Law

50. The increase in length of a wire in tension is directly proportional to the applied load $P$. Write an equation for the length $L$ of a wire that has an initial length of 3.00 m and which stretches 1.00 mm for each 12.5 N. Find the length of the wire with a load of 750 N.

### Straight-Line Depreciation

51. The straight-line method is often used to depreciate a piece of equipment for tax purposes. Starting with the purchase price $P$, the item is assumed to drop in value the same amount each year (the annual depreciation) until the salvage value $S$ is reached (Fig. 25-36). Write an expression for the book value $y$ (the value at any time $t$) as a function of the number of years $t$. For a lathe that cost $15,428 and has a salvage value of $2264, find the book value after 15 years if it is depreciated over a period of 20 years.

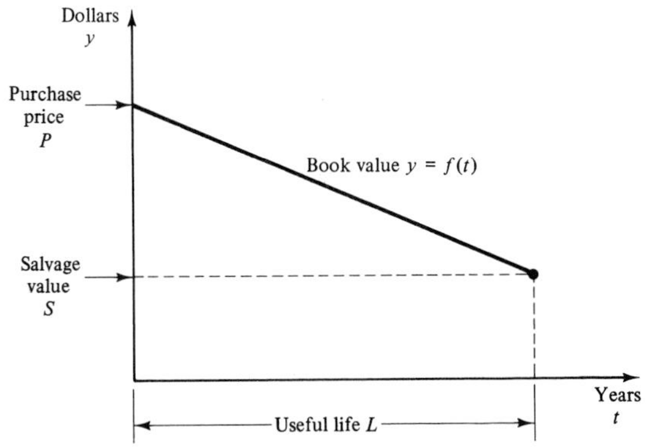


**FIGURE 25-36 Straight-line depreciation.**

### Computer

52. Write a program for the *false position method,* which is used to find the approximate roots to an equation $f(x) = 0$. We seek the value $R$ at which the graph of $y = f(x)$ crosses the $x$ axis (Fig. 25-37). We must make two initial guesses, $x_1$ and $x_2$, which lie on either side of $R$. Using these, we find the ordinates $y_1$ and $y_2$, from which we get the slope $m$ of the line $P_1P_2$ from

$$m = \frac{y_2 - y_1}{x_2 - x_1} = \frac{y_2 - 0}{x_2 - x_3}$$

From this we compute the value $x_3$ at which the line $P_1P_2$ crosses the $x$ axis.

$$x_3 = x_2 - y_2\left(\frac{x_2 - x_1}{y_2 - y_1}\right)$$

The value, $x_3$, becomes our second approximation, replacing either $x_1$ or $x_2$, depending on which side of the root $x_3$ lies, and we repeat the computation until we get as close to $R$ as we wish.

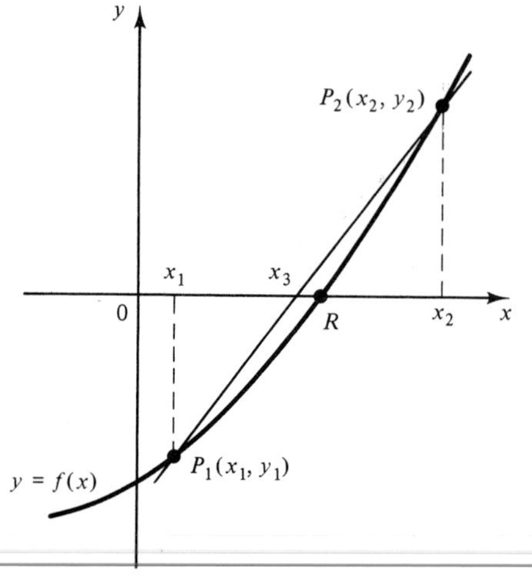


FIGURE 25-37 **False position method.**

## CHAPTER 25 REVIEW PROBLEMS

1. Find the distance between the points (3, 0) and (7, 0).
2. Find the distance between the points (4, −4) and (1, −7).
3. Find the slope of the line perpendicular to a line that has an angle of inclination of 34.8°.
4. Find the angle of inclination in degrees of a line with a slope of −3.
5. Find the angle of inclination of a line perpendicular to a line having a slope of 1.55.
6. Find the angle of inclination of a line passing through (−3, 5) and (5, −6).
7. Find the slope of a line perpendicular to a line having a slope of $a/2b$.
8. Write the equation of a line having a slope of −2 and a $y$ intercept of 5.
9. Find the slope and $y$ intercept of the line $2y - 5 = 3(x - 4)$.
10. Write the equation of a line having a slope of $2p$ and a $y$ intercept of $p - 3q$.
11. Write the equation of the line passing through (−5, −1) and (−2, 6).
12. Write the equation of the line passing through $(-r, s)$ and $(2r, -s)$.
13. Write the equation of the line having a slope of 5 and passing through the point (−4, 7).
14. Write the equation of the line having a slope of $3c$ and passing through the point $(2c, c - 1)$.
15. Write the equation of the line having an $x$ intercept of −3 and a $y$ intercept of 7.
16. Find the acute angle between two lines, if one line has a slope of 1.5, and the other has a slope of 3.4.
17. Write the equation of the line that passes through (2, 5) and is parallel to the $x$ axis.

18. Find the angle of intersection between line $L_1$ having a slope of 2 and line $L_2$ having a slope of 7.
19. Find the directed distance $AB$ between the points $A(-2, 0)$ and $B(-5, 0)$.
20. Find the angle of intersection between line $L_1$ having an angle of inclination of 18° and line $L_2$ having an angle of inclination of 75°.
21. Write the equation of the line that passes through $(-3, 6)$ and is parallel to the $y$ axis.
22. Find the increments in the coordinates of a particle that moves along a curve from (3, 4) to (5, 5).
23. Find the area of a triangle with vertices at (6, 4), (5, −2), and (−3, −4).

*In Problems 24–33, a tangent T, of slope m, and a normal N are drawn to a curve at the point $P(x_1, y_1)$ (Fig. 25-38). Show the following.*

24. The equation of the tangent is $y - y_1 = m(x - x_1)$.
25. The equation of the normal is $x - x_1 + m(y - y_1) = 0$.
26. The $x$ intercept $OA$ of the tangent is $x_1 - y_1/m$.
27. The $y$ intercept $OB$ of the tangent is $y_1 - mx_1$.
28. The length of the tangent from $P$ to the $x$ axis is $PA = \frac{y_1}{m}\sqrt{1 + m^2}$.
29. The length of the tangent from $P$ to the $y$ axis is $PB = x_1\sqrt{1 + m^2}$.
30. The $x$ intercept $OC$ of the normal is $x_1 + my_1$.
31. The $y$ intercept $OD$ of the normal is $y_1 + x_1/m$.

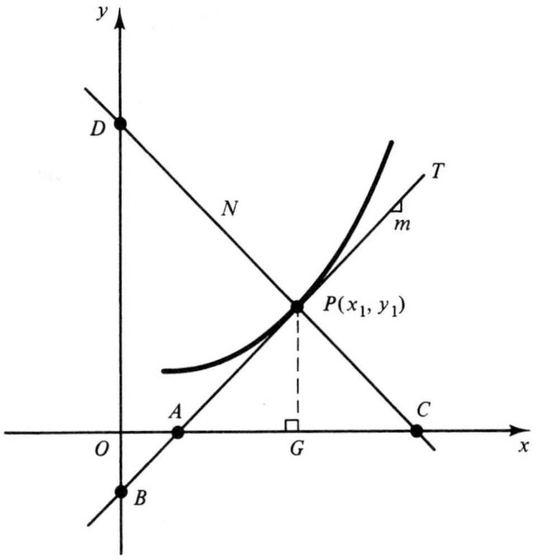


**FIGURE 25-38 Tangent and normal to a curve.**

32. The length of the normal from $P$ to the $x$ axis is $PC = y_1\sqrt{1 + m^2}$.
33. The length of the normal from $P$ to the $y$ axis is $PD = \frac{x_1}{m}\sqrt{1 + m^2}$.

*Writing*

34. Write a short paragraph explaining, in your own words, what is meant by the slope of a curve.

# 26

# THE CONIC SECTIONS

## OBJECTIVES

**When you have completed this chapter, you should be able to:**

- Write the equation of a circle, ellipse, parabola, or hyperbola from given information.
- Write an equation in standard form given the equation of any of the conic sections.
- Determine all the features of interest from the standard equation of any conic section, and make a graph.
- Make a graph of any of the conic sections.
- Tell, by inspection, whether a given second-degree equation represents a circle, ellipse, parabola, or hyperbola.
- Write a new equation for a curve with the axes shifted when given the equation of that curve.
- Solve applied problems involving any of the conic sections.

The extremely useful curves we study in this chapter are called *conic sections,* or just *conics,* because each can be obtained by passing a plane through a right circular cone, as in Fig. 26-1.

When the plane is perpendicular to the cone's axis, it intercepts a *circle*. When the plane is tilted a bit, but not so much as to be parallel to an element (a line on the cone that passes through the vertex) of the cone, we get an *ellipse*. When the plane is parallel to an element of the cone, we get a *parabola*: and when the plane is parallel to the cone's axis, we get the two-branched *hyperbola*. In this section we write equations for each of these curves, and use the equations to solve some interesting problems.

**You might want to try making these shapes by cutting a cone made of modeling clay or damp sand. Also, what position of the plane would you use to make a straight line? A point?**

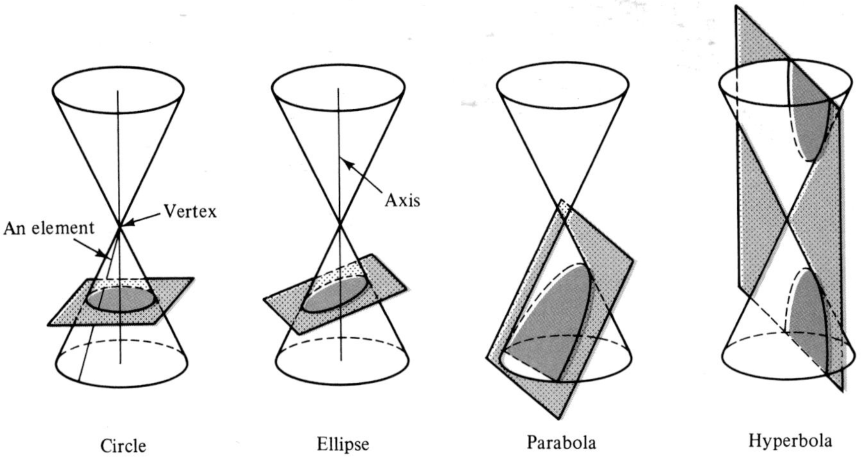


**FIGURE 26-1 Conic sections.**

## 26-1 CIRCLE

| | | |
|---|---|---|
| Definition | A *circle* is a plane curve all points of which are at a fixed distance (the *radius*) from a fixed point (the *center*). | **293** |

### Standard Equation of a Circle with Center at the Origin

Let us write the equation of a circle of radius $r$ whose center is at the origin (Fig. 26-2). Let $x$ and $y$ be the coordinates of any point $P$ on the circle. The equation we develop will give a relationship between $x$ and $y$ that will have two meanings: geometrically, $(x, y)$ will represent a point on the circle, and algebraically, the numbers corresponding to those points will satisfy that equation.

We observe that, by the definition of a circle, the distance $OP$ must be constant and equal to $r$. But by the distance formula (Eq. 273),

$$OP = \sqrt{x^2 + y^2} = r$$

Squaring, we get a *standard equation of a circle* (also called *standard form* of the equation).

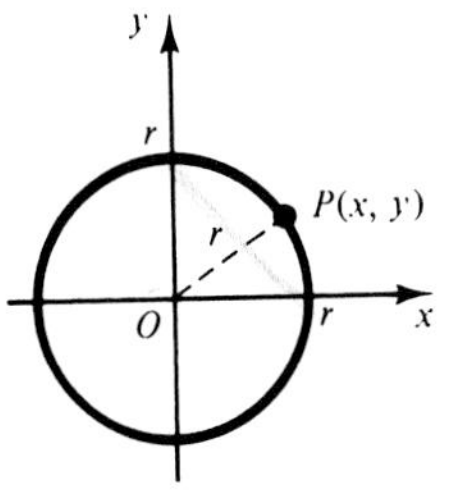


**FIGURE 26-2 Circle with center at origin.**

Note that both $x^2$ and $y^2$ have the same coefficient. Otherwise, the graph is not a circle.

| | | |
|---|---|---|
| Standard Equation, Circle of Radius $r$: Center at Origin | $x^2 + y^2 = r^2$ | **294** |

**EXAMPLE 1:** Write, in standard form, the equation of a circle of radius 3, whose center is at the origin.

**Solution:**

$$x^2 + y^2 = 3^2 = 9$$

## Circle with Center Not at the Origin

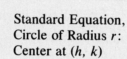

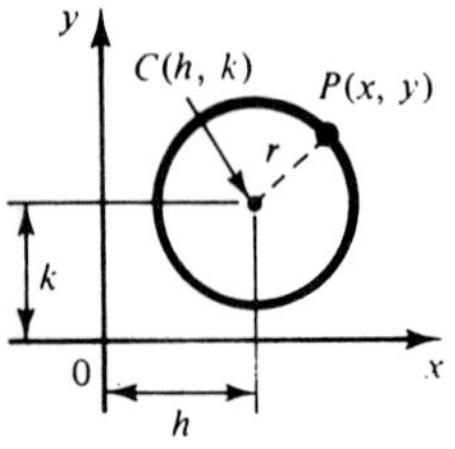

**FIGURE 26-3 Circle with center at $(h, k)$.**

Figure 26-3 shows a circle whose center has the coordinates $(h, k)$. We can think of the difference between this circle and the one in Fig. 26-2 as having its center moved or *translated* $h$ units to the right and $k$ units upward. Our derivation is similar to the preceding one.

$$CP = r = \sqrt{(x - h)^2 + (y - k)^2}$$

Squaring, we get

| | | |
|---|---|---|
| Standard Equation, Circle of Radius $r$: Center at $(h, k)$ | $(x - h)^2 + (y - k)^2 = r^2$ | **295** |

**EXAMPLE 2:** Write, in standard form, the equation of a circle of radius 5 whose center is at $(3, -2)$.

**Solution:** We substitute into Eq. 295 with $r = 5$, $h = 3$, and $k = -2$:

$$(x - 3)^2 + [y - (-2)]^2 = 5^2$$

$$(x - 3)^2 + (y + 2)^2 = 25$$

**EXAMPLE 3:** Find the radius and the coordinates of the center of the circle $(x + 5)^2 + (y - 3)^2 = 16$.

**Solution:** We see that $r^2 = 16$, so the radius $r$ is 4. Also, since

$$x - h = x + 5$$

$$h = -5$$

and since

$$y - k = y - 3$$

$$k = 3$$

so the center is at $(-5, 3)$.

| Common Error | It is easy to get the signs of $h$ and $k$ wrong. In the preceding example, do *not* take $$h = +5 \quad \text{and} \quad k = -3$$ |
|---|---|

### Translation of Axes

We see that Eq. 295 for a circle with its center at $(h, k)$ is almost identical to Eq. 294 for a circle with center at the origin, except that *$x$ has been replaced by $(x - h)$ and $y$ has been replaced by $(y - k)$*. This same substitution will, of course, work for curves other than the circle, and we will use it to translate or *shift* the axes for the other conic sections.

| Translation of Axes | To translate or shift the axes of a curve to the left by a distance $h$ and downward by a distance $k$, replace $x$ by $(x - h)$ and $y$ by $(y - k)$ in the equation of the curve. | **288** |
|---|---|---|

### The General Second-Degree Equation

A second-degree equation in $x$ and $y$ which has all possible terms would have an $x^2$ term, a $y^2$ term, an $xy$ term, and all terms of lesser degree as well. The general second-degree equation is usually written with terms in the following order, where $A$, $B$, $C$, $D$, $E$, and $F$ are constants.

| General Second-Degree Equation | $Ax^2 + Bxy + Cy^2 + Dx + Ey + F = 0$ | **287** |
|---|---|---|

There are six constants in this equation, but only five are independent. We can divide through by any constant and thus eliminate it.

### General Equation of a Circle

If we expand Eq. 295 we get

$$(x - h)^2 + (y - k)^2 = r^2$$

$$x^2 - 2hx + h^2 + y^2 - 2ky + k^2 = r^2$$

Rearranging gives $x^2 + y^2 - 2hx - 2ky + (h^2 + k^2 - r^2) = 0$, or an equation with $D$, $E$, and $F$ as constants.

| General Equation of a Circle | $x^2 + y^2 + Dx + Ey + F = 0$ | **296** |
|---|---|---|

Comparing this with the general second-degree equation, we see that the general second-degree equation

$$Ax^2 + Bxy + Cy^2 + Dx + Ey + F = 0$$

represents a circle if $B = 0$, and $A = C$.

---

**EXAMPLE 4:** Write the equation of Example 3 in general form.

**Solution:** The equation, in standard form, was

$$(x + 5)^2 + (y - 3)^2 = 16$$

Expanding, we obtain

$$x^2 + 10x + 25 + y^2 - 6y + 9 = 16$$

or

$$x^2 + y^2 + 10x - 6y + 18 = 0.$$

---

### Changing from General to Standard Form

When we want to go from general form to standard form, we must *complete the square,* both for $x$ and for $y$.

---

**EXAMPLE 5:** Write the equation $2x^2 + 2y^2 - 18x + 16y + 60 = 0$ in standard form. Find the radius and center, and plot the curve.

**Solution:** We first divide by 2:

$$x^2 + y^2 - 9x + 8y + 30 = 0$$

and separate the $x$ and $y$ terms:

$$(x^2 - 9x \quad ) + (y^2 + 8y \quad ) = -30$$

We complete the square for the $x$ terms with $\frac{81}{4}$ because we take $\frac{1}{2}$ of $(-9)$ and square it. Similarly, for completing the square for the $y$ terms, $(\frac{1}{2} \times 8)^2 = 16$. Completing the square on the left and compensating on the right gives

$$\left(x^2 - 9x + \frac{81}{4}\right) + (y^2 + 8y + 16) = -30 + \frac{81}{4} + 16$$

Factoring each group of terms yields

$$\left(x - \frac{9}{2}\right)^2 + (y + 4)^2 = \left(\frac{5}{2}\right)^2$$

So

$$r = \frac{5}{2}, \qquad h = \frac{9}{2}, \qquad k = -4$$

The circle is shown in Fig. 26-4.

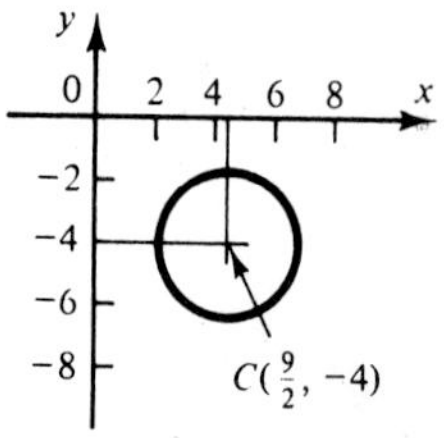


FIGURE 26-4

## Equation of a Circle Found from Three Conditions

Since the equation of a circle either in standard form or in general form has three arbitrary constants, we must have three independent pieces of information in order to write the equation. The given information is substituted into whichever of the two forms of the circle equation is most appropriate, and the equations are solved simultaneously.

---

**EXAMPLE 6:** Write the equation of the circle passing through the points (4, 6), (−2, −2), and (−4, 2).

**Solution:** Since each of these points is on the circle, each must satisfy the general equation of the circle (Eq. 296). So we substitute the coordinates of each in turn into Eq. 296 and get three equations in three unknowns.

At (4, 6): $16 + 36 + 4D + 6E + F = 0$

At (−2, −2): $4 + 4 - 2D - 2E + F = 0$

At (−4, 2): $16 + 4 - 4D + 2E + F = 0$

Collecting terms yields

$$4D + 6E + F = -52 \qquad (1)$$

$$-2D - 2E + F = -8 \qquad (2)$$

$$-4D + 2E + F = -20 \qquad (3)$$

Subtracting (2) from (1) gives

$$6D + 8E = -44 \qquad (4)$$

and subtracting (3) from (2) gives

$$2D - 4E = 12 \qquad (5)$$

Now multiplying (5) by two and adding to (4), we get

$$10D = -20$$

$$D = -2$$

Substituting back gives $E = -4$ and $F = -20$, so our equation is

$$x^2 + y^2 - 2x - 4y - 20 = 0$$

If you now complete the square, you will find that this is a circle whose center is at (1, 2) with a radius of 5.

---

## EXERCISE 1—CIRCLE

### Equation of a Circle

Write the equation of each circle in standard form.

1. center at (0,0), radius = 7
2. center at (0, 0), radius = 4.82
3. center at (2, 3), radius = 5
4. center at (5, 0), diameter = 10
5. center at (5, −3), radius = 4
6. center at (0, −2), radius = 11

Find the center and radius of each circle.

7. $x^2 + y^2 = 49$
8. $x^2 + y^2 = 64.8$
9. $(x - 2)^2 + (y + 4)^2 = 16$
10. $(x + 5)^2 + (y - 2)^2 = 49$
11. $(y + 5)^2 + (x - 3)^2 = 36$
12. $(x - 2.22)^2 + (y + 7.16)^2 = 5.93$
13. $x^2 + y^2 - 8x = 0$
14. $x^2 + y^2 - 2x - 4y = 0$
15. $x^2 + y^2 - 10x + 12y + 25 = 0$

Write the equation of each circle in general form.

16. passes through (0, 0), (0, −1), and (−6, −1)
17. passes through (1, 3), (1, 2), and (2, 5)
18. center at (1, 1), passes through (4, −3)
19. whose diameter joins the points (−2, 5) and (6, −1)
20. passes through origin, $x$ intercept = 4, and $y$ intercept = 6

### Tangent to a Circle

Write the equation of the tangent to each circle at the given point. (*Hint:* The slope of the tangent is the negative reciprocal of the slope of the radius to the given point.)

21. $x^2 + y^2 = 25$ at (4, 3)
22. $(x - 5)^2 + (y - 6)^2 = 100$ at (11, 14)
23. $x^2 + y^2 + 10y = 0$ at (−4, −2)

### Intercepts

Find the $x$ and $y$ intercepts for each circle. (*Hint:* Set $x$ and $y$, in turn, equal to zero to find the intercepts.)

24. $x^2 + y^2 - 6x + 4y + 4 = 0$
25. $x^2 + y^2 - 5x - 7y + 6 = 0$

### Intersecting Circles

Find the point(s) of intersection. (*Hint:* Solve each pair of equations simultaneously as we did in Chapter 13, Exercise 8.)

26. $x^2 + y^2 - 10x = 0$ and $x^2 + y^2 + 2x - 6y = 0$
27. $x^2 + y^2 - 3y - 4 = 0$ and $x^2 + y^2 + 2x - 5y - 2 = 0$
28. $x^2 + y^2 + 2x - 6y + 2 = 0$ and $x^2 + y^2 - 4x + 2 = 0$

### Applications

**Even though you may be able to solve some of these with only the Pythagorean theorem, we suggest that you use analytic geometry for the practice.**

29. Prove that any angle inscribed in a semicircle is a right angle.
30. Write the equation of the circle in Fig. 26-5, taking the axes as shown. Use your equation to find $A$ and $B$.
31. Write the equations for each of the circular arches in Fig. 26-6, taking the axes as shown. Solve simultaneously to get the point of intersection $P$, and compute the height $h$ of the column.
32. Write the equation of the centerline of the circular street shown in Fig. 26-7, taking the origin at the intersection $0$. Use your equation to find the distance $y$.
33. Each side of a Gothic arch is a portion of a circle. Write the equation of one side of the Gothic arch shown in Fig. 26-8. Use the equation to find the width $w$ of the arch at a height of 3.0 ft.

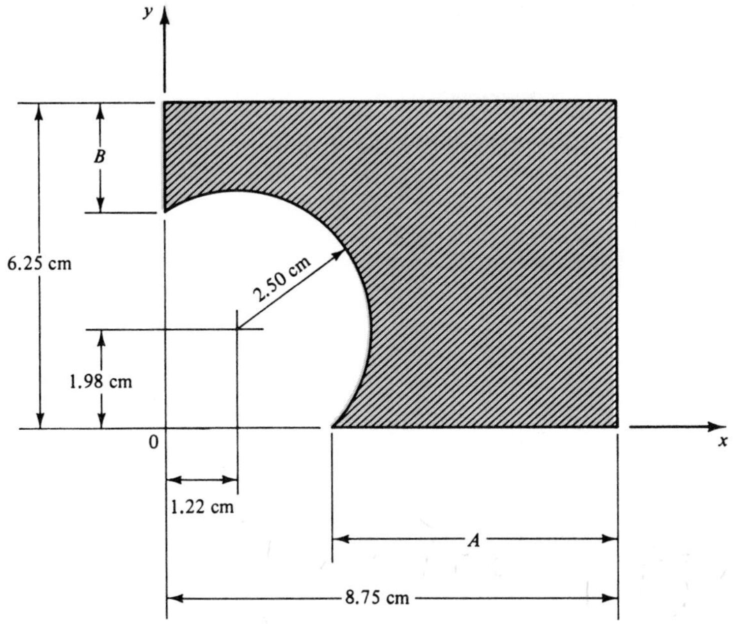


**FIGURE 26-5**

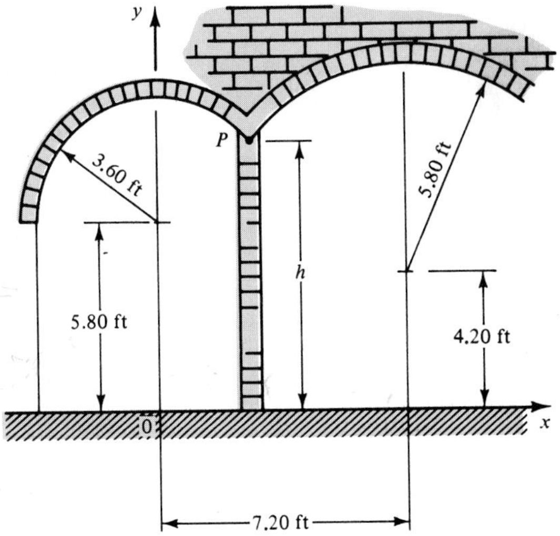


**FIGURE 26-6** **Circular arches.**

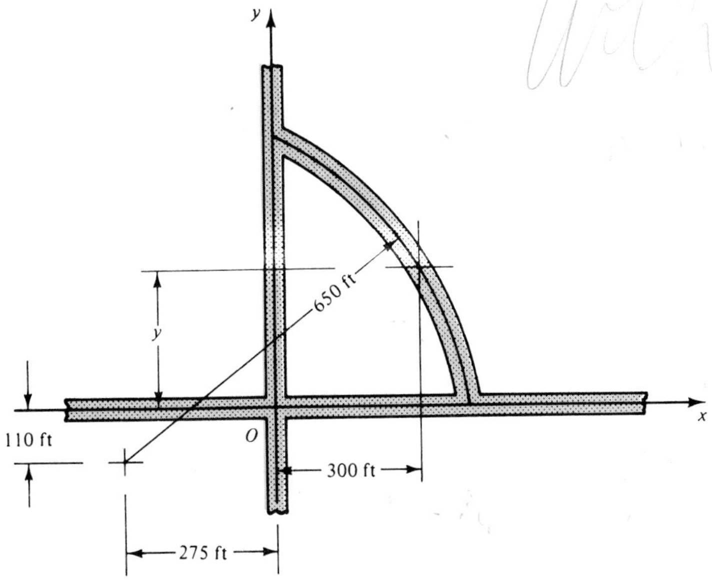


**FIGURE 26-7** **Circular street.**

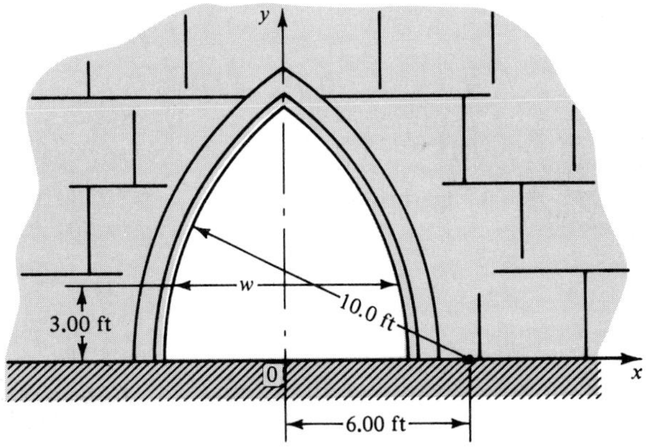


**FIGURE 26-8** **Gothic arch.**

| | | |
|---|---|---|
| Definition | A *parabola* is the set of points in a plane each of which is equidistant from a fixed point, the *focus,* and a fixed line, the *directrix*. | **297** |

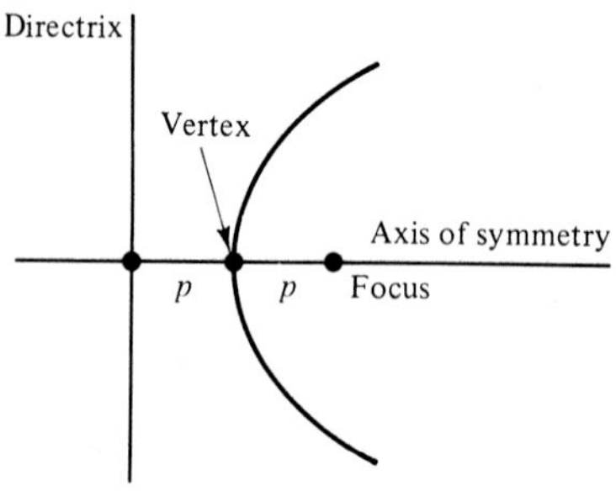


**FIGURE 26-9 Parabola.**

Such a set of points, connected with a smooth curve, is plotted in Fig. 26-9, and shows the typical shape of the parabola. The parabola has an *axis of symmetry* which intersects it at the *vertex*. The distance $p$ from directrix to vertex is equal to the directed distance from the vertex to the focus.

### Standard Equation of a Parabola with Vertex at the Origin

Let us place the parabola on coordinate axes with the vertex at the origin (Fig. 26-10) and with the axis of symmetry along the $x$ axis. Choose any point $P$ on the parabola. Then by the definition of a parabola, $FP = AP$. But in right triangle $FBP$,

$$FP = \sqrt{(x - p)^2 + y^2}$$

and also,

$$AP = p + x$$

But since $FP = AP$,

$$\sqrt{(x - p)^2 + y^2} = p + x$$

Squaring both sides yields

$$(x - p)^2 + y^2 = p^2 + 2px + x^2$$

$$x^2 - 2px + p^2 + y^2 = p^2 + 2px + x^2$$

Collecting terms, we get the standard equation of a parabola with vertex at the origin.

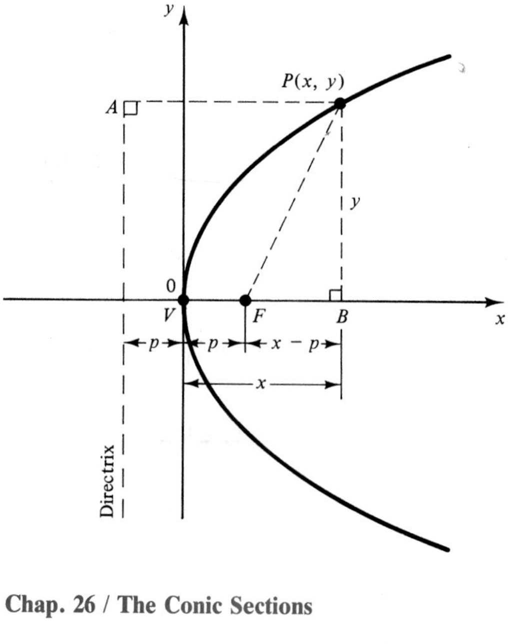


**FIGURE 26-10**

| | | | |
|---|---|---|---|
| Standard Equation of a Parabola: Vertex at Origin, Axis Horizontal | 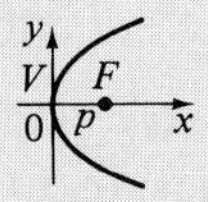  | $y^2 = 4px$ | **298** |

We have defined $p$ as the directed distance $VF$ from the vertex $V$ to the focus $F$. Thus if $p$ is positive, the focus must lie to the right of the vertex, and hence the parabola opens to the right. Conversely, if $p$ is negative, the parabola opens to the left.

---

**EXAMPLE 7:** Find the coordinates of the focus of the parabola $2y^2 + 7x = 0$.

**Solution:** Subtracting $7x$ from both sides and dividing by 2 gives us

$$y^2 = -3.5x$$

Thus $4p = -3.5$ and $p = -0.875$. Since $p$ is negative, the parabola opens to the left. The focus is thus 0.875 unit to the left of the vertex (Fig. 26-11) and so has the coordinates $(-0.875, 0)$.

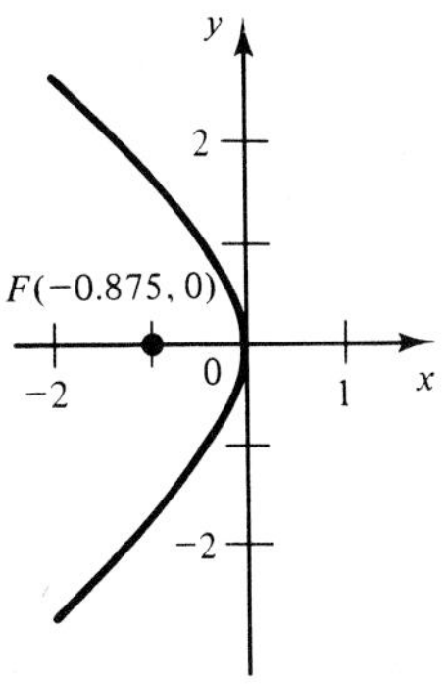


**FIGURE 26-11**

---

**EXAMPLE 8:** Find the coordinates of the focus and write the equation of a parabola whose vertex is at the origin, that has a horizontal axis of symmetry, and that passes through the point $(5, -6)$.

**Solution:** Our given point must satisfy the equation $y^2 = 4px$. Substituting 5 for $x$ and $-6$ for $y$ gives

$$(-6)^2 = 4p(5)$$

So $4p = \frac{36}{5}$ and $p = \frac{9}{5}$. The equation of the parabola is then $y^2 = 36x/5$, or

$$5y^2 = 36x$$

Since the axis is horizontal and $p$ is positive, the focus must be on the $x$ axis and is a distance $p$ $(\frac{9}{5})$ to the right of the origin. The coordinates of $F$ are thus $(\frac{9}{5}, 0)$ (Fig. 26-12).

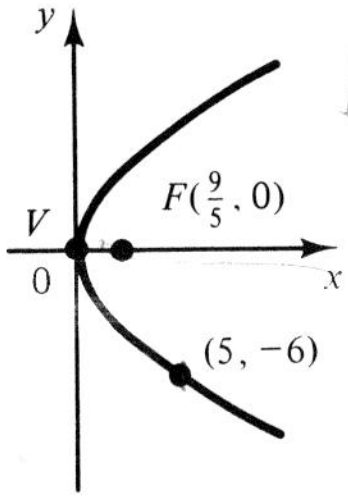


**FIGURE 26-12**

---

## Standard Equation of a Parabola with Vertical Axis

The standard equation for a parabola having a vertical axis is obtained by switching the positions of $x$ and $y$ in Eq. 298,

| | | | |
|---|---|---|---|
| Standard Equation of a Parabola: Vertex at Origin, Axis Vertical | 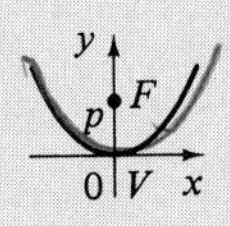  | $x^2 = 4py$ | **299** |

**Note that only one variable is squared in any parabola equation. This gives us the best way to recognize such equations.**

When $p$ is positive, the parabola opens upward; when $p$ is negative, it opens downward.

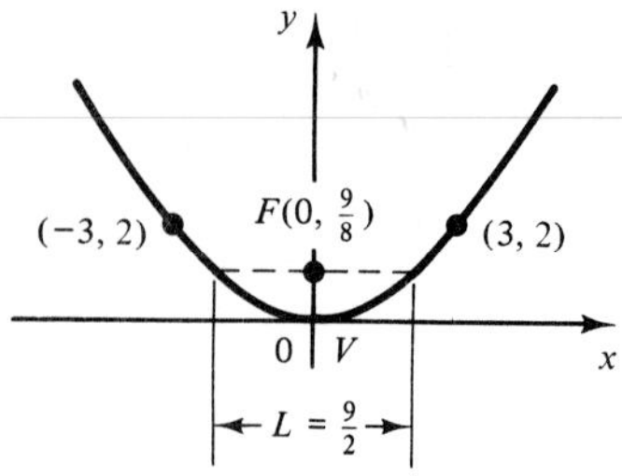


FIGURE 26-13

**EXAMPLE 9:** A parabola has its vertex at the origin and passes through the points (3, 2) and (−3, 2). Write its equation and find the focus.

**Solution:** A sketch (Fig. 26-13) shows that the axis must be vertical, so our equation is of the form $x^2 = 4py$. Substituting 3 for $x$ and 2 for $y$ gives

$$3^2 = 4p(2)$$

from which $4p = \frac{9}{2}$ and $p = \frac{9}{8}$. Our equation is then $x^2 = 9y/2$ or $2x^2 = 9y$. The focus is on the $y$ axis at a distance $p = \frac{9}{8}$ from the origin, so its coordinates are $(0, \frac{9}{8})$.

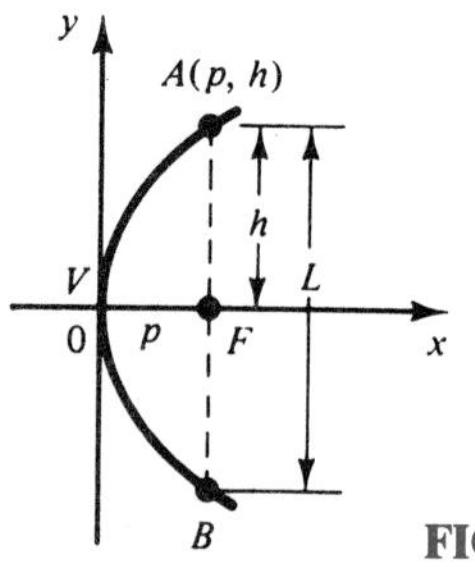


FIGURE 26-14

## Focal Width of a Parabola

The *latus rectum* of a parabola is a line through the focus which is perpendicular to the axis of symmetry, such as line $AB$ in Fig. 26-14. The length of the latus rectum is also called the *focal width.* We will find the focal width, or length $L$ of the latus rectum, by substituting the coordinates $(p, h)$ of point $A$ into Eq. 298:

$$h^2 = 4p(p) = 4p^2$$

$$h = \pm 2p$$

**The focal width is useful for making a quick sketch of the parabola.**

The focal width is twice $h$, so

| | | |
|---|---|---|
| Focal Width of a Parabola | $L = \lvert 4p \rvert$ | **303** |

*The focal width (length of the latus rectum) of a parabola is four times the distance from vertex to focus.*

**EXAMPLE 10:** The focal width for the parabola in Fig. 26-13 is $\frac{9}{2}$, or 4.5 units.

## Shift of Axes

As with the circle, when the vertex of the parabola is not at the origin but at $(h, k)$, our equations will be similar to Eqs. 298 and 299, except that $x$ is replaced with $x - h$ and $y$ is replaced with $y - k$.

| | | | | |
|---|---|---|---|---|
| Standard Equations of a Parabola: Vertex at $(h, k)$ | Axis Horizontal | 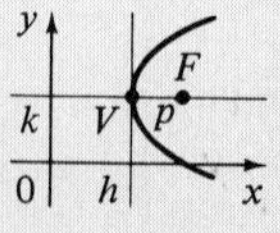 | $(y - k)^2 = 4p(x - h)$ | **300** |
| | Axis Vertical | 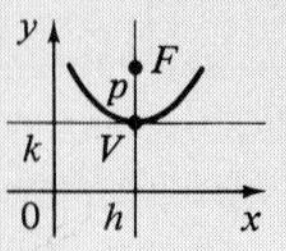 | $(x - h)^2 = 4p(y - k)$ | **301** |

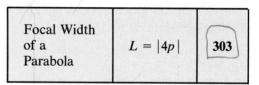

**EXAMPLE 11:** Find the vertex, focus, focal width, and equation of the axis for the parabola $(y - 3)^2 = 8(x + 2)$.

**Solution:** The given equation is of the same form as Eq. 300, so the axis is horizontal. Also, $h = -2$, $k = 3$, and $4p = 8$. So the vertex is at $V(-2, 3)$ (Fig. 26-15). Since $p = \frac{8}{4} = 2$, the focus is 2 units to the right of the vertex, at $F(0, 3)$. The focal width is $4p$, so $L = 8$. The axis is horizontal and 3 units from the $x$ axis, so its equation is $y = 3$.

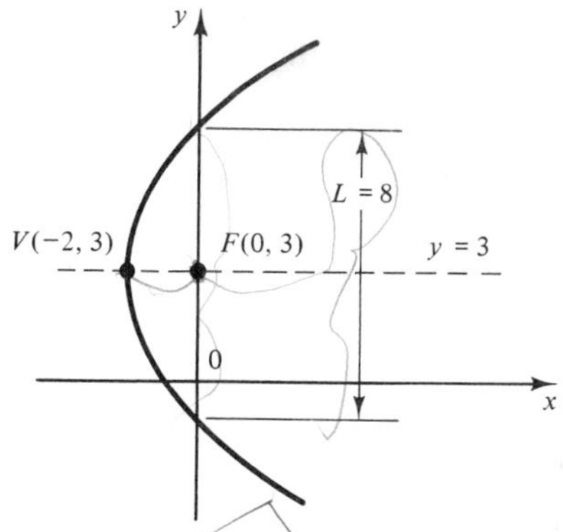


FIGURE 26-15

**EXAMPLE 12:** Write the equation of a parabola that opens upward, with vertex at $(-1, 2)$, and which passes through the point $(1, 3)$ (Fig. 26-16). Find the focus and the focal width.

**Solution:** We substitute $h = -1$ and $k = 2$ into Eq. 301,

$$(x + 1)^2 = 4p(y - 2)$$

Now, since (1, 3) is on the parabola, these coordinates must satisfy our equation. Substituting yields

$$(1 + 1)^2 = 4p(3 - 2)$$

Solving for $p$, we obtain $2^2 = 4p$, or, $p = 1$. So the equation is

$$(x + 1)^2 = 4(y - 2)$$

The focus is $p$ units above the vertex, at $(-1, 3)$. The focal width is, by Eq. 303,

$$L = |4p| = 4(1) = 4 \text{ units}$$

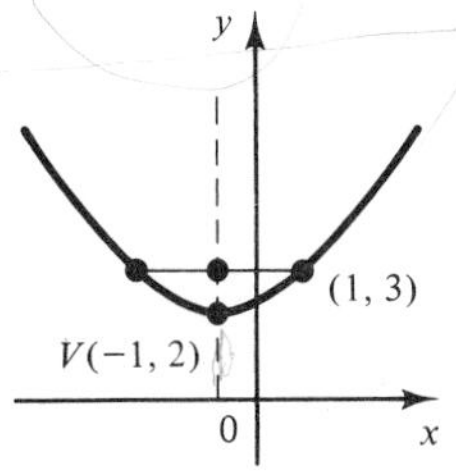


FIGURE 26-16

## General Equation of a Parabola

We get the general equation of the parabola by expanding the standard equation (Eq. 300)

$$(y - k)^2 = 4p(x - h)$$

$$y^2 - 2ky + k^2 = 4px - 4ph$$

or

$$y^2 - 4px - 2ky + (k^2 + 4ph) = 0$$

which is of the following general form ($C$, $D$, $E$, and $F$ are constants).

| | | |
|---|---|---|
| General Equation of a Parabola with Horizontal Axis | $Cy^2 + Dx + Ey + F = 0$ | **302** |

Compare this with the general second-degree equation.

We see that the equation of a parabola having a horizontal axis of symmetry has a $y^2$ term but no $x^2$ term. Conversely, the equation for a parabola with vertical axis has an $x^2$ term but not a $y^2$ term. The parabola is the only conic for which there is only one variable squared.

If the coefficient $B$ of the $xy$ term in the general second-degree equation (Eq. 287) were not zero, it would indicate that the axis of symmetry was *rotated* by some amount and was no longer parallel to a coordinate axis. The presence of an $xy$ term indicates rotation of the ellipse and hyperbola as well.

## Completing the Square

As with the circle, we go from general to standard form by completing the square.

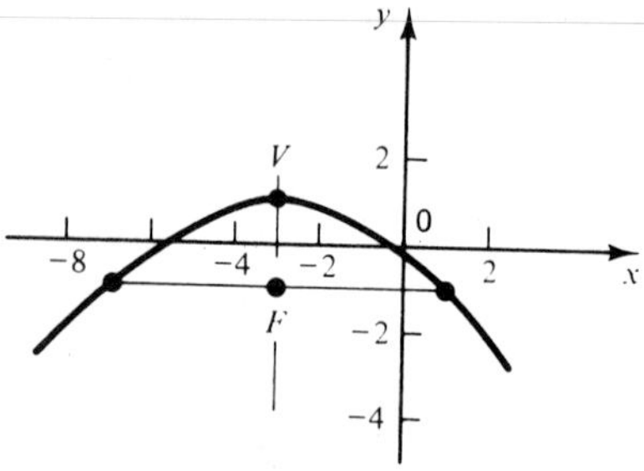


FIGURE 26-17

---

**EXAMPLE 13:** Find the vertex, focus, and focal width for the parabola

$$x^2 + 6x + 8y + 1 = 0$$

**Solution:** Separating the $x$ and $y$ terms, we have

$$x^2 + 6x = -8y - 1$$

Completing the square by adding 9 to both sides, we obtain

$$x^2 + 6x + 9 = -8y - 1 + 9$$

Factoring yields

$$(x + 3)^2 = -8(y - 1)$$

which is the form of Eq. 301, with $h = -3$, $k = 1$, and $p = -2$.

The vertex is $(-3, 1)$. Since the parabola opens downward (Fig. 26-17), the focus is 2 units below the vertex, at $(-3, -1)$. The focal width is $4p$ or 8 units.

---

## EXERCISE 2—PARABOLA

### Parabola with Vertex at Origin

Find the focus and focal width of each parabola.

1. $y^2 = 8x$
2. $x^2 = 16y$
3. $7x^2 + 12y = 0$
4. $3y^2 + 5x = 0$

Write the equation of each parabola.

5. focus at $(0, -2)$
6. passes through $(25, 20)$, axis horizontal
7. passes through $(3, 2)$ and $(3, -2)$
8. passes through $(3, 4)$ and $(-3, 4)$

### Parabola with Vertex Not at Origin

Find the vertex, focus, the focal width, and the equation of the axis for each parabola.

9. $(y - 5)^2 = 12(x - 3)$
10. $(x + 2)^2 = 16(y - 6)$
11. $3x + 2y^2 + 4y - 4 = 0$
12. $y^2 + 8y + 4 - 6x = 0$
13. $y - 3x + x^2 + 1 = 0$
14. $x^2 + 4x - y - 6 = 0$

Write the equation, in general form, for each parabola.

15. vertex at $(1, 2)$, $L = 8$, axis is $y = 2$, opens to the right
16. axis is $y = 3$, passes through $(6, -1)$ and $(3, 1)$
17. passes through $(-3, 3)$, $(-6, 5)$, and $(-11, 7)$, axis horizontal
18. vertex $(0, 2)$, axis is $x = 0$, passes through $(-4, -2)$
19. axis is $y = -1$, passes through $(-4, -2)$ and $(2, 1)$

### Construction of a Parabola

**20.** Use the definition of the parabola (the set of points equidistant from a fixed point and a fixed line) to construct a parabola. Let the distance between the focus and the directrix be 2.0 in. (Fig. 26-18). Then draw a line $L$ parallel to the directrix at some arbitrary distance, say 3.0 in. Then with that same (3.0 in.) distance as radius and $F$ as center, use a compass to draw arcs intersecting $L$ at $P_1$ and $P_2$. Each of these points is now at the same distance (3.0 in.) from $F$ and from the directrix, and is hence a point on the parabola. Repeat the construction with distances other than 3.0 in. to get more points on the parabola.

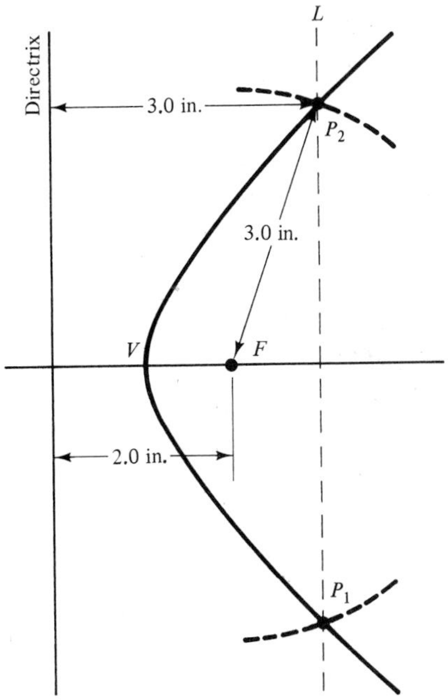


**FIGURE 26-18** **Construction of a parabola.**

### Trajectories

**21.** A ball thrown into the air (Fig. 26-19), will, neglecting air resistance, follow a parabolic path. Write the equation of the path, taking axes as shown. Use your equation to find the height of the ball when it is at a horizontal distance of 95 ft from $O$.

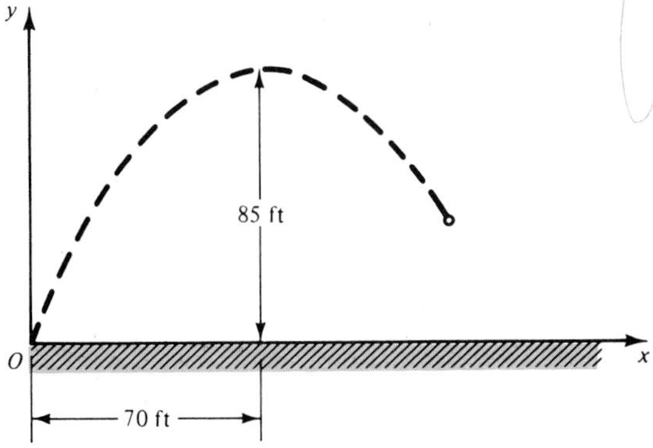


**FIGURE 26-19** **Ball thrown into the air.**

**22.** Some comets follow a parabolic orbit with the sun at the focal point (Fig. 26-20). Taking axes as shown, write the equation of the path if the distance $p$ is 75 million kilometers.

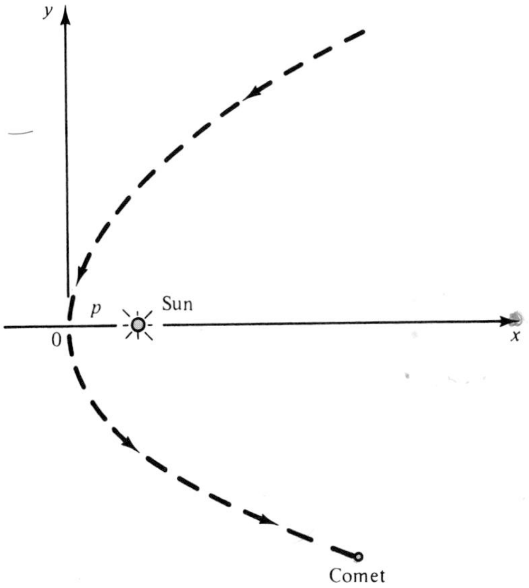


**FIGURE 26-20** **Path of a comet.**

23. An object dropped from a moving aircraft (Fig. 26-21) will follow a parabolic path if air resistance is negligible. A weather instrument released at a height of 3520 m is observed to strike the water at a distance of 2150 m from the point of release. Write the equation of the path, taking axes as shown. Find the height of the instrument when $x$ is 1000 m.

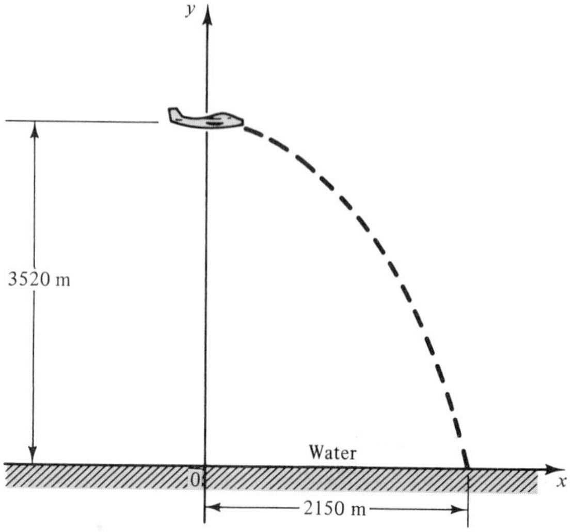


**FIGURE 26-21 Object dropped from an aircraft.**

**Parabolic Arch**

24. A 10-ft-high truck passes under a parabolic arch (Fig. 26-22). Find the maximum distance $x$ that the side of the truck can be from the center of the road.

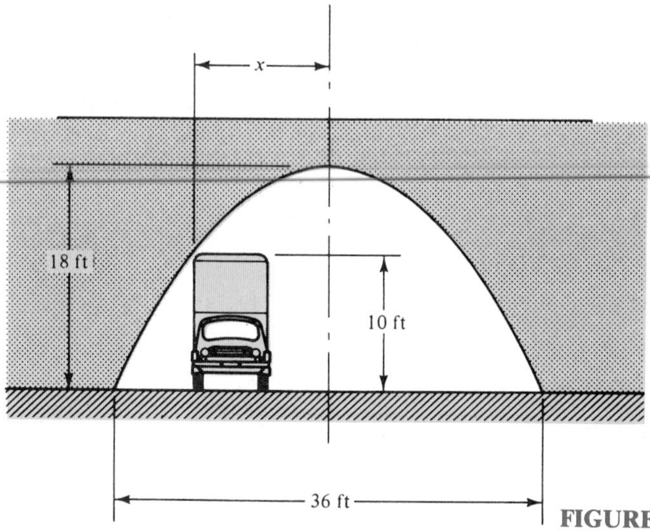


**FIGURE 26-22 Parabolic arch.**

**A cable will hang in the shape of a parabola if the vertical load per horizontal foot is constant.**

25. Assuming the bridge cable $AB$ (Fig. 26-23) to be a parabola, write its equation, taking axes as shown.

**FIGURE 26-23 Parabolic bridge cable.**

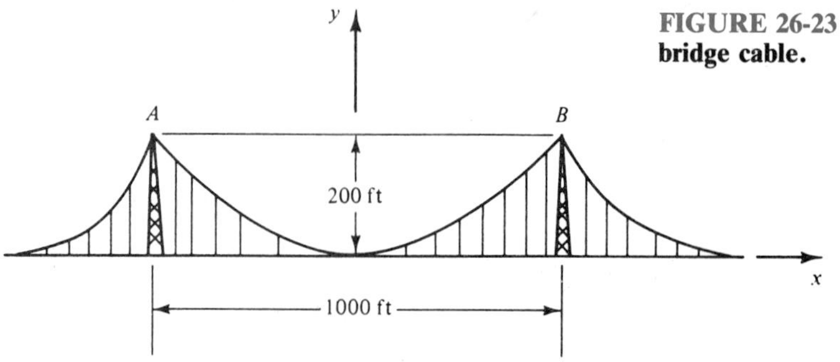

26. A parabolic arch supports a roadway (Fig. 26-24). Write the equation of the arch, taking axes as shown. Use your equation to find the vertical distance from the roadway to the arch at a horizontal distance of 50 m from the center.

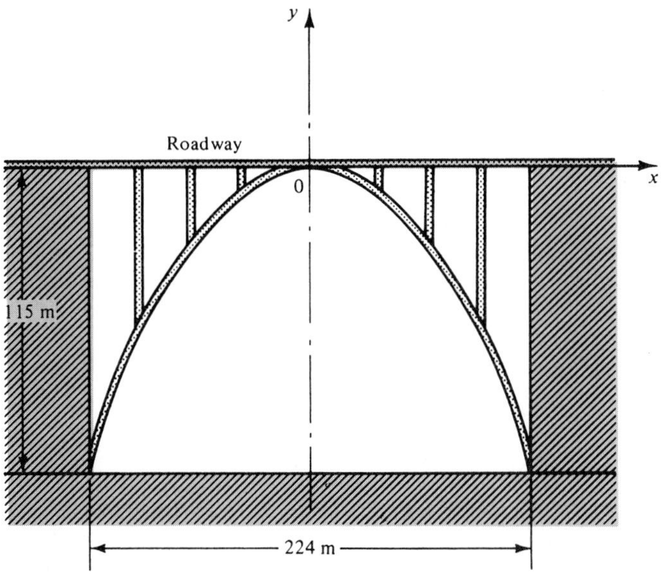


**FIGURE 26-24 Parabolic arch.**

**Parabolic Reflector**

27. A certain solar collector consists of a long panel of polished steel bent into a parabolic shape (Fig. 26-25), which focuses sunlight onto a pipe $P$ at the focal point of the parabola. At what distance $x$ should the pipe be placed?
28. A parabolic collector for receiving television signals from a satellite is shown in Fig. 26-26. The receiver $R$ is at the focus, 1.0 m from the vertex. Find the depth $d$ of the collector.

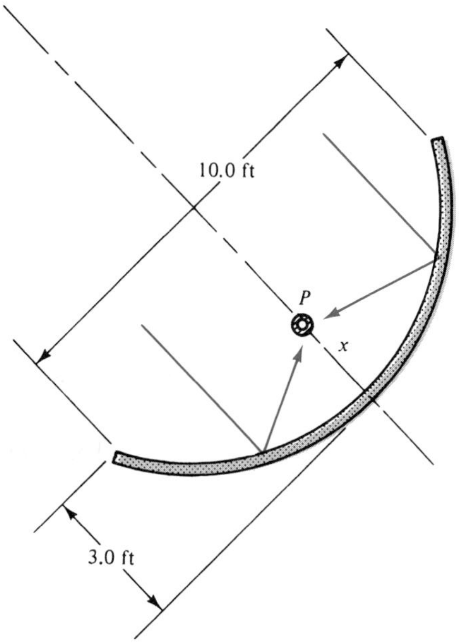


**FIGURE 26-25 Parabolic solar collector.**

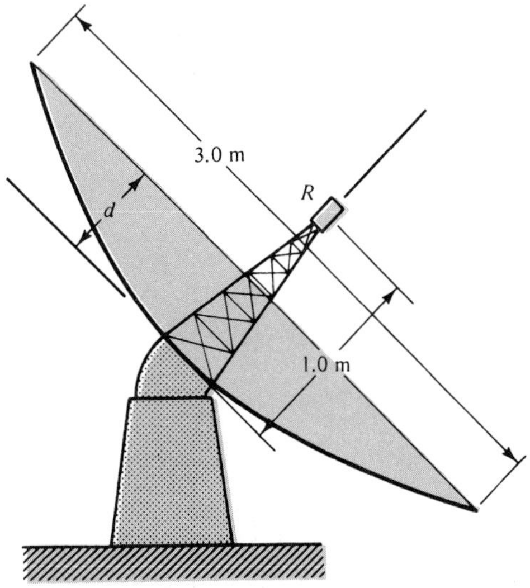


**FIGURE 26-26 Parabolic antenna.**

### Vertical Highway Curves

29. A parabolic curve is to be used at a dip in a highway. The road dips 32.0 m in a horizontal distance of 125 m, and then rises to its previous height in another 125 m. Write the equation of the curve of the roadway, taking the origin at the bottom of the dip and the $y$ axis vertical.
30. Write the equation of the vertical highway curve in Fig. 26-27, taking axes as shown.

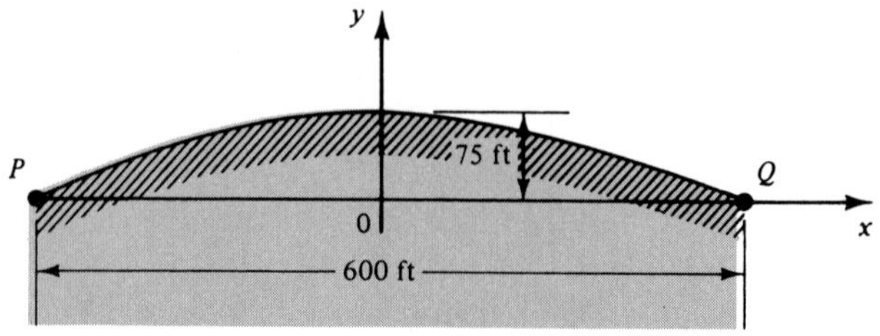


**FIGURE 26-27** **Road over a hill.**

### Beams

31. A simply supported beam with a concentrated load at its midspan (Fig. 26-28) will deflect in the shape of a parabola. If the deflection at the midspan is 1.0 in., write the equation of the parabola (called the elastic curve), taking axes as shown.
32. Using the equation found in Problem 31, find the deflection of the beam in Fig. 26-28 at a distance of 10.0 ft from the left end.

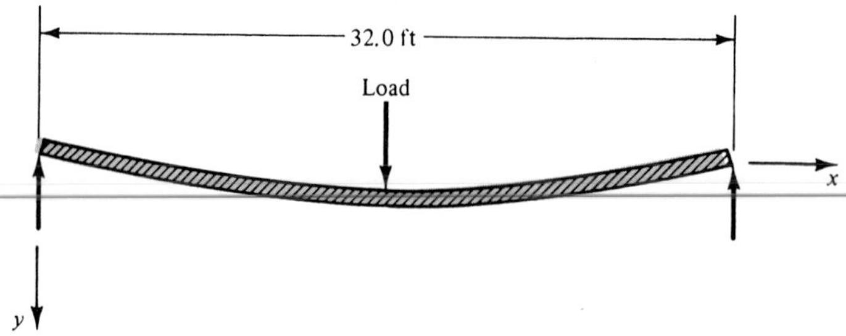


**FIGURE 26-28** **Deflection of a beam.**

## 26-3 ELLIPSE

| | | |
|---|---|---|
| Definition | An *ellipse* is the set of all points in a plane such that the sum of the distances from each point on the ellipse to two fixed points (called the *foci*) is constant. | 305 |

Such a set of points, connected with a smooth curve, is plotted in Fig. 26-29, and shows the typical shape of the ellipse. An ellipse has two axes of symmetry, the *major axis* and the *minor axis,* which intersect at the *center* of the ellipse. A *vertex* is a point where the ellipse crosses the major axis.

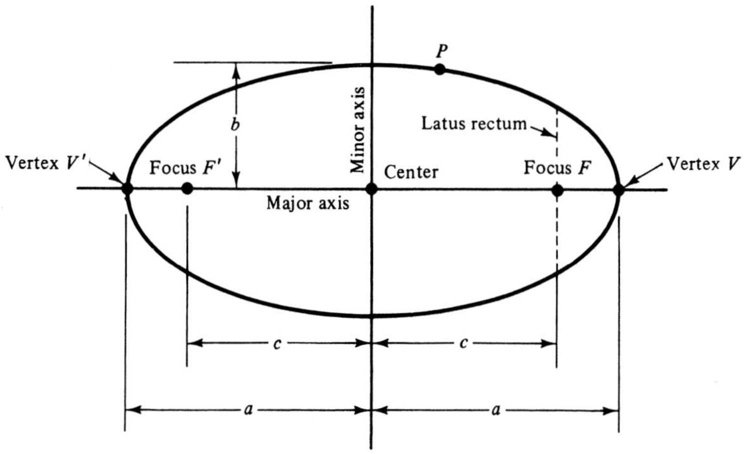


FIGURE 26-29 **Ellipse.**

It is often convenient to speak of *half* the lengths of the major and minor axes, and these are called the *semimajor* and *semiminor* axes, whose lengths we label $a$ and $b$, respectively. The distance from either focus to the center is labeled $c$.

## Distance to Focus

Before deriving an equation for the ellipse let us first write an expression for the distance $c$ from the center to a focus, in terms of the semimajor axis $a$ and the semiminor axis $b$.

From our definition of an ellipse, if $P$ is any point on the ellipse, then

$$PF + PF' = k \tag{1}$$

where $k$ is a constant. If $P$ is taken at a vertex $V$, then (1) becomes

$$VF + VF' = k = 2a \tag{2}$$

because $VF + VF'$ is equal to the length of the major axis. Substituting back into (1) gives us

$$PF + PF' = 2a \tag{3}$$

Figure 26-30 shows our point $P$ moved to the intersection of the ellipse and the minor axis. Here $PF$ and $PF'$ are equal. But since their sum is $2a$, $PF$ and $PF'$ must each equal $a$. By the Pythagorean theorem, $c^2 + b^2 = a^2$, or,

$$c^2 = a^2 - b^2$$

| | | |
|---|---|---|
| Ellipse: Distance from Center to Focus | $c = \sqrt{a^2 - b^2}$ | **311** |

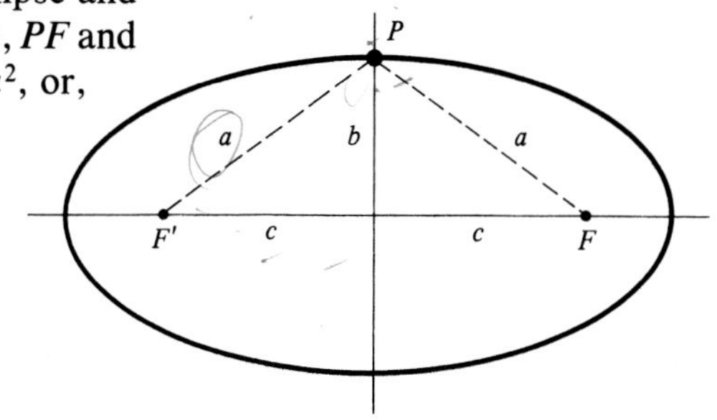


FIGURE 26-30

## Ellipse with Center at the Origin, Major Axis Horizontal

Let us place an ellipse on coordinate axes with its center at the origin and major axis along the $x$ axis, as in Fig. 26-31. If $P(x, y)$ is any point on the ellipse, then by the definition of the ellipse,

$$PF + PF' = 2a$$

**A line from a focus to a point on the ellipse is called a *focal radius*.**

To get $PF$ and $PF'$ in terms of $x$ and $y$, we first drop a perpendicular from $P$ to the $x$ axis. Then in triangle $PQF$,

$$PF = \sqrt{(c - x)^2 + y^2}$$

and in triangle $PQF'$,

$$PF' = \sqrt{(c + x)^2 + y^2}$$

Substituting yields

$$PF + PF' = \sqrt{(c - x)^2 + y^2} + \sqrt{(c + x)^2 + y^2} = 2a$$

Rearranging, we obtain

$$\sqrt{(c + x)^2 + y^2} = 2a - \sqrt{(c - x)^2 + y^2}$$

Squaring, and expanding the binomials,

$$x^2 + 2cx + c^2 + y^2 = 4a^2 - 4a\sqrt{(c - x)^2 + y^2} + c^2 - 2cx + x^2 + y^2$$

Collecting terms, we get

$$cx = a^2 - a\sqrt{(c - x)^2 + y^2}$$

Dividing by $a$ and rearranging,

$$a - \frac{cx}{a} = \sqrt{(c - x)^2 + y^2}$$

Squaring both sides again yields

$$a^2 - 2cx + \frac{c^2x^2}{a^2} = c^2 - 2cx + x^2 + y^2$$

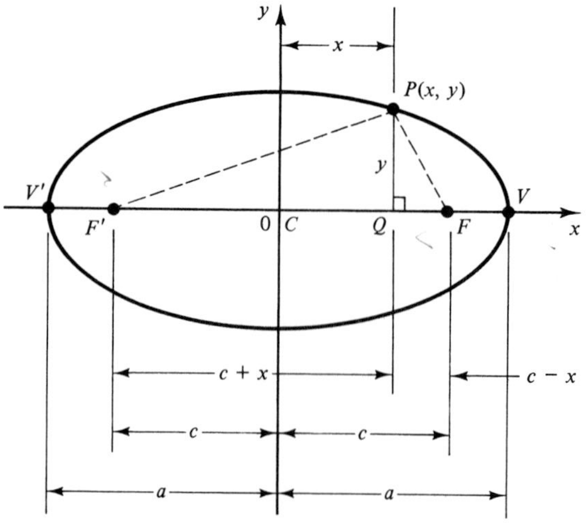


**FIGURE 26-31 Ellipse with center at origin.**

Collecting terms,

$$a^2 + \frac{c^2x^2}{a^2} = c^2 + x^2 + y^2$$

But $c = \sqrt{a^2 - b^2}$. Substituting, we have

$$a^2 + \frac{x^2}{a^2}(a^2 - b^2) = a^2 - b^2 + x^2 + y^2$$

or

$$a^2 + x^2 - \frac{b^2x^2}{a^2} = a^2 - b^2 + x^2 + y^2$$

Collecting terms and rearranging gives us

$$\frac{b^2x^2}{a^2} + y^2 = b^2$$

Finally, dividing through by $b^2$, we get the standard form of the equation of an ellipse with center at origin.

| | | | |
|---|---|---|---|
| Standard Equation of an Ellipse: Center at Origin, Major Axis Horizontal | 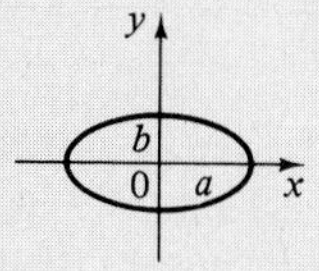 | $\frac{x^2}{a^2} + \frac{y^2}{b^2} = 1$<br>$a > b$ | **306** |

Here $a$ is half the length of the major axis, and $b$ is half the length of the minor axis.

**EXAMPLE 14:** Find the vertices and foci for the ellipse $16x^2 + 36y^2 = 576$.

**Solution:** To be in standard form, our equation must have 1 (unity) on the right side. Dividing by 576 and simplifying gives us

$$\frac{x^2}{36} + \frac{y^2}{16} = 1$$

from which $a = 6$ and $b = 4$. The vertices are then $V(6, 0)$ and $V'(-6, 0)$ (Fig. 26-32). The distance $c$ from the center to a focus is

$$c = \sqrt{6^2 - 4^2} = \sqrt{20} \simeq 4.47$$

so the foci are $F(4.47, 0)$ and $F'(-4.47, 0)$.

$c = \sqrt{a^2 - b^2}$

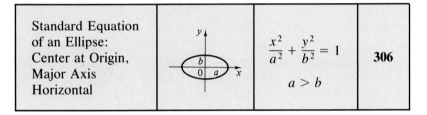

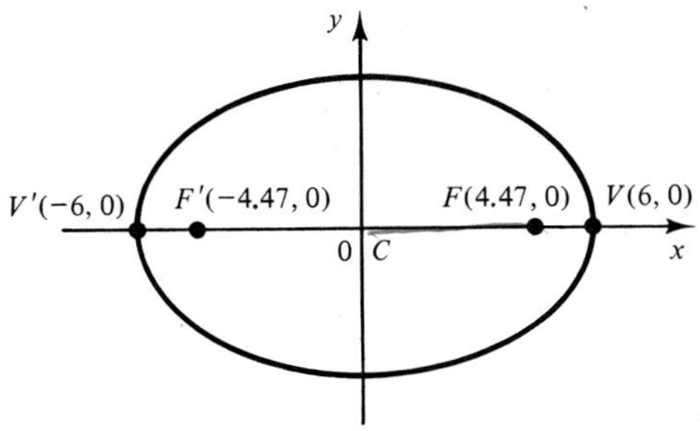


**FIGURE 26-32**

## Ellipse with Center at Origin, Major Axis Vertical

When the major axis is vertical rather than horizontal, the only effect on the standard equation is to interchange the positions of $x$ and $y$.

| | | | |
|---|---|---|---|
| Standard Equation of an Ellipse: Center at Origin, Major Axis Vertical | 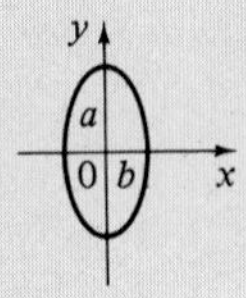 | $\frac{y^2}{a^2} + \frac{x^2}{b^2} = 1$ <br> $a > b$ | **307** |

Notice that the quantities $a$ and $b$ are dimensions of the semimajor and semiminor axes and remain so as the ellipse is turned or shifted. Therefore, for an ellipse in any position, the distance $c$ from center to focus is found the same way as before.

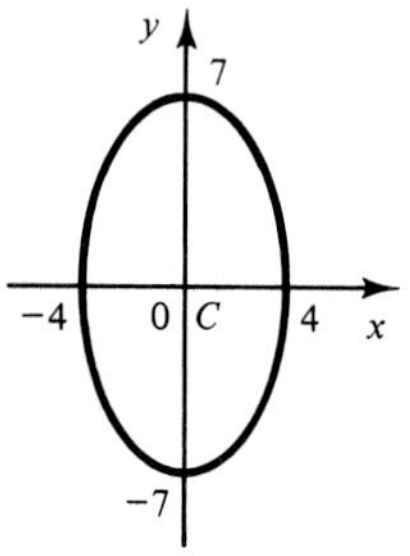

**FIGURE 26-33**

**EXAMPLE 15:** Find the lengths of the major and minor axes and the distance to the focus for the ellipse

$$\frac{x^2}{16} + \frac{y^2}{49} = 1$$

**Solution:** How can we tell which denominator is $a^2$ and which is $b^2$? It is easy: $a$ is always greater than $b$. So $a = 7$ and $b = 4$. Thus the major and minor axes are 14 units and 8 units long (Fig. 26-33). From Eq. 311,

$$c = \sqrt{a^2 - b^2} = \sqrt{49 - 16} = \sqrt{33}$$

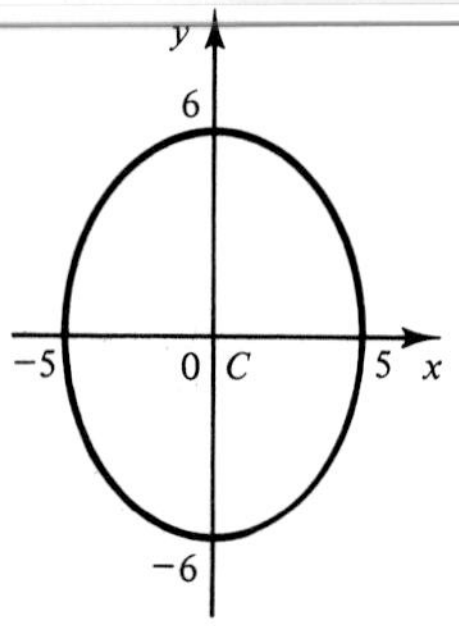

**FIGURE 26-34**

**EXAMPLE 16:** Write the equation of an ellipse with center at the origin, whose major axis is 12 units on the $y$ axis, and whose minor axis is 10 units (Fig. 26-34).

**Solution:** Substituting into Eq. 307, with $a = 6$ and $b = 5$, we obtain

$$\frac{y^2}{36} + \frac{x^2}{25} = 1$$

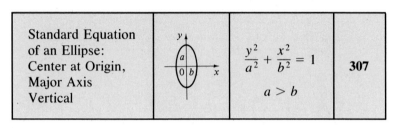

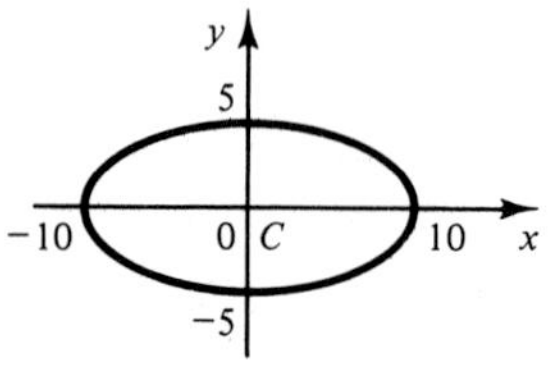

**FIGURE 26-35**

**EXAMPLE 17:** An ellipse whose center is at the origin and whose major axis is on the $x$ axis has a minor axis of 10 units and passes through the point (6, 4) (Fig. 26-35). Write the equation of the ellipse in standard form.

**Solution:** Our equation will have the form of Eq. 306. Substituting, with $b = 5$, we have

$$\frac{x^2}{a^2} + \frac{y^2}{25} = 1$$

Since the ellipse passes through (6, 4), these coordinates must satisfy our equation. Substituting yields

$$\frac{36}{a^2} + \frac{16}{25} = 1$$

Solving for $a^2$, we multiply by the LCD, $25a^2$:

$$36(25) + 16a^2 = 25a^2$$
$$9a^2 = 36(25)$$
$$a^2 = 100$$

So our final equation is

$$\frac{x^2}{100} + \frac{y^2}{25} = 1$$

| Common Error | Do not confuse $a$ and $b$ in the ellipse equations. The *larger* denominator is always $a^2$. Also, the variable ($x$ or $y$) in the same term with $a^2$ tells the direction of the major axis. |
|---|---|

### Shift of Axes

Now consider an ellipse whose center is not at the origin, but at $(h, k)$. The equation for such an ellipse will be the same as before, except that $x$ is replaced by $x - h$ and $y$ is replaced by $y - k$.

| | | | | |
|---|---|---|---|---|
| Standard Equations of an Ellipse: Center at $(h, k)$ | Major Axis Horizontal | 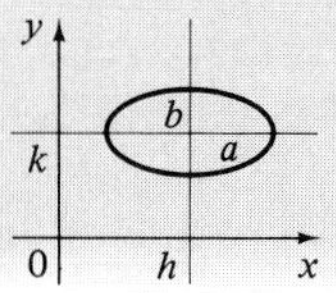 | $\frac{(x-h)^2}{a^2} + \frac{(y-k)^2}{b^2} = 1$ <br> $a > b$ | **308** |
| | Major Axis Vertical | 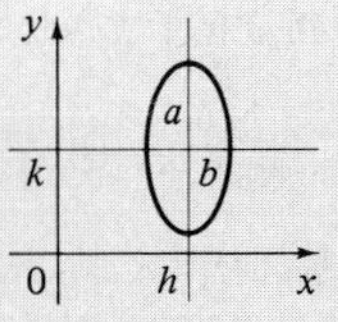 | $\frac{(y-k)^2}{a^2} + \frac{(x-h)^2}{b^2} = 1$ <br> $a > b$ | **309** |

**EXAMPLE 18:** Find the center, vertices, and foci for the ellipse

$$\frac{(x-5)^2}{9} + \frac{(y+3)^2}{16} = 1$$

**Solution:** From the given equation, $h = 5$ and $k = -3$, so the center is $C(5, -3)$ (Fig. 26-36). Also, $a = 4$ and $b = 3$, and the major axis is vertical because $a$ is with the $y$ term. By setting $x = 5$ in the equation of the ellipse, the vertices are $V(5, 1)$ and $V'(5, -7)$. The distance $c$ to the foci is

$$c = \sqrt{4^2 - 3^2} = \sqrt{7} \cong 2.66$$

so the foci are $F(5, -0.34)$ and $F'(5, -5.66)$.

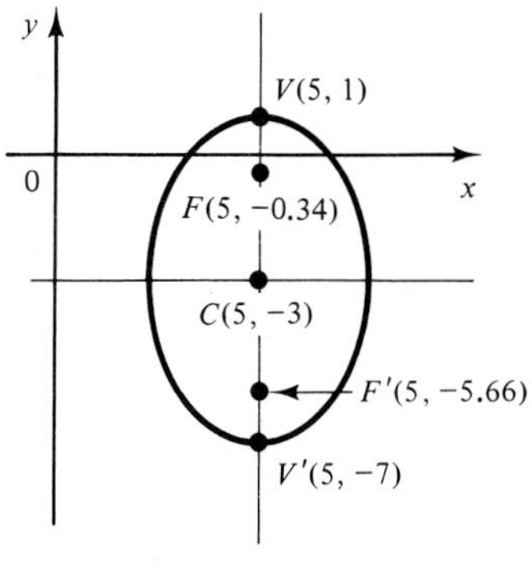


**FIGURE 26-36**

**EXAMPLE 19:** Write the equation in standard form of an ellipse with a vertical major axis 10 units long, center at (3, 5), and whose distance between focal points is 8 units (Fig. 26-37).

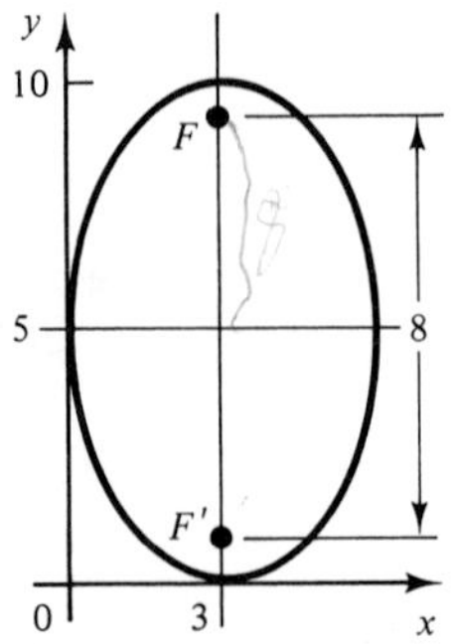


**FIGURE 26-37**

**Yes, Eq. 311 still applies here.**

**Solution:** From the information given, $h = 3$, $k = 5$, $a = 5$, and $c = 4$. By Eq. 311,

$$b = \sqrt{a^2 - c^2} = \sqrt{25 - 16} = 3 \text{ units}$$

Substituting into Eq. 309, we get

$$\frac{(y-5)^2}{25} + \frac{(x-3)^2}{9} = 1$$

## General Equation of an Ellipse

As we did with the circle, we now expand the standard equation for the ellipse to get the general equation. Starting with Eq. 308,

$$\frac{(x-h)^2}{a^2} + \frac{(y-k)^2}{b^2} = 1$$

we multiply through by $a^2b^2$, and expand the binomials:

$$b^2(x^2 - 2hx + h^2) + a^2(y^2 - 2ky + k^2) = a^2b^2$$

$$b^2x^2 - 2b^2hx + b^2h^2 + a^2y^2 - 2a^2ky + a^2k^2 - a^2b^2 = 0$$

or

$$b^2x^2 + a^2y^2 - 2b^2hx - 2a^2ky + (b^2h^2 + a^2k^2 - a^2b^2) = 0$$

which is of the following general form.

| | | |
|---|---|---|
| General Equation of an Ellipse | $Ax^2 + Cy^2 + Dx + Ey + F = 0$ | **310** |

Comparing this with the general equation of the second degree, we see that the general equation represents an ellipse with axes parallel to the coordinate axes if $B = 0$, and if $A$ and $C$ are different but have the same sign.

## Completing the Square

As before, we go from general to standard form by completing the square.

**EXAMPLE 20:** Find the center, foci, vertices, and major and minor axes for the ellipse $9x^2 + 25y^2 + 18x - 50y - 191 = 0$.

**Solution:** Grouping the $x$ terms and the $y$ terms gives us

$$(9x^2 + 18x) + (25y^2 - 50y) = 191$$

Factoring yields

$$9(x^2 + 2x \quad ) + 25(y^2 - 2y \quad ) = 191$$

Completing the square, we obtain

$$9(x^2 + 2x + 1) + 25(y^2 - 2y + 1) = 191 + 9 + 25$$

Factoring gives us

$$9(x + 1)^2 + 25(y - 1)^2 = 225$$

Finally, dividing by 225, we have

$$\frac{(x + 1)^2}{25} + \frac{(y - 1)^2}{9} = 1$$

We see that $h = -1$ and $k = 1$, so the center is at $(-1, 1)$. Also, $a = 5$, so the major axis is 10 units and is horizontal, and $b = 3$, so the minor axis is 6 units. A vertex is located 5 units to the right of the center, at (4, 1), and 5 units to the left, at (−6, 1). From Eq. 311,

$$c = \sqrt{a^2 - b^2}$$

$$= \sqrt{25 - 9}$$

$$= 4$$

So the foci are at (3, 1) and (−5, 1). This ellipse is shown in Fig. 26-38.

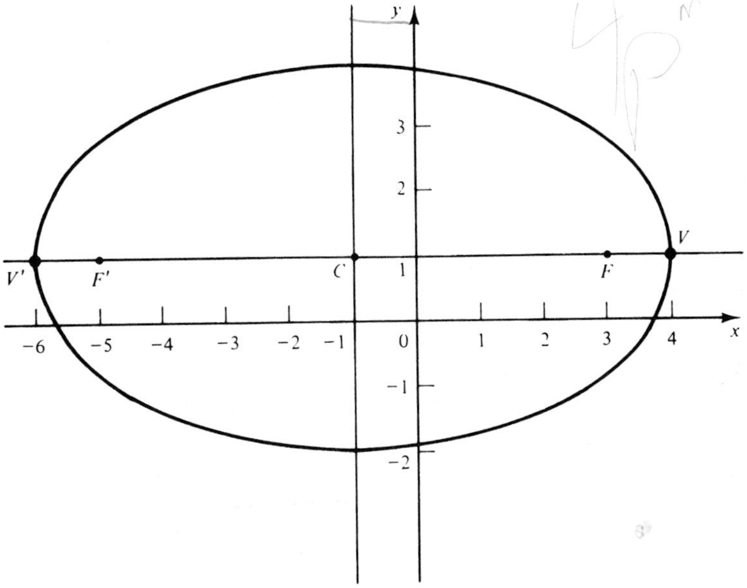


**FIGURE 26-38**

| | |
|---|---|
| Common Error | In Example 20 we needed 1 to complete the square on $x$, but added 9 to the right side, because the expression containing the 1 was multiplied by a factor of 9. Similarly, for $y$, we needed 1 on the left but added 25 to the right. It is very easy to forget to multiply by those factors. |

## Focal Width of an Ellipse

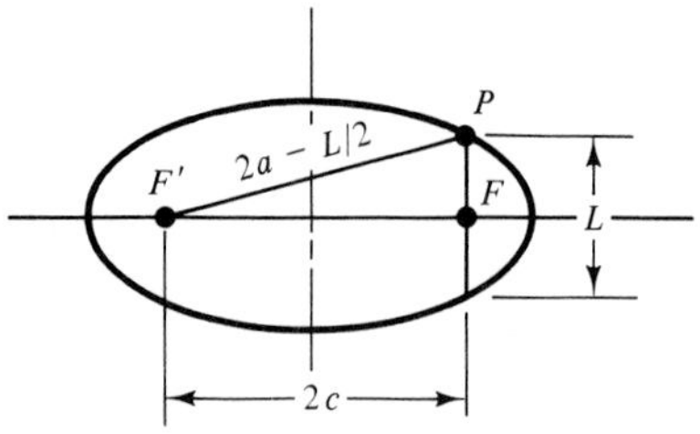


**FIGURE 26-39 Derivation of focal width.**

The focal width $L$, or length of the latus rectum, is the width of the ellipse through the focus (Fig. 26-39). The sum of the focal radii $PF$ and $PF'$ from a point $P$ at one end of the latus rectum must equal $2a$, by the definition of an ellipse. So $PF' = 2a - PF$, or since $L/2 = PF$, $PF' = 2a - L/2$. Squaring both sides,

$$(PF')^2 = 4a^2 - 2aL + \frac{L^2}{4} \tag{1}$$

But in right triangle $PFF'$,

$$(PF')^2 = \left(\frac{L}{2}\right)^2 + (2c)^2 = \frac{L^2}{4} + 4c^2 = \frac{L^2}{4} + 4a^2 - 4b^2 \tag{2}$$

since $c^2 = a^2 - b^2$. Equating (1) and (2) and collecting terms gives $2aL = 4b^2$, or

**As with the parabola, the main use for $L$ is for quick sketching of the ellipse.**

| | | |
|---|---|---|
| Focal Width of an Ellipse | $L = \dfrac{2b^2}{a}$ | **312** |

*The focal width of an ellipse is twice the square of the semiminor axis, divided by the semimajor axis.*

---

**EXAMPLE 21:** The focal width of an ellipse that is 25 m long and 10 m wide is

$$L = \frac{2(5^2)}{12.5} = \frac{50}{12.5} = 4 \text{ m}$$

---

## EXERCISE 3—ELLIPSE

### Ellipse with Center at Origin

Find the coordinates of the vertices and foci for each ellipse.

1. $\dfrac{x^2}{25} + \dfrac{y^2}{16} = 1$
2. $x^2 + 2y^2 = 2$
3. $3x^2 + 4y^2 = 12$
4. $64x^2 + 15y^2 = 960$
5. $4x^2 + 3y^2 = 48$
6. $8x^2 + 25y^2 = 200$

Write the equation of each ellipse.

7. vertices at (±5, 0), foci at (±4, 0)

8. vertical major axis 8 units long, a focus at (0, 2)

9. horizontal major axis 12 units long, passes through $(3, \sqrt{3})$

10. passes through (4, 6) and $(2, 3\sqrt{5})$

11. horizontal major axis 26 units long, distance between foci = 24
12. vertical minor axis 10 units long, distance from focus to vertex = 1
13. passes through (1, 4) and (−6, 1)
14. distance between foci = 18, sum of axes = 54, horizontal major axis

### Ellipse with Center Not at Origin

Find the coordinates of the center, vertices, and foci for each ellipse.

15. $\dfrac{(x-2)^2}{16} + \dfrac{(y+2)^2}{9} = 1$
16. $\dfrac{(x+5)^2}{25} + \dfrac{(y-3)^2}{49} = 1$
17. $5x^2 + 20x + 9y^2 - 54y + 56 = 0$
18. $16x^2 - 128x + 7y^2 + 42y = 129$
19. $7x^2 - 14x + 16y^2 + 32y = 89$
20. $3x^2 - 6x + 4y^2 + 32y + 55 = 0$
21. $25x^2 + 150x + 9y^2 - 36y + 36 = 0$

Write the equation of each ellipse.

22. minor axis = 10, foci at (13, 2) and (−11, 2)
23. center at (0, 3), vertical major axis = 12, length of minor axis = 6
24. center at (2, −1), a vertex at (2, 5), length of minor axis = 3
25. center at (−2, −3), a vertex at (−2, 1), a focus halfway between vertex and center

### Intersections of Curves

Solve simultaneously to find the points of intersection of each ellipse with the given curve.

26. $3x^2 + 6y^2 = 11$ with $y = x + 1$
27. $2x^2 + 3y^2 = 14$ with $y^2 = 4x$
28. $x^2 + 7y^2 = 16$ with $x^2 + y^2 = 10$

### Construction of an Ellipse

29. One way to draw an ellipse is to put two tacks at the focal points (Fig. 26-40) and to adjust a loop of string so that the pencil point passes through the end of the major and minor axes. How far apart must the tacks be placed if the ellipse is to have a major axis of 84 cm and a minor axis of 58 cm?

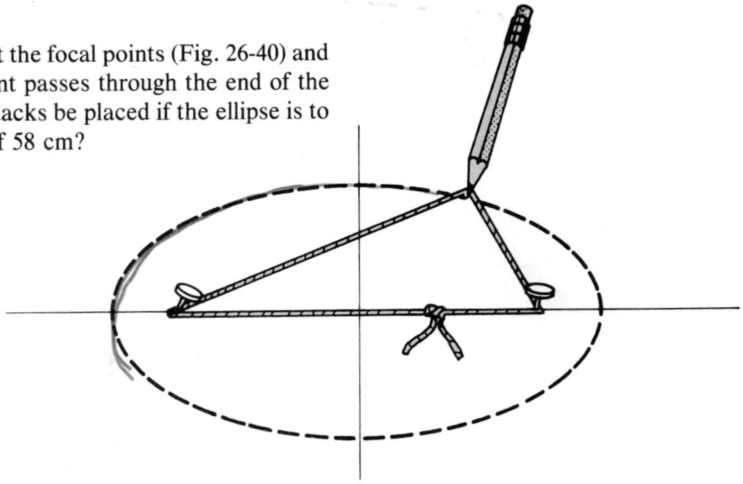

**FIGURE 26-40 Construction of an ellipse.**

## Applications

For Problems 30–33, first write the equation for each ellipse.

**30.** A certain bridge arch is in the shape of half an ellipse 120 ft wide and 30.0 ft high. At what horizontal distance from the center of the arch is the height equal to 15.0 ft?

**31.** A curved mirror in the shape of an ellipse will reflect all rays of light coming from one focus onto the other focus (Fig. 26-41). A certain spot heater is to be made with a heating element at $A$ and the part to be heated at $B$, contained within an ellipsoid (a solid obtained by rotating an ellipse about one axis). Find the width $x$ of the chamber if its length is 25 cm and the distance from $A$ to $B$ is 15 cm.

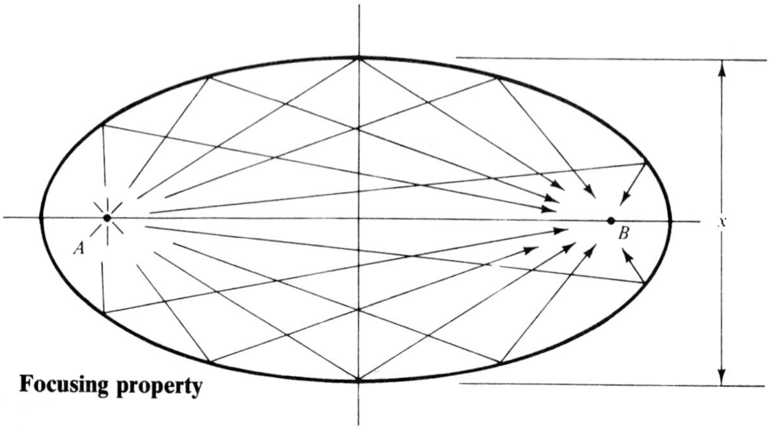


**FIGURE 26-41 Focusing property of the ellipse.**

An *astronomical unit,* AU, is the distance between the earth and the sun, about 92.6 million miles.

**32.** The paths of the planets and certain comets are ellipses, with the sun at one focal point. The path of Halley's comet is an ellipse with a major axis of 36.18 AU and a minor axis of 9.12 AU. What is the greatest distance that Halley's comet gets from the sun?

**33.** An elliptical culvert (Fig. 26-42) is filled with water to a depth of 1.0 ft. Find the width $w$ of the stream.

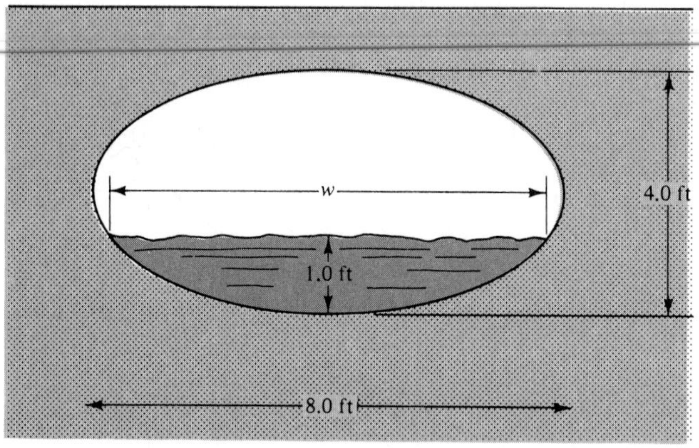


**FIGURE 26-42**

## 26-4 HYPERBOLA

| | | |
|---|---|---|
| Definition | A *hyperbola* is the set of all points in a plane such that the distances from each point to two fixed points, the *foci,* have a constant difference. | **315** |

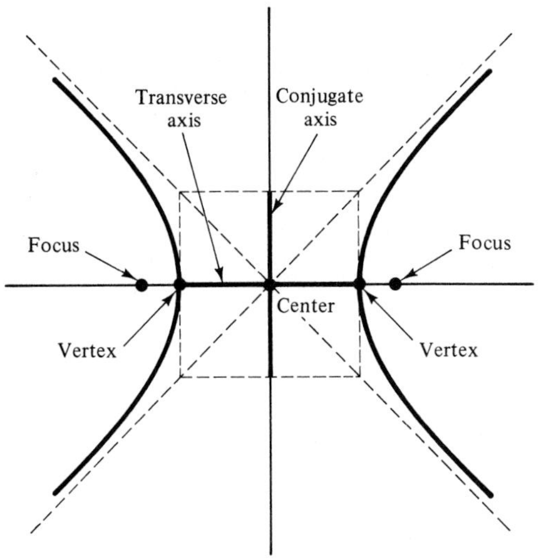


**FIGURE 26-43 Hyperbola.**

Figure 26-43 shows the typical shape of the hyperbola. The point midway between the foci is called the *center* of the hyperbola. The line passing through the foci and the center is one *axis* of the hyperbola. The hyperbola crosses that axis at points called the *vertices*. The line segment connecting the vertices is called the *transverse axis*. A second axis of the hyperbola passes through the center and is perpendicular to the transverse axis. The segment of this axis shown in Fig. 26-43 is called the *conjugate axis*. Half the lengths of the transverse and conjugate axes are the *semitransverse* and *semiconjugate* axes, respectively. They are also referred to as *semiaxes*.

## Standard Equations of a Hyperbola with Center at Origin

We place the hyperbola on coordinate axes, with its center at the origin and its transverse axis on the $x$ axis (Fig. 26-44). Let $a$ be half the transverse axis,

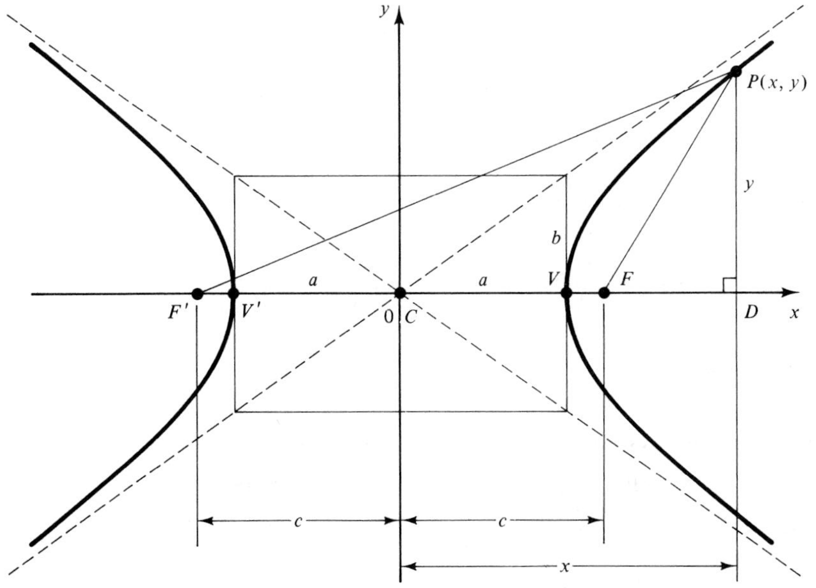


**FIGURE 26-44 Hyperbola with center at origin.**

and let $c$ be half the distance between foci. Now take any point $P$ on the hyperbola and draw the focal radii $PF$ and $PF'$. Then by the definition of the hyperbola, $|PF' - PF| = \text{constant}$. The constant can be found by moving $P$ to the vertex $V$, where

$$\begin{aligned} |PF' - PF| &= VF' - VF \\ &= 2a + V'F' - VF \\ &= 2a \end{aligned}$$

since $V'F'$ and $VF$ are equal. So $|PF' - PF| = 2a$. But in right triangle $PF'D$,

$$PF' = \sqrt{(x + c)^2 + y^2}$$

and in right triangle $PFD$,

$$PF = \sqrt{(x - c)^2 + y^2}$$

so

$$\sqrt{(x + c)^2 + y^2} - \sqrt{(x - c)^2 + y^2} = 2a$$

Notice that this equation is almost identical to the one we had when deriving the equation for the ellipse in Sec. 26-3, and the derivation is almost identical as well. However, to eliminate $c$ from our equation, we define a new quantity $b$, such that $b^2 = c^2 - a^2$. We'll soon give a geometric meaning to the quantity $b$.

**Try to derive these equations. Follow the same steps we used for the ellipse.**

The remainder of the derivation is left as an exercise. The resulting equations are very similar to those for the ellipse.

| | | | | |
|---|---|---|---|---|
| Standard Equations of a Hyperbola: Center at Origin | Transverse Axis Horizontal | 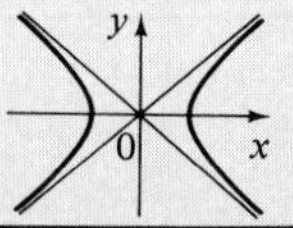 | $\dfrac{x^2}{a^2} - \dfrac{y^2}{b^2} = 1$ | **316** |
| | Transverse Axis Vertical | 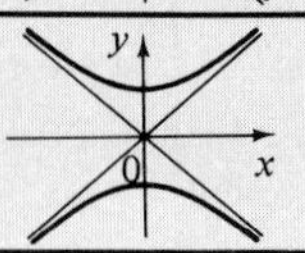 | $\dfrac{y^2}{a^2} - \dfrac{x^2}{b^2} = 1$ | **317** |

## Asymptotes of a Hyperbola

In Fig. 26-44, the dashed diagonal lines are asymptotes of the hyperbola. (The two branches of the hyperbola approach, but do not intersect, the asymptotes.)

We find the slope of these asymptotes from the equation of the ellipse. Solving Eq. 316 for $y$ gives $y^2 = b^2(x^2/a^2 - 1)$, or

$$y = \pm\, b\sqrt{\frac{x^2}{a^2} - 1}$$

As $x$ gets large, the 1 under the radical sign becomes insignificant in comparison with $x^2/a^2$, and the equation for $y$ becomes

$$y = \pm\frac{b}{a}x$$

This is the equation of a straight line having a slope of $b/a$. Thus as $x$ increases, the branches of the hyperbola more closely approach straight lines of slopes $\pm b/a$.

In general, if the distance from a point $P$ on a curve to some line $L$ approaches zero as the distance from $P$ to the origin increases without bound, then $L$ is an *asymptote*. The hyperbola has two such asymptotes, of slopes $\pm b/a$.

For a hyperbola whose transverse axis is *vertical*, the slopes of the asymptotes can be shown to be $\pm a/b$.

| | | | |
|---|---|---|---|
| Slope of the Asymptotes of a Hyperbola | Transverse Axis Horizontal | slope $= \pm\dfrac{b}{a}$ | **322** |
| | Transverse Axis Vertical | slope $= \pm\dfrac{a}{b}$ | **323** |

| | |
|---|---|
| Common Error | The asymptotes are not (usually) perpendicular to each other. The slope of one is **not** the negative reciprocal of the other. |

Looking back at Fig. 26-44, we can now give meaning to the quantity $b$. If an asymptote has a slope $b/a$, it must have a rise of $b$ in a run equal to $a$. Thus $b$ is the distance, perpendicular to the transverse axis, from vertex to asymptote. It is half the length of the conjugate axis.

| | | |
|---|---|---|
| Hyperbola: Distance from Center to Focus | $c = \sqrt{a^2 + b^2}$ | **321** |

**EXAMPLE 22:** Find $a$, $b$, and $c$, the coordinates of the center, the vertices, and the foci, and the slope of the asymptotes for the hyperbola

$$\frac{x^2}{25} - \frac{y^2}{36} = 1$$

**Solution:** This equation is of the same form as Eq. 316, so we know that the center is at the origin and that the transverse axis is on the $x$ axis. Also, $a^2 = 25$ and $b^2 = 36$, so

$$a = 5 \quad \text{and} \quad b = 6$$

The vertices then have the coordinates

$$V(5, 0) \quad \text{and} \quad V'(-5, 0)$$

Then from Eq. 321,

$$c^2 = a^2 + b^2 = 25 + 36 = 61$$

so $c = \sqrt{61} = \approx 7.81$. The coordinates of the foci are then,

$$F(7.81, 0) \quad \text{and} \quad F'(-7.81, 0)$$

The slope of the asymptotes are

$$\pm\frac{b}{a} = \pm\frac{6}{5}$$

This hyperbola is shown in Fig. 26-45. In the following section we show how to make such a graph.

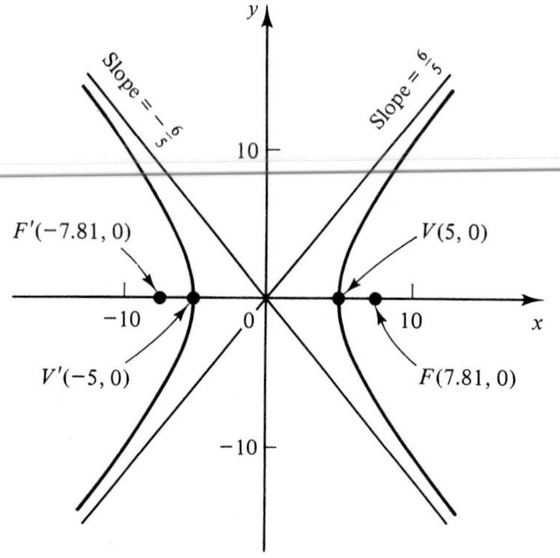


**FIGURE 26-45 Graph of $\frac{x^2}{25} - \frac{y^2}{36} = 1$.**

## Graphing a Hyperbola

A good way to start a sketch of the hyperbola is to draw a rectangle whose dimension along the transverse axis is $2a$ and whose dimension along the conjugate axis is $2b$. The asymptotes are then drawn along the diagonals of this rectangle. Half the diagonal of the rectangle has a length $\sqrt{a^2 + b^2}$, which is equal to $c$, the distance to the foci. Thus an arc of radius $c$ will cut the transverse axes at the focal points.

**EXAMPLE 23:** Find the vertices, foci, semiaxes, slope of the asymptotes, and focal width, and graph the hyperbola $64x^2 - 49y^2 = 3136$.

**As with the parabola and ellipse, a perpendicular through a focus connecting two points on the hyperbola is called a *latus rectum.* Its length, called the focal width, is $2b^2/a$, the same as for the ellipse.**

**Solution:** We write the equation in standard form by dividing by 3136,

$$\frac{x^2}{49} - \frac{y^2}{64} = 1$$

This is the form of Eq. 316, so the transverse axis is horizontal, with $a = 7$ and $b = 8$. We draw a rectangle of width 14 and height 16 (shown shaded in Fig. 26-46), thus locating the vertices at $(\pm 7, 0)$. Diagonals through the rectangle give us the asymptotes of slopes $\pm\frac{8}{7}$. We locate the foci by swinging an arc of radius $c$, equal to half the diagonal of the rectangle. Thus

$$c = \sqrt{7^2 + 8^2} = \sqrt{113} \simeq 10.6$$

The foci then are at $(\pm 10.6, 0)$. We obtain a few more points by computing the focal width,

$$L = \frac{2b^2}{a} = \frac{2(64)}{7} \simeq 18.3$$

This gives us the additional points (10.6, 9.15), (10.6, −9.15), (−10.6, 9.15), and (−10.6, −9.15).

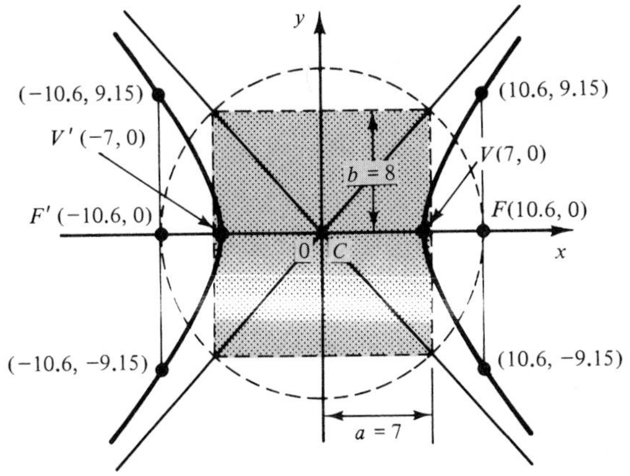


FIGURE 26-46

**EXAMPLE 24:** A hyperbola whose center is at the origin has a focus at (0, −5) and a transverse axis 8 units long (Fig. 26-47). Write the standard equation of the hyperbola and find the slope of the asymptotes.

**Solution:** The transverse axis is 8, so $a = 4$. A focus is 5 units below the origin, so the transverse axis must be vertical, and $c = 5$. From Eq. 321,

$$b = \sqrt{c^2 - a^2} = \sqrt{25 - 16} = 3$$

Substituting into Eq. 317 gives us

$$\frac{y^2}{16} - \frac{x^2}{9} = 1$$

and from Eq. 323,

$$\text{slope of asymptotes} = \pm\frac{a}{b} = \pm\frac{4}{3}$$

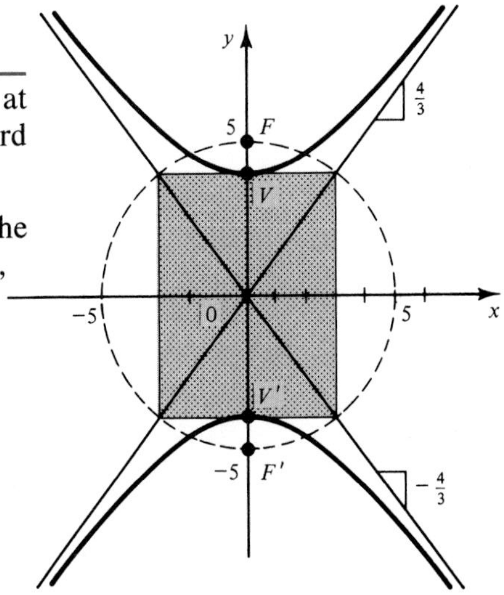


FIGURE 26-47

| | |
|---|---|
| Common Error | Be sure that you know the direction of the transverse axis before computing the slopes of the asymptotes, which are $\pm b/a$ when the axis is horizontal, but $\pm a/b$ when the axis is vertical. |

### Shift of Axes

As with the other conics, we shift the axes by replacing $x$ by $(x - h)$ and $y$ by $(y - k)$ in Eqs. 316 and 317.

| | | | | |
|---|---|---|---|---|
| Standard Equations of a Hyperbola: Center at $(h, k)$ | Transverse Axis Horizontal | 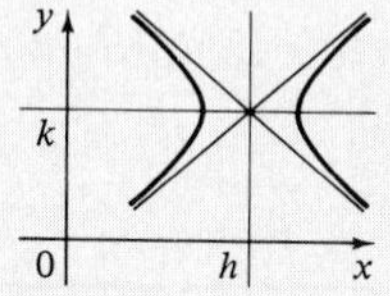 | $$\frac{(x-h)^2}{a^2} - \frac{(y-k)^2}{b^2} = 1$$ | **318** |
| | Transverse Axis Vertical | 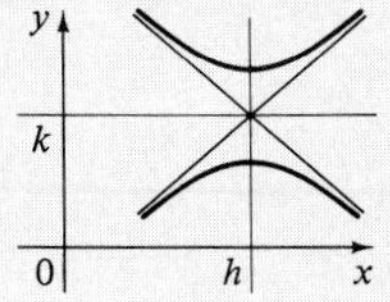 | $$\frac{(y-k)^2}{a^2} - \frac{(x-h)^2}{b^2} = 1$$ | **319** |

**The equations for $c$ and for the slope of the asymptotes are still valid for these cases.**

| | |
|---|---|
| Common Error | Do not confuse these equations with those for the ellipse. Here the terms have **opposite signs.** Also, $a^2$ is always the denominator of the **positive** term, even though it may be smaller than $b^2$. As with the ellipse, the variable in the same term as $a^2$ tells the direction of the transverse axis. |

---

**EXAMPLE 25:** A certain hyperbola has a focus at (1, 0), passes through the origin, has its transverse axis on the $x$ axis, and has a distance of 10 between its focal points. Write its standard equation.

**Solution:** In Fig. 26-48 we plot the given focus $F(1, 0)$. Since the hyperbola passes through the origin, and the transverse axis also passes through the origin, a vertex must be at the origin as well. Then, since $c = 5$, we go 5 units to the left of $F$ and plot the center, $(-4, 0)$. Thus $a = 4$ units, from center to vertex. Then, by Eq. 321,

$$b = \sqrt{c^2 - a^2}$$

$$= \sqrt{25 - 16}$$

$$= 3$$

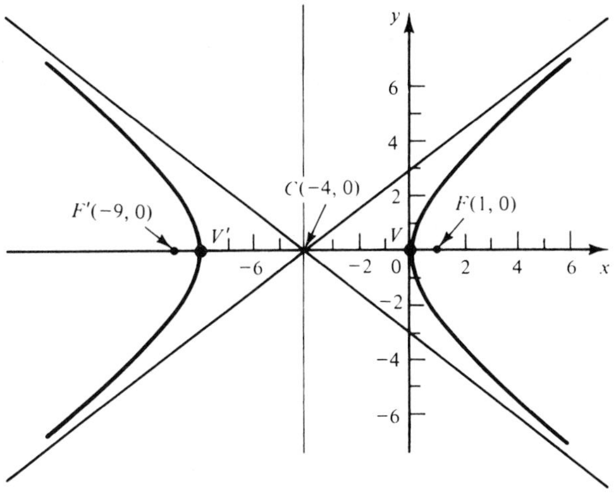


FIGURE 26-48

Substituting into Eq. 318 with

$$h = -4, \quad k = 0, \quad a = 4, \quad b = 3$$

we get

$$\frac{(x + 4)^2}{16} - \frac{y^2}{9} = 1$$

as the required equation.

---

## General Equation of a Hyperbola

Since the standard equations for the ellipse and hyperbola are identical except for the minus sign, it is not surprising that their general equations should be identical except for a minus sign. The general second-degree equation for ellipse or hyperbola is:

**It's easy to feel overwhelmed by the large number of formulas in this chapter. Study the summary of formulas in Appendix A, where you'll find them side by side and can see where they differ and where they are alike.**

| | | |
|---|---|---|
| General Equation of a Hyperbola | $Ax^2 + Cy^2 + Dx + Ey + F = 0$ | **320** |

For the ellipse, $A$ and $C$ have like signs; for the hyperbola, they have opposite signs.

As before, we complete the square to go from general form to standard form.

---

**EXAMPLE 26:** Write the equation $x^2 - 4y^2 - 6x + 8y - 11 = 0$ in standard form. Find the center, vertices, foci, and asymptotes.

**Solution:** Rearranging gives us

$$(x^2 - 6x \qquad) - 4(y^2 - 2y \qquad) = 11$$

Completing the square and factoring yields

$$(x^2 - 6x + 9) - 4(y^2 - 2y + 1) = 11 + 9 - 4$$

$$(x - 3)^2 - 4(y - 1)^2 = 16$$

(Take care, when completing the square, to add −4 to the right side to compensate for 1 times −4 added on the left.) Dividing by 16 gives us

$$\frac{(x-3)^2}{16} - \frac{(y-1)^2}{4} = 1$$

from which $a = 4$, $b = 2$, $h = 3$, and $k = 1$.

From Eq. 321, $c = \sqrt{a^2 + b^2} = \sqrt{16 + 4} \simeq 4.47$, and from Eq. 322,

$$\text{slope of asymptotes} = \pm\frac{b}{a}$$

$$= \pm\frac{1}{2}$$

This hyperbola is plotted in Fig. 26-49.

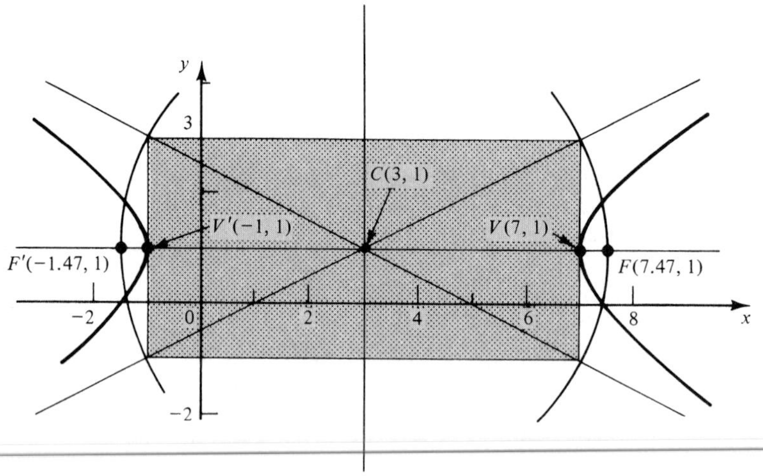


**FIGURE 26-49**

## Hyperbola Whose Asymptotes Are the Coordinate Axes

The graph of an equation of the form

$$xy = k$$

where $k$ is a constant, is a hyperbola similar to those we have studied in this section, but rotated so that the coordinate axes are now the asymptotes, and the transverse and conjugate axes are at 45°.

| | | |
|---|---|---|
| Hyperbola: Axes Rotated 45° | $xy = k$ | **325** |

**EXAMPLE 27:** Plot the equation $xy = -4$.

**Solution:** We select values for $x$ and evaluate $y = -4/x$:

| $x$ | $-4$ | $-3$ | $-2$ | $-1$ | $0$ | $1$ | $2$ | $3$ | $4$ |
|---|---|---|---|---|---|---|---|---|---|
| $y$ | $1$ | $\frac{4}{3}$ | $2$ | $4$ | $-$ | $-4$ | $-2$ | $-\frac{4}{3}$ | $-1$ |

The graph (Fig. 26-50) shows vertices at $(-2, 2)$ and $(2, -2)$. We see from the graph that $a$ and $b$ are equal, and that each is the hypotenuse of a right triangle of side 2. Therefore,

$$a = b = \sqrt{2^2 + 2^2} = \sqrt{8}$$

and from Eq. 321,

$$c = \sqrt{a^2 + b^2} = \sqrt{16} = 4$$

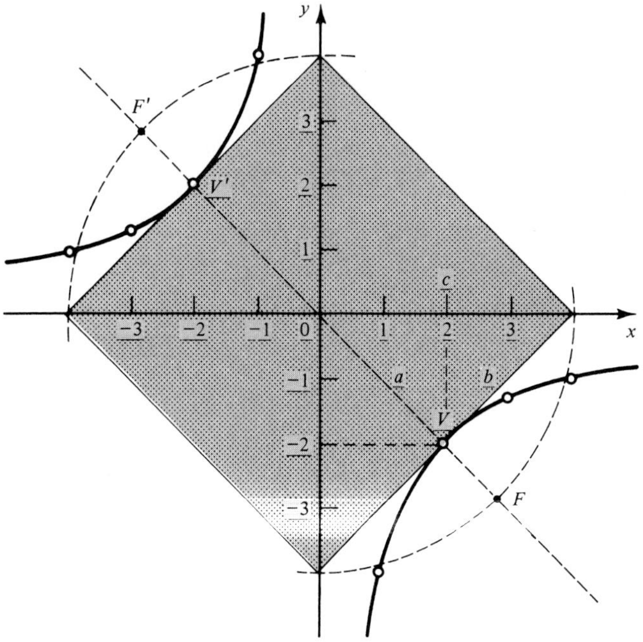


**FIGURE 26-50 Hyperbola whose asymptotes are the coordinate axes.**

When the constant $k$ in Eq. 325 is negative, the hyperbola lies in the second and fourth quadrants, as in Example 27. When $k$ is positive, the hyperbola lies in the first and third quadrants.

## EXERCISE 4—HYPERBOLA

### Hyperbola with Center at Origin

Find the vertices, foci, the lengths $a$ and $b$ of the semiaxes, and the slope of the asymptotes for each hyperbola.

1. $\dfrac{x^2}{16} - \dfrac{y^2}{25} = 1$
2. $\dfrac{y^2}{25} - \dfrac{x^2}{49} = 1$
3. $16x^2 - 9y^2 = 144$
4. $9x^2 - 16y^2 = 144$
5. $x^2 - 4y^2 = 16$
6. $3x^2 - y^2 + 3 = 0$

Write the equation of each hyperbola.

7. vertices at $(\pm 5, 0)$, foci at $(\pm 13, 0)$

8. vertices at $(0, \pm 7)$, foci at $(0, \pm 10)$

9. distance between foci = 8, transverse axis = 6 and is horizontal

10. conjugate axis = 4, foci at $(\pm 2.5, 0)$

11. transverse and conjugate axes equal, passes through (3, 5), transverse axis vertical

12. transverse axis = 16 and is horizontal, conjugate axis = 14

13. transverse axis = 10 and is vertical, curve passes through (8, 10)

14. transverse axis = 8 and is horizontal, curve passes through (5, 6)

### Hyperbola with Center Not at Origin

Find the center, lengths of the semiaxes, foci, slope of the asymptotes, and vertices for each hyperbola.

15. $\dfrac{(x-2)^2}{25} - \dfrac{(y+1)^2}{16} = 1$

16. $\dfrac{(y+3)^2}{25} - \dfrac{(x-2)^2}{49} = 1$

17. $16x^2 - 64x - 9y^2 - 54y = 161$

18. $16x^2 + 32x - 9y^2 + 592 = 0$

19. $4x^2 + 8x - 5y^2 - 10y + 19 = 0$

20. $y^2 - 6y - 3x^2 - 12x = 12$

Write the equation of each hyperbola in standard form.

21. center at (3, 2), length of the transverse axis = 8 and is vertical, length of the conjugate axis = 4

22. foci at $(-1, -1)$ and $(-5, -1)$, $a = b$

23. foci at $(5, -2)$ and $(-3, -2)$, vertex halfway between center and focus

24. length of the conjugate axis = 8 and is horizontal, center at $(-1, -1)$, length of the transverse axis = 16

### Hyperbola Whose Asymptotes Are Coordinate Axes

25. Write the equation of a hyperbola with center at the origin and with asymptotes on the coordinate axes, whose vertex is at (6, 6).
26. Write the equation of a hyperbola with center at the origin and with asymptotes on the coordinate axes, which passes through the point $(-9, 2)$.

### Applications

27. A hyperbolic mirror (Fig. 26-51) has the property that a ray of light directed at one focus will be reflected so as to pass through the other focus. Write the equation of the mirror shown, taking the axes as indicated.

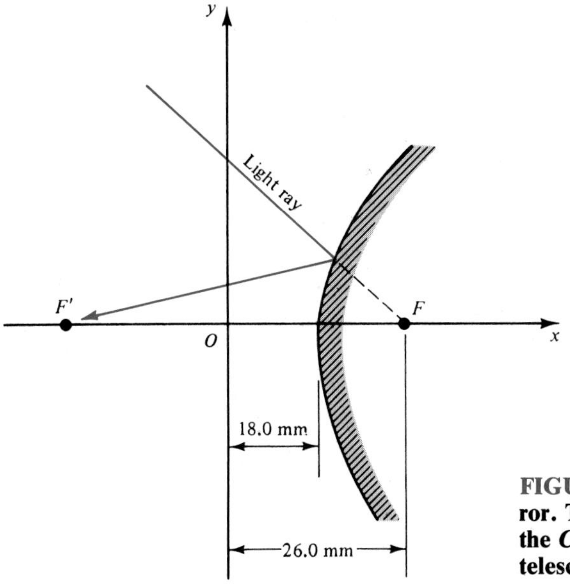


**FIGURE 26-51 Hyperbolic mirror. This type of mirror is used in the *Cassegrain* form of reflecting telescope.**

28. A ship (Fig. 26-52) receives simultaneous radio signals from stations $P$ and $Q$, on the shore. The signal from station $P$ is found to arrive 375 microseconds ($\mu$s) later than that from $Q$. Assuming that the signals travel at a rate of 0.186 mi/$\mu$s, find the distance from the ship to each station. (*Hint:* The ship will be on one branch of a hyperbola $H$ whose foci are at $P$ and $Q$. Write the equation of this hyperbola, taking axes as shown, and then substitute 75.0 mi for $y$ to obtain the distance $x$.)

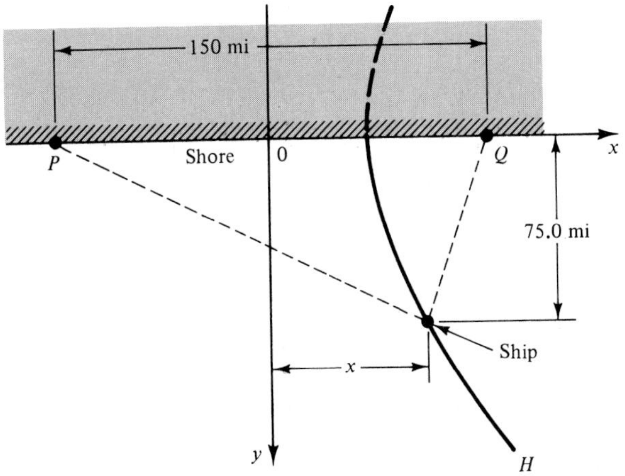


**FIGURE 26-52**

29. Boyle's law states that under certain conditions, the product of the pressure and volume of a gas is constant, or $pv = c$. This equation has the same form as the hyperbola (Eq. 325). If a certain gas has a pressure of 25.0 lb/in$^2$ at a volume of 1000 in$^3$, write Boyle's law for this gas and make a graph of pressure versus volume.

**Computer**

30. Write a program that will accept as input the constants $A$, $B$, and $C$ in the general second-degree equation (Eq. 287), and then identify and print the type of conic represented.

## CHAPTER 26 REVIEW PROBLEMS

*Identify the curve represented by each equation. Find, where applicable, the center, vertices, foci, radius, semi-axes, and so on.*

1. $x^2 - 2x - 4y^2 + 16y = 19$
2. $x^2 + 6x + 4y = 3$
3. $x^2 + y^2 = 8y$
4. $25x^2 - 200x + 9y^2 - 90y = 275$
5. $16x^2 - 9y^2 = 144$
6. $x^2 + y^2 = 9$
7. Write the equation for an ellipse whose center is at the origin, whose major axis ($2a$) is 20 and is horizontal, and whose minor axis ($2b$) equals the distance ($2c$) between the foci.
8. Write an equation for the circle passing through (0, 0), (8, 0), and (0, −6).
9. Write the equation for a parabola whose vertex is at the origin and whose focus is (−4.25, 0).
10. Write an equation for a hyperbola whose transverse axis is horizontal, with center at (1, 1) passing through (6, 2) and (−3, 1).
11. Write the equation of a circle whose center is (−5, 0) and whose radius is 5.
12. Write the equation of a hyperbola whose center is at the origin, whose transverse axis = 8 and is horizontal, passing through (10, 25).
13. Write the equation of an ellipse whose foci are (2, 1) and (−6, 1), and the sum of the focal radii is 10.
14. Write the equation of a parabola whose axis is the line $y = -7$, whose vertex is 3 units to the right of the $y$ axis, and passing through (4, −5).
15. Find the intercepts of the curve $y^2 + 4x - 6y = 16$.
16. Find the points of intersection of $x^2 + y^2 + 2x + 2y = 2$ and $3x^2 + 3y^2 + 5x + 5y = 10$.
17. A stone bridge arch is in the shape of half an ellipse and is 15.0 m wide and 5.00 m high. Find the height of the arch at a distance of 6.00 m from its center.
18. Write the equation of a hyperbola centered at the origin, where the conjugate axis = 12 and is vertical, and the distance between foci is 13.
19. A parabolic arch is 5.00 m high and 6.00 m wide at the base. Find the width of the arch at a height of 2.00 m above the base.
20. A stone thrown in the air follows a parabolic path, and reaches a maximum height of 56.0 ft in a horizontal distance of 48.0 ft. At what horizontal distances from the launch point will the height be 25.0 ft?

*Writing*

21. Suppose that the day before your visit to your former high school math class the teacher unexpectedly asks you to explain how that day's topic, the conic sections, got that name. Write a paragraph on what you will tell the class about how the circle, ellipse, parabola, and hyperbola (and the point and straight line too) can be formed by intersecting a plane and a cone. You may plan to illustrate your talk with a clay model or a piece of cardboard rolled into a cone. You will probably be asked what the conic sections are good for, so write out at least one use for each curve.

# 27

# DERIVATIVES OF ALGEBRAIC FUNCTIONS

## OBJECTIVES

**When you have completed this chapter, you should be able to:**

- Evaluate limits for algebraic, trigonometric, and exponential functions.
- Find the derivative of an algebraic expression using the delta method.
- Find the derivative of an algebraic expression using the rules for derivatives.
- Find the slope of the tangent line to a curve at a given point.
- Find the derivative of an implicit algebraic function.
- Find higher derivatives of an algebraic function.

This chapter begins our study of calculus. We start with the limit concept, which is then applied in finding the derivative of a function. We then develop rules with which we can find derivatives quickly and easily.

The derivative is extremely useful and will enable us to do calculations that we could not do before. But the value of the derivative may not be apparent while we are learning the mechanical details in this chapter, so do not get discouraged. The payoff will come later.

We'll do derivatives of algebraic functions only, in this chapter, and save the trigonometric, logarithmic, and exponential functions for Chapter 30.

## 27-1 LIMITS

### Limit Notation

Suppose that $x$ and $y$ are related by a function, such as $y = 3x$. Then if we are given a value of $x$, we can find a corresponding value for $y$. For example, when $x$ *is* 2, then $y$ *is* 6.

We now want to extend our mathematical language to be able to say what will happen to $y$ *not* when $x$ *is* a certain value, but when $x$ *approaches* a certain value. For example, when $x$ *approaches* 2, then $y$ *approaches* 6. The *notation* we use to say the same thing is

$$\lim_{x\to 2} y = \lim_{x\to 2}(3x) = 6$$

We read this as "the limit, as $x$ approaches 2, of $3x$, is 6."

In general, if the function $f(x)$ approaches some value $L$ as $x$ approaches $a$, we would indicate that with the notation

| | | |
|---|---|---|
| Limit Notation | $\lim_{x\to a} f(x) = L$ | **326** |

Well, why bother with new notation? Why not just say, in the preceding example, that $y$ *is* 6 when $x$ *is* 2? Why is it necessary to creep up on the answer like that?

It is true that limit notation offers no advantage in an example such as the last one. But we really need it when our function *cannot reach* the limit, or is *not even defined* at the limit.

**EXAMPLE 1:** The function $y = 1/x^2$ is graphed in Fig. 27-1. Even though $y$ never reaches 0, we can still write

$$\lim_{x\to\infty} \frac{1}{x^2} = 0$$

Further, even though our function is not even defined at $x = 0$, because it results in division by zero, we can still write

$$\lim_{x\to 0} \frac{1}{x^2} = \infty$$

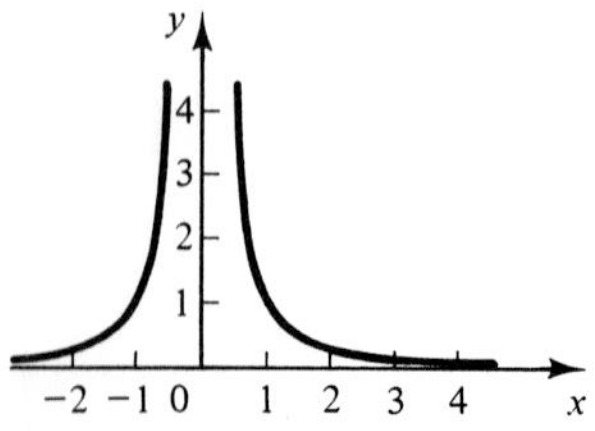


**FIGURE 27-1** **Graph of $y = 1/x^2$.**

**EXAMPLE 2:** In Sec. 25-2 we defined the tangent line at $P$ as the limiting position of the secant line $PQ$, as $Q$ approached $P$. If $\Delta x$ is the horizontal distance between $P$ and $Q$ (Fig. 27-2), we say that $Q$ approaches $P$ if $\Delta x$ approaches zero. We can thus say that the slope $m_t$ of the tangent line is the limit of the slope $m_s$ of the secant line as $\Delta x$ approaches zero. We can now restate this idea in compact form using our new limit notation, as

$$m_t = \lim_{\Delta x \to 0} m_s$$

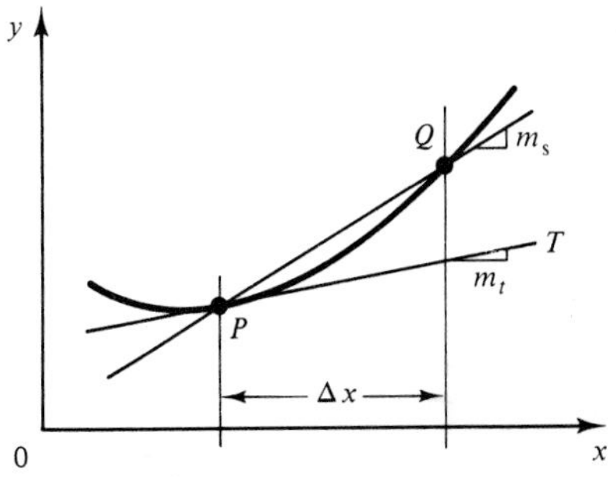


**FIGURE 27-2 Tangent to a curve as the limiting position of the secant.**

### Limits Involving Zero or Infinity

Limits involving zero or infinity can usually be evaluated using the following facts. If $C$ is a nonzero constant, then

| | |
|---|---|
| (1) $\lim_{x \to 0} Cx = 0$ | (4) $\lim_{x \to \infty} Cx = \infty$ |
| (2) $\lim_{x \to 0} \frac{x}{C} = 0$ | (5) $\lim_{x \to \infty} \frac{x}{C} = \infty$ |
| (3) $\lim_{x \to 0} \frac{C}{x} = \infty$ | (6) $\lim_{x \to \infty} \frac{C}{x} = 0$ |

**EXAMPLE 3:**

(a) $\lim_{x \to 0} 5x^3 = 0$ (b) $\lim_{x \to 0} \left(7 + \frac{x}{2}\right) = 7$

(c) $\lim_{x \to 0} \left(3 + \frac{25}{x^2}\right) = \infty$ (d) $\lim_{x \to \infty} (3x - 2) = \infty$

(e) $\lim_{x \to \infty} \left(8 + \frac{x}{4}\right) = \infty$ (f) $\lim_{x \to \infty} \left(\frac{5}{x} - 3\right) = -3$

| Common Error | Be sure to distinguish between a denominator that is zero and one that is **approaching** zero. In the first case we have division by zero, but in the second case we can get a useful answer. |
|---|---|

When we want the limit, as $x$ becomes infinite, of the quotient of two polynomials, such as

$$\lim_{x \to \infty} \frac{4x^3 - 3x^2 + 5}{3x - x^2 - 5x^3}$$

we see that both numerator and denominator become infinite. However, the limit of such an expression can be found if we *divide both numerator and denominator by the highest power of x occurring in either.*

**EXAMPLE 4:** Evaluate $\lim_{x \to \infty} \dfrac{4x^3 - 3x^2 + 5}{3x - x^2 - 5x^3}$.

**Solution:** Dividing both numerator and denominator by $x^3$ gives us

One of the theorems on limits (which we will not prove) is that the limit of the sum of several functions is equal to the sum of their individual limits.

$$\lim_{x \to \infty} \frac{4x^3 - 3x^2 + 5}{3x - x^2 - 5x^3} = \lim_{x \to \infty} \frac{4 - \dfrac{3}{x} + \dfrac{5}{x^3}}{\dfrac{3}{x^2} - \dfrac{1}{x} - 5}$$

$$= \frac{4 - 0 + 0}{0 - 0 - 5}$$

$$= -\frac{4}{5}$$

## Limits of the Form 0/0

This entire discussion of limits is to prepare us for the idea of the derivative. There we will have to find the limit of a fraction in which *both the numerator and the denominator approach zero.*

At first glance, when we see a numerator approaching zero, we expect the entire fraction to approach zero. But when we see a denominator approaching zero, we throw up our hands and cry "division by zero." What then do we make of a fraction in which *both* numerator and denominator approach zero?

First, keep in mind that the denominator is not *equal to* zero; it is only *approaching* zero. Second, even though a shrinking numerator would make a fraction approach zero, in this case the denominator is also shrinking. So, in fact, our fraction will not necessarily approach infinity, or zero, but will often approach some useful finite value.

A useful way to investigate limits is by calculator or computer. We simply substitute $x$ closer and closer to the given value, and see if the expression appears to approach a limit, as shown in the next example.

**EXAMPLE 5:** Evaluate $\lim_{x \to 0} \dfrac{\sqrt{9 + x} - 3}{3x}$.

**Solution:** We see that when $x$ *equals* zero, we get

$$\frac{\sqrt{9 + 0} - 3}{3(0)} = \frac{3 - 3}{0}$$

$$= \frac{0}{0}$$

a result that is indeterminate. But what happens when $x$ *approaches* zero? Let us use the calculator to substitute smaller and smaller values of $x$. Working to five decimal places, we get

This, of course, is not a *proof* that the limit found is the correct one, or even that a limit exists.

| $x$ | 10 | 1 | 0.1 | 0.01 | 0.001 | 0.0001 |
|---|---|---|---|---|---|---|
| | 0.04530 | 0.05409 | 0.05540 | 0.05554 | 0.05555 | 0.05556 |

So as $x$ approaches zero, the given expression appears to approach $0.0555\overline{5}$ as a limit.

**EXAMPLE 6:** Find the limit of $(\sin x)/x$ (where $x$ is in radians) as $x$ approaches zero.

**Solution:** This is another case where both numerator and denominator approach zero. We can find the limit of this expression by calculator, substituting smaller and smaller values of $x$ radians and calculating $(\sin x)/x$ (you can try this yourself), but we will use a different approach.

Let $x$ be a small angle, in radians, in a circle of radius 1 (Fig. 27-3), in which

$$\sin x = \frac{BC}{1} = BC, \qquad \cos x = \frac{OC}{1} = OC, \qquad \tan x = \frac{AD}{OA} = AD$$

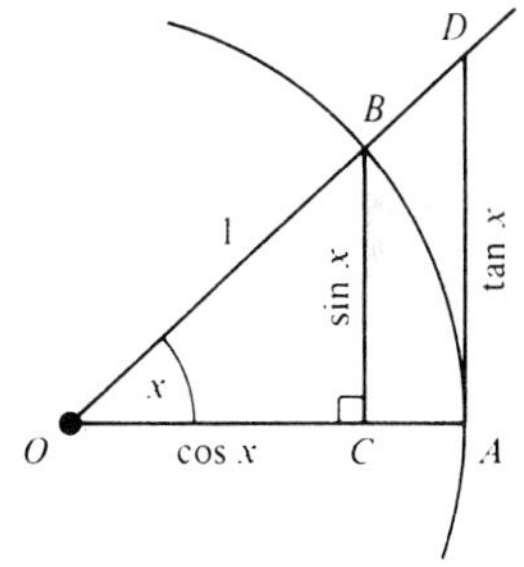


FIGURE 27-3

We see that the area of the sector $OAB$ is greater than the area of triangle $OBC$ but less than the area of triangle $OAD$. But by Eq. 137,

$$\text{area of triangle } OBC = \frac{1}{2} BC \cdot OC$$

$$= \frac{1}{2} \sin x \cos x$$

and

$$\text{area of triangle } OAD = \frac{1}{2} \cdot OA \cdot AD$$

$$= \frac{1}{2} \tan x$$

Also, by Eq. 116,

$$\text{area of sector } OAB = \frac{1}{2} r^2 x$$

$$= \frac{x}{2}$$

So

$$\frac{1}{2} \sin x \cos x < \frac{x}{2} < \frac{1}{2} \tan x$$

Dividing by $\frac{1}{2} \sin x$ gives us

$$\cos x < \frac{x}{\sin x} < \frac{1}{\cos x}$$

Now letting $x$ approach zero, both $\cos x$ and $1/\cos x$ approach 1. Therefore, $x/\sin x$, which is "sandwiched" between them, also approaches 1.

$$\lim_{x \to 0} \frac{x}{\sin x} = 1$$

Taking reciprocals yields

$$\lim_{x \to 0} \frac{\sin x}{x} = 1$$

**We will use this result later when we take the derivative of the sine function.**

Sometimes factoring the given expression will help us find a limit.

**EXAMPLE 7:** Evaluate $\lim_{x\to 3} \frac{x^2 - 9}{x - 3}$.

**Solution:** As $x$ approaches 3, both numerator and denominator approach zero, which tells us nothing about the limit of entire expression. However, if we factor the numerator, we get

$$\lim_{x\to 3} \frac{x^2 - 9}{x - 3} = \lim_{x\to 3} \frac{(x - 3)(x + 3)}{x - 3}$$

$$= \lim_{x\to 3} \left(\frac{x - 3}{x - 3}\right)(x + 3)$$

Now as $x$ approaches 3, both the numerator and the denominator of the fraction $(x - 3)/(x - 3)$ approach zero. But since the numerator and denominator are equal, the fraction will equal 1 for any nonzero value of $x$, no matter how small. So

$$\lim_{x\to 3} \frac{x^2 - 9}{x - 3} = \lim_{x\to 3}(x + 3) = 6$$

**EXAMPLE 8:**

$$\lim_{x\to 2} \frac{x^2 - 3x + 2}{x - 2} = \lim_{x\to 2} \frac{(x - 2)(x - 1)}{x - 2}$$

$$= \lim_{x\to 2} \left(\frac{x - 2}{x - 2}\right)(x - 1)$$

$$= \lim_{x\to 2}(x - 1)$$

$$= 1$$

If you can't find the limit of an expression in its given form, try to **manipulate** the expression into another form in which the limit can be found. This is often done simply by performing the indicated operations.

### Limit Can Depend on Direction of Approach

The limit of $y = 3x$, as $x$ approached 2, was 6. It did not matter whether $x$ approached 2 from below or above (from values less than 2 or from values greater than 2). But sometimes it does matter.

**EXAMPLE 9:** The function $y = 1/x$ is graphed in Fig. 27-4. When $x$ approaches zero from "above" (that is, from the *right* in Fig. 27-4), which we write $x \to 0^+$, we get

$$\lim_{x\to 0^+} \frac{1}{x} = +\infty$$

which means that $y$ grows without bound in the positive direction. But when $x$ approaches zero from "below" (from the *left* in Fig. 27-4), we write

$$\lim_{x\to 0^-} \frac{1}{x} = -\infty$$

which means that $y$ grows without bound in the negative direction.

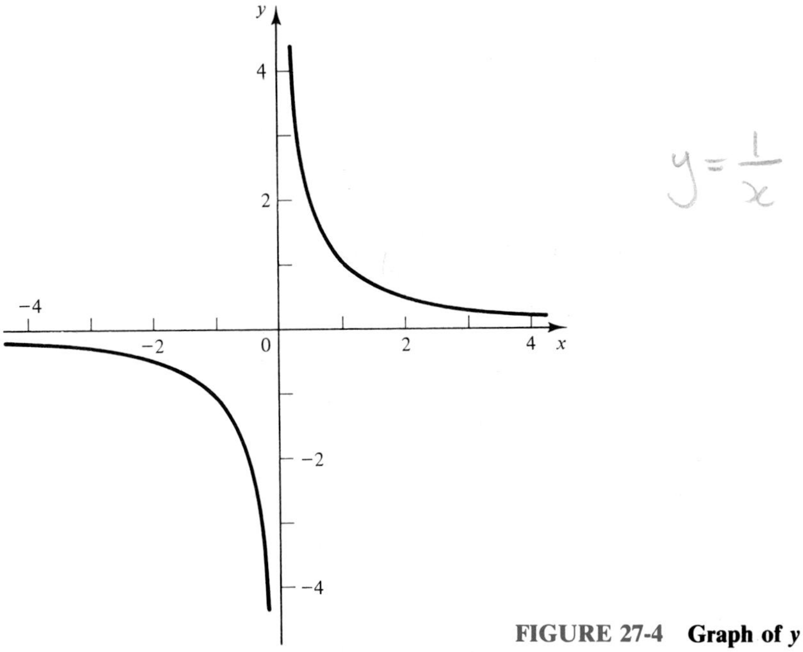


**FIGURE 27-4 Graph of $y = 1/x$.**

## When the Limit Is an Expression

Our main use for limits will be for finding derivatives in the following sections of this chapter. There we will evaluate limits in which both the numerator and denominator approach zero, and the resulting limit is an *expression* rather than a single number. A limit typical of the sort we will have to evaluate later is given in the following example.

**EXAMPLE 10:** Evaluate $\lim_{d\to 0} \dfrac{(x + d)^2 + 5(x + d) - x^2 - 5x}{d}$.

**Solution:** If we try to set $d$ equal to zero,

$$\frac{(x + 0)^2 + 5(x + 0) - x^2 - 5x}{0} = \frac{x^2 + 5x - x^2 - 5x}{0} = \frac{0}{0}$$

we get the indeterminate expression 0/0. Instead, let us remove parentheses in the original expression. We get

$$\lim_{d\to 0} \frac{(x + d)^2 + 5(x + d) - x^2 - 5x}{d}$$

$$= \lim_{d\to 0} \frac{x^2 + 2dx + d^2 + 5x + 5d - x^2 - 5x}{d}$$

$$= \lim_{d\to 0} \frac{2dx + d^2 + 5d}{d} = \lim_{d\to 0}(2x + d + 5)$$

after dividing through by $d$. Now letting $d$ approach zero, we obtain

$$\lim_{d\to 0} \frac{(x + d)^2 + 5(x + d) - x^2 - 5x}{d} = 2x + 5$$

## EXERCISE 1—LIMITS

Evaluate each limit.

1. $\lim_{x\to 2}(x^2 + 2x - 7)$
2. $\lim_{x\to -1}(x^3 - 3x^2 - 5x - 5)$
3. $\lim_{x\to 2}\dfrac{x^2 - x - 1}{x + 3}$
4. $\lim_{x\to 5}\dfrac{5 + 4x - x^2}{5 - x}$
5. $\lim_{x\to 5}\dfrac{x^2 - 25}{x - 5}$
6. $\lim_{x\to -7}\dfrac{49 - x^2}{x + 7}$
7. $\lim_{x\to 1}\dfrac{x^2 + 2x - 3}{x - 1}$
8. $\lim_{x\to 5}\dfrac{x^2 - 12x + 35}{5 - x}$

### Zero in the Limit

**You may want to evaluate some of these by calculator or computer.**

9. $\lim_{x\to 0}(4x^2 - 5x - 8)$
10. $\lim_{x\to 0}\dfrac{3 - 2x}{x + 4}$
11. $\lim_{x\to 0}\dfrac{\sqrt{x} - 4}{\sqrt[3]{x} + 5}$
12. $\lim_{x\to 0}\dfrac{3 + x - x^2}{(x + 3)(5 - x)}$
13. $\lim_{x\to 0}\left(\dfrac{1}{2 + x} - \dfrac{1}{2}\right)\cdot\dfrac{1}{x}$
14. $\lim_{x\to 0} x\cos x$
15. $\lim_{x\to 0}\dfrac{7}{x}$
16. $\lim_{x\to 0}\dfrac{e^x}{x}$
17. $\lim_{x\to 0}\dfrac{\cos x}{x}$
18. $\lim_{x\to 0}\dfrac{\sin x}{\tan x}$
19. $\lim_{x\to 0^+}\dfrac{x + 1}{x}$
20. $\lim_{x\to 0^-}\dfrac{x + 1}{x}$

### Infinity in the Limit

21. $\lim_{x\to\infty}\dfrac{2x + 5}{x - 4}$
22. $\lim_{x\to\infty}\dfrac{5x - x^2}{2x^2 - 3x}$
23. $\lim_{x\to\infty}\dfrac{x^2 + x - 3}{5x^2 + 10}$
24. $\lim_{x\to\infty}\dfrac{4 + 2^x}{3 + 2^x}$
25. $\lim_{x\to -\infty} 10^x$
26. $\lim_{x\to\infty} 10^x$

### When the Limit is an Expression

27. $\lim_{d\to 0} x^2 + 2d + d^2$
28. $\lim_{d\to 0} 2x + d$
29. $\lim_{d\to 0}\dfrac{(x + d)^2 - x^2}{x^2(x + d)}$
30. $\lim_{d\to 0}\dfrac{(x + d)^2 - x^2}{d}$
31. $\lim_{d\to 0}\dfrac{3(x + d) - 3x}{d}$
32. $\lim_{d\to 0}\dfrac{[2(x + d) + 5] - (2x + 5)}{d}$
33. $\lim_{d\to 0} 3x + d - \dfrac{1}{(x + d + 2)(x - 2)}$
34. $\lim_{d\to 0}\dfrac{[(x + d)^2 + 1] - (x^2 + 1)}{d}$
35. $\lim_{d\to 0}\dfrac{(x + d)^3 - x^3}{d}$
36. $\lim_{d\to 0}\dfrac{\sqrt{x + d} - \sqrt{x}}{d}$
37. $\lim_{d\to 0}\dfrac{(x + d)^2 - 2(x + d) - x^2 + 2x}{d}$
38. $\lim_{d\to 0}\dfrac{\dfrac{7}{x + d} - \dfrac{7}{d}}{d}$

### Computer

39. Write a program or use a spreadsheet to compute and print the value of $(\sin x)/x$, where $x$ is in radians. Start with $x = 1$ rad and decrease it each time by a factor of 10, (1, 0.1, 0.01, 0.001, and so forth). Is the value of $(\sin x)/x$ approaching a limit?

## 27-2 THE DERIVATIVE

We have already defined a tangent line $PT$ (Fig. 27-5) as the limiting position of the secant line $PQ$, as $Q$ approached $P$. But we have no way yet to calculate the *slope* of $PT$.

In Chapter 25 we found the slope of a tangent approximately by actually measuring rise and run on a graph of the tangent, and dividing rise by run to find the slope. We'll now derive an equation for finding the slope *exactly*.

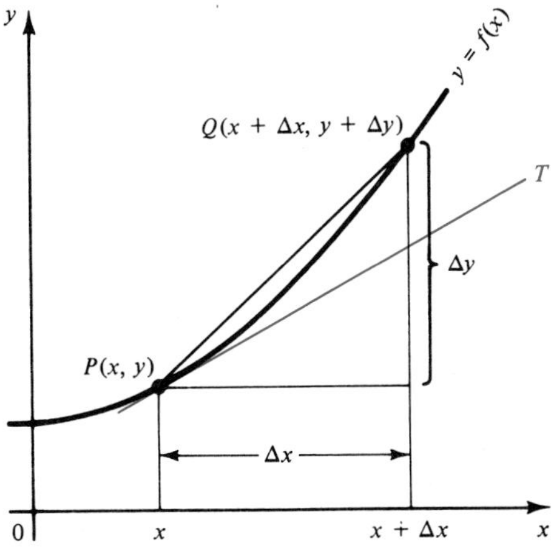


**FIGURE 27-5 Derivation of the derivative.**

Starting from $P(x, y)$ on the curve $y = f(x)$, Fig. 27-5, we locate a second point $Q$ spaced from $P$ by a horizontal increment $\Delta x$. In a run of $\Delta x$ the graph of $y = f(x)$ is seen to rise by an amount that we call $\Delta y$. $PQ$ then has a rise $\Delta y$ in a run of $\Delta x$, so

$$\text{slope of } PQ = \frac{\text{rise}}{\text{run}} = \frac{\Delta y}{\Delta x} = \frac{f(x + \Delta x) - f(x)}{\Delta x}$$

We now let $\Delta x$ approach zero, and point $Q$ will approach $P$ along the curve. Since the tangent $PT$ is the limiting position of the secant $PQ$, the slope of $PQ$ will approach the slope of $PT$.

$$\text{slope of } PT = \lim_{\Delta x \to 0} \frac{\Delta y}{\Delta x} = \lim_{\Delta x \to 0} \frac{f(x + \Delta x) - f(x)}{\Delta x}$$

This important quantity is called the *derivative* of $y$ with respect to $x$, and is given the symbol $\dfrac{dy}{dx}$.

**We will see later that *dy* and *dx* in the symbol *dy/dx* have separate meanings of their own. But for now we will treat *dy/dx* as a single symbol.**

| | | |
|---|---|---|
| The Derivative | $\dfrac{dy}{dx} = \lim_{\Delta x \to 0} \dfrac{\Delta y}{\Delta x}$ or $\dfrac{dy}{dx} = \lim_{\Delta x \to 0} \dfrac{f(x + \Delta x) - f(x)}{\Delta x}$ | **328** |

**EXAMPLE 11:** Find the derivative of $f(x) = x^2$.

**Solution:** By Eq. 328,

$$\frac{dy}{dx} = \lim_{\Delta x \to 0} \frac{f(x + \Delta x) - f(x)}{\Delta x}$$

$$= \lim_{\Delta x \to 0} \frac{(x + \Delta x)^2 - x^2}{\Delta x}$$

$$= \lim_{\Delta x \to 0} \frac{x^2 + 2x\,\Delta x + (\Delta x)^2 - x^2}{\Delta x}$$

$$= \lim_{\Delta x \to 0} \frac{2x\,\Delta x + (\Delta x)^2}{\Delta x}$$

$$= \lim_{\Delta x \to 0} (2x + \Delta x) = 2x$$

| Common Error | The symbol $\Delta x$ is a *single symbol*. It is not $\Delta$ *times* $x$. Thus $x \cdot \Delta x \neq \Delta x^2$ |
|---|---|

## Derivatives by the Delta Method

Since $y = f(x)$, we can also write Eq. 328 using $y$ instead of $f(x)$.

| The Derivative | $\frac{dy}{dx} = \lim_{\Delta x \to 0} \frac{(y + \Delta y) - y}{\Delta x}$ | **328** |
|---|---|---|

Our main use for the delta method will be to derive *rules* with which we can quickly find derivatives.

Instead of substituting directly into Eq. 328, some prefer to apply this equation in a series of steps. This is sometimes referred to as the *delta method*, *delta process*, or *four-step rule*.

**EXAMPLE 12:** (a) Find the derivative of the function $y = 3x^2$ by the delta method and (b) evaluate the derivative at $x = 1$.

**Solution:** (a) Solve using the following steps.

1. Starting from $P(x, y)$, Fig. 27-6a, we locate a second point $Q$, spaced from $P$ by a horizontal distance $\Delta x$ and by a vertical distance $\Delta y$. Since the coordinates of $Q(x + \Delta x, y + \Delta y)$ must satisfy the given function, we may substitute $x + \Delta x$ for $x$ and $y + \Delta y$ for $y$ in the original equation.

$$y + \Delta y = 3(x + \Delta x)^2$$
$$= 3[x^2 + 2x\,\Delta x + (\Delta x)^2]$$
$$= 3x^2 + 6x\,\Delta x + 3(\Delta x)^2$$

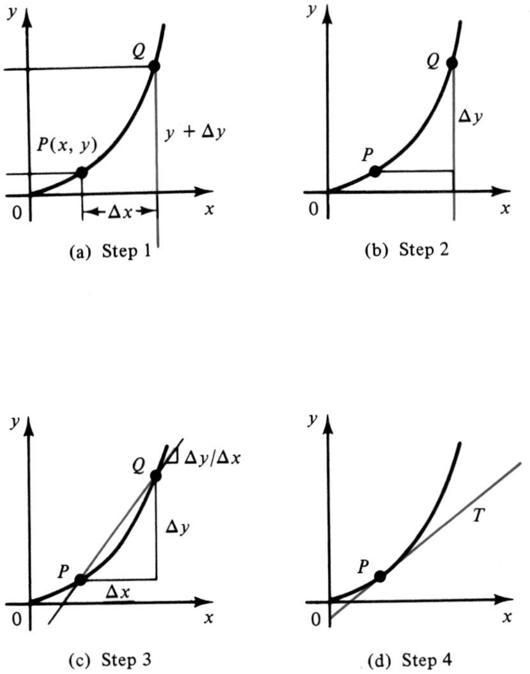


**FIGURE 27-6 Derivatives by the delta method.**

2. Find the rise $\Delta y$ from $P$ to $Q$ (Fig. 27-6b) by subtracting the original function.

$$(y + \Delta y) - y = 3x^2 + 6x\,\Delta x + 3(\Delta x)^2 - 3x^2$$

$$\Delta y = 6x\,\Delta x + 3(\Delta x)^2$$

**Of course, *rise* can be either positive or negative. Thus, a curve could in fact *fall* in a given run, but we still refer to it as the rise.**

3. Find the slope of the secant line $PQ$ (Fig. 27-6c) by dividing the rise $\Delta y$ by the run $\Delta x$.

$$\frac{\Delta y}{\Delta x} = \frac{6x\,\Delta x + 3(\Delta x)^2}{\Delta x}$$

4. Let $\Delta x$ approach zero. This causes $\Delta y$ also to approach zero, and appears to make $\Delta y/\Delta x$ equal to the indeterminate expression 0/0. But recall from our study of limits in the preceding section that a fraction can often have a limit *even when both numerator and denominator approach zero*. To find it, we divide through by $\Delta x$ and get $\Delta y/\Delta x = 6x + 3\,\Delta x$. Then the slope of the tangent $T$ (Fig. 27-6d) is

$$\frac{dy}{dx} = \lim_{\Delta x \to 0} \frac{\Delta y}{\Delta x} = 6x + 3(0) = 6x$$

(b) When $x = 1$,

$$\left.\frac{dy}{dx}\right|_{x=1} = 6(1) = 6$$

The curve $y = 3x^2$ is graphed in Fig. 27-7 with a tangent line of slope 6 drawn at the point (1, 3).

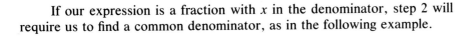

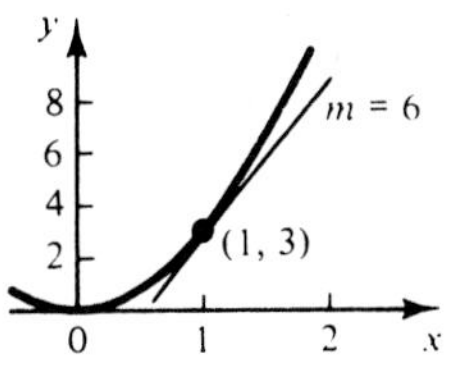


**FIGURE 27-7**

If our expression is a fraction with $x$ in the denominator, step 2 will require us to find a common denominator, as in the following example.

**EXAMPLE 13:** Find the derivative of $y = \dfrac{3}{x^2 + 1}$

**Solution:**

1. Substitute $x + \Delta x$ for $x$ and $y + \Delta y$ for $y$.

$$y + \Delta y = \frac{3}{(x + \Delta x)^2 + 1}$$

2. Subtracting yields

$$y + \Delta y - y = \frac{3}{(x + \Delta x)^2 + 1} - \frac{3}{x^2 + 1}$$

Combining the fractions over a common denominator gives

$$\Delta y = \frac{3(x^2 + 1) - 3[(x + \Delta x)^2 + 1]}{[(x + \Delta x)^2 + 1](x^2 + 1)}$$

which simplifies to

$$\Delta y = \frac{-6x\,\Delta x - 3(\Delta x)^2}{[(x + \Delta x)^2 + 1](x^2 + 1)}$$

3. Dividing by $\Delta x$, we obtain

$$\frac{\Delta y}{\Delta x} = \frac{-6x - 3\,\Delta x}{[(x + \Delta x)^2 + 1](x^2 + 1)}$$

4. Letting $\Delta x$ approach zero gives

$$\frac{dy}{dx} = -\frac{6x}{(x^2 + 1)^2}$$

The following example shows how to use the delta method to differentiate an expression containing a radical.

**EXAMPLE 14:** Find the slope of the tangent to the curve $y = \sqrt{x}$ at the point (4, 2).

**Solution:** We first find the derivative in four steps.

1. $y + \Delta y = \sqrt{x + \Delta x}$
2. $\Delta y = \sqrt{x + \Delta x} - \sqrt{x}$

**When simplifying radicals, we used to rationalize the denominator. Here we rationalize the numerator instead!**

The later steps will be easier if we now write this expression as a fraction with no radicals in the numerator. To do this, we multiply by the conjugate, and get

$$\Delta y = (\sqrt{x + \Delta x} - \sqrt{x}) \cdot \frac{\sqrt{x + \Delta x} + \sqrt{x}}{\sqrt{x + \Delta x} + \sqrt{x}}$$

$$= \frac{(x + \Delta x) - x}{\sqrt{x + \Delta x} + \sqrt{x}}$$

$$= \frac{\Delta x}{\sqrt{x + \Delta x} + \sqrt{x}}$$

**3.** Dividing by $\Delta x$ yields

$$\frac{\Delta y}{\Delta x} = \frac{1}{\sqrt{x + \Delta x} + \sqrt{x}}$$

**4.** Letting $\Delta x$ approach zero gives

$$\frac{dy}{dx} = \frac{1}{2\sqrt{x}}$$

When $x = 4$,

$$\left.\frac{dy}{dx}\right|_{x=4} = \frac{1}{4}$$

which is the slope of the tangent to $y = \sqrt{x}$ at the point (4, 2).

**If the derivation of the derivative was less than convincing to you, you can at least convince yourself that the end result is correct. Make a good graph of this curve, carefully draw the tangent, and measure its slope. It should agree with the result found with the delta method.**

## Other Symbols for the Derivative

There are other symbols which are used instead of $dy/dx$ to indicate a derivative.

A prime (′) symbol is often used. For example, if a certain function is equal to $y$, then the symbol $y'$ can represent the derivative of that function. Further, if $f(x)$ is a certain function, then $f'(x)$ is the derivative of that function. Thus

$$\frac{dy}{dx} = y' = f'(x)$$

The $y'$ or $f'$ notation is handy for specifying the derivative at a particular value of $x$.

**EXAMPLE 15:** To specify a derivative evaluated at $x = 2$, we can write $y'(2)$, or $f'(2)$, instead of the clumsier

$$\left.\frac{dy}{dx}\right|_{x=2}$$

## The Derivative as an Operator

We can think of the derivative as an *operator*: one that operates on a function to produce the derivative of that function. The symbol $d/dx$ or $D$ in front of an expression indicates that the expression is to be differentiated. For example, the symbols

$$\frac{d}{dx}(\ ) \qquad \text{or} \qquad D_x(\ ) \qquad \text{or} \qquad D(\ )$$

mean to find the derivative of the expression enclosed in parentheses. $D(\ )$ means differentiate the function with respect to its independent variable.

**EXAMPLE 16:** We saw in Example 12(a) that if $y = 3x^2$, then $dy/dx = 6x$. This same result can be written

$$\frac{d(3x^2)}{dx} = 6x$$

$$D_x(3x^2) = 6x$$

$$D(3x^2) = 6x$$

**Keep in mind that even though the notation is different, we find the derivative in *exactly the same way.***

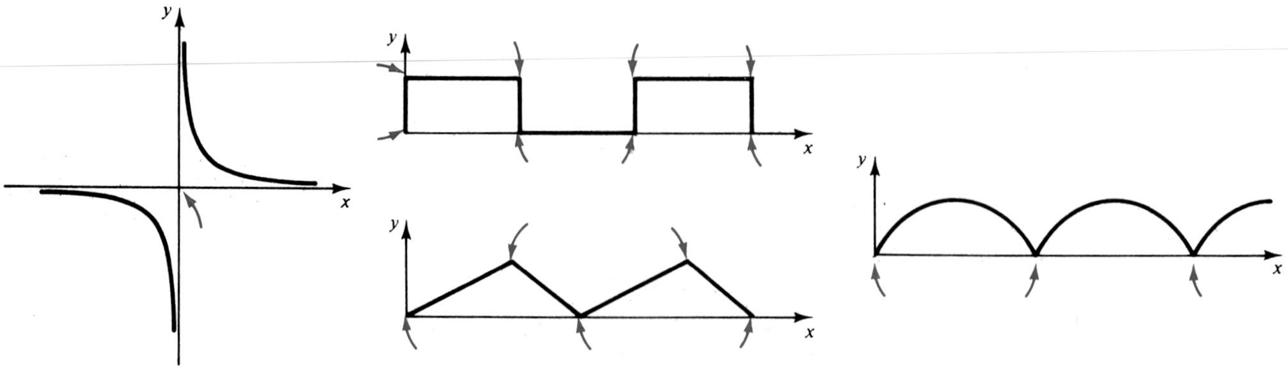

**FIGURE 27-8 The arrows show the points where the derivative does not exist.**

### Points Where the Derivative Does Not Exist

A curve is said to be *discontinuous* at a value of $x$ where there is a break or gap. The derivative does not exist at such points. It also does not exist where the curve has a jump, corner, cusp, or any other feature at which it is not possible to draw a tangent line, as the points shown in Fig. 27-8. At such points, we say that the function is *not differentiable*.

## EXERCISE 2—THE DERIVATIVE

### Delta Method

Find the derivative by the delta method.

1. $y = 2x + 5$
2. $y = 7 - 4x$
3. $y = x^2 + 1$
4. $y = x^2 - 3x + 5$
5. $y = x^3$
6. $y = x^3 - x^2$
7. $y = \dfrac{3}{x}$
8. $y = \dfrac{x}{(x-1)^2}$
9. $y = \sqrt{3 - x}$
10. $y = \dfrac{1}{\sqrt{x}}$

Find the slope of the tangent at the given value of $x$.

11. $y = \dfrac{1}{x^2}$ at $x = 1$
12. $y = \dfrac{1}{x+1}$ at $x = 2$
13. $y = x + \dfrac{1}{x}$ at $x = 2$
14. $y = \dfrac{1}{x}$ at $x = 3$
15. $y = 2x - 3$ at $x = 3$
16. $y = 16x^2$ at $x = 1$
17. $y = 2x^2 - 6$ at $x = 3$
18. $y = x^2 + 2x$ at $x = 1$

### Other Symbols for the Derivative

19. If $y = 2x^2 - 3$, find $y'$.
20. In Problem 19, find $y'(3)$.
21. In Problem 19, find $y'(-1)$.
22. If $f(x) = 7 - 4x^2$, find $f'(x)$.
23. In Problem 22, find $f'(2)$.
24. In Problem 22, find $f'(-3)$.

### Operator Notation

Find the derivative.

25. $\dfrac{d}{dx}(3x + 2)$
26. $\dfrac{d}{dx}(x^2 - 1)$
27. $D_x(7 - 5x)$
28. $D_x(x^2)$
29. $D(3x + 2)$
30. $D(x^2 - 1)$

## Computer

31. Write a program or use a spreadsheet to compute the slope of the secant line drawn to the curve $y = x^2 - 2x$ at the point (3, 3). Start with $\Delta x = 1$, and halve it for each subsequent calculation. Have the computer print the following table.

| DELTA X | X + DELTA X | Y + DELTA Y | DELTA Y | DELTA Y/DELTA X |
|---|---|---|---|---|
| 1 | 4 | 8 | 5 | 5 |
| 0.5 | 3.5 | 5.25 | 2.25 | 4.5 |
| 0.25 | 3.25 | 4.06 | 1.06 | 4.24 |
| 0.125 | 3.125 | 3.516 | 0.516 | 4.13 |
| . | . | . | . | . |
| . | . | . | . | . |
| . | . | . | . | . |

Let the program run until $\Delta y/\Delta x$ stops changing in the second decimal place.

## 27-3 RULES FOR DERIVATIVES

We now apply the delta method to various kinds of simple functions, and derive a rule for each with which we can then find the derivative.

### The Derivative of a Constant

To find the derivative of the function $y = c$, we substitute $c$ for $f(x)$ and $c$ for $f(x + \Delta x)$ in Eq. 328, and get

$$\frac{d(c)}{dx} = \lim_{\Delta x \to 0} \frac{c - c}{\Delta x} = 0$$

| Derivative of a Constant | $\frac{d(c)}{dx} = 0$ | 330 |
|---|---|---|

*The derivative of a constant is zero.*

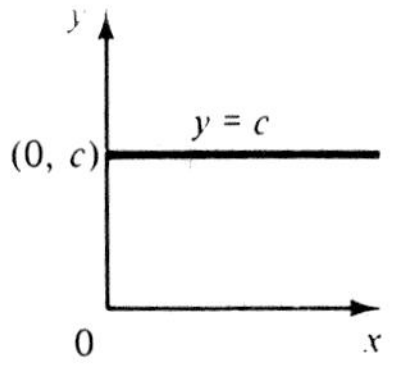

FIGURE 27-9

This is not surprising, because the graph of function $y = c$ is a straight line parallel to the $x$ axis (Fig. 27-9) whose slope is, of course, zero.

---

**EXAMPLE 17:** If $y = 2\pi^2$, then

$$\frac{dy}{dx} = 0$$

---

### Derivative of a Constant Times a Power of $x$

We let $y = cx^n$, where $n$ is a positive integer and $c$ is any constant. Using the delta method:

1. We substitute $x + \Delta x$ for $x$ and $y + \Delta y$ for $y$,

$$y + \Delta y = c(x + \Delta x)^n$$

By the binomial formula, Eq. 240,

**The binomial formula shows the expression we get when a binomial $(a + b)$ is raised to any positive integral power $n$.**

$$y + \Delta y = c\left[x^n + nx^{n-1}(\Delta x) + \frac{n(n-1)}{2}x^{n-2}(\Delta x)^2 + \cdots + (\Delta x)^n\right]$$

2. Subtracting $y = cx^n$, we get

$$\Delta y = c\left[nx^{n-1}(\Delta x) + \frac{n(n-1)}{2}x^{n-2}(\Delta x)^2 + \cdots + (\Delta x)^n\right]$$

3. Dividing by $\Delta x$ yields

$$\frac{\Delta y}{\Delta x} = c\left[nx^{n-1} + \frac{n(n-1)}{2}x^{n-2}\,\Delta x + \cdots + (\Delta x)^{n-1}\right]$$

4. Finally, letting $\Delta x$ approach zero, all terms but the first vanish. Thus

$$\frac{dy}{dx} = cnx^{n-1}$$

or

| | | |
|---|---|---|
| Derivative of a Constant Times a Power of $x$ | $\frac{d}{dx}cx^n = cnx^{n-1}$ | **333** |

*The derivative of a constant times a power of x is equal to the product of the constant, the exponent, and x raised to the exponent reduced by one.*

---

**EXAMPLE 18:**

(a) If $y = x^3$, then by Eq. 333,

$$\frac{dy}{dx} = 3x^{3-1} = 3x^2$$

(b) If $y = 5x^2$, then

$$\frac{dy}{dx} = 5(2)x^{2-1} = 10x$$

---

## Power Function with Negative Exponent

In our derivation of Eq. 333 we had required that the exponent $n$ be a positive integer. We'll now show that the rule is also valid when $n$ is a *negative* integer.

If $n$ is a negative integer, then $m = -n$ is a positive integer. So

$$y = cx^n = cx^{-m} = \frac{c}{x^m}$$

We again use the delta method.

1. We substitute $x + \Delta x$ for $x$ and $y + \Delta y$ for $y$.

$$y = \Delta y = \frac{c}{(x + \Delta x)^m}$$

2. Subtracting gives us

$$\Delta y = \frac{c}{(x + \Delta x)^m} - \frac{c}{x^m} = c\frac{x^m - (x + \Delta x)^m}{x^m(x + \Delta x)^m}$$

$$= c\frac{x^m - (x^m + mx^{m-1}\,\Delta x + kx^{m-2}\,\Delta x^2 + \cdots + \Delta x^m)}{x^m(x + \Delta x)^m}$$

3. Dividing by $\Delta x$ yields

$$\frac{\Delta y}{\Delta x} = -c\frac{mx^{m-1} + kx^{m-2}\,\Delta x + \cdots + \Delta x^{m-1}}{x^m(x + \Delta x)^m}$$

4. Letting $\Delta x$ go to zero, we get

$$\frac{dy}{dx} = -c\frac{mx^{m-1}}{x^{2m}} = -cmx^{m-1-2m} = -cmx^{-m-1} = cnx^{n-1}$$

This shows that Eq. 333 is valid when the exponent $n$ is negative, as well as positive.

---

**EXAMPLE 19:**

(a) If $y = x^{-4}$, then

$$\frac{dy}{dx} = -4x^{-5} = -\frac{4}{x^5}$$

(b) $\dfrac{d(3x^{-2})}{dx} = 3(-2)x^{-3} = -\dfrac{6}{x^3}$

(c) If $y = -3/x^2$, then $y = -3x^{-2}$ and

$$y' = -3(-2)x^{-3} = \frac{6}{x^3}$$

---

### Power Function with Fractional Exponent

We have shown that the exponent $n$ in Eq. 333 can be a positive or a negative integer. The rule is also valid when $n$ is a positive or a negative rational number. We'll prove it in Sec. 27-4.

---

**EXAMPLE 20:** If $y = x^{-5/3}$, then

$$\frac{dy}{dx} = -\frac{5}{3}x^{-8/3}$$

---

To find the derivative of a radical, write it in exponential form, and use Eq. 333.

---

**EXAMPLE 21:** If $y = \sqrt[3]{x^2}$, then

$$\frac{dy}{dx} = \frac{d}{dx}x^{2/3} = \frac{2}{3}x^{-1/3} = \frac{2}{3\sqrt[3]{x}}$$

---

## Derivative of a Sum

We want the derivative of the function

$$y = u + v + w$$

where $u$, $v$, and $w$ are all functions of $x$. These may be visualized as in Fig. 27-10b, c, and d. We use the delta method as before. Starting from $P(x, y)$ on the curve $y = f(x)$, we locate a second point $Q$ spaced from $P$ by a horizontal increment $\Delta x$. In a run of $\Delta x$ the graph of $y = f(x)$ is seen to rise by an amount that we call $\Delta y$ (Fig. 27-10a).

But in a run of $\Delta x$, the graph of $u$ also has a rise, and we call this rise $\Delta u$. Similarly, the graphs of $v$ and $w$ will rise by amounts that we call $\Delta v$ and $\Delta w$.

Thus, at $(x + \Delta x)$, the values of $u$, $v$, $w$, and $y$ are $(u + \Delta u)$, $(v + \Delta v)$, $(w + \Delta w)$, and $(y + \Delta y)$. Substituting these values into the original function $y = u + v + w$ gives

$$y + \Delta y = (u + \Delta u) + (v + \Delta v) + (w + \Delta w)$$

Subtracting the original function yields

$$y + \Delta y - y = u + \Delta u + v + \Delta v + w + \Delta w - u - v - w$$
$$\Delta y = \Delta u + \Delta v + \Delta w$$

Dividing by $\Delta x$ gives us

$$\frac{\Delta y}{\Delta x} = \frac{\Delta u + \Delta v + \Delta w}{\Delta x}$$

So

$$\frac{dy}{dx} = \lim_{\Delta x \to 0} \left( \frac{\Delta u}{\Delta x} + \frac{\Delta v}{\Delta x} + \frac{\Delta w}{\Delta x} \right)$$
$$= \lim_{\Delta x \to 0} \frac{\Delta u}{\Delta x} + \lim_{\Delta x \to 0} \frac{\Delta v}{\Delta x} + \lim_{\Delta x \to 0} \frac{\Delta w}{\Delta x}$$

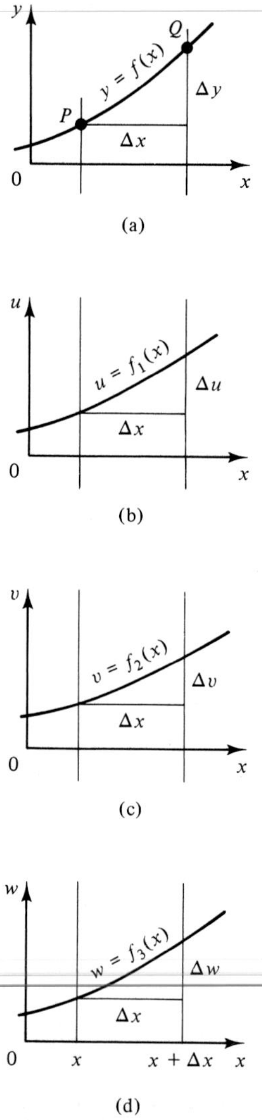


**FIGURE 27-10**

**Here we use the idea that the limit of a sum of several functions is equal to the sum of the individual limits.**

This leads us to the following sum rule.

| | | |
|---|---|---|
| Derivative of a Sum | $\frac{d}{dx}(u + v + w) = \frac{du}{dx} + \frac{dv}{dx} + \frac{dw}{dx}$ | **334** |

*The derivative of the sum of several functions is equal to the sum of the derivatives of those functions.*

**In addition to using the rule for sums here, we also need the rules for a power function and for a constant. We usually have to apply several rules in one problem.**

**EXAMPLE 22:**

$$\frac{d}{dx}(2x^3 - 3x^2 + 5x + 4) = 6x^2 - 6x + 5$$

**EXAMPLE 23:** Find the derivative of $y = \dfrac{x^2 + 3}{x}$.

**Solution:** At first glance it looks as if none of the rules learned so far apply here. But if we rewrite the function as

$$y = \frac{x^2 + 3}{x} = x + \frac{3}{x} = x + 3x^{-1}$$

we may write

$$\frac{dy}{dx} = 1 + 3(-1)x^{-2}$$

$$= 1 - \frac{3}{x^2}$$

| Common Error | Students often forget to simplify an expression as much as possible **before** taking the derivative. |
|---|---|

## Other Variables

So far we have been using $x$ for the independent variable and $y$ for the dependent variable, but now let us get some practice using other variables.

**EXAMPLE 24:**

(a) If $s = 3t^2 - 4t + 5$, then

$$\frac{ds}{dt} = 6t - 4$$

(b) If $y = 7u - 5u^3$, then

$$\frac{dy}{du} = 7 - 15u^2$$

(c) If $u = 9 + x^2 - 2x^4$, then

$$\frac{du}{dx} = 2x - 8x^3$$

**EXAMPLE 25:** Find the derivative of $T = 4z^2 - 2z + 5$.

**Solution:** What are we to find here, $dT/dz$, or $dz/dT$, or $dT/dx$, or something else?

> When we say "derivative," we mean the derivative of the given function **with respect to the independent variable,** unless otherwise noted.

Here we find the derivative of $T$ with respect to the independent variable $z$.

$$\frac{dT}{dz} = 8z - 2$$

## EXERCISE 3—RULES FOR DERIVATIVES

**Here, as usual, the letters *a, b, c, . . . ,* represent constants.**

Find the derivative of each function.

1. $y = a^2$
2. $y = 3b + 7c$
3. $y = x^7$
4. $y = x^4$
5. $y = 3x^2$
6. $y = 5.4x^3$
7. $y = x^{-5}$
8. $y = 2x^{-3}$
9. $y = \dfrac{1}{x}$
10. $y = \dfrac{1}{x^2}$
11. $y = \dfrac{3}{x^3}$
12. $y = \dfrac{3}{2x^2}$
13. $y = 7.5x^{1/3}$
14. $y = 4x^{5/3}$
15. $y = 4\sqrt{x}$
16. $y = 3\sqrt[3]{x}$
17. $y = -17\sqrt{x^3}$
18. $y = -2\sqrt[5]{x^4}$
19. $y = 3 - 2x$
20. $y = 4x^2 + 2x^3$
21. $y = 3x - x^3$
22. $y = x^4 + 3x^2 + 2$
23. $y = 3x^3 + 7x^2 - 2x + 5$
24. $y = x^3 - x^{3/2} + 3x$
25. $y = ax + b$
26. $y = ax^5 - 5bx^3$
27. $y = \dfrac{x^2}{2} - \dfrac{x^7}{7}$
28. $y = \dfrac{x^3}{1.75} + \dfrac{x^2}{2.84}$
29. $y = 2x^{3/4} + 4x^{-1/4}$
30. $y = \dfrac{2}{x} - \dfrac{3}{x^2}$
31. $y = 2x^{4/3} - 3x^{2/3}$
32. $y = x^{2/3} - a^{2/3}$
33. $y = \dfrac{x + 4}{x}$
34. $y = \dfrac{x^3 + 1}{x}$
35. If $y = 2x^3 - 3$, find $y'$.
36. In Problem 35, find $y'(2)$.
37. In Problem 35, find $y'(-3)$.
38. If $f(x) = 7 - 4x^2$, find $f'(x)$.
39. In Problem 38, find $f'(1)$.
40. In Problem 38, find $f'(-3)$.

Find the derivative.

41. $\dfrac{d}{dx}(3x^5 + 2x)$
42. $\dfrac{d}{dx}(2.5x^2 - 1)$
43. $D_x(7.8 - 5.2x^{-2})$
44. $D_x(4x^2 - 1)$
45. $D(3x^2 + 2x)$
46. $D(1.75x^{-2} - 1)$
47. Find the slope of the tangent to the curve $y = x^2 - 2$, where $x$ equals 2.
48. Find the slope of the tangent to the function $y = x - x^2$ at $x = 2$.
49. If $y = x^3 - 5$, find $y'(1)$.
50. If $f(x) = 1/x^2$, find $f'(2)$.
51. If $f(x) = 2.75x^2 - 5.02x$, find $f'(3.36)$.
52. If $y = \sqrt{83.2x^3}$, find $y'(1.74)$.

### Other Variables

Find the derivative with respect to the independent variable.

53. $v = 5t^2 - 3t + 4$
54. $z = 9 - 8w + w^2$
55. $s = 58.3t^3 - 63.8t$
56. $x = 3.82y + 6.25y^4$
57. $y = \sqrt{5w^3}$
58. $w = \dfrac{5}{x} - \dfrac{3}{x^2}$
59. $v = \dfrac{85.3}{t^4}$
60. $T = 3.55\sqrt{1.06w^5}$

## 27-4 DERIVATIVE OF A FUNCTION RAISED TO A POWER

### Composite Functions

You may want to review composite functions in Chapter 4 before going further.

In Sec. 27-3 we derived rule 333 for finding the derivative of $x$ raised to a power. But rule 333 does not apply when we have an *expression* raised to a power, such as

$$y = (2x + 7)^5 \qquad (1)$$

We may consider this function as being made up of two parts. One part is the function, which we shall call $u$, that is being raised to the power.

$$u = 2x + 7 \qquad (2)$$

Then $y$, which is a function of $u$, is seen to be

$$y = u^5 \qquad (3)$$

Our original function (1), which can be obtained by combining (2) and (3), is called a *composite function*.

### The Chain Rule

The chain rule will enable us to take the derivative of a composite function, such as (1). Consider the situation where $y$ is a function of $u$,

$$y = g(u)$$

and $u$, in turn, is a function of $x$,

$$u = h(x)$$

so that $y$ is therefore a function of $x$.

$$y = f(x)$$

The graphs of $u = h(x)$ and $y = f(x)$ are shown in Fig. 27-11. We now start our derivation of the chain rule by recalling our definition of the derivative as the limit of the quotient $\Delta y/\Delta x$ as $\Delta x$ approaches zero.

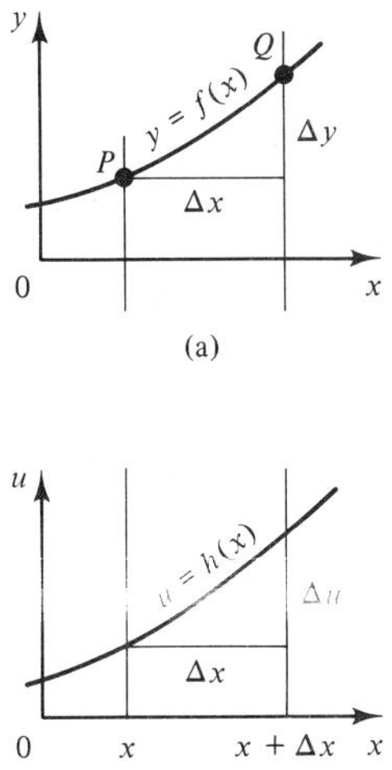

(a) (b)

**FIGURE 27-11**

| | | |
|---|---|---|
| The Derivative | $\dfrac{dy}{dx} = \lim\limits_{\Delta x \to 0} \dfrac{\Delta y}{\Delta x}$ | 328 |

where $\Delta y$ is the rise of the graph of $y = f(x)$ in a run of $\Delta x$ (Fig. 27-11a). But in a run of $\Delta x$, the graph of $u$ also has a rise, and we call this rise $\Delta u$ (Fig. 27-11b). Before taking the limit in Eq. 328, let us multiply the quotient $\Delta y/\Delta x$ by $\Delta u/\Delta u$, assuming here that $\Delta u$ is not zero.

$$\frac{\Delta y}{\Delta x} = \frac{\Delta y}{\Delta x} \cdot \frac{\Delta u}{\Delta u}$$

The chain rule is true even for those rare cases where $\Delta u$ may be zero, but it takes a more complicated proof to show this.

Rearranging gives us

$$\frac{\Delta y}{\Delta x} = \frac{\Delta y}{\Delta u} \cdot \frac{\Delta u}{\Delta x}$$

We now let $\Delta x$ approach zero, and apply a theorem (which we'll use without proof) that the limit of a product of two functions is equal to the product of the limits of the two functions.

$$\lim_{\Delta x \to 0} \frac{\Delta y}{\Delta x} = \lim_{\Delta x \to 0} \frac{\Delta y}{\Delta u} \cdot \frac{\Delta u}{\Delta x} = \left(\lim_{\Delta x \to 0} \frac{\Delta y}{\Delta u}\right)\left(\lim_{\Delta x \to 0} \frac{\Delta u}{\Delta x}\right)$$

But $\Delta u$ also approaches zero as $\Delta x$ approaches zero, so we may write

$$\lim_{\Delta x \to 0} \frac{\Delta y}{\Delta x} = \lim_{\Delta u \to 0} \frac{\Delta y}{\Delta u} \cdot \lim_{\Delta x \to 0} \frac{\Delta u}{\Delta x}$$

Then by our definition of the derivative, Eq. 328,

$$\lim_{\Delta x \to 0} \frac{\Delta y}{\Delta x} = \frac{dy}{dx}$$

Similarly,

$$\lim_{\Delta u \to 0} \frac{\Delta y}{\Delta u} = \frac{dy}{du} \quad \text{and} \quad \lim_{\Delta x \to 0} \frac{\Delta u}{\Delta x} = \frac{du}{dx}$$

So we get

| | | |
|---|---|---|
| Chain Rule | $\dfrac{dy}{dx} = \dfrac{dy}{du} \cdot \dfrac{du}{dx}$ | **329** |

*If y is a function of u, and u is a function of x, then the derivative of y with respect to x is the product of the derivative of y with respect to u and the derivative of u with respect to x.*

### The Power Rule

We now use the chain rule to find the derivative of a function raised to a power, $y = cu^n$. If, in Eq. 333 we replace $x$ by $u$, we get

$$\frac{dy}{du} = \frac{d}{du}(cu^n) = cnu^{n-1}$$

Then by the chain rule, we get $dy/dx$ by multiplying by $du/dx$,

$$\frac{dy}{dx} = \frac{dy}{du} \cdot \frac{du}{dx}$$

**We'll refer to Eq. 335 as the "power rule."**

| | | |
|---|---|---|
| Derivative of a Function Raised to a Power | $\dfrac{d(cu^n)}{dx} = cnu^{n-1}\dfrac{du}{dx}$ | **335** |

*The derivative (with respect to x) of a constant times a function raised to a power is equal to the product of the constant, the power, the function raised to the power less 1, and the derivative of the function (with respect to x).*

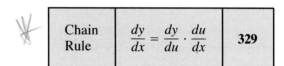

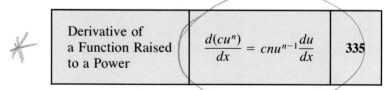

**EXAMPLE 26:** Take the derivative of $y = (x^3 + 1)^5$.

**Solution:** We use the rule for a function raised to a power, with

$$n = 5 \qquad \text{and} \qquad u = x^3 + 1$$

Then

$$\frac{du}{dx} = 3x^2$$

so

$$\frac{dy}{dx} = nu^{n-1}\frac{du}{dx}$$

$$= 5(x^3 + 1)^4\underline{(3x^2)}$$

| Common Error | Don't forget $\frac{du}{dx}$! |
|---|---|

Now simplifying our answer, we get

$$\frac{dy}{dx} = 15x^2(x^3 + 1)^4$$

**EXAMPLE 27:** Take the derivative of $y = \dfrac{3}{x^2 + 2}$.

**Solution:** Rewriting our function as $y = 3(x^2 + 2)^{-1}$ and applying rule 335 with $u = x^2 + 2$, we get

$$\frac{dy}{dx} = 3(-1)(x^2 + 2)^{-2}(2x)$$

$$= -\frac{6x}{(x^2 + 2)^2}$$

We'll now use the power rule for *fractional exponents,* even though we will not prove that the rule is valid for them until later in this chapter.

**EXAMPLE 28:** Differentiate $y = \sqrt[3]{1 + x^2}$.

**Solution:** We rewrite the radical in exponential form,

$$y = (1 + x^2)^{1/3}$$

Then by rule 335, with $u = 1 + x^2$, and $du/dx = 2x$,

$$\frac{dy}{dx} = \frac{1}{3}(1 + x^2)^{-2/3}(2x)$$

Or, returning to radical form,

$$\frac{dy}{dx} = \frac{2x}{3\sqrt[3]{(1 + x^2)^2}}$$

**EXAMPLE 29:** If $f(x) = \dfrac{5}{\sqrt{x^2 + 3}}$, find $f'(1)$.

**Solution:** Rewriting our function without radicals gives us

$$f(x) = 5(x^2 + 3)^{-1/2}$$

Taking the derivative yields

$$f'(x) = 5\left(-\frac{1}{2}\right)(x^2 + 3)^{-3/2}(2x)$$

$$= -\frac{5x}{(x^2 + 3)^{3/2}}$$

Substituting $x = 1$, we obtain

$$f'(1) = -\frac{5}{(1 + 3)^{3/2}} = -\frac{5}{8}$$

## EXERCISE 4—FUNCTION RAISED TO A POWER

**As before, the letters *a, b, c, . . .* , represent constants.**

Find the derivative of each function.

**1.** $y = (2x + 1)^5$

**2.** $y = (2 - 3x^2)^3$

**3.** $y = (3x^2 + 2)^4 - 2x$

**4.** $y = (x^3 + 5x^2 + 7)^2$

**5.** $y = (2 - 5x)^{3/5}$

**6.** $y = -\dfrac{2}{x + 1}$

**7.** $y = \dfrac{2.15}{x^2 + a^2}$

**8.** $y = \dfrac{31.6}{1 - 2x}$

**9.** $y = \dfrac{3}{x^2 + 2}$

**10.** $y = \dfrac{5}{(x^2 - 1)^2}$

**11.** $y = \left(a - \dfrac{b}{x}\right)^2$

**12.** $y = \left(a + \dfrac{b}{x}\right)^3$

**13.** $y = \sqrt{1 - 3x^2}$

**14.** $y = \sqrt{2x^2 - 7x}$

**15.** $y = \sqrt{1 - 2x}$

**16.** $y = \dfrac{b}{a}\sqrt{a^2 - x^2}$

**17.** $y = \sqrt[3]{4 - 9x}$

**18.** $y = \sqrt[3]{a^3 - x^3}$

**19.** $y = \dfrac{1}{\sqrt{x + 1}}$

**20.** $y = \dfrac{1}{\sqrt{x^2 - x}}$

Find the derivative.

**21.** $\dfrac{d}{dx}(3x^5 + 2x)^2$

**22.** $\dfrac{d}{dx}(1.5x^2 - 3)^3$

**23.** $D_x(4.8 - 7.2x^{-2})^2$

**24.** $D(3x^3 - 5)^3$

Find the derivative with respect to the independent variable.

**25.** $v = (5t^2 - 3t + 4)^2$

**26.** $z = (9 - 8w)^3$

**27.** $s = (8.3t^3 - 3.8t)^{-2}$

**28.** $x = \sqrt{3.2y + 6.2y^4}$

29. Find the derivative of the function $y = (4.82x^2 - 8.25x)^3$ when $x = 3.77$.
30. Find the slope of the tangent to the curve $y = 1/(x + 1)$ at $x = 2$.
31. If $y = (x^2 - x)^3$, find $y'(3)$.
32. If $f(x) = \sqrt[3]{2x} + (2x)^{2/3}$, find $f'(4)$.

## 27-5 DERIVATIVES OF PRODUCTS AND QUOTIENTS

### Derivative of a Product

We often need the derivative of the product of two expressions, such as $y = (x^2 + 2)\sqrt{x - 5}$ where each of the expressions is itself a function of $x$. Let us label these expressions $u$ and $v$. So our function is

$$y = uv$$

where $u$ and $v$ are functions of $x$. These may be visualized as in Fig. 27-10a, b, and c. We use the delta method as before. Starting from $P(x, y)$ on the curve $y = f(x)$, we locate a second point $Q$ spaced from $P$ by a horizontal increment $\Delta x$. In a run of $\Delta x$ the graph of $y = f(x)$ is seen to rise by an amount that we call $\Delta y$ (Fig. 27-10a).

But in a run of $\Delta x$, the graph of $u$ also has a rise, and we call this rise $\Delta u$. Similarly, the graph of $v$ will rise by an amount that we call $\Delta v$.

Thus, at $(x + \Delta x)$, the values of $u$ and $v$ and $y$ are $(u + \Delta u)$, $(v + \Delta v)$, and $(y + \Delta y)$. Substituting these values into the original function $y = uv$ gives

$$\begin{aligned} y + \Delta y &= (u + \Delta u)(v + \Delta v) \\ &= uv + u\,\Delta v + v\,\Delta u + \Delta u\,\Delta v \end{aligned}$$

Subtracting $y = uv$ gives us

$$\Delta y = u\,\Delta v + v\,\Delta u + \Delta u\,\Delta v$$

Dividing by $\Delta x$ yields

$$\frac{\Delta y}{\Delta x} = u\frac{\Delta v}{\Delta x} + v\frac{\Delta u}{\Delta x} + \Delta u\frac{\Delta v}{\Delta x}$$

As $\Delta x$ now approaches, 0, $\Delta y$, $\Delta u$, and $\Delta v$ also approach 0, and

$$\lim_{\Delta x \to 0} \frac{\Delta y}{\Delta x} = \frac{dy}{dx}, \qquad \lim_{\Delta x \to 0} \frac{\Delta u}{\Delta x} = \frac{du}{dx}, \qquad \lim_{\Delta x \to 0} \frac{\Delta v}{\Delta x} = \frac{dv}{dx}$$

so

$$\frac{dy}{dx} = u\frac{dv}{dx} + v\frac{du}{dx} + 0\frac{dv}{dx}$$

or

| | | |
|---|---|---|
| Derivative of a Product of Two Factors | $\frac{d(uv)}{dx} = u\frac{dv}{dx} + v\frac{du}{dx}$ | **336** |

*The derivative of a product of two factors is equal to the first factor times the derivative of the second factor, plus the second factor times the derivative of the first.*

**EXAMPLE 30:** Find the derivative of $y = (x^2 + 2)(x - 5)$.

**Solution:** We let the first factor be $u$, and the second be $v$.

$$y = \underbrace{(x^2 + 2)}_{u}\underbrace{(x - 5)}_{v}$$

So

$$\frac{du}{dx} = 2x \quad \text{and} \quad \frac{dv}{dx} = 1$$

Using the product rule,

We could have multiplied the two factors together and taken the derivative term by term. Try it and see if you get the same result. But sometimes this is not possible, as in Example 31.

$$\frac{dy}{dx} = (x^2 + 2)\frac{d}{dx}(x - 5) + (x - 5)\frac{d}{dx}(x^2 + 2)$$

$$= (x^2 + 2)(1) + (x - 5)(2x)$$

$$= x^2 + 2 + 2x^2 - 10x$$

$$= 3x^2 - 10x + 2$$

**EXAMPLE 31:** Differentiate $y = (x + 5)\sqrt{x - 3}$.

**Solution:** By the product rule,

$$\frac{dy}{dx} = (x + 5) \cdot \frac{1}{2}(x - 3)^{-1/2}(1) + \sqrt{x - 3}(1)$$

$$= \frac{x + 5}{2\sqrt{x - 3}} + \sqrt{x - 3}$$

$$\frac{d(uv)}{dx} = u\frac{dv}{dx} + v\frac{dy}{dx}$$

## Products with More Than Two Factors

Our rule for the derivative of a product having two factors can be easily extended. Take an expression with three factors, for example.

$$y = uvw = (uv)w$$

By the product rule,

$$\frac{dy}{dx} = uv\frac{dw}{dx} + w\frac{d(uv)}{dx}$$

$$= uv\frac{dw}{dx} + w\left(u\frac{dv}{dx} + v\frac{du}{dx}\right)$$

so

| Derivative of a Product of Three Factors | $\dfrac{d(uvw)}{dx} = uv\dfrac{dw}{dx} + uw\dfrac{dv}{dx} + vw\dfrac{du}{dx}$ | 337 |
|---|---|---|

*The derivative of the product of three factors is an expression of three terms, each term being the product of two of the factors and the derivative of the third factor.*

**758**

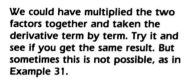

**EXAMPLE 32:** Differentiate $y = x^2(x-2)^5\sqrt{x+3}$.

**Solution:** By Eq. 337,

$$\frac{dy}{dx} = x^2(x-2)^5\left[\frac{1}{2}(x+3)^{-1/2}\right] + x^2(x+3)^{1/2}[5(x-2)^4]$$
$$+ (x-2)^5(x+3)^{1/2}(2x)$$
$$= \frac{x^2(x-2)^5}{2\sqrt{x+3}} + 5x^2(x-2)^4\sqrt{x+3} + 2x(x-2)^5\sqrt{x+3}$$

We now generalize this result (without proof) to cover any number of factors:

| | | |
|---|---|---|
| Derivative of a Product of $n$ Factors | The derivative of the product of $n$ factors is an expression of $n$ terms, each term being the product of $n-1$ of the factors and the derivative of the other factor. | **338** |

### Derivative of a Constant Times a Function

Let us use the product rule for the product $cu$, where $c$ is a constant and $u$ is a function of $x$.

$$\frac{d}{dx}cu = c\frac{du}{dx} + u\frac{dc}{dx} = c\frac{du}{dx}$$

since the derivative $dc/dx$ of a constant is zero. Thus

| | | |
|---|---|---|
| Derivative of a Constant Times a Function | $\frac{d(cu)}{dx} = c\frac{du}{dx}$ | **332** |

*The derivative of the product of a constant and a function is equal to the constant times the derivative of the function.*

**EXAMPLE 33:** If $y = 3(x^2 - 3x)^5$, then

$$\frac{dy}{dx} = 3\frac{d}{dx}(x^2-3x)^5$$
$$= 3(5)(x^2-3x)^4(2x-3)$$
$$= 15(x^2-3x)^4(2x-3)$$

| | |
|---|---|
| Common Error | If one of the factors is a constant, it is much easier to use rule 332 for a constant times a function, rather than the product rule. |

## Derivative of a Quotient

To find the derivative of the function

$$y = \frac{u}{v}$$

where $u$ and $v$ are functions of $x$, we first rewrite it as a product,

$$y = uv^{-1}$$

Now using the rule for products and the rule for a power function,

$$\frac{dy}{dx} = u(-1)v^{-2}\frac{dv}{dx} + v^{-1}\frac{du}{dx}$$

$$= -\frac{u}{v^2}\frac{dv}{dx} + \frac{1}{v}\frac{du}{dx}$$

We combine the two fractions over the LCD, $v^2$, and rearrange.

$$\frac{dy}{dx} = \frac{v}{v^2}\frac{du}{dx} - \frac{u}{v^2}\frac{dv}{dx}$$

or

| | | |
|---|---|---|
| Derivative of a Quotient | $\frac{d}{dx}\left(\frac{u}{v}\right) = \frac{v\frac{du}{dx} - u\frac{dv}{dx}}{v^2}$ | **339** |

*The derivative of a quotient equals the denominator times the derivative of the numerator minus the numerator times the derivative of the denominator, all divided by the square of the denominator.*

**Some prefer to use the product rule to do quotients, treating the quotient $u/v$ as the product $uv^{-1}$.**

**EXAMPLE 34:** Take the derivative of $y = 2x^3/(4x + 1)$.

**Solution:** The numerator is $u = 2x^3$, so

$$\frac{du}{dx} = 6x^2$$

and the denominator is $v = 4x + 1$, so

$$\frac{dv}{dx} = 4$$

| | |
|---|---|
| Common Error | It is very easy, in Eq. 339, to interchange $u$ and $v$, by mistake. |

Applying the quotient rule yields

$$\frac{dy}{dx} = \frac{(4x + 1)(6x^2) - (2x^3)(4)}{(4x + 1)^2}$$

Simplifying, we get

$$\frac{dy}{dx} = \frac{24x^3 + 6x^2 - 8x^3}{(4x + 1)^2} = \frac{16x^3 + 6x^2}{(4x + 1)^2}$$

| Tip | If the numerator or denominator is a constant, it is easier to rewrite the expression as a product and use the power rule. |
|---|---|

**EXAMPLE 35:** Find the derivative of $y = \dfrac{2x^3}{3} + \dfrac{4}{x^2}$.

**Solution:** Rather than use the quotient rule, we rewrite the expression as

$$y = \frac{2}{3}x^3 + 4x^{-2}$$

Using the power rule and the sum rule,

$$\frac{dy}{dx} = \frac{2}{3}(3)x^2 + 4(-2)x^{-3} = 2x^2 - \frac{8}{x^3}$$

**EXAMPLE 36:** Find $s'(3)$ if $s = \dfrac{(t^3 - 3)^2}{\sqrt{t + 1}}$

**Solution:** By the quotient rule,

$$s' = \frac{\sqrt{t+1}\,(2)(t^3 - 3)(3t^2) - (t^3 - 3)^2(\frac{1}{2})(t + 1)^{-1/2}}{t + 1}$$

$$= \frac{6t^2(t^3 - 3)\sqrt{t+1} - \dfrac{(t^3 - 3)^2}{2\sqrt{t+1}}}{t + 1}$$

Then

$$s'(3) = \frac{6(9)(27 - 3)\sqrt{3+1} - \dfrac{(27 - 3)^2}{2\sqrt{3+1}}}{3 + 1} = 612$$

## EXERCISE 5—PRODUCTS AND QUOTIENTS

### Products

Find the derivative of each function.

Do not multiply out, where possible, but use the product rule for the practice.

1. $y = x(x^2 - 3)$
2. $y = x^3(5 - 2x)$
3. $y = (5 + 3x)(3 + 7x)$
4. $y = (7 - 2x)(x + 4)$
5. $y = x(x^2 - 2)^2$
6. $y = x^3(8.24x - 6.24x^3)$
7. $y = (x^2 - 5x)^3(8x - 7)^2$
8. $y = (1 - x^2)(1 + x^2)^2$
9. $y = x\sqrt{1 + 2x}$
10. $y = 3x\sqrt{5 + x^2}$
11. $y = x^2\sqrt{3 - 4x}$
12. $y = 2x^2 - 2x\sqrt{x^2 - 5}$

13. $y = x\sqrt{a + bx}$

14. $y = \sqrt{x}\,(3x^2 + 2x - 3)$

15. $y = (3x + 1)^3\sqrt{4x - 2}$

16. $y = (2x^2 - 3)\sqrt[3]{3x + 5}$

17. $v = (5t^2 - 3t)(t + 4)^2$
18. $z = (6 - 5w)^3(w^2 - 1)$
19. $s = (81.3t^3 - 73.8t)(t - 47.2)^2$
20. $x = (y - 49.3)\sqrt{23.2y + 16.2y^4}$

Find the derivative.

21. $\dfrac{d}{dx}(2x^5 + 5x)^2(x - 3)$

22. $D_x(11.5x + 49.3)(14.8 - 27.2x^{-2})^2$

**Products with More Than Two Factors**

Find the derivative of each function.

23. $y = x(x + 1)^2(x - 2)^3$

24. $y = x\sqrt{x + 1}\sqrt[3]{x}$

25. $y = (x - 1)^{1/2}(x + 1)^{3/2}(x + 2)$

26. $y = x\sqrt{x + 2}\sqrt[3]{x - 1}\sqrt[4]{x + 1}$

**Quotients**

Find the derivative of each function.

27. $y = \dfrac{x}{x + 2}$

28. $y = \dfrac{x}{x^2 + 1}$

29. $y = \dfrac{x^2}{4 - x^2}$

30. $y = \dfrac{x - 1}{x + 1}$

31. $y = \dfrac{x + 2}{x - 3}$

32. $y = \dfrac{2x - 1}{(x - 1)^2}$

33. $y = \dfrac{2x^2 - 1}{(x - 1)^2}$

34. $y = \dfrac{a - x}{n + x}$

35. $y = \dfrac{x^{1/2}}{x^{1/2} + 1}$

36. $s = \sqrt{\dfrac{t - 1}{t + 1}}$

37. $w = \dfrac{z}{\sqrt{z^2 - a^2}}$

38. $v = \sqrt{\dfrac{1 + 2t}{1 - 2t}}$

39. Find the slope of the tangent to the curve $y = \sqrt{16 + 3x}/x$ at $x = 3$.
40. Find the derivative of the function $y = x/(7.42x^2 - 2.75x)$ when $x = 1.47$.
41. If $y = x\sqrt{8 - x^2}$, find $y'(2)$.

42. If $f(x) = x^2/\sqrt{1 + x^3}$, find $f'(2)$.

## 27-6 DERIVATIVES OF IMPLICIT RELATIONS

### Derivatives with Respect to Other Variables

Our rule for the derivative of a function raised to a power is

$$\frac{d(cu^n)}{dx} = cnu^{n-1}\frac{du}{dx} \qquad \textbf{335}$$

Up to now, the variable in the function $u$ has been the *same variable* that we take the derivative with respect to, as in the following example.

**EXAMPLE 37:**

(a) $\frac{d}{dx}(x^3) = 3x^2\frac{dx}{dx} = 3x^2$ (same)

(b) $\frac{d}{dt}(t^5) = 5t^4\frac{dt}{dt} = 5t^4$ (same)

Since $dx/dx = dt/dt = 1$, we have not bothered to write these in. Of course, rule 335 is just as valid when our independent variable is *different* from the variable that we are taking the derivative with respect to. The following example shows the power rule being applied to such cases.

**EXAMPLE 38:**

(a) $\frac{d}{dx}(u^5) = 5u^4\frac{du}{dx}$ (different)

(b) $\frac{d}{dt}(w^4) = 4w^3\frac{dw}{dt}$ (different)

(c) $\frac{d}{dx}(y^6) = 6y^5\frac{dy}{dx}$ (different)

(d) $\frac{d}{dx}(y) = 1y^0\frac{dy}{dx} = \frac{dy}{dx}$ (different)

**Mathematical ideas do not, of course, depend upon which letters of the alphabet we happen to have chosen when doing a derivation. Thus in any of our rules you can replace any letter, say *x*, with any other letter, such as *z* or *t*.**

| Common Error | It is very easy to forget to include the $dy/dx$ in problems such as the following. $\frac{d}{dx}(y^6) = 6y^5\frac{dy}{dx}$ ← don't forget! |
|---|---|

Our other rules for derivatives (for sums, products, quotients, etc.) also work when the independent variable(s) of the function is different from the variable we are taking the derivative with respect to.

**EXAMPLE 39:**

(a) $\frac{d}{dx}(2y + z^3) = 2\frac{dy}{dx} + 3z^2\frac{dz}{dx}$

(b) $\frac{d}{dt}(wz) = w\frac{dz}{dt} + z\frac{dw}{dt}$

(c) $\frac{d}{dx}(x^2y^3) = x^2(3y^2)\frac{dy}{dx} + y^3(2x)\frac{dx}{dx}$

$= 3x^2y^2\frac{dy}{dx} + 2xy^3$

## Implicit Relations

Recall that in an *implicit relation* neither variable is isolated on one side of the equal sign.

**EXAMPLE 40:** $x^2 + y^2 = y^3 - x$ is an implicit relation.

To find the derivative of an implicit relation, take the derivative of both sides of the equation, and then solve for $dy/dx$. When taking the derivatives, keep in mind that the derivative of $x$ with respect to $x$ is 1, and that the derivative of $y$ with respect to $x$ is $dy/dx$.

**We need to be able to differentiate implicit relations because we cannot always solve for one of the variables before differentiating.**

**EXAMPLE 41:** Given the implicit relation in Example 40, find $dy/dx$.

**Solution:** We take the derivative term by term,

$$2x\frac{dx}{dx} + 2y\frac{dy}{dx} = 3y^2\frac{dy}{dx} - \frac{dx}{dx}$$

or

$$2x + 2y\frac{dy}{dx} = 3y^2\frac{dy}{dx} - 1$$

Solving for $dy/dx$ gives us

$$2y\frac{dy}{dx} - 3y^2\frac{dy}{dx} = -2x - 1$$

$$(2y - 3y^2)\frac{dy}{dx} = -2x - 1$$

Note that the derivative, unlike those for explicit functions, contains $y$.

$$\frac{dy}{dx} = \frac{2x + 1}{3y^2 - 2y}$$

When taking implicit derivatives, it is convenient to use the $y'$ notation instead of $dy/dx$.

**EXAMPLE 42:** Find the derivative $dy/dx$ for the relation

$$x^2y^3 = 5$$

**Solution:** Using the product rule, we obtain

$$x^2(3y^2)y' + y^3(2x) = 0$$

$$3x^2y^2y' = -2xy^3$$

$$y' = -\frac{2xy^3}{3x^2y^2}$$

$$= -\frac{2y}{3x}$$

**EXAMPLE 43:** Find $dy/dx$, given $x^2 + 3x^3y + y^2 = 4xy$.

**Solution:** Taking the derivative of each term, we obtain

$$2x + 3x^3y' + y(9x^2) + 2yy' = 4xy' + 4y$$

Moving all terms containing $y'$ to the left side and the other terms to the right side yields

$$3x^3y' + 2yy' - 4xy' = 4y - 2x - 9x^2y$$

Factoring gives us

$$(3x^3 + 2y - 4x)y' = 4y - 2x - 9x^2y$$

So

$$y' = \frac{4y - 2x - 9x^2y}{3x^3 + 2y - 4x}$$

### Power Function with Fractional Exponent

We'll now show that the power rule

| | |
|---|---|
| $\frac{d}{dx}x^n = nx^{n-1}$ | **331** |

and hence Eq. 335, is valid when the exponent $n$ is a fraction.

Let $n = p/q$, where $p$ and $q$ are both integers, positive or negative. Then

$$y = x^n = x^{p/q}$$

Raising both sides to the $q$th power, we have

$$y^q = x^p$$

Using Eq. 333, we take the derivative of each side,

$$qy^{q-1}\frac{dy}{dx} = px^{p-1}\frac{dx}{dx}$$

Solving for $dy/dx$ yields

$$\frac{dy}{dx} = \frac{p}{q}\frac{x^{p-1}}{y^{q-1}} = n\frac{x^{p-1}}{[x^{(p/q)}]^{(q-1)}}$$

since $p/q = n$ and $y = x^{p/q}$. Applying the laws of exponents gives

$$\frac{dy}{dx} = n\frac{x^{p-1}}{x^{p-p/q}} = nx^{p-1-p+n} = nx^{n-1}$$

We have now shown that the power rule works for any rational exponent, positive or negative. It is also valid for an irrational exponent (such as $\pi$), as we will show in Sec. 30-5.

## EXERCISE 6—IMPLICIT RELATIONS

### Derivatives with Respect to Other Variables

1. If $y = 2u^3$, find $dy/dw$.
2. If $z = (w + 3)^2$, find $dz/dy$.
3. If $w = y^2 + u^3$, find $dw/du$.
4. If $y = 3x^2$, find $dy/du$.

Find the derivative.

5. $\frac{d}{dx}(x^3y^2)$
6. $\frac{d}{dx}(w^2 - 3w - 1)$
7. $\frac{d}{dt}\sqrt{3z^2 + 5}$
8. $\frac{d}{dz}(y - 3)\sqrt{y - 2}$

### Derivatives of Implicit Relations

Find $dy/dx$.

9. $5x - 2y = 7$
10. $2x + 3y^2 = 4$
11. $xy = 5$
12. $x^2 + 3xy = 2y$
13. $y^2 = 4ax$
14. $y^2 - 2xy = a^2$
15. $x^3 + y^3 - 3axy = 0$
16. $x^2 + y^2 = r^2$
17. $y + y^3 = x + x^3$
18. $x + 2x^2y = 7$
19. $y^3 - 4x^2y^2 + y^4 = 9$
20. $y^{3/2} + x^{3/2} = 16$

Find the slope of the tangent to each curve at the given point.

21. $x^2 + y^2 = 25$ at $x = 2$ in the first quadrant
22. $x^2 + y^2 = 25$ at (3, 4)
23. $2x^2 + 2y^3 - 9xy = 0$ at (1, 2)
24. $x^2 + xy + y^2 - 3 = 0$ at (1, 1)

## 27-7 HIGHER-ORDER DERIVATIVES

After taking the derivative of a function, we may then take the derivative of the derivative. That is called the *second derivative*. Our original derivative we now call the *first derivative*. The symbols used for the second derivative are

**We will have lots of uses for the second derivative in the next few chapters, but derivatives higher than second order are rarely needed.**

$$\frac{d^2y}{dx^2} \quad \text{or} \quad y'' \quad \text{or} \quad f''(x) \quad \text{or} \quad D^2y$$

We can then go on to find third, fourth, and higher derivatives.

---

**EXAMPLE 44:** Given the function

$$y = x^4 + 2x^3 - 3x^2 + x - 5$$

we get

$$y' = 4x^3 + 6x^2 - 6x + 1$$

and

$$y'' = 12x^2 + 12x - 6$$

$$y''' = 24x + 12$$

$$y^{(iv)} = 24$$

$$y^{(v)} = 0$$

and all higher derivatives will also be zero.

---

**EXAMPLE 45:** Find the second derivative of $y = (x + 2)\sqrt{x - 3}$.

**Solution:** Using the product rule, we obtain

$$y' = (x + 2)\left(\frac{1}{2}\right)(x - 3)^{-1/2} + \sqrt{x - 3}(1)$$

and

$$y'' = \frac{1}{2}\left[(x + 2)\left(-\frac{1}{2}\right)(x - 3)^{-3/2} + (x - 3)^{-1/2}(1)\right] + \frac{1}{2}(x - 3)^{-1/2}$$

$$= -\frac{x + 2}{4(x - 3)^{3/2}} + \frac{1}{\sqrt{x - 3}}$$

| Common Error | Be sure to *simplify* the first derivative before taking the second. |
|---|---|

## EXERCISE 7—HIGHER-ORDER DERIVATIVES

Find the second derivative of each function.

1. $y = 3x^4 - x^3 + 5x$
2. $y = x^3 - 3x^2 + 6$
3. $y = \frac{x^2}{x + 2}$
4. $y = \frac{3 + x}{3 - x}$
5. $y = \sqrt{5 - 4x^2}$
6. $y = \sqrt{x + 2}$
7. $y = (x - 7)(x - 3)^3$
8. $y = x^2\sqrt{2.3x - 5.82}$
9. If $y = x(9 + x^2)^{1/2}$, find $y''(4)$.
10. If $f(x) = 1/\sqrt{x} + \sqrt{x}$, find $f''(1)$.

## CHAPTER 27 REVIEW PROBLEMS

1. Differentiate $y = \sqrt{\frac{x^2 - 1}{x^2 + 1}}$.
2. Evaluate $\lim_{x \to 1} \frac{x^2 + 2 - 3x}{x - 1}$.
3. Find $dy/dx$ if $x^{2/3} + y^{2/3} = 9$.
4. Find $y''$ if $y = \frac{x}{\sqrt{2x + 1}}$.
5. If $f(x) = \sqrt{4x^2 + 9}$, find $f'(2)$.
6. Evaluate $\lim_{x \to \infty} \frac{5x + 3x^2}{x^2 - 1 - 3x}$.
7. Find $dy/dx$ if $\frac{x^2}{a^2} + \frac{y^2}{b^2} = 1$.
8. Find $dy/dx$ if $y = \sqrt[3]{\frac{3x + 2}{2 - 3x}}$.
9. If $f(x) = \sqrt{25 - 3x}$, find $f''(3)$.
10. Find $dy/dx$ if $y = \frac{x^2 + a^2}{a^2 - x^2}$.
11. Find $dy/dx$ by the delta method if $y = 5x - 3x^2$.
12. Evaluate $\lim_{x \to 0} \frac{\sin x}{x + 1}$.
13. Find the slope of the tangent to the curve $y = \frac{1}{\sqrt{25 - x^2}}$ at $x = 3$.
14. Evaluate $\lim_{x \to 0} \frac{e^x}{x}$.
15. If $y = 2.15x^3 - 6.23$, find $y'(5.25)$.
16. Evaluate $\lim_{x \to -7} \frac{x^2 + 6x - 7}{x + 7}$.
17. Find the derivative $\frac{d}{dx}(3x + 2)$.
18. Find the derivative $D(3x^4 + 2)^2$.
19. Find the derivative $D_x(x^2 - 1)(x + 3)^{-4}$.
20. If $v = 5t^2 - 3t + 4$, find $dv/dt$.
21. If $z = 9 - 8w + w^2$, find $dz/dx$.

22. Evaluate $\lim_{x \to 5} \frac{25 - x^2}{x - 5}$.

23. If $f(x) = 7x - 4x^3$, find $f''(x)$.

24. Find the derivative $D_x(21.7x + 19.1)(64.2 - 17.9x^{-2})^2$.

25. If $s = 58.3t^3 - 63.8t$, find $ds/dt$.

26. Find $\frac{d}{dx}(4x^3 - 3x + 2)$

*Find the derivative.*

27. $y = 6x^2 - 2x + 7$

28. $y = (3x + 2)(x^2 - 7)$

29. $y = 16x^3 + 4x^2 - x - 4$

30. $y = \frac{2x}{x^2 - 9}$

31. $y = 2x^{17} + 4x^{12} - 7x + 1/(2x^2)$

32. $y = (2x^3 - 4)^2$

33. $y = \frac{x^2 + 3x}{x - 1}$

34. $y = x^{2/5} + 2x^{1/3}$

*Find $f''(x)$.*

35. $f(x) = 6x^4 + 4x^3 - 7x^2 + 2x - 17$

36. $f(x) = (2x + 1)(5x^2 - 2)$

*Writing.*

37. We find the derivative by the delta method by first finding $\Delta y$ divided by $\Delta x$, and then letting $\Delta x$ approach zero. Explain in your own words why this doesn't give division by zero, causing us to junk the whole calculation.

# 28

# GRAPHICAL APPLICATIONS OF THE DERIVATIVE

## OBJECTIVES

**When you have completed this chapter, you should be able to:**

- Write the equation of the tangent or the normal to a curve.
- Find the angle of intersection between two curves.
- Find the values of $x$ for which a given curve is increasing or decreasing.
- Find maximum and minimum points on a curve.
- Test whether a point is a maximum or minimum point.
- Find points of inflection on a curve.
- Use Newton's method to find roots of an equation.
- Graph curves and regions with the aid of the techniques of this chapter.

We saw in Chapter 27 that the first derivative gives us the slope of the tangent to a curve at any point. This fact leads to several graphical applications of the derivative. In addition to the obvious one of finding the equations of the tangent and normal to a curve at a given point, we will find the "peaks" and "valleys" on a curve by locating the points at which the tangent is parallel to the $x$ axis.

The second derivative is also used here to locate points on a curve where the curvature changes direction, and also to tell whether a point at which the tangent is horizontal is a peak or a valley.

The derivative is further used in Newton's method for finding roots of equations, a technique that works very well on the computer. Finally, the chapter closes with a summary of steps to aid in curve sketching.

## 28-1 TANGENTS AND NORMALS

### Tangent to a Curve

We saw in Chapter 27 that the slope $m_t$ of the tangent to a curve at some point [$(x_1, y_1)$ in Fig. 28-1] is given by the derivative of the equation of the curve evaluated at that point.

$$\text{slope of tangent} = m_t = y'(x_1)$$

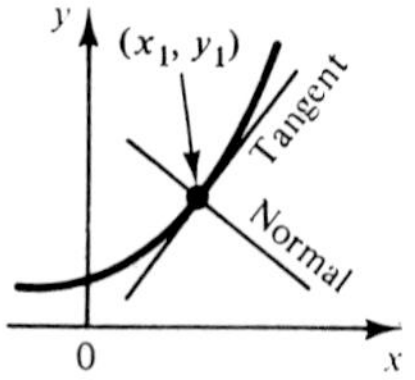


**FIGURE 28-1 Tangent and normal to a curve.**

Knowing the slope and the coordinates of the given point, we then use the point-slope form of the straight-line equation (Eq. 281) to write the equation of the tangent.

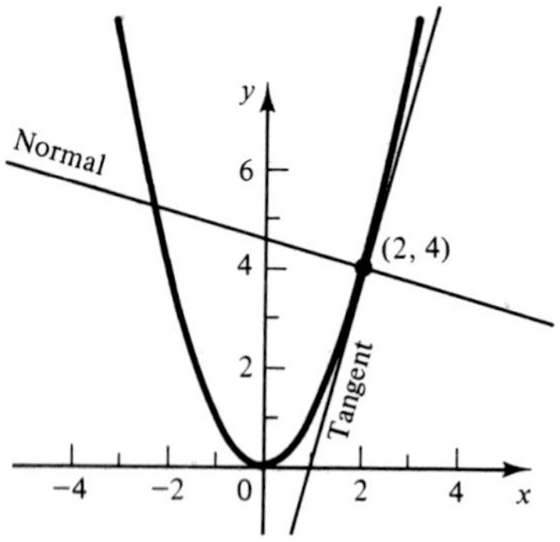


**FIGURE 28-2**

---

**EXAMPLE 1:** Write the equation of the tangent to the curve $y = x^2$, at $x = 2$ (Fig. 28-2).

**Solution:** The derivative is $y' = 2x$. When $x = 2$, $y(2) = 2^2 = 4$ and

$$y'(2) = 2(2) = 4 = m_t$$

Using the point-slope form of the straight-line equation, we obtain

$$\frac{y - 4}{x - 2} = 4$$

or

$$y - 4 = 4x - 8$$

so $4x - y - 4 = 0$ is the equation of the tangent.

---

### Normal to a Curve

The *normal* to a curve at a given point is the line perpendicular to the tangent at that point, as in Fig. 28-1. From Eq. 285, we know that the slope $m_n$ of the normal will be the negative reciprocal of the slope of the tangent.

**EXAMPLE 2:** Find the equation of the normal at the same point as in Example 1.

**Solution:** The slope of the normal is

$$m_n = -\frac{1}{m_t} = -\frac{1}{4}$$

Again using the point-slope form, we have

$$\frac{y - 4}{x - 2} = -\frac{1}{4}$$

$$4y - 16 = -x + 2$$

so $x + 4y - 18 = 0$ is the equation of the normal.

## Implicit Relations

When the equation of the curve is an implicit relation, you may choose to solve for $y$, when possible, before taking the derivative. Often, though, it is easier to take the derivative implicitly, as in the following example.

**EXAMPLE 3:** Find (a) the equation of the tangent to the ellipse $4x^2 + 9y^2 = 40$ at the point $(1, -2)$ and (b) the $x$ intercept of the tangent (Fig. 28-3).

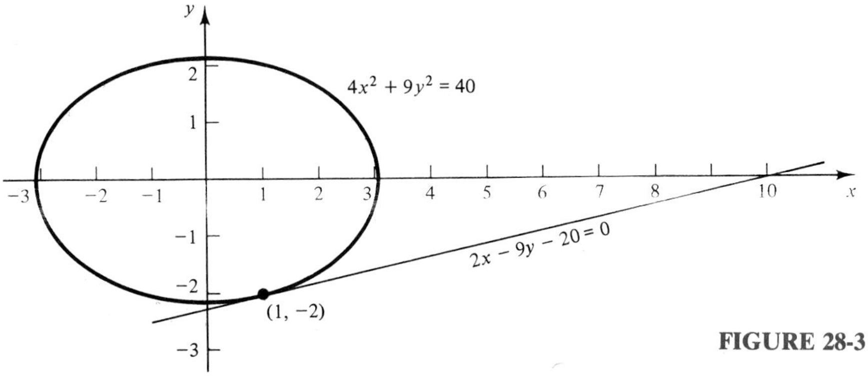


**FIGURE 28-3**

**Solution:** (a) Taking the derivative implicitly, we have $8x + 18yy' = 0$, or

$$y' = -\frac{4x}{9y}$$

At $(1, -2)$,

$$y'(1, -2) = -\frac{4(1)}{9(-2)} = \frac{2}{9} = m_t$$

Using the point-slope form gives us

$$\frac{y - (-2)}{x - 1} = \frac{2}{9}$$

$$9y + 18 = 2x - 2$$

so $2x - 9y - 20 = 0$ is the equation of the tangent.

(b) Setting $y$ equal to zero in the equation of the tangent gives $2x - 20 = 0$, or an $x$ intercept of

$$x = 10$$

## Angle of Intersection of Two Curves

**If the point(s) of intersection are not known, solve the two equations simultaneously.**

The angle between two curves is defined as the angle between their tangents at the point of intersection, which is found from Eq. 286.

**EXAMPLE 4:** Find the angle of intersection between the parabolas

$$(1)\ y^2 = x \qquad \text{and} \qquad (2)\ y = x^2$$

at the point of intersection (1, 1) (Fig. 28-4).

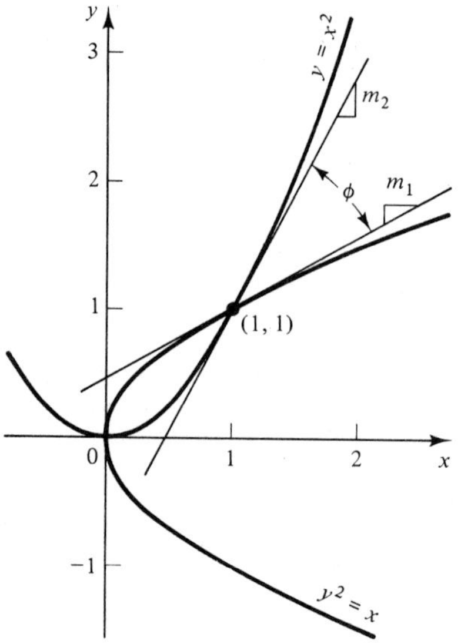


**FIGURE 28-4 Angle of intersection of two curves.**

**Solution:** Taking the derivative of (1), we have $2yy' = 1$, or

$$y' = \frac{1}{2y}$$

At (1, 1),

$$y'(1, 1) = \frac{1}{2} = m_1$$

Now taking the derivative of (2) gives us $y' = 2x$. At (1, 1),

$$y'(1, 1) = 2 = m_2$$

From Eq. 286,

$$\tan\phi = \frac{m_2 - m_1}{1 + m_1 m_2}$$

$$= \frac{2 - \frac{1}{2}}{1 + \frac{1}{2}(2)}$$

$$= \frac{1.5}{2}$$

$$= 0.75$$

$$\phi = 36.9°$$

### EXERCISE 1—TANGENTS AND NORMALS

Write the equation of the tangent and normal at the given point.

1. $y = x^2 + 2$ at $x = 1$
2. $y = x^3 - 3x$ at (2, 2)
3. $y = 3x^2 - 1$ at $x = 2$
4. $y = x^2 - 4x + 5$ at (1, 2)
5. $x^2 + y^2 = 25$ at (3, 4)
6. $16x^2 + 9y^2 = 144$ at (2, 2.98)
7. Find the first quadrant point on the curve $y = x^3 - 3x^2$ at which the tangent to the curve is parallel to $y = 9x + 7$.
8. Write the equation of the tangent to the parabola $y^2 = 4x$ that makes an angle of 45° with the $x$ axis.
9. The curve $y = 2x^3 - 6x^2 - 2x + 1$ has two tangents each of which is parallel to the line $2x + y = 12$. Find their equations.
10. Each of two tangents to the circle $x^2 + y^2 = 25$ has a slope $\frac{3}{4}$. Find the points of contact.
11. Find the equation of the line tangent to the parabola $y = x^2 - 3x + 2$, perpendicular to the line $5x - 2y - 10 = 0$.

**Intercepts**

12. Find the $x$ intercept of the tangent to $y^2 = 18x$ at (2, 6).
13. Find the $x$ intercept of the tangent to the curve $x^2 + y^2 = 50$ at (−5, −5).

**Angles between Curves**

Find the angle(s) of intersection between the given curves.

14. $y = x^2 + x - 2$ and $y = x^2 - 5x + 4$ at (1, 0)
15. $y = -2x$ and $y = x^2(1 - x)$ at (0, 0), (2, −4), and (−1, 2)
16. $y = 2x + 2$ and $x^2 - xy + y^2 = 4$ at (0, 2) and (−2, −2)

## 28-2 MAXIMUM AND MINIMUM POINTS

### Some Definitions

Figure 28-5 shows a path over the mountains from $A$ to $H$. It goes over three peaks, $B$, $D$, and $F$. These are called *maximum points* on the curve from $A$ to $H$. The highest, peak $D$, is called the *absolute maximum* on the curve from $A$ to $H$, while peaks $B$ and $F$ are called *relative maximum* points. Similarly, valley $G$ is called an *absolute minimum,* while valleys $C$ and $E$ are *relative minimums*. All the peaks and valleys are referred to as *extreme values*. Figure 28-6 shows maximum and minimum points on a graph of $y = f(x)$.

**As usual, we limit this discussion to *smooth curves,* without cusps or corners at which there is no derivative.**

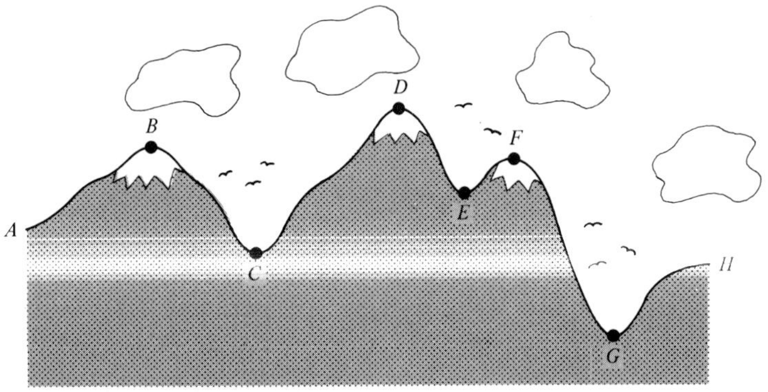


**FIGURE 28-5 Path over the mountains.**

### Increasing and Decreasing Functions

Within the interval from $A$ to $K$ in Fig. 28-6, the function is said to be *increasing* from $A$ to $D$, from $F$ to $H$, and from $J$ to $K$. The function is *decreasing* from $D$ to $F$, and from $H$ to $J$.

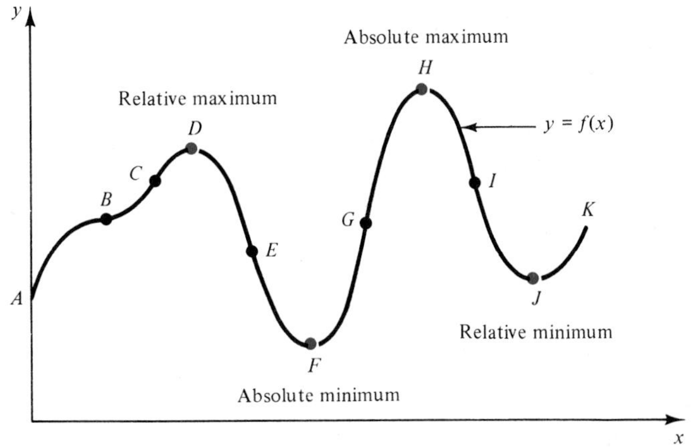


**FIGURE 28-6 Maximum, minimum, and inflection points.**

The slope of the tangent to the curve, and hence the first derivative, is *positive* for an *increasing* function. Conversely, the first derivative is *negative* for a *decreasing* function.

As with the mountains, point $D$ is a relative maximum and $H$ is an absolute maximum. Further, $F$ is an absolute minimum and $J$ is a relative minimum.

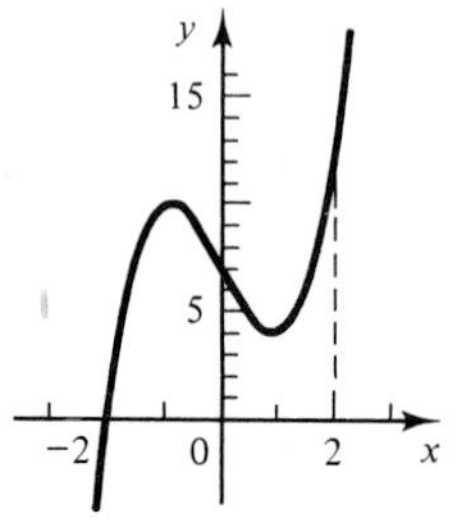


**FIGURE 28-7 Graph of $y = 2x^3 - 5x + 7$.**

**EXAMPLE 5:** Is the function $y = 2x^3 - 5x + 7$ increasing or decreasing at $x = 2$?

**Solution:** The derivative is $y' = 6x^2 - 5$. At $x = 2$,

$$y'(2) = 6(4) - 5 = 19$$

A positive derivative at $x = 2$ means that the given function is *increasing* at $x = 2$ (Fig. 28-7).

**EXAMPLE 6:** For what values of $x$ is the curve $y = 3x^2 - 12x - 2$ rising, and for what values of $x$ is it falling?

**Solution:** The first derivative is $y' = 6x - 12$. This derivative will be equal to zero when $6x - 12 = 0$, or $x = 2$. We see that $y'$ is negative for values of $x$ less than 2. A *negative* derivative tells us that our original function is *falling* in that region. Further, the derivative is positive for $x > 2$, so the given function is rising in that region (Fig. 28-8).

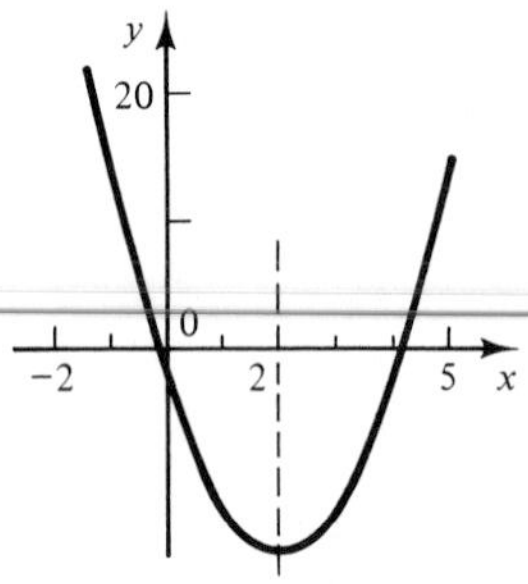


**FIGURE 28-8 Graph of $y = 3x^2 - 12x - 2$.**

## Finding Maximum and Minimum Points

**The idea of maximum and minimum values will lead to some interesting applications in Chapter 29.**

A *stationary point* on a curve is one at which the tangent is horizontal. In Fig. 28-6 they include the maximums, $D$ and $H$, the minimums $F$ and $J$, and point $B$, which is neither a maximum nor a minimum. At all such stationary points *the first derivative is equal to zero*. Since the first derivative is zero at maximum and minimum points (as well as at other stationary points such as point $B$ in Fig. 28-6), we find such points by taking the first derivative of the function and finding the value(s) of $x$ that make the first derivative equal to zero.

| | | |
|---|---|---|
| Maximum and Minimum Points | To find maximum and minimum points (and other stationary points) set the first derivative equal to zero and solve for $x$. | **356** |

**EXAMPLE 7:** Find the maximum and minimum values for the function $y = x^3 - 3x$.

**Solution:** We take the first derivative,

$$y' = 3x^2 - 3$$

We find the value of $x$ that makes the derivative zero by *setting the derivative equal to zero and solving for x.*

$$3x^2 - 3 = 0$$

$$x = \pm 1$$

Solving for $y$,

$$y(1) = 1 - 3 = -2$$

and

$$y(-1) = -1 + 3 = 2$$

So the points of zero slope are $(1, -2)$ and $(-1, 2)$, as in Fig. 28-9.

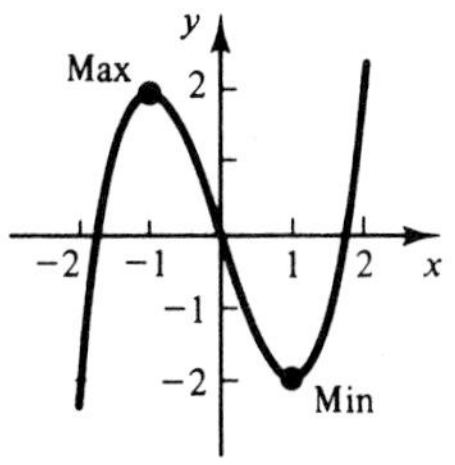


**FIGURE 28-9 Graph of $y = x^3 - 3x$.**

We see from the graph that $(1, -2)$ is a minimum and that $(-1, 2)$ is a maximum. However, we can *test* for whether a point is a maximum or minimum, without having to graph the function, as shown in the following section.

## Testing for Maximum or Minimum

The simplest way to tell whether a stationary point is a maximum, a minimum, or neither is to graph the curve. There are also a few tests that will identify such points.

If we look at the slope of the tangent on either side of a minimum point $B$ (Fig. 28-10), we see that it is negative to the left (at $A$) and positive to the right (at $C$) of that point. The reverse is true for a maximum point. Since the slope is given by the first derivative, we have the following:

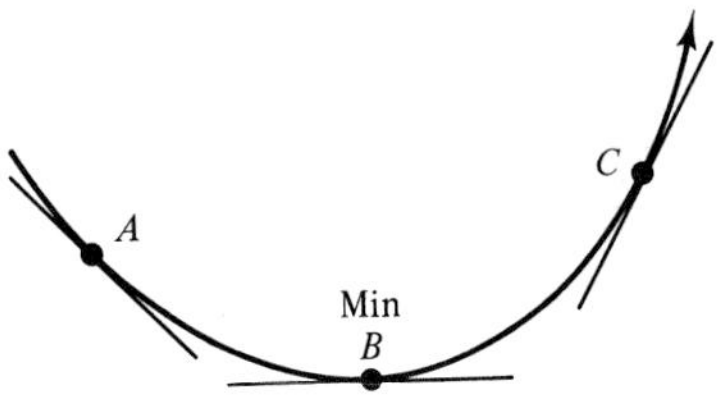


**FIGURE 28-10**

| | | |
|---|---|---|
| First-Derivative Test | The first derivative is negative to the left of, and positive to the right of, a minimum point. The reverse is true for a maximum point. | **357** |

We use this test only on points *close* to the suspected maximum or minimum. If our test points are too far away, the curve might have already changed direction.

Further, a minimum point occurs in a region of a curve (Fig. 28-11a) that is said to be *concave upward,* while a maximum point occurs where a curve is *concave downward* (Fig. 28-11b). Thus a test for concavity at a stationary point will tell whether that point is a maximum or a minimum.

If we look at the slope of the tangent at each of several points on the concave upward curve, we see that the slope is increasing as we proceed in the positive $x$ direction. At point $A$ in Fig. 28-10 the slope is negative, at $B$ it is zero, and at $C$ it is positive. And since the slope of the tangent is increasing, it means that the first derivative (which gives the value of the slope) is also increasing.

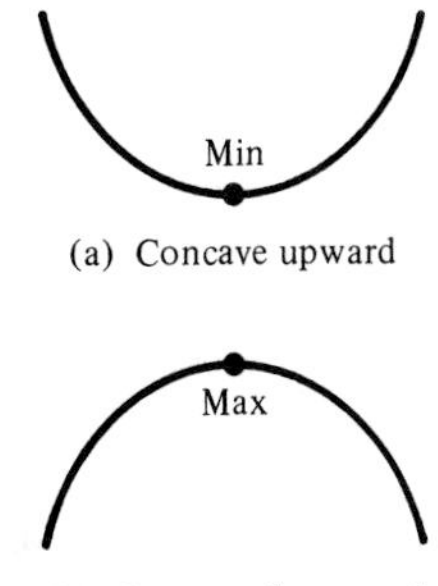


(a) Concave upward

(b) Concave downward

**FIGURE 28-11**

But we have already seen that the derivative is positive for an increasing function. Therefore, if the derivative is increasing, then the derivative of the derivative (that is, the second derivative) must be positive. In other words, *the second derivative is positive where a curve is concave upward.* The opposite is also true, that *the second derivative is negative where a curve is concave downward.* We can also reason that the second derivative is zero when the curve is neither concave upward or downward, such as point $B$ in Fig. 28-6.

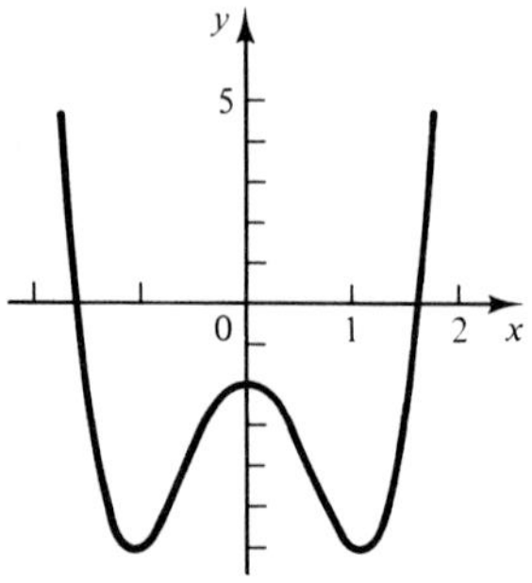


**FIGURE 28-12 Graph of $y = 3x^4 - 7x^2 - 2$.**

**EXAMPLE 8:** Is the curve $y = 3x^4 - 7x^2 - 2$ concave upward or concave downward at (a) $x = 0$ and (b) $x = 1$?

**Solution:** The first derivative is $y' = 12x^3 - 14x$, and the second derivative is $y'' = 36x^2 - 14$.
(a) At $x = 0$,

$$y''(0) = -14$$

A negative second derivative tells that the given curve is *concave downward* at that point, as in Fig. 28-12.
(b) At $x = 1$,

$$y''(1) = 36 - 14 = 22$$

or positive, so the curve is *concave upward* at that point.

Thus the second derivative tells us whether a curve is concave upward or downward at a given point, and provides us with the following test.

This test will not work on rare occasions. For example, the function $y = x^4$ has a minimum point at the origin, but the second derivative there is zero, not a positive number as expected.

| | | |
|---|---|---|
| Second-Derivative Test | If the first derivative at some point is zero, then, if the second derivative is<br>1. Positive, the point is a minimum<br>2. Negative, the point is a maximum<br>3. Zero, the test fails | **358** |

**EXAMPLE 9:** From Example 7, the function $y = x^3 - 3x$ had extreme values of $(1, -2)$ and $(-1, 2)$. Use the second derivative test to decide which is a maximum and which a minimum.

**Solution:** The first derivative is $y' = 3x^2 - 3$ and the second derivative is

$$y'' = 6x$$

At the point $(1, -2)$,

$$y''(1) = 6(1) = 6$$

It might help to think of the curve as a bowl. When it is concave upward it will hold (+) water, and when concave downward it will spill (−) water.

A positive second derivative tells us that $(1, -2)$ is a minimum. Now at $(-1, 2)$,

$$y''(-1) = 6(-1) = -6$$

which is negative, so $(-1, 2)$ is a maximum, as in Fig. 28-9.

| Common Error | It is tempting to group *maximum with positive,* and *minimum with negative*. Remember that they are just the *reverse* of this: |
|---|---|

A third test for distinguishing between a maximum point and a minimum point is called the *ordinate* test.

| Ordinate Test | Find $y$ a small distance to either side of the point to be tested. If $y$ is greater there, we have a minimum; if less, we have a maximum. |
|---|---|

---

**EXAMPLE 10:** Testing the point (1, −2) in Example 9, we compute $y$ at, say, $x = 0.9$ and $x = 1.1$.

$$y(0.9) = (0.9)^3 - 3(0.9) = -1.97$$

$$y(1.1) = (1.1)^3 - 3(1.1) = -1.97$$

Thus the curve is higher a small distance to either side of the extreme value (1, −2), so we conclude that (1, −2) is a minimum point.

---

The procedure for finding maximum and minimum points is no different for an implicit relation, although it usually takes more work.

---

**EXAMPLE 11:** Find any maximum and minimum points on the curve $x^2 + 4y^2 - 6x + 2y + 3 = 0$.

**Solution:** Taking the derivative implicitly gives

$$2x + 8yy' - 6 + 2y' = 0$$

$$y'(8y + 2) = 6 - 2x$$

$$y' = \frac{6 - 2x}{8y + 2}$$

Setting this derivative equal to zero gives $6 - 2x = 0$, or

$$x = 3$$

Substituting $x = 3$ into the original equation, we get

$$9 + 4y^2 - 18 + 2y + 3 = 0$$

Collecting terms yields

$$2y^2 + y - 3 = 0$$

Factoring gives us

$$(y - 1)(2y + 3) = 0$$

$$y = 1, \qquad y = -\frac{3}{2}$$

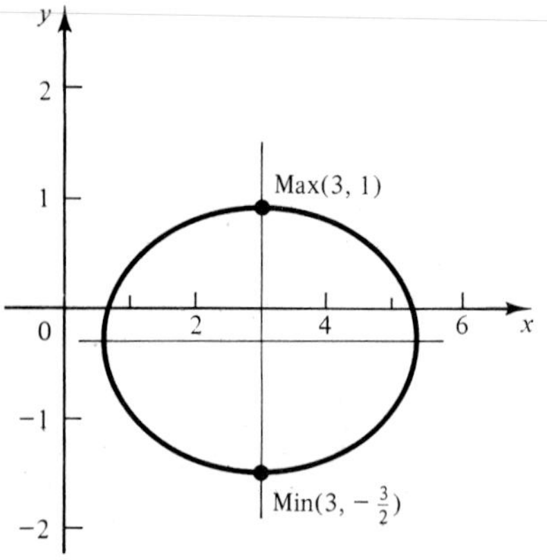


**FIGURE 28-13** **Graph of** $x^2 + 4y^2 - 6x + 2y + 3 = 0$.

So the points of zero slope are (3, 1) and (3, $-\frac{3}{2}$). We now apply the second-derivative test. Using the quotient rule gives

$$y'' = \frac{(8y + 2)(-2) - (6 - 2x)8y'}{(8y + 2)^2}$$

Replacing $y'$ by $(6 - 2x)/(8y + 2)$ and simplifying, we obtain

$$y'' = \frac{(-16y - 4)(8y + 2) - 8(6 - 2x)^2}{(8y + 2)^3}$$

At (3, 1), $y'' = -0.2$. The negative second derivative tells that (3, 1) is a maximum point. At (3, $-\frac{3}{2}$), $y'' = 0.2$, telling that we have a minimum, as shown in Fig. 28-13.

## EXERCISE 2—MAXIMUM AND MINIMUM POINTS

### Increasing and Decreasing Functions

For what values of $x$ is each curve rising, and for what values is each falling?

**1.** $y = 3x + 5$ **2.** $y = 4x^2 + 16x - 7$

Is each function increasing or decreasing at the value indicated?

**3.** $y = 3x^2 - 4$ at $x = 2$
**4.** $y = x^2 - x - 3$ at $x = 0$
**5.** $y = 4x^2 - x$ at $x = -2$
**6.** $y = x^3 + 2x - 4$ at $x = -1$

State whether each curve is concave upward or concave downward at the given value of $x$.

**7.** $y = x^4 + x^2$ at $x = 2$ **8.** $y = 4x^5 - 5x^4$ at $x = 1$
**9.** $y = -2x^3 - 2\sqrt{x + 2}$ at $x = \frac{1}{4}$ **10.** $y = \sqrt{x^2 + 3x}$ at $x = 2$

### Maxima and Minima

Find the maximum and minimum points for each function.

**11.** $y = x^2$ **12.** $y = x^3 + 3x^2 - 2$
**13.** $y = 6x - x^2 + 4$ **14.** $y = x^3 - 3x + 4$
**15.** $y = x^3 - 7x^2 + 36$ **16.** $y = x^4 - 4x^3$
**17.** $y = 2x^2 - x^4$ **18.** $16y = x^2 - 32x$
**19.** $y = x^4 - 4x$ **20.** $2y = x^2 - 4x + 6$
**21.** $y = 3x^4 - 4x^3 - 12x^2$
**22.** $y = 2x^3 - 9x^2 + 12x - 3$
**23.** $y = x^3 + 3x^2 - 9x + 5$
**24.** $y = \dfrac{x^2 - 7x + 6}{x - 10}$
**25.** $y = (x - 1)^4(x + 2)^3$
**26.** $y = (x - 2)^2(2x + 1)$

Find the maximum and minimum points for each implicit relation.

**27.** $4x^2 + 9y^2 = 36$
**28.** $x^2 + y^2 - 2x + 4y = 4$
**29.** $x^2 - x - 2y^2 + 36 = 0$
**30.** $x^2 + y^2 - 8x - 6y = 0$

## 28-3 INFLECTION POINTS

A point where the curvature changes from concave upward to concave downward (or vice versa) is called an *inflection point*, or *point of inflection*. In Fig. 28-6 they are points $B$, $C$, $E$, $G$, and $I$.

We saw that the second derivative was positive where a curve was concave upward and negative where concave downward. In going from positive to negative, the second derivative must somewhere be zero, and this is at the point of inflection.

**Unfortunately, a curve may have a point of inflection where the second derivative does not exist, such as the curve $y = x^{1/3}$ at $x = 0$. Fortunately, these cases are rare.**

| | | |
|---|---|---|
| Inflection Points | To find points of inflection, set the second derivative to zero and solve for $x$. Test by seeing if the second derivative changes sign a small distance to either side of the point. | **359** |

**EXAMPLE 12:** Find any points of inflection on the curve

$$y = x^3 - 3x^2 - 5x + 7.$$

**Solution:** We take the derivative twice,

$$y' = 3x^2 - 6x - 5$$

$$y'' = 6x - 6$$

We now set $y''$ to 0 and solve for $x$.

$$6x - 6 = 0$$

$$x = 1$$

and

$$y(1) = 1 - 3 - 5 + 7 = 0$$

A graph of the given function (Fig. 28-14) shows the point (1, 0) clearly to be a point of inflection. If there were any doubt, we would test it by seeing if the second derivative has opposite signs on either side of the point.

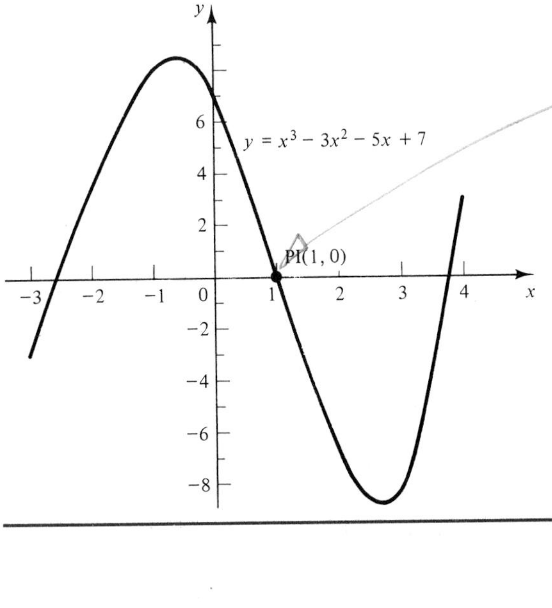


**FIGURE 28-14**

## EXERCISE 3—INFLECTION POINTS

Find the points of inflection for each curve.

1. $y = x^3$
2. $y = x^3 - 3x^2$
3. $y = x^5$
4. $y = 3x^4 - 4x^3 + 1$
5. $y = x^3 - 9x^2 - 24x$
6. $y = 3x - x^3$
7. $y = x^4 - 6x^2$
8. $y = x^2(x^2 + 1)$
9. $y = x^4 + 2x^3 + 10x - 17$
10. $y = (x + 2)(x - 2)(x - 3)$

## 28-4 APPROXIMATE SOLUTION OF EQUATIONS BY NEWTON'S METHOD

**We have had so many laws and methods named "Newton" that you might think there were twenty Newtons all working in different fields. But no. All are from the same Isaac Newton (1642–1727).**

The ability to find the slope of a tangent to a curve simply by taking the derivative is made use of in Newton's method. With this method we can quickly find the root of an equation of the form $y = f(x)$, even though we cannot solve the equation for this root. The result is *approximate,* but this is hardly a drawback since we get as many significant digits as we want.

Figure 28-15 shows a graph of the function $y = f(x)$ and the zero where the curve crosses the $x$ axis. The solution to the equation $f(x) = 0$ is then the value $a$.

**We have already shown an approximate method for finding the roots of an equation. Refer back to the *Midpoint Method* in Sec. 4-5.**

Suppose that our first approximation to $a$ is the value $x_1$ (our first guess is too high). We can *correct* our first guess by subtracting from it the amount $h$, taking for our second approximation the point where the tangent line $T$ crosses the $x$ axis. The slope of the tangent line at $x_1$ is

$$m = f'(x_1) = \frac{\text{rise}}{\text{run}} = \frac{f(x_1)}{h}$$

So

$$h = \frac{f(x_1)}{f'(x_1)}$$

and

$$(\text{second approximation}) = (\text{first guess}) - h$$

or

$$x_2 = x_1 - \frac{f(x_1)}{f'(x_1)}$$

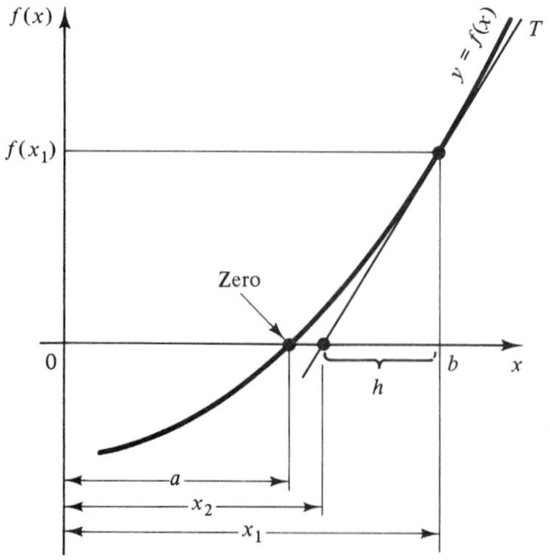


**FIGURE 28-15** **Newton's method.**

You can stop at this point, or repeat the calculation several times to get the accuracy needed, with each new value $x_{n+1}$ obtained from the old value $x_n$ by

| Newton's Method | $x_{n+1} = x_n - \dfrac{f(x_n)}{f'(x_n)}$ | **360** |
|---|---|---|

A good way to get a first guess is to graph the function. Note that Eq. 360 works whether your first guess is to the left or to the right of the zero. Also, the method is the same regardless if the curve is rising or falling where it crosses the $x$ axis.

---

**EXAMPLE 13:** Use Newton's method to find one root of the equation $x^3 - 5x = 5$ to one decimal place. Take $x = 3$ as a first guess.

**Solution:** We first move all terms to one side of the equation and express them as a function of $x$:

$$f(x) = x^3 - 5x - 5$$

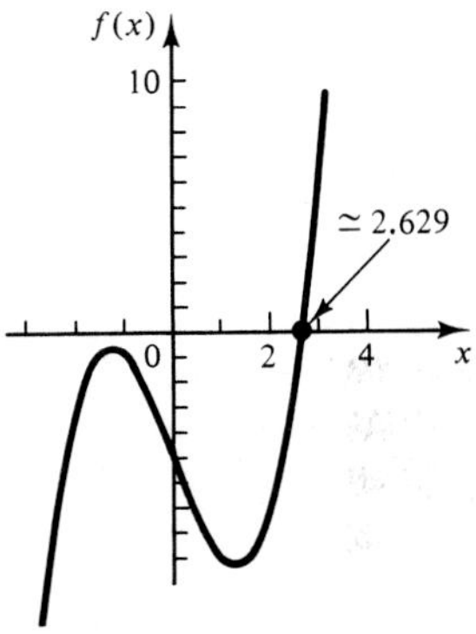


**FIGURE 28-16 A root found by Newton's method.**

This function is graphed in Fig. 28-16, and shows a root near $x = 3$. Taking the derivative yields

$$f'(x) = 3x^2 - 5$$

At the value 3,

$$f'(3) = 3(9) - 5 = 22$$

and

$$f(3) = 27 - 15 - 5 = 7$$

The first correction is then

$$h = \frac{7}{22} = 0.32$$

Our second approximation is then

$$3 - 0.32 = 2.68$$

We now repeat the entire calculation, using 2.68 as our guess.

$$f(2.68) = 0.8488$$

$$f'(2.68) = 16.55$$

$$h = \frac{0.8488}{16.55} = 0.0513$$

$$\text{third approximation} = 2.68 - 0.0513 = 2.629$$

Since $h$ is 0.0513 and getting smaller, we know that it will have no further effect on the first decimal place, so we stop here. Our root, then, to one decimal place, is $x = 2.6$.

---

**Newton's method will usually fail when the root you seek is near a maximum, minimum, or inflection point, or near a discontinuity. If *h* does not shrink rapidly as you repeat the computation, this may be the cause.**

**Can you imagine trying to solve this equation by an exact method?**

**EXAMPLE 14:** Find the root of the equation

$$1.07\sqrt{x^2 + 7.25} = x^3 + 9.84$$

to four decimal places.

**Solution:** Let

$$f(x) = x^3 - 1.07\sqrt{x^2 + 7.25} + 9.84$$

Taking the derivative gives

$$f'(x) = 3x^2 + \frac{1.07x}{\sqrt{x^2 + 7.25}}$$

A plot of our function (Fig. 28-17) shows a zero near $x = -2$, so we use this for our first guess. The calculated values are given in the following computer-generated table.

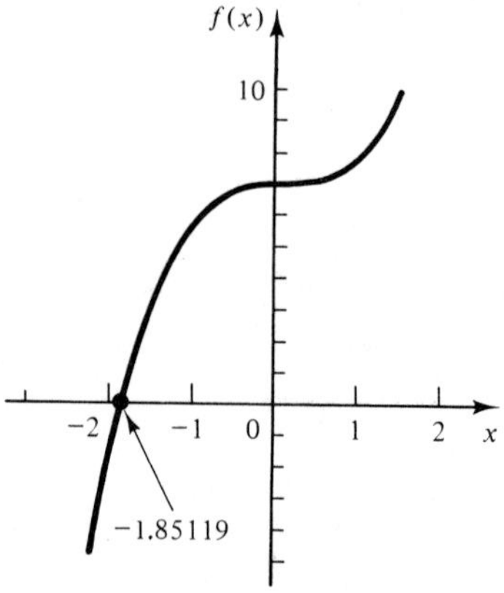


**FIGURE 28-17 A root found by Newton's method.**

| $x$ | $f(x)$ | $f''(x)$ | Correction | New $x$ |
|---|---|---|---|---|
| −2.00000 | 0.05540 | 9.61893 | −0.15392 | −1.84608 |
| −1.84608 | −0.00716 | 9.68154 | 0.00576 | −1.85184 |
| −1.85184 | 0.00089 | 9.67349 | −0.00074 | −1.85110 |
| −1.85110 | −0.00011 | 9.67450 | 0.00009 | −1.85119 |

Our root then, to four decimal places, is −1.8512.

## EXERCISE 4—NEWTON'S METHOD

Each equation has at least one root between $x = -10$ to 10. Find one such root to two decimal places using Newton's method.

1. $x^3 - 40 = 0$
2. $x^3 + 3x^2 - 10 = 0$
3. $x^3 - 3x^2 + 8x - 25 = 0$
4. $x^3 - 6x + 12 = 0$
5. $x^3 + 2x - 8 = 0$
6. $3x^3 - 4x = 1$
7. $x^3 + 4x + 12 = 0$
8. $x^4 + 8x - 12 = 0$
9. A square of side $x$ is cut from each corner of a square sheet of metal 16 in. on a side, and the edges are turned up to form a tray whose volume is 282 in³. Write an equation for the volume as a function of $x$, and use Newton's method to find $x$ to two decimal places.
10. A solid sphere of radius $r$ and specific gravity $S$ will sink in water to a depth $x$, where

$$x^3 - 3rx^2 + 4r^3S = 0$$

Use Newton's method to find $x$ to two decimal places, if $r = 4$ in. and $S = 0.700$.

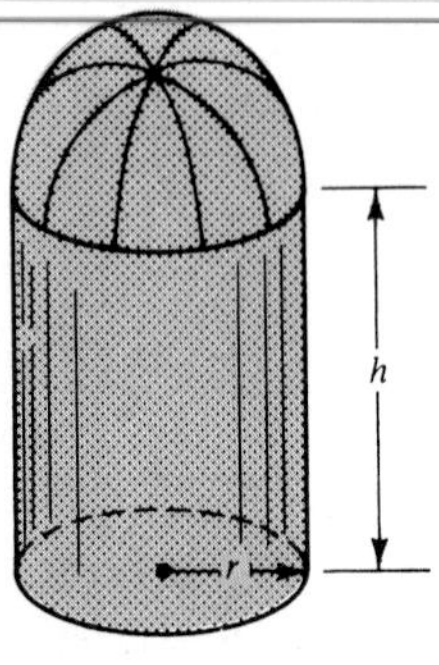


**FIGURE 28-18 Silo.**

11. A certain silo consists of a cylindrical base of radius $r$ and height $h$, topped by a hemisphere (Fig. 28-18). It contains a volume $V$ equal to

$$V = \pi r^2 h + \frac{2}{3}\pi r^3$$

If $h = 20.0$ m and $V = 800$ m³, use Newton's method to find $r$ to two decimal places.

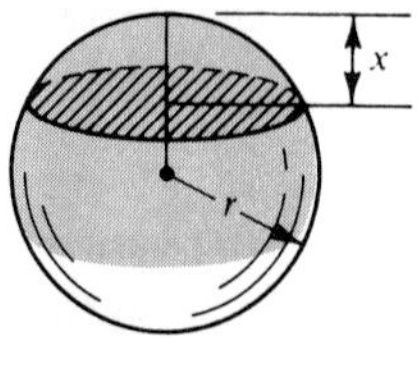


**FIGURE 28-19 Segment cut from a sphere.**

12. A segment of height $x$ cut from a sphere of radius $r$ (Fig. 28-19) has a volume

$$V = \pi\left(rx^2 - \frac{x^3}{3}\right)$$

Find $x$ to two decimal places if $r = 4.0$ cm and $V = 150$ cm³.

### Computer

13. The labor of Newton's method can be reduced by the computer. Write a program or use a spreadsheet that will accept as input the given equation, its derivative, the approximate location of a root, and the degree of accuracy you require. The computer should then locate and print the value of the root. Test your program on any of the problems in this exercise.

## 28-5 CURVE SKETCHING

When we first studied curve sketching, we simply obtained a table of point pairs by substituting into the given function or relation, and then plotted each point. Since then, we have learned methods for finding particular features of interest, such as intercepts or inflection points, which can make curve sketching faster and easier.

Why bother with all this when a computer (and certain calculators) will graph any equation we want?

Because getting the graph is not enough; it must be given meaning. In addition to drawing graphs we also use these ideas to *interpret* graphs and to understand how a function behaves. If challenged could you explain why your computer-drawn curve does strange things, perhaps shooting up to infinity or vanishing at certain values?

Also, with practice, you may be able to look at a function and, with the aid of these ideas, predict the shape of the curve without actually graphing it.

A. *Type of equation:* Does the equation look like any we have studied before—a linear function, power function, quadratic function, trigonometric function, exponential function, logarithmic function, or equation of a conic? For a polynomial, the *degree* will tell the maximum number of extreme points you can expect.

---

**EXAMPLE 15:** The function $y = x^5 - 3x^2 + 4$ is a polynomial of fifth degree, so its graph (Fig. 28-20) may have up to four extreme points and three inflection points. There may, however, be fewer than these.

---

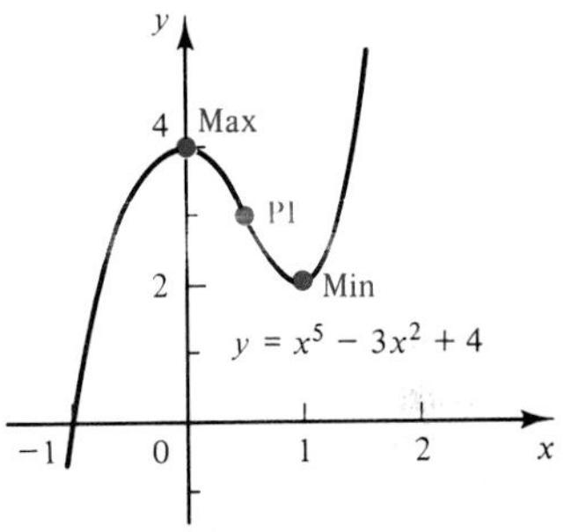


**FIGURE 28-20**

B. *Intercepts:* Find the $x$ intercept(s), or roots, by setting $y$ equal to zero and solving the resulting equation for $x$. Find the $y$ intercept(s) by setting $x$ equal to zero and solving for $y$.

**We look for those aspects of the curve that can easily be found, and look for them in any order. Our goal is to make sketching faster and easier, not to increase our work.**

---

**EXAMPLE 16:**

(a) The function $3x + 2y = 6$ has a $y$ intercept at

$$3(0) + 2y = 6$$

or at $y = 3$, and an $x$ intercept at

$$3x + 2(0) = 6$$

or at $x = 2$, Fig. 28-21.

(b) The function of Example 15 has a $y$ intercept at

$$y = 0 - 0 + 4$$

or at $y = 4$ (Fig. 28-20) but to find the $x$ intercept involves solving a fifth degree equation, so we don't bother.

---

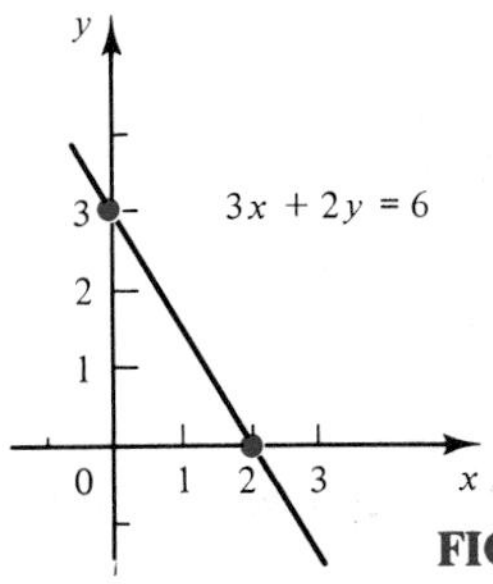


**FIGURE 28-21**

A function whose graph is symmetrical about the y axis is called an *even function.* One symmetrical about the origin is called an *odd function.*

C. *Symmetry:* A curve is symmetrical about the

**(a)** $x$ axis, if the equation does not change when we substitute $-y$ for $y$

**(b)** $y$ axis, if the equation does not change when we substitute $-x$ for $x$

**(c)** Origin, if the equation does not change when we substitute $-x$ for $x$ and $-y$ for $y$

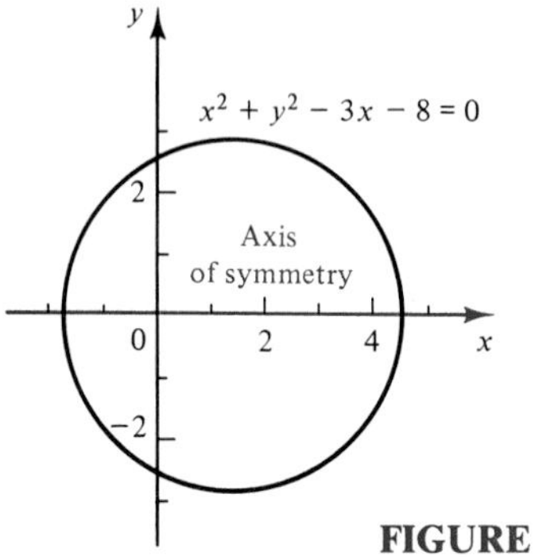


FIGURE 28-22

---

**EXAMPLE 17:** Given the function $x^2 + y^2 - 3x - 8 = 0$,

(a) Substituting $-y$ for $y$ gives $x^2 + (-y)^2 - 3x - 8 = 0$, or $x^2 + y^2 - 3x - 8 = 0$. This is the same as the original, indicating symmetry about the $x$ axis.

(b) Substituting $-x$ for $x$ gives $(-x)^2 + y^2 - 3(-x) - 8 = 0$, or $x^2 + y^2 + 3x - 8 = 0$. This equation is different from the original, indicating no symmetry about the $y$ axis.

(c) Substituting $-x$ for $x$ and $-y$ for $y$ gives $(-x)^2 + (-y)^2 - 3(-x) - 8 = 0$ or $x^2 + y^2 + 3x - 8 = 0$. The equation is different from the original, indicating no symmetry about the origin, as shown in Fig. 28-22.

---

D. *Range:* Look for values of the variables that give division by zero, or that result in negative numbers under a radical sign. The curve will not exist at these values.

---

**EXAMPLE 18:** For the function

$$y = \frac{\sqrt{x + 2}}{x - 5}$$

$y$ is not real for $x < -2$ or for $x = 5$ (Fig. 28-23).

---

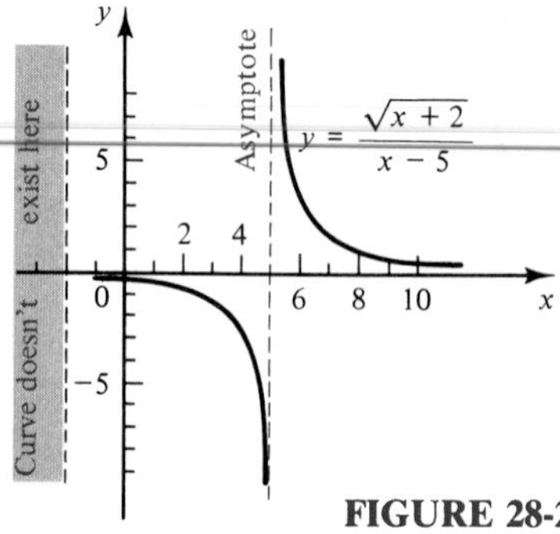


FIGURE 28-23

E. *Asymptotes:* Look for some value of $x$ which, when *approached* by $x$ (from above, or from below) will cause $y$ to become infinite. You will then have found a *vertical asymptote*. Then if $y$ approaches some particular value as $x$ becomes infinite, we will have found a *horizontal asymptote*. Similarly, check what happens to $y$ when $x$ becomes infinite in the negative direction.

---

**EXAMPLE 19:** The function in Example 18 becomes infinite as $x$ approaches 5, so we expect a vertical asymptote at $x = 5$ (Fig. 28-23).

---

F. *Increasing or decreasing function:* The first derivative will be positive in regions where the function is increasing, and negative where the function is decreasing. Thus inspection of the first derivative can show where the curve is rising, and where falling.

---

**EXAMPLE 20:** For the function $y = x^2 - 4x + 2$, the first derivative $y' = 2x - 4$ is positive for $x > 2$ and negative for $x < 2$. Thus we expect the curve to fall in the region to the left of $x = 2$ and rise in the region to the right of $x = 2$ (Fig. 28-24).

---

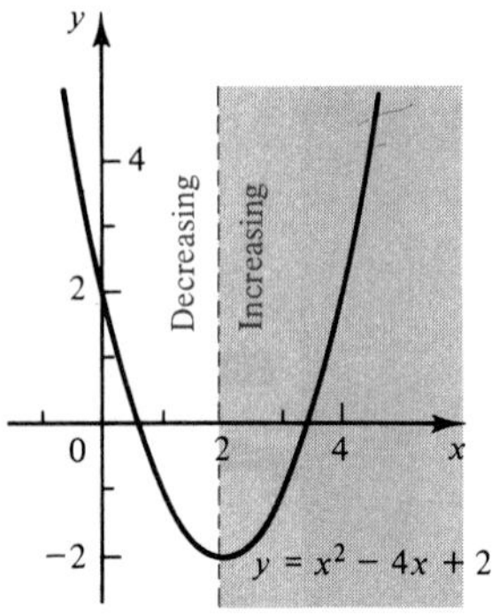


FIGURE 28-24

G. *Maximum and minimum points:* Find the points at which the first derivative is zero. Then test each of these points to decide if it is a maximum, minimum, or neither of these.

In some cases it may be useful to find the values of $y$ that make $dx/dy$ equal to zero, thus finding maximum and minimum points in the *horizontal* direction. Usually, though, we will have enough other information without this step.

**EXAMPLE 21:** Setting the first derivative of the function of Example 20 to zero, $y' = 2x - 4 = 0$, shows an extreme point at $x = 2$. The second derivative, $y'' = 2$, is positive, indicating a minimum point (Fig. 28-24).

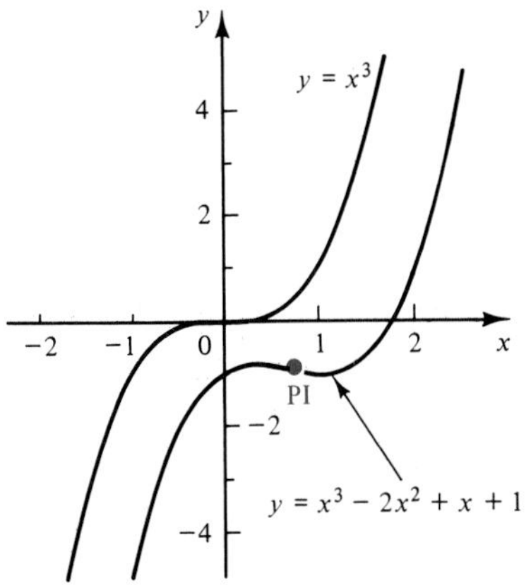


**FIGURE 28-25**

H. *Inflection points:* Find any inflection points by setting the second derivative equal to zero, and solving for $x$. Test by seeing if the second derivative changes sign a small distance to either side of the point.

**EXAMPLE 22:** The function $y = x^3 - 2x^2 + x + 1$ has derivatives $y' = 3x^2 - 4x + 1$ and $y'' = 6x - 4$. An inflection point, found by setting $y''$ to zero and solving for $x$, is at $x = 2/3$ (Fig. 28-25).

I. *Large values:* What happens to $y$ as $x$ gets very large (both in the positive and negative directions)? Does $y$ continue to grow without bound, or approach some asymptote, or is it possible that the curve will turn and perhaps cross the $x$ axis again? Similarly, can you say what happens to $x$ as $y$ gets very large?

**EXAMPLE 23:** As $x$ grows, for the function of Example 22, the second, third, and fourth terms on the right side become less significant compared to the $x^3$ term. So, far from the origin, the function will have the appearance of the function $y = x^3$ (Fig. 28-25).

In graphing any function, we use whichever of these nine items or ideas seem useful and appropriate (seldom all of them), and fill in any gaps in our graph by plotting extra points.

**EXAMPLE 24:** Graph the function $y = \dfrac{x^3}{x - 1}$.

**Solution:** We make a table of point pairs in the usual way:

| $x$ | −4 | −3 | −2 | −1 | 0 | 1 | 2 | 3 | 4 |
|---|---|---|---|---|---|---|---|---|---|
| $y$ | 12.8 | 6.75 | 2.67 | 0.500 | 0 | — | 8 | 13.5 | 21.3 |

These points are plotted in Fig. 28-26. Now if we were not too curious about why we get no value for $y$ at $x = 1$, we might be tempted to connect these

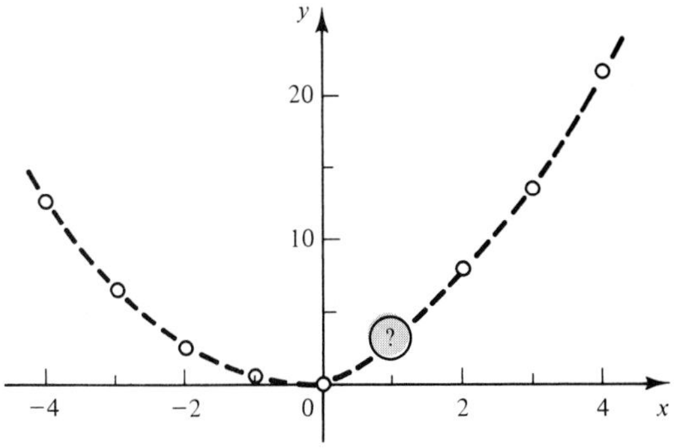


**FIGURE 28-26**

points with the smooth curve shown dashed in Fig. 28-26 which seems to fit the points very well. But we know that something is wrong. The dashed curve shows $y = 3$, approximately, when $x = 1$, while the equation gives division by zero at $x = 1$. Let us now go through the steps for curve sketching, but not necessarily in the order given above, to see what the graph really looks like.

*Range:* We get division by zero when $x = 1$, so the curve cannot exist at that value. On our graph we draw a vertical line at $x = 1$. Our curve cannot cross that line.

*Asymptotes:* As $x$ approaches 1 from above, the denominator of our function gets smaller, but remains positive. Thus $y$ takes on larger and larger positive values. As $x$ approaches 1 from below, the denominator gets smaller but is now negative. Thus $y$ gets very large in the negative direction. Thus the line $x = 1$ is a vertical asymptote.

*Large values:* As $x$ gets large, the $-1$ in the denominator becomes negligible compared to $x$, and our equation is approximately equal to $y = x^3/x$, or $y = x^2$. So, far from the origin, the curve will look like a parabola with vertex at the origin, which opens upward.

*Maximum and minimum points:* Looking at the information already obtained, we know there must be a minimum point somewhere between $x = 1$ and $x = 2$, and perhaps a maximum between $x = 0$ and $x = 1$. To find these points, we take the first derivative,

$$y' = \frac{(x-1)(3x^2) - x^3(1)}{(x-1)^2}$$

$$= \frac{x^2(2x-3)}{(x-1)^2}$$

Setting the first derivative equal to zero, we get $x^2(2x - 3) = 0$, so

$$x = 0 \quad \text{and} \quad x = \frac{3}{2}$$

Solving for $y$ yields $y(0) = 0$, and

$$y\left(\frac{3}{2}\right) = \frac{\left(\frac{3}{2}\right)^3}{\frac{3}{2} - 1}$$

$$= 6.75$$

So the first derivative is zero at the points (0, 0) and (1.5, 6.75). We now need the second derivative to determine what type of point each of these is.

$$y'' = \frac{(x-1)^2(6x^2 - 6x) - x^2(2x-3)(2)(x-1)}{(x-1)^4}$$

which eventually reduces to

$$y'' = \frac{2x(x^3 - 4x^2 + 6x - 3)}{(x-1)^4}$$

Now testing the points found above yields $y''(0) = 0$. So (0, 0) is not a maximum or minimum, but is a stationary point, and also an *inflection point*.

From our graph, we see that the point (1.5, 6.75) can only be a minimum. We verify this with the second derivative test,

$$y''(1.5) = \frac{2(1.5)[(1.5)^3 - 4(1.5)^2 + 6(1.5) - 3]}{(1.5 - 1)^4} = 18$$

The positive second derivative indicates a minimum point, as expected.

*Increasing or decreasing function:* Looking again at the first derivative,

$$y' = \frac{x^2(2x - 3)}{(x - 1)^2}$$

we see that the denominator cannot be negative, and $x^2$ in the numerator cannot be negative, but that $y'$ can be negative when $2x$ is less than 3, or, when $x$ is less than $\frac{3}{2}$. We conclude that the function is decreasing when $x$ is less than $\frac{3}{2}$ and increasing when $x$ is greater than $\frac{3}{2}$.

*Intercepts:* Setting $x = 0$ in our original equation gives $y = 0$. Setting $y = 0$ and solving for $x$, we get $x = 0$. Thus the origin is the only intercept.

*Symmetry:* Substituting $-y$ for $y$ gives

$$-y = \frac{x^3}{x - 1}$$

We get an equation that is not equivalent to the original equation, so there is no symmetry about the $x$ axis. Substituting $-x$ for $x$,

$$y = \frac{(-x)^3}{-x - 1} = \frac{-x^3}{-x - 1} = \frac{x^3}{x + 1}$$

we get a changed equation, which rules out symmetry about the $y$ axis. Finally, substituting $-x$ for $x$ and $-y$ for $y$ yields

$$-y = \frac{(-x)^3}{-x - 1} = \frac{x^3}{x + 1}$$

which is also different than the original. Thus there is no symmetry about the origin. Our final graph is shown in Fig. 28-27.

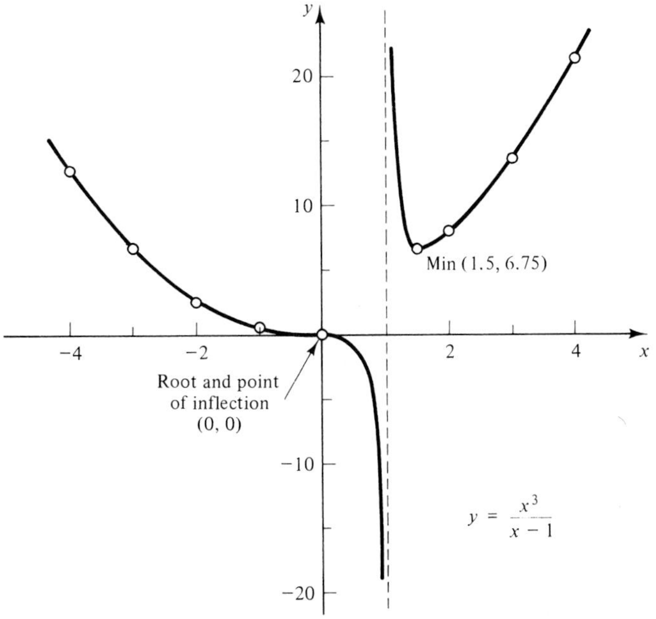


**FIGURE 28-27**

## Graphing Regions

In later applications we will have to identify *regions* that are bounded by two or more curves. Here we get some practice in doing that.

---

**EXAMPLE 25:** Locate the region bounded by the $y$ axis and the curves $y = x^2$, $y = 1/x$, and $y = 4$.

**Solution:** The three curves given are, respectively, a parabola, a hyperbola, and a straight line, and are graphed in Fig. 28-28. We see that three different closed areas are formed, but the shaded area is the only one *bounded by each one of the given curves* and the $y$ axis.

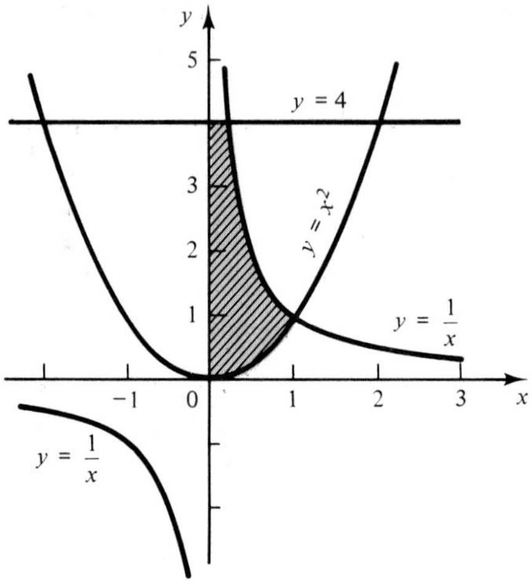


**FIGURE 28-28 Graphing a region bounded by several curves.**

---

## EXERCISE 5—CURVE SKETCHING

Graph each function.

1. $y = 4x^2 - 5$
2. $y = 3x - 2x^2$
3. $y = 5 - \dfrac{1}{x}$
4. $y = \dfrac{3}{x} + x^2$
5. $y = x^4 - 8x^2$
6. $y = \dfrac{1}{x^2 - 1}$
7. $y = x^3 - 9x^2 + 24x - 7$
8. $y = x\sqrt{1 - x}$
9. $y = 5x - x^5$
10. $y = \dfrac{9}{x^2 + 9}$
11. $y = \dfrac{6x}{3 + x^2}$
12. $y = \dfrac{x}{\sqrt{4 - x^2}}$

13. $y = x^3 - 6x^2 + 9x + 3$

14. $y = x^2\sqrt{6 - x^2}$

15. $y = \dfrac{96x - 288}{x^2 + 2x + 1}$

16. $y = \dfrac{\sqrt{x}}{x - 1}$

17. $y = \dfrac{x}{\sqrt{x^2 + 1}}$

18. $y = \dfrac{x^3}{(1 + x^2)^2}$

19. $y = x^2 + 2x$

20. $y = x^2 - 3x + 2$

21. $y = x^3 + 4x^2 - 5$

22. $y = x^4 - x^2$

Graph the region bounded by the given curves.

23. $y = 3x^2$ and $y = 2x$

24. $y^2 = 4x$, $x = 5$, and the $x$ axis, in the first quadrant

25. $y = 5x^2 - 2x$, the $y$ axis, and $y = 4$, in the second quadrant

26. $y = 4/x$, $y = x$, $x = 6$, and the $x$ axis

## CHAPTER 28 REVIEW PROBLEMS

1. Find the roots of $x^3 - 4x + 2 = 0$ to two decimal places.
2. Find any points of inflection on the curve $y = x^3(1 + x^2)$.
3. Find the maximum and minimum points on the curve $y = 3\sqrt[3]{x} - x$.
4. Find any maximum points, minimum points, and points of inflection for the function $3y = x^3 - 3x^2 - 9x + 11$.
5. Write the equations of the tangent and normal to the curve $y = \dfrac{1 + 2x}{3 - x}$ at the point (2, 5).
6. Find the coordinates of the point on the curve $y = \sqrt{13 - x^2}$ at which the slope of the tangent is $-\frac{2}{3}$.

7. Find the $x$ intercept of the tangent to the curve $y = \sqrt{x^2 + 7}$ at the point (3, 4).

8. Graph the function $y = \dfrac{4 - x^2}{\sqrt{1 - x^2}}$ and locate any features of interest (roots, asymptotes, maximum/minimum points, points of inflection).

9. For which values of $x$ is the curve $y = \sqrt{4x}$ rising, and for what values is it falling?

10. Is the function $y = \sqrt{5 - 3x}$ increasing or decreasing at $x = 1$?

11. Is the curve $y = x^2 - x^5$ concave upward or concave downward at $x = 1$?

12. Graph the region bounded by the curves $y = 3x^3$, $x = 1$, and the $x$ axis.

13. Write the equations of the tangent and normal to the curve $y = 3x^3 - 2x + 4$ at $x = 2$.

14. Find the $x$ intercept of the tangent and normal in Problem 13.

15. Find the angle of intersection between the curves $y = x^2/4$, and $y = 2/x$.

16. Graph the function $y = 3x^3\sqrt{9 - x^2}$, and locate any features of interest.

*Writing*

17. In this chapter we gave nine things to look for when graphing a function. Can you list at least seven from memory? Write a paragraph explaining how just one of the nine is useful in graphing.

# 29

# MORE APPLICATIONS OF THE DERIVATIVE

## OBJECTIVES

**When you have completed this chapter, you should be able to:**

- Solve applied problems involving instantaneous rate of change.
- Solve for currents and voltages in capacitors and inductors.
- Compute velocities and accelerations for straight-line or curvilinear motion.
- Solve motion problems given by parametric equations.
- Solve related rate problems.
- Solve applied maximum-minimum problems.
- Use differentials to compute small changes in a function, to find volumes of shells and rings, and to estimate effects of errors in measurement.

What is this stuff good for? This chapter gives the main answer to that question, at least for the derivative.

Most applications for the derivative hinge on two ideas. The first of these is *rate of change*. In preceding chapters we have used the derivative to find the slope of the tangent to a curve at any point. Here we relate derivative and slope to rate of change. We'll show that the derivative of a function gives us the instantaneous rate of change of that function. On a graph of the function, the rate of change at some point will be equal to the slope of the tangent at that point. With this we can find out *how fast* something (say, the pressure in an engine cylinder) is changing. If the changing quantity is a *distance*, the rate of change with respect to time is called *speed*, or *velocity*. In this chapter we find velocities of moving objects, and also the rate of change of velocity, or *acceleration*.

For electrical applications, we use the rate of change of charge to find current, the rate of change of current to find the voltage across an inductor, and the rate of change of voltage to find the current in a capacitor.

We will then go on to use the fact that the derivative is zero at a maximum or minimum point to find maximum and minimum values of a varying quantity.

Finally, we will use the idea of a *differential* to make fast estimations of some quantities that would otherwise be difficult to find.

## 29-1 RATE OF CHANGE

### Average Rate of Change

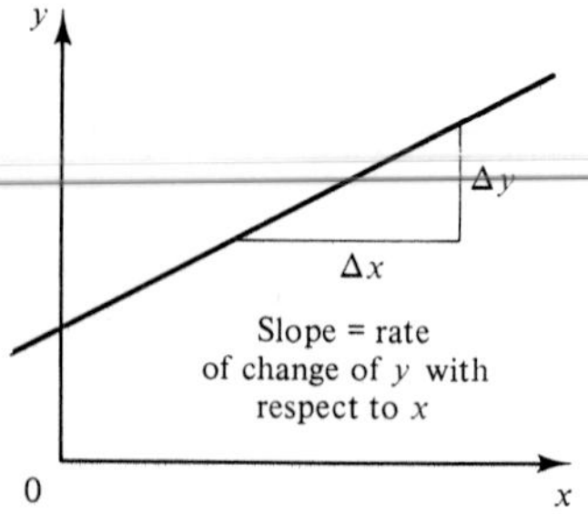

**FIGURE 29-1 Uniform change.**

When a function changes at a uniform rate, its graph is a straight line, Fig. 29-1. Since the slope of the line is constant, any change $\Delta x$ produces the same change $\Delta y$, anywhere on the line. We define the rate of change of this function as the change in $y$ divided by the change in $x$, over any interval. But $\Delta y/\Delta x$ is also the slope of the line, so

$$\text{rate of change} = \frac{\text{change in } y}{\text{change in } x}$$

$$= \frac{\Delta y}{\Delta x}$$

$$= \text{slope of line}$$

---

**EXAMPLE 1:** Find the rate of change of the function $y = 3x - 2$.

**Solution:** This equation represents a straight line with a slope of 3, so

$$\text{rate of change} = 3$$

---

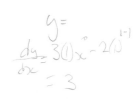

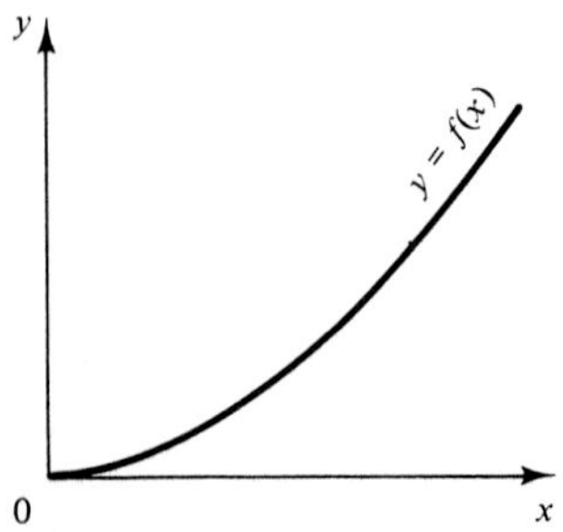

**FIGURE 29-2 Nonuniform change.**

But most real quantities do not change at a constant rate. Their graph is no longer a straight line, but is some curve, as in Fig. 29-2, where the slope changes as $x$ changes.

What can we say about rate of change when a quantity is changing nonuniformly? Two things. We can give either the *average rate of change* over some interval or the *instantaneous rate of change* at some point.

We'll define *average rate of change* here, and the instantaneous rate of change in the following section.

The average rate of change over some interval is defined as the change in $y$ divided by the change in $x$, over that interval

$$\text{average rate of change} = \frac{\text{change in } y}{\text{change in } x} = \frac{\Delta y}{\Delta x} = \text{slope of } PQ$$

For the interval $PQ$ (Fig. 29-3) the average rate of change is the slope of the line connecting $P$ and $Q$. If $x$ were time and $y$ were distance, the average rate of change would be called the *average velocity* over that interval.

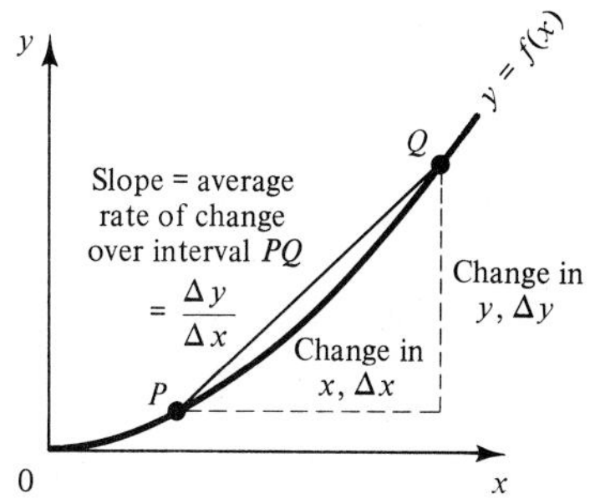


**FIGURE 29-3 Average rate of change.**

---

**EXAMPLE 2:** Find the average rate of change of the function $y = 3x^2 + 1$ over the interval $x = 1$ to $x = 3$.

**Solution:** We compute $y$ at $x = 1$ and $x = 3$,

$$y(1) = 3(1)^2 + 1 = 4$$

and

$$y(3) = 3(3)^2 + 1 = 28$$

The endpoints of the interval are thus $P(1, 4)$ and $Q(3, 28)$. Then

$$\text{average rate of change} = \text{slope of } PQ = \frac{\Delta y}{\Delta x} = \frac{28 - 4}{3 - 1} = 12$$

---

**EXAMPLE 3:** During one day of a certain cross-country journey, a motorist travels 500 mi in 10 h. The average speed is

$$\frac{500 \text{ mi}}{10 \text{ h}} = 50 \text{ mi/h}$$

even though part of the time may have been spent going faster than 50 mi/h, or slower, or even stopped or backing up.

---

## Instantaneous Rate of Change

As the name implies, the instantaneous rate of change of some quantity is the rate of change at a given instant. In the preceding example of the motorist, the instantaneous speed would be given by the speedometer reading at the instant in question (except when backing up).

For some function $y = f(x)$ (Fig. 29-4), we define the instantaneous rate of change at $P$ as the limit of the average rate of change over the interval $PQ$, as the width $\Delta x$ of that interval approaches zero. In symbols,

$$\text{instantaneous rate of change} = \lim_{\Delta x \to 0} \frac{\Delta y}{\Delta x}$$

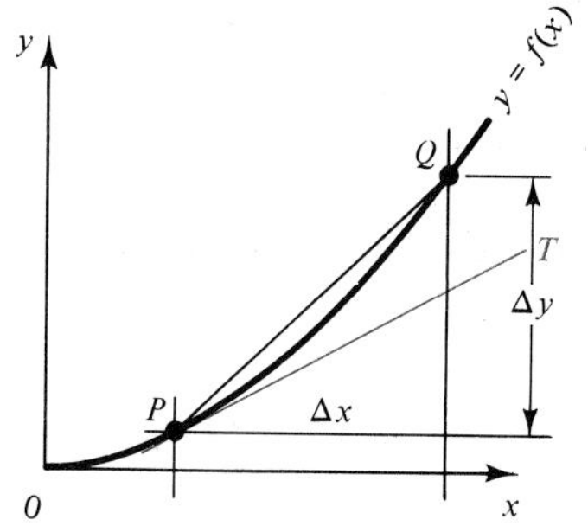


**FIGURE 29-4**

But this expression is none other than our definition of the derivative (Eq. 328), which also gives us the slope of the tangent $PT$. Thus

$$\text{instantaneous rate of change} = \lim_{\Delta x \to 0} \frac{\Delta y}{\Delta x} = \text{slope of } PT = \frac{dy}{dx}$$

*The instantaneous rate of change of a function is given by the derivative of that function.*

**EXAMPLE 4:** Find the instantaneous rate of change of the function $y = 3x^2 + 5$ when $x = 2$.

**Solution:** Taking the derivative,

$$y' = 6x$$

When $x = 2$,

$$y'(2) = 6(2) = 12 = \text{instantaneous rate of change}$$

Of course, rates of change can involve variables other than $x$ and $y$. When two or more related quantities are changing, we often speak about the *rate of change* of one quantity with respect to one of the other quantities. For example, if a steel rod is placed in a furnace, its temperature and its length both increase. Since the length varies with the temperature, we can speak about the *rate of change of length with respect to temperature.* But the length of the bar is also varying with time, so we can speak about the *rate of change of length with respect to time.* Time rates are the most common rates of change with which we have to deal.

**EXAMPLE 5:** The temperature $T$ (°F) in a certain furnace varies with time $t$(s) according to the function $T = 4.85t^3 + 2.96t$. Find the rate of change of temperature at $t = 3.75$ s.

**Solution:** Taking the derivative of $T$ with respect to $t$,

$$\frac{dT}{dt} = 14.55t^2 + 2.96 \qquad \text{°F/s}$$

At $t = 3.75$ s,

$$\frac{dT}{dt} = 14.55(3.75)^2 + 2.96 = 208\text{°F/s}$$

## Electric Current

**Students not specializing in electricity may wish to skip this section.**

The idea of rate of change finds many applications in electrical technology, a few of which we cover here. The *coulomb* (C) is the unit of electrical *charge*. The *current* in amperes (A) is the number of coulombs passing a point in a circuit in 1 second. If the current varies with time, then the *instantaneous current* is given by

| | | |
|---|---|---|
| Current | $i = \dfrac{dq}{dt}$ | **A79** |

*Current is the rate of change of charge.*

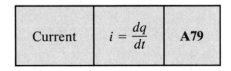

**EXAMPLE 6:** The charge through a 2.85-Ω resistor is given by

$$q = 1.08t^3 - 3.82t \text{ coulomb}$$

Write an expression for (a) the instantaneous current through the resistor, (b) the instantaneous voltage across the resistor, and (c) the instantaneous power in the resistor. (d) Evaluate each at 2.00 s.

**Solution:** (a) $i = dq/dt = 3.24t^2 - 3.82 \quad \text{A}$

(b) By Ohm's law,

$$v = Ri = 2.85(3.24t^2 - 3.82)$$
$$= 9.23t^2 - 10.9 \quad \text{V}$$

(c) Since $P = vi$,

$$P = (9.23t^2 - 10.9)(3.24t^2 - 3.82)$$
$$= 29.9t^4 - 70.5t^2 + 41.6 \quad \text{W}$$

(d) At $t = 2.00$ s,

$$i = 3.24(4.00) - 3.82 = 9.14 \text{ A}$$
$$v = 9.23(4.00) - 10.9 = 26.0 \text{ V}$$
$$P = 29.9(16.0) - 70.5(4.00) + 41.6 = 238 \text{ W}$$

## Current in a Capacitor

If a steady voltage is applied across a capacitor, no current will flow into the capacitor (after the initial transient currents have died down). But if the applied voltage varies with time, the instantaneous current $i$ to the capacitor will be proportional to the rate of change of the voltage. The constant of proportionality is called the *capacitance C*.

| | | |
|---|---|---|
| Current in a Capacitor | $i = C\dfrac{dv}{dt}$ | **A81** |

*The current to a capacitor equals the capacitance times the rate of change of the voltage.*

**The units here are volts for *v*, seconds for *t*, farads for C, and amperes for *i*.**

**EXAMPLE 7:** The voltage applied to a 2.85-microfarad (μF) capacitor is $v = 1.47t^2 + 48.3t - 38.2$ V. Find the current at $t = 2.50$ s.

**Solution:** The derivative of the voltage equation is $dv/dt = 2.94t + 48.3$. Then from Eq. A81,

**A microfarad equals $10^{-6}$ farad.**

$$i = C\frac{dv}{dt} = (2.85 \times 10^{-6})(2.94t + 48.3) \quad \text{A}$$

At $t = 2.50$ s,

$$i = (2.85 \times 10^{-6})[2.94(2.50) + 48.3] = 159 \times 10^{-6} \text{ A}$$
$$= 0.159 \text{ mA}$$

## Voltage across an Inductor

If the current through an inductor (such as a coil of wire) is steady, there will be no voltage drop across the inductor. But if the current varies, a voltage will be induced that is proportional to the rate of change of the current. The constant of proportionality $L$ is called the *inductance* and is measured in henrys (H).

| | | |
|---|---|---|
| Voltage across an Inductor | $v = L\dfrac{di}{dt}$ V | **A86** |

*The voltage across an inductor equals the inductance times the rate of change of the current.*

---

**EXAMPLE 8:** The current in a 8.75-H inductor is given by

$$i = \sqrt{t^2 + 5.83t}$$

Find the voltage across the inductor at $t = 5.00$ s.

**Solution:** By Eq. A86,

$$v = 8.75\frac{di}{dt}$$

$$= 8.75\left(\frac{1}{2}\right)(t^2 + 5.83t)^{-1/2}(2t + 5.83)$$

At $t = 5.00$ s,

$$v = 8.75\left(\frac{1}{2}\right)[25.0 + 5.83(5.00)]^{-1/2}(10.0 + 5.83)$$

$$= 9.41 \text{ V}$$

---

### EXERCISE 1—RATE OF CHANGE

Use Boyle's law, $pv = k$.

1. The air in a certain cylinder is at a pressure of 25.5 lb/in² when its volume is 146 in³. Find the rate of change of the pressure with respect to volume as the piston descends farther.

Use the inverse square law, $I = k/d^2$.

2. A certain light source produces an illumination of 655 lux on a surface at a distance of 2.75 m. Find the rate of change of illumination with respect to distance, and evaluate it at 2.75 m.
3. A spherical balloon starts to shrink as the gas escapes. Find the rate of change of its volume with respect to its radius when the radius is 1.00 m.

Use Eq. A67, $P = I^2R$.

4. The power dissipated in a certain resistor is 865 W at a current of 2.48 A. What is the rate of change of the power with respect to the current as the current starts to increase?
5. The period (in seconds) for a pendulum of length $L$ inches to complete one oscillation is equal to $P = 0.324\sqrt{L}$. Find the rate of change of the period with respect to length when the length is 9.00 in.

The rate of change $dT/dx$ is called the *temperature gradient.*

6. The temperature $T$ at a distance $x$ inches from the end of a certain heated bar is given by $T = 2.24x^3 + 1.85x + 95.4$ (°F). Find the rate of change of temperature with respect to distance at a point 3.75 in. from the end.

7. The cantilever beam (Fig. 29-5) has a deflection $y$ at a distance $x$ from the built-in end,

$$y = \frac{wx^2}{24EI}(x^2 + 6L^2 - 4Lx)$$

where $E$ is the modulus of elasticity and $I$ is the moment of inertia. Write an expression for the rate of change of deflection with respect to the distance $x$. Regard $E$, $I$, $w$, and $L$ as constants.

**The shape taken by the axis of a bent beam is called the *elastic curve*.**

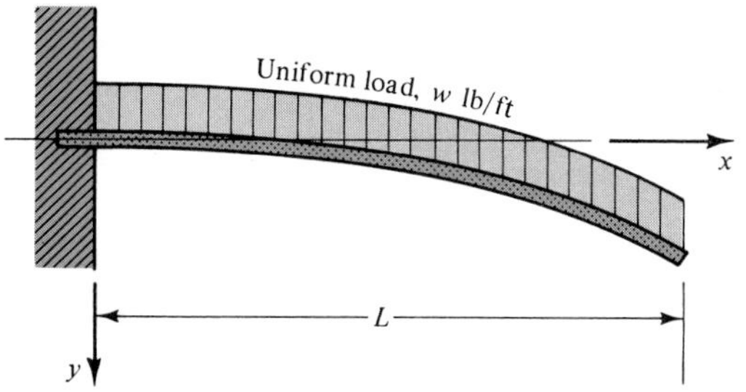


**FIGURE 29-5 Cantilever beam with uniform load.**

8. The equation of the elastic curve for the beam of Fig. 29-6 is

$$y = \frac{wx}{24EI}(L^3 - 2Lx^2 + x^3)$$

Write an expression for the rate of change of deflection (the slope) of the elastic curve at $x = L/4$. Regard $E$, $I$, $w$, and $L$ as constants.

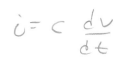

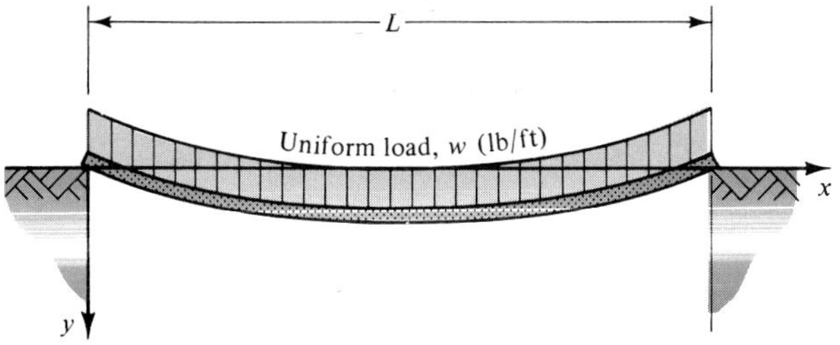


**FIGURE 29-6 Simply supported beam with uniform load.**

9. The charge $q$ (in coulombs) through a 4.82-Ω resistor varies with time according to the function $q = 3.48t^2 - 1.64t$. Write an expression for the instantaneous current through the resistor.
10. Evaluate the current in Problem 9 at $t = 5.92$ s.
11. Find the voltage across the resistor of Problem 9. Evaluate it at $t = 1.75$ s.
12. Find the instantaneous power in the resistor of Problem 9. Evaluate it at $t = 4.88$ s.
13. The charge $q$ (in coulombs) at a resistor varies with time according to the function $q = 22.4t + 41.6t^3$. Write an expression for the instantaneous current through the resistor and evaluate it at 2.50 s.
14. The voltage applied to a 33.5-μF capacitor is $v = 6.27t^2 - 15.3t + 52.2$ V. Find the current at $t = 5.50$ s.
15. The voltage applied to a 1.25-μF capacitor is $v = 3.17 + 28.3t + 29.4t^2$ V. Find the current at $t = 33.2$ s.
16. The current in a 1.44-H inductor is given by $i = 5.22t^2 - 4.02t$. Find the voltage across the inductor at $t = 2.00$ s.
17. The current in a 8.75-H inductor is given by $i = 8.22 + 5.83t^3$. Find the voltage across the inductor at $t = 25.0$ s.

## 29-2 MOTION OF A POINT

### Velocity in Straight-Line Motion

Let us first consider straight-line, or rectilinear motion, and later we will study curvilinear motion. We first distinguish between speed and velocity. As an object moves along some path, the distance traveled *along the path* per unit time is called the *speed*. No account is taken of any change in direction; hence speed is a *scalar* quantity.

*Velocity,* on the other hand, is a *vector* quantity, having both *magnitude* and *direction*. For an object moving along a curved path, the magnitude of the velocity along the path is equal to the speed, and the *direction* of the velocity is the same as that of the *tangent to the curve* at that point. We will also speak of the components of that velocity in directions other than along the path, usually in the $x$ and $y$ directions.

As with average and instantaneous rates of change, we also can have average speed and average velocity, or instantaneous speed or instantaneous velocity. These terms have the same meaning as in Sec. 29-1.

Velocity is the rate of change of displacement, and hence *is given by the derivative* of the displacement. If we give displacement the symbol $s$, then

| | | |
|---|---|---|
| Instantaneous Velocity | $v = \dfrac{ds}{dt}$ | **A23** |

*The velocity is the rate of change of the displacement.*

---

**EXAMPLE 9:** The displacement of an object is given by $s = 2t^2 + 5t + 4$ (in.), where $t$ is the time in seconds. Find the velocity at 1 s.

**Solution:** We take the derivative

$$v = \frac{ds}{dt} = 4t + 5$$

At $t = 1$ s,

$$v(1) = \left.\frac{ds}{dt}\right|_{t=1} = 4(1) + 5 = 9 \text{ in./s}$$

---

### Acceleration in Straight-Line Motion

The acceleration is defined as the time rate of change of velocity. It is also a vector quantity. Since the velocity is itself the derivative of the displacement, the acceleration is the derivative of the derivative of the displacement, or the *second derivative* of displacement, with respect to time. We write the second derivative of $s$ with respect to $t$ as

$$\frac{d^2s}{dt^2}$$

| | | |
|---|---|---|
| Instantaneous Acceleration | $a = \frac{dv}{dt} = \frac{d^2s}{dt^2}$ | **A25** |

*The acceleration is the rate of change of the velocity.*

**EXAMPLE 10:** One point in a certain mechanism moves according to the equation $s = 3t^3 + 5t - 3$ (cm), where $t$ is in seconds. Find the instantaneous velocity and acceleration at $t = 2$ s.

**Solution:** We take the derivative twice, with respect to $t$.

$$v = \frac{ds}{dt} = 9t^2 + 5$$

and

$$a = \frac{dv}{dt} = 18t$$

At $t = 2$ s,

$$v(2) = 9(2)^2 + 5 = 41 \text{ cm/s}$$

and

$$a(2) = 18(2) = 36 \text{ cm/s}^2$$

**Later we will do the reverse of differentiation, called *integration,* to find the velocity from the acceleration, and then the displacement from the velocity.**

## Velocity in Curvilinear Motion

At any instant we may think of a point as moving in a direction *tangent* to the path, as in Fig. 29-7. Thus if the speed is known and the direction of the tangent can be found, the instantaneous velocity (a vector having both magnitude and direction) can be found.

A more useful way of giving the instantaneous velocity, however, is by its $x$ and $y$ components (Fig. 29-8). If the magnitude and direction of the velocity are known, the components can be found by resolving the velocity vector into its $x$ and $y$ components.

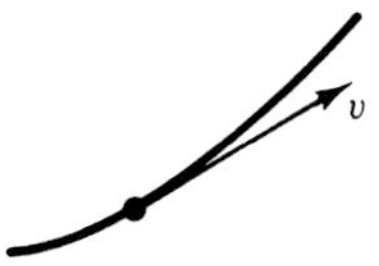


**FIGURE 29-7**

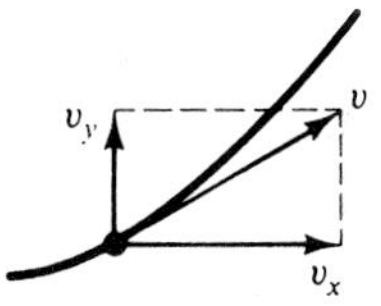


**FIGURE 29-8** ***x* and *y* components of velocity.**

**EXAMPLE 11:** A point moves along the curve $y = 2x^3 - 5x^2 - 1$ (cm). (a) Find the direction of travel at $x = 2.00$ cm. (b) If the speed of the point along the curve is 3.00 cm/s, find the $x$ and $y$ components of the velocity when $x = 2.00$ cm.

**Solution:** (a) Taking the derivative of the given function gives

$$dy/dx = 6x^2 - 10x$$

When $x = 2.00$,

$$y'(2) = 6(2)^2 - 10(2) = 4$$

The slope of the curve at that point is thus 4, and the direction of travel is $\tan^{-1} 4 = 76.0°$.

(b) Resolving the velocity vector into $x$ and $y$ components gives us

$$v_x = 3.00 \cos 76.0° = 0.726 \text{ cm/s}$$

$$v_y = 3.00 \sin 76.0° = 2.91 \text{ cm/s}$$

## Displacement Given by Parametric Equations

If the $x$ displacement and $y$ displacement are each given by a separate function of time (parametric equations) we may find the $x$ and $y$ components directly by taking the derivative of each equation.

| | | | |
|---|---|---|---|
| $x$ and $y$ Components of Velocity | (a) $v_x = \frac{dx}{dt}$ | (b) $v_y = \frac{dy}{dt}$ | **A33** |

Once we have expressions for the $x$ and $y$ components of velocity, we simply have to take the derivative again to get the $x$ and $y$ components of acceleration.

| | | | |
|---|---|---|---|
| $x$ and $y$ Components of Acceleration | (a) $a_x = \frac{dv_x}{dt} = \frac{d^2x}{dt^2}$ | (b) $a_y = \frac{dv_y}{dt} = \frac{d^2y}{dt^2}$ | **A35** |

**EXAMPLE 12:** A point moves along a curve such that its horizontal displacement is

$$x = 2t^3 - 15t \quad \text{cm}$$

and its vertical displacement is

$$y = 3 + t^2 \quad \text{cm}$$

Find (a) the horizontal and vertical components of the instantaneous velocity at $t = 2.00$ s, (b) the magnitude and direction of the instantaneous velocity at $t = 2.00$ s, (c) the $x$ and $y$ components of the instantaneous acceleration at $t = 2.00$ s, and (d) the magnitude and direction of the instantaneous acceleration at the same instant.

**Solution:** (a) Taking derivatives, we get

$$v_x = 6t^2 - 15 \quad \text{and} \quad v_y = 2t$$

At $t = 2.00$ s,

$$v_x = 6(2.00)^2 - 15 = 9.00 \text{ cm/s}$$

and

$$v_y = 2(2.00) = 4.00 \text{ cm/s}$$

(b) The $x$ and $y$ components of the velocity are shown in Fig. 29-9. We find the resultant by vector addition.

$$v = \sqrt{v_x^2 + v_y^2} = \sqrt{81.0 + 16.0} = 9.85 \text{ cm/s}$$

Now finding the angle, we have

$$\tan \theta = \frac{v_y}{v_x} = \frac{4.00}{9.00} = 0.444$$

$$\theta = 24.0°$$

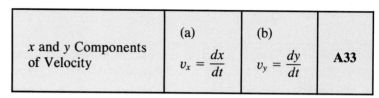

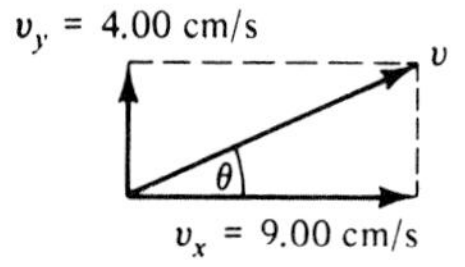

FIGURE 29-9

(c) Again taking derivatives gives us

$$a_x = \frac{dv_x}{dt} = 12t \quad \text{and} \quad a_y = \frac{dv_y}{dt} = 2$$

At $t = 2.00$ s,

$$a_x = 12(2.00) = 24.0 \text{ cm/s}^2 \quad \text{and} \quad a_y = 2.00 \text{ cm/s}^2.$$

(d) Finding the resultant (Fig. 29-10) gives us

$$a = \sqrt{(24.0)^2 + (2.00)^2} = 24.1 \text{cm/s}^2$$

and the angle

$$\tan \phi = \frac{a_y}{a_x} = \frac{2.00}{24.0} = 0.0833$$

$$\phi = 4.76°$$

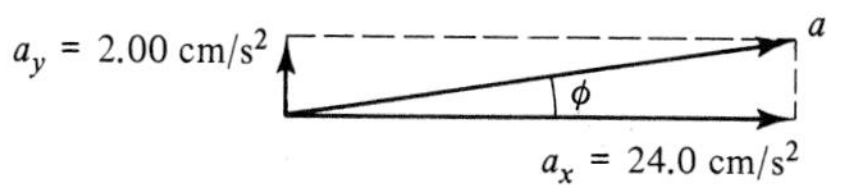


**FIGURE 29-10** ***x* and *y* components of acceleration.**

**Note that the direction of the acceleration is *different* from that of the velocity. The velocity is always in a direction tangent to the path, while the acceleration vector is turned more toward the inside of the curve.**

If the *equation of the path* that the point follows is known, the derivative of that equation will give the slope of the tangent to the curve, and hence the direction of the velocity vector.

**EXAMPLE 13:** A point moves along the curve $y = x^2 + 2$ so that its horizontal displacement is $x = 3t^2 - 2t$ (in.). Find the magnitude and direction of the velocity when $t = 1.50$ s.

**Solution:** The velocity in the $x$ direction is

$$v_x = \frac{dx}{dt} = 6t - 2 \quad \text{in./s}$$

and at $t = 1.50$ s, the horizontal velocity is

$$v_x = 6(1.50) - 2 = 7.00 \text{ in./s}$$

and the horizontal displacement is

$$x = 3(1.50)^2 - 2(1.50) = 3.75 \text{ in.}$$

We now find the slope of the curve:

$$\text{slope} = \frac{dy}{dx} = 2x$$

and at $t = 1.50$, $x = 3.75$ in., so

$$\text{slope} = 2(3.75) = 7.50$$

Thus the angle $\theta$ between $v_x$ and $v$ (Fig. 29-11) is

$$\tan \theta = 7.50$$

$$\theta = 82.4°$$

The magnitude of the velocity vector is then

$$v = \frac{7.00}{\cos 82.4°} = 52.9 \text{ in./s}$$

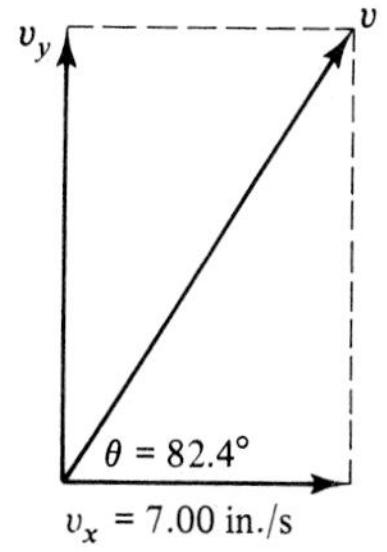


**FIGURE 29-11**

A *trajectory* is the path followed by a projectile, such as a ball thrown into the air. Trajectories are usually described by parametric equations, with the horizontal and vertical motions considered separately.

If air resistance is neglected, a projectile will move horizontally with constant velocity and fall with constant acceleration, just like any other freely falling body. Thus if the initial velocities are $v_{0x}$ and $v_{0y}$ in the horizontal and vertical directions, the parametric equations of motion are

$$x = v_{0x}t \qquad \text{and} \qquad y = v_{0y}t - \frac{gt^2}{2}$$

---

**EXAMPLE 14:** A projectile launched with initial velocity $v_0$ at an angle $\theta_0$ (Fig. 29-12) has horizontal and vertical displacements of

$$x = (v_0 \cos \theta_0)t \qquad \text{and} \qquad y = (v_0 \sin \theta_0)t - \frac{gt^2}{2}$$

Find the horizontal and vertical velocities and accelerations.

**FIGURE 29-12** **A trajectory.**

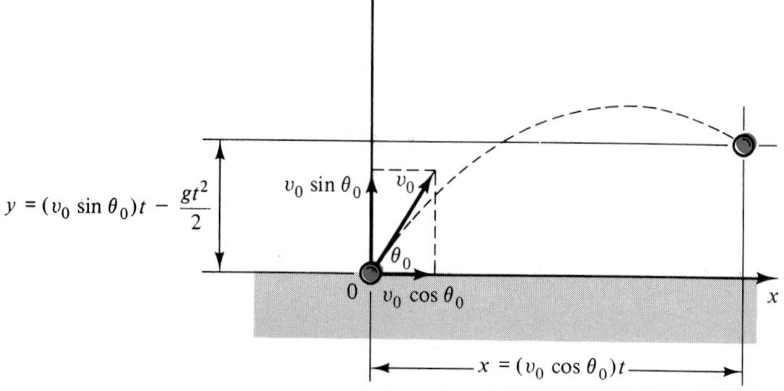


**Solution:** We take derivatives, remembering that $\theta_0$ is a constant.

$$v_x = \frac{dx}{dt} = v_0 \cos \theta_0 \qquad \text{and} \qquad v_y = \frac{dy}{dt} = v_0 \sin \theta_0 - gt$$

We take derivatives again to get the accelerations.

$$a_x = \frac{dv_x}{dt} = 0 \qquad \text{and} \qquad a_y = \frac{dv_y}{dt} = -g$$

As expected, we get a horizontal motion with constant velocity, and a vertical motion with constant acceleration.

---

**EXAMPLE 15:** A projectile is launched at an angle of 62° to the horizontal with an initial velocity of 5850 ft/s. Find (a) the horizontal and vertical position of the projectile and (b) the horizontal and vertical velocity, after 5 s.

**Solution:**

(a) We first resolve the initial velocity into horizontal and vertical components.

$$v_{0x} = v_0 \cos \theta = 5850 \cos 62° = 2746 \text{ ft/s}$$

$$v_{0y} = v_0 \sin \theta = 5850 \sin 62° = 5165 \text{ ft/s}$$

The horizontal and vertical displacements are then

$$x = v_{0x}t = 2746t \qquad \text{ft}$$

and

$$y = v_{0y}t - \frac{gt^2}{2} = 5165t - \frac{gt^2}{2} \qquad \text{ft}$$

At $t = 5$ s, and with $g = 32$ ft/s²,

$$x = 2746(5) = 13{,}730 \text{ ft}$$

and

$$y = 5165(5) - \frac{32(5^2)}{2} = 25{,}425 \text{ ft}$$

(b) The horizontal velocity is constant, so

$$v_x = v_{0x} = 2746 \text{ ft/s}$$

and the vertical velocity is

$$v_y = v_{0y} - gt$$

$$= 5165 - 32(5) = 5005 \text{ ft/s}$$

---

## Rotation

We saw for straight-line motion that velocity is the instantaneous rate of change of displacement. Similarly, for rotation, the *angular velocity*, $\omega$, is the instantaneous rate of change of *angular displacement*, $\theta$.

| | | |
|---|---|---|
| Angular Velocity | $\omega = \frac{d\theta}{dt}$ | **A29** |

*The angular velocity is the instantaneous rate of change of the angular displacement with respect to time.*

Similarly for acceleration,

| | | |
|---|---|---|
| Angular Acceleration | $\alpha = \frac{d\omega}{dt} = \frac{d^2\theta}{dt^2}$ | **A31** |

*The angular acceleration is the instantaneous rate of change of the angular velocity with respect to time.*

**EXAMPLE 16:** The angular displacement of a rotating body is given by $\theta = 1.75t^3 + 2.88t^2 + 4.88$ rad. Find (a) the angular velocity and (b) the angular acceleration, at $t = 2.00$ s.

**Solution:**

(a) From Eq. A29,

$$\omega = \frac{d\theta}{dt} = 5.25t^2 + 5.76t$$

At 2.00 s, $\omega = 5.25(4.00) + 5.76(2.00) = 32.5$ rad/s.

(b) From Eq. A31,

$$\alpha = \frac{d\omega}{dt} = 10.5t + 5.76$$

At 2.00 s, $\alpha = 10.5(2.00) + 5.76 = 26.8$ rad/s$^2$.

## EXERCISE 2—MOTION OF A POINT

### Straight-Line Motion

Find the instantaneous velocity and acceleration at the given time for the straight-line motion described by each equation, where $s$ is in centimeters and $t$ is in seconds.

1. $s = 32t - 8t^2$ at $t = 2$
2. $s = 6t^2 - 2t^3$ at $t = 1$
3. $s = t^2 + t^{-1} + 3$ at $t = \frac{1}{2}$
4. $s = (t + 1)^4 - 3(t + 1)^3$ at $t = -1$
5. $s = 120t - 16t^2$ at $t = 4$
6. $s = 3t - t^4 - 8$ at $t = 1$
7. The distance in feet traveled in time $t$ seconds by a point moving in a straight line is given by the formula $s = 40t + 16t^2$. Find the velocity and the acceleration at the end of 2.0 s.
8. A car moves according to the equation $s = 250t^2 - \frac{5}{4}t^4$, where $t$ is measured in minutes and $s$ in feet. **(a)** How far does the car go in the first 10 min? **(b)** What is the maximum speed? **(c)** How far has the car moved when its maximum speed is reached?
9. If the distance traveled by a ball rolling down an incline in $t$ seconds is $s$ feet, where $s = 6t^2$, find its speed when $t = 5$ s.
10. The height $s$ in feet reached by a ball $t$ seconds after being thrown vertically upward at 320 ft/s is given by $s = 320t - 16t^2$. Find the greatest height reached by the ball, and the velocity with which it reaches the ground.
11. A bullet was fired straight upward so that its height in feet after $t$ seconds was $s = 2000t - 16t^2$. What was its initial velocity? What was its greatest height? What was its velocity at the end of 10 s?
12. If the height $h$ kilometers to which a balloon will rise in $t$ minutes is given by the formula

$$h = \frac{10t}{\sqrt{4000 + t^2}}$$

at what rate is the balloon rising at the end of 30 min?

13. If the equation of motion of a point is $s = 16t^2 - 64t + 64$, find the position and acceleration at which the point first comes to rest.

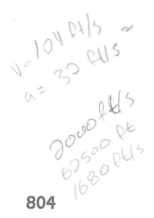

### Motion along a Curve

14. A point moves along the curve $y = 2x^3 - 3x^2 - 3$ with a speed of 14.00 in./s. Find the direction of travel at $x = 1.50$ in. and the $x$ and $y$ components of the velocity.
15. A point has horizontal and vertical displacements (in cm) of $x = 3t^2 + 5t$ and $y = 13 - 3t^2$, respectively. Find the $x$ and $y$ components of the velocity and acceleration at $t = 4.55$ s.
16. A point moves along a curve having the parametric equations $x = 2t$, $y = 2t^2 - 4$. Find the coordinates and the velocity of the point when $t = 1$ s.
17. A point moves on the curve $xy = 2$, such that $x = 2t^2$, where $x$ and $y$ are in feet and $t$ in seconds. Find the magnitude and direction of the velocity of the point when $t = 2$.

8 ft/s
−1.79°

18. A point moves along the curve $y^2 = 8x$ ($x$ and $y$ in feet) in such a way that its $y$ coordinate is $y = t^2 - 4t$, where $y$ is in feet and $t$ in seconds. Find its velocity when $t = 5$ s.
19. A point moves on the path $xy = 8$ such that its abscissa is $x = 4t^2$. Find the $x$ and $y$ components of the velocity and of the acceleration when $t = 2$, and find the total acceleration.
20. A point moves on the curve $y^2 = x$ such that its ordinate is $y = 4t$. Find the magnitude and direction of the acceleration of the point when $t = 1$ s.
21. A point moves on a parabola $y = x^2$ so that its abscissa changes at the constant rate of 2 units/s. How fast is its ordinate changing when the point passes through (3, 9)?
22. A point moves along the parabola $y^2 = 12x$. At what point do the abscissa and ordinate increase at the same rate?
23. A point moves along the parabola $y^2 = 8x$. What is the position of the point when $v_x = v_y$?

rate = 1st deriv with respect to time

### Rotation

24. The angular displacement of a rotating body is given by $\theta = 44.8t^3 + 29.3t^2 + 81.5$ rad. Find the angular velocity at $t = 4.25$ s.
25. Find the angular acceleration in Problem 24 at $t = 22.4$ s.
26. The angular displacement of a rotating body is given by $\theta = 184 + 271t^3$ rad. Find **(a)** the angular velocity and **(b)** the angular acceleration, at $t = 1.25$ s.

## 29-3 RELATED RATES

In *related rate* problems, there are *two* quantities changing with time. The rate of change of one of the quantities is given and the other must be found. A procedure that can be followed is:

1. Locate the *given rate*. Since it is a rate, it can be *expressed as a derivative* with respect to time.
2. Determine the *unknown* rate. Express it also as a derivative with respect to time.
3. Find an *equation* linking the variable in the given rate with that in the unknown rate. If there are other variables in the equation they must be eliminated by means of other relationships.
4. Take the derivative of the equation *with respect to time.*
5. Substitute the given values and solve for the unknown rate.

**You would, of course, study the problem statement and make a diagram, as you would for other word problems.**

**EXAMPLE 17:** A 20-ft ladder (Fig. 29-13) leans against a building. The foot of the ladder is pulled away from the building at a rate of 2.00 ft/s. How fast is the top of the ladder falling when the foot is 10.0 ft from the building?

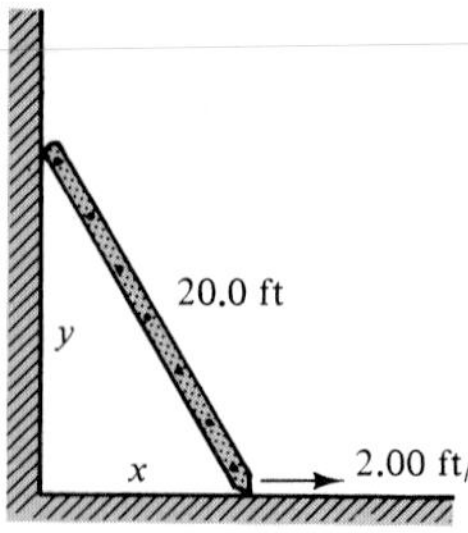


**FIGURE 29-13 The ladder problem.**

**Solution:**

1. If we let $x$ be the distance from the foot of the ladder to the building, we have given

$$\frac{dx}{dt} = 2.00 \text{ ft/s}$$

2. If $y$ is the distance from the ground to the top of the ladder, we are looking for $dy/dt$.
3. The equation linking $x$ and $y$ is the Pythagorean theorem.

$$x^2 + y^2 = 20.0^2$$

**It will often be easiest to take the derivative implicitly rather than first to solve for one of the variables.**

4. We could solve for $y$ before taking the derivative, or do it implicitly as follows:

$$2x\frac{dx}{dt} + 2y\frac{dy}{dt} = 0$$

5. When $x = 10.0$ ft, the height of the top of the ladder is

$$y = \sqrt{400 - (10.0)^2} = 17.3 \text{ ft}$$

We now substitute 2.00 for $dx/dt$, 10.0 for $x$, and 17.3 ft for $y$.

$$2(10.0)(2.00) + 2(17.3)\frac{dy}{dt} = 0$$

$$\frac{dy}{dt} = -1.16 \text{ ft/s}$$

The negative sign indicates that $y$ is decreasing.

**Common Errors**

In step 4 above, when taking the derivative of $x^2$, it is tempting to take the derivative with respect to $x$, rather than $t$. Also, don't forget the $dx/dt$ in the derivative.

$$\frac{d}{dt}x^2 = 2x\boxed{\frac{dx}{dt}}$$

don't forget!

Students often substitute the given values too soon. For example, if we had substituted $x = 10.0$ and $y = 17.3$ before taking the derivative, we would get

$$(10.0)^2 + (17.3)^2 = (20.0)^2$$

Taking the derivative now gives us

$$0 = 0!$$

Do not substitute the given values until *after* you have taken the derivative.

In the next example when we find an equation linking the variables, it is seen to contain *three* variables. In such a case we need a *second equation* with which to eliminate one variable.

---

**EXAMPLE 18:** A conical tank with vertex down has a base radius of 3.00 m and a height of 6.00 m (Fig. 29-14). Water flows in at a rate of 2.00 m³/h. How fast is the water level rising when the depth is 3.00 m?

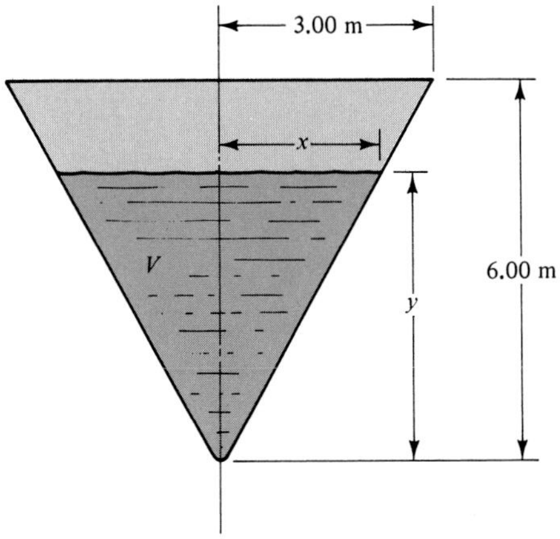


**FIGURE 29-14 Conical tank.**

**Solution:** We let $y$ equal the depth, $x$ the base radius, and $V$ the volume of the water when the tank is partially filled. Then:

1. Given: $dV/dt = 2.00\ \text{m}^3/\text{h}$.
2. Unknown: $dy/dt$ when $y = 3.00$ m.
3. The equation linking $V$ and $y$ is that for the volume of a cone, $V = (\pi/3)x^2y$. But in addition to the two variables in our derivatives, $V$ and $y$, we have the third variable, $x$. We must eliminate $x$ by means of another equation. By similar triangles,

$$\frac{x}{y} = \frac{3}{6}$$

from which $x = \frac{y}{2}$. Substituting yields

$$V = \frac{\pi}{3} \cdot \frac{y^2}{4} \cdot y = \frac{\pi}{12} y^3$$

4. Now we have $V$ as a function of $y$ only. Taking the derivative gives us

$$\frac{dV}{dt} = \frac{\pi}{4} y^2 \frac{dy}{dt}$$

5. Substituting 3.00 for $y$ and 2.00 for $dV/dt$, we obtain

$$2.00 = \frac{\pi}{4}(9.00)\frac{dy}{dt}$$

$$\frac{dy}{dt} = \frac{8.00}{9.00\pi} = 0.283\ \text{m/h}$$

---

| Common Error | You cannot write an equation linking the variables until you have *defined* those variables. Indicate them right on your diagram. Draw axes if needed. |
|---|---|

Some problems have two *independently* moving objects, as in the following example.

---

**EXAMPLE 19:** Ship $A$ leaves a port $P$ and travels west at 11.5 mi/h. After 2.25 h, ship $B$ leaves $P$ and travels north at 19.4 mi/h. How fast are the ships separating 5.00 h after the departure of $A$?

**Solution:** Figure 29-15 shows the ships $t$ hours after $A$ has left. Ship $A$ has gone $11.5t$ miles and $B$ has gone $19.4(t - 2.25)$ miles. The distance $S$ between them is given by

$$S^2 = (11.5t)^2 + [19.4(t - 2.25)]^2$$
$$= 132t^2 + 376(t - 2.25)^2$$

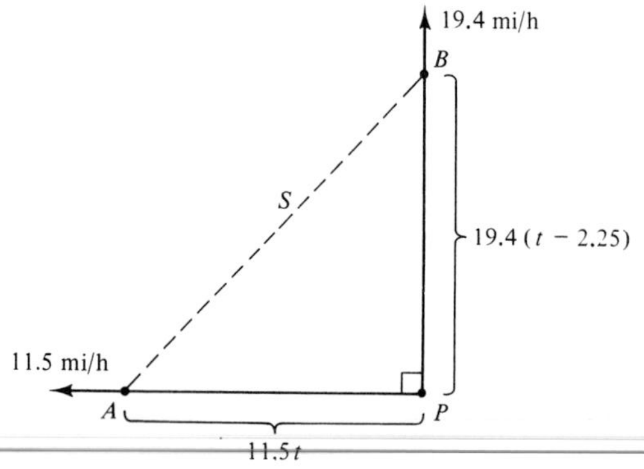


**FIGURE 29-15** **Note that we give the position of each ship after $t$ hours have elapsed.**

Taking the derivative, we have

$$2S\frac{dS}{dt} = 2(132)t + 2(376)(t - 2.25)$$

$$\frac{dS}{dt} = \frac{132t + 376t - 846}{S} = \frac{508t - 846}{S}$$

At $t = 5.00$ h, $A$ has gone $11.5(5.00) = 57.5$ mi, and $B$ has gone $19.4(5.00 - 2.25) = 53.4$ mi. The distance $S$ between them is then

$$S = \sqrt{(57.5)^2 + (53.4)^2} = 78.4 \text{ mi}$$

Substituting gives

$$\frac{dS}{dt} = \frac{508(5.00) - 846}{78.4} = 21.6 \text{ mi/h}$$

---

| Common Error | Be sure to distinguish which quantities in a problem are *constants* and which are *variables*. Represent each variable by a letter, and do not substitute a given numerical value for a variable *until the very last step*. |
|---|---|

15. A lamp is located on the ground 30.0 ft from a building. A person 6.00 ft tall walks from the light toward the building at a rate of 5.00 ft/s. Find the rate at which the person's shadow on the wall is shortening when the person is 15.0 ft from the building.

16. A ball (Fig. 29-20), dropped from a height of 100 ft, is at height $s$ at $t$ seconds, where $s = 100 - gt^2/2$ ft. The sun, at an altitude of 40°, casts a shadow of the ball on the ground. Find the rate $dx/dt$ at which the shadow is traveling along the ground when the ball has fallen 50.0 ft.

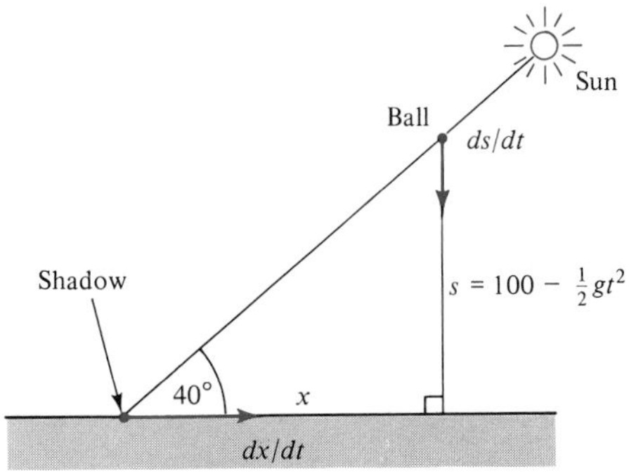

FIGURE 29-20

17. A square sheet of metal 10.0 in. on a side is expanded by increasing its temperature so that each side of the square increases 0.0050 in./s. At what rate is the area of the square increasing at 20.0 s?

18. A circular plate in a furnace is expanding so that its radius is changing 0.010 cm/s. How fast is the area of one face changing when the radius is 5.0 cm?

19. The volume of a cube is increasing at 10.0 in.$^3$/min. At the instant when its volume is 125 in.$^3$, what is the rate of change of its edge?

20. The edge of an expanding cube is changing at the rate of 0.0030 in./s. Find the rate of change of its volume when its edge is 5.0 in. long.

21. At some instant the diameter $x$ of a cylinder (Fig. 29-21) is 10.0 in. and increasing at a rate of 1.00 in./min. At that same instant the height $y$ is 20.0 in. and decreasing at a rate ($dy/dt$) such that the volume is not changing ($dV/dt = 0$). Find $dy/dt$.

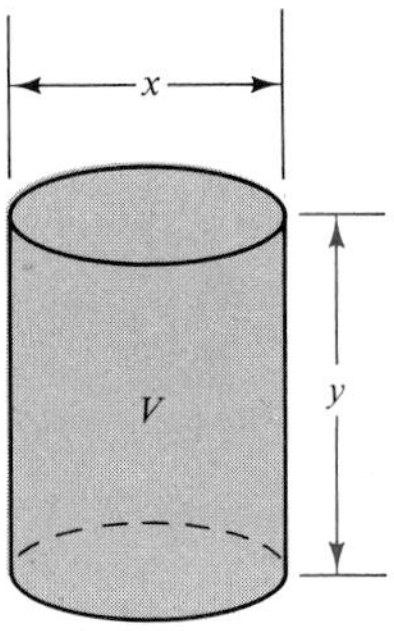

FIGURE 29-21

### Fluid Flow

22. Water is running from a vertical cylindrical tank 3.00 m in diameter at the rate of $3\pi\sqrt{h}$ m$^3$/min, where $h$ is the depth of the water in the tank. How fast is the surface of the water falling when $h = 9.00$ m?

23. Water is flowing into a conical reservoir 20.0 m deep and 10.0 m across the top, at a rate of 15.0 m$^3$/min. How fast is the surface rising when the water is 8.00 m deep?

24. Sand poured on the ground at the rate of 3.00 m$^3$/min forms a conical pile whose height is one-third the diameter of its base. How fast is the altitude of the pile increasing when the radius of its base is 2.00 m?

25. A horizontal trough 10.0 ft long has ends in the shape of an isosceles right triangle (Fig. 29-22). If water is poured into it at the rate of 8.00 ft$^3$/min, at what rate is the surface of the water rising when the water is 2.00 ft deep?

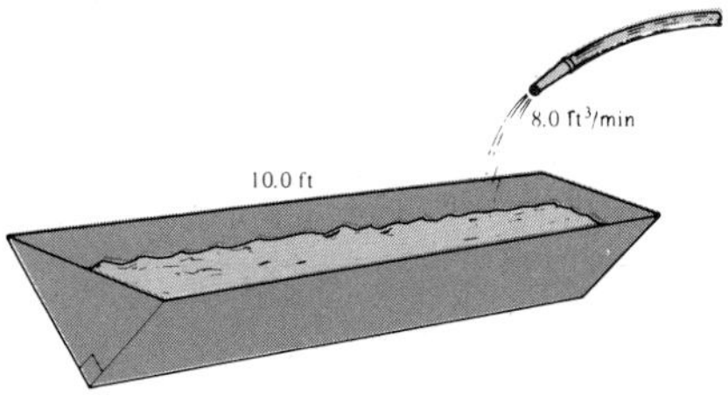

FIGURE 29-22

Use Boyle's law, $pv$ = constant.

26. A tank contains 1000 $ft^3$ of gas at a pressure of 5.00 $lb/in^2$. If the pressure is decreasing at the rate of 0.050 $lb/in^2$ per hour, find the rate of increase of the volume.
27. The adiabatic law for the expansion of air is $pv^{1.4} = C$. If at a given time the volume is observed to be 10.0 $ft^3$, and the pressure is 50.0 $lb/in^2$, at what rate is the pressure changing if the volume is decreasing 1.00 $ft^3/s$?

### Miscellaneous

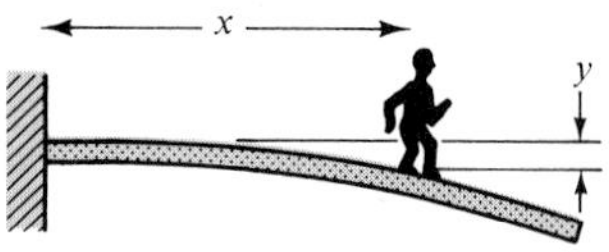

FIGURE 29-23

28. If the speed $v$ ft/s of a certain bullet passing through wood is given by $v = 500\sqrt{1 - 3x}$, where $x$ is the depth in feet, find the rate at which the speed is decreasing after the bullet has penetrated 3.00 in.
29. As a man walks a distance $x$ (ft) along a board (Fig. 29-23), he sinks a distance $y = x^2(x + 2)/150$ in. If he moves at the rate of 2.00 ft/s, how fast is he sinking when $x = 8.00$ ft?
30. A stone dropped into a calm lake causes a series of circular ripples. The radius of the outer one increases at 2.00 ft/s. How rapidly is the disturbed area changing at the end of 3.00 s?

## 29-4 APPLIED MAXIMUM–MINIMUM PROBLEMS

In Chapter 28, we found extreme points, the peaks and valleys on a curve. We did this, you recall, by finding the points where the slope (and hence the first derivative) was zero. We now apply the same idea to problems in which we find, for example, the point of minimum cost, or the point of maximum efficiency, or the point of maximum carrying capacity.

### Suggested Steps for Maximum–Minimum Problems

1. Locate the quantity (which we will call $Q$) to be maximized or minimized, and locate the independent variable, say $x$, which is varied in order to maximize or minimize $Q$.
2. Write an equation linking $Q$ and $x$. If this equation contains another variable as well, that variable must be eliminated by means of a second equation. It is a good idea, although not essential, to sketch the function $Q = f(x)$. The sketch will give the approximate location of any maximum or minimum points.
3. Take the derivative $dQ/dx$.
4. Set the derivative equal to zero. Solve for $x$.
5. Check any extreme points found to see if they are maxima or minima. This can be done simply by looking at your graph of $Q = f(x)$, by your knowledge of the physical problem, or by the first- or second-derivative tests. Also check if the maximum or minimum value you seek is at one of the endpoints. An endpoint can be a maximum or minimum point in the given interval, even though the slope there is not zero.

A list of general suggestions such as these is usually so vague as to be useless without examples. Our first example is one in which the relation between the variables is given verbally in the problem statement.

**EXAMPLE 20:** What two positive numbers whose product is 100 have the least possible sum?

**Solution:**

1. We want to minimize the sum $S$ of the two numbers by varying one of them, which we call $x$. Then

$$\frac{100}{x} = \text{other number}$$

2. The sum of the two numbers is

$$S = x + \frac{100}{x}$$

3. Taking the derivative yields

$$\frac{dS}{dx} = 1 - \frac{100}{x^2}$$

4. Setting the derivative to zero and solving for $x$ gives us

$$x^2 = 100$$

$$x = \pm 10$$

Since we are asked for positive numbers, we discard the $-10$. The other number is $100/10 = 10$.

5. But have we found those numbers that will give a *minimum* sum, as requested, or a *maximum* sum? We can check this by means of the second derivative test. Taking the second derivative, we have

**The sum of two positive numbers whose product is a constant is always a minimum when the numbers are equal as in this example. Can you prove the general case?**

$$S'' = \frac{200}{x^3}$$

When $x = 10$,

$$S''(10) = \frac{200}{1000} = 0.2$$

which is *positive,* indicating a *minimum* point. This is verified by our graph of $S$ versus $x$ (Fig. 29-24), which shows a minimum at $x = 10$.

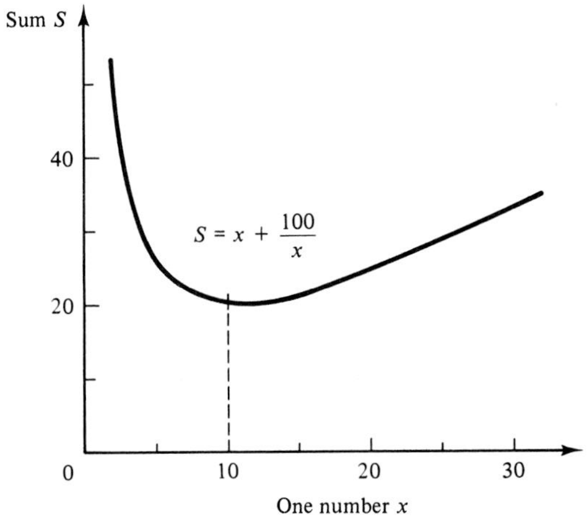


FIGURE 29-24

In the next example, the equation linking the variables is easily written from the geometrical relationships in the problem.

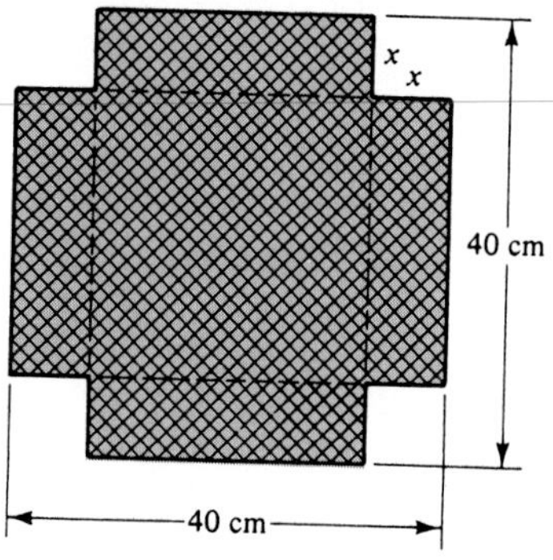


FIGURE 29-25

**EXAMPLE 21:** An open-top box is to be made from a square of sheet metal 40 cm on a side by cutting a square from each corner and bending up the sides along the dashed lines in Fig. 29-25. Find the dimension $x$ of the cutout that will result in a box of the greatest volume.

**Solution:**

1. We want to maximize the volume $V$ by varying $x$.
2. The equation is

$$V = \text{length} \cdot \text{width} \cdot \text{depth}$$
$$= (40 - 2x)(40 - 2x)x$$
$$= x(40 - 2x)^2$$

3. Taking the derivative, we obtain

$$\frac{dV}{dx} = x(2)(40 - 2x)(-2) + (40 - 2x)^2$$
$$= -4x(40 - 2x) + (40 - 2x)^2$$
$$= (40 - 2x)(-4x + 40 - 2x)$$
$$= (40 - 2x)(40 - 6x)$$

4. Setting the derivative to zero and solving for $x$, we have

$$40 - 2x = 0 \qquad\qquad 40 - 6x = 0$$
$$x = 20 \text{ cm} \qquad\qquad x = \frac{20}{3} = 6.67 \text{ cm}$$

We discard $x = 20$ cm, for it is a minimum value, and results in the entire sheet of metal being cut away, and keep $x = 6.67$ cm as our answer.

5. The graph of volume versus $x$ in Fig. 29-26 shows that our point is indeed a maximum, so no further test is needed.

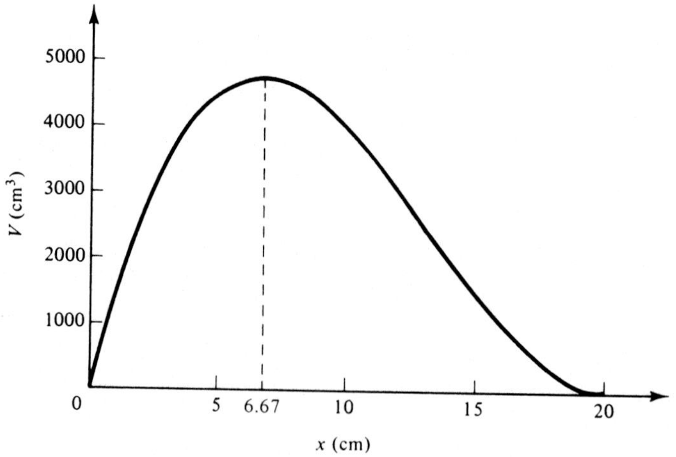


FIGURE 29-26

Our next example is one in which the equation is given.

**EXAMPLE 22:** The deflection $y$ of a simply supported beam with a concentrated load $P$ (Fig. 29-27) at a distance $x$ from the left end of the beam is given by

$$y = \frac{Pbx}{6LEI}(L^2 - x^2 - b^2)$$

where $E$ is the modulus of elasticity, and $I$ is the moment of inertia of the beam's cross section. Find the value of $x$ at which the deflection is a maximum, for a 20.0-ft-long beam with a concentrated load 5.00 ft from the right end.

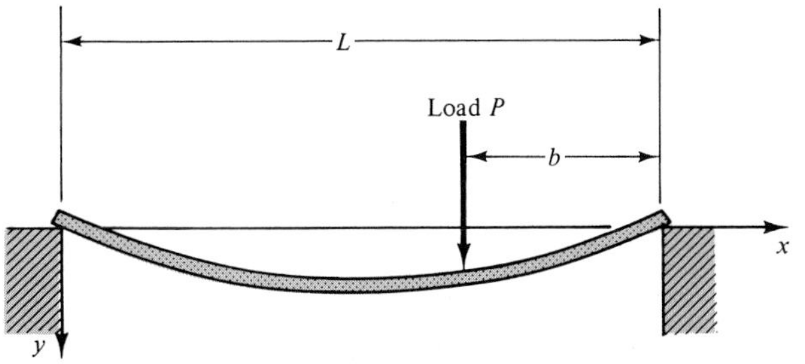


**FIGURE 29-27 Simply supported beam with concentrated load.**

**Solution:**

1. We want to find the distance $x$ from the left end at which the deflection is a maximum.
2. The equation is given in the problem statement.
3. We take the derivative using the product rule, noting that every quantity but $x$ or $y$ is a constant.

$$\frac{dy}{dx} = \frac{Pb}{6LEI}[x(-2x) + (L^2 - x^2 - b^2)(1)]$$

4. We set this derivative equal to zero and solve for $x$:

$$2x^2 = L^2 - x^2 - b^2$$

$$3x^2 = L^2 - b^2$$

$$x = \pm\sqrt{\frac{L^2 - b^2}{3}}$$

We drop the negative value, since $x$ cannot be negative in this problem. Now substituting $L = 20.0$ ft and $b = 5.00$ ft gives us

$$x = \sqrt{\frac{400 - 25}{3}} = 11.2 \text{ ft}$$

Thus the maximum deflection occurs between the load and the midpoint of the beam, as we might expect.

5. It is clear from the physical problem that our point is a maximum, so no test is needed.

The equation linking the variables in the preceding example had only two variables, $x$ and $y$. In the following example, our equation has *three* variables, one of which must be eliminated before we take the derivative. We eliminate the third variable by means of a second equation.

**EXAMPLE 23:** If we assume that the strength of a rectangular beam varies directly as its width and the square of its depth, find the dimensions of the strongest beam that can be cut from a round log 12.0 in. in diameter (Fig. 29-28).

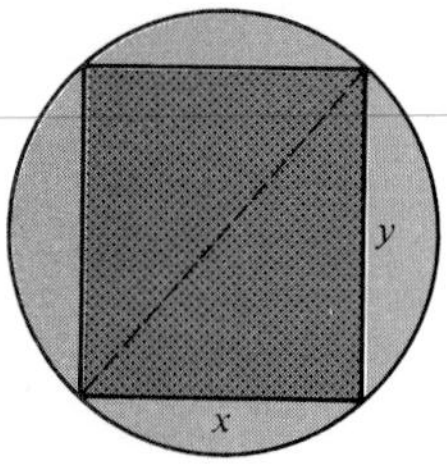


**FIGURE 29-28 Beam cut from a log.**

**Solution:**

1. We want to maximize the strength $S$ by varying the width $x$.
2. The strength $S$ is $S = kxy^2$, where $k$ is a constant of proportionality. Note that we have three variables, $S$, $x$, and $y$. We must eliminate $x$ or $y$. By the Pythagorean theorem,

$$x^2 + y^2 = 12.0^2$$

so

$$y^2 = 144 - x^2$$

Substituting, we obtain

$$S = kx(144 - x^2)$$

3. The derivative is

$$\frac{dS}{dx} = kx(-2x) + k(144 - x^2)$$

$$= -3kx^2 + 144k$$

4. Setting the derivative equal to zero and solving for $x$ gives us $3x^2 = 144$, so

$$x = \pm 6.93 \text{ in.}$$

We discard the negative value, of course. The depth is

$$y = \sqrt{144 - 48} = 9.80 \text{ in.}$$

5. But have we found the dimensions of the beam with *maximum* strength, or *minimum* strength? Let us use the second derivative test to tell us. The second derivative is

$$\frac{d^2S}{dx^2} = -6kx$$

When $x = 6.93$,

$$\frac{d^2S}{dx^2} = -41.6k$$

Since $k$ is positive, the second derivative is negative, which tells us that we have found a *maximum*.

## EXERCISE 4—APPLIED MAXIMUM–MINIMUM PROBLEMS

### Number Problems

1. What number added to half the square of its reciprocal gives the smallest sum?
2. Separate the number 10 into two parts such that their product will be a maximum.

3. Separate the number 20 into two parts such that the product of one part and the square of the other part is a maximum.
4. Separate the number 5 into two parts such that the square of one part times the cube of the other part shall be a maximum.

### Minimum Perimeter

5. A rectangular garden (Fig. 29-29) laid out along your neighbor's lot contains 432 $m^2$. It is to be fenced on all sides. If the neighbor pays for half the shared fence, what should be the dimensions of the garden so that your cost is a minimum?

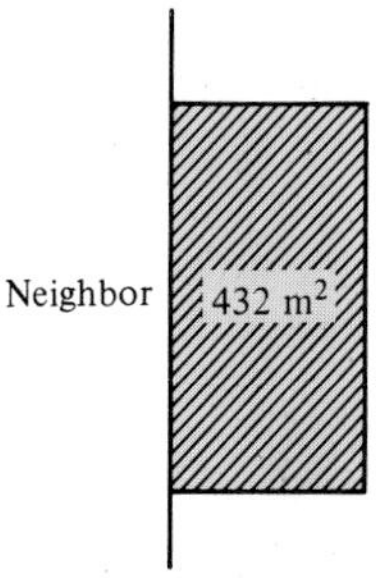


FIGURE 29-29 **Rectangular garden.**

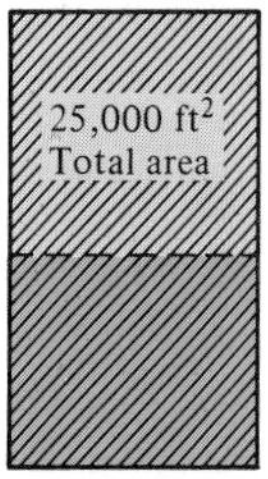


FIGURE 29-30 **Rectangular field.**

6. It is required to enclose a rectangular field by a fence (Fig. 29-30), and then divide it into two lots by a fence parallel to the short sides. If the area of the field is 25,000 $ft^2$, find the lengths of the sides so that the total length of fence will be a minimum.
7. A rectangular pasture 162 $yd^2$ in area is built so that a long, straight wall serves as one side of it. If the length of the fence along the remaining three sides is the least possible, find the dimensions of the pasture.

### Maximum Volume of Containers

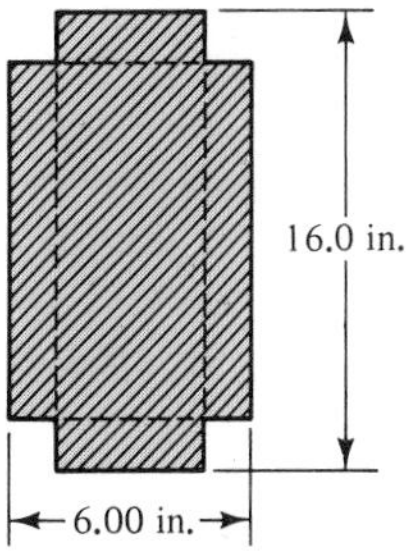


FIGURE 29-31 **Sheet metal for box.**

8. Find the volume of the largest box that can be made from a rectangular sheet of metal 6.00 in. by 16.0 in. (Fig. 29-31) by cutting a square from each corner and turning up the sides.
9. Find the height and base diameter of a cylindrical, topless tin cup of maximum volume if its area (sides and bottom) is 100 $cm^2$.
10. The slant height of a certain cone is 50.0 cm. What cone height will make the volume a maximum?

### Maximum Area of Plane Figures

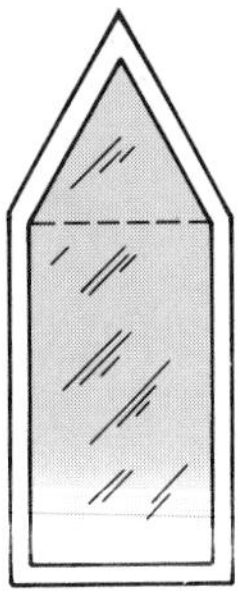

FIGURE 29-32 **Window.**

11. Find the area of the greatest rectangle that has a perimeter of 20.0 in.
12. A window composed of a rectangle surmounted by an equilateral triangle is 15.0 ft in perimeter (Fig. 29-32). Find the dimensions that will make its total area a maximum.
13. Two corners of a rectangle are on the $x$ axis between $x = 0$ and $x = 10$ (Fig. 29-33). The other two corners are on the lines whose equations are $y = 2x$ and $3x + y = 30$. For what value of $y$ will the area of the rectangle be a maximum?

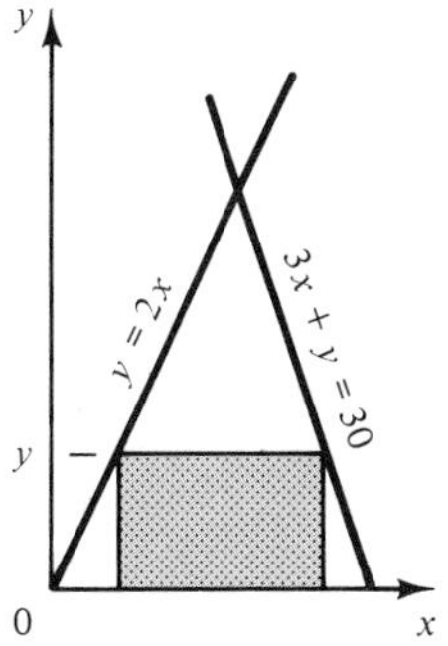


FIGURE 29-33

### Maximum Cross-Sectional Area

14. A trough is to be made of a long rectangular piece of metal by bending up two edges so as to give a rectangular cross section. If the width of the original piece is 14.0 in., how deep should the trough be made in order that its cross-sectional area be a maximum?
15. A gutter is to be made of a strip of metal 12.0 in. wide, the cross section having the form shown in Fig. 29-34. What depth $x$ gives a maximum cross-sectional area?

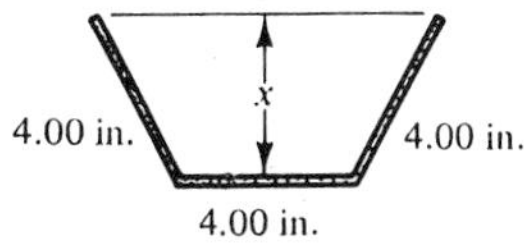


FIGURE 29-34

### Minimum Distance

16. Two railroad tracks intersect at right angles. There is a train on each track, each approaching the crossing at 40.0 mi/h, one being 10.0 mi, the other 20.0 mi from the intersection. What will be their minimum distance apart?
17. Find the point $Q$ on the curve $y = x^2/2$ that is nearest the point (4, 1) (Fig. 29-35).
18. Given one branch of the parabola $y^2 = 8x$ and the point $P(6, 0)$ on the $x$ axis (Fig. 29-36), find the coordinates of point $Q$ so that $PQ$ is a minimum.

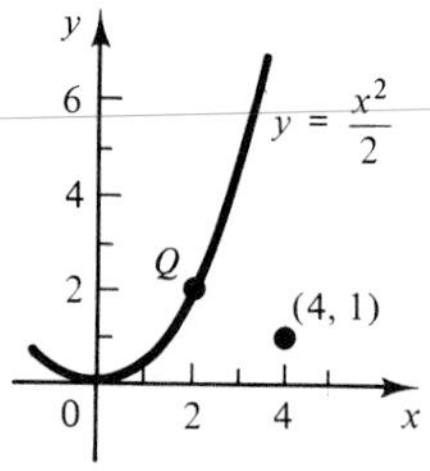


FIGURE 29-35

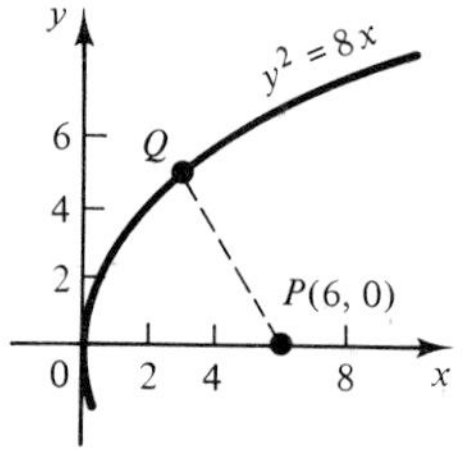


FIGURE 29-36

### Inscribed Plane Figures

19. Find the dimensions of the rectangle of greatest area that can be inscribed in an equilateral triangle each of whose sides is 10.0 in. if one of the sides of the rectangle is on a side of the triangle (Fig. 29-37). (*Hint:* Let the independent variable be the height $x$ of the rectangle.)
20. Find the area of the largest rectangle with sides parallel to the coordinate axes which can be inscribed in the figure bounded by the two parabolas $3y = 12 - x^2$ and $6y = x^2 - 12$ (Fig. 29-38).

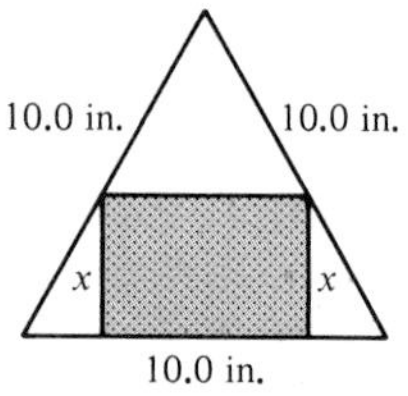


FIGURE 29-37

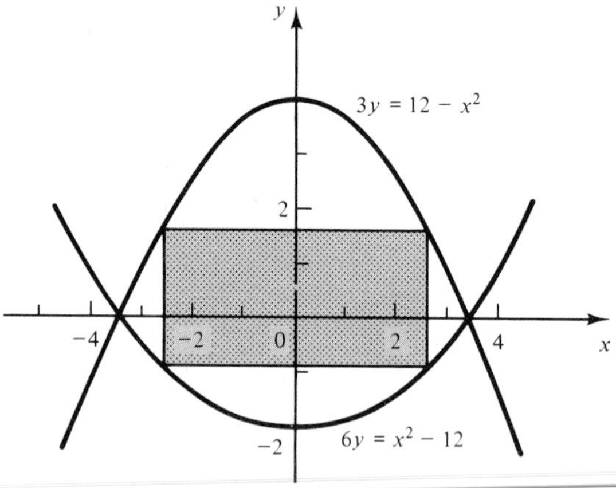


FIGURE 29-38

21. Find the dimensions of the largest rectangle that can be inscribed in an ellipse whose major axis is 20 units and whose minor axis is 14 units (Fig. 29-39).

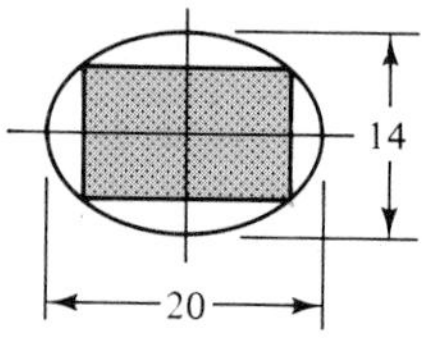


**FIGURE 29-39 Rectangle inscribed in ellipse.**

### Inscribed Volumes

22. Find the dimensions of the largest rectangular parallelepiped with a square base that can be cut from a solid sphere 18.0 in. in diameter (Fig. 29-40).
23. Find the dimensions of the largest right circular cylinder that can be inscribed in a sphere with a diameter of 10.0 cm (Fig. 29-41).

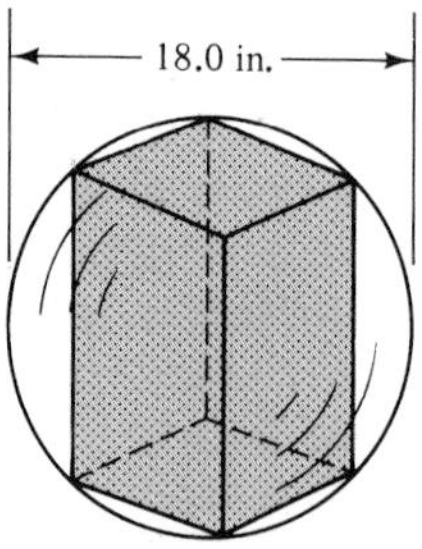


**FIGURE 29-40 Parallelepiped inscribed in sphere.**

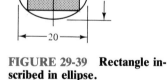

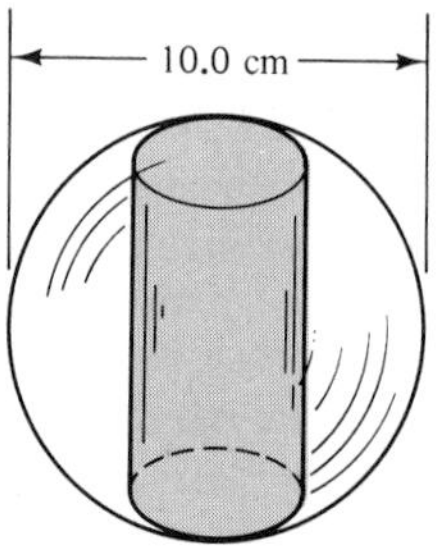


**FIGURE 29-41 Cylinder inscribed in sphere.**

24. Find the height of the cone of minimum volume circumscribed about a sphere of radius 10.0 m (Fig. 29-42).

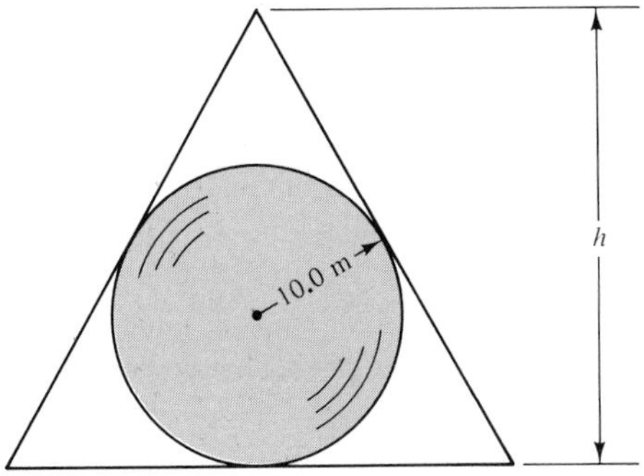


**FIGURE 29-42 Cone circumscribed about sphere.**

25. Find the altitude of the cone of maximum volume that can be inscribed in a sphere of radius 9.00 ft.

### Most Economical Dimensions of Containers

26. What should be the diameter of a can holding 1 qt (58 in$^3$) and requiring the least amount of metal, if the can is open at the top?
27. A silo (Fig. 28-18) has a hemispherical roof, cylindrical sides, and circular floor, all made of steel. Find the dimensions for a silo having a volume of 755 m$^3$ (including the dome) that needs the least steel.

### Minimum Travel Time

28. A man in a rowboat at $P$ (Fig. 29-43) 6.00 mi from shore desires to reach point $Q$ on the shore at a straight-line distance of 10.0 mi from his present position. If he can walk 4.00 mi/h and row 3.00 mi/h, at what point $L$ should he land in order to reach $Q$ in the shortest time?

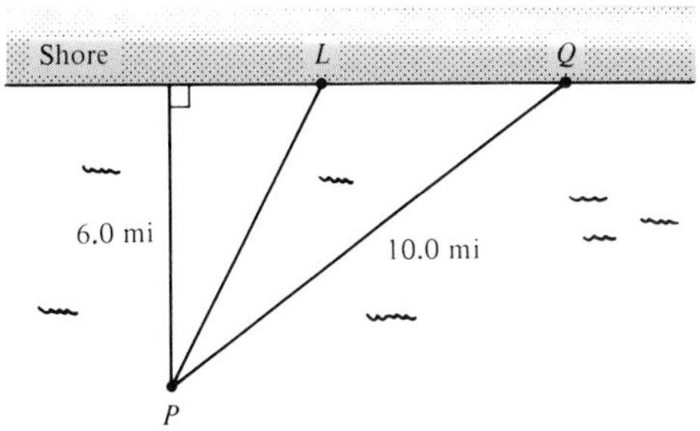


**FIGURE 29-43 Find the fastest route from boat $P$ to shore at $Q$.**

### Beam Problems

29. The strength $S$ of the beam in Fig. 29-28 is given by $S = kxy^2$, where $k$ is a constant. Find $x$ and $y$ for the strongest rectangular beam that can be cut from an 18.0-in.-diameter cylindrical log.
30. The stiffness $Q$ of the beam in Fig. 29-28 is given by $Q = kxy^3$, where $k$ is a constant. Find $x$ and $y$ for the stiffest rectangular beam that can be cut from an 18.0-in.-diameter cylindrical log.

### Light

31. The intensity $E$ of illumination at a point due to a light at a distance $x$ from the point is given by $E = kI/x^2$, where $k$ is a constant and $I$ is the intensity of the source. A light $M$ has an intensity three times that of $N$ (Fig. 29-44). At what distance from $M$ is the illumination a minimum?

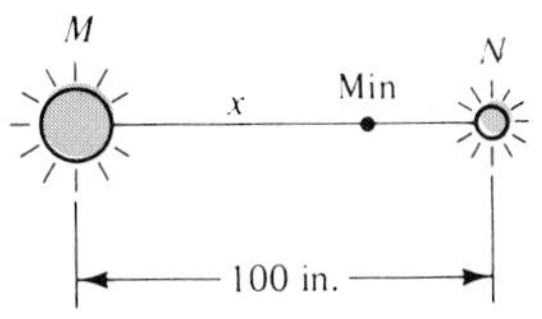


**FIGURE 29-44**

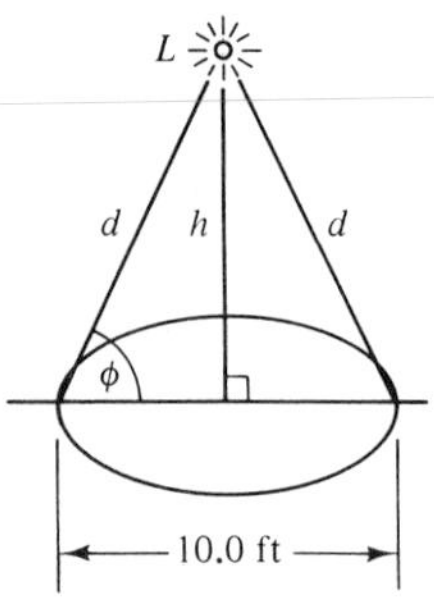


FIGURE 29-45

32. The intensity of illumination $I$ from a given light source is given by

$$I = \frac{k \sin \phi}{d^2}$$

where $k$ is a constant, $\phi$ is the angle at which the rays strike the surface, and $d$ is the distance between the surface and the light (Fig. 29-45). At what height $h$ should a light be suspended directly over the center of a circle 10.0 ft in diameter so that the illumination at the circumference will be a maximum?

### Electrical

33. The power delivered to a load by a 30-V source of internal resistance 2 Ω is $30i - 2i^2$ watts, where $i$ is the current in amperes. For what current will this source deliver the maximum power?
34. When 12 cells, each having an EMF of $e$ and an internal resistance $r$ are connected to a variable load $R$ as in Fig. 29-46 the current in $R$ is

$$i = \frac{3e}{\frac{3r}{4} + R}$$

Show that the maximum power ($i^2R$) delivered to the load is a maximum when the load $R$ is equal to the equivalent internal resistance of the source, $3r/4$.

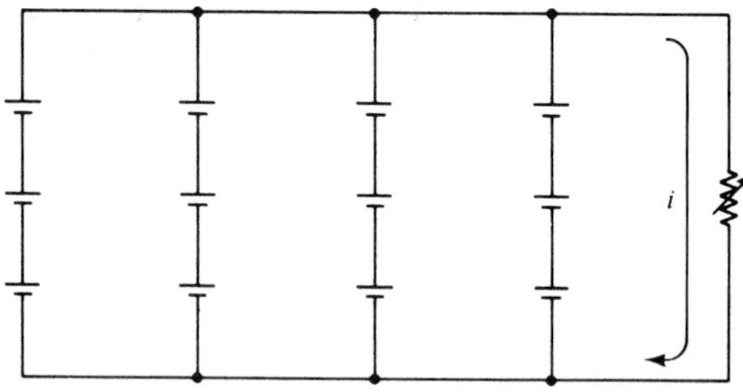


FIGURE 29-46

35. A certain transformer has an efficiency $E$ when delivering a current $i$, where

$$E = \frac{115i - 25 - i^2}{115i}$$

At what current is the efficiency of the transformer a maximum?

### Mechanisms

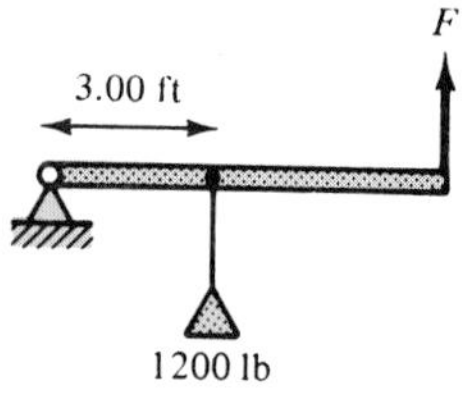


FIGURE 29-47

36. If the lever in Fig. 29-47 weighs 12.0 lb per foot, find its length so as to make the lifting force $F$ a minimum.
37. The efficiency $E$ of a screw is given by

$$E = \frac{x - \mu x^2}{x + \mu}$$

where $\mu$ is the coefficient of friction and $x$ the tangent of the pitch angle of the screw. Find $x$ for maximum efficiency if $\mu = 0.45$.

## 29-5 DIFFERENTIALS

### Definition

Up to now, we have treated the symbol $dy/dx$ as a *whole,* and not as the quotient of two quantities $dy$ and $dx$. Here, we give $dy$ and $dx$ separate names and meanings of their own. The quantity $dy$ is called the *differential of y*, and $dx$ is called the *differential of x*.

These two differentials, $dx$ and $dy$, have a simple geometric interpretation. Figure 29-48 shows a tangent drawn to a curve $y = f(x)$ at some point $P$. The slope of the curve is found by evaluating $dy/dx$ at $P$. *The differential dy is then the rise of the tangent line, in some arbitrary run dx.* Since the rise of a line is equal to the slope of the line times the run:

| | | |
|---|---|---|
| Differential of $y$ | $dy = f'(x)\,dx$ | **361** |

where we have represented the slope by $f'(x)$ instead of $dy/dx$, to avoid confusion.

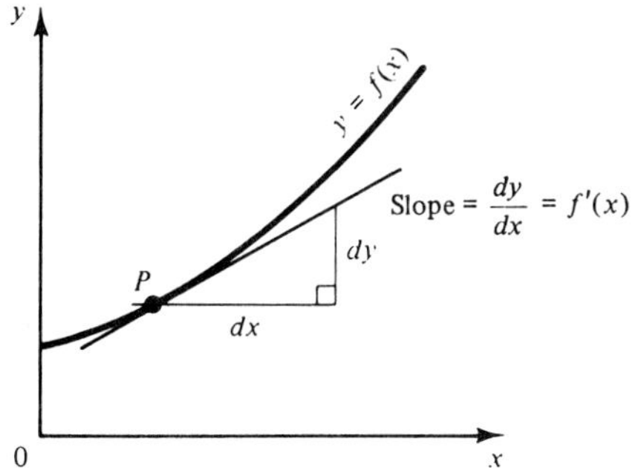


**FIGURE 29-48**

### Differentials and Increments Compared

You may be thinking that the differentials $dx$ and $dy$ look something like the increments $\Delta x$ and $\Delta y$ back in Fig. 27-5. In fact, it is easy to confuse differentials and increments.

Recall from Sec. 27-2 that when we gave a small increment $\Delta x$ to $x$, then the function $f(x)$ changed by an increment $\Delta y$, as shown in Fig. 29-49. We see then that the difference between differentials and increments is that:

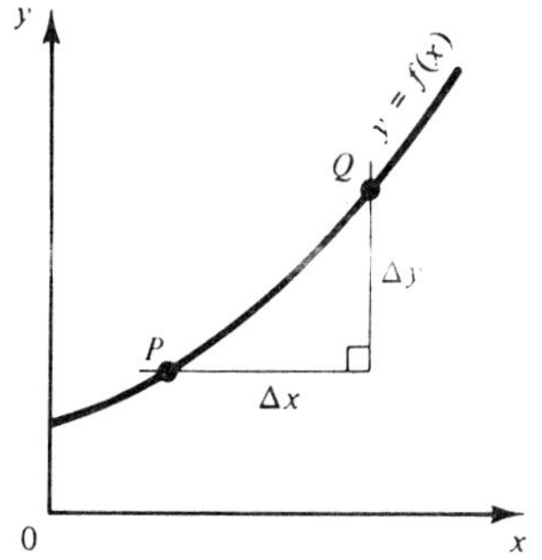


**FIGURE 29-49 Increments.**

1. An increment $\Delta x$ is a *small* change in $x$, while the differential $dx$ can be of any size, large or small (although we will soon see that, for applications, we will choose only small values for differentials).
2. The increment $\Delta y$ is the change in $y$ for the *curve* itself, corresponding to an increment $\Delta x$, whereas the differential $dy$ is the change in $y$ for the *tangent line,* corresponding to a change $dx$.

These quantities can be clearly seen in Fig. 29-50, where we have chosen the differential $dx$ to be equal to the increment $\Delta x$.

Also notice that if we choose a small enough $dx$ (or $\Delta x$) then *dy will be approximately equal to* $\Delta y$. We make use of this fact shortly to find approximate changes in one quantity due to small changes in another quantity.

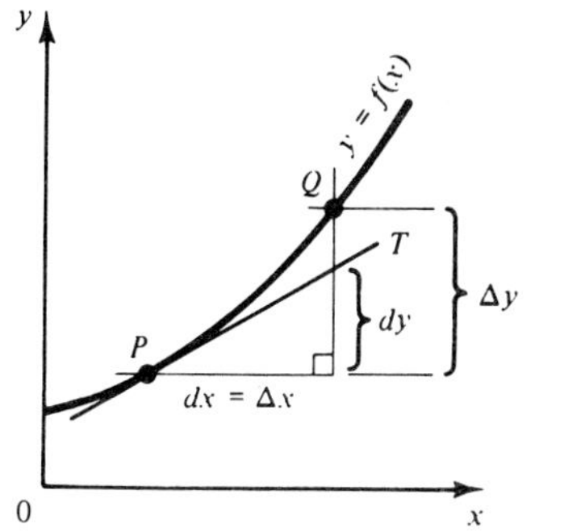


**FIGURE 29-50 Increments and differentials compared.**

## Differential Form

If we take the derivative of some function, say, $y = x^3$ we get

$$\frac{dy}{dx} = 3x^2$$

An equation containing a derivative (called a *differential equation*) is usually written in differential form before it is solved.

Since we may now think of $dy$ and $dx$ as separate quantities (differentials), we can multiply both sides by $dx$,

$$dy = 3x^2\,dx$$

This expression is said to be in *differential form*.

Thus to find $dy$, the differential of $y$, given some function of $x$, simply take the derivative of the function and multiply by $dx$.

---

**EXAMPLE 24:** If $y = 3x^2 - 2x + 5$, find the differential $dy$.

**Solution:** Taking the derivative gives us

$$\frac{dy}{dx} = 6x - 2$$

Multiplying by $dx$, we get the differential of $y$

$$dy = (6x - 2)\,dx$$

---

To find the differential of an *implicit* function, simply find the derivative as in Chap. 27, and then multiply both sides by $dx$.

---

**EXAMPLE 25:** Find the differential of the implicit function

$$x^3 + 3xy - 2y^2 = 8$$

**Solution:** Differentiating term by term gives us

$$3x^2 + 3x\frac{dy}{dx} + 3y - 4y\frac{dy}{dx} = 0$$

$$(3x - 4y)\frac{dy}{dx} = -3x^2 - 3y$$

$$\frac{dy}{dx} = \frac{-3x^2 - 3y}{3x - 4y} = \frac{3x^2 + 3y}{4y - 3x}$$

$$dy = \frac{3x^2 + 3y}{4y - 3x}dx$$

---

## Using Differentials to Estimate Small Changes in a Function

Let us look again at Fig. 29-50. There we took the run equal to the increment $\Delta x$ and also equal to the differential $dx$, and we saw that the tangent line had a rise of $dy$, while the curve itself rose by an amount $\Delta y$.

Now it is clear from the figure that $dy$ and $\Delta y$ are not equal. *But,* if $\Delta x$ is small, *the differential dy can be a close enough approximation to the increment* $\Delta y$ *for many applications.*

We obtain an expression for $\Delta y$ from our definition of the derivative (Eq. 328),

$$\lim_{\Delta x \to 0} \frac{\Delta y}{\Delta x} = \frac{dy}{dx}$$

That is, that $\Delta y/\Delta x$ approaches $dy/dx$ as $\Delta x$ goes to zero. But when $\Delta x$ has some small but finite value, we see that

$$\frac{\Delta y}{\Delta x} \simeq \frac{dy}{dx}$$

Multiplying by $\Delta x$ yields

| | | |
|---|---|---|
| Approximations Using Differentials | $\Delta y \simeq \frac{dy}{dx} \Delta x$ | **362** |

*The approximate change $\Delta y$ in some function due to a small change $\Delta x$ is the change $\Delta x$ multiplied by the derivative of the function.*

---

**EXAMPLE 26:** Using differentials, estimate the change in the function $y = 2x^3$ when $x$ changes from 5 to 5.1. Compare this estimate with the exact value.

**Solution:** The change $\Delta x$ in $x$ is 0.1, and the derivative is

$$\frac{dy}{dx} = 6x^2$$

When $x = 5$,

$$\frac{dy}{dx} = 6(5)^2 = 150$$

Then by Eq. 362,

$$\Delta y \simeq \frac{dy}{dx}\Delta x = 150(0.1) = 15$$

The exact change is

$$2(5.1)^3 - 2(5.0)^3 = 265.302 - 250 = 15.302$$

or about 2% different than our estimate.

---

## Approximate Volumes of Shells and Rings

We can use differentials to derive approximate formulas for the volumes of thin-walled shells, such as the volume of rubber in a spherical balloon of given radius. We do this by thinking of the volume of the wall of the shell as the increase in volume of the sphere as the radius increases by an amount equal to the wall thickness.

**EXAMPLE 27:** Derive a formula for the volume of a hollow sphere of inside radius $r$, having a wall thickness $t$.

**Solution:** The volume of a sphere is, by Eq. 128, $V = \frac{4}{3}\pi r^3$. Taking the derivative, we have

$$\frac{dV}{dr} = 4\pi r^2$$

The increment $\Delta r$ as $r$ increases from the inside of the wall to the outside is equal to the wall thickness $t$, so by Eq. 362,

$$\Delta V \simeq \frac{dV}{dr}\Delta r \simeq 4\pi r^2 t$$

which is, then, the formula for the volume of a hollow sphere.

## Estimating the Effect of Small Errors in Measurement

Equation 362 tells us how much $y$ will depart from its original value when $x$ departs from its original value by an amount $\Delta x$. In this section we let $\Delta x$ represent some small error in measurement or manufacture, and compute the resulting error in some quantity $y$ that is related to $x$.

**EXAMPLE 28:** In order to calculate the power dissipated in a certain 1565-Ω resistor, the current through the resistor is measured and found to be 2.84 A. What would be the error in the calculated value of the power if the error in the current measurement is 5%?

**Solution:** The power to a resistor is, by Eq. A67, $P = I^2R$. Taking the derivative, we have

$$\frac{dP}{dI} = 2IR$$

The increment in $I$ is equal to 5% of the measurement,

$$\Delta I = 0.05(2.84) = 0.142 \text{ A}$$

Then by Eq. 362,

$$\Delta P \simeq \frac{dP}{dI}\Delta I = 2IR(0.142)$$

Substituting $I = 2.84$ A and $R = 1565\ \Omega$ gives us

$$\Delta P \simeq 2(2.84)(1565)(0.142)$$
$$= 1260 \text{ W}$$

## EXERCISE 5—DIFFERENTIALS

Write the differential $dy$ for each function.

1. $y = x^3$
2. $y = x^2 + 2x$
3. $y = \dfrac{x-1}{x+1}$
4. $y = (2 - 3x^2)^3$

5. $y = (x + 1)^2(2x + 3)^3$

6. $y = \sqrt{1 - 2x}$

7. $y = \dfrac{\sqrt{x - 4}}{3 - 2x}$

Write the differential $dy$ in terms of $x$, $y$, and $dx$ for each implicit relation.

8. $3x^2 - 2xy + 2y^2 = 3$

9. $x^3 + 2y^3 = 5$

10. $2x^2 + 3xy + 4y^2 = 20$

11. $2\sqrt{x} + 3\sqrt{y} = 4$

**Estimating Small Changes**

Find the approximate change in $y$ when $x$ changes as indicated.

12. $y = 5x^2$, and $x$ changes from 8 to 8.01
13. $y = \sqrt{x}$, and $x$ changes from 10 to 10.1
14. $y = \sqrt[3]{3x}$, and $x$ changes from 4 to 3.99
15. $y = 2\sqrt{x^2 - 3x}$, and $x$ changes from 15 to 15.3
16. The height $h$ of a certain projectile at time $t$ is given by $h = 100 + 160t - 16t^2$ (ft). Find the approximate change in height during the interval when $t$ changes from 15 s to 15.1 s.
17. The approximate time $t$ for one oscillation of a pendulum of length $L$ ft is $t = 2\pi\sqrt{L/g}$ (s) where $g = 32.2$ ft/s approximately. By how much will $t$ change if a 13-ft-long pendulum is lengthened by 0.01 ft?
18. A cube of metal 10.00 cm on a side is heated until each side has increased by 0.01 cm. Find the approximate increase in volume.

**Shells and Rings**

Use differentials to derive an approximate formula for the volume of each figure.

19. a thin cylindrical shell of thickness $t$, height $h$, and radius $r$
20. a cubical shell of edge $x$ and thickness $t$
21. a cylindrical shell, including top, but no bottom, of radius $r$, thickness $t$, and height equal to the radius
22. Write an approximate formula for the area of a circular ring of radius $r$ and width $w$ (Fig. 29-51).

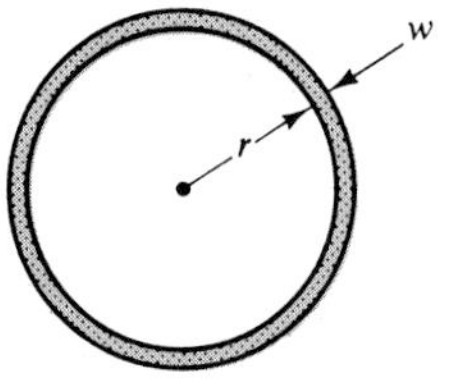


FIGURE 29-51

**Effects of Errors in Measurement**

23. If $x$ and $y$ are related by the function $y = x^{2/3}$, what is the approximate error in $y$ when $x$ is measured at 27.9 instead of the correct value of 27?
24. What is the approximate error in the volume of a cube of edge 6 cm if an error of 0.2 mm is made when measuring the edge?
25. What is the approximate error in the surface area of a sphere of radius 3 ft if an error of 0.01 ft is made when measuring the radius?
26. How accurately must the edge of a cubical box be made so that the volume will not differ from the intended value of 1000 ft$^3$ by more than 3 ft$^3$?
27. To what percent accuracy must the diameter of a circle be measured so that the calculated area will be correct to 1%?

## CHAPTER 29 REVIEW PROBLEMS

1. Airplane A is flying south at a speed of 120 ft/s. It passes over a bridge 12 min before another airplane, B, which is flying east at the same height at a speed of 160 ft/s. How fast are the airplanes separating 12 min after B passes over the bridge?
2. Find the approximate change in the function $y = 3x^2 - 2x + 5$, when $x$ changes from 5 to 5.01.
3. A person walks towards the base of a 60-m-high tower at the rate of 5.0 km/h. At what rate does the person approach the top of the tower when 80 m from the base?
4. A point moves along the hyperbola $x^2 - y^2 = 144$ with a horizontal velocity $v_x = 15$ cm/s. Find the total velocity when the point is at (13, 5).
5. A conical tank with vertex down has a vertex angle of 60°. Water flows from the tank at a rate of 5.0 cm³/min. At what rate is the inner surface of the tank being exposed when the water is 6.0 cm deep?
6. Find the instantaneous velocity and acceleration at $t =$ 2.0 s for a point moving in a straight line according to the equation $s = 4t^2 - 6t$.
7. A turbine blade (Fig. 29-52) is driven by a jet of water having a speed $s$. The power output from the turbine is given by $P = k(sv - v^2)$, where $v$ is the blade speed and $k$ is a constant. Find the blade speed $v$ for maximum power output.

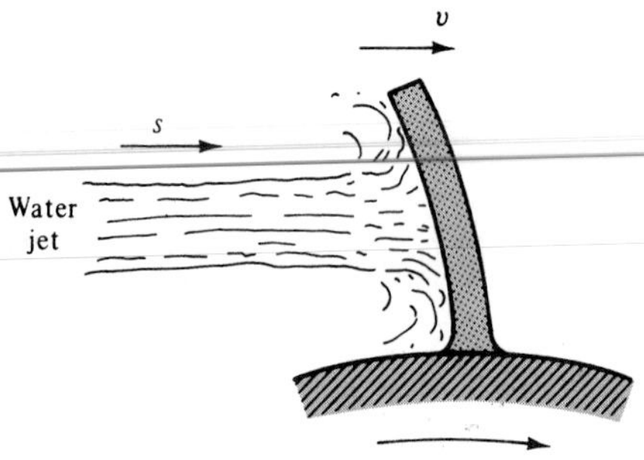


**FIGURE 29-52 Turbine blade.**

8. A pole (Fig. 29-53), is braced by a cable 24.0 ft long. Find the distance $x$ from the foot of the pole to the cable anchor so that the moment produced by the tension in the cable about the foot of the pole is a maximum. Assume that the tension in the cable does not change as the anchor point is changed.

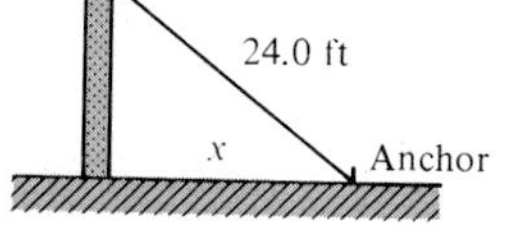


**FIGURE 29-53 Pole braced by a cable.**

9. Find the height of a right circular cylinder of maximum volume that can be inscribed in a sphere of radius 6.
10. Three sides of a trapezoid each have a length of 10 units. What length must the fourth side be to make the area a maximum?
11. The air in a certain balloon has a pressure of 40.0 lb/in², a volume of 5.0 ft³, and is expanding at the rate of 0.20 ft³/s. If the pressure and volume are related by the equation $pv^{1.41} =$ constant, find the rate at which the pressure is changing.
12. A certain item costs \$10 to make, and the number that can be sold is estimated to be inversely proportional to the cube of the selling price. What selling price will give the greatest net profit?
13. The distance $s$ of a point moving in a straight line is given by

$$s = -t^3 + 3t^2 + 24t + 28$$

At what times and at what distances is the point at rest?
14. A stone dropped into water produces a circular wave which increases in radius at the rate of 5.0 ft/s. How fast is the area within the ripple increasing when its diameter is 20.0 ft?
15. What is the area of the largest rectangle that can be drawn with one side on the $x$ axis and with two corners on the curve $y = 8(x^2 + 4)$?
16. The power $P$ delivered to a load by a 120-V source having an internal resistance of 5 Ω is $P = 120I - 5I^2$, where $I$ is the current to the load. At what current will the power be a maximum?
17. Separate the number 10 into two parts so that the product of the square of one part and the cube of the other part is a maximum.
18. The radius of a circular metal plate is increasing at the rate of 0.01 m/s. At what rate is the area increasing when the radius is 2.0 m?
19. Find the dimensions of the largest rectangular box with square base and open top that can be made from 300 in² of metal.
20. Use differentials to show that a 1% increase in the side of a square gives a 2% increase in its area.
21. Use differentials to show that a 1% increase in the edge of a cube gives a 3% increase in its volume.
22. The charge through a 8.24-Ω resistor varies with time according to the function $q = 2.26t^3 - 8.28$ coulomb. Write an expression for the instantaneous current through the resistor.
23. The charge at a resistor varies with time according to the function $q = 2.84t^2 + 6.25t^3$ coulomb. Write an expression for the instantaneous current through the resistor and evaluate it at 1.25 s.

**24.** The voltage applied to a 3.25 $\mu$F capacitor is $v = 1.03t^2 + 1.33t + 2.52$ V. Find the current at $t = 15.0$ s.

**25.** The angular displacement of a rotating body is given by $\theta = 18.5t^2 + 12.8t + 14.8$ rad. Find **(a)** the angular velocity and **(b)** the angular acceleration, at $t = 3.50$ s.

*Writing*

**26.** Once again our writing assignment is to make up a word problem, but now one that leads to a max/min or a related rate problem. As before, swap with a classmate, solve each other's problems, note anything unclear, unrealistic, or ambiguous, and then rewrite your problem if needed.

# 30

# DERIVATIVES OF TRIGONOMETRIC, LOGARITHMIC, AND EXPONENTIAL FUNCTIONS

## OBJECTIVES

**When you have completed this chapter, you should be able to:**

- Find the derivative of expressions containing trigonometric functions.
- Solve applied problems requiring derivatives of the trigonometric functions.
- Find derivatives of the inverse trigonometric functions.
- Find derivatives of logarithmic and exponential functions.
- Solve applied problems involving derivatives of logarithmic and exponential functions.
- Use logarithmic differentiation to find derivatives of certain algebraic expressions.

In this chapter we extend our ability to take derivatives to include the trigonometric, logarithmic, and exponential functions. This will enable us to solve a larger range of problems than was possible before. After learning the rules for taking derivatives of these functions, we apply them to problems quite similar to those in Chapters 28 and 29, that is, tangents, related rates, maximum–minimum, and the rest.

If you have not used the trigonometric, logarithmic, or exponential functions for a while, you might want to thumb quickly through that material before starting here.

## 30-1 DERIVATIVES OF THE SINE AND COSINE FUNCTIONS

### Derivative of sin *u*

We want the derivative of the function $y = \sin u$ where $u$ is a function of $x$ [such as $y = \sin(x^2 + 3x)$]. We will use the delta method to derive a rule for finding this derivative. We give an increment $\Delta u$ to $u$, and $y$ thus changes by an amount $\Delta y$,

**You might want to glance back at the delta method in Sec. 27-2.**

$$y + \Delta y = \sin(u + \Delta u)$$

Subtracting the original equation gives

$$\Delta y = \sin(u + \Delta u) - \sin u$$

We now make use of the identity, Eq. 177,

$$\sin \alpha - \sin \beta = 2 \cos \frac{\alpha + \beta}{2} \sin \frac{\alpha - \beta}{2}$$

to transform our equation for $\Delta y$ into a more useful form. We let

$$\alpha = u + \Delta u \qquad \text{and} \qquad \beta = u$$

so

$$\Delta y = 2 \cos \frac{u + \Delta u + u}{2} \sin \frac{u + \Delta u - u}{2}$$

$$= 2 \cos\left(u + \frac{\Delta u}{2}\right) \sin \frac{\Delta u}{2}$$

Now dividing by $\Delta u$,

$$\frac{\Delta y}{\Delta u} = 2\left[\cos\left(u + \frac{\Delta u}{2}\right)\right] \frac{\sin(\Delta u/2)}{\Delta u}$$

$$= \left[\cos\left(u + \frac{\Delta u}{2}\right)\right] \frac{\sin(\Delta u/2)}{\Delta u/2}$$

If we now let $\Delta u$ approach zero, the quantity

$$\frac{\sin(\Delta u/2)}{\Delta u/2}$$

approaches 1, as we saw in Sec. 27-1. Also, the quantity $\Delta u/2$ will approach zero, leaving us with

**Remember that when we evaluated this limit in Sec. 27-1 we required the angle ($\Delta u/2$ in this case) to be *in radians.* Thus the formulas we derive here also require the angle to be in radians.**

$$\frac{dy}{du} = \cos u$$

Now by the chain rule,

$$\frac{dy}{dx} = \frac{dy}{du} \cdot \frac{du}{dx}$$

So

| | | |
|---|---|---|
| Derivative of the Sine | $\dfrac{d(\sin u)}{dx} = \cos u \dfrac{du}{dx}$ | **340** |

*The derivative of the sine of some function is the cosine of that function, multiplied by the derivative of that function.*

---

**EXAMPLE 1:**

(a) If $y = \sin 3x$, then

$$\frac{dy}{dx} = (\cos 3x) \cdot \frac{d}{dx} 3x$$
$$= 3 \cos 3x$$

(b) If $y = \sin(x^3 + 2x^2)$, then

$$y' = \cos(x^3 + 2x^2) \cdot \frac{d}{dx}(x^3 + 2x^2)$$
$$= \cos(x^3 + 2x^2)(3x^2 + 4x)$$
$$= (3x^2 + 4x) \cos(x^3 + 2x^2)$$

**Recall that $\sin^3 x$ is the same as $(\sin x)^3$.**

(c) If $y = \sin^3 x$, then by the power rule

$$y' = 3(\sin x)^2 \cdot \frac{d}{dx}(\sin x)$$
$$= 3 \sin^2 x \cos x$$

---

**EXAMPLE 2:** Find the slope of the tangent to the curve $y = x^2 - \sin^2 x$, at $x = 2$ (Fig. 30-1).

**Solution:** Taking the derivative yields

$$y' = 2x - 2 \sin x \cos x$$

At $x = 2$ rad,

$$y'(2) = 4 - 2 \sin 2 \cos 2$$
$$= 4 - 2(0.909)(-0.416)$$
$$= 4.757$$

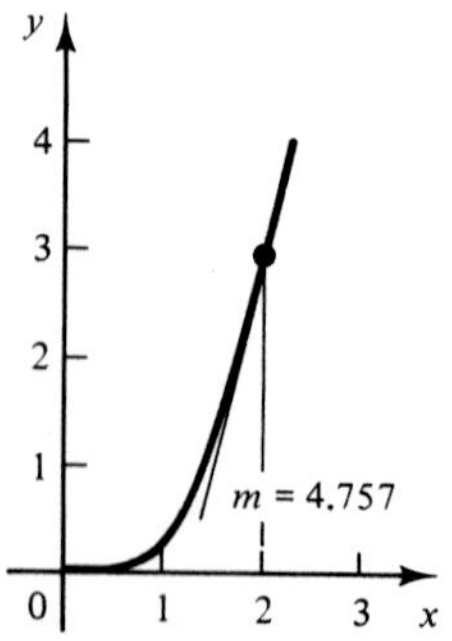


**FIGURE 30-1 Graph of $y = x^2 - \sin^2 x$.**

---

| | |
|---|---|
| Common Error | Remember that $x$ is in *radians*, unless otherwise specified. Be sure that your calculator is in radian mode. |

## Derivative of cos *u*

We now take the derivative of $y = \cos u$.

We do not need the delta method again, because we can relate the cosine to the sine with Eq. 154b, $\cos A = \sin B$, where $A$ and $B$ are complementary angles ($B = \pi/2 - A$; see Fig. 30-2). So

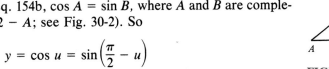

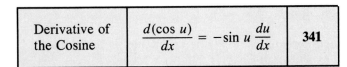

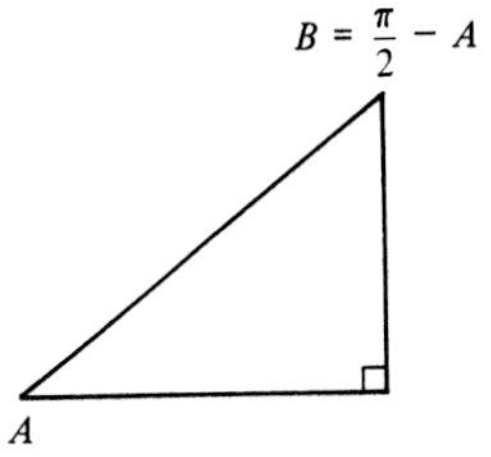

FIGURE 30-2

$$y = \cos u = \sin\left(\frac{\pi}{2} - u\right)$$

Then by Eq. 340,

$$\frac{dy}{dx} = \cos\left(\frac{\pi}{2} - u\right)\left(-\frac{du}{dx}\right)$$

But $\cos(\pi/2 - u) = \sin u$, so

| | | |
|---|---|---|
| Derivative of the Cosine | $\dfrac{d(\cos u)}{dx} = -\sin u \dfrac{du}{dx}$ | **341** |

*The derivative of the cosine of some function is the negative of the sine of that function, multiplied by the derivative of that function.*

---

**EXAMPLE 3:** If $y = \cos 3x^2$, then

$$y' = (-\sin 3x^2)\frac{d}{dx}(3x^2)$$

$$= (-\sin 3x^2)6x = -6x \sin 3x^2$$

---

| | |
|---|---|
| Common Error | $\cos 3x^2$ *is not the same* as $(\cos 3x)^2$.<br>$\cos 3x^2$ *is* the same as $\cos(3x^2)$. |

---

**EXAMPLE 4:** Differentiate $y = \sin 3x \cos 5x$.

**Solution:** Using the product rule, we have

$$\frac{dy}{dx} = (\sin 3x)(-\sin 5x)(5) + (\cos 5x)(\cos 3x)(3)$$

$$= -5 \sin 3x \sin 5x + 3 \cos 5x \cos 3x$$

---

**EXAMPLE 5:** Differentiate $y = \dfrac{2 \cos x}{\sin 3x}$.

**Solution:** Using the quotient rule gives us

$$y' = \frac{(\sin 3x)(-2 \sin x) - (2 \cos x)(3 \cos 3x)}{(\sin 3x)^2}$$

$$= \frac{-2 \sin 3x \sin x - 6 \cos x \cos 3x}{\sin^2 3x}$$

---

**EXAMPLE 6:** Find the maximum and minimum points on the curve $y = 3 \cos x$.

**Solution:** We take the derivative and set it equal to zero.

$$y' = -3 \sin x = 0$$

$$\sin x = 0$$

$$x = 0, \pm\pi, \pm 2\pi, \pm 3\pi, \ldots$$

The second derivative is

$$y'' = -3 \cos x$$

which is negative when $x$ equals $0, \pm 2\pi, \pm 4\pi, \ldots$, so these are the locations of the maximum points. The others, where the second derivative is positive, are minimum points. This, of course, agrees with our plot of the cosine curve in Fig. 30-3.

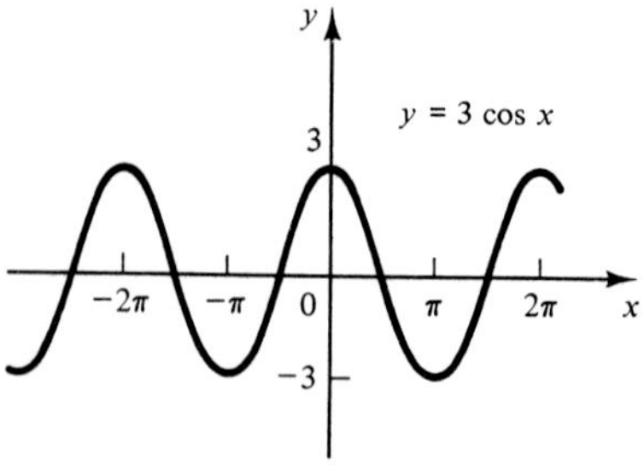


**FIGURE 30-3**

*Implicit* derivatives involving trigonometric functions are handled the same way as shown in Chapter 27.

**EXAMPLE 7:** Find $dy/dx$ if $\sin xy = x^2 y$.

**Solution:**

$$(\cos xy)(xy' + y) = x^2 y' + 2xy$$

$$xy' \cos xy + y \cos xy = x^2 y' + 2xy$$

$$xy' \cos xy - x^2 y' = 2xy - y \cos xy$$

$$(x \cos xy - x^2)y' = 2xy - y \cos xy$$

$$y' = \frac{2xy - y \cos xy}{x \cos xy - x^2}$$

## EXERCISE 1—SINE AND COSINE FUNCTIONS

### First Derivatives

Find the derivative.

1. $y = \sin x$
2. $y = 3 \cos 2x$
3. $y = \cos^3 x$
4. $y = \sin x^2$
5. $y = \sin 3x$
6. $y = \cos 6x$
7. $y = \sin x \cos x$
8. $y = 15.4 \cos^5 x$
9. $y = 3.75\, x \cos x$
10. $y = \sin(\theta + \pi) \cos(\theta - \pi)$
11. $y = \sin^2(\pi - x)$
12. $y = \dfrac{\sin \theta}{\theta}$
13. $y = \sin 2x \cos x$
14. $y = \sin^5 x$
15. $y = \sin^2 x \cos x$
16. $y = \sin 2x \cos 3x$
17. $y = 1.23 \sin^2 x \cos 3x$
18. $y = \frac{1}{2} \sin^2 x$
19. $y = \sqrt{\cos 2t}$
20. $y = 42.7 \sin^2 x$

### Second Derivatives

Find the second derivative of each function.

21. $y = \sin x$
22. $y = \frac{1}{4} \cos 2\theta$
23. $y = x \cos x$
24. If $f(x) = x^2 \cos^3 x$, find $f''(0)$.
25. If $f(x) = x \sin(\pi/2)x$, find $f''(1)$.

### Implicit Functions

Find $dy/dx$ for each implicit function.

26. $y \sin x = 1$
27. $xy - y \sin x - x \cos y = 0$
28. $y = \cos(x - y)$
29. $x = \sin(x + y)$
30. $x \sin y - y \sin x = 0$

### Tangents

Find the slope of the tangent at the given value of $x$.

31. $y = \sin x$ at $x = 2$ rad
32. $y = x - \cos x$ at $x = 1$ rad
33. $y = x \sin \frac{x}{2}$ at $x = 2$ rad
34. $y = \sin x \cos 2x$ at $x = 1$ rad

### Extreme Values and Inflection Points

For each curve, find the maximum, minimum, and inflection points between $x = 0$ and $x = 2\pi$.

35. $y = \sin x$
36. $y = \frac{x}{2} - \sin x$
37. $y = 3 \sin x - 4 \cos x$

### Newton's Method

Sketch each function from $x = 0$ to 10. Calculate a root to three decimal places by Newton's method. If there is more than one root, find only the smallest positive root.

38. $\cos x - x = 0$
39. $\cos 2x - x = 0$
40. $3 \sin x - x = 0$
41. $2 \sin x - x^2 = 0$
42. $\cos x - 2x^2 = 0$

## 30-2 DERIVATIVES OF THE TANGENT, COTANGENT, SECANT, AND COSECANT

### Derivative of tan u

We seek the derivative of $y = \tan u$. By Eq. 162,

$$y = \frac{\sin u}{\cos u}$$

Using the rule for the derivative of a quotient (Eq. 339),

$$\frac{dy}{dx} = \frac{\cos^2 u \frac{du}{dx} + \sin^2 u \frac{du}{dx}}{\cos^2 u}$$

$$= \frac{(\sin^2 u + \cos^2 u) \frac{du}{dx}}{\cos^2 u}$$

and since $\sin^2 u + \cos^2 u = 1$,

$$\frac{dy}{dx} = \frac{1}{\cos^2 u} \frac{du}{dx}$$

and since $1/\cos^2 u = \sec^2 u$,

$$\frac{dy}{dx} = \sec^2 u \frac{du}{dx}$$

So

| | | |
|---|---|---|
| Derivative of the Tangent | $\frac{d(\tan u)}{dx} = \sec^2 u \frac{du}{dx}$ | **342** |

*The derivative of the tangent of some function is the secant squared of that function, multiplied by the derivative of that function.*

---

**EXAMPLE 8:**

(a) If $y = 3 \tan x^2$, then

$$y' = 3(\sec^2 x^2)(2x) = 6x \sec^2 x^2$$

(b) If $y = 2 \sin 3x \tan 3x$, then, by the product rule,

$$\begin{aligned} y' &= 2[\sin 3x(\sec^2 3x)(3) + \tan 3x(\cos 3x)(3)] \\ &= 6 \sin 3x \sec^2 3x + 6 \cos 3x \tan 3x \end{aligned}$$

---

## Derivatives of cot *u*, sec *u*, and csc *u*

Each of these derivatives can be obtained using the rules for the sine, cosine, and tangent already derived, and the following identities

$$\cot u = \frac{\cos u}{\sin u} \qquad \sec u = \frac{1}{\cos u} \qquad \csc u = \frac{1}{\sin u}$$

We list those derivatives here, together with those already found.

**You should memorize at least the first three of these. Notice how the signs alternate. The derivative of each cofunction is negative.**

| | | |
|---|---|---|
| Derivatives of the Trigonometric Functions | $\frac{d(\sin u)}{dx} = \cos u \frac{du}{dx}$ | **340** |
| | $\frac{d(\cos u)}{dx} = -\sin u \frac{du}{dx}$ | **341** |
| | $\frac{d(\tan u)}{dx} = \sec^2 u \frac{du}{dx}$ | **342** |
| | $\frac{d(\cot u)}{dx} = -\csc^2 u \frac{du}{dx}$ | **343** |
| | $\frac{d(\sec u)}{dx} = \sec u \tan u \frac{du}{dx}$ | **344** |
| | $\frac{d(\csc u)}{dx} = -\csc u \cot u \frac{du}{dx}$ | **345** |

**EXAMPLE 9:**

(a) If $y = \sec(x^3 - 2x)$, then

$$y' = [\sec(x^3 - 2x)\tan(x^3 - 2x)](3x^2 - 2)$$
$$= (3x^2 - 2)\ \sec(x^3 - 2x)\tan(x^3 - 2x)$$

(b) If $y = \cot^3 5x$, then by the power rule,

$$y' = 3(\cot 5x)^2(-\csc^2 5x)(5)$$
$$= -15 \cot^2 5x \csc^2 5x$$

## EXERCISE 2—TANGENT, COTANGENT, SECANT, AND COSECANT

### First Derivative

Find the derivative.

1. $y = \tan 2x$
2. $y = \sec 4x$
3. $y = 5 \csc 3x$
4. $y = 9 \cot 8x$
5. $y = 3.25 \tan x^2$
6. $y = 5.14 \sec 2.11x^2$
7. $y = 7 \csc x^3$
8. $y = 9 \cot 3x^3$
9. $y = x \tan x$
10. $y = x \sec x^2$
11. $y = 5x \csc 6x$
12. $y = 9x^2 \cot 2x$
13. $w = \sin \theta \tan 2\theta$
14. $s = \cos t \sec 4t$
15. $v = 5 \tan t \csc 3t$
16. $z = 2 \sin 2\theta \cot 8\theta$
17. If $y = 5.83 \tan^2 2x$, find $y'(1)$.
18. If $f(x) = \sec^3 x$, find $f'(3)$.
19. If $f(x) = 3 \csc^3 3x$, find $f'(3)$.
20. If $y = 9.55x \cot^2 8x$, find $y'(1)$.

### Second Derivative

Find the second derivative.

21. $y = 3 \tan x$
22. $y = 2 \sec 5\theta$
23. If $y = 3 \csc 2\theta$, find $y''(1)$.
24. If $f(x) = 6 \cot 4x$, find $f''(3)$.

### Implicit Functions

Find $dy/dx$ for each implicit function.

25. $y \tan x = 2$
26. $xy + y \cot x = 0$
27. $\sec(x + y) = 7$
28. $x \cot y = y \sec x$

### Tangents

29. Find the tangent to the curve $y = \tan x$, at $x = 1$ rad.
30. Find the tangent to the curve $y = \sec 2x$, at $x = 2$ rad.

### Extreme Values and Inflection Points

For each function, find any maximum, minimum, or inflection points between 0 and $\pi$.

31. $y = 2x - \tan x$
32. $y = \tan x - 4x$

### Rate of Change

33. An object moves with simple harmonic motion so that its displacement $y$ at time $t$ is $y = 6 \sin 4t$ cm. Find the velocity and acceleration of the object when $t = 0.05$ s.

## Related Rates

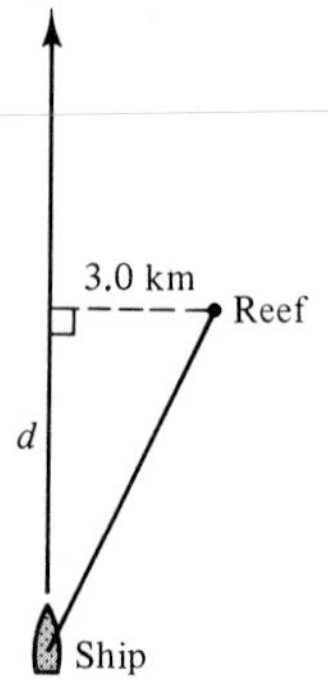

FIGURE 30-4

34. A ship is sailing at 10.0 km/h in a straight line (Fig. 30-4). It keeps its searchlight trained on a reef which is at a perpendicular distance of 3.0 km from the path of the ship. How fast (rad/h) is the light turning when the distance $d$ is 5.0 km?
35. Two cables pass over fixed pulleys $A$ and $B$ (Fig. 30-5) forming an isosceles triangle $ABP$. Point $P$ is being raised at the rate of 3.0 in./min. How fast is $\theta$ changing when $h$ is 4.0 ft?
36. The illumination at a point $P$ on the ground (Fig. 30-6) due to a flare $F$ is $I = k \sin \theta / d^2$ lux, where $k$ is a constant. Find the rate of change of $I$ when the flare is 100 ft above the ground and falling at a rate of 1.0 ft/s, if the illumination at $P$ at that instant is 65 lux.

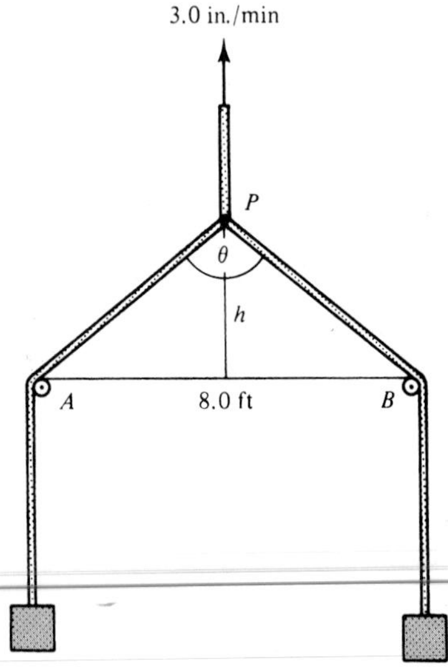

FIGURE 30-5

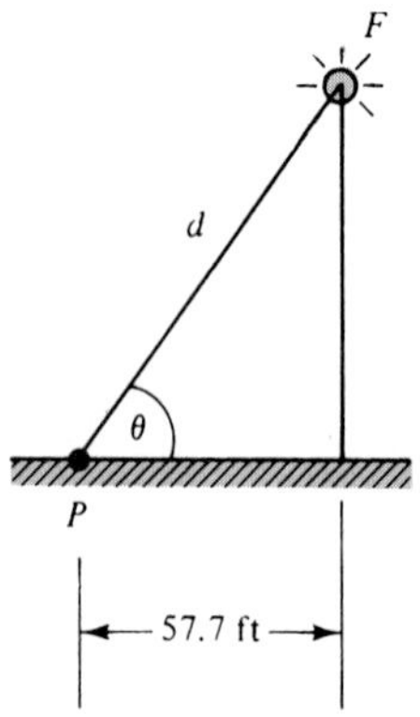

FIGURE 30-6 **Falling flare.**

## Maximum–Minimum Problems

37. Find the length of the shortest ladder (Fig. 30-7) that will touch the ground, the wall, and the house.

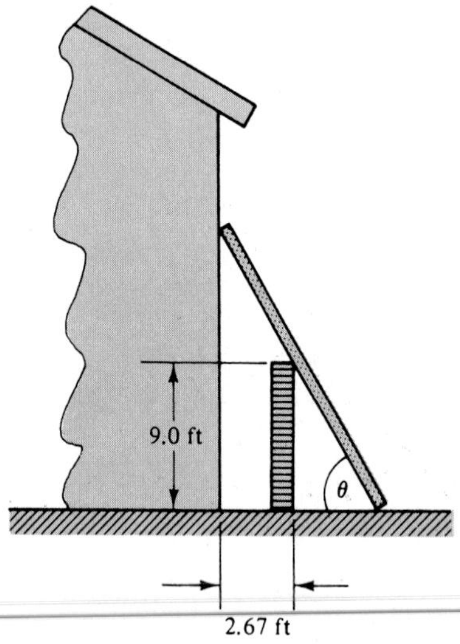

FIGURE 30-7

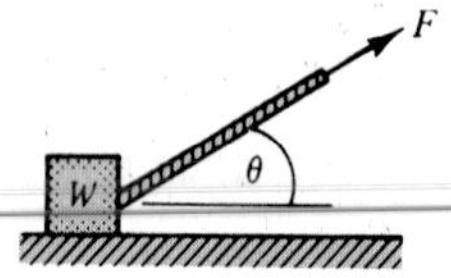

FIGURE 30-8

38. The range $x$ of a projectile fired at an angle $\theta$ with the horizontal at a velocity $v$ is $x = (v^2/g) \sin 2\theta$, where $g$ is the acceleration due to gravity. Find $\theta$ for maximum range.
39. A force $F$ (Fig. 30-8) pulls the weight along a horizontal surface. If $f$ is the coefficient of friction, then

$$F = \frac{fW}{f \sin \theta + \cos \theta}$$

Find $\theta$ for a minimum force when $f = 0.60$.
40. A 20-ft-long steel girder is dragged along a corridor 10.0 ft wide, and then around a corner into another corridor at right angles to the first (Fig. 30-9). Neglecting the thickness of the girder, what must be the width of the second corridor to allow the girder to turn the corner?

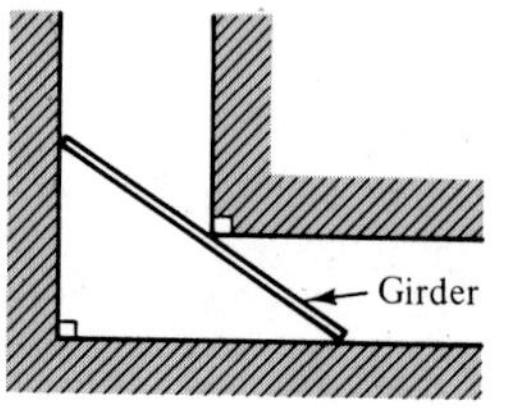

FIGURE 30-9 **Top view of a girder dragged along a corridor.**

41. If a girder (Fig. 30-9) is to be dragged from a 12.8-ft-wide corridor into another 5.4-ft-wide corridor, find the length of the longest girder that will fit around the corner. (Neglect the thickness of the girder.)

## 30-3 DERIVATIVES OF THE INVERSE TRIGONOMETRIC FUNCTIONS

### Derivative of Arcsin *u*

We now seek the derivative of $y = \text{Sin}^{-1}u$, where $y$ is some angle whose sine is $u$, as in Fig. 30-10, whose value we restrict to the range $-\pi/2$ to $\pi/2$. We can then write

$$\sin y = u$$

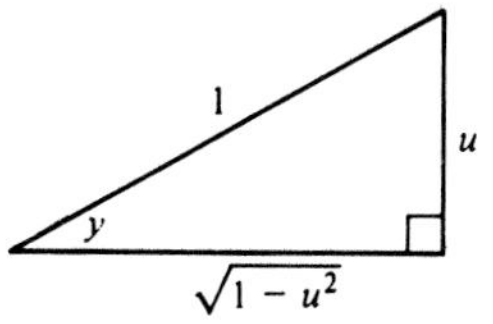

FIGURE 30-10

Taking the derivative yields

$$\cos y \frac{dy}{dx} = \frac{du}{dx}$$

so

$$\frac{dy}{dx} = \frac{1}{\cos y}\frac{du}{dx} \tag{1}$$

But from Eq. 164,

$$\cos^2 y = 1 - \sin^2 y$$

$$\cos y = \pm\sqrt{1 - \sin^2 y}$$

But since $y$ is restricted to values between $-\pi/2$ and $\pi/2$, $\cos y$ cannot be negative. So

$$\cos y = +\sqrt{1 - \sin^2 y} = \sqrt{1 - u^2}$$

Substituting into (1), we have

| | | |
|---|---|---|
| Derivative of the Arcsin | $\dfrac{d(\text{Sin}^{-1}u)}{dx} = \dfrac{1}{\sqrt{1 - u^2}}\dfrac{du}{dx}$<br>$-1 < u < 1$ | **346** |

**EXAMPLE 10:** If $y = \text{Sin}^{-1}3x$, then

$$y' = \frac{1}{\sqrt{1 - (3x)^2}}(3)$$

$$= \frac{3}{\sqrt{1 - 9x^2}}$$

### Derivatives of Arccos, Arctan, Arccot, Arcsec, and Arccsc

The rules for taking derivatives of the remaining inverse trigonometric functions, and the Arcsin as well, are

Try to derive one or more of these. Follow steps similar to those we used for the derivative of Arcsin.

| | | |
|---|---|---|
| Derivatives of the Inverse Trigonometric Functions | $\frac{d(\text{Sin}^{-1}u)}{dx} = \frac{1}{\sqrt{1-u^2}}\frac{du}{dx}$ <br> $-1 < u < 1$ | **346** |
| | $\frac{d(\text{Cos}^{-1}u)}{dx} = \frac{-1}{\sqrt{1-u^2}}\frac{du}{dx}$ <br> $-1 < u < 1$ | **347** |
| | $\frac{d(\text{Tan}^{-1}u)}{dx} = \frac{1}{1+u^2}\frac{du}{dx}$ | **348** |
| | $\frac{d(\text{Cot}^{-1}u)}{dx} = \frac{-1}{1+u^2}\frac{du}{dx}$ | **349** |
| | $\frac{d(\text{Sec}^{-1}u)}{dx} = \frac{1}{u\sqrt{u^2-1}}\frac{du}{dx}$ <br> $\|u\| > 1$ | **350** |
| | $\frac{d(\text{Csc}^{-1}u)}{dx} = \frac{-1}{u\sqrt{u^2-1}}\frac{du}{dx}$ <br> $\|u\| > 1$ | **351** |

**EXAMPLE 11:**

(a) If $y = \text{Cot}^{-1}(x^2 + 1)$, then

$$y' = \frac{-1}{1 + (x^2 + 1)^2}(2x)$$

$$= \frac{-2x}{2 + 2x^2 + x^4}$$

(b) If $y = \text{Cos}^{-1}\sqrt{1 - x}$, then

$$y' = \frac{-1}{\sqrt{1 - (1 - x)}} \cdot \frac{-1}{2\sqrt{1 - x}}$$

$$= \frac{1}{2\sqrt{x - x^2}}$$

## EXERCISE 3—INVERSE TRIGONOMETRIC FUNCTIONS

Find the derivative.

1. $y = x\,\text{Sin}^{-1}x$

2. $y = \text{Sin}^{-1}\frac{x}{a}$

3. $y = \operatorname{Cos}^{-1}\dfrac{x}{a}$

4. $y = \operatorname{Tan}^{-1}(\sec x + \tan x)$

5. $y = \operatorname{Sin}^{-1}\dfrac{\sin x - \cos x}{\sqrt{2}}$

6. $y = \sqrt{2ax - x^2} + a \operatorname{Cos}^{-1}\dfrac{\sqrt{2ax - x^2}}{a}$

7. $y = t^2 \operatorname{Cos}^{-1}t$

8. $y = \operatorname{Arcsin} 2x$

9. $y = \operatorname{Arctan}(1 + 2x)$

10. $y = \operatorname{Arccot}(2x + 5)^2$

11. $y = \operatorname{Arccot} \dfrac{x}{a}$

12. $y = \operatorname{Arcsec} \dfrac{1}{x}$

13. $y = \operatorname{Arccsc} 2x$

14. $y = \operatorname{Arcsin}\sqrt{x}$

15. $y = t^2 \operatorname{Arcsin}\dfrac{t}{2}$

16. $y = \operatorname{Sin}^{-1}\dfrac{x}{\sqrt{1 + x^2}}$

17. $y = \operatorname{Sec}^{-1}\dfrac{a}{\sqrt{a^2 - x^2}}$

Find the slope of the tangent to each curve.

18. $y = x \operatorname{Arcsin} x$ at $x = \frac{1}{2}$

19. $y = \dfrac{\operatorname{Arctan} x}{x}$ at $x = 1$

20. $y = x^2 \operatorname{Arccsc}\sqrt{x}$ at $x = 2$

21. $y = \sqrt{x} \operatorname{Arccot}\dfrac{x}{4}$ at $x = 4$

22. Find the equations of the tangents to the curve $y = \operatorname{Arctan} x$ having a slope of $\frac{1}{4}$.

23. Use differentials to find the approximate amount by which Arcsin $x$ will change (in degrees) when $x$ changes from 0.50 to 0.52.

## 30-4 DERIVATIVES OF LOGARITHMIC FUNCTIONS

### Derivative of $\log_b u$

Let us now use the delta method again to find the derivative of the logarithmic function $y = \log_b u$. We first let $u$ take on an increment $\Delta u$ and $y$ an increment $\Delta y$,

$$y + \Delta y = \log_b(u + \Delta u)$$

Subtracting gives us

$$\Delta y = \log_b(u + \Delta u) - \log_b u = \log_b\frac{u + \Delta u}{u}$$

by the law of logarithms for quotients. Now dividing by $\Delta u$ yields

$$\frac{\Delta y}{\Delta u} = \frac{1}{\Delta u}\log_b\frac{u + \Delta u}{u}$$

We now do some manipulation to get our expression into a form that will be easier to evaluate. We start by multiplying the right side by $u/u$.

$$\begin{aligned}\frac{\Delta y}{\Delta u} &= \frac{u}{u} \cdot \frac{1}{\Delta u}\log_b\frac{u + \Delta u}{u} \\ &= \frac{1}{u} \cdot \frac{u}{\Delta u}\log_b\frac{u + \Delta u}{u}\end{aligned}$$

Then using the law of logarithms for powers (Eq. 188), we have

$$\frac{\Delta y}{\Delta u} = \frac{1}{u}\log_b\left(\frac{u + \Delta u}{u}\right)^{u/\Delta u}$$

We now let $\Delta u$ approach zero,

$$\lim_{\Delta u \to 0} \frac{\Delta y}{\Delta u} = \frac{dy}{du} = \lim_{\Delta u \to 0} \frac{1}{u}\log_b\left(\frac{u + \Delta u}{u}\right)^{u/\Delta u}$$

$$= \frac{1}{u}\log_b\left[\lim_{\Delta u \to 0}\left(1 + \frac{\Delta u}{u}\right)^{u/\Delta u}\right]$$

Let us simplify the expression inside the brackets by making the substitution

$$k = \frac{u}{\Delta u}$$

Then $k$ will approach infinity as $\Delta u$ approaches zero, and the expression inside the brackets becomes

$$\lim_{\Delta u \to 0}\left(1 + \frac{\Delta u}{u}\right)^{u/\Delta u} = \lim_{k \to \infty}\left(1 + \frac{1}{k}\right)^k$$

This limit defines the number $e$, the familiar base of natural logarithms.

Glance back at Sec. 19-2 where we developed this expression

$$e \equiv \lim_{k \to \infty}\left(1 + \frac{1}{k}\right)^k$$

Our derivative thus becomes

$$\frac{dy}{du} = \frac{1}{u}\log_b e$$

We are nearly finished now. Using the chain rule, we get $dy/dx$ by multiplying $dy/du$ by $du/dx$, so

| | | |
|---|---|---|
| Derivative of $\log_b u$ | $\frac{d(\log_b u)}{dx} = \frac{1}{u}\log_b e\frac{du}{dx}$ | **352a** |

or, since $\log_b e = 1/\ln b$,

This form is more useful when the base $b$ is a number other than 10.

| | | |
|---|---|---|
| Derivative of $\log_b u$ | $\frac{d(\log_b u)}{dx} = \frac{1}{u \ln b}\frac{du}{dx}$ | **352b** |

**EXAMPLE 12:** Take the derivative of $y = \log(x^2 - 3x)$.

**Solution:** The base $b$, when not otherwise indicated, is taken as 10. By Eq. 352a,

$$\frac{dy}{dx} = \frac{1}{x^2 - 3x}(\log e)(2x - 3)$$

$$= \frac{2x - 3}{x^2 - 3x}\log e$$

Or, since $\log e \simeq 0.4343$,

$$\frac{dy}{dx} = 0.4343\left(\frac{2x - 3}{x^2 - 3x}\right)$$

---

**EXAMPLE 13:** Find the derivative of $y = x \log_3 x^2$.

**Solution:** By the product rule,

$$y' = x\left(\frac{1}{x^2 \ln 3}\right)(2x) + \log_3 x^2$$

$$= \frac{2}{\ln 3} + \log_3 x^2 \simeq 1.82 + \log_3 x^2$$

---

### Derivative of ln *u*

Our efforts in deriving Eqs. 352 will now pay off, because we can use that result to find the derivative of the natural logarithm of a function, as well as the derivatives of exponential functions in the following sections. To find the derivative of $y = \ln u$ we use Eq. 352a:

$$\frac{dy}{dx} = \frac{1}{u}\ln e\frac{du}{dx}$$

But by Eq. 191, $\ln e = 1$, so

| | | |
|---|---|---|
| Derivative of ln *u* | $\dfrac{d(\ln u)}{dx} = \dfrac{1}{u}\dfrac{du}{dx}$ | **353** |

*The derivative of the natural logarithm of a function is the reciprocal of that function multiplied by its derivative.*

---

**EXAMPLE 14:** Differentiate $y = \ln(2x^3 + 5x)$.

**Solution:** By Eq. 353,

$$\frac{dy}{dx} = \frac{1}{2x^3 + 5x}(6x^2 + 5)$$

$$= \frac{6x^2 + 5}{2x^3 + 5x}$$

---

The rule for derivatives of the logarithmic function is often used with our former rules for derivatives.

**EXAMPLE 15:** Take the derivative of $y = x^3 \ln(5x + 2)$.

**Solution:** Using the product rule together with our rule for logarithms gives us

$$\frac{dy}{dx} = x^3\left(\frac{1}{5x + 2}\right)5 + [\ln(5x + 2)]3x^2 = \frac{5x^3}{5x + 2} + 3x^2 \ln(5x + 2)$$

Our work is sometimes made easier if we first use the laws of logarithms to simplify a given expression.

**EXAMPLE 16:** Take the derivative of $y = \ln\dfrac{x\sqrt{2x - 3}}{\sqrt[3]{4x + 1}}$.

**Solution:** We first rewrite the given expression using the laws of logarithms,

$$y = \ln x + \frac{1}{2}\ln(2x - 3) - \frac{1}{3}\ln(4x + 1)$$

We now take the derivative term by term,

$$\frac{dy}{dx} = \frac{1}{x} + \frac{1}{2}\left(\frac{1}{2x - 3}\right)2 - \frac{1}{3}\left(\frac{1}{4x + 1}\right)4$$

$$= \frac{1}{x} + \frac{1}{2x - 3} - \frac{4}{3(4x + 1)}$$

## Logarithmic Differentiation

Here we use logarithms to aid in differentiating nonlogarithmic expressions. Derivatives of some complicated expressions can be found more easily if we first take the logarithm of both sides of the given expression and simplify by means of the laws of logarithms. (These operations will not change the meaning of the original expression.) We then take the derivative.

**EXAMPLE 17:** Differentiate $y = \dfrac{\sqrt{x - 2}\sqrt[3]{x + 3}}{\sqrt[4]{x + 1}}$.

**We could instead take the common log of both sides, but the natural log has a simpler derivative.**

**Solution:** We will use logarithmic differentiation here. Instead of proceeding in the usual way, we first take the natural log of both sides of the equation, and apply the laws of logarithms:

$$\ln y = \frac{1}{2}\ln(x - 2) + \frac{1}{3}\ln(x + 3) - \frac{1}{4}\ln(x + 1)$$

Taking the derivative, we have

$$\frac{1}{y}\frac{dy}{dx} = \frac{1}{2}\left(\frac{1}{x - 2}\right) + \frac{1}{3}\left(\frac{1}{x + 3}\right) - \frac{1}{4}\left(\frac{1}{x + 1}\right)$$

Finally, multiplying by $y$ to solve for $dy/dx$ gives us

**We could now use the original expression to replace the *y* in the answer.**

$$\frac{dy}{dx} = y\left[\frac{1}{2(x - 2)} + \frac{1}{3(x + 3)} - \frac{1}{4(x + 1)}\right]$$

In other cases, this method will allow us to take derivatives not possible with our other rules.

**EXAMPLE 18:** Find the derivative of $y = x^{2x}$.

**Solution:** This is not a power function, because the exponent is not a constant, nor is it an exponential function, because the base is not a constant. So neither rule 335 nor 354 applies. But let us use logarithmic differentiation by first taking the natural logarithm of both sides:

$$\ln y = \ln x^{2x} = 2x \ln x$$

Now taking the derivative by means of the product rule yields

$$\frac{1}{y}\frac{dy}{dx} = 2x\left(\frac{1}{x}\right) + (\ln x)2$$

$$= 2 + 2 \ln x$$

Multiplying by $y$, we get

$$\frac{dy}{dx} = 2(1 + \ln x)y$$

Replacing $y$ by $x^{2x}$ gives us

$$\frac{dy}{dx} = 2(1 + \ln x)x^{2x}$$

## EXERCISE 4—LOGARITHMIC FUNCTIONS

### Derivative of $\log_b u$

Differentiate.

1. $y = \log 7x$
2. $y = \log x^{-2}$
3. $y = \log_b x^3$
4. $y = \log_a(x^2 - 3x)$
5. $y = \log(x\sqrt{5 + 6x})$
6. $y = \log_a\left(\dfrac{1}{2x + 5}\right)$
7. $y = x \log\dfrac{2}{x}$
8. $y = \log\dfrac{(1 + 3x)}{x^2}$

### Derivative of ln u

Differentiate.

9. $y = \ln 3x$
10. $y = \ln x^3$
11. $y = \ln(x^2 - 3x)$
12. $y = \ln(4x - x^3)$
13. $y = 2.75x \ln 1.02x^3$
14. $w = z^2 \ln(1 - z^2)$
15. $y = \dfrac{\ln(x + 5)}{x^2}$
16. $y = \dfrac{\ln x^2}{3 \ln(x - 4)}$
17. $s = \ln\sqrt{t - 5}$
18. $y = 5.06 \ln\sqrt{x^2 - 3.25x}$

### With Trigonometric Functions

Differentiate.

19. $y = \ln \sin x$
20. $y = \ln \sec x$
21. $y = \sin x \ln \sin x$
22. $y = \ln(\sec x + \tan x)$

### Implicit Relations

Find $dy/dx$.

23. $y \ln y + \cos x = 0$

24. $\ln x^2 - 2x \sin y = 0$

25. $x - y = \ln(x + y)$

26. $xy = a^2 \ln\frac{x}{a}$

27. $\ln y + x = 10$

### Logarithmic Differentiation

**Remember to start these by taking the logarithm of both sides.**

Differentiate.

28. $y = \dfrac{\sqrt{x+2}}{\sqrt[3]{2-x}}$

29. $y = \dfrac{\sqrt{a^2 - x^2}}{x}$

30. $y = x^x$

31. $y = x^{\sin x}$

32. $y = (\cot x)^{\sin x}$

33. $y = (\cos^{-1} x)^x$

### Tangent to a Curve

Find the slope of the tangent at the point indicated.

34. $y = \log x$ at $x = 1$

35. $y = \ln x$ where $y = 0$

36. $y = \ln(x^2 + 2)$ at $x = 4$

37. $y = \log(4x - 3)$ at $x = 2$

38. Find the equation of the tangent to the curve $y = \ln x$ at $y = 0$.

### Angle of Intersection

Find the angle of intersection of each pair of curves.

39. $y = \ln(x + 1)$ and $y = \ln(7 - 2x)$ at $x = 2$

40. $y = x \ln x$ and $y = x \ln(1 - x)$ at $x = \frac{1}{2}$

### Extreme Values and Points of Inflection

Find the maximum, minimum, and inflection points for each curve.

41. $y = x \ln x$

42. $y = x^3 \ln x$

43. $y = \dfrac{x}{\ln x}$

44. $y = \ln(8x - x^2)$

### Newton's Method

Find the smallest positive root between $x = 0$ and $x = 10$, to two decimal places.

45. $x - 10 \log x = 0$

46. $\tan x - \log x = 0$

### Applications

47. A certain underwater cable has a core of copper wires covered by insulation. The speed of transmission of a signal along the cable is

$$S = x^2 \ln\frac{1}{x}$$

where $x$ is the ratio of the radius of the core to the thickness of the insulation. What value of $x$ gives the greatest signal speed?

48. The heat loss $q$ per foot of cylindrical pipe insulation (Fig. 30-11) having an inside radius $r_1$ and outside radius $r_2$ is given by the logarithmic equation

$$q = \frac{2\pi k(t_1 - t_2)}{\ln(r_2/r_1)} \quad \text{Btu/h}$$

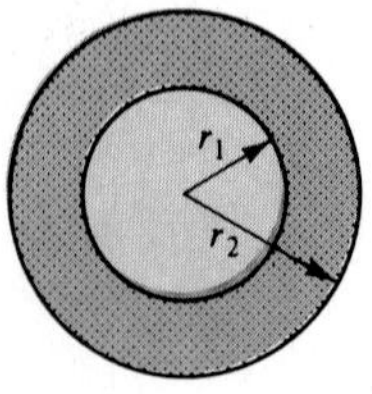

FIGURE 30-11 **Insulated pipe.**

where $t_1$ and $t_2$ are the inside and outside temperatures (°F) and $k$ is the conductivity of the insulation. A 4-in.-thick insulation having a conductivity of 0.036 is wrapped around a 9-in.-diameter pipe at 550°F, and the surroundings are at 90°F. Find the rate of change of heat loss $q$ if the insulation thickness is decreasing at the rate of 0.1 in./h.

49. The pH value of a solution having a concentration $C$ of hydrogen ions is $\text{pH} = -\log_{10}C$. Use differentials to estimate the pH change if the concentration of hydrogen ions is increased from $2.0 \times 10^{-5}$ to $2.1 \times 10^{-5}$.

50. The difference in elevation, in feet, between two locations having barometric readings of $B_1$ and $B_2$ inches of mercury is given by $h = 60{,}470 \log \dfrac{B_2}{B_1}$, where $B_1$ is the pressure at the upper location. At what rate is an airplane changing in elevation when the barometric pressure outside the airplane is 21.5 in. of mercury and decreasing at the rate of 0.50 in. per minute?

**Hint:** Treat $B_2$ as a *constant.*

51. The power input to a certain amplifier is 2.0 W; the power output is 400 W, but due to a defective component is dropping at the rate of 0.5 W per day. Use Eq. A102 to find the rate (decibels per day) at which the decibels are decreasing.

## 30-5 DERIVATIVE OF THE EXPONENTIAL FUNCTION

### Derivative of $b^u$

We seek the derivative of the exponential function $y = b^u$, where $b$ is a constant and $u$ is a function of $x$. We can get the derivative without having to use the delta method by using the rule we derived for the logarithmic function (Eq. 353). We first take the natural logarithm of both sides,

$$\ln y = \ln b^u = u \ln b$$

by Eq. 188. We now take the derivative of both sides, remembering that $\ln b$ is a constant,

$$\frac{1}{y}\frac{dy}{dx} = \frac{du}{dx}\ln b$$

Multiplying by $y$, we have

$$\frac{dy}{dx} = y\frac{du}{dx}\ln b$$

Finally, replacing $y$ by $b^u$ gives us

| | | |
|---|---|---|
| Derivative of $b^u$ | $\dfrac{d(b^u)}{dx} = b^u\dfrac{du}{dx}\ln b$ | **354** |

*The derivative of a base b raised to a function u is the product of $b^u$, the derivative of the function, and the natural log of the base.*

---

**EXAMPLE 19:** Find the derivative of $y = 10^{x^2+2}$.

**Solution:** By Eq. 354,

$$\frac{dy}{dx} = 10^{x^2+2}(2x)\ln 10 = 2x(\ln 10)10^{x^2+2}$$

---

| Common Error | Do not confuse the exponential function $y = b^x$ with the power function $y = x^n$. The derivative of $b^x$ **is not** $xb^{x-1}$! The derivative of $b^x$ *is* $b^x \ln b$. |
|---|---|

## Derivative of $e^u$

We will use Eq. 354 mostly when the base $b$ is the base of natural logarithms, $e$.

$$y = e^u$$

Taking the derivative by Eq. 354 gives us

$$\frac{dy}{dx} = e^u \frac{du}{dx} \ln e$$

But since $\ln e = 1$, we get

| Derivative of $e^u$ | $\frac{d}{dx}e^u = e^u \frac{du}{dx}$ | **355** |
|---|---|---|

*The derivative of an exponential function $e^u$ is the same exponential function, multiplied by the derivative of the exponent.*

Note that $y = c_1 e^{mx}$ and its derivatives $y' = c_2 e^{mx}$, $y'' = c_3 e^{mx}$, etc., are all *like terms,* since $c_1$, $c_2$, and $c_3$ are constants. We use this fact when solving differential equations in Chapter 35.

**EXAMPLE 20:** Find the first, second, and third derivatives of $y = 2e^{3x}$.

**Solution:** By rule 355,

$$y' = 6e^{3x} \qquad y'' = 18e^{3x} \qquad y''' = 54e^{3x}$$

**EXAMPLE 21:** Find the derivative of $y = e^{x^3+5x^2}$.

**Solution:**

$$\frac{dy}{dx} = e^{x^3+5x^2}(3x^2 + 10x)$$

**EXAMPLE 22:** Find the derivative of $y = x^3 e^{x^2}$.

**Solution:** Using the product rule together with Eq. 355, we have

$$\frac{dy}{dx} = x^3(e^{x^2})(2x) + e^{x^2}(3x^2)$$

$$= 2x^4e^{x^2} + 3x^2e^{x^2} = x^2e^{x^2}(2x^2 + 3)$$

## Derivative of $y = x^n$, Where $n$ Is Any Real Number

We now return to some unfinished business regarding the power rule. We have already shown that the derivative of the power function $x^n$ is $nx^{n-1}$, when $n$ is any rational number, positive or negative. Now we show that $n$ can also be an irrational number, such as $e$ or $x$. Making use of the fact that $e^{\ln z} = z$, we let

$$x = e^{\ln x}$$

Raising to the $n$th power gives

$$x^n = (e^{\ln x})^n = e^{n \ln x}$$

Then

$$\frac{d}{dx}x^n = \frac{d}{dx}e^{n \ln x}$$

By Eq. 355,

$$\frac{d}{dx}x^n = \frac{n}{x}e^{n \ln x}$$

Substituting $x^n$ for $e^{n \ln x}$ gives

$$\frac{d}{dx}x^n = \frac{n}{x}x^n = nx^{n-1}$$

Thus the power rule holds when the exponent $n$ is any real number.

---

**EXAMPLE 23:** The derivative of $3x^\pi$ is

$$\frac{d}{dx}3x^\pi = 3\pi x^{\pi-1}$$

---

## EXERCISE 5—EXPONENTIAL FUNCTION

### Derivative of $b^u$

Differentiate.

1. $y = 3^{2x}$
2. $y = 10^{2x+3}$
3. $y = (x)(10^{2x+3})$
4. $y = 10^{3x}$
5. $y = 2^{x^2}$
6. $y = 7^{2x}$

### Derivative of $e^u$

Differentiate.

7. $y = e^{2x}$
8. $y = e^{x^2}$
9. $y = e^{e^x}$
10. $y = e^{3x^2+4}$
11. $y = e^{\sqrt{1-x^2}}$
12. $y = xe^x$
13. $y = \dfrac{2}{e^x}$
14. $y = (3x + 2)e^{-x^2}$
15. $y = x^2e^{3x}$
16. $y = xe^{-x}$
17. $y = \dfrac{e^x}{x}$
18. $y = \dfrac{x^2 - 2}{e^{3x}}$
19. $y = \dfrac{e^x - e^{-x}}{x^2}$
20. $y = \dfrac{e^x - x}{e^{-x} + x^2}$
21. $y = (x + e^x)^2$
22. $y = (e^x + 2x)^3$

23. $y = \dfrac{(1 + e^x)^2}{x}$

24. $y = \left(\dfrac{e^x + 1}{e^x - 1}\right)^2$

### With Trigonometric Functions

Differentiate.

25. $y = \sin^3 e^x$
26. $y = e^x \sin x$
27. $y = e^\theta \cos 2\theta$
28. $y = e^x(\cos bx + \sin bx)$

### Implicit Relations

Find $dy/dx$.

29. $e^x + e^y = 1$
30. $e^x \sin y = 0$

31. $e^y = \sin(x + y)$

Evaluate each expression.

32. $f'(2)$ where $f(x) = e^{\sin(\pi x/2)}$

33. $f''(0)$ where $f(t) = e^{\sin t} \cos t$
34. $f'(1)$ and $f''(1)$ where $f(x) = e^{x-1} \sin \pi x$

### With Logarithmic Functions

Differentiate.

35. $y = e^x \ln x$

36. $y = \ln(x^2e^x)$

37. $y = \ln x^{e^x}$

38. $y = \ln e^{2x}$

Find the second derivative.

39. $y = e^t \cos t$

40. $y = e^{-t} \sin 2t$

41. $y = e^x \sin x$

42. $y = \frac{1}{2}(e^x + e^{-x})$

### Newton's Method

Find the smallest root that is greater than zero, to two decimal places.

43. $e^x + x - 3 = 0$

44. $xe^{-0.02x} = 1$

45. $5e^{-x} + x - 5 = 0$

46. $e^x = \tan x$

### Maximum, Minimum, and Inflection Points

47. Find the minimum point of $y = e^{2x} + 5e^{-2x}$.
48. Find the maximum point and the points of inflection of $y = e^{-x^2}$.
49. Find the maximum and minimum points for one cycle of $y = 10e^{-x} \sin x$.

### Applications

50. If \$10,000 is invested for $t$ years at an annual interest rate of 10%, compounded continuously, it will accumulate to an amount $y$, where $y = 10{,}000e^{0.1t}$. At what rate, in dollars per year, is the balance growing when **(a)** $t = 0$ years; **(b)** $t = 10$ years?

51. If we assume that the price of an automobile is increasing or "inflating" exponentially at an annual rate of 8%, at what rate in dollars per year is the price of a car that initially cost \$9000 increasing after 3 years?

52. When a certain object is placed in an oven at 1000°F, its temperature $T$ rises according to the equation $T = 1000(1 - e^{-0.1t})$, where $t$ is the elapsed time (minutes). How fast is the temperature rising (in degrees per minute) when **(a)** $t = 0$; **(b)** $t = 10$ min?

53. A catenary has the equation $y = \frac{1}{2}(e^x + e^{-x})$. Find the slope of the catenary when $x = 5$.

54. Verify that the minimum point on the catenary described in Problem 53 occurs at $x = 0$.

55. The speed $N$ of a certain flywheel is decaying exponentially according to the equation $N = 1855e^{-0.5t}$ (rev/min), where $t$ is the time (minutes) after the power is disconnected. Find the angular acceleration (the rate of change of $N$) when $t = 1$ min.

56. The height $y$ of a certain pendulum released from a height of 50.0 cm is $y = 50.0e^{-0.75t}$ cm, where $t$ is the time after release, in seconds. Find the vertical component of the velocity of the pendulum when $t = 1.00$ s.

57. The number of bacteria in a certain culture is growing exponentially. The number $N$ of bacteria at any time $t$ (hours) is $N = 10{,}000e^{0.1t}$. At what rate (number of bacteria per hour) is the population increasing when **(a)** $t = 0$; **(b)** $t = 100$ h?

58. A capacitor (Fig. 30-12) is charged to 300 V. When the switch is closed, the voltage across $R$ is initially 300 V but then drops according to the equation $V_1 = 300e^{-t/RC}$, where $t$ is the time (in seconds), $R$ is the resistance, and $C$ is the capacitance. If the voltage $V_2$ also starts to rise at the instant of switch closure so that $V_2 = 100t$, the total voltage $V$ will be $V = 300e^{-t/RC} + 100t$. Find $t$ when $V$ is a minimum.

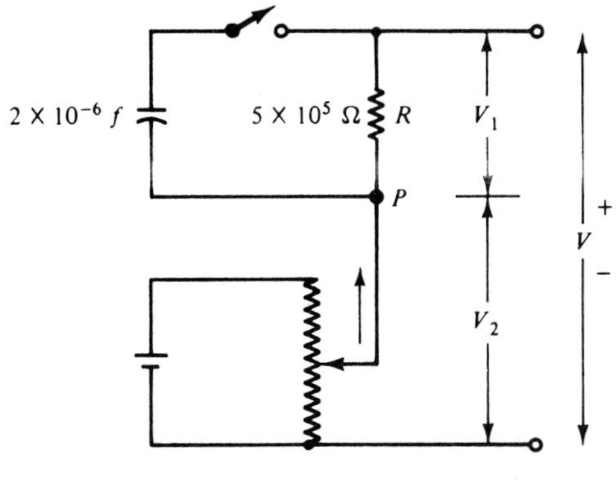


FIGURE 30-12

59. The atmospheric pressure at a height of $h$ miles above the earth's surface is given by $p = 29.92e^{-h/5}$ inches of mercury. Find the rate of change of the pressure on a rocket which is at 18 mi and climbing at a rate of 1500 mi/h.

60. The equation in Problem 59 becomes $p = 2121e^{-0.000037h}$ when $h$ is in feet and $p$ is in pounds per square foot. Find the rate of change of pressure on an aircraft at 5000 ft, climbing at a rate of 10 ft/s.

61. The approximate density of seawater at a depth of $h$ miles is $d = 64.0e^{0.00676h}$ lb/ft$^3$. Use differentials to estimate the change in density when going from a depth of 1.0 mi to 1.1 mi.

62. The force $F$ needed to hold a weight $W$ (Fig. 30-13) is $F = We^{-\mu\theta}$. For a certain beam with $\mu = 0.15$, an angle of wrap of 4.62 rad is needed to hold a weight of 200 lb with a force of 100 lb. Find the rate of change of $F$ if the rope is unwrapping at a rate of 15°/s.

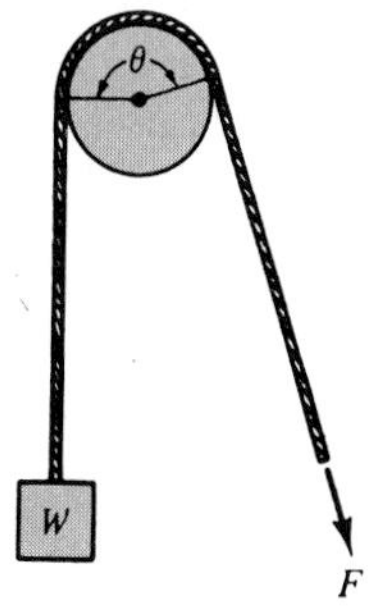


FIGURE 30-13

## CHAPTER 30 REVIEW PROBLEMS

*Find dy/dx.*

1. $y = \frac{a}{2}(e^{x/a} - e^{-x/a})$
2. $y = 5^{2x+3}$
3. $y = 8 \tan \sqrt{x}$
4. $y = \sec^2 x$
5. $y = x \operatorname{Arctan} 4x$
6. $y = \frac{1}{\sqrt{\operatorname{Arcsin} 2x}}$
7. $y^2 = \sin 2x$
8. $y = xe^{2x}$
9. $y = x \sin x$
10. $y = x^2 \sin x$
11. $y = x^3 \cos x$
12. $y = \ln \sin(x^2 + 3x)$
13. $y = \frac{\sin x}{x}$
14. $y = (\log x)^2$
15. $y = \log x(1 + x^2)$
16. $\cos(x - y) = 2x$
17. $y = \ln(x + \sqrt{x^2 + a^2})$
18. $y = \ln(x + 10)$
19. $y = \csc 3x$
20. $y = \ln(x^2 + 3x)$
21. $y = \frac{\sin x}{\cos x}$
22. $y = \frac{1}{\cos^2 x}$
23. $y = \ln(2x^3 + x)$
24. $y = x \operatorname{Arcsin} 2x$
25. $y = x^2 \operatorname{Arccos} x$
26. Find the minimum point of the curve $y = \ln(x^2 - 2x + 3)$.
27. Find the points of inflection of the curve $xy = 4 \log(x/2)$.
28. Find a minimum point and a point of inflection on the curve $y \ln x = x$. Write the equation of the tangent at the point of inflection.
29. At what $x$ between $-\pi/2$ and $\pi/2$ is there a maximum on the curve $y = 2 \tan x - \tan^2 x$?

*Find the value of dy/dx for the given value of x.*

30. $y = x \operatorname{Arccos} x$ at $x = -\frac{1}{2}$
31. $y = \frac{\operatorname{Arcsec} 2x}{\sqrt{x}}$ at $x = 1$
32. If $x^2 + y^2 = \ln y + 2$, find $y'$ and $y''$ at the point (1, 1).

*Find the equation of the tangent to each curve.*

33. $y = \sin x$ at $x = \pi/6$
34. $y = x \ln x$ parallel to the line $3x - 2y = 5$.
35. At what $x$ is the tangent to the curve $y = \tan x$ parallel to the line $y = 2x + 5$?

*Find the smallest positive root between x = 0 and x = 10, to three decimal places.*

36. $\sin 3x - \cos 2x = 0$
37. $2 \sin \frac{1}{2}x - \cos 2x = 0$
38. Find the angle of intersection between $y = \ln(x^3/8 - 1)$ and $y = \ln(3x - x^2/4 - 1)$ at $x = 4$.
39. A casting is taken from one oven at 1500°F and placed in another oven whose temperature is 0°F and rising at a linear rate of 100° per hour. The temperature $T$ of the casting after $t$ hours is then $T = 100t + 1500e^{-0.2t}$. Find the minimum temperature reached by the casting and the time at which it occurs.
40. A statue 11 ft tall is on a pedestal, so that the bottom of the statue is 25 ft above eye level. How far from the statue (measured horizontally) should an observer stand so that the statue will subtend the greatest angle at the observer's eye?

*Writing*

41. Examine this chapter and locate any points that are not clear. Write a letter to the author in which you give specific examples of what you think can be explained better. Point out where you would like more examples, and be specific about the type. Send your letter to the author at Vermont Technical College, Randolph Center, VT 05061. We promise an answer. (You may, of course, do the same for anything in this text.)

# 31

# INTEGRATION

## OBJECTIVES

**When you have completed this chapter, you should be able to:**

- Find the integral of certain algebraic functions.
- Check an integral by differentiating.
- Solve simple differential equations.
- Evaluate the constant of integration, given boundary conditions.
- Perform successive integration.
- Use a table of integrals.
- Use integration to find displacement and velocity of a body in straight-line or curvilinear motion.
- Use integration to find the charge, voltage, and current in capacitors and inductors.

Every operation in mathematics has its inverse; we reverse the squaring operation by taking the square root, the arcsin is the inverse of the sine, and so on. In this chapter we learn how to reverse the process of differentiation with the process of *integration*.

The material in this and the following three chapters usually falls under the heading of *integral calculus*, in contrast to the differential calculus already introduced. We will present some applications (accelerated motion and electric circuits) in this chapter, but most applications of integration will be in Chapters 32 and 33.

## 31-1 THE INDEFINITE INTEGRAL

### Reversing the Process of Differentiation

An *antiderivative* of an expression is a new expression which, if differentiated, gives the original expression.

---

**EXAMPLE 1:** The derivative of $x^3$ is $3x^2$, so an antiderivative of $3x^2$ is $x^3$.

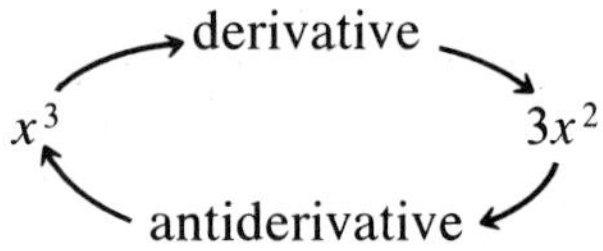


---

The derivative of $x^3$ is $3x^2$, but the derivative $x^3 + 6$ is also $3x^2$. The derivatives of $x^3 - 99$, and $x^3$ + *any constant* are also $3x^2$. This constant, called the *constant of integration*, must be included when we find *the* antiderivative of a function.

---

We will learn how to evaluate the constant of integration in Sec. 31-3.

**EXAMPLE 2:** The derivative of $x^3$ + (any constant) is $3x^2$, so the antiderivative of $3x^2$ is $x^3$ + (a constant). We use $C$ to stand for the constant.

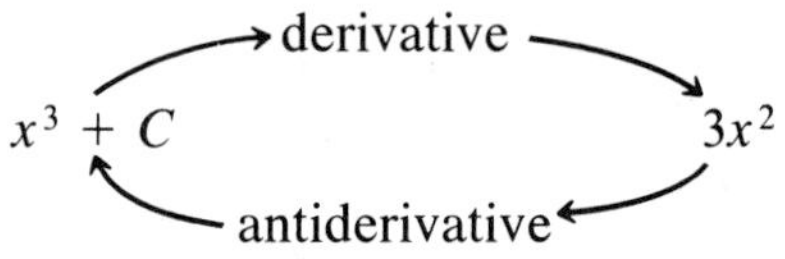


---

The antiderivative is also called the *indefinite integral*, and we will use both names. The process of finding the integral or antiderivative is called *integration*.

### The Integral Sign

We have seen above that the derivative of $x^3 + C$ is $3x^2$, so the antiderivative of $3x^2$ is $x^3 + C$. Let us now state this same idea more formally.

Let there be a function $F(x) + C$. Its derivative is

$$\frac{d}{dx}[F(x) + C] = F'(x) + 0 = F'(x)$$

where we have used the familiar prime (′) notation to indicate a derivative. Going to differential form, we have

$$d[F(x) + C] = F'(x)\,dx$$

Thus the differential of $F(x) + C$ is $F'(x)\,dx$. Conversely, the antiderivative of $F'(x)\,dx$ is $F(x) + C$:

$$\text{antiderivative of } F'(x)\,dx = F(x) + C$$

Instead of writing "antiderivative of" we use the *integral sign* $\int$ to indicate the antiderivative.

**It is no accident that the integral sign looks like "S." We will see in Chapter 32 that it stands for *summation.***

***The integral is called indefinite* because of the unknown constant. In Chapter 32 we study the definite integral, which has no unknown constant.**

| | | |
|---|---|---|
| Indefinite Integral or Antiderivative | $\int F'(x)\,dx = F(x) + C$ | **363** |

*The integral of the differential of a function is equal to the function itself, plus a constant.*

We read this as "the integral of $F'(x)\,dx$ with respect to $x$ is $F(x) + C$." The expression $F'(x)\,dx$ to be integrated is called the *integrand,* and $C$ is the constant of integration.

---

**EXAMPLE 3:** What is the indefinite integral of $3x^2\,dx$?

**Solution:** The function $x^3$ has $3x^2\,dx$ as a differential, so the integral of $3x^2\,dx$ is $x^3 + C$.

$$\int 3x^2\,dx = x^3 + C$$

---

If we let $F'(x)$ be denoted by $f(x)$, Eq. 363 becomes

| | | |
|---|---|---|
| Indefinite Integral or Antiderivative | $\int f(x)\,dx = F(x) + C$ where $f(x) = F'(x)$. | **364** |

**We use both capital and lowercase *F* in this section. Be careful not to mix them up.**

## Some Rules for Finding Integrals

We can obtain rules for integration by reversing the rules we had previously derived for differentiation. Such a list of rules, called a *table of integrals,* is given in Appendix D. This is a very short table of integrals; some fill entire books.

Let there be a function $u = F(x)$ whose derivative is

$$\frac{du}{dx} = F'(x)$$

or

$$du = F'(x)\,dx$$

Just as we can take the derivative of both sides of an equation, we can take the integral of both sides. This gives

$$\int du = \int F'(x)\,dx = F(x) + C$$

by Eq. 363. Substituting $F(x) = u$, we get our first rule for finding integrals.

| Rule 1. | $\int du = u + C$ |
|---|---|

*The integral of the differential of a function is equal to the function itself, plus a constant of integration.*

---

**EXAMPLE 4:** We use Rule 1 to integrate the following expressions.

(a) $\int dy = y + C$

(b) $\int dz = z + C$

(c) $\int d(x^3) = x^3 + C$

(d) $\int d(x^2 + 2x) = x^2 + 2x + C$

(e) $\int d\left(\dfrac{u^{n+1}}{n+1}\right) = \dfrac{u^{n+1}}{n+1} + C$

---

For our second rule, we use the fact that the derivative of a constant times a function equals the constant times the derivative of the function. Reversing the process, we have

| Rule 2. | $\int a\,f(x)\,dx = a \int f(x)\,dx = a\,F(x) + C$ |
|---|---|

*The integral of a constant times a function is equal to the constant times the integral of the function, plus a constant of integration.*

Rule 2 says that we may move constants to the left of the integral sign, as in the following example.

---

**EXAMPLE 5:** Find $\int 3kx^2\,dx$.

**Solution:** From Example 3 we know that $\int 3x^2\,dx = x^3 + C$, so if $k$ is some constant,

$$\int 3kx^2\,dx = k \int 3x^2\,dx = k(x^3 + C) = kx^3 + kC_1$$
$$= kx^3 + C$$

where we have replaced the constant $kC_1$, by another constant $C$.

---

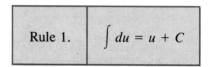

**EXAMPLE 6:** Find $\int 5\,dx$.

**Solution:** By Rule 2,

$$\int 5\,dx = 5\int dx$$

Then by Rule 1,

$$5\int dx = 5(x + C_1)$$
$$= 5x + C$$

where we have replaced $5C_1$ by $C$.

We get our third rule by noting that the derivative of a function with several terms is the sum of the derivatives of each term. Reversing gives

| Rule 3. | $\int [f(x) + g(x) + h(x) \cdots]\,dx = \int f(x)\,dx + \int g(x)\,dx + \int h(x)\,dx + \cdots + C$ |
|---|---|

**The constant $C$ comes from combining the individual constants $C_1, C_2, \ldots$ into a single term.**

*The integral of a function with several terms equals the sum of the integrals of those terms, plus a constant.*

Rule 3 says that when integrating an expression having several terms, we may integrate each term separately.

**EXAMPLE 7:** Integrate $\int (3x^2 + 5)\,dx$.

**Solution:**

By Rule 3:
$$\int (3x^2 + 5)\,dx = \int 3x^2\,dx + \int 5\,dx$$

By Rule 2:
$$= \int 3x^2\,dx + 5\int dx$$
$$= x^3 + C_1 + 5x + C_2$$
$$= x^3 + 5x + C$$

where we have combined $C_1$ and $C_2$ into a single constant $C$.

**From now on, we will not bother writing $C_1$ and $C_2$, but will combine them immediately into a single constant $C$.**

Our fourth rule is for a power of $x$. We know that the derivative of $x^{n+1}/(n + 1)$ is

$$\frac{d}{dx}\left(\frac{x^{n+1}}{n+1}\right) = (n + 1)\frac{x^n}{n+1} = x^n$$

or in differential form

$$d\left(\frac{x^{n+1}}{n+1}\right) = x^n\,dx$$

Taking the integral of both sides gives

$$\int d\left(\frac{x^{n+1}}{n+1}\right) = \int x^n \, dx$$

This gives us Rule 4,

| Rule 4. | $\int x^n \, dx = \dfrac{x^{n+1}}{n+1} + C \qquad (n \neq -1)$ |
|---|---|

*The integral of x raised to a power is x raised to that power increased by one, divided by the new power, plus a constant.*

Note that Rule 4 is not valid for $n = -1$, for this would give division by zero. We'll use a later rule (7) when $n$ is $-1$.

We'll now give some examples of these rules for integration.

---

**EXAMPLE 8:** Integrate $\int x^3 \, dx$.

**Solution:** We use Rule 4, with $n = 3$.

$$\int x^3 \, dx = \frac{x^{3+1}}{3+1} + C = \frac{x^4}{4} + C$$

---

**EXAMPLE 9:** Integrate $\int x^5 \, dx$.

**Solution:** This is similar to Example 8, except that the exponent is 5. By Rule 4,

$$\int x^5 \, dx = \frac{x^{5+1}}{5+1} + C = \frac{x^6}{6} + C$$

---

**EXAMPLE 10:** Integrate $\int 7x^3 \, dx$.

**Solution:** This is similar to Example 8, except that here our function is multiplied by 7. By Rule 2, we may move the constant factor 7 to the left of the integral sign.

$$\begin{aligned}\int 7x^3 \, dx &= 7 \int x^3 \, dx \\ &= 7\left(\frac{x^4}{4} + C_1\right) \\ &= \frac{7x^4}{4} + 7C_1 \\ &= \frac{7x^4}{4} + C\end{aligned}$$

where $C = 7C_1$. Since $C_1$ is an unknown constant, $7C_1$ is also an unknown constant that can be more simply represented by $C$. From now on we will not even bother with individual constants such as $C_1$ but will write the final constant $C$ directly.

---

**EXAMPLE 11:** Integrate $\int (x^3 + x^5)\,dx$.

**Solution:** By Rule 3, we can integrate each term separately.

$$\int (x^3 + x^5)dx = \int x^3\,dx + \int x^5\,dx$$

$$= \frac{x^4}{4} + \frac{x^6}{6} + C$$

Even though each of the two integrals has produced its own constant of integration, we have combined them immediately into the single constant $C$.

**EXAMPLE 12:** Integrate $\int (5x^2 + 2x - 3)\,dx$.

**Solution:** Integrating term by term yields

$$\int (5x^2 + 2x - 3)dx = \frac{5x^3}{3} + x^2 - 3x + C$$

Rule 4 is also used when the exponent $n$ is not an integer.

**EXAMPLE 13:** Integrate $\int \sqrt[3]{x}\,dx$.

**Solution:** We write the radical in exponential form, with $n = \frac{1}{3}$.

$$\int \sqrt[3]{x}\,dx = \int x^{1/3}\,dx$$

By Rule 4:

$$= \frac{x^{4/3}}{4/3} + C$$

$$= \frac{3x^{4/3}}{4} + C$$

The exponent $n$ can also be negative (with the exception of $-1$, which would result in division by zero).

**EXAMPLE 14:** Integrate $\int \frac{1}{t^3}\,dt$.

The variable of integration can, of course, be any letter other than $x$, as in this example.

**Solution:**

$$\int \frac{1}{t^3}dt = \int t^{-3}\,dt$$

$$= \frac{t^{-2}}{-2} + C = -\frac{1}{2t^2} + C$$

## Checking an Integral by Differentiating

Many rules for integration are presented without derivation or proof, so you would be correct in being suspicious of them. However, you can convince yourself that a rule works (and that you have used it correctly) simply by taking the derivative of your result. You should get back the original expression.

Handwritten margin notes (left side):

$(x^2+3)(x^2+3)$

$x^4 + 3x^2 + 3x^2 + 9$
$= (x^4 + 6x^2 + 9)$

$\int (x^2+3)^2 \, dx$

$= \int (x^4 + 6x^2 + 9) \, dx$

$= \int x^4 \, dx + \int 6x^2 \, dx + \int 9 \, dx$

$= \int x^4 \, dx + 6\int x^2 \, dx + 9\int dx$

$= \dfrac{x^{4+1}}{4+1} + \dfrac{6x^{2+1}}{2+1} + 9x + C$

$= \dfrac{x^5}{5} + \dfrac{\cancel{6}x^3}{\cancel{3}} + 9x + C$

$= x^5 \dfrac{}{5} + 2x^3 + 9x + C$

$= \dfrac{x^5}{5} + 2x^3 + 9x + C$

$\int x^5 - 2x^3 + 5x \, dx$

$\int \dfrac{x^5 - 2x^3 + 5x}{x} \, dx$

$\int (x^4 - 2x^2 + 5) \, dx$

$\int x^4 \, dx - 2\int x^2 \, dx + 5\int dx$

$\int x^4 \, dx - \dfrac{2x^3}{3} + 5x + C$

$= \dfrac{x^5}{5} - \dfrac{2x^3}{3}$

$= \dfrac{x^5}{5}$

---

**EXAMPLE 15:** Taking the derivative of the expression obtained in Example 14, we have

$$\frac{d}{dt}\left(-\frac{t^{-2}}{2} + C\right) = -\frac{1}{2}(-2t^{-3}) + 0 = t^{-3} = \frac{1}{t^3}$$

which is the expression we started with, so our integration was correct.

### Simplify before Integrating

If an expression does not seem to fit any given rule at first, try changing its form by performing the indicated operations (squaring, removing parentheses, and so on).

---

**EXAMPLE 16:** Integrate $\int (x^2 + 3)^2 \, dx$.

**Solution:** None of our rules (so far) seem to fit. Rule 3, for example, is for $x$ raised to a power, *not* for $(x^2 + 3)$ raised to a power. However, if we square $x^2 + 3$, we get

$$\int (x^2 + 3)^2 \, dx = \int (x^4 + 6x^2 + 9) \, dx$$

$$= \frac{x^5}{5} + \frac{6x^3}{3} + 9x + C$$

$$= \frac{x^5}{5} + 2x^3 + 9x + C$$

---

**EXAMPLE 17:** Integrate $\int \dfrac{x^5 - 2x^3 + 5x}{x} \, dx$.

**Solution:** This looks complicated at first, but let us perform the indicated division first:

$$\int \frac{x^5 - 2x^3 + 5x}{x} \, dx = \int (x^4 - 2x^2 + 5) \, dx$$

Now Rules 3 and 4 can be used:

$$= \frac{x^5}{5} - \frac{2x^3}{3} + 5x + C$$

---

**EXAMPLE 18:** Evaluate $\int \dfrac{x^3 - x^2 + 5x - 5}{x - 1} \, dx$.

We can also simplify here by factoring the numerator and canceling.

**Solution:** Again, no rule seems to fit, so we try to simplify the given expression by long division.

$$
\begin{array}{r}
x^2 + 5 \phantom{000} \\
x - 1 \overline{\smash{\big)}\, x^3 - x^2 + 5x - 5} \\
\underline{x^3 - x^2 \phantom{0000000}} \\
0 + 5x - 5 \\
\underline{5x - 5} \\
\end{array}
$$

So our quotient is $x^2 + 5$, which we now integrate,

$$\int \frac{x^3 - x^2 + 5x - 5}{x - 1} \, dx = \int (x^2 + 5) \, dx = \frac{x^3}{3} + 5x + C$$

## Simple Differential Equations

Suppose that we have the derivative of a function and we wish to find the original function from its derivative. Let's use our familiar

**This equation, as well as any other that contains a derivative, is called a *differential equation.* In this example we are solving a differential equation.**

$$\frac{dy}{dx} = 3x^2$$

We now seek the equation, $y = F(x)$, of which $3x^2$ is the derivative. Proceed as follows. First write the given differential equation in differential form by multiplying both sides by $dx$.

$$dy = 3x^2\,dx$$

We now take the integral of both sides of the equation,

$$\int dy = \int 3x^2\,dx$$

so

$$y + C_1 = x^3 + C_2$$

We now combine the two arbitrary constants, $C_1$ and $C_2$.

$$y = x^3 + C_2 - C_1 = x^3 + C$$

This is the function whose derivative is $3x^2$. This function is also called the *solution* to the differential equation $dy/dx = 3x^2$.

### EXERCISE 1—THE INDEFINITE INTEGRAL

Find the indefinite integral.

1. $\int dx$
2. $\int dy$
3. $\int x\,dx$
4. $\int x^4\,dx$
5. $\int \frac{dx}{x^2}$
6. $\int x^{2/3}\,dx$
7. $\int 3x\,dx$
8. $\int 6x\,dx$
9. $\int 3x^3\,dx$
10. $\int x^5\,dx$
11. $\int x^n\,dx$
12. $\int x^{1/2}\,dx$
13. $\int \frac{dx}{\sqrt{x}}$
14. $\int 3x^2\,dx$
15. $\int 4x^3\,dx$
16. $\int \left(\frac{x^2}{2} - \frac{2}{x^2}\right) dx$
17. $\int \frac{7}{2}x^{5/2}\,dx$
18. $\int 5x^4\,dx$
19. $\int 2(x + 1)dx$
20. $\int x^{4/3}\,dx$
21. $\int \frac{4}{3}x^{1/3}\,dx$
22. $\int (x^{3/2} - 2x^{2/3} + 5\sqrt{x} - 3)dx$
23. $\int 2u\,du$
24. $\int \sqrt[3]{t}\,dt$
25. $\int 4s^{1/2}\,ds$
26. $\int 3\sqrt[3]{y}\,dy$

Simplify and integrate.

27. $\int \sqrt{x}(3x - 2)\,dx$

28. $\int (x + 1)^2\,dx$

29. $\int \frac{4x^2 - 2\sqrt{x}}{x}dx$

30. $\int (t + 2)(t - 3)dt$

31. $\int (1 - s)^3\,ds$

32. $\int \frac{v^3 + 2v^2 - 3v - 6}{v + 2}dv$

Find the function whose derivative is given (solve each differential equation).

33. $\frac{dy}{dx} = 4x^2$

34. $\frac{dy}{dx} = 2x(x^2 + 6)$

35. $\frac{dy}{dx} = x^{-3}$

36. $\frac{ds}{dt} = 10t^{-6}$

37. $\frac{ds}{dt} = \frac{1}{2}t^{-2/3}$

38. $\frac{dv}{dt} = 6t^3 - 3t^{-2}$

## 31-2 INTEGRAL OF A POWER FUNCTION

Our next rule is for a function $u$ raised to a power $n$. If $u$ is some function of $x$, the derivative of $u^{n+1}/(n + 1)$ is

$$\frac{d}{dx}\left(\frac{u^{n+1}}{n + 1}\right) = (n + 1)\frac{u^n}{n + 1}\frac{du}{dx} = u^n\frac{du}{dx}$$

or

$$d\left(\frac{u^{n+1}}{n + 1}\right) = u^n\,du$$

Taking the integral of both sides yields

$$\int d\left(\frac{u^{n+1}}{n + 1}\right) = \int u^n\,du$$

Since the integral of the differential of a function is the function itself, we get

| | |
|---|---|
| Rule 5. | $\int u^n\,du = \frac{u^{n+1}}{n + 1} + C \quad (n \neq -1)$ |

The expression $u$ in Rule 5 can be any function of $x$, say, $x^3 + 3$. However, in order to use Rule 5, *the quantity $u^n$ must be followed by the derivative of $u$.* Thus if $u = x^3 + 3$, then $du = 3x^2\,dx$, as in the following example.

**EXAMPLE 19:** Find the integral $\int (x^3 + 3)^6(3x^2)\,dx$.

**Solution:** If we let

$$u = x^3 + 3$$

we see that the derivative of $u$ is

$$\frac{du}{dx} = 3x^2$$

or, in differential form,

$$du = 3x^2\,dx$$

Notice now that our given integral exactly matches Rule 5.

$$\int \underbrace{(x^3 + 3)}_{u}{}^{\textcircled{n}\,6}\underbrace{(3x^2)\,dx}_{du}$$

We now apply Rule 5.

$$\int (x^3 + 3)^6(3x^2)\,dx = \frac{(x^3 + 3)^7}{7} + C$$

---

Students are often puzzled at the "disappearance" of the $3x^2\,dx$ in Example 19.

$$\int (x^3 + 3)^6 \boxed{(3x^2)\,dx} = \frac{(x^3 + 3)^7}{7} + C$$

Where did this go?

The $3x^2\,dx$ is the differential of $x^3$, and does not remain after integration. Do not be alarmed when it vanishes.

Very often an integral will not exactly match the form of Rule 5 or any other rule in the table. However, if all we lack is a constant factor, it can usually be supplied as in the following examples.

---

**EXAMPLE 20:** Integrate $\int (x^3 + 3)^6 x^2 dx$.

**Solution:** This is almost identical to Example 19, except that the factor 3 is missing. Realize that we cannot use Rule 5 *yet,* because if

$$u = x^3 + 3$$

then

$$du = 3x^2\,dx$$

Our integral contains $x^2\,dx$ but not $3x^2\,dx$, which we need in order to use Rule 5. But we can insert a factor of 3 into our integrand, as long as we compensate for it by multiplying the whole integral by $\frac{1}{3}$.

$$\frac{1}{3}\int (x^3 + 3)^6(3x^2)\,dx$$

compensate (the $\frac{1}{3}$) — insert (the 3)

Integrating gives us

$$\frac{1}{3}\int (x^3 + 3)^6(3x^2)\,dx = \left(\frac{1}{3}\right)\frac{(x^3 + 3)^7}{7} + C$$

$$= \frac{(x^3 + 3)^7}{21} + C$$

---

**EXAMPLE 21:** Evaluate $\int \frac{5z\,dz}{\sqrt{3z^2 - 7}}$.

**Solution:** We rewrite the given expression in exponential form and move the 5 outside the integral.

$$\int \frac{5z\,dz}{\sqrt{3z^2 - 7}} = 5 \int (3z^2 - 7)^{-1/2}\, z\,dz$$

Here $n = -\frac{1}{2}$. Let $u = 3z^2 - 7$, so $du$ is $6z\,dz$. Comparing with our actual integral, we see that we need to insert a 6 before the $z$ and compensate with $\frac{1}{6}$ outside the integral sign.

$$5 \int (3z^2 - 7)^{-1/2}\, z\,dz = 5\left(\frac{1}{6}\right) \int (3z^2 - 7)^{-1/2}(6z\,dz)$$

$$= \frac{5}{6}\,\frac{(3z^2 - 7)^{1/2}}{\frac{1}{2}} + C = \frac{5}{3}\sqrt{3z^2 - 7} + C$$

| | |
|---|---|
| Common Error | Rule 2 allows us to move *only constant factors* to the left of the integral sign. You cannot use this same procedure with variables. |

Sometimes Rule 5 can be used in unexpected places, as in the following example.

**EXAMPLE 22:** Evaluate $\int \sin^3 2x \cos 2x\,dx$.

**Solution:** Remember that $\sin^3 2x = (\sin 2x)^3$. We try to use Rule 5, with $n = 3$. If we let

$$u = \sin 2x$$

then

$$du = 2 \cos 2x\,dx$$

We insert the factor 2, compensate with $\frac{1}{2}$ to the left of the integral sign, and integrate using Rule 5.

$$\frac{1}{2}\int (\sin 2x)^3(2\cos 2x\,dx) = \frac{1}{2}\left(\frac{\sin^4 2x}{4}\right) + C = \frac{1}{8}\sin^4 2x + C$$

## EXERCISE 2—INTEGRAL OF A POWER FUNCTION

Integrate.

1. $\int (x^4 + 1)^3\, 4x^3\,dx$
2. $\int (2x^2 - 6)^3\, 4x\,dx$
3. $\int 2(x^2 + 2x)(2x + 2)dx$
4. $\int (1 - 2x)^5\,dx$
5. $\int \frac{dx}{(1 - x)^2}$
6. $\int 6(x^2 - 1)^2\, x\,dx$
7. $\int 9(x^3 + 1)^2\, x^2\,dx$
8. $\int 3x\sqrt{x^2 - 1}\,dx$

9. $\int \frac{y^2\,dy}{\sqrt{1 - y^3}}$

10. $\int (x + 1)(x^2 + 2x + 6)^2\,dx$

11. $\int \frac{4x\,dx}{(9 - x^2)^2}$

12. $\int z\sqrt{z^2 - 2}\,dz$

13. $\int \tan^2 x \sec^2 x\,dx$

14. $\int \cos^3 2x \sin 2x\,dx$

Find the function whose derivative is given.

15. $\frac{dy}{dx} = x(1 - x^2)^3$

16. $\frac{dy}{dx} = x^2(x^3 - 2)^{1/2}$

17. $\frac{dy}{dx} = (3x^2 + x)(2x^3 + x^2)^3$

18. $\frac{ds}{dt} = \frac{t^2}{\sqrt{5t^3 + 7}}$

19. $\frac{dv}{dt} = 5t\sqrt{7 - t^2}$

20. $\frac{dy}{dx} = \frac{x}{(9 - 4x^2)^2}$

## 31-3 CONSTANT OF INTEGRATION

### Families of Curves

We have seen in the preceding sections that we get a constant of integration whenever we find an indefinite integral. We now give a geometric meaning to this constant.

---

**EXAMPLE 23:** Find the function whose derivative is $dy/dx = 2x$, and make a graph of the function.

**Solution:** Proceeding as in Sec. 31-1, we write the given differential equation in differential form, $dy = 2x\,dx$. Integrating, we get

$$y = \int 2x\,dx = 2\left(\frac{x^2}{2}\right) + C = x^2 + C$$

From Sec. 26-2 we recognize this as a parabola opening upward. Further, its vertex is on the $y$ axis at a distance of $C$ units from the origin. When we graph this function (Fig. 31-1) we get not a single curve, but a *family of curves,* each of which has a different value for $C$. (Although we show integer values of $C$ here, $C$ need not be an integer.)

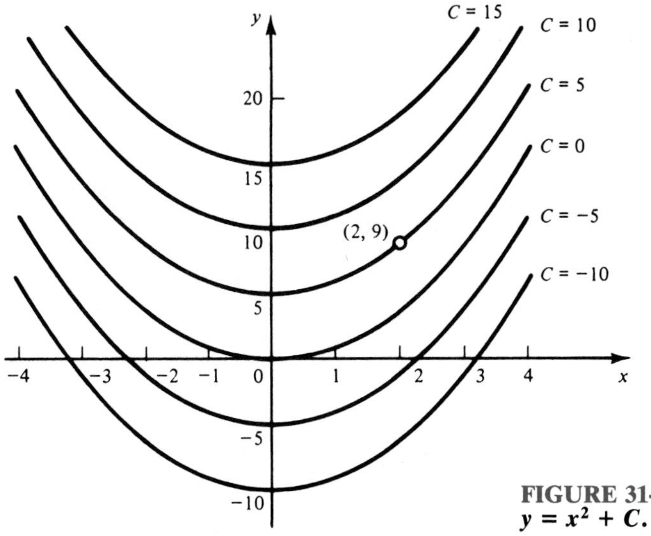


**FIGURE 31-1 Family of curves, $y = x^2 + C$.**

---

## Boundary Conditions

The term *initial* is more appropriate when our variable is *time*.

In order for us to evaluate the constant of integration, more information must be given. Such additional information is called a *boundary condition* or *initial condition*.

---

**EXAMPLE 24:** Find the constant of integration in Example 23 if the curve whose derivative is $2x$ is to pass through the point (2, 9).

**Solution:** The equation of the curve was found to be $y = x^2 + C$. Letting $x = 2$ and $y = 9$, we have

$$C = 9 - 2^2 = 5$$

as can be verified from Fig. 31-1.

---

## Successive Integration

If we are given the *second* derivative of some function and wish to find the function itself, we simply have to *integrate twice*. However, each time we integrate we get another constant of integration. These constants can be found if we are given enough additional information, as in the following example.

---

**EXAMPLE 25:** Find the equation of a curve that has a second derivative $y'' = 4x$, and that has a slope of 10 at the point (2, 9).

**Solution:** We can write the second derivative as

$$\frac{d}{dx}(y') = 4x$$

or, in differential form,

$$d(y') = 4x\,dx$$

Integrating gives us

$$y' = \int 4x\,dx = 2x^2 + C_1 \qquad (1)$$

But the slope, and hence $y'$, is 10 when $x = 2$, so

$$C_1 = 10 - 2(2)^2 = 2$$

So (1) becomes $y' = 2x^2 + 2$ or, in differential form,

$$dy = (2x^2 + 2)\,dx$$

Integrating again, we obtain

$$y = \int (2x^2 + 2)dx = \frac{2x^3}{3} + 2x + C_2 \qquad (2)$$

But $y = 9$ when $x = 2$, so

$$C_2 = 9 - \frac{2(2)^3}{3} - 2(2) = -\frac{1}{3}$$

Substituting into (2), we get

$$y = \frac{2x^3}{3} + 2x - \frac{1}{3}$$

as our final equation, with all constants evaluated.

---

## EXERCISE 3—CONSTANT OF INTEGRATION

### Families of Curves

Write the function that has the given derivative. Then graph that function for $C = -1$, $C = 0$, and $C = 1$.

1. $\dfrac{dy}{dx} = 3$
2. $\dfrac{dy}{dx} = 5x$
3. $\dfrac{dy}{dx} = 3x^2$

### Constant of Integration

Write the function that has the given derivative and passes through the given point.

4. $y' = 3x$, passes through (2, 6)
5. $y' = x^2$, passes through (1, 1)
6. $y' = \sqrt{x}$, passes through (2, 4)
7. If $dy/dx = 2x + 1$, and $y = 7$ when $x = 1$, find the value of $y$ when $x = 3$.
8. If $dy/dx = \sqrt{2x}$ and $y = \frac{1}{3}$ when $x = \frac{1}{2}$, find the value of $y$ when $x = 2$.

### Successive Integration

9. Find the equation of a curve that passes through the point (3, 0), has the slope $\frac{7}{2}$ at that point, and has a second derivative $y'' = x$.
10. Find the equation of the curve for which $y'' = 4$ if the curve is tangent to the line $y = 3x$ at (2, 6).
11. Find the equation of a curve that passes through the point (1, 0), is tangent to the line $6x + y = 6$ at that point, and has a second derivative $y'' = 12/x^3$.
12. The second derivative of a curve is $y'' = 12x^2 - 6$. The tangent to this curve at (2, 4) is perpendicular to the line $x + 20y - 40 = 0$. Find the equation of the curve.

## 31-4 USE OF A TABLE OF INTEGRALS

A very short table of integrals is given in Appendix D, but much longer ones are available. With an extensive table we can often find the integral of a given expression without first having to greatly modify it. It is usually necessary, however, to insert constant factors, as we did before, and place a suitable compensating factor to the left of the integral sign.

**This is a good time to review the rules already covered.**

### Integral of *du/u*

The derivative of $\ln |u|$ (Fig. 31-2a) is

$$\frac{d(\ln |u|)}{dx} = \frac{1}{u}\frac{du}{dx} \qquad \textbf{353}$$

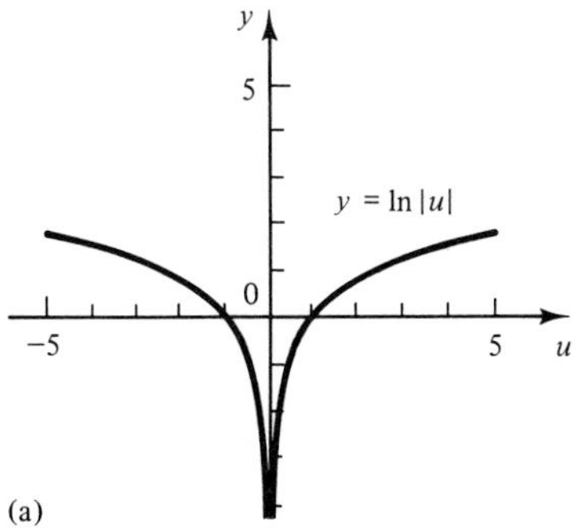


FIGURE 31-2(a)

or, in differential form, $d(\ln |u|) = du/u$. Thus the integral of $du/u$ is $\ln |u|$.

| Rule 7. | $\displaystyle\int \frac{du}{u} = \ln |u| + C \qquad (u \neq 0)$ |
|---|---|

**When we took derivatives of ln $u$, we didn't care about negative values because $u$ had to be positive for the logarithm to exist. But we sometimes want the integral of $y = 1/u$ at negative values of $u$.**

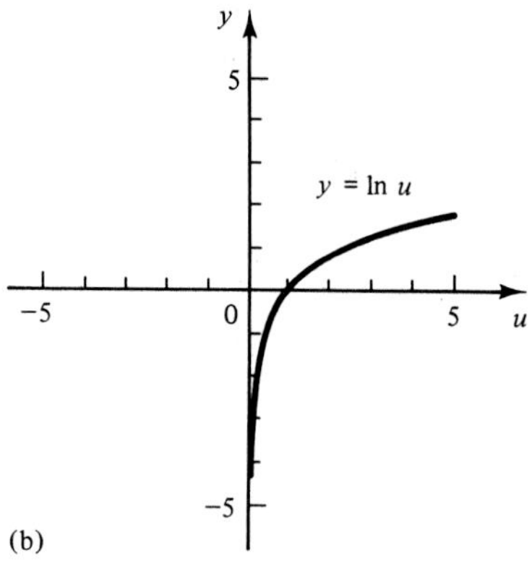


**FIGURE 31-2(b)**

We used ln $|u|$ rather than ln $u$ in our derivation because ln $|u|$ exists for negative values of $u$, while ln $u$ does not (Fig. 31-2b). Had we used ln $u$, the rule derived would not have been valid for negative values of $u$.

To use Rule 7, look for a function $u$ in the denominator of the integral.

---

**EXAMPLE 26:** Integrate $\int \frac{5}{x} dx$.

**Solution:** Here we see that $u = x$.

$$\int \frac{5}{x} dx = 5 \int \frac{dx}{x} = 5 \ln |x| + C$$

by Rule 7.

---

**EXAMPLE 27:** Integrate $\int \frac{x\, dx}{3 - x^2}$.

**Solution:** Let $u = 3 - x^2$. Then $du$ is $-2x\, dx$. Our integral does not have the $-2$ factor, so we insert a factor of $-2$ and compensate with a factor of $-\frac{1}{2}$,

$$\int \frac{x\, dx}{3 - x^2} = -\frac{1}{2} \int \frac{-2x\, dx}{3 - x^2} = -\frac{1}{2} \ln |3 - x^2| + C$$

by Rule 7.

---

**EXAMPLE 28:** Find the equation of the curve whose derivative is $6x/(x^2 - 4)$ and which passes through the point (0, 5).

**Solution:** Our derivative, in differential form, is

$$dy = \frac{6x\, dx}{x^2 - 4}$$

To integrate, we let $u = x^2 - 4$ and $du = 2x\, dx$. Then, by Rule 7,

$$y = 3 \int \frac{2x\, dx}{x^2 - 4} = 3 \ln |x^2 - 4| + C$$

At (0, 5) we get $C = 5 - 3 \ln |0 - 4| = 0.841$, so

$$y = 3 \ln |x^2 - 4| + 0.841$$

This function (graphed as a solid line in Fig. 31-3) has a domain from $x = -2$ to $x = 2$.

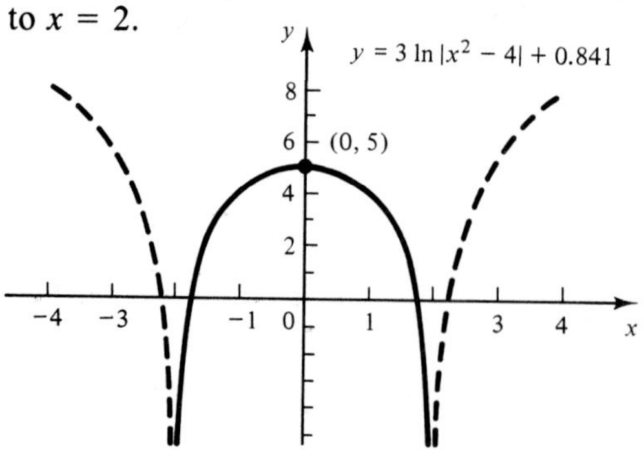


**FIGURE 31-3**

Do not try to use our function $y = 3 \ln |x^2 - 4| + 0.841$ for $x < -2$ or $x > 2$. Those parts of the curve (shown dashed) *are not continuous* with the section for which we had the known value, so the function we found may not apply there. Our function only applies to $-2 < x < 2$.

---

### Integral of $e^u\,du$

Since the derivative of $e^u$ is

$$\frac{d(e^u)}{dx} = e^u \frac{du}{dx}$$

or $d(e^u) = e^u\,du$, then integrating gives

$$\int e^u\,du = \int d(e^u) = e^u + C$$

or

| Rule 8. | $\int e^u\,du = e^u + C$ |
|---|---|

---

**EXAMPLE 29:** Integrate $\int e^{6x}\,dx$.

**Solution:** To match the form $\int e^u\,du$, let

$$u = 6x$$

$$du = 6\,dx$$

We insert a factor of 6, and compensate with $\frac{1}{6}$:

$$\int e^{6x}\,dx = \frac{1}{6}\int e^{6x}(6\,dx) = \frac{1}{6}e^{6x} + C$$

by Rule 8.

---

**EXAMPLE 30:** Integrate $\displaystyle\int \frac{6e^{\sqrt{3x}}}{\sqrt{3x}}\,dx$.

**Solution:** Since the derivative of $\sqrt{3x}$ is $\frac{3}{2}(3x)^{-1/2}$, we insert a factor of $\frac{3}{2}$ and compensate:

$$\int \frac{6e^{\sqrt{3x}}}{\sqrt{3x}}\,dx = 6\int e^{\sqrt{3x}}(3x)^{-1/2}\,dx$$

$$= 6\left(\frac{2}{3}\right)\int e^{\sqrt{3x}}\left[\frac{3}{2}(3x)^{-1/2}\,dx\right]$$

$$= 4e^{\sqrt{3x}} + C$$

by Rule 8.

---

### Integral of $b^u\,du$

The derivative of $b^u/\ln b$ is

$$\frac{d}{dx}\left(\frac{b^u}{\ln b}\right) = \frac{1}{\ln b}(b^u)(\ln b)\frac{du}{dx} = b^u\frac{du}{dx}$$

or, in differential form, $d(b^u/\ln b) = b^u\,du$. Thus the integral of $b^u\,du$ is

| Rule 9. | $\int b^u\,du = \dfrac{b^u}{\ln b} + C \qquad (b > 0,\ b \neq 1)$ |
|---|---|

**EXAMPLE 31:** Integrate $\int 3xa^{2x^2}\,dx$.

**Solution:** We take 3 out of the integral sign. Let

$$u = 2x^2$$

Then $du = 4x\,dx$.

$$\int 3xa^{2x^2}\,dx = 3\int a^{2x^2}x\,dx$$

We insert 4 and compensate with $\frac{1}{4}$.

$$= 3\left(\frac{1}{4}\right)\int a^{2x^2}(4x\,dx)$$

$$= \frac{3a^{2x^2}}{4\ln a} + C$$

by Rule 9.

### Integrals of the Trigonometric Functions

To our growing list of rules we add those for the six trigonometric functions. By Eq. 341,

$$\frac{d(-\cos u)}{dx} = \sin u\frac{du}{dx}$$

or $d(-\cos u) = \sin u\,du$. Taking the integral of both sides gives

$$\int \sin u\,du = -\cos u + C$$

The integrals of the other trigonometric functions are found in the same way. We thus get the following rules,

| | |
|---|---|
| Rule 10. | $\int \sin u\,du = -\cos u + C$ |
| Rule 11. | $\int \cos u\,du = \sin u + C$ |
| Rule 12. | $\int \tan u\,du = -\ln\lvert\cos u\rvert + C$ |
| Rule 13. | $\int \cot u\,du = \ln\lvert\sin u\rvert + C$ |
| Rule 14. | $\int \sec u\,du = \ln\lvert\sec u + \tan u\rvert + C$ |
| Rule 15. | $\int \csc u\,du = \ln\lvert\csc u - \cot u\rvert + C$ |

We use these rules just as we did the preceding ones. Match the given integral *exactly* with one of the rules, inserting a factor and compensating when necessary, and then copy off the integral in the rule.

**EXAMPLE 32:** Integrate $\int x \sin x^2\,dx$.

**Solution:**

$$\int x \sin x^2\,dx = \int \sin x^2(x\,dx)$$

$$= \frac{1}{2}\int \sin x^2(2x\,dx) = -\frac{1}{2}\cos x^2 + C$$

from Rule 10.

Sometimes the trigonometric identities can be used to simplify an expression before integrating.

**EXAMPLE 33:** Integrate $\int \frac{\cot 5x}{\cos 5x}\,dx$.

**Solution:** We replace $\cot 5x$ by $\cos 5x/\sin 5x$:

$$\int \frac{\cot 5x}{\cos 5x}\,dx = \int \frac{\cos 5x}{\sin 5x \cos 5x}\,dx = \int \frac{1}{\sin 5x}\,dx$$

$$= \int \csc 5x\,dx$$

$$= \frac{1}{5}\int \csc 5x(5\,dx) = \frac{1}{5}\ln|\csc 5x - \cot 5x| + C$$

by Rule 15.

### Miscellaneous Rules from the Table

Now that you can use Rules 1 to 15, you should find it no harder to use any rule from the table of integrals.

**EXAMPLE 34:** Integrate $\int e^{3x} \cos 2x\,dx$.

**Solution:** We search the table for a similar form and find

| | |
|---|---|
| Rule 42. | $\int e^{au} \cos bu\,du = \frac{e^{au}}{a^2 + b^2}(a \cos bu + b \sin bu) + C$ |

This matches our integral if we set

$$a = 3 \qquad b = 2 \qquad u = x \qquad du = dx$$

so

$$\int e^{3x} \cos 2x\,dx = \frac{e^{3x}}{3^2 + 2^2}(3 \cos 2x + 2 \sin 2x) + C$$

$$= \frac{e^{3x}}{13}(3 \cos 2x + 2 \sin 2x) + C$$

**EXAMPLE 35:** Integrate $\int \frac{dx}{4x^2 + 25}$.

**Solution:** From the table we find

| Rule 56. | $\int \frac{du}{a^2 + b^2u^2} = \frac{1}{ab} \tan^{-1} \frac{bu}{a} + C$ |
|---|---|

Letting $a = 5$, $b = 2$, $u = x$, and $du = dx$, our integral matches Rule 56, if we rearrange the denominator.

$$\int \frac{dx}{25 + 4x^2} = \frac{1}{10} \tan^{-1} \frac{2x}{5} + C$$

**EXAMPLE 36:** Integrate $\int \frac{dx}{(4x^2 + 9)2x}$.

**Solution:** We match this with

| Rule 60. | $\int \frac{du}{u(u^2 + a^2)} = \frac{1}{2a^2} \ln \left\| \frac{u^2}{u^2 + a^2} \right\| + C$ |
|---|---|

with $a = 3$, $u = 2x$, and $du = 2dx$, so

$$\int \frac{dx}{(4x^2 + 9)2x} = \frac{1}{2} \int \frac{2\,dx}{(2x)[(2x)^2 + 3^2]} = \frac{1}{36} \ln \left| \frac{4x^2}{4x^2 + 9} \right| + C$$

| Common Error | When using the table of integrals, be sure that the integral chosen completely matches the given integral, and especially that *all factors of du are present*. |
|---|---|

**EXAMPLE 37:** Integrate $\int \frac{dx}{x\sqrt{x^2 + 16}}$.

**Solution:** We use

| Rule 64. | $\int \frac{du}{u\sqrt{u^2 + a^2}} = \frac{1}{a} \ln \left\| \frac{u}{a + \sqrt{u^2 + a^2}} \right\| + C$ |
|---|---|

with $u = x$, $a = 4$, and $du = dx$, so

$$\int \frac{dx}{x\sqrt{x^2 + 16}} = \frac{1}{4} \ln \left| \frac{x}{4 + \sqrt{x^2 + 16}} \right| + C$$

## EXERCISE 4—USE OF A TABLE OF INTEGRALS

Integrate.

### Integral of *du/u*

1. $\int \frac{3}{x}\,dx$
2. $\int \frac{dx}{x-1}$
3. $\int \frac{5z^2}{z^3-3}\,dz$
4. $\int \frac{t\,dt}{6-t^2}$
5. $\int \frac{x+1}{x}\,dx$
6. $\int \frac{w^2+5}{w}\,dw$

### Exponential Functions

**Each of these problems has been contrived to match a table entry. But what if you had one that did not match? We'll learn how to deal with those in Chapter 34.**

7. $\int a^{5x}\,dx$
8. $\int a^{9x}\,dx$
9. $\int 5^{7x}\,dx$
10. $\int 10^{x}\,dx$
11. $\int a^{3y}\,dy$
12. $\int x a^{3x^2}\,dx$
13. $\int 4e^{x}\,dx$
14. $\int e^{2x}\,dx$
15. $\int x e^{x^2}\,dx$
16. $\int x^2 e^{x^3}\,dx$
17. $\int x e^{3x^2}\,dx$
18. $\int \sqrt{e^t}\,dt$
19. $\int \frac{e^{\sqrt{x}}\,dx}{\sqrt{x}}$
20. $\int (x+3)e^{x^2+6x-2}\,dx$
21. $\int (e^{x}-1)^2\,dx$
22. $\int x e^{-x^2}\,dx$
23. $\int \frac{e^{\sqrt{x-2}}}{\sqrt{x-2}}\,dx$
24. $\int \frac{(e^{x/2}-e^{-x/2})^2}{4}\,dx$
25. $\int (e^{x/a}+e^{-x/a})\,dx$
26. $\int (e^{x/a}-e^{-x/a})^2\,dx$

### Trigonometric Functions

27. $\int \sin 3x\,dx$
28. $\int \cos 7x\,dx$
29. $\int \tan 5\theta\,d\theta$
30. $\int \sec 2\theta\,d\theta$
31. $\int \sec 4x\,dx$
32. $\int \cot 8x\,dx$

33. $\int 3 \tan 9\theta \, d\theta$

34. $\int 7 \sec 3\theta \, d\theta$

35. $\int x \sin x^2 \, dx$

36. $\int 5x \cos 2x^2 \, dx$

37. $\int \theta^2 \tan \theta^3 \, d\theta$

38. $\int \theta \sec 2\theta^2 \, d\theta$

39. $\int \sin(x + 1) \, dx$

40. $\int \cos(7x - 3) \, dx$

41. $\int \tan(4 - 5\theta) \, d\theta$

42. $\int \sec(2\theta + 3) \, d\theta$

43. $\int x \sec(4x^2 - 3) \, dx$

44. $\int 3x^2 \cot(8x^3 + 3) \, dx$

**Miscellaneous Integrals from the Table**

45. $\int \frac{dt}{4 - 9t^2}$

46. $\int \frac{ds}{\sqrt{s^2 - 16}}$

47. $\int \sqrt{25 - 9x^2} \, dx$

48. $\int \sqrt{4x^2 + 9} \, dx$

49. $\int \frac{dx}{x^2 + 9}$

50. $\int \frac{dx}{x^2 + 2x}$

51. $\int \frac{dx}{16x^2 + 9}$

52. $\int x\sqrt{1 + 3x} \, dx$

53. $\int \frac{dx}{9 - 4x^2}$

54. $\int \sqrt{1 + 9x^2} \, dx$

55. $\int \frac{dx}{x^2 - 4}$

56. $\int \frac{dy}{\sqrt{25 - y^2}}$

57. $\int \sqrt{\frac{x^2}{4} - 1} \, dx$

58. $\int \frac{x^2 \, dx}{\sqrt{3x + 5}}$

59. $\int \frac{5x \, dx}{\sqrt{1 - x^4}}$

## 31-5 APPLICATIONS TO MOTION

Our main applications for the integral come in the next several chapters, but here we give two types of applications for the indefinite integral: motion and, in the next section, electric circuits.

### Displacement and Velocity

In Sec. 29-2 we saw that the velocity $v$ of a moving point was defined as the rate of change of the displacement $s$ of the point. The velocity was thus equal to the derivative of the displacement, or $v = ds/dt$. We now reverse the process and find the displacement when given the velocity. Since $ds = v\,dt$, integrating gives

| | | |
|---|---|---|
| Displacement | $s = \int v\,dt$ | **A22** |

Similarly, the acceleration $a$ is given by $dv/dt$, so $dv = a\,dt$. Integrating we get

| | | |
|---|---|---|
| Instantaneous Velocity | $v = \int a\,dt$ | **A24** |

Thus, if given an equation for acceleration, we can integrate to get an equation for velocity, and integrate again to get displacement. We evaluate the constants of integration by substituting boundary conditions, as shown in the following example.

---

**EXAMPLE 38:** A particle moves with a constant acceleration of 4 ft/s². It has an initial velocity of 6 ft/s and an initial displacement of 2 ft. Find the equations for velocity and displacement, and graph the displacement, velocity, and acceleration for $t = 0$ to 10 s.

**Solution:** We are given $a = dv/dt = 4$, so,

$$v = \int 4\,dt = 4t + C_1$$

Since $v = 6$ ft/s when $t = 0$, we get $C_1 = 6$, so

$$v = 4t + 6 \qquad \text{ft/s}$$

is our equation for the velocity of the particle. Now since $v = ds/dt$,

$$ds = (4t + 6)\,dt$$

Integrating again gives us

$$s = \frac{4t^2}{2} + 6t + C_2$$

Since the initial displacement is 2 ft when $t = 0$, we get $C_2 = 2$. The complete equation for the displacement of the particle is then

$$s = 2t^2 + 6t + 2 \qquad \text{ft}$$

The three curves are graphed one above the other in Fig. 31-4 with the same scale used for the horizontal axis of each. Note that each curve is the derivative of the one below it, and conversely, each curve is the integral of the one above it.

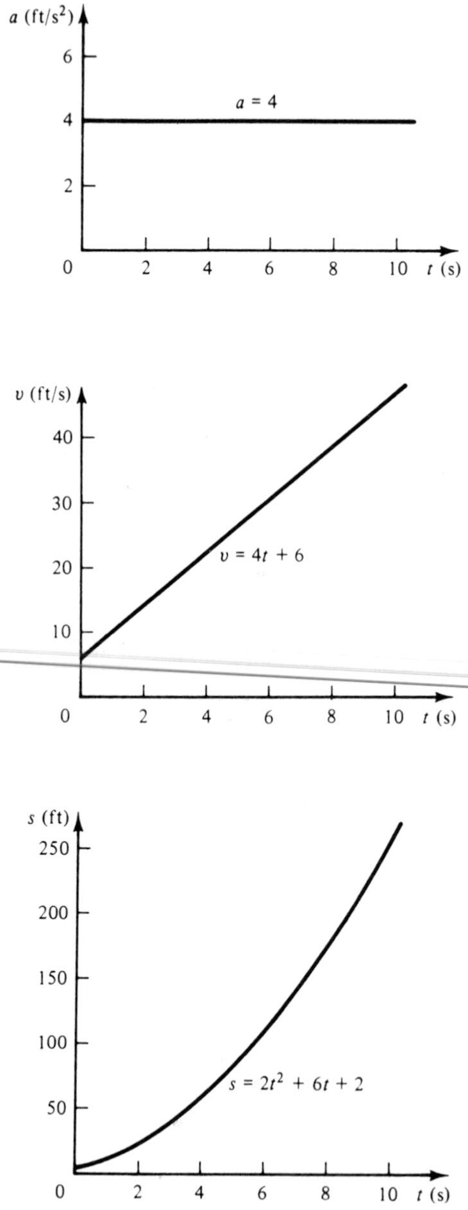

**FIGURE 31-4**

## Freely Falling Body

Integration provides us with a slick way to derive the equations for the displacement and velocity of a freely falling body.

**EXAMPLE 39:** An object falls with constant acceleration, $g$, due to gravity. Write the equations for the displacement and velocity of the body at any time.

**Solution:** We are given that $a = dv/dt = g$, so

$$dv = g\,dt$$

Integrating, we find that

$$v = \int g\,dt = gt + C_1$$

When $t = 0$, $v = C_1$, so $C_1$ is the initial velocity. Let us relabel the constant $v_0$.

| $v = v_0 + gt$ | **A19** |
|---|---|

But $v = ds/dt$, so

$$ds = (v_0 + gt)\,dt$$

$$s = \int (v_0 + gt)\,dt = v_0 t + \frac{gt^2}{2} + C_2$$

These two formulas are slightly different from Eqs. A18 and A19 given in Appendix A. There $a$ is used instead of $g$ to stand for any constant acceleration, and in Eq. A18, the initial displacement $s_0$ is assumed to be zero.

When $t = 0$, $s = C_2$, so we interpret $C_2$ as the initial displacement. Let us call it $s_0$. So the displacement is

| $s = s_0 + v_0 t + \dfrac{gt^2}{2}$ | **A18** |
|---|---|

## Motion along a Curve

In Sec. 29-2 the motion of a point along a curve was described by parametric equations, with the $x$ and $y$ displacements each given by a separate function of time. We saw that $dx/dt$ gave the velocity $v_x$ in the $x$ direction and that $dy/dt$ gave the velocity $v_y$ in the $y$ direction. Now, given the velocities, we integrate to get the displacements.

| Displacement in $x$ and $y$ Directions | (a) $x = \int v_x\,dt$ | (b) $y = \int v_y\,dt$ | **A32** |
|---|---|---|---|

Similarly, if we have parametric equations for the accelerations in the $x$ and $y$ directions, we integrate to get the velocities.

| Velocity in $x$ and $y$ Directions | (a) $v_x = \int a_x\,dt$ | (b) $v_y = \int a_y\,dt$ | **A34** |
|---|---|---|---|

**EXAMPLE 40:** An object starts from (2, 4) with initial velocities $v_x = 7$ cm/s and $v_y = 5$ cm/s, and moves along a curved path. It has $x$ and $y$ accelerations of $a_x = 3t$ cm/s$^2$ and $a_y = 5$ cm/s$^2$. Write expressions for the $x$ and $y$ components of velocity and displacement.

**Solution:** We integrate to find the velocities,

$$v_x = \int 3t\,dt = \frac{3t^2}{2} + C_1 \qquad \text{and} \qquad v_y = \int 5\,dt = 5t + C_2$$

At $t = 0$, $v_x = 7$ and $v_y = 5$, so

$$v_x = \frac{3t^2}{2} + 7 \quad \text{cm/s} \qquad \text{and} \qquad v_y = 5t + 5 \quad \text{cm/s}$$

Integrating again gives the displacements.

$$x = \int \left(\frac{3t^2}{2} + 7\right) dt \qquad \text{and} \qquad y = \int (5t + 5)\,dt$$

$$= \frac{t^3}{2} + 7t + C_3 \qquad\qquad = \frac{5t^2}{2} + 5t + C_4$$

At $t = 0$, $x = 2$ and $y = 4$, so our complete equations for the displacements are

$$x = \frac{t^3}{2} + 7t + 2 \quad \text{cm} \qquad \text{and} \qquad y = \frac{5t^2}{2} + 5t + 4 \quad \text{cm}$$

## Rotation

In Sec. 29-2 we saw that the angular velocity $\omega$ of a rotating body was given by the derivative $d\theta/dt$ of the angular displacement $\theta$. Thus $\theta$ is the integral of the angular velocity,

| | | |
|---|---|---|
| Angular Displacement | $\theta = \int \omega\,dt$ | **A28** |

Similarly, the angular velocity is the integral of the angular acceleration $\alpha$,

| | | |
|---|---|---|
| Angular Velocity | $\omega = \int \alpha\,dt$ | **A30** |

**EXAMPLE 41:** A flywheel in a machine starts from rest and accelerates at $3.85t$ rad/s$^2$. Find the angular velocity and the total number of revolutions after 10.0 s.

**Solution:** We integrate to get the angular velocity,

$$\omega = \int 3.85t\, dt = \frac{3.85t^2}{2} + C_1 \qquad \text{rad/s}$$

Since the flywheel starts from rest, $\omega = 0$ at $t = 0$, so $C_1 = 0$. Integrating again gives the angular displacement.

$$\theta = \int \frac{3.85t^2}{2} dt = \frac{3.85t^3}{6} + C_2 \qquad \text{rad}$$

Since $\theta$ is 0 at $t = 0$, we get $C_2 = 0$. Now evaluating $\omega$ and $\theta$ at $t = 10.0$ s,

$$\omega = \frac{3.85(10.0)^2}{2} = 193 \text{ rad/s}$$

and

$$\theta = \frac{3.85(10.0)^3}{6} = 642 \text{ rad} = 102 \text{ revolutions}$$

## EXERCISE 5—APPLICATIONS TO MOTION

### Displacement and Velocity

1. The acceleration of a point is given by $a = 4 - t^2$ m/s$^2$. Write an equation for the velocity, if $v = 2$ m/s when $t = 3$ s.
2. A car starts from rest and continues at a rate of $v = \frac{1}{8}t^2$ ft/s. Find the function that relates distance $s$ the car has traveled to the time $t$, in seconds. How far will the car go in 4 s?
3. A particle starts from the origin. Its $x$ component of velocity is $t^2 - 4$ and its $y$ component is $4t$ (cm/s). **(a)** Write equations for $x$ and $y$ and **(b)** find the distance between the particle and the origin when $t = 2$ s.
4. A body is moving at the rate $v = \frac{3}{2}t^2$ (m/s). Find the distance it will move in $t$ seconds if $s = 0$ when $t = 0$.
5. The acceleration of a falling body is $a = -32$ ft/s$^2$. Find the relation between $s$ and $t$ if $s = 0$ and $v = 20$ ft/s when $t = 0$.

### Motion along a Curve

6. A point starts from rest at the origin and moves along a curved path with $x$ and $y$ accelerations of $a_x = 2$ cm/s$^2$ and $a_y = 8t$ cm/s$^2$. Write expressions for the $x$ and $y$ components of velocity.
7. A point starts from rest at the origin and moves along a curved path with $x$ and $y$ accelerations of $a_x = 5t^2$ cm/s$^2$ and $a_y = 2t$ cm/s$^2$. Find the $x$ and $y$ components of velocity at $t = 10$ s.
8. A point starts from (5, 2) with initial velocities of $v_x = 2$ cm/s and $v_y = 4$ cm/s and moves along a curved path. It has $x$ and $y$ accelerations of $a_x = 7t$ and $a_y = 2$. Find the $x$ and $y$ displacements at $t = 5$ s.
9. A point starts from (9, 1) with initial velocities of $v_x = 6$ cm/s and $v_y = 2$ cm/s and moves along a curved path. It has $x$ and $y$ accelerations of $a_x = 3t$ and $a_y = 2t^2$. Find the $x$ and $y$ components of velocity at $t = 15$ s.

10. A wheel starts from rest and accelerates at 3.00 rad/s$^2$. Find the angular velocity after 12.0 s.
11. A certain gear starts from rest and accelerates at $8.5t^2$ rad/s$^2$. Find the total number of revolutions after 20.0 s.
12. A link in a mechanism rotating with an angular velocity of 3.00 rad/s is given an acceleration of 5.00 rad/s$^2$ at $t = 0$. Find the angular velocity after 20.0 s.
13. A pulley in a magnetic tape drive is rotating at 1.25 rad/s when it is given an acceleration of 7.24 rad/s$^2$ at $t = 0$. Find the angular velocity at 2.00 s.

## 31-6 APPLICATIONS TO ELECTRIC CIRCUITS

### Charge

We stated in Sec. 29-1 that the current $i$ (amperes) at some point in a conductor was equal to the time rate of change of the charge $q$ (coulombs) passing that point, or $i = dq/dt$. We can now solve this equation for $q$. Multiplying by $dt$ gives $dq = i\,dt$. Integrating, we get

| | | |
|---|---|---|
| Charge | $q = \int i\,dt$ coulombs | **A80** |

**EXAMPLE 42:** The current to a certain capacitor is given by $i = 2t^3 + t^2 + 3$. The initial charge on the capacitor is 6.83 coulombs. Find (a) an expression for the charge on the capacitor and (b) the charge when $t = 5.00$ *s*.

**Solution:**

(a) Integrating the expression for current, we obtain

$$q = \int i\,dt = \int (2t^3 + t^2 + 3)\,dt = \frac{t^4}{2} + \frac{t^3}{3} + 3t + C \qquad \text{coulombs}$$

We find the constant of integration by substituting the initial conditions, $q = 6.83$ coulombs at $t = 0$. So $C = 6.83$ coulombs. Our complete equation is then

$$q = \frac{t^4}{2} + \frac{t^3}{3} + 3t + 6.83 \qquad \text{coulombs}$$

(b) When $t = 5.00$ s,

$$q = \frac{(5.00)^4}{2} + \frac{(5.00)^3}{3} + 3(5.00) + 6.83 = 376 \text{ coulombs}$$

### Voltage across a Capacitor

The current in a capacitor has already been given by Eq. A81, $i = C\,dv/dt$, where $i$ is in amperes, $C$ in farads, $v$ in volts, and $t$ in seconds. We now integrate to find the voltage across the capacitor.

$$dv = \frac{1}{C}i\,dt$$

| | | | |
|---|---|---|---|
| Voltage across a Capacitor | $v = \frac{1}{C}\int i\,dt$ | volts | **A82** |

**EXAMPLE 43:** A 1.25-F capacitor that has an initial voltage of 25.0 V is charged with a current that varies with time according to the equation $i = t\sqrt{t^2 + 6.83}$. Find the voltage across the capacitor at 1.00 s.

**Solution:** By Eq. A82,

$$v = \frac{1}{1.25}\int t\sqrt{t^2 + 6.83}\,dt = 0.80\left(\frac{1}{2}\right)\int (t^2 + 6.38)^{1/2}(2t\,dt)$$

$$= \frac{0.40(t^2 + 6.83)^{3/2}}{3/2} + k = 0.267(t^2 + 6.83)^{3/2} + k$$

where we have used $k$ for the constant of integration to avoid confusion with the symbol for capacitance. Since $v = 25.0$ V when $t = 0$, we get

$$k = 25.0 - 0.267(6.83)^{3/2} = 20.2 \text{ V}$$

When $t = 1.00$ s,

$$v = 0.267(1.00^2 + 6.83)^{3/2} + 20.2 = 26.0 \text{ V}$$

### Current in an Inductor

The voltage across an inductor was given by Eq. A86 as $v = L\,di/dt$, where $L$ is the inductance in henrys. From this we get

| | | | |
|---|---|---|---|
| Current in an Inductor | $i = \frac{1}{L}\int v\,dt$ | amperes | **A85** |

**EXAMPLE 44:** The voltage across a 10.6-H inductor is $v = \sqrt{3t + 25.4}$ V. Find the current in the inductor at 5.25 s if the initial current is 6.15 A.

**Solution:** From Eq. A85,

$$i = \frac{1}{10.6}\int \sqrt{3t + 25.4}\,dt = 0.0943\left(\frac{1}{3}\right)\int (3t + 25.4)^{1/2}(3\,dt)$$

$$= \frac{0.0314(3t + 25.4)^{3/2}}{3/2} + C = 0.0210(3t + 25.4)^{3/2} + C$$

When $t = 0$, $i = 6.15$ A, so

$$C = 6.15 - 0.0210(25.4)^{3/2} = 3.46 \text{ A}$$

When $t = 5.25$ s,

$$i = 0.0314[3(5.25) + 25.4]^{3/2} + 3.46 = 11.8 \text{ A}$$

## EXERCISE 6—APPLICATIONS TO ELECTRIC CIRCUITS

1. The current to a capacitor is given by $i = 2t + 3$. The initial charge on the capacitor is 8.13 C. Find the charge when $t = 1.00$ s.
2. The current to a certain circuit is given by $i = t^2 + 4$. If the initial charge is zero, find the charge at 2.50 s.
3. The current to a certain capacitor is $i = 3.25 + t^3$. If the initial charge on the capacitor is 16.8 C, find the charge when $t = 3.75$ s.
4. A 21.5-F capacitor with zero initial voltage has a charging current of $i = \sqrt{t}$. Find the voltage across the capacitor at 2.00 s.
5. A 15.2-F capacitor has an initial voltage of 2.00 V. It is charged with a current given by $i = t\sqrt{5 + t^2}$. Find the voltage across the capacitor at 1.75 s.
6. A 75.0-$\mu$F capacitor has an initial voltage of 125 V and is charged with a current equal to $i = \sqrt{t} + 16.3$. Find the voltage across the capacitor at 4.00 s.
7. The voltage across a 1.05-H inductor is $v = \sqrt{23t}$ V. Find the current in the inductor at 1.25 s if the initial current is zero.
8. The voltage across a 52.0-H inductor is $v = t^2 - 3t$ V. If the initial current is 2.00 A, find the current in the inductor at 1.00 s.
9. The voltage across a 15.0-H inductor is given by $v = 28.5 + \sqrt{6t}$ V. Find the current in the inductor at 2.50 s if the initial current is 15.0 A.

## CHAPTER 31 REVIEW PROBLEMS

1. $\int \tan 3\theta \, d\theta$
2. $\int \cot 5\theta \, d\theta$
3. $\int a^{2x} \, dx$
4. $\int \frac{dx}{\sqrt[3]{x}}$
5. $\int (e^{5x} + a^{5x}) \, dx$
6. $\int (e^x + 4)e^{-x} \, dx$
7. $\int (e^{2x+1} + x) \, dx$
8. $\int 6e^{3x} \, dx$
9. $\int \frac{x^4 + x^3 + 1}{x^3} \, dx$
10. $\int e^{5x} \, dx$
11. $\int \frac{dx}{e^x}$
12. $\int \csc^2(3x + 2) \, dx$
13. $\int \csc^2 3x \, dx$
14. $\int 3.1 \, y^2 \, dy$
15. $\int \frac{2dt}{t^2}$
16. $\int \sqrt{4x} \, dx$
17. $\int \sec 2\theta \, d\theta$
18. $\int \tan^2 5\theta \, d\theta$
19. $\int x^2(x^3 - 4)^2 \, dx$
20. $\int (x^4 - 2x^3)(2x^3 - 3x^2) \, dx$
21. $\int \frac{x \, dx}{x^2 + 3}$
22. $\int \frac{dx}{x + 5}$
23. Find the equation of a curve that passes through the point (3, 0), has a slope of 9 at that point, and has a second derivative $y'' = x$.
24. The rate of growth of the number $N$ of bacteria in a culture is $dN/dt = 0.5N$. If $N = 100$ when $t = 0$, derive the formula for $N$ at any time.
25. The acceleration of an object that starts from rest is given by $a = 3t$. Write equations for the velocity and displacement of the object.
26. The voltage across a 25.0-H inductor is given by $v = 8.9 + \sqrt{3t}$ V. Find the current in the inductor at 5.00 s if the initial current is 1.00 A.

27. A flywheel starts from rest and accelerates at $7.25t^2$ rad/s$^2$. Find the angular velocity and the total number of revolutions after 20.0 s.

28. The current to a certain capacitor is $i = t^3 + 18.5$. If the initial charge on the capacitor is 6.84 coulombs, find the charge when $t = 5.25$ s.

29. A point starts from (1, 1) with initial velocities of $v_x = 4$ cm/s and $v_y = 15$ cm/s, and moves along a curved path. It has $x$ and $y$ components of acceleration of $a_x = t$ and $a_y = 5t$. Write expressions for the $x$ and $y$ components of velocity and displacement.

30. A 15.0-F capacitor has an initial voltage of 25 V and is charged with a current equal to $i = \sqrt{4t + 21.6}$. Find the voltage across the capacitor at 14.00 s.

*Writing*

31. Integration is the inverse of differentiation. List as many other pairs of inverse mathematical operations as you can. Describe in your own words when the inverse of each operation gives an indefinite result.

# 32

# THE DEFINITE INTEGRAL

## OBJECTIVES

**When you have completed this chapter, you should be able to:**

- Evaluate definite integrals.
- Determine approximate areas under curves by the midpoint method.
- Apply the Fundamental Theorem of Calculus to determine exact areas under curves.
- Find the area between two curves using horizontal or vertical elements.
- Determine areas involving more than one region.

In this chapter we first define the *definite integral,* and then show how to evaluate it. Next we discuss the problem of finding the area bounded by a curve and the $x$ axis, between two given values of $x$. We find such areas, first approximately by the *midpoint method,* and then exactly by means of the definite integral. In the process we develop the *fundamental theorem of calculus,* which ties together the derivative, the integral, and the area under a curve. Finally, we learn a fast way to set up the integral for finding areas, which will be of great use later for finding quantities, such as volumes, by integration.

## 32-1 THE DEFINITE INTEGRAL

In Chapter 31 we learned how to find the indefinite integral, or antiderivative, of a function. For example,

$$\int x^2\,dx = \frac{x^3}{3} + C \tag{1}$$

We can, of course, evaluate the antiderivative at some particular value, say, $x = 6$. Substituting into (1) gives us

$$\int x^2\,dx \bigg|_{x=6} = \frac{6^3}{3} + C = 72 + C$$

Similarly, we can evaluate the same integral at, say, $x = 3$. Again substituting into (1) yields

$$\int x^2\,dx \bigg|_{x=3} = \frac{3^3}{3} + C = 9 + C$$

Suppose, now, that we subtract the second antiderivative from the first. We get

$$72 + C - 9 - C = 63$$

Although we do not know the value of the constant $C$, we do know that it has the same value in both expressions, since both were obtained from (1), so $C$ will drop out when we subtract.

We now introduce new notation. We let

$$\int x^2\,dx \bigg|_{x=6} - \int x^2\,dx \bigg|_{x=3} = \int_3^6 x^2\,dx$$

The expression on the right is called a *definite integral.* Here 6 is called the *upper limit* and 3 is the *lower limit.* This notation tells us to evaluate the antiderivative at the upper limit, and from that, subtract the antiderivative evaluated at the lower limit. Notice that a definite integral (unlike the indefinite integral) has a *numerical* value, in this case, 63.

**But what is this number, and what is it good for? We'll soon see that it gives us the area under the curve $y = x^2$, from $x = 3$ to $x = 6$, and that the definite integral has lots of applications.**

In general, if

$$\int f(x)\,dx = F(x) + C$$

then

$$\int f(x)\,dx \bigg|_{x=a} = F(a) + C$$

$$\int f(x)\,dx \Big|_{x=b} = F(b) + C$$

and

$$\int_a^b f(x)\,dx = F(b) + C - F(a) - C$$

**We require, as usual, that the function $f(x)$ be continuous in the interval under consideration.**

The constants drop out, leaving

| | | |
|---|---|---|
| Definite Integral | $\int_a^b f(x)\,dx = F(b) - F(a)$ | **365** |

*The definite integral of a function is equal to the antiderivative of that function evaluated at the upper limit b minus the antiderivative evaluated at the lower limit a.*

### Evaluating a Definite Integral

To *evaluate* a definite integral, first integrate the expression (omitting the constant of integration), and write the upper and lower limits on a vertical bar or bracket to the right of the integral. Then substitute the upper limit, then the lower limit, and subtract.

---

**EXAMPLE 1:** Evaluate $\int_2^4 x^2\,dx$.

**Solution:**

$$\int_2^4 x^2\,dx = \frac{x^3}{3}\Big|_2^4$$

$$= \frac{4^3}{3} - \frac{2^3}{3} = \frac{56}{3}$$

---

**EXAMPLE 2:** Evaluate $\int_0^{\pi/2} \sin 2x\,dx$.

**Solution:**

$$\int_0^{\pi/2} \sin 2x\,dx = \frac{1}{2}\int_0^{\pi/2} \sin 2x(2\,dx)$$

By Rule 10:

$$= -\frac{1}{2}\cos 2x \Big|_0^{\pi/2}$$

$$= -\frac{1}{2}\left[\cos 2\left(\frac{\pi}{2}\right) - \cos 0\right] = 1$$

---

| | |
|---|---|
| Common Error | Don't assume that an integral is zero when either the upper or lower limit is zero. Work it out. |

**EXAMPLE 3:** Evaluate $\int_1^2 e^{3x}\,dx$.

**Solution:**

$$\int_1^2 e^{3x}\,dx = \frac{1}{3}\int_1^2 e^{3x}\,(3\,dx)$$

$$= \frac{1}{3}e^{3x}\Big|_1^2$$

$$= \frac{1}{3}[e^6 - e^3] \simeq 128$$

by Rule 8.

Integrals will sometimes produce expressions with absolute value signs, as in the following example.

**EXAMPLE 4:**

$$\int_{-3}^{-2}\frac{dx}{x} = \ln|x|\Big|_{-3}^{-2} = \ln|-2| - \ln|-3|$$

The logarithm of a negative number is not defined. But here we are taking the logarithm of the *absolute value* of a negative number. Thus

$$\ln|-2| - \ln|-3| = \ln 2 - \ln 3 \simeq -0.405$$

## Continuity

If a function is *discontinuous* between two limits $a$ and $b$, the definite integral is not defined over that interval.

**EXAMPLE 5:** The integral $\int_{-3}^{2}\frac{dx}{x}$ is not defined, because the function $y = 1/x$ is discontinuous at $x = 0$.

| | |
|---|---|
| Common Error | Be sure that your function is continuous between the given limits before evaluating a definite integral. |

## EXERCISE 1—THE DEFINITE INTEGRAL

Evaluate each definite integral.

1. $\int_1^2 x\,dx$
2. $\int_{-2}^2 x^2\,dx$
3. $\int_1^3 7x^2\,dx$
4. $\int_{-2}^2 3x^4\,dx$
5. $\int_0^4 (x^2 + 2x)\,dx$
6. $\int_{-2}^2 x^2(x + 2)\,dx$
7. $\int_2^4 (x + 3)^2\,dx$
8. $\int_0^a (a^2x - x^3)\,dx$
9. $\int_1^{10}\frac{dx}{x}$
10. $\int_1^e \frac{dx}{x}$

11. $\int_0^1 \frac{x\,dx}{4 + x^2}$

12. $\int_{-2}^{-3} \frac{2t\,dt}{1 + t^2}$

13. $\int_0^1 \frac{dx}{\sqrt{3 - 2x}}$

14. $\int_0^1 \frac{x\,dx}{\sqrt{2 - x^2}}$

15. $\int_0^1 xe^{x^2}\,dx$

16. $\int_0^1 \frac{dx}{e^{3x}}$

17. $\int_0^{\pi} \sin\phi\,d\phi$

18. $\int_0^{\pi/2} \cos\phi\,d\phi$

19. $\int_0^{\pi} \cos\frac{\theta}{2}\,d\theta$

20. $\int_{\pi/3}^{\pi/2} \sin^2 x \cos x\,dx$

## 32-2 AREA UNDER A CURVE

### Summation Notation

Before we derive an expression for the area under a curve, we must learn some new notation to express the sum of a string of terms. We use the capital Greek sigma $\Sigma$ to stand for summation, or adding up. Thus

$$\Sigma\, n$$

means to sum a string of $n$'s. Of course, we must indicate a starting and ending value for $n$, and these values are placed on the sigma symbol. Thus

$$\sum_{n=1}^{5} n$$

means to add up the $n$'s starting with $n = 1$ and ending with $n = 5$.

$$\sum_{n=1}^{5} n = 1 + 2 + 3 + 4 + 5 = 15$$

---

**EXAMPLE 6:** Evaluate $\sum_{n=1}^{4} (n^2 - 1)$.

**Solution:**

$$\sum_{n=1}^{4} (n^2 - 1) = (1^2 - 1) + (2^2 - 1) + (3^2 - 1) + (4^2 - 1)$$

$$= 0 + 3 + 8 + 15 = 26$$

---

**EXAMPLE 7:** Evaluate the sums indicated.

(a) $\sum_{k=2}^{5} k^2 = 2^2 + 3^2 + 4^2 + 5^2 = 54$

(b) $\sum_{x=1}^{4} f(x) = f(1) + f(2) + f(3) + f(4)$

(c) $\sum_{i=1}^{n} f(x_i) = f(x_1) + f(x_2) + f(x_3) + \cdots + f(x_n)$

---

In the following section, we use the sigma notation for expressions similar to Example 7(c).

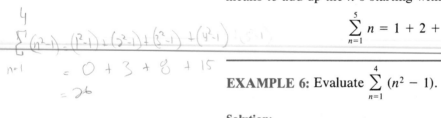

## Approximate Area under a Curve

Figure 32-1 shows a graph of some function $f(x)$. Our problem is to find the area (shown lightly shaded) bounded by that curve, the $x$ axis, and the lines $x = a$ and $x = b$.

We start by subdividing that area into $n$ vertical strips, called *panels*, by drawing vertical lines at $x_0, x_1, x_2, \ldots, x_n$. The panels do not have to be of equal width, but we make them equal for simplicity. Let the width of each panel be $\Delta x$.

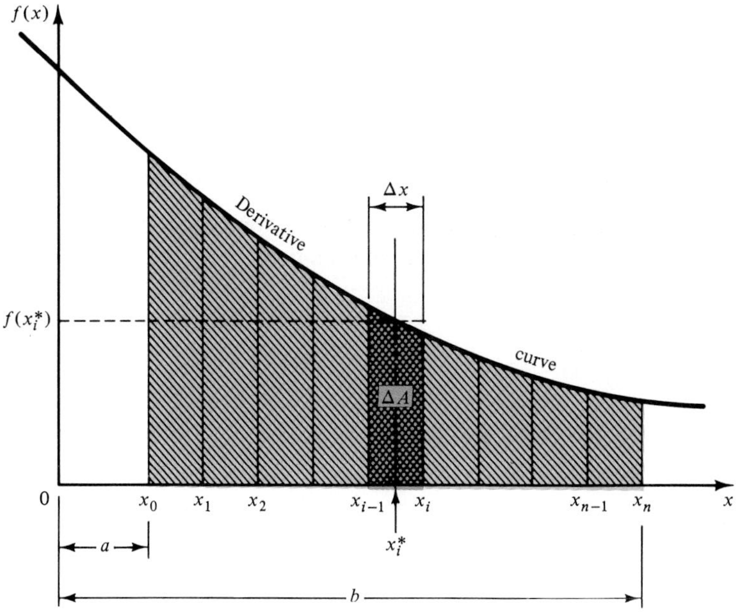


FIGURE 32-1

Now look at one particular panel, the one lying between $x_{i-1}$ and $x_i$ (shown shaded darkly). Anywhere within this panel we choose a point $x_i^*$. The height of the curve at this value of $x$ is then $f(x_i^*)$. The area $\Delta A$ of the dark panel is then *approximately* equal to the area of a rectangle of width $\Delta x$ and height $f(x_i^*)$.

$$\Delta A \cong f(x_i^*)\ \Delta x$$

Our approximate area may be greater than or less than the actual area $\Delta A$, depending on where we chose $x_i^*$. (Later we show that $x_i^*$ is chosen in a particular place, but for now, consider it to be anywhere between $x_{i-1}$ and $x_i$.)

The area of the first panel is, similarly, $f(x_1^*)\ \Delta x$; of the second panel, $f(x_2^*)\ \Delta x$; and so on. To get an approximate value for the total area, we add up the areas of each panel.

$$A \cong f(x_1^*)\ \Delta x + f(x_2^*)\ \Delta x + f(x_3^*)\ \Delta x + \cdots + f(x_n^*)\ \Delta x$$

Rewriting this expression using our sigma notation gives

**These are called *Riemann sums,* after Georg Friedrich Bernhard Riemann (1826–1866).**

$$A \simeq \sum_{i=1}^{n} f(x_i^*)\ \Delta x$$

## Midpoint Method

If we approximate the area under a curve by rectangular panels of width $\Delta x$ and choose our values of $x^*$ at the *midpoints* of these panels, we are using what is often called the *midpoint method*.

**In this chapter we use the midpoint method mainly as a lead-in to finding areas by integration. But it is also a valuable method for finding areas for functions that cannot be exactly integrated. Other standard approximation methods are given in Sec. 34-6.**

| | | |
|---|---|---|
| Midpoint Method | $A \simeq \sum_{i=1}^{n} f(x_i^*)\,\Delta x$ where $f(x_i^*)$ is the height of the $i$th panel at its midpoint. | **371** |

$y = 3x^2$ $x = 0$ to $x = 10$

**EXAMPLE 8:** Use the midpoint method to calculate the approximate area under the curve $f(x) = 3x^2$, from $x = 0$ to $x = 10$, taking panels of width 2.

**Solution:** Our graph (Fig. 32-2) shows the panels, with midpoints at 1, 3, 5, 7, and 9. At each midpoint $x^*$ we compute the height $f(x^*)$ of the curve.

| $x^*$ | 1 | 3 | 5 | 7 | 9 |
|---|---|---|---|---|---|
| $f(x^*)$ | 3 | 27 | 75 | 147 | 243 |

The approximate area is then the sum of the areas of each panel.

**We'll see later that the exact area is 1000 square units.**

$$A \simeq 3(2) + 27(2) + 75(2) + 147(2) + 243(2)$$
$$= 495(2) = 990 \text{ square units}$$

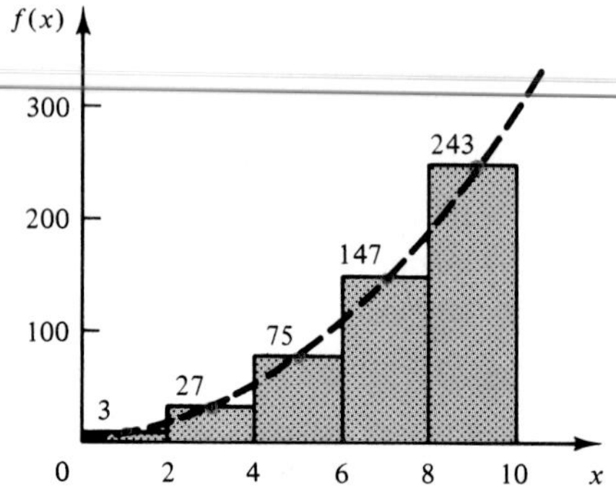


**FIGURE 32-2 Area by midpoint method.**

## Exact Area under a Curve

We can obtain greater accuracy in computing the area under a curve simply by reducing the width of each rectangular panel. Clearly, the panels in Fig. 32-3b are a better fit to the curve than those in Fig. 32-3a. Thus, as the panel width $\Delta x$ approaches zero (and the number of panels approaches infinity) the sum of the areas of the panels approaches the exact area $A$ under the curve.

| | | |
|---|---|---|
| Exact Area under a Curve | $A = \lim_{\Delta x \to 0} \sum_{i=1}^{n} f(x_i^*)\,\Delta x$ | **366** |

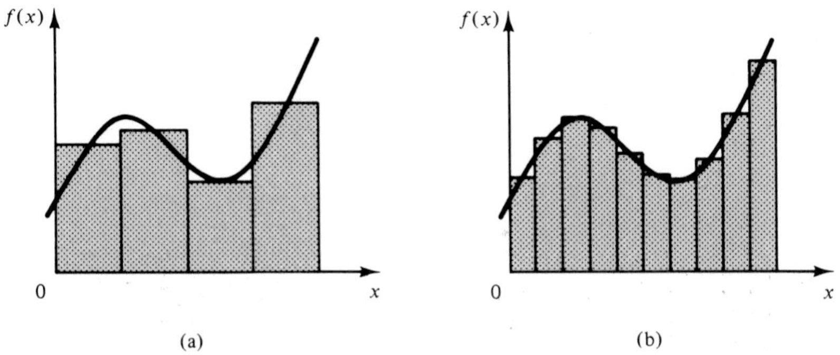


**FIGURE 32-3 More panels give greater accuracy.**

**EXAMPLE 9:** Compute the area under the curve in Example 8 by the midpoint method, using panel widths of 2, 1, $\frac{1}{2}$, $\frac{1}{4}$, and so on.

**Solution:** We compute the approximate area just as in Example 8. We omit the tedious computations (which were done by computer) and show only the results.

| *Panel Width* | *Area* |
|---|---|
| 2.0000 | 990.0000 |
| 1.0000 | 997.5000 |
| 0.5000 | 999.3750 |
| 0.2500 | 999.8438 |
| 0.1250 | 999.9609 |
| 0.0625 | 999.9902 |
| 0.0313 | 999.9968 |
| 0.0156 | 999.9996 |

Notice that as the panel width decreases, the computed area seems to be approaching a limit of 1000. We'll see in the next section that 1000 is the exact area under the curve.

## The Fundamental Theorem of Calculus

How, then, do we find the exact area under a curve? Must we compute the areas of hundreds of rectangular panels and add them up, as in Example 9? No. We can let the process of integration add them up for us, as we'll now see.

Let us return to Fig. 32-1, but now we draw above it the curve $F(x)$ (Fig. 32-4). Thus the lower curve $f(x)$ is the derivative of the upper curve $F(x)$, and conversely, the upper curve $F(x)$ is the integral of the lower.

For the midpoint method, we arbitrarily selected $x_i^*$ at the midpoint of each panel. We now do it differently. *We select $x_i^*$ so that the slope at point $Q$ on the integral (upper) curve is equal to the slope of the straight line PR* (Fig. 32-5).

There is a theorem, called the *mean value theorem*, that says there must be at least one point $Q$, between $P$ and $R$, at which the slope is equal to the slope of $PR$. We won't prove it, but can you see intuitively that it must be so?

The slope at $Q$ is equal to $f(x_i^*)$, and the slope of $PR$ is equal to

$$f(x_i^*) = \frac{\text{rise}}{\text{run}} = \frac{F(x_i) - F(x_{i-1})}{\Delta x}$$

or

$$f(x_i^*)\,\Delta x = F(x_i) - F(x_{i-1})$$

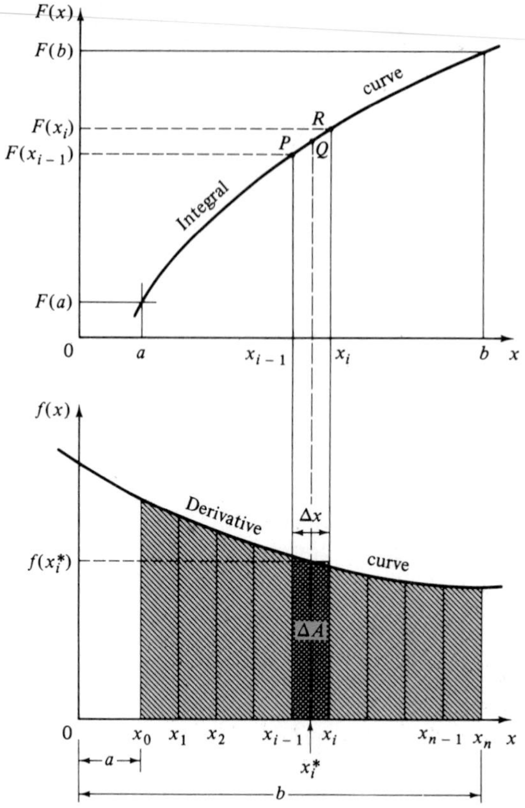


**FIGURE 32-4 Area as the limit of a sum.**

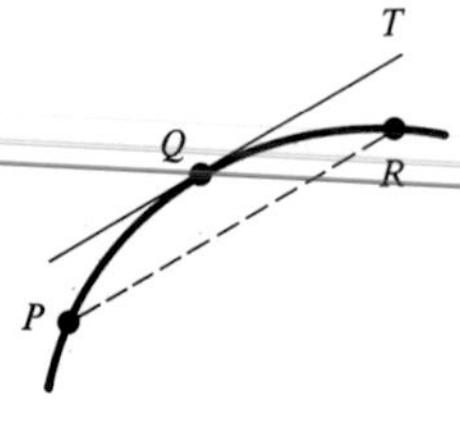


**FIGURE 32-5**

If we write this expression for each panel, we get

$$f(x_1^*)\ \Delta x = F(x_1) - F(a)$$

$$f(x_2^*)\ \Delta x = F(x_2) - F(x_1)$$

$$f(x_3^*)\ \Delta x = F(x_3) - F(x_2)$$

$$\vdots \qquad\qquad \vdots$$

$$f(x_n^*)\ \Delta x = F(b) - F(x_{n-1})$$

If we add all these equations, every term on the right drops out except $F(a)$ and $F(b)$.

$$f(x_1^*)\ \Delta x + f(x_2^*)\ \Delta x + f(x_3^*)\ \Delta x + \cdots + f(x_n^*)\ \Delta x = F(b) - F(a)$$

$$\sum_{i=1}^{n} f(x_i^*)\ \Delta x = F(b) - F(a)$$

As before, we let $\Delta x$ approach zero.

$$\lim_{\Delta x \to 0} \sum_{i=1}^{n} f(x_i^*)\ \Delta x = F(b) - F(a)$$

The left side of this equation is equal to the exact area $A$ under the curve (Eq. 366). The right side is equal to the definite integral from $a$ to $b$ of the function $f(x)$. Thus we get

| | | |
|---|---|---|
| Fundamental Theorem of Calculus | $A = \int_a^b f(x)\,dx = F(b) - F(a)$ | **365** |

This result is so important that it is called the *Fundamental Theorem of Calculus*. It gives the amazingly simple result that the area under a curve between two limits is equal to the change in the integral between the same limits, as shown graphically in Fig. 32-6.

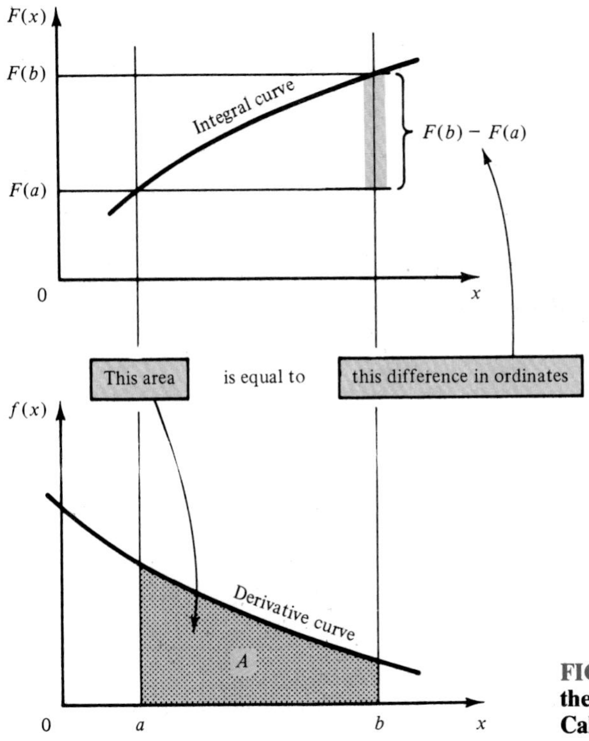


**FIGURE 32-6 Illustration of the Fundamental Theorem of Calculus.**

---

**EXAMPLE 10:** Find the area bounded by the curve $y = 3x^2 + x + 1$, the $x$ axis, and the lines $x = 2$ and $x = 5$.

**Solution:** By Eq. 365,

$$A = \int_2^5 (3x^2 + x + 1)\,dx = x^3 + \frac{x^2}{2} + x \Big|_2^5$$

$$= \left(5^3 + \frac{5^2}{2} + 5\right) - \left(2^3 + \frac{2^2}{2} + 2\right)$$

$$\simeq 130.5 \text{ square units}$$

---

## EXERCISE 2—Area under a Curve

### Sigma Notation

Evaluate each expression.

1. $\sum_{n=1}^{5} n$

2. $\sum_{r=1}^{9} r^2$

3. $\sum_{n=1}^{7} 3n$

4. $\sum_{m=1}^{4} \frac{1}{m}$

5. $\sum_{n=1}^{5} n(n-1)$

6. $\sum_{q=1}^{6} \frac{q}{q+1}$

### Approximate Areas by Midpoint Method

Find the approximate area (in square units) under each curve by the midpoint method, using panels 2 units wide.

7. $y = x^2 + 1$ from $x = 0$ to 8.
8. $y = x^2 + 3$ from $x = -4$ to 4.
9. $y = \frac{1}{x}$ from $x = 2$ to 10.
10. $y = 2 + x^4$ from $x = -10$ to 0.

### Exact Areas by Integration

Find the area (in square units) bounded by each curve, the given lines, and the $x$ axis.

11. $y = 2x$ from $x = 0$ to $x = 10$
12. $y = x^2 + 1$ from $x = 1$ to 20
13. $y = 3 + x^2$ from $x = -5$ to 5
14. $y = x^4 + 4$ from $x = -10$ to $-2$
15. $y = x^3$ from $x = 0$ to $x = 4$
16. $y = 9 - x^2$ from $x = 0$ to $x = 3$
17. $y = 1/\sqrt{x}$ from $x = \frac{1}{2}$ to $x = 8$
18. $y = x^3 + 3x^2 + 2x + 10$ from $x = -3$ to $x = 3$
19. $y = x^2 + x + 1$ from $x = 2$ to $x = 3$
20. $y = \sqrt{3x}$ from $x = 2$ to $x = 8$
21. $y = 2x + \frac{1}{x^2}$ from $x = 1$ to $x = 4$
22. $y = \frac{10}{\sqrt{x+4}}$ from $x = 0$ to $x = 5$

### Computer

23. Write a program or use a spreadsheet to compute areas using the midpoint method, and test it on any of the problems in this exercise. Have the program accept any panel width as input, and print the panel width and the approximate area.

## 32-3 FINDING AREAS BY MEANS OF THE DEFINITE INTEGRAL

### A Fast Way to Set Up the Integral

In Sec. 32-2 we have found areas bounded by simple curves and the $x$ axis. In this section we go on to find areas bounded by the $y$, rather than the $x$ axis, and areas *between* curves. Later, we find volumes, surface areas, and so on, where Eq. 365 cannot be used directly. We will have to set up a *different integral each time*. Looking back at the work we had to do to get Eq. 365 makes us wish for an easier way to set up an integral. We have shown in Sec. 32-2 that the definite integral can be thought of as the sum of many small elements of area. This provides us with an intuitive shortcut for setting up an integral, without having to go through a long derivation each time.

Think of the integral sign as an "S," standing for *sum*. It indicates that we are to *add up* the elements that are written after the integral sign. Thus

$$A = \int_a^b f(x)\,dx$$

can be read: "The area $A$ is the sum of all the elements having a height $f(x)$ and a width $dx$, between the limits of $a$ and $b$," as shown in Fig. 32-7.

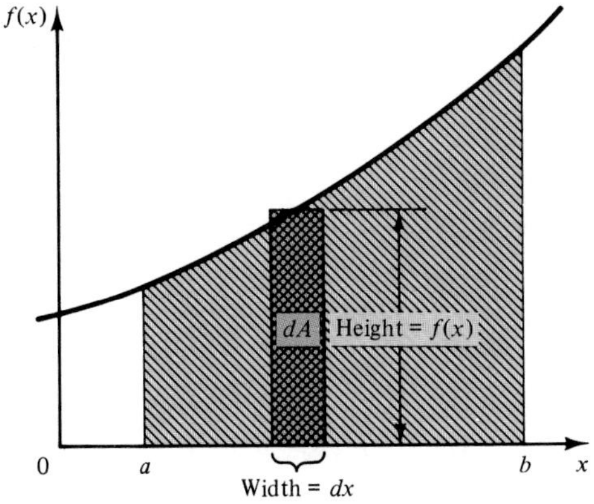


FIGURE 32-7

---

**EXAMPLE 11:** Find the area bounded by the curve $y = x^2 + 3$, the $x$ axis, and the lines $x = 1$ and $x = 4$:

Solution: The usual steps are as follows:

1. Make a sketch showing the bounded area, as in Fig. 32-8. Locate a point $(x, y)$ on the curve.
2. Through $(x, y)$ draw a rectangular element of area, which we call $dA$. Give the rectangle dimensions. We call the width $dx$, because it is measured in the $x$ direction, and we call the height $y$. The area of the element is thus

$$dA = y\,dx = (x^2 + 3)\,dx$$

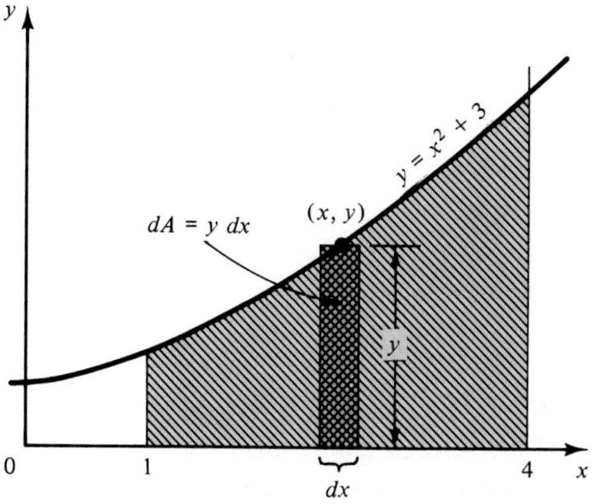


FIGURE 32-8

This looks like a long procedure, but it will save us a great deal of time later.

3. We think of $A$ as the sum of all the small $dA$'s. We accomplish the summation by integration.

$$A = \int dA = \int (x^2 + 3)\,dx$$

4. We locate the limits of integration from the figure. Since we are summing the elements in the $x$ direction, our limits must be on $x$. It is clear that we start the summing at $x = 1$ and end at $x = 4$. So

$$A = \int_1^4 (x^2 + 3)\,dx$$

5. Check that all parts of the integral, including the integrand, the differential, and the limits of integration, are in terms of the *same variable*. In our example, everything is in terms of $x$, and the limits are on $x$, so we can proceed. If, however, our integral contained both $x$ and $y$, one of the variables would have to be eliminated.
6. Our integral is now set up. Integrating gives

$$A = \int_1^4 (x^2 + 3)\,dx = \frac{x^3}{3} + 3x \Big|_1^4$$

$$= \frac{4^3}{3} + 3(4) - \left[\frac{1^3}{3} + 3(1)\right] = 30 \text{ square units}$$

## Area between Two Curves

Either of the curves $f(x)$ or $g(x)$ can, of course, be the $x$ axis.

Suppose that we want the area $A$ bounded by an upper curve $y = f(x)$ and a lower curve $y = g(x)$, between the limits $a$ and $b$ (Fig. 32-9). We draw a vertical element whose width is $dx$, whose height is $f(x) - g(x)$, and whose area $dA$ is

$$dA = [f(x) - g(x)]\,dx$$

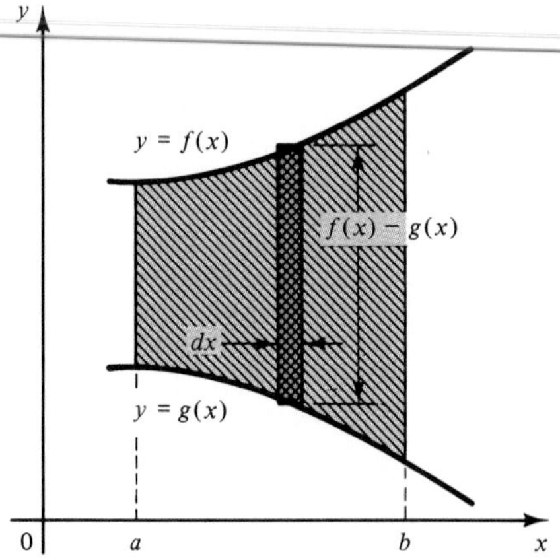


**FIGURE 32-9** **Area between two curves.**

Integrating from $a$ to $b$ gives the total area

$$A = \int_a^b [f(x) - g(x)]\,dx$$

It is important to get *positive* lengths for the elements. To do this, be sure to properly identify the "upper curve" for the region and subtract the values on the "lower curve" from it.

**EXAMPLE 12:** Find the first-quadrant area bounded by the curves

$$y = x^2 + 3 \qquad y = 3x - x^2 \qquad x = 0 \qquad x = 3$$

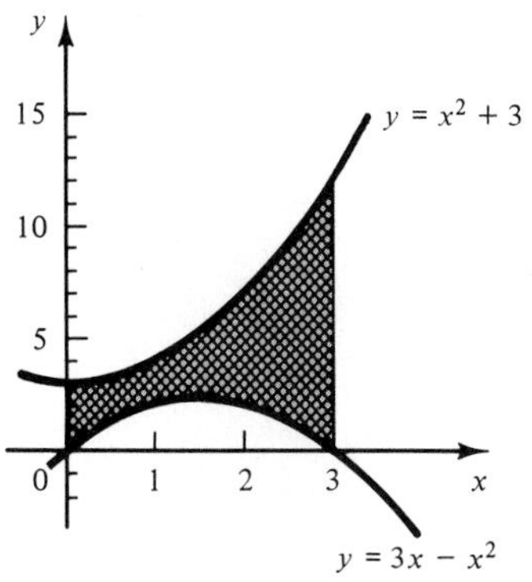

**FIGURE 32-10**

**Solution:** We make a sketch (Fig. 32-10). Letting $f(x) = x^2 + 3$ and $g(x) = 3x - x^2$, we get

$$f(x) - g(x) = (x^2 + 3) - (3x - x^2) = 2x^2 - 3x + 3$$

The area bounded by the curves is then

$$A = \int_0^3 (2x^2 - 3x + 3)\,dx = \frac{2x^3}{3} - \frac{3x^2}{2} + 3x \Big|_0^3$$

$$= 18 - \frac{27}{2} + 9 = \frac{27}{2} \text{ square units}$$

Don't worry if all or part of the desired area is below the $x$ axis. Just follow the same procedure as when the area is above the axis, and the signs will work out by themselves. Be sure, however, that the lengths of the elements are *positive* by subtracting the lower curve from the upper curve.

**EXAMPLE 13:** Find the area bounded by the curves $y = \sqrt{x}$ and $y = x - 3$, between $x = 1$ and $x = 4$ (Fig. 32-11).

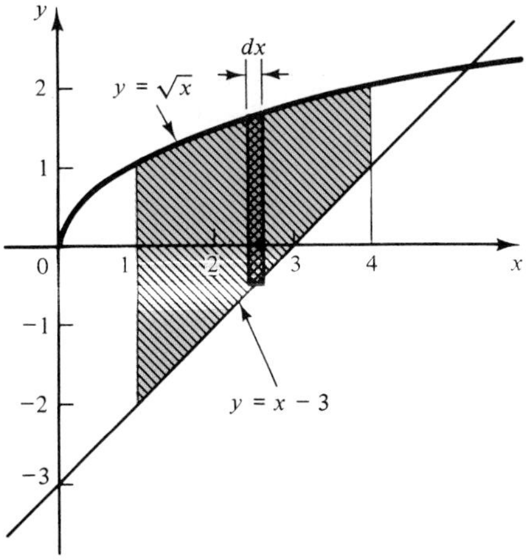

**FIGURE 32-11**

**Solution:** Letting $f(x) = \sqrt{x}$ and $g(x) = x - 3$, we get

$$A = \int_a^b [f(x) - g(x)]\,dx$$

$$= \int_1^4 [\sqrt{x} - x + 3]\,dx$$

$$= \frac{2x^{3/2}}{3} - \frac{x^2}{2} + 3x \Big|_1^4$$

$$= 6\tfrac{1}{6}$$

## Limits Not Given

If we must find the area bounded by two curves and the limits $a$ and $b$ are not given, we must solve the given equations simultaneously to find their points of intersection. These points will provide the limits we need.

**EXAMPLE 14:** Find the area bounded by the parabola $y = 2 - x^2$ and the straight line $y = x$.

**Solution:** Our sketch (Fig. 32-12) shows two points of intersection. To find them, let us solve the equations simultaneously by setting one equation equal to the other.

$$2 - x^2 = x$$

or $x^2 + x - 2 = 0$. We solve this quadratic by factoring.

$$(x + 2)(x - 1) = 0$$

so $x = -2$ and $x = 1$. Substituting back gives the points of intersection (1, 1) and (−2, −2).

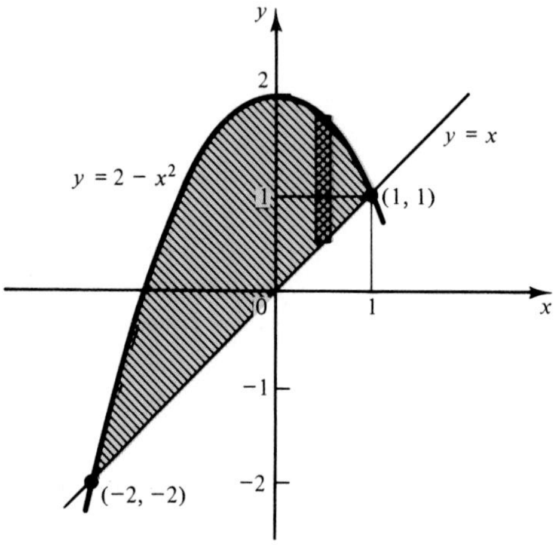


FIGURE 32-12

We draw a vertical element of width $dx$ and whose height is the upper curve minus the lower, or

$$2 - x^2 - x$$

The area $dA$ of the strip is then

$$dA = (2 - x^2 - x)\,dx$$

We integrate, taking as limits the values of $x$ (−2 and 1) found earlier by simultaneous solution of the given equations.

$$A = \int_{-2}^{1} (2 - x^2 - x)\,dx$$

$$= 2x - \frac{x^3}{3} - \frac{x^2}{2}\Big|_{-2}^{1}$$

$$= 4\tfrac{1}{2}$$

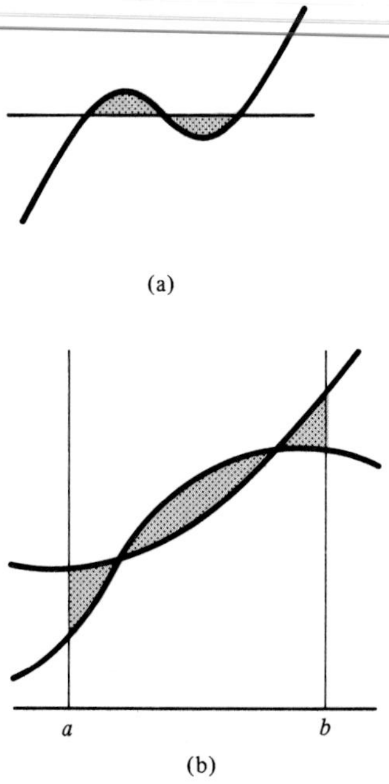


**FIGURE 32-13 Intersecting curves may bound more than one region.**

## Several Regions

Sometimes the given curves may cross in several places, or the limits may be specified in such a way as to define *more than one region,* as in Fig. 32-13. In such cases we find the area of each region separately, and then add them. Note that one of the given curves may be "upper" for one region and "lower" for another.

**EXAMPLE 15:** Find the area bounded by the curve $y = x^2 - 4$ and the $x$ axis, between $x = 1$ and $x = 3$.

**Solution:** Our sketch (Fig. 32-14) shows two regions. For the first region the upper curve is $y = 0$ and the lower is $y = x^2 - 4$. The reverse is true for the second region. We set up two separate integrals,

$$A_1 = \int_1^2 [0 - (x^2 - 4)]\, dx = 4x - \frac{x^3}{3}\Big|_1^2 = \frac{5}{3}$$

$$A_2 = \int_2^3 [(x^2 - 4) - 0]\, dx = \frac{x^3}{3} - 4x\Big|_2^3 = \frac{7}{3}$$

Adding gives us

$$A = A_1 + A_2 = \frac{5}{3} + \frac{7}{3} = 4 \text{ square units}$$

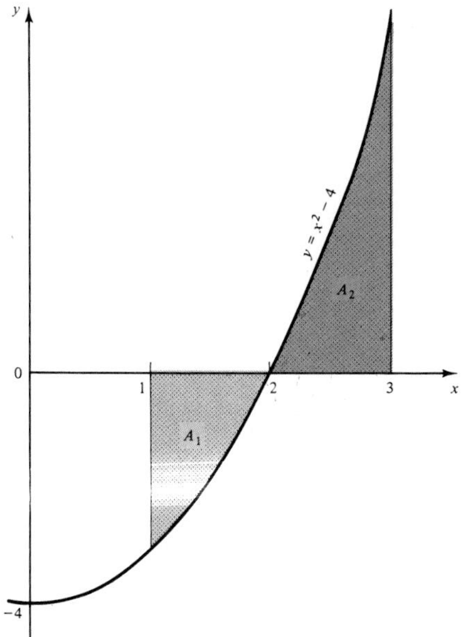


FIGURE 32-14

**EXAMPLE 16:** Find the area bounded by the curve $y = 1/x$ and the $x$ axis (a) from $x = 1$ to 4 and (b) from $x = -1$ to 4.

**Solution:** Integrating, we obtain

$$A = \int_a^b \frac{dx}{x} = \ln |x|\Big|_a^b = \ln |b| - \ln |a|$$

(a) For the limits 1 to 4,

$$A = \ln 4 - \ln 1 = 1.386$$

(b) For the limits $-1$ to 4,

$$A = \ln |4| - \ln |-1| = \ln 4 - \ln 1 = 1.386 \text{ (?)}$$

We appear to get the same area between the limits $-1$ and 4 as we did for the limits 1 and 4. However, a graph (Fig. 32-15) shows that the curve $y = 1/x$ *is discontinuous* at $x = 0$, so we cannot integrate over the interval $-1$ to 4.

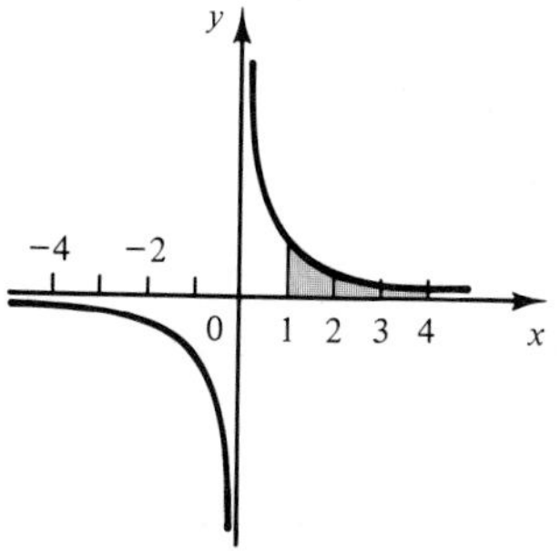


FIGURE 32-15

| | |
|---|---|
| Common Error | Don't try to set up these area problems without making a sketch. Don't try to integrate across a discontinuity. |

### Horizontal Elements

So far we have used vertical elements of area to set up our problems. Often, however, the use of horizontal elements will make our work easier, as in the following example.

---

**EXAMPLE 17:** Find the first-quadrant area bounded by the curve $y = x^2 + 3$, the $y$ axis, and the lines $y = 7$ and $y = 12$ (Fig. 32-16).

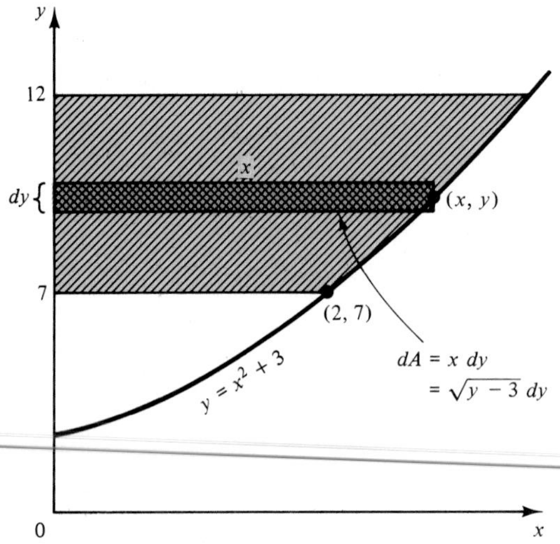


**FIGURE 32-16**

**Solution:** We locate a point $(x, y)$ on the curve. If we were to draw a vertical element through $(x, y)$, its height would be 5 units in the region to the left of the point (2, 7), and $12 - y$ units in the region to the right of (2, 7). Thus we would need two different expressions for the height of the element. To avoid this complication we choose a *horizontal* element, whose length is simply $x$, and whose width is $dy$. So

$$dA = x\,dy = (y - 3)^{1/2}\,dy$$

Integrating, we have

$$A = \int_7^{12} (y - 3)^{1/2}\,dy$$

$$= \left.\frac{2(y - 3)^{3/2}}{3}\right|_7^{12}$$

$$= \frac{2}{3}(9)^{3/2} - \frac{2}{3}(4)^{3/2} = \frac{38}{3} \text{ square units}$$

---

**EXAMPLE 18:** Find the area bounded by the curves $y^2 = 12x$ and $y^2 = 24x - 36$.

**Solution:** We first plot the two curves (Fig. 32-17) which we recognize from their equations to be parabolas opening to the right. We find their points of intersection by solving simultaneously,

$$24x - 36 = 12x$$

$$x = 3$$

$$y = \pm\sqrt{12x} = \pm 6$$

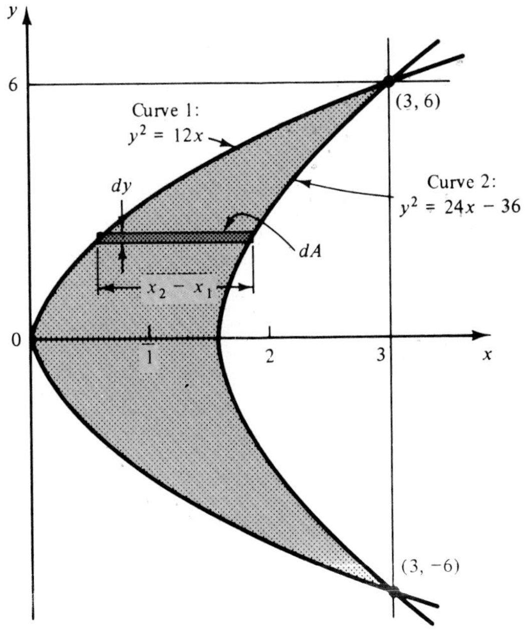


**FIGURE 32-17**

Since we have symmetry about the $x$ axis, let us solve for the first quadrant area, and later double it. We draw a horizontal strip of width $dy$ and length $x_2 - x_1$. The area $dA$ is then $dA = (x_2 - x_1)\,dy$. Integrating, we obtain

$$A = \int_0^6 (x_2 - x_1)\,dy$$

$$= \int_0^6 \left(\frac{y^2}{24} + \frac{36}{24} - \frac{y^2}{12}\right) dy$$

$$= \int_0^6 \left(\frac{3}{2} - \frac{y^2}{24}\right) dy$$

$$= \frac{3y}{2} - \frac{y^3}{72}\bigg|_0^6$$

$$= \frac{3(6)}{2} - \frac{(6)^3}{72} = 6$$

By symmetry, the total area between the two curves is twice this, or 12 square units.

| | |
|---|---|
| Common Error | Don't always assume that a vertical element is the best choice. Try setting up the integral in Example 17 or 18 using a vertical element. What problems arise? |

## EXERCISE 3—FINDING AREAS

### Areas Bounded by the y Axis

Find the first quadrant area bounded by the given curve, the $y$ axis, and the given lines.

1. $y = x^2 + 2$ from $y = 3$ to 5
2. $8y^2 = x$ from $y = 0$ to 10
3. $y^3 = 4x$ from $y = 0$ to $y = 4$
4. $y = 4 - x^2$ from $y = 0$ to $y = 3$

### Areas Bounded by Both Axes

Find the first-quadrant area bounded by each curve and both coordinate axes.

5. $y^2 = 16 - x$
6. $y = x^3 - 8x^2 + 15x$
7. $x + y + y^2 = 2$
8. $\sqrt{x} + \sqrt{y} = 1$

### Areas Bounded by Two Branches of a Curve

Find the area bounded by the given curve and given line.

9. $y^2 = x$ and $x = 4$
10. $y^2 = 2x$ and $x = 4$
11. $4y^2 = x^3$ and $x = 8$
12. $y^2 = x^2(x^2 - 1)$ and $x = 2$

### Areas above and below the x Axis

13. Find the area bounded by the curve $10y = x^2 - 80$, the $x$ axis, and the lines $x = 1$ and $x = 6$.
14. Find the area bounded by the curve $y = x^3$, the $x$ axis, and the lines $x = -3$ and $x = 0$.

Find only the portion of the area below the $x$ axis.

15. $y = x^3 - 4x^2 + 3x$
16. $y = x^2 - 4x + 3$

### Periodic Functions

Find the area under each curve.

17. $y = \sin x$, from $x = 0$ to $\pi$
18. $y = 2 \cos x$, from $x = -\pi/2$ to $\pi/2$
19. $y = 2 \sin \frac{1}{2}\pi x$, from $x = 0$ to 2 rad
20. Find the area between the curve $y = \sin x$ and the $x$ axis from $x = 1$ rad to $x = 3$ rad.
21. Find the area between the curve $y = \cos x$, the $x$ axis, and the ordinates at $x = 0$ and $x = \frac{3}{2}\pi$.

### Areas between Curves

22. Find the area bounded by the lines $y = 3x$, $y = 15 - 3x$, and the $x$ axis.
23. Find the area bounded by the curves $y^2 = 4x$ and $2x - y = 4$.
24. Find the area bounded by the parabola $y = 6 + 4x - x^2$ and the chord joining $(-2, -6)$ and $(4, 6)$.
25. Find the area bounded by the curve $y^2 = x^3$ and the line $x = 4$.

26. Find the area bounded by the curve $y^3 = x^2$ and the chord joining $(-1, 1)$ and $(8, 4)$.
27. Find the area between the parabolas $y^2 = 4x$ and $x^2 = 4y$.
28. Find the area between the parabolas $y^2 = 2x$ and $x^2 = 2y$.

### Areas of Geometric Figures

Use integration to verify the formula for the area of each figure.

**Hint: Integrate using Rule 69 for Problem 33 and several of those to follow.**

29. square of side $a$
30. rectangle with sides $a$ and $b$
31. right triangle (Fig. 32-18a)
32. triangle (Fig. 32-18b)
33. circle of radius $r$
34. segment of circle (Fig. 32-18c)
35. ellipse (Fig. 32-18d)
36. parabola (Fig. 32-18e)

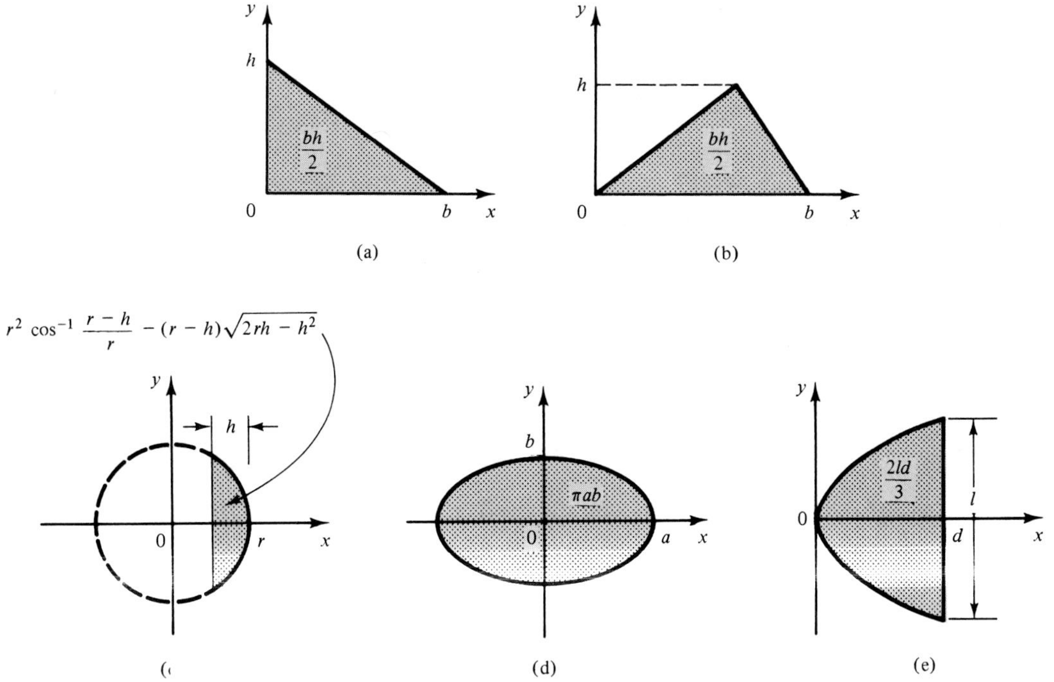


FIGURE 32-18 **Areas of some geometric figures.**

### Applications

37. An elliptical culvert is partly full of water (Fig. 32-19). Find, by integration, the cross-sectional area of the water.

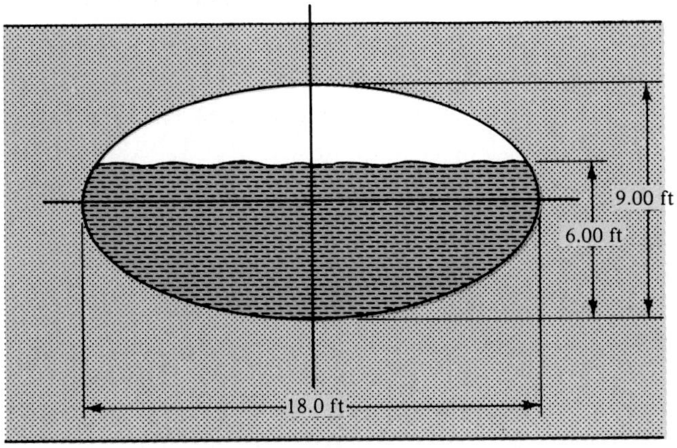


FIGURE 32-19 **Elliptical culvert.**

38. A mirror (Fig. 32-20) has a parabolic face. Find the volume of glass in the mirror.

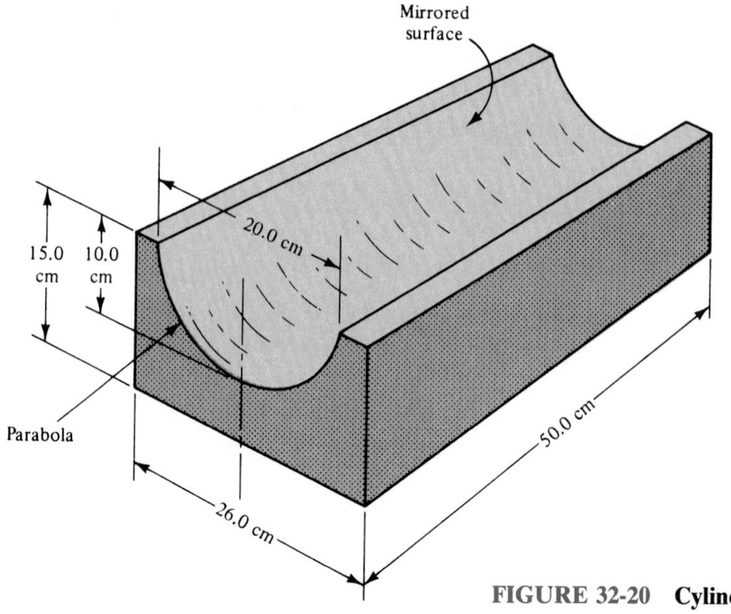


**FIGURE 32-20 Cylindrical mirror.**

39. Figure 32-21 shows a concrete column which has an elliptical cross section. Find the volume of concrete in the column.

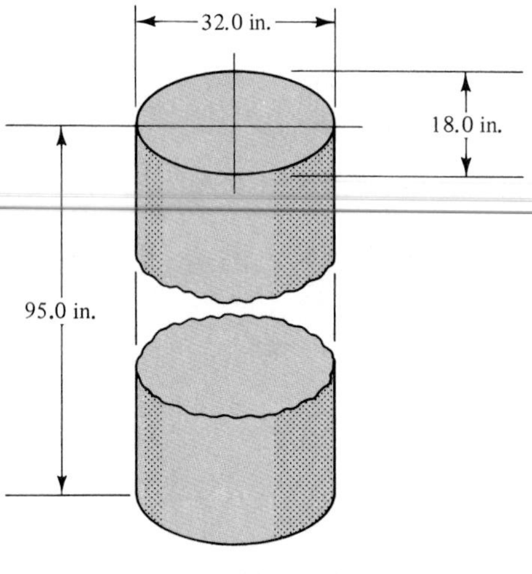


**FIGURE 32-21 Concrete column.**

Note that no equations are given for the curves in these applications. Thus before setting up each integral you must draw axes, chosen to put the equation in simplest form, and then write the equation of the curve.

40. A concrete roof beam for an auditorium has a straight top edge and a parabolic lower edge (Fig. 32-22). Find the volume of concrete in the beam.

**FIGURE 32-22 Roof beam.**

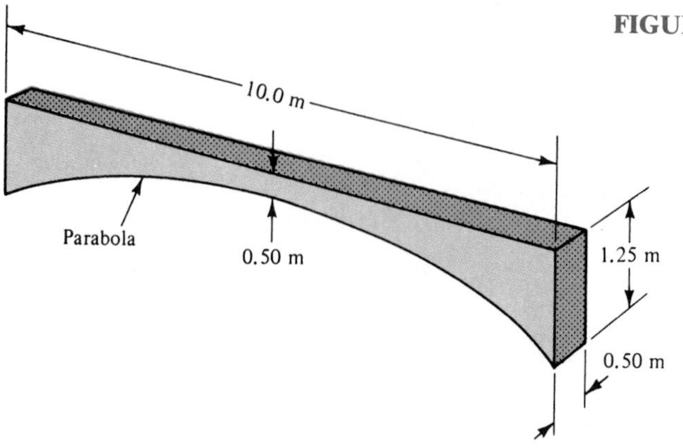

41. The deck of a certain ship has the shape of two intersecting parabolic curves (Fig. 32-23). Find the area of the deck.

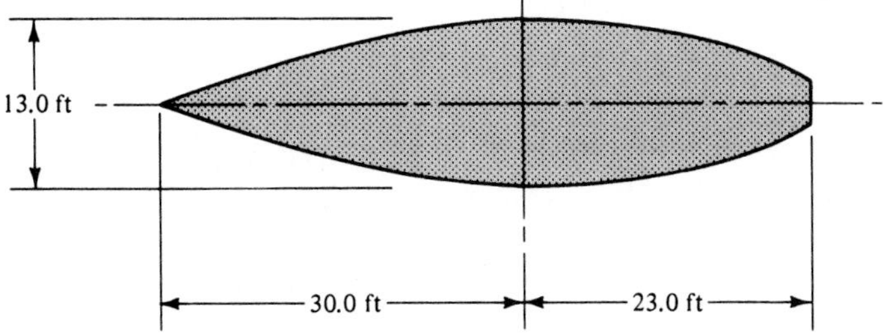


FIGURE 32-23 **Top view of a ship's deck.**

42. A lens (Fig. 32-24) has a cross section formed by two intersecting circular arcs. Find by integration the cross-sectional area of the lens.

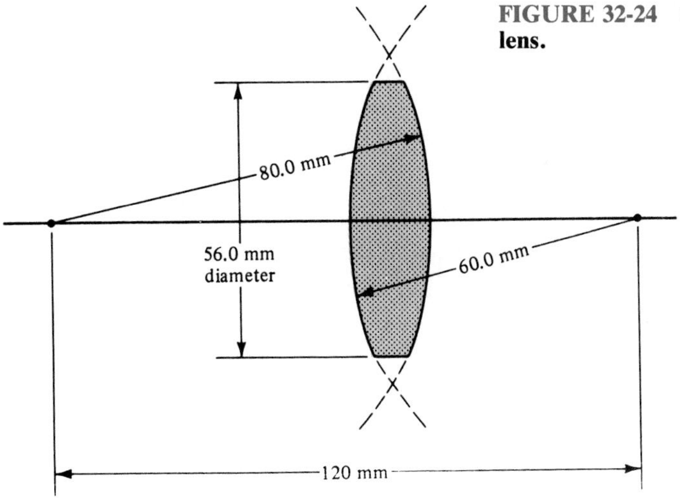


FIGURE 32-24 **Cross section of a lens.**

43. A window (Fig. 32-25) has the shape of a parabola above and a circular arc below. Find the area of the window.

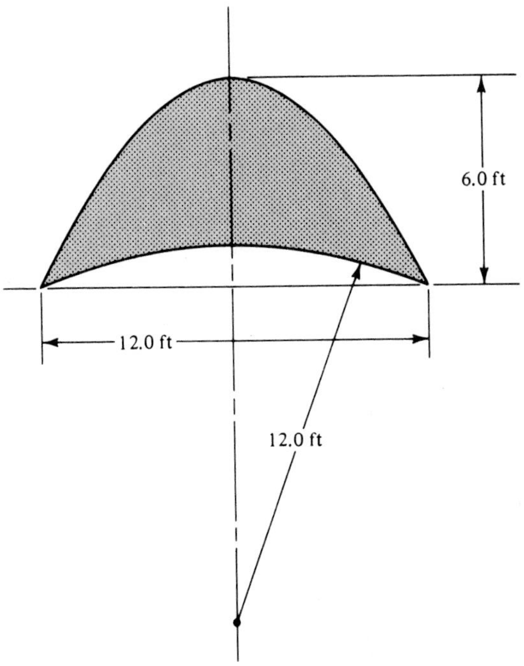


FIGURE 32-25 **Window.**

## CHAPTER 32 REVIEW PROBLEMS

*Evaluate each definite integral.*

1. $\int_{-1}^{0} \frac{dx}{1 - x}$
2. $\int_{0}^{2} \frac{x\,dx}{4 + x^2}$
3. $\int_{0}^{\ln 3} e^{2x}\,dx$
4. $\int_{0}^{a} (\sqrt{a} - \sqrt{x})^2\,dx$
5. $\int_{2}^{7} (x^2 - 2x + 3)\,dx$
6. $\int_{0}^{\pi} \cos 3x\,dx$
7. $\int_{-\pi}^{0} x \sin 2x^2\,dx$
8. $\int_{2}^{1} \sqrt{7 - 3x}\,dx$
9. Evaluate $\sum_{n=1}^{5} n^2(n - 1)$.
10. Evaluate $\sum_{d=1}^{4} \frac{d^2}{d - 1}$.
11. Find the approximate area under the curve $y = 5 + x^2$ from $x = 1$ to 9, by the midpoint method. Use panels 2 units wide.
12. Find the area bounded by the parabola $y^2 = 8x$, the $x$ axis, and the line $x = 2$.
13. Find the entire area of the ellipse

$$\frac{x^2}{16} + \frac{y^2}{9} = 1$$

(Use Rule 69.)

14. Find the area bounded by the coordinate axes and the curve $x^{1/2} + y^{1/2} = 2$.
15. Find the area between the curve $x^2 = 8y$ and $y = \frac{64}{x^2 + 16}$. (Use Rule 56.)
16. Find the area bounded by the curves $y^2 = 8x$ and $x^2 = 8y$.
17. Find the area bounded by the curve $y = x^{3/2}$ and the line $x = 4$.
18. Find the area bounded by the curve $y = 1/x$ and the $x$ axis, between the limits $x = 1$ and $x = 3$.
19. Find the area bounded by the curve $y^2 = x^3 - x^2$ and the line $x = 2$.
20. Find the area bounded by $xy = 6$, the lines $x = 1$ and $x = 6$, and the $x$ axis.

*Writing*

21. List the steps needed to find the area between two curves, and give a short description of each step.

# 33

# APPLICATIONS OF THE DEFINITE INTEGRAL

## OBJECTIVES

**When you have completed this chapter, you should be able to:**

- Find the volume of a solid of revolution using the disk or shell method.
- Determine a volume rotated about the $x$ axis, the $y$ axis, or a noncoordinate axis.
- Calculate the length of a curve.
- Determine the surface of revolution about the $x$ axis or about the $y$ axis.
- Find the average value of a function.
- Determine the root-mean-square of a function.
- Find centroids of areas and of volumes of revolution.
- Calculate fluid pressure.
- Determine the work done by a variable force.
- Find the moment of inertia of plane areas.
- Calculate polar moments of inertia using the disk method or the shell method.

We have defined the definite integral in terms of the area under a curve, and in Chapter 32 have used the definite integral to compute plane areas. From this you might think that the definite integral is used only for finding areas, but nothing could be further from the truth. Here we show that there is a wide range of applications for the definite integral.

In each application we set up our integral using the shortcut method of Sec. 32-3, that is, we define a small element of the quantity we want to compute, and then sum up all such elements by integration. We can thus find important quantities such as volume, surface area, length of arc, average value, and root-mean-square value.

We continue our applications of the definite integral by finding moments and centroids of areas. We start with simple shapes, go on to those that require integration, and then compute centroids of solids of revolution. This is followed by two applications involving centroids: finding the force due to fluid pressure, and computing the work required for various tasks.

We conclude by computing moment of inertia for an area, and the polar moment of inertia of solids of revolution, both useful quantities when studying strength of materials, or the motion of rigid bodies.

## 33-1 VOLUMES BY INTEGRATION

### Solids of Revolution

When an area is rotated about some axis $L$, it sweeps out a *solid of revolution*. It is clear from Figure 33-1 that every cross section of that solid of revolution at right angles to the axis of rotation is a circle.

When the area $A$ is rotated about an axis $L$ located at some fixed distance from the area, we get a solid of revolution with a cylindrical hole down its center (Fig. 33-2). The cross section of this solid, at right angles to the axis of rotation, consists of a ring bounded by an outer circle and an inner

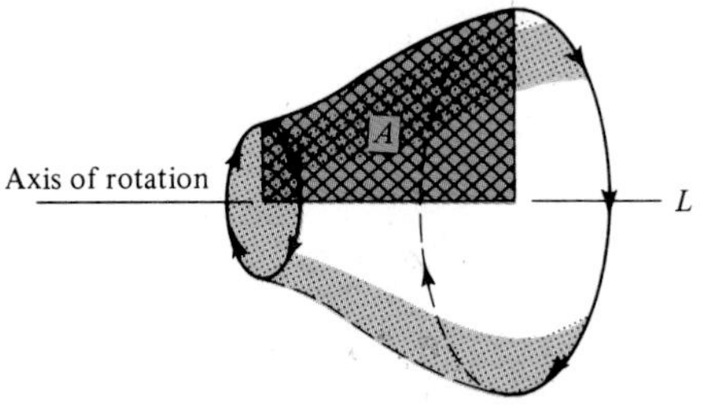


**FIGURE 33-1 Solid of revolution.**

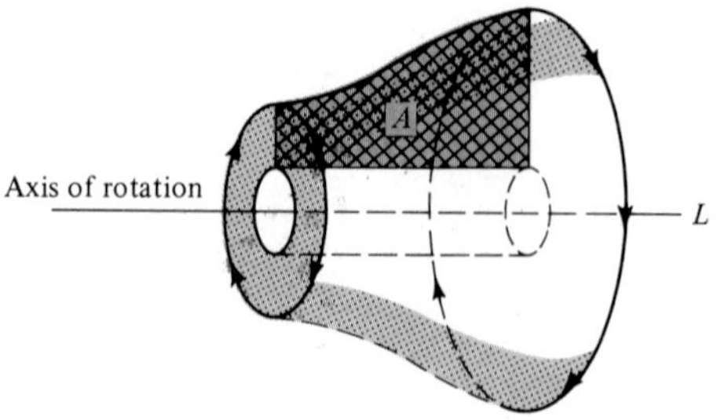


**FIGURE 33-2 Solid of revolution with an axial hole.**

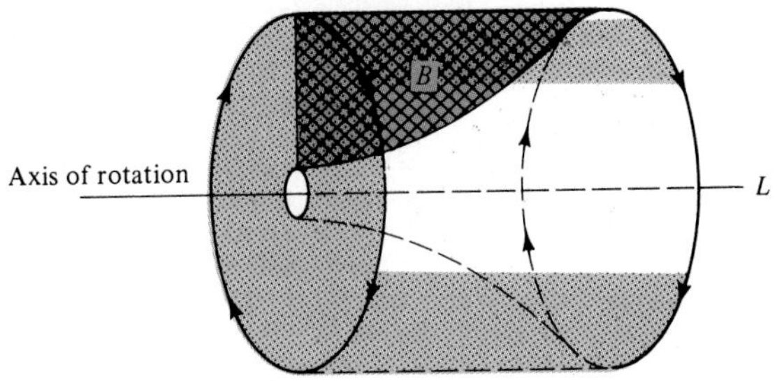


FIGURE 33-3

concentric circle. When the area $B$ is rotated about axis $L$ (Fig. 33-3), we get a solid of revolution with a hole of varying diameter down its center. We first learn how to calculate the volume of a solid with no hole, and then cover "hollow" solids of revolution.

### Volumes by the Disk Method—Rotation about the $x$ Axis

We may think of a solid of revolution as being made up of a stack of thin disks, like a stack of coins of different sizes (Fig. 33-4). Each disk is called an *element* of the total volume. We let the radius of one such disk be $r$ (which varies with the disk's position in the stack), and let the thickness be equal to $dh$. Since a disk (Fig. 33-5) is a cylinder, we calculate its volume $dV$ by Eq. 377,

$$dV = \pi r^2 \, dh$$

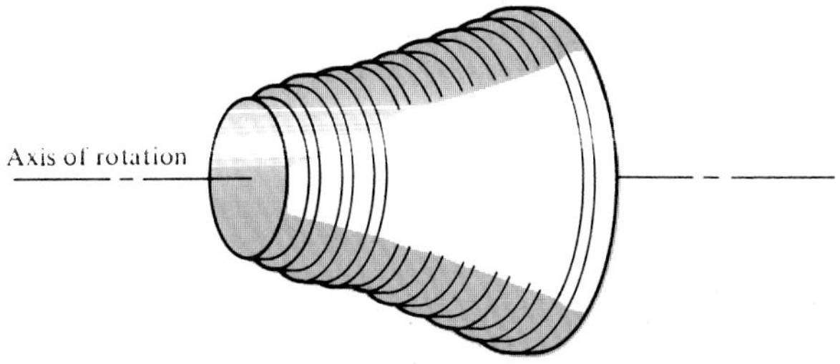


**FIGURE 33-4 Solid of revolution approximated by a stack of thin disks.**

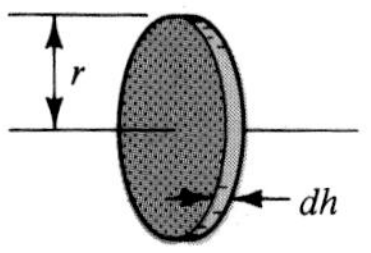


**FIGURE 33-5 One disk.**

Now using the shortcut method of Sec. 32-3 for setting up a definite integral, we "sum" the volumes of all such disk-shaped elements by integrating from one end of the solid to the other.

| | | |
|---|---|---|
| Volumes by the Disk Method | $V = \pi \int_a^b r^2 \, dh$ | **378** |

In an actual problem, we must express $r$ and $h$ in terms of $x$ and $y$, as in the following example.

**EXAMPLE 1:** The area bounded by the curve $y = 8/x$, the $x$ axis, and the lines $x = 1$ and $x = 8$, is rotated about the $x$ axis. Find the volume generated.

**Solution:** We sketch the solid (Fig. 33-6) and a typical disk, touching the curve at some point $(x, y)$. The radius $r$ of the disk is equal to $y$, and the thickness $dh$ of the disk is $dx$. So by Eq. 378,

$$V = \pi \int_a^b y^2\, dx$$

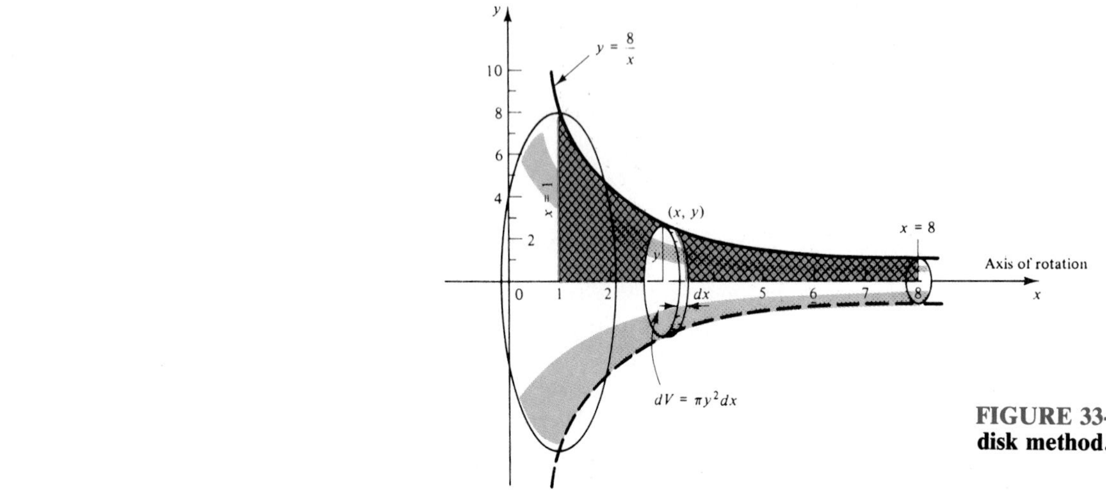


**FIGURE 33-6 Volumes by the disk method.**

For $y$ we substitute $8/x$, and for $a$ and $b$, the limits 1 and 8,

$$V = \pi \int_1^8 \left(\frac{8}{x}\right)^2 dx = 64\pi \int_1^8 x^{-2}\, dx$$

Integrate: $$= -64\pi[x^{-1}]_1^8$$

$$= -64\pi\left(\frac{1}{8} - 1\right) = 56\pi \text{ cubic units}$$

| Common Error | Remember that all parts of the integral, including the limits, must be expressed in terms of the *same variable*. |
|---|---|

### Volumes by the Disk Method—Rotation about the *y* Axis

When our area is rotated about the $y$ rather than the $x$ axis, we take our element of area as a horizontal disk rather than a vertical one.

**EXAMPLE 2:** Find the volume generated when the first-quadrant area bounded by $y = x^2$, the $y$ axis, and the lines $y = 1$ and $y = 4$, is rotated about the $y$ axis.

**Solution:** Our disk-shaped element of volume (Fig. 33-7) now has a radius $x$ and a thickness $dy$. So by Eq. 378,

$$V = \pi \int_a^b x^2\, dy$$

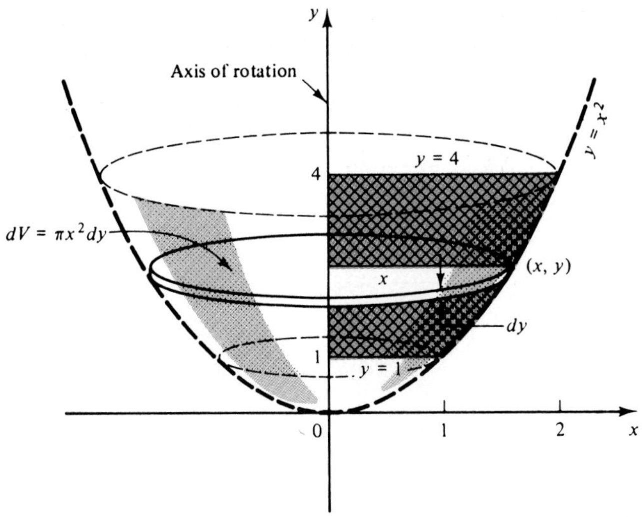


**FIGURE 33-7**

Substituting $y$ for $x^2$, and inserting the limits 1 and 4, we have

$$V = \pi \int_1^4 y\, dy = \pi \left[\frac{y^2}{2}\right]_1^4$$

$$= \frac{\pi}{2}(4^2 - 1^2) = \frac{15\pi}{2} \text{ cubic units}$$

---

## Volumes by the Shell Method

Instead of using a thin disk for our element of volume, it is sometimes easier to use a *thin-walled shell* (Fig. 33-8). To visualize such shells, imagine a solid of revolution to be turned from a log, with the axis of revolution along the centerline of the log. Each annual growth ring would have the shape of a thin-walled shell, with the solid of revolution being made up of many such shells nested one inside the other.

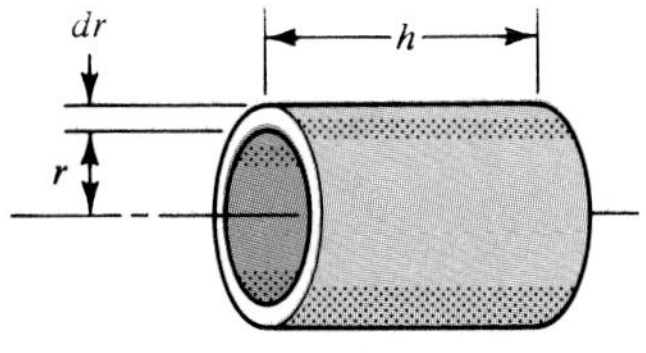


**FIGURE 33-8 Thin-walled shell.**

The volume $dV$ of a thin-walled cylindrical shell of radius $r$, height $h$, and wall thickness $dr$ is

$$dV = \text{circumference} \times \text{height} \times \text{wall thickness}$$

$$= 2\pi rh\, dr$$

Integrating gives

| | | |
|---|---|---|
| Volumes by the Shell Method | $V = 2\pi \int_a^b rh\, dr$ | **382** |

Notice that here the integration limits are in terms of $r$. As with the disk method, $r$ and $h$ must be expressed in terms of $x$ and $y$ in a particular problem.

**EXAMPLE 3:** The first-quadrant area bounded by the curve $y = x^2$, the $y$ axis, and the line $y = 4$ (Fig. 33-9) is rotated about the $y$ axis. Find the volume generated, by the shell method.

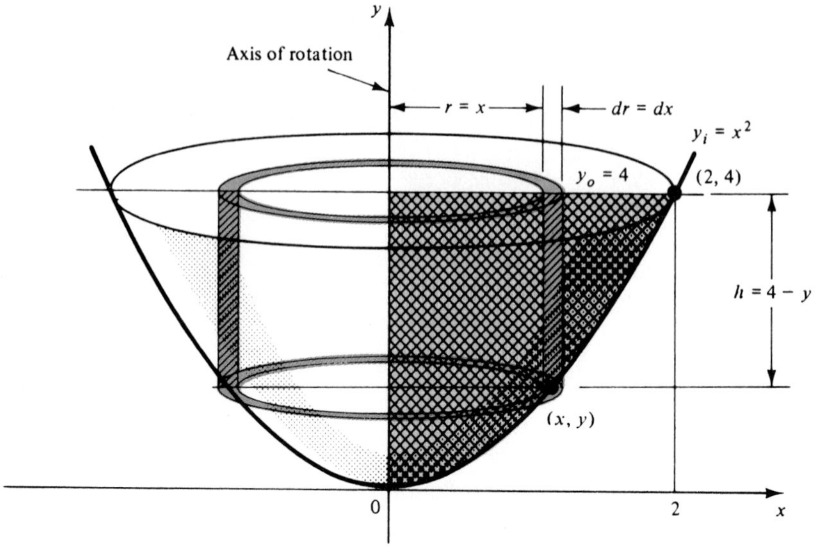


**FIGURE 33-9 Volumes by the shell method.**

**Solution:** Through a point $(x, y)$ on the curve we sketch an element of area of height $h = (4 - y)$ and thickness $dx$. As the first-quadrant area rotates about the $y$ axis, generating a solid of revolution, our element of area sweeps out a shell of radius $r = x$ and thickness $dr = dx$. The volume $dV$ of the shell is then

$$\begin{aligned} dV &= 2\pi rh\ dr = 2\pi x(4 - y)\ dx \\ &= 2\pi x(4 - x^2)\ dx \end{aligned}$$

since $y = x^2$. Integrating gives the total volume,

$$\begin{aligned} V &= 2\pi \int_0^2 x(4 - x^2)\ dx \\ &= 2\pi \int_0^2 (4x - x^3)\ dx \\ &= 2\pi \left[2x^2 - \frac{x^4}{4}\right]_0^2 = 2\pi \left[2(2)^2 - \frac{2^4}{4}\right] \\ &= 8\pi \text{ cubic units} \end{aligned}$$

## Solid of Revolution with Hole

To find the volume of a solid of revolution with an axial hole, such as in Figs. 33-2 and 33-3, we can first find the volume of hole and solid separately, and then subtract. Or we can find the volume of the solid of revolution directly by either the ring or shell method, as in the following example.

**EXAMPLE 4:** The first-quadrant area bounded by the curve $y^2 = 4x$, the $x$ axis, and the line $x = 4$ is rotated about the $y$ axis. Find the volume generated (a) by the ring method and (b) by the shell method.

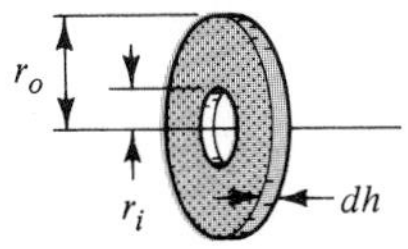


**FIGURE 33-10 Ring or washer.**

**Solution:**

(a) *Ring method:* The volume $dV$ of a thin ring (Fig. 33-10) is given by

$$dV = \pi(r_o^2 - r_i^2)\,dh$$

where $r_o$ is the outer radius and $r_i$ is the inner radius. Integrating gives

**The ring method is also called the *washer* method.**

| | | |
|---|---|---|
| Volume by the Ring Method | $V = \pi \int_a^b (r_o^2 - r_i^2)\,dh$ | **380** |

On our given solid (Fig. 33-11a) we show an element of volume in the shape of a ring, centered on the $y$ axis. Its outer and inner radii are $x_o$ and $x_i$, respectively, and its thickness $dh$ is here $dy$. Then by Eq. 380,

$$V = \pi \int_a^b (x_o^2 - x_i^2)\,dy$$

In our problem, $x_o = 4$ and $x_i = y^2/4$. Substituting these values and placing the limits on $y$ of 0 and 4 gives

$$V = \pi \int_0^4 \left(16 - \frac{y^4}{16}\right) dy$$

Integrating, we obtain

$$V = \pi\left[16y - \frac{y^5}{80}\right]_0^4 = \pi\left[16(4) - \frac{4^5}{80}\right] = \frac{256}{5}\pi \text{ cubic units}$$

(b) *Shell method:* On the given solid we indicate an element of volume in the shape of a shell (Fig. 33-11b). Its inner radius $r$ is $x$, its thickness $dr$ is $dx$, and its height $h$ is $y$. Then by Eq. 381,

$$dV = 2\pi xy\,dx$$

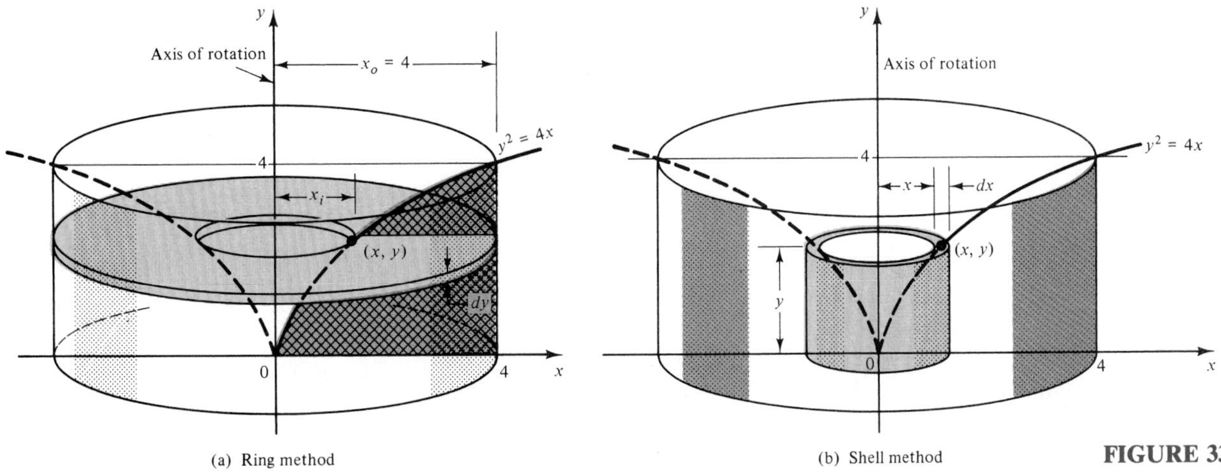


**FIGURE 33-11**

So the total volume is

$$V = 2\pi \int xy\, dx$$

Replacing $y$ with $2\sqrt{x}$, we have

$$V = 2\pi \int_0^4 x(2x^{1/2})\, dx$$

$$= 4\pi \int_0^4 x^{3/2}\, dx$$

$$= 4\pi \left[\frac{x^{5/2}}{5/2}\right]_0^4$$

$$= \frac{8\pi}{5}(4)^{5/2} = \frac{256}{5}\pi \text{ cubic units}$$

as by the ring method.

| Common Error | In Example 4 the radius $x$ in the shell method varies from 0 to 4. The limits of integration are therefore 0 to 4, not $-4$ to 4. |
|---|---|

## Rotation about a Noncoordinate Axis

We can, of course, get a volume of revolution by rotating a given area about some axis other than a coordinate axis. This often results in a solid with a hole in it. As with the preceding hollow figure, these can usually be set up with shells, or, as in the following example, with rings.

**EXAMPLE 5:** The first-quadrant area bounded by the curve $y = x^2$, the $y$ axis, and the line $y = 4$, is rotated about the line $x = 3$. Find the volume generated, by the ring method.

**Solution:** Through the point $(x, y)$ on the curve (Fig. 33-12) we draw a ring-shaped element of volume, with outside radius of 3 units, inside radius of $(3 - x)$ units, and thickness $dy$. The volume $dV$ of the element is then, by Eq. 379,

$$dV = \pi[3^2 - (3 - x)^2]\, dy$$

$$= \pi(9 - 9 + 6x - x^2)\, dy$$

$$= \pi(6x - x^2)\, dy$$

Substituting $\sqrt{y}$ for $x$ gives

$$dV = \pi(6\sqrt{y} - y)\, dy$$

Integrating, we get

$$V = \int_0^4 (6y^{1/2} - y)\, dy = \pi\left[\frac{12y^{3/2}}{3} - \frac{y^2}{2}\right]_0^4$$

$$= \pi\left[4(4)^{3/2} - \frac{16}{2}\right] = 24\,\pi \text{ cubic units}$$

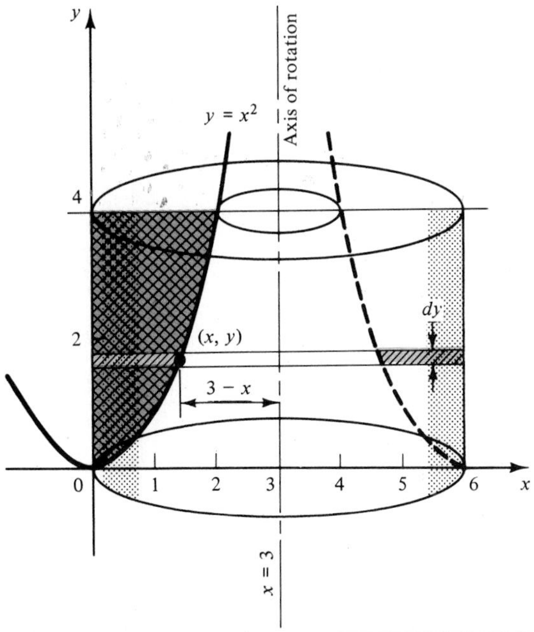


FIGURE 33-12

## EXERCISE 1—VOLUMES BY INTEGRATION

### Rotation about the x Axis

Find the volume generated by rotating the first-quadrant area bounded by each set of curves about the $x$ axis. Use either the disk or the shell method.

1. $y = x^3$ and $x = 2$
2. $y = \dfrac{x^2}{4}$ and $x = 4$
3. $y = \dfrac{x^{3/2}}{2}$ and $x = 2$
4. $y^2 = x^3 - 3x^2 + 2x$ and $x = 1$
5. $y^2(2 - x) = x^3$ and $x = 1$
6. $\sqrt{x} + \sqrt{y} = 1$
7. $x^{2/3} + y^{2/3} = 1$ from $x = 0$ to $x = 1$
8. one arch of $y = \sin x$
9. the catenary $y = \frac{1}{2}(e^x + e^{-x})$ from $x = 0$ to $x = 1$.
10. $y^2 = x\left(\dfrac{x-3}{x-4}\right)$

### Rotation about the y Axis

Find the volume generated by rotating about the $y$ axis the first-quadrant area bounded by each set of curves.

11. $y = x^3$, the $y$ axis, and $y = 8$
12. $2y^2 = x^3$, $x = 0$, and $y = 2$
13. $9x^2 + 16y^2 = 144$
14. $\left(\dfrac{x}{2}\right)^2 + \left(\dfrac{y}{3}\right)^{2/3} = 1$
15. one arch of the sine curve (use the shell method and Rule 31)
16. $y^2 = 4x$, and $y = 4$

### Rotation about a Noncoordinate Axis

Find the volume generated by rotating about the indicated axis the first-quadrant area bounded by each set of curves.

17. $x = 4$ and $y^2 = x^3$, about $x = 4$
18. $y = e^x$ and $x = 1$, about $x = 1$ (use Rule 37)

19. above $y = 3$ and below $y = 4x - x^2$, about $y = 3$
20. $y^2 = x^3$ and $y = 8$, about $y = 9$
21. $y = 4$ and $y = 4 + 6x - 2x^2$, about $y = 4$

A ***paraboloid of revolution*** **is the figure formed by revolving a parabola about its axis.**

## Applications

22. The nose cone of a certain rocket is a paraboloid of revolution (Fig. 33-13). Find its volume.

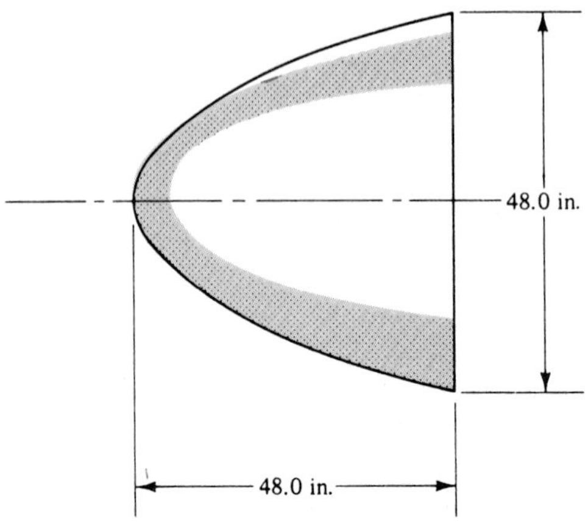


FIGURE 33-13 **Rocket nose cone.**

23. A wing tank for an airplane is a solid of revolution formed by rotating the curve $8y = 4x - x^2$, from $x = 0$ to $x = 3.5$, about the $x$ axis (Fig. 33-14). Find the volume of the tank.

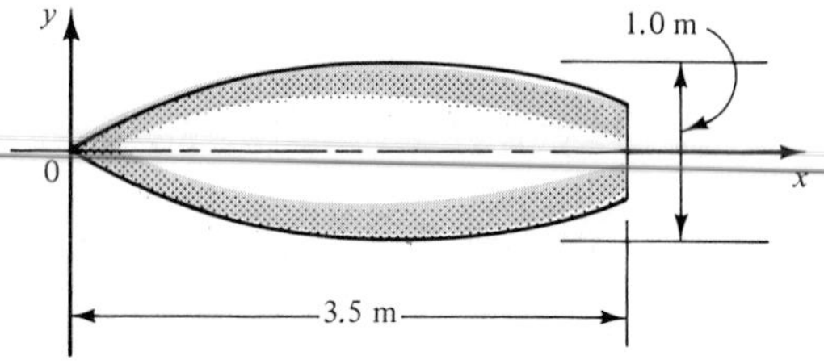


FIGURE 33-14 **Airplane wing tank.**

24. The bullet (Fig. 33-15) consists of a cylinder and a paraboloid of revolution, and is made of lead having a density of 11.3 g/cm³. Find its weight.

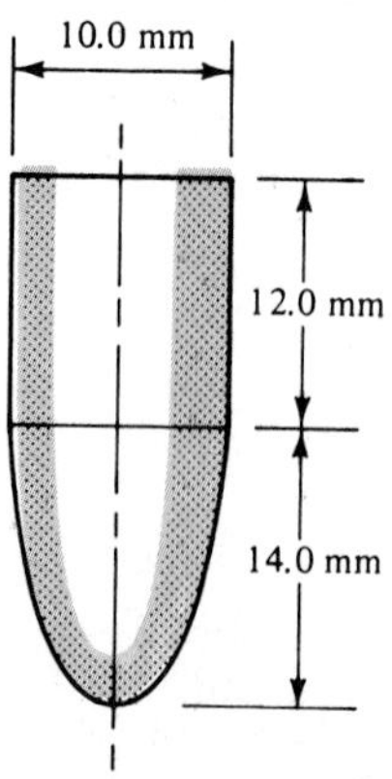


FIGURE 33-15 **Bullet.**

25. The telescope mirror, shown in cross section in Fig. 33-16, is formed by rotating the area under the hyperbola $y^2/100 - x^2/1225 = 1$ about the $y$ axis, and has a 20-cm-diameter hole at its center. Find the volume of glass in the mirror.

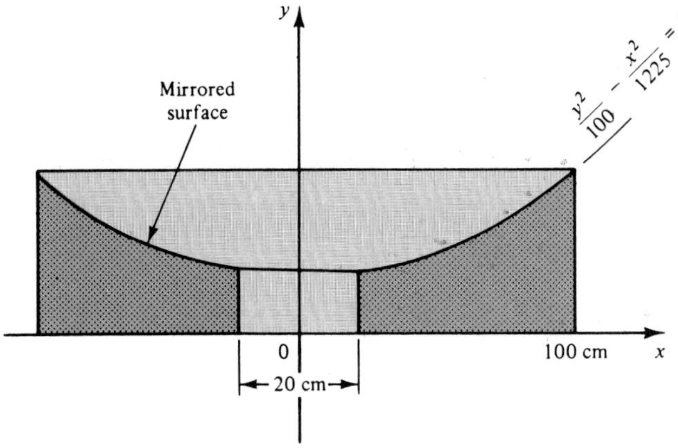


**FIGURE 33-16 Telescope mirror.**

## 33-2 LENGTH OF ARC

In this section we develop a way to find the length of a curve between two endpoints, such as the distance $PQ$ in Fig. 33-17. We will get the distance measured *along the curve,* as if the curve were stretched out straight, and then measured.

Again we use our intuitive method to set up the integral. We think of the curve as being made up of many short sections, each of length $\Delta s$. By the Pythagorean theorem,

$$(\Delta s)^2 \simeq (\Delta x)^2 + (\Delta y)^2$$

Dividing by $(\Delta x)^2$ gives us

$$\frac{(\Delta s)^2}{(\Delta x)^2} \simeq 1 + \frac{(\Delta y)^2}{(\Delta x)^2}$$

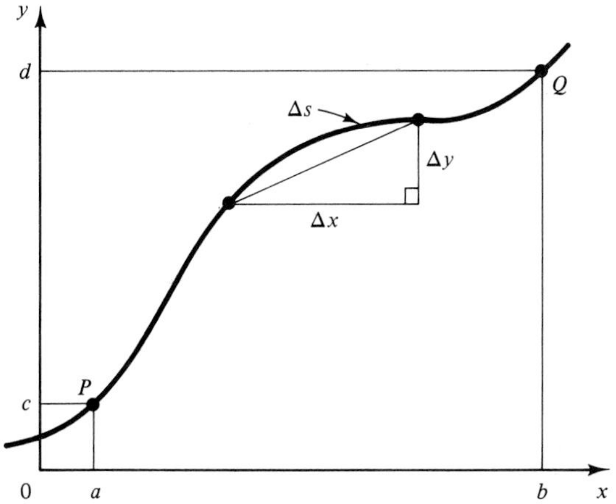


**FIGURE 33-17 Length of arc. We are considering here (as usual) only smooth continuous curves.**

Taking the square root yields

$$\frac{\Delta s}{\Delta x} \simeq \sqrt{1 + \left(\frac{\Delta y}{\Delta x}\right)^2}$$

We now let the number of these short sections of curve approach infinity as we let $\Delta x$ approach zero.

$$\frac{ds}{dx} = \lim_{\Delta x \to 0} \frac{\Delta s}{\Delta x} = \sqrt{1 + \left(\frac{dy}{dx}\right)^2}$$

or

$$ds = \sqrt{1 + \left(\frac{dy}{dx}\right)^2}\, dx$$

Now thinking of integration as a summing process, we use it to add up all the small segments of length $ds$:

| | | |
|---|---|---|
| Length of Arc | $s = \int_a^b \sqrt{1 + \left(\frac{dy}{dx}\right)^2}\, dx$ | **383** |

Finding the length of arc is sometimes referred to as *rectifying* a curve.

**EXAMPLE 6:** Find the first quadrant length of the curve $y^2 = x^3$ from $x = 0$ to $x = 4$.

**Solution:** Let us first find $(dy/dx)^2$.

$$y = x^{3/2}$$

$$\frac{dy}{dx} = \frac{3}{2}x^{1/2}$$

$$\left(\frac{dy}{dx}\right)^2 = \frac{9x}{4}$$

Then by Eq. 383,

$$\begin{aligned} s &= \int_0^4 \sqrt{1 + \left(\frac{dy}{dx}\right)^2}\, dx \\ &= \int_0^4 \sqrt{1 + \frac{9x}{4}}\, dx \\ &= \frac{4}{9}\int_0^4 \left(1 + \frac{9x}{4}\right)^{1/2}\left(\frac{9}{4}dx\right) \\ &= \frac{8}{27}\left[\left(1 + \frac{9x}{4}\right)^{3/2}\right]_0^4 \\ &= \frac{8}{27}(10^{3/2} - 1) \simeq 9.07 \end{aligned}$$

## Another Form of the Arc Length Equation

Another form of the equation for arc length, which can be derived in a similar way to Eq. 383, is

| | | |
|---|---|---|
| Length of Arc | $s = \int_c^d \sqrt{1 + \left(\frac{dx}{dy}\right)^2}\, dy$ | **384** |

This equation is more useful when the equation of the curve is given in the form $x = f(y)$, instead of the more usual form $y = f(x)$.

---

**EXAMPLE 7:** Find the length of the curve $x = 4y^2$, between $y = 0$ and $y = 4$.

**Solution:** Taking the derivative, we have

$$\frac{dx}{dy} = 8y$$

so $(dx/dy)^2 = 64y^2$. Substituting into Eq. 384,

$$s = \int_0^4 \sqrt{1 + 64y^2}\, dy = \frac{1}{8}\int_0^4 \sqrt{1 + (8y)^2}\,(8\, dy)$$

Using Rule 66, with $u = 8y$ and $a = 1$,

When finding length of arc we usually wind up with messy integrals. Often they can be evaluated using Rule 66.

$$s = \frac{1}{8}\left[\frac{8y}{2}\sqrt{1 + 64y^2} + \frac{1}{2}\ln\left|8y + \sqrt{1 + 64y^2}\right|\right]_0^4$$

$$= \frac{1}{8}\left[4(4)\sqrt{1 + 64(16)} + \frac{1}{2}\ln\left|32 + \sqrt{1 + 64(16)}\right| - 0\right] = 64.3$$

---

## EXERCISE 2—LENGTH OF ARC

Find the length of each curve.

Many of these will require the use of Rule 66.

1. $y^2 = x^3$ in the first quadrant from $x = 0$ to $x = \frac{5}{9}$
2. $6y = x^2$ from the origin to the point $(4, \frac{8}{3})$
3. $y = \ln \sec x$ from the origin to the point $(\pi/3, \ln 2)$
4. the circle $x^2 + y^2 = 36$ (use Rule 61)
5. $x^{2/3} + y^{2/3} = 1$ from $x = 0$ to $x = 1$
6. $4y = x^2$ from $x = 0$ to $x = 4$
7. $2y^2 = x^3$ from the origin to $x = 10$
8. the arch of the parabola $y = 4x - x^2$ that lies above the $x$ axis
9. the length in one quadrant of the curve $\left(\frac{x}{2}\right)^{2/3} + \left(\frac{y}{3}\right)^{2/3} = 1$
10. the catenary $y = \dfrac{a(e^{x/a} + e^{-x/a})}{2}$ from $x = 0$ to $x = 6$; use $a = 3$
11. $y^3 = x^2$ between points (0, 0) and (8, 4)
12. $y^2 = 8x$ from its vertex to one end of its latus rectum

Hint: Use Eq. 384 for Problems 11 and 12.

## Applications

13. Find the length of the cable $AB$ (Fig. 33-18) that is in the shape of a parabola.

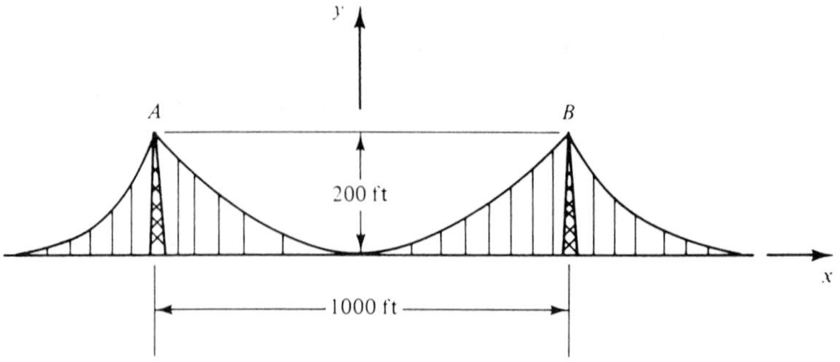


**FIGURE 33-18 Suspension bridge.**

14. A roadway has a parabolic shape at the top of a hill (Fig. 33-19). If the road is 30 ft wide, find the cost of paving from $P$ to $Q$, at the rate of \$35 per square foot.

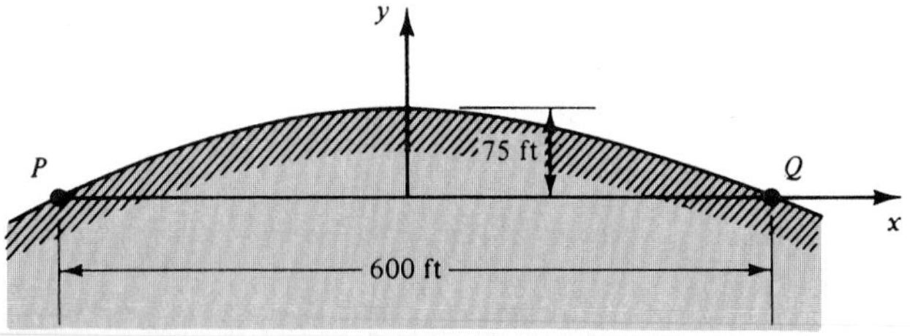


**FIGURE 33-19 Road over a hill.**

15. The equation of the bridge arch in Fig. 33-20 is $y = 0.0625x^2 - 5x + 100$. Find its length.

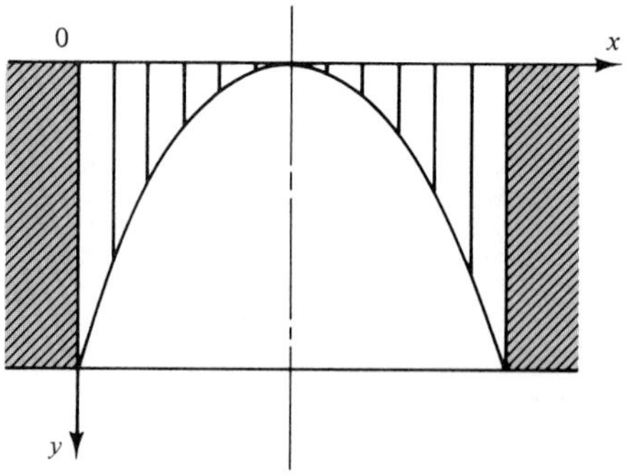


**FIGURE 33-20 Parabolic bridge arch. Note that the $y$ axis is positive downward.**

16. Find the surface area of the curved portion of the mirror in Fig. 32-20.
17. Find the perimeter of the window in Fig. 32-25.

## 33-3 AREA OF SURFACE OF REVOLUTION

### Surface of Revolution

In Sec. 33-1 we rotated an *area* about an axis and got a solid of revolution. We now find the *area* of the surface of such a solid. Alternatively, we can think of a surface of revolution as being generated when a curve rotates about some axis. Let us take the curve of Fig. 33-21 and rotate it about the $x$ axis. The arc $PQ$ sweeps out a surface of revolution while the small section $ds$ sweeps out a hoop-shaped element of that surface.

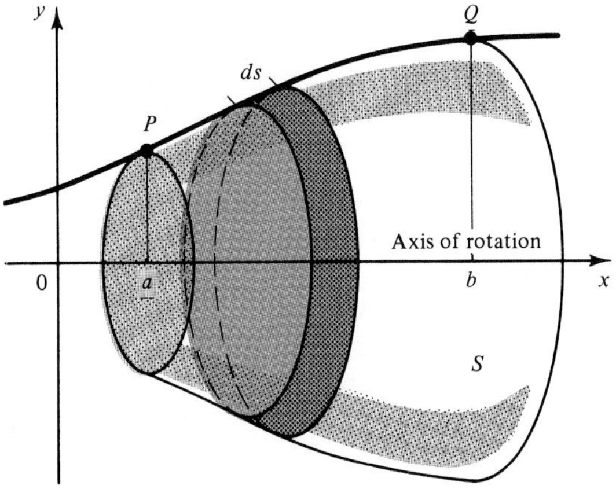


**FIGURE 33-21 Surface of revolution.**

### Surface Area—Rotation about the *x* Axis

The area of a hoop-shaped or circular band is equal to the length of the edge times the average circumference of the hoop. Our element $ds$ is at a radius $y$ from the $x$ axis, so the area $dS$ of the hoop is

$$dS = 2\pi y \, ds$$

But from Sec. 33-2, we saw that

$$ds = \sqrt{1 + \left(\frac{dy}{dx}\right)^2} \, dx$$

So

**We are using capital *S* for surface area and small *s* for arc length. Be careful not to confuse the two.**

$$dS = 2\pi y \sqrt{1 + \left(\frac{dy}{dx}\right)^2} \, dx$$

We then integrate from $a$ to $b$, to sum up the areas of all such elements:

| | | |
|---|---|---|
| Area of Surface of Revolution about $x$ Axis | $S = 2\pi \int_a^b y \sqrt{1 + \left(\frac{dy}{dx}\right)^2} \, dx$ | **385** |

**EXAMPLE 8:** The portion of the parabola $y^2 = 16x$ between $x = 0$ and $x = 16$ is rotated about the $x$ axis. Find the area of the surface of revolution generated.

**Solution:** We first find the derivative. Solving for $y$ gives us $y = 4x^{1/2}$. Then

$$\frac{dy}{dx} = 2x^{-1/2}$$

and

$$\left(\frac{dy}{dx}\right)^2 = 4x^{-1}$$

**Surface area problems, like those of length of arc, usually result in difficult integrals.**

Substituting into Eq. 385, we obtain

$$S = 2\pi \int_0^{16} 4x^{1/2}\sqrt{1 + 4x^{-1}}\, dx$$

$$= 8\pi \int_0^{16} (x + 4)^{1/2}\, dx$$

$$= \frac{16\pi}{3}(20^{3/2} - 4^{3/2}) \approx 1365 \text{ square units}$$

## Surface Area—Rotation about the *y* Axis

The equation for the area of a surface of revolution whose axis of revolution is the $y$ axis can be derived in a similar way. It is

| | | |
|---|---|---|
| Area of Surface of Revolution about $y$ Axis | $S = 2\pi \int_a^b x\sqrt{1 + \left(\frac{dy}{dx}\right)^2}\, dx$ | **386** |

**EXAMPLE 9:** The portion of the curve $y = x^2$ lying between the points (0, 0) and (2, 4) is rotated about the $y$ axis. Find the area of the surface generated.

**Solution:** Taking the derivative gives $dy/dx = 2x$, so

$$\left(\frac{dy}{dx}\right)^2 = 4x^2$$

Substituting into Eq. 386, we have

$$S = 2\pi \int_0^2 x\sqrt{1 + 4x^2}\, dx$$

$$= \frac{2\pi}{8} \int_0^2 (1 + 4x^2)^{1/2}(8x\, dx)$$

$$= \frac{\pi}{4} \cdot \frac{2(1 + 4x^2)^{3/2}}{3}\Bigg|_0^2 = \frac{\pi}{6}(17^{3/2} - 1^{3/2}) = 36.2 \text{ square units}$$

## EXERCISE 3—AREA OF SURFACE OF REVOLUTION

Find the area of the surface generated by rotating each curve about the $x$ axis.

1. $y = \dfrac{x^2}{9}$ from $x = 0$ to $x = 3$
2. $y^2 = 2x$ from $x = 0$ to $x = 4$
3. $y^2 = 9x$ from $x = 0$ to $x = 4$
4. $y^2 = 4x$ from $x = 0$ to $x = 1$
5. $y^2 = 4 - x$ in the first quadrant
6. one arch of the curve $y = \sin x$
7. $y^2 = 24 - 4x$ from $x = 3$ to $x = 6$
8. $x^{2/3} + y^{2/3} = 1$ in the first quadrant
9. $y = \dfrac{x^3}{6} + \dfrac{1}{2x}$ from $x = 1$ to $x = 3$
10. $y = e^{-x}$ from $x = 0$ to $x = 100$

Find the area of the surface generated by rotating each curve about the $y$ axis.

11. $y = 3x^2$ from $x = 0$ to 5
12. $y = 5x^2$ from $x = 2$ to 4
13. $y = 4 - x^2$ from $x = 0$ to 2
14. $y = 24 - x^2$ from $x = 2$ to 4

### Geometric Figures

15. Find the surface area of a sphere by rotating the curve $x^2 + y^2 = r^2$ about a diameter.
16. Find the area of the curved surface of a cone by rotating about the $x$ axis the line connecting the origin and the point $(a, b)$.

### Applications

17. Find the surface area of the nose cone shown in Fig. 33-13.
18. Find the cost of copper plating 10,000 bullets (Fig. 33-15) at the rate of \$15 per square meter.

# 33-4 AVERAGE AND ROOT-MEAN-SQUARE VALUES

## Average Value of a Function

The area $A$ under the curve $y = f(x)$ (Fig. 33-22) between $x = a$ and $x = b$ is, by Eq. 365,

$$A = \int_a^b f(x)\, dx$$

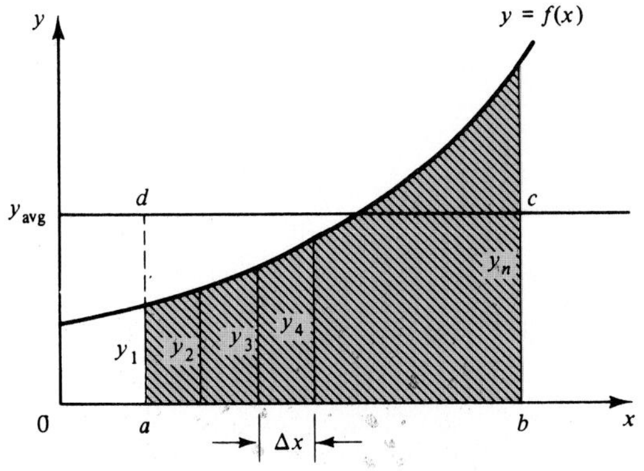


**FIGURE 33-22 Average ordinate.**

Within the same interval, the *average ordinate* of that function, $y_{avg}$, is that value of $y$ which will cause the rectangle *abcd* to have the same area as that under the curve, or

$$(b - a)y_{avg} = A = \int_a^b f(x)\,dx$$

so

| | | |
|---|---|---|
| Average Ordinate | $y_{avg} = \dfrac{1}{b - a}\displaystyle\int_a^b f(x)\,dx$ | **404** |

---

**EXAMPLE 10:** Find the average ordinate of a half-cycle of the sinusoidal voltage

$$v = V \sin \theta \qquad \text{volts}$$

**Solution:** By Eq. 404, with $a = 0$ and $b = \pi$,

$$\begin{aligned} V_{avg} &= \frac{V}{\pi - 0}\int_0^{\pi} \sin \theta\, d\theta \\ &= \frac{V}{\pi}[-\cos \theta]_0^{\pi} \\ &= \frac{V}{\pi}(-\cos \pi + \cos 0) = \frac{2}{\pi}V = 0.637\,V \qquad \text{volts} \end{aligned}$$

---

## Root-Mean-Square Value of a Function

The *root-mean-square* (rms) value of a function is the square root of the average of the squares of the ordinates. If, in Fig. 33-22 we take $n$ values of $y$ spaced apart by a distance $\Delta x$, the rms value is approximately

$$\text{rms} \simeq \sqrt{\frac{y_1^2 + y_2^2 + y_3^2 + \cdots + y_n^2}{n}}$$

or, using summation notation,

$$\text{rms} \simeq \sqrt{\frac{\sum_{i=1}^{n} y_i^2}{n}}$$

Multiplying numerator and denominator of the fraction under the radical by $\Delta x$,

$$\text{rms} \simeq \sqrt{\frac{\sum_{i=1}^{n} y_i^2\,\Delta x}{n\,\Delta x}}$$

But $n\,\Delta x$ is simply the width $(b - a)$ of the interval. If we now let $n$ approach infinity, we get

$$\lim_{n\to\infty} \sum_{i=1}^{n} y_i^2\,\Delta x = \int_a^b [f(x)]^2\,dx$$

Therefore,

| | | |
|---|---|---|
| Root-Mean-Square Value | $\text{rms} = \sqrt{\frac{1}{b-a}\int_a^b [f(x)]^2\,dx}$ | **405** |

**EXAMPLE 11:** Find the rms value for the sinusoidal voltage of Example 10.

**Solution:** We substitute into Eq. 405, with $a = 0$ and $b = \pi$,

$$\text{rms} = \sqrt{\frac{1}{\pi - 0}\int_0^\pi V^2 \sin^2\theta\,d\theta} = \sqrt{\frac{V^2}{\pi}\int_0^\pi \sin^2\theta\,d\theta}$$

But by Rule 16,

$$\int_0^\pi \sin^2\theta\,d\theta = \frac{\theta}{2} - \frac{\sin 2\theta}{4}\Big|_0^\pi$$

$$= \frac{\pi}{2} - \frac{\sin 2\pi}{4} = \frac{\pi}{2}$$

So

$$\text{rms} = \sqrt{\frac{V^2}{\pi}\cdot\frac{\pi}{2}} = \frac{V}{\sqrt{2}} = 0.707\,V$$

**In electrical work, the rms value of an alternating current of voltage is also called the *effective* value. We see then that the effective value is 0.707 of the peak value.**

## EXERCISE 4—AVERAGE AND ROOT-MEAN-SQUARE VALUES

Find the average ordinate for each function in the given interval.

1. $y = x^2$ from 0 to 6
2. $y = x^3$ from $-5$ to 5
3. $y = \sqrt{1 + 2x}$ from 4 to 12
4. $y = \dfrac{x}{\sqrt{9 + x^2}}$ from 0 to 4
5. $y = \sin^2 x$ from 0 to $\pi/2$
6. $2y = \cos 2x + 1$ from 0 to $\pi$

Find the rms value for each function in the given interval.

7. $y = 2x + 1$ from 0 to 6
8. $y = \sin 2x$ from 0 to $\pi/2$
9. $y = x + 2x^2$ from 1 to 4
10. $y = 3\tan x$ from 0 to $\pi/4$
11. $y = 2\cos x$ from $\pi/6$ to $\pi/2$
12. $y = 5\sin 2x$ from 0 to $\pi/6$

## 33-5 CENTROIDS

### Center of Gravity and Centroid

The *center of gravity* (or *center of mass*) of a body is the point where all the mass can be thought to be concentrated, without altering the effect that the earth's gravity has upon it. For simple shapes such as a sphere, cube, or cylinder, the center of gravity is exactly where you would expect it to be, at the center of the object. Figure 33-23 shows how the center of gravity of an irregular figure would be located.

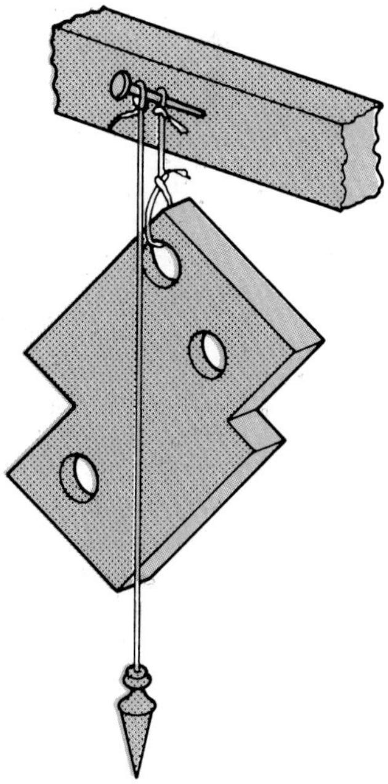

FIGURE 33-23

Any object hung from a point will swing to where its center of gravity is directly below the point of suspension.

A plane area has no thickness, and hence has no weight or mass. Since it makes no sense to speak of center of mass or center of gravity for a weightless figure, we use instead the word *centroid*.

### First Moment

We will learn about second moments (moments of inertia) in Sec. 33-8.

To calculate the position of the centroid in figures of various shapes, we need the idea of the *first moment*, often referred to as *the moment*. It is not new to us. We know from our previous work that the *moment of a force* about some point $a$ is the product of the force $F$ and the perpendicular distance $d$ from the point to the line of action of the force. In a similar way, we speak of the *moment of an area* about some axis (Fig. 33-24a):

$$\text{moment of an area} = \text{area} \times \text{distance to the centroid}$$

or *moment of a volume* (Figure 33-24b):

$$\text{moment of a volume} = \text{volume} \times \text{distance to the centroid}$$

or *moment of a mass* (Fig. 33-24c):

$$\text{moment of a mass} = \text{mass} \times \text{distance to the centroid}$$

In each case, the distance is that from the axis about which we take the moment, measured to some point on the area, volume, or mass. But to which point on the figure shall we measure? To the *centroid* or *center of gravity*.

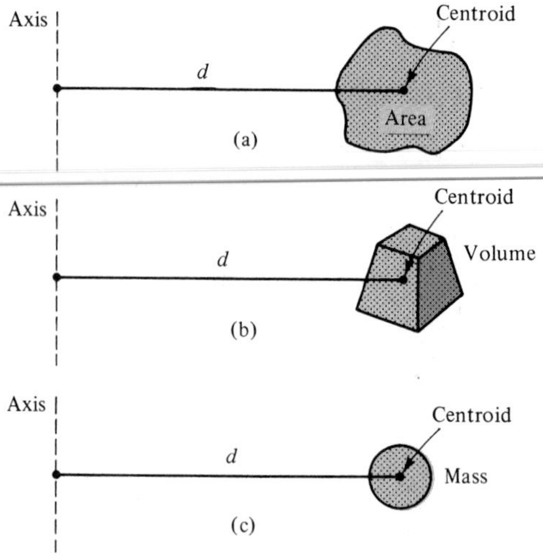


**FIGURE 33-24 Moment of (a) an area, (b) a volume, and (c) a mass.**

### Centroids of Simple Shapes

If a shape has an axis of symmetry, the centroid will be located on that axis. If the centroid is on the $x$ axis, its coordinates will be $(\bar{x}, 0)$; if on the $y$ axis, $(0, \bar{y})$. If there are two or more axes of symmetry, the centroid is found at the intersection of those axes.

We can easily find the centroid of an area that can be subdivided into simple regions, each of whose centroid location is known by symmetry. We first find the moment of each region about some convenient axis by multiplying the area of that region by the distance of its centroid from the axis. We then make use of the following fact, which we give without proof: that for an area subdivided into smaller regions, *the moment of that area about a given axis is equal to the sum of the moments of the individual regions about that same axis.*

**EXAMPLE 12:** Find the location of the centroid of the shape in Fig. 33-25a.

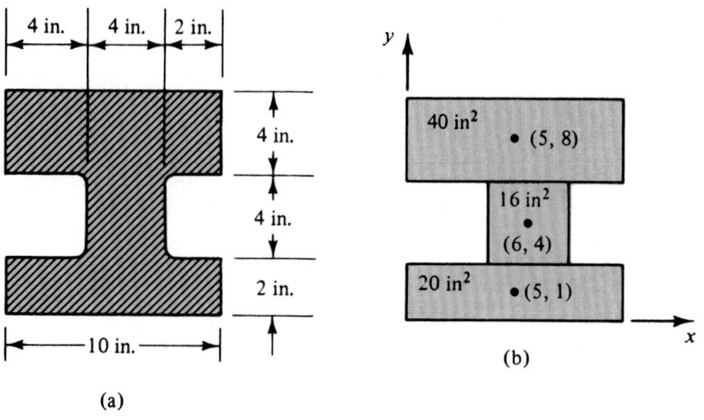


**FIGURE 33-25**

**Solution:** We subdivide the area into rectangles (Fig. 33-25b), locate the centroid, and compute the area of each. We also choose axes from which we will measure the coordinates of the centroid, $\bar{x}$ and $\bar{y}$. The total area is

$$40 + 16 + 20 = 76 \text{ in}^2$$

The moment $M_y$ about the $y$ axis of the entire region is the sum of the moments of the individual regions.

$$M_y = 40(5) + 16(6) + 20(5) = 396 \text{ in}^3$$

Then, since

area × distance to centroid = moment

$$\bar{x} = \frac{\text{moment about } y \text{ axis, } M_y}{\text{area}}$$

$$= \frac{396 \text{ in}^3}{76 \text{ in}^2} = 5.21 \text{ in.}$$

The moment $M_x$ about the $x$ axis is

$$M_x = 40(8) + 16(4) + 20(1) = 404 \text{ in}^3$$

So

$$\bar{y} = \frac{\text{moment about } x \text{ axis, } M_x}{\text{area}}$$

$$= \frac{404 \text{ in}^3}{76 \text{ in}^2} = 5.32 \text{ in.}$$

### Centroids by Integration

If an area does not have axes of symmetry whose intersection gives us the location of the centroid, we can often find it by integration. We subdivide the area into thin strips, compute the first moment of each, sum these moments by integration, and then divide by the total area to get the distance to the centroid.

Consider the area bounded by the curves $y_1 = f_1(x)$ and $y_2 = f_2(x)$ and the lines $x = a$ and $x = b$ (Fig. 33-26). We draw a vertical element of area of width $dx$ and height $(y_2 - y_1)$. Since the strip is narrow, all points on it may be considered to be the same distance $x$ from the $y$ axis. The moment $dM_y$ of that strip about the $y$ axis is then

$$dM_y = x\,dA = x(y_2 - y_1)\,dx$$

since $dA = (y_2 - y_1)\,dx$. We get the total moment $M_y$ by integrating,

$$M_y = \int_a^b x(y_2 - y_1)\,dx$$

But since the moment $M_y$ is equal to the area $A$ times the distance $\bar{x}$ to the centroid, we get $\bar{x}$ by dividing the moment by the area. So

| | | |
|---|---|---|
| Horizontal Distance to Centroid | $\bar{x} = \dfrac{1}{A}\displaystyle\int_a^b x(y_2 - y_1)\,dx$ | **387** |

To find $\bar{x}$ we must have the area $A$. It can be found by integration as in Chapter 32.

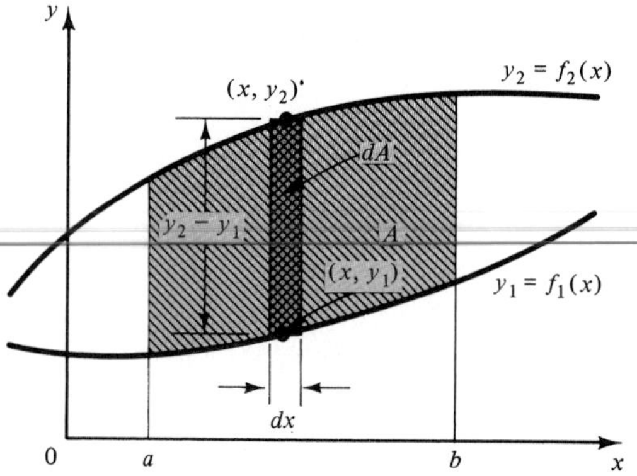

**FIGURE 33-26** **Centroid of irregular area found by integration.**

We now find the moment about the $x$ axis. The centroid of the vertical element is at its midpoint, which is at a distance of $(y_1 + y_2)/2$ from the $x$ axis. Thus the moment of the element about the $x$ axis is

$$dM_x = \frac{y_1 + y_2}{2}(y_2 - y_1)\,dx$$

**Equations 387 and 388 apply only for *vertical elements.* When using horizontal elements, interchange *x* and *y* in these equations.**

Integrating and dividing by the area gives us $\bar{y}$.

| | | |
|---|---|---|
| Vertical Distance to Centroid | $\bar{y} = \dfrac{1}{2A}\displaystyle\int_a^b (y_1 + y_2)(y_2 - y_1)\,dx$ | **388** |

Our first example is for an area bounded by one curve and the $x$ axis.

**EXAMPLE 13:** Find the centroid of the area bounded by the parabola $y^2 = 4x$, the $x$ axis, and the line $x = 1$ (Fig. 33-27).

**Solution:** We need the area, so we find that first. We draw a vertical strip having an area $y\,dx$ and integrate.

$$A = \int y\,dx = 2\int_0^1 x^{1/2}\,dx = 2\left[\frac{x^{3/2}}{3/2}\right]_0^1 = \frac{4}{3}\text{ ft}^2$$

Then by Eq. 387,

$$\bar{x} = \frac{1}{A}\int_0^1 x(2x^{1/2} - 0)\,dx$$

$$= \frac{3}{4}\int_0^1 2x^{3/2}\,dx = \frac{3}{2}\frac{x^{5/2}}{(5/2)}\bigg|_0^1 = \frac{3}{5}\text{ ft}$$

and by Eq. 388,

$$\bar{y} = \frac{1}{2A}\int_0^1 (2x^{1/2} + 0)(2x^{1/2} - 0)\,dx$$

$$= \frac{3}{8}\int_0^1 4x\,dx = \frac{3}{8}\frac{4x^2}{2}\bigg|_0^1 = \frac{3}{4}\text{ ft}$$

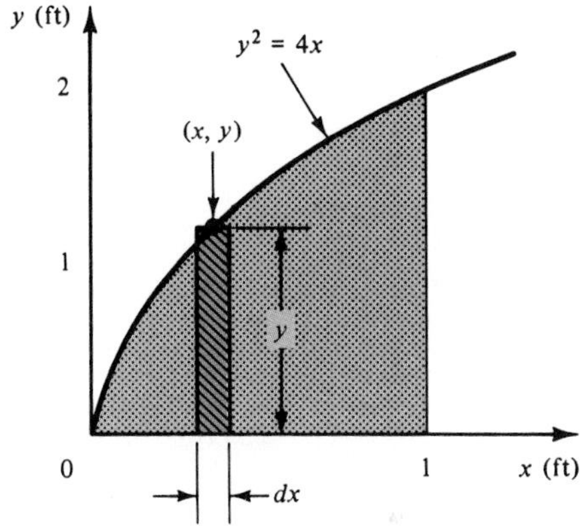

FIGURE 33-27

For areas that are not bounded by a curve and a coordinate axis, but are instead bounded by two curves, the work is only slightly more complicated, as shown in the following example.

**EXAMPLE 14:** Find the coordinates of the centroid of the area bounded by the curves $6y = x^2 - 4x + 4$ and $3y = 16 - x^2$.

**Solution:** We plot the curves (Fig. 33-28) and find their points of intersection by solving simultaneously. Multiplying the second equation by $-2$ and adding the resulting equation to the first gives

$$3x^2 - 4x - 28 = 0$$

Solving by quadratic formula (work not shown) we find that the points of intersection are at $x = -2.46$ and $x = 3.79$. We take a vertical strip whose width is $dx$ and whose height is $y_2 - y_1$, where

$$y_2 - y_1 = \frac{16}{3} - \frac{x^2}{3} - \frac{x^2}{6} + \frac{4x}{6} - \frac{4}{6} = \frac{28 + 4x - 3x^2}{6}$$

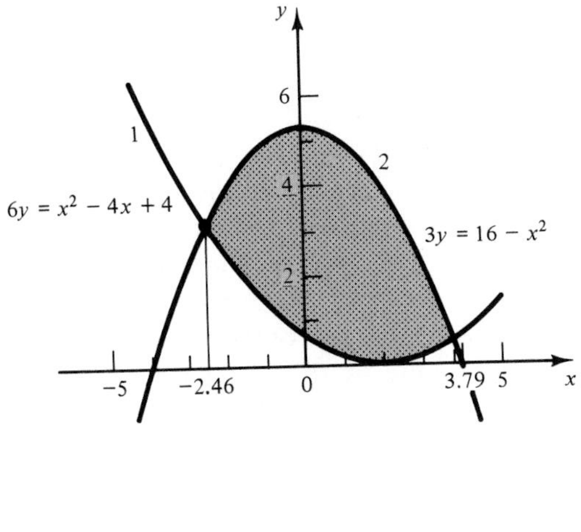

FIGURE 33-28

The area is then

$$A = \int_{-2.46}^{3.79} (y_2 - y_1)\, dx$$

$$= \frac{1}{6}\int_{-2.46}^{3.79} (28 + 4x - 3x^2)\, dx$$

$$= 20.4$$

Then from Eq. 387,

$$M_y = A\bar{x} = \int_{-2.46}^{3.79} x(y_2 - y_1)\, dx$$

$$= \frac{1}{6}\int_{-2.46}^{3.79} (28x + 4x^2 - 3x^3)\, dx$$

$$= 13.6$$

Then

$$\bar{x} = \frac{M_y}{A} = \frac{13.6}{20.4} = 0.667$$

We now substitute into Eq. 388, with

$$y_1 + y_2 = \frac{36 - 4x - x^2}{6}$$

Thus

$$M_x = A\bar{y} = \frac{1}{72}\int_{-2.46}^{3.79} (36 - 4x - x^2)(28 + 4x - 3x^2)\, dx$$

$$= \frac{1}{72}\int_{-2.46}^{3.79} (3x^4 + 8x^3 - 152x^2 + 32x + 1008)\, dx$$

$$= \frac{1}{72}\left|\frac{3x^5}{5} + \frac{8x^4}{4} - \frac{152x^3}{3} + \frac{32x^2}{2} + 1008x\right|_{-2.46}^{3.79}$$

$$= 52.5$$

so

$$\bar{y} = \frac{M_x}{A} = \frac{52.5}{20.4} = 2.57$$

---

## Centroids of Solids of Revolution by Integration

A solid of revolution is, of course, symmetrical about the axis of revolution, so the centroid must be on that axis. We only have to find the position of the centroid along that axis. The procedure is similar to that for an area. We think of the solid as being subdivided into many small elements of volume, find the sum of the moments for each element by integration, and set this equal to the moment of the entire solid (the product of its total volume and the distance to the centroid). We then divide by the volume to obtain the distance to the centroid.

**EXAMPLE 15:** Find the centroid of a hemisphere of radius $r$.

**Solution:** We place the hemisphere on coordinate axes (Fig. 33-29), and consider it as the solid obtained by rotating the first-quadrant portion of the curve $x^2 + y^2 = r^2$ about the $x$ axis. Through the point $(x, y)$ we draw an element of volume of radius $y$ and thickness $dx$, at a distance $x$ from the base of the hemisphere. Its volume is thus

$$dV = \pi y^2\,dx$$

and its moment about the base of the hemisphere (the $y$ axis) is

$$dM_y = \pi x y^2\,dx = \pi x(r^2 - x^2)\,dx$$

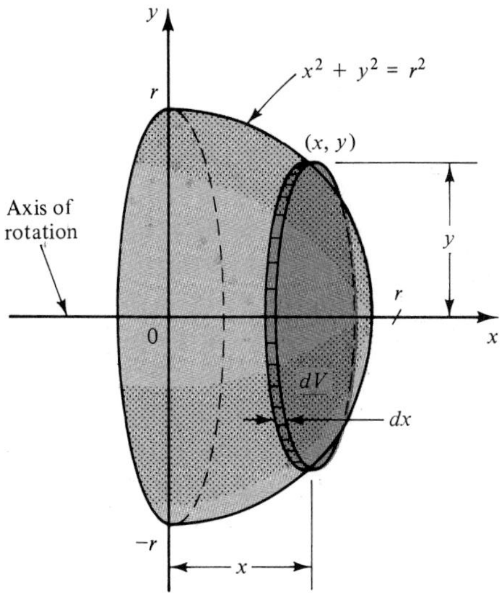


**FIGURE 33-29 Finding the centroid of a hemisphere.**

Integrating gives us the total moment,

$$M_y = \pi \int_0^r x(r^2 - x^2)\,dx = \pi \int_0^r (r^2x - x^3)\,dx$$

$$= \pi \left[\frac{r^2x^2}{2} - \frac{x^4}{4}\right]_0^r = \frac{\pi r^4}{4}$$

Of course, these methods work only for a solid that is homogeneous; that is, one whose density is the same throughout.

The total moment also equals the volume ($\frac{2}{3}\pi r^3$ for a hemisphere) times the distance $\bar{x}$ to the centroid, so

$$\bar{x} = \frac{\pi r^4/4}{2\pi r^3/3} = \frac{3r}{8}$$

Thus the centroid is located at $(3r/8, 0)$.

As with centroids of areas, we can write a formula for finding the centroid of a solid of revolution. For the volume $V$ (Fig. 33-30) formed by rotating the curve $y = f(x)$ about the $x$ axis, the distance to the centroid is:

| | | |
|---|---|---|
| Distance to the Centroid of Volume of Revolution about $x$ Axis | $\bar{x} = \frac{\pi}{V}\int_a^b xy^2\,dx$ | **389** |

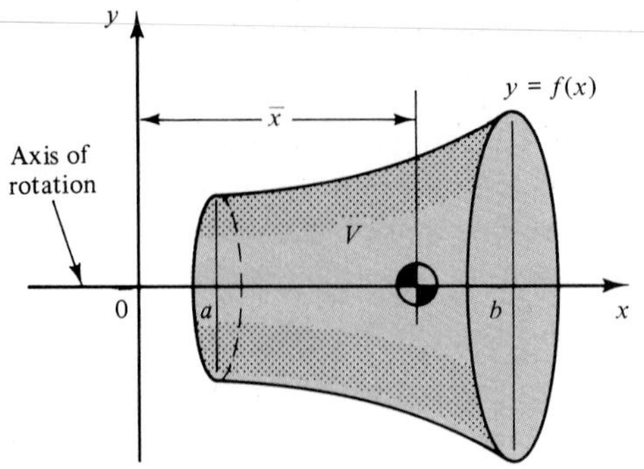


**FIGURE 33-30 Centroid of a solid of revolution.**

For volumes formed by rotation about the $y$ axis, we simply interchange $x$ and $y$ in Eq. 389 and get

| | | |
|---|---|---|
| Distance to the Centroid of Volume of Revolution about $y$ Axis | $\bar{y} = \frac{\pi}{V}\int_c^d yx^2\,dy$ | **390** |

These formulas work for solid figures only, not for those that have holes down their center.

## EXERCISE 5—CENTROIDS

### Without Integration

1. Find the centroid of four particles of equal mass located at (0, 0), (4, 2), (3, −5), and (−2, −3).
2. Find the centroid in Fig. 33-31a.
3. Find the centroid in Fig. 33-31b.
4. Find the centroid in Fig. 33-31c.

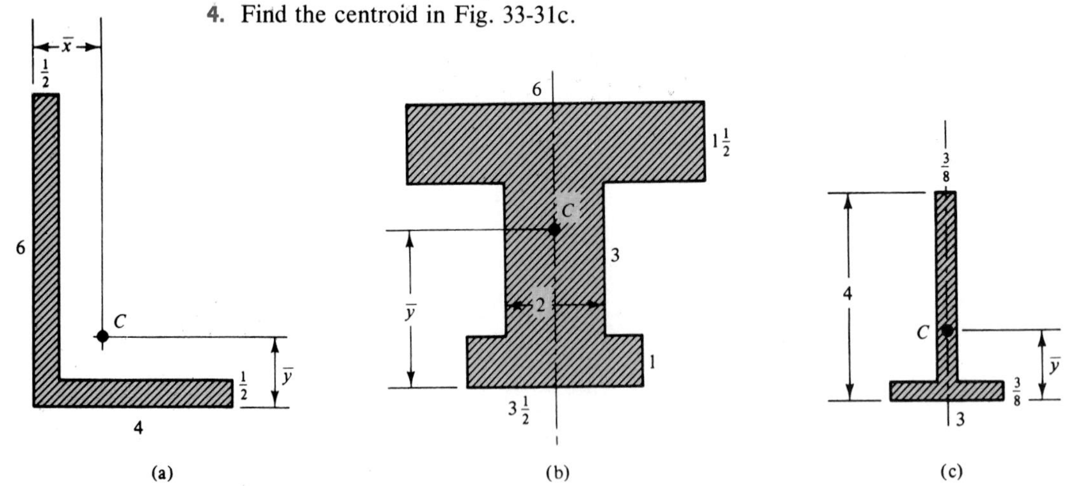


**FIGURE 33-31**

### Centroids of Areas by Integration

Find the specified coordinate of the centroid of each area.

5. bounded by $y^2 = 4x$ and $x = 4$; find $\bar{x}$ and $\bar{y}$
6. bounded by $y^2 = 2x$ and $x = 5$; find $\bar{x}$ and $\bar{y}$

7. bounded by $y = \frac{1}{2}(e^x + e^{-x})$, the coordinate axes, and the line $x = 1$; find $\bar{x}$
8. bounded by $y = x^2$, the $x$ axis, and $x = 3$; find $\bar{y}$
9. bounded by $y = e^x$, $x = 0$, and $x = 1$; find $\bar{y}$
10. bounded by $\sqrt{x} + \sqrt{y} = 1$ and the coordinate axes (area $= \frac{1}{6}$); find $\bar{x}$ and $\bar{y}$

### Areas Bounded by Two Curves

Find the specified coordinate of the centroid of each area.

11. bounded by $y = x^3$ and $y = 4x$, in the first quadrant; find $\bar{x}$
12. bounded by $x = 4y - y^2$ and $y = x$; find $\bar{y}$
13. bounded by $y^2 = x$ and $x^2 = y$; find $\bar{x}$ and $\bar{y}$
14. bounded by $y^2 = 4x$ and $y = 2x - 4$; find $\bar{y}$
15. bounded by $2y = x^2$ and $y = x^3$; find $\bar{x}$
16. bounded by $y = x^2 - 2x - 3$ and $y = 6x - x^2 - 3$ (area = 21.33); find $\bar{x}$ and $\bar{y}$
17. bounded by $y = x^2$ and $y = 2x + 3$, in the first quadrant; find $\bar{x}$

### Centroids of Volumes of Revolution

Find the distance from the origin to the centroid of each volume.

18. formed by rotating the area bounded by $x^2 + y^2 = 4$, $x = 0$, $x = 1$, and the $x$ axis about the $x$ axis
19. formed by rotating the area bounded by $6y = x^2$, the line $x = 6$, and the $x$ axis about the $x$ axis
20. formed by rotating the first quadrant area under the curve $y^2 = 4x$, from $x = 0$ to $x = 1$, about the $x$ axis
21. formed by rotating the area bounded by $y^2 = 4x$, $y = 6$, and the $y$ axis about the $y$ axis
22. formed by rotating the area bounded by $y = e^x$, $x = 0$ and $x = 1$ about the $x$ axis
23. a paraboloid of revolution bounded by a plane through the focus perpendicular to the axis of symmetry
24. formed by rotating the first-quadrant portion of the ellipse $x^2/64 + y^2/36 = 1$ about the $x$ axis
25. a right circular cone of height $h$, measured from its base

### Applications

26. A certain airplane rudder (Fig. 33-32) consists of one quadrant of an ellipse, and a quadrant of a circle. Find the coordinates of the centroid.

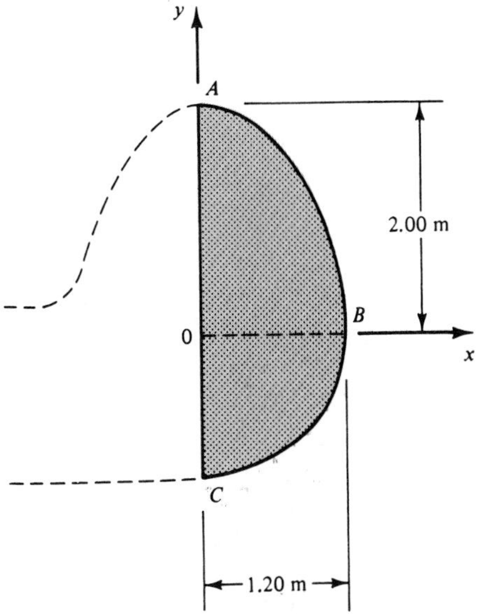


**FIGURE 33-32 Airplane rudder.**

**27.** The vane on a certain wind generator has the shape of a semicircle attached to a trapezoid (Fig. 33-33). Find the distance $\bar{x}$ to the centroid.

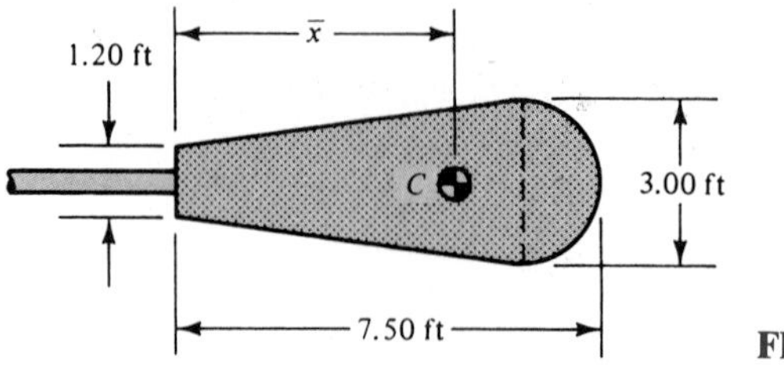


**FIGURE 33-33 Wind vane.**

**28.** A certain rocket (Fig. 33-34) consists of a cylinder attached to a paraboloid of revolution. Find the distance from the nose to the centroid of the total volume of the rocket.

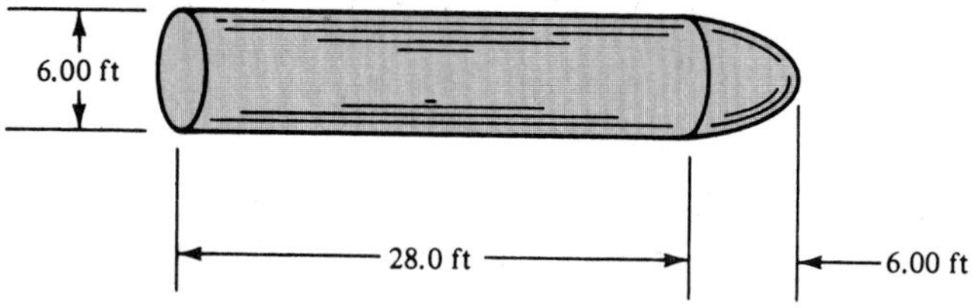


**FIGURE 33-34 Rocket.**

**29.** An optical instrument contains a mirror in the shape of a paraboloid of revolution (Fig. 33-35) hollowed out of a cylindrical block of glass. Find the distance from the flat bottom of the mirror to the centroid of the mirror.

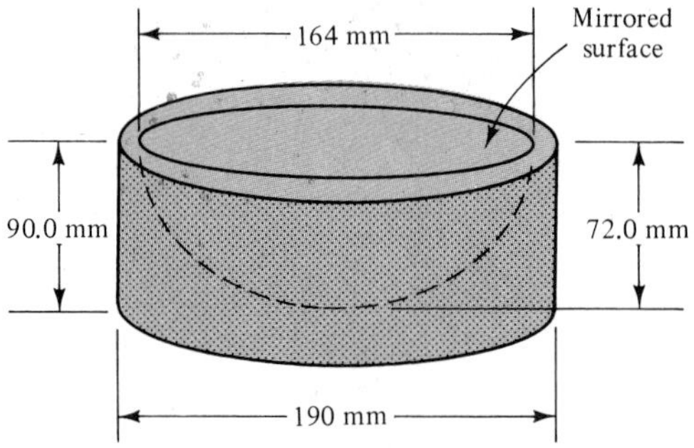


**FIGURE 33-35 Paraboloidal mirror.**

## 33-6 FLUID PRESSURE

The pressure at any point on a submerged surface varies directly with the depth of that point below the surface. Thus the pressure on a diver at a depth of 50 ft will be twice that at 25 ft.

The pressure on a submerged area is equal to the weight of the column of fluid above that area. Thus a square foot of area at a depth of 20 ft. supports a column of water having a volume of $20 \times 1^2 = 20$ ft$^3$. Since the

Recall that
weight = volume × density
(Eq. A43).

density of water is 62.4 lb/ft$^3$, the weight of this column is $20 \times 62.4 = 1248$ lb, so the pressure is 1248 lb/ft$^2$ at a depth of 20 ft. Further, Pascal's law says that the pressure is the same in *all directions,* so the same 1248 lb/ft$^2$ will be felt by a surface that is horizontal, vertical, or at any angle.

The *force* exerted by the fluid (*fluid pressure*) can be found by multiplying the pressure per unit area at a given depth by the total area.

| | | |
|---|---|---|
| Total Force on a Surface | force = pressure × area | **A46** |

A complication arises from an area that has points at various depths, and hence has different pressures over its surface. To compute the force on such a surface, we first compute the force on a narrow horizontal strip of area, assuming that the pressure is the same everywhere on that strip, and then add up the forces on all such strips by integration.

**EXAMPLE 16:** Find an expression for the force on the vertical area $A$ submerged in a fluid of density $\delta$ (Fig. 33-36).

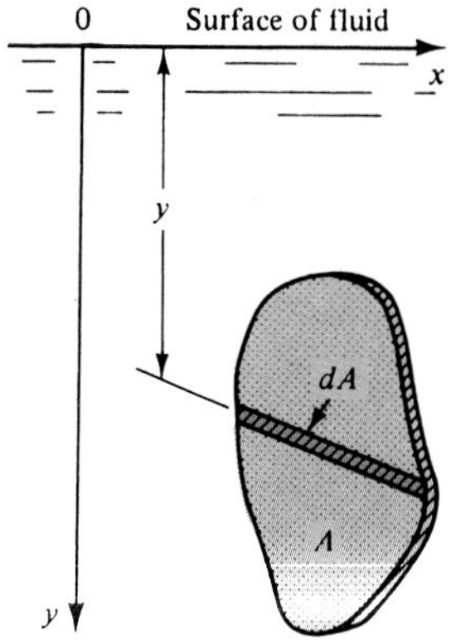


**FIGURE 33-36 Force on a submerged vertical surface.**

**Solution:** Let us take our origin at the surface of the fluid, with the $y$ axis downward. We draw a horizontal strip whose area is $dA$, located at a depth $y$ below the surface. The pressure at depth $y$ is $y\delta$, so the force $dF$ on the strip is, by Eq. A46,

$$dF = y\delta\, dA = \delta(y\, dA)$$

Integrating, we get

| | | |
|---|---|---|
| Force of Pressure | $F = \delta \int y\, dA$ | **A47** |

But the product $y\,dA$ is nothing but *the first moment of the area dA about the x axis*. Thus integration will give us the moment $M_x$ of the entire area, about the $x$ axis, multiplied by the density,

$$F = \delta \int y\, dA = \delta M_x$$

But the moment $M_x$ is also equal to the area $A$ times the distance $\bar{y}$ to the centroid, so

| | | |
|---|---|---|
| Force of Pressure | $F = \delta \bar{y} A$ | **A48** |

Thus *the force of pressure on a submerged, vertical surface is equal to the product of its area, the distance to its centroid, and the density of the fluid.*

We can compute the force on a submerged surface by using either Eq. A47 or A48. We give now an example of each method.

**EXAMPLE 17:** A vertical wall in a dam (Fig. 33-37) holds back water whose level is at the top of the wall. Find the total force of pressure on the wall by integration.

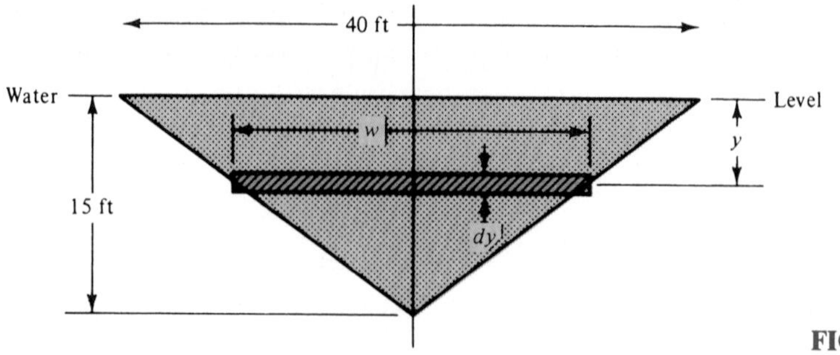


FIGURE 33-37

**Solution:** We sketch an element of area, with width $w$ and height $dy$. So

$$dA = w\,dy \tag{1}$$

We wish to express $w$ in terms of $y$. By similar triangles

$$\frac{w}{40} = \frac{15 - y}{15}$$

$$w = \frac{40}{15}(15 - y)$$

Substituting into (1) yields

$$dA = \frac{40}{15}(15 - y)\,dy$$

Then by Eq. A47,

$$F = \delta \int y\,dA = \frac{40\delta}{15}\int_0^{15} y(15 - y)\,dy$$

$$= \frac{40\delta}{15}\int_0^{15} (15y - y^2)\,dy$$

$$= \frac{40\delta}{15}\left[\frac{15y^2}{2} - \frac{y^3}{3}\right]_0^{15}$$

$$= \frac{40(62.4)}{15}\left[\frac{15(15)^2}{2} - \frac{(15)^3}{3}\right] = 93{,}600 \text{ lb}$$

**EXAMPLE 18:** The area shown in Fig. 33-27 is submerged in water (density = 62.4 lb/ft$^3$) so that the origin is 10.0 ft below the surface. Find the force on the area using Eq. A48.

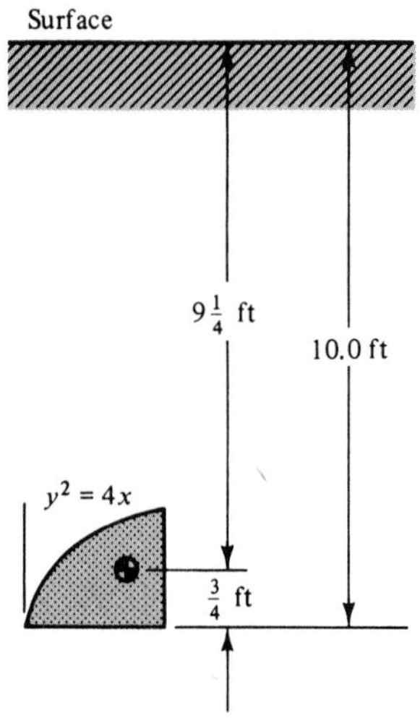


FIGURE 33-38

**Solution:** The area and the distance to the centroid have already been found in Example 13 of Sec. 33-5: area $= \frac{4}{3}$ ft$^2$ and $\bar{y} = \frac{3}{4}$ ft up from the origin, as shown in Fig. 33-38. The depth of the centroid below the surface is then

$$10.0 - \frac{3}{4} = 9\frac{1}{4} \text{ ft}$$

Thus by Eq. A48,

$$\text{force of pressure} = \frac{62.4 \text{ lb}}{\text{ft}^3} \cdot 9\frac{1}{4} \text{ ft} \cdot \frac{4}{3} \text{ ft}^3 = 770 \text{ lb}$$

## EXERCISE 6—FLUID PRESSURE

1. A circular plate 6.0 ft in diameter is placed vertically in a dam with its center 50 ft below the surface of the water. Find the force of pressure on the plate.
2. A vertical rectangular plate in the wall of a reservoir has its upper edge 20.0 ft below the surface of the water. It is 6.0 ft wide and 4.0 ft high. Find the force of pressure on the plate.
3. A vertical rectangular gate in a dam is 10.0 ft wide and 6.0 ft high. Find the force on the gate when the water level is 8.0 ft above the top of the gate.
4. A vertical cylindrical tank has a diameter of 30.0 ft and a height of 50.0 ft. Find the total force on the curved surface when the tank is full of water.
5. A trough, whose cross section is an equilateral triangle with a vertex down, has sides 2.0 ft long. Find the total force on one end when the trough is full of water.
6. A horizontal cylindrical boiler 4.0 ft in diameter is half full of water. Find the force on one end.
7. A horizontal tank of oil (density = 60 lb/ft$^3$) has ends in the shape of an ellipse with horizontal axis 12.0 ft long and vertical axis 6.0 ft long (Fig. 33-39). Find the force on one end when the tank is half full.

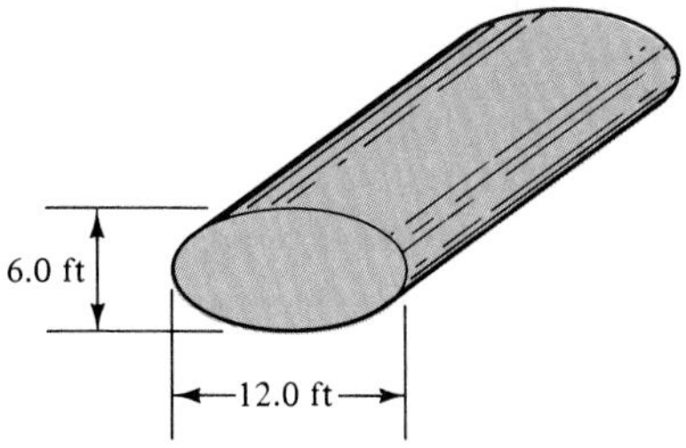


**FIGURE 33-39 Oil tank.**

8. The cross section of a certain trough is a parabola with vertex down. It is 2.0 ft deep and 4.0 ft wide at the top. Find the force on one end when the trough is full of water.

## 33-7 WORK

### Definition

When a constant force acts on an object which moves in the direction of the force, the *work* done by the force is defined as the product of the force and the distance moved by the object.

| | | |
|---|---|---|
| Work Done by a Constant Force | work = force × distance | **A6** |

---

**EXAMPLE 19:** The work needed to lift a 100-lb weight a distance of 2 ft is 200 ft-lb.

---

### Variable Force

Equation A6 applies when the force is *constant,* but this is not always the case. For example, the force needed to stretch or compress a spring increases as the spring gets extended (Fig. 33-40). Or, as another example, the expanding gases in an automobile cylinder exert a variable force on the piston.

If we let the variable force be represented by $F(x)$, acting in the $x$ direction from $x = a$ to $x = b$, the work done by this force may be defined as

| Work Done by a Variable Force | $W = \int_a^b F(x)\, dx$ | **A7** |
|---|---|---|

We first apply this formula to find the work done in stretching or compressing a spring.

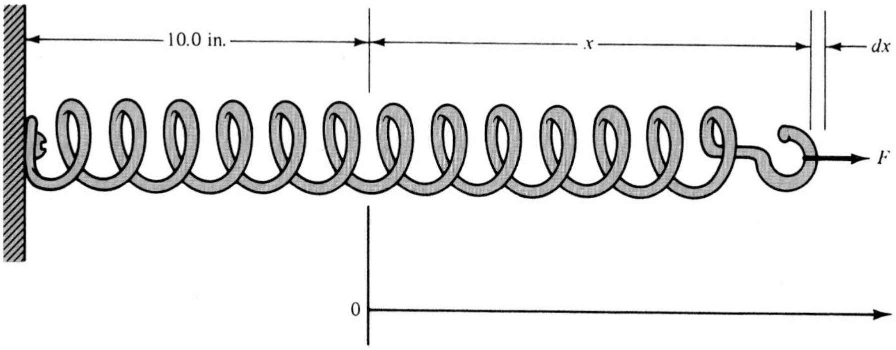


**FIGURE 33-40**

---

**EXAMPLE 20:** A certain spring (Fig. 33-40) has a free length (when no force is applied) of 10.0 in. The spring constant is 12.0 lb/in. Find the work needed to stretch the spring from a length of 12.0 in. to a length of 14.0 in.

**Solution:** We draw the spring partly stretched, as shown, taking our $x$ axis in the direction of movement of the force, with zero at the free position of the end of the spring. The force needed to hold the spring in this position is equal to the spring constant $k$ times the deflection $x$:

| $F = kx$ | **A61** |
|---|---|

If we assume that the force does not change when stretching the spring an additional small amount $dx$, the work done is

$$dW = F\, dx = kx\, dx$$

We get the total work by integrating:

$$W = \int F\, dx = k \int_2^4 x\, dx = \frac{kx^2}{2}\Big|_2^4$$

$$= \frac{12.0}{2}(4^2 - 2^2) = 72 \text{ in. lb}$$

---

| | |
|---|---|
| Common Error | Be sure to measure spring deflections from the *free* end of the spring, not from the fixed end. |

Another typical problem is that of finding the work needed to pump the fluid out of a tank. Such problems may be solved by noting that the work required is equal to the weight of the fluid times the distance which the centroid of the fluid (when still in the tank) must be raised.

---

**EXAMPLE 21:** A hemispherical tank (Fig. 33-41) is filled with water having a density of 62.4 lb/ft$^3$. Find the work needed to pump all the water to a height of 10.0 ft above the top of the tank.

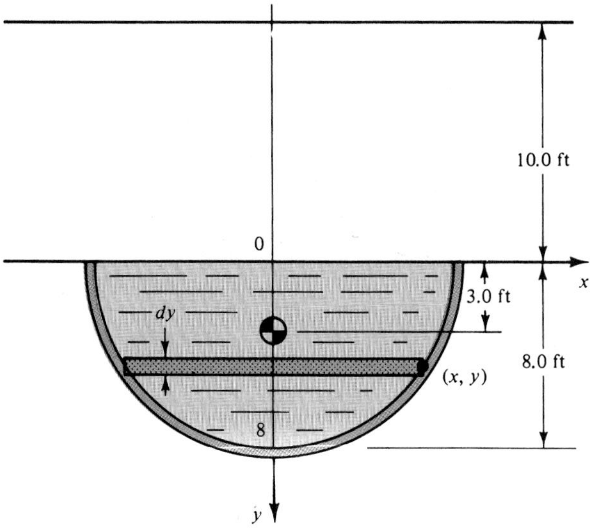


**FIGURE 33-41 Hemispherical tank.**

**Solution:** The distance to the centroid of a hemisphere of radius $r$ was found in Example 15 to be $3r/8$, so the centroid of our hemisphere is at a distance

$$\bar{y} = \frac{3}{8} \cdot 8 = 3.0 \text{ ft}$$

as shown. It must therefore be raised a distance of 13.0 ft. We find the weight of the tankful of water by multiplying its volume times its density:

$$\begin{aligned} \text{weight} &= \text{volume} \times \text{density} \\ &= \frac{2}{3}\pi(8^3)(62.4) = 66{,}900 \text{ lb} \end{aligned}$$

The work done is then

$$\text{work} = 66{,}900 \text{ lb}(13.0 \text{ ft}) = 870{,}000 \text{ ft lb}$$

---

If we do not know the location of the centroid, we can find the work directly, by integration.

**EXAMPLE 22:** Repeat Example 21 by integration, assuming that the location of the centroid is not known.

**Solution:** We choose coordinate axes as shown in Fig. 33-41. Through some point $(x, y)$ on the curve we draw an element of volume whose volume $dV$ is

$$dV = \pi x^2\, dy$$

and whose weight is

$$62.4\, dV = 62.4\pi x^2\, dy$$

Since this element must be lifted a distance of $(10 + y)$ ft, the work required is

$$dW = (10 + y)(62.4\pi x^2\, dy)$$

Integrating gives us

$$W = 62.4\pi \int (10 + y)x^2\, dy$$

But using the equation for a circle (Eq. 294), we have

$$x^2 = r^2 - y^2 = 64 - y^2$$

We substitute, to get the expression in terms of $y$,

$$W = 62.4\pi \int_0^8 (10 + y)(64 - y^2)\, dy$$

$$= 62.4\pi \int_0^8 (640 + 64y - 10y^2 - y^3)\, dy$$

$$= 62.4\pi \left[640y + 32y^2 - \frac{10y^3}{3} - \frac{y^4}{4}\right]_0^8 = 870{,}000 \text{ ft lb}$$

as by the method of Example 21.

## EXERCISE 7—WORK

### Springs

1. A force of 50.0 lb will stretch a spring that has a free length of 12.0 in. to a length of 13.0 in. How much work is needed to stretch the spring from a length of 14.0 in. to 16.0 in.?
2. A spring whose free length is 10.0 in. has a spring constant of 12.0 lb/in. Find the work needed to stretch this spring from 12.0 in. to 15.0 in.
3. A spring has a spring constant of 8.0 lb/in. and a free length of 5.0 in. Find the work required to stretch it from 6.0 in. to 8.0 in.

### Tanks

4. Find the work required to pump all the water to the top and out of a vertical cylindrical tank, 16.0 ft in diameter and 20.0 ft deep, that is completely filled at the start.
5. A hemispherical tank 12.0 ft in diameter is filled with water to a depth of 4.0 ft. How much work is needed to pump the water to the top of the tank?
6. A conical tank 20.0 ft deep and 20.0 ft across the top is full of water. Find the work needed to pump the water to a height of 15.0 ft above the top of the tank.

7. A tank has the shape of a frustum of a cone, with a top diameter of 8.0 ft, a bottom diameter of 12.0 ft, and a height of 10.0 ft. How much work is needed to pump the contents to a height of 10.0 ft above the tank, if it is filled with oil of density 50.0 $lb/ft^3$?

### Gas Laws

8. Find the work needed to compress air initially at a pressure of 15.0 $lb/in^2$ from a volume of 200 $ft^3$ to 50.0 $ft^3$. (*Hint:* The work $dW$ done in moving the piston (Fig. 33-42) is $dW = F\,dx$. Express both $F$ and $dx$ in terms of $v$ by means of Eqs. 126 and A46, and integrate.)
9. Air is compressed (Fig. 33-42) from an initial pressure of 15.0 $lb/in^2$ and volume of 200 $ft^3$ to a pressure of 80.0 $lb/in^2$. How much work was needed to compress the air?
10. If the pressure and volume of air are related by the equation $pv^{1.4} = k$, find the work needed to compress air initially at 14.5 $lb/in^2$ and 10.0 $ft^3$ to a pressure of 100 $lb/in^2$.

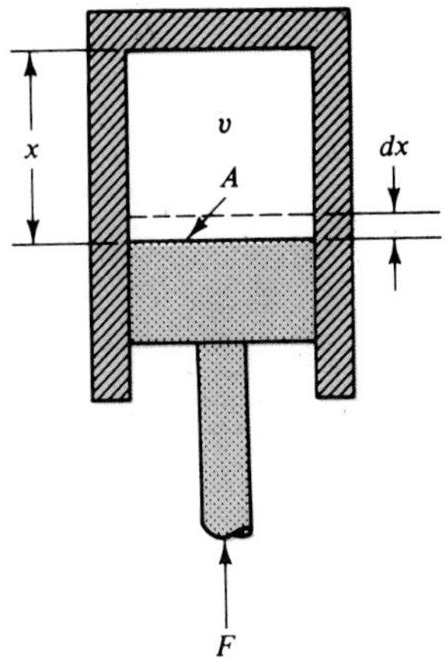


**FIGURE 33-42 Piston and cylinder. Assume that the pressure and volume are related by the equation $pv$ = constant.**

### Miscellaneous

11. The force of attraction (in pounds) between two masses separated by a distance $d$ is equal to $k/d^2$, where $k$ is a constant. If two masses are 50 ft apart, find the work needed to separate them another 50 ft. Express your answer in terms of $k$.
12. Find the work needed to wind up a vertical cable 100 ft long, weighing 3.0 lb/ft.
13. A 500-ft-long cable weighs 1.0 lb/ft and is hanging from a tower with a 200-lb weight at its end. How much work is needed to raise the weight and the cable a distance of 20.0 ft?

## 33-8 MOMENT OF INERTIA

### Moment of Inertia of an Area

Figure 33-43a and b each show a small area at a distance $r$ from a line $L$ in the same plane. In each case the dimensions of the area are such that we may consider all points on the area as being at the same distance $r$ from the line $L$.

In Sec. 33-5 we defined the *first moment* of the area about $L$ as being the product of the area times the distance to the line. We now define the *second moment*, or *moment of inertia I*, as the product of the area times the *square* of the distance to the line.

**The moment of inertia is very important in beam design. It is not only the cross-sectional area of a beam that determines how well the beam will resist bending, but *how far* that area is located from the axis of the beam. The moment of inertia, taking into account both the area and its location with respect to the beam axis, is a measure of the resistance to bending.**

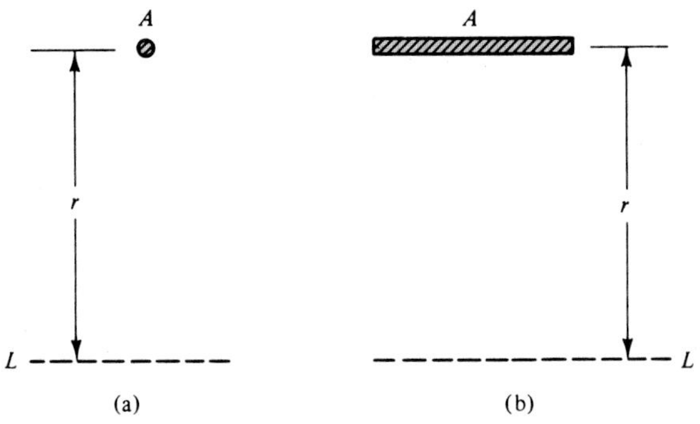


**FIGURE 33-43**

| Moment of Inertia of an Area | $I = Ar^2$ | **391** |
|---|---|---|

The distance $r$, which is the distance to the line, is called the *radius of gyration*. We write $I_x$ to denote the moment of inertia about the $x$ axis, and $I_y$ for the moment of inertia about the $y$ axis.

---

**EXAMPLE 23:** Find the moment of inertia of the thin strip in Fig. 33-43b if it has a length of 8.0 cm, a width of 0.20 cm, and is 7.0 cm from axis $L$.

**Solution:** The area of the strip is $8.0(0.20) = 1.6\ \text{cm}^2$, so the moment of inertia is, by Eq. 391,

$$I = 1.6(7.0)^2 = 78.4\ \text{cm}^4$$

---

## Moment of Inertia of a Rectangle

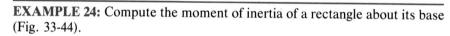

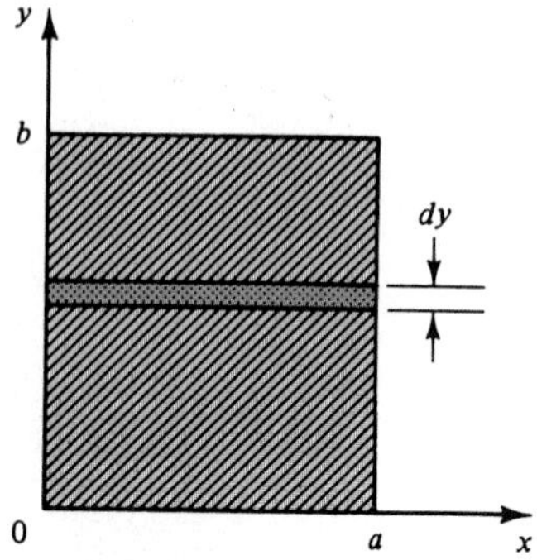

**FIGURE 33-44 Moment of inertia of a rectangle.**

In Example 23 our area was a thin strip parallel to the axis, with all points on the area at the same distance $r$ from the axis. But what shall we use for $r$ when dealing with an extended area, such as the rectangle in Fig. 33-44? Again calculus comes to our aid. Since we can easily compute the moment of inertia of a thin strip, we slice our area into many thin strips and add up their individual moments of inertia by integration.

---

**EXAMPLE 24:** Compute the moment of inertia of a rectangle about its base (Fig. 33-44).

**Solution:** We draw a single strip of area parallel to the axis about which we are taking the moment. This strip has a width $dy$ and a length $a$. Its area is then

$$dA = a\,dy$$

All points on the strip are at a distance $y$ from the $x$ axis, so the moment of inertia, $dI_x$, of the single strip is

$$dI_x = y^2(a\,dy)$$

We add up the moments of all such strips by integrating from $y = 0$ to $y = b$,

$$I_x = a\int_0^b y^2\,dy = \left.\frac{ay^3}{3}\right|_0^b = \frac{ab^3}{3}$$

---

## Radius of Gyration

If all the area in a plane figure were squeezed into a single thin strip of equal area and placed parallel to the $x$ axis at such a distance that it had the *same moment of inertia* as the original rectangle, it would be at a distance $r$ that we call the *radius of gyration*. If $I$ is the moment of inertia of some area $A$ about some axis, then

$$Ar^2 = I$$

So

| | | |
|---|---|---|
| Radius of Gyration | $r = \sqrt{\dfrac{I}{A}}$ | **395** |

We use $r_x$ and $r_y$ to denote the radius of gyration about the $x$ and $y$ axes, respectively.

---

**EXAMPLE 25:** Find the radius of gyration for the rectangle in Example 24.

**Solution:** The area of the rectangle is $ab$, so by Eq. 395,

$$r_x = \sqrt{\frac{I}{A}} = \sqrt{\frac{ab^3}{3ab}} = \frac{b}{\sqrt{3}} \simeq 0.577b$$

Note that the *centroid* is at a distance of $0.5b$ from the edge of the rectangle. This shows that *the radius of gyration is not equal to the distance to the centroid.*

---

| | |
|---|---|
| Common Error | Do *not* use the distance to the *centroid* when computing moment of inertia. |

### Moment of Inertia of an Area by Integration

Since we are now able to write the moment of inertia of a rectangular area we can now derive formulas for the moment of inertia of other areas, such as in Fig. 33-45. The area of the vertical strip shown in Fig. 33-45 is $dA$ and its distance from the $y$ axis is $x$, so its moment of inertia about the $y$ axis is, by Eq. 391,

$$dI_y = x^2\, dA = x^2 y\, dx$$

since $dA = y\, dx$. The total moment is then found by integrating:

| | | |
|---|---|---|
| Moment of Inertia of an Area about $y$ Axis | $I_y = \int x^2 y\, dx$ | **393** |

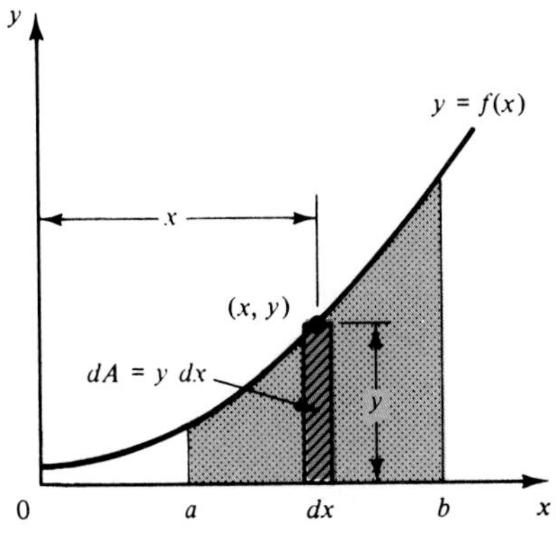


FIGURE 33-45

To find the moment of inertia about the $x$ axis, we use the result of Example 24, that the moment of inertia of a rectangle about its base is $ab^3/3$. Thus the moment of inertia of the rectangle in Fig. 33-45 is

$$dI_x = \frac{1}{3}y^3\,dx$$

As before, the total moment of inertia is found by integrating:

| Moment of Inertia of an Area about $x$ Axis | $I_x = \frac{1}{3}\int y^3\,dx$ | **392** |
|---|---|---|

**EXAMPLE 26:** Find the moment of inertia about the $x$ and $y$ axes for the area under the curve $y = x^2$, from $x = 1$ to $x = 3$.

**Solution:** From Eq. 393,

$$I_y = \int_1^3 x^2(x^2)\,dx$$

$$= \int_1^3 x^4\,dx$$

$$= \left.\frac{x^5}{5}\right|_1^3 \simeq 48.4$$

Now using Eq. 392, with $y^3 = (x^2)^3 = x^6$,

$$I_x = \frac{1}{3}\int_1^3 x^6\,dx = \frac{1}{3}\left[\frac{x^7}{7}\right]_1^3 = \frac{1}{21}(3^7 - 1^7) = \frac{2186}{21} \simeq 104.1$$

**EXAMPLE 27:** Find the radius of gyration of the area in Example 26, about the $x$ axis.

**Solution:** We first find the area of the figure. By Eq. 365,

$$A = \int_1^3 x^2\,dx = \left.\frac{x^3}{3}\right|_1^3 = \frac{3^3}{3} - \frac{1^3}{3} \simeq 8.667$$

From Example 26, $I_x = 104.1$. So by Eq. 395,

$$r_x = \sqrt{\frac{I_x}{A}} = \sqrt{\frac{104.1}{8.667}} = 3.466$$

## Polar Moment of Inertia

**Polar moment of inertia is needed when studying rotation of rigid bodies.**

In the preceding sections we had found the moment of inertia of an area about some line in the plane of that area. Now we find moment of inertia of a *solid of revolution* about its axis of revolution. We call this the *polar moment of inertia*. We first find the polar moment of inertia for a thin-walled shell, then use that result to find the polar moment of inertia for any solid of revolution.

## Polar Moment of Inertia by the Shell Method

A thin-walled cylindrical shell (Fig. 33-46) has a volume equal to the product of its circumference, wall thickness, and height,

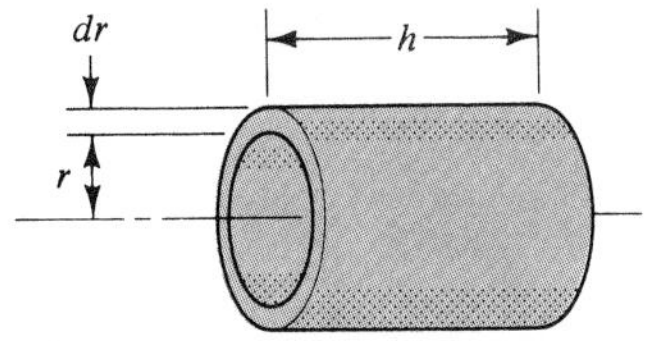


FIGURE 33-46

| | |
|---|---|
| $dV = 2\pi rh\ dr$ | **381** |

If we let $m$ represent the mass per unit volume, the mass of the shell is then

$$dM = 2\pi mrh\ dr$$

Since we may consider all particles in this shell to be at a distance $r$ from the axis of revolution, we obtain the moment of inertia of the shell by multiplying the mass by $r^2$,

| | | |
|---|---|---|
| Polar Moment of Inertia of a Shell | $dI = 2\pi mr^3h\ dr$ | **400** |

We then think of the entire solid of revolution as being formed by concentric shells. The polar moment of inertia of an entire solid of revolution is then found by integration.

| | | |
|---|---|---|
| Polar Moment of Inertia by the Shell Method | $I = 2\pi m \int r^3h\ dr$ | **402** |

---

**EXAMPLE 28:** Find the polar moment of inertia of a solid cylinder (Fig. 33-47) about its axis.

**Solution:** We draw elements of volume in the form of concentric shells. The moment of inertia of each is

$$dI_x = 2\pi my^3h\ dy$$

Integrating yields

$$I_x = 2\pi mh \int_0^r y^3\ dy = 2\pi mh\left[\frac{y^4}{4}\right]_0^r$$

$$= \frac{\pi mhr^4}{2}$$

But since the volume of the cylinder is $\pi r^2 h$, we get

$$I_x = \frac{1}{2}mVr^2$$

In other words, *the polar moment of inertia of a cylindrical solid is equal to half the product of its density, its volume, and the square of its radius.*

---

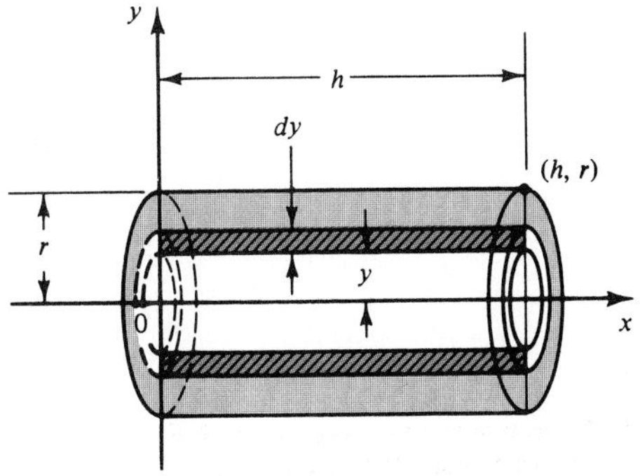


FIGURE 33-47 **Polar moment of inertia of a solid cylinder.**

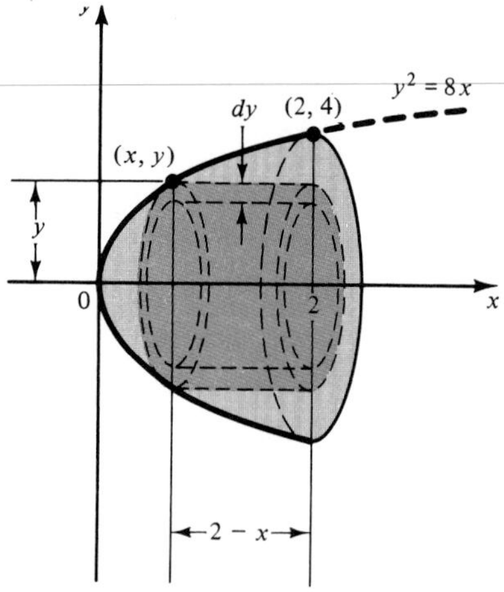


**FIGURE 33-48 Polar moment of inertia by shell method.**

**EXAMPLE 29:** The first-quadrant area under the curve $y^2 = 8x$, from $x = 0$ to $x = 2$, is rotated about the $x$ axis. Use the shell method to find the polar moment of inertia of the paraboloid generated (Fig. 33-48).

**Solution:** Our shell element has a radius $y$, a length $2 - x$, and a thickness $dy$. By Eq. 400,

$$dI_x = 2\pi m y^3(2 - x)\,dy$$

Replacing $x$ by $y^2/8$ gives

$$dI_x = 2\pi m y^3\left(2 - \frac{y^2}{8}\right) dy$$

Integrating gives us

$$I_x = 2\pi m \int_0^4 \left(2y^3 - \frac{y^5}{8}\right) dy = 2\pi m \left| \frac{2y^4}{4} - \frac{y^6}{48} \right|_0^4 = 268m$$

### Polar Moment of Inertia by the Disk Method

Sometimes the disk method will result in an integral that is easier to evaluate than that obtained by the shell method.

For the solid of revolution in Fig. 33-49, we choose a disk-shaped element of volume of radius $r$ and thickness $dh$. Since it is a cylinder, we use the moment of inertia of a cylinder found in Example 28,

$$dI = \frac{m\pi r^4\,dh}{2}$$

Integrating gives

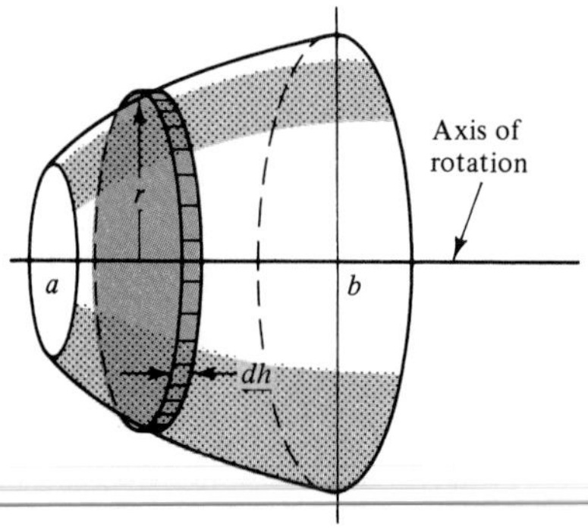


**FIGURE 33-49 Polar moment of inertia by disk method.**

| | | |
|---|---|---|
| Polar Moment of Inertia by the Disk Method | $I = \frac{m\pi}{2} \int_a^b r^4\,dh$ | **401** |

**EXAMPLE 30:** Repeat Example 29 by the disk method.

**Solution:** We use Eq. 401, with $r = y$ and $dh = dx$.

$$I_x = \frac{m\pi}{2} \int y^4\,dx$$

But $y^2 = 8x$, so $y^4 = 64x^2$. Substituting yields

$$I_x = 32m\pi \int_0^2 x^2\,dx$$

$$= 32m\pi \left[\frac{x^3}{3}\right]_0^2$$

$$= 32m\pi \left(\frac{2^3}{3}\right) \simeq 268m$$

as before.

## EXERCISE 8—MOMENT OF INERTIA

### Moment of Inertia of Plane Areas

1. Find the moment of inertia about the $x$ axis of the area bounded by $y = x$, $y = 0$, and $x = 1$.
2. Find the radius of gyration about the $x$ axis of the first-quadrant area bounded by $y^2 = 4x$, $x = 4$, and $y = 0$.
3. Find the radius of gyration about the $y$ axis of the area in Problem 2.
4. Find the moment of inertia about the $y$ axis of the area bounded by $x + y = 3$, $x = 0$, and $y = 0$.
5. Find the moment of inertia about the $x$ axis of the first-quadrant area bounded by the curve $y = 4 - x^2$ and the coordinate axes.
6. Find the moment of inertia about the $y$ axis of the area in Problem 5.
7. Find the moment of inertia about the $x$ axis of the first-quadrant area bounded by the curve $y^3 = 1 - x^2$.
8. Find the moment of inertia about the $y$ axis of the area in Problem 1.
9. Find the radius of gyration of the area under one arch of the sine curve $y = \sin x$ with respect to the $x$ axis.
10. Find the moment of inertia of the area bounded by the curve $y = e^x$, the line $x = 1$, and the coordinate axes with respect to the $x$ axis.

### Polar Moment of Inertia

Find the polar moment of inertia of the volume formed when a first-quadrant area with the following boundaries is rotated about the $x$ axis.

11. bounded by $y = x$, $x = 2$, and the $x$ axis
12. bounded by $y = x + 1$, from $x = 1$ to $x = 2$, and the $x$ axis
13. bounded by the curve $y = x^2$, the line $x = 2$, and the $x$ axis
14. bounded by $\sqrt{x} + \sqrt{y} = 2$ and the coordinate axes

Find the polar moment of inertia of each solid with respect to its axis in terms of the total mass $M$ of the solid.

15. a right circular cone of height $h$ and base radius $r$
16. a sphere of radius $r$
17. a paraboloid of revolution bounded by a plane through the focus perpendicular to the axis of symmetry

## CHAPTER 33 REVIEW PROBLEMS

1. Find the volume generated when the area bounded by the parabolas $y^2 = 4x$ and $y^2 = 5 - x$ is rotated about the $x$ axis.
2. Find the length of the curve $y = \frac{4}{5}x^2$ from the origin to $x = 4$.
3. The area bounded by the parabolas $y^2 = 4x$ and $y^2 = x + 3$ is rotated about the $x$ axis. Find the surface area of the solid generated.
4. The area bounded by the parabola $y^2 = 4x$, from $x = 0$ to $x = 8$, is rotated about the $x$ axis. Find the volume generated.
5. Find the volume of the solid generated by rotating the ellipse $x^2/16 + y^2/9 = 1$ about the $x$ axis.
6. Find the length of the parabola $y^2 = 8x$ from the vertex to one end of the latus rectum.
7. The area bounded by the curves $x^2 = 4y$, $x - 2y + 4 = 0$, and the $y$ axis is rotated about the $y$ axis. Find the surface area of the volume generated.
8. Find the volume generated when the area bounded by the curve $y^2 = 16x$, from $x = 0$ to $x = 4$, is rotated about the $x$ axis.
9. Find the volume generated when the area bounded by

$y^2 = x^3$, the $y$ axis, and $y = 8$ is rotated about the line $x = 4$.

10. The cables on a certain suspension bridge hang in the shape of a parabola. The towers are 100 m apart and the cables dip 10 m below the tops of the towers. Find the length of the cables.

11. The first-quadrant area bounded by the curves $y = x^3$ and $y = 4x$ is rotated about the $x$ axis. Find the surface area of the solid generated.

12. Find the volume generated when the area bounded by $y^2 = x^3$, $x = 4$, and the $x$ axis is rotated about the line $y = 8$.

13. Find the average ordinate for the function $y = \sin^2 x$, for $x = 0$ to $2\pi$.

14. Find the rms value for the function $y = 2x + x^2$, for the interval $x = -1$ to $x = 3$.

15. A cylindrical, horizontal tank is 6.0 ft in diameter and is half full of water. Find the force on one end of the tank.

16. Find the moment of inertia of a sphere of radius $r$ and density $m$ with respect to a diameter.

17. A spring has a free length of 12.0 in. and a spring constant of 5.45 lb/in. Find the work needed to compress it from a length of 11.0 in. to 9.0 in.

18. Find the coordinates of the centroid of the area bounded by the curve $y^2 = 2x$ and the line $y = x$.

19. A horizontal cylindrical tank 8.0 ft in diameter is half full of oil (60 lb/ft$^3$). Find the force on one end.

20. Find the moment of inertia of a right circular cone of base radius $r$, height $h$, and mass $m$ with respect to its axis.

21. Find the centroid of a semicircle of radius 10.

22. A bucket that weighs 3.0 lb and has a volume of 2.0 ft$^3$ is filled with water. It is being raised at a rate of 5.0 ft/s while water leaks from the bucket at a rate of 0.01 ft$^3$/s. Find the work done in raising the bucket 100 ft.

23. Find the centroid of the area bounded by one quadrant of the ellipse $x^2/16 + y^2/9 = 1$.

24. Find the radius of gyration of the area bounded by the parabola $y^2 = 16x$ and its latus rectum, with respect to its latus rectum.

25. A conical tank is 8.0 ft in diameter at the top and 12.0 ft deep, and is filled with a liquid having a density of 80.0 lb/ft$^3$. How much work is required to pump all the liquid to the top of the tank?

26. Find the coordinates of the centroid of the area bounded by the $x$ axis and a half-cycle of the sine curve, $y = \sin x$.

*Writing*

27. Suppose you have designed a tank in the shape of a solid of revolution, and have found its centroid by integration. Your manager insists that it is impossible to find the center of gravity of something that isn't even built yet. Write a memo to your manager explaining how it is done.

# 34

# METHODS OF INTEGRATION

## OBJECTIVES

**When you have completed this chapter, you should be able to:**

- Evaluate integrals using integration by parts.
- Integrate rational fractions with denominators having nonrepeated linear factors, repeated linear factors, or quadratic factors.
- Use algebraic substitution to evaluate an integral.
- Integrate using trigonometric substitution.
- Evaluate improper integrals containing infinite limits or discontinuous integrands.
- Do approximate integration using the average ordinate method, the trapezoid rule, or Simpson's rule.

All practical methods of integration depend eventually on the use of a table of integrals. However, a given expression may not match any of the forms in the table, and may even look completely different. The methods given in this chapter will help us to manipulate a given expression into one of the listed forms.

But even with all our new methods there will always be expressions we cannot integrate, so we conclude this chapter with a few "approximate methods" of integration, that is, by methods that find the value of an integral by numerical approximations. These methods have taken on added importance because they can be used on the computer.

## 34-1 INTEGRATION BY PARTS

### Derivation of the Rule

Integration by parts is a useful method that enables us to split products into two parts for easier integration. We start with the rule for the derivative of a product (Eq. 336), in differential form,

$$d(uv) = u\,dv + v\,du$$

Rearranging, we get

$$u\,dv = d(uv) - v\,du$$

Integrating gives us

$$\int u\,dv = \int d(uv) - \int v\,du$$

or

| Rule 6. | $\int u\,dv = uv - \int v\,du$ |
|---|---|

This rule is listed as number 6 in our table of integrals.

---

**EXAMPLE 1:** Integrate $\int x \cos 3x\,dx$.

**Solution:** The integrand is a product to which none of our previous rules apply. We use integration by parts. We separate $x \cos 3x\,dx$ into two parts, one of which we call $u$ and the other $dv$. But which part shall we call $u$ and which $dv$?

A good rule of thumb is to take $dv$ as *the most complicated part,* but one that you can still *easily integrate*. So we let

$$u = x \qquad \text{and} \qquad dv = \cos 3x\,dx$$

Now our Rule 6 requires that we know $du$ and $v$ in addition to $u$ and $dv$. We obtain $du$ by differentiating $u$, and find $v$ by integrating $dv$.

$$du = dx \qquad \text{and} \qquad v = \int \cos 3x\, dx$$

$$= \frac{1}{3}\int \cos 3x(3\, dx)$$

$$= \frac{1}{3}\sin 3x + C_1$$

Applying Rule 6, we have

$$\int \underbrace{x}_{u} \underbrace{\cos x\, dx}_{dv} = \underbrace{x}_{u}\underbrace{\left(\frac{1}{3}\sin 3x + C_1\right)}_{v} - \int \underbrace{\left(\frac{1}{3}\sin 3x + C_1\right)}_{v} \underbrace{dx}_{du}$$

$$= \frac{x}{3}\sin 3x + C_1 x - \frac{1}{9}\int \sin 3x(3\, dx) - \int C_1\, dx$$

$$= \frac{x}{3}\sin 3x + C_1 x + \frac{1}{9}\cos 3x - C_1 x + C$$

$$= \frac{x}{3}\sin 3x + \frac{1}{9}\cos 3x + C$$

---

Note that the constant $C_1$ obtained when integrating $dv$ *does not appear in the final integral,* and in fact, the final result would be the same as if we had not introduced $C_1$ in the first place. So in the following examples we will follow the usual practice of dropping the constant $C_1$ when integrating $dv$.

---

**EXAMPLE 2:** Integrate $\int x \ln x\, dx$.

**Solution:** Let us choose

$$u = \ln x \qquad \text{and} \qquad dv = x\, dx$$

Then

$$du = \frac{1}{x}\, dx \qquad \text{and} \qquad v = \frac{x^2}{2} \quad \text{(plus a constant (which we drop)}$$

By Rule 6,

$$\int \underbrace{\ln x}_{u}\, \underbrace{(x\, dx)}_{dv} = \underbrace{(\ln x)}_{u}\underbrace{\left(\frac{x^2}{2}\right)}_{v} - \int \underbrace{\frac{x^2}{2}}_{v}\underbrace{\left(\frac{dx}{x}\right)}_{du}$$

$$= \frac{x^2}{2}\ln x - \frac{1}{2}\int x\, dx$$

$$= \frac{x^2}{2}\ln x - \frac{x^2}{4} + C$$

---

Sometimes our first choice of $u$ and $dv$ will result in an integral no easier to evaluate than our original.

**EXAMPLE 3:** Integrate $\int xe^x\,dx$.

**Solution:** *First try:* Let

$$u = e^x \quad \text{and} \quad dv = x\,dx$$

Then

$$du = e^x\,dx \quad \text{and} \quad v = \frac{x^2}{2}$$

By Rule 6,

$$\int xe^x\,dx = \frac{x^2e^x}{2} - \frac{1}{2}\int x^2e^x\,dx$$

we get an integral which is more difficult than the one we started with.

*Second try:* Let

$$u = x \quad \text{and} \quad dv = e^x\,dx$$

Then

$$du = dx \quad \text{and} \quad v = e^x$$

By Rule 6,

$$\int xe^x\,dx = xe^x - \int e^x\,dx$$
$$= xe^x - e^x + C$$

Sometimes we may have to integrate by parts *twice* to get the final result.

**EXAMPLE 4:** Integrate $\int x^2 \sin x\,dx$.

**Solution:** Let

$$u = x^2 \quad \text{and} \quad dv = \sin x\,dx$$

Then

$$du = 2x\,dx \quad \text{and} \quad v = -\cos x$$

By Rule 6,

$$\int x^2 \sin x\,dx = -x^2 \cos x + 2\int x \cos x\,dx \tag{1}$$

Our new integral in (1) is similar to the original, but we note that the power of $x$ has been reduced from 2 to 1. This suggests that integrating by parts *again* may reduce the exponent to zero. Let's try

$$u = x \quad \text{and} \quad dv = \cos x\,dx$$

Then

$$du = dx \quad \text{and} \quad v = \sin x$$

By Rule 6,

$$\int x \cos x\,dx = x \sin x - \int \sin x\,dx$$
$$= x \sin x + \cos x + C_1$$

Substituting back into (1) gives us

$$\int x^2 \sin x\,dx = -x^2 \cos x + 2(x \sin x + \cos x + C_1)$$
$$= -x^2 \cos x + 2x \sin x + 2 \cos x + C$$

Once in a while, integrating by parts *twice* will result in an integral *which, except for its coefficient, is identical to the original one.* When this happens, we merely have to move these identical integrals to the same side of the equation, and combine them algebraically.

**EXAMPLE 5:** Integrate $\int e^x \cos x\, dx$.

**Solution:** We choose

$$u = e^x \qquad \text{and} \qquad dv = \cos x\, dx$$

Then

$$du = e^x\, dx \qquad \text{and} \qquad v = \sin x$$

So

$$\int e^x \cos x\, dx = e^x \sin x - \int e^x \sin x\, dx \tag{1}$$

The integral here is similar to the original, except that now we have sine instead of cosine. Integrating by parts again should give us the cosine again. We let

$$u = e^x \qquad \text{and} \qquad dv = \sin x\, dx$$

Then

$$du = e^x\, dx \qquad \text{and} \qquad v = -\cos x$$

and

$$\int e^x \sin x\, dx = -e^x \cos x + \int e^x \cos x\, dx$$

The integral here is now the same as the original. Substituting back into (1) yields

$$\begin{aligned}\int e^x \cos x\, dx &= e^x \sin x - \left(-e^x \cos x + \int e^x \cos x\, dx\right)\\ &= e^x \sin x + e^x \cos x - \int e^x \cos x\, dx\end{aligned}$$

Moving the integral to the left side gives us

Two identical integrals can be added just like any other identical quantities: by combining coefficients.

$$2\int e^x \cos x\, dx = e^x \sin x + e^x \cos x$$

Finally, dividing by 2, we obtain

$$\int e^x \cos x\, dx = \frac{e^x}{2}(\sin x + \cos x) + C$$

## EXERCISE 1—INTEGRATION BY PARTS

Integrate by parts.

Integrate by parts for the practice, even though you find a rule that fits.

1. $\int x \sin x\, dx$
2. $\int x\sqrt{1-x}\, dx$
3. $\int x \sec^2 x\, dx$
4. $\int_0^2 x \sin \frac{x}{2}\, dx$
5. $\int x \cos x\, dx$
6. $\int x^2 \ln x\, dx$
7. $\int x \sin^2 3x\, dx$
8. $\int \cos^3 x\, dx$
9. $\int_1^3 \frac{\ln(x+1)\, dx}{\sqrt{x+1}}$
10. $\int \frac{x^3\, dx}{\sqrt{1-x^2}}$

11. $\int \frac{\ln(x + 1)\,dx}{(x + 1)^2}$

12. $\int x^3 \ln x\,dx$

13. $\int_0^4 xe^{2x}\,dx$

14. $\int x \tan^2 x\,dx$

15. $\int x^3\sqrt{1 - x^2}\,dx$

16. $\int \frac{x^2\,dx}{(1 + x^2)^2}$

17. $\int \frac{x^2\,dx}{(1 - x^2)^{3/2}}$

18. $\int_{-2}^{2} \frac{xe^x\,dx}{(1 + x)^2}$

19. $\int \cos x \ln \sin x\,dx$

Integrate by parts twice.

20. $\int x^2 e^{2x}\,dx$

21. $\int x^2 e^{-x}\,dx$

22. $\int x^2 e^x\,dx$

23. $\int e^x \sin x\,dx$

24. $\int x^2 \cos x\,dx$

25. $\int_1^3 e^{-x} \sin 4x\,dx$

26. $\int e^x \cos x\,dx$

27. $\int e^{-x} \cos \pi x\,dx$

## 34-2 INTEGRATING RATIONAL FRACTIONS

### Rational Algebraic Fractions

**Recall that in a polynomial, all the powers of *x* are positive integers.**

A rational algebraic fraction is one in which both numerator and denominator are polynomials.

---

**EXAMPLE 6:**

(a) $\frac{x^2}{x^3 - 2}$ is a *proper* rational fraction because the numerator is of lower degree than the denominator.

(b) $\frac{x^3 - 2}{x^2}$ is an *improper* rational fraction.

(c) $\frac{\sqrt{x}}{x^3 - 2}$ is *not* a rational fraction.

---

### Integrating Improper Rational Fractions

Perform the indicated division and integrate term by term.

---

**EXAMPLE 7:** Integrate $\int \frac{x^3 - 2x^2 - 5x - 2}{x + 1}\,dx$.

**Solution:** When we carry out the long division (work not shown), we get a quotient of $x^2 - 3x - 2$, so

$$\int \frac{x^3 - 2x^2 - 5x - 2}{x + 1}\,dx = \int (x^2 - 3x - 2)dx$$

$$= \frac{x^3}{3} - \frac{3x^2}{2} - 2x + C$$

---

| Common Error | It is easy to overlook the simple operation of dividing the numerator by the denominator of the given expression. Be sure to consider it whenever you have to integrate an algebraic fraction. |
|---|---|

Long division works fine if there is *no remainder,* as in Example 7. When there is a remainder, it will be a *proper* rational fraction. We now learn how to integrate such a proper rational fraction.

### Fractions with a Quadratic Denominator

We know that by *completing the square,* a quadratic trinomial $Ax^2 + Bx + C$ can be written in the form $u^2 \pm a^2$. This sometimes allows us to use Rules 56 to 60, as in the following example.

---

**EXAMPLE 8:** Integrate $\int \frac{4\,dx}{x^2 + 3x - 1}$.

**Solution:** We start by writing the denominator in the form $u^2 \pm a^2$ by completing the square.

$$x^2 + 3x - 1 = \left(x^2 + 3x + \frac{9}{4}\right) - \frac{9}{4} - 1$$

$$= \left(x + \frac{3}{2}\right)^2 - \frac{13}{4}$$

$$= \left(x + \frac{3}{2}\right)^2 - \left(\frac{\sqrt{13}}{2}\right)^2$$

Our integral is then

$$\int \frac{4\,dx}{\left(x + \frac{3}{2}\right)^2 - \left(\frac{\sqrt{13}}{2}\right)^2}$$

Using Rule 57,

$$\int \frac{du}{u^2 - a^2} = \frac{1}{2a} \ln \left|\frac{u - a}{u + a}\right| + C$$

with $u = x + 3/2$ and $a = \sqrt{13}/2$, we get

$$\int \frac{4\,dx}{x^2 + 3x - 1} = 4 \cdot \frac{1}{\frac{2\sqrt{13}}{2}} \ln \left|\frac{x + \frac{3}{2} - \frac{\sqrt{13}}{2}}{x + \frac{3}{2} + \frac{\sqrt{13}}{2}}\right| + C$$

or, in decimal form,

$$= 1.109 \ln \left|\frac{x - 0.303}{x + 3.303}\right| + C$$

---

If the numerator in Example 8 had contained an $x$ term, the integral would not have matched any rule in our table. Sometimes, however, the numerator can be separated into two parts, one of which is the derivative of the denominator.

**EXAMPLE 9:** Integrate

$$\int \frac{2x + 7}{x^2 + 3x - 1}\, dx$$

**Solution:** The derivative of the denominator is $2x + 3$. We can get this in the numerator by splitting the 7 into 3 and 4,

$$\int \frac{(2x + 3) + 4}{x^2 + 3x - 1}\, dx$$

We can now separate this integral into two,

$$\int \frac{2x + 3}{x^2 + 3x - 1}\, dx + \int \frac{4}{x^2 + 3x - 1}\, dx$$

The first of these two integrals is, by Rule 7,

$$\int \frac{2x + 3}{x^2 + 3x - 1}\, dx = \ln |x^2 + 3x - 1| + C_1$$

The second integral is the same as that in Example 8. Our complete integral is then

$$\int \frac{2x + 7}{x^2 + 3x - 1}\, dx = \ln |x^2 + 3x - 1| + 1.109 \ln \left|\frac{x - 0.303}{x + 3.303}\right| + C$$

## Partial Fractions

When studying algebra, we learned how to combine several fractions into a single fraction by writing all fractions with a common denominator, and then added or subtracted the numerators as indicated. Now we do the *reverse*. Given a proper rational fraction, we *separate it* into several simpler fractions whose sum is the original fraction. These simpler fractions are called *partial fractions*.

**We'll need partial fractions again when we study the Laplace transform.**

**EXAMPLE 10:** The sum of

$$\frac{2}{x + 1} \quad \text{and} \quad \frac{3}{x - 2}$$

is

$$\frac{2(x - 2) + 3(x + 1)}{(x + 1)(x - 2)} = \frac{5x - 1}{x^2 - x - 2}$$

Therefore,

$$\frac{2}{x + 1} \quad \text{and} \quad \frac{3}{x - 2}$$

are the partial fractions of

$$\frac{5x - 1}{x^2 - x - 2}$$

The *first step* in finding partial fractions is to *factor the denominator* of the given fraction to find the set of all possible factors. The remaining steps then depend on the nature of those factors.

1. For each linear factor $ax + b$ in the denominator, there will be a partial fraction $A/(ax + b)$.
2. For the repeated linear factors $(ax + b)^n$, there will be $n$ partial fractions,

$$\frac{A_1}{ax + b} + \frac{A_2}{(ax + b)^2} + \cdots + \frac{A_n}{(ax + b)^n}$$

3. For each quadratic factor $ax^2 + bx + c$, there will be a partial fraction $(Ax + B)/(ax^2 + bx + c)$.
4. For the repeated quadratic factors $(ax^2 + bx + c)^n$, there will be $n$ partial fractions,

$$\frac{A_1x + B_1}{ax^2 + bx + c} + \frac{A_2x + B_2}{(ax^2 + bx + c)^2} + \cdots + \frac{A_nx + B_n}{(ax^2 + bx + c)^n}$$

**The *A*'s and *B*'s are constants here.**

### Denominator with Nonrepeated Linear Factors

This is the simplest case, and the method of partial fractions is best shown by an example.

---

**EXAMPLE 11:** Separate $\frac{x + 2}{x^3 - x}$ into partial fractions.

**Solution:** We first factor the denominator into

$$x \qquad x + 1 \qquad x - 1$$

each a nonrepeated linear factor. Next we assume that each of these factors will be the denominator of a partial fraction, whose numerator is yet to be found, so let

$$\frac{x + 2}{x^3 - x} = \frac{A}{x} + \frac{B}{x + 1} + \frac{C}{x - 1} \tag{1}$$

where $A$, $B$, and $C$ are the unknown numerators. If our partial fractions are to be *proper,* the numerator of each must be of lower degree than the denominator. Since each denominator is of first degree, $A$, $B$, and $C$ *must be constants.*

We now clear fractions by multiplying both sides of (1) by $x^3 - x$,

$$x + 2 = A(x^2 - 1) + Bx(x - 1) + Cx(x + 1)$$

or

$$x + 2 = (A + B + C)x^2 + (-B + C)x - A$$

For this equation to be true, the coefficients of like powers of $x$ on both sides of the equation must be equal. The coefficient of $x$ is 1, on the left side, and $(-B + C)$ on the right side. Therefore,

$$1 = -B + C \tag{2}$$

Similarly for the $x^2$ term,

**This is called the method of undetermined coefficients.**

$$0 = A + B + C \tag{3}$$

and for the constant term,

$$2 = -A \tag{4}$$

Solving (2), (3), and (4) simultaneously gives

$$A = -2 \qquad B = \frac{1}{2} \qquad \text{and} \qquad C = \frac{3}{2}$$

You can check your work by recombining the partial fractions.

Substituting back into (1) gives us

$$\frac{x+2}{x^3-x} = -\frac{2}{x} + \frac{1}{2(x+1)} + \frac{3}{2(x-1)}$$

---

**EXAMPLE 12:** Integrate $\int \frac{x+2}{x^3-x}\,dx$.

**Solution:** From Example 11,

$$\int \frac{x+2}{x^3-x}\,dx = \int \left[-\frac{2}{x} + \frac{1}{2(x+1)} + \frac{3}{2(x-1)}\right] dx$$

$$= -2 \ln|x| + \frac{1}{2} \ln|x+1| + \frac{3}{2} \ln|x-1| + C$$

---

### Denominator with Repeated Linear Factors

We now consider fractions where a linear factor appears more than once in the denominator. Of course, there may also be one or more nonrepeated linear factors, as in the following example.

---

**EXAMPLE 13:** Separate the fraction $\frac{x^3+1}{x(x-1)^3}$ into partial fractions.

**Solution:** We assume that

$$\frac{x^3+1}{x(x-1)^3} = \frac{A}{x} + \frac{B}{(x-1)} + \frac{C}{(x-1)^2} + \frac{D}{(x-1)^3}$$

because $x(x-1)^3$ has $x$, $(x-1)$, $(x-1)^2$, and $(x-1)^3$ as its set of possible factors. Multiplying both sides by $x(x-1)^3$ yields

$$\begin{aligned} x^3 + 1 &= A(x-1)^3 + Bx(x-1)^2 + Cx(x-1) + Dx \\ &= Ax^3 - 3Ax^2 + 3Ax - A + Bx^3 - 2Bx^2 + Bx + Cx^2 - Cx + Dx \\ &= (A+B)x^3 + (-3A - 2B + C)x^2 + (3A + B - C + D)x - A \end{aligned}$$

Equating coefficients of like powers of $x$ gives

$$\begin{aligned} A + B &= 1 \\ -3A - 2B + C &= 0 \\ 3A + B - C + D &= 0 \\ -A &= 1 \end{aligned}$$

Solving this set of equations simultaneously gives

$$A = -1 \qquad B = 2 \qquad C = 1 \qquad D = 2$$

so

$$\frac{x^3+1}{x(x-1)^3} = -\frac{1}{x} + \frac{2}{x-1} + \frac{1}{(x-1)^2} + \frac{2}{(x-1)^3}$$

---

If a denominator has a factor $x^n$, treat it as having the linear factor $x$ repeated:

$$x(x)(x)(x)\ \ldots$$

**EXAMPLE 14:** Separate into partial fractions,

$$\frac{2}{s^3(3s-2)}$$

**Solution:** We have the linear factor $(3s - 2)$ and the repeated linear factors, $s$, $s$, and $s$. So

$$\frac{2}{s^3(3s-2)} = \frac{A}{3s-2} + \frac{B}{s} + \frac{C}{s^2} + \frac{D}{s^3}$$

Multiplying through by $s^3(3s - 2)$ gives

$$2 = As^3 + Bs^2(3s-2) + Cs(3s-2) + D(3s-2)$$

Removing parentheses and collecting terms, we obtain

$$2 = (A + 3B)s^3 + (-2B + 3C)s^2 + (3D - 2C)s - 2D$$

Equating coefficients gives

$$A + 3B = 0 \qquad (1)$$

$$3C - 2B = 0 \qquad (2)$$

$$3D - 2C = 0 \qquad (3)$$

$$-2D = 2 \qquad (4)$$

From (4) we get $D = -1$. Then, from (3),

$$2C = 3D = -3$$

or $C = -\frac{3}{2}$. Then, from (2),

$$2B = 3C = -\frac{9}{2}$$

or $B = -\frac{9}{4}$. Finally, from (1),

$$A = -3B = \frac{27}{4}$$

Our partial fraction is then

$$\frac{2}{s^3(3s-2)} = \frac{27}{4(3s-2)} - \frac{9}{4s} - \frac{3}{2s^2} - \frac{1}{s^3}$$

### Denominator with Quadratic Factors

We now deal with fractions whose denominators have one or more quadratic factors, in addition to any linear factors.

**EXAMPLE 15:** Separate the fraction $\dfrac{x+2}{x^3+3x^2-x}$ into partial fractions.

**Solution:** The factors of the denominator are $x$ and $x^2 + 3x - 1$, so we assume that

$$\frac{x+2}{(x^2+3x-1)x} = \frac{Ax+B}{x^2+3x-1} + \frac{C}{x}$$

Multiplying through by $(x^2 + 3x - 1)x$, we have

$$\begin{aligned} x + 2 &= Ax^2 + Bx + Cx^2 + 3Cx - C \\ &= (A + C)x^2 + (B + 3C)x - C \end{aligned}$$

Equating the coefficients of $x^2$ gives

$$A + C = 0$$

and of $x$,

$$B + 3C = 1$$

and the constant terms,

$$-C = 2$$

Solving simultaneously gives

$$A = 2 \qquad B = 7 \qquad C = -2$$

So our partial fractions are

$$\frac{x+2}{x^3+3x^2-x} = \frac{2x+7}{x^2+3x-1} - \frac{2}{x}$$

**EXAMPLE 16:** Integrate $\displaystyle\int \frac{x+2}{x^3+3x^2-x}\,dx$.

**Solution:** Using the results of Examples 15 and 9, we have

$$\begin{aligned} \int \frac{x+2}{x^3+3x^2-x} &= \int \frac{2x+7}{x^2+3x-1}\,dx - \int \frac{2}{x}\,dx \\ &= \ln|x^3 - 3x - 1| + 1.109 \ln\left|\frac{x-0.303}{x+3.303}\right| - 2\ln|x| + C \end{aligned}$$

## EXERCISE 2—INTEGRATING RATIONAL FRACTIONS

Integrate.

### Nonrepeated Linear Factors

1. $\displaystyle\int \frac{(4x-2)\,dx}{x^3-x^2-2x}$

2. $\displaystyle\int_2^3 \frac{(3x-1)\,dx}{x^3-x}$

3. $\displaystyle\int_1^3 \frac{(2-x^2)\,dx}{x^3+3x^2+2x}$

4. $\displaystyle\int \frac{x^2\,dx}{x^2-4x+5}$

5. $\displaystyle\int \frac{x+1}{x^3+2x^2-3x}\,dx$

**Repeated Linear Factors**

6. $\int_2^4 \frac{(x^3 - 2)\,dx}{x^3 - x^2}$

7. $\int_0^1 \frac{x - 5}{(x - 3)^2}\,dx$

8. $\int_2^3 \frac{x^2\,dx}{(x - 1)^3}$

9. $\int \frac{(2x - 5)\,dx}{(x - 2)^3}$

10. $\int \frac{dx}{x^3 + x^2}$

11. $\int_2^3 \frac{dx}{(x^2 + x)(x - 1)^2}$

12. $\int_0^2 \frac{x - 2}{(x + 1)^3}\,dx$

**Quadratic Factors**

13. $\int_1^2 \frac{dx}{x^4 + x^2}$

14. $\int_0^1 \frac{(x^2 - 3)\,dx}{(x + 2)(x^2 + 1)}$

15. $\int_1^2 \frac{(4x^2 + 6)\,dx}{x^3 + 3x}$

16. $\int_1^4 \frac{(5x^2 + 4)\,dx}{x^3 + 4x}$

17. $\int \frac{x^2\,dx}{1 - x^4}$

18. $\int_1^2 \frac{(x - 3)\,dx}{x^3 + x^2}$

19. $\int_0^1 \frac{5x\,dx}{(x + 2)(x^2 + 1)}$

20. $\int_3^4 \frac{(5x^2 - 4x)\,dx}{x^4 - 16}$

21. $\int \frac{x^5\,dx}{(x^2 + 4)^2}$

## 34-3 INTEGRATING BY ALGEBRAIC SUBSTITUTION

Expressions containing radicals are usually harder to integrate than those that do not. In this section we learn how to remove all radicals and fractional exponents from an expression by means of a suitable algebraic substitution.

**This is also called *integration by rationalization.***

### Expressions Containing Fractional Powers of *x*

If the expression to be integrated contains a factor of, say, $x^{1/3}$, we would substitute,

$$z = x^{1/3}$$

and the fractional exponents would vanish. If the same problem should also contain $x^{1/2}$, for example, we would substitute

$$z = x^{1/6}$$

because 6 is the lowest common denominator of the fractional exponents.

---

**EXAMPLE 17:** Integrate $\int \frac{x^{1/2}\,dx}{1 + x^{3/4}}$.

**Solution:** Here $x$ is raised to the $\frac{1}{2}$ power in one place, and to the $\frac{3}{4}$ power in another. The LCD of $\frac{1}{2}$ and $\frac{3}{4}$ is 4, so we substitute

$$z = x^{1/4}$$

So

$$x^{1/2} = (x^{1/4})^2 = z^2$$

and

$$x^{3/4} = (x^{1/4})^3 = z^3$$

and since

$$x = z^4$$

$$dx = 4z^3\,dz$$

Substituting into our given integral gives us

$$\int \frac{x^{1/2}\,dx}{1 + x^{3/4}} = \int \frac{z^2}{1 + z^3}(4z^3\,dz)$$

$$= 4\int \frac{z^5}{1 + z^3}\,dz$$

The integrand is now a rational fraction, such as those we studied in Sec. 34-2. As before, we divide numerator by denominator:

$$4\int \frac{z^5}{1 + z^3}dx = 4\int \left(z^2 - \frac{z^2}{1 + z^3}\right) dz$$

$$= \frac{4z^3}{3} - \frac{4}{3}\ln|1 + z^3| + C$$

Finally, we substitute back, $z = x^{1/4}$.

$$\int \frac{x^{1/2}\,dx}{1 + x^{3/4}} = \frac{4x^{3/4}}{3} - \frac{4}{3}\ln|1 + x^{3/4}| + C$$

---

### Expressions Containing Fractional Powers of a Binomial

The procedure here is very similar to that of the preceding case, where only $x$ was raised to a fractional power.

---

**EXAMPLE 18:** Integrate $\int x\sqrt{3 + 2x}\,dx$.

**Solution:** Let $z = \sqrt{3 + 2x}$. Then

$$x = \frac{z^2 - 3}{2}$$

So

$$dx = \frac{1}{2}(2z)\,dz = z\,dz$$

Substituting, we get

$$\int x\sqrt{3 + 2x}\,dx = \int \frac{z^2 - 3}{2} \cdot z \cdot z\,dz$$

$$= \frac{1}{2}\int (z^4 - 3z^2)\,dz$$

$$= \frac{z^5}{10} - \frac{3z^3}{6} + C$$

Now substituting $\sqrt{3 + 2x}$ for $z$, we obtain

$$\int x\sqrt{3 + 2x}\,dz = \frac{1}{10}(3 + 2x)^{5/2} - \frac{1}{2}(3 + 2x)^{3/2} + C$$

---

## Definite Integrals

When using the method of substitution with a definite integral, we may also substitute in the limits of integration. This eliminates the need to substitute back to the original variable.

---

**EXAMPLE 19:** Evaluate $\int_0^3 \frac{x\,dx}{\sqrt{1+x}}$.

**Solution:** We let $z = \sqrt{1+x}$. So

$$x = z^2 - 1$$

and

$$dx = 2z\,dz$$

We now compute the limits on $z$.
When $x = 0$,

$$z = \sqrt{1+0} = 1$$

and when $x = 3$,

$$z = \sqrt{1+3} = 2$$

Making the substitution yields

$$\int_0^3 \frac{x\,dx}{\sqrt{1+x}} = \int_1^2 \frac{z^2-1}{z} \cdot 2z\,dz$$

$$= 2\int_1^2 (z^2-1)\,dz = \left[\frac{2z^3}{3} - 2z\right]_1^2 = \frac{8}{3}$$

---

## EXERCISE 3—INTEGRATING BY ALGEBRAIC FRACTIONS

Integrate.

1. $\int \frac{dx}{1+\sqrt{x}}$
2. $\int \frac{dx}{x-\sqrt{x}}$
3. $\int \frac{dx}{x-\sqrt[3]{x}}$
4. $\int \frac{dx}{\sqrt{x}+\sqrt[4]{x^3}}$
5. $\int \frac{x\,dx}{\sqrt[3]{1+x}}$
6. $\int \frac{x\,dx}{\sqrt{x-1}}$
7. $\int \frac{x\,dx}{\sqrt{2-7x}}$
8. $\int \frac{\sqrt{x^2-1}\,dx}{x}$
9. $\int \frac{x^2\,dx}{(4x+1)^{5/2}}$
10. $\int \frac{x\,dx}{(1+x)^{3/2}}$

11. $\int \frac{(x + 5)\,dx}{(x + 4)\sqrt{x + 2}}$

12. $\int \frac{dx}{x^{5/8} + x^{3/4}}$

13. $\int \frac{dx}{x\sqrt{1 - x^2}}$

14. $\int \frac{(x^{3/2} - x^{1/3})\,dx}{6x^{1/4}}$

15. $\int x\sqrt[3]{1 + x}\,dx$

Evaluate.

16. $\int_0^3 \frac{dx}{(x + 2)\sqrt{x + 1}}$

17. $\int_0^1 \frac{x^{3/2}\,dx}{x + 1}$

18. $\int_0^4 \frac{dx}{1 + \sqrt{x}}$

19. $\int_0^{1/2} \frac{dx}{\sqrt{2x}(9 + \sqrt[3]{2x})}$

20. $\int_1^4 \frac{x\,dx}{\sqrt{4x + 2}}$

## 34-4 INTEGRATING BY TRIGONOMETRIC SUBSTITUTION

The two legs $a$ and $b$ of a right triangle are related to the hypotenuse $c$ by the familiar Pythagorean theorem

$$a^2 + b^2 = c^2$$

Thus when we have an expression of the form $\sqrt{u^2 + a^2}$, we can consider it as the hypotenuse of a right triangle whose legs are $u$ and $a$. Doing so enables us to integrate expressions involving the sum or difference of two squares, especially when they are under a radical sign. A similar substitution will work for other radical expressions, as given in the following table.

| Expression | Substitution |
|---|---|
| $\sqrt{u^2 + a^2}$ | let $u = a \tan \theta$ |
| $\sqrt{a^2 - u^2}$ | let $u = a \sin \theta$ |
| $\sqrt{u^2 - a^2}$ | let $u = a \sec \theta$ |

These are not the only possible substitutions, but they work well, as you will see in the following examples.

**EXAMPLE 20:** Integrate $\int \frac{dx}{\sqrt{x^2 + 9}}$.

**Solution:** Our substitution, from the table above, is

$$x = 3 \tan \theta \tag{1}$$

We sketch a right triangle (Fig. 34-1) with $x$ as the leg opposite to $\theta$ and 3 as the leg adjacent to $\theta$. The hypotenuse is then $\sqrt{x^2 + 9}$.

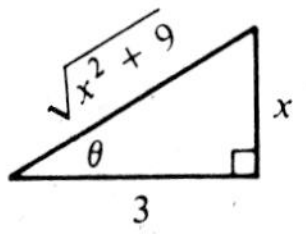


**FIGURE 34-1**

We now write our integral in terms of $\theta$. Since $\cos \theta = 3/\sqrt{x^2 + 9}$,

$$\sqrt{x^2 + 9} = \frac{3}{\cos \theta} = 3 \sec \theta \tag{2}$$

For the numerator, we find $dx$ by taking the derivative of (1).

$$dx = 3 \sec^2 \theta \, d\theta \tag{3}$$

Substituting (2) and (3) into our original integral, we obtain

$$\int \frac{dx}{\sqrt{x^2 + 9}} = \int \frac{3 \sec^2 \theta \, d\theta}{3 \sec \theta}$$

$$= \int \sec \theta \, d\theta$$

By Rule 14:

$$= \ln |\sec \theta + \tan \theta| + C_1$$

We now return to our original variable, $x$. From Fig. 34-1,

$$\sec \theta = \frac{\sqrt{x^2 + 9}}{3} \quad \text{and} \quad \tan \theta = \frac{x}{3}$$

So

$$\int \frac{dx}{\sqrt{x^2 + 9}} = \ln \left| \frac{\sqrt{x^2 + 9}}{3} + \frac{x}{3} \right| + C_1$$

$$= \ln |\sqrt{x^2 + 9} + x| - \ln 3 + C_1$$

$$= \ln |\sqrt{x^2 + 9} + x| + C$$

where $C_1$ and $-\ln 3$ have been combined into a single constant $C$.

**EXAMPLE 21:** Integrate $\int \sqrt{9 - 4x^2}\, dx$.

**Solution:** Our radical is of the form $\sqrt{a^2 - u^2}$, with $a = 3$ and $u = 2x$. Our substitution is then

$$2x = 3 \sin \theta$$

or

$$x = \frac{3}{2}\sin \theta$$

Then

$$dx = \frac{3}{2}\cos \theta \, d\theta$$

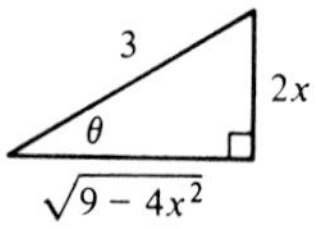

**FIGURE 34-2**

We sketch a right triangle with hypotenuse 3 and one leg $2x$, as in Fig. 34-2. Then

$$\cos \theta = \frac{\sqrt{9 - 4x^2}}{3}$$

so

$$\sqrt{9 - 4x^2} = 3 \cos \theta$$

Substituting into our given integral gives us

$$\int \sqrt{9 - 4x^2}\, dx = \int (3 \cos \theta)\left(\frac{3}{2} \cos \theta \, d\theta\right)$$

$$= \frac{9}{2} \int \cos^2 \theta \, d\theta$$

By Rule 17:

$$= \frac{9}{2}\left(\frac{\theta}{2} + \frac{\sin 2\theta}{4}\right) + C$$

We now use Eq. 170 to replace $\sin 2\theta$ with $2 \sin \theta \cos \theta$,

$$\int \sqrt{9 - 4x^2}\, dx = \frac{9}{4}(\theta + \sin \theta \cos \theta) + C$$

We now substitute back to return to the variable $x$. From Fig. 34-2 we see that

$$\sin \theta = \frac{2x}{3}$$

and that

$$\theta = \sin^{-1}\frac{2x}{3}$$

and

$$\cos \theta = \frac{\sqrt{9 - 4x^2}}{3}$$

So

$$\int \sqrt{9 - 4x^2}\, dx = \frac{9}{4}\left(\sin^{-1}\frac{2x}{3} + \frac{2x}{3} \cdot \frac{\sqrt{9 - 4x^2}}{3}\right) + C$$

$$= \frac{9}{4}\left(\sin^{-1}\frac{2x}{3} + \frac{2x\sqrt{9 - 4x^2}}{9}\right) + C$$

## Definite Integrals

When evaluating a *definite* integral, we may use the limits on $x$ after substituting back into $x$, or we may *change the limits* when we change variables, as shown in the following example.

---

**EXAMPLE 22:** Integrate $\int_2^4 \sqrt{x^2 - 4}\, dx$.

**Solution:** We substitute $x = 2 \sec \theta$, so

$$dx = 2 \sec \theta \tan \theta \, d\theta$$

We sketch the triangle (Fig. 34-3) from which

$$\sqrt{x^2 - 4} = 2 \tan \theta$$

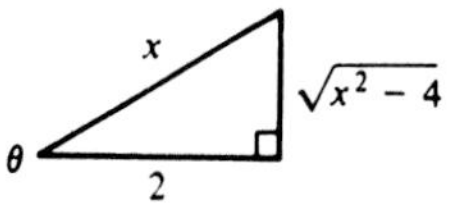


FIGURE 34-3

Our integral is, then,

$$\int_2^4 \sqrt{x^2 - 4}\, dx = \int (2 \tan \theta)(2 \sec \theta \tan \theta \, d\theta)$$

$$= 4 \int \tan^2\theta \sec \theta \, d\theta$$

$$= 4 \int (\sec^2\theta - 1) \sec \theta \, d\theta$$

$$= 4 \int (\sec^3\theta - \sec \theta) \, d\theta$$

which we can integrate by Rules 14 and 26. But what about limits? What happens to $\theta$ when $x$ goes from 2 to 4? From Fig. 34-3 we see that

$$\cos \theta = \frac{2}{x}$$

| when $x = 2$: | when $x = 4$: |
|---|---|
| $\cos \theta = 1$ | $\cos \theta = \frac{1}{2}$ |
| $\theta = 0°$ | $\theta = 60°$ |

**Note that $\theta$ is always acute, so when determining $\theta$ we need only consider angles less than 90°.**

So our integral is then

$$\int_{x=2}^{x=4} \sqrt{x^2 - 4}\, dx = 4 \int_{\theta=0°}^{\theta=60°} (\sec^3\theta - \sec \theta) \, d\theta$$

By Rules 14 and 26,

$$= 4\left[\frac{1}{2} \sec \theta \tan \theta + \frac{1}{2} |\ln \sec \theta + \tan \theta| - \ln|\sec \theta + \tan \theta|\right]_{0°}^{60°}$$

$$= 2\left[\sec \theta \tan \theta - \ln|\sec \theta + \tan \theta|\right]_{0°}^{60°}$$

$$= 2[2(1.732) - \ln|2 + 1.732|] - 2[0 - \ln|1 + 0|]$$

$$= 4.294$$

---

## EXERCISE 4—INTEGRATING BY TRIGONOMETRIC SUBSTITUTION

Integrate.

1. $\int \frac{5\,dx}{(5-x^2)^{3/2}}$

2. $\int \frac{\sqrt{9-x^2}\,dx}{x^4}$

3. $\int \frac{dx}{x\sqrt{x^2+4}}$

4. $\int_3^4 \frac{dx}{x^3\sqrt{x^2-9}}$

5. $\int \frac{dx}{x^2\sqrt{4-x^2}}$

6. $\int \frac{dx}{(x^2+2)^{3/2}}$

7. $\int \frac{x^2\,dx}{\sqrt{4-x^2}}$

8. $\int_1^5 \frac{dx}{x\sqrt{25-x^2}}$

9. $\int \frac{\sqrt{16-x^2}\,dx}{x^2}$

10. $\int \frac{dx}{\sqrt{x^2+2}}$

11. $\int_3^6 \frac{x^2\,dx}{\sqrt{x^2-6}}$

12. $\int_3^5 \frac{x^2\,dx}{(x^2+8)^{3/2}}$

13. $\int \frac{dx}{x^2\sqrt{x^2-7}}$

## 34-5 IMPROPER INTEGRALS

Up to now, we have assumed that the limits in the definite integral

$$\int_a^b f(x)\,dx \qquad (1)$$

were finite, and that the integrand $f(x)$ was continuous for all values of $x$ in the interval $a \le x \le b$. If either or both of these conditions is not met, the integral is called an *improper integral*.

Thus an integral is called improper if

1. one or both limits is infinite, or
2. the integrand is discontinuous at some $x$ within the interval $a \le x \le b$.

---

**EXAMPLE 23:**

(a) The integral

$$\int_2^\infty x^2\,dx$$

is improper because one limit is infinite.

(b) The integral

$$\int_0^5 \frac{x^2}{x-3}\,dx$$

is improper because the integrand, $f(x) = x^2/(x-3)$, is discontinuous at $x = 3$.

---

We now treat each type separately.

### Infinite Limit

The value of a definite integral with one or both limits infinite is defined by the following relationships:

$$\int_a^{\infty} f(x)\,dx = \lim_{b\to\infty} \int_a^b f(x)\,dx \tag{2}$$

$$\int_{-\infty}^{b} f(x)\,dx = \lim_{a\to-\infty} \int_a^b f(x)\,dx \tag{3}$$

$$\int_{-\infty}^{\infty} f(x)\,dx = \lim_{\substack{b\to\infty \\ a\to-\infty}} \int_a^b f(x)\,dx \tag{4}$$

When the limit has a finite value, we say that the integral *converges*, or *exists*. When the limit is infinite, we say that the integral *diverges*, or *does not exist*.

---

**EXAMPLE 24:** Does the improper integral

$$\int_1^{\infty} \frac{dx}{x}$$

converge or diverge?

**Solution:** From (2),

$$\int_1^{\infty} \frac{dx}{x} = \lim_{b\to\infty} \int_1^b \frac{dx}{x}$$

$$= \lim_{b\to\infty} \Big[\ln x\Big]_1^b$$

$$= \lim_{b\to\infty} [\ln b] = \infty$$

Since there is no limiting value, we say that the given integral diverges, or does not exist.

---

**EXAMPLE 25:** Does the following integral converge or diverge?

**Solution:**

$$\int_{-\infty}^{0} e^x\,dx = \lim_{a\to-\infty} \Big[e^x\Big]_a^0$$

$$= \lim_{a\to-\infty} [1 - e^a]$$

$$= 1$$

Thus the given limit converges and has a value of 1.

---

**EXAMPLE 26:** Find the area under the curve $y = 1/x^2$ (Fig. 34-4) from $x = 2$ to infinity.

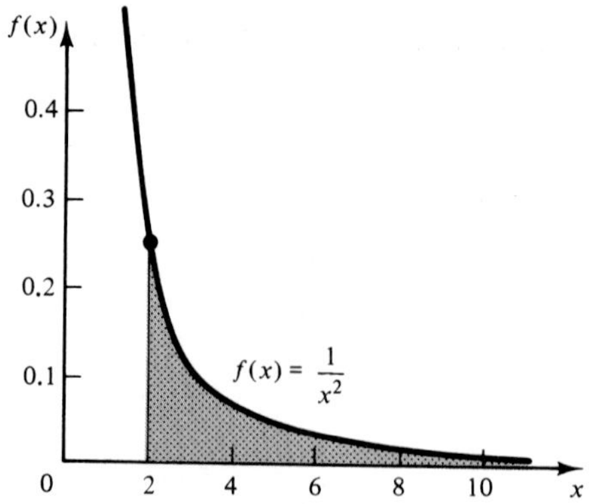


**FIGURE 34-4**

**Solution:**

$$A = \int_2^{\infty} \frac{dx}{x^2} = \lim_{b\to\infty} \int_2^b \frac{dx}{x^2}$$

$$= \lim_{b\to\infty} \left| -\frac{1}{x} \right|_2^b$$

$$= \lim_{b\to\infty} \left| -\frac{1}{b} + \frac{1}{2} \right|$$

$$= \frac{1}{2}$$

Thus the area under the curve is not infinite, as might be expected, but has a finite value of 1/2.

In many cases we need not carry out the limiting process as in Example 26, but can treat the infinite limit as if it were a very large number. Thus, repeating Example 26, we have

$$\int_2^{\infty} \frac{dx}{x^2} = \left| -\frac{1}{x} \right|_2^{\infty} = \left| -\frac{1}{\infty} + \frac{1}{2} \right| = \frac{1}{2}$$

## Discontinuous Integrand

We now cover the case where the integrand is discontinuous at one of the limits. If the integrand $f(x)$ in (1) is continuous for all values of $x$ in the interval $a \le x \le b$ except at $b$, then the integral (1) is defined by

$$\int_a^b f(x)\,dx = \lim_{x\to b^-} \int_a^x f(x)\,dx \qquad (5)$$

where the notation $x \to b^-$ means that $x$ approaches $b$ from values that are less than $b$. Similarly, if the integrand is discontinuous only at the lower limit $a$, we have

$$\int_a^b f(x)\,dx = \lim_{x\to a^+} \int_x^b f(x)\,dx \qquad (6)$$

where $x \to a^+$ means that $x$ approaches $a$ from values greater than $a$.

**EXAMPLE 27:** Evaluate the improper integral $\int_0^4 \frac{dx}{\sqrt{x}}$.

**Solution:** Here our function $f(x) = 1/\sqrt{x}$ is discontinuous at the lower limit, 0. Thus from (6),

$$\int_0^4 \frac{dx}{\sqrt{x}} = \lim_{x \to 0^+} \int_x^4 \frac{dx}{\sqrt{x}} = \lim_{x \to 0^+} \left[2\sqrt{x}\right]_x^4$$
$$= 4 - \lim_{x \to 0^+} 2\sqrt{x} = 4$$

## EXERCISE 5—IMPROPER INTEGRALS

Evaluate each improper integral that converges.

### Infinite Limit

1. $\int_1^\infty \frac{dx}{x^3}$
2. $\int_3^\infty \frac{dx}{(x-2)^3}$
3. $\int_{-\infty}^1 e^x\, dx$
4. $\int_5^\infty \frac{dx}{\sqrt{x-1}}$
5. $\int_{-\infty}^\infty \frac{dx}{1+x^2}$
6. $\int_0^\infty \frac{dx}{(x^2+4)^{3/2}}$

### Discontinuous Integrand

7. $\int_0^1 \frac{dx}{\sqrt{1-x}}$
8. $\int_0^1 \frac{dx}{x^3}$
9. $\int_0^{\pi/2} \tan x\, dx$
10. $\int_1^2 \frac{dx}{x\sqrt{x^2-1}}$
11. $\int_{-1}^1 \frac{dx}{\sqrt{1-x^2}}$
12. $\int_1^2 \frac{dx}{(x-1)^{2/3}}$

## 34-6 APPROXIMATE INTEGRATION

### The Need for Approximate Methods

We have studied many rules and methods for evaluating a definite integral. But unfortunately many functions cannot be integrated by these methods. Further, all our methods so far require that the function to be integrated be known in the form of an equation. But in practice, our function may simply be described in a graph, or perhaps a table of point pairs. In these cases we need a method to approximate the integration.

All our approximate methods are based on the fact that the definite integral can be interpreted as the area bounded by the function and the $x$ axis, between the given limits. Thus, even though the problem we are solving may have nothing to do with area (we could be finding work, or a volume, for example), we attack it by finding the area under a curve.

There are many methods for finding the area under a curve, and even mechanical devices such as the planimeter for doing the same. We have already described the *midpoint method* in Sec. 32-2 and here we cover the *average ordinate method*, the *trapezoid rule*, *prismoidal formula*, and *Simpson's rule*.

### Average Ordinate Method

In Sec. 33-4 we showed how to compute the average ordinate, by dividing the area under a curve by the total interval $b - a$.

| | |
|---|---|
| $$y_{avg} = \frac{1}{b-a}\int_a^b f(x)\,dx$$ | **404** |

Here we reverse the procedure. We compute the average ordinate (simply by averaging the ordinates of points on the curve), and use it to compute the approximate area. Thus

| | | |
|---|---|---|
| Average Ordinate Method | $A \simeq y_{avg}(b-a)$ | **372** |

---

**EXAMPLE 28:** A curve passes through the points

**Notice that the data points do not have to be equally spaced, although accuracy is usually better if they are.**

| $x$ | 1 | 2 | 4 | 5 | 7 | 7.5 | 9 |
|---|---|---|---|---|---|---|---|
| $y$ | 3.51 | 4.23 | 6.29 | 7.81 | 7.96 | 8.91 | 9.25 |

Find the area under the curve.

**Solution:** The average of the seven ordinates is

$$y_{avg} \simeq \frac{3.51 + 4.23 + 6.29 + 7.81 + 7.96 + 8.91 + 9.25}{7} = 6.85$$

So $A \simeq 6.85(9-1) = 54.8$ square units.

---

## Trapezoid Rule

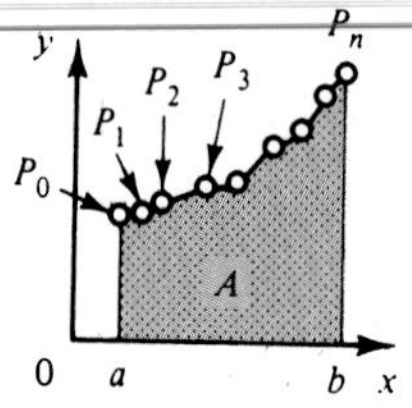


FIGURE 34-5

Let us suppose that our function is in the form of a table of point pairs, and that we plot each of these points as in Fig. 34-5. Joining these points with a smooth curve would give us a graph of the function, and it is the area under this curve that we seek.

We subdivide this area into a number of vertical *panels* by connecting the points with straight lines, and by dropping a perpendicular from each point to the $x$ axis (Fig. 34-6). Each panel then has the shape of a *trapezoid*—hence the name of the rule. If our first data point has the coordinates $(x_0, y_0)$ and the second point is $(x_1, y_1)$, the area of the first panel is, by Eq. 111,

$$\frac{1}{2}(x_1 - x_0)(y_1 + y_0)$$

and the sum of all the panels gives us the approximate area under the curve:

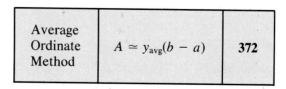

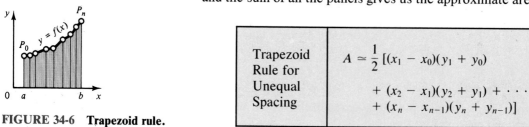

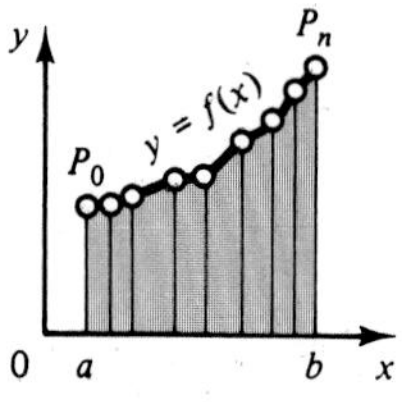


FIGURE 34-6 **Trapezoid rule.**

| | | |
|---|---|---|
| Trapezoid Rule for Unequal Spacing | $$A \simeq \frac{1}{2}[(x_1 - x_0)(y_1 + y_0) + (x_2 - x_1)(y_2 + y_1) + \cdots + (x_n - x_{n-1})(y_n + y_{n-1})]$$ | **373** |

If the points are equally spaced in the horizontal direction at a distance of

$$h = x_1 - x_0 = x_2 - x_1 = \cdots = x_n - x_{n-1}$$

Equation 373 reduces to:

| | | |
|---|---|---|
| Trapezoid Rule for Equal Spacing | $A \simeq h\left[\frac{1}{2}(y_0 + y_n) + y_1 + y_2 + \cdots + y_{n-1}\right]$ | **374** |

---

**EXAMPLE 29:** Find the area under the graph of the function given by the following table of ordered pairs:

| $x$ | 0 | 1 | 2 | 3 | 4 | 5 | 6 | 7 | 8 | 9 | 10 |
|---|---|---|---|---|---|---|---|---|---|---|---|
| $y$ | 4 | 3 | 4 | 7 | 12 | 19 | 28 | 39 | 52 | 67 | 84 |

**Solution:** By Eq. 374 with $h = 1$,

$$A \simeq 1\left[\frac{1}{2}(4 + 84) + 3 + 4 + 7 + 12 + 19 + 28 + 39 + 52 + 67\right] = 275$$

As a check, we note that the points given are actually points on the curve $y = x^2 - 2x + 4$. Integrating gives us

$$\int_0^{10} (x^2 - 2x + 4)\, dx = \frac{x^3}{3} - x^2 + 4x\Big|_0^{10} = 273\frac{1}{3}$$

Our approximate area thus agrees with the exact value within about 0.6%.

---

When the function is given in the form of an equation or a graph, we are free to select as many points as we want, for use in Eq. 374. The more points selected, the greater will be the accuracy, and the greater will be the labor.

---

**EXAMPLE 30:** Evaluate the integral

$$\int_0^{10} \left(\frac{x^2}{10} + 2\right) dx$$

taking points spaced 1 unit apart.

**Solution:** Making a table of point pairs for the function

$$y = \frac{x^2}{10} + 2$$

| $x$ | 0 | 1 | 2 | 3 | 4 | 5 | 6 | 7 | 8 | 9 | 10 |
|---|---|---|---|---|---|---|---|---|---|---|---|
| $y$ | 2 | 2.1 | 2.4 | 2.9 | 3.6 | 4.5 | 5.6 | 6.9 | 8.4 | 11.1 | 12 |

sum = 47.5 (of the $y$ values from 2.1 to 11.1)

Then Eq. 374,

$$\int_0^{10} \left(\frac{x^2}{10} + 2\right) dx \simeq 1\left[\frac{1}{2}(2 + 12) + 47.5\right] \simeq 54.5$$

---

## Prismoidal Formula

The prismoidal formula gives us the area under a curve that is a parabola if we are given three equally spaced points on the curve. Let us find the area under a parabola (Fig. 34-7a). Since the area does not depend on the location of the origin, let us place the $y$ axis halfway between $a$ and $b$ (Fig. 34-7b) so that it divides our area into two panels of width $h$. Let $y_0$, $y_1$, and $y_2$ be the heights of the curve at $x = -h$, 0, and $h$, respectively.

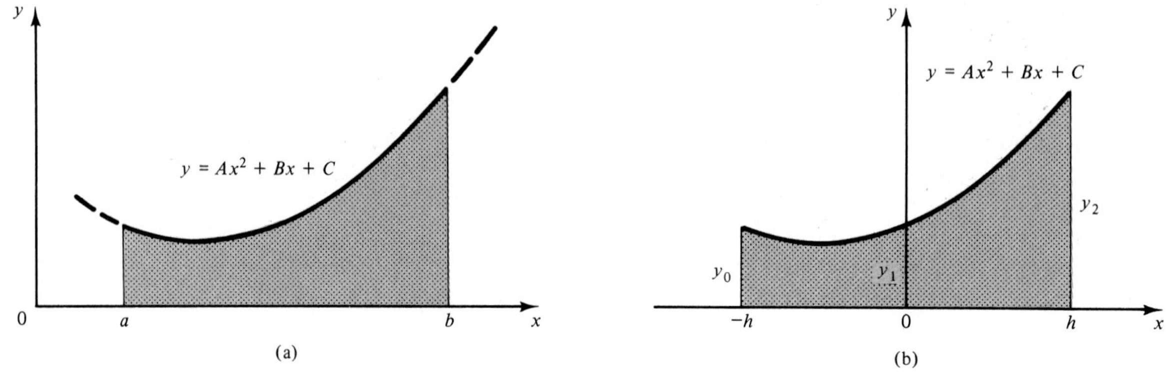


**FIGURE 34-7 Areas by prismoidal formula.**

Then

$$\text{Area} = \int_{-h}^{h} y\,dx = \int_{-h}^{h} (Ax^2 + Bx + C)\,dx = \frac{Ax^3}{3} + \frac{Bx^2}{2} + Cx\Big|_{-h}^{h}$$

$$= \frac{2}{3}Ah^3 + 2Ch = \frac{h}{3}(2Ah^2 + 6C) \qquad (1)$$

We now get $A$ and $C$ in terms of $y_0$, $y_1$, and $y_2$ by substituting coordinates of the three points on the curve, one by one, into the equation of the parabola, $y = Ax^2 + Bx + C$:

$$y(-h) = y_0 = A(-h)^2 + B(-h) + C = Ah^2 - Bh + C$$

$$y(0) = y_1 = A(0) + B(0) + C = C$$

$$y(h) = y_2 = Ah^2 + Bh + C$$

Adding $y_0$ and $y_2$ gives

$$y_0 + y_2 = 2Ah^2 + 2C = 2Ah^2 + 2y_1$$

from which $A = (y_0 - 2y_1 + y_2)/2h^2$. Substituting into (1) yields

$$\text{Area} = \frac{h}{3}(2Ah^2 + 6C) = \frac{h}{3}\left(2h^2\,\frac{y_0 - 2y_1 + y_2}{2h_2} + 6y_1\right)$$

or

**Our main use for the prismoidal formula will be in the derivation of Simpson's rule.**

| | | |
|---|---|---|
| Prismoidal Formula | $\text{Area} = \frac{h}{3}(y_0 + 4y_1 + y_2)$ | **375** |

**EXAMPLE 31:** A curve passes through the points (1, 2.05), (3, 5.83), and (5, 17.9). Assuming the curve to be a parabola, find the area bounded by the curve and the $x$ axis, from $x = 1$ to 5.

**Solution:** Substituting into the prismoidal formula with $h = 2$, $y_0 = 2.05$, $y_1 = 5.83$, and $y_2 = 17.9$,

$$A \simeq \frac{2}{3}[2.05 + 4(5.83) + 17.9] = 28.8 \text{ square units}$$

### Simpson's Rule

The prismoidal formula gives us the area under a curve that is a parabola, or *the approximate area under a curve that is assumed to be a parabola,* if we know (or can find) the coordinates of three equally spaced points on the curve. When there are more than three points on the curve, as in Fig. 34-5, we apply the prismoidal formula to each group of three, in turn.

Named for the English mathematician Thomas Simpson (1710–1761).

Thus the area of each pair of panels is

Panels 1 and 2: $\dfrac{h(y_0 + 4y_1 + y_2)}{3}$

Panels 3 and 4: $\dfrac{h(y_2 + 4y_3 + y_4)}{3}$

⋮

Panels $n - 1$ and $n$: $\dfrac{h(y_{n-2} + 4y_{n-1} + y_n)}{3}$

Adding these areas gives the approximate area under the entire curve, assuming it to be made up of parabolic segments.

| | | |
|---|---|---|
| Simpson's Rule | $A \simeq \frac{h}{3}(y_0 + 4y_1 + 2y_2 + 4y_3 + \cdots + 4y_{n-1} + y_n)$ | **376** |

**EXAMPLE 32:** A curve passes through the following data points.

| $x$ | 0 | 1 | 2 | 3 | 4 | 5 | 6 | 7 | 8 | 9 | 10 |
|---|---|---|---|---|---|---|---|---|---|---|---|
| $y$ | 4.02 | 5.23 | 5.66 | 6.05 | 5.81 | 5.62 | 5.53 | 5.71 | 6.32 | 7.55 | 8.91 |

If we had an equation rather than a table of data, we would start by choosing an even number of intervals and then make a table of point pairs.

Find the approximate area under the curve using Simpson's rule.

**Solution:** Substituting, with $h = 1$, we obtain

$$A \simeq \frac{4.02 + 4(5.23) + 2(5.66) + 4(6.05) + 2(5.81) + 4(5.62) + 2(5.53) + 4(5.71) + 2(6.32) + 4(7.55) + 8.91}{3}$$

$$= 60.1 \text{ square units}$$

## EXERCISE 6—APPROXIMATE INTEGRATION

Evaluate each integral using either the average ordinate method, the trapezoid rule, or Simpson's rule. Choose 10 panels of equal width. Check your answer by integrating.

1. $\int_1^3 \frac{1 + 2x}{x + x^2}\, dx$

2. $\int_1^{10} \frac{dx}{x}$

3. $\int_1^8 (4x^{1/3} + 3)\, dx$

4. $\int_0^{\pi} \sin x\, dx$

Find the area bounded by each curve and the $x$ axis, by any method. Use 10 panels.

5. $y = \frac{1}{\sqrt{x}}$ from $x = 4$ to 9

6. $y = \frac{2x}{1 + x^2}$ from $x = 2$ to 3

7. $y = x\sqrt{1 - x^2}$ from $x = 0$ to 1

8. $y = \frac{x + 6}{\sqrt{x + 4}}$ from $x = 0$ to 5

Find the area bounded by the given data points and the $x$ axis. Use any method.

9.

| $x$ | 0 | 1 | 2 | 3 | 4 | 5 | 6 | 7 | 8 | 9 | 10 |
|---|---|---|---|---|---|---|---|---|---|---|---|
| $y$ | 415 | 593 | 577 | 615 | 511 | 552 | 559 | 541 | 612 | 745 | 893 |

10.

| $x$ | 0 | 1 | 2 | 3 | 4 | 5 | 6 | 7 | 8 | 9 | 10 |
|---|---|---|---|---|---|---|---|---|---|---|---|
| $y$ | 3.02 | 4.63 | 4.76 | 5.08 | 6.31 | 6.60 | 6.23 | 6.48 | 7.27 | 8.93 | 9.11 |

11.

| $x$ | 0 | 1 | 2 | 3 | 4 | 5 | 6 | 7 | 8 | 9 | 10 |
|---|---|---|---|---|---|---|---|---|---|---|---|
| $y$ | 24.0 | 25.2 | 25.6 | 26.0 | 25.8 | 25.6 | 25.5 | 25.7 | 26.3 | 27.5 | 28.9 |

12.

| $x$ | 0 | 1 | 2 | 3 | 4 | 5 | 6 | 7 | 8 | 9 | 10 |
|---|---|---|---|---|---|---|---|---|---|---|---|
| $y$ | 462 | 533 | 576 | 625 | 591 | 522 | 563 | 511 | 602 | 745 | 821 |

### Applications

13. A ship's deck is measured every 30.0 ft, and has the following widths, in feet:

5.5 34.0 53.5 57.2 61.0 59.4 59.0 56.8 45.2 12.0

Find the area of the deck.

14. The dimensions, in inches, of a streamlined strut are shown in Fig. 34-8. Find the cross-sectional area.

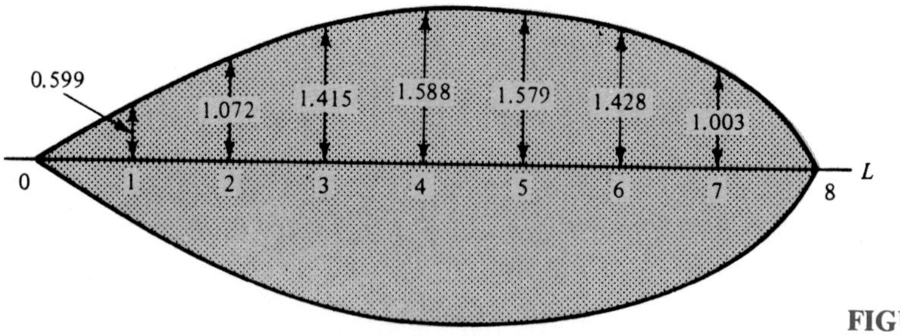


FIGURE 34-8

15. The cross-sectional areas $A$ of a certain ship at various depths $D$ below the waterline are:

| $D(ft)$ | 0 | 3 | 6 | 9 | 12 | 15 | 18 |
|---|---|---|---|---|---|---|---|
| $A(ft^2)$ | 7500 | 7150 | 6640 | 5680 | 4225 | 2430 | 260 |

Find the volume of the ship's hull below the waterline.

16. The pressure and volume of an expanding gas in a cylinder are measured and found to be

| *Volume* ($in^3$) | 20 | 40 | 60 | 80 | 100 |
|---|---|---|---|---|---|
| *Pressure* ($lb/in^2$) | 68.7 | 31.4 | 19.8 | 14.3 | 11.3 |

Find the work done by the gas by finding the area under the $p$–$v$ curve.

17. The light output of a certain lamp (W/nm) as a function of the wavelength of the light (nm) is given as

**One nanometer (nm) equals $10^{-9}$ meter.**

| *Wavelength* | 460 | 480 | 500 | 520 | 540 | 560 | 580 | 600 | 620 |
|---|---|---|---|---|---|---|---|---|---|
| *Output* | 0 | 21.5 | 347 | 1260 | 1360 | 939 | 338 | 64.6 | 0 |

Find the total output wattage of the lamp by finding the area under the output curve.

18. A certain tank has the following cross-sectional areas at a distance $x$ from one end:

| $x(ft)$ | 0 | 0.3 | 0.6 | 0.9 | 1.2 | 1.5 | 1.8 |
|---|---|---|---|---|---|---|---|
| *Area* ($ft^2$) | 28.3 | 27.3 | 26.4 | 24.2 | 21.6 | 17.3 | 13.9 |

Find the volume of the tank.

**Computer**

19. Write a program or use a spreadsheet that will accept as input the coordinates of any number of data points, and then compute and print the area below those points, using either the average ordinate method, the trapezoid rule, or Simpson's rule.

## CHAPTER 34 REVIEW PROBLEMS

*Integrate. Try some of the definite integrals by an approximate method.*

1. $\int \cot^2 x \csc^4 x\, dx$
2. $\int \sqrt{5 - 3x^2}\, dx$
3. $\int \frac{dx}{9x^2 - 4}$
4. $\int \sin x \sec^2 x\, dx$
5. $\int \frac{(3x - 1)\, dx}{x^2 + 9}$
6. $\int \frac{x^2\, dx}{(9 - x^2)^{3/2}}$
7. $\int \frac{dx}{x\sqrt{1 + x^2}}$
8. $\int \frac{(5x^2 - 3)\, dx}{x^3 - x}$
9. $\int \frac{dx}{2x^2 - 2x + 1}$
10. $\int \frac{dx}{\sqrt{1 + x^2}}$
11. $\int \frac{x\, dx}{\cos^2 x}$
12. $\int \sqrt{1 - 4x^2}\, dx$
13. $\int \frac{(x + 2)\, dx}{x^2 + x + 1}$
14. $\int \frac{\sqrt{x^2 - 1}\, dx}{x^3}$
15. $\int_0^5 \frac{(x^2 - 3)\, dx}{(x + 2)(x + 1)^2}$
16. $\int \frac{dx}{x^2\sqrt{5 - x^2}}$
17. $\int_0^4 \frac{9x^2\, dx}{(2x + 1)(x + 2)^2}$
18. $\int \frac{dx}{\sqrt{4 - (x + 3)^2}}$
19. $\int \frac{dx}{9x^2 - 1}$
20. $\int \frac{(3x - 7)\, dx}{(x - 2)(x - 3)}$
21. $\int \frac{x\, dx}{(1 - x)^5}$
22. $\int \frac{dx}{1 - \sqrt{x}}$
23. $\int \cot^3 x \sin x\, dx$
24. $\int \frac{3x\, dx}{\sqrt[3]{x + 1}}$

25. $\int_0^1 \frac{(3x^2 + 7x)\,dx}{(x+1)(x+2)(x+3)}$

26. $\int \frac{(5x + 9)\,dx}{(x - 9)^{3/2}}$

27. $\int_0^1 \frac{(2x^2 + x + 3)\,dx}{(x+1)(x^2+1)}$

28. $\int_2^\infty \frac{dx}{x\sqrt{x^2 - 4}}$

29. $\int_0^2 \frac{dx}{(x-2)^{2/3}}$

30. $\int_0^\infty e^{-x} \sin x\,dx$

31. $\int_0^1 \frac{dx}{x^2\sqrt{x^2+1}}$

*Writing*

32. Assume that there is just one week of class left, time enough to study either exact methods of integration (Sections 34-1 to 34-5) or approximate methods (Section 34-6). Your instructor asks your opinion. Write a letter explaining which you think is more important, and why.

# 35

# DIFFERENTIAL EQUATIONS

## OBJECTIVES

**When you have completed this chapter, you should be able to:**

- Solve first-order differential equations that have separable variables or that are exact.
- Solve homogeneous first-order differential equations.
- Solve first-order linear differential equations and Bernoulli's equation.
- Solve applications of first-order differential equations.
- Solve second-order differential equations that have separable variables.
- Use the auxiliary equation to determine the general solution to second-order differential equations with right side equal to zero.
- Use the method of undetermined coefficients to solve a second-order differential equation with right side not equal to zero.
- Use second-order differential equations to solve RLC circuits.

A *differential equation* is one that contains one or more derivatives. We have already solved some simple differential equations in Chapter 31 and here we go on to solve more difficult types. We begin with first-order differential equations. Our main applications of these will be in exponential growth and decay, motion, and electrical circuits. We then continue our study with second-order differential equations, those that have second derivatives. They may, of course, also have first derivatives, but no third or higher derivatives. Here we solve types of second-order equations that are fairly simple but of great practical importance just the same. This will be borne out by the applications to electrical circuits presented in this chapter.

## 35-1 DIFFERENTIAL EQUATIONS

A *differential equation* is one that contains one or more derivatives. We sometimes refer to "differential equations" by the abbreviation DE.

---

**EXAMPLE 1:** Some differential equations, using different notation for the derivative, are

(a) $\frac{dy}{dx} + 5 = 2xy$

(b) $y'' - 4y' + xy = 0$

(c) $Dy + 5 = 2xy$

(d) $D^2y - 4Dy + xy = 0$

---

### Differential Form

A differential equation containing the derivative $dy/dx$ is put into *differential form* simply by multiplying through by $dx$.

---

**EXAMPLE 2:** Example 1(a), in differential form, is

$$dy + 5\,dx = 2xy\,dx$$

---

### Ordinary versus Partial Differential Equations

We get an *ordinary* differential equation when our differential equation contains only two variables, and "ordinary" derivatives, as in Examples 1 and 2. When the equation contains *partial* derivatives, due to the presence of three or more variables, we have a *partial differential equation*.

---

**EXAMPLE 3:** $\frac{\partial x}{\partial t} = 5\frac{\partial y}{\partial t}$ is a partial differential equation.

The symbol $\partial$ is used for partial derivatives.

---

We cover only *ordinary* differential equations in this book.

### Order

The *order* of a differential equation is the order of the highest-order derivative in the equation.

**EXAMPLE 4:**

(a) $\dfrac{dy}{dx} - x = 2y$ is of *first* order.

(b) $\dfrac{d^2y}{dx^2} - \dfrac{dy}{dx} = 3x$ is of *second* order.

(c) $5y''' - 3y'' = xy$ is of *third* order.

## Degree

The *degree of a derivative* is the power to which that derivative is raised.

It's important to recognize the type of DE here for the same reason as for other equations. It lets us pick the right method of solution.

**EXAMPLE 5:** $(y')^2$ is of second degree.

The *degree of a differential equation* is the degree of the highest-order derivative in the equation. The equation must be rationalized and cleared of fractions before its degree can be determined.

**EXAMPLE 6:**

(a) $(y'')^3 - 5(y')^4 = 7$ is a second-order equation of *third* degree.

(b) To find the degree of the differential equation

$$\frac{x}{\sqrt{y' - 2}} = 1$$

we clear fractions and square both sides, getting

$$\sqrt{y' - 2} = x$$

or

$$y' - 2 = x^2$$

which is of first degree.

| Common Error | Don't confuse order and degree. The symbols $\dfrac{d^2y}{dx^2}$ and $\left(\dfrac{dy}{dx}\right)^2$ have different meanings. |
| --- | --- |

## Solving a Simple Differential Equation

We have already solved some simple differential equations by multiplying both sides of the equation by $dx$ and integrating.

**EXAMPLE 7:** Solve the differential equation $dy/dx = x^2 - 3$.

**Solution:** We first put our equation into differential form, getting $dy = (x^2 - 3)\,dx$. Integrating, we get

$$y = \int (x^2 - 3)\,dx = \frac{x^3}{3} - 3x + C$$

### Checking a Solution

Any *function* that satisfies a differential equation is called a *solution* of that equation. Thus, to check a solution, we substitute it and its derivatives into the original equation and see if an identity is obtained.

---

**EXAMPLE 8:** Is the function $y = e^{2x}$ a solution of the differential equation $y'' - 3y' + 2y = 0$?

**Solution:** Taking the first and second derivatives of the function gives

$$y' = 2e^{2x} \qquad \text{and} \qquad y'' = 4e^{2x}$$

Substituting the function and its derivatives into the differential equation gives

$$4e^{2x} - 6e^{2x} + 2e^{2x} = 0$$

which is an identity. Thus $y = e^{2x}$ is a solution to the differential equation. We will see shortly that it is one of many solutions to the given equation.

---

### General and Particular Solutions

We have just seen that the function $y = e^{2x}$ is a solution of the differential equation $y'' - 3y' + 2y = 0$. But is it the *only* solution? Try the following functions yourself, and you will see that they are also solutions to the given equation.

$$y = 4e^{2x} \tag{1}$$

$$y = Ce^{2x} \tag{2}$$

$$y = C_1e^{2x} + C_2e^{x} \tag{3}$$

where $C$, $C_1$, and $C_2$ are *arbitrary constants*.

There is a simple relation, which we state without proof, between the order of a differential equation and the number of constants in the solution.

> The solution of an $n$th-order differential equation can have at most $n$ arbitrary constants. A solution having the maximum number of constants is called the *general solution* or *complete solution*.

The differential equation in Example 8 is of second order, so the solution can have up to two arbitrary constants. Thus (3) is the general solution, while (1) and (2) are called *particular solutions*. When we later solve a differential equation, we will first obtain the general solution. Then, by using other given information, we will evaluate the arbitrary constants to obtain a particular solution. The "other given information" is referred to as *boundary conditions* or *initial conditions*.

## EXERCISE 1—DIFFERENTIAL EQUATIONS

Give the order and degree of each equation, and state whether it is an ordinary or partial differential equation.

1. $\frac{dy}{dx} + 3xy = 5$
2. $y'' + 3y' = 5x$
3. $D^3y - 4Dy = 2xy$
4. $\frac{\partial^2 y}{\partial x^2} + 4y = 7$
5. $3(y'')^4 - 5y' = 3y$
6. $4\frac{dy}{dx} - 3\left(\frac{d^2y}{dx^2}\right)^3 = x^2y$

Solve each differential equation.

7. $\frac{dy}{dx} = 7x$
8. $2y' = x^2$
9. $4x - 3y' = 5$
10. $3Dy = 5x + 2$
11. $dy = x^2\,dx$
12. $dy - 4x\,dx = 0$

Show that each function is a solution to the given differential equation.

13. $y' = \frac{2y}{x}$, $\quad y = Cx^2$
14. $\frac{dy}{dx} = \frac{x^2}{y^3}$, $\quad 4x^3 - 3y^4 = C$
15. $Dy = \frac{2y}{x}$ $\quad y = Cx^2$
16. $y' \cot x + 3 + y = 0$, $\quad y = C \cos x - 3$

## 35-2 FIRST-ORDER DIFFERENTIAL EQUATION, VARIABLES SEPARABLE

Given a differential equation of first order, $dy/dx = f(x, y)$, it is sometimes possible to *separate* the variables $x$ and $y$. This means that when we multiply both sides by $dx$ the resulting equation can be put into a form that has $dy$ multiplied by a function of $y$ only and $dx$ multiplied by a function of $x$ only.

| | | |
|---|---|---|
| Form of First-Order DE, Variables Separable | $f(y)\,dy = g(x)\,dx$ | **406** |

If this is possible, we can obtain a solution simply by integrating term by term.

**EXAMPLE 9:** Solve the differential equation $y' = x^2/y$.

**Solution:**

1. Rewrite the equation in differential form. We do this by replacing $y'$ with $dy/dx$ and multiplying by $dx$.

$$dy = \frac{x^2\,dx}{y}$$

2. Separate the variables. We do this here by multiplying both sides by $y$, and get $y\,dy = x^2\,dx$. The variables are now separated, and our equation is in the form of Eq. 406.
3. Integrate:

$$\int y\,dy = \int x^2\,dx$$

$$\frac{y^2}{2} = \frac{x^3}{3} + C_1$$

where $C_1$ is an arbitrary constant.

4. Simplify the answer. We can do this by multiplying by the LCD 6.

$$3y^2 = 2x^3 + C$$

where we have replaced $6C_1$ by $C$.

We'll often leave our answer in implicit form, as we have done here.

## Simplifying the Solution

Often the simplification of a solution to a differential equation will involve several steps.

**EXAMPLE 10:** Solve the differential equation $dy/dx = 4xy$.

**Solution:** Multiplying both sides by $dx/y$ and integrating gives us

$$\frac{dy}{y} = 4x\,dx$$

$$\ln|y| = 2x^2 + C$$

We can leave our solution in this implicit form or solve for $y$.

$$|y| = e^{2x^2+C} = e^{2x^2}e^C$$

Let us replace $e^C$ by another constant $k$. Since $k$ can be positive or negative, we can remove the absolute value symbols from $y$, getting

$$y = ke^{2x^2}$$

The solution to a DE can take on different forms, depending on how you simplify it. Don't be discouraged if your solution does not at first match the one in the answer key.

> We will often interchange one arbitrary constant with another of a different form, such as replacing $C_1$ by $8C$ or by $\ln C$ or by $\sin C$ or by $e^C$, or any other form that will help us to simplify an expression.

The laws of exponents or of logarithms will often be helpful in simplifying an answer, as in the following example.

**EXAMPLE 11:** Solve the differential equation $dy/dx = y/(5 - x)$.

**Solution:**

1. Going to differential form, $dy = y\,dx/(5 - x)$.

2. Separating the variables yields

$$\frac{dy}{y} = \frac{dx}{5 - x}$$

3. Integrating gives $\ln y = -\ln(5 - x) + C_1$.

4. Simplifying, we have $\ln y + \ln(5 - x) = C_1$. Then we use our laws of logarithms and obtain

$$\ln y(5 - x) = C_1$$

Here it is convenient to replace $C_1$ with $\ln C$, so $y(5 - x) = C$, where $C = e^{C_1}$. So

$$y = \frac{C}{5 - x}$$

where $x \neq 5$.

## Logarithmic, Exponential, or Trigonometric Equations

To separate the variables in certain equations, it may be necessary to use our laws of exponents or logarithms, or the trigonometric identities.

**EXAMPLE 12:** Solve the equation $4y'e^{4y} \cos x = e^{2y} \sin x$.

**Solution:** We replace $y'$ with $dy/dx$ and multiply through by $dx$, getting $4e^{4y} \cos x\,dy = e^{2y} \sin x\,dx$. We can eliminate $x$ on the left side by dividing through by $\cos x$. Similarly, we can eliminate $y$ from the right by dividing by $e^{2y}$.

$$\frac{4e^{4y}}{e^{2y}}\,dy = \frac{\sin x}{\cos x}\,dx$$

or $4e^{2y}\,dy = \tan x\,dx$. Integrating gives the solution:

$$2e^{2y} = -\ln|\cos x| + C$$

## Particular Solution

We can evaluate the constants in our solution to a differential equation when given suitable *boundary conditions,* as in the following example.

**EXAMPLE 13:** Solve the equation $2y(1 + x^2)y' + x(1 + y^2) = 0$, subject to the condition that $y = 2$ when $x = 0$.

**Solution:** Separating variables and integrating, we obtain

$$\frac{2y\,dy}{1 + y^2} + \frac{x\,dx}{1 + x^2} = 0$$

$$\int \frac{2y\,dy}{1 + y^2} + \frac{1}{2}\int \frac{2x\,dx}{1 + x^2} = 0$$

$$\ln|1 + y^2| + \frac{1}{2}\ln|1 + x^2| = C$$

Simplifying, we drop the absolute value signs since $(1 + x^2)$ and $(1 + y^2)$ cannot be negative. Then we multiply by 2 and apply the laws of logarithms:

$$2\ln(1 + y^2) + \ln(1 + x^2) = 2C$$

$$\ln(1 + x^2)(1 + y^2)^2 = 2C$$

$$(1 + x^2)(1 + y^2)^2 = e^{2C}$$

Applying the boundary conditions that $y = 2$ when $x = 0$ gives

$$e^{2C} = (1 + 0)(1 + 2^2)^2 = 25$$

Our particular solution is then $(1 + x^2)(1 + y^2)^2 = 25$.

## EXERCISE 2—FIRST-ORDER DIFFERENTIAL EQUATION, VARIABLES SEPARABLE

### General Solution

We'll have applications later in this chapter.

Find the general solution to each differential equation.

1. $y' = \frac{x}{y}$
2. $\frac{dy}{dx} = \frac{2y}{x}$
3. $dy = x^2y\,dx$
4. $y' = xy^3$
5. $y' = \frac{x^2}{y^3}$
6. $y' = \frac{x^2 + x}{y - y^2}$
7. $xy\,dx - (x^2 + 1)\,dy = 0$
8. $y' = x^3y^5$
9. $(1 + x^2)\,dy + (y^2 + 1)\,dx = 0$
10. $y' = x^2e^{-3y}$
11. $\sqrt{1 + x^2}\,dy + xy\,dx = 0$
12. $y^2\,dx = (1 - x)\,dy$
13. $(y^2 + 1)\,dx = (x^2 + 1)\,dy$
14. $y^3\,dx = x^3\,dy$
15. $(2 + y)\,dx + (x - 2)\,dy = 0$
16. $y' = \frac{e^{x-y}}{e^x + 1}$
17. $(x - xy^2)\,dx = -(x^2y + y)\,dy$

### With Exponential Functions

18. $dy = e^{-x}\,dx$
19. $ye^{2x} = (1 + e^{2x})y'$
20. $e^{y}(y' + 1) = 1$
21. $e^{x-y}\,dx + e^{y-x}\,dy = 0$

### With Trigonometric Functions

22. $(3 + y)\,dx + \cot x\,dy = 0$
23. $\tan y\,dx + (1 + x)\,dy = 0$
24. $\tan y\,dx + \tan x\,dy = 0$
25. $\cos x \sin y\,dy + \sin x \cos y\,dx = 0$
26. $\sin x \cos^2 y\,dx + \cos^2 x\,dy = 0$
27. $4 \sin x \sec y\,dx = \sec x\,dy$

### Particular Solution

Find the particular solution to each differential equation, using the given boundary condition.

28. $x\,dx = 4y\,dy, \quad x = 5$ when $y = 2$
29. $y^2y' = x^2, \quad x = 0$ when $y = 1$
30. $\sqrt{x^2 + 1}\,y' + 3xy^2 = 0, \quad x = 1$ when $y = 1$
31. $y' \sin y = \cos x, \quad x = \pi/2$ when $y = 0$
32. $x(y + 1)y' = y(1 + x), \quad x = 1$ when $y = 1$

## 35-3 EXACT FIRST-ORDER DIFFERENTIAL EQUATIONS

When the left side of a first-order differential equation is the exact differential of some function, we call that equation an *exact differential equation*. Even if we cannot separate the variables in such a differential equation, it might still be solved by integrating a *combination* of terms.

---

**EXAMPLE 14:** Solve $y\,dx + x\,dy = x\,dx$.

**Solution:** The variables are not now separated, and we see that no amount of manipulation will separate them. However, the combination of terms $y\,dx + x\,dy$ on the left side may ring a bell. In fact, it is the derivative of the product of $x$ and $y$ (Eq. 336):

$$\frac{d(xy)}{dx} = x\frac{dy}{dx} + y\frac{dx}{dx}$$

or

$$d(xy) = x\,dy + y\,dx$$

This, then, is the left side of our given equation. The right side contains only $x$'s, so we integrate:

$$\int (y\,dx + x\,dy) = \int x\,dx$$

$$\int d(xy) = \int x\,dx$$

$$xy = \frac{x^2}{2} + C$$

or

$$y = \frac{x}{2} + \frac{C}{x}$$

---

## Integrable Combinations

The expression $y\,dx + x\,dy$ from Example 14 is called an *integrable combination*. Some of the most frequently used combinations are as follows.

| | | |
|---|---|---|
| Integrable Combinations | $x\,dy + y\,dx = d(xy)$ | **407** |
| | $\dfrac{x\,dy - y\,dx}{x^2} = d\left(\dfrac{y}{x}\right)$ | **408** |
| | $\dfrac{y\,dx - x\,dy}{y^2} = d\left(\dfrac{x}{y}\right)$ | **409** |
| | $\dfrac{x\,dy - y\,dx}{x^2 + y^2} = d\left(\tan^{-1}\dfrac{y}{x}\right)$ | **410** |

---

**EXAMPLE 15:** Solve $dy/dx = y(1 - xy)/x$.

**Solution:** We go to differential form and clear denominators by multiplying through by $x\,dx$, getting $x\,dy = y(1 - xy)\,dx$. Removing parentheses, we obtain

$$x\,dy = y\,dx - xy^2\,dx$$

On the lookout for an integrable combination, we move the $y\,dx$ term to the left side.

$$x\,dy - y\,dx = -xy^2\,dx$$

**Any expression that we multiply by (such as $-1/y^2$ here) to make our equation exact, is called an *integrating factor.***

We see now that the left side will be the differential of $x/y$ if we multiply by $-1/y^2$.

$$\frac{y\,dx - x\,dy}{y^2} = x\,dx$$

Integrating gives $x/y = x^2/2 + C_1$, or

$$y = \frac{2x}{x^2 + C}$$

---

## Particular Solution

As before, we substitute boundary conditions to evaluate the constant of integration.

**EXAMPLE 16:** Solve $2xy\,dy - 4x\,dx + y^2\,dx = 0$, such that $x = 1$ when $y = 2$.

**Solution:** It looks as though the first and third terms might be an integrable combination, so we transpose the $-4x\,dx$.

**A certain amount of trial-and-error work may be needed to get the DE into a suitable form.**

$$2xy\,dy + y^2\,dx = 4x\,dx$$

The left side is the derivative of the product $xy^2$,

$$d(xy^2) = 4x\,dx$$

Integrating gives $xy^2 = 2x^2 + C$. Substituting the boundary conditions, we get $C = xy^2 - 2x^2 = 1(2)^2 - 2(1)^2 = 2$, so our particular solution is

$$xy^2 = 2x^2 + 2$$

## EXERCISE 3—EXACT FIRST-ORDER DIFFERENTIAL EQUATIONS

### Integrable Combinations

Find the general solution of each differential equation.

**1.** $y\,dx + x\,dy = 7\,dx$

**2.** $x\,dy = (4 - y)\,dx$

**3.** $x\dfrac{dy}{dx} = 3 - y$

**4.** $y + xy' = 9$

**5.** $2xy' = x - 2y$

**6.** $x\dfrac{dy}{dx} = 2x - y$

**7.** $x\,dy = (3x^2 + y)\,dx$

**8.** $(x + y)\,dx + x\,dy = 0$

**9.** $3x^2 + 2y + 2xy' = 0$

**10.** $(1 - 2x^2y)\dfrac{dy}{dx} = 2xy^2$

**11.** $(2x - y)y' = x - 2y$

**12.** $\dfrac{y\,dx - x\,dy}{y^2} = x\,dx$

**13.** $y\,dx - x\,dy = 2y^2\,dx$

**14.** $(x - 2x^2y)\,dy = y\,dx$

**15.** $(4y^3 + x)\dfrac{dy}{dx} = y$

**16.** $(y - x)y' + 2xy^2 + y = 0$

**17.** $3x - 2y^2 - 4xyy' = 0$

**18.** $\dfrac{x\,dy - y\,dx}{x^2 + y^2} = 5\,dy$

Find the particular solution to each differential equation, using the given boundary condition.

**19.** $4x = y + xy'$, $x = 3$ when $y = 1$

**20.** $y\,dx = (x - 2x^2y)\,dy$, $x = 1$ when $y = 2$

**21.** $y = (3y^3 + x)\dfrac{dy}{dx}$, $x = 1$ when $y = 1$

**22.** $4x^2 = -2y - 2xy'$, $x = 5$ when $y = 2$

**23.** $3x - 2y = (2x - 3y)\dfrac{dy}{dx}$, $x = 2$ when $y = 2$

**24.** $5x = 2y^2 + 4xyy'$, $x = 1$ when $y = 4$

## 35-4 FIRST-ORDER HOMOGENEOUS DIFFERENTIAL EQUATIONS

### Recognizing a Homogeneous Differential Equation

If each variable in a function is replaced by $t$ times the variable, and a power of $t$ can be factored out, we say that the function is *homogeneous*. The power of $t$ that can be factored out of the function is the *degree of homogeneity* of the function.

---

**EXAMPLE 17:** Is $\sqrt{x^4 + xy^3}$ a homogeneous function?

**Solution:** We replace $x$ with $tx$ and $y$ with $ty$ and get

$$\sqrt{(tx)^4 + (tx)(ty)^3} = \sqrt{t^4x^4 + t^4xy^3} = t^2\sqrt{x^4 + xy^3}$$

Since we were able to factor out a $t^2$, we say that our function is homogeneous to the second degree.

---

A *homogeneous polynomial* is one in which every term is of the same degree.

---

**EXAMPLE 18:**

(a) $x^2 + xy - y^2$ is a homogeneous polynomial of degree 2.

(b) $x^2 + x - y^2$ is not homogeneous.

---

A first-order *homogeneous differential equation* is one of the following form.

| | | |
|---|---|---|
| Form of First-Order Homogeneous DE | $M\,dx + N\,dy = 0$ | **411** |

Here $M$ and $N$ are functions of $x$ and $y$, and are homogeneous functions of the same degree.

---

**EXAMPLE 19:**

(a) $(x^2 + y^2)\,dx + xy\,dy = 0$ is a first-order homogeneous differential equation.

(b) $(x^2 + y^2)\,dx + x\,dy = 0$ is not homogeneous.

---

### Solving a First-Order Homogeneous Differential Equation

Sometimes we can transform a homogeneous differential equation whose variables cannot be separated into one whose variables can be separated by making the substitution

$$y = vx$$

as explained in the following example.

**EXAMPLE 20:** Solve

$$x\frac{dy}{dx} - y = \sqrt{x^2 - y^2}$$

**Solution:** We first check if the given equation is homogeneous of first degree. We put the equation into the form of Eq. 411 by multiplying by $dx$ and rearranging.

$$x\,dy - y\,dx = \sqrt{x^2 - y^2}\,dx$$

$$(\sqrt{x^2 - y^2} + y)\,dx - x\,dy = 0$$

This is now in the form of Eq. 411,

$$M\,dx + N\,dy = 0$$

with $M = \sqrt{x^2 - y^2} + y$ and $N = -x$. To test if $M$ is homogeneous, we replace $x$ by $tx$ and $y$ by $ty$.

$$\sqrt{(tx)^2 - (ty)^2} + ty = \sqrt{t^2x^2 - t^2y^2} + ty$$

$$= t\sqrt{x^2 - y^2} + ty$$

$$= t[\sqrt{x^2 - y^2} + y]$$

We see that $t$ can be factored out, and is of first degree. Thus $M$ is homogeneous of first degree, as is $N$, so the given differential equation is homogeneous. To solve, we make the substitution $y = vx$ to transform the given equation into one whose variables can be separated. However, when we substitute for $y$ we must also substitute for $dy/dx$. Let

$$y = vx$$

Then by the product rule

$$\frac{dy}{dx} = v + x\frac{dv}{dx}$$

Substituting into the given equation yields

$$x\left(v + x\frac{dv}{dx}\right) - vx = \sqrt{x^2 + v^2x^2}$$

which simplifies to

$$x\frac{dv}{dx} = \sqrt{1 + v^2}$$

Separating variables, we have

$$\frac{dv}{\sqrt{1 + v^2}} = \frac{dx}{x}$$

Integrating by Rule 62, we obtain

$$\ln |v + \sqrt{1 + v^2}| = \ln |x| + C_1$$

$$\ln \left| \frac{v + \sqrt{1 + v^2}}{x} \right| = C_1$$

$$\frac{v + \sqrt{1 + v^2}}{x} = e^{C_1}$$

$$v + \sqrt{1 + v^2} = Cx$$

where $C = e^{C_1}$. Now subtracting $v$ from both sides, squaring, and simplifying gives

$$C^2x^2 - 2Cvx = 1$$

Finally, substituting back, $v = y/x$, we get

$$C^2x^2 - 2Cy = 1$$

---

## EXERCISE 4—FIRST-ORDER HOMOGENEOUS DIFFERENTIAL EQUATIONS

Find the general solution to each differential equation.

These are not easy.

1. $(x - y)\,dx - 2x\,dy = 0$
2. $(3y - x)\,dx = (x + y)\,dy$
3. $(x^2 - xy)y' + y^2 = 0$
4. $(x^2 - xy)\,dy + (x^2 - xy + y^2)\,dx = 0$
5. $xy^2\,dy - (x^3 + y^3)\,dx = 0$
6. $2x^3y' + y^3 - x^2y = 0$

Find the particular solution, using the given boundary condition.

7. $x - y = 2xy', \quad x = 1, y = 1$
8. $3xy^2\,dy = (3y^3 - x^3)\,dx, \quad x = 1, y = 2$
9. $(2x + y)\,dx = y\,dy, \quad x = 2, y = 1$
10. $(x^3 + y^3)\,dx - xy^2\,dy = 0, \quad x = 1, y = 0$

## 35-5 FIRST-ORDER LINEAR DIFFERENTIAL EQUATIONS

When describing the degree of a term, we usually add the degrees of each variable in the term. Thus $x^2y^3$ is of fifth degree.

Sometimes, however, we want to describe the degree of a term with regard to just one of the variables. Thus we say that $x^2y^3$ is of second degree in $x$, and of third degree in $y$.

In determining the degree of a term, we must also consider any derivatives in that term. We consider $dy/dx$ to contribute one "degree" to $y$ in this computation; $d^2y/dx^2$ to contribute two, and so on. Thus the term $xy\,dy/dx$ is of first degree in $x$ and of second degree in $y$.

A first-order differential equation is called *linear* if each term is of first degree or less *in the dependent variable y*.

**EXAMPLE 21:**

(a) $y' + x^2y = e^x$ is linear.

(b) $y' + xy^2 = e^x$ is not linear because $y$ is squared in the second term.

(c) $y\, dy/dx - xy = 5$ is not linear because we must add the exponents of $y$ and $dy/dx$, making the first term of second degree.

A first-order linear differential equation can always be written in the standard form

| Form of First-Order Linear DE | $\dfrac{dy}{dx} + Py = Q$ | **412** |
|---|---|---|

where $P$ and $Q$ are functions of $x$ only.

**EXAMPLE 22:** Write the equation $xy' - e^x + y = xy$ in the form of Eq. 412.

**Solution:** Rearranging gives $xy' + y - xy = e^x$. Factoring, we have

$$xy' + (1 - x)y = e^x$$

Dividing by $x$ gives us the standard form,

$$y' + \frac{1 - x}{x}y = \frac{e^x}{x}$$

where $P = (1 - x)/x$ and $Q = e^x/x$.

## Integrating Factor

The left side of a first-order linear differential equation can always be made into an integrable combination by multiplying by an *integrating factor R*. We now find such a factor.

Multiplying Eq. 412 by $R$, the left side becomes

$$R\frac{dy}{dx} + yRP \qquad (1)$$

Let us try to make the left side *the exact derivative of the product Ry of y and the integrating factor R*. The derivative of $Ry$ is

$$R\frac{dy}{dx} + y\frac{dR}{dx} \qquad (2)$$

But (1) and (2) will be equal if $dR/dx = RP$. Since $P$ is a function of $x$ only, we can separate variables,

$$\frac{dR}{R} = P\, dx$$

We omit the constant of integration because we seek only one integrating factor.

Integrating, $\ln R = \int P\,dx$, or

| | | |
|---|---|---|
| Integrating Factor | $R = e^{\int P\,dx}$ | **413** |

Thus, *multiplying a given first-order linear equation by the integrating factor* $R = e^{\int P\,dx}$ *will make the left side of the equation the exact derivative of* $Ry$. We can then proceed to integrate to get the solution of the given differential equation.

---

**EXAMPLE 23:** Solve $\dfrac{dy}{dx} + \dfrac{4y}{x} = 3.$

**Solution:** Our equation is in standard form with $P = 4/x$ and $Q = 3$. Then

$$\int P\,dx = 4\int \frac{dx}{x} = 4\ln x = \ln x^4$$

Our integrating factor $R$ is then

Can you show why $e^{\ln x^4} = x^4$?

$$R = e^{\int P\,dx} = e^{\ln x^4} = x^4$$

Multiplying our given equation by $x^4$ and going to differential form gives

$$x^4\,dy + 4x^3y\,dx = 3x^4\,dx$$

Notice that the left side is now the derivative of $y$ times the integrating factor, $d(x^4y)$. Integrating, we obtain

$$x^4y = \frac{3x^5}{5} + C$$

or

$$y = \frac{3x}{5} + \frac{C}{x^4}$$

---

In summary, to solve the first-order linear equation

$$y' + Py = Q$$

multiply by an integrating factor $R = e^{\int P\,dx}$, and the solution to the equation becomes

$$yR = \int QR\,dx$$

or

| | | |
|---|---|---|
| Solution to First-Order Linear DE | $ye^{\int P\,dx} = \int Qe^{\int P\,dx}\,dx$ | **414** |

Next we try an equation having trigonometric functions.

**EXAMPLE 24:** Solve $y' + y \cot x = \csc x$.

**Solution:** This is a first-order differential equation in standard form, with $P = \cot x$ and $Q = \csc x$. Thus $\int P\,dx = \int \cot x\,dx = \ln|\sin x|$. The integrating factor $R$ is $e$ raised to the $\ln|\sin x|$, or simply $\sin x$. Then, by Eq. 414, the solution to the differential equation is

$$y \sin x = \int \csc x \sin x\,dx$$

Since $\csc x \sin x = 1$, this simplifies to

$$y \sin x = \int dx$$

so

$$y \sin x = x + C$$

## Particular Solution

As before, we find the general solution and then substitute the boundary conditions to evaluate $C$.

**EXAMPLE 25:** Solve $y' + \left(\dfrac{1-2x}{x^2}\right)y = 1$ given that $y = 2$ when $x = 1$.

**Solution:** We are in standard form with $P = (1 - 2x)/x^2$. We find the integrating factor by first integrating $\int P\,dx$. This gives

$$P = x^{-2} - 2x^{-1}$$

$$\int P\,dx = -\frac{1}{x} - 2\ln x$$

$$= \frac{1}{x} - \ln x^2$$

Our integrating factor is then

$$R = e^{\int P\,dx} = e^{-1/x - \ln x^2} = \frac{e^{-1/x}}{e^{\ln x^2}} = \frac{1}{x^2 e^{1/x}}$$

Substituting into Eq. 414, the general solution is

$$\frac{y}{x^2 e^{1/x}} = \int \frac{dx}{x^2 e^{1/x}}$$

$$= \int e^{-1/x} x^{-2}\,dx = e^{-1/x} + C$$

When $x = 1$ and $y = 2$, $C = 2/e - 1/e = 1/e$, so the particular solution is

$$\frac{y}{x^2 e^{1/x}} = e^{-1/x} + \frac{1}{e}$$

or

$$y = x^2\left(1 + \frac{e^{1/x}}{e}\right)$$

## Equations Reducible to Linear Form

Sometimes a first-order differential equation that is nonlinear can be expressed in linear form. One type of nonlinear equation easily reducible to linear form is one in which the right side contains a power of $y$ as a factor, an equation known as a Bernoulli equation.

After James Bernoulli (1654–1705), the most famous of this family of Swiss mathematicians.

| | | |
|---|---|---|
| Form of Bernoulli's Equation | $\frac{dy}{dx} + Py = Qy^n$ | **422** |

Here $P$ and $Q$ are functions of $x$, as before. We solve such an equation by making the substitution

$$z = y^{1-n}$$

---

**EXAMPLE 26:** Solve the equation $y' + y/x = x^2y^6$.

**Solution:** This is a Bernoulli equation with $n = 6$. We first divide through by $y^6$ and get

$$\frac{1}{y^6}\frac{dy}{dx} + \frac{y^{-5}}{x} = x^2 \tag{1}$$

We let $z = y^{1-6} = y^{-5}$. The derivative of $z$ with respect to $x$ is

$$\frac{dz}{dx} = -\frac{5}{y^6}\frac{dy}{dx}$$

or

$$\frac{dy}{dx} = -\frac{y^6}{5}\frac{dz}{dx}$$

Substituting for $y$ and $dy/dx$ into (1) gives

$$-\frac{1}{5}\frac{dz}{dx} + \frac{1}{x}z = x^2$$

or

$$\frac{dz}{dx} - \frac{5}{x}z = -5x^2 \tag{2}$$

This equation is now linear in $z$. We proceed to solve it as before, with $P = -5/x$ and $Q = -5x^2$. We now find the integrating factor. The integral of $P$ is $\int (-5/x)\,dx = -5 \ln x = \ln x^{-5}$, so the integrating factor is

$$R = e^{\int P\,dx} = x^{-5} = \frac{1}{x^5}$$

From Eq. 414,

$$\frac{z}{x^5} = -5\int x^{-3}\,dx = \frac{-5x^{-2}}{-2} + C$$

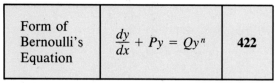

This is the solution of the differential equation (2). We now solve for $z$ and then substitute back $z = 1/y^5$,

$$z = \frac{5}{2}x^3 + Cx^5 = \frac{1}{y^5}$$

Our solution of (1) is then

$$y^5 = \frac{1}{5x^3/2 + Cx^5}$$

or

$$y^5 = \frac{2}{5x^3 + C_1x^5}$$

---

## Summary

To solve a first-order differential equation:

1. If you can *separate the variables,* integrate term by term.
2. If the DE is exact (contains an *integrable combination*), isolate that combination on one side of the equation and take the integral of both sides.
3. If the DE is *homogeneous,* start by substituting $y = vx$, and $dy/dx = v + x\, dv/dx$.
4. If the DE is *linear,* $y' + Py = Q$, multiply by the integrating factor $R = e^{\int P\,dx}$, and the solution will be $y = (1/R) \int QR\, dx$.
5. If the DE is a *Bernoulli's equation,* $y' + Py = Qy^n$, substitute $z = y^{1-n}$ to make it first-order linear, and proceed as in step 4.
6. If *boundary conditions* are given, substitute them to evaluate the constant of integration.

## EXERCISE 5—FIRST-ORDER LINEAR DIFFERENTIAL EQUATIONS

Find the general solution to each differential equation.

1. $\dfrac{dy}{dx} + \dfrac{y}{x} = 4$
2. $y' + \dfrac{y}{x} = 3x$
3. $xy' = 4x^3 - y$
4. $\dfrac{dy}{dx} + xy = 2x$
5. $y' = x^2 - x^2y$
6. $y' - \dfrac{y}{x} = \dfrac{-1}{x^2}$
7. $y' = \dfrac{3 - xy}{2x^2}$
8. $y' = x + \dfrac{2y}{x}$
9. $xy' = 2y - x$
10. $(x + 1)y' - 2y = (x + 1)^4$
11. $y' = \dfrac{2 - 4x^2y}{x + x^3}$
12. $(x + 1)y' = 2(x + y + 1)$
13. $xy' + x^2y + y = 0$
14. $(1 + x^3)\, dy = (1 - 3x^2y)\, dx$

### With Exponential and Logarithmic Expressions

15. $y' + y = e^x$

16. $y' = e^{2x} + y$

17. $y' = 2y + 4e^{2x}$

18. $xy' - e^x + y + xy = 0$

19. $y' = \dfrac{4 \ln x - 2x^2y}{x^3}$

### With Trigonometric Expressions

20. $\dfrac{dy}{dx} + y \sin x = 3 \sin x$

21. $y' + y = \sin x$

22. $y' + 2xy = 2x \cos x^2$

23. $y' = 2 \cos x - y$

24. $y' = \sec x - y \cot x$

### Bernoulli's Equation

25. $y' + \dfrac{y}{x} = 3x^2y^2$

26. $xy' + x^2y^2 + y = 0$

27. $y' = y - xy^2(x + 2)$

28. $y' + 2xy = xe^{-x^2}y^3$

### Particular Solution

Find the particular solution to each differential equation, using the given boundary condition.

29. $xy' + y = 4x, \quad x = 1$ when $y = 5$

30. $\dfrac{dy}{dx} + 5x = x - xy, \quad x = 2$ when $y = 1$

31. $y' + \dfrac{y}{x} = 5, \quad x = 1$ when $y = 2$

32. $y' = 2 + \dfrac{3y}{x}, \quad x = 2$ when $y = 6$

33. $y' = \tan^2 x + y \cot x, \quad x = \dfrac{\pi}{4}$ when $y = 2$

34. $\dfrac{dy}{dx} + 5y = 3e^x, \quad x = 1$ when $y = 1$

## 35-6 GEOMETRIC APPLICATIONS OF FIRST-ORDER DIFFERENTIAL EQUATIONS

Now that we are able to solve some simple differential equations of first order, we turn to applications. Here we must not only solve the equation, but must first *set up* the differential equation. The geometric problems we do first will help to prepare us for the physical applications that follow.

### Setting Up a Differential Equation

When reading the problem statement, look for the words "slope" or "rate of change." Each of these can be represented by the first derivative.

---

**EXAMPLE 27:**

(a) The statement *"the slope of a curve at every point is equal to twice the ordinate"* is represented by the differential equation

$$\frac{dy}{dx} = 2y$$

(b) The statement *"the ratio of abscissa to ordinate at each point on a curve is proportional to the rate of change at that point"* can be written

$$\frac{x}{y} = k\frac{dy}{dx}$$

(c) The statement *"the slope of a curve at every point is inversely proportional to the square of the ordinate at that point"* can be described by the equation

$$\frac{dy}{dx} = \frac{k}{y^2}$$

---

### Finding an Equation Whose Slope Is Specified

Once the equation is written, it is solved by the methods of the preceding sections.

---

**EXAMPLE 28:** The slope of a curve at each point is one-tenth the product of the ordinate and the square of the abscissa, and the curve passes through the point (2, 3). Find the equation of the curve.

**Solution:** The differential equation is

$$\frac{dy}{dx} = \frac{x^2y}{10}$$

In solving a differential equation, we first see if the variables can be separated. In this case they can be,

$$\frac{10\,dy}{y} = x^2\,dx$$

Integrating gives us

$$10 \ln y = \frac{x^3}{3} + C$$

At (2, 3),

$$C = 10 \ln 3 - \frac{8}{3} = 8.32$$

Our curve thus has the equation $\ln y = x^3/30 + 0.832$.

---

### Tangents and Normals to Curves

In setting up problems involving tangents and normals to curves, recall that

$$\text{slope of the tangent} = \frac{dy}{dx} = y'$$

$$\text{slope of the normal} = -\frac{1}{y'}$$

**EXAMPLE 29:** A curve (Fig. 35-1) passes through the point (4, 2). If from any point $P$ on the curve, the line $OP$ and the tangent $PT$ is drawn, the triangle $OPT$ is isosceles. Find the equation of the curve.

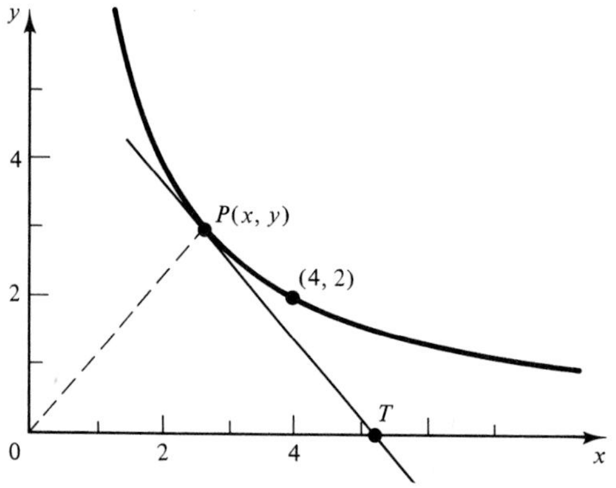


FIGURE 35-1

**Solution:** The slope of the tangent is $dy/dx$, and the slope of $OP$ is $y/x$. Since the triangle is isosceles, these slopes must be equal but of opposite signs.

$$\frac{dy}{dx} = -\frac{y}{x}$$

We can solve this equation by separation of variables or as the integrable combination $x\,dy + y\,dx = 0$. Either way gives the hyperbola

$$xy = C$$

(see Eq. 325). At the point (4, 2), $C = 4(2) = 8$. So our equation is $xy = 8$.

### Orthogonal Trajectories

In Sec. 31-3 we graphed a relation that had an arbitrary constant and got a *family of curves*. For example, the relation $x^2 + y^2 = C^2$ represents a family of circles of radius $C$, whose center is at the origin. Another curve that cuts each curve of the family at right angles is called an *orthogonal trajectory* to that family.

To find the orthogonal trajectory to a family of curves:

1. Differentiate the equation of the family to get the slope.
2. Eliminate the constant contained in the original equation.
3. Take the negative reciprocal of that slope to get the slope of the orthogonal trajectory.
4. Solve the resulting differential equation to get the equation of the orthogonal trajectory.

**EXAMPLE 30:** Find the equation of the orthogonal trajectories to the parabolas $y^2 = px$.

**Solution:**

1. The derivative is $2yy' = p$, so

$$y' = \frac{p}{2y} \tag{1}$$

2. The constant $p$, from the original equation, is $y^2/x$. Substituting $y^2/x$ for $p$ in (1) gives $y' = y/2x$.

| Common Error | Be sure to eliminate the constant ($p$ in this example) before continuing. |
|---|---|

3. The slope of the orthogonal trajectory is, by Eq. 285, the negative reciprocal of the slope of the given family,

$$y' = \frac{-2x}{y}$$

4. Separating variables yields

$$y\,dy = -2x\,dx$$

Integrating gives the solution,

$$\frac{y^2}{2} = \frac{-2x^2}{2} + C_1$$

which is a family of ellipses, $2x^2 + y^2 = C$ (Fig. 35-2).

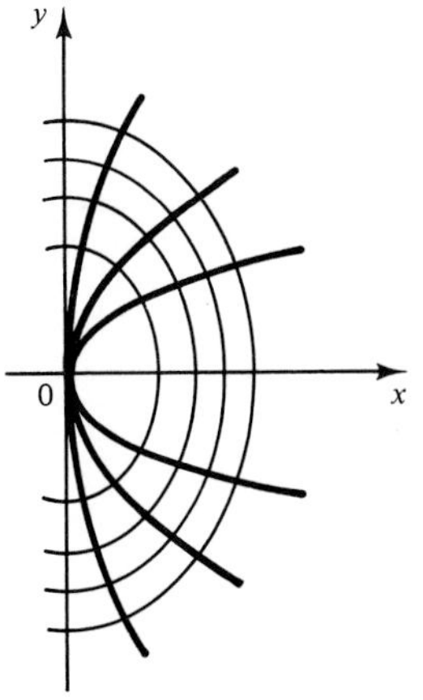


**FIGURE 35-2 Orthogonal ellipses and parabolas. Each intersection is at 90°.**

## EXERCISE 6—GEOMETRIC APPLICATIONS

### Slope of Curves

1. Find the equation of the curve that passes through the point (2, 9) and whose slope is $y' = x + 1/x + y/x$.
2. The slope of a certain curve at any point is equal to the reciprocal of the ordinate at that point. Write the equation of the curve if it passes through the point (1, 3).
3. Find the equation of a curve whose slope at any point is equal to the abscissa of that point divided by the ordinate, and which passes through the point (3, 4).
4. Find the equation of a curve that passes through (1, 1) and whose slope at any point is equal to the product of the ordinate and abscissa.
5. A curve passes through the point (2, 3) and has a slope equal to the sum of the abscissa and ordinate at each point. Find its equation.

### Tangents and Normals

6. The distance to the $x$ intercept $C$ of the normal (Fig. 35-3) is given by $OC = x + y'y$. Write the equation of the curve passing through (1, 2) for which $OC$ is equal to three times the abscissa of $P$.

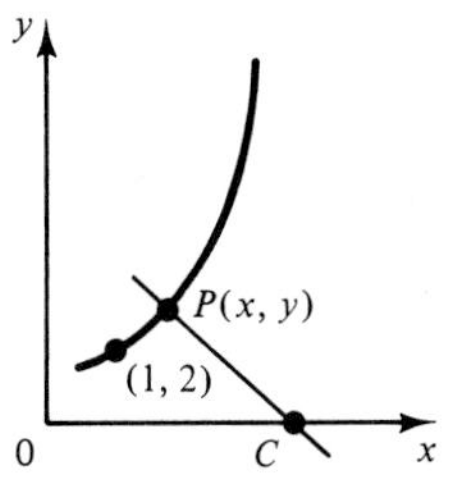


**FIGURE 35-3**

The equations for the various distances associated with tangents and normals were given in Chapter 25 Review Problems.

7. A certain first-quadrant curve (Fig. 35-4) passes through the point (4, 1). If a tangent is drawn through any point $P$, the portion $AB$ of the tangent that lies between the coordinate axes is bisected by $P$. Find the equation of the curve, given that

$$AP = -\left(\frac{y}{y'}\right)\sqrt{1 + (y')^2} \quad \text{and} \quad BP = x\sqrt{1 + (y')^2}$$

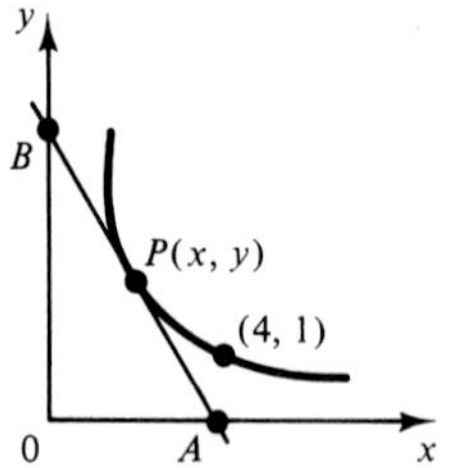

FIGURE 35-4

8. A tangent $PT$ is drawn to a curve at a point $P$ (Fig. 35-5). The distance $OT$ from the origin to a tangent through $P$ is given by

$$OT = \frac{xy' - y}{\sqrt{1 + (y')^2}}$$

Find the equation of the curve passing through the point (2, 4) so that $OT$ is equal to the abscissa of $P$.

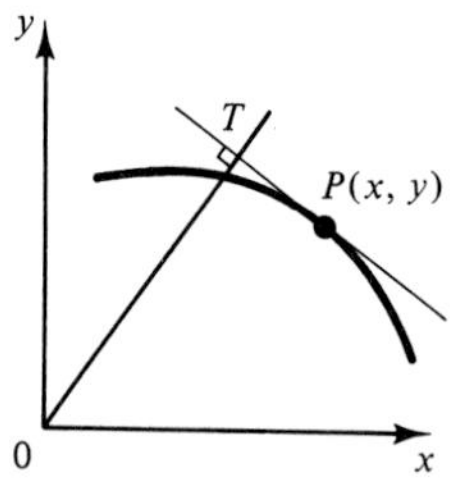

FIGURE 35-5

9. Find the equation of the curve passing through (0, 1) for which the length $PC$ of a normal through $P$ equals the square of the ordinate of $P$ (Fig. 35-6), where

$$PC = y\sqrt{1 + (y')^2}$$

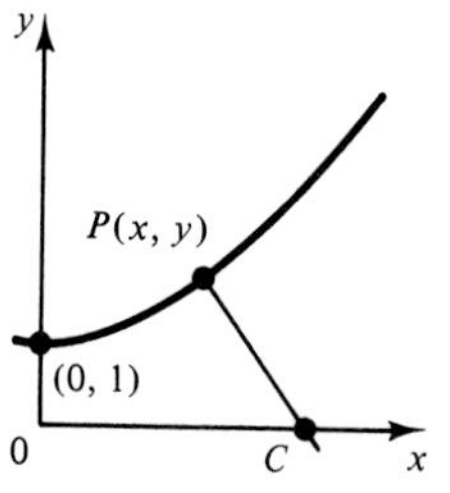

FIGURE 35-6

10. Find the equation of the curve passing through (4, 4) such that the distance OT (Fig. 35-5) is equal to the ordinate of $P$.

**Orthogonal Trajectories**

Write the equation of the orthogonal trajectories to each family of curves.

11. the circles, $x^2 + y^2 = r^2$
12. the parabolas, $x^2 = ay$
13. the hyperbolas $x^2 - y^2 = Cy$

## 35-7 EXPONENTIAL GROWTH AND DECAY

In Chapter 19 we derived Eqs. 197, 199, and 200 for exponential growth and decay and exponential growth to an upper limit by means of compound interest formulas. Here we derive the equation for exponential growth to an upper limit by solving the differential equation that describes such growth. The derivation of equations for exponential growth and decay are left as an exercise.

---

**EXAMPLE 31:** A quantity starts from zero and grows with time such that its rate of growth is proportional to the difference between the final amount $a$ and the present amount $y$. Find an equation for $y$ as a function of time.

**Solution:** The amount present at time $t$ is $y$, and the rate of growth of $y$ we write as $dy/dt$. Since the rate of growth is proportional to $(a - y)$, we write the differential equation

$$\frac{dy}{dt} = n(a - y)$$

where $n$ is a constant of proportionality. Separating variables, we obtain

$$\frac{dy}{a - y} = n\,dt$$

Integrating gives us

$$-\ln(a - y) = nt + C$$

Going to exponential form and simplifying yields

$$a - y = e^{-nt-C} = e^{-nt}e^{-C} = C_1e^{-nt}$$

where $C_1 = e^{-C}$. Applying the initial condition that $y = 0$ when $t = 0$ gives $C_1 = a$, so our equation becomes $a - y = ae^{-nt}$, or

| | | |
|---|---|---|
| Exponential Growth to an Upper Limit | $y = a(1 - e^{-nt})$ | **200** |

---

### Motion in a Resisting Fluid

Here we continue our study of motion. In Chapter 29 we showed that the instantaneous velocity is given by the derivative of the displacement, and that the instantaneous acceleration is given by the derivative of the velocity (or by the second derivative of the displacement). In Chapter 31 we solved simple differential equations to find displacement given the velocity or acceleration. Here we do a type of problem that we were not able to solve in Chapter 31.

In this type of problem, an object falls through a fluid (usually air or water) which exerts a *resisting force* that is proportional to the velocity of the object and in the opposite direction. We set up these problems using Newton's second law, $F = ma$, and we'll see that the motion follows the law for exponential growth described by Eq. 200.

**EXAMPLE 32:** A crate falls from rest from an airplane. The air resistance is proportional to the crate's velocity, and the crate reaches a limiting speed of 218 ft/s. (a) Write an equation for the crate's velocity. (b) Find the crate's velocity after 0.75 s.

**Solution:** (a) By Newton's second law,

$$F = ma = \frac{W}{g}\frac{dv}{dt}$$

where $W$ is the weight of the crate, $dv/dt$ is the acceleration, and $g = 32.2$ ft/s$^2$. Taking the downward direction as positive, the resultant force $F$ is equal to $W - kv$, where $k$ is a constant of proportionality. So

$$W - kv = \frac{W}{32.2}\frac{dv}{dt}$$

We can find $k$ by noting that the acceleration must be zero when the limiting speed (218 ft/s) is reached. Thus

$$W - 218k = 0$$

Notice that the weight $W$ has dropped out.

So $k = W/218$. Our differential equation, after multiplying by $218/W$, is then $218 - v = 6.77\, dv/dt$. Separating variables and integrating, we have

$$\frac{dv}{218 - v} = 0.148\, dt$$

$$\ln|218 - v| = -0.148t + C$$

$$218 - v = e^{-0.148t+C} = e^{-0.148t}e^{C}$$

$$= C_1 e^{-0.148t}$$

Our equation for $v$ is of the same form as Eq. 200, for exponential growth to an upper limit.

Since $v = 0$ when $t = 0$, $C_1 = 218 - 0 = 218$. Then

$$v = 218(1 - e^{-0.148t})$$

(b) When $t = 0.75$ s,

$$v = 218(1 - e^{-0.111}) = 22.9 \text{ ft/s}$$

## EXERCISE 7—EXPONENTIAL GROWTH AND DECAY

### Exponential Growth

1. A quantity grows with time such that its rate of growth $dy/dt$ is proportional to the present amount $y$. Use this statement to derive the equation for exponential growth, $y = ae^{nt}$.
2. A biomedical company finds that a certain bacteria used for crop insect control will grow exponentially at the rate of 12% per hour. Starting with 1000 bacteria, how many will they have after 10 h?
3. If the U.S. energy consumption in 1979 was 37 million barrels (bbl) per day oil equivalent and is growing exponentially at a rate of 6.9% per year, estimate the daily oil consumption in the year 2000.

We'll have more problems involving exponential growth and decay in the following section on electric circuits.

### Exponential Decay

4. A quantity decreases with time such that its rate of decrease $dy/dt$ is proportional to the present amount $y$. Use this statement to derive the equation for exponential decay, $y = ae^{-nt}$.

5. An iron ingot is 1850°F above room temperature. If it cools exponentially at 3.5% per minute, find its temperature (above room temperature) after 2.50 h.
6. A certain pulley in a tape drive is rotating at 2550 rev/min. After the power is shut off, its speed decreases exponentially at a rate of 12.5% per second. Find the pulley's speed after 5.00 s.

### Exponential Growth to an Upper Limit

7. A forging, initially at 0°F, is placed in a furnace at 1550°F, where its temperature rises exponentially at the rate of 6.5% per minute. Find its temperature after 25.0 min.
8. If we assume that the compressive strength of concrete increases exponentially with time to an upper limit of 4000 lb/in$^2$, and that the rate of increase is 52.5% per week, find the strength after 2 weeks.

### Motion in a Resisting Medium

9. A 45.5-lb carton is initially at rest. It is then pulled horizontally by a 19.4-lb force in the direction of motion, and resisted by a frictional force which is equal (in pounds) to four times the carton's velocity (in ft/s). Show that the differential equation of motion is $dv/dt = 13.7 - 2.83v$.
10. For the carton in Problem 9, find the velocity after 1.25 s.
11. An instrument package is dropped from an airplane. It falls from rest through air whose resisting force is proportional to the speed of the package. The terminal speed is 155 ft/s. Show that the acceleration is given by the differential equation $a = dv/dt = g - gv/155$.
12. Find the speed of the instrument package in Problem 11 after 0.50 s.
13. Find the displacement of the instrument package in Problem 11 after 1.00 s.
14. A 157-lb stone falls from rest from a cliff. If the air resistance is proportional to the square of the stone's speed, and the limiting speed of the stone is 125 ft/s, show that the differential equation of motion is $dv/dt = g - gv^2/15{,}625$.
15. Find the time for the velocity of the stone in Problem 14 to be 60 ft/s.
16. A 15.0-lb ball is thrown downward from an airplane with a speed of 21.0 ft/s. If we assume the air resistance to be proportional to the ball's speed, and the limiting speed is 135 ft/s, show that the velocity of the ball is given by $v = 135 - 114e^{-t/4.19}$ ft/s.
17. Find the time at which the ball in Problem 16 is going at a speed of 70.0 ft/s.
18. A box falls from rest and encounters air resistance proportional to the cube of the speed. The limiting speed is 12.5 ft/s. Show that the acceleration is given by the differential equation $60.7\ dv/dt = 1953 - v^3$.

## 35-8 SERIES *RL* AND *RC* CIRCUITS

### Series *RL* Circuit

Figure 35-7 shows a resistance of $R$ ohms in series with an inductance of $L$ henrys. The switch can either connect these elements to a battery of voltage $E$ (position 1, charge) or to a short circuit (position 2, discharge). In either case our object is to find the current $i$ in the circuit. We will see that it is composed of two parts; a *steady-state* current that flows long after the switch has been thrown, and a *transient* current that dies down shortly after the switch is thrown.

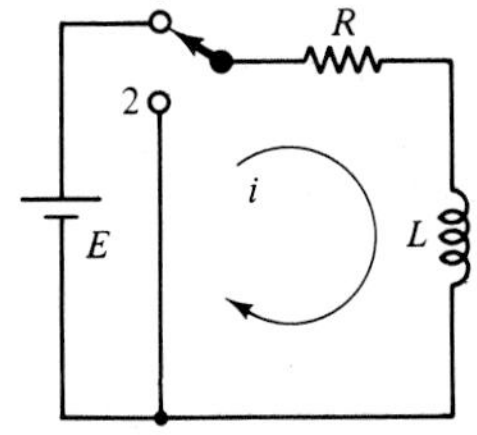

**FIGURE 35-7** ***RL* circuit.**

**EXAMPLE 33:** After being in position 2 for a long time, the switch (Fig. 35-7) is thrown into position 1 at $t = 0$. Write an expression for (a) the current $i$ and (b) the voltage across the inductor.

**Solution:** (a) The voltage $V_L$ across an inductance $L$ is given by Eq. A86, $v_L = L\,di/dt$. Using Kirchhoff's voltage law (Eq. A68) gives

$$Ri + L\frac{di}{dt} = E \tag{1}$$

or

$$\frac{di}{dt} + \frac{R}{L}i = \frac{E}{L} \tag{2}$$

We recognize this as a first-order linear differential equation. Our integrating factor is

$$e^{\int (R/L)dt} = e^{Rt/L}$$

In electrical problems, we will use $k$ for the constant of integration, saving $C$ for capacitance. Similarly, $R$ here is resistance, not the integrating constant.

Multiplying (2) by the integrating factor and integrating, we obtain

$$ie^{Rt/L} = \frac{E}{L}\int e^{Rt/L}\,dt = \frac{E}{R}e^{Rt/L} + k$$

Dividing by $e^{Rt/L}$ gives us

$$i = \frac{E}{R} + ke^{-Rt/L} \tag{3}$$

We now evaluate $k$ by noting that $i = 0$ at $t = 0$. So $ke^0 = k = -E/R$. Substituting into (3) yields

| | | |
|---|---|---|
| Current in a Charging Inductor | $i = \frac{E}{R}(1 - e^{-Rt/L})$ | **A87** |

Note that the equation for the voltage across the inductor is of the same form as for exponential decay (Eq. 199), and that the equation for the current is the same as that for exponential growth to an upper limit (Eq. 200).

The first term in this expression $(E/R)$ is the steady-state current, and the second term $(E/R)(e^{-Rt/L})$ is the transient current.

(b) From (1), the voltage across the inductor is

$$v_L = L\frac{di}{dt} = E - Ri$$

Using Eq. A87, we see that $Ri = E - Ee^{-Rt/L}$. Then

$$v_L = E - E + Ee^{-Rt/L}$$

So

| | | |
|---|---|---|
| Voltage across a Charging Inductor | $v_L = Ee^{-Rt/L}$ | **A88** |

### Series *RC* Circuit

We now analyze the *RC* circuit as we did the *RL* circuit.

---

**EXAMPLE 34:** A fully charged capacitor (Fig. 35-8) is discharged by throwing the switch from 1 to 2 at $t = 0$. Write an expression for (a) the voltage across the capacitor and (b) the current $i$.

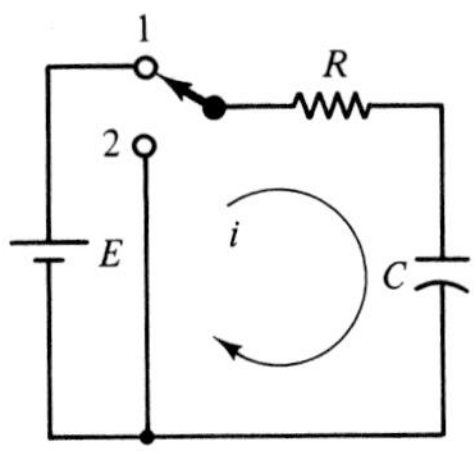


**FIGURE 35-8** ***RC* circuit.**

**Solution:**

(a) By Kirchhoff's law, the voltage $v_R$ across the resistor at any instant must be equal to the voltage $v$ across the capacitor, but of opposite sign. Further, the current through the resistor, $v_R/R$, or $-v/R$, must be equal to the current $C\,dv/dt$ in the capacitor.

$$-\frac{v}{R} = C\frac{dv}{dt}$$

Separating variables and integrating gives

$$\frac{dv}{v} = -\frac{dt}{RC}$$

$$\ln v = -\frac{t}{RC} + k$$

But at $t = 0$ the voltage across the capacitor is the battery voltage $E$, so $k = \ln E$. Substituting, we obtain

$$\ln v - \ln E = \ln\frac{v}{E} = -\frac{t}{RC}$$

Or, in exponential form, $v/E = e^{-t/RC}$, or

| | | |
|---|---|---|
| Voltage across a Discharging Capacitor | $v = Ee^{-t/RC}$ | **A84** |

(b) We get the current through the resistor (and the capacitor) by dividing the voltage $v$ by $R$,

**Equations A83 and A84 are both for exponential decay.**

| | | |
|---|---|---|
| Current in a Discharging Capacitor | $i = \frac{E}{R}e^{-t/RC}$ | **A83** |

---

**EXAMPLE 35:** For the circuit of Fig. 35-8, $R = 1540\ \Omega$, $C = 125\ \mu F$, and $E = 115$ V. If the switch is thrown from position 1 to position 2 at $t = 0$, find the current and the voltage across the capacitor at $t = 60$ ms.

**Solution:** We first compute $1/RC$.

$$\frac{1}{RC} = \frac{1}{1540 \times 125 \times 10^{-6}} = 5.19$$

Then from Eqs. A83 and A84,

$$i = \frac{E}{R} e^{-t/RC} = \frac{115}{1540} e^{-5.19t}$$

and

$$v = Ee^{-t/RC} = 115\ e^{-5.19t}$$

At $t = 0.060$ s, $e^{-5.19t} = 0.732$, so

$$i = \frac{115}{1540}(0.732) = 0.0547 \text{ A} = 54.7 \text{ mA}$$

and

$$v = 115(0.732) = 84.2 \text{ V}$$

## Alternating Source

We now consider the case where the *RL* circuit or *RC* circuit is connected to an alternating rather than a direct source of voltage.

**EXAMPLE 36:** A switch (Fig. 35-9) is closed at $t = 0$, thus applying an alternating voltage of amplitude $E$ to a resistor and inductor in series. Write an expression for the current.

**Solution:** By Kirchhoff's voltage law, $Ri + L\ di/dt = E \sin \omega t$, or

$$\frac{di}{dt} + \frac{R}{L} i = \frac{E}{L} \sin \omega t$$

This is a first-order linear differential equation. Our integrating factor is $e^{\int R/L dt} = e^{Rt/L}$. Thus

$$ie^{Rt/L} = \frac{E}{L}\int e^{Rt/L} \sin \omega t\ dt$$

$$= \frac{E}{L}\frac{e^{Rt/L}}{(R^2/L^2 + \omega^2)}\left(\frac{R}{L} \sin \omega t - \omega \cos \omega t\right) + k$$

by Rule 41. We now divide through by $e^{Rt/L}$, and after some manipulation get

$$i = E\frac{R \sin \omega t - \omega L \cos \omega t}{R^2 + \omega^2 L^2} + ke^{-Rt/L}$$

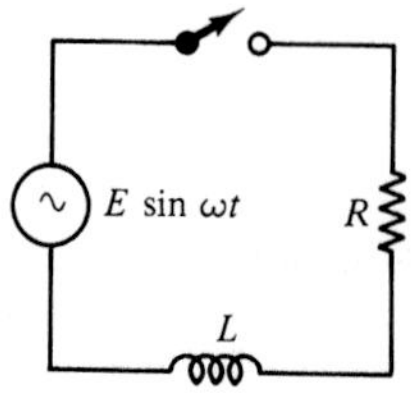


**FIGURE 35-9** ***RL* circuit with ac source.**

From the impedance triangle (Fig. 35-10) we see that $R^2 + \omega^2 L^2 = Z^2$, the square of the impedance. Further, by Ohm's law for ac, $E/Z = I$, the amplitude of the current wave. Thus

$$i = \frac{E}{Z}\left(\frac{R}{Z}\sin \omega t - \frac{\omega L}{Z}\cos \omega t\right) + ke^{-Rt/L}$$

$$= I\left(\frac{R}{Z}\sin \omega t - \frac{\omega L}{Z}\cos \omega t\right) + ke^{-Rt/L}$$

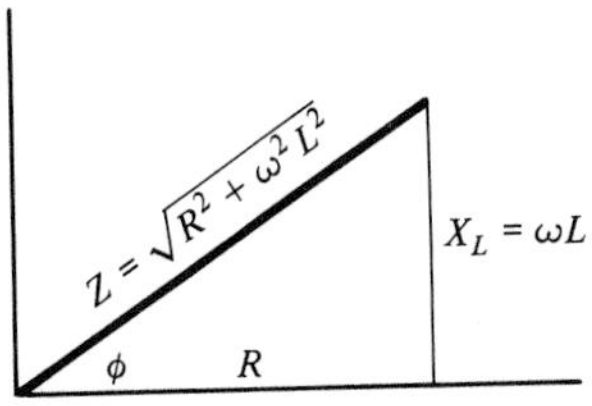


**FIGURE 35-10 Impedance triangle.**

Again from the impedance triangle, $R/Z = \cos \phi$ and $\omega L/Z = \sin \phi$. Substituting yields

$$i = I(\sin \omega t \cos \phi - \cos \omega t \sin \phi) + ke^{-Rt/L}$$

$$= I \sin(\omega t - \phi) + ke^{-Rt/L}$$

which we get by means of the trigonometric identity for the sine of the difference of two angles (Eq. 167). Evaluating $k$, we note that $i = 0$ when $t = 0$, so

$$k = -I \sin(-\phi) = I \sin \phi = \frac{IX_L}{Z}$$

where, from the impedance triangle, $\sin \phi = X_L/Z$. Substituting, we obtain

$$i = \underbrace{I \sin(\omega t - \phi)}_{\text{steady-state current}} + \underbrace{\frac{IX_L}{Z} e^{-Rt/L}}_{\text{transient current}}$$

Our current thus has two parts; (1) a steady-state alternating current of magnitude $I$, out of phase with the applied voltage by an angle $\phi$, and (2) a transient current with an initial value of $IX_L/Z$, which decays exponentially.

## EXERCISE 8—SERIES *RL* AND *RC* CIRCUITS

### Series RL Circuit

1. If the inductor in Fig. 35-7 is discharged by throwing the switch from position 1 to 2, show that the current decays exponentially according to the function $i = (E/R)e^{-Rt/L}$.
2. The voltage across an inductor is equal to $L\, di/dt$. Show that the magnitude of the voltage across the inductance in Problem 1 decays exponentially according to the function $v = Ee^{-Rt/L}$.
3. When the switch in Fig. 35-7 is thrown from 2 to 1, (charging) we showed that the current grows exponentially to an upper limit and is given by $i = (E/R)(1 - e^{-Rt/L})$. Show that the voltage across the inductance ($L\, di/dt$) decays exponentially and is given by $v = Ee^{-Rt/L}$.
4. For the circuit of Fig. 35-7, $R = 382\ \Omega$, $L = 4.75$ H, and $E = 125$ V. If the switch is thrown from 2 to 1 (charging), find the current and the voltage across the inductance at $t = 2.00$ ms.

### Series RC Circuit

5. The voltage $v$ across the capacitor in Fig. 35-8 during charging is described by the differential equation $(E - v)/R = C\, dv/dt$. Solve this differential equation to show that the voltage is given by $v = E(1 - e^{-t/RC})$. Hint: Write the given

equation in the form of a first-order linear differential equation and solve using Eq. 414.

6. Show that, in problem 5, the current through the resistor (and hence through the capacitor) is given by $i = (E/R)e^{-t/RC}$.
7. For the circuit of Fig. 35-8, $R = 538\ \Omega$, $C = 525\ \mu F$, and $E = 125$ V. If the switch is thrown from 2 to 1 (charging), find the current and the voltage across the capacitor at $t = 2.00$ ms.

### Circuits in Which R, L, and C Are Not Constant

8. For the circuit of Fig. 35-7, $L = 2.00\ H$, $E = 60.0$ V, and the resistance decreases with time according to the expression $R = 4.00/(t + 1)$. Show that the current $i$ is given by $i = 10(t + 1) - 10(t + 1)^{-2}$.
9. For the circuit in Problem 8, find the current at $t = 1.55$ ms.
10. For the circuit of Fig. 35-7, $R = 10.0\ \Omega$, $E = 100$ V, and the inductance varies with time according to the expression $L = 5.00t + 2.00$. Show that the current $i$ is given by the expression $i = 10.0 - 40/(5t + 2)^2$.
11. For the circuit in Problem 10, find the current at $t = 4.82$ ms.
12. For the circuit of Fig. 35-7, $E = 300$ V, the resistance varies with time according to the expression $R = 4.00t$, and the inductance varies with time according to the expression $L = t^2 + 4.00$. Show that the current $i$ as a function of time is given by $i = 100t(t^2 + 12)/(t^2 + 4)^2$.
13. For the circuit in Problem 12, find the current at $t = 1.85$ ms.
14. For the circuit of Fig. 35-8, $C = 2.55\ \mu F$, $E = 625$ V, and the resistance varies with current according to the expression $R = 493 + 372i$. Show that the differential equation for current is $di/dt + 1.51i\ di/dt + 795i = 0$.

### Series RL or RC Circuit with Alternating Current

15. For the circuit of Fig. 35-9, $R = 233\ \Omega$, $L = 5.82$ H, and $e = 58.0 \sin 377t$ volts. If the switch is closed when $e$ is zero and increasing, show that the current is given by $i = 26.3 \sin(377t - 83.9°) + 26.1e^{-40t}$ mA.
16. For the circuit in Problem 15, find the current at $t = 2.00$ ms.
17. For the circuit in Fig. 35-8, the applied voltage is alternating and is given by $e = E \sin \omega t$. If the switch is thrown from 2 to 1 (charging) when $e$ is zero and increasing, show that the current is given by $i = (E/Z)[\sin(\omega t + \phi) - e^{-t/RC} \sin \phi]$. In your derivation, follow the steps used for the $RL$ circuit with an ac source.
18. For the circuit in Problem 17, $R = 837\ \Omega$, $C = 2.96\ \mu F$, and $e = 58.0 \sin 377t$. Find the current at $t = 1.00$ ms.

## 35-9 SECOND-ORDER DIFFERENTIAL EQUATIONS

### The General Second-Order Linear Differential Equation

A linear differential equation of second order can be written in the form

$$Py'' + Qy' + Ry = S$$

where $P$, $Q$, $R$, and $S$ are constants or functions of $x$.

A second-order linear differential equation *with constant coefficients is one where* $P$, $Q$, and $R$ are *constants,* although $S$ can be a function of $x$, such as

| | | |
|---|---|---|
| Form of Second-Order Linear DE, Right Side Not Zero | $ay'' + by' + cy = f(x)$ | **420** |

where $a$, $b$, and $c$ are constants. This is the type of equation we will solve in this chapter.

Equation 420 is said to be *homogeneous* if $f(x) = 0$, and is called *nonhomogeneous* if $f(x)$ is not zero. We will, instead, usually refer to these equations as "right-hand side zero" and "right-hand side not zero."

### Operator Notation

Differential equations are often written using the $D$ operator we first introduced in Chapter 27, where

**We'll usually use the more familiar $y'$ notation rather than the $D$ operator.**

$$Dy = y' \qquad D^2y = y'' \qquad D^3y = y''' \qquad \text{etc.}$$

Thus Eq. 420 can be written

$$aD^2y + bDy + c = f(x)$$

### Second-Order Differential Equations with Variables Separable

We develop methods for solving the general second-order equation in Sec. 35-10. However, simple differential equations of second-order which are lacking a first derivative term can be solved by separation of variables, as in the following example.

---

**EXAMPLE 37:** Solve the equation $y'' = 3 \cos x$ (where $x$ is in radians) if $y' = 1$ at the point (2, 1).

**Solution:** Replacing $y''$ by $d(y')/dx$ and multiplying both sides by $dx$ to separate variables,

$$d(y') = 3 \cos x \, dx$$

Integrating gives us

$$y' = 3 \sin x + C_1$$

Since $y' = 1$ when $x = 2$ rad, $C_1 = 1 - 3 \sin 2 = -1.73$, so

$$y' = 3 \sin x - 1.73$$

**Notice that we have to integrate twice to solve a second-order DE and that two constants of integration have to be evaluated.**

or $dy = 3 \sin x \, dx - 1.73 \, dx$. Integrating again, we have

$$y = -3 \cos x - 1.73x + C_2$$

At the point (2, 1), $C_2 = 1 + 3 \cos 2 + 1.73(2) = 3.21$. Our solution is then

$$y = -3 \cos x - 1.73x + 3.21$$

---

### EXERCISE 9—SECOND-ORDER DIFFERENTIAL EQUATIONS

Solve each equation for $y$.

**1.** $y'' = 5$

**2.** $y'' = x$

**3.** $y'' = 3e^x$

**4.** $y'' = \sin 2x$

**5.** $y'' - x^2 = 0$, where $y' = 1$ at the point (0, 0)

**6.** $xy'' = 1$, where $y' = 2$ at the point (1, 1)

## 35-10 SECOND-ORDER DIFFERENTIAL EQUATIONS WITH CONSTANT COEFFICIENTS AND RIGHT SIDE ZERO

### Solving a Second-Order Equation with Right Side Equal to Zero

If the right side, $f(x)$, in Eq. 420 is zero, and $a$, $b$, and $c$ are constants, we have the (homogeneous) equation

| | | |
|---|---|---|
| Form of Second-Order DE, Right Side Zero | $ay'' + by' + cy = 0$ | **415** |

To solve this equation, we note that the sum of the three terms on the left side must equal zero. So a solution must be a value of $y$ that will make these terms alike, so they may be combined to give a sum of zero. Also note that each term on the left contains $y$ or its first or second derivative. Thus *a possible solution is a function such that it and its derivatives are like terms.* Recall from Chapter 30 that one such function was the exponential function

$$y = e^{mx}$$

It has derivatives $y' = me^{mx}$ and $y'' = m^2e^{mx}$, which are all like terms. We thus try this for our solution. Substituting $y = e^{mx}$ and its derivatives into Eq. 415 gives

$$am^2e^{mx} + bme^{mx} + ce^{mx} = 0$$

Factoring, we obtain

$$e^{mx}(am^2 + bm + c) = 0$$

Since $e^{mx}$ can never be zero, this equation is satisfied only when $am^2 + bm + c = 0$. This is called the *auxiliary* or *characteristic equation.*

**Note that the coefficients in the auxiliary equation are the same as those in the original DE. We can thus get the auxiliary equation by inspection of the DE.**

| | | |
|---|---|---|
| Auxiliary Equation of Second-Order DE, Right Side Zero | $am^2 + bm + c = 0$ | **416** |

The auxiliary equation is a quadratic, and has two roots, which we call $m_1$ and $m_2$. Either of the two values of $m$ makes the auxiliary equation equal to zero. Thus we get *two solutions* to Eq. 415, $y = e^{m_1x}$ and $y = e^{m_2x}$.

### General Solution

We now show that if $y_1$ and $y_2$ are each solutions to $ay'' + by' + cy = 0$, then $y = c_1y_1 + c_2y_2$ is also a solution.

We first observe that if $y = c_1y_1 + c_2y_2$, then

$$y' = c_1y_1' + c_2y_2' \qquad \text{and} \qquad y'' = c_1y_1'' + c_2y_2''$$

Substituting into the differential equation (415), we have

$$a(c_1y_1'' + c_2y_2'') + b(c_1y_1' + c_2y_2') + c(c_1y_1 + c_2y_2) = 0$$

This simplifies to

$$c_1(ay_1'' + by_1' + cy_1) + c_2(ay_2'' + by_2' + cy_2) = 0$$

But $ay_1'' + by_1' + cy_1 = 0$ and $ay_2'' + by_2' + cy_2 = 0$ from Eq. 415, so

$$c_1(0) + c_2(0) = 0$$

showing that if $y_l$ and $y_2$ are solutions to the differential equation, then $y = c_1y_1 + c_2y_2$ is also a solution to the differential equation. Since $e^{m_1x}$ and $y = e^{m_2x}$ are solutions to Eq. 415, the complete solution is then of the form $y = c_1e^{m_1x} + c_2e^{m_2x}$.

| | | |
|---|---|---|
| General Solution of Second-Order DE, Right Side Zero | $y = c_1e^{m_1x} + c_2e^{m_2x}$ | **417** |

We see that the solution to a second-order differential equation depends on the nature of the roots of the auxiliary equation. We thus look at the roots of the auxiliary equation in order to quickly write the solution to the differential equation.

The roots of a quadratic can be real and unequal, real and equal, or nonreal. We now give examples of each case.

## Roots Real and Unequal

---

**EXAMPLE 38:** Solve the equation $y'' - 3y' + 2y = 0$.

**Solution:** We get the auxiliary equation by inspection.

$$m^2 - 3m + 2 = 0$$

It factors into $(m - 1)$ and $(m - 2)$. Setting each factor equal to zero gives $m = 1$ and $m = 2$. These roots are real and unequal; our solution, by Eq. 417 is then

$$y = c_1e^x + c_2e^{2x}$$

---

Sometimes one root of the auxiliary equation will be zero, as in the next example.

---

**EXAMPLE 39:** Solve the equation $y'' - 5y' = 0$.

**Solution:** The auxiliary equation is $m^2 - 5m = 0$, which factors into $m(m - 5) = 0$. Setting each factor equal to zero gives

$$m = 0 \quad \text{and} \quad m = 5$$

Our solution is then

$$y = c_1 + c_2e^{5x}$$

---

If the auxiliary equation cannot be factored, we use the quadratic formula to find its roots.

**EXAMPLE 40:** Solve $4.82y'' + 5.85y' - 7.26y = 0$.

**Solution:** The auxiliary equation is $4.82m^2 + 5.85m - 7.26 = 0$. By the quadratic formula,

$$m = \frac{-5.85 \pm \sqrt{(5.85)^2 - 4(4.82)(-7.26)}}{2(4.82)} = 0.762 \quad \text{or} \quad -1.98$$

Our solution is then

$$y = c_1e^{0.762x} + c_2e^{-1.98x}$$

### Roots Real and Equal

If $b^2 - 4ac$ is zero, the auxiliary equation has the double root

$$m = \frac{-b \pm \sqrt{b^2 - 4ac}}{2a} = -\frac{b}{2a}$$

Our solution $y = c_1e^{mx} + c_2e^{mx}$ seems to contain two arbitrary constants, but actually does not. Factoring gives

$$y = (c_1 + c_2)e^{mx} = c_3e^{mx}$$

where $c_3 = c_1 + c_2$. A solution with one constant cannot be a complete solution for a second-order equation.

Let us *assume* that there is *a second solution* $y = ue^{mx}$, where $u$ is some function of $x$ that we are free to choose. Differentiating, we have

$$y' = mue^{mx} + u'e^{mx}$$

and

$$y'' = m^2ue^{mx} + mu'e^{mx} + mu'e^{mx} + u''e^{mx}$$

Substituting into $ay'' + by' + cy = 0$ gives

$$a(m^2ue^{mx} + 2mu'e^{mx} + u''e^{mx}) + b(mue^{mx} + u'e^{mx}) + cue^{mx} = 0$$

which simplifies to

$$e^{mx}[u(am^2 + bm + c) + u'(2am + b) + au''] = 0$$

But $am^2 + bm + c = 0$. Also, $m = -b/2a$, so $2am + b = 0$. Our equation then becomes $e^{mx}(u'') = 0$. Since $e^{mx}$ cannot be zero, we have $u'' = 0$. Thus any $u$ that has a zero second derivative will make $ue^{mx}$ a solution to the differential equation. The simplest $u$ (not a constant) for which $u'' = 0$ is $u = x$. Thus $xe^{mx}$ is a solution to the differential equation, and the complete solution to Eq. 415 is

| | | |
|---|---|---|
| Equal Roots | $y = c_1e^{mx} + c_2xe^{mx}$ | **418** |

**EXAMPLE 41:** Solve $y'' - 6y' + 9y = 0$.

**Solution:** The auxiliary equation $m^2 - 6m + 9 = 0$ has the double root $m = 3$. Our solution is then

$$y = c_1e^{3x} + c_2xe^{3x}$$

### Euler's Formula

When the roots of the auxiliary equation are nonreal, our solution will contain expressions of the form $e^{jbx}$. In the following section we will want to simplify such expressions using *Euler's formula,* which we derive here.

We have already used Euler's formula in Sec. 20-4.

Let $z = \cos\theta + j\sin\theta$, where $j = \sqrt{-1}$ and $\theta$ is in radians. Then

$$\frac{dz}{d\theta} = -\sin\theta + j\cos\theta$$

Multiplying by $j$ (and recalling that $j^2 = -1$) gives

$$j\frac{dz}{d\theta} = -j\sin\theta - \cos\theta = -z$$

Multiplying by $-j$ we get $dz/d\theta = jz$. We now separate variables and integrate

$$\frac{dz}{z} = j\,d\theta$$

Integrating, we obtain $\ln z = j\theta + c$

When $\theta = 0$, $z = \cos 0 + j\sin 0 = 1$. So $c = \ln z - j\theta = \ln 1 - 0 = 0$. Thus $\ln z = j\theta$, or, in exponential form, $z = e^{j\theta}$. But $z = \cos\theta + j\sin\theta$, so we arrive at Euler's formula:

| Euler's Formula | $e^{j\theta} = \cos\theta + j\sin\theta$ | **227** |
|---|---|---|

We can get two more useful forms of Eq. 227 for $e^{jbx}$ and $e^{-jbx}$. First we set $\theta = bx$. Thus

$$e^{jbx} = \cos bx + j\sin bx \tag{1}$$

Further,

$$e^{-jbx} = \cos(-bx) + j\sin(-bx) = \cos bx - j\sin bx \tag{2}$$

since $\cos(-A) = \cos A$, and $\sin(-A) = -\sin A$.

### Roots Not Real

We return to our second-order differential equation whose solution we are finding by means of the auxiliary equation. We now see that if the auxiliary equation has the nonreal roots $a + jb$ and $a - jb$, our solution becomes

$$\begin{aligned} y &= c_1e^{(a+jb)x} + c_2e^{(a-jb)x} \\ &= c_1e^{ax}e^{jbx} + c_2e^{ax}e^{-jbx} = e^{ax}(c_1e^{jbx} + c_2e^{-jbx}) \end{aligned}$$

Using Euler's Formula gives

$$\begin{aligned} y &= e^{ax}[c_1(\cos bx + j\sin bx) + c_2(\cos bx - j\sin bx)] \\ &= e^{ax}[(c_1 + c_2)\cos bx + j(c_1 - c_2)\sin bx]. \end{aligned}$$

Replacing $c_1 + c_2$ by $C_1$, and $j(c_1 - c_2)$ by $C_2$ gives

| Nonreal Roots | $y = e^{ax}(C_1\cos bx + C_2\sin bx)$ | **419a** |
|---|---|---|

A more compact form of the solution may be obtained by using the equation for the sum of a sine wave and a cosine wave of the same frequency. In Sec. 17-2 we derived the formula

$$A \sin \omega t + B \cos \omega t = R \sin(\omega t + \phi)$$

where $$R = \sqrt{A^2 + B^2} \quad \text{and} \quad \phi = \arctan\frac{B}{A}$$

Thus the solution to the differential equation can take the alternative form

We'll find this form handy for applications.

| | | |
|---|---|---|
| Nonreal Roots | $y = Ce^{ax} \sin(bx + \phi)$ | **419b** |

where $C = \sqrt{C_1^2 + C_2^2}$ and $\phi = \arctan C_1/C_2$.

---

**EXAMPLE 42:** Solve $y'' - 4y' + 13y = 0$.

**Solution:** The auxiliary equation $m^2 - 4m + 13 = 0$ has the roots $m = 2 \pm j3$, two nonreal roots. Substituting into Eq. 419a with $a = 2$ and $b = 3$ gives

$$y = e^{2x}(C_1 \cos 3x + C_2 \sin 3x)$$

or

$$y = Ce^{2x} \sin(3x + \phi)$$

in the alternative form of Eq. 419b.

---

### Summary

The types of solutions to a second-order differential equation with right side zero and with constant coefficients are summarized here.

| Second-Order DE with Right Side Zero $ay'' + by' + cy = 0$ | | |
|---|---|---|
| If the roots of auxiliary equation $am^2 + bm + c = 0$ are: | Then the solution to $ay'' + by' + cy = 0$ is: | |
| Real and Unequal | $y = c_1e^{m_1x} + c_2e^{m_2x}$ | **417** |
| Real and Equal | $y = c_1e^{mx} + c_2xe^{mx}$ | **418** |
| Nonreal | $y = e^{ax}(C_1 \cos bx + C_2 \sin bx)$ | **419a** |
| | $y = Ce^{ax} \sin(bx + \phi)$ | **419b** |

### Particular Solution

As before, we use the boundary conditions to find the two constants in the solution.

---

**EXAMPLE 43:** Solve $y'' - 4y' + 3y = 0$ if $y' = 5$ at (1, 2).

**Solution:** The auxiliary equation $m^2 - 4m + 3 = 0$ has roots $m = 1$ and $m = 3$. Our solution is then

$$y = c_1e^x + c_2e^{3x}$$

At (1, 2) we get

$$2 = c_1e + c_2e^3 \tag{1}$$

Here we have one equation and two unknowns. We get a second equation by taking the derivative of $y$.

$$y' = c_1e^x + 3c_2e^{3x}$$

Substituting the boundary condition $y' = 5$ when $x = 1$ gives

$$5 = c_1e + 3c_2e^3 \tag{2}$$

We solve (1) and (2) simultaneously. Subtracting (1) from (2) gives $2c_2e^3 = 3$, or

$$c_2 = \frac{3}{2e^3}$$

$$= 0.0747$$

Then, from (1),

$$c_1 = \frac{2 - (0.0747)e^3}{e}$$

$$= 0.184$$

Our particular solution is then

$$y = 0.184e^x + 0.0747e^{3x}$$

---

### Third-Order Differential Equations

We now show (without proof) how the method of the preceding sections can be extended to simple third-order equations that can be easily factored.

---

**EXAMPLE 44:** Solve $y''' - 4y'' - 11y' + 30y = 0$.

**Solution:** We write the auxiliary equation by inspection,

$$m^3 - 4m^2 - 11m + 30 = 0$$

which factors, by trial and error, into

$$(m - 2)(m - 5)(m + 3) = 0$$

giving roots of 2, 5, and $-3$. The solution to the given equation is then

$$y = C_1e^{2x} + C_2e^{5x} + C_3e^{-3x}$$

---

## EXERCISE 10—SECOND-ORDER DIFFERENTIAL EQUATIONS WITH CONSTANT COEFFICIENTS AND RIGHT SIDE ZERO

Find the general solution to each differential equation.

### Second-Order DE, Roots of Auxiliary Equation Real and Unequal

1. $y'' - 6y' + 5y = 0$
2. $2y'' - 5y' - 3y = 0$
3. $y'' - 3y' + 2y = 0$
4. $y'' + 4y' - 5y = 0$
5. $y'' - y' - 6y = 0$
6. $y'' + 5y' + 6y = 0$
7. $5y'' - 2y' = 0$
8. $y'' + 4y' + 3y = 0$
9. $6y'' + 5y' - 6y = 0$
10. $y'' - 4y' + y = 0$

### Second-Order DE, Roots of Auxiliary Equation Real and Equal

11. $y'' - 4y' + 4y = 0$
12. $y'' - 6y' + 9y = 0$
13. $y'' - 2y' + y = 0$
14. $9y'' - 6y' + y = 0$
15. $y'' + 4y' + 4y = 0$
16. $4y'' + 4y' + y = 0$
17. $y'' + 2y' + y = 0$
18. $y'' - 10y' + 25y = 0$

### Second-Order DE, Roots of Auxiliary Equation Not Real

19. $y'' + 4y' + 13y = 0$
20. $y'' - 2y' + 2y = 0$
21. $y'' - 6y' + 25y = 0$
22. $y'' + 2y' + 2y = 0$
23. $y'' + 4y = 0$
24. $y'' + 2y = 0$
25. $y'' - 4y' + 5y = 0$
26. $y'' + 10y' + 425y = 0$

### Particular Solution

Solve each differential equation. Use the given boundary conditions to find the constants of integration.

27. $y'' + 6y' + 9y = 0,\quad y = 0$ and $y' = 3$ when $x = 0$
28. $y'' + 3y' + 2y = 0,\quad y = 0$ and $y' = 1$ when $x = 0$
29. $y'' - 2y' + y = 0,\quad y = 5$ and $y' = -9$ when $x = 0$
30. $y'' + 3y' - 4y = 0,\quad y = 4$ and $y' = -2$ when $x = 0$
31. $y'' - 2y' = 0,\quad y = 1 + e^2$ and $y' = 2e^2$ when $x = 1$
32. $y'' + 2y' + y = 0,\quad y = 1$ and $y' = -1$ when $x = 0$
33. $y'' - 4y = 0,\quad y = 1$ and $y' = -1$ when $x = 0$
34. $y'' + 9y = 0,\quad y = 2$ and $y' = 0$ when $x = \pi/6$
35. $y'' + 2y' + 2y = 0,\quad y = 0$ and $y' = 1$ when $x = 0$
36. $y'' + 4y' + 13y = 0,\quad y = 0$ and $y' = 12$ when $x = 0$

### Third-Order DE

Solve each differential equation.

37. $y''' - 2y'' - y' + 2y = 0$

38. $y''' - y' = 0$
39. $y''' - 6y'' + 11y' - 6y = 0$
40. $y''' + y'' - 4y' - 4y = 0$
41. $y''' - 3y'' - y' + 3y = 0$
42. $y''' - 7y' + 6y = 0$
43. $4y''' - 3y' + y = 0$
44. $y''' - y'' = 0$

## 35-11 SECOND-ORDER DIFFERENTIAL EQUATIONS WITH RIGHT SIDE NOT ZERO

### Complementary Function and Particular Integral

We'll now see that the solution to a differential equation is made up of *two parts*, the *complementary function* and the *particular integral*. We show this now for a first-order equation, and later, for a second-order equation.

The solution to a first-order differential equation, say

$$y' + \frac{y}{x} = 4 \tag{1}$$

can be found by the methods of Secs. 35-1 to 35-9. The solution to (1) is

$$y = c/x + 2x$$

Note that the solution has two parts. Let us label one part $y_c$ and the other $y_p$. Thus $y = y_c + y_p$, where $y_c = c/x$ and $y_p = 2x$.

Let us substitute for $y$ only $y_c = c/x$ into the left side of (1), and for $y'$ the derivative $-c/x^2$.

$$-\frac{c}{x^2} + \frac{c}{x^2} = 0$$

We get zero instead of the required 4 on the right-hand side. Thus $y_c$ *does not satisfy (1)*. It does, however, satisfy what we call the *reduced equation,* obtained by setting the right side equal to zero. We call $y_c$ the *complementary function*.

We now substitute only $y_p = 2x$ as $y$ in the left side of (1). Similarly, for $y'$, we substitute 2. We now have

$$2 + \frac{2x}{x} = 4$$

We see that $y_p$ *does* satisfy (1), and is hence a solution. But it cannot be a complete solution because it has no arbitrary constant. We call $y_p$ a *particular integral*. The quantity $y_c$ had the required constant but did not, by itself, satisfy (1).

**Don't confuse *particular integral* with *particular solution.***

However, *the sum of $y_c$ and $y_p$ satisfies (1) and has the required number of constants, and is hence the complete solution.*

| | | |
|---|---|---|
| Complete Solution | $y = y_c + y_p$ <br> $= \begin{pmatrix}\text{complementary}\\ \text{function}\end{pmatrix} + \begin{pmatrix}\text{particular}\\ \text{integral}\end{pmatrix}$ | **421** |

## Second-Order Differential Equation

We have seen that the solution to a first-order equation is made up of a complementary function and a particular integral. But is the same true of the second-order equation?

$$ay'' + by' + cy = f(x) \tag{420}$$

If a particular integral $y_p$ is a solution to Eq. 420, we get, on substituting,

$$ay_p'' + by_p' + cy_p = f(x) \tag{2}$$

If the complementary function $y_c$ is a complete solution to the reduced equation

$$ay'' + by' + cy = 0 \tag{415}$$

we get

$$ay_c'' + by_c' + cy_c = 0 \tag{3}$$

Adding (2) and (3) gives us

$$a(y_p'' + y_c'') + b(y_p' + y_c') + c(y_p + y_c) = f(x)$$

Since the sum of the two derivatives is the derivative of the sum, we get

$$a(y_p + y_c)'' + b(y_p + y_c)' + c(y_p + y_c) = f(x)$$

This shows that $y_p + y_c$ is a solution to Eq. 420.

---

**EXAMPLE 45:** Given that the solution to $y'' - 5y' + 6y = 3x$ is

$$y = \underbrace{c_1e^{3x} + c_2e^{2x}}_{\text{complementary function}} + \underbrace{\frac{x}{2} + \frac{5}{12}}_{\text{particular integral}}$$

prove by substitution that (a) the complementary function will make the left side of the given equation equal to zero, and that (b) the particular integral will make the left side equal to $3x$.

**Solution:** Given $y'' - 5y + 6y = 3x$

(a)

$$\text{Let } y_c = c_1e^{3x} + c_2e^{2x}$$

$$y_c' = 3c_1e^{3x} + 2c_2e^{2x}$$

$$y_c'' = 9c_1e^{3x} + 4c_2e^{2x}$$

Substituting on the left gives

$$9c_1e^{3x} + 4c_2e^{2x} - 5(3c_1e^{3x} + 2c_2e^{2x}) + 6(c_1e^{3x} + c_2e^{2x})$$

which equals zero.

(b)

$$\text{Let } y_p = \frac{x}{2} + \frac{5}{12}$$

$$y_p' = \frac{1}{2}$$

$$y_p'' = 0$$

Substituting gives

$$0 - 5\left(\frac{1}{2}\right) + 6\left(\frac{x}{2} + \frac{5}{12}\right)$$

or $3x$.

---

## Finding the Particular Integral

We already know how to find the complementary function $y_c$. We set the right side of the given equation to zero and then solve that (reduced) equation just as we did in the preceding section.

To see how to find the particular integral $y_p$, we start with Eq. 420 and isolate $y$. We get

$$y = \frac{1}{c}[f(x) - ay'' - by']$$

For this equation to balance, $y$ must contain terms similar to those in $f(x)$. Further, $y$ must contain terms similar to those in its own first and second derivatives. Thus it seems reasonable to try a solution consisting of the sum of $f(x)$, $f'(x)$, and $f''(x)$, each with an (as yet) undetermined (constant) coefficient.

---

**EXAMPLE 46:** Find $y_p$ for the equation $y'' - 5y' + 6y = 3x$.

**Solution:** Here $f(x) = 3x$, $f'(x) = 3$, and $f''(x) = 0$. Then

$$y_p = Ax + B$$

where $A$ and $B$ are constants yet to be found.

This is sometimes called the ***trial function.***

---

## Method of Undetermined Coefficients

To find the constants in $y_p$, we use the method of undetermined coefficients. We substitute $y_p$ and its first and second derivatives into the differential equation. We then equate coefficients of like terms and solve for the constants.

**We used the method of undetermined coefficients before when finding partial fractions.**

---

**EXAMPLE 47:** Determine the constants in $y_p$ in Example 46, and find the complete solution to the given equation given the complementary function $y_c = c_1e^{3x} + c_2e^{2x}$.

**Solution:** In Example 46 we had $y_p = Ax + B$. Taking derivatives gives us

$$y_p' = A \qquad \text{and} \qquad y_p'' = 0$$

Substituting into the original equation, we obtain

$$0 - 5A + 6(Ax + B) = 3x$$

$$6Ax + (6B - 5A) = 3x$$

Equating the coefficients on both sides of this equation, we see that the equation is satisfied when $6A = 3$, or $A = \frac{1}{2}$, and $6B - 5A = 0$. Thus $B = 5A/6 = \frac{5}{12}$. Our particular integral is then $y_p = x/2 + \frac{5}{12}$, and the complete solution is

$$y = y_c + y_p = c_1e^{3x} + c_2e^{2x} + \frac{x}{2} + \frac{5}{12}$$

---

## General Procedure

Thus to solve a second-order linear differential equation with constant coefficients (right side not zero):

| $ay'' + by' + cy = f(x)$ | 420 |
|---|---|

When we say "duplicate," we mean terms that are alike, regardless of numerical coefficient. These are discussed in the following section.

1. Find the *complementary function* $y_c$ by solving the auxiliary equation.
2. Write the *particular integral* $y_p$. It should contain each term from the right side $f(x)$ (less coefficients) as well as the first and second derivatives of each term of $f(x)$ (less coefficients). Discard any duplicates.
3. If a term in $y_p$ is a duplicate of one in $y_c$, multiply that term in $y_p$ by $x^n$, using the lowest $n$ that *eliminates that duplication* and any new duplication with other terms in $y_p$.
4. *Write* $y_p$, each term with an undetermined coefficient.
5. *Substitute* $y_p$ and its first and second derivatives into the differential equation.
6. *Evaluate the coefficients* by the method of undetermined coefficients.
7. *Combine* $y_c$ and $y_p$ to obtain the complete solution.

We illustrate these steps in the following example.

---

**EXAMPLE 48:** Solve $y'' - y' - 6y = 36x + 50 \sin x$.

**Solution:**

1. The auxiliary equation $m^2 - m - 6 = 0$ has the roots $m = 3$ and $m = -2$, so the complementary function is $y_c = c_1 e^{3x} + c_2 e^{-2x}$.
2. The terms in $f(x)$ and their derivatives (less coefficients) are

| | | | |
|---|---|---|---|
| Term in $f(x)$: | $x$ | $\sin x$ | duplicates |
| First derivative: | Constant | $\cos x$ | |
| Second derivative: | 0 | $\sin x$ | |

We discard the duplicate $\sin x$ term.

3. Comparing the terms from $y_c$ and these possible terms that will form $y_p$, we see that no term in $y_p$ is a duplicate of one in $y_c$.
4. Our particular integral, $y_p$, is then

$$y_p = A + Bx + C \sin x + D \cos x$$

5. The derivatives of $y_p$ are

$$y_p' = B + C \cos x - D \sin x$$

and

$$y_p'' = -C \sin x - D \cos x$$

Substituting into the differential equation gives

$$y_p'' - y_p' - 6y_p = 36x + 50 \sin x$$

$$(-C \sin x - D \cos x) - (B + C \cos x - D \sin x) - 6(A + Bx + C \sin x + D \cos x) = 36x + 50 \sin x$$

6. Collecting terms and equating coefficients of like terms from the left and the right sides of this equation gives the equations

$$-6B = 36 \qquad -B - 6A = 0 \qquad D - 7C = 50 \qquad -7D - C = 0$$

from which $A = 1$, $B = -6$, $C = -7$, and $D = 1$. Our particular integral is, then, $y_p = 1 - 6x - 7 \sin x + \cos x$.

7. The complete solution is thus $y_c + y_p$, or

$$y = c_1 e^{3x} + c_2 e^{-2x} + 1 - 6x - 7 \sin x + \cos x$$

## Duplicate Terms in the Solution

The terms in the particular integral $y_p$ must be *independent*. If $y_p$ has duplicate terms, only one should be kept. However, if a term in $y_p$ is a duplicate of one in the complementary function $y_c$ (except for the coefficient), *multiply that term in $y_p$ by $x^n$, using the lowest n that will eliminate that duplication* and any new duplication with other terms in $y_p$.

**EXAMPLE 49:** Solve the equation $y'' - 4y' + 4y = e^{2x}$.

**Solution:**

1. The complementary function (work not shown) is

$$y_c = c_1 e^{2x} + c_2 x e^{2x}.$$

2. Our particular integral should contain $e^{2x}$ and its derivatives, which are also of the form $e^{2x}$. But since these are duplicates, we need $e^{2x}$ only once.

3. But $e^{2x}$ is a duplicate of the first term, $c_1 e^{2x}$, in $y_c$. If we multiply by $x$, we see that $xe^{2x}$ is now a duplicate of the *second* term in $y_c$. We thus need $x^2 e^{2x}$.

   We now see that in the process of eliminating duplicates with $y_c$, we may have created a new duplicate within $y_p$. If so, we again multiply by $x^n$, using the lowest $n$ that would eliminate that new duplication as well. Here, $x^2 e^{2x}$ does not duplicate any term in $y_p$, so we proceed.

4. Our particular integral is thus $y_p = Ax^2 e^{2x}$.

5. Taking derivatives

$$y_p' = 2Ax^2 e^{2x} + 2Axe^{2x}$$

and

$$y_p'' = 4Ax^2 e^{2x} + 8Axe^{2x} + 2Ae^{2x}$$

Substituting into the differential equation gives

$$4Ax^2 e^{2x} + 8Axe^{2x} + 2Ae^{2x} - 4(2Ax^2 e^{2x} + 2Axe^{2x}) + 4(Ax^2 e^{2x}) = e^{2x}$$

6. Collecting terms and solving for $A$ gives $A = \frac{1}{2}$.

7. Our complete solution is then

$$y = c_1 e^{2x} + c_2 x e^{2x} + \frac{1}{2} x^2 e^{2x}$$

## EXERCISE 11—SECOND-ORDER DIFFERENTIAL EQUATIONS WITH RIGHT SIDE NOT ZERO

Solve each second-order differential equation.

**With Algebraic Expressions**

1. $y'' - 4y = 12$
2. $y'' + y' - 2y = 3 - 6x$
3. $y'' - y' - 2y = 4x$
4. $y'' + y' = x + 2$
5. $y'' - 4y = x^3 + x$

**With Exponential Expressions**

6. $y'' + 2y' - 3y = 42e^{4x}$
7. $y'' - y' - 2y = 6e^x$
8. $y'' + y' = 6e^x + 3$
9. $y'' - 4y = 4x - 3e^x$
10. $y'' - y = e^x + 2e^{2x}$
11. $y'' + 4y' + 4y = 8e^{2x} + x$
12. $y'' - y = xe^x$

**With Trigonometric Expressions**

13. $y'' + 4y = \sin 2x$
14. $y'' + y' = 6 \sin 2x$
15. $y'' + 2y' + y = \cos x$
16. $y'' + 4y' + 4y = \cos x$
17. $y'' + y = 2 \cos x - 3 \cos 2x$
18. $y'' + y = \sin x + 1$

**With Exponential and Trigonometric Expressions**

19. $y'' + y = e^x \sin x$
20. $y'' + y = 10e^x \cos x$
21. $y'' - 4y' + 5y = e^{2x} \sin x$
22. $y'' + 2y' + 5y = 3e^{-x} \sin x - 10$

**Particular Solution**

Find the particular solution to each differential equation, using the given boundary conditions.

23. $y'' - 4y' = 8, \quad y = y' = 0$ when $x = 0$
24. $y'' + 2y' - 3y = 6, \quad y = 0$ and $y' = 2$ when $x = 0$
25. $y'' + 4y = 2, \quad y = 0$ when $x = 0$ and $y = \frac{1}{2}$ when $x = \pi/4$
26. $y'' + 4y' + 3y = 4e^{-x}, \quad y = 0$ and $y' = 2$ when $x = 0$
27. $y'' - 2y' + y = 2e^x, \quad y' = 2e$ at $(1, 0)$
28. $y'' - 9y = 18 \cos 3x + 9, \quad y = -1$ and $y' = 3$ when $x = 0$
29. $y'' + y = -2 \sin x, \quad y = 0$ at $x = 0$ and $x = \pi/2$

## 35-12 *RLC* CIRCUITS

In Sec. 35-8 we studied the *RL* circuit and the *RC* circuit. Each gave rise to a first-order differential equation. We'll now see that the *RLC* circuit will result in a second-order differential equation.

A switch (Fig. 35-11) is closed at $t = 0$. The sum of the voltage drops must equal the applied voltage, so

$$Ri + L\frac{di}{dt} + \frac{q}{C} = E \tag{1}$$

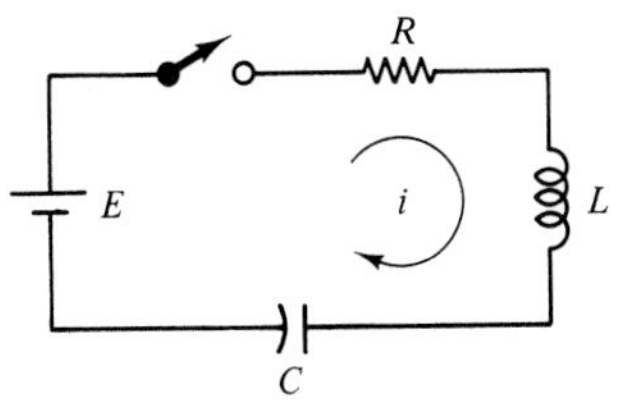


**FIGURE 35-11** ***RLC*** **circuit with dc source.**

Replacing $q$ by $\int i\,dt$ and differentiating gives

$$R\frac{di}{dt} + L\frac{d^2i}{dt^2} + \frac{i}{C} = 0$$

or

$$Li'' + Ri' + \left(\frac{1}{C}\right)i = 0 \tag{2}$$

This is a second-order linear differential equation, which we now solve as we did before. The auxiliary equation $Lm^2 + Rm + 1/C = 0$ has the roots

$$m = \frac{-R \pm \sqrt{R^2 - 4L/C}}{2L}$$

$$= -\frac{R}{2L} \pm \sqrt{\frac{R^2}{4L^2} - \frac{1}{LC}}$$

We now let

$$a^2 = \frac{R^2}{4L^2} \quad \text{and} \quad \omega_n^2 = \frac{1}{LC}$$

so that

| | | |
|---|---|---|
| Resonant Frequency of *RLC* Circuit with DC Source | $\omega_n = \sqrt{\frac{1}{LC}}$ | **A89** |

and our roots become

$$m = -a \pm \sqrt{a^2 - \omega_n^2}$$

or

$$m = -a \pm j\omega_d$$

where $\omega_d^2 = \omega_n^2 - a^2$. We have three possible cases.

**_RLC_ Circuit with DC Source: Second-Order Differential Equation**

| *Roots of Auxiliary Equation* | *Type of Solution* | |
|---|---|---|
| Nonreal | Underdamped | $a < \omega_n$ |
| | No damping (series *LC* circuit) | $a = 0$ |
| Real, equal | Critically damped | $a = \omega_n$ |
| Real, unequal | Overdamped | $a > \omega_n$ |

We consider the underdamped and overdamped cases with a dc source, and the underdamped case with an ac source.

### Underdamped, with DC Source, $a < \omega_n$

We first consider the case where $R$ is not zero but is low enough so that $a < \omega_n$. The roots of the auxiliary equation are then nonreal and the current is

$$i = e^{-at}(k_1 \sin \omega_d t + k_2 \cos \omega_d t)$$

Since $i$ is zero at $t = 0$ we get $k_2 = 0$. The current is then

$$i = k_1 e^{-at} \sin \omega_d t$$

Taking the derivative yields

$$\frac{di}{dt} = k_1 \omega_d e^{-at} \cos \omega_d t - a k_1 e^{-at} \sin \omega_d t$$

At $t = 0$ the capacitor behaves as a short circuit, and there is also no voltage drop across the resistor (since $i = 0$). The entire voltage $E$ then appears across the inductor. Since $E = L\, di/dt$, then $di/dt = E/L$, so

$$k_1 = \frac{E}{\omega_d L}$$

The current is then

| | | |
|---|---|---|
| Underdamped *RLC* Circuit | $i = \frac{E}{\omega_d L} e^{-at} \sin \omega_d t$ | **A91** |

where, from our previous substitution,

| | | |
|---|---|---|
| Underdamped *RLC* Circuit | $\omega_d = \sqrt{\omega_n^2 - a^2} = \sqrt{\omega_n^2 - \frac{R^2}{4L^2}}$ | **A92** |

We get a damped sine wave whose amplitude decreases exponentially with time.

---

**EXAMPLE 50:** A switch (Fig. 35-11) is closed at $t = 0$. If $R = 225\ \Omega$, $L = 1.50$ H, $C = 4.75\ \mu$F, and $E = 75.4$ V, write an expression for the instantaneous current.

**Solution:** We first compute $LC$,

$$LC = 1.50(4.75 \times 10^{-6}) = 7.13 \times 10^{-6}$$

Then, by Eq. A89,

$$\omega_n = \sqrt{\frac{1}{LC}} = \sqrt{\frac{10^6}{7.13}} = 375 \text{ rad/s}$$

and

$$a = \frac{R}{2L} = \frac{225}{2(1.50)} = 75.0 \text{ rad/s}$$

Then, by Eq. A92,

$$\omega_d = \sqrt{\omega_n^2 - a^2} = \sqrt{(375)^2 - (75.0)^2} = 367 \text{ rad/s}$$

The instantaneous current is then

$$i = \frac{E}{\omega_d L} e^{-at} \sin \omega_d t = \frac{75.4}{367(1.50)} e^{-75t} \sin 367t \quad \text{A}$$

$$= 137e^{-75t} \sin 367t \quad \text{mA}$$

This curve is plotted in Fig. 35-12 showing the damped sine wave.

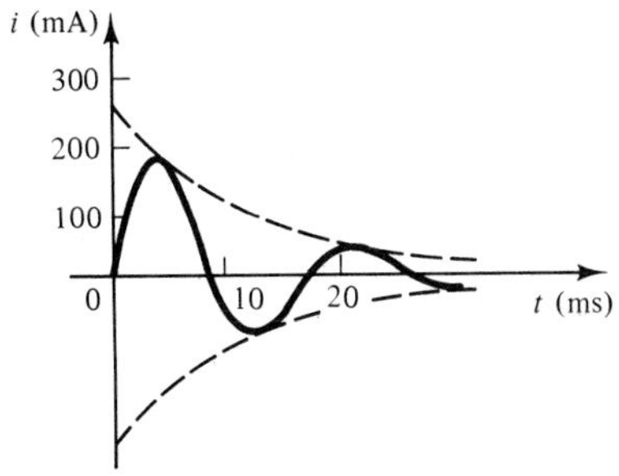


FIGURE 35-12

## No Damping: The Series *LC* Circuit

When the resistance $R$ is zero, $a = 0$ and $\omega_d = \omega_n$. From Eq. A91 we get

This, of course, is a theoretical case, for a real circuit always has some resistance.

| | | |
|---|---|---|
| Series *LC* Circuit | $i = \dfrac{E}{\omega_n L} \sin \omega_n t$ | **A90** |

which represents a sine wave with amplitude $E/\omega_n L$.

**EXAMPLE 51:** Repeat Example 50 with $R = 0$.

**Solution:** The value of $\omega_n$, from before, is 375 Hz. The amplitude of the current wave is

$$\frac{E}{\omega_n L} = \frac{75.4}{375(1.50)} = 0.134 \quad \text{A}$$

so the instantaneous current is

$$i = 134 \sin 375t \text{ mA}$$

## Overdamped, with DC Source, $a > \omega_n$

If the resistance is relatively large, so that $a > \omega_n$, the auxiliary equation has the real and unequal roots

$$m_1 = -a + j\omega_d \qquad \text{and} \qquad m_2 = -a - j\omega_d$$

The current is then

$$i = k_1 e^{m_1 t} + k_2 e^{m_2 t} \tag{1}$$

Since $i(0) = 0$ we have

$$k_1 + k_2 = 0 \tag{2}$$

Taking the derivative of (1), we obtain

$$\frac{di}{dt} = m_1 k_1 e^{m_1 t} + m_2 k_2 e^{m_2 t}$$

Since $di/dt = E/L$ at $t = 0$,

$$\frac{E}{L} = m_1 k_1 + m_2 k_2 \tag{3}$$

Solving (2) and (3) simultaneously gives

$$k_1 = \frac{E}{(m_1 - m_2)L} \quad \text{and} \quad k_2 = -\frac{E}{(m_1 - m_2)L}$$

where $m_1 - m_2 = -a + j\omega_d + a + j\omega_d = 2j\omega_d$. The current is then

| Overdamped *RLC* Circuit | $i = \dfrac{E}{2j\omega_d L}[e^{(-a+j\omega_d)t} - e^{(-a-j\omega_d)t}]$ | **A93** |
|---|---|---|

**EXAMPLE 52:** For the circuit of Examples 50 and 51, let $R = 2550\ \Omega$ and compute the instantaneous current.

**Solution:** $a = \dfrac{R}{2L} = \dfrac{2550}{2(1.50)} = 850$ rad/s. Since $\omega_n = 375$ rad/s, we have

$$\omega_d = \sqrt{(375)^2 - (850)^2} = j763 \text{ rad/s}$$

Then $-a - j\omega_d = -87.0$ and $-a + j\omega_d = -1613$. From Eq. A93,

$$i = \frac{75.4}{2(-763)(1.50)}(e^{-1613t} - e^{-87.0t})$$

$$= 32.9(e^{-87.0t} - e^{-1613t}) \quad \text{mA}$$

This equation is graphed in Fig. 35-13.

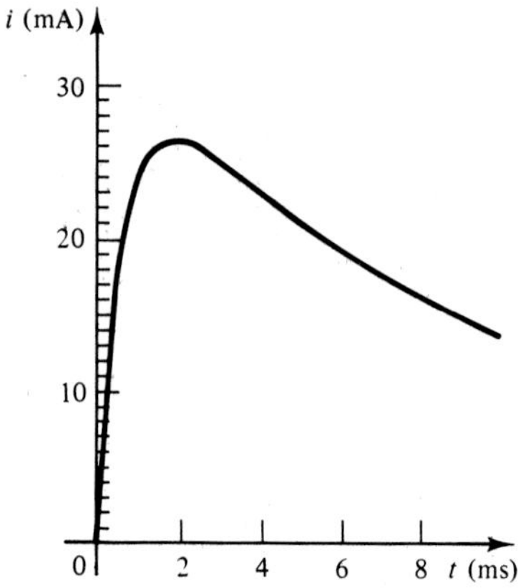


**FIGURE 35-13 Current in an overdamped circuit.**

### Underdamped, with AC Source

Up to now we have considered only a dc source. We now repeat the underdamped (low-resistance) case with an alternating voltage $E \sin \omega t$.

A switch (Fig. 35-14) is closed at $t = 0$. The sum of the voltage drops must equal the applied voltage, so

$$Ri + L\frac{di}{dt} + \frac{q}{C} = E \sin \omega t \tag{1}$$

Replacing $q$ by $\int i\, dt$ and differentiating gives

$$R\frac{di}{dt} + L\frac{d^2i}{dt^2} + \frac{i}{C} = \omega E \cos \omega t$$

or

$$Li'' + Ri' + \left(\frac{1}{C}\right)i = \omega E \cos \omega t \tag{2}$$

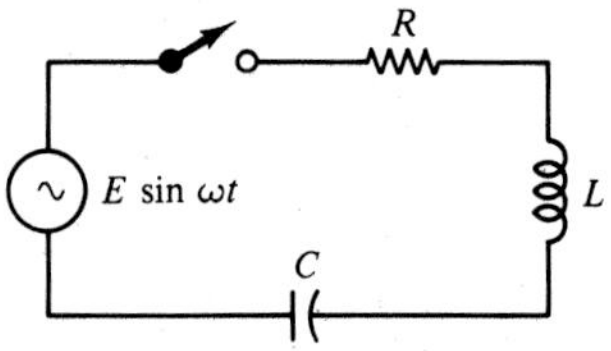


**FIGURE 35-14** ***RLC*** **circuit with ac source.**

The complementary function is the same as was calculated for the dc case,

$$i_c = e^{-at}(k_1 \sin \omega_d t + k_2 \cos \omega_d t)$$

The particular integral $i_p$ will have a sine term and a cosine term,

$$i_p = A \sin \omega t + B \cos \omega t$$

Taking first and second derivatives yields

In Chapter 36 we'll do this type of problem by the Laplace transform.

$$i' = \omega A \cos \omega t - \omega B \sin \omega t$$

$$i'' = -\omega^2 A \sin \omega t - \omega^2 B \cos \omega t$$

Substituting into (2), we get

$$-L\omega^2 A \sin \omega t - L\omega^2 B \cos \omega t - R\omega B \sin \omega t + R\omega A \cos \omega t + \frac{A}{C} \sin \omega t$$

$$+ \frac{B}{C} \cos \omega t = \omega E \cos \omega t$$

Equating the coefficients of the sine terms gives

$$-L\omega^2 A - R\omega B + \frac{A}{C} = 0$$

from which

$$-RB = A\left(\omega L - \frac{1}{\omega C}\right) = AX \quad (3)$$

where $X$ is the reactance of the circuit. Equating the coefficients of the cosine terms gives

$$-L\omega^2 B + R\omega A + \frac{B}{C} = \omega E$$

or

$$RA - E = B\left(\omega L - \frac{1}{\omega C}\right) = BX$$

Solving (3) and (4) simultaneously gives

$$A = -\frac{RE}{R_2 + X^2} = \frac{RE}{Z^2}$$

where $Z$ is the impedance of the circuit, and

$$B = -\frac{EX}{R^2 + X^2} = -\frac{EX}{Z^2}$$

Our particular integral thus becomes

$$y_p = -\frac{RE}{Z^2} \sin \omega t - \frac{EX}{Z^2} \cos \omega t$$

$$= -\frac{E}{Z^2}(R \sin \omega t - X \cos \omega t)$$

From the impedance triangle (Fig. 35-15)

$$R = Z \cos \phi \qquad \text{and} \qquad X = Z \sin \phi$$

where $\phi$ is the phase angle. Thus

$$R \sin \omega t - X \cos \omega t = Z \sin \omega t \cos \phi - Z \cos \omega t \sin \phi$$

$$= Z \sin(\omega t - \phi)$$

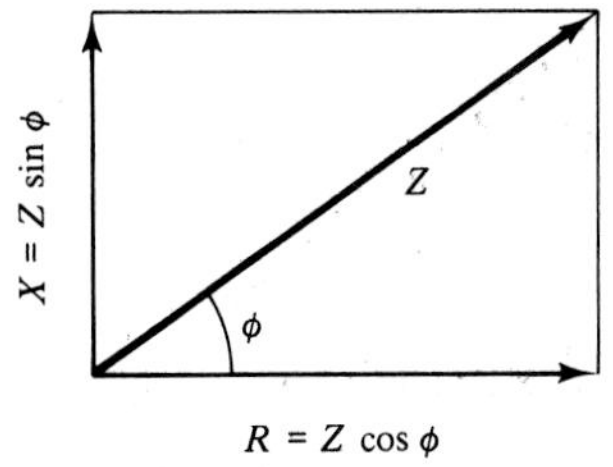


**FIGURE 35-15 Impedance triangle.**

by the trigonometric identity (Eq. 167). Thus $i_p$ becomes $E/Z \sin(\omega t - \phi)$, or

$$i_p = I_{\max} \sin(\omega t - \phi)$$

since $E/Z$ gives the maximum current $I_{\max}$. The total current is the sum of the complementary function $i_c$ and the particular integral $i_p$,

$$i = \underbrace{e^{-at}(k_1 \sin \omega_d t + k_2 \cos \omega_d t)}_{\substack{\text{transient} \\ \text{current}}} + \underbrace{I_{\max} \sin(\omega t - \phi)}_{\substack{\text{steady-state} \\ \text{current}}}$$

Thus the current is made up of two parts; a *transient* part that dies quickly with time, and a *steady-state* part that continues as long as the ac source is connected. We are usually interested only in the steady-state current.

| | | |
|---|---|---|
| Steady-State Current for *RLC* Circuit, AC Source, Underdamped Case | $i_{ss} = \dfrac{E}{Z} \sin(\omega t - \phi)$ | **A100** |

---

**EXAMPLE 53:** Find the steady-state current for an *RLC* circuit if $R = 345\ \Omega$, $L = 0.726$ H, $C = 41.4\ \mu$F, and $e = 155 \sin 285t$.

**Solution:** By Eqs. A94 and A95.

$$X_L = \omega L = 285(0.726) = 207\ \Omega$$

and

$$X_C = \frac{1}{\omega C} = \frac{1}{285(41.4 \times 10^{-6})} = 84.8\ \Omega$$

The total reactance is, by Eq. A96,

$$X = X_L - X_C = 207 - 84.8 = 122\ \Omega$$

The impedance $Z$ is found from Eq. A97:

$$Z = \sqrt{R^2 + X^2} = \sqrt{(345)^2 + (122)^2} = 366\ \Omega$$

The phase angle, from Eq. A98, is

$$\phi = \tan^{-1}\frac{X}{R} = \tan^{-1}\frac{122}{345} = 19.5°$$

The steady-state current is then

$$\begin{aligned} i_{ss} &= \frac{E}{Z}\sin(\omega t - \phi) \\ &= \frac{155}{366}\sin(285t - 19.5°) \\ &= 0.423 \sin(285t - 19.5°) \quad \text{A} \end{aligned}$$

Figure 35-16 shows the applied voltage and the steady-state current, with a phase difference of 19.5°, or 0.239 ms.

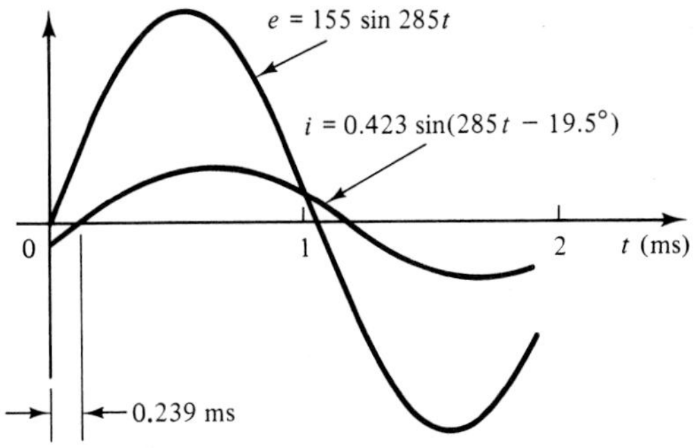

FIGURE 35-16

### Resonance

The current in a series *RLC* circuit will be a maximum when the impedance $Z$ is zero. This will occur when the reactance $X$ is zero, so

$$X = \omega L - \frac{1}{\omega C} = 0$$

Solving for $\omega$, we get $\omega^2 = 1/LC$ or $\omega_n^2$. Thus

| | | |
|---|---|---|
| Resonant Frequency | $\omega = \dfrac{1}{\sqrt{LC}} = \omega_n$ | **A89** |

**EXAMPLE 54:** Find the resonant frequency for the circuit of Example 53, and write an expression for the steady-state current at that frequency.

*Solution:* The resonant frequency is

$$\omega_n = \frac{1}{\sqrt{LC}} = \frac{1}{\sqrt{0.726(41.4 \times 10^{-6})}} = 182 \text{ rad/s}$$

Since $X_L = X_C$, then $X = 0$, $Z = R$, and $\phi = 0$. Thus $I_{max}$ is 155/345 = 0.448 A and the steady-state current is then

$$i_{ss} = 448 \sin 182t \quad \text{mA}$$

## EXERCISE 12—*RLC* Circuits

### Series LC Circuit with DC Source

1. In an *LC* circuit, $C = 1.00\ \mu F$, $L = 1.00$ H, and $E = 100$ V. At $t = 0$ the charge and the current are both zero. Using Eq. A90, show that $i = 0.1 \sin 1000t$ A.
2. For Problem 1, take the integral of $i$ to show that $q = 10^{-4}(1 - \cos 1000t)$ coulomb (C).
3. For the circuit of Problem 1, with $E = 0$ and an initial charge of $255 \times 10^{-6}$ C ($i_0$ is still 0), the differential equation, in terms of charge, is $Lq'' + q/C = 0$. Solve this DE for $q$ and show that $q = 255 \cos 1000t\ \mu C$.
4. By differentiating the expression for charge in Problem 3, show that the current is $i = -255 \sin 1000t$ mA.

### Series RLC Circuit with DC Source

5. In an *RLC* circuit, $R = 1.55\ \Omega$, $C = 250\ \mu F$, $L = 0.125$ H, and $E = 100$ V. The current and charge are 0 when $t = 0$. **(a)** Show that the circuit is underdamped. **(b)** Using Eq. A91, show that $i = 4.47\ e^{-6.2t} \sin 179t$ A.
6. Integrate $i$ in Problem 5 to show that $q = -e^{-6.2t}(0.866 \sin 179t + 25.0 \cos 179t) + 25.0$ mC.
7. In an *RLC* circuit, $R = 1.75\ \Omega$, $C = 4.25$ F, $L = 1.50$ H, and $E = 100$ V. The current and charge are 0 when $t = 0$. **(a)** Show that the circuit is overdamped. **(b)** Using Eq. A93, show that $i = -77.8e^{-1.01t} + 77.8e^{-0.155t}$ A.
8. Integrate the expression for $i$ for the circuit in Problem 7 to show that $q = 76.8e^{-1.01t} - 502e^{-0.155t} + 425$ C.

### Series RLC Circuit with AC Source

9. For an *RLC* circuit, $R = 10.5\ \Omega$, $L = 0.125$ H, $C = 225\ \mu F$, and $e = 100 \sin 175t$. Using Eq. A100, show that the steady-state current is $i = 9.03 \sin(175t + 18.4°)$.
10. Find the resonant frequency for Problem 9.
11. In an *RLC* circuit, $R = 1550\ \Omega$, $L = 0.350$ H, $C = 20.0\ \mu F$, and $e = 250 \sin 377t$. Using Eq. A100, show that the steady-state current is $i = 161 \sin 377t$ mA.
12. If $R = 10.5\ \Omega$, $L = 0.175$ H, $C = 1.50 \times 10^{-3}$ F, and $e = 175 \sin 55t$, find the amplitude of the steady-state current.
13. Find the resonant frequency for the circuit in Problem 12.
14. Find the amplitude of the steady-state current at resonance for the circuit in Problem 12.
15. For the circuit of Fig. 35-19 $R = 110\ \Omega$, $L = 5.60 \times 10^{-6}$ H, and $\omega = 6.00 \times 10^5$ rad/s. What value of $C$ will produce resonance?
16. If an *RLC* circuit has the values $R = 10\ \Omega$, $L = 0.200$ H, $C = 500\ \mu F$, and $e = 100 \sin \omega t$, find the resonant frequency.
17. For the circuit in Problem 16, find the amplitude of the steady-state current at resonance.

## CHAPTER 35 REVIEW PROBLEMS

*Find the general solution to each first-order differential equation.*

1. $xy + y + xy' = e^x$
2. $y + xy' = 4x^3$
3. $y' + y - 2 \cos x = 0$
4. $y + x^2y^2 + xy' = 0$
5. $2y + 3x^2 + 2xy' = 0$
6. $y' - 3x^2y^2 + \dfrac{y}{x} = 0$
7. $y^2 + (x^2 - xy)y' = 0$
8. $y' = e^{-y} - 1$
9. $(1 - x)\dfrac{dy}{dx} = y^2$
10. $y' \tan x + \tan y = 0$
11. $y + 2xy^2 + (y - x)y' = 0$

*Find the particular solution to each differential equation, using the given boundary condition.*

12. $x\,dx = 2y\,dy$, $\quad x = 3$ when $y = 1$
13. $y' \sin y = \cos x$, $\quad x = \dfrac{\pi}{4}$ when $y = 0$
14. $xy' + y = 4x$, $\quad x = 2$ when $y = 1$
15. $y\,dx = (x - 2x^2y)\,dy$, $\quad x = 2$ when $y = 1$
16. $3xy^2\,dy = (3y^3 - x^3)\,dx$, $\quad x = 3$ when $y = 1$
17. A gear is rotating at 1550 rev/min. Its speed decreases exponentially at a rate of 9.5% per second, after the power is shut off. Find the gear's speed after 6.00 s.
18. For the circuit of Fig. 35-7, $R = 1350\ \Omega$, $L = 7.25$ H, and $E = 225$ V. If the switch is thrown from position 2 to position 1, find the current and the voltage across the inductance at $t = 3.00$ ms.
19. Write the equation of the orthagonal trajectories to each family of parabolas, $x^2 = 4y$.

20. For the circuit of Fig. 35-8, $R = 2550\ \Omega$, $C = 145\ \mu F$, and $E = 95$ V. If the switch is thrown from position 2 to position 1, find the current and the voltage across the capacitor at $t = 5.00$ ms.

21. Find the equation of the curve that passes through the point (1, 2) and whose slope is $y' = 2 + y/x$.

22. An object is dropped and falls from rest through air whose resisting force is proportional to the speed of the package. The terminal speed is 275 ft/s. Show that the acceleration is given by the differential equation $a = dv/dt = g - gv/275$.

23. A certain yeast is found to grow exponentially at the rate of 15% per hour. Starting with 500 g of yeast, how many grams will there be after 15 h?

*Solve each differential equation.*

24. $y'' + 2y' - 3y = 7, \quad y = 1$ and $y' = 2$ when $x = 0$
25. $y'' - y' - 2y = 5x$
26. $y'' - 2y = 2x^3 + 3x$
27. $y'' - 4y' - 5y = e^{4x}$
28. $y''' - 2y' = 0$
29. $y'' - 5y' = 6, \quad y = y' = 1$ when $x = 0$
30. $y'' + 2y' - 3y = 0$
31. $y'' - 7y' - 18y = 0$
32. $y'' + 6y' + 9y = 0, \quad y = 0$ and $y' = 2$ when $x = 0$
33. $y''' - 6y'' + 11y' - 6y = 0$
34. $y'' - 2y = 3x^3e^x$
35. $y'' + 2y = 3 \sin 2x$
36. $y'' + y' = 5 \sin 3x$
37. $y'' + 3y' + 2y = 0, \quad y = 1$ and $y' = 2$ when $x = 0$
38. $y'' + 9y = 0$
39. $y'' + 3y' - 4y = 0$
40. $y'' - 2y' + y = 0, \quad y = 0$ and $y' = 2$ when $x = 0$
41. $y'' + 5y' - y = 0$
42. $y'' - 4y' + 4y = 0$
43. $y'' + 3y' - 4y = 0, \quad y = 0$ and $y' = 2$ when $x = 0$
44. $y'' + 4y' + 4y = 0$
45. $y'' + 6y' + 9y = 0$
46. $y'' - 2y' + y = 0$
47. $9y'' - 6y' + y = 0$
48. $y'' - x^2 = 0$, where $y' = 4$ at the point (1, 2)

49. For an *RLC* circuit, $R = 3.75\ \Omega$, $C = 150\ \mu F$, $L = 0.100$ H, and $E = 120$ V. The current and charge are 0 when $t = 0$. Find $i$ as a function of time.

50. In Problem 49, find an expression for the charge.

51. For an *LC* circuit, $C = 1.75\ \mu F$, $L = 2.20$ H, and $E = 110$ V. At $t = 0$ the charge and the current are both zero. Find $i$ as a function of time.

52. In Problem 51, find the charge as a function of time.

*Writing*

53. We have given seven steps for the solution of a second-order linear differential equation with constant coefficients. List as many of these steps as you can and write a one-sentence explanation of each.

# 36

# Solving Differential Equations by the Laplace Transform and by Numerical Methods

## OBJECTIVES

**When you have completed this chapter, you should be able to:**

- Find the Laplace transform of a function by direct integration or by using a table.
- Find the Laplace transform of an expression containing derivatives, and substitute initial conditions.
- Determine the inverse of a Laplace transform by completing the square or by partial fractions.
- Solve first-order and second-order differential equations using Laplace transforms.
- Solve electrical applications using Laplace transforms.
- Solve first-order differential equations using Euler's method, modified Euler's method, or the Runge-Kutta method.
- Solve second-order differential equations using the modified Euler's method or the Runge-Kutta method.

The methods for solving differential equations that we learned in Chapter 35 are often called *classical methods*. In this chapter we learn other powerful techniques: the *Laplace transform* and *numerical methods*. The Laplace transform enables us to transform a differential equation into an algebraic one. We can then solve the algebraic equation and apply an inverse transform to obtain the solution to the differential equation.

The Laplace transform is good for finding particular solutions of differential equations. For example, it enables us to solve *initial value* problems, that is, when the value of the function is known at $t = 0$. Thus the equations we deal with here are functions of time, such as electric circuit problems, rather than functions of $x$.

We first use the Laplace transform to solve first- and second-order differential equations with constant coefficients. We then do some applications and compare our results with those obtained by classical methods in Chapter 35. We go on to solve differential equations using *numerical methods*. These are methods that use successive approximations to give a solution in the form of *numbers*, rather than equations. One great advantage to numerical methods is that they can be programmed on a computer.

## 36-1 THE LAPLACE TRANSFORM OF A FUNCTION

### Laplace Transform of Some Simple Functions

If $y$ is some function of time, so that $y = f(t)$, the Laplace transform of that function is defined by the improper integral

| Laplace Transform | $\mathscr{L}[f(t)] = \int_0^\infty f(t)e^{-st}\,dt$ | **423** |
|---|---|---|

After the French analyst, probabilist, astronomer, and physicist Pierre Laplace (1749–1827).

The transformed expression is *a function of s*, which we call $F(s)$. Thus

$$\mathscr{L}[f(t)] = F(s)$$

We can write the transform of an expression by direct integration, that is, by writing an integral using Eq. 423, and then evaluating it.

Glance back at Sec. 34-5 if you have forgotten how to evaluate an improper integral.

**EXAMPLE 1:** If $y = f(t) = 1$, then

$$\mathscr{L}[f(t)] = \mathscr{L}[1] = \int_0^\infty (1)e^{-st}\,dt$$

To evaluate the integral, we multiply by $-s$ and compensate with $-1/s$:

$$\mathscr{L}[1] = -\frac{1}{s}\int_0^\infty e^{-st}(-s\,dt) = -\frac{1}{s}e^{-st}\Big|_0^\infty$$

Figure 36-1 shows a graph of $y = e^{-u}$, which will help us to evaluate the limits. Note that as $u$ approaches infinity, $e^{-u}$ approaches zero, and as $u$ approaches zero, $e^{-u}$ approaches 1. So we take $e^{-st}$ equal to 0 at the upper limit $t = \infty$ and $e^{-st} = 1$ at the lower limit $t = 0$. Thus

$$\mathscr{L}[1] = -\frac{1}{s}(0 - 1) = \frac{1}{s}$$

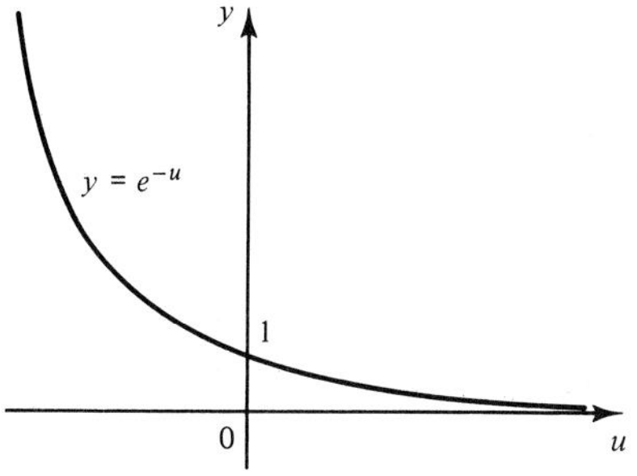

FIGURE 36-1

Rule 2 from our table of integrals in Appendix C says that $\int af(x)\,dx = a \int f(x)\,dx$. It follows then that

$$\mathcal{L}[af(t)] = a\mathcal{L}[f(t)]$$

*The Laplace transform of a constant times a function is equal to the constant times the transform of the function.*

---

**EXAMPLE 2:** If $\mathcal{L}[1] = 1/s$, as above, then

$$\mathcal{L}[5] = 5\mathcal{L}[1] = \frac{5}{s}$$

---

**EXAMPLE 3:** If $y = f(t) = t$, then

$$\mathcal{L}[f(t)] = \mathcal{L}[t] = \int_0^\infty te^{-st}\,dt$$

We evaluate the right side using Rule 37, with $u = t$ and $a = -s$,

$$\mathcal{L}[t] = \frac{e^{-st}}{s^2}(-st - 1)\Big|_0^\infty = 0 - \left(-\frac{1}{s^2}\right) = \frac{1}{s^2}$$

---

Note that in each case the transformed expression is a function of $s$ only.

**EXAMPLE 4:** Find $\mathcal{L}[\sin at]$.

**Solution:** Using Eq. 423, we find that

$$\mathcal{L}[\sin at] = \int_0^\infty e^{-st} \sin at\,dt$$

$$= \frac{e^{-st}}{s^2 + a^2}(-s \sin at - a \cos at)\Big|_0^\infty$$

$$= 0 - \frac{1}{s^2 + a^2}(-a) = \frac{a}{s^2 + a^2}$$

---

### Transform of a Sum

If we have the sum of several terms,

$$f(t) = a\,g(t) + b\,h(t) \cdots$$

then

$$\mathcal{L}[f(t)] = \int_0^\infty [a\,g(t) + b\,h(t) + \cdots]e^{-st}\,dt$$

$$= a\int_0^\infty g(t)e^{-st}\,dt + b\int_0^\infty h(t)e^{-st}\,dt + \cdots$$

so

$$\mathcal{L}[a\,g(t) + b\,h(t) \cdots] = a\mathcal{L}[g(t)] + b\mathcal{L}[h(t)] + \cdots$$

*The Laplace transform of a sum of several terms is the sum of the transforms of each term.*

This property allows us to work term by term in writing the transform of an expression made up of sums or differences.

**EXAMPLE 5:** If $y = 4 + 5t - 2 \sin 3t$, then

$$\mathscr{L}[y] = \frac{4}{s} + \frac{5}{s^2} - 2\frac{3}{s^2 + 9}$$

### Transform of a Derivative

To solve differential equations, we must be able to take the transform of a derivative. Let $f'(t)$ be the derivative of some function $f(t)$. Then

When we write the Laplace transform of a function, we often say that we are "taking the transform" of the function.

$$\mathscr{L}[f'(t)] = \int_0^\infty f'(t)e^{-st}\,dt$$

Integrating by parts, we let $u = e^{-st}$ and $dv = f'(t)\,dt$. So $du = -se^{-st}\,dt$ and $v = \int f'(t)\,dt = f(t)$. Then $\mathscr{L}[f'(t)] = u\,dv - v\,du$, or

$$\mathscr{L}[f'(t)] = f(t)e^{-st}\Big|_0^\infty + s\int_0^\infty f(t)e^{-st}\,dt$$

$$= f(t)e^{-st}\Big|_0^\infty + s\mathscr{L}[f(t)]$$

$$= 0 - f(0) + s\mathscr{L}[f(t)]$$

so

| | |
|---|---|
| Transform of a Derivative | $\mathscr{L}[f'(t)] = s\mathscr{L}[f(t)] - f(0)$ |

*The transform of a derivative of a function equals s times the transform of the function minus the function evaluated at $t = 0$.*

Thus if we have a function $y = f(t)$, then

$$\mathscr{L}[y'] = s\mathscr{L}[y] - y(0)$$

Note that the transform of a derivative contains $f(0)$ or $y(0)$, the value of the function at $t = 0$. Thus *to evaluate the Laplace transform of a derivative, we must know the initial conditions.*

**EXAMPLE 6:** Take the Laplace transform of both sides of the differential equation $y' = 6t$, with the initial condition that $y(0) = 5$.

**Solution:**

If two quantities are equal, their transforms are equal. This means that we can take the transform of both sides of an equation without changing the meaning of the equation.

$$y' = 6t$$

$$\mathscr{L}[y'] = \mathscr{L}[6t]$$

$$s\mathscr{L}[y] - y(0) = \frac{6}{s^2}$$

or, substituting,

$$s\mathscr{L}[y] - 5 = \frac{6}{s^2}$$

since $y(0) = 5$. Later we will solve such an equation for $\mathscr{L}[y]$ and then find the inverse of the transform to obtain $y$.

Similarly, we can find the transform of the second derivative, since the second derivative of $f(t)$ is the derivative of $f'(t)$.

$$\mathcal{L}[f''(t)] = s\mathcal{L}[f'(t)] - f'(0)$$

Substituting, we get

$$\mathcal{L}[f''(t)] = s\{s\mathcal{L}[f(t)] - f(0)\} - f'(0)$$

| Transform of the Second Derivative | $\mathcal{L}[f''(t)] = s^2\mathcal{L}[f(t)] - sf(0) - f'(0)$ |
|---|---|

Another way of expressing this, if $y = f(t)$, is

$$\mathcal{L}[y''] = s^2\mathcal{L}[y] - sy(0) - y'(0)$$

**EXAMPLE 7:** Transform both sides of the second-order differential equation $y'' - 3y' + 4y = 5t$ if $y(0) = 6$ and $y'(0) = 7$.

**Solution:**

$$y'' - 3y' + 4y = 5t$$

$$\mathcal{L}[y''] - 3\mathcal{L}[y'] + 4\mathcal{L}[y] = \mathcal{L}[5t]$$

$$s^2\mathcal{L}[y] - sy(0) - y'(0) - 3\{s\mathcal{L}[y] - y(0)\} + 4\mathcal{L}[y] = \frac{5}{s^2}$$

Substituting $y(0) = 6$ and $y'(0) = 7$, we obtain

$$s^2\mathcal{L}[y] - 6s - 7 - 3s\mathcal{L}[y] - 3(6) + 4\mathcal{L}[y] = \frac{5}{s^2}$$

or

$$s^2\mathcal{L}[y] - 3s\mathcal{L}[y] + 4\mathcal{L}[y] - 6s - 25 = \frac{5}{s^2}$$

## Table of Laplace Transforms

The transforms of some common functions are given in Table 36-1. Instead of transforming a function step by step, we simply look it up in the table.

**TABLE 36-1 Short Table of Laplace Transforms**

| | $f(t)$ | $\mathcal{L}[f(t)] = F(s)$ |
|---|---|---|
| 1 | $f(t)$ | $F(s) = \int_0^\infty e^{-st}f(t)\,dt$ |
| 2 | $f'(t)$ | $s\mathcal{L}[f(t)] - f(0)$ |
| 3 | $f''(t)$ | $s^2\mathcal{L}[f(t)] - sf(0) - f'(0)$ |
| 4 | $a\,g(t) + b\,h(t) + \cdots$ | $a\mathcal{L}[g(t)] + b\mathcal{L}[h(t)] + \cdots$ |
| 5 | $1$ | $\frac{1}{s}$ |
| 6 | $t$ | $\frac{1}{s^2}$ |
| 7 | $t^n$ | $\frac{n!}{s^{n+1}}$ |
| 8 | $\frac{t^{n-1}}{(n-1)!}$ | $\frac{1}{s^n}$ |

**Here $n$ is a positive integer.**

TABLE 36-1 (*Continued*)

| | $f(t)$ | $\mathscr{L}[f(t)] = F(s)$ |
|---|---|---|
| 9 | $e^{at}$ | $\dfrac{1}{s-a}$ |
| 10 | $1 - e^{-at}$ | $\dfrac{a}{s(s+a)}$ |
| 11 | $te^{at}$ | $\dfrac{1}{(s-a)^2}$ |
| 12 | $e^{at}(1 + at)$ | $\dfrac{s}{(s-a)^2}$ |
| 13 | $t^n e^{at}$ | $\dfrac{n!}{(s-a)^{n+1}}$ |
| 14 | $i^{n-1}e^{-at}$ | $\dfrac{(n-1)!}{(s+a)^n}$ |
| 15 | $e^{-at} - e^{-bt}$ | $\dfrac{b-a}{(s+a)(s+b)}$ |
| 16 | $ae^{-at} - be^{-bt}$ | $\dfrac{s(a-b)}{(s+a)(s+b)}$ |
| 17 | $\sin at$ | $\dfrac{a}{s^2+a^2}$ |
| 18 | $\cos at$ | $\dfrac{s}{s^2+a^2}$ |
| 19 | $t \sin at$ | $\dfrac{2as}{(s^2+a^2)^2}$ |
| 20 | $t \cos at$ | $\dfrac{s^2-a^2}{(s^2+a^2)^2}$ |
| 21 | $1 - \cos at$ | $\dfrac{a^2}{s(s^2+a^2)}$ |
| 22 | $at - \sin at$ | $\dfrac{a^3}{s^2(s^2+a^2)}$ |
| 23 | $e^{-at} \sin bt$ | $\dfrac{b}{(s+a)^2+b^2}$ |
| 24 | $e^{-at} \cos bt$ | $\dfrac{s+a}{(s+a)^2+b^2}$ |
| 25 | $\sin at - at \cos at$ | $\dfrac{2a^3}{(s^2+a^2)^2}$ |
| 26 | $\sin at + at \cos at$ | $\dfrac{2as^2}{(s^2+a^2)^2}$ |
| 27 | $\cos at - \frac{1}{2} at \sin at$ | $\dfrac{s^3}{(s^2+a^2)^2}$ |
| 28 | $\dfrac{b}{a^2}(e^{-at} + at - 1)$ | $\dfrac{b}{s^2(s+a)}$ |
| 29 | $\int_0^t f(t)\, dt$ | $\dfrac{\mathscr{L}[f(t)]}{s}$ |

**EXAMPLE 8:** Find the Laplace transform of $t^3e^{2t}$ by using Table 36-1.

**Solution:** Our function matches Transform 13, with $n = 3$ and $a = 2$, so

$$\mathcal{L}[t^3e^{2t}] = \frac{3!}{(s-2)^{3+1}} = \frac{6}{(s-2)^4}$$

## EXERCISE 1—THE LAPLACE TRANSFORM OF A FUNCTION

### Transforms by Direct Integration

Find the Laplace transform of each function by direct integration.

1. $f(t) = 6$
2. $f(t) = t$
3. $f(t) = t^2$
4. $f(t) = 2t^2$
5. $f(t) = \cos 5t$
6. $f(t) = e^t \sin t$

### Transforms by Table

Use Table 36-1 to find the Laplace transform of each function.

7. $f(t) = t^2 + 4$
8. $f(t) = t^3 - 2t^2 + 3t - 4$
9. $f(t) = 3e^t + 2e^{-t}$
10. $f(t) = 5te^{3t}$
11. $f(t) = \sin 2t + \cos 3t$
12. $f(t) = 3 + e^{4t}$
13. $f(t) = 5e^{3t} \cos 5t$
14. $f(t) = 2e^{-4t} - t^2$
15. $f(t) = 5e^t - t \sin 3t$
16. $f(t) = t^3 + 4t^2 - 3e^t$

### Transforms of Derivatives

Find the Laplace transform of each expression and substitute the given initial conditions.

17. $y' + 2y, \quad y(0) = 1$
18. $3y' + 2y, \quad y(0) = 3$
19. $y' - 4y, \quad y(0) = 0$
20. $5y' - 3y, \quad y(0) = 2$
21. $y'' + 3y' - y, \quad y(0) = 1, y'(0) = 3$
22. $y'' - y' + 2y, \quad y(0) = 1, y'(0) = 0$
23. $2y'' + 3y' + y, \quad y(0) = 2, y'(0) = 3$
24. $3y'' - y' + 2y, \quad y(0) = 2, y'(0) = 1$

## 36-2 INVERSE TRANSFORMS

Before we can use the Laplace transform to solve differential equations, we must be able to transform a function of $s$ back to a function of $t$. The *inverse Laplace transform* is denoted by $\mathcal{L}^{-1}$. Thus, if $\mathcal{L}[f(t)] = F(s)$, then

**We'll see that finding the inverse is often harder than finding the transform itself.**

| | | |
|---|---|---|
| Inverse Laplace Transform | $\mathcal{L}^{-1}[F(s)] = f(t)$ | **424** |

We use Table 36-1 to find the inverse of some Laplace transforms. We put our given expression in a form that matches a right-hand entry of Table 36-1 and then read the corresponding entry at the left.

**EXAMPLE 9:** If $F(s) = 4/(s^2 + 16)$, find $f(t)$.

**Solution:** We search Table 36-1 in the right-hand column for an expression of similar form and find Transform 17,

$$\mathscr{L}[\sin at] = \frac{a}{s^2 + a^2}$$

which matches our expression if $a = 4$. Thus

$$f(t) = \sin 4t$$

**EXAMPLE 10:** Find $y$ if $\mathscr{L}[y] = 5/(s - 7)^4$.

**Solution:** From the table we find Transform 13,

$$\mathscr{L}[t^n e^{at}] = \frac{n!}{(s - a)^{n+1}}$$

In order to match our function, we must have $a = 7$ and $n = 3$. The numerator then must be $3! = 3(2) = 6$. So we insert 6 in the numerator and compensate with 6 in the denominator. We then rewrite our function as

$$\mathscr{L}[y] = \frac{5}{6}\frac{6}{(s - 7)^4}$$

which now is the same as Transform 13 with a coefficient of $\frac{5}{6}$. The inverse is then

$$y = \frac{5}{6}t^3e^{7t}$$

## Completing the Square

To get our given function to match a right-hand table entry, we may have to do some algebra. Sometimes we must complete the square to make our function match one in the table.

**EXAMPLE 11:** Find $y$ if $\mathscr{L}[y] = \dfrac{s - 1}{s^2 + 4s + 20}$.

**Solution:** This does not now match any functions in our table, but it will if we complete the square on the denominator.

This is no different from the method we used earlier when we completed the square.

$$s^2 + 4s + 20 = (s^2 + 4s \qquad) + 20$$
$$= (s^2 + 4s + 4) + 20 - 4 = (s + 2)^2 + 4^2$$

The denominator is now of the same form as in Transform 24. However, to use Transform 24, the numerator must be $s + 2$, not $s - 1$. So let us add 3 and subtract 3 from the numerator, so that $s - 1 = (s + 2) - 3$. Then

$$\mathscr{L}[y] = \frac{s - 1}{s^2 + 4s + 20} = \frac{(s + 2) - 3}{(s + 2)^2 + 4^2} = \frac{s + 2}{(s + 2)^2 + 4^2} - \frac{3}{(s + 2)^2 + 4^2}$$
$$= \frac{s + 2}{(s + 2)^2 + 4^2} - \frac{3}{4} \cdot \frac{4}{(s + 2)^2 + 4^2}$$

Now the first expression matches Transform 24 and the second matches Transform 23. We find the inverse of these transforms to be

$$y = e^{-2t}\cos 4t - \frac{3}{4}e^{-2t}\sin 4t = e^{-2t}\left(\cos 4t - \frac{3}{4}\sin 4t\right)$$

## Partial Fractions

We used partial fractions in Sec. 34-2 to make a given expression match one listed in the table of integrals. Now we use partial fractions to make an expression match one listed in our table of Laplace transforms.

---

**EXAMPLE 12:** Find $y$ if $\mathscr{L}[y] = 12/(s^2 - 2s - 8)$.

**Solution:** We separate $\mathscr{L}[y]$ into partial fractions,

$$\mathscr{L}[y] = \frac{12}{s^2 - 2s - 8}$$

$$= \frac{A}{s - 4} + \frac{B}{s + 2}$$

We can also complete the square here, but would find that the resulting expression would not match a table entry. Some trial-and-error work in algebra may be necessary to get the function to match a table entry.

so

$$12 = A(s + 2) + B(s - 4)$$

$$= (A + B)s + (2A + 4B)$$

from which $A + B = 0$ and $2A + 4B = 12$. Solving simultaneously gives $A = 2$ and $B = -2$, so

$$\mathscr{L}[y] = \frac{12}{s^2 - 2s - 8}$$

$$= \frac{2}{s - 4} - \frac{2}{s + 2}$$

Using Transform 9 twice, we get

$$y = 2e^{4t} - 2e^{-2t}$$

---

## EXERCISE 2—INVERSE TRANSFORMS

Find the inverse of each transform.

1. $\dfrac{1}{s}$
2. $\dfrac{3}{s^2}$
3. $\dfrac{2}{s^3}$
4. $\dfrac{1}{s^2 + 3s}$
5. $\dfrac{4}{4 + s^2}$
6. $\dfrac{s}{(s - 6)^2}$
7. $\dfrac{3s}{s^2 + 2}$
8. $\dfrac{4}{s^3 + 9s}$
9. $\dfrac{s + 4}{(s - 9)^2}$
10. $\dfrac{3s}{(s^2 + 4)^2}$
11. $\dfrac{5}{(s + 2)^2 + 9}$
12. $\dfrac{s + 2}{s^2 - 6s + 8}$
13. $\dfrac{2s^2 + 1}{s(s^2 + 1)}$
14. $\dfrac{s}{s^2 + 2s + 1}$
15. $\dfrac{5s - 2}{s^2(s - 1)(s + 2)}$
16. $\dfrac{1}{(s + 1)(s + 2)^2}$
17. $\dfrac{2}{s^2 + s - 2}$
18. $\dfrac{s + 1}{s^2 + 2s}$
19. $\dfrac{2s}{s^2 + 5s + 6}$
20. $\dfrac{3s}{(s^2 + 4)(s^2 + 1)}$

## 36-3 SOLVING DIFFERENTIAL EQUATIONS BY THE LAPLACE TRANSFORM

To solve a differential equation with the Laplace transform:

1. Take the transform of each side of the equation.
2. Solve for $\mathscr{L}[y] = F(s)$.
3. Manipulate $F(s)$ until it matches one or more table entries.
4. Take the inverse transform to find $y = f(t)$.

We start with a very simple example.

---

**EXAMPLE 13:** Solve the first-order differential equation $y' + y = 2$ if $y(0) = 0$.

**Solution:**

1. We write the Laplace transform of both sides and get

$$\mathscr{L}[y'] + \mathscr{L}[y] = \mathscr{L}[2]$$

$$s\mathscr{L}[y] - y(0) + \mathscr{L}[y] = \frac{2}{s}$$

2. We substitute 0 for $y(0)$ and solve for $\mathscr{L}[y]$.

$$(s + 1)\mathscr{L}[y] = \frac{2}{s}$$

$$\mathscr{L}[y] = \frac{2}{s(s + 1)}$$

3. Our function will match Transform 10 if we write it as

$$\mathscr{L}[y] = 2\frac{1}{s(s + 1)}$$

4. Taking the inverse gives

$$y = 2(1 - e^{-t})$$

---

In our next example we express $F(s)$ as three partial fractions in order to take the inverse transform.

---

**EXAMPLE 14:** Solve the first-order differential equation $y' - 3y + 4 = 9t$ if $y(0) = 2$.

**Solution:**

1. We take the Laplace transform of both sides and get

$$\mathscr{L}[y'] - \mathscr{L}[3y] + \mathscr{L}[4] = \mathscr{L}[9t]$$

$$s\mathscr{L}[y] - y(0) - 3\mathscr{L}[y] + \frac{4}{s} = \frac{9}{s^2}$$

We substitute 2 for $y(0)$ and solve for $\mathscr{L}[y]$.

$$(s-3)\mathscr{L}[y] = \frac{9}{s^2} - \frac{4}{s} + 2$$

$$\mathscr{L}[y] = \frac{9}{s^2(s-3)} - \frac{4}{s(s-3)} + \frac{2}{s-3}$$

3. We match our expressions with Transforms 28, 10, and 9:

$$\mathscr{L}[y] = \frac{9}{s^2(s-3)} + \frac{4}{3}\frac{-3}{s(s-3)} + 2\frac{1}{s-3}$$

4. Taking the inverse gives

$$y = e^{3t} - 3t - 1 + \frac{4}{3}(1 - e^{3t}) + 2e^{3t} = \frac{5}{3}e^{3t} - 3t + \frac{1}{3}$$

---

We now use the Laplace transform to solve a second-order differential equation.

---

**EXAMPLE 15:** Solve the second-order differential equation $y'' + 4y - 3 = 0$, where $y$ is a function of $t$, if $y$ and $y'$ are both zero at $t = 0$.

**Solution:**

1. Taking the Laplace transform of both sides,

$$\mathscr{L}[y''] + \mathscr{L}[4y] - \mathscr{L}[3] = 0$$

$$s^2\mathscr{L}[y] - sy(0) - y'(0) + 4\mathscr{L}[y] - \frac{3}{s} = 0$$

Substituting $y(0) = 0$ and $y'(0) = 0$ gives

$$s^2\mathscr{L}[y] - 0 - 0 + 4\mathscr{L}[y] - \frac{3}{s} = 0$$

2. Solving for $\mathscr{L}[y]$, we have

$$(s^2 + 4)\mathscr{L}[y] = \frac{3}{s}$$

$$\mathscr{L}[y] = \frac{3}{s(s^2 + 4)}$$

3. We match it to Transform 21:

$$\mathscr{L}[y] = \frac{3}{s(s^2+4)} = \frac{3}{4}\frac{4}{s(s^2 + 2^2)}$$

4. Taking the inverse gives

$$y = \frac{3}{4}(1 - \cos 2t)$$

---

We have seen that the hardest part in solving a differential equation by Laplace transform is often in doing the algebra necessary to get the given function to match a table entry. We must often use partial fractions or completing the square, and sometimes both, as in the following example.

**EXAMPLE 16:** Solve $y'' + 4y' + 5y = t$, where $y(0) = 1$ and $y'(0) = 2$.

**Solution:**

1. Taking the Laplace transform of both sides and substituting initial conditions, we find that

$$s^2\mathscr{L}[y] - sy(0) - y'(0) + 4s\mathscr{L}[y] - 4y(0) + 5\mathscr{L}[y] = \frac{1}{s^2}$$

$$\mathscr{L}[y](s^2 + 4s + 5) = \frac{1}{s^2} + s + 6$$

2. Solving for $\mathscr{L}[y]$ gives

$$\mathscr{L}[y] = \frac{1}{s^2(s^2 + 4s + 5)} + \frac{s + 6}{s^2 + 4s + 5} \qquad (1)$$

3. Taking for now just the first fraction in (1), which we'll call $F_1(s)$, we obtain

$$F_1(s) = \frac{1}{s^2(s^2 + 4s + 5)} = \frac{As + B}{s^2} + \frac{Cs + D}{s^2 + 4s + 5}$$

Multiplying by $s^2(s^2 + 4s + 5)$ gives

$$1 = As^3 + 4As^2 + 5As + Bs^2 + 4Bs + 5B + Cs^3 + Ds^2$$

$$= (A + C)s^3 + (4A + B + D)s^2 + (5A + 4B)s + 5B$$

Equating coefficients gives

$$A = -\frac{4}{25} \qquad B = \frac{1}{5} \qquad C = \frac{4}{25} \qquad D = \frac{11}{25}$$

So

$$F_1(s) = \frac{-4s/25 + 1/5}{s^2} + \frac{4s/25 + 11/25}{s^2 + 4s + 5}$$

$$= \frac{1}{25}\left(\frac{5}{s^2} - \frac{4}{s} + \frac{4s + 11}{s^2 + 4s + 5}\right)$$

We can use Transforms 5 and 6 for $4/s$ and $5/s^2$ but there is no match for the third term in the parentheses. However, by completing the square we can get the denominator to match those in Transforms 23 and 24. Thus

$$s^2 + 4s + 5 = s^2 + 4s + 4 - 4 + 5$$

$$= (s + 2)^2 + 1^2$$

We now manipulate that third term into the form of Transforms 23 and 24.

$$\frac{4s + 11}{s^2 + 4s + 5} = \frac{4s + 11}{(s + 2)^2 + 1^2} = (4)\frac{s + 11/4 + 2 - 2}{(s + 2)^2 + 1^2}$$

$$= (4)\frac{s + 2 + 3/4}{(s + 2)^2 + 1^2} = (4)\frac{s + 2}{(s + 2)^2 + 1^2} + (3)\frac{1}{(s + 2)^2 + 1^2}$$

Combining the third term with the first and second gives

$$F_1(s) = \frac{1}{25}\left[\frac{5}{s^2} - \frac{4}{s} + (4)\frac{s+2}{(s+2)^2+1^2} + (3)\frac{1}{(s+2)^2+1^2}\right]$$

Returning to the remaining fraction in (1), which we call $F_2(s)$, we complete the square in the denominator and use partial fractions to get

$$F_2(s) = \frac{s+6}{(s+2)^2+1^2} = \frac{s+2-2+6}{(s+2)^2+1^2}$$

$$= \frac{s+2}{(s+2)^2+1^2} + \frac{4}{(s+2)^2+1^2}$$

4. Taking the inverse transform of $F_1(s)$ and $F_2(s)$, we obtain

$$y = \frac{1}{25}[5t - 4 + 4e^{-2t}\cos t + 3e^{-2t}\sin t] + e^{-2t}\cos t + 4e^{-2t}\sin t$$

which simplifies to

$$y = \frac{1}{25}(5t - 4 + 29e^{-2t}\cos t + 103e^{-2t}\sin t)$$

## EXERCISE 3—SOLVING DIFFERENTIAL EQUATIONS BY THE LAPLACE TRANSFORM

Solve each differential equation by the Laplace transform.

### First-Order Equations

1. $y' - 3y = 0,\quad y(0) = 1$
2. $2y' + y = 1,\quad y(0) = 3$
3. $4y' - 2y = t,\quad y(0) = 0$
4. $y' + 5y = e^{2t},\quad y(0) = 2$
5. $3y' - 2y = t^2,\quad y(0) = 3$
6. $y' - 3y = \sin t,\quad y(0) = 0$
7. $y' + 2y = \cos 2t,\quad y(0) = 0$
8. $4y' - y = 3t^3,\quad y(0) = 0$

### Second-Order Equations

9. $y'' + 2y' - 3y = 0,\quad y(0) = 0,\ y'(0) = 2$
10. $y'' + y' + y = 1,\quad y(0) = 0,\ y'(0) = 0$
11. $y'' + 3y' = 3,\quad y(0) = 1,\ y'(0) = 2$
12. $y'' + 2y = 2,\quad y(0) = 0,\ y'(0) = 3$
13. $2y'' + y = 4t,\quad y(0) = 3,\ y'(0) = 0$
14. $y'' + y' + 5y = t,\quad y(0) = 1,\ y'(0) = 2$
15. $y'' + 4y' + 3y = t,\quad y(0) = 2,\ y'(0) = 2$
16. $y'' + 4y' = 8t^3,\quad y(0) = 0,\ y'(0) = 0$
17. $3y'' + y' = \sin t,\quad y(0) = 2,\ y'(0) = 3$
18. $2y'' + y' + 2y = 3,\quad y(0) = 2,\ y'(0) = 1$
19. $y'' - 2y' + y = e^t,\quad y(0) = 0,\ y'(0) = 0$
20. $2y'' + 32y = \cos 2t,\quad y(0) = 0,\ y'(0) = 1$
21. $y'' + 2y' + 3y = te^t,\quad y(0) = 0,\ y'(0) = 0$
22. $3y'' + 2y' - y = \sin 3t,\quad y(0) = 0,\ y'(0) = 0$

### Motion in a Resisting Medium

23. A 15.0-kg object is dropped from rest through air whose resisting force is equal to 1.85 times the object's speed, in m/s. Find its speed after 1.00 s.

### Exponential Growth and Decay

24. A certain steel ingot is 1900°F and cools at a rate (in °F/min) equal to 1.25 times its present temperature (°F). Find its temperature after 5.00 min.
25. The rate of growth (bacteria/h) of a colony of bacteria is equal to 2.5 times the number present at any instant. How many bacteria are there after 24 h if there are 5000 at first?

### Mechanical Vibrations

26. A weight that hangs from a spring is pulled down 1.00 in. at $t = 0$ and released from rest. The differential equation of motion is $x'' + 6.25x' + 25.8x = 0$. Write the equation for $x$ as a function of time.
27. An alternating force is applied to a weight such that the equation of motion is $x'' + 6.25x' = 45.3 \cos 2.25t$. If $v$ and $x$ are zero at $t = 0$, write an equation for $x$ as a function of time.

## 36-4 ELECTRICAL APPLICATIONS

Many of the differential equations for motion and electric circuits covered in Chapter 35 are nicely handled by the Laplace transform. However, since the Laplace transform is used mainly for electric circuits, we emphasize that application here. The method is illustrated by examples of several types of circuits, with dc and ac sources.

### Series *RC* Circuit with DC Source

**EXAMPLE 17:** A capacitor (Fig. 36-2) is discharged through a resistor by throwing the switch from position 1 to position 2 at $t = 0$. Find the current $i$.

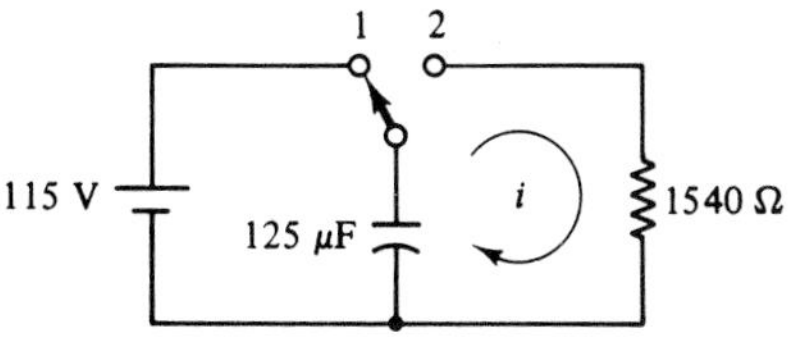


**FIGURE 36-2** ***RC* circuit.**

**Solution:** Summing the voltages around the loop gives

$$Ri + \frac{1}{C}\int_0^t i\,dt - v_c = 0$$

Substituting values and rearranging yields

$$1540i + \frac{1}{125 \times 10^{-6}}\int_0^t i\,dt = 115$$

Taking the transform of each term, we obtain

$$1540\mathcal{L}[i] + 8000\frac{\mathcal{L}[i]}{s} = \frac{115}{s}$$

We solve for $\mathcal{L}[i]$ and rewrite our expression in the form of a table entry.

$$\mathcal{L}[i]\left(1540 + \frac{8000}{s}\right) = \frac{115}{s}$$

$$\mathcal{L}[i]\left(\frac{1540s + 8000}{s}\right) = \frac{115}{s}$$

$$\mathcal{L}[i] = \frac{115}{1540s + 8000} = \frac{0.0747}{s + 5.19}$$

Taking the inverse transform using Transform 9 gives us

$$i = 0.0747e^{-5.19t}$$

This is the same result obtained by classical methods in Chapter 35, Example 35.

---

**EXAMPLE 18:** A capacitor (Fig. 36-3) has an initial charge of 35 V, with the polarity as marked. Find the current if the switch is closed at $t = 0$.

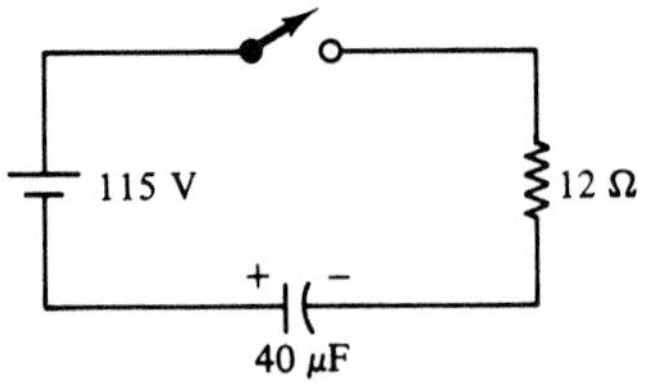


**FIGURE 36-3** ***RC* circuit.**

**Solution:** Summing the voltages around the loops gives

$$Ri + \frac{1}{C}\int_0^t i\,dt - v_c - E = 0$$

Substituting values and rearranging, we see that

$$12i + \frac{1}{40 \times 10^{-6}}\int_0^t i\,dt = 150$$

Taking the transform of each term, we have

$$12\mathcal{L}[i] + 25{,}000\frac{\mathcal{L}[i]}{s} = \frac{150}{s}$$

Solving for $\mathcal{L}[i]$ yields

$$\mathcal{L}[i]\left(12 + \frac{25{,}000}{s}\right) = \frac{150}{s}$$

$$\mathcal{L}[i]\left(\frac{12s + 25{,}000}{s}\right) = \frac{150}{s}$$

$$\mathcal{L}[i] = \frac{150}{12s + 25{,}000} = \frac{12.5}{s + 2080}$$

Taking the inverse transform using Transform 9, we obtain

$$i = 12.5e^{-2080t}$$

---

## Series *RL* Circuit with DC Source

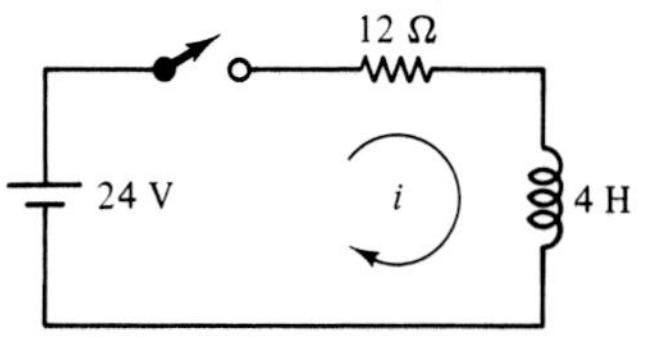


FIGURE 36-4 ***RL* circuit.**

**EXAMPLE 19:** The initial current in an *RL* circuit (Fig. 36-4) is 1 A in the direction shown. The switch is closed at $t = 0$. Find the current.

**Solution:** Summing the voltages around the loop gives

$$E = Ri + L\frac{di}{dt}$$

Substituting values and rearranging, we find that

$$24 = 12i + 4i'$$

or

$$6 = 3i + i'$$

Taking the transform of each term yields

$$\frac{6}{s} = 3\mathscr{L}[i] + s\mathscr{L}[i] - i(0)$$

Substituting 1 for $i(0)$ and solving for $\mathscr{L}[i]$, we have

$$\frac{6}{s} = \mathscr{L}[i](s + 3) - 1$$

$$\mathscr{L}[i] = \frac{6}{s(s + 3)} + \frac{1}{s + 3}$$

Taking the inverse transform using Transforms 9 and 10 gives

$$i = 2(1 - e^{-3t}) + e^{-3t}$$

or

$$i = 2 - e^{-3t}$$

The current is graphed in Fig. 36-5.

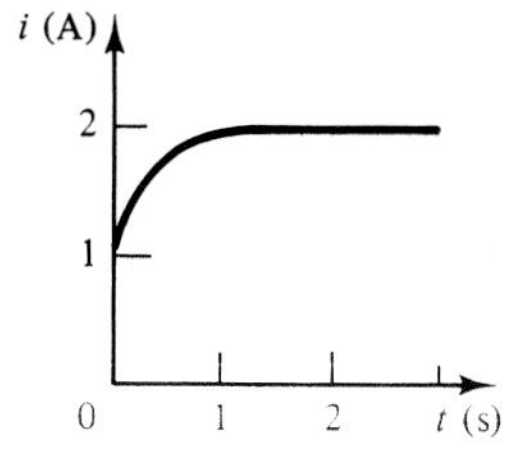


**FIGURE 36-5**

## Series *RL* Circuit with AC Source

**EXAMPLE 20:** A switch (Fig. 36-6) is closed at $t = 0$ when applied voltage is zero and increasing. Find the current.

**Solution:** Summing the voltages around the loop gives

$$Ri + L\frac{di}{dt} = v$$

Substituting values and rearranging, we find that

$$4i + 0.02\frac{di}{dt} = 80 \sin 400t$$

Taking the transform of each term yields

$$4\mathscr{L}[i] + 0.02[s\mathscr{L}[i] - i(0)] = \frac{80(400)}{s^2 + (400)^2}$$

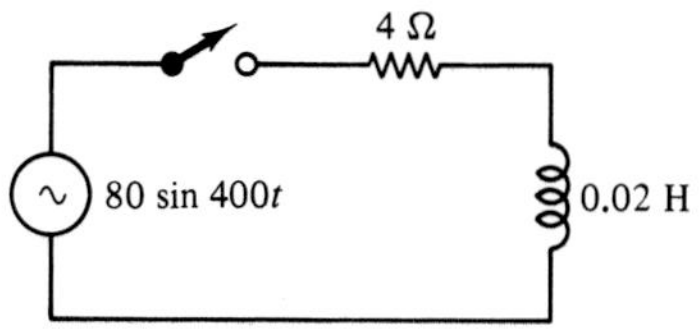


**FIGURE 36-6 *RL* circuit with ac source.**

Substituting 0 for $i(0)$ and solving for $\mathscr{L}[i]$, we have

$$\mathscr{L}[i](0.02s + 4) = \frac{32{,}000}{s^2 + (400)^2}$$

$$\mathscr{L}[i] = \frac{32{,}000}{[s^2 + (400)^2](0.02s + 4)} = \frac{1{,}600{,}000}{[s^2 + (400)^2](s + 200)}$$

$$= \frac{1{,}600{,}000}{(s - j400)(s + j400)(s + 200)}$$

Splitting this expression into partial fractions yields

$$\frac{1{,}600{,}000}{(s - j400)(s + j400)(s + 200)} = \frac{A}{s - j400} + \frac{B}{s + j400} + \frac{C}{s + 200}$$

Multiplying both sides by $(s - j400)$ gives

$$\frac{1{,}600{,}000}{(s + j400)(s + 200)} = A + \frac{B(s - j400)}{s + j400} + \frac{C(s - j400)}{s + 200}$$

Letting $s = j400$ causes the partial fractions containing $B$ and $C$ to vanish. Solving for $A$, we get

$$A = \frac{1{,}600{,}000}{(j400 + j400)(j400 + 200)} = 2(-2 - j)$$

after simplification (not shown). We find $B$ and $C$ by the same method, and get $B = 2(-2 + j)$, and $C = 8$. Substituting back, we have

$$\mathscr{L}[y] = \frac{2(-2 - j)}{s - j400} + \frac{2(-2 + j)}{s + j400} + \frac{8}{s + 200}$$

We multiply the numerator and denominator of the first two fractions each by the conjugate of its denominator,

$$\mathscr{L}[i] = \frac{2(-2 - j)(s + j400)}{(s - j400)(s + j400)} + \frac{2(-2 + j)(s - j400)}{(s + j400)(s - j400)} + \frac{8}{s + 200}$$

We combine the first two terms over a common denominator and collect terms, getting

$$\mathscr{L}[i] = \frac{1600 - 8s}{s^2 + (400)^2} + \frac{8}{s + 200}$$

$$= 4\frac{400}{s^2 + (400)^2} - 8\frac{s}{s^2 + (400)^2} + 8\frac{1}{s + 200}$$

Taking the inverse transform using Transforms 9, 17, and 18, we find that

$$i = 4 \sin 400t - 8 \cos 400t + 8e^{-200t}$$

We now combine the first two terms in the form $I \sin(400t + \theta)$, where

$$I = \sqrt{4^2 + 8^2} = 8.94 \quad \text{and} \quad \theta = \tan^{-1}\left(-\frac{8}{4}\right) = -63.4° = -1.107 \text{ rad}$$

so

$$i = 8.94 \sin(400t - 1.107) + 8e^{-200t}$$

A graph of this wave, as well as the applied voltage wave, is shown in Fig. 36-7.

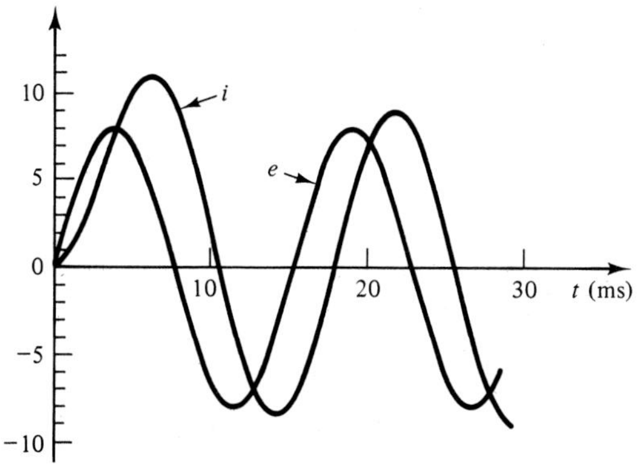


FIGURE 36-7

## Series *RLC* Circuit with DC Source

**EXAMPLE 21:** A switch (Fig. 36-8) is closed at $t = 0$, and there is no initial charge on the capacitor. Write an expression for the current.

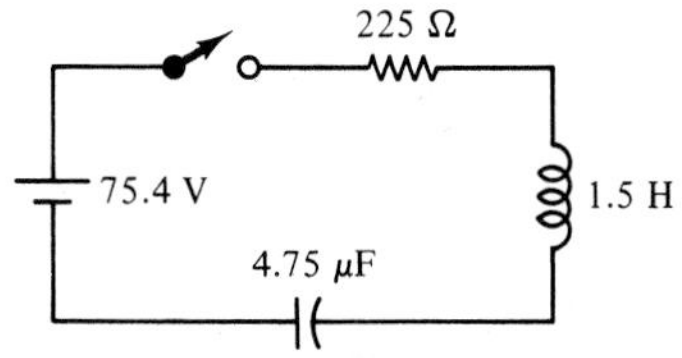


**FIGURE 36-8** ***RLC* circuit.**

**Solution:**

1. The differential equation for this circuit is

$$Ri + L\frac{di}{dt} + \frac{1}{C}\int_0^t i\,dt = E$$

2. Transforming each term and substituting values gives

$$225\mathscr{L}[i] + 1.5\{s\,\mathscr{L}[i] - i(0)\} + \frac{10^6\mathscr{L}[i]}{4.75s} = \frac{75.4}{s}$$

3. Setting $i(0)$ to 0 and solving for $\mathscr{L}[i]$ yields

$$\mathscr{L}[i] = \frac{50.3}{s^2 + 150s + 140{,}000}$$

4. We now try to match this expression with those in the table. Let us factor the denominator.

$$\begin{aligned} s^2 + 150s + 140{,}000 &= (s^2 + 150s + 5625) + 140{,}000 - 5625 \\ &= (s + 75)^2 - (-134{,}000) \\ &= (s + 75)^2 - j^2(367)^2 \\ &= (s + 75 + j367)(s + 75 - j367) \end{aligned}$$

after replacing $-1$ by $j^2$, and factoring the difference of two squares. So

$$\begin{aligned} \mathscr{L}[i] &= \frac{50.3}{s^2 + 150s + 140{,}000} \\ &= \frac{50.3}{(s + 75 + j367)(s + 75 - j367)} \end{aligned}$$

We can now find partial fractions,

$$\frac{50.3}{(s + 75 + j367)(s + 75 - j367)} = \frac{A}{s + 75 + j367} + \frac{B}{s + 75 - j367}$$

Multiplying through by $s + 75 + j367$ gives

$$\frac{50.3}{s + 75 - j367} = A + \frac{B(s + 75 + j367)}{s + 75 - j367}$$

Setting $s = -75 - j367$, we get

$$A = \frac{50.3}{-75 - j367 + 75 - j367} = \frac{50.3}{-j734} = j0.0685$$

We see that most of the work when using the Laplace transform is in algebra, rather than in calculus.

We find $B$ by multiplying through by $s + 75 - j367$ and then setting $s = j367 - 75$, and we get $B = -j0.0625$ (work not shown). Substituting gives us

$$\mathcal{L}[i] = \frac{j0.0685}{s + 75 + j367} - \frac{j0.0685}{s + 75 - j367}$$

This still doesn't match any table entry. Let us now combine the two fractions over a common denominator.

$$\mathcal{L}[i] = \frac{j0.0685}{s + 75 + j367} - \frac{j0.0685}{s + 75 - j367}$$

$$= \frac{j0.0685(s + 75 - j367) - j0.0685(s + 75 + j367)}{(s + 75 + j367)(s + 75 - j367)}$$

$$= \frac{-j^2 50.4}{s^2 + 150s + 140{,}000} = 0.137 \frac{367}{(s + 75)^2 + (367)^2}$$

5. We finally have a match, with Transform 23. Taking the inverse transform gives

$$i = 137e^{-75t} \sin 367t \qquad \text{mA}$$

which agrees with the result found in Chapter 35, Example 50.

## EXERCISE 4—ELECTRICAL APPLICATIONS

1. The current in a certain $RC$ circuit satisfies the equation $172i' + 2750i = 115$. If $i$ is zero at $t = 0$, write an equation for $i$.
2. In an $RL$ circuit, $R = 3750\ \Omega$ and $L = 0.150$ H. It is connected to a dc source of 250 V at $t = 0$. If $i$ is zero at $t = 0$, write an equation for the instantaneous current.
3. The current in an $RLC$ circuit satisfies the equation $0.55i'' + 482i' + 7350i = 120$. If $i$ and $i'$ are zero at $t = 0$, write an equation for $i$.
4. In an $RLC$ circuit, $R = 2950\ \Omega$, $L = 0.120$ H, and $C = 1.55\ \mu$F. It is connected to a dc source of 115 V at $t = 0$. If $i$ is zero at $t = 0$, write an equation for the instantaneous current.
5. The current in a certain $RLC$ circuit satisfies the equation $3.15i + 0.0223i' + 1680 \int_0^t i\, dt = 1$. If $i$ is zero at $t = 0$, write an equation for the instantaneous current.
6. In an $RL$ circuit, $R = 4150\ \Omega$ and $L = 0.127$ H. It is connected to a dc source of 28.4 V at $t = 0$. If $i$ is zero at $t = 0$, write an equation for the instantaneous current.
7. In an $RL$ circuit, $R = 8.37\ \Omega$ and $L = 0.250$ H. It is connected to a dc source of 50.5 V at $t = 0$. If $i$ is zero at $t = 0$, write an equation for the instantaneous current.

## 36-5 NUMERICAL SOLUTION OF FIRST-ORDER DIFFERENTIAL EQUATIONS

There are many differential equations that cannot be solved by the "classical" methods of Chapter 35 or even by the Laplace transform shown earlier in this chapter. Further, even if an equation can be solved analytically, we may want to use a computer, for which the preceding methods are not useful. In these cases we can use a numerical method.

We start with Euler's method, the earliest and simplest numerical method for solving differential equations, but not the most accurate. It does, however, serve as a good introduction to other methods. We show a modified Euler's method that includes predictor-corrector steps, and the Runge–Kutta method. We apply these methods to both first- and second-order equations.

### Euler's Method

Suppose that we have a first-order differential equation, which we write in the form

$$y' = f(x, y) \tag{1}$$

and a boundary condition that $x = x_p$ when $y = y_p$. We seek a solution $y = F(x)$ such that the graph of this function (Fig. 36-9) passes through $P(x_p, y_p)$ and has a slope at $P$ given by (1), $m_p = f(x_p, y_p)$.

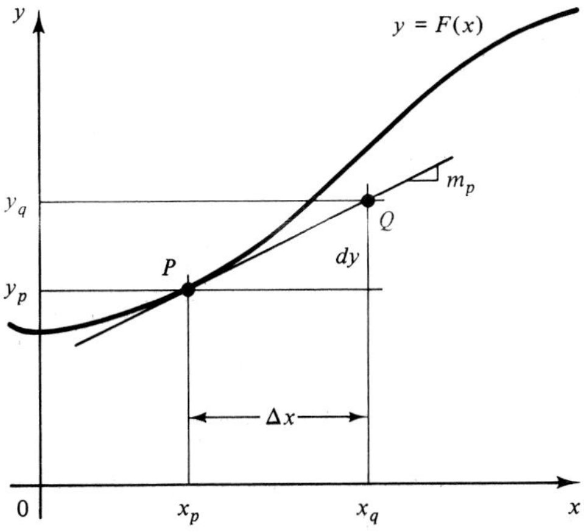


**FIGURE 36-9** **Euler's method.**

Having the slope at $P$, we then step a distance $\Delta x$ to the right. The rise $dy$ of the tangent line is then

$$\text{rise} = (\text{slope})(\text{run}) = m_p\,\Delta x$$

Thus the point $Q$ has the coordinates $(x_p + \Delta x,\ y_p + m_p\,\Delta x)$. *Point $Q$ is probably not on the curve $y = F(x)$, but may be close enough to use as an approximation to the curve.*

**This is no different than when, in Sec. 29-5, we used differentials to approximate a small change in a function.**

From $Q$, we repeat the process enough times to reconstruct as much of the curve $y = F(x)$ as needed. Thus having the coordinates $(x_p, y_p)$ at $P$ and the slope $m_p$ at $P$, we find the coordinates $(x_q, y_q)$ of $Q$ by the following iteration formulas.

After a Swiss mathematician, Leonhard Euler (1707–1783).

| | | |
|---|---|---|
| Euler's Method | $x_q = x_p + \Delta x$<br>$y_q = y_p + m_p\,\Delta x$ | **425** |

**EXAMPLE 22:** Find an approximate solution to $y' = x^2/y$, with the boundary condition that $y = 2$ when $x = 3$. Calculate $y$ for $x = 3$ to 10, in steps of 1.

**Solution:** The slope at (3, 2) is

$$m = y'(3, 2) = \frac{9}{2} = 4.5$$

If $\Delta x = 1$, the rise is

$$dy = m\,\Delta x = 4.5(1) = 4.5$$

The ordinate of our next point is then $2 + 4.5 = 6.5$. The abscissa of the next point is $3 + 1 = 4$. So the coordinates of our next point are (4, 6.5). Repeating the process using (4, 6.5) as $(x_p, y_p)$, we calculate the next point by first getting $m$ and $dy$:

$$m = y'(4, 6.5) = \frac{16}{6.5} = 2.462$$

$$dy = 2.452(1) = 2.462$$

So the next point is (5, 8.962). The remaining values are given in the following computer-generated table.

| $x$ | *Approximate y* | *Exact y* | *Error* |
|---|---|---|---|
| 3 | 2.00000 | 2.00000 | 0.00000 |
| 4 | 6.50000 | 5.35413 | 1.14587 |
| 5 | 8.96154 | 8.32666 | 0.63487 |
| 6 | 11.75124 | 11.40176 | 0.34948 |
| 7 | 14.81475 | 14.65151 | 0.16324 |
| 8 | 18.12226 | 18.09236 | 0.02991 |
| 9 | 21.65383 | 21.72556 | 0.07173 |
| 10 | 25.39451 | 25.54734 | 0.15283 |

The solution to our differential equation, then, is in the form of a set of $(x, y)$ pairs, the first two columns in the table above. We do not get an equation for our solution.

Reducing the step size with this method and those to follow will result in better accuracy (and more work).

We normally use numerical methods for a differential equation whose exact solution cannot be found. Here we have used an equation whose solution is known so that we may check our answer. The solution is $3y^2 = 2x^3 - 42$, which we use to compute the values shown in the third column of the table, with the difference between exact and approximate values in the third column. Note that the error at $x = 10$ is about 0.6%.

### Modified Euler's Method with Predictor-Corrector Steps

The various *predictor–corrector* methods all have two main steps. First, the *predictor* step tries to predict the next point on the curve from preceding values, and then the *corrector* step tries to improve the predicted values.

Here we modify Euler's method of the preceding section as the predictor, using the coordinates and slope at the present point $P$ to estimate the next point $Q$. The corrector step then recomputes $Q$ *using the average of the slopes at $P$ and $Q$*. Our iteration formula is then

| Modified Euler's Method | $x_q = x_p + \Delta x$<br>$y_q = y_p + \dfrac{m_p + m_q}{2}\Delta x$ | **426** |
|---|---|---|

---

**EXAMPLE 23:** Repeat Example 22 using the modified Euler's method.

**Solution:** Using data from before, the slope $m_1$ at (3, 2) was 4.5, the second point was predicted to be (4, 6.5), and the slope $m_2$ at (4, 6.5) was 2.462. The recomputed ordinate at the second point is then, by Eq. 426,

$$y_2 = 2 + \frac{4.5 + 2.462}{2}(1)$$

$$= 5.481$$

so our corrected second point is (4, 5.481). The slope at this point is

$$m_2 = y'(4, 5.481) = 2.919$$

We use this to predict $y_3$.

$$y_3(\text{predicted}) = y_2 + m_2\,\Delta x$$

$$= 5.481 + 2.919(1)$$

$$= 8.400$$

The slope at (5, 8.400) is

$$m_3 = y'(5, 8.400) = 2.976$$

The corrected $y_3$ is then

$$y_3(\text{corrected}) = 5.481 + \frac{2.919 + 2.976}{2}(1) = 8.429$$

The remaining values are given in the following table.

| $x$ | *Approximate y* | *Exact y* | *Error* |
|---|---|---|---|
| 3 | 2.00000 | 2.00000 | 0.00000 |
| 4 | 5.48077 | 5.35413 | 0.12664 |
| 5 | 8.42850 | 8.32666 | 0.10184 |
| 6 | 11.49126 | 11.40176 | 0.08950 |
| 7 | 14.73299 | 14.65151 | 0.08148 |
| 8 | 18.16790 | 18.09236 | 0.07555 |
| 9 | 21.79642 | 21.72556 | 0.07086 |
| 10 | 25.61434 | 25.54734 | 0.06700 |

Note that the error at $x = 10$ is about 0.26%, compared with 0.6% before. We have gained in accuracy at the cost of added computation.

---

## Runge–Kutta Method

The Runge–Kutta method, like the modified Euler's, is a predictor-corrector method.

*Prediction:* Starting at $P$ (Fig. 36-10), we use

the slope $m_p$ to predict $R$,
the slope $m_r$ to predict $S$,
the slope $m_s$ to predict $Q$.

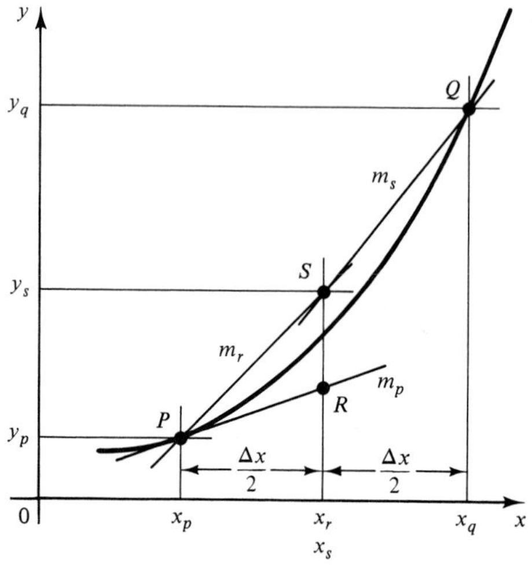


**FIGURE 36-10 Runge-Kutta method.**

*Correction:* We then take a weighted average of the slopes at $P$, $Q$, $R$, and $S$, giving twice the weight to the slopes at the midpoints $R$ and $S$ than at the endpoints $P$ and $Q$.

$$m_{avg} = \frac{1}{6}(m_p + 2m_r + 2m_s + m_q)$$

and use the average slope and the coordinates of $P$ to find $Q$.

$$y_q = y_p + m_{avg}\,\Delta x$$

After two German mathematicians, Carl Runge (1856–1927) and Wilhelm Kutta (1867–1944).

| | | |
|---|---|---|
| Runge–Kutta Method | $x_q = x_p + \Delta x$ <br> $y_q = y_p + m_{avg}\,\Delta x$ <br> where $m_{avg} = \frac{1}{6}(m_p + 2m_r + 2m_s + m_q)$ <br> and $m_p = f'(x_p, y_p)$ <br> $m_r = f'\left(x_p + \frac{\Delta x}{2}, y_p + m_p \frac{\Delta x}{2}\right)$ <br> $m_s = f'\left(x_p + \frac{\Delta x}{2}, y_p + m_r \frac{\Delta x}{2}\right)$ <br> $m_q = f'(x_p + \Delta x, y_p + m_s\,\Delta x)$ | **427** |

**EXAMPLE 24:** Repeat Example 23 using the Runge–Kutta method and a step of 1.

**Solution:** Starting at $P(3, 2)$, with $m_p = 4.5$,

$$y_r = 2 + 4.5(0.5) = 4.25$$

$$m_r = f'(3.5, 4.25) = \frac{(3.5)^2}{4.25} = 2.882$$

$$y_s = 2 + 2.882(0.5) = 3.441$$

$$m_s = f'(3.5, 3.441) = \frac{(3.5)^2}{3.441} = 3.560$$

$$y_q = 2 + 3.560(1) = 5.560$$

$$m_q = f'(4, 5.560) = \frac{4^2}{5.560} = 2.878$$

$$m_{avg} = \frac{4.5 + 2(2.882) + 2(3.560) + 2.878}{6} = 3.377$$

$$y_q(\text{corrected}) = y_p + m_{avg}\,\Delta x = 2 + 3.377(1) = 5.377$$

So our second point is (3, 5.377). The rest of the calculation is given in the following table.

| $x$ | *Approximate y* | *Exact y* | *Error* |
|---|---|---|---|
| 3 | 2.00000 | 2.000000 | 0.00000 |
| 4 | 5.37703 | 5.354126 | 0.02290 |
| 5 | 8.34141 | 8.326664 | 0.01475 |
| 6 | 11.41255 | 11.401760 | 0.01079 |
| 7 | 14.65992 | 14.651510 | 0.00841 |
| 8 | 18.09918 | 18.092360 | 0.00682 |
| 9 | 21.73125 | 21.725560 | 0.00569 |
| 10 | 25.55219 | 25.547340 | 0.00484 |

The error here at $x = 10$ with the Runge–Kutta method is about 0.02%, compared with 0.26% for the modified Euler method.

## Comparison of the Three Methods

The following table shows the errors obtained when solving the same differential equation by our three different methods, each with a step size of 1.

| $x$ | *Euler* | *Modified Euler* | *Runge–Kutta* |
|---|---|---|---|
| 3 | 0.00000 | 0.00000 | 0.00000 |
| 4 | 1.14587 | 0.12664 | 0.02290 |
| 5 | 0.63487 | 0.10184 | 0.01475 |
| 6 | 0.34948 | 0.08950 | 0.01079 |
| 7 | 0.16324 | 0.08148 | 0.00841 |
| 8 | 0.02991 | 0.07555 | 0.00682 |
| 9 | 0.07173 | 0.07086 | 0.00569 |
| 10 | 0.15283 | 0.06700 | 0.00484 |

### Divergence

As with most numerical methods, the computation can diverge, with our answers getting further and further from the true values. The computations shown here can diverge if either the step size $\Delta x$ is too large, or the range of $x$ over which we compute is too large. Numerical methods can also diverge if the points used in the computation are near a discontinuity, a maximum or minimum point, or a point of inflection.

When doing the computations on the computer (the only practical way) it is easy to change the step size. A good practice is to reduce the step size by a factor of 10 each time (1, 0.1, 0.01, . . .) until the answers do not change in a given decimal place. Further reduction of the step size will eventually result in numbers too small for the computer to handle and produce unreliable results.

## EXERCISE 5—NUMERICAL SOLUTION OF FIRST-ORDER DIFFERENTIAL EQUATIONS

Solve each differential equation and find the value of $y$ asked for. Use either Euler's method, the modified Euler method, or the Runge–Kutta method, with a step size of 0.1.

1. $y' + xy = 3$; $y(4) = 2$. Find $y(5)$.
2. $y' + 5y^2 = 3x$; $y(0) = 3$. Find $y(1)$.
3. $x^4y' + ye^x = xy^3$; $y(2) = 2$. Find $y(3)$.
4. $xy' + y^2 = 3 \ln x$; $y(0) = 3$. Find $y(1)$.
5. $1.94x^2y' + 8.23x \sin y - 2.99y = 0$; $y(0) = 3$. Find $y(1)$.
6. $y' \ln x + 3.99(x - y)^2 = 4.28y$; $y(2) = 1.76$. Find $y(3)$.
7. $(x + y)y' + 2.76x^2y^2 = 5.33y$; $y(0) = 1$. Find $y(1)$.
8. $y' + 297x\, e^x + 192x \cos y = 0$; $y(2) = 1$. Find $y(3)$.
9. $xy' + xy^2 = 8y \ln |xy|$; $y(1) = 3$. Find $y(2)$.
10. $x^2y' - 88.5x^2 \ln y + 32.7x = 0$; $y(1) = 23.9$. Find $y(2)$.

### Computer

11. Write a program or use a spreadsheet to solve a first-order differential equation by Euler's method. Try it on any of the equations above.
12. Write a program or use a spreadsheet to solve a first-order differential equation by the modified Euler's method. Try it on any of the equations above.
13. Write a program or use a spreadsheet to solve a first-order differential equation by the Runge–Kutta method. Try it on any of the equations above.
14. If you have done one of the calculations above using a spreadsheet which has a graphing utility, use that utility to make a graph of $x$ versus $y$ over the given range.

## 36-6 NUMERICAL SOLUTION OF SECOND-ORDER DIFFERENTIAL EQUATIONS

To solve a second-order differential equation numerically, we first transform the given equation into two first-order equations by substituting another variable, $m$, for $y'$.

**EXAMPLE 25:** By making the substitution $y' = m$, we can transform the second-order equation

$$y'' + 3y' + 4xy = 0$$

into the two first-order equations

$$y' = m$$

and

$$m' = -3m - 4xy$$
$$= f(x, y, m)$$

---

As before, we start with the given point $P(x_p, y_p)$, at which we must also know the slope $m_p$. We proceed from $P$ to the next point $Q(x_q, y_q)$ as follows:

1. Compute the average first derivative, $m_{avg}$, over the interval $\Delta x$.
2. Compute the average second derivative $m'_{avg}$ over that interval.
3. Find the coordinates $x_q$ and $y_q$ at $Q$ by using the equations

$$x_q = x_p + \Delta x$$

and

$$y_q = y_p + m_{avg}\,\Delta x$$

4. Find the first derivative $m_q$ at $Q$ by using

$$m_q = m_p + m'_{avg}\,\Delta x$$

5. Go to step 1 and repeat the sequence for the next interval.

We can find $m_{avg}$ and $m'_{avg}$ in steps 1 and 2 either by the modified Euler's method or by the Runge–Kutta method. The latter method, as before, requires us to find values at the intermediate points $R$ and $S$. The equations for second-order equations are similar to those we had for first-order equations.

At $P$: $x_p$, $y_p$, and $m_p$ are given, $m'_p = f(x_p, y_p, m_p)$.

| | | |
|---|---|---|
| At $R$: | $x_r = x_p + \dfrac{\Delta x}{2}$ | $y_r = y_p + m_p \dfrac{\Delta x}{2}$ |
| | $m_r = m_p + m'_p \dfrac{\Delta x}{2}$ | $m'_r = f(x_r, y_r, m_r)$ |
| At $S$: | $x_s = x_p + \dfrac{\Delta x}{2}$ | $y_s = y_p + m_r \dfrac{\Delta x}{2}$ |
| | $m_s = m_p + m'_r \dfrac{\Delta x}{2}$ | $m'_s = f(x_s, y_s, m_s)$ |
| At $Q$: | $x_q = x_p + \Delta x$ | $y_q = y_p + m_s\,\Delta x$ |
| | $m_q = m_p + m'_s\,\Delta x$ | $m'_q = f(x_q, y_q, m_q)$ |

**EXAMPLE 26:** Find an approximate solution to the equation

$$y'' - 2y' + y = 2e^x$$

with boundary conditions $y' = 2e$ at (1, 0). Find points from 1 to 3 in steps of 0.1.

**Solution:** Replacing $y'$ by $m$ and solving for $m'$ gives

$$m' = f(x, y, m)$$
$$= 2e^x - y + 2m$$

At *P*: The boundary values are $x_p = 1$, $y_p = 0$, and $m_p = 5.4366$. Then

$$m_p' = f(x_p, y_p, m_p) = 2e^1 - 0 + 2(5.4366) = 16.3097$$

At *R*:

$$x_r = x_p + \frac{\Delta x}{2} = 1 + 0.05 = 1.05$$

$$y_r = y_p + m_p \frac{\Delta x}{2} = 0 + 5.4366(0.05) = 0.2718$$

$$m_r = m_p + m_p' \frac{\Delta x}{2} = 5.4366 + 16.3097(0.05) = 6.2521$$

$$m_r' = 2e^{1.05} - 0.2718 + 2(6.2521) = 17.9477$$

At *S*:

$$x_s = 1.05$$

$$y_s = y_p + m_r \frac{\Delta x}{2} = 0 + 6.2521(0.05) = 0.3126$$

$$m_s = m_p + m_r' \frac{\Delta x}{2} = 5.4366 + 17.9477(0.05) = 6.3360$$

$$m_s' = 2e^{1.05} - 0.3126 + 2(6.3360) = 18.0707$$

At *Q*:

$$x_q = x_p + \Delta x = 1.1$$

$$y_q = y_p + m_s \, \Delta x = 0 + 6.3360(0.1) = 0.6336$$

$$m_q = m_p + m_s' \, \Delta x = 5.4366 + 18.0707(0.1) = 7.2437$$

$$m_q' = 2e^{1.1} - 0.6336 + 2(7.2437) = 19.8621$$

Then

$$m_{avg} = \frac{m_p + 2m_r + 2m_s + m_q}{6}$$

$$= \frac{5.4366 + 2(6.2521) + 2(6.3360) + 7.2437}{6} = 6.3094$$

and

$$m'_{avg} = \frac{m_p' + 2m_r' + 2m_s' + m_q'}{6}$$

$$= \frac{16.3097 + 2(17.9477) + 2(18.0707) + 19.8621}{6} = 18.0348$$

So

$$y_q = y_p + m_{avg}\,\Delta x = 0 + 6.3094(0.1) = 0.63094$$

and

$$m_q = m_p + m'_{avg}\,\Delta x = 5.4366 + 18.0348(0.1) = 7.2401$$

The computation is then repeated. The remaining values are given in the following table, along with the exact $y$ and exact slope, found by analytical solution of the given equation.

| $x$ | $y$ | *Exact y* | *Slope* | *Exact Slope* |
|---|---|---|---|---|
| 1.0 | 0.000 | 0.000 | 5.437 | 5.437 |
| 1.1 | 0.631 | 0.631 | 7.240 | 7.240 |
| 1.2 | 1.461 | 1.461 | 9.429 | 9.429 |
| 1.3 | 2.532 | 2.532 | 12.072 | 12.072 |
| 1.4 | 3.893 | 3.893 | 15.248 | 15.248 |
| 1.5 | 5.602 | 5.602 | 19.047 | 19.047 |
| 1.6 | 7.727 | 7.727 | 23.576 | 23.576 |
| 1.7 | 10.346 | 10.346 | 28.957 | 28.957 |
| 1.8 | 13.551 | 13.551 | 35.330 | 35.330 |
| 1.9 | 17.450 | 17.450 | 42.856 | 42.857 |
| 2.0 | 22.167 | 22.167 | 51.723 | 51.723 |
| 2.1 | 27.846 | 27.847 | 62.144 | 62.145 |
| 2.2 | 34.656 | 34.656 | 74.366 | 74.366 |
| 2.3 | 42.789 | 42.789 | 88.670 | 88.670 |
| 2.4 | 52.470 | 52.470 | 105.381 | 105.382 |
| 2.5 | 63.958 | 63.958 | 124.870 | 124.870 |
| 2.6 | 77.551 | 77.551 | 147.562 | 147.562 |
| 2.7 | 93.593 | 93.593 | 173.943 | 173.944 |
| 2.8 | 112.480 | 112.481 | 204.570 | 204.571 |
| 2.9 | 134.669 | 134.670 | 240.079 | 240.080 |
| 3.0 | 160.683 | 160.684 | 281.196 | 281.197 |

## EXERCISE 6—NUMERICAL SOLUTION OF SECOND-ORDER DIFFERENTIAL EQUATIONS

Solve each equation by the Runge–Kutta method and find the value of $y$ at $x = 2$. Take a step size of 0.1 unit.

1. $y'' + y' + xy = 3, \quad y'(1, 1) = 2$
2. $y'' - xy' + 3xy^2 = 2y, \quad y'(1, 1) = 2$
3. $3y'' + y' + ye^x = xy^3, \quad y'(1, 2) = 2$
4. $2y'' - xy' + y^2 = 5 \ln x, \quad y'(1, 0) = 1$
5. $xy'' - 2.4x^2y' + 5.3x \sin y - 7.4y = 0, \quad y'(1, 1) = 0.1$
6. $y'' + y' \ln x + 2.69(x - y)^2 = 6.26y, \quad y'(1, 3.27) = 8.26$
7. $x^2y'' - (x + y)y' + 8.26x^2y^2 = 1.83y, \quad y'(1, 1) = 2$
8. $y'' - y' + 183xye^x + 826x \cos y = 0, \quad y'(1, 73.4) = 273$
9. $y'' - xy' + 59.2xy = 74.1y \ln xy, \quad y'(1, 1) = 2$
10. $xy'' - x^2y' + 11.4x^2 \ln y + 62.2 = 0, \quad y'(1, 8) = 52.1$

### Computer

11. Write a program or use a spreadsheet to solve a second-order differential equation by the Runge–Kutta method. Try your program on any of the equations above. If your spreadsheet has a graphing utility, use it to graph $x$ versus $y$ over the given range.

## CHAPTER 36 REVIEW PROBLEMS

*Find the Laplace transform by direct integration.*

1. $f(t) = 2t$

2. $f(t) = 3t^2$

3. $f(t) = \cos 3t$

*Find the Laplace transform by Table 36-1.*

4. $f(t) = 2t^3 + 3t$

5. $f(t) = 3te^{2t}$

6. $f(t) = 2t + 3e^{2t}$

7. $f(t) = 3e^{-t} - 2t^2$

8. $f(t) = 3t^3 + 5t^2 + 4e^t$

9. $y' + 3y, \quad y(0) = 1$

10. $y' - 6y, \quad y(0) = 1$

11. $3y' + 2y, \quad y(0) = 0$

12. $2y'' + y' - 3y, \quad y(0) = 1, y'(0) = 3$

13. $y'' + 3y' + 4y, \quad y(0) = 1, y'(0) = 3$

*Find the inverse transform.*

14. $F(s) = \dfrac{6}{4 + s^2}$

15. $F(s) = \dfrac{s}{(s - 3)^2}$

16. $F(s) = \dfrac{s + 6}{(s - 2)^2}$

17. $F(s) = \dfrac{2s}{(s^2 + 5)^2}$

18. $F(s) = \dfrac{3}{(s + 4)^2 + 5}$

19. $F(s) = \dfrac{s + 1}{s^2 - 6s + 8}$

20. $F(s) = \dfrac{4s^2 + 2}{s(s^2 + 3)}$

21. $F(s) = \dfrac{3s}{s^2 + 2s + 1}$

22. $F(s) = \dfrac{3}{s^2 + s + 4}$

23. $F(s) = \dfrac{4s}{s^2 + 4s + 4}$

*Solve each differential equation by the Laplace transform.*

24. $y' + 2y = 0, \quad y(0) = 1$

25. $y' - 3y = t^2, \quad y(0) = 2$

26. $3y' - y = 2t, \quad y(0) = 1$

27. $y'' + 2y' + y = 2, \quad y(0) = 0, y'(0) = 0$

28. $y'' + 3y' + 2y = t, \quad y(0) = 1, y'(0) = 2$

29. $y'' + 2y = 4t, \quad y(0) = 3, y'(0) = 0$

30. $y'' + 3y' + 2y = e^{2t}, \quad y(0) = 2, y'(0) = 1$

31. A 21.5-lb ball is dropped from rest through air whose resisting force is equal to 2.75 times the ball's speed, in ft/s. Write an expression for the speed of the ball.

32. A weight that hangs from a spring is pulled down 0.50 cm at $t = 0$ and released from rest. The differential equation of motion is $x'' + 3.22x' + 18.5 = 0$. Write the equation for $x$ as a function of time.

33. The current in a certain $RC$ circuit satisfies the equation $8.24i' + 149i = 100$. If $i$ is zero at $t = 0$, write an equation for $i$.

34. The current in an $RLC$ circuit satisfies the equation $5.45i'' + 2820i' + 9730i = 10$. If $i$ and $i'$ are zero at $t = 0$, write an equation for $i$.

35. In an $RLC$ circuit, $R = 3750\ \Omega$, $L = 0.150$ H, and $C = 1.25\ \mu$F. It is connected to a dc source of 150 V at $t = 0$. If $i$ is zero at $t = 0$, write an equation for the instantaneous current.

36. In an $RL$ circuit, $R = 3750\ \Omega$ and $L = 0.150$ H. It is connected to an ac source of $28.4 \sin 84t$ V at $t = 0$. If $i$ is zero at $t = 0$, write an equation for the instantaneous current.

*Solve each differential equation. Use any numerical method with a step size of 0.1 unit.*

37. $y' + xy = 4y$; $y(2) = 2$. Find $y(3)$.
38. $xy' + 3y^2 = 5xy$; $y(1) = 5$. Find $y(2)$.
39. $3y'' - y' + xy^2 = 2 \ln y$; $y'(1, 2) = 8$. Find $y(2)$.
40. $yy'' - 7.2yy' + 2.8x \sin x - 2.2y = 0$; $y'(1, 5) = 8.2$. Find $y(2)$.
41. $y'' + y' \ln y + 5.27(x - 2y)^2 = 2.84x$; $y'(1, 1) = 3$. Find $y(2)$.
42. $x^2y'' - (1 + y)y' + 3.67x^2 = 5.28x$; $y'(1, 2) = 3$. Find $y(2)$.

**43.** $y'' - xy' + 6e^x + 126 \cos y = 0$; $y'(1, 2) = 5$. Find $y(2)$.

**44.** $y' + xy^2 = 8 \ln y$; $y(1) = 7$. Find $y(2)$.

**45.** $7.35y' + 2.85x \sin y - 7.34x = 0$; $y(2) = 4.3$. Find $y(3)$.

**46.** $yy'' - y' + 77.2y^2 = 28.4x \ln |xy|$; $y'(3, 5) = 3$. Find $y(4)$.

*Writing*

**47.** Suppose your company gets a new project requiring you to solve many differential equations. You see this as a chance to get that computer you've always wanted. Write a memo to your boss explaining how the computer can be used to solve those differential equations.

# 37

# INFINITE SERIES

## OBJECTIVES

**When you have completed this chapter, you should be able to:**

- Determine if a series converges using the partial sum test, the limit test, or the ratio test.
- Determine the interval of convergence of a power series using the ratio test.
- Expand a function using a Maclaurin series or a Taylor series.
- Compute numerical values using a Maclaurin series or a Taylor series.
- Add, subtract, multiply, divide, differentiate, and integrate series.
- Describe waveforms using a Fourier series.

To find the sine of an angle or the logarithm of a number, you would probably use a calculator, computer, or a table. But where did the table come from? And how can the chips in a calculator or computer find sines or logs when all they can do are the four arithmetic operations of addition, subtraction, multiplication, and division?

We can, in fact, find the sine of an angle $x$ (in radians) from a polynomial, such as

$$\sin x \simeq x - \frac{x^3}{6} + \frac{x^5}{120}$$

and the natural logarithm from

$$\ln x \simeq (x - 1) - \frac{(x - 1)^2}{2} + \frac{2(x - 1)^3}{6}$$

But where did these formulas come from? How accurate are they, and can the accuracy be improved? Are they good for all values of $x$ or for just a small range of $x$? Can similar formulas be derived for other functions, such as $e^x$? In this chapter we answer such questions.

The technique of representing a function by a polynomial is called *polynomial approximation,* and each of these formulas is really the first several terms of an infinitely long series. We learn how to write Maclaurin and Taylor series for such functions, and how to estimate and improve the accuracy of computation.

Why bother with infinite series when we can get these functions so easily using a calculator or computer? First, we may want to understand how a computer finds certain functions, even if we may never have to design the logic to do it. Perhaps more important, many numerical methods rely on polynomial approximation of functions. Finally, some limits, derivatives, or integrals are best found by this method.

**Look back at Chapter 23, where we examined infinite series by computer. Here we treat the same ideas analytically.**

We first give some tests to see if a series converges or diverges, for a series that diverges is of no use for computation. We then write polynomial approximations for some common functions and use them for computation, and look at other uses for series. We conclude with infinite series of sine and cosine terms, called *Fourier series*. These enable us to write series to represent periodic functions, such as waveforms, and are extremely useful in electronics and mechanical vibrations.

## 37-1 CONVERGENCE AND DIVERGENCE OF INFINITE SERIES

In many applications, we first replace a given function by an infinite series and then do any required operations on the series rather than the original function. But we cannot work with an infinite number of terms, nor do we want to work with many terms of the series. For practical applications, we want to represent our original function *by only the first few terms* of an infinite series. Thus we need a series in which the terms decrease in magnitude rapidly, and for which the sum of the first several terms is not too different from the sum of all the terms of the series. Such a series is said to *converge*. A series that does not converge is said to *diverge*. In this section we give three tests for convergence.

**Convergence of a series is an extensive topic. We're only scratching the surface here.**

### Partial Sum Test

In Chapter 23 we said that an infinite series

| | |
|---|---|
| $u_1 + u_2 + u_3 + \cdots + u_n + \cdots$ | **428** |

is the indicated sum of an infinite number of terms.

Now let $S_n$ stand for the sum of the first $n$ terms of an infinite series, so that

$$S_1 = u_1 \qquad S_2 = u_1 + u_2 \qquad S_3 = u_1 + u_2 + u_3 \qquad \text{etc.}$$

These are called *partial sums*. The infinite sequence of partial sums, $S_1$, $S_2$, $S_3$, . . . may or may not have a definite limit $S$ as $n$ becomes infinite. If a limit $S$ exists, we call it the sum of the series. If an infinite series has a sum $S$, the series is said to converge. Otherwise, the series diverges.

| | | |
|---|---|---|
| Partial Sum Test | $\lim_{n\to\infty} S_n = S$ | **430** |

*If the partial sums of an infinite series have a limit S, the series converges. Otherwise, the series diverges.*

It is not hard to see that a finite series will have a finite sum. But is it possible that the sum of an infinite number of terms can be finite? Yes; however, we must look at the infinite sequence of partial sums $S_1$, $S_2$, $S_3$, . . . , $S_n$, to determine if this sequence approaches a specific limit $S$, as in the following example.

---

**EXAMPLE 1:** Does the following geometric series converge?

$$1 + \frac{1}{2} + \frac{1}{2^2} + \frac{1}{2^3} + \cdots + \frac{1}{2^{n-1}} + \cdots$$

**Solution:** The sum of $n$ terms of a geometric series is given by Eq. 237. We substitute, with common ratio $r = \frac{1}{2}$ and $a = 1$.

$$S_n = \frac{a(1 - r^n)}{1 - r}$$

$$= \frac{1 - (1/2)^n}{1 - 1/2} = 2 - \frac{1}{2^{n-1}}$$

Since the limit of this sum is 2 as $n$ approaches infinity, the given series converges.

---

### Limit Test

**You may want to review limits (Sec. 27-1) before going further.**

We have defined a convergent infinite series as one for which the partial sums approach zero. For this to happen, it is clear that the terms themselves must get smaller as $n$ increases, and approach zero as $n$ becomes infinite. We may use this fact, which we call the *limit test,* as our second test for convergence.

| | | |
|---|---|---|
| Limit Test | $\lim_{n\to\infty} u_n = 0$ | **429** |

*If the terms of an infinite series do not approach zero as n approaches infinity, the series diverges.*

However, if the terms do approach zero, we cannot say for sure that the series converges. That the terms should approach zero is a *necessary* condition for convergence, but not a *sufficient* condition for convergence.

---

**EXAMPLE 2:** Does the infinite series

$$1 + \frac{2}{3} + \frac{3}{5} + \cdots + \frac{n}{2n - 1} + \cdots$$

converge or diverge?

**Solution:** The series diverges because the general term does not approach zero as $n$ approaches infinity.

$$\lim_{n\to\infty} \frac{n}{2n - 1} = \lim_{n\to\infty} \frac{1}{2 - 1/n} = \frac{1}{2} \neq 0$$

---

**EXAMPLE 3:** The harmonic series

$$1 + \frac{1}{2} + \frac{1}{3} + \cdots + \frac{1}{n} + \cdots$$

also has terms that approach zero as $n$ approaches infinity. However, we cannot tell from this information alone whether this series converges. In fact, it does *not* converge, as we had found by computer in Chapter 23, Example 8.

---

### Ratio Test

A third important test for convergence is the ratio test. We look at the ratio of two successive terms of the series as $n$ increases. If the limit of that ratio has an absolute value less than 1, the series converges. If the limit is greater than 1, the series diverges. However, a limit of 1 gives insufficient information.

| | | |
|---|---|---|
| Ratio Test | If $\lim_{n\to\infty} \left\lvert \frac{u_{n+1}}{u_n} \right\rvert$<br>(a) is less than 1, the series converges.<br>(b) is greater than 1, the series diverges.<br>(c) is equal to 1, the test fails to give conclusive information. | **431** |

**Also called the *Cauchy ratio test.* There are many tests for convergence which we cannot give here, but the ratio test will be the most useful for our later work.**

**EXAMPLE 4:** Use the ratio test on

$$\frac{1}{e} + \frac{2}{e^2} + \frac{3}{e^3} + \cdots + \frac{n}{e^n} + \cdots$$

to determine if this series converges or diverges.

**Solution:** By the ratio test,

$$\frac{u_{n+1}}{u_n} = \frac{n+1}{e^{n+1}} \div \frac{n}{e^n}$$

$$= \frac{n+1}{e^{n+1}} \cdot \frac{e^n}{n}$$

$$= \frac{n+1}{ne} = \frac{1 + 1/n}{e}$$

or

$$\lim_{n\to\infty} \left| \frac{u_{n-1}}{u_n} \right| = \frac{1}{e}$$

Since $1/e$ is less than 1, the series converges.

## EXERCISE 1—CONVERGENCE AND DIVERGENCE

### Partial Sum Test

Use the partial sum test to determine if each series converges.

1. The geometric series $9 + 3 + 1 + \cdots$
2. The geometric series $100 + 10 + 1 + \cdots$
3. The geometric series $2 + 2.02 + 2.04 + \cdots$
4. The geometric series $8.00 + 7.20 + 6.48 + \cdots$

### Limit Test

Use the limit test to decide, if possible, whether each series converges or diverges.

5. The geometric series $16 + 8 + 4 + 2 + \cdots$
6. The geometric series $1 + 1.05 + 1.1025 + \cdots$
7. The geometric series $100 + 50 + 25 + \cdots$
8. The harmonic series $1/10 + 1/8 + 1/6 + \cdots$

### Ratio Test

Use the ratio test to determine, if possible, if each series converges or diverges.

9. $\frac{1}{2!} + \frac{2}{3!} + \frac{3}{4!} + \cdots + \frac{n}{(n+1)!} + \cdots$
10. $1 + \frac{3}{2!} + \frac{5}{3!} + \frac{7}{4!} + \cdots + \frac{2n-1}{n!} + \cdots$
11. $1 + \frac{9}{8} + \frac{27}{16} + \frac{81}{128} + \cdots + \frac{3^n}{n \cdot 2^n} + \cdots$
12. $3 - \frac{3^3}{3 \cdot 1!} + \frac{3^5}{5 \cdot 2!} - \frac{3^7}{7 \cdot 3!} + \cdots + \frac{3^{2n-1}}{(2n-1)(n-1)!} + \cdots$
13. $1 + \frac{1}{2} + \frac{1}{3} + \frac{1}{4} + \cdots + \frac{1}{n} + \cdots$
14. $1 + \frac{4}{7} + \frac{9}{49} + \frac{16}{343} + \cdots + \frac{n^2}{7^{n-1}} + \cdots$

## 37-2 MACLAURIN SERIES

### Power Series

An infinite series of the form

$$c_0 + c_1x + c_2x^2 + \cdots + c_nx^n + \cdots$$

in which the $c$'s are constants and $x$ is a variable, is called a *power series* in $x$. Note that this series is a function of $x$, whereas in our previous examples the terms did not contain $x$.

---

**EXAMPLE 5:** Is the series $x - \frac{x^2}{2} + \frac{x^3}{3} - \cdots + (-1)^{n-1}\frac{x^n}{n} + \cdots$ a power series?

**Solution:** Since it matches the form of a power series with $c_0 = 0$, $c_1 = 1$, $c_2 = -\frac{1}{2}$, and so on, the given series is a power series.

---

Another way of writing a power series is

$$c_0 + c_1(x - a) + c_2(x - a)^2 + \cdots + c_n(x - a)^n + \cdots$$

where $a$ and the $c$'s are constants.

### Interval of Convergence

A power series, if it converges, does so only for specific values of $x$. The range of $x$ for which a particular power series converges is called the *interval of convergence*. We find this interval by means of the ratio test.

---

**EXAMPLE 6:** Find the interval of convergence of the series given in Example 5.

**Solution:** Applying the ratio test,

$$\lim_{n\to\infty}\left|\frac{u_{n+1}}{u_n}\right| = \lim_{n\to\infty}\left|\frac{x^{n+1}}{n+1}\left(\frac{n}{x_n}\right)\right| = \lim_{n\to\infty}\left|\frac{n}{n+1}(x)\right| = |x|$$

Thus the series converges when $x$ lies between 1 and $-1$ and diverges when $x$ is greater than 1 or less than $-1$. Our interval of convergence is thus $-1 < x < 1$.

---

It is usual to test for convergence at the endpoints of the interval as well, but this requires tests that we have not covered. So in Example 6 we do not test for $x = 1$ or $x = -1$.

### Representing a Function by a Series

We know that it is possible to represent certain functions by an infinite power series. Take the function

$$f(x) = \frac{1}{1 + x}$$

By ordinary division (try it) we get the infinite power series

$$(1 + x)^{-1} = 1 - x + x^2 - x^3 + \cdots$$

Although we do not show it here, this power series can be shown to converge when $x$ is less than 1. Can we also write a power series to represent other functions? Yes, we can represent many (but not all) functions by a power series. We show some useful power series in the following section.

## Writing a Maclaurin Series

We require that the function being represented is continuous and differentiable over the interval of interest. Later in this chapter we use Fourier series to represent discontinuous functions.

Let us assume that a certain function $f(x)$ can be represented by an infinite power series, so that

$$f(x) = c_0 + c_1x + c_2x^2 + c_3x^3 + \cdots + c_{n-1}x^{n-1} + \cdots \quad (1)$$

We now evaluate the constants in this equation. We get $c_0$ by setting $x$ to zero. Thus

$$f(0) = c_0$$

Next, taking the derivative of (1) gives

$$f'(x) = c_1 + 2c_2x + 3c_3x^2 + 4c_4x^3 + \cdots$$

from which we find $c_1$.

$$f'(0) = c_1$$

We continue taking derivatives and evaluating each at $x = 0$. Thus

$$f''(x) = 2c_2 + 3(2)c_3x + 4(3)c_4x^2 + \cdots$$

$$f''(0) = 2c_2$$

$$f'''(x) = 3(2)c_3 + 4(3)(2)c_4x + \cdots$$

$$f'''(0) = 3(2)c_3$$

and so on. The constants in (1) are thus

$$c_0 = f(0), \qquad c_1 = f'(0), \qquad c_2 = \frac{f''(0)}{2}, \qquad c_3 = \frac{f'''(0)}{3(2)}$$

and, in general,

$$c_n = \frac{f^{(n)}(0)}{n!}$$

Substituting the $c$'s back into (1) gives us what is called a Maclaurin series.

Named for a Scottish mathematician and physicist, Colin Maclaurin (1698–1746).

| Maclaurin Series | $f(x) = f(0) + f'(0)x + \frac{f''(0)}{2!}x^2 + \cdots + \frac{f^{(n)}(0)}{n!}x^n + \cdots$ | 432 |
|---|---|---|

The Maclaurin series expresses a function $f(x)$ in terms of an infinite power series whose $n$th coefficient is the $n$th derivative of $f(x)$, evaluated at $x = 0$, divided by $n!$

Notice that in order to write a Maclaurin series for a function, the function and its derivatives must exist at $x = 0$.

We now apply the Maclaurin series expansion to two important functions mentioned previously, the sine and the natural logarithm.

**EXAMPLE 7:** Write a Maclaurin series for the function

$$f(x) = \sin x$$

**Solution:** We take successive derivatives and evaluate each at $x = 0$.

$$\begin{aligned} f(x) &= \sin x & f(0) &= 0 \\ f'(x) &= \cos x & f'(0) &= 1 \\ f''(x) &= -\sin x & f''(0) &= 0 \\ f'''(x) &= -\cos x & f'''(0) &= -1 \\ f^{(iv)}(x) &= \sin x & f^{(iv)}(0) &= 0 \\ &\text{etc.} & &\text{etc.} \end{aligned}$$

Substituting into Eq. 432, we have

| | |
|---|---|
| $\sin x = x - \frac{x^3}{3!} + \frac{x^5}{5!} - \frac{x^7}{7!} + \cdots$ | **210** |

This series is the first listed in Table 37-1. Put a bookmark at this table, for we'll be referring to it often in this chapter.

**TABLE 37-1 Some Taylor Series***

| | *Expanded About:* |
|---|---|
| 1. $\sin x = x - \frac{x^3}{3!} + \frac{x^5}{5!} - \frac{x^7}{7!} + \cdots$ (Eq. 210) | |
| 2. $\sin x = \frac{1}{2} + \frac{\sqrt{3}}{2}\left(x - \frac{\pi}{6}\right) - \frac{1}{4}\left(x - \frac{\pi}{6}\right)^2 - \frac{\sqrt{3}}{12}\left(x - \frac{\pi}{6}\right)^3 + \cdots$ | $a = \frac{\pi}{6}$ |
| 3. $\cos x = 1 - \frac{x^2}{2!} + \frac{x^4}{4!} - \frac{x^6}{6!} + \cdots$ (Eq. 211) | |
| 4. $\cos x = \frac{\sqrt{2}}{2}\left[1 - \left(x - \frac{\pi}{4}\right) - \frac{(x - \pi/4)^2}{2} + \cdots\right]$ | $a = \frac{\pi}{4}$ |
| 5. $\tan x = x + \frac{x^3}{3} + \frac{2x^5}{15} + \frac{17x^7}{315} + \cdots$ | |
| 6. $\tan x = 1 + 2\left(x - \frac{\pi}{4}\right) + 2\left(x - \frac{\pi}{4}\right)^2 + \cdots$ | $a = \frac{\pi}{4}$ |
| 7. $\sec x = 1 + \frac{x^2}{2} + \frac{5x^4}{24} + \cdots$ | |
| 8. $\sec x = 2 + 2\sqrt{3}\left(x - \frac{\pi}{3}\right) + 7\left(x - \frac{\pi}{3}\right)^2 + \cdots$ | $a = \frac{\pi}{3}$ |
| 9. $\sin^2 x = x^2 - \frac{x^4}{3} + \frac{2x^6}{45} - \frac{x^8}{315} + \cdots$ | |
| 10. $\cos^2 x = 1 - x^2 + \frac{x^4}{3} - \frac{2x^6}{45} \cdots$ | |

* Given are Taylor series expansions for some common functions. The column "Expanded About" gives the value about which the series is expanded. If no value is given, the series has been expanded about $a = 0$, and the series is also called a Maclaurin series.

| | *Expanded About:* |
|---|---|
| 11. $\sin 2x = 2x - \frac{2^3x^3}{3!} + \frac{2^5x^5}{5!} - \frac{2^7x^7}{7!} + \cdots$ | |
| 12. $\cos 2x = 1 - 2x^2 + \frac{2^4x^4}{4!} - \frac{2^6x^6}{6!} + \cdots$ | |
| 13. $\sin x^2 = x^2 - \frac{x^6}{3!} + \frac{x^{10}}{5!} - \cdots$ | |
| 14. $\cos x^2 = 1 - \frac{x^4}{2!} + \frac{x^8}{4!} - \frac{x^{12}}{6!} + \cdots$ | |
| 15. $\sec^2 x = 1 + x^2 + 2x^4/3 + \cdots$ | |
| 16. $\sinh x = \frac{e^x - e^{-x}}{2} = x + \frac{x^3}{3!} + \frac{x^5}{5!} + \frac{x^7}{7!} + \cdots$ | |
| 17. $\cosh x = \frac{e^x + x^{-x}}{2} = 1 + \frac{x^2}{2!} + \frac{x^4}{4!} + \frac{x^6}{6!} + \cdots$ | |
| 18. $\frac{1}{x} = 1 - (x-1) + (x-1)^2 - (x-1)^3 + \cdots$ | $a = 1$ |
| 19. $\frac{1}{x} = \frac{1}{2} - \frac{x-2}{4} + \frac{(x-2)^2}{16} + \cdots$ | $a = 2$ |
| 20. $\frac{1}{1+x} = 1 - x + x^2 - x^3 + \cdots$ | |
| 21. $\sqrt{x} = 1 + \frac{1}{2}(x-1) - \frac{1}{8}(x-1)^2 + \cdots$ | $a = 1$ |
| 22. $\sqrt[3]{x} = -1 + \frac{x+1}{3} + \frac{(x+1)^2}{9} + \frac{5(x+1)^3}{81} + \cdots$ | $a = -1$ |
| 23. $\sqrt{1+x} = 1 + \frac{x}{2} - \frac{x^2}{8} + \frac{x^3}{16} - \cdots$ | |
| 24. $\sqrt{x^2+1} = \sqrt{2} + \frac{(x-1)}{\sqrt{2}} + \frac{(x-1)^2}{4\sqrt{2}} + \cdots$ | $a = 1$ |
| 25. $e^x = 1 + x + \frac{x^2}{2!} + \frac{x^3}{3!} + \cdots$ (Eq. 203) | |
| 26. $e^x = e\left[1 + (x-1) + \frac{(x-1)^2}{2!} + \cdots\right]$ | $a = 1$ |
| 27. $e^{-x} = 1 - x + \frac{x^2}{2!} - \frac{x^3}{3!} + \cdots$ | |
| 28. $e^{-x} = e\left[1 - (x+1) + \frac{(x+1)^2}{2!} + \cdots\right]$ | $a = -1$ |
| 29. $e^{2x} = 1 + 2x + 2x^2 + \frac{4x^3}{3} + \cdots$ | |
| 30. $e^{-2x} = 1 - 2x + 2x^2 - \frac{4x^3}{3} + \cdots$ | |
| 31. $e^{-x^2} = 1 - x^2 + \frac{x^4}{2!} - \frac{x^6}{3!} + \cdots$ | |
| 32. $xe^{2x} = x + 2x^2 + 2x^3 + \cdots$ | |
| 33. $xe^{-2x} = x - 2x^2 + 2x^3 - \cdots$ | |
| 34. $x^2e^{2x} = x^2 + 2x^3 + 2x^4 + \cdots$ | |
| 35. $e^{jx} = 1 + jx - \frac{x^2}{2!} - j\frac{x^3}{3!} + \frac{x^4}{4!} + \cdots$ | |

| | Expanded About: |
|---|---|
| 36. $e^x \sin x = x + x^2 + \frac{x^3}{3} - \cdots$ | |
| 37. $e^x \cos x = 1 + x - \frac{x^3}{3} - \frac{x^4}{6} + \cdots$ | |
| 38. $e^{-x} \cos 2x = 1 - x - \frac{3x^2}{2} + \frac{11x^3}{6} + \cdots$ | |
| 39. $e^{\sin x} = 1 + \sin x + \frac{\sin^2 x}{2!} + \frac{\sin^3 x}{3!} + \cdots$ | |
| 40. $\ln(1 + x) = x - \frac{x^2}{2} + \frac{x^3}{3} - \frac{x^4}{4} + \cdots$ | |
| 41. $\ln(1 - x) = -x - \frac{x^2}{2} - \frac{x^3}{3} - \frac{x^4}{4} - \cdots$ | |
| 42. $\ln(1 - x^2) = -x^2 - \frac{x^4}{2} - \frac{x^6}{3} - \frac{x^8}{4} - \cdots$ | |
| 43. $\ln x = (x - 1) - \frac{(x-1)^2}{2!} + \frac{2(x-1)^3}{3!} - \frac{6(x-1)^4}{4!} + \cdots$ | $a = 1$ |
| 44. $\ln \cos x = -\frac{x^2}{2} - \frac{x^4}{12} - \frac{x^6}{45} - \cdots$ | |

Now we turn to the natural logarithm.

**EXAMPLE 8:** Write a Maclaurin series for $f(x) = \ln x$.

**Solution:** We proceed as before,

$$f(x) = \ln x \qquad \text{but } f(0) \text{ is undefined!}$$

Therefore, we cannot write a Maclaurin series for ln $x$. In order to write a Maclaurin series for a function, *the function and all its derivatives must exist at $x = 0$*. In the next section, we'll see how to write an expression for ln $x$ using a Taylor series.

## Computing with Maclaurin Series

Maclaurin series are useful in computations because after we have represented a function by a Maclaurin series, we can find the value of the function simply by substituting into the series.

**EXAMPLE 9:** Find the sine of 0.5 rad using the first three terms of the Maclaurin series. Work to seven decimal places.

**Solution:** Substituting into Eq. 210, we find that

$$\sin 0.5 = 0.5 - \frac{(0.5)^3}{6} + \frac{(0.5)^5}{120}$$

$$= 0.5 - 0.0208333 + 0.0002604 = 0.4794271$$

By calculator (which we assume to be correct to seven decimal places) the sine of 0.5 rad is 0.4794255. Our answer is thus too high by 0.0000016, or about 0.0003%.

The error caused by discarding terms of a series is called the *truncation error*.

### Accuracy

In Example 9 we could estimate the truncation error by finding the exact value by calculator. But what if we were designing a computer to calculate sines, and must know *in advance* what the accuracy is?

**TABLE 37-2 Sin $x$ Calculated by Maclaurin Series**

| *Angle* | *1 Term* | *2 Terms* | *3 Terms* | *4 Terms* | *By Computer* |
|---|---|---|---|---|---|
| 0 | 0.000000 | 0.000000 | 0.000000 | 0.000000 | 0.000000 |
| 10 | 0.174533 | 0.173647 | 0.173648 | 0.173648 | 0.173648 |
| 20 | 0.349066 | 0.341977 | 0.342020 | 0.342020 | 0.342020 |
| 30 | 0.523599 | 0.499674 | 0.500002 | 0.500000 | 0.500000 |
| 40 | 0.698132 | 0.641422 | 0.642804 | 0.642788 | 0.642788 |
| 50 | 0.872665 | 0.761903 | 0.766120 | 0.766044 | 0.766045 |
| 60 | 1.047198 | 0.855801 | 0.866295 | 0.866021 | 0.866025 |
| 70 | 1.221731 | 0.917800 | 0.940482 | 0.939676 | 0.939693 |
| 80 | 1.396263 | 0.942582 | 0.986806 | 0.984753 | 0.984808 |
| 90 | 1.570796 | 0.924832 | 1.004525 | 0.999843 | 1.000000 |

**Because the sine function is periodic, we need only calculate sines of angles up to 90°. Sines of larger angles can be found from these.**

Table 37-2 shows the sines of the angles from 0 to 90°, computed using one, two, three, and four terms of the Maclaurin series (Eq. 210). In the last column is the sine, given by the computer, which we assume to be accurate to the number of places given. Notice that the values obtained by series differ from the computer values. The absolute value of the percent error of each is given in Table 37-3 and plotted in Fig. 37-1.

**TABLE 37-3 Percent Error**

| *Angle* | *1 Term* | *2 Terms* | *3 Terms* | *4 Terms* |
|---|---|---|---|---|
| 0 | 0.0000 | 0.0000 | 0.0000 | 0.0000 |
| 10 | 0.5095 | 0.0008 | 0.0000 | 0.0000 |
| 20 | 2.0600 | 0.0126 | 0.0000 | 0.0000 |
| 30 | 4.7198 | 0.0652 | 0.0004 | 0.0000 |
| 40 | 8.6100 | 0.2125 | 0.0025 | 0.0000 |
| 50 | 13.9183 | 0.5407 | 0.0099 | 0.0001 |
| 60 | 20.9200 | 1.1806 | 0.0312 | 0.0005 |
| 70 | 30.0138 | 2.3298 | 0.0840 | 0.0018 |
| 80 | 41.7803 | 4.2877 | 0.2029 | 0.0056 |
| 90 | 57.0796 | 7.5158 | 0.4525 | 0.0157 |

Note that the error decreases rapidly as more terms of the series are used. Also note that the series gives the exact answer at $x = 0$ regardless of the number of terms, and that the error increases as we go farther from zero. With a Maclaurin series, we say that the function is *expanded about* $x = 0$.

In the next section we'll be able to expand a function about any point $x = a$ using a Taylor series, and that the best accuracy, as here, will be near that point.

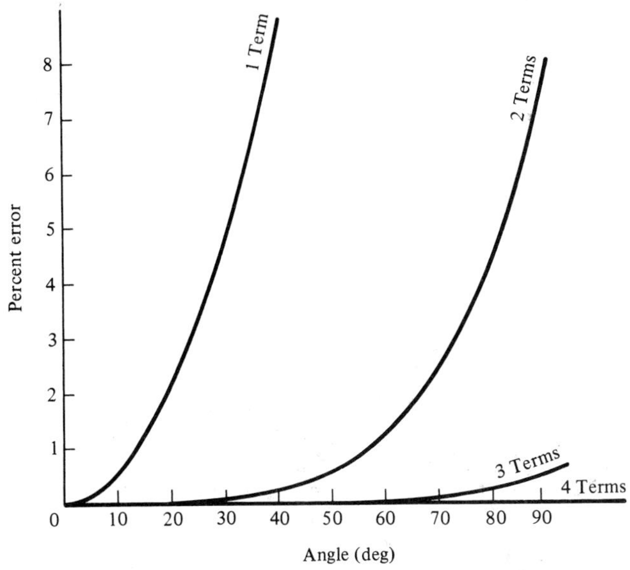


**FIGURE 37-1 Percent error in the sine of an angle, using one, two, three, and four terms of a Maclaurin series.**

## EXERCISE 2—MACLAURIN SERIES

Use the ratio test to find the interval of convergence of each power series.

1. $1 + x + x^2 + \cdots + x^{n-1} + \cdots$
2. $1 + \dfrac{2x}{5} + \dfrac{3x^2}{5^2} + \cdots + \dfrac{nx^{n-1}}{5^{n-1}} + \cdots$
3. $x + 2!x^2 + 3!x^3 + \cdots + n!x^n + \cdots$
4. $2x + 4x^2 + 8x^3 + \cdots + 2^n x^n + \cdots$
5. $x - \dfrac{x^3}{3} + \dfrac{x^5}{5} - \dfrac{x^7}{7} + \cdots$
6. $\dfrac{x}{1!} + \dfrac{x^2}{2!} + \cdots + \dfrac{x^n}{n!} + \cdots$

Verify each Maclaurin series expansion from Table 37-1. Starting from $f(x)$, derive the Maclaurin series. Compare your answer with the series in Table 37-1.

7. Series 20
8. Series 23
9. Series 25
10. Series 27
11. Series 31
12. Series 32
13. Series 40
14. Series 11
15. Series 3
16. Series 5
17. Series 7
18. Series 14
19. Series 9
20. Series 44

Compute each number using three terms of the appropriate Maclaurin's series.

21. $\sqrt{e}$
22. $\sin 1$
23. $\ln 1.2$
24. $\cos 8°$
25. $\sqrt{1.1}$
26. $\sec \dfrac{1}{4}$

## 37-3 TAYLOR SERIES

### Writing a Taylor Series

We have seen that certain functions, such as $f(x) = \ln x$, cannot be represented by a Maclaurin series. Further, a Maclaurin series may converge too slowly to be useful. However, if we cannot represent a function by a power

**Named for Brook Taylor (1685–1731), an English analyst, geometer, and philosopher.**

series in $x$ (a Maclaurin series), we can often represent the function by a power series in $(x - a)$, called a *Taylor series*.

Let us assume that a function $f(x)$ can be represented by a power series in $(x - a)$, where $a$ is some constant.

$$f(x) = c_0 + c_1(x - a) + c_2(x - a)^2 + \cdots + c_n(x - a)^n + \cdots \quad (1)$$

**Since the Maclaurin series is a special case of the Taylor series, they are both referred to as Taylor series.**

As with the Maclaurin series, we take successive derivatives. We find the $c$'s by evaluating each derivative at $x = a$.

$$f(x) = c_0 + c_1(x - a) + c_2(x - a)^2 + \cdots \qquad f(a) = c_0$$

$$f'(x) = c_1 + 2c_2(x - a) + \cdots \qquad f'(a) = c_1$$

$$f''(x) = 2c_2 + 3(2)c_3(x - a) + \cdots \qquad f''(a) = 2c_2$$

$$f'''(x) = 3(2)c_3 + 4(3)(2)c_4(x - a) + \cdots \qquad f'''(a) = 3 \cdot 2 \cdot c_3$$

$$\vdots \qquad \qquad \vdots$$

$$f^{(n)}(x) = n!c_n + \cdots \qquad f^{(n)}(a) = n!c_n$$

The values of $c$ obtained are substituted back into (1) and we get the following standard form of the Taylor series:

| Taylor Series | $f(x) = f(a) + f'(a)(x - a) + \dfrac{f''(a)}{2!}(x - a)^2 + \cdots + \dfrac{f^{(n)}(a)}{n!}(x - a)^n + \cdots$ | **433** |
|---|---|---|

**EXAMPLE 10:** Write four terms of a Taylor series for $f(x) = \ln x$, expanded about $a = 1$.

**Solution:** We take derivatives and evaluate each at $x = 1$.

$$f(x) = \ln x \qquad f(1) = 0$$

$$f'(x) = \frac{1}{x} \qquad f'(1) = 1$$

$$f''(x) = -x^{-2} \qquad f''(1) = -1$$

$$f'''(x) = 2x^{-3} \qquad f'''(1) = 2$$

$$f^{(iv)}(x) = -6x^{-4} \qquad f^{(iv)}(1) = -6$$

Substituting into Eq. 433 gives

**This is Series 43 in Table 37-1. Recall that we were not able to write a Maclaurin series for ln *x*.**

$$\ln x = 0 + 1(x - 1) + \frac{-1}{2}(x - 1)^2 + \frac{2}{6}(x - 1)^3 + \frac{-6}{24}(x - 1)^4 + \cdots$$

$$= (x - 1) - \frac{(x - 1)^2}{2} + \frac{(x - 1)^3}{3} - \frac{(x - 1)^4}{4} + \cdots$$

## Computing with Taylor Series

As with the Maclaurin series, we simply substitute the required $x$ into the proper Taylor series. For best accuracy, we expand the Taylor series about a point that is close to the value we wish to compute. Of course, $a$ must be chosen such that $f(a)$ is known.

---

**EXAMPLE 11:** Evaluate ln 0.9 using four terms of a Taylor series. Work to seven decimal places.

**Solution:** We know from Example 10 that

$$\ln x = (x-1) - \frac{(x-1)^2}{2} + \frac{(x-1)^3}{3} - \frac{(x-1)^4}{4} + \cdots$$

We choose $a = 1$, since 1 is close to 0.9 and ln 1 is known. Substituting into the series with $x = 0.9$ and with

$$x - 1 = 0.9 - 1 = -0.1$$

gives

$$\begin{aligned}\ln 0.9 &= -0.1 - \frac{(-0.1)^2}{2} + \frac{(-0.1)^3}{3} - \frac{(-0.1)^4}{4} + \cdots \\ &= -0.1 - \frac{0.01}{2} + \frac{-0.001}{3} - \frac{0.0001}{4} + \cdots \\ &= -0.1 - 0.005 - 0.0003333 - 0.0000250 = -0.1053583\end{aligned}$$

So $\ln 0.9 \simeq -0.1053583$. By calculator, $\ln 0.9 = -0.1053605$, to seven decimal places, a difference of about 0.002% from our answer.

---

**EXAMPLE 12:** Calculate the sines of the angles in 10° intervals from 0 to 90°, using two, three, and four terms of a Taylor series expanded about 45°.

**Solution:** We take derivatives and evaluate each at $x = \pi/4$, remembering to work in radians,

$$\begin{aligned} f(x) &= \sin x & f\left(\frac{\pi}{4}\right) &= \frac{\sqrt{2}}{2} \\ f'(x) &= \cos x & f'\left(\frac{\pi}{4}\right) &= \frac{\sqrt{2}}{2} \\ f''(x) &= -\sin x & f''\left(\frac{\pi}{4}\right) &= -\frac{\sqrt{2}}{2} \\ f'''(x) &= -\cos x & f'''\left(\frac{\pi}{4}\right) &= -\frac{\sqrt{2}}{2} \end{aligned}$$

Substituting into Eq. 433 gives

$$\sin x = \frac{\sqrt{2}}{2} + \frac{\sqrt{2}}{2}\left(x - \frac{\pi}{4}\right) - \frac{\sqrt{2}}{2(2!)}\left(x - \frac{\pi}{4}\right)^2 - \frac{\sqrt{2}}{2(3!)}\left(x - \frac{\pi}{4}\right)^3 \cdots$$

A computer calculation of sin $x$, using two, three, and four terms of this series, is given in Table 37-4. The absolute value of the truncation error, as a percent, is shown in Table 37-5 and plotted in Fig. 37-2.

**TABLE 37-4 Sin *x* Calculated by Taylor Series**

| *Angle* | *2 Terms* | *3 Terms* | *4 Terms* | *By Computer* |
|---|---|---|---|---|
| 0 | 0.151746 | −0.066343 | −0.009247 | 0.000000 |
| 10 | 0.275160 | 0.143229 | 0.170093 | 0.173648 |
| 20 | 0.398573 | 0.331262 | 0.341052 | 0.342020 |
| 30 | 0.521987 | 0.497755 | 0.499869 | 0.500000 |
| 40 | 0.645400 | 0.642708 | 0.642786 | 0.642788 |
| 50 | 0.768814 | 0.766121 | 0.766043 | 0.766045 |
| 60 | 0.892227 | 0.867995 | 0.865880 | 0.866025 |
| 70 | 1.015640 | 0.948329 | 0.938539 | 0.939693 |
| 80 | 1.139054 | 1.007123 | 0.980259 | 0.984808 |
| 90 | 1.262467 | 1.044378 | 0.987282 | 1.000000 |

**TABLE 37-5 Percent Error**

| *Angle* | *2 Terms* | *3 Terms* | *4 Terms* |
|---|---|---|---|
| 10 | 58.4582 | 17.5176 | 2.0473 |
| 20 | 16.5350 | 3.1456 | 0.2832 |
| 30 | 4.3973 | 0.4491 | 0.0262 |
| 40 | 0.4064 | 0.0124 | 0.0003 |
| 50 | 0.3615 | 0.0100 | 0.0002 |
| 60 | 3.0255 | 0.2274 | 0.0168 |
| 70 | 8.0822 | 0.9190 | 0.1228 |
| 80 | 15.6625 | 2.2659 | 0.4619 |
| 90 | 26.2467 | 4.4378 | 1.2718 |

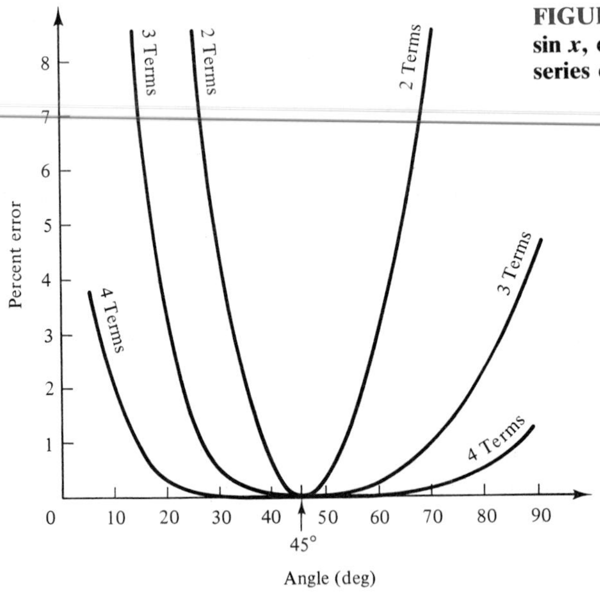


**FIGURE 37-2 Percent error in sin *x*, computed with a Taylor series expanded about 45°.**

Note that we get the best accuracy near the point about which the series is expanded ($x = 45°$). Also note that at other values the Taylor series gives *less accurate* results than the Maclaurin series with the same number of terms (compare Fig. 37-1 and Fig. 37-2). That is because the Maclaurin series converges much faster than the Taylor series.

## EXERCISE 3—TAYLOR SERIES

Verify each Taylor series from Table 37-1. Starting from $f(x)$, derive the Taylor series. Compare your answer with the series in Table 37-1.

1. Series 18
2. Series 2
3. Series 4
4. Series 6
5. Series 8
6. Series 21
7. Series 22
8. Series 26
9. Series 28
10. Series 43

Compute each value using three terms of the appropriate Taylor series. Compare your answer with that from a calculator.

11. $\dfrac{1}{1.1}$
12. $\sin 32°$
13. $\cos 46.5°$ (use $\cos 45° = 0.707107$)
14. $\tan 44.2°$
15. $\sec 61.4°$
16. $\sqrt{1.24}$
17. $\sqrt[3]{-0.8}$
18. $e^{1.1}$
19. $\ln (1.25)$
20. Calculate $\sin 65°$ using **(a)** three terms of a Maclaurin series (Series 1), **(b)** three terms of a Taylor series expanded about $\pi/6$ (Series 2), and **(c)** a calculator. Compare the results of each.

## 37-4 OPERATIONS WITH POWER SERIES

To expand a function in a power series, we can write a Maclaurin or Taylor series by the methods of Sec. 37-2 or 37-3. A faster way to get a new series is to modify or combine, when possible, series already derived.

### Substitution

Given a series for $f(x)$, a new series can be obtained simply by substituting a different expression for $x$.

---

**EXAMPLE 13:** We get Series 39 for $e^{\sin x}$ from the series for $e^x$:

$$e^x = 1 + x + \frac{x^2}{2!} + \frac{x^3}{3!} + \cdots \tag{25}$$

Substituting $\sin x$ for $x$ gives

$$e^{\sin x} = 1 + \sin x + \frac{\sin^2 x}{2!} + \frac{\sin^3 x}{3!} + \cdots \tag{39}$$

---

### Addition and Subtraction

*Two power series may be added or subtracted, term by term, for those values of x for which both series converge.*

**EXAMPLE 14:** Given the series

$$40. \qquad \ln(1 + x) = x - \frac{x^2}{2} + \frac{x^3}{3} - \frac{x^4}{4} + \cdots$$

and

$$41. \qquad \ln(1 - x) = -x - \frac{x^2}{2} - \frac{x^3}{3} - \frac{x^4}{4} - \cdots$$

Adding these term by term gives

$$42. \qquad \ln(1 - x^2) = -x^2 - \frac{x^4}{2} - \cdots$$

since $\ln(1 + x) + \ln(1 - x) = \ln[(1 + x)(1 - x)] = \ln(1 - x^2)$.

## Multiplication and Division

**A series can also be multiplied by a constant or by a polynomial. Thus a series for $x \ln x$ can be found by multiplying the series for $\ln x$ by $x$.**

*One power series may be multiplied or divided by another, term by term, for those values of x for which both converge, provided that division by zero does not occur.*

**EXAMPLE 15:** Find the power series for $e^x \sin x$. Discard any terms of fifth degree or higher.

**Solution:** We will multiply series 25 for $e^x$ and Series 1 for $\sin x$ to get Series 36 for $e^x \sin x$.

We trim each series to terms of fourth degree or less and multiply term by term, discarding any products of fifth degree or higher.

$$e^x \sin x = \left(1 + x + \frac{x^2}{2!} + \frac{x^3}{3!} + \frac{x^4}{4!} + \cdots\right)\left(x - \frac{x^3}{3!} \cdots\right) = x - \frac{x^3}{3!} + x^2 + \frac{x^3}{2!} \cdots$$

$$36. \qquad e^x \sin x = x + x^2 + \frac{x^3}{3} - \cdots$$

## Differentiation and Integration

*A power series may be differentiated or integrated term by term for values of x within its interval of convergence (but not, usually, at the endpoints).*

**EXAMPLE 16:** Derive a power series for $\ln(1 + x)$.

**Solution:** Series 20,

$$\frac{1}{1 + x} = 1 - x + x^2 - x^3 + \cdots$$

converges for $|x| < 1$. Integrating term by term between the limits 0 and $x$ gives

$$\int_0^x \frac{dx}{1 + x} = \left[x - \frac{x^2}{2} + \frac{x^3}{3} - \frac{x^4}{4} + \cdots\right]_0^x$$

But, from Rule 7,

$$\int_0^x \frac{dx}{1 + x} = \ln(1 + x)$$

So we have derived Series 40,

$$40. \qquad \ln(1 + x) = x - \frac{x^2}{2} + \frac{x^3}{3} - \frac{x^4}{4} + \cdots$$

### Evaluating Definite Integrals

If an expression cannot be integrated by our former methods, it still may be possible to evaluate its definite integral. We replace the function by a power series and then integrate term by term.

---

**EXAMPLE 17:** Find the approximate area under the curve $y = (\sin x)/x$ from $x = 0$ to $x = 1$.

**Solution:** The area is given by the definite integral

$$A = \int_0^1 \frac{\sin x}{x} dx$$

This expression is not in our table of integrals, nor have we any earlier method of dealing with it. Instead, let us take the series for sin $x$ (Eq. 210) and divide it by $x$, getting

$$\frac{\sin x}{x} = 1 - \frac{x^2}{3!} + \frac{x^4}{5!} - \cdots$$

We integrate the power series term by term, getting

$$A = \int_0^1 \frac{\sin x}{x} dx$$

$$= \left[x - \frac{x^3}{3(3!)} + \frac{x^5}{5(5!)} - \frac{x^7}{7(7!)} + \cdots\right]_0^1$$

$$\simeq 1 - 0.05556 + 0.00167 - 0.00003 + \cdots$$

$$= 0.94608$$

---

### Euler's Formula

Recall that we had used differential equations to derive Euler's formula in Sec. 35-10.

Here we show how Euler's formula (Eq. 227) may be derived by substituting into the power series for $e^x$. If we replace $x$ by $jx$ in Series 25 we get

$$e^{jx} = 1 + jx + \frac{(jx)^2}{2!} + \frac{(jx)^3}{3!} + \frac{(jx)^4}{4!} + \cdots$$

$$= 1 + jx + \frac{j^2x^2}{2!} + \frac{j^3x^3}{3!} + \frac{j^4x^4}{4!} + \cdots$$

But by Eq. 212, $j^2 = -1, j^3 = -j, j^4 = j$, and so on. So

$$e^{jx} = 1 + jx - \frac{x^2}{2!} - j\frac{x^3}{3!} + \frac{x^4}{4!} + \cdots$$

$$= \left[1 - \frac{x^2}{2!} + \frac{x^4}{4!} - \cdots\right] + j\left[x - \frac{x^3}{3!} + \frac{x^5}{5!} - \cdots\right]$$

But the series in the brackets are those for sin $x$ and cos $x$ (Series 1 and 3), so

| | | |
|---|---|---|
| Euler's Formula | $e^{jx} = \cos x - j \sin x$ | **227** |

### Substitution

By substituting into Series 25 for $e^x$, verify the following.

1. $e^{-x} = 1 - x + \frac{x^2}{2!} - \frac{x^3}{3!} + \cdots$
2. $e^{2x} = 1 + 2x + 2x^2 + \frac{4x^3}{3} + \cdots$
3. $e^{-x^2} = 1 - x^2 + \frac{x^4}{2!} - \frac{x^6}{3!} + \cdots$

By substituting into Series 1 for $\sin x$, verify the following.

4. $\sin 2x = 2x - \frac{2^3x^3}{3!} + \frac{2^5x^5}{5!} - \frac{2^7x^7}{7!} + \cdots$
5. $\sin x^2 = x^2 - \frac{x^6}{3!} + \frac{x^{10}}{5!} - \cdots$

### Addition and Subtraction

Add or subtract two earlier series to verify the following.

6. $\sec x + \tan x = 1 + x + \frac{x^2}{2} + \frac{x^3}{x} + \frac{5x^4}{24} + \cdots$ (Use Series 5 and 7)
7. $\sin x + \cos x = 1 + x - \frac{x^2}{2!} - \frac{x^3}{3!} + \frac{x^4}{4!} + \cdots$ (Use Series 1 and 3)
8. $\sinh x = \frac{e^x - e^{-x}}{2} = x + \frac{x^3}{3!} + \frac{x^5}{5!} + \frac{x^7}{7!} + \cdots$ (Use Series 25 and 27)
9. $\cosh x = \frac{e^x + e^{-x}}{2} = 1 + \frac{x^2}{2!} + \frac{x^4}{4!} + \frac{x^6}{6!} + \cdots$ (Use Series 25 and 27)
10. Using Series 12 for $\cos 2x$ and the identity $\sin^2 x = (1 - \cos 2x)/2$, show that
$\sin^2 x = x^2 - \frac{x^4}{3} + \frac{2x^6}{45} - \frac{x^8}{315} + \cdots$
11. Using Series 12 for $\cos 2x$ and the identity $\cos^2 x = (1 + \cos 2x)/2$, show that
$\cos^2 x = 1 - x^2 + \frac{x^4}{3} - \frac{2x^6}{45} \cdots$

### Multiplication and Division

Multiply or divide earlier series to obtain the following.

12. $xe^{2x} = x + 2x^2 + 2x^3 + \cdots$ (Use Series 29)
13. $xe^{-2x} = x - 2x^2 + 2x^3 - \cdots$ (Use Series 30)
14. $e^x \cos x = 1 + x - \frac{x^3}{3} - \frac{x^4}{6} + \cdots$ (Use Series 3 and 25)
15. $e^{-x} \cos 2x = 1 - x - \frac{3x^2}{2} + \frac{11x^3}{6} + \cdots$ (Use Series 12 and 27)
16. $\frac{\sin x}{x} = 1 - \frac{x^2}{3!} + \frac{x^4}{5!} - \frac{x^6}{7!} + \cdots$ (Use Series 1)
17. $\frac{e^x - 1}{x} = 1 + \frac{x}{2!} + \frac{x^2}{3!} + \frac{x^3}{4!} + \cdots$ (Use Series 25)
18. $\tan x = \frac{\sin x}{\cos x} = x + \frac{x^3}{3} + \frac{2x^5}{15} + \frac{17x^7}{315} + \cdots$ (Use Series 1 and 3)
19. Square the first four terms of Series 1 for $\sin x$ to obtain the series for $\sin^2 x$ given in Problem 10.
20. Cube the first three terms of Series 3 for $\cos x$ to obtain the series for $\cos^3 x$.

### Differentiation and Integration

21. Find the series for $\cos x$ by differentiating the series for $\sin x$.
22. Differentiate the series for $\tan x$ to show that Series 15 for $\sec^2 x$ is

$\sec^2 x = 1 + x^2 + 2x^4/3 + \cdots$.

23. Find the series for $1/(1 + x)$ by differentiating the series for $\ln(1 + x)$.
24. Find the series for $\ln \cos x$ by integrating the series for $\tan x$.
25. Find the series for $\ln(1 - x)$ by integrating the series for $1/(1 - x)$.
26. Find the series for $\ln(\sec x + \tan x)$ by integrating the series for $\sec x$.

Evaluate each integral by integrating the first three terms of each series.

27. $\int_1^2 \frac{e^x}{\sqrt{x}} dx$
28. $\int_0^1 \sin x^2 \, dx$
29. $\int_0^{1/4} e^x \ln(x + 1) \, dx$

## 37-5 FOURIER SERIES

Thus far we have used power series to approximate continuous functions. But a power series cannot be written for a *discontinuous function* such as a square wave that is defined by a different expression in different parts of the interval. For these, we use a Fourier series. In fact, most periodic functions can be represented by an infinite series of sine and cosine terms if the amplitude and frequency of each term are chosen properly. Here we learn how to write such a series. Then we show how to use any symmetry in the waveform to reduce the work of computation, and conclude with a method for writing a Fourier series for a waveform for which we have no equation, such as from an oscillogram trace.

### Definition of Fourier Series

A Fourier series is an infinite series of sine and cosine terms of the form

| | | |
|---|---|---|
| Fourier Series | $f(x) = \frac{a_0}{2} + a_1 \cos x + a_2 \cos 2x + a_3 \cos 3x + \cdots + a_n \cos nx + \cdots$ $+ b_1 \sin x + b_2 \sin 2x + b_3 \sin 3x + \cdots + b_n \sin nx + \cdots$ | 435 |

Named for a French analyst and mathematical physicist, Jean Baptiste Joseph Fourier (1768–1830).

Here $a_0/2$ is the *constant term*, $a_1 \cos x$ and $b_1 \sin x$ are called the *fundamental* components, and the remaining terms are called the *harmonics:* second harmonic, third harmonic, and so on. (We have let the constant term be $a_0/2$ instead of just $a_0$ because it will lead to a simpler expression for $a_0$ in a later derivation.)

The terms of the Fourier series are made up of sine and cosine functions. The fundamental terms ($a_1 \cos x$ and $b_1 \sin x$) have a period of $2\pi$; that is, the values repeat every $2\pi$ radians. The values of the harmonics, having a higher frequency than the fundamental, will repeat more often than $2\pi$. Thus the fifth harmonic has a period of $2\pi/5$. Therefore, the values of the entire series repeat every $2\pi$ radians. Thus the behavior of the entire series can be found by studying only one cycle, from 0 to $2\pi$, or from $-\pi$ to $\pi$, or any other interval containing a full cycle. To write a Fourier series for a periodic function, we observe that the period of the Fourier series is the same as that of the waveform for which it is written.

The Fourier series is an infinite series, but as with Taylor series, we use only a few terms for a given computation.

## Finding the Coefficients

To write a Fourier series, we must evaluate the coefficients: the $a$'s and $b$'s. We first find $a_0$. To do this, let us integrate both sides of the Fourier series over one cycle. We can choose any two limits of integration that are spaced $2\pi$ radians apart, such as $-\pi$ and $\pi$.

$$\int_{-\pi}^{\pi} f(x)\,dx = \frac{a_0}{2}\int_{-\pi}^{\pi} dx + a_1\int_{-\pi}^{\pi} \cos x\,dx + a_2\int_{-\pi}^{\pi} \cos 2x\,dx + \cdots + a_n\int_{-\pi}^{\pi} \cos nx\,dx + \cdots + b_1\int_{-\pi}^{\pi} \sin x\,dx + b_2\int_{-\pi}^{\pi} \sin 2x\,dx + \cdots + b_n\int_{-\pi}^{\pi} \sin nx\,dx + \cdots \tag{1}$$

But when we evaluate a definite integral of a sine wave or a cosine wave over one cycle the result is zero, so (1) reduces to

$$\int_{-\pi}^{\pi} f(x)\,dx = \frac{a_0}{2}\int_{-\pi}^{\pi} dx = \frac{a_0}{2}x\Big|_0^{2\pi} = a_0\pi$$

from which

| | |
|---|---|
| $a_0 = \dfrac{1}{\pi}\displaystyle\int_{-\pi}^{\pi} f(x)\,dx$ | **436** |

Next we find the general $a$ coefficient, $a_n$. Multiplying (1) by $\cos nx$ and integrating gives

$$\int_{-\pi}^{\pi} f(x)\cos nx\,dx = \frac{a_0}{2}\int_{-\pi}^{\pi} \cos nx\,dx + a_1\int_{-\pi}^{\pi} \cos x\cos nx\,dx + a_2\int_{-\pi}^{\pi} \cos 2x\cos nx\,dx + \cdots + a_n\int_{-\pi}^{\pi} \cos nx\cos nx\,dx + \cdots + b_1\int_{-\pi}^{\pi} \sin x\cos nx\,dx + b_2\int_{-\pi}^{\pi} \sin 2x\cos nx\,dx + \cdots + b_n\int_{-\pi}^{\pi} \sin nx\cos nx\,dx + \cdots \tag{2}$$

As before, when we evaluate the definite integral of the cosine function over one cycle the result is zero. Further, the trigonometric identity (Eq. 181)

$$2\cos A\cos B = \cos(A - B) + \cos(A + B)$$

shows that the product of two cosine functions can be written as the sum of two cosine functions. Thus the integral of the product of two cosines is equal to the sum of the integrals of two cosines, each of which is zero when integrated over one cycle. Thus (2) reduces to

$$\int_{-\pi}^{\pi} f(x)\cos nx\,dx = a_n\int_{-\pi}^{\pi}\cos nx\cos nx\,dx + \cdots$$
$$+ b_1\int_{-\pi}^{\pi}\sin x\cos nx\,dx$$
$$+ b_2\int_{-\pi}^{\pi}\sin 2x\cos nx\,dx + \cdots$$
$$+ b_n\int_{-\pi}^{\pi}\sin nx\cos nx\,dx + \cdots \quad (3)$$

In a similar way, the identity (Eq. 182)

$$2\sin A\cos B = \sin(A + B) + \sin(A - B)$$

shows that those terms containing the integral of the sine times the cosine will vanish. This leaves

$$\int_{-\pi}^{\pi} f(x)\cos nx\,dx = a_n\int_{-\pi}^{\pi}\cos^2 nx\,dx$$
$$= \frac{a_n}{n}\left[\frac{nx}{2} + \frac{\sin 2nx}{4}\right]_{-\pi}^{\pi} = \pi a_n$$

by Rule 17, so

| $$a_n = \frac{1}{\pi}\int_{-\pi}^{\pi} f(x)\cos nx\,dx$$ | **437** |
|---|---|

We now use a similar method to find the $b$ coefficients. We multiply (1) by $\sin nx$ and get

$$\int_{-\pi}^{\pi} f(x)\sin nx\,dx = \frac{a_0}{2}\int_{-\pi}^{\pi}\sin nx\,dx + a_1\int_{-\pi}^{\pi}\cos x\sin nx\,dx$$
$$+ a_2\int_{-\pi}^{\pi}\cos 2x\sin nx\,dx + \cdots + a_n\int_{-\pi}^{\pi}\cos nx\sin nx\,dx + \cdots$$
$$+ b_1\int_{-\pi}^{\pi}\sin x\sin nx\,dx + b_2\int_{-\pi}^{\pi}\sin 2x\sin nx\,dx + \cdots$$
$$+ b_n\int_{-\pi}^{\pi}\sin nx\sin nx\,dx + \cdots \quad (4)$$

The trigonometric identity (Eq. 180)

$$2\sin A\sin B = \cos(A - B) - \cos(A + B)$$

states that the product of two sines is equal to the difference of two cosines, each of which has an integral that is zero over a full cycle. Other integrals vanish for reasons given before, so (4) reduces to

$$\begin{aligned}\int_{-\pi}^{\pi} f(x) \sin nx\, dx &= b_n \int_{-\pi}^{\pi} \sin^2 nx\, dx \\ &= b_n \left[ \frac{nx}{2} - \frac{\sin 2nx}{4} \right]_{-\pi}^{\pi} \\ &= \pi b_n\end{aligned}$$

by Rule 16, so

| | |
|---|---|
| $b_n = \frac{1}{\pi} \int_{-\pi}^{\pi} f(x) \sin nx\, dx$ | **438** |

In summary,

| | | | |
|---|---|---|---|
| Fourier Series | A periodic function $f(x)$ can be replaced with the Fourier series $f(x) = \frac{a_0}{2} + a_1 \cos x + a_2 \cos 2x + a_3 \cos 3x + \cdots + b_1 \sin x + b_2 \sin 2x + b_3 \sin 3x + \cdots$ | | **435** |
| | where | $a_0 = \frac{1}{\pi} \int_{-\pi}^{\pi} f(x)\, dx$ | **436** |
| | | $a_n = \frac{1}{\pi} \int_{-\pi}^{\pi} f(x) \cos nx\, dx$ | **437** |
| | | $b_n = \frac{1}{\pi} \int_{-\pi}^{\pi} f(x) \sin nx\, dx$ | **438** |

**Keep in mind that the limits can be any two values spaced $2\pi$ apart. We'll sometimes use limits of 0 and $2\pi$. In Sec. 37-7 we'll rewrite the series for functions having any period.**

When we represent a wave by a Fourier series we say we *synthesize* the wave by the series. We get a better synthesis of the original waveform by using more terms of the series.

**EXAMPLE 18:** Write a Fourier series to represent a square wave (Waveform 1 in Table 37-6).

**Solution:** The first step is to write one or more equations that describe the waveform over a full cycle. For Waveform 1, these are

$$f(x) = \begin{cases} -1 & \text{for } -\pi \le x < 0 \\ 1 & \text{for } 0 \le x < \pi \end{cases}$$

**Square wave**

1. 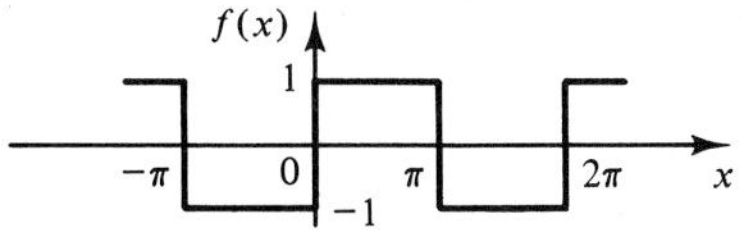


$$f(x) = \frac{4}{\pi}\left(\sin x + \frac{1}{3}\sin 3x + \frac{1}{5}\sin 5x + \frac{1}{7}\sin 7x + \cdots\right)$$

2. 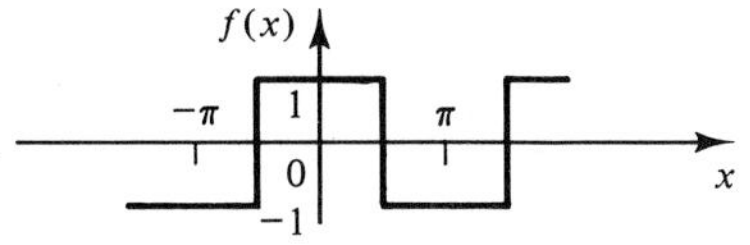


$$f(x) = \frac{4}{\pi}\left(\cos x - \frac{1}{3}\cos 3x + \frac{1}{5}\cos 5x - \cdots\right)$$

3. 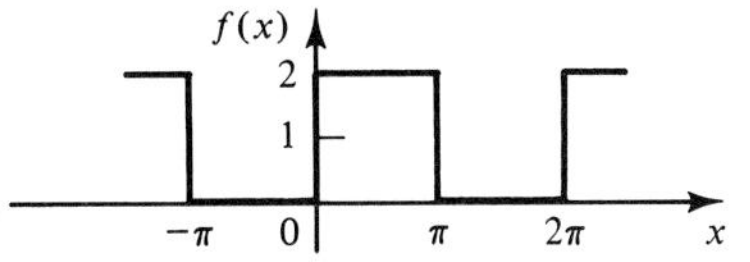


$$f(x) = \frac{4}{\pi}\left(\frac{\pi}{4} + \sin x + \frac{1}{3}\sin 3x + \frac{1}{5}\sin 5x + \cdots\right)$$

**Triangular wave**

4. 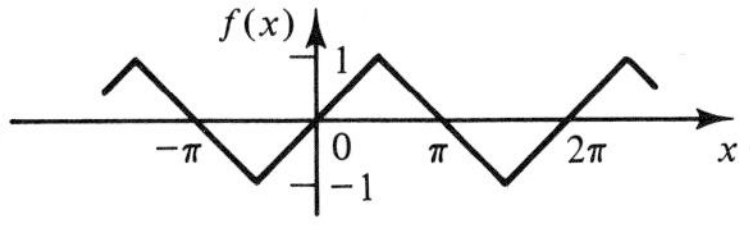


$$f(x) = \frac{8}{\pi^2}\left(\sin x - \frac{1}{3^2}\sin 3x + \frac{1}{5^2}\sin 5x - \cdots\right)$$

5. 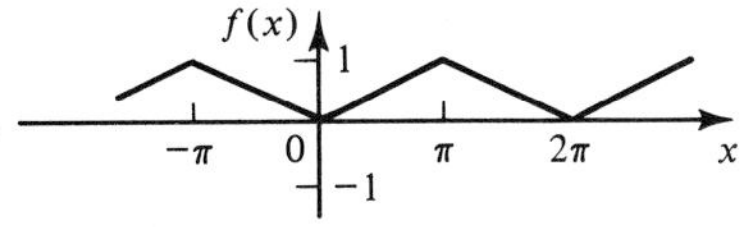


$$f(x) = \frac{1}{2} - \frac{4}{\pi^2}\left(\cos x + \frac{1}{3^2}\cos 3x + \frac{1}{5^2}\cos 5x + \cdots\right)$$

**Sawtooth wave**

6. 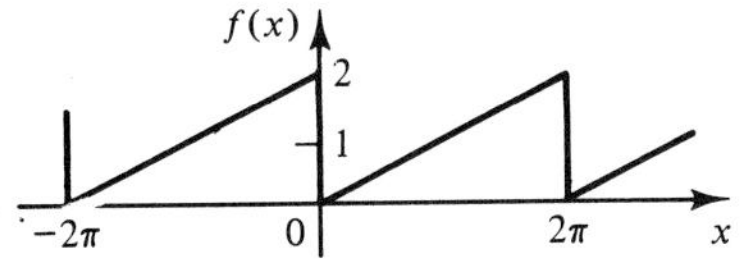


$$f(x) = 1 - \frac{2}{\pi}\left(\sin x + \frac{1}{2}\sin 2x + \frac{1}{3}\sin 3x + \cdots\right)$$

7. 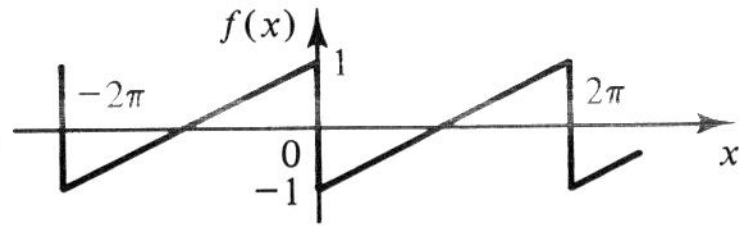


$$f(x) = -\frac{2}{\pi}\left(\sin x + \frac{1}{2}\sin 2x + \frac{1}{3}\sin 3x + \cdots\right)$$

8. 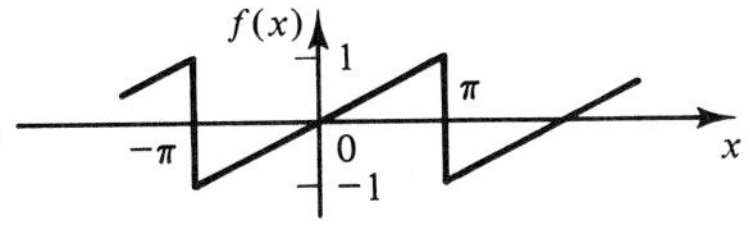


$$f(x) = \frac{2}{\pi}\left(\sin x - \frac{1}{2}\sin 2x + \frac{1}{3}\sin 3x - \frac{1}{4}\sin 4x + \cdots\right)$$

9. 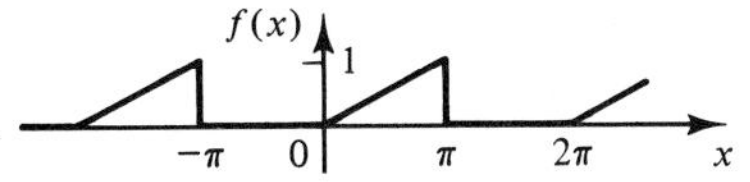


$$f(x) = \frac{1}{4} - \frac{2}{\pi^2}\left(\cos x + \frac{1}{3^2}\cos 3x + \cdots\right) + \frac{1}{\pi}\left(\sin x - \frac{1}{2}\sin 2x + \cdots\right)$$

**Half-wave rectifier**

10. 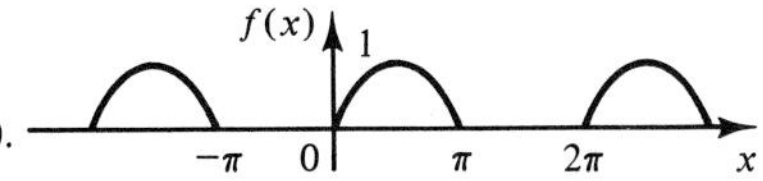

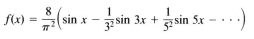

$$f(x) = \frac{1}{\pi}\left(1 + \frac{\pi}{2}\sin x - \frac{2}{3}\cos 2x - \frac{2}{15}\cos 4x - \frac{2}{35}\cos 6x + \cdots\right)$$

11. 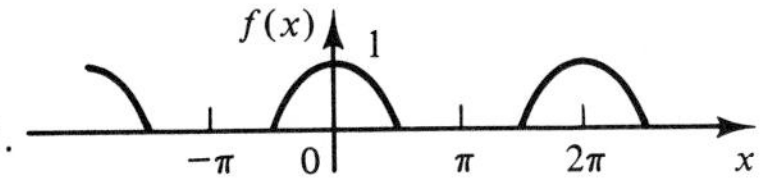

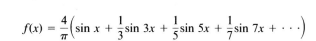

$$f(x) = \frac{1}{\pi}\left(1 + \frac{\pi}{2}\cos x + \frac{2}{3}\cos 2x - \frac{2}{15}\cos 4x + \frac{2}{35}\cos 6x + \cdots\right)$$

**Full-wave rectifier**

12. 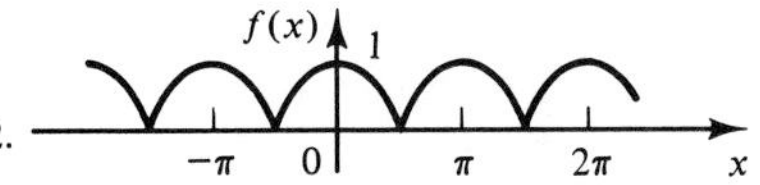


$$f(x) = \frac{2}{\pi}\left(1 + \frac{2}{3}\cos 2x - \frac{2}{15}\cos 4x + \frac{2}{35}\cos 6x + \cdots\right)$$

Since two equations describe $f(x)$ over a full cycle, we need two separate integrals to find each $a$ or $b$: one with limits $-\pi$ to 0, and another from 0 to $\pi$. We first find $a_0$ from Eq. 436,

$$a_0 = \frac{1}{\pi}\int_{-\pi}^{\pi} f(x)\,dx = \frac{1}{\pi}\int_{-\pi}^{0} (-1)\,dx + \frac{1}{\pi}\int_{0}^{\pi} (1)\,dx$$

$$= -\frac{x}{\pi}\Big|_{-\pi}^{0} + \frac{x}{\pi}\Big|_{0}^{\pi} = 0$$

We then get $a_n$ from Eq. 437,

$$a_n = \frac{1}{\pi}\int_{-\pi}^{0} (-1)\cos nx\,dx + \frac{1}{\pi}\int_{0}^{\pi} (1)\cos nx\,dx$$

$$= -\frac{\sin nx}{n\pi}\Big|_{-\pi}^{0} + \frac{\sin nx}{n\pi}\Big|_{0}^{\pi} = 0$$

**In Sec. 37-6 we'll learn how to tell in advance if all the sine or cosine terms will vanish.**

since the sine is zero for any angle that is a multiple of $\pi$. Thus all the $a$ coefficients are zero and our series contains no cosine terms. Then, by Eq. 438,

$$b_1 = \frac{1}{\pi}\int_{-\pi}^{0} (-1)\sin x\,dx + \frac{1}{\pi}\int_{0}^{\pi} (1)\sin x\,dx$$

$$= \frac{\cos x}{\pi}\Big|_{-\pi}^{0} - \frac{\cos x}{\pi}\Big|_{0}^{\pi} = \frac{4}{\pi}$$

$$b_2 = \frac{1}{\pi}\int_{-\pi}^{0} (-1)\sin 2x\,dx + \frac{1}{\pi}\int_{0}^{\pi} (1)\sin 2x\,dx$$

$$= \frac{\cos 2x}{2\pi}\Big|_{-\pi}^{0} - \frac{\cos 2x}{2\pi}\Big|_{0}^{\pi} = \frac{1}{2\pi}[(1-1)-(1-1)] = 0$$

Note that $b$ would be zero for $n = 4$ or, in fact, for all even values of $n$, since $\cos 2x = \cos 4x = \cos 6x$, and so on. Our series, then, contains no even harmonics.

$$b_3 = \frac{1}{\pi}\int_{-\pi}^{0} (-1)\sin 3x\,dx + \frac{1}{\pi}\int_{0}^{\pi} (1)\sin 3x\,dx$$

$$= \frac{\cos 3x}{3\pi}\Big|_{-\pi}^{0} - \frac{\cos 3x}{3\pi}\Big|_{0}^{\pi}$$

$$= \frac{4}{3\pi}$$

Continuing the same way gives values of $b$ of $4/\pi$, $4/3\pi$, $4/5\,\pi$, and so on. Our final Fourier series is then

$$f(x) = \frac{4}{\pi}\sin x + \frac{4}{3\pi}\sin 3x + \frac{4}{5\pi}\sin 5x + \frac{4}{7\pi}\sin 7x + \cdots$$

as in Table 37-6. Figure 37-3b shows the first two terms (the fundamental and the third harmonic) of our Fourier series. Figure 37-3c shows the first four terms (fundamental plus third, fifth, and seventh harmonics), and Fig. 37-3d shows the first six terms (fundamental plus third, fifth, seventh, ninth, and eleventh harmonics).

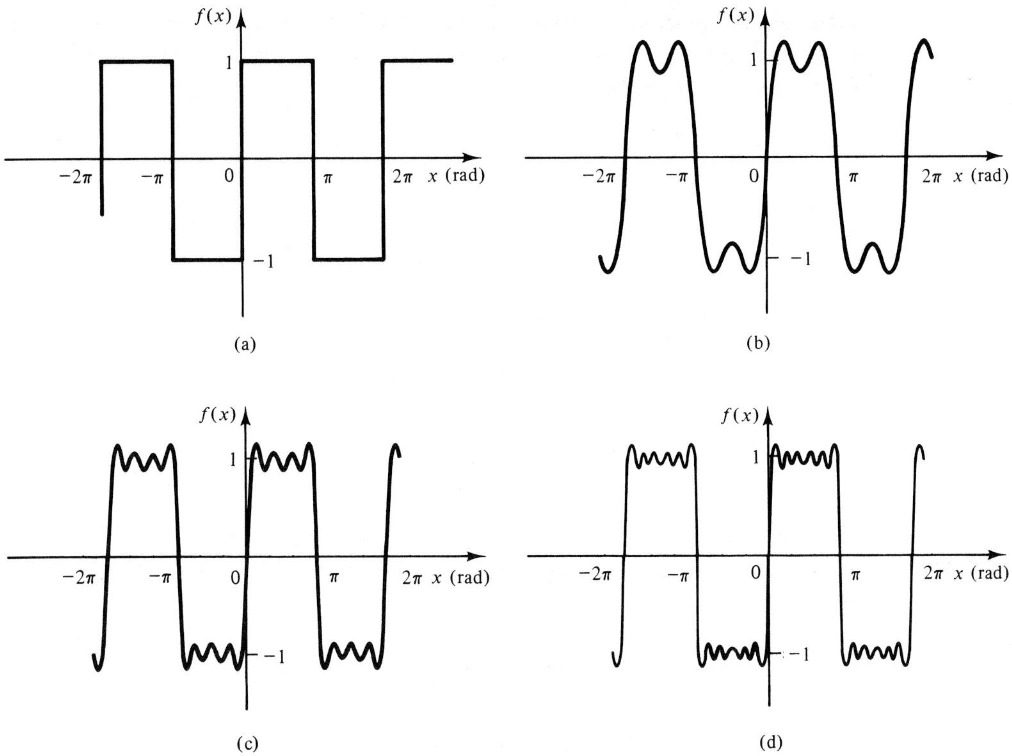


**FIGURE 37-3 A square wave, synthesized using two, four, and six terms of a Fourier series.**

## EXERCISE 5—FOURIER SERIES

Write seven terms of the Fourier series given the following coefficients.

1. $a_0 = 4$, $a_1 = 3$, $a_2 = 2$, $a_3 = 1$; $b_1 = 4$, $b_2 = 3$, $b_3 = 2$
2. $a_0 = 1.6$, $a_1 = 5.2$, $a_2 = 3.1$, $a_3 = 1.4$; $b_1 = 7.5$, $b_2 = 5.3$, $b_3 = 2.8$

Write a Fourier series for the indicated waveforms in Table 37-6, first writing the functions that describe the waveform. Check your Fourier series against those given in the table.

3. Waveform 3
4. Waveform 4
5. Waveform 6
6. Waveform 7
7. Waveform 8
8. Waveform 9
9. Waveform 10

### Computer

10. Write a program to synthesize a function of Fourier series. The program should accept as input the first seven Fourier coefficients and print the value of the function at values of $x$ from 0 to 360°.

## 37-6 WAVEFORM SYMMETRIES

### Kinds of Symmetry

We have seen that for the square wave, all the $a_n$ coefficients of the Fourier series were zero, and thus the final series had no cosine terms. If we could tell in advance that all the cosine terms would vanish, we could simplify our work by not even concerning ourselves with them. And, in fact, we can tell in advance which terms vanish by noticing the kind of symmetry the given function has. The three kinds of symmetry that are most useful here are:

1. If a function is *symmetric about the origin,* it is called an *odd* function, and its Fourier series has only sine terms.
2. If a function is *symmetric about the y axis,* it is called an *even* function, and its Fourier series has no sine terms.
3. If a function has *half-wave symmetry,* its Fourier series has only odd harmonics.

We now cover each type of symmetry.

## Odd Functions

We noted in Chapter 28 that a function $f(x)$ is symmetric about the origin if it remained unchanged when we substitute both $-x$ for $x$ and $-y$ for $y$. Another way of saying this is that $f(x) = -f(-x)$. Such functions are called *odd functions* (Fig. 37-4a).

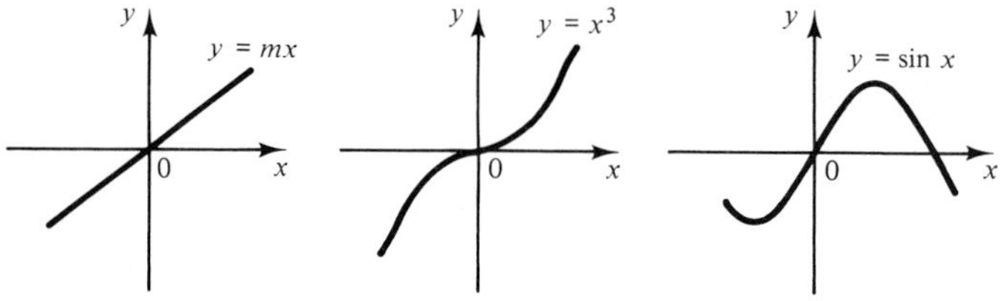


(a) Odd functions

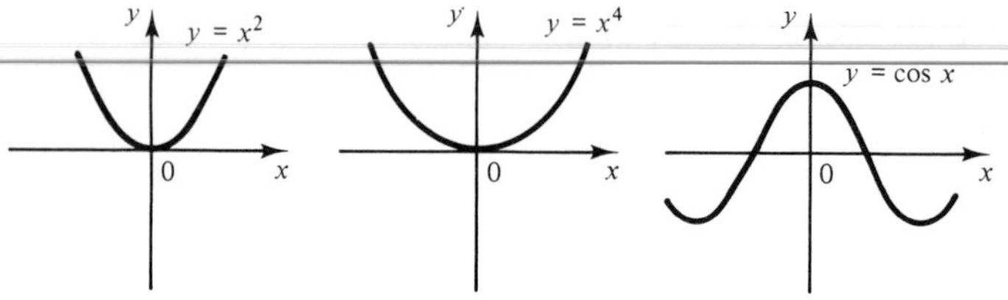


(b) Even functions

**FIGURE 37-4**

---

**EXAMPLE 19:** The function $y = x^3$ is odd because $(-x)^3 = -x^3$. Similarly, $x, x^5, x^7, \ldots$ are odd, with the name "odd" coinciding with whether the exponent is odd or even.

---

## Even Functions

We noted earlier that a function is symmetric about the $y$ axis (Fig. 37-4b) if it remains unchanged when we substitute $-x$ for $x$. In other words, $f(x) = f(-x)$. These are called *even functions.*

---

**EXAMPLE 20:** The function $y = x^2$ is even because $(-x)^2 = x^2$. Similarly, other even powers of $x$, such as $x^4, x^6, \ldots$, are even.

---

### Fourier Expansion of Odd and Even Functions

In the power series expansion for the sine function

| | |
|---|---|
| $\sin x = x - \frac{x^3}{6} + \frac{x^5}{120} - \cdots$ | **210** |

every term is an odd function; so that sin $x$ itself is an odd function. In the power series expansion for the cosine function

| | |
|---|---|
| $\cos x = 1 - \frac{x^2}{2} + \frac{x^4}{24} - \cdots$ | **211** |

every term is an even function, so cos $x$, made up of all even functions, is itself an even function. Thus we conclude the following:

| | | |
|---|---|---|
| Odd and Even Functions | (a) Odd functions have Fourier series with only sine terms (and no constant term).<br>(b) Even functions have Fourier series with only cosine terms (and may have a constant term). | **443** |

**EXAMPLE 21:** In Table 37-6 Waveforms 1, 4, 7, and 8 are odd. When deriving their Fourier series from scratch, we would save work by not even looking for cosine terms. Waveforms 2, 11, and 12 are even, while the others are neither odd nor even.

### Shift of Axes

A typical problem is one in which a periodic waveform is given and we are to write a Fourier series for it. When writing a Fourier series for a periodic waveform we are often free to choose the position of the $y$ axis. Our choice of axis can change the function from even to odd, or vice versa.

**EXAMPLE 22:** In Table 37-6 notice that Waveforms 1 and 2 are the same waveform but with the axis shifted by $\pi/2$ radians. Thus Waveform 1 is an odd function, while Waveform 2 is even. A similar shift is seen with the half-wave rectifier waveforms (10 and 11).

A *vertical* shift of the $x$ axis will affect the constant term in the series.

**EXAMPLE 23:** Waveform 3 is the same as Waveform 1, except for a vertical shift of 1 unit. Thus the series for Waveform 3 has a constant term of 1 unit. The same vertical shift can be seen with Waveforms 6 and 7.

## Half-Wave Symmetry

When the negative half-cycle of a periodic wave has the same shape as the positive half-cycle (except that it is, of course, inverted), we say that the wave has *half-wave symmetry*. The sine wave, the cosine wave, as well as Waveforms 1, 2, and 4 have half-wave symmetry.

A quick graphical way to test for half-wave symmetry is to measure the ordinates on the wave at two points spaced half a cycle apart. The two ordinates should be equal in magnitude but opposite in sign.

Figure 37-5 shows a sine wave and second, third, and fourth harmonics. For each, one arrow shows the ordinate at an arbitrarily chosen point, and the other arrow shows the ordinate half a cycle away ($\pi$ radians for the sine wave). For half-wave symmetry, the arrows on a given waveform should be equal in length but opposite in direction. Note that the fundamental and the odd harmonics have half-wave symmetry, while the even harmonics do not. This is also true for harmonics higher than those shown. Thus for the Fourier series representing a waveform, we conclude:

| | | |
|---|---|---|
| Half-Wave Symmetry | A waveform that has half-wave symmetry has only odd harmonics in its Fourier series | **444** |

| | |
|---|---|
| Common Error | Don't confuse odd functions with odd harmonics. Thus, sin $4x$ is an even harmonic but an odd function. |

By seeing if a waveform is an odd or even function or if it has half-wave symmetry, we can reduce the work of computation. Or, when possible, we can choose our axes to create symmetry, as in the following example.

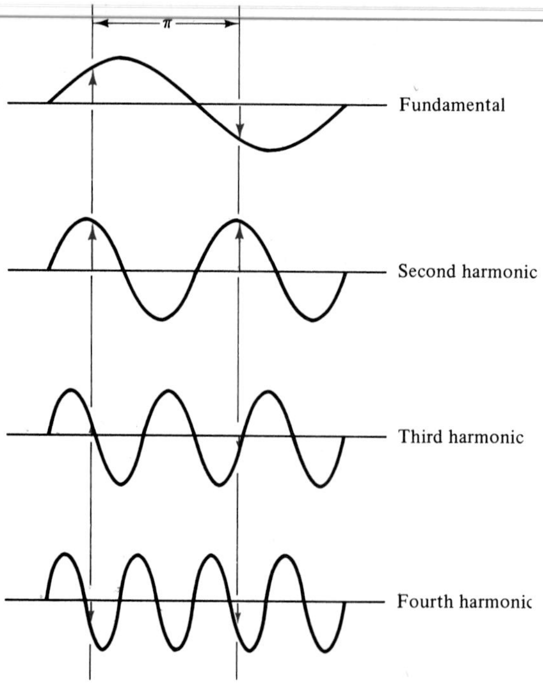


**FIGURE 37-5 The fundamental and the third harmonic have half-wave symmetry.**

**EXAMPLE 24:** Choose the $y$ axis and write a Fourier series for the triangular wave in Fig. 37-6a.

**Solution:** The given waveform has no symmetry, so let us temporarily shift the $x$ axis up 2 units, which will eliminate the constant term from our series. We will add 2 to our final series to account for this shift. We are free to locate the $y$ axis, so let us place it as in Fig. 37-6b. Our waveform is now an odd function and has half-wave symmetry, so we expect that the series we write will be composed of odd-harmonic sine terms,

$$y = b_1 \sin x + b_3 \sin 3x + b_5 \sin 5x + \cdots.$$

From Eq. 438,

$$b_n = \frac{1}{\pi} \int_{-\pi}^{\pi} f(x) \sin nx \, dx$$

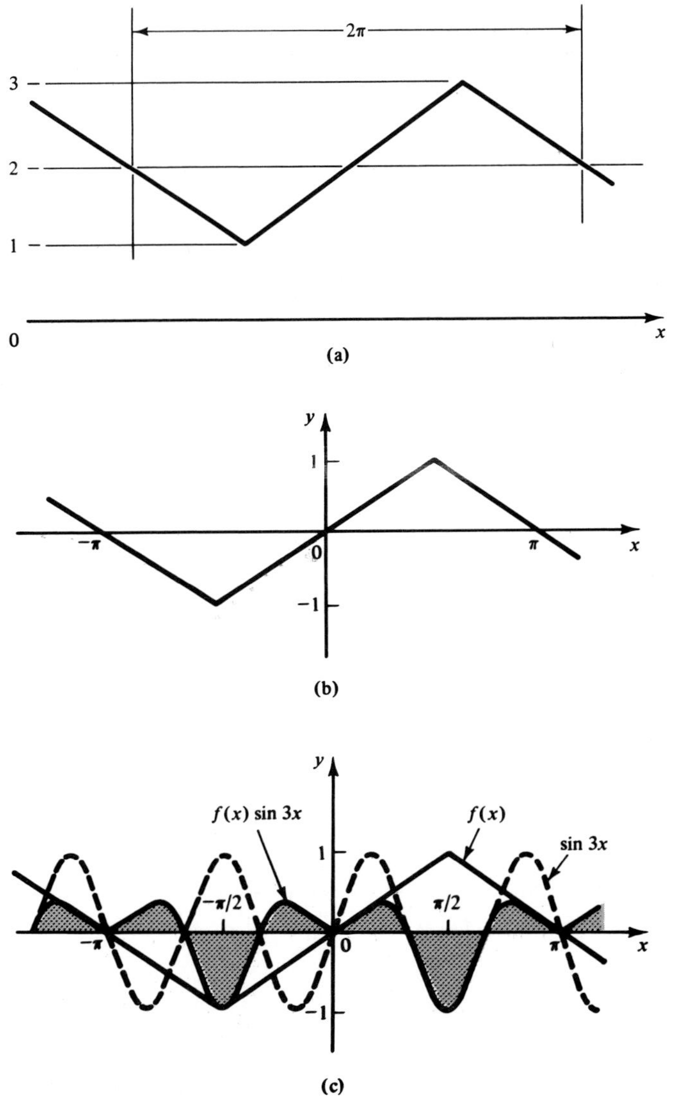


**FIGURE 37-6 Synthesis of a triangular wave.**

From 0 to $\pi/2$ our waveform is a straight line through the origin with a slope of $2/\pi$, so

$$f(x) = \frac{2}{\pi}x \qquad \left(0 < x < \frac{\pi}{2}\right)$$

The equation of the waveform is different elsewhere in the interval $-\pi$ to $\pi$, but we'll see that we need only the portion from 0 to $\pi/2$.

If we graph $f(x)\sin nx$, for any odd value of $n$, say 3, we get a curve such as in Fig. 37-6c. The integral of $f(x)\sin nx$ corresponds to the shaded portion under the curve. Note that this area repeats, and that the area under the curve between 0 and $\pi/2$ is one-fourth the area from $-\pi$ to $\pi$. Thus we need only integrate from 0 to $\pi/2$ and multiply the result by 4. Thus

$$b_n = \frac{1}{\pi}\int_{-\pi}^{\pi} f(x)\sin nx\,dx = \frac{4}{\pi}\int_0^{\pi/2} f(x)\sin nx\,dx$$

$$= \frac{4}{\pi}\int_0^{\pi/2} \frac{2}{\pi}x\sin nx\,dx = \frac{8}{\pi^2}\int_0^{\pi/2} x\sin nx\,dx$$

$$= \frac{8}{\pi^2}\left[\frac{1}{n^2}\sin nx - \frac{x}{n}\cos nx\right]_0^{\pi/2}$$

$$= \frac{8}{\pi^2}\left[\frac{1}{n^2}\sin\frac{n\pi}{2} - \frac{\pi}{2n}\cos\frac{n\pi}{2}\right]$$

We have only odd harmonics, so $n = 1, 3, 5, 7, \ldots$. For these values of $n$, $\cos(n\pi/2) = 0$, so

$$b_n = \frac{8}{\pi^2}\left(\frac{1}{n^2}\sin\frac{n\pi}{2}\right)$$

from which

$$b_1 = \frac{8}{\pi^2}\left(\frac{1}{1^2}\right) = \frac{8}{\pi^2} \qquad b_3 = \frac{8}{\pi^2}\left(\frac{1}{3^2}\right)(-1) = -\frac{8}{9\pi^2}$$

$$b_5 = \frac{8}{\pi^2}\left(\frac{1}{5^2}\right)(1) = \frac{8}{25\pi^2} \qquad b_7 = \frac{8}{\pi^2}\left(\frac{1}{7^2}\right)(-1) = -\frac{8}{49\pi^2}$$

Our final series, remembering to add a constant term of 2 units for the vertical shift of axis, is then

$$y = 2 + \frac{8}{\pi^2}\left(\sin x - \frac{1}{9}\sin 3x + \frac{1}{25}\sin 5x - \frac{1}{49}\sin 7x + \cdots\right)$$

This is the same as Waveform 4, except for the 2-unit shift.

## EXERCISE 6—WAVEFORM SYMMETRIES

Label each function as odd, even, or neither.

1. Fig. 37-7a
2. Fig. 37-7b
3. Fig. 37-7c
4. Fig. 37-7d
5. Fig. 37-7e
6. Fig. 37-7f

Which functions have half-wave symmetry?

7. Fig. 37-7a
8. Fig. 37-7b
9. Fig. 37-7c
10. Fig. 37-7d
11. Fig. 37-7e
12. Fig. 37-7f

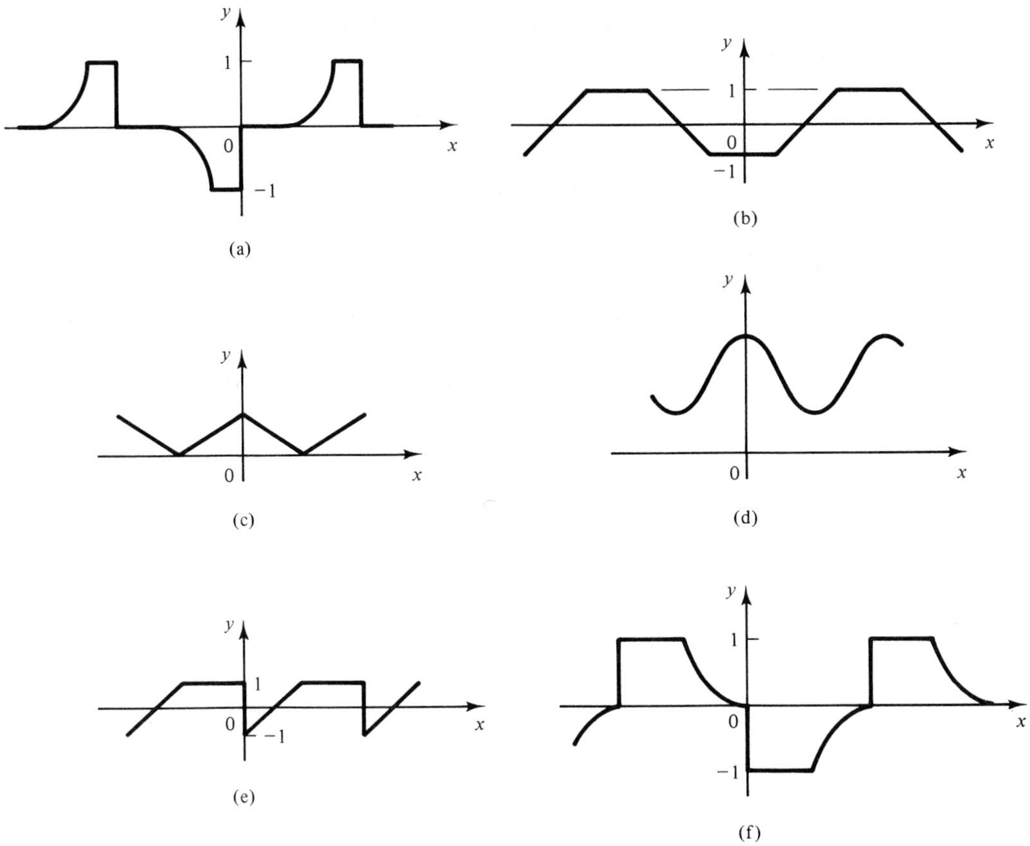


**FIGURE 37-7**

Using symmetry to simplify your work, write a Fourier series for the indicated waveforms in Table 37-6. As before, start by writing the functions that describe the waveform. Check your series against those given in the table.

**13.** Waveform 2
**14.** Waveform 5
**15.** Waveform 11
**16.** Waveform 12

## 37-7 WAVEFORMS WITH PERIOD OF 2*L*

So far we have written Fourier series only for functions with a period of $2\pi$. But a waveform could have some other period, say 8. Here we modify our formulas so that functions with any period can be represented.

Figure 37-8a shows a function $y = g(z)$ with a period of $2\pi$. The Fourier coefficients for this function are

$$a_0 = \frac{1}{\pi} \int_{-\pi}^{\pi} g(z)\, dz \tag{1}$$

$$a_n = \frac{1}{\pi} \int_{-\pi}^{\pi} g(z) \cos nz\, dz \tag{2}$$

and

$$b_n = \frac{1}{\pi} \int_{-\pi}^{\pi} g(z) \sin nz\, dz \tag{3}$$

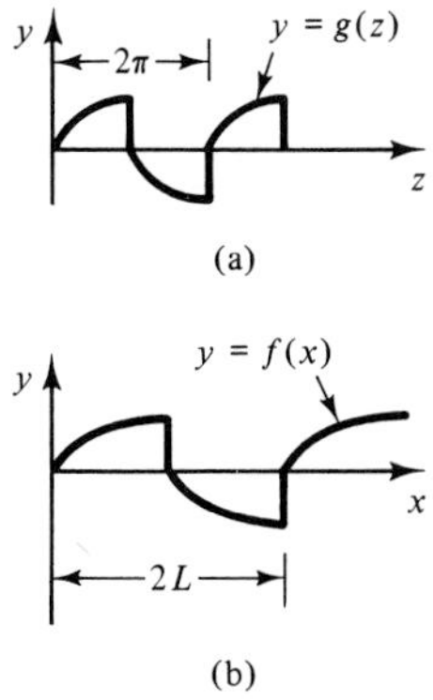


**FIGURE 37-8**

Figure 37-8b shows a function $y = f(x)$ with a period $2L$, where $x$ and $z$ are related by the proportion

$$\frac{x}{2L} = \frac{z}{2\pi}$$

or $z = \pi x/L$. Substituting into (1), with $g(z) = y = f(x)$ and $dz = d(\pi x/L) = (\pi/L)\,dx$, we obtain

$$a_0 = \frac{1}{\pi}\int_{-\pi}^{\pi} g(z)\,dz = \frac{1}{\pi}\int_{-L}^{L} f(x)\frac{\pi}{L}dx = \frac{1}{L}\int_{-L}^{L} f(x)\,dx$$

Next, from (2) we get

$$a_n = \frac{1}{\pi}\int_{-\pi}^{\pi} g(z)\cos nz\,dz$$

$$= \frac{1}{\pi}\int_{-L}^{L} f(x)\left(\cos\frac{n\pi x}{L}\right)\frac{\pi}{L}dx$$

$$= \frac{1}{L}\int_{-L}^{L} f(x)\cos\frac{n\pi x}{L}dx$$

Similarly, from (3),

$$b_n = \frac{1}{\pi}\int_{-\pi}^{\pi} g(z)\sin nz\,dz = \frac{1}{\pi}\int_{-L}^{L} f(x)\left(\sin\frac{n\pi x}{L}\right)\frac{\pi}{L}dx$$

$$= \frac{1}{L}\int_{-L}^{L} f(x)\sin\frac{n\pi x}{L}dx$$

In summary,

| | | | |
|---|---|---|---|
| Waveforms with Period of $2L$ | A periodic function can be represented by the Fourier series $f(x) = \frac{a_0}{2} + a_1\cos\frac{\pi x}{L} + a_2\cos\frac{2\pi x}{L} + a_3\cos\frac{3\pi x}{L} + \cdots + b_1\sin\frac{\pi x}{L} + b_2\sin\frac{2\pi x}{L} + b_3\sin\frac{3\pi x}{L} + \cdots$ | | **439** |
| | where | $a_0 = \frac{1}{L}\int_{-L}^{L} f(x)\,dx$ | **440** |
| | | $a_n = \frac{1}{L}\int_{-L}^{L} f(x)\cos\frac{n\pi x}{L}\,dx$ | **441** |
| | | $b_n = \frac{1}{L}\int_{-L}^{L} f(x)\sin\frac{n\pi x}{L}\,dx$ | **442** |

**EXAMPLE 25:** Write the Fourier series for the waveform in Fig. 37-9a.

**Solution:** Since the function is odd, we expect only sine terms, and no constant term. The equation of the waveform from −6 to 6 is $f(x) = x/6$. From Eq. 442, with $L = 6$,

$$b_n = \frac{1}{L}\int_{-L}^{L} f(x)\sin\frac{n\pi x}{L}\,dx$$

$$= \frac{1}{6}\int_{-6}^{6} \frac{x}{6}\sin\frac{n\pi x}{6}\,dx$$

Integrating by Rule 31 and evaluating at the limits gives

$$b_n = \frac{1}{(n\pi)^2}\left[\sin\frac{n\pi x}{6} - \frac{n\pi x}{6}\cos\frac{n\pi x}{6}\right]_{-6}^{6}$$

$$= -\frac{2}{n\pi}\cos n\pi$$

from which

$$b_1 = \frac{2}{\pi} \qquad b_2 = \frac{-1}{\pi} \qquad b_3 = \frac{2}{3\pi} \qquad b_4 = \frac{-1}{2\pi} \qquad \text{etc.}$$

Our series is then

$$y = \frac{2}{\pi}\left(\sin\frac{\pi x}{6} - \frac{1}{2}\sin\frac{\pi x}{3} + \frac{1}{3}\sin\frac{\pi x}{2} - \frac{1}{4}\sin\frac{2\pi x}{3} + \cdots\right)$$

Figure 37-9b shows a graph of the first five terms of this series.

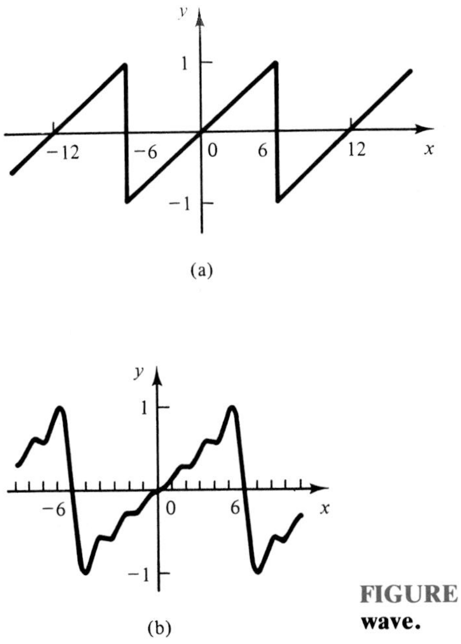


**FIGURE 37-9 Synthesis of a sawtooth wave.**

## EXERCISE 7—WAVEFORMS WITH PERIOD OF $2L$

Write a Fourier series for each waveform in Fig. 37-10. Check your series against those given.

1. Fig. 37-10a
2. Fig. 37-10b
3. Fig. 37-10c
4. Fig. 37-10d
5. Fig. 37-10e
6. Fig. 37-10f

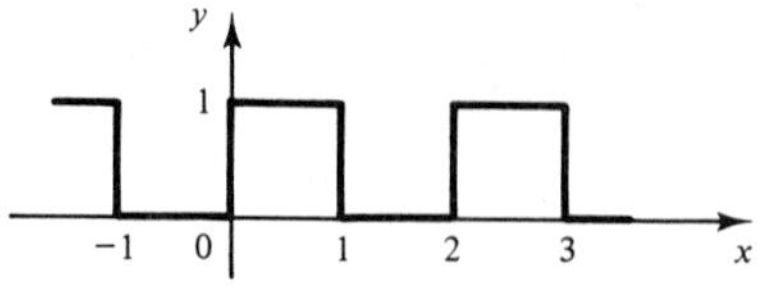


$$y = \frac{1}{2} + \frac{2}{\pi}\left(\sin \pi x + \frac{1}{3}\sin 3\pi x + \cdots\right)$$

(a)

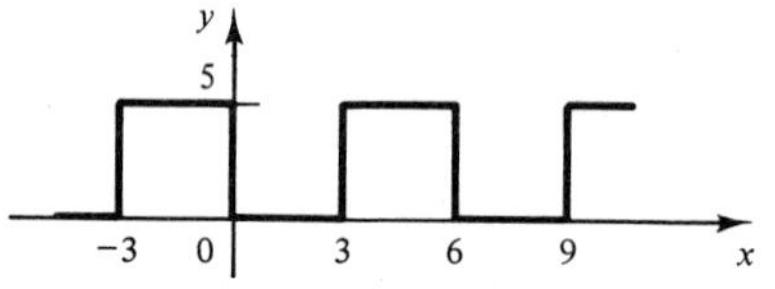


$$y = \frac{5}{2} - \frac{10}{\pi}\left(\sin \frac{\pi x}{3} + \frac{1}{3}\sin 3\pi x - \cdots\right)$$

(b)

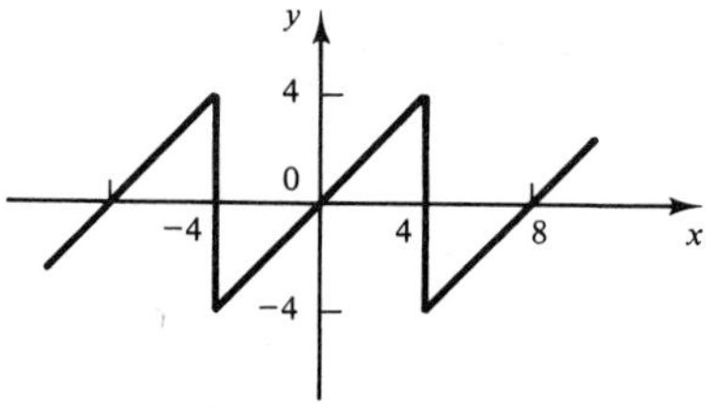


$$y = \frac{8}{\pi}\left(\sin \frac{\pi x}{4} - \frac{1}{2}\sin \frac{2\pi x}{4} + \frac{1}{3}\sin \frac{3\pi x}{4} - \cdots\right)$$

(c)

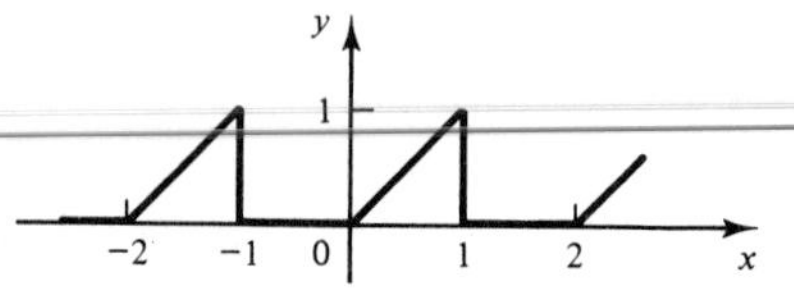


$$y = \frac{1}{4} - \frac{2}{\pi^2}\left(\cos \pi x + \frac{1}{9}\cos 3\pi x + \frac{1}{25}\cos 5\pi x + \cdots\right)$$
$$+ \frac{1}{\pi}\left(\sin \pi x - \frac{1}{2}\sin 2\pi x + \frac{1}{3}\sin 3\pi x - \cdots\right)$$

(d)

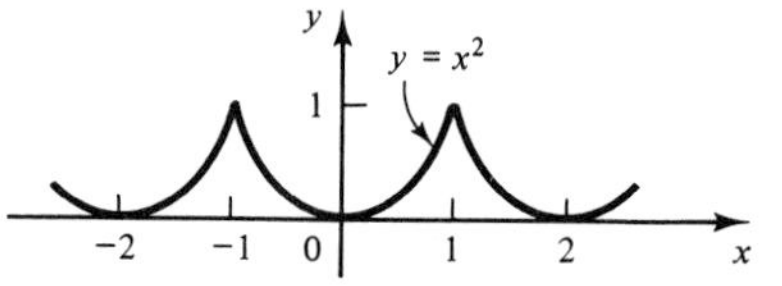


$$y = \frac{1}{3} - \frac{4}{\pi^2}\left(\cos \pi x - \frac{1}{4}\cos 2\pi x + \frac{1}{9}\cos 3\pi x - \cdots\right)$$

(e)

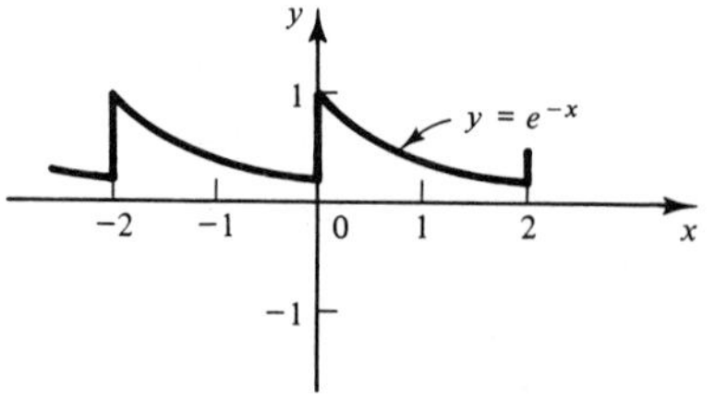


$$y = 0.4324 + 0.0795 \cos \pi x + 0.0214 \cos 2\pi x$$
$$+ 0.00963 \cos 3\pi x + \cdots$$
$$+ 0.2499 \sin \pi x + 0.1342 \sin 2\pi x$$
$$+ 0.0907 \sin 3\pi x + \cdots$$

(f)

**FIGURE 37-10**

## 37-8 A NUMERICAL METHOD FOR FINDING FOURIER SERIES

So far we have written Fourier series for waveforms for which we have equations. But what about a wave whose shape we know from an oscilloscope trace or from a set of meter readings but for which we have no equation? Here we give a simple numerical method, easily programmed for the computer. It is based on numerical methods for finding the area under a curve covered in Sec. 34-6.

---

**EXAMPLE 26:** Find the first eight terms of the Fourier series for the waveform in Fig. 37-11.

**Solution:** *Using symmetry:* We first note that the wave does not appear to be odd or even, so we expect our series to contain both sine and cosine terms. There is half-wave symmetry, however, so we look only for odd harmonics. The first eight terms, then, will be

$$y = a_1 \cos x + a_3 \cos 3x + a_5 \cos 5x + a_7 \cos 7x$$
$$+ b_1 \sin x + b_3 \sin 3x + b_5 \sin 5x + b_7 \sin 7x$$

*Finding the coefficients:* From Eq. 437 the first coefficient is

$$a_1 = \frac{1}{\pi} \int_0^{2\pi} y \cos x \, dx$$

But the value of the integral $\int y \cos x \, dx$ is equal to the area under the $y \cos x$ curve, which we find by numerical integration. We use the average ordinate method here. Taking advantage of half-wave symmetry, we integrate over only a half-cycle, and double the result.

If $y_{avg}$ is the average ordinate of the $y \cos x$ curve over a half-cycle (0 to $\pi$), then by Eq. 404,

$$\int_0^{2\pi} y \cos x \, dx = 2\pi y_{avg}$$

so

$$a_1 = \frac{1}{\pi} \int_0^{2\pi} y \cos x \, dx = 2y_{avg}$$

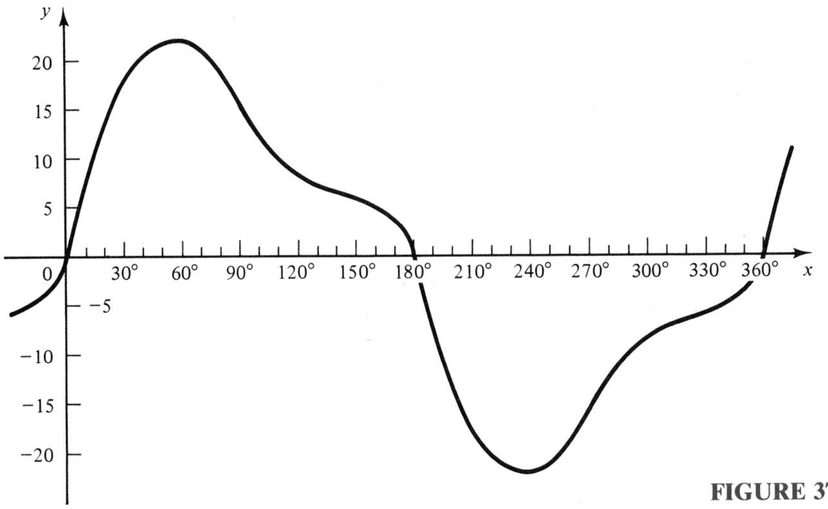


FIGURE 37-11

TABLE 37-7

| $x$ | $y$ | $\sin x$ | $y \sin x$ | $\cos x$ | $y \cos x$ | $\sin 3x$ | $y \sin 3x$ | $\cos 3x$ | $y \cos 3x$ |
|---|---|---|---|---|---|---|---|---|---|
| 0 | 0.0 | 0.0000 | 0.000 | 1.0000 | 0.000 | 0.0000 | 0.000 | 1.0000 | 0.000 |
| 10 | 10.0 | 0.1736 | 1.736 | 0.9848 | 9.848 | 0.5000 | 5.000 | 0.8660 | 8.660 |
| 20 | 14.8 | 0.3420 | 5.062 | 0.9397 | 13.907 | 0.8660 | 12.817 | 0.5000 | 7.400 |
| 30 | 16.1 | 0.5000 | 8.050 | 0.8660 | 13.943 | 1.0000 | 16.100 | 0.0000 | 0.000 |
| 40 | 20.0 | 0.6428 | 12.856 | 0.7660 | 15.321 | 0.8660 | 17.320 | −0.5000 | −10.000 |
| 50 | 21.0 | 0.7660 | 16.087 | 0.6428 | 13.499 | 0.5000 | 10.500 | −0.8660 | −18.187 |
| 60 | 21.2 | 0.8660 | 18.360 | 0.5000 | 10.600 | 0.0000 | −0.000 | −1.0000 | −21.200 |
| 70 | 20.3 | 0.9397 | 19.076 | 0.3420 | 6.943 | −0.5000 | −10.150 | −0.8660 | −17.580 |
| 80 | 17.7 | 0.9848 | 17.431 | 0.1736 | 3.074 | −0.8660 | −15.329 | −0.5000 | −8.850 |
| 90 | 14.1 | 1.0000 | 14.100 | 0.0000 | 0.000 | −1.0000 | −14.100 | 0.0000 | 0.000 |
| 100 | 11.2 | 0.9848 | 11.030 | −0.1737 | −1.945 | −0.8660 | −9.699 | 0.5000 | 5.600 |
| 110 | 8.8 | 0.9397 | 8.269 | −0.3420 | −3.010 | −0.5000 | −4.400 | 0.8660 | 7.621 |
| 120 | 7.5 | 0.8660 | 6.495 | −0.5000 | −3.750 | 0.0000 | 0.000 | 1.0000 | 7.500 |
| 130 | 6.6 | 0.7660 | 5.056 | −0.6428 | −4.242 | 0.5000 | 3.300 | 0.8660 | 5.716 |
| 140 | 6.1 | 0.6428 | 3.921 | −0.7660 | −4.673 | 0.8660 | 5.283 | 0.5000 | 3.050 |
| 150 | 5.5 | 0.5000 | 2.750 | −0.8660 | −4.763 | 1.000 | 5.500 | −0.0000 | −0.000 |
| 160 | 4.6 | 0.3420 | 1.573 | −0.9397 | −4.323 | 0.8660 | 3.984 | −0.5000 | −2.300 |
| 170 | 3.6 | 0.1736 | 0.625 | −0.9848 | −3.545 | 0.5000 | 1.800 | −0.8660 | −3.118 |
| 180 | 0.0 | 0.0000 | 0.000 | −1.0000 | 0.000 | −0.0000 | 0.000 | −1.0000 | 0.000 |
| | | Sum: | 152.477 | | 56.883 | | 27.926 | | −35.688 |
| | | $y_{avg}$: | 8.025 | | 2.994 | | 1.470 | | −1.878 |
| | | | $b_1 = 16.050$ | | $a_1 = 5.988$ | | $b_3 = 2.940$ | | $a_3 = -3.756$ |

We divide the half-cycle into a number of intervals, say at every 10°. At each $x$ we measure the ordinate $y$ and calculate $y \cos x$. For example, at $x = 20°$, $y$ is measured at 14.8, so

$$y \cos x = 14.8 \cos 20° = 14.8(0.9397) = 13.907$$

The computed values of $y \cos x$ for the remaining angles are shown in Table 37-7.

We then add all the values of $y \cos x$ over the half-cycle and get an average value by dividing that sum (56.883) by the total number of data points, 19. Thus

$$y_{avg} = \frac{56.883}{19} = 2.994$$

Our first coefficient is then

$$a_1 = 2y_{avg} = 2(2.994) = 5.988$$

The remainder of the computation is similar, and the results are shown in Table 37-7. Our final Fourier series is then

$$y = 5.988 \cos x - 3.756 \cos 3x - 0.176 \cos 5x - 0.512 \cos 7x$$
$$+ 16.050 \sin x + 2.940 \sin 3x + 0.866 \sin 5x + 0.750 \sin 7x$$

*Check:* To test our result we compute $y$ by series and get the values listed in Table 37-8 and graphed in Fig. 37-12. For better accuracy we could compute more terms of the series or use finer spacing when finding $y_{avg}$. This would not be much more work when done by computer.

| sin 5x | y sin 5x | cos 5x | y cos 5x | sin 7x | y sin 7x | cos 7x | y cos 7x |
|---|---|---|---|---|---|---|---|
| 0.0000 | 0.000 | 1.0000 | 0.000 | 0.0000 | 0.000 | 1.0000 | 0.000 |
| 0.7660 | 7.660 | 0.6428 | 6.428 | 0.9397 | 9.397 | 0.3420 | 3.420 |
| 0.9848 | 14.575 | −0.1737 | −2.570 | 0.6428 | 9.513 | −0.7660 | −11.338 |
| 0.5000 | 8.050 | −0.8660 | −13.943 | −0.5000 | −8.050 | −0.8660 | −13.943 |
| −0.3420 | −6.841 | −0.9397 | −18.794 | −0.9848 | −19.696 | 0.1737 | 3.473 |
| −0.9397 | −19.734 | −0.3420 | −7.182 | −0.1736 | −3.646 | 0.9848 | 20.681 |
| −0.8660 | −18.360 | 0.5000 | 10.600 | 0.8660 | 18.360 | 0.5000 | 10.600 |
| −0.1736 | −3.525 | 0.9848 | 19.992 | 0.7660 | 15.550 | −0.6428 | −13.049 |
| 0.6428 | 11.378 | 0.7660 | 13.559 | −0.3420 | −6.054 | −0.9397 | −16.632 |
| 1.0000 | 14.100 | −0.0000 | −0.000 | −1.0000 | −14.100 | 0.0000 | 0.000 |
| 0.6428 | 7.199 | −0.7661 | −8.580 | −0.3420 | −3.830 | 0.9397 | 10.525 |
| −0.1737 | −1.528 | −0.9848 | −8.666 | 0.7661 | 6.741 | 0.6428 | 5.656 |
| −0.8660 | −6.495 | −0.5000 | −3.750 | 0.8660 | 6.495 | −0.5000 | −3.750 |
| −0.9397 | −6.202 | 0.3420 | 2.257 | −0.1737 | −1.146 | −0.9848 | −6.500 |
| −0.3420 | −2.086 | 0.9397 | 5.732 | −0.9848 | −6.007 | −0.1736 | −1.059 |
| 0.5000 | 2.750 | 0.8660 | 4.763 | −0.5000 | −2.750 | 0.8660 | 4.763 |
| 0.9848 | 4.530 | 0.1736 | 0.799 | 0.6428 | 2.957 | 0.7660 | 3.524 |
| 0.7660 | 2.758 | −0.6428 | −2.314 | 0.9397 | 3.383 | −0.3421 | −1.231 |
| −0.0000 | 0.000 | −1.0000 | 0.000 | −0.0001 | 0.000 | −1.0000 | 0.000 |
| | 8.230 | | −1.670 | | 7.116 | | −4.860 |
| | 0.433 | | −0.088 | | 0.375 | | −0.256 |
| | $b_5$ = 0.866 | | $a_5$ = −0.176 | | $b_7$ = 0.750 | | $a_7$ = −0.512 |

TABLE 37-8

| x | Original y | y from Series |
|---|---|---|
| 0 | 0.0 | 1.5 |
| 10 | 10.0 | 8.0 |
| 20 | 14.8 | 13.5 |
| 30 | 16.1 | 16.8 |
| 40 | 20.0 | 18.4 |
| 50 | 21.0 | 19.5 |
| 60 | 21.2 | 20.2 |
| 70 | 20.3 | 19.5 |
| 80 | 17.7 | 16.8 |
| 90 | 14.1 | 13.2 |
| 100 | 11.2 | 10.3 |
| 110 | 8.8 | 8.6 |
| 120 | 7.5 | 7.4 |
| 130 | 6.6 | 6.2 |
| 140 | 6.1 | 5.3 |
| 150 | 5.5 | 5.2 |
| 160 | 4.6 | 5.2 |
| 170 | 3.6 | 3.3 |
| 180 | 0.0 | −1.5 |

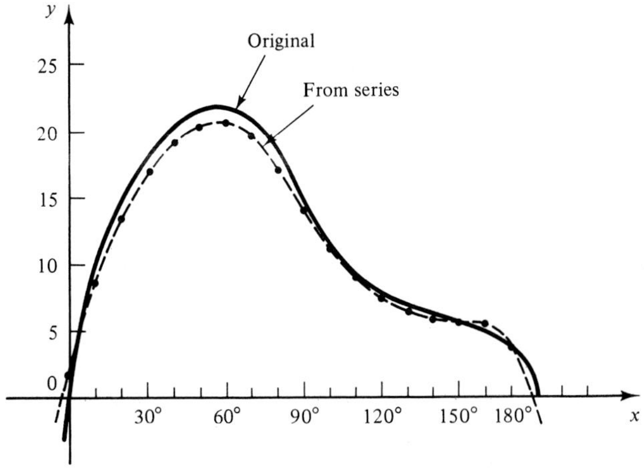


**FIGURE 37-12**

## EXERCISE 8—A NUMERICAL METHOD FOR FINDING FOURIER SERIES

Find the first six terms of the Fourier series for each waveform. Assume half-wave symmetry.

**1.**

| | | | | | | | | | | |
|---|---|---|---|---|---|---|---|---|---|---|
| $x$ | 0° | 20° | 40° | 60° | 80° | 100° | 120° | 140° | 160° | 180° |
| $y$ | 0 | 2.1 | 4.2 | 4.5 | 7.0 | 10.1 | 14.3 | 14.8 | 13.9 | 0 |

**2.**

| | | | | | | | | | | |
|---|---|---|---|---|---|---|---|---|---|---|
| $x$ | 0° | 20° | 40° | 60° | 80° | 100° | 120° | 140° | 160° | 180° |
| $y$ | 0 | 31 | 52 | 55 | 80 | 111 | 153 | 108 | 49 | 0 |

**3.**

| | | | | | | | | | | |
|---|---|---|---|---|---|---|---|---|---|---|
| $x$ | 0° | 20° | 40° | 60° | 80° | 100° | 120° | 140° | 160° | 180° |
| $y$ | 0 | 12.1 | 24.2 | 34.5 | 47.0 | 30.1 | 24.3 | 11.8 | 9.9 | 0 |

**4.**

| | | | | | | | | | | |
|---|---|---|---|---|---|---|---|---|---|---|
| $x$ | 0° | 20° | 40° | 60° | 80° | 100° | 120° | 140° | 160° | 180° |
| $y$ | 0 | 4.6 | 9.4 | 15.4 | 12.3 | 10.8 | 14.1 | 16.8 | 10.9 | 0 |

### Computer

**5.** Write a program or use a spreadsheet to compute the coefficients of the first eight terms of a Fourier series for a given set of experimental data. Test your program on the problems above.

## CHAPTER 37 REVIEW PROBLEMS

*Compute each number using three terms of the appropriate Maclaurin series.*

**1.** $e^3$

**2.** $\sin 0.5$

*Compute each value using three terms of the appropriate Taylor series from Table 37-1.*

**3.** $\dfrac{1}{2.1}$

**4.** $\sin 31.5°$

**5.** Use the ratio test to find the interval of convergence of the power series

$$1 + x + x^2 + x^3 + \cdots + x^n + \cdots$$

**6.** Use the ratio test to determine if the following series converges or diverges.

$$1 + \frac{3}{3!} + \frac{5}{5!} + \frac{7}{7!} + \cdots$$

*Write the first five terms of each series, given the general term.*

**7.** $u_n = 2n^2$

**8.** $u_n = 3n - 1$

*Multiply or divide earlier series to verify the following.*

**9.** $x^2e^{2x} = x^2 + 2x^3 + 2x^4 + \cdots$ (Use Series 29)

**10.** $\dfrac{\sin x}{x^2} = \dfrac{1}{x} - \dfrac{x}{3!} + \dfrac{x^3}{5!} - \cdots$

*Deduce the general term of each series. Use it to predict the next two terms.*

**11.** $-1 + 2 + 7 + 14 + 23 + \cdots$

**12.** $4 + 7 + 10 + 13 + 16 + \cdots$

**13.** Evaluate using series, $\displaystyle\int_0^1 \frac{e^x - 1}{x}\,dx$

*Deduce a recursion relation for each series. Use it to predict the next two terms.*

**14.** $3 + 5 + 9 + 17 + \cdots$

**15.** $3 + 7 + 47 + \cdots$

**16.** By substituting into Series 3 for $\cos x$ verify the series

14. $$\cos x^2 = 1 - \frac{x^4}{2!} + \frac{x^8}{4!} - \frac{x^{12}}{6!} + \cdots$$

17. Verify the Maclaurin expansion

$$(1 + x)^3 = 1 + 3x + 3x^2 + x^3$$

18. Verify the following Taylor series, expanded about $a = 1$,

24. $$\sqrt{x^2 + 1} = \sqrt{2} + \frac{(x - 1)}{\sqrt{2}} + \frac{(x - 1)^2}{4\sqrt{2}} + \cdots$$

*Label each function as odd, even, or neither.*

19. Fig. 37-13a
20. Fig. 37-13b
21. Fig. 37-13c
22. Fig. 37-13d

*Which functions have half-wave symmetry?*

23. Fig. 37-13a
24. Fig. 37-13b
25. Fig. 37-13c
26. Fig. 37-13d

*Write a Fourier series for each waveform.*

27. Fig. 37-13a
28. Fig. 37-13b
29. Fig. 37-13c
30. Fig. 37-13d
31. Find the first six terms of the Fourier series for the given waveform. Assume half-wave symmetry.

| $x$ | 0° | 20° | 40° | 60° | 80° | 100° | 120° | 140° | 160° | 180° |
|---|---|---|---|---|---|---|---|---|---|---|
| $y$ | 0 | 3.2 | 5.5 | 6.2 | 8.3 | 11.7 | 15.1 | 14.3 | 12.6 | 0 |

*Writing*

32. Your friend writes saying that you are crazy to study infinite series. "They're useless," he says, "because there is no way to compute an infinite number of terms. What can you do with a string of numbers that goes on forever?" Reply in writing to your friend and either accept or reject his statement. Give specific reasons for your choice.

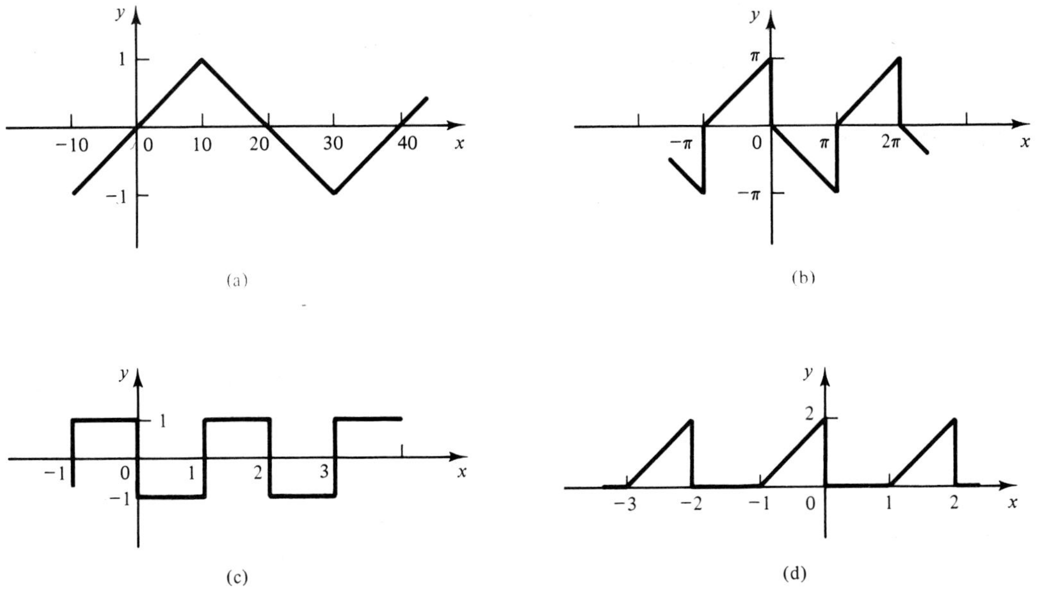


FIGURE 37-13

# A

# Summary of Facts and Formulas

**Some of the formulas in this Summary are included for reference even though they may not appear elsewhere in the text.**

| | No. | | | | Page |
|---|---|---|---|---|---|
| ALGEBRAIC LAWS | 1 | Commutative Law | Addition | $a + b = b + a$ | 8 |
| | 2 | | Multiplication | $ab = ba$ | 11, 59 |
| | 3 | Associative Law | Addition | $a + (b + c) = (a + b) + c = (a + c) + b$ | 8 |
| | 4 | | Multiplication | $a(bc) = (ab)c = (ac)b = abc$ | 11 |
| | 5 | Distributive Law | | $a(b + c) = ab + ac$ | 12, 60, 184 |
| RULES OF SIGNS | 6 | Addition and Subtraction | | $a + (-b) = a - (+b) = a - b$ | 7 |
| | 7 | | | $a + (+b) = a - (-b) = a + b$ | 7 |
| | 8 | Multiplication | | $(+a)(+b) = (-a)(-b) = +ab$ | 12, 59 |
| | 9 | | | $(+a)(-b) = (-a)(+b) = -(+a)(+b) = -ab$ | 12, 59 |
| | 10 | Division | | $\frac{+a}{+b} = \frac{-a}{-b} = -\frac{-a}{+b} = -\frac{+a}{-b} = \frac{a}{b}$ | 16, 68, 209 |
| | 11 | | | $\frac{+a}{-b} = \frac{-a}{+b} = -\frac{-a}{-b} = -\frac{a}{b}$ | 16, 68, 209 |
| PERCENTAGE | 12 | Amount = base × rate, $A = BP$ | | | 36 |
| | 13 | $\text{Percent change} = \frac{\text{new value} - \text{original value}}{\text{original value}} \times 100\%$ | | | 38 |
| | 14 | $\text{Percent error} = \frac{\text{measured value} - \text{known value}}{\text{known value}} \times 100\%$ | | | 39 |
| | 15 | $\text{Percent concentration of ingredient } A = \frac{\text{amount of } A}{\text{amount of mixture}} \times 100\%$ | | | 39, 90 |
| | 16 | $\text{Percent efficiency} = \frac{\text{output}}{\text{input}} \times 100\%$ | | | 38 |

| | No. | | | | Page |
|---|---|---|---|---|---|
| BINARY NUMBERS | 17 | Largest $n$-Bit Binary Number | | $2^n - 1$ | 578 |
| | 18 | Addition | | $0 + 0 = 0$<br>$0 + 1 = 1$<br>$1 + 0 = 1$<br>$1 + 1 = 0$ with carry to next column | |
| | 19 | Subtraction | | $0 - 0 = 0$<br>$0 - 1 = 1$ with borrow from next column<br>$1 - 0 = 1$<br>$1 - 1 = 0$<br>$10 - 1 = 1$ | |
| | 20 | | | Subtracting $B$ from $A$ is equivalent to adding the two's complement of $B$ to $A$ | |
| | 21 | Multiplication | | $0 \times 0 = 0$<br>$0 \times 1 = 0$<br>$1 \times 0 = 0$<br>$1 \times 1 = 1$ | |
| | 22 | Division | | $0 \div 0$ is not defined<br>$1 \div 0$ is not defined<br>$0 \div 1 = 0$<br>$1 \div 1 = 1$ | |
| | 23 | Complements | $n$-Bit One's Complement of $x$ | $(2^n - 1) - x$ | |
| | 24 | | | The one's complement of a number can be written by changing the 1's to 0's and the 0's to 1's | |
| | 25 | | $n$-Bit Two's Complement of $x$ | $2^n - x$ | |
| | 26 | | | To find the two's complement of a number first write the one's complement, and add 1 | |
| | 27 | Negative Binary Numbers | | If $M$ is a positive binary number, then $-M$ is the two's complement of $M$ | |

| | No. | | | Formula | Page |
|---|---|---|---|---|---|
| EXPONENTS | 28 | Definition | | $a^n = \underbrace{a \cdot a \cdot a \cdot \ldots \cdot a}_{n \text{ factors}}$ | 52 |
| | 29 | Laws of Exponents | Products | $x^a \cdot x^b = x^{a+b}$ | 53<br>60<br>319 |
| | 30 | | Quotients | $\frac{x^a}{x^b} = x^{a-b} \quad (x \neq 0)$ | 54<br>68<br>320 |
| | 31 | | Powers | $(x^a)^b = x^{ab} = (x^b)^a$ | 55<br>187<br>319 |
| | 32 | | Product Raised to a Power | $(xy)^n = x^n \cdot y^n$ | 55<br>319 |
| | 33 | | Quotient Raised to a Power | $\left(\frac{x}{y}\right)^n = \frac{x^n}{y^n} \quad (y \neq 0)$ | 56<br>320 |
| | 34 | | Zero Exponent | $x^0 = 1 \quad (x \neq 0)$ | 56<br>318 |
| | 35 | | Negative Exponent | $x^{-a} = \frac{1}{x^a} \quad (x \neq 0)$ | 57<br>318 |
| | 36 | Fractional Exponents | | $a^{1/n} = \sqrt[n]{a}$ | 20<br>323<br>525 |
| | 37 | | | $a^{m/n} = \sqrt[n]{a^m} = (\sqrt[n]{a})^m$ | 324 |
| RADICALS | 38 | Rules of Radicals | Root of a Product | $\sqrt[n]{ab} = \sqrt[n]{a}\,\sqrt[n]{b}$ | 325 |
| | 39 | | Root of a Quotient | $\sqrt[n]{\frac{a}{b}} = \frac{\sqrt[n]{a}}{\sqrt[n]{b}}$ | 326 |
| | 40 | | Root of a Power | $\sqrt[n]{a^m} = (\sqrt[n]{a})^m$ | |
| SPECIAL PRODUCTS AND FACTORING | 41 | Binomials | Difference of Two Squares | $a^2 - b^2 = (a - b)(a + b)$ | 63<br>186 |
| | 42 | | Sum of Two Cubes | $a^3 + b^3 = (a + b)(a^2 - ab + b^2)$ | 201 |
| | 43 | | Difference of Two Cubes | $a^3 - b^3 = (a - b)(a^2 + ab + b^2)$ | 201 |
| | 44 | Trinomials | Test for Factorability | $ax^2 + bx + c$ is factorable if $b^2 - 4ac$ is a perfect square | 189 |
| | 45 | | Leading Coefficient = 1 | $x^2 + (a + b)x + ab = (x + a)(x + b)$ | 63<br>190 |
| | 46 | | General Quadratic Trinomial | $acx^2 + (ad + bc)x + bd = (ax + b)(cx + d)$ | 63<br>196 |
| | 47 | | Perfect Square Trinomials | $a^2 + 2ab + b^2 = (a + b)^2$ | 64<br>199 |
| | 48 | | | $a^2 - 2ab + b^2 = (a - b)^2$ | 64<br>199 |
| | 49 | Factoring by Grouping | | $ac + ad + bc + bd = (a + b)(c + d)$ | |

| | No. | | | | Page |
|---|---|---|---|---|---|
| FRACTIONS | 50 | Simplifying | | $\frac{ad}{bd} = \frac{a}{b}$ | 207 |
| | 51 | Multiplication | | $\frac{a}{b} \cdot \frac{c}{d} = \frac{ac}{bd}$ | 210 |
| | 52 | Division | | $\frac{a}{b} \div \frac{c}{d} = \frac{a}{b} \cdot \frac{d}{c} = \frac{ad}{bc}$ | 211 |
| | 53 | Addition and Subtraction | Same Denominators | $\frac{a}{b} \pm \frac{c}{b} = \frac{a \pm c}{b}$ | 215 |
| | 54 | | Different Denominators | $\frac{a}{b} \pm \frac{c}{d} = \frac{ad}{bd} \pm \frac{bc}{bd} = \frac{ad \pm bc}{bd}$ | 216 |
| PROPORTION | 55 | In the Proportion $a:b = c:d$ | The product of the means equals the product of the extremes | $ad = bc$ | |
| | 56 | | The extremes may be interchanged | $d:b = c:a$ | |
| | 57 | | The means may be interchanged | $a:c = b:d$ | |
| | 58 | | The means may be interchanged with the extremes | $b:a = d:c$ | |
| | 59 | Mean Proportional | In the Proportion | $a:b = b:c$ | |
| | | | Geometric Mean | $b = \pm\sqrt{ac}$ | 478<br>622 |
| VARIATION | 60 | $k$ = Constant of Proportionality | Direct | $y \propto x$ or $y = kx$ | 480 |
| | 61 | | Inverse | $y \propto \frac{1}{x}$ or $y = \frac{k}{x}$ | 492 |
| | 62 | | Joint | $y \propto xw$ or $y = kxw$ | 497 |

| Topic | No. | Subtopic | | | Formula | Page |
|---|---|---|---|---|---|---|
| SYSTEMS OF LINEAR EQUATIONS | 63 | Algebraic solution | | $a_1x + b_1y = c_1$<br>$a_2x + b_2y = c_2$ | $x = \dfrac{b_2c_1 - b_1c_2}{a_1b_2 - a_2b_1}$, and $y = \dfrac{a_1c_2 - a_2c_1}{a_1b_2 - a_2b_1}$ where $a_1b_2 - a_2b_1 \neq 0$ | 252<br>308<br>313 |
| SYSTEMS OF LINEAR EQUATIONS | 64 | Algebraic solution | | $a_1x + b_1y + c_1z = k_1$<br>$a_2x + b_2y + c_2z = k_2$<br>$a_3x + b_3y + c_3z = k_3$ | $x = \dfrac{b_2c_3k_1 + b_1c_2k_3 + b_3c_1k_2 - b_2c_1k_3 - b_3c_2k_1 - b_1c_3k_2}{a_1b_2c_3 + a_3b_1c_2 + a_2b_3c_1 - a_3b_2c_1 - a_1b_3c_2 - a_2b_1c_3}$<br>$y = \dfrac{a_1c_3k_2 + a_3c_2k_1 + a_2c_1k_3 - a_3c_1k_2 - a_1c_2k_3 - a_2c_3k_1}{a_1b_2c_3 + a_3b_1c_2 + a_2b_3c_1 - a_3b_2c_1 - a_1b_3c_2 - a_2b_1c_3}$<br>$z = \dfrac{a_1b_2k_3 + a_3b_1k_2 + a_2b_3k_1 - a_3b_2k_1 - a_1b_3k_2 - a_2b_1k_3}{a_1b_2c_3 + a_3b_1c_2 + a_2b_3c_1 - a_3b_2c_1 - a_1b_3c_2 - a_2b_1c_3}$ | 274 |
| SYSTEMS OF LINEAR EQUATIONS | 65 | Determinants | Value of a Determinant | Second Order | $\begin{vmatrix} a_1 & b_1 \\ a_2 & b_2 \end{vmatrix} = a_1b_2 - a_2b_1$ | 268 |
| SYSTEMS OF LINEAR EQUATIONS | 66 | Determinants | Value of a Determinant | Third Order | $\begin{vmatrix} a_1 & b_1 & c_1 \\ a_2 & b_2 & c_2 \\ a_3 & b_3 & c_3 \end{vmatrix} = a_1b_2c_3 + a_3b_1c_2 + a_2b_3c_1 - (a_3b_2c_1 + a_1b_3c_2 + a_2b_1c_3)$ | 274 |
| SYSTEMS OF LINEAR EQUATIONS | 67 | Determinants | Value of a Determinant | Minors | The signed minor of element $b$ in the determinant $\begin{vmatrix} a & b & c \\ d & e & f \\ g & h & i \end{vmatrix}$ is $-\begin{vmatrix} d & f \\ g & i \end{vmatrix}$ | 272 |
| SYSTEMS OF LINEAR EQUATIONS | 68 | Determinants | Value of a Determinant | Minors | To find the value of a determinant:<br>1. Choose any row or any column to develop by minors.<br>2. Write the product of every element in that row or column and its signed minor.<br>3. Add these products to get the value of the determinant. | 273 |
| SYSTEMS OF LINEAR EQUATIONS | 69 | Determinants | Cramer's Rule | | The solution for any variable is a fraction whose denominator is the determinant of the coefficients, and whose numerator is also the determinant of the coefficients, except that the column of coefficients of the variable being solved for is replaced by the column of constants | 275 |
| SYSTEMS OF LINEAR EQUATIONS | 70 | Determinants | Cramer's Rule | Two Equations | $x = \dfrac{\begin{vmatrix} c_1 & b_1 \\ c_2 & b_2 \end{vmatrix}}{\begin{vmatrix} a_1 & b_1 \\ a_2 & b_2 \end{vmatrix}}$ and $y = \dfrac{\begin{vmatrix} a_1 & c_1 \\ a_2 & c_2 \end{vmatrix}}{\begin{vmatrix} a_1 & b_1 \\ a_2 & b_2 \end{vmatrix}}$ | 269 |
| SYSTEMS OF LINEAR EQUATIONS | 71 | Determinants | Cramer's Rule | Three Equations | $x = \dfrac{\begin{vmatrix} k_1 & b_1 & c_1 \\ k_2 & b_2 & c_2 \\ k_3 & b_3 & c_3 \end{vmatrix}}{\Delta}$, $y = \dfrac{\begin{vmatrix} a_1 & k_1 & c_1 \\ a_2 & k_2 & c_2 \\ a_3 & k_3 & c_3 \end{vmatrix}}{\Delta}$, $z = \dfrac{\begin{vmatrix} a_1 & b_1 & k_1 \\ a_2 & b_2 & k_2 \\ a_3 & b_3 & k_3 \end{vmatrix}}{\Delta}$<br>Where $\Delta = \begin{vmatrix} a_1 & b_1 & c_1 \\ a_2 & b_2 & c_2 \\ a_3 & b_3 & c_3 \end{vmatrix} \neq 0$ | 275 |

| Category | No. | Subcategory | Topic | Fact / Formula | Page |
|---|---|---|---|---|---|
| SYSTEMS OF LINEAR EQUATIONS (Continued) | 72 | Properties of Determinants | Zero Row or Column | If all elements in a row (or column) are zero, the value of the determinant is zero | 279 |
| | 73 | | Identical Rows or Columns | The value of a determinant is zero if two rows (or columns) are identical | 279 |
| | 74 | | Zeros below the Principal Diagonal | If all elements below the principal diagonal are zeros, then the value of the determinant is the product of the elements along the principal diagonal | 280 |
| | 75 | | Interchanging Rows with Columns | The value of a determinant is unchanged if we change the rows to columns and the columns to rows | 280 |
| | 76 | | Interchange of Rows or Columns | A determinant will change sign when we interchange two rows (or columns) | 280 |
| | 77 | | Multiplying by a Constant | If each element in a row (or column) is multiplied by some constant, the value of the determinant is multiplied by that constant | 281 |
| | 78 | | Multiples of One Row or Column Added to Another | The value of a determinant is unchanged when the elements of a row (or column) are multiplied by some factor, and then added to the corresponding elements of another row or column | 282 |
| MATRICES | 79 | Addition | Commutative Law | $\mathbf{A} + \mathbf{B} = \mathbf{B} + \mathbf{A}$ | 292 |
| | 80 | | Associative Law | $\mathbf{A} + (\mathbf{B} + \mathbf{C}) = (\mathbf{A} + \mathbf{B}) + \mathbf{C} = (\mathbf{A} + \mathbf{C}) + \mathbf{B}$ | 292 |
| | 81 | | Addition and Subtraction | $\begin{pmatrix} a & b \\ c & d \end{pmatrix} + \begin{pmatrix} w & x \\ y & z \end{pmatrix} = \begin{pmatrix} a + w & b + x \\ c + y & d + z \end{pmatrix}$ | 292 |
| | 82 | Multiplication | Conformable Matrices | The product **AB** of two matrices **A** and **B** is defined only when the number of columns in **A** equals the number of rows in **B** | 293 |
| | 83 | | Commutative Law | $\mathbf{AB} \neq \mathbf{BA}$ Matrix multiplication is *not* commutative | 298 |
| | 84 | | Associative Law | $\mathbf{A}(\mathbf{BC}) = (\mathbf{AB})\mathbf{C} = \mathbf{ABC}$ | 293 |
| | 85 | | Distributive Law | $\mathbf{A}(\mathbf{B} + \mathbf{C}) = \mathbf{AB} + \mathbf{AC}$ | 293 |
| | 86 | | Dimensions of the Product | $(m \times p)(p \times n) = (m \times n)$ | 294 |
| | 87 | | Product of a Scalar and a Matrix | $k\begin{pmatrix} a & b \\ c & d \end{pmatrix} = \begin{pmatrix} ka & kb \\ kc & kd \end{pmatrix}$ | 293 |
| | 88 | | Scalar Product of a Row Vector and a Column Vector | $\underset{(1 \times 2)}{(a \quad b)} \underset{(2 \times 1)}{\begin{pmatrix} x \\ y \end{pmatrix}} = \underset{(1 \times 1)}{(ax + by)}$ | 294 |
| | 89 | | Tensor Product of a Column Vector and a Row Vector | $\underset{(2 \times 1)}{\begin{pmatrix} a \\ b \end{pmatrix}} \underset{(1 \times 2)}{(x \quad y)} = \underset{(2 \times 2)}{\begin{pmatrix} ax & ay \\ bx & by \end{pmatrix}}$ | 299 |
| | 90 | | Product of a Row Vector and a Matrix | $\underset{(1 \times 2)}{(a \quad b)} \underset{(2 \times 3)}{\begin{pmatrix} u & v & w \\ x & y & z \end{pmatrix}} = \underset{(1 \times 3)}{(au + bx \quad av + by \quad aw + bz)}$ | 295 |

| Category | No. | Subheading | Topic | Fact or Formula | Page |
|---|---|---|---|---|---|
| MATRICES (Continued) | 91 | Matrix Multiplication (Continued) | Product of a Matrix and a Column Vector | (2 X 2) (2 X 1) (2 X 1)<br>$\begin{pmatrix} a & b \\ c & d \end{pmatrix} \begin{pmatrix} x \\ y \end{pmatrix} = \begin{pmatrix} ax + by \\ cx + dy \end{pmatrix}$ | 298 |
| MATRICES (Continued) | 92 | Matrix Multiplication (Continued) | Product of Two Matrices | (2 X 3) (3 X 2) (2 X 2)<br>$\begin{pmatrix} a & b & c \\ d & e & f \end{pmatrix} \begin{pmatrix} u & x \\ v & y \\ w & z \end{pmatrix} = \begin{pmatrix} au + bv + cw & ax + by + cz \\ du + ev + fw & dx + ey + fz \end{pmatrix}$ | 297 |
| MATRICES (Continued) | 93 | Matrix Multiplication (Continued) | Product of a Matrix and Its Inverse | $\mathbf{AA^{-1} = A^{-1}A = I}$ | 308 |
| MATRICES (Continued) | 94 | Matrix Multiplication (Continued) | Multiplying by the Unit Matrix | $\mathbf{AI = IA = A}$ | 303 |
| MATRICES (Continued) | 95 | Solving Systems of Equations | Matrix Form for a System of Equations | $\mathbf{AX = B}$ | 304<br>306 |
| MATRICES (Continued) | 96 | Solving Systems of Equations | Elementary Transformations of a Matrix | 1. Interchange any rows<br>2. Multiply a row by a nonzero constant<br>3. Add a constant multiple of one row to another row | 304 |
| MATRICES (Continued) | 97 | Solving Systems of Equations | Unit Matrix Method | When we transform the coefficient matrix **A** into the unit matrix **I**, the column vector **B** gets transformed into the solution set | 307 |
| MATRICES (Continued) | 98 | Solving Systems of Equations | Solving a Set of Equations Using the Inverse | $\mathbf{X = A^{-1}B}$ | 311 |
| QUADRATICS | 99 | | General Form | $ax^2 + bx + c = 0$ | 340 |
| QUADRATICS | 100 | | Quadratic Formula | $x = \dfrac{-b \pm \sqrt{b^2 - 4ac}}{2a}$ | 349 |
| QUADRATICS | 101 | | The Discriminant | If $a$, $b$, and $c$ are real, and<br>if $b^2 - 4ac > 0$ the roots are real and unequal<br>if $b^2 - 4ac = 0$ the roots are real and equal<br>if $b^2 - 4ac < 0$ the roots are not real | 351 |
| QUADRATICS | 102 | | Polynomial of Degree $n$ | $a_0x^n + a_1x^{n-1} + \cdots + a_{n-1}x + a_n$ | 363<br>485 |
| QUADRATICS | 103 | | Factor Theorem | If a polynomial equation $f(x) = 0$ has a root $r$, then $(x - r)$ is a factor of the polynomial $f(x)$; conversely, if $(x - r)$ is a factor of a polynomial $f(x)$, then $r$ is a root of $f(x) = 0$ | 363 |

| | No. | | | | Page |
|---|---|---|---|---|---|
| INTERSECTING LINES | 104 | Opposite angles of two intersecting straight lines are equal | | | 131 |
| | 105 | If two parallel straight lines are cut by a transversal, corresponding angles are equal and alternate interior angles are equal | | | 132 |
| | 106 | If two lines are cut by a number of parallels, the corresponding segments are proportional | | | 132 |
| QUADRILATERALS | 107 | 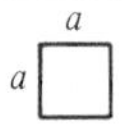 | Square | Area = $a^2$ | 144 |
| | 108 | 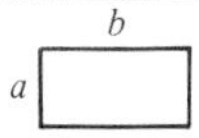 | Rectangle | Area = $ab$ | 144 |
| | 109 | 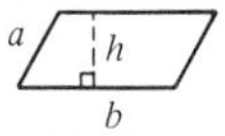 | Parallelogram: Diagonals bisect each other | Area = $bh$ | 144 |
| | 110 | 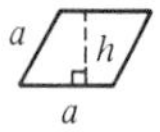 | Rhombus: Diagonals intersect at right angles | Area = $ah$ | 144 |
| | 111 | 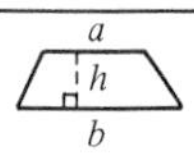 | Trapezoid | Area = $\frac{(a + b)h}{2}$ | 144 |
| POLYGON | 112 | 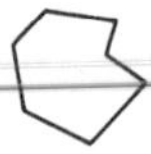 $n$ sides | Sum of Angles = $(n - 2)$ 180° | | 135 |
| CIRCLES | 113 | 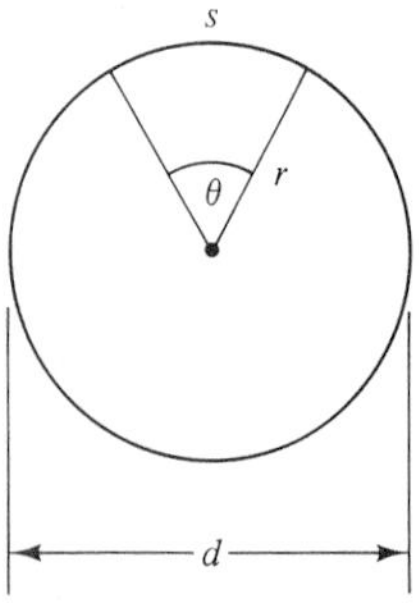 | Circumference = $2\pi r = \pi d$ | | 145 |
| | 114 | | Area = $\pi r^2 = \frac{\pi d^2}{4}$ | | 25<br>145 |
| | 115 | | Central angle $\theta$ (radians) = $\frac{s}{r}$ | | 404 |
| | 116 | | Area of sector = $\frac{rs}{2} = \frac{r^2\theta}{2}$ | | 403 |
| | 117 | | 1 revolution = $2\pi$ radians = 360° 1° = 60 min 1 min = 60 sec | | 156<br>399 |
| | 118 | | Any angle inscribed in a semicircle is a right angle | | 147 |
| | 119 | 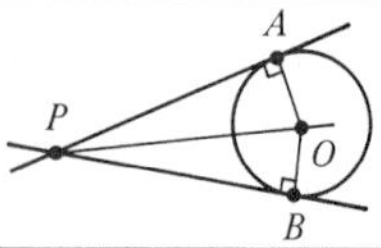 | Tangents to a Circle | Tangent $AP$ is perpendicular to radius $OA$ | 146 |
| | 120 | | | Tangent $AP$ = tangent $BP$<br>$OP$ bisects angle $APB$ | 147 |
| | 121 | 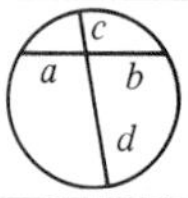 | Intersecting Chords | $ab = cd$ | 146 |

| | No. | | | | Page |
|---|---|---|---|---|---|
| SOLIDS | 122 | 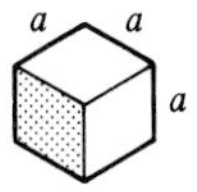 | Cube | Volume = $a^3$ | 150 |
| | 123 | | | Surface area = $6a^2$ | 150 |
| | 124 | 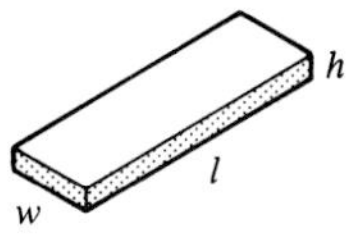 | Rectangular Parallel-epiped | Volume = $lwh$ | 150 |
| | 125 | | | Surface area = $2(lw + hw + lh)$ | 150 |
| | 126 | 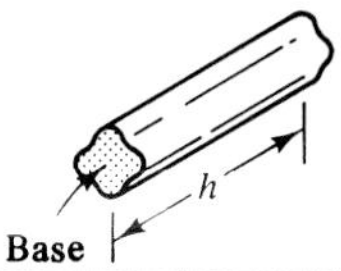 | Any Cylinder or Prism | Volume = (area of base)(altitude) | 150 |
| | 127 | | Right Cylinder or Prism | Lateral area (not incl. bases) = (perimeter of base)(altitude) | 150 |
| | 128 | 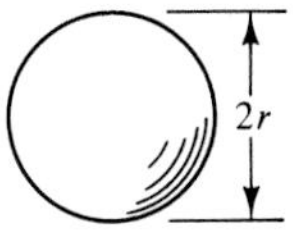 | Sphere | Volume = $\frac{4}{3}\pi r^3$ | 150 |
| | 129 | | | Surface area = $4\pi r^2$ | 150 |
| | 130 | 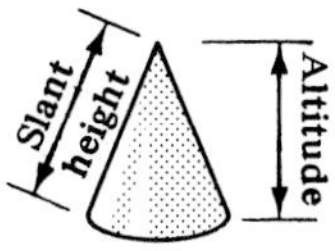 | Any Cone or Pyramid | Volume = $\frac{1}{3}$(area of base)(altitude) | 150 |
| | 131 | | Right Circular Cone or Regular Pyramid | Lateral area = $\frac{1}{2}$(perimeter of base) × (slant height) | 150 |
| | 132 | 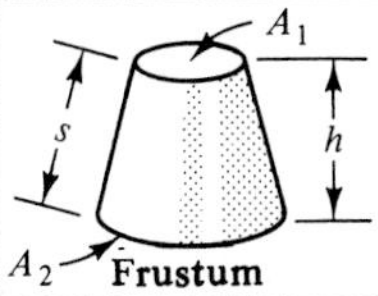 | Any Cone or Pyramid | Volume = $\frac{h}{3}(A_1 + A_2 + \sqrt{A_1A_2})$ | 150 |
| | 133 | | Right Circular Cone or Regular Pyramid | Lateral area = $\frac{s}{2}$(sum of base perimeters) = $\frac{s}{2}(P_1 + P_2)$ | 150 |
| SIMILAR FIGURES | 134 | 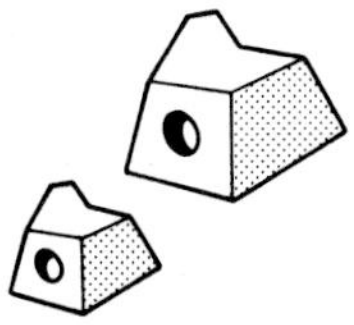 | Corresponding dimensions of plane or solid similar figures are in proportion | | 488 |
| | 135 | | Areas of similar plane or solid figures are proportional to the squares of any two corresponding dimensions | | 489 |
| | 136 | | Volumes of similar solid figures are proportional to the cubes of any two corresponding dimensions | | 489 |

| | No. | | | | Page |
|---|---|---|---|---|---|
| ANY TRIANGLE | 137 | Areas 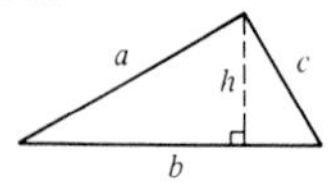 | | Area $= \frac{1}{2}bh$ | 136 |
| | 138 | | Hero's Formula: | Area $= \sqrt{s(s-a)(s-b)(s-c)}$ where $s = \frac{1}{2}(a+b+c)$ | 137 |
| | 139 | 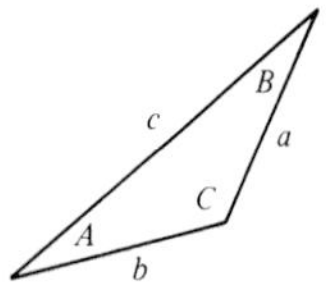 | Sum of the Angles | $A + B + C = 180°$ | 137 |
| | 140 | | Law of Sines | $\frac{a}{\sin A} = \frac{b}{\sin B} = \frac{c}{\sin C}$ | 380 |
| | 141 | | Law of Cosines | $a^2 = b^2 + c^2 - 2bc \cos A$<br>$b^2 = a^2 + c^2 - 2ac \cos B$<br>$c^2 = a^2 + b^2 - 2ab \cos C$ | 385 |
| | 142 |  | Exterior Angle | $\theta = A + B$ | 138 |
| SIMILAR TRIANGLES | 143 | If two angles of a triangle equal two angles of another triangle, the triangles are similar | | | 138 |
| | 144 | Corresponding sides of similar triangles are in proportion | | | 138 |
| RIGHT TRIANGLES | 145 | 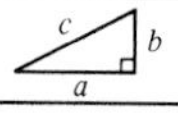 | Pythagorean Theorem | $a^2 + b^2 = c^2$ | 139<br>166 |
| | 146 | Trigonometric Ratios 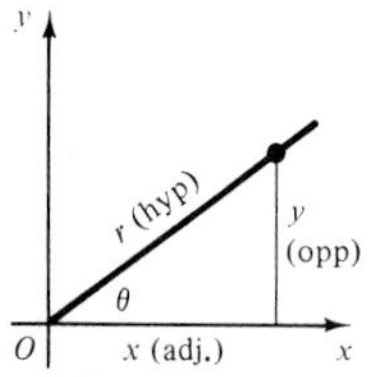 | Sine | $\sin\theta = \frac{y}{r} = \frac{\text{opposite side}}{\text{hypotenuse}}$ | 158<br>166 |
| | 147 | | Cosine | $\cos\theta = \frac{x}{r} = \frac{\text{adjacent side}}{\text{hypotenuse}}$ | 158<br>166 |
| | 148 | | Tangent | $\tan\theta = \frac{y}{x} = \frac{\text{opposite side}}{\text{adjacent side}}$ | 158<br>166 |
| | 149 | | Cotangent | $\cot\theta = \frac{x}{y} = \frac{\text{adjacent side}}{\text{opposite side}}$ | 158 |
| | 150 | | Secant | $\sec\theta = \frac{r}{x} = \frac{\text{hypotenuse}}{\text{adjacent side}}$ | 158 |
| | 151 | | Cosecant | $\csc\theta = \frac{r}{y} = \frac{\text{hypotenuse}}{\text{opposite side}}$ | 158 |
| | 152 | Reciprocal Relations | (a) $\sin\theta = \frac{1}{\csc\theta}$ | (b) $\cos\theta = \frac{1}{\sec\theta}$ (c) $\tan\theta = \frac{1}{\cot\theta}$ | 159<br>374<br>448 |
| | 153 |  | In a right triangle, the altitude to the hypotenuse forms two right triangles which are similar to each other and to the original triangle | | 137 |
| | 154 | $A$ and $B$ Are Complementary Angles 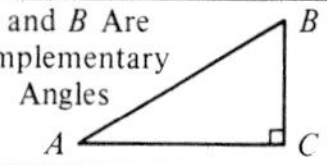 | Cofunctions | (a) $\sin A = \cos B$<br>(b) $\cos A = \sin B$<br>(c) $\tan A = \cot B$<br>(d) $\cot A = \tan B$<br>(e) $\sec A = \csc B$<br>(f) $\csc A = \sec B$ | 170 |
| CONGRUENT TRIANGLES | 155 | Two Triangles Are Congruent If | Two angles and a side of one are equal to two angles and the corresponding side of the other (ASA), (AAS) | | 138 |
| | 156 | | Two sides and the included angle of one are equal, respectively to two sides and the included angle of the other (SAS) | | 138 |
| | 157 | | Three sides of one are equal to the three sides of the other (SSS) | | 138 |
| COORDINATE SYSTEMS | 158 | 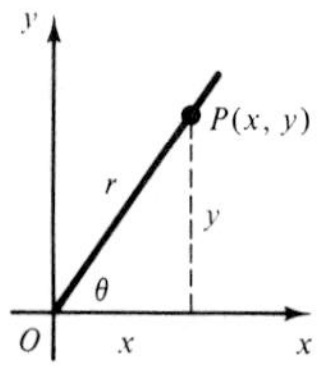 | Rectangular | $x = r\cos\theta$ | 441 |
| | 159 | | | $y = r\sin\theta$ | 441 |
| | 160 | | Polar | $r = \sqrt{x^2 + y^2}$ | 441 |
| | 161 | | | $\theta = \arctan \frac{y}{x}$ | 441 |

| | No. | | | Page |
|---|---|---|---|---|
| TRIGONOMETRIC IDENTITIES | 162 | Quotient Relations | $\tan\theta = \dfrac{\sin\theta}{\cos\theta}$ | 449 |
| | 163 | | $\cot\theta = \dfrac{\cos\theta}{\sin\theta}$ | 449 |
| | 164 | Pythagorean Relations | $\sin^2\theta + \cos^2\theta = 1$ | 449 |
| | 165 | | $1 + \tan^2\theta = \sec^2\theta$ | 449 |
| | 166 | | $1 + \cot^2\theta = \csc^2\theta$ | 450 |
| | 167 | Sum or Difference of Two Angles | $\sin(\alpha \pm \beta) = \sin\alpha\cos\beta \pm \cos\alpha\sin\beta$ | 455 |
| | 168 | | $\cos(\alpha \pm \beta) = \cos\alpha\cos\beta \mp \sin\alpha\sin\beta$ | 456 |
| | 169 | | $\tan(\alpha \pm \beta) = \dfrac{\tan\alpha \pm \tan\beta}{1 \mp \tan\alpha\tan\beta}$ | 457 |
| | 170 | Double-Angle Relations | $\sin 2\alpha = 2\sin\alpha\cos\alpha$ | 460 |
| | 171 | | (a) $\cos 2\alpha = \cos^2\alpha - \sin^2\alpha$ (b) $\cos 2\alpha = 1 - 2\sin^2\alpha$ (c) $\cos 2\alpha = 2\cos^2\alpha - 1$ | 460 461 |
| | 172 | | $\tan 2\alpha = \dfrac{2\tan\alpha}{1 - \tan^2\alpha}$ | 462 |
| | 173 | Half-Angle Relations | $\sin\dfrac{\alpha}{2} = \pm\sqrt{\dfrac{1 - \cos\alpha}{2}}$ | 463 |
| | 174 | | $\cos\dfrac{\alpha}{2} = \mp\sqrt{\dfrac{1 + \cos\alpha}{2}}$ | 463 |
| | 175 | | (a) $\tan\dfrac{\alpha}{2} = \dfrac{1 - \cos\alpha}{\sin\alpha}$ (b) $\tan\dfrac{\alpha}{2} = \dfrac{\sin\alpha}{1 + \cos\alpha}$ (c) $\tan\dfrac{\alpha}{2} = \pm\sqrt{\dfrac{1 - \cos\alpha}{1 + \cos\alpha}}$ | 464 |
| | 176 | Sum or Difference of Two Functions | $\sin\alpha + \sin\beta = 2\sin\frac{1}{2}(\alpha + \beta)\cos\frac{1}{2}(\alpha - \beta)$ | |
| | 177 | | $\sin\alpha - \sin\beta = 2\cos\frac{1}{2}(\alpha + \beta)\sin\frac{1}{2}(\alpha - \beta)$ | |
| | 178 | | $\cos\alpha + \cos\beta = 2\cos\frac{1}{2}(\alpha + \beta)\cos\frac{1}{2}(\alpha - \beta)$ | |
| | 179 | | $\cos\alpha - \cos\beta = 2\sin\frac{1}{2}(\alpha + \beta)\sin\frac{1}{2}(\alpha - \beta)$ | |
| | 180 | Product of Two Functions | $\sin\alpha\sin\beta = \frac{1}{2}\cos(\alpha - \beta) - \frac{1}{2}\cos(\alpha + \beta)$ | |
| | 181 | | $\cos\alpha\cos\beta = \frac{1}{2}\cos(\alpha - \beta) + \frac{1}{2}\cos(\alpha + \beta)$ | |
| | 182 | | $\sin\alpha\cos\beta = \frac{1}{2}\sin(\alpha + \beta) + \frac{1}{2}\sin(\alpha - \beta)$ | |
| | 183 | Inverse Trigonometric Functions | $\theta = \text{arcsine } C = \arctan\dfrac{C}{\sqrt{1 - C^2}}$ | 164 |
| | 184 | | $\theta = \arccos D = \arctan\dfrac{\sqrt{1 - D^2}}{D}$ | 165 |

| | No. | | | | Page |
|---|---|---|---|---|---|
| LOGARITHMS | 185 | Exponential to Logarithmic Form | | If $b^x = y$ then $x = \log_b y$ $(y > 0, b > 0, b \neq 1)$ | 516 |
| | 186 | Laws of Logarithms | Products | $\log_b MN = \log_b M + \log_b N$ | 521 |
| | 187 | | Quotients | $\log_b \frac{M}{N} = \log_b M - \log_b N$ | 522 |
| | 188 | | Powers | $\log_b M^P = p \log_b M$ | 523 |
| | 189 | | Roots | $\log_b \sqrt[q]{M} = \frac{1}{q} \log_b M$ | 525 |
| | 190 | | Log of 1 | $\log_b 1 = 0$ | 526 |
| | 191 | | Log of the Base | $\log_b b = 1$ | 526 |
| | 192 | Log of the Base Raised to a Power | | $\log_b b^n = n$ | 526 |
| | 193 | Change of Base | | $\log N = \frac{\ln N}{\ln 10} \cong \frac{\ln N}{2.3026}$ | 528 |

SOME USEFUL FUNCTIONS

| No. | | | | Page |
|---|---|---|---|---|
| 194 | Power Function | 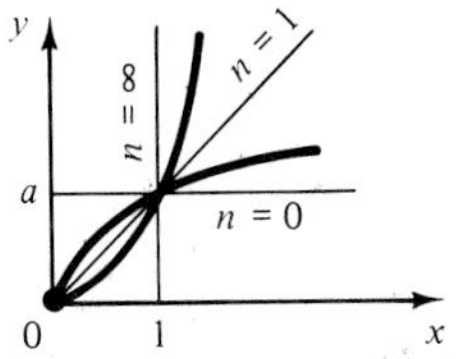 | $y = ax^n$ | 484<br>541 |
| 195 | Exponential Function | 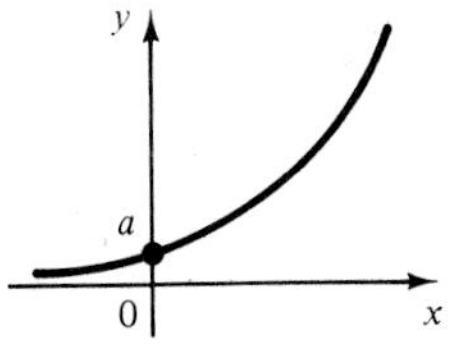 | $y = a(b)^{nx}$ | 543 |
| 196 | Series Approximation | $b^x = 1 + x \ln b + \frac{(x \ln b)^2}{2!} + \frac{(x \ln b)^3}{3!} + \cdots \quad (b > 0)$ | | 520 |
| 197 | Exponential Growth | 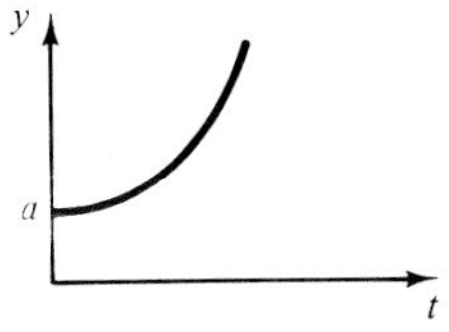 | $y = ae^{nt}$ | 510<br>532<br>623 |
| 198 | | | Doubling Time or Half-Life: $t = \frac{\ln 2}{n}$ | 532<br>533 |
| 199 | Exponential Decay |  | $y = ae^{-nt}$ | 511<br>533 |
| 200 | Exponential Growth to an Upper Limit | 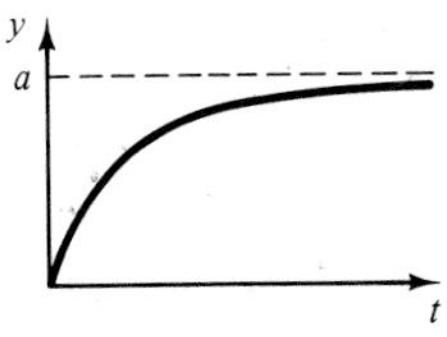 | $y = a(1 - e^{-nt})$ | 512<br>1001 |
| 201 | Time Constant | $T = \frac{1}{\text{growth rate } n}$ | | 512 |
| 202 | Series Approximations | $e = 2 + \frac{1}{2!} + \frac{1}{3!} + \frac{1}{4!} + \cdots$ | | 515 |
| 203 | | $e^x = 1 + x + \frac{x^2}{2!} + \frac{x^3}{3!} + \frac{x^4}{4!} + \cdots$ | | 515 |

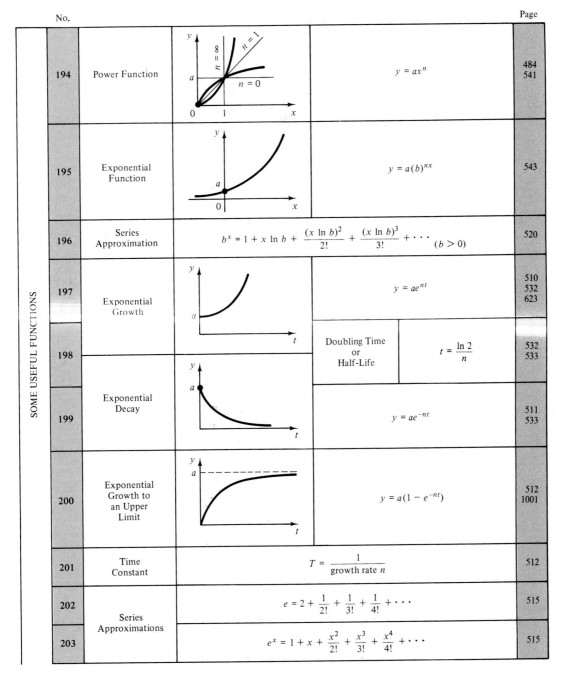

SOME USEFUL FUNCTIONS (Continued)

| No. | | | | Page |
|---|---|---|---|---|
| 204 | Logarithmic Function | 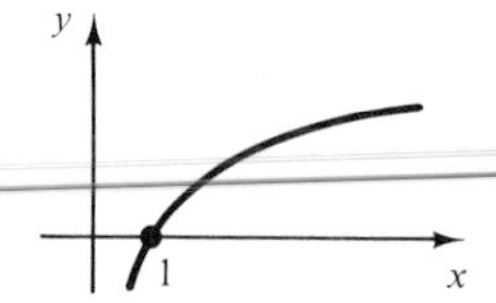 | $y = \log_b x$<br>$(x > 0,\ b > 0,\ b \neq 1)$ | 518 |
| 205 | Series Approximation | $\ln x = 2a + \frac{2a^3}{3} + \frac{2a^5}{5} + \frac{2a^7}{7} + \cdots$<br>where<br>$a = \frac{x-1}{x+1}$ | | 520 |
| 206 | Sine Wave of Amplitude $a$ | 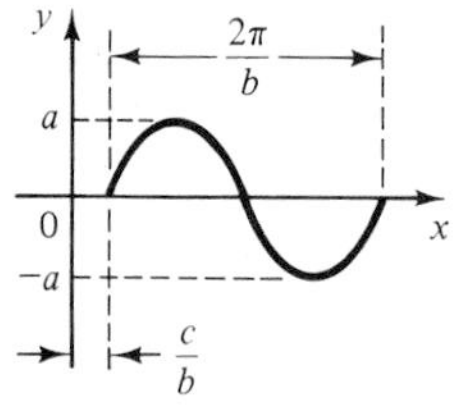 | $y = a \sin (bx + c)$ | 413 |
| 207 | | | Period = $\frac{360}{b}$ deg/cycle = $\frac{2\pi}{b}$ rad/cycle | 416 |
| 208 | | | Frequency = $\frac{b}{360}$ cycle/deg = $\frac{b}{2\pi}$ cycle/rad | 416 |
| 209 | | | Phase shift = $-\frac{c}{b}$ | 418 |
| 210 | Series Approximations | $\sin x = x - \frac{x^3}{3!} + \frac{x^5}{5!} - \frac{x^7}{7!} + \cdots$ | | 166<br>1069<br>1089 |
| 211 | | $\cos x = 1 - \frac{x^2}{2!} + \frac{x^4}{4!} - \frac{x^6}{6!} + \cdots$ | | 166<br>1089 |

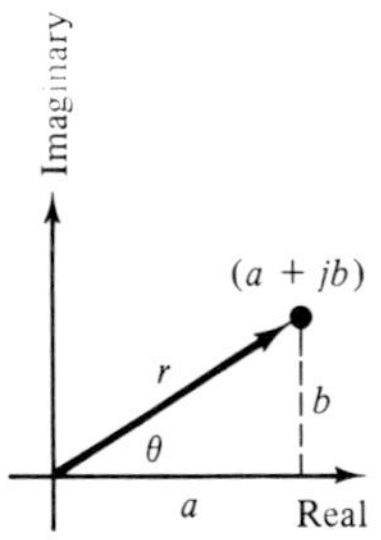


| | No. | | | | | Page |
|---|---|---|---|---|---|---|
| COMPLEX NUMBERS | 212 | Powers of $j$ | | | $j=\sqrt{-1}$, $j^2=-1$, $j^3=-j$, $j^4=1$, $j^5=j$, etc. | 551 |
| | 213 | | Rectangular Form | | $a+jb$ | 551 |
| | 214 | | | Sums | $(a+jb)+(c+jd)=(a+c)+j(b+d)$ | 551 |
| | 215 | | | Differences | $(a+jb)-(c+jd)=(a-c)+j(b-d)$ | 551 |
| | 216 | | | Products | $(a+jb)(c+jd)=(ac-bd)+j(ad+bc)$ | 552 |
| | 217 | | | Quotients | $\dfrac{a+jb}{c+jd}=\dfrac{ac+bd}{c^2+d^2}+j\,\dfrac{bc-ad}{c^2+d^2}$ | 554 |
| | 218 | | Trigonometric Form | | $a+jb=r(\cos\theta+j\sin\theta)$ | 558 |
| | 219 | | | where | $a=r\cos\theta$ | 557 |
| | 220 | | | | $b=r\sin\theta$ | 557 |
| | 221 | | | | $r=\sqrt{a^2+b^2}$ | 557 |
| | 222 | | | | $\theta=\arctan\dfrac{b}{a}$ | 557 |
| | 223 | | Polar Form | | $r\underline{/\theta}=a+jb$ | 557 |
| | 224 | | | Products | $r\underline{/\theta}\cdot r'\underline{/\theta'}=rr'\underline{/\theta+\theta'}$ | 559 |
| | 225 | | | Quotients | $r\underline{/\theta}\div r'\underline{/\theta'}=\dfrac{r}{r'}\underline{/\theta-\theta'}$ | 560 |
| | 226 | | | Roots and Powers | DeMoivre's Theorem: $(r\underline{/\theta})^n=r^n\underline{/n\theta}$ | 560 |
| | 227 | | Exponential Form | Euler's Formula | $re^{j\theta}=r(\cos\theta+j\sin\theta)$ | 563 1013 1079 |
| | 228 | | | Products | $r_1e^{j\theta_1}\cdot r_2e^{j\theta_2}=r_1r_2e^{j(\theta_1+\theta_2)}$ | 564 |
| | 229 | | | Quotients | $\dfrac{r_1e^{j\theta_1}}{r_2e^{j\theta_2}}=\dfrac{r_1}{r_2}\,e^{j(\theta_1-\theta_2)}$ | 565 |
| | 230 | | | Powers and Roots | $(re^{j\theta})^n=r^ne^{jn\theta}$ | 565 |

|  | No. |  |  |  | Page |
|---|---|---|---|---|---|
| PROGRESSIONS | 231 | Arithmetic Progression<br>Common Difference = $d$ | Recursion Formula | $a_n = a_{n-1} + d$ | 613 |
|  | 232 |  | General Term | $a_n = a + (n-1)d$ | 614 |
|  | 233 |  | Sum of $n$ Terms | $s_n = \dfrac{n(a + a_n)}{2}$ | 615 |
|  | 234 |  |  | $s_n = \dfrac{n[2a + (n-1)d]}{2}$ | 615 |
|  | 235 | Geometric Progression<br>Common Ratio = $r$ | Recursion Formula | $a_n = ra_{n-1}$ | 620 |
|  | 236 |  | General Term | $a_n = ar^{n-1}$ | 620 |
|  | 237 |  | Sum of $n$ Terms | $s_n = \dfrac{a(1 - r^n)}{1 - r}$ | 621 |
|  | 238 |  |  | $s_n = \dfrac{a - ra_n}{1 - r}$ | 621 |
|  | 239 |  | Sum to Infinity | $S = \dfrac{a}{1 - r}$ where $\lvert r \rvert < 1$ | 625 |
| BINOMIAL THEOREM | 240 | Binomial Expansion |  | $(a + b)^n = a^n + na^{n-1}b + \dfrac{n(n-1)}{2!}a^{n-2}b^2 + \dfrac{n(n-1)(n-2)}{3!}a^{n-3}b^3 + \cdots + b^n$ | 628 |
|  | 241 | General Term |  | $r\text{th term} = \dfrac{n!}{(r-1)!\,(n-r+1)!}a^{n-r+1}b^{r-1}$ | 631 |
|  | 242 | Binomial Series |  | $(a + b)^n = a^n + na^{n-1}b + \dfrac{n(n-1)}{2!}a^{n-2}b^2 + \dfrac{n(n-1)(n-2)}{3!}a^{n-3}b^3 + \cdots$<br>where $\lvert a \rvert > \lvert b \rvert$ | 634 |
|  | 243 |  |  | $(1 + x)^n = 1 + nx + \dfrac{n(n-1)}{2!}x^2 + \dfrac{n(n-1)(n-2)}{3!}x^3 + \cdots$<br>where $\lvert x \rvert < 1$ | 634 |
| STATISTICS AND PROBABILITY | 244 | Measures of Central Tendency | Arithmetic Mean | $\bar{x} = \dfrac{\Sigma x}{n}$ | 649 |
|  | 245 |  | Median | The median of a set of numbers arranged in order of magnitude is the middle value, or the mean of the two middle values. | 649 |
|  | 246 |  | Mode | The mode of a set of numbers is the measurement(s) that occur most often in the set. | 650 |
|  | 247 | Measures of Dispersion | Range | The range of a set of numbers is the difference between the largest and the smallest number in the set. | 650 |
|  | 248 |  | Variance $s^2$ | $s^2 = \dfrac{\Sigma (x - \bar{x})^2}{n}$ | 652 |
|  | 249 |  | Standard Deviation $s$ | The standard deviation of a set of numbers is the positive square root of the variance. | 653 |
|  | 250 | Probability | Of a Single Event | $P(A) = \dfrac{\text{number of ways in which } A \text{ can happen}}{\text{number of equally likely ways}}$ | 655 |
|  | 251 |  | Of Two Events Both Occurring | $P(A, B) = P(A)P(B)$ | 656 |
|  | 252 |  | Of Several Events All Occurring | $P(A, B, C, \ldots) = P(A)P(B)P(C) \cdots$ | 656 |
|  | 253 |  | Of Either of Two Events Occurring | $P(A + B) = P(A) + P(B) - P(A, B)$ | 656 |
|  | 254 |  | Of Two Mutually Exclusive Events Occurring | $P(A + B) = P(A) + P(B)$ | 657 |
|  | 255 | Gaussian Distribution | 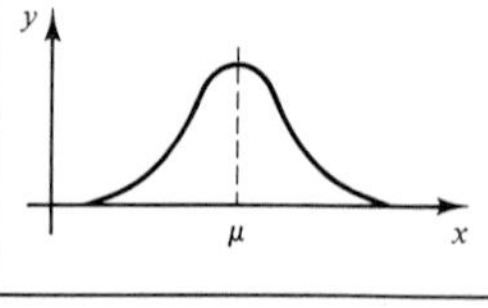 | $y = \dfrac{1}{\sigma\sqrt{2\pi}}e^{-(x-\mu)^2/2\sigma^2}$ | 660 |
|  | 256 | Standard Error | Of the Mean | $SE_{\bar{x}} = \dfrac{\sigma}{\sqrt{n}} = \dfrac{s}{\sqrt{n}}$ | 668 |
|  | 257 |  | Of the Standard Deviation | $SE_s = \dfrac{\sigma}{\sqrt{2n}}$ | 669 |

| BOOLEAN ALGEBRA AND SETS | No. | Boolean Operators | | Truth Table | Venn Diagram | Switch Diagram | Logic Gate | Page |
|---|---|---|---|---|---|---|---|---|
| | 258 | | AND | $A$ $B$ $A \cdot B$<br>0 0 0<br>0 1 0<br>1 0 0<br>1 1 1 | 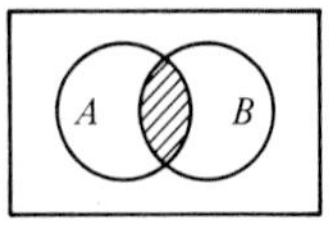<br>Intersection<br>$A \cap B = \{x : x \in A, x \in B\}$ | 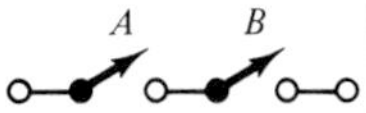<br>In Series | 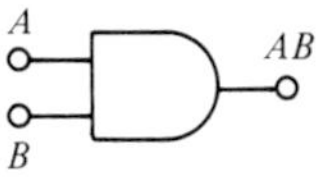<br>AND Gate | |
| | 259 | | OR | $A$ $B$ $A + B$<br>0 0 0<br>0 1 1<br>1 0 1<br>1 1 1 | 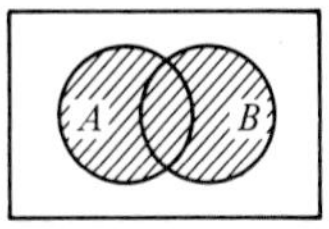<br>Union<br>$A \cup B = \{x : x \in A$ or $x \in B\}$ | 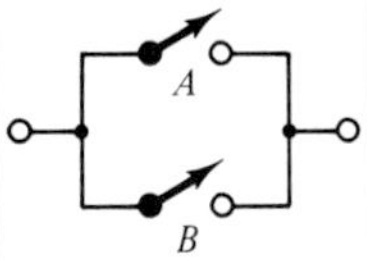<br>In Parallel | 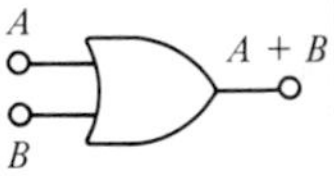<br>OR Gate | |
| | 260 | | NOT | $A$ $\overline{A}$<br>0 1<br>1 0 | 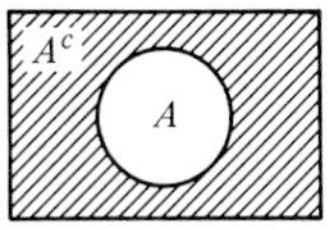<br>Complement<br>$A^c = \{x : x \in U, x \notin A\}$ | 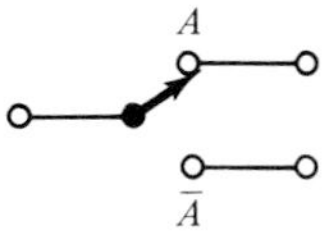 | 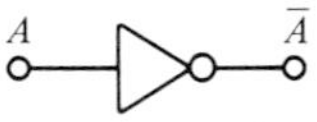<br>Inverter | |
| | 261 | | Exclusive OR | $A$ $B$ $A \oplus B$<br>0 0 0<br>0 1 1<br>1 0 1<br>1 1 0 | 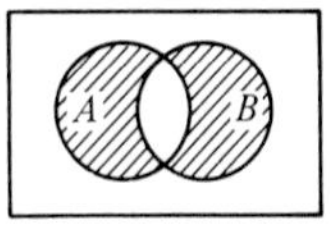<br>$A \oplus B = \{x : x \in A$ or $x \in B, x \notin A \cap B\}$ | 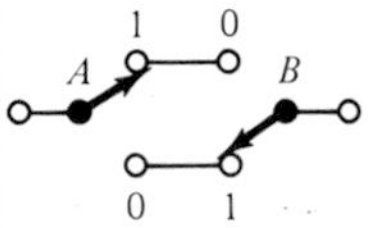 | 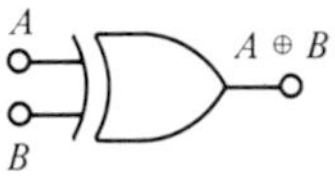<br>Exclusive OR Gate | |

| No. | | | | AND | OR | Page |
|---|---|---|---|---|---|---|
| | BOOLEAN ALGEBRA (Continued) | Laws of Boolean Algebra | | | | |
| 262 | | | Commutative Laws | (a) $AB \equiv BA$ | (b) $A + B \equiv B + A$ | |
| 263 | | | Boundedness Laws | (a) $A \cdot 0 \equiv 0$ | (b) $A + 1 \equiv 1$ | |
| 264 | | | Identity Laws | (a) $A \cdot 1 \equiv A$ | (b) $A + 0 \equiv A$ | |
| 265 | | | Idempotent Laws | (a) $AA \equiv A$ | (b) $A + A \equiv A$ | |
| 266 | | | Complement Laws | (a) $A \cdot \overline{A} \equiv 0$ | (b) $A + \overline{A} \equiv 1$ | |
| 267 | | | Associative Laws | (a) $A(BC) \equiv (AB)C$ | (b) $A + (B + C) \equiv (A + B) + C$ | |
| 268 | | | Distributive Laws | (a) $A(B + C) \equiv AB + AC$ | (b) $A + BC \equiv (A + B)(A + C)$ | |
| 269 | | | | (a) $A(\overline{A} + B) \equiv AB$ | (b) $A + \overline{A} \cdot B \equiv A + B$ | |
| 270 | | | Absorption Laws | (a) $A(A + B) \equiv A$ | (b) $A + (AB) \equiv A$ | |
| 271 | | | DeMorgan's Laws | (a) $\overline{A \cdot B} \equiv \overline{A} + \overline{B}$ | (b) $\overline{A + B} \equiv \overline{A} \cdot \overline{B}$ | |
| 272 | | | Involution Law | $\overline{\overline{A}} \equiv A$ | | |

| Section | No. | Figure | Topic | Subtopic | Formula | Page |
|---|---|---|---|---|---|---|
| THE STRAIGHT LINE | 273 | | Distance Formula | | $d = \sqrt{(\Delta x)^2 + (\Delta y)^2} = \sqrt{(x_2 - x_1)^2 + (y_2 - y_1)^2}$ | 676 |
| | 274 | | Slope $m$ | | $m = \frac{\text{rise}}{\text{run}} = \frac{\Delta y}{\Delta x} = \frac{y_2 - y_1}{x_2 - x_1}$ | 678 |
| | 275 | | | | $m = \tan$ (angle of inclination) $= \tan \theta$ <br> $0 \leq \theta < 180°$ | 679 |
| | 276 | 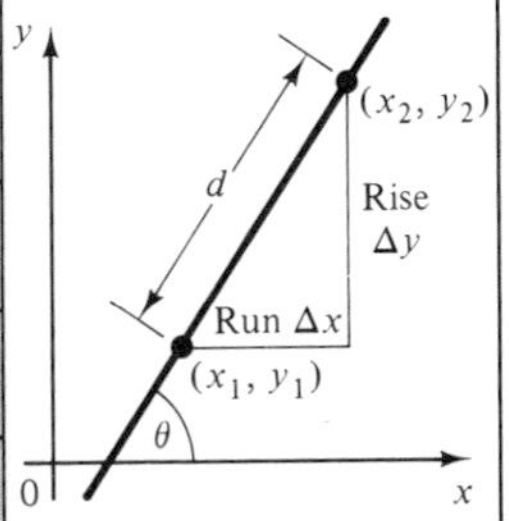  | Equation of Straight Line | General Form | $Ax + By + C = 0$ | 686 |
| | 277 | | | Parallel to $x$ axis | $y = b$ | 688 |
| | 278 | | | Parallel to $y$ axis | $x = a$ | 688 |
| | 279 | | | Slope-Intercept Form | $y = mx + b$ | 685 |
| | 280 | | | Two-Point Form | $\frac{y - y_1}{x - x_1} = \frac{y_2 - y_1}{x_2 - x_1}$ | 687 |
| | 281 | | | Point-Slope Form | $m = \frac{y - y_1}{x - x_1}$ | 686 |
| | 282 | | | Intercept Form | $\frac{x}{a} + \frac{y}{b} = 1$ | |
| | 283 | | | Polar Form | $r \cos(\theta - \beta) = p$ | |
| | 284 | 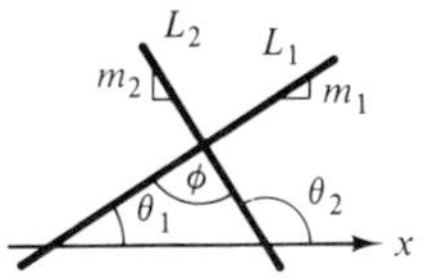  | If $L_1$ and $L_2$ are parallel, then | | $m_1 = m_2$ | 680 |
| | 285 | | If $L_1$ and $L_2$ are perpendicular, then | | $m_1 = -\frac{1}{m_2}$ | 681 |
| | 286 | | Angle of Intersection | | $\tan \phi = \frac{m_2 - m_1}{1 + m_1 m_2}$ | 682 |
| CONIC SECTIONS | 287 | Any Conic | General Second-Degree Equation | | $Ax^2 + Bxy + Cy^2 + Dx + Ey + F = 0$ | 697 |
| | 288 | | Translation of Axes | | To translate or shift the axes of a curve to the left by a distance $h$ and downward by a distance $k$, replace $x$ by $(x - h)$ and $y$ by $(y - k)$ in the equation of the curve. | 697 |
| | 289 | | Eccentricity 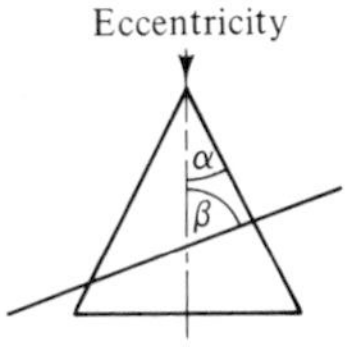  | | $e = \frac{\cos \beta}{\cos \alpha}$ | |
| | 290 | | | | $e = 0$ for the Circle <br> $0 < e < 1$ for the Ellipse <br> $e = 1$ for the Parabola <br> $e > 1$ for the Hyperbola | |
| | 291 | | Definition of a Conic | | $PF = e \cdot PD$ | |
| | 292 | | Polar Equation for the Conics | | $r = \frac{ke}{1 - e \cos \theta}$ | |

CONIC SECTIONS (Continued)

| No. | | | | | Page |
|---|---|---|---|---|---|
| 293 | Circle of Radius $r$ | The set of points in a plane equidistant from a fixed point. | | | 695 |
| 294 | Circle of Radius $r$ | 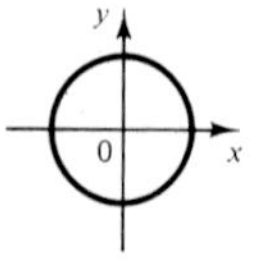  | Standard Form | $x^2 + y^2 = r^2$ | 696 |
| 295 | Circle of Radius $r$ | 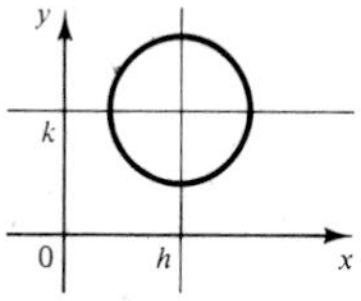  | Standard Form | $(x - h)^2 + (y - k)^2 = r^2$ | 696 |
| 296 | Circle of Radius $r$ | General Form | | $x^2 + y^2 + Dx + Ey + F = 0$ | 697 |
| 297 | Parabola | The set of points in a plane such that the distance $PF$ from each point to a fixed point (the focus) is equal to the distance $PD$ to a fixed line (the directrix). $PF = PD$ | | | 702 |
| 298 | Parabola | 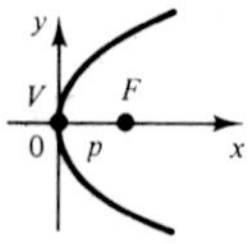  | Standard Form | $y^2 = 4px$ | 703 |
| 299 | Parabola | 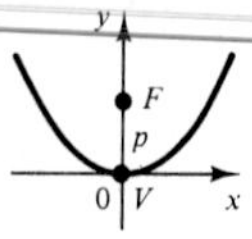  | Standard Form | $x^2 = 4py$ | 703 |
| 300 | Parabola | 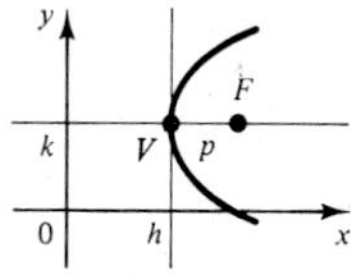  | Standard Form | $(y - k)^2 = 4p(x - h)$ | 704 |
| 301 | Parabola | 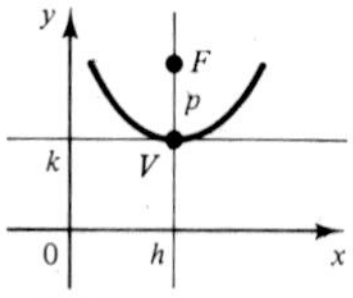  | Standard Form | $(x - h)^2 = 4p(y - k)$ | 704 |
| 302 | Parabola | General Form | | $Cy^2 + Dx + Ey + F = 0$ or $Ax^2 + Dx + Ey + F = 0$ | 705 |
| 303 | Parabola | Focal Width | | $L = \| 4p \|$ | 704 |
| 304 | Parabola | Area 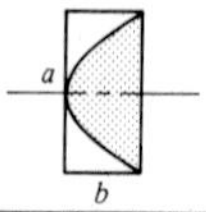  | | Area $= \frac{2}{3}ab$ | |

| No. | | | Page |
|---|---|---|---|
| 305 | The set of points in a plane such that the sum of the distances $PF$ and $PF'$ from each point to two fixed points (the foci) is constant, and equal to the length of the major axis. $PF + PF' = 2a$ | | 710 |
| 306 | 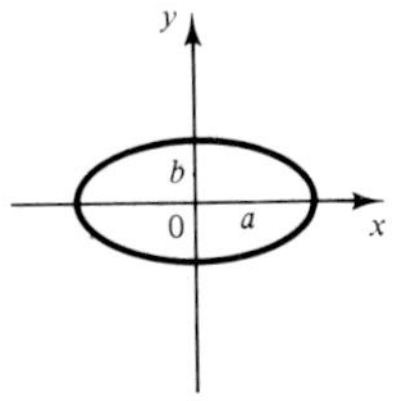 | $\dfrac{x^2}{a^2} + \dfrac{y^2}{b^2} = 1$ $a > b$ | 713 |
| 307 | 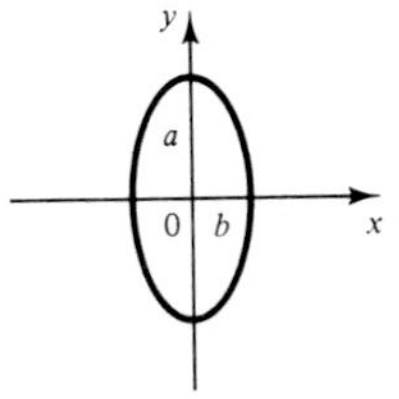 | $\dfrac{y^2}{a^2} + \dfrac{x^2}{b^2} = 1$ $a > b$ | 714 |
| 308 | 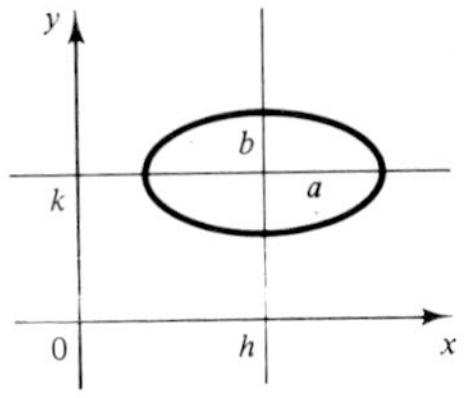 | $\dfrac{(x-h)^2}{a^2} + \dfrac{(y-k)^2}{b^2} = 1$ $a > b$ | 715 |
| 309 | 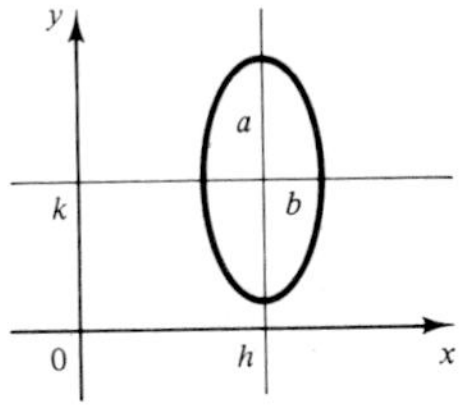 | $\dfrac{(y-k)^2}{a^2} + \dfrac{(x-h)^2}{b^2} = 1$ $a > b$ | 715 |
| 310 | General Form | $Ax^2 + Cy^2 + Dx + Ey + F = 0$ $A \neq C$, but have same signs | 716 |
| 311 | Distance from Center to Focus | $c = \sqrt{a^2 - b^2}$ | 711 |
| 312 | Focal Width | $L = \dfrac{2b^2}{a}$ | 718 |
| 313 | 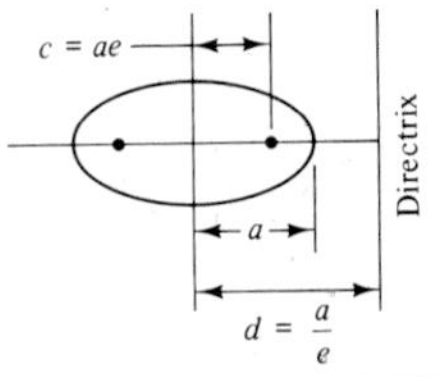  | Eccentricity $e = \dfrac{a}{d} = \dfrac{c}{a}$ | |
| 314 | Area $= \pi ab$ | | |

CONIC SECTIONS (Continued) — Hyperbola

| No. | | | | Page |
|---|---|---|---|---|
| 315 | The set of points in a plane such that the difference of the distances $PF$ and $PF'$ from each point to two fixed points (the foci) is constant, and equal to the distance between the vertices. $PF' - PF = 2a$ | | | 720 |
| 316 | 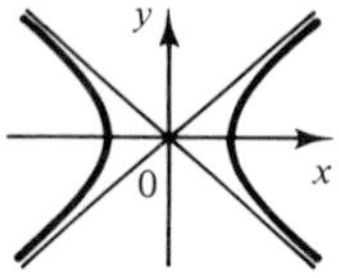 | Standard Form | $\dfrac{x^2}{a^2} - \dfrac{y^2}{b^2} = 1$ | 722 |
| 317 | 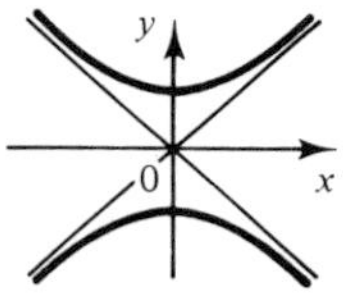 | Standard Form | $\dfrac{y^2}{a^2} - \dfrac{x^2}{b^2} = 1$ | 722 |
| 318 | 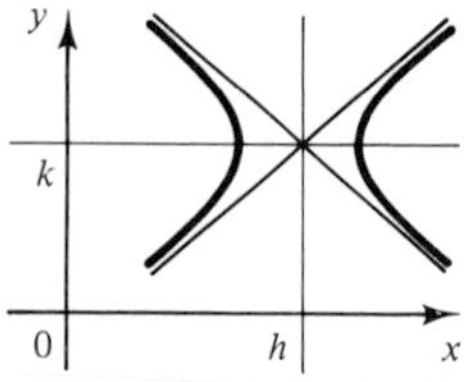 | Standard Form | $\dfrac{(x-h)^2}{a^2} - \dfrac{(y-k)^2}{b^2} = 1$ | 726 |
| 319 | 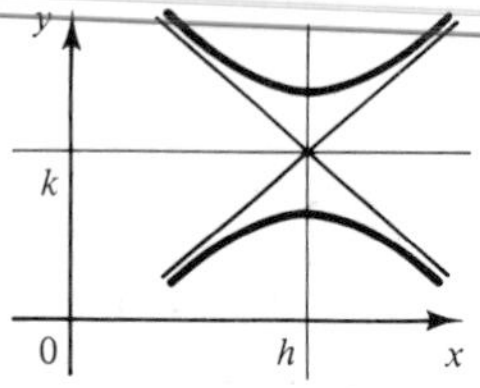 | Standard Form | $\dfrac{(y-k)^2}{a^2} - \dfrac{(x-h)^2}{b^2} = 1$ | 726 |
| 320 | General Form | | $Ax^2 + Cy^2 + Dx + Ey + F = 0$ <br> $A \neq C$, and have opposite signs | 727 |
| 321 | Distance from Center to Focus | | $c = \sqrt{a^2 + b^2}$ | 723 |
| 322 | Slope of Asymptotes | Transverse Axis Horizontal | Slope $= \pm \dfrac{b}{a}$ | 723 |
| 323 | Slope of Asymptotes | Transverse Axis Vertical | Slope $= \pm \dfrac{a}{b}$ | 723 |
| 324 | Length of Latus Rectum | | $L = \dfrac{2b^2}{a}$ | |
| 325 | 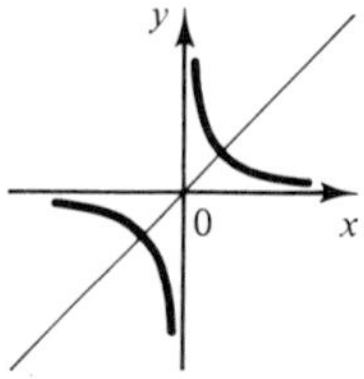 | | Axes Rotated 45°: $xy = k$ | 728 |

| | No. | | | | Page |
|---|---|---|---|---|---|
| DIFFERENTIAL CALCULUS | 326 | | Limit Notation | $\lim_{x \to a} f(x) = L$ | 625 |
| DIFFERENTIAL CALCULUS | 327 | | Increments | $\Delta x = x_2 - x_1, \quad \Delta y = y_2 - y_1$ | 675 |
| DIFFERENTIAL CALCULUS | 328 | | Definition of the Derivative | $\frac{dy}{dx} = \lim_{\Delta x \to 0} \frac{\Delta y}{\Delta x} = \lim_{\Delta x \to 0} \frac{f(x + \Delta x) - f(x)}{\Delta x}$ <br> $= \lim_{\Delta x \to 0} \frac{(y + \Delta y) - y}{\Delta x}$ | 741 742 753 |
| DIFFERENTIAL CALCULUS | 329 | | The Chain Rule | $\frac{dy}{dx} = \frac{dy}{du} \cdot \frac{du}{dx}$ | 754 |
| DIFFERENTIAL CALCULUS | 330 | Rules for Derivatives | Of a Constant | $\frac{d(c)}{dx} = 0$ | |
| DIFFERENTIAL CALCULUS | 331 | Rules for Derivatives | Of a Power Function | $\frac{d}{dx} x^n = nx^{n-1}$ | 765 |
| DIFFERENTIAL CALCULUS | 332 | Rules for Derivatives | Of a Constant Times a Function | $\frac{d(cu)}{dx} = c \frac{du}{dx}$ | 759 |
| DIFFERENTIAL CALCULUS | 333 | Rules for Derivatives | Of a Constant Times a Power of $x$ | $\frac{d}{dx} cx^n = cnx^{n-1}$ | 748 |
| DIFFERENTIAL CALCULUS | 334 | Rules for Derivatives | Of a Sum | $\frac{d}{dx}(u + v + w) = \frac{du}{dx} + \frac{dv}{dx} + \frac{dw}{dx}$ | 750 |
| DIFFERENTIAL CALCULUS | 335 | Rules for Derivatives | Of a Function Raised to a Power | $\frac{d(cu^n)}{dx} = cnu^{n-1} \frac{du}{dx}$ | 754 762 |
| DIFFERENTIAL CALCULUS | 336 | Rules for Derivatives | Of a Product | $\frac{d(uv)}{dx} = u \frac{dv}{dx} + v \frac{du}{dx}$ | 757 |
| DIFFERENTIAL CALCULUS | 337 | Rules for Derivatives | Of a Product of Three Factors | $\frac{d(uvw)}{dx} = uv \frac{dw}{dx} + uw \frac{dv}{dx} + vw \frac{du}{dx}$ | 758 |
| DIFFERENTIAL CALCULUS | 338 | Rules for Derivatives | Of a Product of $n$ Factors | The derivative is an expression of $n$ terms, each term being the product of $n - 1$ of the factors and the derivative of the other factor | 759 |
| DIFFERENTIAL CALCULUS | 339 | Rules for Derivatives | Of a Quotient | $\frac{d}{dx}\left(\frac{u}{v}\right) = \frac{v \frac{du}{dx} - u \frac{dv}{dx}}{v^2}$ | 760 |

| No. | | | | | Page |
|---|---|---|---|---|---|
| 340 | DIFFERENTIAL CALCULUS (Continued) — Rules for Derivatives (cont.) | Of the Trigonometric Functions | $\frac{d(\sin u)}{dx} = \cos u \frac{du}{dx}$ | | 830 834 |
| 341 | | | $\frac{d(\cos u)}{dx} = -\sin u \frac{du}{dx}$ | | 831 834 |
| 342 | | | $\frac{d(\tan u)}{dx} = \sec^2 u \frac{du}{dx}$ | | 834 |
| 343 | | | $\frac{d(\cot u)}{dx} = -\csc^2 u \frac{du}{dx}$ | | 834 |
| 344 | | | $\frac{d(\sec u)}{dx} = \sec u \tan u \frac{du}{dx}$ | | 834 |
| 345 | | | $\frac{d(\csc u)}{dx} = -\csc u \cot u \frac{du}{dx}$ | | 834 |
| 346 | | Of the Inverse Trigonometric Functions | $\frac{d(\mathrm{Sin}^{-1} u)}{dx} = \frac{1}{\sqrt{1-u^2}} \frac{du}{dx}$ | $-1 < u < 1$ | 837 838 |
| 347 | | | $\frac{d(\mathrm{Cos}^{-1} u)}{dx} = \frac{-1}{\sqrt{1-u^2}} \frac{du}{dx}$ | $-1 < u < 1$ | 838 |
| 348 | | | $\frac{d(\mathrm{Tan}^{-1} u)}{dx} = \frac{1}{1+u^2} \frac{du}{dx}$ | | 838 |
| 349 | | | $\frac{d(\mathrm{Cot}^{-1} u)}{dx} = \frac{-1}{1+u^2} \frac{du}{dx}$ | | 838 |
| 350 | | | $\frac{d(\mathrm{Sec}^{-1} u)}{dx} = \frac{1}{u\sqrt{u^2-1}} \frac{du}{dx}$ | $\lvert u \rvert > 1$ | 838 |
| 351 | | | $\frac{d(\mathrm{Csc}^{-1} u)}{dx} = \frac{-1}{u\sqrt{u^2-1}} \frac{du}{dx}$ | $\lvert u \rvert > 1$ | 838 |
| 352 | | Of Logarithmic and Exponential Functions | (a) $\frac{d}{dx}(\log_b u) = \frac{1}{u} \log_b e \frac{du}{dx}$ | (b) $\frac{d}{dx}(\log_b u) = \frac{1}{u \ln b} \frac{du}{dx}$ | 840 |
| 353 | | | $\frac{d}{dx}(\ln u) = \frac{1}{u} \frac{du}{dx}$ | | 841 865 |
| 354 | | | $\frac{d}{dx} b^u = b^u \frac{du}{dx} \ln b$ | | 845 |
| 355 | | | $\frac{d}{dx} e^u = e^u \frac{du}{dx}$ | | 846 |
| 356 | Graphical Applications | Maximum and Minimum Points | To find maximum and minimum points (and other stationary points) set the first derivative equal to zero and solve for $x$. | | 774 |
| 357 | | First-Derivative Test | The first derivative is negative to the left of, and positive to the right of, a minimum point. The reverse is true for a maximum point. | | 775 |
| 358 | | Second-Derivative Test | If the first derivative at some point is zero, then, if the second derivative is<br>1. Positive, the point is a minimum<br>2. Negative, the point is a maximum<br>3. Zero, the test fails | | 776 |
| 359 | | Inflection Points | To find points of inflection, set the second derivative to zero and solve for $x$. Test by seeing if the second derivative changes sign a small distance to either side of the point. | | 779 |
| 360 | | Newton's Method | $x_{n+1} = x_n - \frac{f(x_n)}{f'(x_n)}$ | | 781 |
| 361 | Differentials | Differential of $y$ | $dy = f'(x)\,dx$ | | 821 |
| 362 | | Approximations Using Differentials | $\Delta y \cong \frac{dy}{dx} \Delta x$ | | 823 |

| | No. | | | | | Page |
|---|---|---|---|---|---|---|
| INTEGRAL CALCULUS | 363 | The Indefinite Integral | | | $\int F'(x)\,dx = F(x) + C$ | 853 |
| | 364 | | | | $\int f(x)\,dx = F(x) + C \qquad$ where $f(x) = F'(x)$ | 853 |
| | 365 | The Definite Integral | The Fundamental Theorem 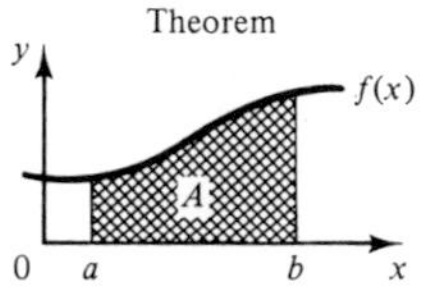  | | $A = \int_a^b f(x)\,dx = F(b) - F(a)$ | 884<br>891 |
| | 366 | | Defined by Riemann Sums | | $A = \lim_{\Delta x \to 0} \sum_{i=1}^{n} f(x_i^*)\,\Delta x = \int_a^b f(x)\,dx$ | 888 |
| | 367 | | Properties of the Definite Integral | | $\int_a^b c\,f(x)\,dx = c\int_a^b f(x)\,dx$ | |
| | 368 | | | | $\int_a^b [f(x) + g(x)]\,dx = \int_a^b f(x)\,dx + \int_a^b g(x)\,dx$ | |
| | 369 | | | | $\int_a^b f(x)\,dx = -\int_b^a f(x)\,dx$ | |
| | 370 | | | | $\int_a^b f(x)\,dx = \int_a^c f(x)\,dx + \int_c^b f(x)\,dx$ | |
| | 371 | Approximate Integration | 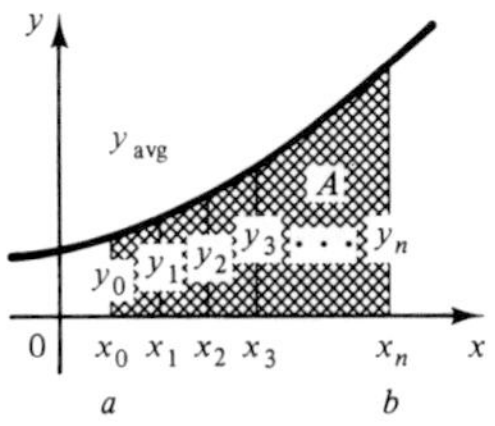  | Midpoint Method | $A \cong \sum_{i=1}^{n} f(x_i^*)\,\Delta x$<br>where $f(x_i^*)$ is the height of the $i$th panel at its midpoint | 888 |
| | 372 | | | Average Ordinate Method | $A \cong y_{\text{avg}}\,(b - a)$ | 970 |
| | 373 | | | Trapezoid Rule | With unequal spacing<br>$A \cong \frac{1}{2}[(x_1 - x_0)(y_1 + y_0) + (x_2 - x_1)(y_2 - y_1) + \cdots + (x_n - x_{n-1})(y_n + y_{n-1})]$ | 970 |
| | 374 | | | | With equal spacing, $h = x_1 - x_0$:<br>$A \cong h[\frac{1}{2}(y_0 + y_n) + y_1 + y_2 + \cdots + y_{n-1}]$ | 971 |
| | 375 | | | Prismoidal Formula | $A = \frac{h}{3}(y_0 + 4y_1 + y_2)$ | 972 |
| | 376 | | | Simpson's Rule | $A \cong \frac{h}{3}(y_0 + 4y_1 + 2y_2 + 4y_3 + \cdots + 4y_{n-1} + y_n)$ | 973 |

| | No. | | | | | Page |
|---|---|---|---|---|---|---|
| APPLICATIONS OF THE DEFINITE INTEGRAL | 377 | Volumes of Solids of Revolution | Disk Method | 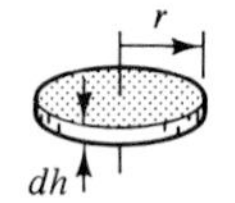 | Volume $= dV = \pi r^2\, dh$ | |
| | 378 | | | 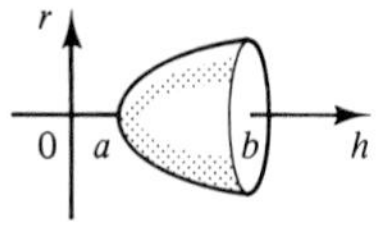 | $V = \pi \int_a^b r^2\, dh$ | 907 |
| | 379 | | Ring Method | 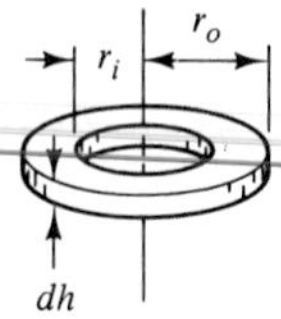 | $dV = \pi(r_o^2 - r_i^2)\, dh$ | |
| | 380 | | | 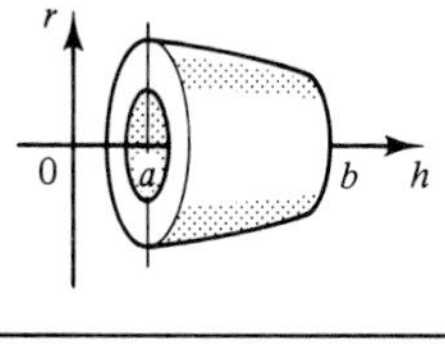 | $V = \pi \int_a^b (r_o^2 - r_i^2)\, dh$ | 911 |
| | 381 | | Shell Method | 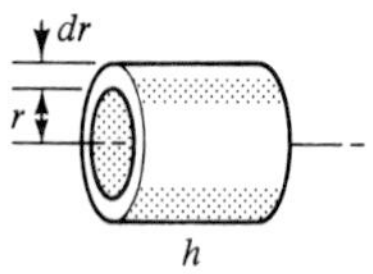 | $dV = 2\pi rh\, dr$ | 943 |
| | 382 | | | 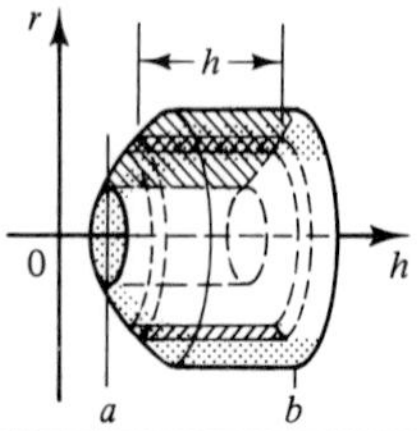 | $V = 2\pi \int_a^b rh\, dr$ | 909 |

APPLICATIONS OF THE DEFINITE INTEGRAL

| No. | | | | Page |
|---|---|---|---|---|
| 383 | Length of Arc | 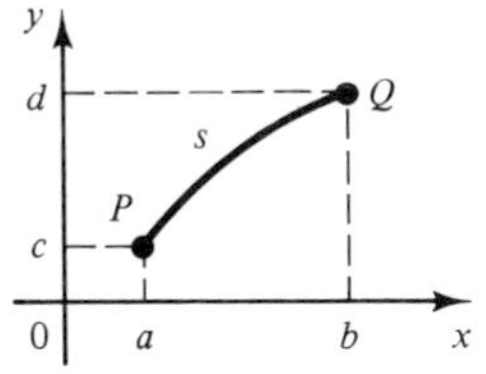  | $s = \int_a^b \sqrt{1 + \left(\frac{dy}{dx}\right)^2}\, dx$ | 916 |
| 384 | Length of Arc | | $s = \int_c^d \sqrt{1 + \left(\frac{dx}{dy}\right)^2}\, dy$ | 917 |
| 385 | Surface Area | 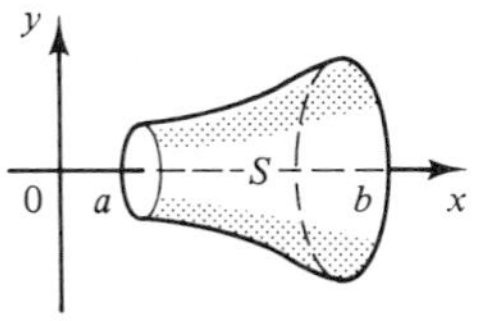  | About $x$ Axis: $S = 2\pi \int_a^b y \sqrt{1 + \left(\frac{dy}{dx}\right)^2}\, dx$ | 919 |
| 386 | Surface Area | 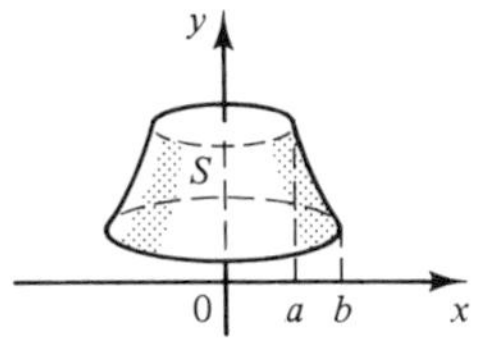  | About $y$ Axis: $S = 2\pi \int_a^b x \sqrt{1 + \left(\frac{dy}{dx}\right)^2}\, dx$ | 920 |
| 387 | Centroids | Of Plane Area: 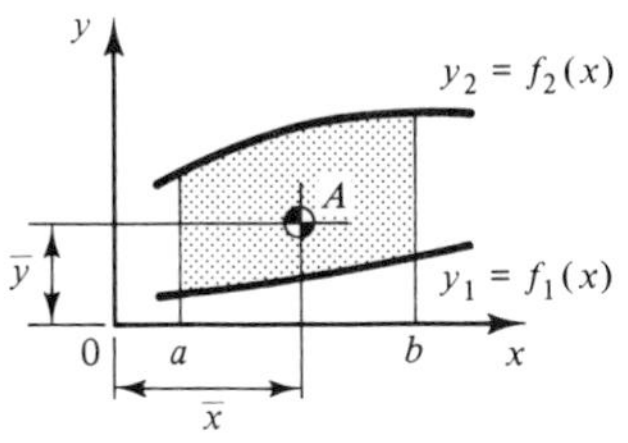  | $\bar{x} = \frac{1}{A} \int_a^b x(y_2 - y_1)\, dx$ | 926 |
| 388 | Centroids | | $\bar{y} = \frac{1}{2A} \int_a^b (y_1 + y_2)(y_2 - y_1)\, dx$ | 926 |
| 389 | Centroids | Of Solid of Revolution of Volume $V$: 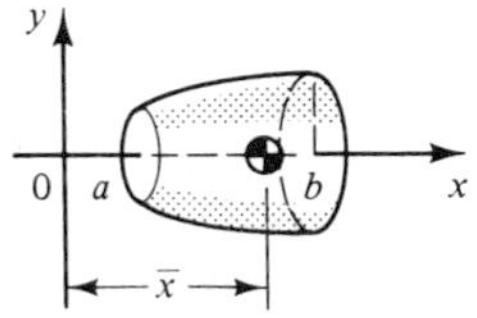  | About $x$ Axis: $\bar{x} = \frac{\pi}{V} \int_a^b xy^2\, dx$ | 929 |
| 390 | Centroids | 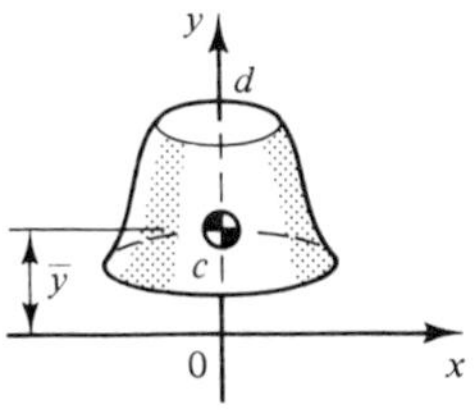  | About $y$ Axis: $\bar{y} = \frac{\pi}{V} \int_c^d yx^2\, dy$ | 930 |

## APPLICATIONS OF THE DEFINITE INTEGRAL (Continued)

| No. | | | | | Page |
|---|---|---|---|---|---|
| 391 | Moment of Inertia | Of Areas | Thin Strip: 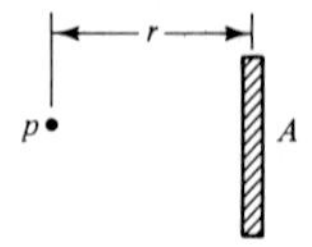  | $I_p = Ar^2$ | 940 |
| 392 | | | Extended Area: 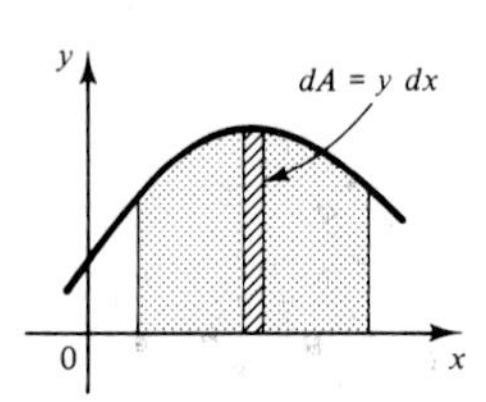  | $I_x = \frac{1}{3}\int y^3\,dx$ | 942 |
| 393 | | | | $I_y = \int x^2 y\,dx$ | 941 |
| 394 | | | | Polar $I_0 = I_x + I_y$ | |
| 395 | | | | Radius of Gyration: $r = \sqrt{\frac{I}{A}}$ | 941 |
| 396 | | | 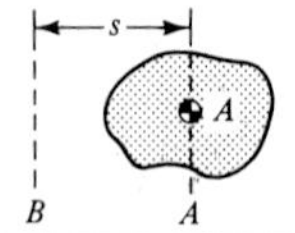  | Parallel Axis Theorem: $I_B = I_A + As^2$ | |
| 397 | | Of Masses | 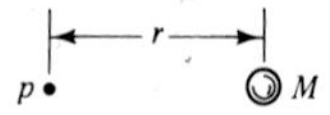  | $I_p = Mr^2$ | |
| 398 | | | Disk: 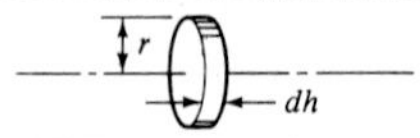  | About Axis of Revolution (Polar Moment of Inertia): $dI = \frac{m\pi}{2} r^4\,dh$ | |
| 399 | | | Ring: 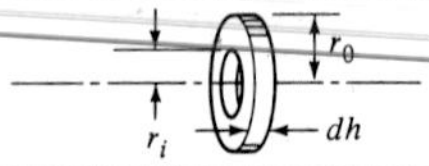  | $dI = \frac{m\pi}{2}(r_0^4 - r_i^4)\,dh$ | |
| 400 | | | Shell: 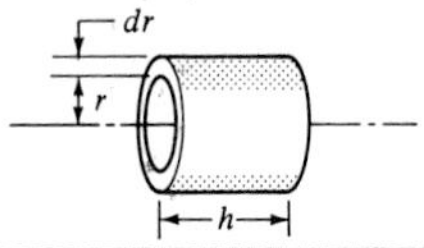  | $dI = 2\pi m r^3 h\,dr$ | 943 |
| 401 | | | Solid of Revolution: 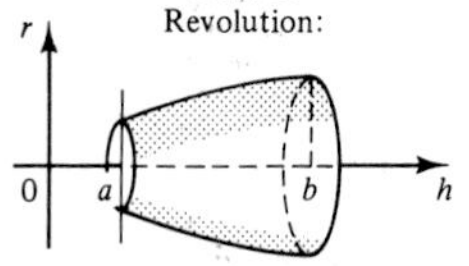  | Disk Method: $I = \frac{m\pi}{2}\int_a^b r^4\,dh$ | 944 |
| 402 | | | | Shell Method: $I = 2\pi m \int r^3 h\,dr$ | 943 |
| 403 | | | 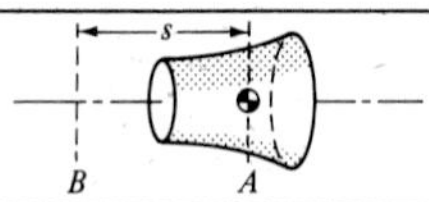  | Parallel Axis Theorem: $I_B = I_A + Ms^2$ | |
| 404 | Average and rms Values | | 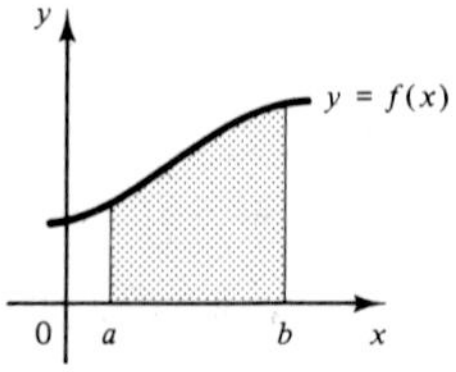  | Average Ordinate: $y_{avg} = \frac{1}{b-a}\int_a^b f(x)\,dx$ | 922<br>970 |
| 405 | | | | Root-Mean-Square Value: $\text{rms} = \sqrt{\frac{1}{b-a}\int_a^b [f(x)]^2\,dx}$ | 923 |

| No. | | | | | | Page |
|---|---|---|---|---|---|---|
| **DIFFERENTIAL EQUATIONS** | | | | | | |
| 406 | First-Order | Variables Separable | | | $f(y)\,dy = g(x)\,dx$ | 981 |
| 407 | | Integrable Combinations | | | $x\,dy + y\,dx = d(xy)$ | 986 |
| 408 | | | | | $\dfrac{x\,dy - y\,dx}{x^2} = d\left(\dfrac{y}{x}\right)$ | 986 |
| 409 | | | | | $\dfrac{y\,dx - x\,dy}{y^2} = d\left(\dfrac{x}{y}\right)$ | 986 |
| 410 | | | | | $\dfrac{x\,dy - y\,dx}{x^2 + y^2} = d\left(\tan^{-1}\dfrac{y}{x}\right)$ | 986 |
| 411 | | Homogeneous | | | $M\,dx + N\,dy = 0$ (Substitute $y = vx$) | 988 |
| 412 | | First-Order Linear | Form | | $y' + Py = Q$ | 991 |
| 413 | | | Integrating Factor | | $R = e^{\int P dx}$ | 992 |
| 414 | | | Solution | | $ye^{\int P dx} = \displaystyle\int Qe^{\int P dx}\,dx$ | 992 |
| 415 | Second-Order | Right Side Zero | Form | | $ay'' + by' + cy = 0$ | 1010 |
| 416 | | | Auxiliary Equation | | $am^2 + bm + c = 0$ | 1010 |
| | | | Form of Solution | Roots of Auxiliary Equation | Solution | |
| 417 | | | | Real and Unequal | $y = c_1 e^{m_1 x} + c_2 e^{m_2 x}$ | 1011 1014 |
| 418 | | | | Real and Equal | $y = c_1 e^{mx} + c_2 x e^{mx}$ | 1012 1014 |
| 419 | | | | Nonreal | (a) $y = e^{ax}(C_1 \cos bx + C_2 \sin bx)$ or (b) $y = Ce^{ax}\sin(bx + \phi)$ | 1013 1014 / 1014 |
| 420 | | Right Side Not Zero | Form | | $ay'' + by' + cy = f(x)$ | 1008 1020 |
| 421 | | | Complete Solution | | $y = \quad y_c \quad + \quad y_p$ complementary function $\qquad$ particular integral | 1017 |
| 422 | | Bernoulli's Equation | | | $\dfrac{dy}{dx} + Py = Qy^n$ (Substitute $z = y^{1-n}$) | 994 |

| | No. | | | | Page |
|---|---|---|---|---|---|
| DIFFERENTIAL EQUATIONS (Continued) | 423 | Laplace Transform | Definition | $\mathcal{L}[f(t)] = \int_0^\infty f(t)e^{-st}\,dt$ | 1033 |
| | 424 | | Inverse Transform | $\mathcal{L}^{-1}[F(s)] = f(t)$ | 1038 |
| | 425 | Numerical Solution | Euler's Method | $x_q = x_p + \Delta x$<br>$y_q = y_p + m_p \Delta x$ | 1052 |
| | 426 | | Modified Euler's Method | $x_q = x_p + \Delta x$<br>$y_q = y_p + \left(\frac{m_p + m_q}{2}\right)\Delta x$ | 1053 |
| | 427 | | Runge-Kutta Method | $x_q = x_p + \Delta x$<br>$y_q = y_p + m_{\text{avg}}\,\Delta x$<br>where $m_{\text{avg}} = \left(\frac{1}{6}\right)(m_p + 2m_r + 2m_s + m_q)$<br>and $m_p = f'(x_p, y_p)$<br>$m_r = f'\left(x_p + \frac{\Delta x}{2}, y_p + m_p \frac{\Delta x}{2}\right)$<br>$m_s = f'\left(x_p + \frac{\Delta x}{2}, y_p + m_r \frac{\Delta x}{2}\right)$<br>$m_q = f'(x_p + \Delta x, y_p + m_s\,\Delta x)$ | 1054 |

| | No. | | | | | Page |
|---|---|---|---|---|---|---|
| INFINITE SERIES | 428 | Power Series | Notation | | $u_1 + u_2 + u_3 + \cdots + u_n + \cdots$ | 608<br>1064 |
| | 429 | | Tests for Convergence | Limit Test | $\lim\limits_{n \approx \infty} u_n = 0$ | 1065 |
| | 430 | | | Partial Sum Test | $\lim\limits_{n \to \infty} S_n = S$ | 1064 |
| | 431 | | | Ratio Test | If $\lim\limits_{n \to \infty} \left\lvert \dfrac{u_{n+1}}{u_n} \right\rvert$<br>(a) is less than 1, the series converges.<br>(b) is greater than 1, the series diverges.<br>(c) is equal to 1, the test fails. | 1065 |
| | 432 | | Maclaurin's Series | | $f(x) = f(0) + f'(0)x + \dfrac{f''(0)}{2!} x^2 + \cdots + \dfrac{f^{(n)}(0)}{n!} x^n + \cdots$ | 1068 |
| | 433 | | Taylor's Series | | $f(x) = f(a) + f'(a)(x-a) + \dfrac{f''(a)}{2!}(x-a)^2 + \cdots + \dfrac{f^{(n)}(a)}{n!}(x-a)^n + \cdots$ | 1074 |
| | 434 | | | Remainder after $n$ Terms | $R_n = \dfrac{(x-a)^n}{n!} f^{(n)}(c)$<br>where $c$ lies between $a$ and $x$. | |
| | 435 | Fourier Series | Period of $2\pi$ | | $f(x) = a_0/2 + a_1 \cos x + a_2 \cos 2x + a_3 \cos 3x + \cdots + a_n \cos nx + \cdots$<br>$+ b_1 \sin x + b_2 \sin 2x + b_3 \sin 3x + \cdots + b_n \sin nx + \cdots$ | 1081<br>1084 |
| | 436 | | | where | $a_0 = \dfrac{1}{\pi} \displaystyle\int_{-\pi}^{\pi} f(x)\, dx$ | 1082<br>1084 |
| | 437 | | | | $a_n = \dfrac{1}{\pi} \displaystyle\int_{-\pi}^{\pi} f(x) \cos nx\, dx$ | 1083<br>1084 |
| | 438 | | | | $b_n = \dfrac{1}{\pi} \displaystyle\int_{-\pi}^{\pi} f(x) \sin nx\, dx$ | 1084 |
| | 439 | | Period of $2L$ | | $f(x) = \dfrac{a_0}{2} + a_1 \cos \dfrac{\pi x}{L} + a_2 \cos \dfrac{2\pi x}{L} + a_3 \cos \dfrac{3\pi x}{L} + \cdots$<br>$+ b_1 \sin \dfrac{\pi x}{L} + b_2 \sin \dfrac{2\pi x}{L} + b_3 \sin \dfrac{3\pi x}{L} + \cdots$ | 1094 |
| | 440 | | | where | $a_0 = \dfrac{1}{L} \displaystyle\int_{-L}^{L} f(x)\, dx$ | 1094 |
| | 441 | | | | $a_n = \dfrac{1}{L} \displaystyle\int_{-L}^{L} f(x) \cos \dfrac{n\pi x}{L}\, dx$ | 1094 |
| | 442 | | | | $b_n = \dfrac{1}{L} \displaystyle\int_{-L}^{L} f(x) \sin \dfrac{n\pi x}{L}\, dx$ | 1094 |
| | 443 | | Waveform Symmetries | Odd and Even Functions | (a) Odd functions have Fourier Series with only sine terms (and no constant term)<br>(b) Even functions have Fourier Series with only cosine terms (and may have a constant term). | 1089 |
| | 444 | | | Half-Wave Symmetry | A waveform that has half-wave symmetry has only odd harmonics in its Fourier Series. | 1090 |

## APPLICATIONS

| | No. | | | | Page |
|---|---|---|---|---|---|
| MIXTURES | A1 | Mixture Containing Ingredients $A, B, C, \ldots$ | | Total amount of mixture = amount of $A$ + amount of $B$ + $\cdots$ | 90 |
| | A2 | | | Final amount of each ingredient = initial amount + amount added − amount removed | 90 |
| | A3 | Combination of Two Mixtures | | Final amount of $A$ = amount of $A$ from mixture 1 + amount of $A$ from mixture 2 | 91 |
| | A4 | Fluid Flow | | Amount of flow = flow rate × elapsed time<br>$A = QT$ | |
| WORK | A5 | | | Amount done = rate of work × time worked | 229 |
| | A6 | 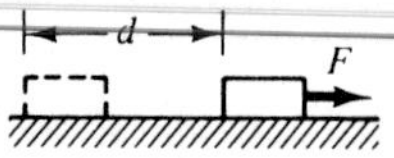 | Constant Force | Work = force × distance = $Fd$ | 935 |
| | A7 | | Variable Force | Work = $\int_a^b F(x)\,dx$ | 936 |
| FINANCIAL | A8 | Unit Cost | | Unit cost = $\dfrac{\text{total cost}}{\text{number of units}}$ | |
| | A9 | Interest: Principal $a$ Invested at Rate $n$ for $t$ years Accumulates to Amount $y$ | Simple | $y = a(1 + nt)$ | |
| | A10 | | Compounded Annually | $y = a(1 + n)^t$ | 507 |
| | A11 | | Compounded $m$ times/yr | $y = a\left(1 + \dfrac{n}{m}\right)^{mt}$ | 508 |
| STATICS | A12 | 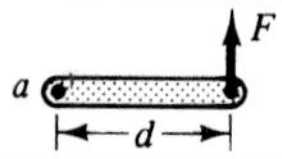 | Moment about Point $a$ | $M_a = Fd$ | 93 |
| | A13 | Equations of Equilibrium (Newton's First Law) | | The sum of all horizontal forces = 0 | 93 |
| | A14 | | | The sum of all vertical forces = 0 | 93 |
| | A15 | | | The sum of all moments about any point = 0 | 93 |
| | A16 |  | Coefficient of Friction | $\mu = \dfrac{f}{N}$ | |

| | No. | | | | | Page |
|---|---|---|---|---|---|---|
| MOTION | A17 | Linear Motion | Uniform Motion (Constant Speed) | | Distance = rate × time<br>$D = Rt$ | 228 |
| | A18 | | Uniformly Accelerated (Constant Acceleration $a$, Initial Velocity $v_0$)<br>For free fall, $a = g = 9.807 \text{ m/s}^2$ $= 32.2 \text{ ft/s}^2$ | Displacement at Time $t$ | $s = v_0 t + \frac{at^2}{2}$ | 875 |
| | A19 | | | Velocity at Time $t$ | $v = v_0 + at$ | 875 |
| | A20 | | | Newton's Second Law | $F = ma$ | 232 |
| | A21 | | Nonuniform Motion | Average Speed | Average speed = $\frac{\text{total distance traveled}}{\text{total time elapsed}}$ | |
| | A22 | | | Displacement | $s = \int v\, dt$ | 873 |
| | A23 | | | Instantaneous Velocity | $v = \frac{ds}{dt}$ | 798 |
| | A24 | | | | $v = \int a\, dt$ | 873 |
| | A25 | | | Instantaneous Acceleration | $a = \frac{dv}{dt} = \frac{d^2 s}{dt^2}$ | 799 |
| | A26 | Rotation | Uniform Motion | Angular Displacement | $\theta = \omega t$ | 407 |
| | A27 | | | Linear Speed of Point at Radius $r$ | $v = \omega r$ | 408 |
| | A28 | | Nonuniform Motion | Angular Displacement | $\theta = \int \omega\, dt$ | 876 |
| | A29 | | | Angular Velocity | $\omega = \frac{d\theta}{dt}$ | 803 |
| | A30 | | | | $\omega = \int \alpha\, dt$ | 876 |
| | A31 | | | Angular Acceleration | $\alpha = \frac{d\omega}{dt} = \frac{d^2\theta}{dt^2}$ | 803 |
| | A32 | Curvilinear Motion: | $x$ and $y$ Components | Displacement | (a) $x = \int v_x\, dt$ (b) $y = \int v_y\, dt$ | 875 |
| | A33 | | | Velocity | (a) $v_x = \frac{dx}{dt}$ (b) $v_y = \frac{dy}{dt}$ | 800 |
| | A34 | | | | (a) $v_x = \int a_x\, dt$ (b) $v_y = \int a_y\, dt$ | 875 |
| | A35 | | | Acceleration | (a) $a_x = \frac{dv_x}{dt} = \frac{d^2 x}{dt^2}$ (b) $a_y = \frac{dv_y}{dt} = \frac{d^2 y}{dt^2}$ | 800 |

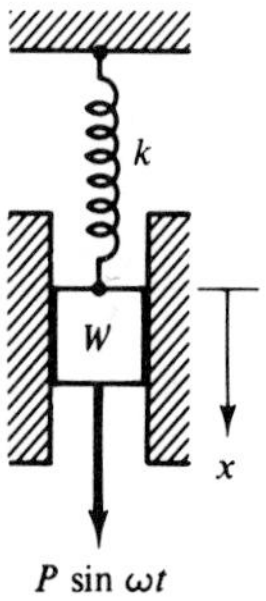


| | No. | | | | | Page |
|---|---|---|---|---|---|---|
| MECHANICAL VIBRATIONS | A36 | | Free Vibrations ($P = 0$) | Simple Harmonic Motion (No Damping) | $x = x_0 \cos \omega_n t$ | |
| | A37 | | | | Undamped Angular Velocity $\omega_n = \sqrt{\dfrac{kg}{W}}$ | |
| | A38 | | | | Natural Frequency $f_n = \dfrac{\omega_n}{2\pi}$ | |
| | A39 | | | Under-damped | $x = x_0 e^{-at} \cos \omega_d t$ | |
| | A40 | $P \sin \omega t$ | | | Damped Angular Velocity $\omega_d = \sqrt{\omega_n^2 - \dfrac{c^2 g^2}{\omega^2}}$ | |
| | A41 | Coefficient of friction = $c$ | | Overdamped | $x = C_1 e^{m_1 t} + C_2 e^{m_2 t}$ | |
| | A42 | | Forced Vibrations | Maximum Deflection | $x_0 = \dfrac{Pg}{W\sqrt{4a^2\omega^2 + (\omega_n^2 - \omega^2)^2}}$ | |

| | No. | | | | Page |
|---|---|---|---|---|---|
| MATERIAL PROPERTIES | A43 | Density | | Density = $\frac{\text{weight}}{\text{volume}}$ or $\frac{\text{mass}}{\text{volume}}$ | |
| | A44 | Mass | | Mass = $\frac{\text{weight}}{\text{acceleration due to gravity}}$ | |
| | A45 | Specific Gravity | | SG = $\frac{\text{density of substance}}{\text{density of water}}$ | 503 |
| | A46 |   Pressure | Total Force on a Surface | Force = pressure × area | 933 |
| | A47 | 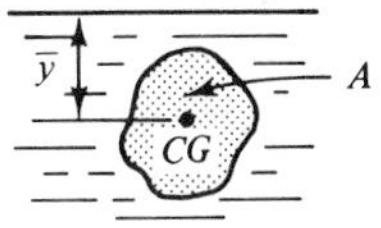  | Force on a Submerged Surface | $F = \delta \int y\, dA$ | 933 |
| | A48 | | | $F = \delta \bar{y} A$ | 933 |
| | A49 | pH | | pH = −10 log concentration | 539 |
| TEMPERATURE | A50 | Conversions between Degrees Celsuis ($C$) and Degrees Fahrenheit ($F$) | | $C = \frac{5}{9}(F - 32)$ | |
| | A51 | | | $F = \frac{9}{5}C + 32$ | |
| STRENGTH OF MATERIALS | A52 | 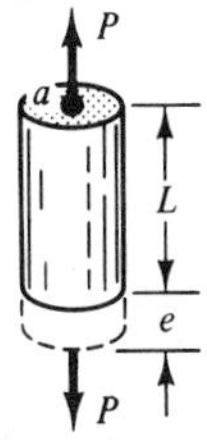  Tension or Compression | Normal Stress | $\sigma = \frac{P}{a}$ | |
| | A53 | | Strain | $\epsilon = \frac{e}{L}$ | |
| | A54 | | Modulus of Elasticity and Hooke's Law | $E = \frac{PL}{ae}$ | |
| | A55 | | | $E = \frac{\sigma}{\epsilon}$ | |
| | A56 | Thermal Expansion 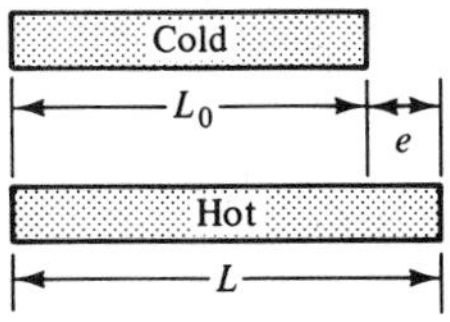  Temperature change = $\Delta t$ Coefficient of thermal expansion = $\alpha$ | Elongation | $e = \alpha L\ \Delta t$ | |
| | A57 | | New Length | $L = L_0(1 + \alpha\Delta t)$ | |
| | A58 | | Strain | $\epsilon = \frac{e}{L} = \alpha\Delta t$ | |
| | A59 | | Stress, if Restrained | $\sigma = E\epsilon = E\alpha\ \Delta t$ | |
| | A60 | | Force, if Restrained | $P = a\sigma = aE\alpha\ \Delta t$ | |
| | A61 | 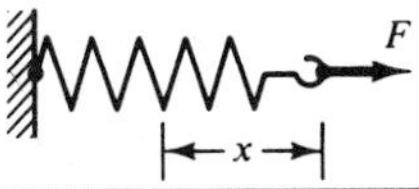  | Force needed to Deform a Spring | $F$ = spring constant × distance = $kx$ | 480 936 |

| | No. | | | | Page |
|---|---|---|---|---|---|
| ELECTRICAL TECHNOLOGY | A62 | Ohm's Law | | Current = $\dfrac{\text{voltage}}{\text{resistance}}$ $\qquad I = \dfrac{V}{R}$ | 502 |
| | A63 | Combinations of Resistors | In Series | $R = R_1 + R_2 + R_3 + \cdots$ | 11, 33 |
| | A64 | | In Parallel | $\dfrac{1}{R} = \dfrac{1}{R_1} + \dfrac{1}{R_2} + \dfrac{1}{R_3} + \cdots$ | 33, 237 |
| | A65 | Power Dissipated in a Resistor 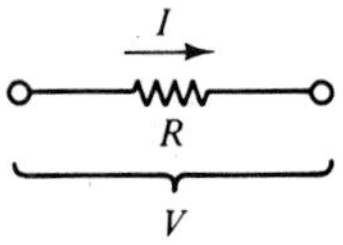  | | Power = $P = VI$ | 112 |
| | A66 | | | $P = \dfrac{V^2}{R}$ | 203 |
| | A67 | | | $P = I^2R$ | 112 |
| | A68 | Kirchhoff's Laws | Loops | The sum of the voltage rises and drops around any closed loop is zero | 237, 259, 264 |
| | A69 | | Nodes | The sum of the currents entering and leaving any node is zero | |
| | A70 | Resistance Change with Temperature | | $R = R_1[1 + \alpha(t - t_1)]$ | 114, 237 |
| | A71 | Resistance of a Wire 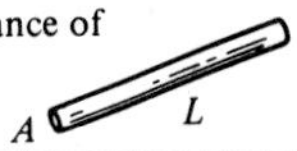  | | $R = \dfrac{\rho L}{A}$ | 224 |
| | A72 | Combinations of Capacitors | In Series | $\dfrac{1}{C} = \dfrac{1}{C_1} + \dfrac{1}{C_2} + \dfrac{1}{C_3} + \cdots$ | |
| | A73 | | In Parallel | $C = C_1 + C_2 + C_3 + \cdots$ | 33 |
| | A74 | Charge on a Capacitor at Voltage $V$ | | $Q = CV$ | |

ELECTRICAL TECHNOLOGY (Continued)

| No. | | | Sinusoidal Form | Complex Form | Page |
|---|---|---|---|---|---|
| A75 | Alternating Voltage | | $v = V_m \cos(\omega t + \phi_1)$ | $\mathbf{V} = V_m \underline{/\phi_1}$ | 435<br>570 |
| A76 | Alternating Current | | $i = I_m \cos(\omega t + \phi_2)$ | $\mathbf{I} = I_m \underline{/\phi_2}$ | 435<br>570 |
| A77 | Period | | $P = \dfrac{2\pi}{\omega}$ seconds | | 430 |
| A78 | Frequency | | $f = \dfrac{1}{P} = \dfrac{\omega}{2\pi}$ hertz | | 431 |
| A79 | Current | | $i = \dfrac{dq}{dt}$ | | 794 |
| A80 | Charge | | $q = \int i\, dt$ coulombs | | 878 |
| A81 | Capacitor | Instantaneous Current | $i = C \dfrac{dv}{dt}$ | | 795 |
| A82 | Capacitor | Instantaneous Voltage | $v = \dfrac{1}{C} \int i\, dt$ volts | | 879 |
| A83 | Capacitor | Current when Charging or Discharging | Series *RC* Circuit 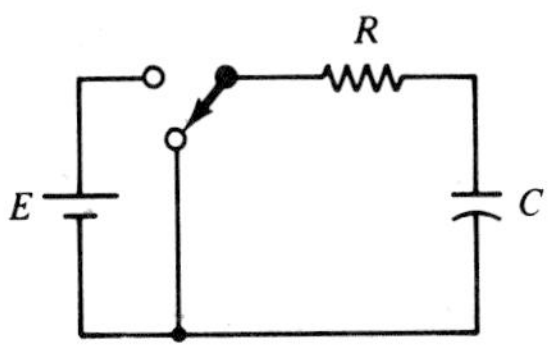 | $i = \dfrac{E}{R} e^{-t/RC}$ | 514<br>1005 |
| A84 | Capacitor | Voltage when Discharging | | $v = Ee^{-t/RC}$ | 1005 |
| A85 | Inductor | Instantaneous Current | $i = \dfrac{1}{L} \int v\, dt$ amperes | | 879 |
| A86 | Inductor | Instantaneous Voltage | $v = L \dfrac{di}{dt}$ | | 796 |
| A87 | Inductor | Current when Charging | Series *RL* Circuit 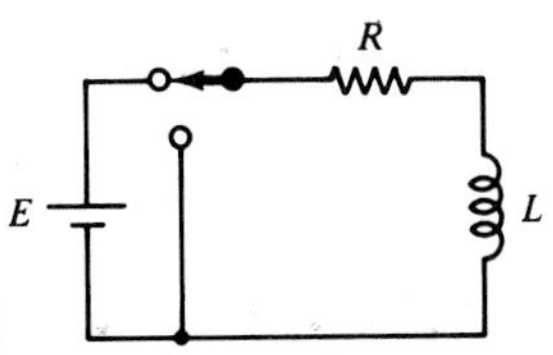 | $i = \dfrac{E}{R}(1 - e^{-Rt/L})$ | 514<br>1004 |
| A88 | Inductor | Voltage when Charging or Discharging | | $v = Ee^{-Rt/L}$ | 1004 |

ELECTRICAL TECHNOLOGY (Continued)

| No. | | | | | Page |
|---|---|---|---|---|---|
| A89 | Series $RLC$ Circuit | DC Source | Resonant Frequency | $\omega_n = \sqrt{\frac{1}{LC}}$ | 1023 1029 |
| A90 | | | No Resistance: The Series $L$ $C$ Circuit: | $i = \frac{E}{\omega_n L} \sin \omega_n t$ | 1025 |
| A91 | | | Underdamped | $i = \frac{E}{\omega_d L} e^{-at} \sin \omega_d t$ | 1024 |
| A92 | 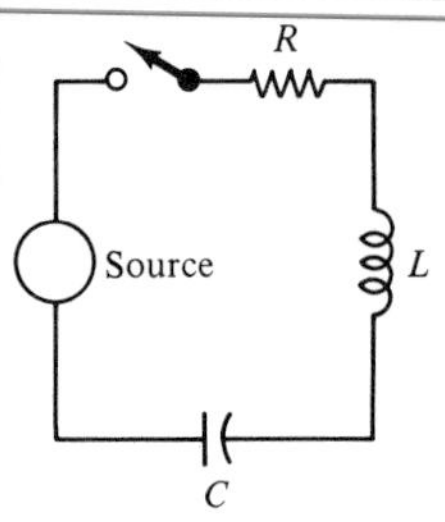 | | | where $\omega_d = \sqrt{\omega_n^2 - \frac{R^2}{4L^2}}$ | 1024 |
| A93 | | | Overdamped | $i = \frac{E}{2 j\omega_d L} \left[e^{(-a + j\omega_d)t} - e^{(-a - j\omega_d)t}\right]$ | 1026 |
| A94 | | AC Source | Inductive Reactance | $X_L = \omega L$ | |
| A95 | | | Capacitive Reactance | $X_C = \frac{1}{\omega C}$ | |
| A96 | | | Total Reactance | $X = X_L - X_C$ | 178 |
| A97 | | | Magnitude of Impedance | $\mid Z \mid = \sqrt{R^2 + X^2} = \sqrt{R^2 + \left(\omega L - \frac{1}{\omega C}\right)^2}$ | 178 |
| A98 | | | Phase Angle | $\phi = \arctan \frac{X}{R}$ | 178 |
| A99 | | | Complex Impedance | $Z = R + jX = Z\angle\phi = Ze^{j\phi}$ | 571 |
| A100 | | | Steady-State Current | $i_{ss} = \frac{E}{Z} \sin(\omega t - \phi)$ | 1028 |
| A101 | Ohm's Law for AC | | | **V = ZI** | 571 |
| A102 | Decibels Gained or Lost | | | $G = 10 \log_{10} \frac{P_2}{P_1}$ dB | 539 |

# B

# Conversion Factors

| Unit | Equals | |
|---|---|---|
| | *Length* | |
| 1 angstrom | $1 \times 10^{-10}$ | meter |
| | $1 \times 10^{-4}$ | micrometer (micron) |
| 1 centimeter | $10^{-2}$ | meter |
| | 0.3937 | inch |
| 1 foot | 12 | inches |
| | 0.3048 | meter |
| 1 inch | 25.4 | millimeters |
| | 2.54 | centimeters |
| 1 kilometer | 3281 | feet |
| | 0.5400 | nautical mile |
| | 0.6214 | statute mile |
| | 1094 | yards |
| 1 light-year | $9.461 \times 10^{12}$ | kilometers |
| | $5.879 \times 10^{12}$ | statute miles |
| 1 meter | $10^{10}$ | angstroms |
| | 3.281 | feet |
| | 39.37 | inches |
| | 1.094 | yards |
| 1 micron | $10^{4}$ | angstroms |
| | $10^{-4}$ | centimeter |
| | $10^{-6}$ | meter |
| 1 nautical mile (International) | 8.439 | cables |
| | 6076 | feet |
| | 1852 | meters |
| | 1.151 | statute miles |
| 1 statute mile | 5280 | feet |
| | 8 | furlongs |
| | 1.609 | kilometers |
| | 0.8690 | nautical mile |
| 1 yard | 3 | feet |
| | 0.9144 | meter |
| | *Angles* | |
| 1 degree | 60 | minutes |
| | 0.01745 | radian |
| | 3600 | seconds |
| | $2.778 \times 10^{-3}$ | revolution |
| 1 minute of arc | 0.01667 | degree |
| | $2.909 \times 10^{-4}$ | radian |
| | 60 | seconds |
| 1 radian | 0.1592 | revolution |
| | 57.296 | degrees |
| | 3438 | minutes |
| 1 second of arc | $2.778 \times 10^{-4}$ | degree |
| | 0.01667 | minute |

*Source:* Adapted from P. Calter, *Schaum's Outline of Technical Mathematics*, McGraw-Hill Book Company, New York, 1979.

| Unit | Equals | |
|---|---|---|
| | *Area* | |
| 1 acre | 4047 | square meters |
| | 43 560 | square feet |
| 1 are | 0.024 71 | acre |
| | 1 | square dekameter |
| | 100 | square meters |
| 1 hectare | 2.471 | acres |
| | 100 | ares |
| | 10 000 | square meters |
| 1 square foot | 144 | square inches |
| | 0.092 90 | square meter |
| 1 square inch | 6.452 | square centimeters |
| 1 square kilometer | 247.1 | acres |
| 1 square meter | 10.76 | square feet |
| 1 square mile | 640 | acres |
| | $2.788 \times 10^7$ | square feet |
| | 2.590 | square kilometers |
| | *Volume* | |
| 1 board-foot | 144 | cubic inches |
| 1 bushel (U.S.) | 1.244 | cubic feet |
| | 35.24 | liters |
| 1 cord | 128 | cubic feet |
| | 3.625 | cubic meters |
| 1 cubic foot | 7.481 | gallons (U.S. liquid) |
| | 28.32 | liters |
| 1 cubic inch | 0.01639 | liter |
| | 16.39 | milliliters |
| 1 cubic meter | 35.31 | cubic feet |
| | $10^6$ | cubic centimeter |
| 1 cubic millimeter | $6.102 \times 10^{-5}$ | cubic inch |
| 1 cubic yard | 27 | cubic feet |
| | 0.7646 | cubic meter |
| 1 gallon (imperial) | 277.4 | cubic inches |
| | 4.546 | liters |
| 1 gallon (U.S. liquid) | 231 | cubic inches |
| | 3.785 | liters |
| 1 kiloliter | 35.31 | cubic feet |
| | 1.000 | cubic meter |
| | 1.308 | cubic yards |
| | 220 | imperial gallons |
| 1 liter | $10^3$ | cubic centimeters |
| | $10^6$ | cubic millimeters |
| | $10^{-3}$ | cubic meter |
| | 61.02 | cubic inches |

| Unit | Equals | |
|---|---|---|
| | *Mass* | |
| 1 gram | $10^{-3}$ | kilogram |
| | $6.854 \times 10^{-5}$ | slug |
| 1 kilogram | 1000 | grams |
| | 0.06854 | slug |
| 1 slug | 14.59 | kilograms |
| | 14,590 | grams |
| 1 metric ton | 1000 | kilograms |
| | *Force* | |
| 1 dyne | $10^{-5}$ | newton |
| 1 newton | $10^{5}$ | dynes |
| | 0.2248 | pound |
| | 3.597 | ounces |
| 1 pound | 4.448 | newtons |
| | 16 | ounces |
| 1 ton | 2000 | pounds |
| | *Velocity* | |
| 1 foot/minute | 0.3048 | meter/minute |
| | 0.011 364 | mile/hour |
| 1 foot/second | 1097 | kilometers/hour |
| | 18.29 | meters/minute |
| | 0.6818 | mile/hour |
| 1 kilometer/hour | 3281 | feet/hour |
| | 54.68 | feet/minute |
| | 0.6214 | mile/hour |
| 1 kilometer/minute | 3281 | feet/minute |
| | 37.28 | miles/hour |
| 1 knot | 6076 | feet/hour |
| | 101.3 | feet/minute |
| | 1.852 | kilometers/hour |
| | 30.87 | meters/minute |
| | 1.151 | miles/hour |
| 1 meter/hour | 3.281 | feet/hour |
| 1 mile/hour | 1.467 | feet/second |
| | 1.609 | kilometers/hour |
| | *Power* | |
| 1 British thermal unit/hour | 0.2929 | watt |
| 1 Btu/pound | 2.324 | joules/gram |
| 1 Btu-second | 1.414 | horsepower |
| | 1.054 | kilowatts |
| | 1054 | watts |

| Unit | Equals | |
|---|---|---|
| | ***Power (continued)*** | |
| 1 horsepower | 42.44 | Btu/minute |
| | 550 | footpounds/second |
| | 746 | watts |
| 1 kilowatt | 3414 | Btu/hour |
| | 737.6 | footpounds/second |
| | 1.341 | horsepower |
| | $10^3$ | joules/second |
| | 999.8 | international watt |
| 1 watt | 44.25 | footpounds/minute |
| | 1 | joule/second |
| | ***Pressure*** | |
| 1 atmosphere | 1.013 | bars |
| | 14.70 | pounds/square inch |
| | 760 | torrs |
| | 101 | kilopascals |
| 1 bar | $10^6$ | baryes |
| | 14.50 | pounds-force/square inch |
| 1 barye | $10^{-6}$ | bar |
| 1 inch of mercury | 0.033 86 | bar |
| | 70.73 | pounds/square foot |
| 1 pascal | 1 | newton/square meter |
| 1 pound/square inch | 0.068 03 | atmosphere |
| | ***Energy*** | |
| 1 British thermal unit | 1054 | joules |
| | 1054 | wattseconds |
| 1 foot-pound | 1.356 | joules |
| | 1.356 | newtonmeters |
| 1 joule | 0.7376 | foot-pound |
| | 1 | wattsecond |
| | 0.2391 | calories |
| 1 kilowatthour | 3410 | British thermal units |
| | 1.341 | horsepowerhours |
| 1 newtonmeter | 0.7376 | footpounds |
| 1 watthour | 3.414 | British thermal units |
| | 2655 | footpounds |
| | 3600 | joules |

# C

# SUMMARY OF BASIC

This brief listing contains commands, statements, and functions. *Commands*, such as RUN, or LIST are not part of a program, but tell the computer what to do with a program. Commands are typed without line numbers and are executed immediately. Program *statements*, such as PRINT or GO TO, tell the computer what to do during a run of the program. *Functions*, such as COS( ) or EXP( ), tell the computer what operation to perform on the quantity enclosed in parentheses (called the *argument*).

ABS( )

A function which returns the absolute value of the expression in parentheses.

*Example:* The lines

```
10 LET X = -5
20 PRINT ABS(X)
```

will cause the value 5 to be printed.

ATN( )

A function which returns the angle (in radians) whose tangent is specified.

*Example:* The statement PRINT ATN(2) will cause the arc-tangent of 2 (1.1071 radians) to be printed.

CONT

A command which causes a program to continue running after a STOP or CTRL-C (interrupt) has been executed.

COS( )

A function which returns the cosine of the angle (in radians).

*Example:* The statement PRINT COS(1.2) will cause the number 0.362358 (the cosine of 1.2 radians) to be printed.

DATA

A statement used to store numbers and strings, for later access by the READ statement.

*Example:* The statement DATA 5, 2, 9, 3 stores the numbers 5, 2, 9, and 3 for later access, and the statement

DATA JOHN, MARY, BILL

stores the given three names for later access. See READ.

DEF FN

A statement which defines a new function written by the user.

*Examples:*

```
DEF FNA(X) = 3*X + 2
DEF FNG(X,Y,Z) = (5*X - 2*Y^2 )*Z
```

DIM

A statement which reserves space for lists or tables.

*Example:* DIM A(50), DIM P$(20), B(25,30)

END

A statement which stops the run and closes all files. It is the last line in a program.

EXP( )

A function which raises e (the base of natural logarithms) to the power specified.

*Example:* The statement PRINT EXP(2.5) will compute and print the value of $e^{2.5}$, (12.182).

FOR . . . NEXT

A pair of statements used to set up a loop.

*Examples:* The lines

```
10 FOR N = 1 TO 10
20 PRINT "HELLO"
30 NEXT N
```

will cause the word "HELLO" to be printed ten times.

The lines

```
10 FOR X = 20 TO 30 STEP .5
20 PRINT X,
30 NEXT X
```

will cause the numbers 20, 20.5, 21, . . . 30 to be printed.

GOSUB . . . RETURN

A pair of statements used to branch to and return from a subroutine.
*Example:* The line GOSUB 500 causes a branching to the subroutine starting on line 500. The RETURN statement placed at the end of the subroutine returns us to the line immediately following the GOSUB statement.

GOTO

Causes a branch to the specified line number.
*Example:* GOTO 150

IF . . . THEN

A statement which causes a branching to another line if the specified condition is met.
*Example:* The line

```
40 IF A > B THEN 200
```

will cause branching to line 200 if A is greater than B. If not, we go to the next line.

INPUT

A statement which allows input of data from the keyboard during a run.
*Example:* The statement INPUT X will cause the run to stop, and a question mark will be printed. The operator must then type a number, which will be accepted as X by the program.

*Example:* The statement
20 INPUT "WHAT ARE THE TWO NAMES"; A$, B$
will print the *prompt string* WHAT ARE THE TWO NAMES, then a question mark, and then stop the run until you enter two strings, separated by a comma.

INT( )

A function which returns the next lowest whole number.
*Examples:* The statement PRINT INT(5.995) will cause the integer 5 to be printed. Also, INT(−7.1) returns −8.

KILL

A command which deletes a file from the disk.
*Example:* KILL "CIRCLE.BAS"

LEFT$( )

A function which returns the leftmost characters of the specified string.
*Example:* The line 50 PRINT LEFT$(A$,5) will cause the five leftmost characters of A$ to be printed.

LEN( )

A function which returns the number of characters, incuding blanks, in the specified string.
*Example:* 10 PRINT LEN(B$)

| | |
|---|---|
| LET | A statement used to assign a value to a variable.<br>*Examples:* 30 LET X = 5<br>40 LET Y = 3*X−2<br>50 LET SUM = X + Y<br>(The LET statement is optional in many versions of BASIC.) |
| LIST | A command to list all or part of a program.<br>*Examples:* LIST lists the entire program.<br>LIST 100–150 lists lines 100 to 150 only.<br>LIST–200 lists up to line 200.<br>LIST 300– lists from line 300 to the end. |
| LLIST | A command to list the program on the printer. |
| LOAD | A command to load a program or file from the disk.<br>*Example:* LOAD "B:QUADRATIC" will load the program QUADRATIC from drive B. |
| LOG( ) | A function which returns the natural logarithm of the argument. |
| LPRINT | See PRINT. |
| MID$( ) | A function which returns a string of characters from the middle of the specified string.<br>*Example:* In the program<br>10 A$ = "COMPUTER"<br>20 PRINT MID$(A$,4,3)<br>line 20 will extract and print a string of length 3 from A$, starting with the 4th character. Thus, "PUT" will be printed. |
| NEW | A command used to clear the memory before entering a new program. |
| ON . . . GO TO | A statement which allows branching to one of several line numbers.<br>*Example:* Given the line<br>10 ON X GO TO 150, 300, 450<br>we branch to line 150 if X = 1, to line 300 if X = 2, and to line 450 if X = 3. |
| PRINT | A statement which causes information to be printed.<br>*Examples:*<br>20 PRINT X causes the value of X to be printed.<br>30 PRINT A, B, C causes the values of A, B, and C to be printed, each in a separate column.<br>40 PRINT A; B; C causes the values of A, B, and C to be printed with just a single space between them.<br>50 PRINT "HELLO" causes the word "HELLO" to be printed.<br>The statement LPRINT will cause the printing to occur at the printer, rather than at the video terminal. |
| PRINT USING | A statement which specifies a format for the printing of strings or numbers. |

*Example:* 20 PRINT USING "###.##"; X
will cause the number X to be printed with three digits before the decimal point and two digits following the decimal point. There are many other ways to format numbers and strings. Consult your users' manual.

RANDOMIZE

A statement which allows you to obtain a different list of random numbers each time you use RND( ). Put the line 10 RANDOMIZE at the start of each program that uses random numbers. Then when you RUN, you will be asked to enter a number. Each such number will cause a different set of random numbers to be generated.

READ

A statement which reads values from a DATA statement and assigns them to variables.
*Example:* 10 DATA JOHN, 24
20 READ N$, A
Line 20 will read the string "JOHN" and assign it to N$, and read the number 24 and assign it to N.

REM

This statement allows us to place remarks in a program. REM's are ignored during a RUN.

*Example:* 10 REM THIS PROGRAM COMPUTES AVERAGES

RENUM

A command to renumber a program.
*Example:* RENUM renumbers a program with the line numbers, 10, 20, 30, . . .
RENUM 100, 300, 20 will start renumbering your program at the old line 300, which it changes to 100, and continues in increments of 20.

RESTORE

A statement which allows data to be reread.

RETURN

see GOSUB

RIGHT$( )

A statement which returns the rightmost characters from the specified string. Similar to LEFT$( ).

RND

A function which returns a random number between 0 and 1.
*Example:* The program
```
10 FOR N = 1 TO 10
20 PRINT RND
30 NEXT N
```
will cause 10 random numbers to be printed.

RUN

A command to run the program in memory.

SAVE

A command to save a program on disk.
*Example:* SAVE "B:ROOTS"
will save the program "ROOTS" on the disk in drive B.

| | |
|---|---|
| SGN( ) | A function which returns a value of +1, 0, or −1, depending on whether the argument is positive, zero, or negative, respectively. |
| SIN( ) | A function which returns the sine of the angle (in radians). |
| SQR( ) | A function which returns the square root of the argument. |
| STOP | A statement which stops a run, but does not close files. To resume the run, type CONT. |
| SYSTEM | A command to exit BASIC and return to the operating system. |
| TAB( ) | A function which tabs to the print position specified. *Example:* The line 30 PRINT TAB(25); "HELLO" will cause the word HELLO to be printed starting at a position 25 characters from the left edge of the screen or paper. |
| TAN( ) | A function which returns the tangent of the angle (in radians). |

D

# TABLE OF INTEGRALS

**Note: Many integrals have alternate forms that are not shown here. Don't be surprised if another table of integrals gives an expression that looks very different than one listed here.**

| | |
|---|---|
| *Basic forms* | 1. $\int du = u + C$ |
| | 2. $\int af(x)\,dx = a\int f(x)\,dx = aF(x) + C$ |
| | 3. $\int [f(x) + g(x) + h(x) + \cdots]\,dx = \int f(x)\,dx + \int g(x)\,dx + \int h(x)\,dx + \cdots + C$ |
| | 4. $\int x^n\,dx = \frac{x^{n+1}}{n+1} + C \qquad (n \neq -1)$ |
| | 5. $\int u^n\,du = \frac{u^{n+1}}{n+1} + C \qquad (n \neq -1)$ |
| | 6. $\int u\,dv = uv - \int v\,du$ |
| | 7. $\int \frac{du}{u} = \ln\|u\| + C$ |
| | 8. $\int e^u\,du = e^u + C$ |
| | 9. $\int b^u\,du = \frac{b^u}{\ln b} + C \qquad (b > 0, b \neq 1)$ |
| *Trigonometric functions* | 10. $\int \sin u\,du = -\cos u + C$ |
| | 11. $\int \cos u\,du = \sin u + C$ |
| | 12. $\int \tan u\,du = -\ln\|\cos u\| + C$ |
| | 13. $\int \cot u\,du = \ln\|\sin u\| + C$ |
| | 14. $\int \sec u\,du = \ln\|\sec u + \tan u\| + C$ |
| | 15. $\int \csc u\,du = \ln\|\csc u - \cot u\| + C$ |
| *Squares of the trigonometric functions* | 16. $\int \sin^2 u\,du = \frac{u}{2} - \frac{\sin 2u}{4} + C$ |
| | 17. $\int \cos^2 u\,du = \frac{u}{2} + \frac{\sin 2u}{4} + C$ |
| | 18. $\int \tan^2 u\,du = \tan u - u + C$ |
| | 19. $\int \cot^2 u\,du = -\cot u - u + C$ |
| | 20. $\int \sec^2 u\,du = \tan u + C$ |
| | 21. $\int \csc^2 u\,du = -\cot u + C$ |

| | |
|---|---|
| *Cubes of the trigonometric functions* | 22. $\int \sin^3 u\, du = \frac{\cos^3 u}{3} - \cos u + C$ |
| | 23. $\int \cos^3 u\, du = \sin u - \frac{\sin^3 u}{3} + C$ |
| | 24. $\int \tan^3 u\, dx = \frac{1}{2}\tan^2 u + \ln\lvert\cos u\rvert + C$ |
| | 25. $\int \cot^3 u\, dx = -\frac{1}{2}\cot^2 u - \ln\lvert\sin u\rvert + C$ |
| | 26. $\int \sec^3 u\, du = \frac{1}{2}\sec u \tan u + \frac{1}{2}\ln\lvert\sec u + \tan u\rvert + C$ |
| | 27. $\int \csc^3 u\, du = -\frac{1}{2}\csc u \cot u + \frac{1}{2}\ln\lvert\csc u - \cot\rvert u + C$ |
| *Miscellaneous trigonometric forms* | 28. $\int \sec u \tan u\, du = \sec u + C$ |
| | 29. $\int \csc u \cot u\, du = -\csc u + C$ |
| | 30. $\int \sin^2 u \cos^2 u\, du = \frac{u}{8} - \frac{1}{32}\sin 4u + C$ |
| | 31. $\int u \sin u\, du = \sin u - u\cos u + C$ |
| | 32. $\int u \cos u\, du = \cos u + u \sin u + C$ |
| | 33. $\int u^2 \sin u\, du = 2u \sin u - (u^2 - 2)\cos u + C$ |
| | 34. $\int u^2 \cos u\, du = 2u\cos u + (u^2 - 2)\sin u + C$ |
| | 35. $\int \text{Sin}^{-1} u\, du = u\,\text{Sin}^{-1} u + \sqrt{1 - u^2} + C$ |
| | 36. $\int \text{Tan}^{-1} u\, du = u\,\text{Tan}^{-1} u - \ln\sqrt{1 + u^2} + C$ |
| *Exponential and logarithmic forms* | 37. $\int ue^{au}\, du = \frac{e^{au}}{a^2}(au - 1) + C$ |
| | 38. $\int u^2 e^{au}\, du = \frac{e^{au}}{a^3}(a^2u^2 - 2au + 2) + C$ |
| | 39. $\int u^n e^u\, du = u^n e^u - n\int u^{n-1} e^u\, du$ |
| | 40. $\int \frac{e^u\, du}{u^n} = \frac{-e^u}{(n-1)u^{n-1}} + \frac{1}{n-1}\int \frac{e^u\, du}{u^{n-1}} \quad (n \neq 1)$ |
| | 41. $\int e^{au}\sin bu\, du = \frac{e^{au}}{a^2 + b^2}(a \sin bu - b\cos bu) + C$ |
| | 42. $\int e^{au}\cos bu\, du = \frac{e^{au}}{a^2 + b^2}(a\cos bu + b\sin bu) + C$ |
| | 43. $\int \ln u\, du = u(\log u - 1) + C$ |
| | 44. $\int u^n \ln\lvert u\rvert\, du = u^{n+1}\left[\frac{\ln\lvert u\rvert}{n+1} - \frac{1}{(n+1)^2}\right] + C \quad (n \neq -1)$ |
| *Forms involving* $a + bu$ | 45. $\int \frac{u\, du}{a + bu} = \frac{1}{b^2}[a + bu - a\ln\lvert a + bu\rvert] + C$ |
| | 46. $\int \frac{u^2\, du}{a + bu} = \frac{1}{b^3}\left[\frac{1}{2}(a + bu)^2 - 2a(a + bu) + a^2 \ln\lvert a + bu\rvert\right] + C$ |
| | 47. $\int \frac{u\, du}{(a + bu)^2} = \frac{1}{b^2}\left[\frac{a}{a + bu} + \ln\lvert a + bu\rvert\right] + C$ |

***Forms involving*** $a + bu$

48. $$\int \frac{u^2\,du}{(a+bu)^2} = \frac{1}{b^3}\left[a + bu - \frac{a^2}{a+bu} - 2a\ln|a+bu|\right] + C$$

49. $$\int \frac{du}{u(a+bu)} = \frac{-1}{a}\ln\left|\frac{a+bu}{u}\right| + C$$

50. $$\int \frac{du}{u^2(a+bu)} = \frac{-1}{au} + \frac{b}{a^2}\ln\left|\frac{a+bu}{u}\right| + C$$

51. $$\int \frac{du}{u(a+bu)^2} = \frac{1}{a(a+bu)} - \frac{1}{a^2}\ln\left|\frac{a+bu}{u}\right| + C$$

52. $$\int u\sqrt{a+bu}\,du = \frac{2(3bu-2a)}{15b^2}(a+bu)^{3/2} + C$$

53. $$\int u^2\sqrt{a+bu}\,du = \frac{2(15b^2u^2 - 12abu + 8a^2)}{105b^3}(a+bu)^{3/2} + C$$

54. $$\int \frac{u\,du}{\sqrt{a+bu}} = \frac{2(bu-2a)}{3b^2}\sqrt{a+bu} + C$$

55. $$\int \frac{u^2\,du}{\sqrt{a+bu}} = \frac{2(3b^2u^2 - 4abu + 8a^2)}{15b^3}\sqrt{a+bu} + C$$

---

***Forms involving*** $u^2 \pm a^2$ $(a > 0)$

56. $$\int \frac{du}{a^2 + b^2u^2} = \frac{1}{ab}\tan^{-1}\frac{bu}{a} + C$$

57. $$\int \frac{du}{u^2 - a^2} = \frac{1}{2a}\ln\left|\frac{u-a}{u+a}\right| + C$$

58. $$\int \frac{u^2\,du}{u^2 - a^2} = u + \frac{a}{2}\ln\left|\frac{u-a}{u+a}\right| + C$$

59. $$\int \frac{u^2\,du}{u^2 + a^2} = u - a\,\mathrm{Tan}^{-1}\frac{u}{a} + C$$

60. $$\int \frac{du}{u(u^2 \pm a^2)} = \frac{\pm 1}{2a^2}\ln\left|\frac{u^2}{u^2 \pm a^2}\right| + C$$

---

*Forms involving* $\sqrt{a^2 \pm u^2}$ *and* $\sqrt{u^2 \pm a^2}$ $(a > 0)$

61. $$\int \frac{du}{\sqrt{a^2 - u^2}} = \mathrm{Sin}^{-1}\frac{u}{a} + C$$

62. $$\int \frac{du}{\sqrt{u^2 \pm a^2}} = \ln|u + \sqrt{u^2 \pm a^2}| + C$$

63. $$\int \frac{u^2\,du}{\sqrt{u^2 \pm a^2}} = \frac{u}{2}\sqrt{u^2 \pm a^2} \mp \frac{a^2}{2}\ln|u + \sqrt{u^2 \pm a^2}| + C$$

64. $$\int \frac{du}{u\sqrt{u^2 + a^2}} = \frac{1}{a}\ln\left|\frac{u}{a + \sqrt{u^2 + a^2}}\right| + C$$

65. $$\int \frac{du}{u\sqrt{u^2 - a^2}} = \frac{1}{a}\mathrm{Sec}^{-1}\frac{u}{a} + C$$

66. $$\int \sqrt{u^2 \pm a^2}\,du = \frac{u}{2}\sqrt{u^2 \pm a^2} \pm \frac{a^2}{2}\ln|u + \sqrt{u^2 \pm a^2}| + C$$

67. $$\int \frac{\sqrt{u^2 + a^2}\,du}{u} = \sqrt{u^2 + a^2} - a\ln\left|\frac{a + \sqrt{u^2 + a^2}}{u}\right| + C$$

68. $$\int \frac{\sqrt{u^2 - a^2}\,du}{u} = \sqrt{u^2 - a^2} - a\,\mathrm{Sec}^{-1}\frac{u}{a} + C$$

69. $$\int \sqrt{a^2 - u^2}\,du = \frac{u}{2}\sqrt{a^2 - u^2} + \frac{a^2}{2}\mathrm{Sin}^{-1}\frac{u}{a} + C$$

70. $$\int u^2\sqrt{a^2 - u^2}\,du = \frac{-u}{4}(a^2 - u^2)^{3/2} + \frac{a^2u}{8}\sqrt{a^2 - u^2} + \frac{a^4}{8}\mathrm{Sin}^{-1}\frac{u}{a} + C$$

71. $$\int \frac{\sqrt{a^2 - u^2}\,du}{u} = \sqrt{a^2 - u^2} - a\ln\left|\frac{a + \sqrt{a^2 - u^2}}{u}\right| + C$$

72. $$\int \frac{\sqrt{a^2 - u^2}\,du}{u^2} = \frac{-\sqrt{a^2 - u^2}}{u} - \mathrm{Sin}^{-1}\frac{u}{a} + C$$

73. $$\int \frac{u^2\,du}{\sqrt{a^2 - u^2}} = \frac{-u}{2}\sqrt{a^2 - u^2} + \frac{a^2}{2}\mathrm{Sin}^{-1}\frac{u}{a} + C$$

---

# E

# Answers to Selected Problems

## CHAPTER ONE

### Exercise 1, Page 6

**1.** $7 < 10$ **3.** $-3 < 4$ **5.** $\frac{3}{4} = 0.75$ **7.** $-18$ **9.** 13 **11.** 4 **13.** 2 **15.** 5 **17.** 5 **19.** 2.00 **21.** 55.86 **23.** 2.96 **25.** 278.38 **27.** 745.6 **29.** 0.5 **31.** 34.9 **33.** 0.8 **35.** 7600 **37.** 274,800 **39.** 2860 **41.** 484,000 **43.** 29.6 **45.** 8370

### Exercise 2, Page 10

**1.** $-1090$ **3.** $-1116$ **5.** 105,233 **7.** 1789 **9.** $-1129$ **11.** $-850$ **13.** 1827 **15.** 4931 **17.** 593.44 **19.** $-0.00031$ **21.** 78,388 $mi^2$ **23.** 35.0 cm **25.** 41.1 Ω

### Exercise 3, Page 14

**1.** $73\bar{0}0$ **3.** 0.525 **5.** $-17,800$ **7.** 22.9 **9.** \$3320 **11.** \$1400 **13.** \$1,570,000 **15.** 17,180 rev **17.** 980.03 cm

### Exercise 4, Page 17

**1.** 163 **3.** $-0.347$ **5.** 0.7062 **7.** 70,840 **9.** 17 **11.** 371.708 m **13.** 10 **15.** 0.00144 **17.** $-0.00000253$ **19.** $-175$ **21.** 0.2003 **23.** 0.0901 **25.** 0.9930 **27.** 313 Ω **29.** 0.279

### Exercise 5, Page 21

**1.** 8 **3.** $-8$ **5.** 1 **7.** 1 **9.** 1 **11.** $-1$ **13.** 100 **15.** 1 **17.** 1000 **19.** 0.01 **21.** 10,000 **23.** 0.00001 **25.** 1.035 **27.** $-112$ **29.** 0.0146 **31.** 0.0279 **33.** 59.8 **35.** 0.0000193 **37.** 125 W **39.** 878,000 $cm^3$ **41.** 5 **43.** 7 **45.** $-2$ **47.** 7.01 **49.** 4.45 **51.** 62.25 **53.** $-7.28$ **55.** $-1.405$ **57.** 4480 Ω

### Exercise 6, Page 25

**1.** 56.4 m **3.** 391 in. **5.** 321,000 $m^2$ **7.** 13.7 $in^2$ **9.** 68.1 slugs **11.** 278 oz **13.** 1243 lb **15.** 43 mi/h **17.** 89.1 km/h

### Exercise 7, Page 26

**1.** 17 **3.** $-37$ **5.** 14 **7.** 14.1 **9.** 233 **11.** 8.00 **13.** \$3975 **15.** 53.3°C **17.** \$13,266

### Exercise 8, Page 32

**1.** $10^2$ **3.** $10^{-4}$ **5.** $10^8$ **7.** 0.01 **9.** 0.1 **11.** $1.86 \times 10^5$ **13.** $2.5742 \times 10^4$ **15.** $9.83 \times 10^4$ **17.** 2850 **19.** 90,000 **21.** 0.003667 **23.** 10 **25.** $10^3$ **27.** $10^3$ **29.** $10^7$ **31.** $10^2$ **33.** 400 **35.** $2.1 \times 10^9$ **37.** $2 \times 10^6$ **39.** $2 \times 10^{-9}$ **41.** $7 \times 10^5$ **43.** $10.7 \times 10^3$ **45.** $3.1 \times 10^{-2}$ **47.** $1.55 \times 10^6$ **49.** $2.11 \times 10^4$ **51.** $1.7 \times 10^2$ **53.** $9.79 \times 10^6$ Ω **55.** $2.72 \times 10^{-6}$ W **57.** $6.77 \times 10^{-5}$ F **59.** 120 h **61.** 174 km **63.** 39.4 kg **65.** $1.45 \times 10^3$ MΩ **67.** $2.584 \times 10^{-3}$ μF

### Exercise 9, Page 40

**1.** 372% **3.** 0.55% **5.** 40.0% **7.** 70.0% **9.** 0.23 **11.** 2.875 **13.** $\frac{3}{8}$ **15.** $1\frac{1}{2}$ **17.** 100 tons **19.** 220 kg **21.** 120 liters **23.** 1090 Ω **25.** \$1400 **27.** 519 **29.** $28\frac{1}{8}$ **31.** 100 **33.** 107 km **35.** \$500 **37.** 50% **39.** 20% **41.** $33\frac{1}{3}$% **43.** 11.6% **45.** 1.5% **47.** 500 **49.** 600 **51.** 100 **53.** 640 **55.** 96.6% **57.** 31.3% **59.** 11% **61.** 5.99% **63.** 67% **65.** 73% **67.** 2.34% **69.** 112.5 and 312.5 V **71.** 37.5% **73.** 25 liters

### Review Problems, Page 42

**1.** 83.35 **3.** 88.1 **5.** 0.346 **7.** 94.7 **9.** 5.46 **11.** 6.76 **13.** 30.6 **15.** 17.4 **17(a)** 179, **(b)** 1.08, **(c)** 4.86, **(d)** 45,700 **19.** 70.2% **21.** $3.63 \times 10^6$ **23.** 12,000 kW **25.** 3.4% **27.** $7 \times 10^{11}$ bbl **29.** 10,200 **31.** $4.16 \times 10^{-10}$ **33.** 2.42 **35.** $-\frac{2}{3} < -0.660$ **37.** 2370 **39.** $-1.72$ **41.** 64.5% **43.** 219 N **45.** 525 ft **47.** 22% **49.** 7216 **51.** 4.939 **53.** 109

**55.** 0.207 **57.** 14.7 **59.** 6.20 **61.** 2 **63.** 83.4 **65.** 93.52 cm **67.** 9.07 **69.** 75.2% **71.** 0.0737 **73.** 7.239 **75.** 121 **77.** 1.21 **79.** 0.30

## CHAPTER 2

### Exercise 1, Page 49

**1.** 2 **3.** 1 **5.** 5 **7.** $b/4$ **9.** $3/2a$

### Exercise 2, Page 51

**1.** $12x$ **3.** $-10ab$ **5.** $2m - c$ **7.** $4a - 2b - 3c - d - 17$ **9.** $11x + 10ax$ **11.** $-26x^3 + 2x^2 + x - 31$ **13.** 0 **15.** $20 - 2ay^3$ **17.** $18b^2 - 3ac - 2d$ **19.** $7m^2 - 5m^3 + 15ab - 4q - z$ **21.** $5.83(a + b)$ **23.** $3a^2 - 3a + x - b$ **25.** $-1.1 - 4.4x$ **27.** $4a^2 - 5a + y + 1$ **29.** $2a + 2b - 2m$ **31.** $6m - 3z$ **33.** $5w + 2z + 6x$ **35.** $a + 12$

### Exercise 3, Page 57

**1.** 16 **3.** $-27$ **5.** 0.0001 **7.** $a^8$ **9.** $y^{2a-2}$ **11.** $10^{a+b}$ **13.** $y^3$ **15.** $x$ **17.** 100 **19.** $x$ **21.** $x^{12}$ **23.** $a^{xy}$ **25.** $x^{2a+2}$ **27.** $8x^3$ **29.** $27a^3b^3c^3$ **31.** $-1/27$ **33.** $8a^3/27b^6$ **35.** $1/a^2$ **37.** $y^3/27$ **39.** $9b^6/4a^2$ **41.** $2/x^2 + 3/y^3$ **43.** $x^{-1}$ **45.** $x^2y^{-2}$ **47.** $a^{-3}b^{-2}$ **49.** 1 **51.** 1/9 **53.** 1

### Exercise 4, Page 65

**1.** $x^5$ **3.** $10a^3b^4$ **5.** $36a^6b^5$ **7.** $-8p^4q^4$ **9.** $2a^2 - 10a$ **11.** $x^2y - xy^2 + x^2y^2$ **13.** $-12p^3q - 8p^2q^2 + 12pq^3$ **15.** $6a^3b^3 - 12a^3b^2 + 9a^2b^3 + 3a^2b^2$ **17.** $a^2 + ac + ab + bc$ **19.** $x^3 - xy + 3x^2y - 3y^2$ **21.** $9m^2 - 9n^2$ **23.** $49c^2d^4 - 16y^6z^2$ **25.** $x^4 - y^4$ **27.** $a^2 - ac - c - 1$ **29.** $m^3 - 5m^2 - m + 14$ **31.** $x^8 - x^2$ **33.** $-z^3 + 2z^2 - 3z + 2$ **35.** $a^3 + y^3$ **37.** $a^3 - y^3$ **39.** $x^3 - px^2 - mx^2 + mpx + nx^2 - npx - mnx + mnp$ **41.** $m^5 - 1$ **43.** $9a^2x - 6abx + b^2x$ **45.** $x^3 - 3x - 2$ **47.** $x^3 - cx^2 + bx^2 - bcx + ax^2 - acx + abx - abc$ **49.** $-c^5 - c^4 + c + 1$ **51.** $4a^6 - 10a^4c^2 + 14a^4 - 25a^2c^2 + 10a^2$ **53.** $a^3 + a^2c - ab^2 + b^2c - b^3 + 2abc - ac^2 + bc^2 - c^3 + a^2b$ **55.** $n^5 - 34n^3 + 57n^2 - 20$ **57.** $a^2 + 2ac + c^2$ **59.** $m^2 - 2mn + n^2$ **61.** $A^2 + 2AB + B^2$ **63.** $14.4a^2 - 16.7ax + 4.84x^2$ **65.** $4c^2 - 12cd + 9d^2$ **67.** $a^2 - 2a + 1$ **69.** $y^4 - 40y^2 + 400$ **71.** $a^2 + 2ab + b^2 + 2ac + 2bc + c^2$ **73.** $x^2 + 2x - 2xy - 2y + y^2 + 1$ **75.** $x^4 + 2x^3y + 3x^2y^2 + 2xy^3 + y^4$ **77.** $a^3 + 3a^2b + 3ab^2 + b^3$ **79.** $p^3 - 9p^2q + 27pq^2 - 27q^3$ **81.** $a^3 - 3a^2b + 3ab^2 - b^3$ **83.** $3a + 3b + 3c$ **85.** $p - q$ **87.** $31y - 5z$ **89.** $2x - 4$ **91.** $7 - 5x$

### Exercise 5, Page 73

**1.** $z^2$ **3.** $-5xyz$ **5.** $2f$ **7.** $-2xz$ **9.** $-5xy^2$ **11.** $3c$ **13.** $-5y$ **15.** $-3w$ **17.** $7n$ **19.** $4b$ **21.** $b^2$ **23.** $x^{m-n}$ **25.** $4s$ **27.** $-24$ **29.** $-5z/xy$ **31.** $-5n^2/m^2x$ **33.** $3ab$ **35.** $3x^5$ **37.** $42/y$ **39.** $-4x$ **41.** $2a^2 - a$ **43.** $7x^2 - 1$ **45.** $3x^4 - 5x^2$ **47.** $2x^2 - 3x$ **49.** $a + 2c$ **51.** $ax - 2y$ **53.** $2x + y$ **55.** $ab - c$ **57.** $xy - x^2 - y^2$ **59.** $1 - a - b$ **61.** $a - 3b + c^2$ **63.** $x - 2y + y^2/x$ **65.** $m + 2 - 3m/n$ **67.** $a - b - c$ **69.** $ab - 2 - 3b^2$ **71.** $5a - 10x - 2$ **73.** $x + 8$ **75.** $x - 6 + 15/(x + 2)$ **77.** $a + 5$ **79.** $9a^2 + 6ab + 4b^2$ **81.** $5x + 13 - 35/(3 - x)$ **83.** $-3a - 2$ **85.** $x^4 + x + 1 + R(x^2 + x + 1)$ **87.** $x^2 + 3x + 1$ **89.** $a + b - c$ **91.** $c^2 + c + 2$ **93.** $x^3 + 2x^2 - 6x + 5$ **95.** $4x^3 - 3x^2 - x + 6$ **97.** $2x^3 - x^2 + 5$

### Review Problems, Page 75

**1.** $b^6 - b^4x^2 + b^4x^3 - b^2x^5 + b^2x^4 - x^6$ **3.** $3.86 \times 10^{14}$ **5.** $9x^4 - 6mx^3 - 6m^3x + 10m^2x^2 + m^4$ **7.** $8x^3 + 12x^2 + 6x + 1$ **9.** $6a^2x^5$ **11.** $16a^2 - 24ab + 9b^2$ **13.** $x^2y^2 - 6xy + 8$ **15.** $4a^2 - 12ab + 9b^2$ **17.** $ab - b^4 - a^2b$ **19.** $16m^4 - c^4$ **21.** $2a^2$ **23.** $24 - 8y$ **25.** $6x^3 + 9x^2y - 3xy^2 - 6y^3$ **27.** $13w - 6$ **29.** $a^5 + 32c^5$ **31.** $(a - c)^{m-2}$ **33.** $x + y$ **35.** $b^3 - 9b^2 + 27b - 27$ **37.** $x^2/2y^3$ **39.** $x^2 + 2x - 8$ **41.** $8.01 \times 10^6$ **43.** $10x^3y^4$ **45.** $x^3 + 3x^2 - 4x$ **47.** $1.77 \times 10^8$ **49.** $6a^3b^3$

## CHAPTER 3

### Exercise 1, Page 82

**1.** 33 **3.** 4/5 **5.** 1/2 **7.** 1/2 **9.** 4 **11.** 3 **13.** −1.4 **15.** 1 **17.** −1/2 **19.** 2 **21.** 5 **23.** 28 **25.** 3 **27.** 3 **29.** 35 **31.** −5/23 **33.** 5/6 **35.** 19/5 **37.** 1 **39.** 3/25 **41.** 0 **43.** 0

### Exercise 2, Page 87

**1.** $5x + 8$ **3.** $x$ and $83 - x$ **5.** $x$ and $6 - x$ **7.** $A$, $2A$, and $180 - 3A$ **9.** $82x$ km **11.** 5 **13.** 57, 58, 59 **15.** 14 **17.** 6 **19.** 4 **21.** 20 **23.** 14 **25.** 81

### Exercise 3, Page 89

**1.** 8 technicians **3.** $130,137 **5.** 57,000 gal **7.** $64.29 to brother, $128.57 to uncle, $257.14 to bank **9.** $94,000 **11.** $12,000

### Exercise 4, Page 92

**1.** 333 gal **3.** 485 kg of 18% mixture, 215 kg of 31% mixture **5.** 1375 kg **7.** 1.3 liters **9.** 60 lb **11.** 59 lb

### Exercise 5, Page 95

**1.** 4.97 ft **3.** $R_1$ = 753 lb downwards, $R_2$ = 3100 lb upwards **5.** 339 cm **7.** 6445 lb at right, 8425 lb at left

### Review Problems, Page 96

**1.** 12 **3.** 5 **5.** 0 **7.** $10\frac{1}{2}$ **9.** 2 **11.** 4 **13.** 4 **15.** 6 **17.** −5/7 **19.** −6 **21.** 9.3 **23.** 7 **25.** 6 **27.** 47 cm, 53 cm **29.** 12 kg **31.** $13,500 **33.** 15,000 ft **35.** −13/5 **37.** 13 **39.** −5/7 **41.** −5/2 **43.** −7/3 **45.** $R_1$ = 608 lb, $R_2$ = 605 lb

## CHAPTER 4

### Exercise 1, Page 104

**1.** Is a function **3.** Not a function **5.** Not a function **7.** Yes **9.** $y = x^3$ **11.** $y = x + 2x^2$ **13.** $y = (2/3)(x - 4)$ **15.** The amount by which 5 exceeds $x$ **17.** Twice the cube root of $x$ **19.** $A = f(b, h) = bh/2$ **21.** $V = 4\pi r^3/3$ **23.** $d = 55t$ mi **25.** $H = 125t - gt^2/2$ **27.** Domain: −10, −7, 0, 5, 10; Range: 3, 7, 10, 20 **29.** $x \geq 7$, $y \geq 0$ **31.** $x \neq 0$, all $y$ **33.** $x \neq 9$, $y \neq 1$ **35.** $x < 1$, $y > 0$ **37.** $x \geq 1$, $y \geq 0$ **39.** $-5 \leq y \leq 70$

### Exercise 2, Page 112

**1.** Explicit **3.** Implicit **5.** $x$ is independent, $y$ is dependent **7.** $x$ and $y$ are independent, $w$ is dependent **9.** $x$ and $y$ are independent, $z$ is dependent **11.** $y = 2/x + 3$ **13.** $y = x(2x + 1)/3$ **15.** $e = PL/aE$ **17.** $x^2 + 2$ **19.** $(4 - 3x)^3$ **21.** −125 **23.** 6 **25.** −21 **27.** 12.5 **29.** −5 **31.** −15 **33.** 5/4 **35.** $2a^2 + 4$ **37.** $5a + 5b + 1$ **39.** $9\frac{1}{2}$ **41.** 142 **43.** 20 **45.** −52 **47.** 18 **49.** $y = (x + 15)/14$ **51.** $y = -x - 18$ **53.** $1.063 \times 10^4\ \Omega$, $1.084 \times 10^4\ \Omega$, $1.105 \times 10^4\ \Omega$

### Exercise 3, Page 115

**1.** Fourth **3.** Second **5.** Fourth **7.** First and fourth **9.** $x = 7$ **11.** $E(-1.8, -0.7)$, $F(-1.4, -1.4)$, $G(1.4, -0.6)$, $H(2.5, -1.8)$

13.

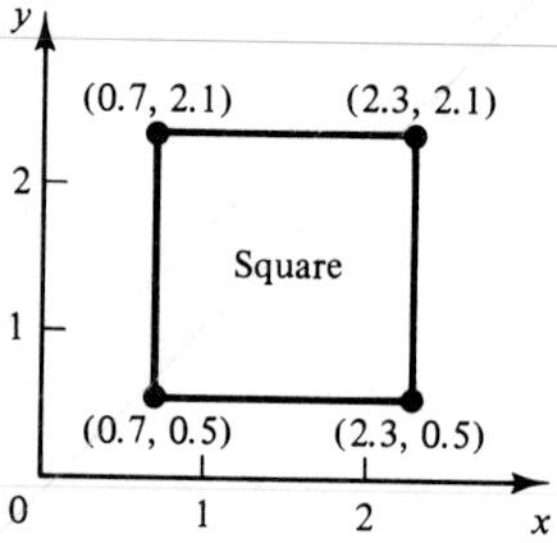


15.

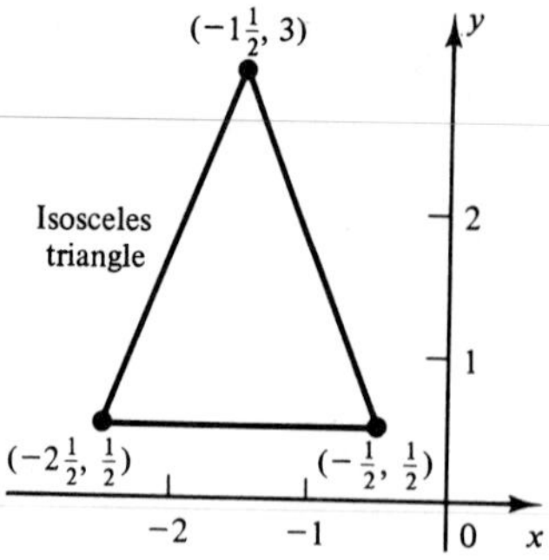


17. (4, −5)

## Exercise 4, Page 121

1.

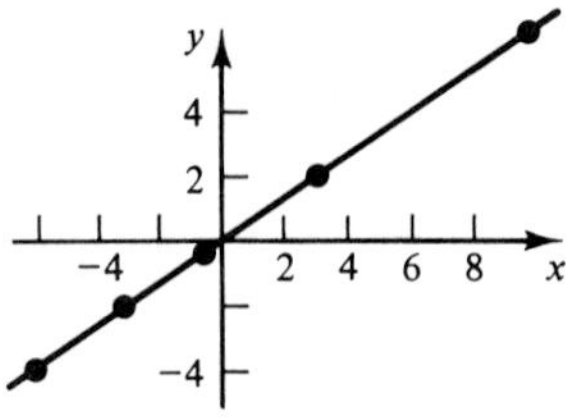

3.

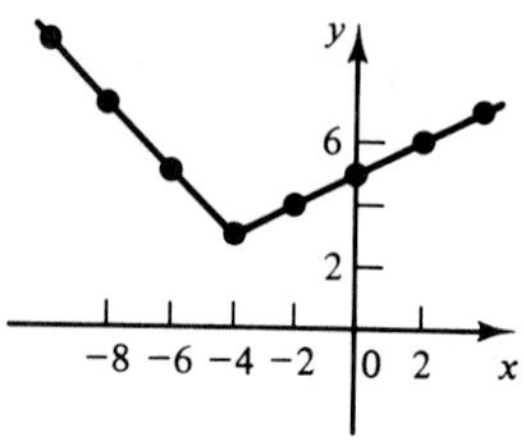

5.

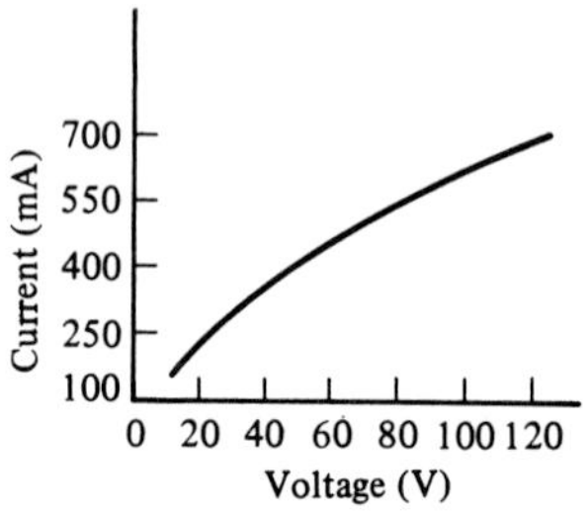


7.

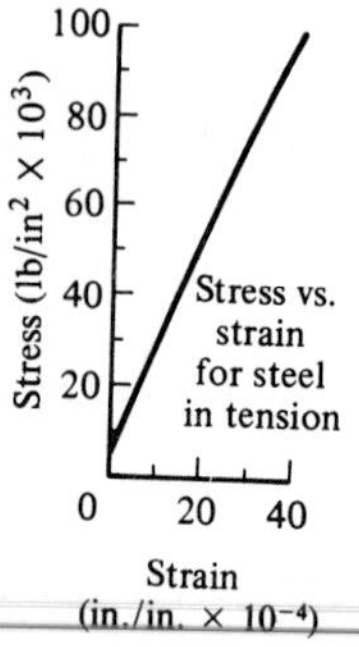


9. (a) and (c)

11.

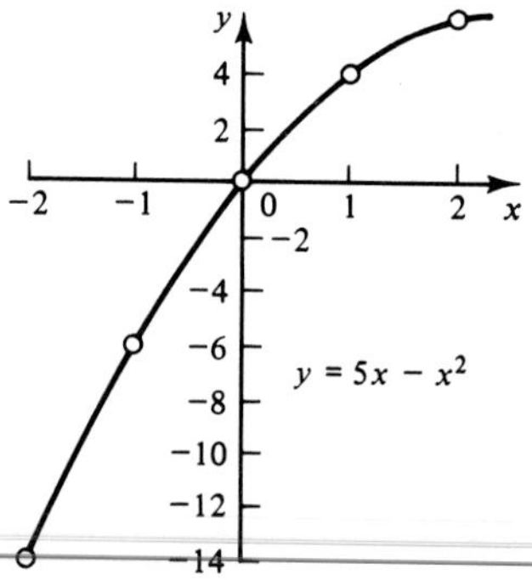


13.

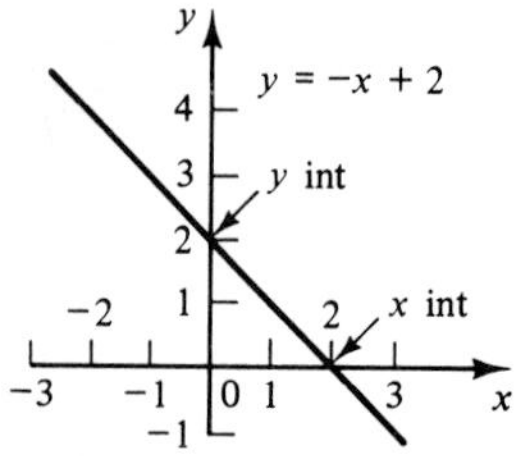


15.

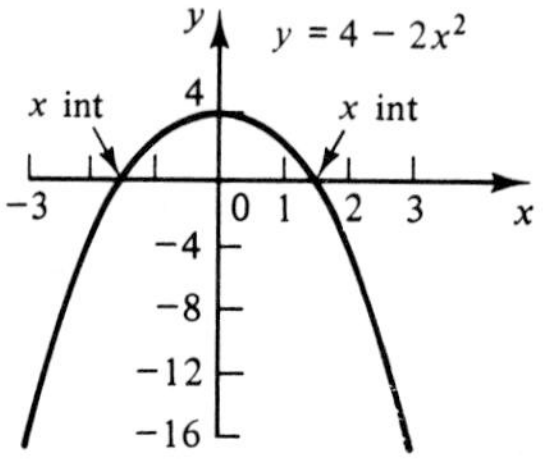


17.

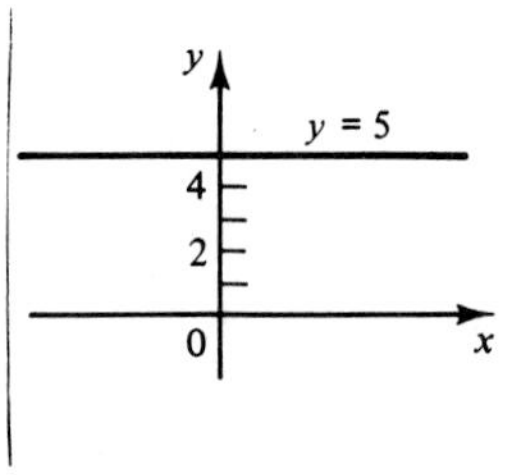


19.

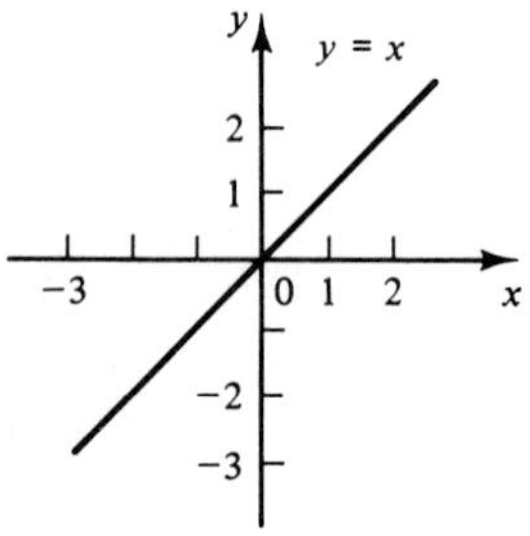


21. a: Domain: $x \geq 0$, Range: all $y$; b: Domain: all $x$, Range: all $y$

23.

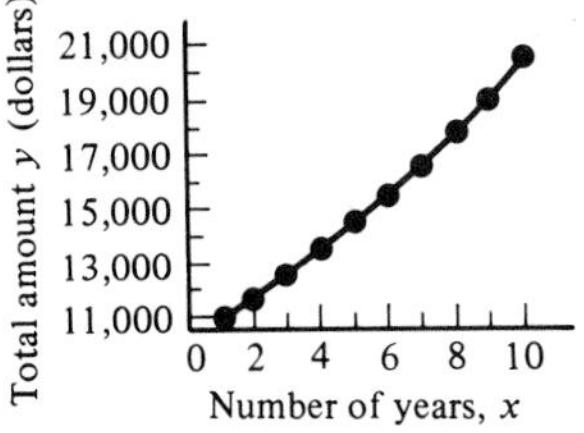


25.

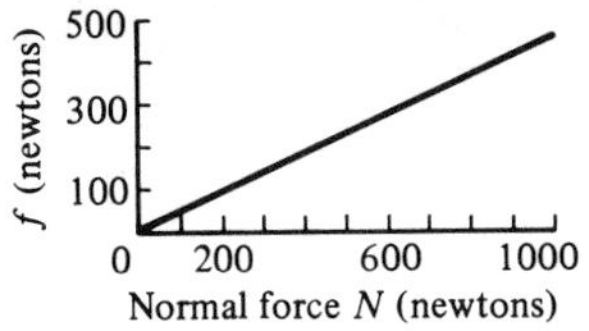


27.

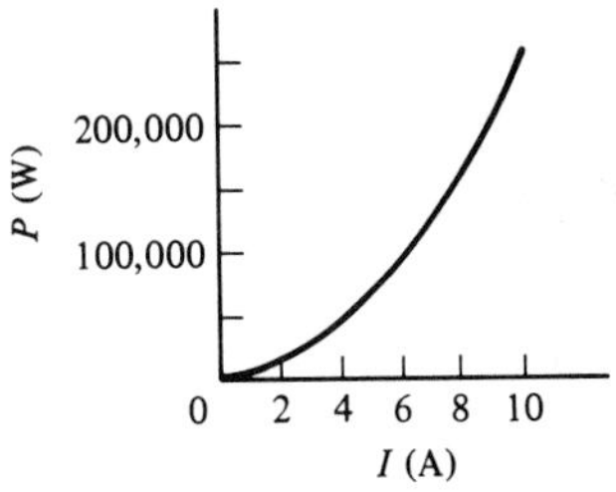


29.

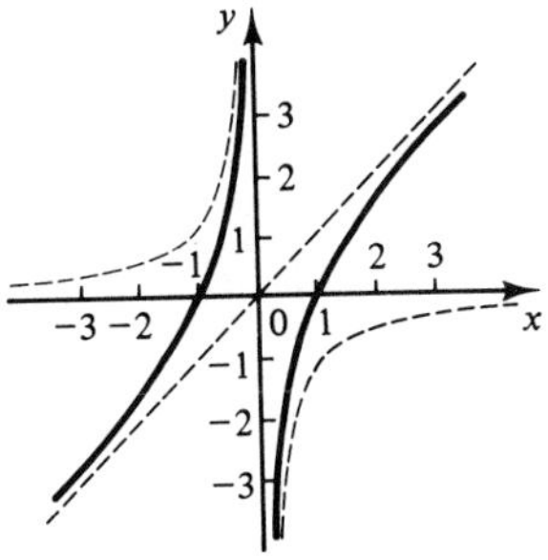

31.

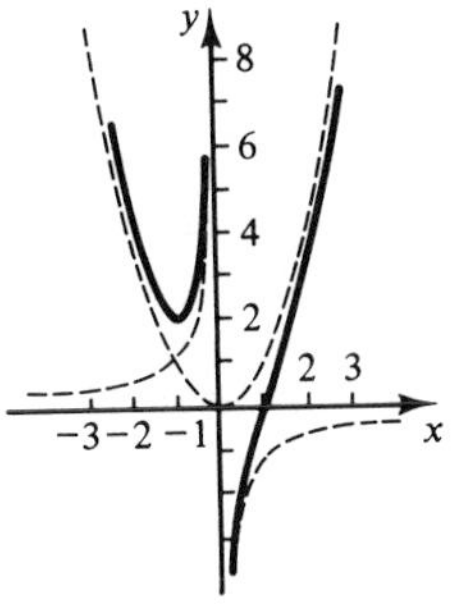

33.

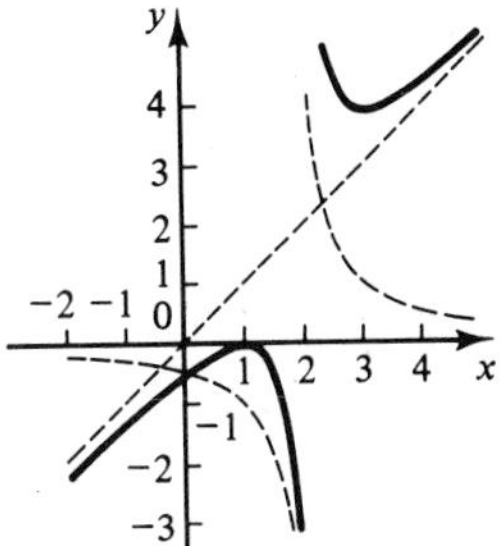

35.

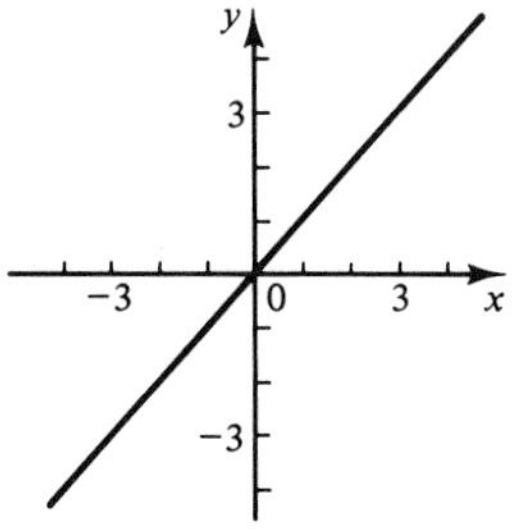

37.

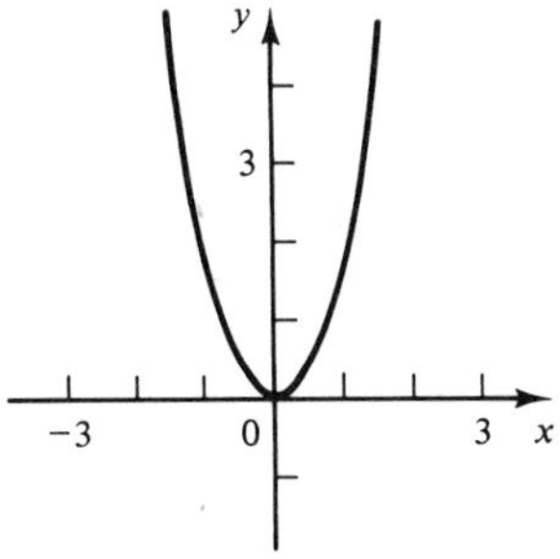

Exercise 5, Page 126

1.

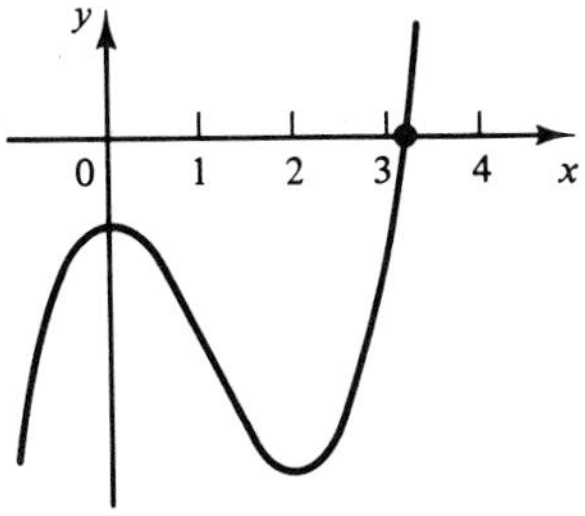

3.

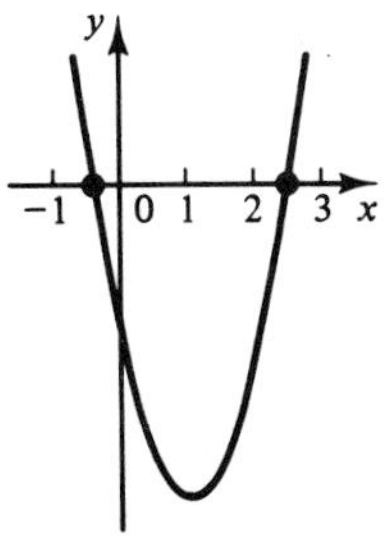

5.

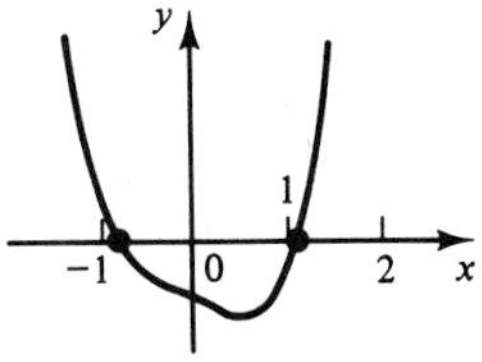

7.

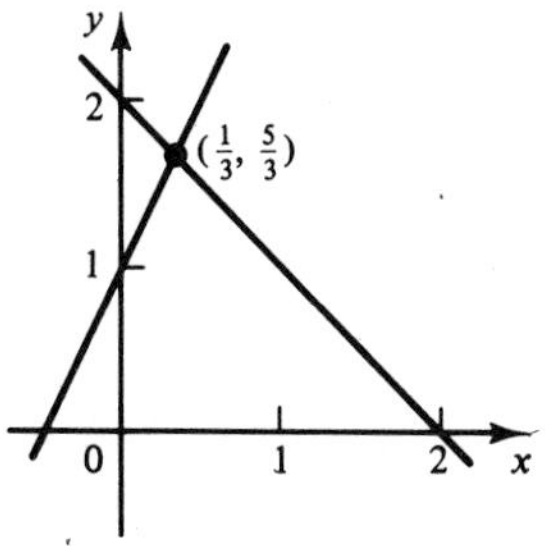


9.

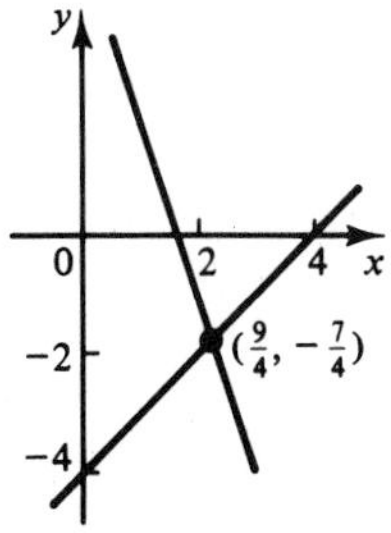

### Review Problems, Page 127

**1(a)** Is a function, **(b)** Not a function, **(c)** Not a function  **3.** $S = 4\pi r^2$
**5(a)** Explicit, $y$ independent, $w$ dependent, **(b)** Implicit  **7.** $w = (3 - x^2 - y^2)/2$  **9.** $7x^2$  **11.** 6

**13.**

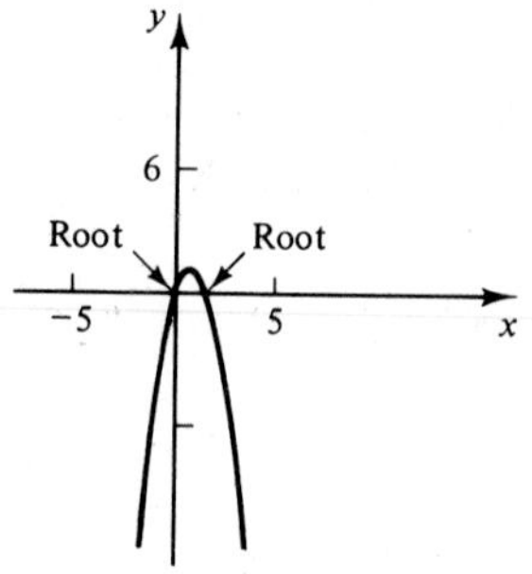


**15.** 13/90  **17.** $y$ is 7 less than five times the cube of $x$  **19.** $x = \pm(6 - y)/3$  **21.** $8u^2 - 29$
**23.** 28  **25.** $y = (x + 5)/9$

**27.**

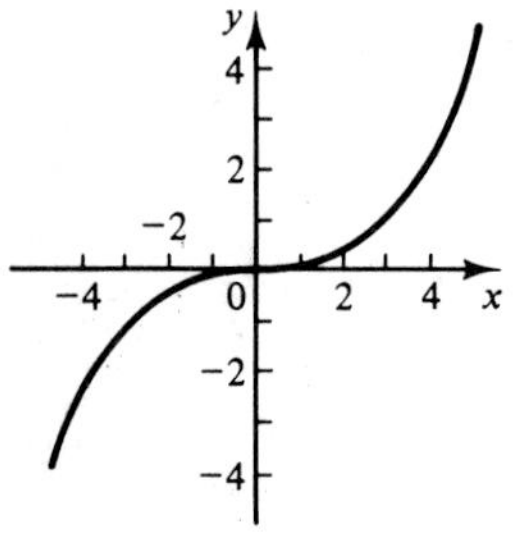


## CHAPTER 5

### Exercise 1, Page 133

**1(a)** 62.8°, **(b)** 64.6°, **(c)** 37.7°  **3.** 5.05  **5.** $A = C = 46.3°$, $B = D = 134°$

### Exercise 2, Page 140

**1.** \$4405  **3.** 605 cm  **5.** 75.0 ft, 117 ft, 205 ft  **7.** 35.4 m  **9.** 35.4 m  **11.** 15.5 ft
**13.** 35.4 ft  **15.** 0.577 in.  **17.** 39.0 in.  **19.** 43.5 acres  **21.** 2.62 m  **23.** 5.2 mm

### Exercise 3, Page 145

**1.** 423 square units  **3.** 248 square units  **5.** \$3042  **7.** \$348  **9.** \$3220  **11.** \$861

### Exercise 4, Page 147

**1.** 7.16 m  **3.** 15.8 in.  **5.** 104 m  **7.** 319 acres  **9.** 0.74 unit  **11.** 44.7 units
**13.** 218 cm  **15.** 69.1 $m^2$  **17.** 247 cm

### Exercise 5, Page 150

**1.** $6.34 \times 10^7$ $in^3$  **3(a)** 4.2 m, **(b)** 150 $m^3$  **5.** 176 $in^3$  **7.** 0.64 in.  **9.** 1.0 in.  **11.** 760 $in^2$
**13.** 3800 lb  **15.** 38.0 $ft^3$  **17.** $1.72 \times 10^9$, $6.96 \times 10^6$  **19.** 15.8, 1.12  **21.** 213 lb

### Review Problems, Page 152

**1.** 1440 mi/h  **3.** 2.89 m  **5.** 13 m  **7.** 176,000 square units  **9.** 43.1 in.  **11.** 17.5 m
**13.** 161°  **15.** 213 $cm^3$  **17.** 1030 $in^2$  **19.** 1150 $m^2$  **21.** 87,800 $cm^3$  **23.** 11.0 $ft^3$  **25.** 114 $cm^3$

## CHAPTER 6

### Exercise 1, Page 157

**1.** 0.485 rad **3.** 0.615 rad **5.** 3.5 rad **7.** 0.7595 rev **9.** 0.191 rev **11.** 0.215 rev **13.** 162° **15.** 171° **17.** 29.45° **19.** 244°57′45″ **21.** 161°54′36″ **23.** 185°58′19″

### Exercise 2, Page 165

**5.** 53° **7.** 63° **9.** 0.528, 0.849, 0.622, 1.61, 1.18, 1.89 **11.** 0.447, 0.894, 0.500 **13.** 0.832, 0.555, 1.50 **15.** 0.491, 0.871, 0.564 **17.** 0.9794 **19.** 3.0415 **21.** 0.9319 **23.** 1.0235 **25.** 0.955, 0.296, 3.230 **27.** 0.546, 0.838, 0.652 **29.** 0.677, 0.736, 0.920 **31.** 30.0° **33.** 31.9° **35.** 28.9° **37.** 3/5, 4/5, 3/4 **39.** 5/13, 12/13, 5/12 **41.** 75.0° **43.** 40.5° **45.** 13.6°

### Exercise 3, Page 170

**1.** $B = 47.1°$, $b = 167$, $c = 228$ **3.** $b = 1.08$, $A = 58.1°$, $c = 2.05$ **5.** $B = 0.441$ rad, $b = 134$, $c = 314$ **7.** $c = 470$, $A = 54.3°$, $B = 35.7°$ **9.** $a = 2.80$, $A = 35.2°$, $B = 54.8°$ **11.** $b = 25.6$, $A = 46.9°$, $B = 43.1°$ **13.** cos 52° **15.** cot 71° **17.** tan 26.8° **19.** cot 54°46′ **21.** cos 0.696 rad

### Exercise 4, Page 171

**1.** 402 m **3.** 285 ft **5.** 64.9 m **7.** 40.0 yd **9.** 128 m **11.** 21.4 km, 14.7 km **13.** 432 mi, S58°31′W **15.** 156 mi N., 162 mi E. **17.** 30.2°, 20.5 ft **19.** 37.6° **21.** 139, 112 **23.** 63°, 117° **25.** 125 cm **27.** 2.55 in. **29.** 9.398 cm **31.** 7.91 cm **33.** 3.366 in.

### Exercise 5, Page 177

**1.** 3.28, 3.68 **3.** 0.9917, 1.602 **5.** 616, 51.7° **7.** 8811, 56.70° **9.** 2.42, 31.1° **11.** 8.36, 35.3° **13.** 4.57, 60.2°

### Exercise 6, Page 179

**1.** 12.3 N **3.** 1522 N **5.** 25.6° **7.** 4.94 tons **9.** 119 km/h **11.** −224 cm/s, 472 cm/s **13.** 4270 m/min, 6230 m/min **15.** $R = 2690\ \Omega$, $Z = 3774\ \Omega$ **17.** $R = 861\ \Omega$, $X = 458\ \Omega$ **19.** 4427 Ω, 37.2°

### Review Problems, Page 180

**1.** 38°12′, 0.667 rad, 0.106 rev **3.** 157.3 deg, 157°18′, 0.4369 rev **5.** 0.919, 0.394, 2.33, 0.429, 2.54, 1.09, 66.8° **7.** 0.600, 0.800, 0.750, 1.33, 1.25, 1.67, 36.9° **9.** 0.9558, 0.2940, 3.2506, 0.3076, 3.4009, 1.0463 **11.** 0.8674, 0.4976, 1.7433, 0.5736, 2.0098, 1.1528 **13.** 34.5° **15.** 70.7° **17.** 65.88° **19.** $B = 61.5°$, $a = 2.02$, $c = 4.23$ **21.** 356, 810 **23.** 473, 35.5° **25.** 7.29 ft **27.** 0.5120 **29.** 1.3175 **31.** 2.1742 **33.** 60.1° **35.** 46.5° **37.** 46.5°

## Chapter 7

### Exercise 1, Page 185

**1.** $y^2(3 + y)$ **3.** $x^3(x^2 - 2x + 3)$ **5.** $a(3 + a - 3a^2)$ **7.** $5(x + y)[1 + 3(x + y)]$ **9.** $(1/x)(3 + 2/x - 5/x^2)$ **11.** $a^2(5b + 6c)$ **13.** $xy(4x + cy + 3y^2)$ **15.** $3ay(a^2 - 2ay + 3y^2)$ **17.** $cd(5a - 2cd + b)$ **19.** $4x^2(2y^2 + 3z^2)$ **21.** $ab(3a + c - d)$ **23.** $L_0(1 + \alpha t)$ **25.** $R_1[1 + \alpha(t - t_1)]$ **27.** $t(v_0 + at/2)$ **29.** $(4\pi D/3)(r_2^3 - r_1^3)$

### Exercise 2, Page 188

**1.** $(2 - x)(2 + x)$ **3.** $(3a - x)(3a + x)$ **5.** $4(x - y)(x + y)$ **7.** $(x - 3y)(x + 3y)$ **9.** $(3c - 4d)(3c + 4d)$ **11.** $(3y - 1)(3y + 1)$ **13.** $(m - n)(m + n)(m^2 + n^2)$ **15.** $(m^n - n^m)(m^n + n^m)$ **17.** $(a^2 - b)(a^2 + b)(a^4 + b^2)(a^8 + b^4)$ **19.** $(5x^2 - 4y^3)(5x^2 + 4y^3)$ **21.** $(4a^2 - 11)(4a^2 + 11)$ **23.** $(5a^2b^2 - 3)(5a^2b^2 + 3)$ **25.** $\pi(r_2 - r_1)(r_2 + r_1)$ **27.** $4\pi(r_1 - r_2)(r_1 + r_2)$ **29.** $m(v_1 - v_2)(v_1 + v_2)/2$ **31.** $\pi h(R - r)(R + r)$

### Exercise 3, Page 193

**1.** Not factorable **3.** Factorable **5.** Factorable **7.** $(x - 7)(x - 3)$ **9.** $(x - 9)(x - 1)$ **11.** $(x + 10)(x - 3)$ **13.** $(x + 4)(x + 3)$ **15.** $(x - 7)(x + 3)$ **17.** $(x + 4)(x + 2)$ **19.** $(b - 5)(b - 3)$ **21.** $(b - 4)(b + 3)$ **23.** $(x + 7y)(x + 12y)$ **25.** $(a + 3b)(a - 2b)$ **27.** $(a + 9x)(a - 7x)$ **29.** $(a - 24bc)(a + 4bc)$ **31.** $(a + 48bc)(a + bc)$ **33.** $(a + b + 1)(a + b - 8)$ **35.** $(x^2 - 5y^2)(x - 2y)(x + 2y)$ **37.** $(t - 2)(t - 12)$ **39.** $(m - 500)(m - 500)$ **41.** $(R - 300)(R - 100)$

### Exercise 4, Page 195

**1.** $(a^2 + 4)(a + 3)$ **3.** $(x - 1)(x^2 + 1)$ **5.** $(x - b)(x + 3)$ **7.** $(x - 2)(3 + y)$ **9.** $(x + y - 2)(x + y + 2)$ **11.** $(m - n + 2)(m + n - 2)$ **13.** $(x + 3)(a + b)$ **15.** $(a + b)(a - c)$ **17.** $(2 + x^2)(a + b)$ **19.** $(2a - c)(3a + b)$ **21.** $(a + b)(b - c)$ **23.** $(x - a - b)(x + a + b)$

### Exercise 5, Page 198

**1.** $(4x - 1)(x - 3)$ **3.** $(5x + 1)(x + 2)$ **5.** $(3b + 2)(4b - 3)$ **7.** $(2a - 3)(a + 2)$ **9.** $(x - 7)(5x - 3)$ **11.** $3(x + 1)(x + 1)$ **13.** $(3x + 2)(x - 1)$ **15.** $2(2x - 3)(x - 1)$ **17.** $(2a - 1)(2a + 3)$ **19.** $(3a - 7)(3a + 2)$ **21.** $(7x^3 - 3y)(7x^3 + 5y)$ **23.** $(4x^3 + 1)(x^3 + 3)$ **25.** $(5x^{2n} + 1)(x^{2n} + 2)$ **27.** $3[(a + x)^n + 1][(a + x)^n - 2]$ **29.** $(2t - 9)(8t - 5)$

### Exercise 6, Page 200

**1.** $(x + 2)^2$ **3.** $(y - 1)^2$ **5.** $2(y - 3)^2$ **7.** $(3 + x)^2$ **9.** $(3x + 1)^2$ **11.** $9(y - 1)^2$ **13.** $4(2 + a)^2$ **15.** $(7a - 2)^2$ **17.** $(x + y)^2$ **19.** $(aw + b)^2$ **21.** $(x + 5a)^2$ **23.** $(x + 4y)^2$ **25.** $(z^3 + 8)^2$ **27.** $(7 - x^3)^2$ **29.** $(a - b)^2(a + b)^2$ **31.** $(2a^n + 3b^n)^2$

### Exercise 7, Page 202

**1.** $(4 + x)(16 - 4x + x^2)$ **3.** $2(a - 2)(a^2 + 2a + 4)$ **5.** $(x - 1)(x^2 + x + 1)$ **7.** $(x + 1)(x^2 - x + 1)$ **9.** $(a + 4)(a^2 - 4a + 16)$ **11.** $(x + 5)(x^2 - 5x + 25)$ **13.** $8(3 - a)(9 + 3a + a^2)$ **15.** $(7 + 4x)(49 - 28x + 16x^2)$ **17.** $(y^3 + 4x)(y^6 - 4xy^3 + 16x^2)$ **19.** $(3x^5 + 2a^2)(9x^{10} - 6x^5a^2 + 4a^4)$ **21.** $(2a^{2x} - 5b^x)(4a^{4x} + 10a^{2x}b^x + 25b^{2x})$ **23.** $(4x^n - y^{3n})(16x^{2n} + 4x^ny^{3n} + y^{6n})$ **25.** $(ab + 3x)(a^2b^2 - 3abx + 9x^2)$ **27.** $(xyz - 2)(x^2y^2z^2 + 2xyz + 4)$ **29.** $(3x + y - z)[9x^2 - 3x(y - z) + (y - z)^2]$ **31.** $(4\pi/3)(r_2 - r_1)(r_2^2 + r_2r_1 + r_1^2)$

### Review Problems, Page 203

**1.** $(x - 5)(x + 3)$ **3.** $(x^3 - y^2)(x^3 + y^2)$ **5.** $(2x - 1)(x + 2)$ **7.** $(a - b + c + d)[(a - b)^2 - (a - b)(c + d) + (c + d)^2]$ **9.** $(x/y)(x - 1)$ **11.** $(y + 5)(x - 2)$ **13.** $(x - y)(1 - b)$ **15.** $(y + 2 - z)(y + 2 + z)$ **17.** $(x - 3)(x - 4)$ **19.** $(4x^{2n} + 9y^{4n})(2x^n + 3y^{2n})(2x^n - 3y^{2n})$ **21.** $(3a + 2z^2)^2$ **23.** $(x^m + 1)^2$ **25.** $(x - y - z)(x - y + z)$ **27.** $(1 + 4x)(1 - 4x)$ **29.** $x^2(3x + 1)(3x - 1)$ **31.** $(p - q - 3)[(p - q)^2 + 3(p - q) + 9]$ **33.** $(3a - 2w)(9a^2 + 6aw + 4w^2)$ **35.** $4(2x - y)^2$ **37.** $(3b + y)(2a + x)$ **39.** $2(y^2 - 3)(y^2 + 3)$ **41.** $(1/R)(V_2 - V_1)(V_2 + V_1)$

## CHAPTER 8

### Exercise 1, Page 209

**1.** $x \neq 0$ **3.** $x \neq 5$ **5.** $x \neq 2, x \neq 1$ **7.** 0.5833 **9.** 0.9375 **11.** 3.6667 **13.** 7/16 **15.** 11/16 **17.** 7/9 **19.** $-1$ **21.** $d - c$ **23.** $a/3$ **25.** $3m/4p^2$ **27.** $(2a - 3b)/2a$ **29.** $x/(x - 1)$ **31.** $\dfrac{x + 2}{x^2 + 2x + 4}$ **33.** $\dfrac{n(m - 4)}{3(m - 2)}$ **35.** $\dfrac{2(a + 1)}{a - 1}$ **37.** $\dfrac{x + z}{x^2 + xz + z^2}$ **39.** $\dfrac{a - 2}{a - 3}$ **41.** $(x - 1)/2y$ **43.** $\dfrac{2}{3(x^4y^4 - 1)}$ **45.** $\dfrac{x + y}{x - y}$ **47.** $(b - 3 + a)/5$

### Exercise 2, Page 213

**1.** 2/15 **3.** 6/7 **5.** $2\frac{2}{15}$ **7.** 75/8 **9.** $1\frac{1}{2}$ **11.** $a^4b^4/2y^{2n}$ **13.** $7p^2/4xz$ **15.** $x - a$ **17.** $ab/(x^2 - y^2)$ **19.** $cd/(x^2 - y^2)^2$ **21.** $\dfrac{(x - 1)(x + 4)}{(x + 3)(x - 5)}$ **23.** $\dfrac{(x + 1)(x + 2)}{(x - 4)(x - 3)}$ **25.** $\dfrac{(2x - 3)(x + 1)}{(x - 1)(x - 6)}$

**27.** $1\frac{15}{16}$ **29.** 9/128 **31.** 5/18 **33.** $5\frac{1}{4}$ **35.** $b^2$ **37.** $8a^3/3dxy$ **39.** $(3an + cm)/(x^4 - y^4)$
**41.** $1/(c - d)$ **43.** $1/(x + 2)$ **45.** $\dfrac{6(x-1)}{x+2}$ **47.** $\dfrac{(p+2)^2}{(p-1)^2}$ **49.** $\dfrac{(z-1)(z+4)}{(z+3)(z-5)}$ **51.** $1\frac{2}{3}$ **53.** $2\frac{5}{12}$
**55.** $9\frac{2}{5}$ **57.** $x + 1/x$ **59.** $\dfrac{2}{x} - \dfrac{1}{2}$ **61.** $3m/4\pi r^3$ **63.** $2P/(a + b)h$ **65.** $4\pi d^3/3$

## Exercise 3, Page 218

**1.** 1 **3.** 1/7 **5.** 5/3 **7.** 7/6 **9.** 19/16 **11.** 7/18 **13.** 13/5 **15.** −92/15 **17.** 67/35
**19.** 8/3 **21.** 13/4 **23.** 91/16 **25.** 6/a **27.** $(a + 3)/y$ **29.** $x$ **31.** $x/(a - b)$
**33.** $19/(a + 1)$ **35.** $19a/10x$ **37.** $\dfrac{2ax - 3x + 12}{6x}$ **39.** $(5b - a)/6$ **41.** $\dfrac{9-x}{x^2-1}$ **43.** $\dfrac{2b(2a-b)}{a^2-b^2}$
**45.** $\dfrac{2x^2-1}{x^4-x^2}$ **47.** $\dfrac{2(3x^2+2x-3)}{x^3-7x-6}$ **49.** $\dfrac{x^2-2x}{(x-3)(x+3)}$ **51.** $\dfrac{5x^2+7x+16}{x^3-8}$ **53.** $-\dfrac{d}{(x+1)(x+d+1)}$
**55.** $\dfrac{d(x^2+dx-1)}{x(x+d)}$ **57.** $(x^2 + 1)/x$ **59.** $2/(x^2 - 1)$ **61.** $(3a - a^2 - 2)/a$ **63.** $(abx + a + b)/ax$
**65.** $4\frac{7}{12}$ mi **67.** 2/25 min **69.** 11/8 machines **71.** $\dfrac{2h(a+b) - \pi d^2}{4}$ **73.** $\dfrac{V_1V_2d + VV_2d_1 + VV_1d_2}{VV_1V_2}$

## Exercise 4, Page 222

**1.** 85/12 **3.** 65/132 **5.** 69/95 **7.** $\dfrac{3(4x+y)}{4(3x-y)}$ **9.** $y/(y - x)$ **11.** $\dfrac{5(3a^2+x)}{3(20+x)}$
**13.** $\dfrac{2(3acx+2d)}{3(2acx+3d)}$ **15.** $\dfrac{2x^2-y^2}{x-3y}$ **17.** $\dfrac{(x+2)(x-1)}{(x+1)(x-2)}$ **19.** 1 **21.** $\dfrac{4}{3(a+1)}$ **23.** 1
**25.** $\dfrac{x+y+z}{x-y+z}$ **27.** $\dfrac{\rho L_1L_2}{A_2L_1 + A_1L_2}$

## Exercise 5, Page 226

**1.** 12 **3.** 20 **5.** 14 **7.** 72 **9.** 24 **11.** 7 **13.** 24 **15.** 24 **17.** 5/2 **19.** 12
**21.** 17 **23.** 3/13 **25.** −2/3 **27.** −11/4 **29.** 1/2 **31.** No solution **33.** −2 **35.** 4
**37.** 8 **39.** 2

## Exercise 6, Page 230

**1.** 150 **3.** 5 and 6 **5.** 15 and 35 **7.** 26 h **9.** 240 mi **11.** 2 days **13.** 5.8 days
**15.** 2.4 h **17.** 4 h **19.** 12.4 h **21.** 7.8 weeks **23.** 5.6 winters **25.** 6.13 h **27.** $11,360
**29.** $1500, $2000

## Exercise 7, Page 236

**1.** $bc/2a$ **3.** $\dfrac{bz-ay}{a-b}$ **5.** $\dfrac{a^2d+3d^2}{4ac+d^2}$ **7.** $\dfrac{b+cd}{a^2+a-d}$ **9.** $2z/a + 3w$ **11.** $(b - m)/3$
**13.** $(c - m)/(a - b - d)$ **15.** $b$ **17.** $\dfrac{2c+bc+3b}{3-2b+c}$ **19.** $\dfrac{w}{w(w+y)-1}$ **21.** $(p - q)/3p$
**23.** $(5b - 2a)/3$ **25.** $\dfrac{a^2(c+1)(c-a)}{c^2}$ **27.** $\dfrac{ab}{a+b}$ **29.** $\dfrac{ab}{b-a}$ **31.** $md/(m + n)$ and $nd/(m + n)$
**33.** $\dfrac{anm}{nm-m-n}$ **35.** $\dfrac{L-L_0}{L_0\,\alpha}$ **37.** $PL/Ee$ **39.** $(kAt_1 - qL)/kA$ **41.** $y/(1 + nt)$ **43.** $L/(1 + \alpha\,\Delta t)$
**45.** $F/(m_1 + m_2 + m_3)$ **47.** $\dfrac{m_2(25-x)}{x-10}$ **49.** $E/(gy + \frac{1}{2}v^2)$

## Review Problems, Page 238

**1.** $2ab/(a - b)$ **3.** $1/(3m - 1)$ **5.** $a^2 + 1 + 1/a^2$ **7.** $a/c$ **9.** $4/(4 - x^2)$ **11.** 0
**13.** $\dfrac{a^2+ab+b^2}{a+b}$ **15.** $(2a - 3c)/2a$ **17.** $3a/(a + 2)$ **19.** $\dfrac{x-y-z}{x+y-z}$ **21.** $\dfrac{a+b-c-d}{a-b+c-d}$
**23.** $(b + 5)/(c - a)$ **25.** 1 **27.** 11 **29.** $(r - pq)/(p - q)$ **31.** $q(p + 1)/(p^2 + r)$ **33.** 7
**35.** 24 **37.** 3 **39.** 6.35 days **41.** $19\frac{22}{45}$ **43.** $2ax^2y/7w$

## CHAPTER 9

### Exercise 1, Page 248

**1.** (2, −1) **3.** (1, 2) **5.** (−0.24, 0.90) **7.** (3, 5) **9.** (−3, 3) **11.** (1, 2) **13.** (3, 2) **15.** (15, 6) **17.** (3, 4) **19.** (2, 3) **21.** (1.63, 0.0967) **23.** $m = 2, n = 3$ **25.** $w = 6, z = 1$

### Exercise 2, Page 253

**1.** (60, 36) **3.** (87/7, 108/7) **5.** (15, 12) **7.** $m = 4, n = 3$ **9.** $r = 3.77, s = 1.23$ **11.** (1/2, 1/3) **13.** (1/3, 1/2) **15.** (1/10, 1/12) **17.** $w = 1/36, z = 1/60$ **19.** $[(a + 2b)/7, (3b - 2a)/7]$ **21.** $\left(\frac{c(n-d)}{an-dm}, \frac{c(m-a)}{an-dm}\right)$

### Exercise 3, Page 256

**1.** 8, 16 **3.** 4/21 **5.** 82 **7.** 7, 41 **9.** 9 × 16 **11.** 7 ft/s, $5\frac{1}{4}$ ft/s **13.** 5 mi/h, $11\frac{1}{4}$ mi **15.** \$2500 at 4% **17.** \$6000, \$4000 **19.** 35 acres, 65 acres **21.** 18.6 kg zinc, 2.86 kg lead **23.** 125 gal of 5% mixture, 375 gal of 11% mixture **25.** $F_1 = 395$ lb, $F_2 = 345$ lb **27.** 87.7 cm, 65.3 cm **29.** 500 parts/h, 300 parts/h **31.** 9.33 h, 16.8 h **33.** Wind: 26.9 kW/h, hydro: 12.2 kW/h **35.** $I_1 = 109$ mA, $I_2 = 87$ mA

### Exercise 4, Page 263

**1.** (15, 20, 25) **3.** (1, 2, 3) **5.** (5, 6, 7) **7.** (1, 2, 3) **9.** (3, 4, 5) **11.** (6.20, 13.7, 18.1) **13.** (2, 3, 1) **15.** (1/2, 1/3, 1/4) **17.** (1, −1, 1/2) **19.** $(a/11, 5a/11, 7a/11)$ **21.** $[(a + b - c), (a - b + c), (-a + b + c)]$ **23.** 876 **25.** 361 **27.** \$4000 in checking, \$6000 in MMA, \$3000 in CD **29.** 2/3 A, 7/3 A, 2/3 A

### Review Problems, Page 265

**1.** (3, 5) **3.** (5, −2) **5.** (2, −1, 1) **7.** (1, 2, −3) **9.** (8, 10) **11.** (2, 3) **13.** $[(a + b - c)/2, (a - b + c)/2, (b - a + c)/2]$ **15.** (7, 5) **17.** (6, 8, 10) **19.** (2, 3, 1) **21.** (13, 17) **23.** (2, 9) **25.** 8 days

## CHAPTER 10

### Exercise 1, Page 270

**1.** −14 **3.** 15 **5.** 0 **7.** 17.3 **9.** −2/5 **11.** $ad - bc$ **13.** $\sin^2\theta - 3\tan\theta$ **15.** $3i_1 \cos\theta - i_2 \sin\theta$ **17.** (3, 5) **19.** (−3, 3) **21.** (1, 2) **23.** (3, 2) **25.** (15, 6) **27.** (3, 4) **29.** (2, 3) **31.** (1.63, 0.0967) **33.** $m = 2, n = 3$ **35.** $w = 6, z = 1$ **37.** (−6/13, 30/13) **39.** $\left(\frac{dp - bq}{ad - bc}, \frac{aq - cp}{ad - bc}\right)$ **41.** (3, 4)

### Exercise 2, Page 278

**1.** 11 **3.** 45 **5.** 48 **7.** −28 **9.** (5, 6, 7) **11.** (15, 20, 25) **13.** (1, 2, 3) **15.** (3, 4, 5) **17.** (6, 8, 10) **19.** (3, 6, 9) **21.** (2, −1, 5)

### Exercise 3, Page 285

**1.** 2 **3.** 18 **5.** −66 **7.** $x = 2, y = 3, z = 4, w = 5$ **9.** $x = -4, y = -3, z = 2, w = 5$ **11.** $x = a - c, y = b + c, z = 0, w = a - b$ **13.** $x = 4, y = 5, z = 6, w = 7, u = 8$ **15.** $v = 3, w = 2, x = 4, y = 5, z = 6$ **17.** \$40,000, \$30,000, \$24,000, \$26,000 **19.** 140 kg, 100 kg, 200 kg, 160 kg

### Review Problems, Page 287

**1.** 20 **3.** 0 **5.** −3 **7.** 0 **9.** 18 **11.** 15 **13.** 0 **15.** −29 **17.** 133 **19.** $x = -4, y = -3, z = 2, w = 1$ **21.** (3, 5) **23.** (4, 3) **25.** (5, 1) **27.** (2, 5) **29.** (2, 1) **31.** (3, 2) **33.** (9/7, 37/7)

## CHAPTER 11

### Exercise 1, Page 291

**1.** A, B, D, E, F, I **3.** C, H **5.** I **7.** B, I **9.** F **11.** 6 **13.** $4 \times 3$ **15.** $2 \times 4$

**17.** $\begin{pmatrix} 0 & 0 \\ 0 & 0 \\ 0 & 0 \\ 0 & 0 \end{pmatrix}$ **19.** If $y = 6$, $x = 2$, $w = 7$, and $z = 8$

### Exercise 2, Page 300

**1.** (12 13 5 8) **3.** $\begin{pmatrix} 14 \\ -10 \\ 6 \\ 14 \end{pmatrix}$ **5.** $\begin{pmatrix} 7 & 1 & -4 & -7 \\ -9 & -4 & 9 & 2 \\ 3 & 7 & -3 & -9 \\ -1 & -4 & -5 & 6 \end{pmatrix}$ **7.** (−15 −27 6 −21 −9)

**9.** $\begin{pmatrix} -12 & -6 & 9 & 0 \\ 3 & -18 & -9 & 12 \end{pmatrix}$ **11.** $7\begin{pmatrix} 3 & 1 & -2 \\ 7 & 9 & 4 \\ 2 & -3 & 8 \\ -6 & 10 & 12 \end{pmatrix}$ **13.** $\begin{pmatrix} 6 & 16 & 3 \\ -16 & 8 & 19 \\ 31 & -14 & 15 \end{pmatrix}$ **15.** 2**A** + 2**B** **17.** 35

**19.** −61 **21.** (10 27 −14 15 −7) **23.** (−60 18 10) **25.** (15 2 40 2 34)

**27.** $\begin{pmatrix} 19 & -13 \\ 17 & -43 \end{pmatrix}$ **29.** $\begin{pmatrix} -20 & 10 \\ 8 & -2 \\ -4 & 16 \end{pmatrix}$ **31.** $\begin{pmatrix} 6 & -10 \\ 3 & -5 \end{pmatrix}$ **33.** $\begin{pmatrix} 0 & 2 \\ 2 & 0 \\ -10 & -6 \\ -5 & 0 \end{pmatrix}$ **35.** $\begin{pmatrix} 2 & 0 & -2 & 1 \\ 8 & 8 & -1 & 2 \\ 2 & 2 & 2 & 6 \end{pmatrix}$

**37.** $\mathbf{AB} = \begin{pmatrix} 1 & 2 & -1 \\ -2 & 9 & 2 \\ -1 & -2 & 1 \end{pmatrix}$, $\mathbf{BA} = \begin{pmatrix} 5 & -2 & 2 \\ 4 & -2 & 0 \\ -2 & 3 & 8 \end{pmatrix}$ **39.** $\begin{pmatrix} 778 & 1050 \\ 1225 & 1653 \end{pmatrix}$ **41.** $\begin{pmatrix} 5 \\ 1 \end{pmatrix}$ **43.** $\begin{pmatrix} 13 \\ -1 \\ 10 \end{pmatrix}$

**45.** $\begin{pmatrix} 6 & 4 & -8 \\ 0 & 0 & 0 \\ 9 & 6 & -12 \\ -3 & -2 & 4 \end{pmatrix}$

### Exercise 4, Page 315

**1.** $\begin{pmatrix} 0 & -0.2 \\ 0.125 & 0.1 \end{pmatrix}$ **3.** $\begin{pmatrix} 0.118 & 0.059 \\ -0.2 & 0.157 \end{pmatrix}$ **5.** $\begin{pmatrix} 0.074 & 0.168 & 0.057 \\ 0.148 & -0.164 & 0.115 \\ 0.090 & 0.094 & 0.042 \end{pmatrix}$

**7.** $\begin{pmatrix} 0.500 & 0 & -0.500 & 0 \\ 10.273 & -0.545 & -13.091 & 1.364 \\ -4.727 & 0.455 & 5.909 & -0.636 \\ -4.591 & 0.182 & 5.864 & -0.455 \end{pmatrix}$

### Review Problems, Page 315

**1.** (2 12 10 −2) **3.** 86 **5.** $\begin{pmatrix} 0.063 & 0.250 \\ -0.188 & 0.250 \end{pmatrix}$ **7.** $\begin{pmatrix} 7 & 2 & 6 \\ 9 & 0 & 3 \end{pmatrix}$ **9.** $x = -4$, $y = -3$, $z = 2$, $w = 1$

**11.** (2, −1, 1) **13.** (3, 5) **15.** −0.557 mA, 0.188 mA, −0.973 mA, −0.271 mA

## CHAPTER 12

### Exercise 1, Page 322

**1.** $3/x$ **3.** 2 **5.** $a/4b^2$ **7.** $a^3/b^2$ **9.** $p^3/q$ **11.** 1 **13.** $1/(16a^6b^4c^{12})$

**15.** $1/(x + y)$ **17.** $\dfrac{m^4}{(1 - 6m^2n)^2}$ **19.** $2/x + 1/y^2$ **21.** $1/(3m)^3 - 2/n^2$ **23.** $x^{2n} + 2x^ny^m + y^{2m}$

**25.** $x/2y^7$ **27.** $p$ **29.** $x^{m^2}$ **31.** $b/3$ **33.** $9y^2/4x^2$ **35.** $27q^3y^3/8p^3x^3$ **37.** $9a^8b^6/25x^4y^2$

**39.** $4^na^{6n}x^{6n}/9^nb^{4n}y^{2n}$ **41.** $3p^2/2q^2x^4z^3$ **43.** $5^pw^{2p}/2^pz^p$ **45.** $25n^4x^8/9m^6y^6$

**47.** $(1 + a)^2(b - 1)/(1 - a + a^2)$ **49.** $x^7(x - y)^2$ **51.** $9a^{2n} + 6a^nb^n + 4b^{2n}$ **53.** $\dfrac{(x + 1)^2(y - 1)}{(y + 1)^2(x^2 - x + 1)}$
**55.** $(7/15)a^{n-m+2}x^{2-n}y^{n-3}$ **57.** $a^3(5a^2 - 16b^2)$ **59.** $a^{4-m}b^{n-3}/5$ **61.** $x^3(x^2 + y^2)$
**63.** $(a + b)^4/(a^2 + b^2)^2$

### Exercise 2, Page 328

**1.** $\sqrt[4]{a}$ **3.** $\sqrt[4]{z^3}$ **5.** $\sqrt{m - n}$ **7.** $\sqrt[3]{y/x}$ **9.** $b^{1/2}$ **11.** $y$ **13.** $(a + b)^{1/n}$ **15.** $xy$
**17.** $3\sqrt{2}$ **19.** $3\sqrt{7}$ **21.** $-2\sqrt[3]{7}$ **23.** $a\sqrt{a}$ **25.** $6x\sqrt{y}$ **27.** $2x^2\sqrt[3]{2y}$ **29.** $6y^2\sqrt[5]{xy}$
**31.** $a\sqrt{a - b}$ **33.** $3\sqrt{m^3 + 2n}$ **35.** $x\sqrt[3]{x - a^2}$ **37.** $(a - b)(a + b)\sqrt{a}$ **39.** $\sqrt{21}/7$ **41.** $\sqrt[3]{2}/2$
**43.** $\sqrt[3]{6}/3$ **45.** $\sqrt{2x}/2x$ **47.** $a\sqrt{15ab}/5b$ **49.** $\sqrt[3]{x}/x$ **51.** $x\sqrt[3]{18}/3$ **53.** $(x - y)\sqrt{x^2 + xy}$
**55.** $\dfrac{(m + n)\sqrt{m^2 - mn}}{m - n}$ **57.** $\dfrac{(2a - 6)\sqrt{6a}}{3a}$

### Exercise 3, Page 333

**1.** $\sqrt{6}$ **3.** $\sqrt{6}$ **5.** $25\sqrt{2}$ **7.** $-7\sqrt[3]{2}$ **9.** $-5\sqrt[3]{5}$ **11.** $-\sqrt[4]{3}$ **13.** $10x\sqrt{2y}$
**15.** $11\sqrt{6}/10$ **17.** $(a + b)\sqrt{x}$ **19.** $(x - 2a)\sqrt{y}$ **21.** $-4a\sqrt{3x}$ **23.** $(5a + 3c)\sqrt[3]{10b}$
**25.** $(a^2b^2c^2 - 2a + bc)\sqrt[5]{a^3bc^2}$ **27.** $6\sqrt{2x}$ **29.** $\dfrac{25}{18y}\sqrt{3x}$ **31.** 45 **33.** $16\sqrt{5}$ **35.** $24\sqrt[3]{5}$
**37.** $6\sqrt[6]{72}$ **39.** $10\sqrt[6]{72}$ **41.** $6x\sqrt{3ay}$ **43.** $2\sqrt[5]{x^3y^3}$ **45.** $\sqrt[4]{a^2b}$ **47.** $2xy\sqrt[6]{x^4yz^3}$ **49.** $2\sqrt{15} - 6$
**51.** $x^2\sqrt{1 - xy}$ **53.** $a^2 - b$ **55.** $16x - 12\sqrt{xy} - 10y$ **57.** $9y$ **59.** $9x^3\sqrt[3]{4x}$ **61.** $9 - 30\sqrt{a} + 25a$
**63.** $250x\sqrt{2x}$ **65.** $250ax$ **67.** 3/4 **69.** $\sqrt[6]{18}/4$ **71.** $3\sqrt[3]{2a}/a$ **73.** $\dfrac{4\sqrt[3]{18}}{3} + \sqrt[3]{4} + 3$ **75.** 3/2
**77.** $2\sqrt[6]{a^5b^2c^3}/ac$ **79.** $\dfrac{1}{2c}\sqrt[12]{a^2bc^{10}}$ **81.** $\dfrac{8 + 5\sqrt{2}}{2}$ **83.** $4\sqrt{ax}/a$ **85.** $6\sqrt[3]{2x}/x$ **87.** $-\dfrac{30 + 17\sqrt{30}}{28}$
**89.** $\dfrac{a^2 - a\sqrt{b}}{a^2 - b}$ **91.** $\dfrac{x - 2\sqrt{xy} + y}{x - y}$ **93.** $\dfrac{5\sqrt{3}x - 5\sqrt{xy} + \sqrt{6xy} - \sqrt{2}y}{6x - 2y}$
**95.** $\dfrac{3\sqrt{6a} - 4\sqrt{3ab} + 6\sqrt{10ab} - 8\sqrt{5b}}{9a - 8b}$

### Exercise 4, Page 337

**1.** 36 **3.** 4 **5.** 8 **7.** 8 **9.** 7 **11.** 4.79 **13.** 3 **15.** 8 **17.** 1/3 **19.** 7
**21.** $C = 1/[\omega^2L \pm \omega\sqrt{Z^2 - R^2}]$

### Review Problems, Page 337

**1.** $2\sqrt{13}$ **3.** $3\sqrt[3]{6}$ **5.** $\sqrt[3]{2}$ **7.** $9x\sqrt[4]{x}$ **9.** $(a - b)x\sqrt[3]{(a - b)^2x}$ **11.** $\dfrac{\sqrt{a^2 - 4}}{a + 2}$
**13.** $x^2\sqrt{1 - xy}$ **15.** $-\dfrac{1}{46}(6 + 15\sqrt{2} - 4\sqrt{3} - 10\sqrt{6})$ **17.** $\dfrac{2x^2 + 2x\sqrt{x^2 - y^2} - y^2}{y^2}$ **19.** $26\sqrt[3]{x^2}$
**21.** $\sqrt[6]{a^3b^2}$ **23.** $9 + 12\sqrt{x} + 4x$ **25.** $7\sqrt{2}$ **27.** $\dfrac{25\sqrt{2}}{12}$ **29.** $\sqrt{2b}/2b$ **31.** $72\sqrt{2}$
**33.** 10 **35.** 14 **37.** 15.2 **39.** $x^{2n-1} + (xy)^{n-1} + x^ny^{n-2} + y^{2n-3}$ **41.** $27x^3y^6/8$ **43.** $8x^9y^6$
**45.** $3/w^2$ **47.** $1/3x$ **49.** $1/x - 2/y^2$ **51.** $1/9x^2 + y^8/4x^4$ **53.** $p^{2a-1} + (pq)^{a-1} + p^aq^{a-2} + q^{2a-3}$
**55.** $p^2/q$ **57.** $r^2/s^3$ **59.** 1

## CHAPTER 13

### Exercise 1, Page 345

**1.** 0, 2/5 **3.** 0, 9/2 **5.** ±3 **7.** ±1 **9.** 0, 0.355 **11.** 3, −5 **13.** 5, −4
**15.** 2, −1 **17.** −1, −2 **19.** 9, −2 **21.** −4, −8 **23.** 16, −4 **25.** 5/2, −1 **27.** 3/2, −4
**29.** 1/5, −3 **31.** 2, −1 **33.** 1/6, −5/4 **35.** 6, −3 **37.** 1/2, −2/3 **39.** 9, −4 **41.** 1, −1/6
**43.** 4, −2/3 **45.** 5, −17/3 **47.** ±6 **49.** 1, −13/6 **51.** $-a$, $-3a$ **53.** $-2b$, $-4b/3$

**55.** $-3y^2/7, -2y^2/3$ **57.** $x^2 - 11x + 28 = 0$ **59.** $15x^2 - 19x + 6 = 0$ **61.** $6x^2 + 31x + 35 = 0$
**63.** $x^2 - 2x - 4 = 0$ **65.** 6, 15/4 **67.** 4, (−21 doesn't check) **69.** 2 (−2 doesn't check)

### Exercise 2, Page 348

**1.** $4 \pm \sqrt{14}$ **3.** $(-7 \pm \sqrt{61})/2$ **5.** $(3 \pm \sqrt{89})/8$ **7.** $(-3 \pm \sqrt{21})/3$ **9.** $(-7 \pm \sqrt{129})/8$

### Exercise 3, Page 351

**1.** 3.17, 8.83 **3.** 3.89, −4.87 **5.** 1.96, −5.96 **7.** 0.401, −0.485 **9.** 0.170, −0.599
**11.** 2.87, 0.465 **13.** 0.907, −2.57 **15.** 5.87, −1.87 **17.** 2.31, 0.623 **19.** 5.08, 10.9
**21.** 0.132, −15.1 **23.** 12.0, −14.0 **25.** Real, unequal **27.** Not real **29.** Not real

### Exercise 4, Page 354

**1.** 2/3 or 3/2 **3.** 4 and 11 **5.** 5 and 15 **7.** 4 × 6 m **9.** 15 × 30 cm **11.** 162 × 200 m
**13.** 32 × 48 ft **15.** 6.47 in. **17.** 146 mi/h **19.** 80 km/h **21.** 9 h and 12 h **23.** 7 m/day
**25.** At each end **27.** 0.630 s and 8.385 s **29.** 125 Ω and 665 Ω **31.** 0.381 A and 0.769 A
**33.** −0.5 A and 0.1 A **35.** −1.1 V

### Exercise 5, Page 359

**1.**

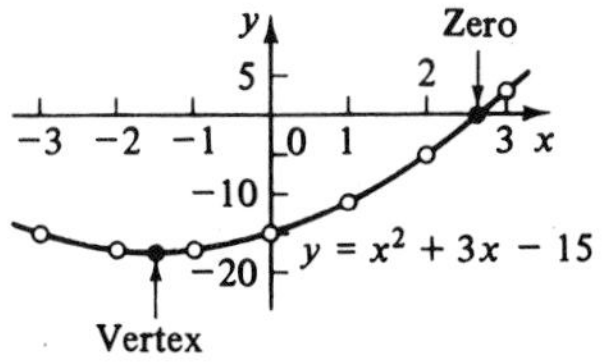


**3.**

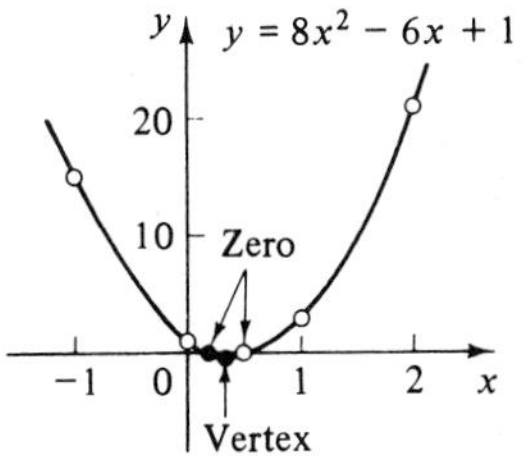


**7.**

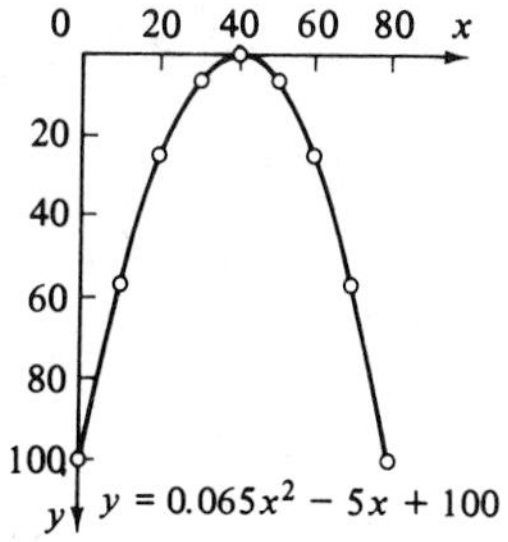


**9.** 20 cm

**11.** 6.33 ft **13.**

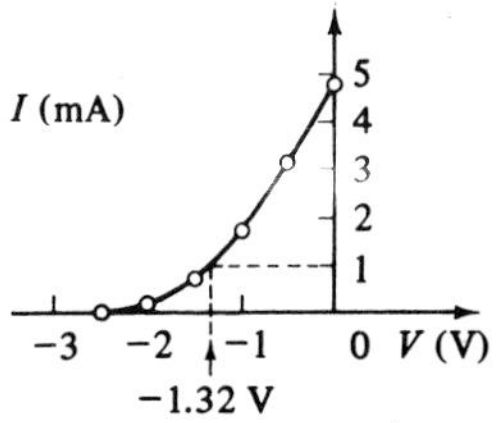


### Exercise 6, Page 362

**1.** $\sqrt[3]{2}, \sqrt[3]{4}$ **3.** 1 **5.** 1 (0 doesn't check) **7.** −1, −1/27 **9.** 1, −64 **11.** $\sqrt[3]{2}/8$
**13.** 625 (256 doesn't check) **15.** 496 (243 doesn't check)

### Exercise 7, Page 365

**1.** 3, −5 **3.** 2, −1 **5.** 3/2, −4 **7.** 4/3, −8/3

### Exercise 8, Page 369

**1.** (6, 2) **3.** (3, 2), (−5, 6) **5.** (2, 3), (−46, 15) **7.** (2.61, −1.2), (−2.61, −1.2)
**9.** (4.86, 0.845), (4.86, −0.845) **11.** (4, −7), (7, −4)

### Review Problems, Page 369

**1.** 6, −1 **3.** ±2, ±3 **5.** 0, 5 **7.** 0, −2 **9.** 5, −2/3 **11.** 1, 1/2 **13.** ±3
**15.** 1.79, −2.79 **17.** 4, 9 **19.** 0.692, −0.803 **21.** $\pm\sqrt{10}$ **23.** 1/2, −2 **25.** 0, 9
**27.** 1, 512 **29.** ±5 **31.** 1, −3 **33.** 202 bags **35.** 15 × 30 ft **37.** 2, 4, −3
**39.** (7.41, −0.41), (−0.41, 7.41) **41.** 3 mi/h **43.** $1\frac{1}{2}$ s **45.** 5 × 16 in. or 8 × 10 in. **47.** 3.56 km/h
**49.** 12 ft × 12 ft

## CHAPTER 14

### Exercise 1, Page 378

**1.** II **3.** IV **5.** I **7.** II **9.** II or III **11.** IV **13.** III **15.** pos **17.** pos

**19.** pos **21.** neg

| | sin | cos | tan |
|---|---|---|---|
| **23.** | + | − | − |
| **25.** | − | + | − |
| **27.** | + | − | − |

| | r | sin | cos | tan | cot | sec | csc |
|---|---|---|---|---|---|---|---|
| **29.** | 12.65 | 0.947 | −0.316 | −3 | −0.333 | −3.16 | 1.05 |
| **31.** | 17 | −0.471 | −0.882 | 0.533 | 1.88 | −1.13 | −2.13 |
| **33.** | 5 | 0 | 1 | 0 | — | 1 | — |

| | sin | cos | tan |
|---|---|---|---|
| **35.** | −3/5 | −4/5 | 3/4 |
| **37.** | $-2/\sqrt{5}$ | $-1/\sqrt{5}$ | 2 |
| **39.** | 2/3 | $-\sqrt{5}/3$ | $-2/\sqrt{5}$ |

| | sin | cos | tan |
|---|---|---|---|
| **41.** | 0.9816 | −0.1908 | −5.145 |
| **43.** | −0.4848 | 0.8746 | −0.5543 |
| **45.** | −0.8898 | 0.4563 | −1.950 |

| | sin | cos | tan |
|---|---|---|---|
| **47.** | 0.8090 | −0.5878 | −1.376 |
| **49.** | 0.9108 | −0.4128 | −2.206 |
| **51.** | −0.2264 | −0.9740 | 0.2324 |

| | sin | cos | tan |
|---|---|---|---|
| **53.** | −0.1959 | 0.9806 | −0.1998 |
| **55.** | −0.8481 | −0.5299 | 1.600 |

**57.** 0.8192 **59.** 1.906 **61.** 1.711 **63.** 1.111 **65.** 135°, 315°
**67.** 33.2°, 326.8° **69.** 81.1°, 261.1° **71.** 90°, 270° **73.** 227.4°, 312.6° **75.** 34.2°, 325.8°
**77.** 102.6°, 282.6° **79.** 1/2 **81.** 0 **83.** 1 **85.** 0 **87.** 0 **89.** −1 **91.** 0 **93.** 1/2
**95.** 1 **97.** 0

### Exercise 2, Page 383

**1.** $A = 32.8°$, $C = 101.0°$, $c = 413$ **3.** $A = 107.8°$, $B = 29.2°$, $a = 29$
**5.** $B = 71.1°$, $b = 8.20$, $c = 6.34$ **7.** $A = 103.6°$, $a = 21.7$, $c = 15.7$ **9.** $A = 23°$, $a = 43$, $b = 90$
**11.** $A = 106.4°$, $a = 90.74$, $c = 29.55$ **13.** $C = 150°$, $b = 34.1$, $c = 82.0$

### Exercise 3, Page 388

**1.** $A = 44.2°$, $B = 29.8°$, $c = 21.7$ **3.** $B = 48.7°$, $C = 63.0°$, $a = 22.6$
**5.** $A = 56.9°$, $B = 95.8°$, $c = 70.1$ **7.** $B = 25.5°$, $C = 19.5°$, $a = 452$
**9.** $A = 45.6°$, $B = 80.4°$, $C = 54.0°$ **11.** $B = 30.7°$, $C = 34.0°$, $a = 82.9$
**13.** $A = 69.0°$, $B = 44.0°$, $c = 8.95$ **15.** $B = 10.3°$, $C = 11.7°$, $a = 3.69$
**17.** $A = 66.21°$, $B = 72.02°$, $c = 1052$ **19.** $A = 69.7°$, $B = 51.7°$, $C = 58.6°$

### Exercise 4, Page 390

**1.** 30.8 m and 85.6 m **3.** 32.3°, 60.3°, 87.4° **5.** N48°48′W **7.** S59, 4°E **9.** 598 km
**11.** 28.3 m **13.** 77.3 m, 131 m **15.** 107 ft **17.** 337 mm **19.** 33.7 cm **21.** 73.4 in.
**23.** 53.8 mm and 78.2 mm **25.** 419 **27.** $A = 45°$, $B = 60°$, $C = 75°$, $AC = 1220$, $AB = 1360$
**29.** 21.9 ft

### Exercise 5, Page 395

| | Resultant | Angle |
|---|---|---|
| **1.** | 521 | 10.0° |
| **3.** | 87.1 | 31.9° |
| **5.** | 6708 | 41.2° |

**7.** 37.3 N at 20.5° **9.** 121 N at N 59.9° W **11.** 1715 N at 29.8°
**13.** 44.2° and 20.7° **15.** Wind: S 37.4° E, Plane: S 84.8° W **17.** 413 km/h at N 41.2° E
**19.** 632 km/h, 3.09° **21.** 26.9 A, 32.3° **23.** 39.8 at 26.2°

### Review Problems, Page 396

**1.** $A = 24.1°$, $B = 20.9°$, $c = 77.6$ **3.** $A = 61.6°$, $C = 80.0°$, $b = 1.30$
**5.** $B = 20.2°$, $C = 27.8°$, $a = 82.4$ **7.** IV **9.** II **11.** neg **13.** neg **15.** neg
**17.** −0.800 −0.600 1.33 **19.** 1200 at 36.3° **21.** 22.1 at 121° **23.** 1.11 km

| | sin | cos | tan |
|---|---|---|---|
| **25.** | 0.0872 | −0.9962 | −0.0875 |
| **27.** | 0.7948 | −0.6069 | −1.3095 |

**29.** 0.4698 **31.** 1.469 **33.** S 3.5° E **35.** 130.8°, 310.8°
**37.** 47.5°, 132.5° **39.** 80.0°, 280.0° **41.** 695 lb, 17.0°

## CHAPTER 15

### Exercise 1, Page 402

**1.** 0.834 rad **3.** 0.615 rad **5.** 9.74 rad **7.** 0.279 rev **9.** 0.497 rev **11.** 0.178 rev **13.** 162° **15.** 21.3° **17.** 65.3° **19.** $\pi/3$ **21.** $11\pi/30$ **23.** $7\pi/10$ **25.** $13\pi/30$ **27.** $20\pi/9$ **29.** $9\pi/20$ **31.** 22.5° **33.** 147° **35.** 20.0° **37.** 158° **39.** 24.0° **41.** 15.0° **43.** 0.8660 **45.** 0.4863 **47.** −0.3090 **49.** 1.067 **51.** −2.747 **53.** −0.8090 **55.** 1.116 **57.** 0.5000 **59.** 0.1585 **61.** 280 **63.** 3.86 in.

### Exercise 2, Page 405

**1.** 6.07 **3.** 230 **5.** 43.5 in. **7.** 2.21 rad **9.** 1.24 rad **11.** 0.824 **13.** 125 ft **15.** 3790 mi **17.** 589 m **19.** 24.5 cm **21.** $2 \times 10^8$ mi **23.** 0.251 m **25.** $r = 345$ mm, $R = 700$ mm, $\theta = 61.8°$ **27.** 5880 mi **29.** 14.1 cm **31.** 87.6064 ft

### Exercise 3, Page 409

**1.** 194 rad/s, 11,100°/s **3.** 12.9 rev/min, 1.35 rad/s **5.** 8.02 rev/min, 0.840 rad/s **7.** 4350° **9.** 0.00749 s **11.** 353 rev/min **13.** 1.62 m/s **15.** 1037 mi/h **17.** 57.6 rad/s

### Review Problems, Page 410

**1.** 77.1° **3.** 20° **5.** 165° **7.** 2.42 rad/s **9.** $5\pi/3$ **11.** $23\pi/18$ **13.** 0.3420 **15.** 1.000 **17.** −0.01827 **19.** 155 rev/min **21.** 336 mi/h **23.** 1.1089 **25.** −0.8812 **27.** −1.0291 **29.** 0.9777 **31.** 0.9097 **33.** 222 mm

## CHAPTER 16

### Exercise 1, Page 422

**1.** $P = 1$ s, $f = 1$ cycle/s, amp. = 7

**3.**

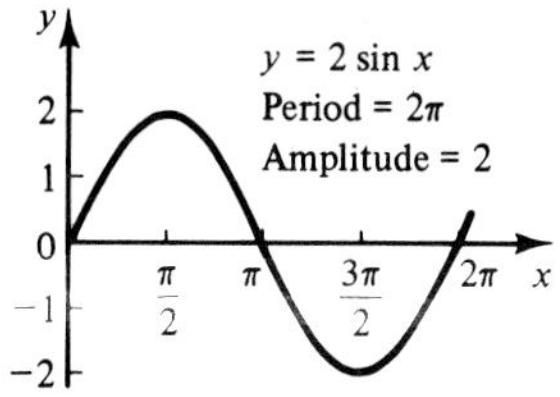


**5.**

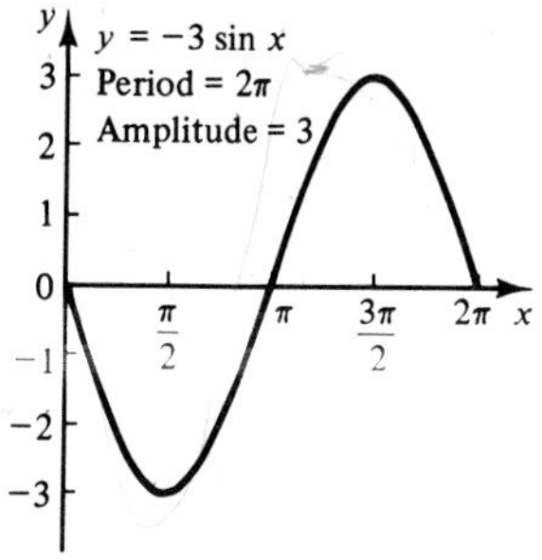


**7.**

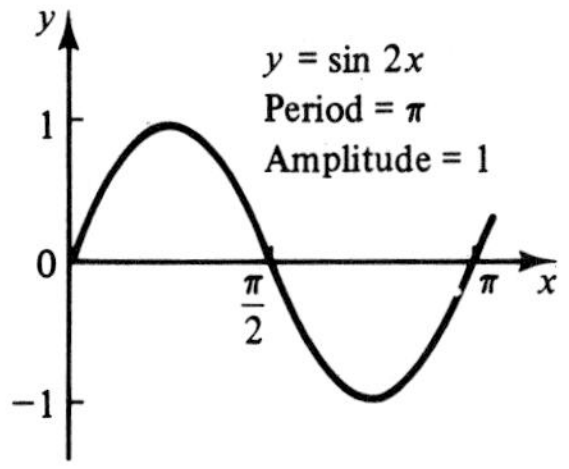


**9.**

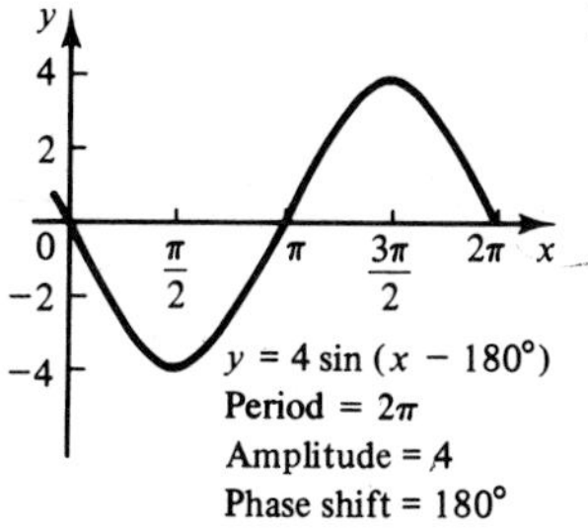


**11.**

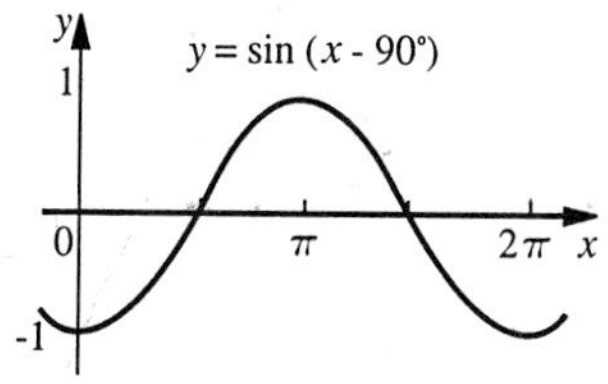


**13.**

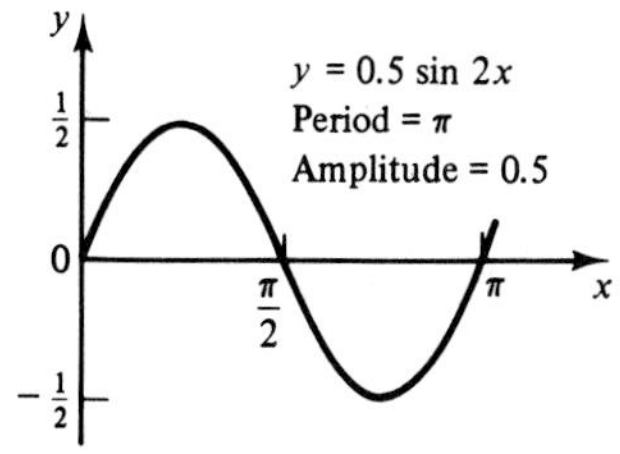


**15.**

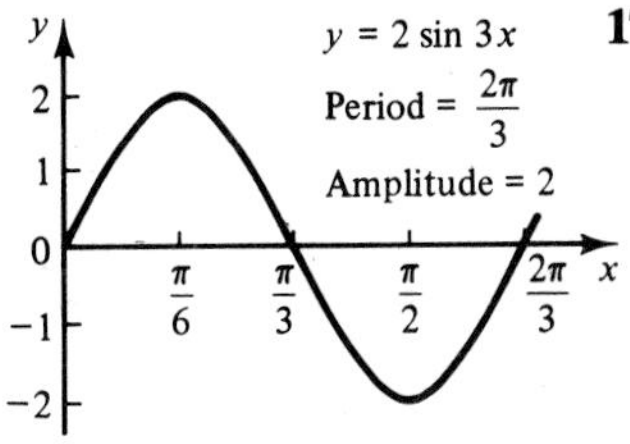


**17.**

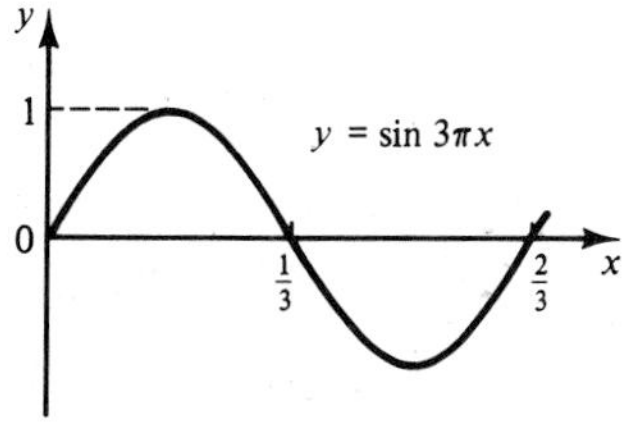

**19.**

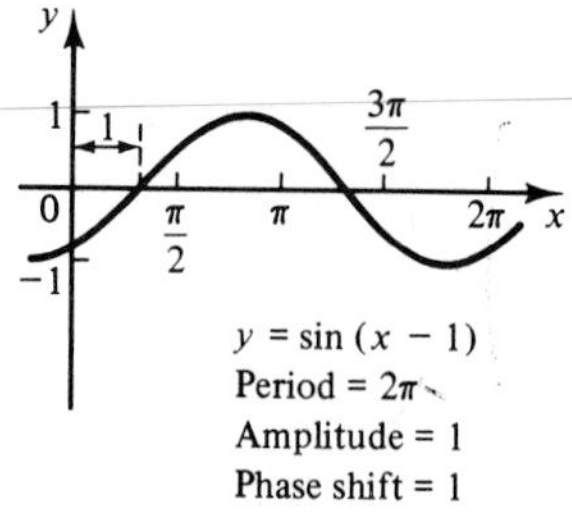

$$y = \sin(x - 1)$$
Period $= 2\pi$
Amplitude $= 1$
Phase shift $= 1$

**21.** $y = \sin\left(x + \dfrac{\pi}{2}\right)$

Period $= 2\pi$
Amplitude $= 1$
Phase shift $= -\dfrac{\pi}{2}$

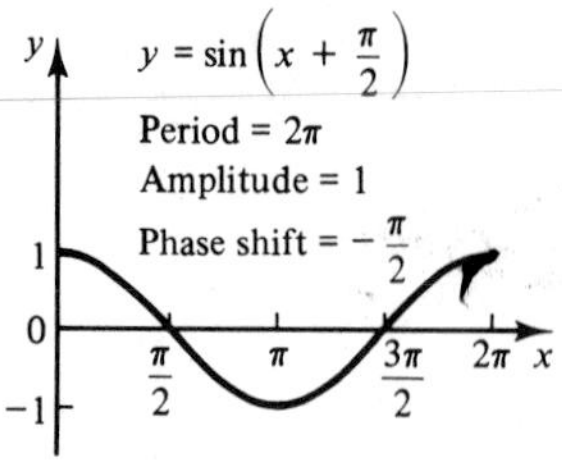

**23(a)** $y = \sin 2x$, **(b)** $y = \sin(x - \pi/2)$, **(c)** $y = 2\sin(x - \pi)$    **25.** $y = -2\sin(x/3 + \pi/12)$

**Exercise 2, Page 428**

**1.** $y = 3\cos x$
Amplitude $= 3$
Period $= 2\pi$
Phase shift $= 0$

**3.** $y = 2\cos 3x$
Amplitude $= 2$
Period $= \dfrac{2\pi}{3}$
Phase shift $= 0$

**5.** $y = 3\cos\left(x - \dfrac{\pi}{4}\right)$
Amplitude $= 3$
Period $= 2\pi$
Phase shift $= \dfrac{\pi}{4}$

**7.** $y = 2\tan x$

**9.** $y = 3\tan 2x$

**11.** $y = 2\tan(3x - 2)$

**13.** $y = x + \sin x$, $y = x$, $y = \sin x$

**15.** $y = \dfrac{2}{x}$, $y = \dfrac{2}{x} - \sin x$, $y = \sin x$

**17.** $y = \sin x$, $y = \cos 2x$, $y = \sin x + \cos 2x$

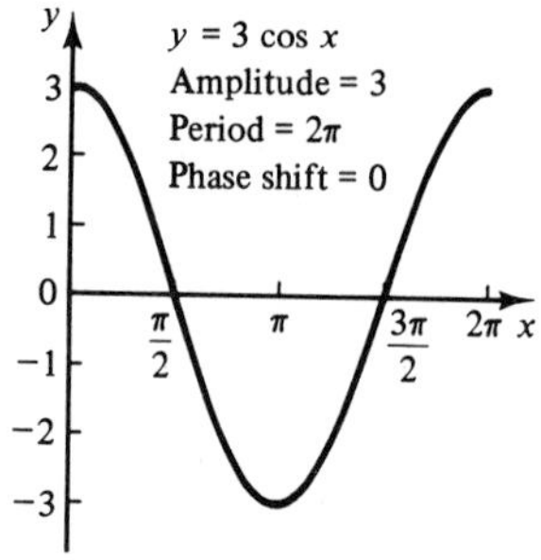

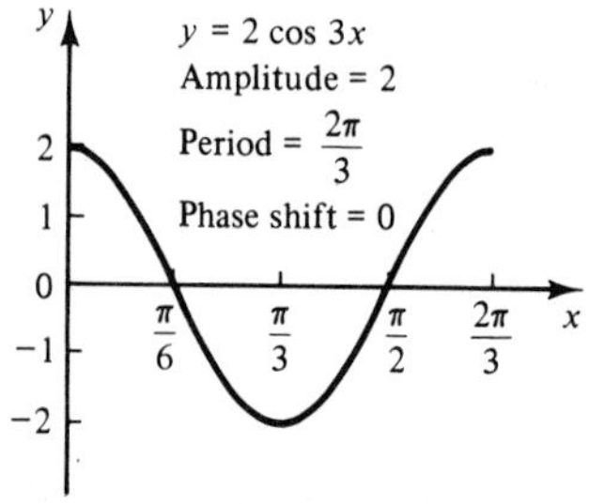

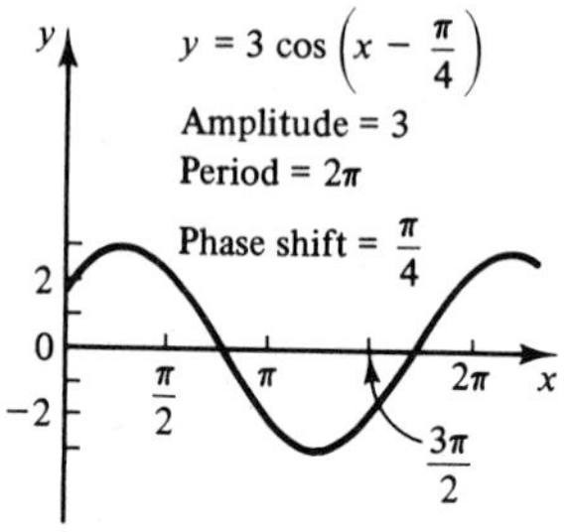

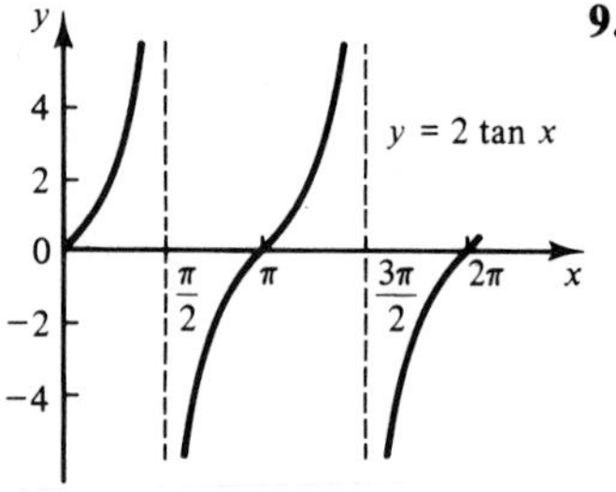

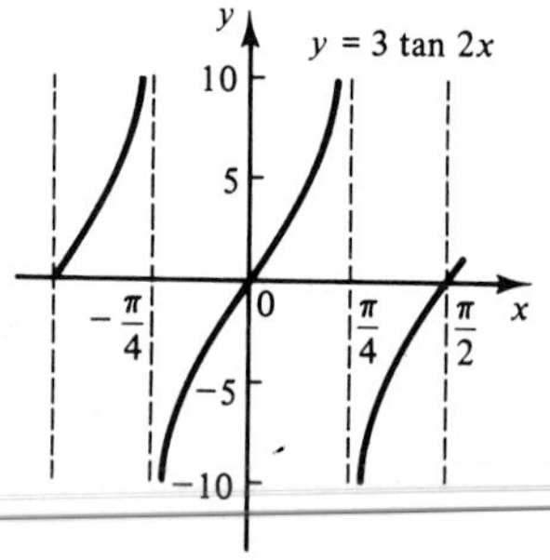

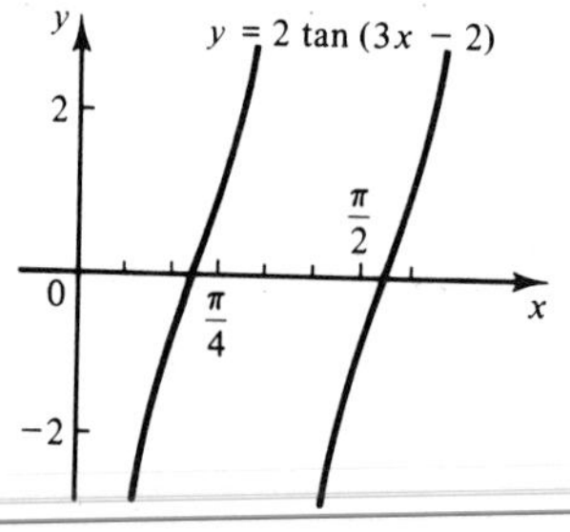

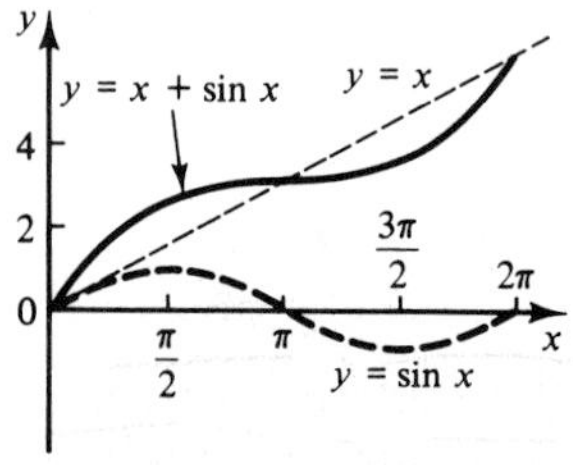

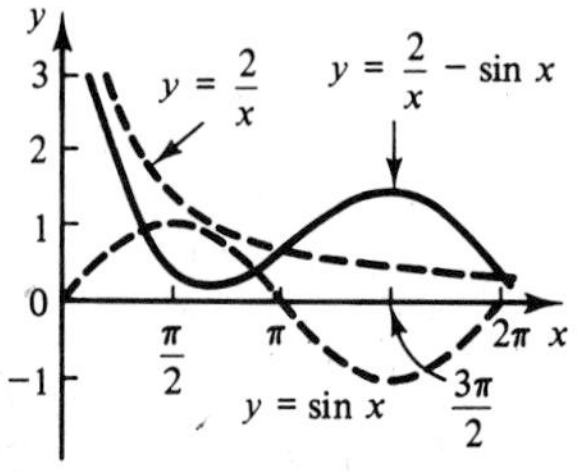

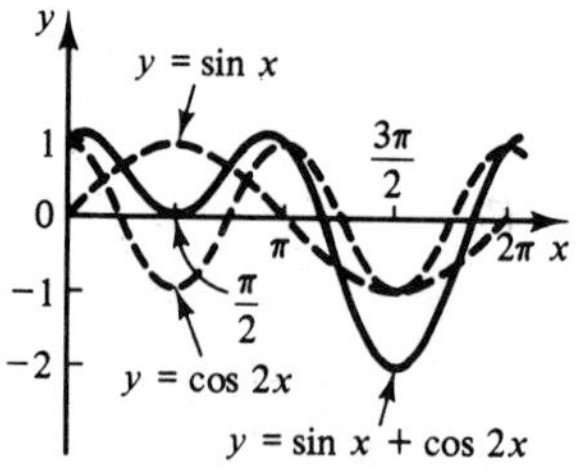

**Exercise 3, Page 437**

**1.** $P = 0.0147$ s, $\omega = 427$ rad/s    **3.** $P = 0.0002$ s, $\omega = 31{,}400$ rad/s    **5.** $f = 8$ Hz, $\omega = 50.3$ rad/s
**7.** $3.33$ s    **9.** $P = 0.0138$ s, $f = 72.4$ Hz    **11.** $P = 0.0126$ s, $f = 79.6$ Hz
**13.** $P = 400$ ms, $a = 10$, $\phi = 1.1$ rad    **15.** $y = 5\sin(750t + 15°)$
**17.** $y = 3\sin 377t$

**19.** $y = 375\sin\left(55t + \dfrac{\pi}{4}\right)$    **21.** $y = 35.7\sin(\omega t + 45.4°)$

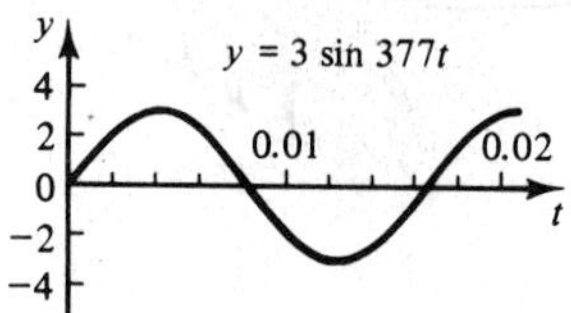

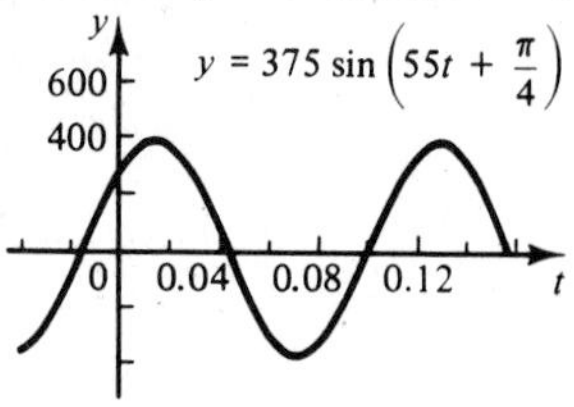

**23.** $y = 843 \sin(\omega t - 28.2°)$    **25.** $y = 9.40 \sin(\omega t + 51.7°)$    **27.** $y = 660 \sin(\omega t + 56.5°)$
**29.** $v_{max} = 4.27$ V, $P = 13.6$ ms, $f = 73.7$ Hz, $\phi = 27° = 0.471$ rad, $v(0.12) = -2.11$ V
**31.** $i = 49.2 \sin(220t + 63.2°)$ mA

### Exercise 4, Page 442

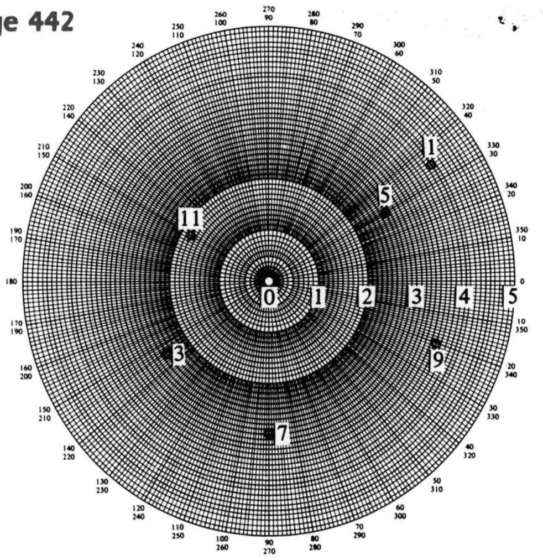

**27.** $(6.71, 63.4°)$    **29.** $(5, 143°)$    **31.** $(7.61, 231°)$    **33.** $(597, 238°)$    **35.** $(3.41, 3.66)$
**37.** $(298, -331)$    **39.** $(2.83, -2.83)$    **41.** $(12.3, -8.60)$    **43.** $(-7.93, 5.76)$    **45.** $x^2 + y^2 = 2y$
**47.** $x^2 + y^2 = 1 - y/x$    **49.** $x^2 + y^2 = 4 - x$    **51.** $r \sin \theta + 3 = 0$    **53.** $r = 1$    **55.** $\sin \theta = r \cos^2 \theta$

### Review Problems, Page 445

**1.**

$y = 3 \sin 2x$
Period $= \pi$
Amplitude $= 3$

**3.**

$y = 1.5 \sin\left(3x + \dfrac{\pi}{2}\right)$
Period $= \dfrac{2\pi}{3}$
Amplitude $= 1.5$
Phase shift $= -\dfrac{\pi}{6}$

**5.**

$y = 2.5 \sin\left(4x + \dfrac{2\pi}{9}\right)$
Period $= \dfrac{\pi}{2}$
Amplitude $= 2.5$
Phase shift $= -\dfrac{\pi}{18}$

**7.** $y = 5 \sin(2x/3 + \pi/9)$

**9.**

$y = x$
$y = x + \cos x$
$y = \cos x$

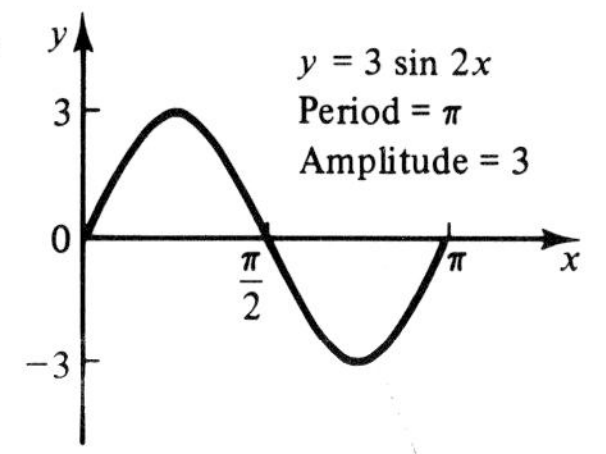

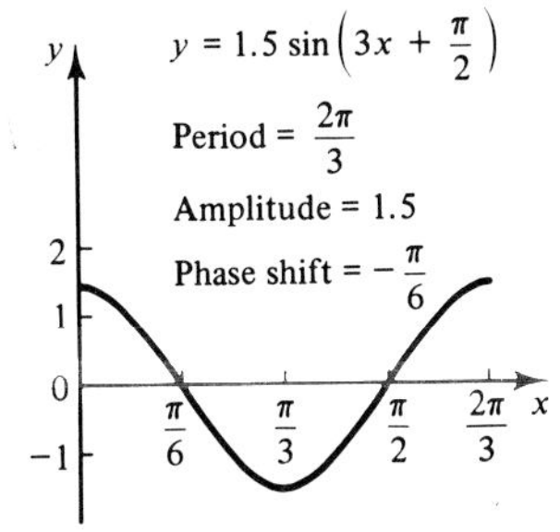

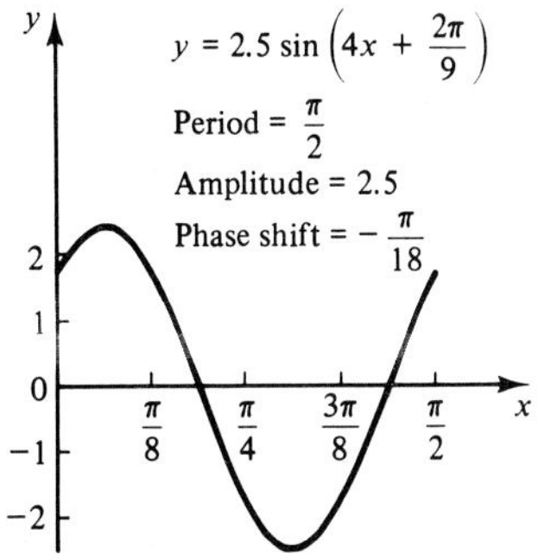

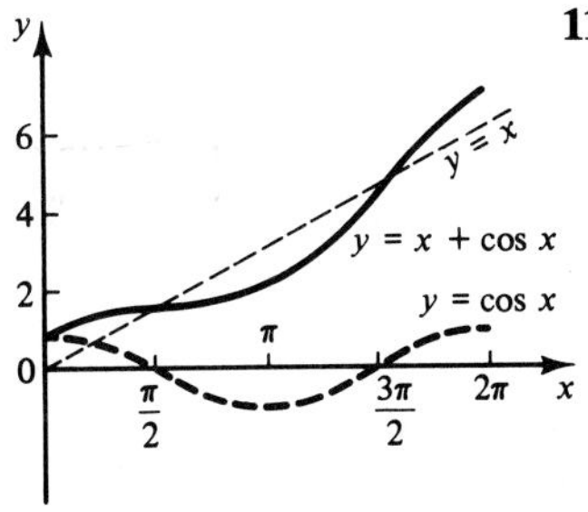

**11 and 13.**

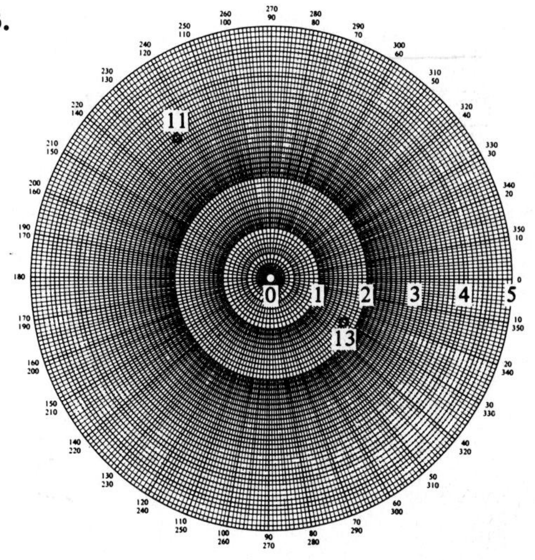

15.

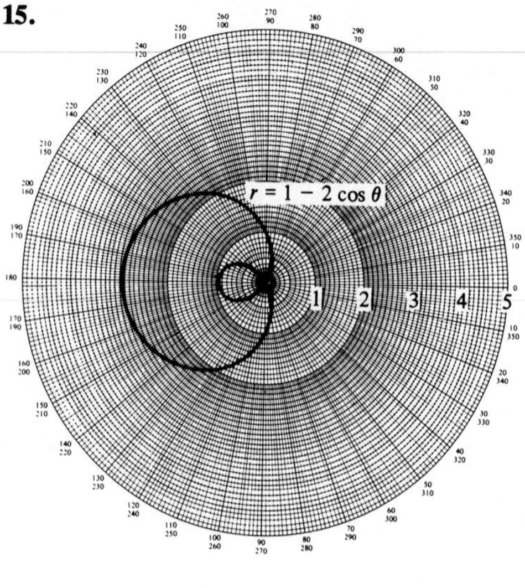


17. (6.52, 144°) 19. (2.54, 2.82) 21. (2.73, −2.64)

23. $x^2 + y^2 = 2x$ 25. $r = 1/(5 \cos \theta + 2 \sin \theta)$ 27. $P = 0.00833$ s, $\omega = 754$ rad/s 29. 3.33 s

31.

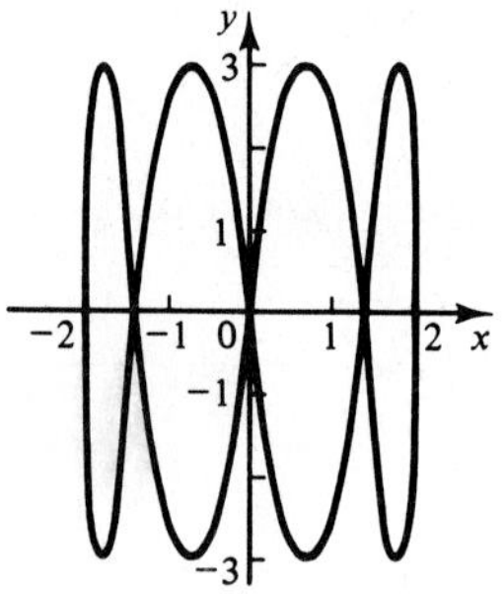


33(a) $y = 101 \sin(\omega t - 15.2°)$, (b) $y = 3.16 \sin(\omega t + 56.8°)$

## CHAPTER 17

### Exercise 1, Page 452

1. $(\sin x - 1)/\cos x$ 3. $1/\cos \theta$ 5. $1/\cos \theta$ 7. $-\tan^2 x$ 9. $\sin \theta$ 11. $\sec \theta$ 13. $\tan x$ 15. $\cos \theta$ 17. $\sin \theta$ 19. 1 21. 1 23. $\sin x$ 25. $\sec x$ 27. $\tan^2 x$ 29. $-\tan^2 \theta$

### Exercise 2, Page 459

1. $\sin \theta \cos 30° + \cos \theta \sin 30°$ 3. $\sin x \cos 60° + \cos x \sin 60°$ 5. $-\sin x$ 7. $\sin \theta \cos 2\phi + \cos \theta \sin 2\phi$ 9. $(\tan 2\theta - \tan 3\alpha)/(1 + \tan 2\theta \tan 3\alpha)$ 11. $-\cos \theta - \sin \theta$ 13. $-\sin x$ 33. $y = 11{,}200 \sin(\omega t + 41.0°)$ 35. $y = 112 \sin(\omega t + 41.4°)$

### Exercise 3, Page 462

1. 1 3. $\sin 2x$ 5. 2

### Exercise 5, Page 470

1. 60°, 120°, 240°, 300° 3. 15°, 135° 5. 30°, 150°, 210°, 330° 7. 45°, 135°, 225°, 315° 9. 0°, 45°, 180°, 225° 11. 60°, 120°, 240°, 300° 13. 60°, 180°, 300° 15. 120°, 240° 17. 0°, 60°, 120°, 180°, 240°, 300° 19. 135°, 315° 21. 0°, 60°, 180°, 300°

### Exercise 6, Page 473

1. 22.0° 3. 281.0° 5. 81.8° 7. 75.4° 9. 33.2°

## Review Problems, Page 474

**13.** 1 **15.** 2 $\csc^2\theta$ **17.** 1 **19.** 1 **21.** 1 **23.** $-\tan\theta$ **25.** 120°, 240° **27.** 45°, 225° **29.** 15°, 45°, 75°, 135°, 195°, 225°, 255°, 315° **31.** 90°, 306.9° **33.** 189.3°, 350.7° **35.** 60°, 300° **37.** 55.6° **39.** 74.3° **41.** 78.2 $\sin(\omega t + 52.5°)$

# CHAPTER 18

## Exercise 1, Page 479

**1.** 4.5 **3.** 2.67 **5.** 8 **7.** 8 **9.** $3x$ **11.** $-5a$ **13.** $(x + 2)/x$ **15.** $\pm 12$ **17.** $\pm 15$ **19.** 30, 42 **21.** 24, 32 **23.** Elder = $40,000, younger = $35,000 **25.** 9, 11 **27.** 300 turns

## Exercise 2, Page 483

**1.** 197 **3.** 71.5

**5.**

| $x$ | 9 | 11 | 15 |
|---|---|---|---|
| $y$ | 45 | 55 | 75 |

**7.**

| $x$ | 115 | 125 | 138 | 154 |
|---|---|---|---|---|
| $y$ | 140 | 152 | 167 | 187 |

**9.** 438 km **11.** 137 **13.** 67 hp **15.** 911 m **17.** 1800 $cm^3$

## Exercise 3, Page 490

**1.** 2254 **3.** 66.7

**5.**

| $x$ | 18.2 | 75.6 | 27.5 |
|---|---|---|---|
| $y$ | 29.7 | 8842 | 154 |

**7.**

| $x$ | 315 | 122 | 782 |
|---|---|---|---|
| $y$ | 203 | 148 | 275 |

**9.**

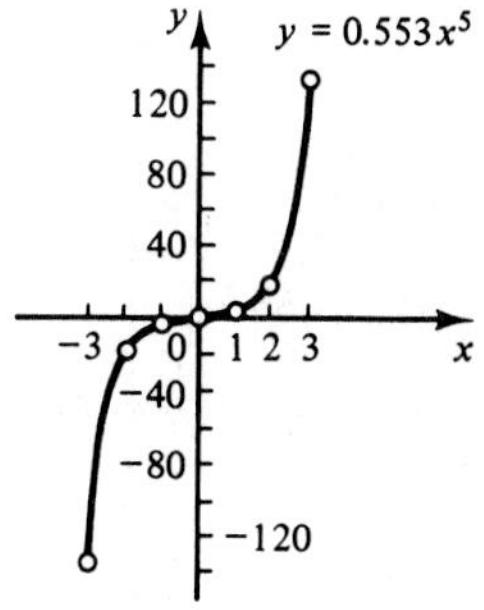


**11.** 396 m **13.** 4.03 s **15.** 765 W **17.** 1.73 **19.** f8 **21.** f13 **23.** 8.30 $ft^3$ **25.** 236 metric tons **27.** 640 $ft^2$ **29.** 3.67 in. **31.** 28,900 $m^2$

## Exercise 4, Page [illegible]

**1.** 1417 **3.** $y$ is halved **5.** $x = 91.5$, $y = 61.5$ **7.** $x = 9236$, $y = 1443$ **9.** 41.4%

**11.**

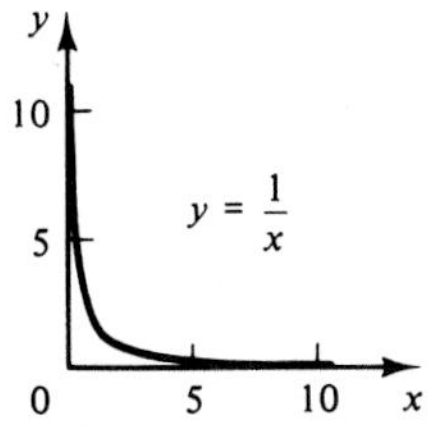


**13.** 103 lb/in$^2$ **15.** $4.49 \times 10^{-6}$ dyn **17.** 78.9 lb **19.** 200 lux **21.** 5.31 m **23.** 0.909 **25.** $33\frac{1}{3}$% increase

## Exercise 5, Page 499

**1.** 352 **3.** 4.2% increase **5.** $y = 121$, $w = 116$, $x = 43.5$ **7.** 13.9 **9.** $y = 4.08\left(\frac{x^{3/2}}{w}\right)$ **11.** 13.8% decrease **13.** 1/3 **15.** 9/4 **17.** 169 W **19.** 3.58 $m^3$ **21.** 3.0 weeks **23.** 12,600 lb **25.** 394 times/s **27.** 2.21

## Review Problems, Page 501

**1.** 1038 **3.** 10% increase **5.** 121 liters/min **7.** 1.73 **9.** 29.3 yr **11.** shortened by 0.31 in. **13.** 8.5 **15.** 22.9 h **17.** 215,000 mi **19.** 42 kg **21.** 2.8 **23.** 45% increase **25.** 469 gal **27.** 102 $m^2$ **29.** 1.12 in. **31.** 17.6 $m^2$ **33.** 12.7 tons **35.** 144 oz **37.** $610 **39.** 1.5 **41.** 7613 $cm^2$

## CHAPTER 19

### Exercise 1, Page 508

**1.**

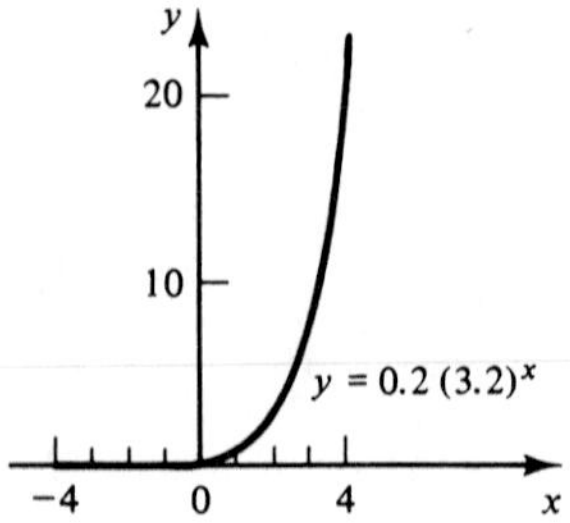


**3.**

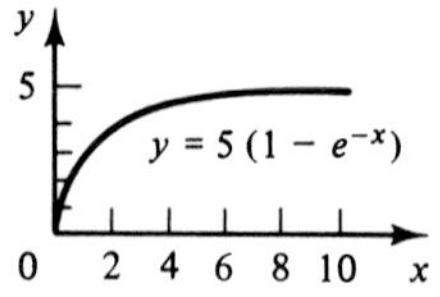


**5.**

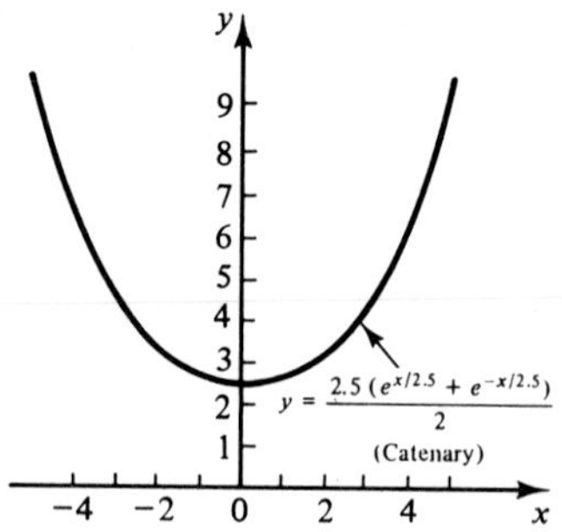


**7.** \$4851.94 **9(a)** \$6.73, **(b)** \$7.33, **(c)** \$7.39 **11.** \$1822.67

### Exercise 2, Page 513

**1.** 284 units **3.** 5.16 billion **5.** 80 million bbl/day **7.** 506°C **9.** 0.101 A **11.** 1048°F **13.** 9.604 in. Hg **15.** 72 mA **17.** 30%

### Exercise 3, Page 519

**1.** $\log_3 81 = 4$ **3.** $\log_4 4096 = 6$ **5.** $\log_x 995 = 5$ **7.** $10^2 = 100$ **9.** $5^3 = 125$ **11.** $3^{57} = x$ **13.** $5^y = x$ **15.**

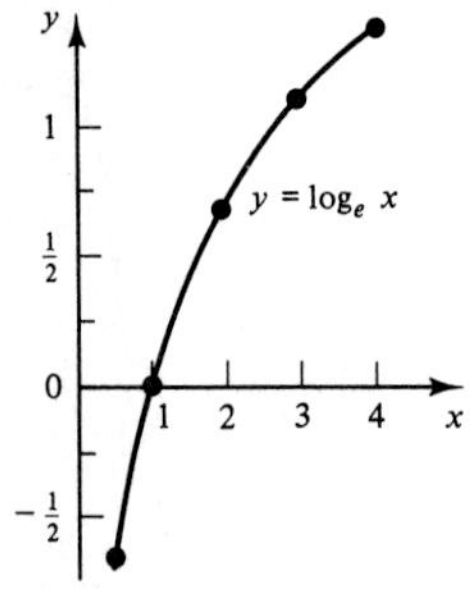


**17.** 1.4409 **19.** 0.7731 **21.** 1.6839 **23.** 2.9222 **25.** 1.4378 **27.** −2.1325 **29.** 4.5437 **31.** 38.4 **33.** 187,000 **35.** $1.39 \times 10^{-3}$ **37.** $5.93 \times 10^{-3}$ **39.** 9530 **41.** $1.07 \times 10^{-4}$ **43.** 3.8774 **45.** 7.7685 **47.** −0.1684 **49.** −4.7843 **51.** −9.9825 **53.** 10.5558 **55.** 17.22 **57.** 2.408 **59.** 0.6942 **61.** 2.550 **63.** $7.572 \times 10^6$ **65.** $1.063 \times 10^{-8}$

### Exercise 4, Page 528

**1.** $\log 2 - \log 3$ **3.** $\log a + \log b$ **5.** $\log x + \log y + \log z$ **7.** $\log 3 + \log x - \log 4$ **9.** $-\log 2 - \log x$ **11.** $\log a + \log b + \log c - \log d$ **13.** log 12 **15.** log 3/2 **17.** log 27 **19.** $\log a^3c^4/b^2$ **21.** $\log xy^2z^3/ab^2c^3$ **23.** $\log[\sqrt{x+2}(x-2)]$ **25.** $2^x = xy^2$ **27.** $a^2 + 2b^2 = 0$ **29.** $p - q = 100$ **31.** 1 **33.** 2 **35.** $x$ **37.** $3y$ **39.** 549.0 **41.** 322 **43.** 2.735 **45.** 12.58 **47.** 3.634 **49.** 3.63 **51.** 1.639 **53.** 2.28 **55.** 195 **57.** −8.796 **59.** 5.46

### Exercise 5, Page 534

**1.** 2.81 **3.** 16.1 **5.** 2.46 **7.** 1.17 **9.** 5.10 **11.** 1.49 **13.** 0.239 **15.** −3.44 **17.** 1.39 **19.** 0 **21.** $3e^{4.159x}$ **23.** 0.0147 s **25.** 1.5 s **27.** 28,600 ft **29.** 4.62 rad **31.** 2.41 yr **33.** 7.95 yr **35.** 9.9 yr **37.** $N < 3.5\%$

### Exercise 6, Page 538

**1.** 2 **3.** 1/3 **5.** 2/3 **7.** 2/3 **9.** 2 **11.** 3 **13.** 25 **15.** 6 **17.** 3/2 **19.** 47.5 **21.** −11.0, 9.05 **23.** 10/3 **25.** 22 **27.** 0.928 **29.** 0.916 **31.** 12 **33.** 101 **35.** 50,000 **37.** 7 yr **39.** 3.5 **41.** 27.36 in. Hg **43.** 1300 kW **45.** −3.01 dB **47.** 17.5 yr **49.** $10^{-7}$

## Exercise 7, Page 545

**1.**

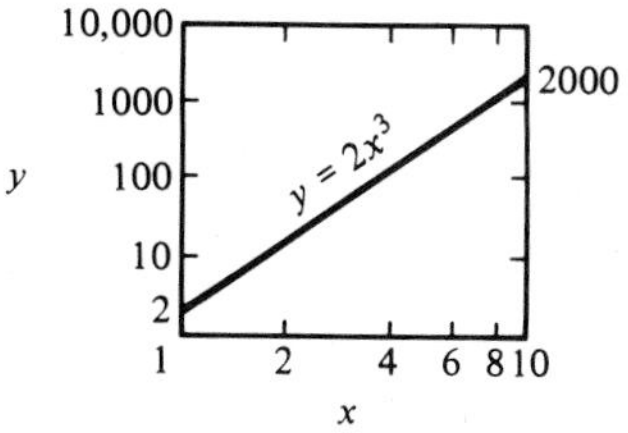


**3.**

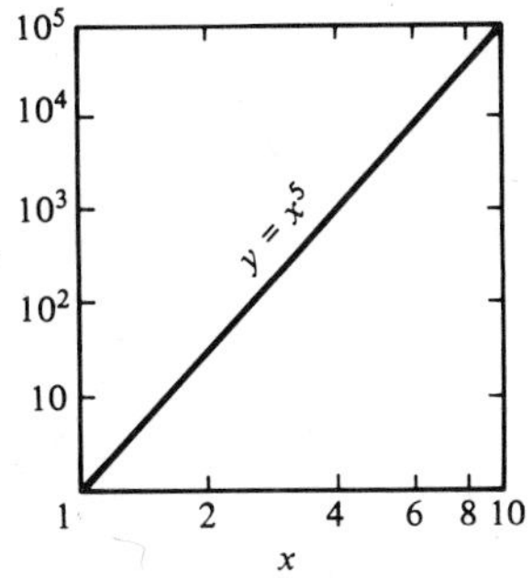


**5.**

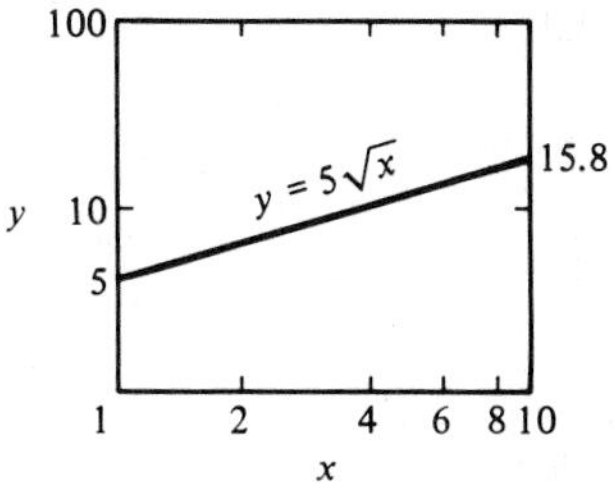


**7.**

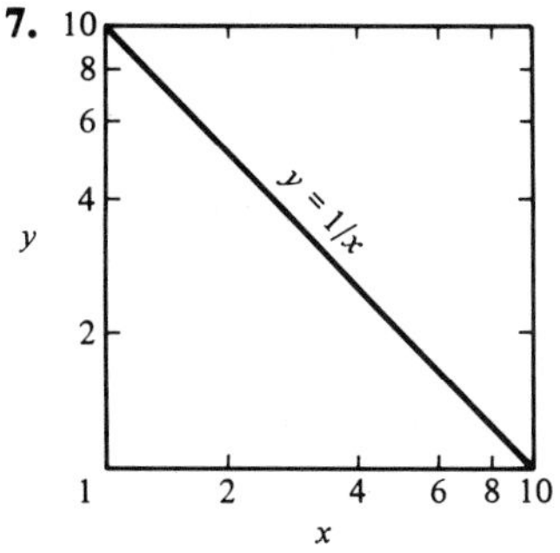


**9.**

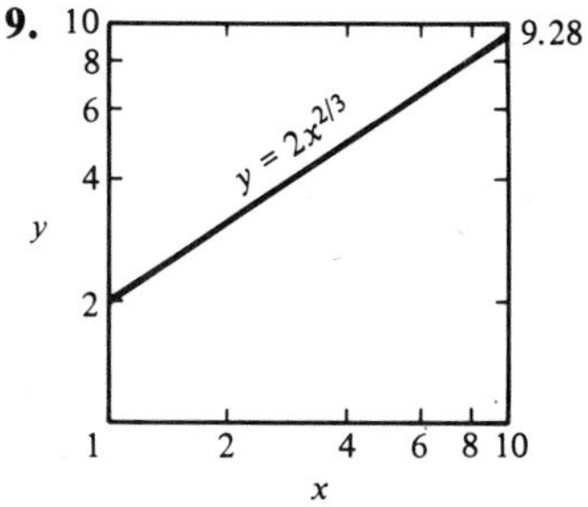


**11.** $y = 2000x^{-2.97}$

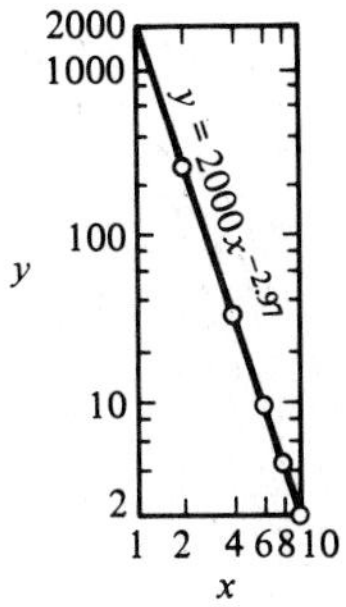


**13.**

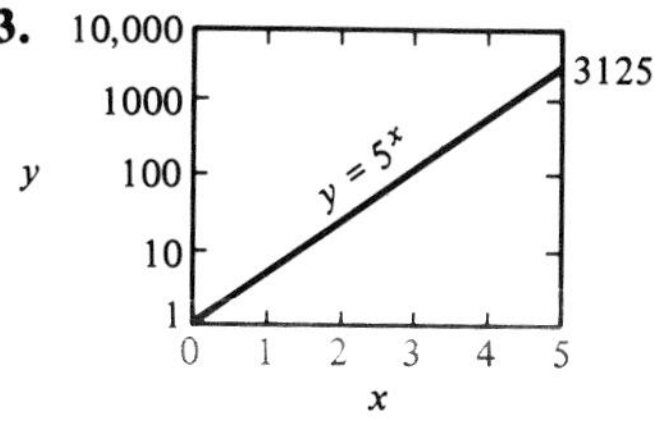


**15.**

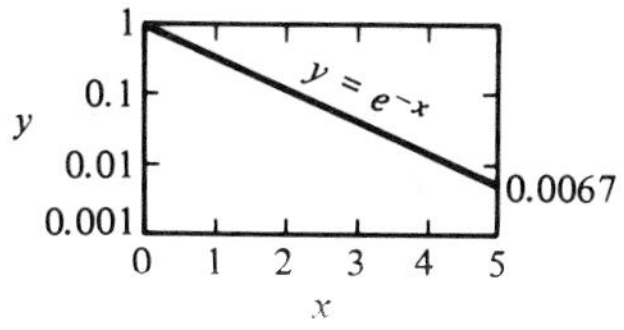


**17.**

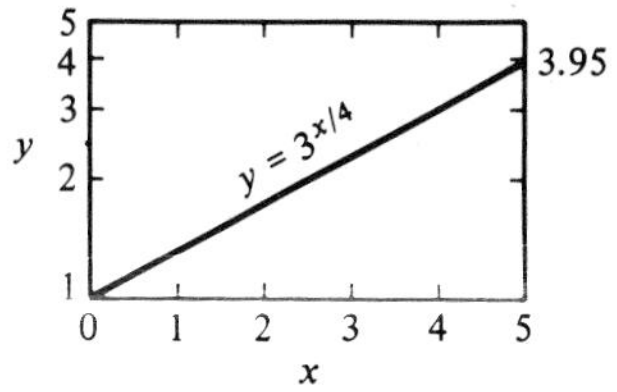


**19.**

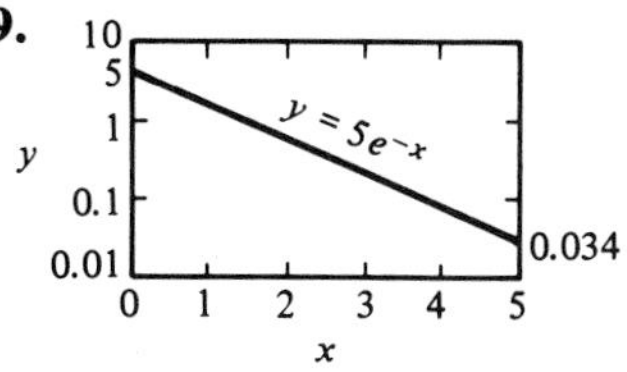


**21.** $y = 2.5^x$

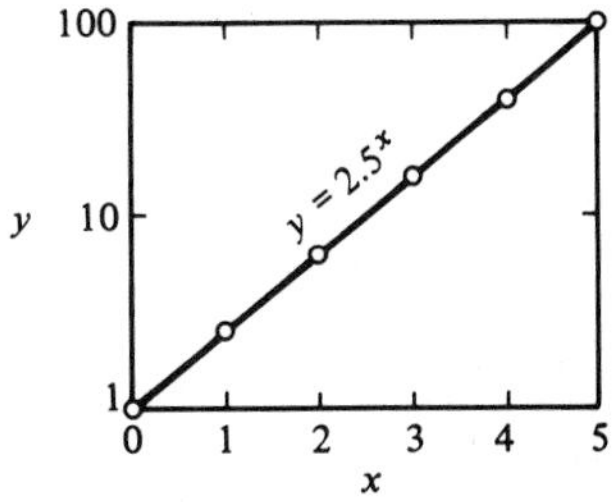


**23.** $i = 16.5v^{0.598}$

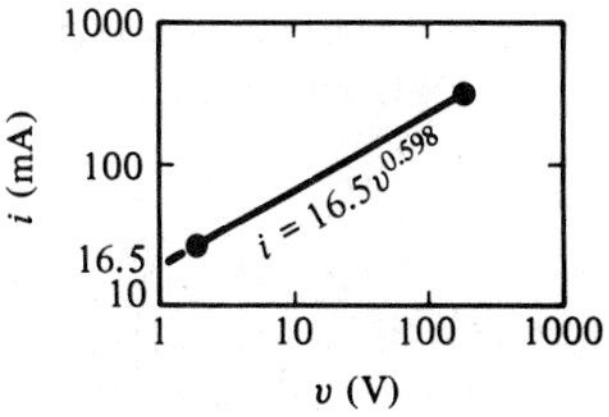


**25.** $p = 460v^{-1.07}$

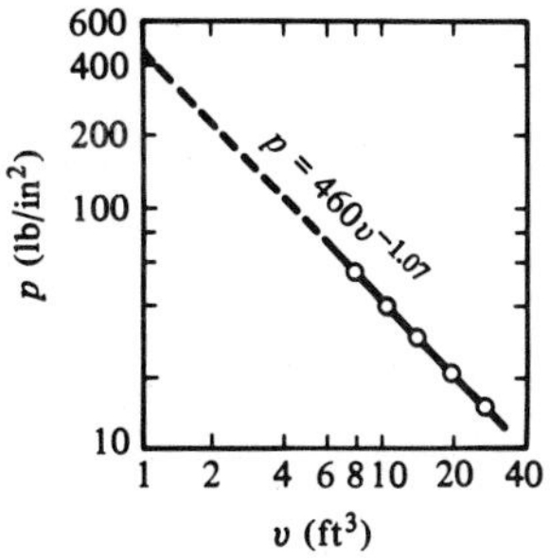

## Review Problems, Page 547

**1.** $\log_x 352 = 5.2$ **3.** $\log_{24} x = 1.4$ **5.** $x^{124} = 5.2$ **7.** 3/4 **9.** 1/128
**11.** $\log 3 + \log x - \log z$ **13.** log 10 **15.** $\log(\sqrt{p}/\sqrt[4]{q})$ **17.** 2.5611 **19.** 701.5 **21.** 0.3365
**23.** 2.943 **25.** 4.4394 **27.** −4.7410 **29.** 4.362 **31.** 2.101 **33.** 13.44 **35.** 3.17
**37.** $10^{-3}$ **39.** \$2070.63 **41.** 828 rev/min **43.** 23 yr **45.**

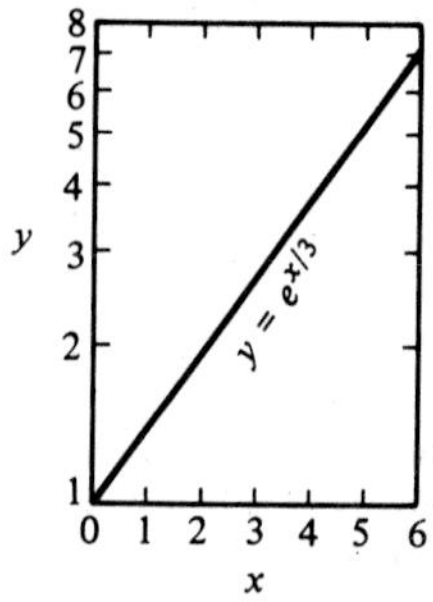


# CHAPTER 20

## Exercise 1, Page 555

**1.** $j3$ **3.** $j/2$ **5.** $4 + j2$ **7.** $-5 + j3/2$ **9.** $j5$ **11.** $-1 + j$ **13.** $2a + j2$
**15.** $3/4 + j/6$ **17.** $4.03 + j1.20$ **19.** $j$ **21.** $j$ **23.** $j14$ **25.** −15 **27.** $-j96$ **29.** −25
**31.** $6 - j8$ **33.** $8 + j20$ **35.** $36 + j8$ **37.** $42 - j39$ **39.** $21 - j20$ **41.** $2 + j3$
**43.** $p - jq$ **45.** $n + jm$ **47.** $j2$ **49.** 2 **51.** $2 + j$ **53.** $-5/2 + j/2$ **55.** $11/34 - j41/34$
**57.** $0.833 \pm j1.28$ **59.** $1.00 \pm j2.24$ **61.** $(x + j3)(x - j3)$ **63.** $(2y + jz)(2y - jz)$

## Exercise 2, Page 556

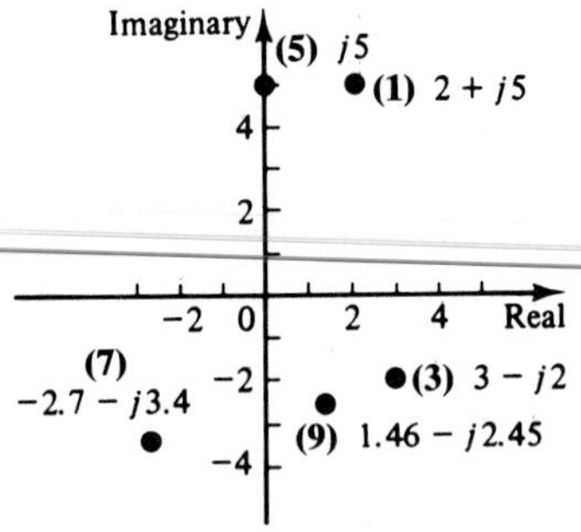


## Exercise 3, Page 562

**1.** $6.4\underline{/38.7^\circ}$, $6.4(\cos 38.7^\circ + j \sin 38.7^\circ)$ **3.** $5\underline{/323^\circ}$, $5(\cos 323^\circ + j \sin 323^\circ)$
**5.** $5.39\underline{/202^\circ}$, $5.39(\cos 202^\circ + j \sin 202^\circ)$ **7.** $10.3\underline{/209^\circ}$, $10.3(\cos 209^\circ + j \sin 209^\circ)$
**9.** $8.06\underline{/240^\circ}$, $8.06(\cos 240^\circ + j \sin 240^\circ)$ **11.** $1.63 + j2.52$, $3\underline{/57^\circ}$ **13.** $-7.79 + j4.50$, $9\underline{/150^\circ}$
**15.** $3.70 + j4.01$, $5.46\underline{/47.3^\circ}$ **17.** $4.64 + j7.71$, $9(\cos 59^\circ + j \sin 59^\circ)$
**19.** $4.21 - j5.59$, $7(\cos 307^\circ + j \sin 307^\circ)$ **21.** $15(\cos 40^\circ + j \sin 40^\circ)$ **23.** $16.1(\cos 54.9^\circ + j \sin 54.9^\circ)$
**25.** $56\underline{/60^\circ}$ **27.** $2(\cos 25^\circ + j \sin 25^\circ)$ **29.** $4.70(\cos 50.2^\circ + j \sin 50.2^\circ)$ **31.** $10\underline{/60^\circ}$
**33.** $8(\cos 45^\circ + j \sin 45^\circ)$ **35.** $49\underline{/40^\circ}$ **37.** $7.55\underline{/26^\circ}$, $7.55\underline{/206^\circ}$
**39.** $3.36\underline{/24.3^\circ}$, $3.36\underline{/144.3^\circ}$, $3.36\underline{/264.3^\circ}$ **41.** $13.8 + j7.44$ and $-13.8 - j7.44$

## Exercise 4, Page 565

**1.** $3.61e^{j0.983}$ **3.** $3e^{j0.873}$ **5.** $2.5e^{j\pi/6}$ **7.** $5.4e^{j\pi/12}$
**9.** $-4.95 + j0.706$, $5\underline{/172^\circ}$, $5(\cos 172^\circ + j \sin 172^\circ)$ **11.** $0.156 + j2.19$, $2.2\underline{/85.9^\circ}$, $2.2(\cos 85.9 + j \sin 85.9^\circ)$
**13.** $18e^{j6}$ **15.** $21e^{j4}$ **17.** $3.57e^{j7}$ **19.** $3e^{j3}$ **21.** $11e^{j3}$ **23.** $3e^{j}$ **25.** $9e^{j10}$ **27.** $8e^{j3}$

## Exercise 5, Page 568

**1.** $4.59 + j5.28$, $7\underline{/49^\circ}$ **3.** $54.8 + j185$, $193\underline{/73.5^\circ}$ **5.** $-31.2 + j23.3$, $39\underline{/143^\circ}$ **7.** $8 - j2$
**9.** $38.3 + j60.3$ **11.** $8.67 + j6.30$ **13.** $29 - j29$ **15.** $12(\cos 32^\circ - j \sin 32^\circ)$ **17.** $3.08\underline{/32^\circ}$

### Exercise 6, Page 572

**1.** $250\angle 25^\circ$ **3.** $57\angle 270^\circ$ **5.** $144\angle 0^\circ$ **7.** $150 \cos \omega t$ **9.** $300 \cos(\omega t - 90^\circ)$ **11.** $7.5 \cos \omega t$
**13.** $155 + j0 = 155\angle 0^\circ$ **15.** $0 - j18 = 18\angle 270^\circ$ **17.** $72 - j42$, $83.4\angle -30.3^\circ$
**19.** $552 - j572$, $795\angle -46.0^\circ$ **21(a)** $\mathbf{V} = 603 \cos(\omega t + 85.3^\circ)$, **(b)** $\mathbf{V} = 293 \cos(\omega t - 75.5^\circ)$

### Review Problems, Page 574

**1.** $3 + j2$, $3.61\angle 33.7^\circ$, $3.61(\cos 33.7^\circ + j \sin 33.7^\circ)$, $3.61e^{j0.588}$
**3.** $-2 + j7$, $7.28\angle 106^\circ$, $7.28(\cos 106^\circ + j \sin 106^\circ)$, $7.28e^{j1.85}$ **5.** $j$ **7.** $9 + j2$ **9.** $60.8 + j45.6$
**11.** $11 + j7$ **13.** $12(\cos 38^\circ + j \sin 38^\circ)$ **15.** $33/17 - j21/17$ **17.** $2(\cos 45^\circ + j \sin 45^\circ)$
**19 and 21.** **23.** $7 - j24$ **25.** $125(\cos 30^\circ + j \sin 30^\circ)$ **27.** $45\angle 32^\circ$

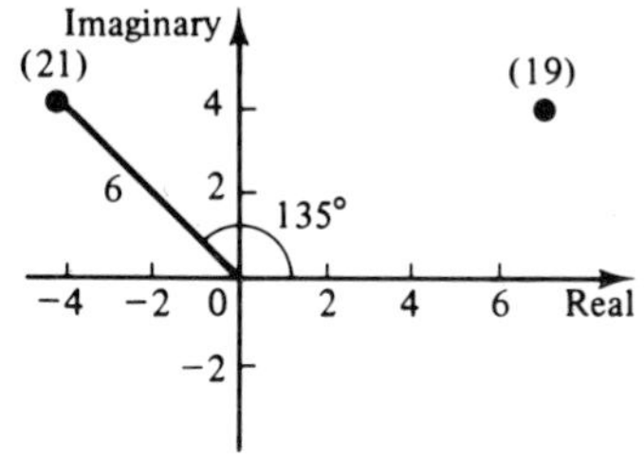


**29.** $1.25\angle -45^\circ$

## CHAPTER 21

### Exercise 1, Page 581

**1.** 2 **3.** 3 **5.** 6 **7.** 13 **9.** 12 **11.** 5 **13.** 103 **15.** 119 **17.** 101 **19.** 10
**21.** 0100 1000 **23.** 0101 1101 **25.** 1 0001 0010 **27.** 0111 0110 **29.** 10 0000 1011 0111
**31.** 1 0100 0011 0011 0100 **33.** 0.1 **35.** 0.11 **37.** 0.0100 1100 **39.** 0.1000 1100 **41.** 0.111
**43.** 0.0110 0011 **45.** 0.5 **47.** 0.25 **49.** 0.5625 **51.** 101.1 **53.** 100.011
**55.** 11 1011 0100.0111 1000 **57.** 1.5 **59.** 2.25 **61.** 25.40625

### Exercise 2, Page 585

**1.** D **3.** 9 **5.** 93 **7.** D8 **9.** 92A6 **11.** 9.3 **13.** 1.38 **15.** 0110 1111
**17.** 0100 1010 **19.** 0010 1111 0011 0101 **21.** 0100 0111 1010 0010 **23.** 0101.1111
**25.** 1001.1010 1010 **27.** 242 **29.** 51 **31.** 14,244 **33.** 62,068 **35.** 3.9375
**37.** 2748.8671875 **39.** 27 **41.** 399 **43.** AB5 **45.** 6C8

### Exercise 3, Page 586

**1.** 6 **3.** 7 **5.** 63 **7.** 155 **9.** 111 **11.** 01 0110 **13.** 1 1001 0011
**15.** 1010 1010 0011 **17.** 0110 0110 0110 1000

### Exercise 4, Page 588

**1.** 0110 0010 **3.** 0010 0111 0100 **5.** 0100 0010.1001 0001 **7.** 9 **9.** 61 **11.** 36.8

### Review Problems, Page 588

**1.** 6.125 **3.** 0100 0101 **5.** 1264 **7.** 1 1010.111 **9.** 348 **11.** BE **13.** 20AA
**15.** 134 **17.** 0100 1110 0101 1101 **19.** 0100 1010

## CHAPTER 22

### Exercise 1, Page 593

**1.**

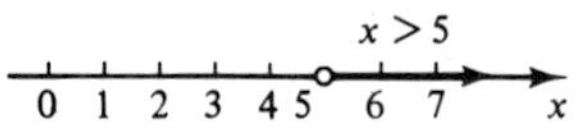


**3.**

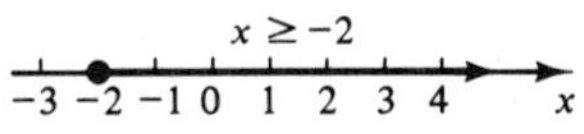


**5.**

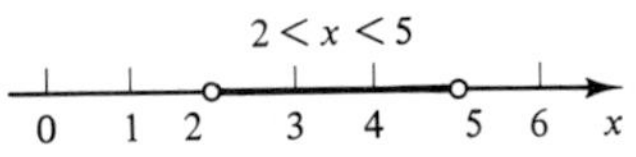

7.
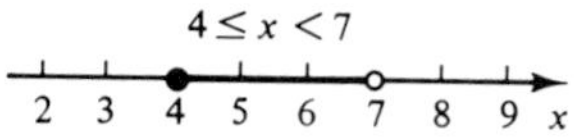


9.
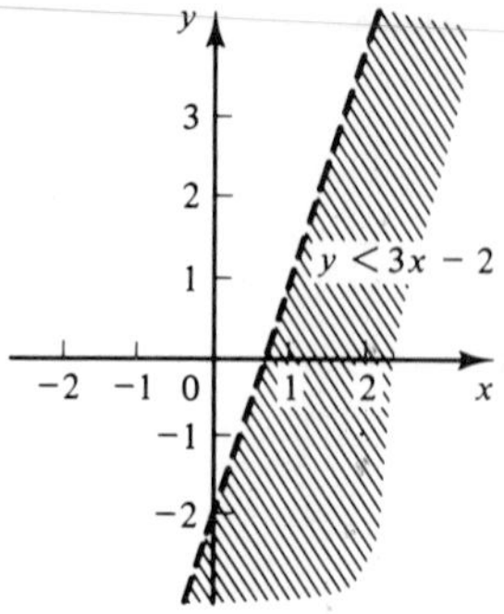


11.
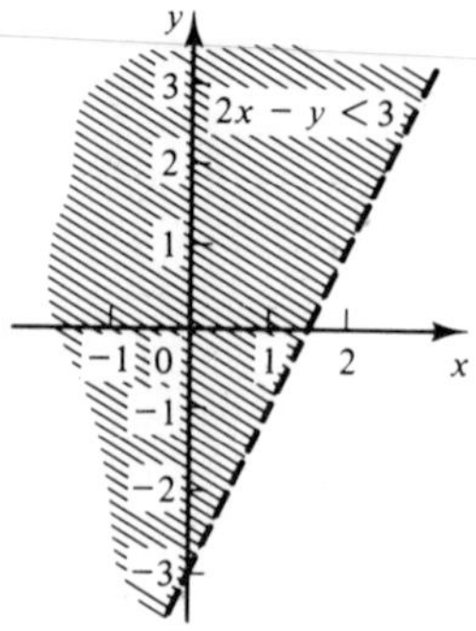


13.
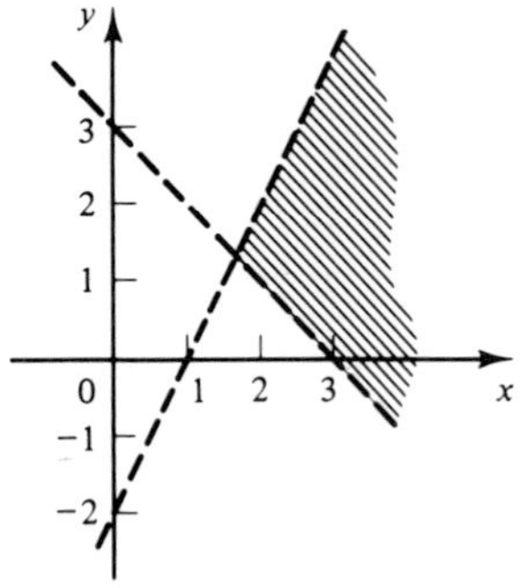

15.
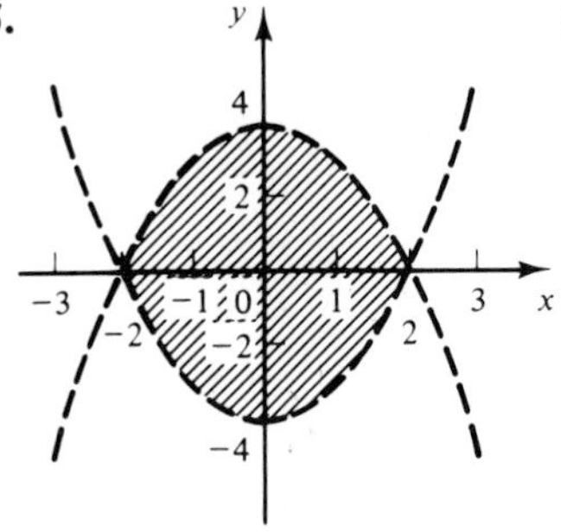

17.
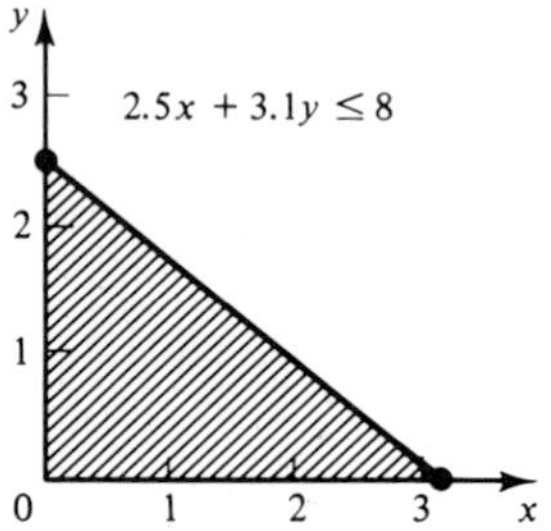


## Exercise 2, Page 599

**1.** $x > 9$ **3.** $-3 \leq x \leq 3$ **5.** $x < 1$ or $x > 3/2$ **7.** $x \geq 40/7$ **9.** $x \leq -5$ or $x \geq 1$
**11.** $x \leq -2$ or $x \geq 3$ **13.** $x > 3$ or $x < -3$ **15.** $x > 8/5$ or $x < -8/5$ **17.** $x \geq 10/3$ or $x \leq -2$
**19.** $x < 5$ **21.** $x \leq -1/2$ or $x \geq 7/8$ **23.** $x \leq 7.8$ h

## Exercise 3, Page 605

**1.** $x = 1.5$, $y = 3.5$ **3.** $x = 4.59$, $y = 0.724$ **5.** 69 pulleys, 68 sprockets
**7.** 8 disk drives, 4 monitors

## Review Problems, Page 605

1.
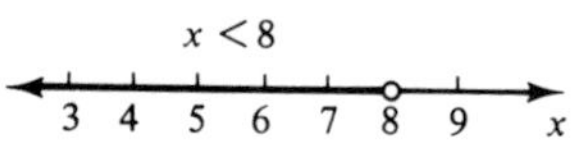


3.
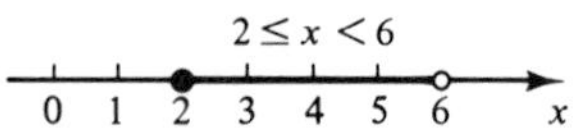


5.
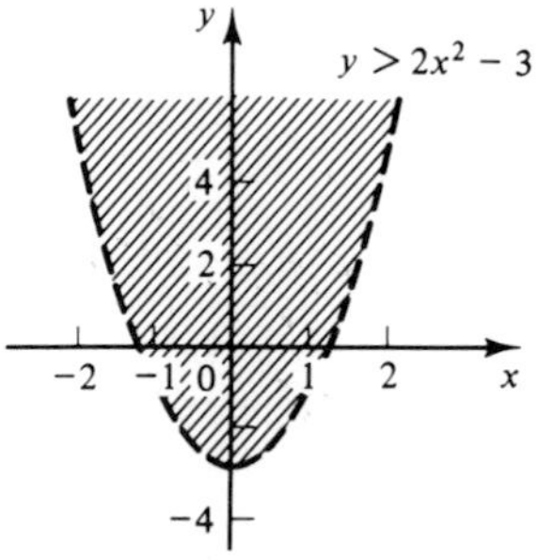


**7.** $x > 8$ **9.** $x < -3$ or $x > 11$ **11.** $x = 2.5$, $y = 0$

**13.** $x \leq -5$ or $x \geq -3$ **15.** $x \leq -7$ or $x \geq 2$

# CHAPTER 23

## Exercise 1, Page 612

**1.** $3 + 6 + 9 + 12 + 15 + \cdots + 3n + \cdots$ **3.** $2 + 3/4 + 4/9 + 5/16 + 6/25 + \cdots + (n + 1)/n^2 + \cdots$
**5.** $u_n = 2n$; 8, 10 **7.** $u_n = 2^n/(n + 3)$; 4, 64/9 **9.** $u_n = u_{n-1} + 4$; 13, 17
**11.** $u_n = (u_{n-1})^2$; 6561, 43,046,721 **13(a)** Diverges, **(b)** Diverges, **(c)** Converges

## Exercise 2, Page 618

**1.** 46 **3.** 43 **5.** 49 **7.** $x + 24y$ **9.** $3, 7\frac{1}{3}, 11\frac{2}{3}, 16, 20\frac{1}{3}, \ldots$ **11.** 5, 11, 17, 23, 29 . . .
**13.** −7 **15.** 234 **17.** 225 **19.** 15 **21.** 10, 15 **23.** $-6\frac{3}{5}, -7\frac{1}{5}, -7\frac{4}{5}, -8\frac{2}{5}$ **25.** 3/14
**27.** 7/11, 7/13 **29(a)** $I = 800 - 80(n - 1)$, **(b)** \$4400 **31.** \$94,000 **33.** 1610 ft

### Exercise 3, Page 623

**1.** 80 **3.** −9375 **5.** 5115 **7.** −292,968 **9.** ±15 **11.** ±30 **13.** 24, 72 **15.** ±20, 80, ±320 **17.** $e^{1/2}$ or 1.649 **19.** 3.23°F **21.** 9.3 yr **23.** 28.75 ft **25.** 62 **27.** 2.01 **29.** 1.97 **31.** \$184,201 **33.** \$7776

### Exercise 4, Page 627

**1.** $5 - c$ **3.** $3/(c + 4)$ **5.** 288 **7.** 12.5 **9.** 19/33 **11.** $7\frac{15}{22}$ **13.** 45.5 in. **15.** Sum = 1

### Exercise 5, Page 635

**1.** 720 **3.** 42 **5.** 35 **7.** $x^7 + 7x^6y + 21x^5y^2 + 35x^4y^3 + 35x^3y^4 + 21x^2y^5 + 7xy^6 + y^7$
**9.** $81a^4 - 216a^3b + 216a^2b^2 - 96ab^3 + 16b^4$ **11.** $x^{5/2} + 5x^2y^{2/3} + 10x^{3/2}y^{4/3} + 10xy^2 + 5x^{1/2}y^{8/3} + y^{10/3}$
**13.** $(a/b)^6 - 6(a/b)^4 + 15(a/b)^2 - 20 + 15(b/a)^2 - 6(b/a)^4 + (b/a)^6$
**15.** $32a^{10} + 80a^8b^{1/2} + 80a^6b + 40a^4b^{3/2} + 10a^2b^2 + b^{5/2}$ **17.** $x^{16} + 8x^{14}y^3 + 28x^{12}y^6 + 56x^{10}y^9 + \cdots$
**19.** $a^9 - 9a^8b^4 + 36a^7b^8 - 84a^6b^{12} + \cdots$ **21.** $8a^6 + 12a^5 + 42a^4 + 37a^3 + 63a^2 + 27a + 27$
**23.** $16a^8 + 32a^7 - 104a^6 - 184a^5 + 289a^4 + 368a^3 - 416a^2 - 256a + 256$ **25.** $512{,}512x^{10}$ **27.** $330x^4a^7$
**29.** $6435x^{14}$ **31.** $1 + a/2 - a^2/8 + a^3/16 - \cdots$ **33.** $1 - 3a + 6a^2 - 10a^3 + \cdots$

### Review Problems, Page 636

**1.** 35 **3.** 6, 9, 12, 15 **5.** $2\frac{2}{5}$, 3, 4, 6 **7.** 85 **9.** $a^5 - 10a^4 + 40a^3 - 80a^2 + 80a - 32$
**11.** 48 **13.** $9 - x$ **15.** 8 **17.** 5/7 **19.** 49/90 **21.** 1215 **23.** −1024 **25.** 20, 50
**27.** 1440 **29.** 252 **31.** $128x^7/y^{14} - 448x^{13/2}/y^{11} + 672x^6/y^8 - 560x^{11/2}/y^5 + \cdots$
**33.** $1 - a/2 + 3a^2/8 - 5a^3/16 + \cdots$ **35.** $-26{,}730a^4b^7$

## CHAPTER 24

### Exercise 1, Page 642

**1.** Discrete **3.** Categorical **5.** Categorical **7.**

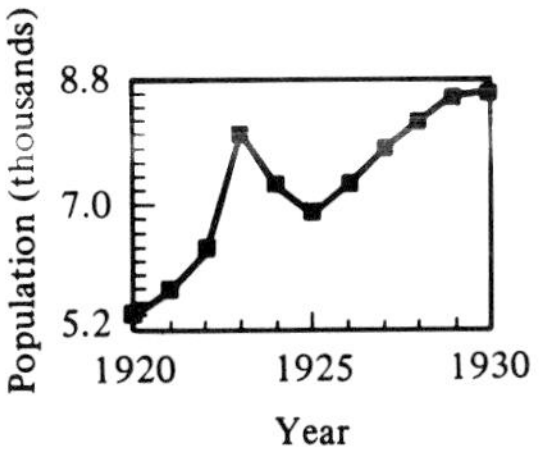


**9.**

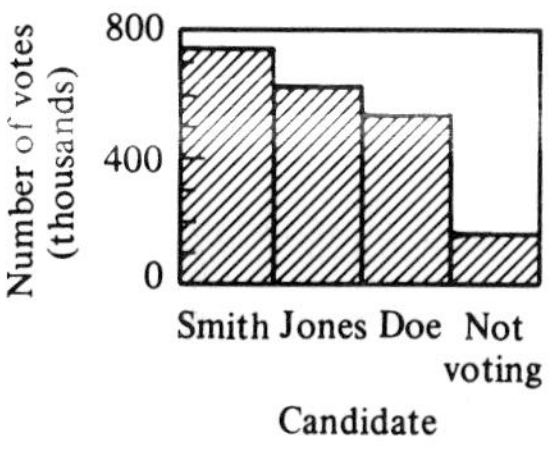


### Exercise 2, Page 647

**1(a)** Range = 172 − 111 = 61

| Class Midpt. | Class Limits | | 1(b) Abs. Freq. | 1(c) Rel. Freq. (%) |
|---|---|---|---|---|
| 113 | 110.5 | 115.5 | 3 | 7.5 |
| 118 | 115.5 | 120.5 | 4 | 10.0 |
| 123 | 120.5 | 125.5 | 1 | 2.5 |
| 128 | 125.5 | 130.5 | 3 | 7.5 |
| 133 | 130.5 | 135.5 | 0 | 0.0 |
| 138 | 135.5 | 140.5 | 2 | 5.0 |
| 143 | 140.5 | 145.5 | 2 | 5.0 |
| 148 | 145.5 | 150.5 | 6 | 15.0 |
| 153 | 150.5 | 155.5 | 7 | 17.5 |
| 158 | 155.5 | 160.5 | 2 | 5.0 |
| 163 | 160.5 | 165.5 | 5 | 12.5 |
| 168 | 165.5 | 170.5 | 3 | 7.5 |
| 173 | 170.5 | 175.5 | 2 | 5.0 |

**3(a)** Range = 99.2 − 48.4 = 50.8

| Class Midpt. | Class Limits | | 3(b) Abs. Freq. | 3(c) Rel. Freq. (%) |
|---|---|---|---|---|
| 525 | 500.1 | 550 | 0 | 0.0 |
| 575 | 550.1 | 600 | 1 | 3.3 |
| 625 | 600.1 | 650 | 2 | 6.7 |
| 675 | 650.1 | 700 | 4 | 13.3 |
| 725 | 700.1 | 750 | 3 | 10.0 |
| 775 | 750.1 | 800 | 7 | 23.3 |
| 825 | 800.1 | 850 | 1 | 3.3 |
| 875 | 850.1 | 900 | 7 | 23.3 |
| 925 | 900.1 | 950 | 4 | 13.3 |
| 975 | 950.1 | 1000 | 1 | 3.3 |

**5.**

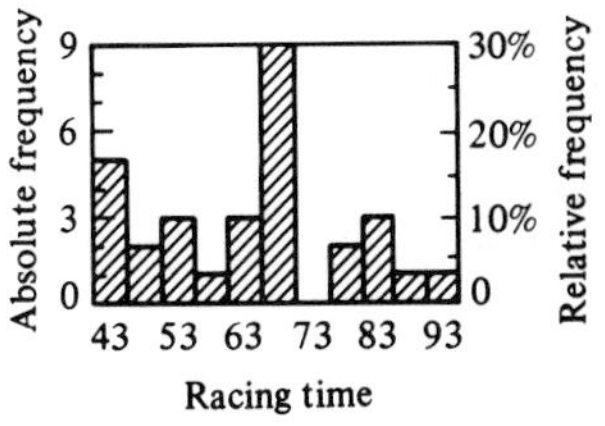

**7.**

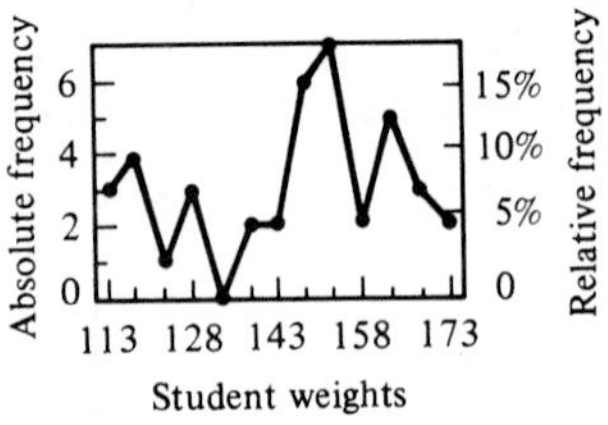


**9.**

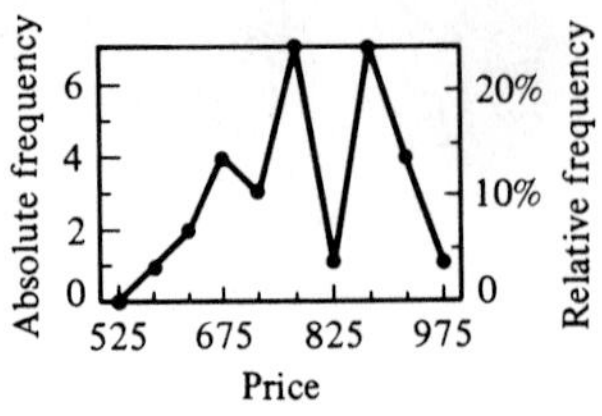


| | 11(a) | | 11(b) | |
|---|---|---|---|---|
| | | | *Cumulative Freq.* | |
| *Class Midpt.* | *Class Limits* | | *Abs.* | *Rel. (%)* |
| 43 | 40.05 | 45.05 | 5 | 16.7 |
| 48 | 45.05 | 50.05 | 7 | 23.3 |
| 53 | 50.05 | 55.05 | 10 | 33.3 |
| 58 | 55.05 | 60.05 | 11 | 36.7 |
| 63 | 60.05 | 65.05 | 14 | 46.7 |
| 68 | 65.05 | 70.05 | 23 | 76.7 |
| 73 | 70.05 | 75.05 | 23 | 76.7 |
| 78 | 75.05 | 80.05 | 25 | 83.3 |
| 83 | 80.05 | 85.05 | 28 | 93.3 |
| 88 | 85.05 | 90.05 | 29 | 96.7 |
| 93 | 90.05 | 95.05 | 30 | 100.0 |

**13.**

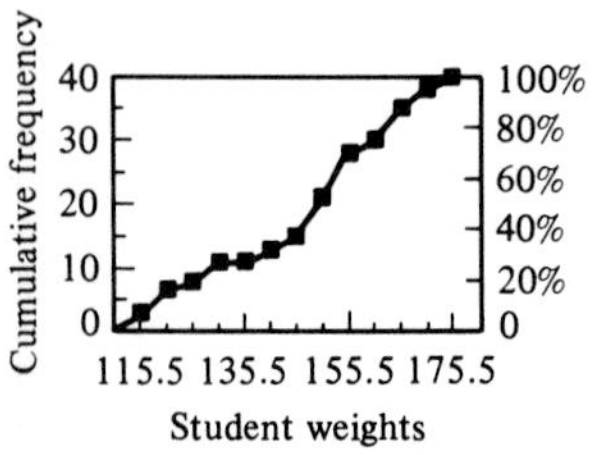


**15.**

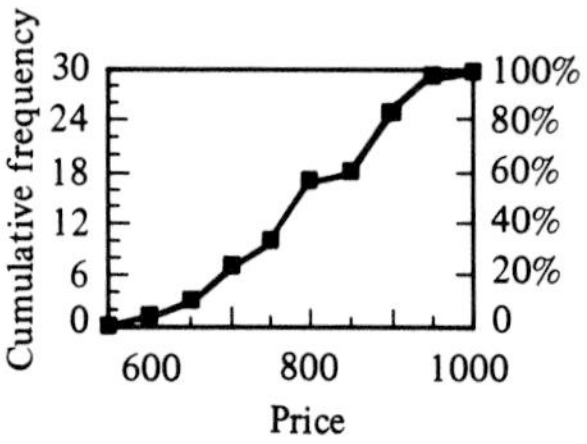


### Exercise 3, Page 653

**1.** $\bar{x} = 77$ **3.** $\bar{x} = 145$ lb **5.** $\bar{x} = \$796$ **7.** 158 **9.** 71.3 min **11.** 76, 96
**13.** 114, 117, 127, 147, 149, 153, 154, 155, 158, 162, 166, and 172 lb **15.** \$864 **17.** 83
**19.** $Q_1 = 4.28$, $Q_2 = 7.40$, $Q_3 = 10.02$; Quartile range = 5.74 **21.** $s^2 = 610$, $s = 24.7$
**23.** $s^2 = 201$, $s = 14.2$

### Exercise 4, Page 663

**1.** 8/15 **3.** 0.375 **5.** 1/9 **7.** 1/36 **9.** 0.63 **11.** 5/18 **13.** 0.4332 **15.** 0.2119
**17.** 620 students **19.** 36 students **21.** 1

### Exercise 5, Page 670

**1.** 69.47 ± 0.34 in. **3.** 2.35 ± 0.24 in. **5.** 164 ± 2.88 **7.** 16.31 ± 2.04

### Review Problems, Page 670

**1.** Continuous **3.** Categorical **5.**

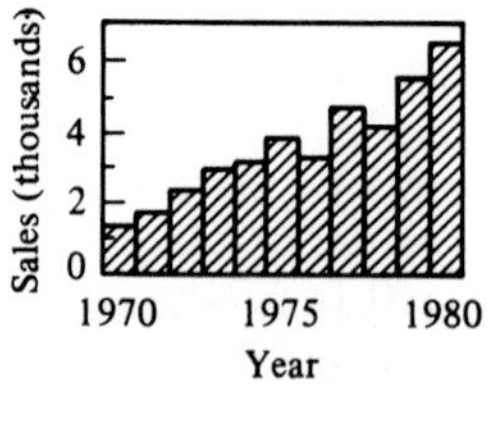


**7.** 5/36 **9.** 0.75 **11.** 0.1915

**13.** 0.3446 **15.** 81 **17.**

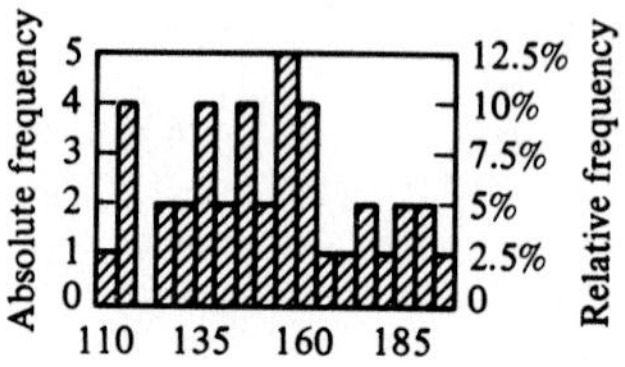

19.

| Data | Cumulative Absolute Frequency | Cumulative Relative Frequency |
|---|---|---|
| Under 112.5 | 1 | 1/40 (2.5%) |
| Under 117.5 | 5 | 5/40 (12.5%) |
| Under 122.5 | 5 | 5/40 (12.5%) |
| Under 127.5 | 7 | 7/40 (17.5%) |
| Under 132.5 | 9 | 9/40 (22.5%) |
| Under 137.5 | 13 | 13/40 (32.5%) |
| Under 142.5 | 15 | 15/40 (37.5%) |
| Under 147.5 | 19 | 19/40 (47.5%) |
| Under 152.5 | 21 | 21/40 (52.5%) |
| Under 157.5 | 26 | 26/40 (65%) |
| Under 162.5 | 30 | 30/40 (75%) |
| Under 167.5 | 31 | 31/40 (77.5%) |
| Under 172.5 | 32 | 32/40 (80%) |
| Under 177.5 | 34 | 34/40 (85%) |
| Under 182.5 | 35 | 35/40 (87.5%) |
| Under 187.5 | 37 | 37/40 (92.5%) |
| Under 192.5 | 39 | 39/40 (97.5%) |
| Under 197.5 | 40 | 40/40 (100%) |

**21.** $\bar{x} = 150$ **23.** 137 and 153 **25.** $s = 22.1$ **27.** $150 \pm 7.0$ **29.** $22.1 \pm 4.94$ **31.** 13 **33.** $Q_1 = 407$, $Q_2 = 718$, $Q_3 = 1055$; Quartile range = 648

## CHAPTER 25

### Exercise 1, Page 676

**1.** 3 **3.** 9 **5.** 6.6 **7.** 4 **9.** 1 **11.** 3.34 **13.** 7.21 **15.** 12.5 **17.** 131 **19.** 2 **21.** 3 **23.** 1 **25.** 26.4 ft **27.** 5.10, 8.06, 9.22, area = 20.6

### Exercise 2, Page 683

**1.** $-1/10$ **3.** $(3 - a)/(a + 2)$ **5.** 4.04 **7.** 0.787 **9.** $-3.77$ **11.** $-0.439$ **13.** 71.6° **15.** 104° **17.** 0° **19.** $-1/5$ **21.** $-0.21$ **23.** 0.351 **25.** $-0.620$ **27.** 1.062 **29.** 41.3° **31.** 45° **33.** 45° **35.** 64° **37.**

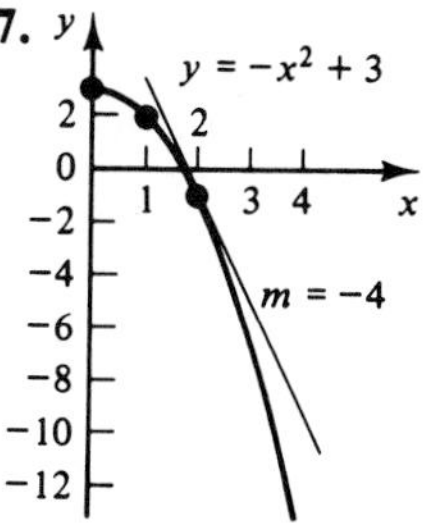


**39.** 2008 ft **41.** 1260 m **43.** 0.911° **45.** 33.7°

### Exercise 3, Page 688

**1.** $m = 3, b = -5$ **3.** $m = -1/2, b = -1/4$ **5.** $m = 0, b = 6$ **7.** $m = -1/2, b = 3/2$ **9.** $m = 2/3, b = 11/3$ **11.** $4x - y - 3 = 0$ **13.** $2.25x - y - 1.48 = 0$ **15.** $3x - 4y + 11 = 0$ **17.** $4x + y + 3 = 0$ **19.** $3x - 5y - 15 = 0$ **21.** $5x - y + 3 = 0$ **23.** $x + 3y + 15 = 0$ **25.** $5x - y + 15 = 0$ **27.** $x + 5y - 6 = 0$ **29.** $1.11x - y + 4 = 0$ **31.** $x - 2.38y - 6.38 = 0$ **33.** $y = 2$ **35.** $x - 2y = 0$ **37.** $x + 2 = 0$ **39.** $R = R_1[1 + \alpha(t - t_1)]$ **41.** 57.4°C **43.** 58.0 lb **47.** $P = 20.6 + 0.432x$, 21.8 ft **49.** $t = -0.789x + 25$, $x = 31.7$ cm, $m = -0.789$ **51.** $y = P + t(S - P)/L$, \$5555

### Review Problems, Page 692

**1.** 4 **3.** $-1.44$ **5.** 147.2° **7.** $-2b/a$ **9.** $m = 3/2, b = -7/2$ **11.** $7x - 3y + 32 = 0$ **13.** $5x - y + 27 = 0$ **15.** $7x - 3y + 21 = 0$ **17.** $y = 5$ **19.** $-3$ **21.** $x + 3 = 0$ **23.** 23.2

## CHAPTER 26

### Exercise 1, Page 699

**1.** $x^2 + y^2 = 49$ **3.** $(x - 2)^2 + (y - 3)^2 = 25$ **5.** $(x - 5)^2 + (y + 3)^2 = 16$ **7.** $C(0, 0)$, $r = 7$
**9.** $C(2, -4)$, $r = 4$ **11.** $C(3, -5)$, $r = 6$ **13.** $C(4, 0)$, $r = 4$ **15.** $C(5, -6)$, $r = 6$
**17.** $x^2 + y^2 - 9x - 5y + 14 = 0$ **19.** $x^2 + y^2 - 4x - 4y = 17$ **21.** $4x + 3y = 25$
**23.** $4x - 3y + 10 = 0$ **25.** (3, 0), (2, 0), (0, 1), (0, 6) **27.** (2, 3) and (−3/2, −1/2) **31.** 8.0 ft
**33.** 7.08 ft

### Exercise 2, Page 706

**1.** $F(2, 0)$, $L = 8$ **3.** $F(0, -3/7)$, $L = 12/7$ **5.** $x^2 + 8y = 0$ **7.** $3y^2 = 4x$
**9.** $V(3, 5)$, $F(6, 5)$, $L = 12$, $y = 5$ **11.** $V(2, -1)$, $F(13/8, -1)$, $L = 3/2$, $y = -1$
**13.** $V(3/2, 5/4)$, $F(3/2, 1)$, $L = 1$, $x = 3/2$ **15.** $y^2 - 8x - 4y + 12 = 0$ **17.** $y^2 + 4x - 2y + 9 = 0$
**19.** $2y^2 + 4y - x - 4 = 0$ **21.** 74.1 ft **23.** 2760 m **25.** $x^2 = 1250y$ **27.** 2.08 ft
**29.** $x^2 = 488y$ **31.** $(x - 16)^2 = -3072(y - 1/12)$

### Exercise 3, Page 718

**1.** $V(\pm 5, 0)$, $F(\pm 3, 0)$ **3.** $V(\pm 2, 0)$, $F(\pm 1, 0)$ **5.** $V(0, \pm 4)$, $F(0, \pm 2)$ **7.** $\dfrac{x^2}{25} + \dfrac{y^2}{9} = 1$
**9.** $\dfrac{x^2}{36} + \dfrac{y^2}{4} = 1$ **11.** $25x^2 + 169y^2 = 4{,}225$ **13.** $3x^2 + 7y^2 = 115$
**15.** $C(2, -2)$, $V(6, -2)$, $(-2, -2)$, $F(4.65, -2)$, $(-0.65, -2)$ **17.** $C(-2, 3)$, $V(1, 3)$, $(-5, 3)$, $F(0, 3)$, $(-4, 3)$
**19.** $C(1, -1)$, $V(5, -1)$, $(-3, -1)$, $F(4, -1)$, $(-2, -1)$ **21.** $C(-3, 2)$, $V(-3, 7)$, $(-3, -3)$, $F(-3, 6)$, $(-3, -2)$
**23.** $\dfrac{x^2}{9} + \dfrac{(y - 3)^2}{36} = 1$ **25.** $\dfrac{(x + 2)^2}{12} + \dfrac{(y + 3)^2}{16} = 1$ **27.** (1, 2), (1, −2) **29.** 60.8 cm **31.** 20 cm
**33.** 6.93 ft

### Exercise 4, Page 729

**1.** $V(\pm 4, 0)$, $F(\pm 6.40, 0)$, slope $= \pm 5/4$; $a = 4$, $b = 5$ **3.** $V(\pm 3, 0)$, $F(\pm 5, 0)$, slope $= \pm 4/3$; $a = 3$, $b = 4$
**5.** $V(\pm 4, 0)$, $F(\pm 4.47, 0)$, slope $= \pm 1/2$; $a = 4$, $b = 2$ **7.** $\dfrac{x^2}{25} - \dfrac{y^2}{144} = 1$ **9.** $\dfrac{x^2}{9} - \dfrac{y^2}{7} = 1$
**11.** $\dfrac{y^2}{16} - \dfrac{x^2}{16} = 1$ **13.** $\dfrac{y^2}{25} - \dfrac{3x^2}{64} = 1$
**15.** $C(2, -1)$, $a = 5$, $b = 4$, $F(8.40, -1)$, $F'(-4.40, -1)$, $V(7, -1)$, $V'(-3, -1)$, slope $= \pm 4/5$
**17.** $C(2, -3)$, $a = 3$, $b = 4$, $F(7, -3)$, $F'(-3, -3)$, $V(5, -3)$, $V'(-1, -3)$, slope $= \pm 4/3$
**19.** $C(-1, -1)$, $a = 2$, $b = \sqrt{5}$, $F(-1, 2)$, $F'(-1, -4)$, $V(-1, 1)$, $V'(-1, -3)$, slope $= \pm 2/\sqrt{5}$
**21.** $\dfrac{(y - 2)^2}{16} - \dfrac{(x - 3)^2}{4} = 1$ **23.** $\dfrac{(x - 1)^2}{4} - \dfrac{(y + 2)^2}{12} = 1$ **25.** $xy = 36$ **27.** $\dfrac{x^2}{324} - \dfrac{y^2}{352} = 1$
**29.** $pv = 25{,}000$

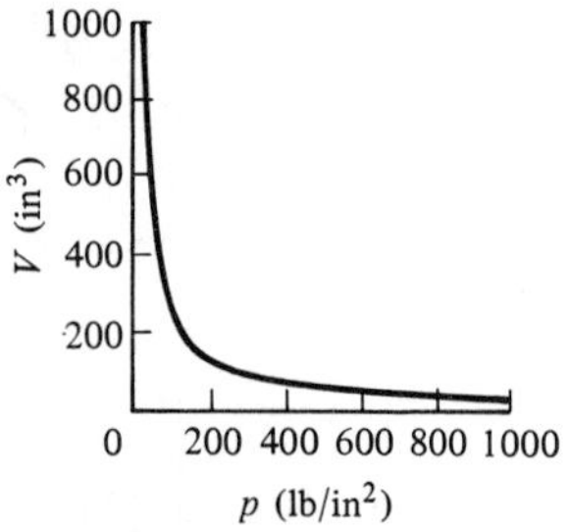


### Review Problems, Page 732

**1.** Hyperbola, $C(1, 2)$, $a = 2$, $b = 1$, $F(3.24, 2)$, $(-1.24, 2)$, $V(3, 2)$, $(-1, 2)$ **3.** Circle, $C(0, 4)$, $r = 4$
**5.** Hyperbola, $C(0, 0)$, $F(\pm 5, 0)$, $a = 3$, $b = 4$, $V(\pm 3, 0)$ **7.** $x^2 + 2y^2 = 100$ **9.** $y^2 = -17x$
**11.** $(x + 5)^2 + y^2 = 25$ **13.** $\dfrac{(x + 2)^2}{25} + \dfrac{(y - 1)^2}{9} = 1$ **15.** $x$ int. = 4, $y$ int. = 8 and −2 **17.** 3.0 m
**19.** 4.64 m

## CHAPTER 27

### Exercise 1, Page 740

**1.** 1 **3.** 1/5 **5.** 10 **7.** 4 **9.** −8 **11.** −4/5 **13.** −1/4 **15.** ∞ **17.** ∞ **19.** +∞ **21.** 2 **23.** 1/5 **25.** 0 **27.** $x^2$ **29.** 0 **31.** 3 **33.** $3x - \dfrac{1}{x^2 - 4}$ **35.** $3x^2$ **37.** $2x - 2$

### Exercise 2, Page 746

**1.** 2 **3.** $2x$ **5.** $3x^2$ **7.** $-3/x^2$ **9.** $-\dfrac{1}{2\sqrt{3 - x}}$ **11.** −2 **13.** 3/4 **15.** 2 **17.** 12 **19.** $4x$ **21.** −4 **23.** −16 **25.** 3 **27.** −5 **29.** 3

### Exercise 3, Page 752

**1.** 0 **3.** $7x^6$ **5.** $6x$ **7.** $-5x^{-6}$ **9.** $-1/x^2$ **11.** $-9x^{-4}$ **13.** $2.5x^{-2/3}$ **15.** $2x^{-1/2}$ **17.** $-51\sqrt{x}/2$ **19.** −2 **21.** $3 - 3x^2$ **23.** $9x^2 + 14x - 2$ **25.** $a$ **27.** $x - x^6$ **29.** $(3/2)x^{-1/4} - x^{-5/4}$ **31.** $(8/3)x^{1/3} - 2x^{-1/3}$ **33.** $-4x^{-2}$ **35.** $6x^2$ **37.** 54 **39.** −8 **41.** $15x^4 + 2$ **43.** $10.4/x^3$ **45.** $6x + 2$ **47.** 4 **49.** 3 **51.** 13.5 **53.** $10t - 3$ **55.** $175t^2 - 63.8$ **57.** $3\sqrt{5w}/2$ **59.** $-341/t^5$

### Exercise 4, Page 756

**1.** $10(2x + 1)^4$ **3.** $24x(3x^2 + 2)^3 - 2$ **5.** $\dfrac{-3}{(2 - 5x)^{2/5}}$ **7.** $\dfrac{-4.30x}{(x^2 + a^2)^2}$ **9.** $-\dfrac{6x}{(x^2 + 2)^2}$ **11.** $\dfrac{2b}{x^2}\left(a - \dfrac{b}{x}\right)$ **13.** $\dfrac{-3x}{\sqrt{1 - 3x^2}}$ **15.** $-\dfrac{1}{\sqrt{1 - 2x}}$ **17.** $\dfrac{-3}{(4 - 9x)^{2/3}}$ **19.** $\dfrac{-1}{2\sqrt{(x + 1)^3}}$ **21.** $2(3x^5 + 2x)(15x^4 + 2)$ **23.** $\dfrac{28.8(4.8 - 7.2x^{-2})}{x^3}$ **25.** $2(5t^2 - 3t + 4)(10t - 3)$ **27.** $\dfrac{7.6 - 49.8t^2}{(8.3t^3 - 3.8t)^3}$ **29.** 118,000 **31.** 540

### Exercise 5, Page 761

**1.** $3x^2 - 3$ **3.** $42x + 44$ **5.** $5x^4 - 12x^2 + 4$ **7.** $(x^2 - 5x)^2(8x - 7)(64x^2 - 242x + 105)$ **9.** $\dfrac{x}{\sqrt{1 + 2x}} + \sqrt{1 + 2x}$ **11.** $\dfrac{-2x^2}{\sqrt{3 - 4x}} + 2x\sqrt{3 - 4x}$ **13.** $\dfrac{bx}{2\sqrt{a + bx}} + \sqrt{a + bx}$ **15.** $\dfrac{2(3x + 1)^3}{\sqrt{4x - 2}} + 9(3x + 1)^2\sqrt{4x - 2}$ **17.** $(t + 4)(20t^2 + 31t - 12)$ **19.** $2(81.3t^3 - 73.8t)(t - 47.2) + (t - 47.2)^2(244t^2 - 73.8)$ **21.** $(2x^5 + 5x)^2 + (2x - 6)(2x^5 + 5x)(10x^4 + 5)$ **23.** $3x(x + 1)^2(x - 2)^2 + (2x^2 + 2x)(x - 2)^3 + (x + 1)^2(x - 2)^3$ **25.** $(x - 1)^{1/2}(x + 1)^{3/2} + 3/2(x^2 - 1)^{1/2}(x + 2) + \dfrac{(x + 1)^{3/2}(x + 2)}{2\sqrt{x - 1}}$ **27.** $\dfrac{2}{(x + 2)^2}$ **29.** $\dfrac{8x}{(4 - x^2)^2}$ **31.** $\dfrac{-5}{(x - 3)^2}$ **33.** $\dfrac{2 - 4x}{(x - 1)^3}$ **35.** $\dfrac{1}{2\sqrt{x}(\sqrt{x} + 1)^2}$ **37.** $\dfrac{-a^2}{(z^2 - a^2)^{3/2}}$ **39.** −0.456 **41.** 0

### Exercise 6, Page 765

**1.** $6u^2\dfrac{du}{dw}$ **3.** $2y\dfrac{dy}{du} + 3u^2$ **5.** $2x^3y\dfrac{dy}{dx} + 3x^2y^2$ **7.** $\dfrac{3z}{\sqrt{3z^2 + 5}}\dfrac{dz}{dt}$ **9.** 5/2 **11.** $-y/x$ **13.** $2a/y$ **15.** $\dfrac{ay - x^2}{y^2 - ax}$ **17.** $\dfrac{1 + 3x^2}{1 + 3y^2}$ **19.** $\dfrac{8xy}{3y - 8x^2 + 4y^2}$ **21.** −0.436 **23.** 14/15

### Exercise 7, Page 767

**1.** $36x^2 - 6x$ **3.** $\dfrac{8}{(x + 2)^3}$ **5.** $\dfrac{-20}{(5 - 4x^2)^{3/2}}$ **7.** $4(x - 3)^2 + 8(x - 6)(x - 3)$ **9.** 1.888

## Review Problems, Page 767

**1.** $\dfrac{2x}{(x^2+1)\sqrt{x^4-1}}$ **3.** $-\sqrt[3]{y/x}$ **5.** 8/5 **7.** $\dfrac{-b^2x}{a^2y}$ **9.** $-0.03516$ **11.** $5-6x$

**13.** 3/64 **15.** 178 **17.** 3 **19.** $-4(x^2-1)(x+3)^{-5}+2x(x+3)^{-4}$ **21.** $-8\dfrac{dw}{dx}+2w\dfrac{dw}{dx}$

**23.** $-24x$ **25.** $174.9t^2-63.8$ **27.** $12x-2$ **29.** $48x^2+8x-1$ **31.** $34x^{16}+48x^{11}-7-\dfrac{1}{x^3}$

**33.** $\dfrac{x^2-2x-3}{(x-1)^2}$ **35.** $72x^2+24x-14$

## CHAPTER 28

### Exercise 1, Page 772

**1.** $2x-y+1=0,\ x+2y-7=0$ **3.** $12x-y-13=0,\ x+12y-134=0$
**5.** $3x+4y=25,\ 4x-3y=0$ **7.** (3, 0) **9.** $2x+y=1$ and $2x+y+7=0$
**11.** $2x+5y=1.55$ **13.** $-10$ **15.** 63.4°, 19.4°, 15.3°

### Exercise 2, Page 778

**1.** Rising for all $x$ **3.** Increasing **5.** Decreasing **7.** Upward **9.** Downward
**11.** min(0, 0) **13.** max(3, 13) **15.** max(0, 36), min(4.67, −14.8) **17.** max(1, 1), (−1, 1); min(0, 0)
**19.** min(1, −3) **21.** max(0, 0); min(−1, −5), (2, −32) **23.** max(−3, 32), min(1, 0)
**25.** max(−5/7, 18.36), min(1, 0) **27.** max(0, 2), min(0, −2) **29.** max(1/2, −4.23), min(1/2, 4.23)

### Exercise 3, Page 780

**1.** (0, 0) **3.** (0, 0) **5.** (3, −126) **7.** (−1, −5) and (1, −5) **9.** (0, −17) and (−1, −28)

### Exercise 4, Page 782

**1.** 3.42 **3.** 3.06 **5.** 1.67 **7.** −1.72 **9.** 1.89 in. **11.** 3.38 m

### Exercise 5, Page 788

**1.**

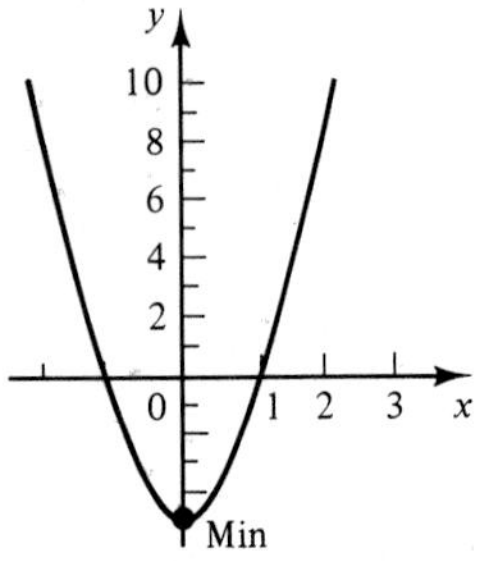


**3.**

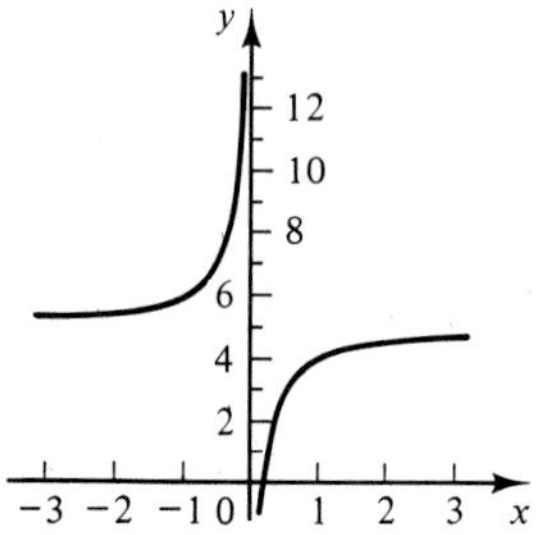


**5.**

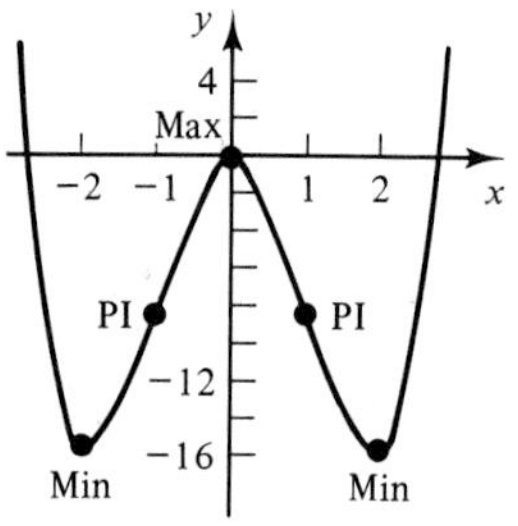


**7.**

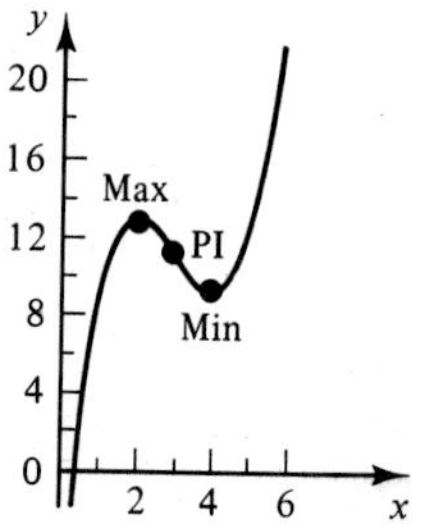


**9.**

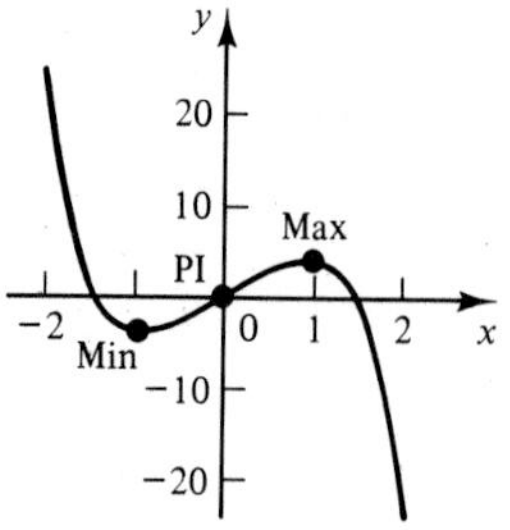


**11.**

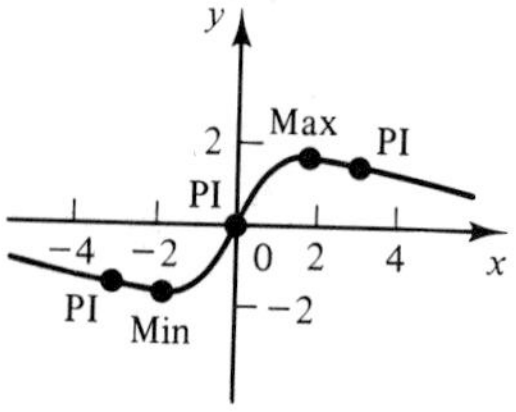

13.

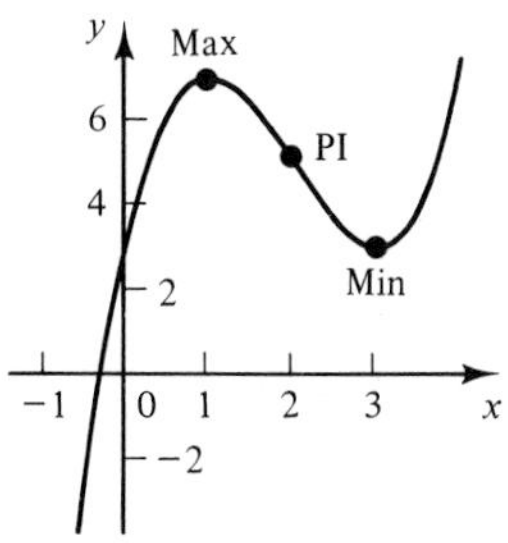


15.

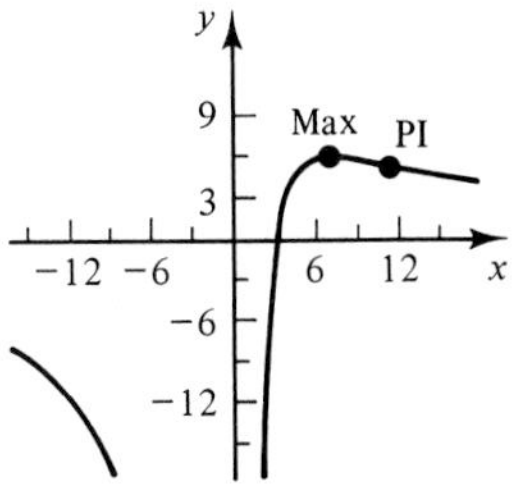


17.

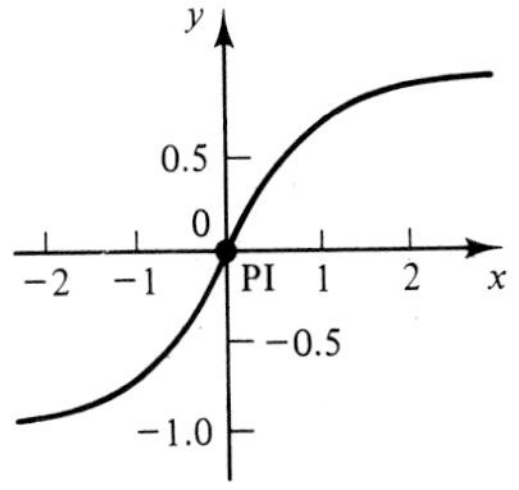


19.

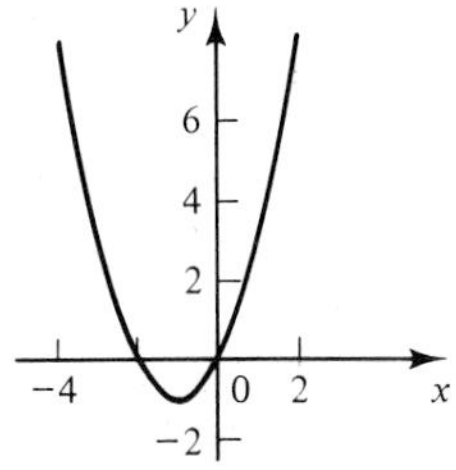


21.

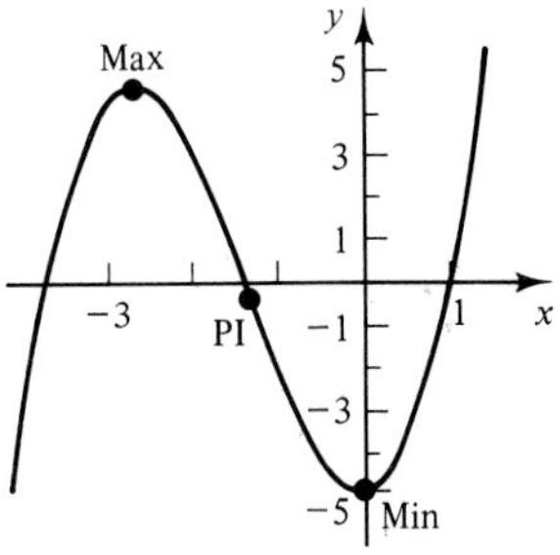


23.

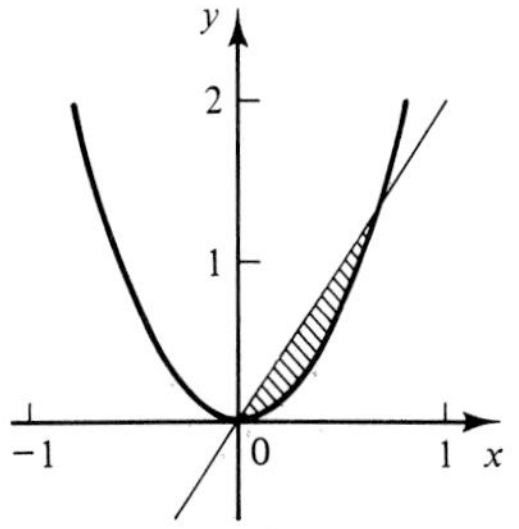


25.

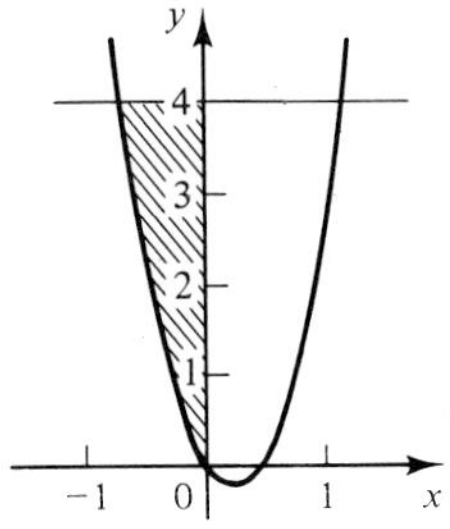


### Review Problems, Page 789

**1.** −2.22, 0.54, 1.68 **3.** max(1, 2) **5.** $7x - y - 9 = 0$, $x + 7y - 37 = 0$ **7.** −7/3
**9.** Rising for $x > 0$, never falls **11.** Downward **13.** $34x - y = 44$, $x + 34y = 818$ **15.** 71.6°

## CHAPTER 29

### Exercise 1, Page 796

**1.** −0.175 lb/in² per cubic inch **3.** 12.6 m³/m **5.** 0.054 s/in. **7.** $\dfrac{dy}{dx} = \dfrac{wx}{6EI}(x^2 + 3L^2 - 3Lx)$
**9.** $i = dq/dt = 6.96t - 1.64$ A **11.** 50.9 V **13.** 804 A **15.** 2.48 mA **17.** 95.6 kV

### Exercise 2, Page 804

**1.** $v = 0$, $a = -16$ **3.** $v = -3$, $a = 18$ **5.** $v = -8$, $a = -32$ **7.** $v = 104$ ft/s, $a = 32$ ft/s²
**9.** 60 ft/s **11.** 2000 ft/s, 62,500 ft, 1680 ft/s **13.** $s = 0$, $a = 32$ units/s²
**15.** $v_x = 6t + 5$, $v_y = -6t$, $a_x = 6$, $a_y = -6$ **17.** 8.00 ft/s at −1.79°
**19.** $v_x = 16$ units/s, $v_y = -1/2$ units/s, $a_x = 8$ units/s², $a_y = 3/4$ units/s², $a = 8.04$ units/s² **21.** 12 units/s
**23.** (2, 4) **25.** 6000 rad/s²

### Exercise 3, Page 809

**1.** 60 m/s **3.** 3.58 mi/h **5.** 2.40 m/s **7.** 4.47 ft/s **9.** 0.83 ft/s **11.** 170 km/h
**13.** 8.33 ft/s **15.** 4 ft/s **17.** 0.101 in²/s **19.** 0.133 in./min **21.** −4 in./min **23.** 1.19 m/min
**25.** 0.2 ft/min **27.** 7.0 lb/in²/s **29.** 2.99 in./s

### Exercise 4, Page 816

**1.** 1 **3.** $6\frac{2}{3}$ and $13\frac{1}{3}$ **5.** 18 m × 24 m **7.** 9 × 18 yd **9.** $h = 3.26$ cm, $d = 6.51$ cm **11.** 25 in$^2$ **13.** 6 **15.** 3.46 in. **17.** (2, 2) **19.** 5 × 4.33 in. **21.** 9.9 × 14.1 units **23.** $r = 4.08$ cm, $h = 5.77$ cm **25.** 12 ft **27.** $r = 5.24$ m, $h = 5.24$ m **29.** 10.4 × 14.7 in. **31.** 59.0 in. **33.** 7.5 A **35.** 5.0 A **37.** 0.65

### Exercise 5, Page 824

**1.** $3x^2\,dx$ **3.** $\dfrac{2\,dx}{(x+1)^2}$ **5.** $2(x+1)(2x+3)^2(5x+6)\,dx$ **7.** $\dfrac{2x-13}{2\sqrt{x-4}(3-2x)^2}dx$ **9.** $\dfrac{-x^2}{2y^2}dx$ **11.** $\dfrac{-2\sqrt{y}}{3\sqrt{x}}dx$ **13.** 0.016 **15.** 0.60 **17.** 0.00153 s **19.** $V = 2\pi rht$ **21.** $V = 3\pi r^2t$ **23.** 0.2 **25.** 0.75 ft$^2$ **27.** 1/2%

### Review Problems, Page 826

**1.** 188 ft/s **3.** 4.0 km/h **5.** 3.33 cm$^2$/min **7.** $s/2$ **9.** 6.93 **11.** −2.26 lb/in$^2$ per second **13.** $t = 4$, $s = 108$; $t = -2$, $s = 0$ **15.** 4 **17.** 4 and 6 **19.** 10 in. × 10 in. × 5 in. **23.** 36.4 A **25(a)** 142 rad/s, **(b)** 37.0 rad/s$^2$

## CHAPTER 30

### Exercise 1, Page 832

**1.** $\cos x$ **3.** $-3\sin x\cos^2 x$ **5.** $3\cos 3x$ **7.** $\cos^2 x - \sin^2 x$ **9.** $3.75(\cos x - x\sin x)$ **11.** $-2\sin(\pi - x)\cos(\pi - x)$ **13.** $2\cos 2x\cos x - \sin 2x\sin x$ **15.** $2\sin x\cos^2 x - \sin^3 x$ **17.** $1.23(2\cos 3x\sin x\cos x - 3\sin^2 x\sin 3x)$ **19.** $-\sin 2t/\sqrt{\cos 2t}$ **21.** $-\sin x$ **23.** $-2\sin x - x\cos x$ **25.** $-\pi^2/4$ **27.** $\dfrac{y\cos x + \cos y - y}{x\sin y - \sin x + x}$ **29.** $\sec(x+y) - 1$ **31.** −0.4161 **33.** 1.381 **35.** max($\pi/2$, 1), min($3\pi/2$, −1), PI($\pi$, 0) **37.** max(2.50, 5), min(5.64, −5), PI(0.927, 0), (4.07, 0) **39.** 0.515 **41.** 1.404

### Exercise 2, Page 835

**1.** $2\sec^2 2x$ **3.** $-15\csc 3x\cot 3x$ **5.** $6.50x\sec^2 x^2$ **7.** $-21x^2\csc x^3\cot x^3$ **9.** $x\sec^2 x + 1$ **11.** $5\csc 6x - 30x\csc 6x\cot 6x$ **13.** $2\sin\theta\sec^2 2\theta + \tan 2\theta\cos\theta$ **15.** $5\csc 3t\sec^2 t - 15\tan t\csc 3t\cot 3t$ **17.** −294 **19.** 853 **21.** $6\sec^2 x\tan x$ **23.** 18.72 **25.** $-y\sec x\csc x$ **27.** −1 **29.** $3.43x - y = 1.87$ **31.** max($\pi/4$, 0.571), min($3\pi/4$, 5.71), PI(0, 0), ($\pi$, $2\pi$) **33.** $v = 2.35$ cm/s, $a = -19.1$ cm/s$^2$ **35.** −3.58 deg/min **37.** 15.6 ft **39.** 31° **41.** 25 ft

### Exercise 3, Page 838

**1.** $\text{Sin}^{-1}x + \dfrac{x}{\sqrt{1-x^2}}$ **3.** $\dfrac{-1}{\sqrt{a^2-x^2}}$ **5.** $\dfrac{\cos x + \sin x}{\sqrt{1+\sin 2x}}$ **7.** $-\dfrac{t^2}{\sqrt{1-t^2}} + 2t\,\text{Cos}^{-1}t$ **9.** $\dfrac{1}{1+2x+2x^2}$ **11.** $\dfrac{-a}{a^2+x^2}$ **13.** $\dfrac{-1}{x\sqrt{4x^2-1}}$ **15.** $2t\,\text{Arcsin}\dfrac{t}{2} + \dfrac{t^2}{\sqrt{4-t^2}}$ **17.** $\dfrac{1}{\sqrt{a^2-x^2}}$ **19.** −0.285 **21.** −0.054 **23.** 1.3°

### Exercise 4, Page 843

**1.** $\dfrac{1}{x}\log e$ **3.** $\dfrac{3}{x\ln b}$ **5.** $\dfrac{(5+9x)\log e}{5x+6x^2}$ **7.** $\left(\dfrac{x^2}{2}\right)\log e + \log\left(\dfrac{2}{x}\right)$ **9.** $\dfrac{1}{x}$ **11.** $\dfrac{2x-3}{x^2-3x}$ **13.** $8.25 + 2.75\ln 1.02x^3$ **15.** $\dfrac{1}{x^2(x+5)} - \dfrac{2\ln(x+5)}{x^3}$ **17.** $\dfrac{1}{2t-10}$ **19.** $\cot x$

**21.** $\cos x(1 + \ln \sin x)$ **23.** $\dfrac{\sin x}{1 + \ln y}$ **25.** $\dfrac{x + y - 1}{x + y + 1}$ **27.** $-y$ **29.** $-\dfrac{a^2}{x^2\sqrt{a^2 - x^2}}$

**31.** $x^{\sin x}[(\sin x)/x + \cos x \ln x]$ **33.** $(\text{Arccos } x)^x\left(\ln \text{Arccos } x - \dfrac{x}{\sqrt{1 - x^2}\,\text{Arccos } x}\right)$ **35.** 1

**37.** 0.3474 **39.** 128° **41.** min$(1/e, -1/e)$ **43.** min$(e, e)$, PI$(e^2, \frac{1}{2}e^2)$ **45.** 1.37 **47.** 0.607
**49.** −0.022 **51.** −0.0054 dB/day

### Exercise 5, Page 847

**1.** $2(3^{2x}) \ln 3$ **3.** $10^{2x+3}(1 + 2x \ln 10)$ **5.** $2x(2^{x^2}) \ln 2$ **7.** $2e^{2x}$ **9.** $e^{x+e^x}$ **11.** $-\dfrac{xe^{\sqrt{1-x^2}}}{\sqrt{1 - x^2}}$

**13.** $-\dfrac{2}{e^x}$ **15.** $xe^{3x}(3x + 2)$ **17.** $\dfrac{e^x(x - 1)}{x^2}$ **19.** $\dfrac{(x - 2)e^x + (x + 2)e^{-x}}{x^3}$ **21.** $2(x + xe^x + e^x + e^{2x})$

**23.** $\dfrac{(1 + e^x)(2xe^x - e^x - 1)}{x^2}$ **25.** $3e^x \sin^2 e^x \cos e^x$ **27.** $e^\theta(\cos 2\theta - 2 \sin 2\theta)$ **29.** $-e^{(x-y)}$

**31.** $\dfrac{\cos(x + y)}{e^y - \cos(x + y)}$ **33.** 0 **35.** $e^x\left(\ln x + \dfrac{1}{x}\right)$ **37.** $e^x\left(\dfrac{1}{x} + \ln x\right)$ **39.** $-2e^t \sin t$

**41.** $2e^x \cos x$ **43.** 0.79 **45.** 4.97 **47.** min(0.402, 4.474) **49.** max($\pi/4$, 3.224), min($5\pi/4$, −0.139)
**51.** \$915.30 per year **53.** 74.2 **55.** −563 rev/min$^2$ **57(a)** 1000 bacteria/h, **(b)** $22 \times 10^6$ bacteria/h
**59.** −245 in. Hg/h **61.** 0.044 lb/ft$^3$

### Review Problems, Page 850

**1.** $\dfrac{1}{2}(e^{x/a} + e^{-x/a})$ **3.** $\dfrac{4}{\sqrt{x}} \sec^2\sqrt{x}$ **5.** $\dfrac{4x}{16x^2 + 1} + \text{Arctan } 4x$ **7.** $\dfrac{\cos 2x}{y}$ **9.** $x \cos x + \sin x$

**11.** $3x^2 \cos x - x^3 \sin x$ **13.** $\dfrac{x \cos x - \sin x}{x^2}$ **15.** $\dfrac{1 + 3x^2}{x^3 + x}\log e$ **17.** $\dfrac{1}{\sqrt{x^2 + a^2}}$

**19.** $-3 \csc 3x \cot 3x$ **21.** $\sec^2 x$ **23.** $\dfrac{6x^2 + 1}{2x^3 + x}$ **25.** $2x \text{ Arccos } x - \dfrac{x^2}{\sqrt{1 - x^2}}$ **27.** (8.96, 0.291)

**29.** $x = \pi/4$ **31.** 0.053 **33.** $y - \dfrac{1}{2} = \dfrac{\sqrt{3}}{2}\left(x - \dfrac{\pi}{6}\right)$ **35.** $x = \pi/4$ **37.** 0.517

**39.** 1049°F at 5.49 h

## CHAPTER 31

### Exercise 1, Page 859

**1.** $x + C$ **3.** $\dfrac{1}{2}x^2 + C$ **5.** $-\dfrac{1}{x} + C$ **7.** $\dfrac{3x^2}{2} + C$ **9.** $\dfrac{3}{4}x^4 + C$ **11.** $\dfrac{x^{n+1}}{n + 1} + C$

**13.** $2\sqrt{x} + C$ **15.** $x^4 + C$ **17.** $x^{7/2} + C$ **19.** $x^2 + 2x + C$ **21.** $x^{4/3} + C$ **23.** $u^2 + C$

**25.** $\dfrac{8}{3}s^{3/2} + C$ **27.** $\dfrac{6x^{5/2}}{5} - \dfrac{4x^{3/2}}{3} + C$ **29.** $2x^2 - 4\sqrt{x} + C$ **31.** $s - \dfrac{3}{2}s^2 + s^3 - \dfrac{s^4}{4} + C$

**33.** $y = \dfrac{4x^3}{3} + C$ **35.** $y = -\dfrac{1}{2x^2} + C$ **37.** $s = \dfrac{3t^{1/3}}{2} + C$

### Exercise 2, Page 862

**1.** $\dfrac{1}{4}(x^4 + 1)^4 + C$ **3.** $(x^2 + 2x)^2 + C$ **5.** $\dfrac{1}{1 - x} + C$ **7.** $(x^3 + 1)^3 + C$ **9.** $-\dfrac{2}{3}(1 - y^3)^{1/2} + C$

**11.** $\dfrac{2}{9 - x^2} + C$ **13.** $\dfrac{1}{3}\tan^3 x + C$ **15.** $y = -\dfrac{1}{8}(1 - x^2)^4 + C$

**17.** $y = \dfrac{1}{8}(2x^3 + x^2)^4 + C$ **19.** $v = \left(-\dfrac{5}{3}\right)(7 - t^2)^{3/2} + C$

### Exercise 3, Page 865

**1.** $y = 3x + C$

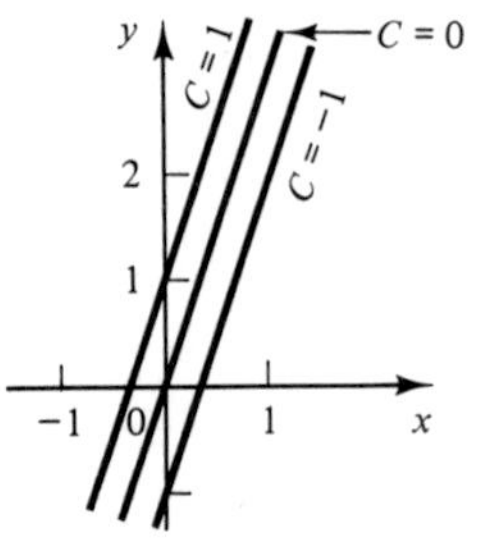


**3.** $y = x^3 + C$

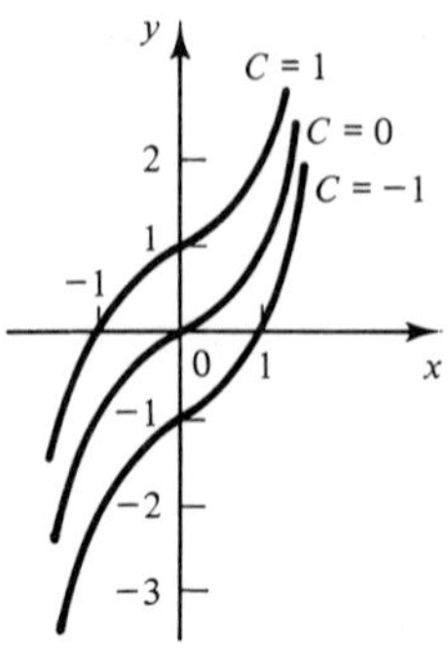


**5.** $x^3 - 3y + 2 = 0$

**7.** 17 **9.** $6y = x^3 - 6x - 9$ **11.** $6x + xy = 6$

### Exercise 4, Page 871

**1.** $3 \ln|x| + C$ **3.** $\frac{5}{3}\ln|z^3 - 3| + C$ **5.** $x + \ln|x| + C$ **7.** $\frac{a^{5x}}{5 \ln a} + C$ **9.** $\frac{5^{7x}}{7 \ln 5} + C$

**11.** $\frac{a^{3y}}{3 \ln a} + C$ **13.** $4e^x + C$ **15.** $\frac{e^{x^2}}{2} + C$ **17.** $\frac{1}{6}e^{3x^2} + C$ **19.** $2e^{\sqrt{x}} + C$

**21.** $\frac{1}{2}e^{2x} - 2e^x + x + C$ **23.** $2e^{\sqrt{x-2}} + C$ **25.** $ae^{x/a} - ae^{-x/a} + C$ **27.** $-\frac{1}{3}\cos 3x + C$

**29.** $-\frac{1}{5}\ln|\cos 5\theta| + C$ **31.** $\frac{1}{4}\ln|\sec 4x + \tan 4x| + C$ **33.** $-\frac{1}{3}\ln|\cos 9\theta| + C$ **35.** $-\frac{1}{2}\cos x^2 + C$

**37.** $-\frac{1}{3}\ln|\cos \theta^3| + C$ **39.** $-\cos(x + 1) + C$ **41.** $\frac{1}{5}\ln|\cos(4 - 5\theta)| + C$

**43.** $\frac{1}{8}\ln|\sec(4x^2 - 3) + \tan(4x^2 - 3)| + C$ **45.** $\frac{1}{12}\ln\left|\frac{3t + 2}{3t - 2}\right| + C$ **47.** $\frac{x}{2}\sqrt{25 - 9x^2} + \frac{25}{6}\text{Sin}^{-1}\left(\frac{3x}{5}\right) + C$

**49.** $\frac{1}{3}\text{Tan}^{-1}\frac{x}{3} + C$ **51.** $\frac{1}{12}\text{Tan}^{-1}\left(\frac{4x}{3}\right) + C$ **53.** $\frac{1}{12}\ln\left|\frac{2x + 3}{2x - 3}\right| + C$ **55.** $\frac{1}{4}\ln\left|\frac{x - 2}{x + 2}\right| + C$

**57.** $\frac{x}{4}\sqrt{x^2 - 4} - \ln|x + \sqrt{x^2 - 4}| + C$ **59.** $\frac{5}{2}\text{Arcsin } x^2 + C$

### Exercise 5, Page 877

**1.** $v = 4t - \frac{1}{3}t^3 - 1$ **3(a)** $x = \frac{1}{3}t^3 - 4t$, $y = 2t^2$; **(b)** 9.6 cm **5.** $s = 20t - 16t^2$

**7.** $v_x = 1667$ cm/s, $v_y = 100$ cm/s **9.** $v_x = 344$ cm/s, $v_y = 2250$ cm/s **11.** 18,000 rev

**13.** 15.7 rad/s

### Exercise 6, Page 880

**1.** 12.1 coulombs **3.** 78.4 coulombs **5.** 2.26 V **7.** 4.25 A **9.** 20.2 A

### Review Problems, Page 880

**1.** $-\frac{1}{3}\ln|\cos 3\theta| + C$ **3.** $\frac{a^{2x}}{2 \ln a} + C$ **5.** $\frac{1}{5}\left(e^{5x} + \frac{a^{5x}}{\ln a}\right) + C$ **7.** $\frac{1}{2}(e^{2x+1} + x^2) + C$

**9.** $\frac{x^2}{2} + x - \frac{1}{2x^2} + C$ **11.** $-\frac{1}{e^x} + C$ **13.** $-\frac{1}{3}\cot 3x + C$ **15.** $-\frac{2}{t} + C$

**17.** $\frac{1}{2}\ln|\sec 2\theta + \tan 2\theta| + C$ **19.** $\frac{1}{9}(x^3 - 4)^3 + C$ **21.** $\frac{1}{2}\ln|x^2 + 3| + C$ **23.** $y = \frac{x^3}{6} + \frac{9x}{2} - 18$

**25.** $v = \frac{3t^2}{2}$, $s = \frac{t^3}{2} + s_0$ **27.** 19,400 rad/s, 15,400 rev **29.** $s_x = \frac{t^3}{6} + 4t + 1$, $s_y = \frac{5t^3}{6} + 15t + 1$

## CHAPTER 32

### Exercise 1, Page 885

**1.** 3/2 **3.** 60.7 **5.** 112/3 **7.** 218/3 **9.** 2.30 **11.** 0.112 **13.** 0.732 **15.** 0.859
**17.** 2 **19.** 2

### Exercise 2, Page 892

**1.** 15 **3.** 84 **5.** 40 **7.** 176 square units **9.** 1.575 square units **11.** 100 square units
**13.** $113\frac{1}{3}$ square units **15.** 64 square units **17.** 4.24 square units **19.** 9.83 square units
**21.** 15.75 square units

### Exercise 3, Page 900

**1.** 2.797 **3.** 16 **5.** $42\frac{2}{3}$ **7.** $1\frac{1}{6}$ **9.** $10\frac{2}{3}$ **11.** 72.4 **13.** 32.83 **15.** $2\frac{2}{3}$ **17.** 2
**19.** $8/\pi$ **21.** 3 **23.** 9 **25.** 25.6 **27.** $5\frac{1}{3}$ **37.** 90.1 $ft^2$ **39.** 24.9 $ft^3$ **41.** 500 $ft^2$
**43.** 35 $ft^2$

### Review Problems, Page 904

**1.** 0.693 **3.** 4 **5.** $81\frac{2}{3}$ **7.** $-1/2$ **9.** 170 **11.** 280 **13.** $12\pi$ **15.** 19.8
**17.** 25.6 **19.** 2.13

## CHAPTER 33

### Exercise 1, Page 913

**1.** 57.4 **3.** $\pi$ **5.** 0.666 **7.** 0.479 **9.** 4.42 **11.** 60.3 **13.** $32\pi$ **15.** 19.7
**17.** 91.9 **19.** 3.35 **21.** 102 **23.** 1.65 $m^3$ **25.** 0.683 $m^3$

### Exercise 2, Page 917

**1.** 0.704 **3.** 1.32 **5.** 3/2 **7.** 24.8 **9.** 19/5 **11.** 9.07 **13.** 1096 ft **15.** 223 ft
**17.** 30.3 ft

### Exercise 3, Page 921

**1.** 32.1 **3.** 154 **5.** 36.2 **7.** 58.6 **9.** 72.6 **11.** 1570 **13.** 36.2 **15.** $4\pi r^2$
**17.** 36.2 $ft^2$

### Exercise 4, Page 923

**1.** 12 **3.** 4.08 **5.** 1/2 **7.** 7.81 **9.** 19.1 **11.** 1.08

### Exercise 5, Page 930

**1.** $(5/4, -3/2)$ **3.** 3.22 **5.** (2.4, 0) **7.** $\bar{x} = 0.538$ **9.** $\bar{y} = 0.93$ **11.** $\bar{x} = 1.067$
**13.** $\bar{x} = 9/20, \bar{y} = 9/20$ **15.** $\bar{x} = 0.299$ **17.** $\bar{x} = 1.25$ **19.** 5 **21.** 5 **23.** $2p/3$ **25.** $h/4$
**27.** 4.13 ft **29.** 36.1 mm

### Exercise 6, Page 935

**1.** 88,200 lb **3.** 41,200 lb **5.** 62.4 lb **7.** 2160 lb

### Exercise 7, Page 938

**1.** 25.0 ft-lb **3.** 32 in.-lb **5.** 50,200 ft-lb **7.** $5.71 \times 10^5$ ft-lb **9.** 723,000 ft-lb
**11.** $k/100$ ft-lb **13.** 13,800 ft-lb

### Exercise 8, Page 945

**1.** 1/12 **3.** 2.62 **5.** 19.5 **7.** 2/9 **9.** 0.471 **11.** $10.1m$ **13.** $89.4m$ **15.** $3Mr^2/10$
**17.** $4Mp^2/3$

### Review Problems, Page 945

**1.** 31.4 **3.** 51.5 **5.** 151 **7.** 197 **9.** 310 **11.** 407 **13.** 1/2 **15.** 1123 lb
**17.** 21.8 in.-lb **19.** 2560 lb **21.** (4.24, 0) **23.** (1.70, 1.27) **25.** 48,300 ft-lb

## CHAPTER 34

### Exercise 1, Page 951

**1.** $\sin x - x\cos x + C$ **3.** $x\tan x + \ln|\cos x| + C$ **5.** $x\sin x + \cos x + C$
**7.** $\frac{x^2}{4} - \frac{x}{12}\sin 6x - \frac{1}{72}\cos 6x + C$ **9.** 1.24 **11.** $-\frac{1}{x+1}(\ln|x+1| + 1) + C$ **13.** 5217
**15.** $\frac{-x^2}{3}(1-x^2)^{3/2} - \frac{2}{15}(1-x^2)^{5/2} + C$ or $-\frac{1}{15}(1-x^2)^{3/2}(3x^2+2) + C$ **17.** $\frac{x}{\sqrt{1-x^2}} - \sin^{-1}x + C$
**19.** $\sin x(\ln|\sin x| - 1) + C$ **21.** $-x^2e^{-x} - 2xe^{-x} - 2e^{-x} + C$ **23.** $\frac{e^x}{2}(\sin x - \cos x) + C$ **25.** −0.081
**27.** $\frac{e^{-x}}{\pi^2+1}(\pi\sin\pi x - \cos\pi x)$

### Exercise 2, Page 958

**1.** $\ln\left|\frac{x(x-2)}{(x+1)^2}\right| + C$ **3.** −0.1054 **5.** $\ln\left|\frac{(x-1)^{1/2}}{x^{1/3}(x+3)^{1/6}}\right| + C$ **7.** −0.739 **9.** $\frac{9-4x}{2(x-2)^2} + C$
**11.** 0.0637 **13.** 0.1784 **15.** 1.946 **17.** $\frac{1}{4}\ln\left|\frac{x+1}{x-1}\right| - \frac{1}{2}\text{Arctan } x + C$ **19.** 0.667
**21.** $\frac{x^2}{2} - 4\ln|x^2+4| - \frac{8}{x^2+4} + C$

### Exercise 3, Page 961

**1.** $2\sqrt{x} - 2\ln|1+\sqrt{x}| + C$ **3.** $\left(\frac{3}{2}\right)\ln|x^{2/3} - 1| + C$
**5.** $\frac{3}{5}\sqrt[3]{(1+x)^5} - \frac{3}{2}\sqrt[3]{(1+x)^2} + C$ or $\frac{3}{10}(1+x)^{2/3}(2x-3) + C$ **7.** $-\frac{2}{147}\sqrt{2-7x}(7x+4) + C$
**9.** $\frac{6x^2+6x+1}{12(4x+1)^{3/2}} + C$ **11.** $2\sqrt{x+2} + \sqrt{2}\text{ Arctan}\sqrt{\frac{x+2}{2}} + C$ **13.** $\ln\left|\frac{1-\sqrt{1-x^2}}{x}\right| + C$
**15.** $\frac{3}{28}(1+x)^{4/3}(4x-3) + C$ **17.** 0.2375 **19.** 0.1042

### Exercise 4, Page 966

**1.** $\frac{x}{\sqrt{5-x^2}} + C$ **3.** $\frac{1}{2}\ln\left|\frac{x}{2+\sqrt{x^2+4}}\right| + C$ or $\frac{1}{2}\ln\left|\frac{\sqrt{x^2-4}-2}{x}\right| + C$ **5.** $\frac{-\sqrt{4-x^2}}{4x} + C$
**7.** $2\text{ Arcsin}\frac{x}{2} - \frac{x}{2}\sqrt{4-x^2} + C$ **9.** $-\frac{\sqrt{16-x^2}}{x} - \text{Arcsin}\frac{x}{4} + C$ **11.** 16.49 **13.** $\frac{\sqrt{x^2-7}}{7x} + C$

### Exercise 5, Page 969

**1.** 1/2 **3.** $e$ **5.** $\pi$ **7.** 2 **9.** Diverges **11.** $\pi$

### Exercise 6, Page 974

**1.** 1.79 **3.** 66.0 **5.** 2.0 **7.** 0.323 **9.** 5960 **11.** 260 **13.** 13,000 ft$^2$
**15.** 90,000 ft$^3$ **17.** 86.6 kW

### Review Problems, Page 975

**1.** $-\frac{1}{5}\cot^5 x - \frac{1}{3}\cot^3 x + C$ **3.** $\frac{1}{12}\ln\left|\frac{3x-2}{3x+2}\right| + C$ **5.** $\frac{3}{2}\ln|x^2+9| - \frac{1}{3}\text{ Arctan}\frac{x}{3} + C$

**7.** $\ln\left|\frac{\sqrt{1+x^2}-1}{x}\right| + C$ **9.** $\tan^{-1}(2x-1) + C$ **11.** $x \tan x + \ln|\cos x| + C$

**13.** $\frac{1}{2}\ln|x^2+x+1| + \sqrt{3}\,\text{Arctan}\left(\frac{2x+1}{\sqrt{3}}\right) + C$ **15.** $-0.4139$ **17.** 1.493 **19.** $\frac{1}{6}\ln\left|\frac{3x-1}{3x+1}\right| + C$

**21.** $\frac{4x-1}{12(1-x)^4} + C$ **23.** $-\csc x - \sin x + C$ **25.** 0.2877 **27.** 2.171 **29.** $3\sqrt[3]{2}$ **31.** Diverges

## CHAPTER 35

### Exercise 1, Page 981

**1.** First order, first degree, ordinary **3.** Third order, first degree, ordinary

**5.** Second order, fourth degree, ordinary **7.** $y = \frac{7x^2}{2} + C$ **9.** $y = \frac{2x^2}{3} - \frac{5x}{3} + C$ **11.** $y = \frac{x^3}{3} + C$

### Exercise 2, Page 984

**1.** $y = \pm\sqrt{x^2 + C}$ **3.** $\ln|y^3| = x^3 + C$ **5.** $4x^3 - 3y^4 = C$ **7.** $y = \frac{\sqrt{x^2+1}}{C}$

**9.** $\arctan y + \arctan x = C$ **11.** $\sqrt{1+x^2} + \ln y = C$ **13.** $\arctan x = \arctan y + C$

**15.** $2x + xy - 2y = C$ **17.** $x^2 + Cy^2 = C - 1$ **19.** $y = C\sqrt{1+e^{2x}}$ **21.** $e^{2x} + e^{2y} = C$

**23.** $(1+x)\sin y = C$ **25.** $\cos x \cos y = C$ **27.** $2\sin^2 x - \sin y = C$ **29.** $y^3 - x^3 = 1$

**31.** $\sin x + \cos y = 2$

### Exercise 3, Page 987

**1.** $xy - 7x = C$ **3.** $C = xy - 3x$ **5.** $4xy - x^2 = C$ **7.** $C = y/x - 3x$ **9.** $x^3 + 2xy = C$

**11.** $C = x^2 + y^2 - 4xy$ **13.** $x/y = 2x + C$ **15.** $2y^3 = x + Cy$ **17.** $4xy^2 - 3x^2 = C$

**19.** $2x^2 - xy = 15$ **21.** $3y^3 - y = 2x$ **23.** $3x^2 + 3y^2 - 4xy = 8$

### Exercise 4, Page 990

**1.** $x(x-3y)^2 = C$ **3.** $x \ln y - y = Cx$ **5.** $y^3 = x^3(3\ln x + C)$ **7.** $x(x-3y)^2 = 4$

**9.** $(y-2x)^2(x+y) = 27$

### Exercise 5, Page 995

**1.** $y = 2x + \frac{C}{x}$ **3.** $y = x^3 + \frac{C}{x}$ **5.** $y = 1 + Ce^{-x^3/3}$ **7.** $y = \frac{C}{\sqrt{x}} - \frac{3}{x}$ **9.** $y = x + Cx^2$

**11.** $(1+x^2)^2 y = 2\ln x + x^2 + C$ **13.** $y = \frac{C}{xe^{x^2/2}}$ **15.** $y = \frac{e^x}{2} + \frac{C}{e^x}$ **17.** $y = (4x + C)e^{2x}$

**19.** $x^2 y = 2\ln^2 x + C$ **21.** $y = \frac{1}{2}(\sin x - \cos x) + Ce^{-x}$ **23.** $y = \sin x + \cos x + Ce^{-x}$

**25.** $xy\left(C - \frac{3x^2}{2}\right) = 1$ **27.** $y = \frac{1}{x^2 + Ce^{-x}}$ **29.** $xy - 2x^2 = 3$ **31.** $y = \frac{5x}{2} - \frac{1}{2x}$

**33.** $y = \tan x + \sqrt{2}\sin x$

### Exercise 6, Page 999

**1.** $y = x^2 + 3x - 1$ **3.** $y^2 = x^2 + 7$ **5.** $y = 0.812e^x - x - 1$ **7.** $xy = 4$ **9.** $y = \frac{e^x + e^{-x}}{2}$

**11.** $y = Cx$ **13.** $3xy^2 + x^3 = k$

### Exercise 7, Page 1002

**3.** 158 million bbl/day **5.** 9.71°F **7.** 1240°F **13.** 15.0 ft **15.** 2.03 s **17.** 2.36 s

### Exercise 8, Page 1007

**7.** 231 mA, 0.882 V **9.** 46.4 mA **11.** 237 mA **13.** 139 mA

## Exercise 9, Page 1009

**1.** $y = \frac{5}{2}x^2 + C_1x + C_2$ **3.** $y = 3e^x + C_1x + C_2$ **5.** $y = \frac{x^4}{12} + x$

## Exercise 10, Page 1016

**1.** $y = C_1e^x + C_2e^{5x}$ **3.** $y = C_1e^x + C_2e^{2x}$ **5.** $y = C_1e^{3x} + C_2e^{-2x}$ **7.** $y = C_1 + C_2e^{2x/5}$
**9.** $y = C_1e^{2x/3} + C_2e^{-3x/2}$ **11.** $y = (C_1 + C_2x)e^{2x}$ **13.** $y = C_1e^x + C_2xe^x$ **15.** $y = (C_1 + C_2x)e^{-2x}$
**17.** $y = C_1e^{-x} + C_2xe^{-x}$ **19.** $y = e^{-2x}(C_1 \cos 3x + C_2 \sin 3x)$ **21.** $y = e^{3x}(C_1 \cos 4x + C_2 \sin 4x)$
**23.** $y = C_1 \cos 2x + C_2 \sin 2x$ **25.** $y = e^{2x}(C_1 \cos x + C_2 \sin x)$ **27.** $y = 3xe^{-3x}$ **29.** $y = 5e^x - 14xe^x$
**31.** $y = 1 + e^{2x}$ **33.** $y = \frac{1}{4}e^{2x} + \frac{3}{4}e^{-2x}$ **35.** $y = e^{-x} \sin x$ **37.** $y = C_1e^x + C_2e^{-x} + C_3e^{2x}$
**39.** $y = C_1e^x + C_2e^{2x} + C_3e^{3x}$ **41.** $y = C_1e^x + C_2e^{-x} + C_3e^{3x}$ **43.** $y = (C_1 + C_2)e^{x/2} + C_3e^{-x}$

## Exercise 11, Page 1022

**1.** $y = C_1e^{2x} + C_2e^{-2x} - 3$ **3.** $y = C_1e^{2x} + C_2e^{-x} - 2x + 1$ **5.** $y = C_1e^{2x} + C_2e^{-2x} - \frac{1}{4}x^3 - \frac{5}{8}x$
**7.** $y = C_1e^{2x} + C_2e^{-x} - 3e^x$ **9.** $y = C_1e^{2x} + C_2e^{-2x} + e^x - x$
**11.** $y = e^{-2x}(C_1 + C_2x) + (2e^{2x} + x - 1)/4$ **13.** $y = C_1 \cos 2x + C_2 \sin 2x - \frac{x \cos 2x}{4}$
**15.** $y = (C_1 + C_2x)e^{-x} + (1/2) \sin x$ **17.** $y = C_1 \cos x + C_2 \sin x + \cos 2x + x \sin x$
**19.** $y = C_1 \cos x + C_2 \sin x + (e^x/5)(\sin x - 2 \cos x)$ **21.** $y = e^{2x}[C_1 \sin x + C_2 \cos x - (x/2) \cos x]$
**23.** $y = \frac{e^{4x}}{2} - 2x - \frac{1}{2}$ **25.** $y = \frac{1 - \cos 2x}{2}$ **27.** $y = e^x(x^2 - 1)$ **29.** $y = x \cos x$

## Exercise 12, Page 1029

**13.** 9.84 Hz **15.** 0.496 μF **17.** 10 A

## Review Problems, Page 1030

**1.** $y = \frac{e^x}{2x} + \frac{C}{xe^x}$ **3.** $y = \sin x + \cos x + Ce^{-x}$ **5.** $x^3 + 2xy = C$ **7.** $x \ln y - y = Cx$
**9.** $y \ln|1 - x| + Cy = 1$ **11.** $x^2 + \frac{x}{y} + \ln y = C$ **13.** $\sin x + \cos x = 1.707$ **15.** $x + 2y = 2xy^2$
**17.** 14.6 rev/s **19.** $y = -2 \ln|x| + C$ **21.** $y = 2x \ln x + 2x$ **23.** 4740 g
**25.** $y = C_1e^{2x} + C_2e^{-x} - \frac{5}{2}x$ **27.** $y = C_1e^{5x} + C_2e^{-x} - \frac{1}{5}e^{4x}$ **29.** $y = (11e^{5x} - 30x + 14)/25$
**31.** $y = C_1e^{9x} + C_2e^{-2x}$ **33.** $y = C_1e^{2x} + C_2e^{3x} + C_3e^x$ **35.** $y = C_1 \cos \sqrt{2}x + C_2 \sin \sqrt{2}x - \frac{3}{2}\sin 2x$
**37.** $y = 4e^{-x} - 3e^{-2x}$ **39.** $y = C_1e^x + C_2e^{-4x}$ **41.** $y = C_1e^{-0.519x} + C_2e^{0.192x}$ **43.** $5y = 2(e^{-4x} - e^x)$
**45.** $y = C_1e^{-3x} + C_2xe^{-3x}$ **47.** $y = C_1e^{x/3} + C_2xe^{x/3}$ **49.** $i = 4.66e^{-18.75t} \sin 258t$ A
**51.** $i = 98.1 \sin 510t$ mA

# CHAPTER 36

## Exercise 1, Page 1038

**1.** $\frac{6}{s}$ **3.** $\frac{2}{s^3}$ **5.** $\frac{s}{s^2 + 25}$ **7.** $\frac{2}{s^3} + \frac{4}{s}$ **9.** $\frac{3}{s - 1} + \frac{2}{s + 1}$ **11.** $\frac{2}{s^2 + 4} + \frac{s}{s^2 + 9}$
**13.** $\frac{5(s - 3)}{(s - 3)^2 + 25}$ **15.** $\frac{5}{s - 1} - \frac{6s}{(s^2 + 9)^2}$ **17.** $s\mathcal{L}[y] - 1 + 2\mathcal{L}[y]$ **19.** $s\mathcal{L}[y] - 4\mathcal{L}[y]$
**21.** $s^2\mathcal{L}[y] + 3s\mathcal{L}[y] - \mathcal{L}[y] - s - 6$ **23.** $2s^2\mathcal{L}[y] + 3s\mathcal{L}[y] + \mathcal{L}[y] - 4s - 12$

## Exercise 2, Page 1040

**1.** 1 **3.** $t^2$ **5.** $2 \sin 2t$ **7.** $3 \cos \sqrt{2}t$ **9.** $e^{9t}(1 + 13t)$ **11.** $\frac{5}{3}e^{-2t} \sin 3t$

**13.** $\cos t + 1$ **15.** $e^t + e^{-2t} + t - 2$ **17.** $\frac{2}{3}e^t - \frac{2}{3}e^{-2t}$ **19.** $2(3e^{-3t} - 2e^{-2t})$

### Exercise 3, Page 1044

**1.** $y = e^{3t}$ **3.** $y = -1 - \frac{t}{2} + e^{t/2}$ **5.** $y = \frac{21}{4}e^{2t/3} - \frac{9}{4} - \frac{3}{2}t - \frac{1}{2}t^2$ **7.** $y = \frac{1}{4}\sin 2t + \frac{1}{4}\cos 2t - \frac{1}{4}e^{-2t}$

**9.** $y = \frac{1}{2}e^t - \frac{1}{2}e^{-3t}$ **11.** $y = \frac{4}{3} + t - \frac{1}{3}e^{-3t}$ **13.** $y = 4t + 3\cos 0.707t - 5.66\sin 0.707t$

**15.** $y = \frac{t}{3} - \frac{4}{9} + \frac{9}{2}e^{-t} - \frac{37}{18}e^{-3t}$ **17.** $y = 12 - 9.9e^{-t/3} - 0.1\cos t - 0.3\sin t$ **19.** $y = \frac{1}{2}t^2e^{2t}$

**21.** $y = 0.111e^{-t}\cos\sqrt{2}t + 0.0393e^{-t}\sin\sqrt{2}t - 0.111e^t + 0.167te^t$ **23.** 5.68 m/s **25.** $5.7 \times 10^{29}$ bacteria

**27.** $x = 1.02e^{-6.25t} - 1.02\cos 2.25t + 2.84\sin 2.25t$

### Exercise 4, Page 1050

**1.** $i = 41.8(1 - e^{-16t})$ mA **3.** $i = 16.4(1 - e^{-16t})$ mA **5.** $i = 169e^{-70.5t}\sin 265t$ mA

**7.** $i = 6.03(1 - e^{-33.5t})$ A

### Exercise 5, Page 1056

**1.** 0.644 **3.** 1.899 **5.** 5.824 **7.** 3.123 **9.** 13.751

### Exercise 6, Page 1059

**1.** 2.502 **3.** 12.090 **5.** 9.033 **7.** −1.267 **9.** 3.267

### Review Problems, Page 1060

**1.** $\frac{2}{s^2}$ **3.** $\frac{s}{s^2 + 9}$ **5.** $\frac{3}{(s - 2)^2}$ **7.** $\frac{3}{s + 1} - \frac{4}{s^3}$ **9.** $(s + 3)\mathscr{L}[y] - 1$ **11.** $(3s + 2)\mathscr{L}[y]$

**13.** $(s^2 + 3s + 4)\mathscr{L}[y] - s - 6$ **15.** $e^{3t}(1 - 3t)$ **17.** $\left(\frac{1}{\sqrt{5}}\right)t\sin\sqrt{5}t$ **19.** $\frac{5e^{4t} - 3e^{2t}}{2}$ **21.** $3e^{-t}(1 - t)$

**23.** $4e^{-2t}(1 - 2t)$ **25.** $y = \frac{-2}{27} - \frac{2}{9}t - \frac{1}{3}t^2 + \frac{56}{27}e^{3t}$ **27.** $y = 2 - 2e^{-t}(1 + t)$

**29.** $y = 2t + 3\cos\sqrt{2}t - \sqrt{2}\sin\sqrt{2}t$ **31.** $v = 7.82(1 - e^{-4.12t})$ ft/s **33.** $i = 0.671(1 - e^{-18.1t})$ A

**35.** $i = 40.7(e^{-200t} - e^{-24,800t})$ mA **37.** 8.913 **39.** 6.167 **41.** 0.844 **43.** 5.184 **45.** 7.065

## CHAPTER 37

### Exercise 1, Page 1066

**1.** Converges **3.** Diverges **5.** Cannot tell **7.** Cannot tell **9.** Converges **11.** Diverges

**13.** Test fails

### Exercise 2, Page 1073

**1.** $-1 < x < 1$ **3.** $x = 0$ **5.** $-1 < x < 1$ **21.** 1.625 **23.** 0.1827 **25.** 1.0488

### Exercise 3, Page 1077

**11.** 0.91 **13.** 0.688353 **15.** 2.0888 **17.** −0.92889 **19.** 0.2240

### Exercise 4, Page 1080

**21.** $1 - \frac{x^2}{2!} + \frac{x^4}{4!} - \frac{x^6}{6!} + \cdots$ **23.** $1 - x + x^2 - x^3 + \cdots$ **25.** $x + \frac{x^2}{2} + \frac{x^3}{3} + \frac{x^4}{4} + \cdots$ **27.** 2.98

**29.** 0.0308

### Exercise 5, Page 1087

**1.** $2 + 3\cos x + 2\cos 2x + \cos 3x + 4\sin x + 3\sin 2x + 2\sin 3x$

## Exercise 6, Page 1092

**1.** Neither **3.** Even **5.** Neither **7.** Yes **9.** No **11.** No

## Exercise 8, Page 1100

**1.** $y = -4.92 \cos x + 2.16 \cos 3x + 0.95 \cos 5x + 10.17 \sin x + 3.09 \sin 3x + 0.78 \sin 5x$

**3.** $y = 3.94 \cos x - 4.76 \cos 3x + 1.21 \cos 5x + 31.51 \sin x - 3.31 \sin 3x + 1.59 \sin 5x$

## Review Problems, Page 1160

**1.** 8.5 **3.** 0.4756 **5.** $-1 < x < 1$ **7.** $2 + 8 + 18 + 32 + 50 + \cdots$

**11.** $\cdots + 34 + 47 + \cdots + (n^2 - 2) + \cdots$ **13.** 1.316

**15.** $\cdots + 2207 + 4{,}870{,}847 + \cdots + (u^2 - 2) + \cdots$ **19.** Odd **21.** Odd **23.** Yes **25.** Yes

**27.** $y = \frac{8}{\pi^2}\left(\sin\frac{\pi x}{20} - \frac{1}{9}\sin\frac{3\pi x}{20} + \frac{1}{25}\sin\frac{5\pi x}{20} - \cdots\right)$ **29.** $y = -\frac{4}{\pi}\left(\sin \pi x + \frac{1}{3}\sin 3\pi x + \frac{1}{5}\sin 5\pi x + \cdots\right)$

**31.** $y = -4.11 \cos x + 2.07 \cos 3x + 0.57 \cos 5x + 11.26 \sin x + 2.69 \sin 3x + 0.63 \sin 5x$

# INDEX TO APPLICATIONS

**FINANCIAL**



**FLUIDS**



**GEODESY**



**LIGHT, OPTICS**



**MACHINE SHOP**



**MATERIALS**



**MECHANISMS**



**NAVIGATION**



**STATICS**



**STRENGTHS OF MATERIALS**



**STRUCTURES**



**SURVEYING**



**THERMAL**

# GENERAL INDEX

498

1-10

19-20